D0151891

MULTIVARIABLE CALCULUS

MULTIVARIABLE CALCULUS

EIGHTH EDITION

JAMES STEWART

McMASTER UNIVERSITY

AND

UNIVERSITY OF TORONTO

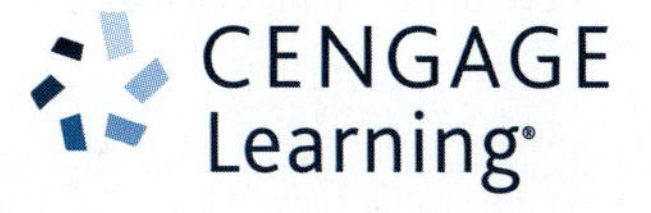

Australia • Brazil • Mexico • Singapore • United Kingdom • United States

***Multivariable Calculus*, Eighth Edition**
James Stewart

Product Manager: Neha Taleja
Senior Content Developer: Stacy Green
Associate Content Developer: Samantha Lugtu
Product Assistant: Stephanie Kreuz
Media Developer: Lynh Pham
Marketing Manager: Ryan Ahern
Content Project Manager: Cheryll Linthicum
Art Director: Vernon Boes
Manufacturing Planner: Becky Cross
Production Service: TECHarts
Photo and Text Researcher: Lumina Datamatics
Copy Editor: Kathi Townes, TECHarts
Illustrator: TECHarts
Text Designer: Diane Beasley
Cover Designer: Irene Morris, Morris Design
Compositor: Stephanie Kuhns, TECHarts
Cover Image: elisanth/123RF; tharrison/Getty Images

Library of Congress Control Number: 2015936919

Student Edition:
ISBN: 978-1-305-26664-3

Loose-leaf Edition:
ISBN: 978-1-305-65423-5

Cengage Learning
20 Channel Center Street
Boston, MA 02210
USA

K05T15

Printed in the United States of America
Print Number: 01 Print Year: 2015

To my family

Contents

17 Second-Order Differential Equations 1193

Appendixes A1

Index A45

Preface

> A great discovery solves a great problem but there is a grain of discovery in the solution of any problem. Your problem may be modest; but if it challenges your curiosity and brings into play your inventive faculties, and if you solve it by your own means, you may experience the tension and enjoy the triumph of discovery.
>
> GEORGE POLYA

The art of teaching, Mark Van Doren said, is the art of assisting discovery. I have tried to write a book that assists students in discovering calculus—both for its practical power and its surprising beauty. In this edition, as in the first seven editions, I aim to convey to the student a sense of the utility of calculus and develop technical competence, but I also strive to give some appreciation for the intrinsic beauty of the subject. Newton undoubtedly experienced a sense of triumph when he made his great discoveries. I want students to share some of that excitement.

The emphasis is on understanding concepts. I think that nearly everybody agrees that this should be the primary goal of calculus instruction. In fact, the impetus for the current calculus reform movement came from the Tulane Conference in 1986, which formulated as their first recommendation:

Focus on conceptual understanding.

I have tried to implement this goal through the *Rule of Three:* "Topics should be presented geometrically, numerically, and algebraically." Visualization, numerical and graphical experimentation, and other approaches have changed how we teach conceptual reasoning in fundamental ways. More recently, the Rule of Three has been expanded to become the *Rule of Four* by emphasizing the verbal, or descriptive, point of view as well.

In writing the eighth edition my premise has been that it is possible to achieve conceptual understanding and still retain the best traditions of traditional calculus. The book contains elements of reform, but within the context of a traditional curriculum.

Alternate Versions

I have written several other calculus textbooks that might be preferable for some instructors. Most of them also come in single variable and multivariable versions.

- *Calculus: Early Transcendentals,* Eighth Edition, is similar to the present textbook except that the exponential, logarithmic, and inverse trigonometric functions are covered in the first semester.
- *Essential Calculus*, Second Edition, is a much briefer book (840 pages), though it contains almost all of the topics in *Calculus,* Eighth Edition. The relative brevity is achieved through briefer exposition of some topics and putting some features on the website.
- *Essential Calculus: Early Transcendentals,* Second Edition, resembles *Essential Calculus,* but the exponential, logarithmic, and inverse trigonometric functions are covered in Chapter 3.

- *Calculus: Concepts and Contexts*, Fourth Edition, emphasizes conceptual understanding even more strongly than this book. The coverage of topics is not encyclopedic and the material on transcendental functions and on parametric equations is woven throughout the book instead of being treated in separate chapters.
- *Calculus: Early Vectors* introduces vectors and vector functions in the first semester and integrates them throughout the book. It is suitable for students taking engineering and physics courses concurrently with calculus.
- *Brief Applied Calculus* is intended for students in business, the social sciences, and the life sciences.
- *Biocalculus: Calculus for the Life Sciences* is intended to show students in the life sciences how calculus relates to biology.
- *Biocalculus: Calculus, Probability, and Statistics for the Life Sciences* contains all the content of *Biocalculus: Calculus for the Life Sciences* as well as three additional chapters covering probability and statistics.

What's New in the Eighth Edition?

The changes have resulted from talking with my colleagues and students at the University of Toronto and from reading journals, as well as suggestions from users and reviewers. Here are some of the many improvements that I've incorporated into this edition:

- The data in examples and exercises have been updated to be more timely.
- New examples have been added (see Examples 11.2.5 and 14.3.3, for instance). And the solutions to some of the existing examples have been amplified.
- One new project has been added: In the project *The Speedo LZR Racer* (page 976) it is explained that this suit reduces drag in the water and, as a result, many swimming records were broken. Students are asked why a small decrease in drag can have a big effect on performance.
- I have streamlined Chapter 15 (Multiple Integrals) by combining the first two sections so that iterated integrals are treated earlier.
- More than 20% of the exercises in each chapter are new. Here are some of my favorites: 12.5.81, 12.6.29–30, 14.6.65–66. In addition, there are some good new Problems Plus. (See Problem 8 on page 1026.)

Features

Conceptual Exercises

The most important way to foster conceptual understanding is through the problems that we assign. To that end I have devised various types of problems. Some exercise sets begin with requests to explain the meanings of the basic concepts of the section. (See, for instance, the first few exercises in Sections 11.2, 14.2, and 14.3.) Similarly, all the review sections begin with a Concept Check and a True-False Quiz. Other exercises test conceptual understanding through graphs or tables (see Exercises 10.1.24–27, 11.10.2, 13.2.1–2, 13.3.33–39, 14.1.1–2, 14.1.32–38, 14.1.41–44, 14.3.3–10, 14.6.1–2, 14.7.3–4, 15.1.6–8, 16.1.11–18, 16.2.17–18, and 16.3.1–2).

Another type of exercise uses verbal description to test conceptual understanding. I particularly value problems that combine and compare graphical, numerical, and algebraic approaches.

Graded Exercise Sets

Each exercise set is carefully graded, progressing from basic conceptual exercises and skill-development problems to more challenging problems involving applications and proofs.

Real-World Data

My assistants and I spent a great deal of time looking in libraries, contacting companies and government agencies, and searching the Internet for interesting real-world data to introduce, motivate, and illustrate the concepts of calculus. As a result, many of the examples and exercises deal with functions defined by such numerical data or graphs. Functions of two variables are illustrated by a table of values of the wind-chill index as a function of air temperature and wind speed (Example 14.1.2). Partial derivatives are introduced in Section 14.3 by examining a column in a table of values of the heat index (perceived air temperature) as a function of the actual temperature and the relative humidity. This example is pursued further in connection with linear approximations (Example 14.4.3). Directional derivatives are introduced in Section 14.6 by using a temperature contour map to estimate the rate of change of temperature at Reno in the direction of Las Vegas. Double integrals are used to estimate the average snowfall in Colorado on December 20–21, 2006 (Example 15.1.9). Vector fields are introduced in Section 16.1 by depictions of actual velocity vector fields showing San Francisco Bay wind patterns.

Projects

One way of involving students and making them active learners is to have them work (perhaps in groups) on extended projects that give a feeling of substantial accomplishment when completed. I have included four kinds of projects: *Applied Projects* involve applications that are designed to appeal to the imagination of students. The project after Section 14.8 uses Lagrange multipliers to determine the masses of the three stages of a rocket so as to minimize the total mass while enabling the rocket to reach a desired velocity. *Laboratory Projects* involve technology; the one following Section 10.2 shows how to use Bézier curves to design shapes that represent letters for a laser printer. *Discovery Projects* explore aspects of geometry: tetrahedra (after Section 12.4), hyperspheres (after Section 15.6), and intersections of three cylinders (after Section 15.7). The *Writing Project* after Section 17.8 explores the historical and physical origins of Green's Theorem and Stokes' Theorem and the interactions of the three men involved. Many additional projects can be found in the *Instructor's Guide*.

Tools for Enriching Calculus

TEC is a companion to the text and is intended to enrich and complement its contents. (It is now accessible in the eBook via CourseMate and Enhanced WebAssign. Selected Visuals and Modules are available at www.stewartcalculus.com.) Developed by Harvey Keynes, Dan Clegg, Hubert Hohn, and myself, TEC uses a discovery and exploratory approach. In sections of the book where technology is particularly appropriate, marginal icons direct students to TEC Modules that provide a laboratory environment in which they can explore the topic in different ways and at different levels. **Visuals are animations of figures in text; Modules are more elaborate activities and include exercises.** Instructors can choose to become involved at several different levels, ranging from simply encouraging students to use the Visuals and Modules for independent exploration, to assigning specific exercises from those included with each Module, or to creating additional exercises, labs, and projects that make use of the Visuals and Modules.

TEC also includes Homework Hints for representative exercises (usually odd-numbered) in every section of the text, indicated by printing the exercise number in red. These hints are usually presented in the form of questions and try to imitate an effective teaching assistant by functioning as a silent tutor. They are constructed so as not to reveal any more of the actual solution than is minimally necessary to make further progress.

Enhanced WebAssign

Technology is having an impact on the way homework is assigned to students, particularly in large classes. The use of online homework is growing and its appeal depends on ease of use, grading precision, and reliability. With the Eighth Edition we have been working with the calculus community and WebAssign to develop an online homework system. Up to 70% of the exercises in each section are assignable as online homework, including free response, multiple choice, and multi-part formats.

The system also includes Active Examples, in which students are guided in step-by-step tutorials through text examples, with links to the textbook and to video solutions.

Website

Visit CengageBrain.com or stewartcalculus.com for these additional materials:

- Homework Hints
- Algebra Review
- Lies My Calculator and Computer Told Me
- History of Mathematics, with links to the better historical websites
- Additional Topics (complete with exercise sets): Fourier Series, Formulas for the Remainder Term in Taylor Series, Rotation of Axes
- Archived Problems (drill exercises that appeared in previous editions, together with their solutions)
- Challenge Problems (some from the Problems Plus sections from prior editions)
- Links, for particular topics, to outside Web resources
- Selected Visuals and Modules from Tools for Enriching Calculus (TEC)

Content

10 Parametric Equations and Polar Coordinates

This chapter introduces parametric and polar curves and applies the methods of calculus to them. Parametric curves are well suited to laboratory projects; the two presented here involve families of curves and Bézier curves. A brief treatment of conic sections in polar coordinates prepares the way for Kepler's Laws in Chapter 13.

11 Infinite Sequences and Series

The convergence tests have intuitive justifications (see page 759) as well as formal proofs. Numerical estimates of sums of series are based on which test was used to prove convergence. The emphasis is on Taylor series and polynomials and their applications to physics. Error estimates include those from graphing devices.

12 Vectors and the Geometry of Space The material on three-dimensional analytic geometry and vectors is divided into two chapters. Chapter 12 deals with vectors, the dot and cross products, lines, planes, and surfaces.

13 Vector Functions This chapter covers vector-valued functions, their derivatives and integrals, the length and curvature of space curves, and velocity and acceleration along space curves, culminating in Kepler's laws.

14 Partial Derivatives Functions of two or more variables are studied from verbal, numerical, visual, and algebraic points of view. In particular, I introduce partial derivatives by looking at a specific column in a table of values of the heat index (perceived air temperature) as a function of the actual temperature and the relative humidity.

15 Multiple Integrals Contour maps and the Midpoint Rule are used to estimate the average snowfall and average temperature in given regions. Double and triple integrals are used to compute probabilities, surface areas, and (in projects) volumes of hyperspheres and volumes of intersections of three cylinders. Cylindrical and spherical coordinates are introduced in the context of evaluating triple integrals.

16 Vector Calculus Vector fields are introduced through pictures of velocity fields showing San Francisco Bay wind patterns. The similarities among the Fundamental Theorem for line integrals, Green's Theorem, Stokes' Theorem, and the Divergence Theorem are emphasized.

17 Second-Order Differential Equations Since first-order differential equations are covered in Chapter 9, this final chapter deals with second-order linear differential equations, their application to vibrating springs and electric circuits, and series solutions.

Ancillaries

Multivariable Calculus, Eighth Edition, is supported by a complete set of ancillaries developed under my direction. Each piece has been designed to enhance student understanding and to facilitate creative instruction. The tables on pages xx–xxi describe each of these ancillaries.

Acknowledgments

The preparation of this and previous editions has involved much time spent reading the reasoned (but sometimes contradictory) advice from a large number of astute reviewers. I greatly appreciate the time they spent to understand my motivation for the approach taken. I have learned something from each of them.

Eighth Edition Reviewers

Jay Abramson, *Arizona State University*
Adam Bowers, *University of California San Diego*
Neena Chopra, *The Pennsylvania State University*
Edward Dobson, *Mississippi State University*
Isaac Goldbring, *University of Illinois at Chicago*
Lea Jenkins, *Clemson University*
Rebecca Wahl, *Butler University*

Technology Reviewers

Maria Andersen, *Muskegon Community College*
Eric Aurand, *Eastfield College*
Joy Becker, *University of Wisconsin–Stout*
Przemyslaw Bogacki, *Old Dominion University*
Amy Elizabeth Bowman, *University of Alabama in Huntsville*
Monica Brown, *University of Missouri–St. Louis*
Roxanne Byrne, *University of Colorado at Denver and Health Sciences Center*
Teri Christiansen, *University of Missouri–Columbia*
Bobby Dale Daniel, *Lamar University*
Jennifer Daniel, *Lamar University*
Andras Domokos, *California State University, Sacramento*
Timothy Flaherty, *Carnegie Mellon University*
Lee Gibson, *University of Louisville*
Jane Golden, *Hillsborough Community College*
Semion Gutman, *University of Oklahoma*
Diane Hoffoss, *University of San Diego*
Lorraine Hughes, *Mississippi State University*
Jay Jahangiri, *Kent State University*
John Jernigan, *Community College of Philadelphia*
Brian Karasek, *South Mountain Community College*
Jason Kozinski, *University of Florida*
Carole Krueger, *The University of Texas at Arlington*
Ken Kubota, *University of Kentucky*
John Mitchell, *Clark College*
Donald Paul, *Tulsa Community College*
Chad Pierson, *University of Minnesota, Duluth*
Lanita Presson, *University of Alabama in Huntsville*
Karin Reinhold, *State University of New York at Albany*
Thomas Riedel, *University of Louisville*
Christopher Schroeder, *Morehead State University*
Angela Sharp, *University of Minnesota, Duluth*
Patricia Shaw, *Mississippi State University*
Carl Spitznagel, *John Carroll University*
Mohammad Tabanjeh, *Virginia State University*
Capt. Koichi Takagi, *United States Naval Academy*
Lorna TenEyck, *Chemeketa Community College*
Roger Werbylo, *Pima Community College*
David Williams, *Clayton State University*
Zhuan Ye, *Northern Illinois University*

Previous Edition Reviewers

B. D. Aggarwala, *University of Calgary*
John Alberghini, *Manchester Community College*
Michael Albert, *Carnegie-Mellon University*
Daniel Anderson, *University of Iowa*
Amy Austin, *Texas A&M University*
Donna J. Bailey, *Northeast Missouri State University*
Wayne Barber, *Chemeketa Community College*
Marilyn Belkin, *Villanova University*
Neil Berger, *University of Illinois, Chicago*
David Berman, *University of New Orleans*
Anthony J. Bevelacqua, *University of North Dakota*
Richard Biggs, *University of Western Ontario*
Robert Blumenthal, *Oglethorpe University*
Martina Bode, *Northwestern University*
Barbara Bohannon, *Hofstra University*
Jay Bourland, *Colorado State University*
Philip L. Bowers, *Florida State University*
Amy Elizabeth Bowman, *University of Alabama in Huntsville*
Stephen W. Brady, *Wichita State University*
Michael Breen, *Tennessee Technological University*
Robert N. Bryan, *University of Western Ontario*
David Buchthal, *University of Akron*
Jenna Carpenter, *Louisiana Tech University*
Jorge Cassio, *Miami-Dade Community College*
Jack Ceder, *University of California, Santa Barbara*
Scott Chapman, *Trinity University*
Zhen-Qing Chen, *University of Washington—Seattle*
James Choike, *Oklahoma State University*
Barbara Cortzen, *DePaul University*
Carl Cowen, *Purdue University*
Philip S. Crooke, *Vanderbilt University*

Charles N. Curtis, *Missouri Southern State College*
Daniel Cyphert, *Armstrong State College*
Robert Dahlin
M. Hilary Davies, *University of Alaska Anchorage*
Gregory J. Davis, *University of Wisconsin–Green Bay*
Elias Deeba, *University of Houston–Downtown*
Daniel DiMaria, *Suffolk Community College*
Seymour Ditor, *University of Western Ontario*
Greg Dresden, *Washington and Lee University*
Daniel Drucker, *Wayne State University*
Kenn Dunn, *Dalhousie University*
Dennis Dunninger, *Michigan State University*
Bruce Edwards, *University of Florida*
David Ellis, *San Francisco State University*
John Ellison, *Grove City College*
Martin Erickson, *Truman State University*
Garret Etgen, *University of Houston*
Theodore G. Faticoni, *Fordham University*
Laurene V. Fausett, *Georgia Southern University*
Norman Feldman, *Sonoma State University*
Le Baron O. Ferguson, *University of California—Riverside*
Newman Fisher, *San Francisco State University*
José D. Flores, *The University of South Dakota*
William Francis, *Michigan Technological University*
James T. Franklin, *Valencia Community College, East*
Stanley Friedlander, *Bronx Community College*
Patrick Gallagher, *Columbia University–New York*
Paul Garrett, *University of Minnesota–Minneapolis*
Frederick Gass, *Miami University of Ohio*
Bruce Gilligan, *University of Regina*
Matthias K. Gobbert, *University of Maryland, Baltimore County*
Gerald Goff, *Oklahoma State University*
Stuart Goldenberg, *California Polytechnic State University*
John A. Graham, *Buckingham Browne & Nichols School*
Richard Grassl, *University of New Mexico*
Michael Gregory, *University of North Dakota*
Charles Groetsch, *University of Cincinnati*
Paul Triantafilos Hadavas, *Armstrong Atlantic State University*
Salim M. Haïdar, *Grand Valley State University*
D. W. Hall, *Michigan State University*
Robert L. Hall, *University of Wisconsin–Milwaukee*
Howard B. Hamilton, *California State University, Sacramento*
Darel Hardy, *Colorado State University*
Shari Harris, *John Wood Community College*
Gary W. Harrison, *College of Charleston*
Melvin Hausner, *New York University/Courant Institute*
Curtis Herink, *Mercer University*
Russell Herman, *University of North Carolina at Wilmington*
Allen Hesse, *Rochester Community College*
Randall R. Holmes, *Auburn University*
James F. Hurley, *University of Connecticut*
Amer Iqbal, *University of Washington—Seattle*
Matthew A. Isom, *Arizona State University*
Gerald Janusz, *University of Illinois at Urbana-Champaign*
John H. Jenkins, *Embry-Riddle Aeronautical University, Prescott Campus*
Clement Jeske, *University of Wisconsin, Platteville*
Carl Jockusch, *University of Illinois at Urbana-Champaign*
Jan E. H. Johansson, *University of Vermont*
Jerry Johnson, *Oklahoma State University*
Zsuzsanna M. Kadas, *St. Michael's College*
Nets Katz, *Indiana University Bloomington*
Matt Kaufman
Matthias Kawski, *Arizona State University*
Frederick W. Keene, *Pasadena City College*
Robert L. Kelley, *University of Miami*
Akhtar Khan, *Rochester Institute of Technology*
Marianne Korten, *Kansas State University*
Virgil Kowalik, *Texas A&I University*
Kevin Kreider, *University of Akron*
Leonard Krop, *DePaul University*
Mark Krusemeyer, *Carleton College*
John C. Lawlor, *University of Vermont*
Christopher C. Leary, *State University of New York at Geneseo*
David Leeming, *University of Victoria*
Sam Lesseig, *Northeast Missouri State University*
Phil Locke, *University of Maine*
Joyce Longman, *Villanova University*

Joan McCarter, *Arizona State University*
Phil McCartney, *Northern Kentucky University*
Igor Malyshev, *San Jose State University*
Larry Mansfield, *Queens College*
Mary Martin, *Colgate University*
Nathaniel F. G. Martin, *University of Virginia*
Gerald Y. Matsumoto, *American River College*
James McKinney, *California State Polytechnic University, Pomona*
Tom Metzger, *University of Pittsburgh*
Richard Millspaugh, *University of North Dakota*
Lon H. Mitchell, *Virginia Commonwealth University*
Michael Montaño, *Riverside Community College*
Teri Jo Murphy, *University of Oklahoma*
Martin Nakashima, *California State Polytechnic University, Pomona*
Ho Kuen Ng, *San Jose State University*
Richard Nowakowski, *Dalhousie University*
Hussain S. Nur, *California State University, Fresno*
Norma Ortiz-Robinson, *Virginia Commonwealth University*
Wayne N. Palmer, *Utica College*
Vincent Panico, *University of the Pacific*
F. J. Papp, *University of Michigan–Dearborn*
Mike Penna, *Indiana University–Purdue University Indianapolis*
Mark Pinsky, *Northwestern University*
Lothar Redlin, *The Pennsylvania State University*
Joel W. Robbin, *University of Wisconsin–Madison*
Lila Roberts, *Georgia College and State University*
E. Arthur Robinson, Jr., *The George Washington University*
Richard Rockwell, *Pacific Union College*
Rob Root, *Lafayette College*
Richard Ruedemann, *Arizona State University*
David Ryeburn, *Simon Fraser University*
Richard St. Andre, *Central Michigan University*
Ricardo Salinas, *San Antonio College*
Robert Schmidt, *South Dakota State University*
Eric Schreiner, *Western Michigan University*
Mihr J. Shah, *Kent State University–Trumbull*
Qin Sheng, *Baylor University*
Theodore Shifrin, *University of Georgia*
Wayne Skrapek, *University of Saskatchewan*
Larry Small, *Los Angeles Pierce College*
Teresa Morgan Smith, *Blinn College*
William Smith, *University of North Carolina*
Donald W. Solomon, *University of Wisconsin Milwaukee*
Edward Spitznagel, *Washington University*
Joseph Stampfli, *Indiana University*
Kristin Stoley, *Blinn College*
M. B. Tavakoli, *Chaffey College*
Magdalena Toda, *Texas Tech University*
Ruth Trygstad, *Salt Lake Community College*
Paul Xavier Uhlig, *St. Mary's University, San Antonio*
Stan Ver Nooy, *University of Oregon*
Andrei Verona, *California State University–Los Angeles*
Klaus Volpert, *Villanova University*
Russell C. Walker, *Carnegie Mellon University*
William L. Walton, *McCallie School*
Peiyong Wang, *Wayne State University*
Jack Weiner, *University of Guelph*
Alan Weinstein, *University of California, Berkeley*
Theodore W. Wilcox, *Rochester Institute of Technology*
Steven Willard, *University of Alberta*
Robert Wilson, *University of Wisconsin–Madison*
Jerome Wolbert, *University of Michigan–Ann Arbor*
Dennis H. Wortman, *University of Massachusetts, Boston*
Mary Wright, *Southern Illinois University–Carbondale*
Paul M. Wright, *Austin Community College*
Xian Wu, *University of South Carolina*

In addition, I would like to thank R. B. Burckel, Bruce Colletti, David Behrman, John Dersch, Gove Effinger, Bill Emerson, Dan Kalman, Quyan Khan, Alfonso Gracia-Saz, Allan MacIsaac, Tami Martin, Monica Nitsche, Lamia Raffo, Norton Starr, and Jim Trefzger for their suggestions; Al Shenk and Dennis Zill for permission to use exercises from their calculus texts; COMAP for permission to use project material; George Bergman, David Bleecker, Dan Clegg, Victor Kaftal, Anthony Lam, Jamie Lawson, Ira Rosenholtz, Paul Sally, Lowell Smylie, and Larry Wallen for ideas for exercises; Dan Drucker for the roller derby project; Thomas Banchoff, Tom Farmer, Fred Gass, John Ramsay, Larry Riddle, Philip Straffin, and Klaus Volpert for ideas for projects; Dan Anderson, Dan Clegg, Jeff Cole, Dan Drucker, and Barbara Frank for solving the new exercises and suggesting ways to improve them; Marv Riedesel and Mary Johnson for accuracy in proofreading; Andy Bulman-Fleming, Lothar Redlin, Gina Sanders, and Saleem Watson for additional proofreading; and Jeff Cole and Dan Clegg for their careful preparation and proofreading of the answer manuscript.

In addition, I thank those who have contributed to past editions: Ed Barbeau, Jordan Bell, George Bergman, Fred Brauer, Andy Bulman-Fleming, Bob Burton, David Cusick, Tom DiCiccio, Garret Etgen, Chris Fisher, Leon Gerber, Stuart Goldenberg, Arnold Good, Gene Hecht, Harvey Keynes, E. L. Koh, Zdislav Kovarik, Kevin Kreider, Emile LeBlanc, David Leep, Gerald Leibowitz, Larry Peterson, Mary Pugh, Lothar Redlin, Carl Riehm, John Ringland, Peter Rosenthal, Dusty Sabo, Doug Shaw, Dan Silver, Simon Smith, Norton Starr, Saleem Watson, Alan Weinstein, and Gail Wolkowicz.

I also thank Kathi Townes, Stephanie Kuhns, Kristina Elliott, and Kira Abdallah of TECHarts for their production services and the following Cengage Learning staff: Cheryll Linthicum, content project manager; Stacy Green, senior content developer; Samantha Lugtu, associate content developer; Stephanie Kreuz, product assistant; Lynh Pham, media developer; Ryan Ahern, marketing manager; and Vernon Boes, art director. They have all done an outstanding job.

I have been very fortunate to have worked with some of the best mathematics editors in the business over the past three decades: Ron Munro, Harry Campbell, Craig Barth, Jeremy Hayhurst, Gary Ostedt, Bob Pirtle, Richard Stratton, Liz Covello, and now Neha Taleja. All of them have contributed greatly to the success of this book.

JAMES STEWART

Ancillaries for Instructors

Instructor's Guide

by Douglas Shaw

ISBN 978-1-305-27178-4

Each section of the text is discussed from several viewpoints. The Instructor's Guide contains suggested time to allot, points to stress, text discussion topics, core materials for lecture, workshop/discussion suggestions, group work exercises in a form suitable for handout, and suggested homework assignments.

Complete Solutions Manual

Multivariable

By Dan Clegg and Barbara Frank

ISBN 978-1-305-27611-6

Includes worked-out solutions to all exercises in the text.

Printed Test Bank

By William Steven Harmon

ISBN 978-1-305-27180-7

Contains text-specific multiple-choice and free response test items.

Cengage Learning Testing Powered by Cognero

(login.cengage.com)

This flexible online system allows you to author, edit, and manage test bank content from multiple Cengage Learning solutions; create multiple test versions in an instant; and deliver tests from your LMS, your classroom, or wherever you want.

Ancillaries for Instructors and Students

Stewart Website

www.stewartcalculus.com

Contents: *Homework Hints* ▪ *Algebra Review* ▪ *Additional Topics* ▪ *Drill exercises* ▪ *Challenge Problems* ▪ *Web Links* ▪ *History of Mathematics* ▪ *Tools for Enriching Calculus (TEC)*

TEC TOOLS FOR ENRICHING™ CALCULUS

By James Stewart, Harvey Keynes, Dan Clegg, and developer Hubert Hohn

Tools for Enriching Calculus (TEC) functions as both a powerful tool for instructors and as a tutorial environment in which students can explore and review selected topics. The Flash simulation modules in TEC include instructions, written and audio explanations of the concepts, and exercises. TEC is accessible in the eBook via CourseMate and Enhanced WebAssign. Selected Visuals and Modules are available at www.stewartcalculus.com.

ENHANCED WebAssign Enhanced WebAssign®

www.webassign.net

Printed Access Code: ISBN 978-1-285-85826-5

Instant Access Code ISBN: 978-1-285-85825-8

Exclusively from Cengage Learning, Enhanced WebAssign offers an extensive online program for Stewart's Calculus *to encourage the practice that is so critical for concept mastery. The meticulously crafted pedagogy and exercises in our proven texts become even more effective in Enhanced WebAssign, supplemented by multimedia tutorial support and immediate feedback as students complete their assignments. Key features include:*

- *Thousands of homework problems that match your textbook's end-of-section exercises*
- *Opportunities for students to review prerequisite skills and content both at the start of the course and at the beginning of each section*
- Read It *eBook pages,* Watch It *videos,* Master It *tutorials, and* Chat About It *links*
- *A customizable Cengage* YouBook *with highlighting, note-taking, and search features, as well as links to multimedia resources*
- Personal Study Plans *(based on diagnostic quizzing) that identify chapter topics that students will need to master*
- *A WebAssign* Answer Evaluator *that recognizes and accepts equivalent mathematical responses in the same way an instructor grades*
- *A* Show My Work *feature that gives instructors the option of seeing students' detailed solutions*
- Visualizing Calculus Animations, Lecture Videos, *and more*

■ Electronic items ■ Printed items

Cengage Customizable YouBook

YouBook *is an eBook that is both interactive and customizable. Containing all the content from Stewart's* Calculus, YouBook *features a text edit tool that allows instructors to modify the textbook narrative as needed. With* YouBook, *instructors can quickly reorder entire sections and chapters or hide any content they don't teach to create an eBook that perfectly matches their syllabus. Instructors can further customize the text by adding instructor-created or YouTube video links. Additional media assets include animated figures, video clips, highlighting and note-taking features, and more.* YouBook *is available within Enhanced WebAssign.*

CourseMate

CourseMate is a perfect self-study tool for students, and requires no set up from instructors. CourseMate brings course concepts to life with interactive learning, study, and exam preparation tools that support the printed textbook. CourseMate for Stewart's Calculus *includes an interactive eBook, Tools for Enriching Calculus, videos, quizzes, flashcards, and more. For instructors, CourseMate includes Engagement Tracker, a first-of-its-kind tool that monitors student engagement.*

CengageBrain.com

To access additional course materials, please visit www.cengagebrain.com. *At the CengageBrain.com home page, search for the ISBN of your title (from the back cover of your book) using the search box at the top of the page. This will take you to the product page where these resources can be found.*

Ancillaries for Students

Student Solutions Manual

Multivariable

By Dan Clegg and Barbara Frank

ISBN 978-1-305-27182-1

Provides completely worked-out solutions to all odd-numbered exercises in the text, giving students a chance to check their answer and ensure they took the correct steps to arrive at the answer. The Student Solutions Manual *can be ordered or accessed online as an eBook at* www.cengagebrain.com *by searching the ISBN.*

Study Guide

Multivariable

By Richard St. Andre

ISBN 978-1-305-27184-5

For each section of the text, the Study Guide *provides students with a brief introduction, a short list of concepts to master, and summary and focus questions with explained answers. The* Study Guide *also contains "Technology Plus" questions and multiple-choice "On Your Own" exam-style questions. The* Study Guide *can be ordered or accessed online as an eBook at* www.cengagebrain.com *by searching the ISBN.*

A Companion to Calculus

By Dennis Ebersole, Doris Schattschneider, Alicia Sevilla, and Kay Somers

ISBN 978-0-495-01124-8

Written to improve algebra and problem-solving skills of students taking a calculus course, every chapter in this companion is keyed to a calculus topic, providing conceptual background and specific algebra techniques needed to understand and solve calculus problems related to that topic. It is designed for calculus courses that integrate the review of precalculus concepts or for individual use. Order a copy of the text or access the eBook online at www.cengagebrain.com *by searching the ISBN.*

Linear Algebra for Calculus

by Konrad J. Heuvers, William P. Francis, John H. Kuisti, Deborah F. Lockhart, Daniel S. Moak, and Gene M. Ortner

ISBN 978-0-534-25248-9

This comprehensive book, designed to supplement the calculus course, provides an introduction to and review of the basic ideas of linear algebra. Order a copy of the text or access the eBook online at www.cengagebrain.com *by searching the ISBN.*

■ Electronic items ■ Printed items

To the Student

Reading a calculus textbook is different from reading a newspaper or a novel, or even a physics book. Don't be discouraged if you have to read a passage more than once in order to understand it. You should have pencil and paper and calculator at hand to sketch a diagram or make a calculation.

Some students start by trying their homework problems and read the text only if they get stuck on an exercise. I suggest that a far better plan is to read and understand a section of the text before attempting the exercises. In particular, you should look at the definitions to see the exact meanings of the terms. And before you read each example, I suggest that you cover up the solution and try solving the problem yourself. You'll get a lot more from looking at the solution if you do so.

Part of the aim of this course is to train you to think logically. Learn to write the solutions of the exercises in a connected, step-by-step fashion with explanatory sentences—not just a string of disconnected equations or formulas.

The answers to the odd-numbered exercises appear at the back of the book, in Appendix H. Some exercises ask for a verbal explanation or interpretation or description. In such cases there is no single correct way of expressing the answer, so don't worry that you haven't found the definitive answer. In addition, there are often several different forms in which to express a numerical or algebraic answer, so if your answer differs from mine, don't immediately assume you're wrong. For example, if the answer given in the back of the book is $\sqrt{2} - 1$ and you obtain $1/(1 + \sqrt{2})$, then you're right and rationalizing the denominator will show that the answers are equivalent.

The icon indicates an exercise that definitely requires the use of either a graphing calculator or a computer with graphing software. But that doesn't mean that graphing devices can't be used to check your work on the other exercises as well. The symbol CAS is reserved for problems in which the full resources of a computer algebra system (like Maple, Mathematica, or the TI-89) are required.

You will also encounter the symbol ⊘, which warns you against committing an error. I have placed this symbol in the margin in situations where I have observed that a large proportion of my students tend to make the same mistake.

Tools for Enriching Calculus, which is a companion to this text, is referred to by means of the symbol TEC and can be accessed in the *eBook* via Enhanced WebAssign and CourseMate (selected Visuals and Modules are available at stewartcalculus.com). It directs you to modules in which you can explore aspects of calculus for which the computer is particularly useful.

You will notice that some exercise numbers are printed in red: 5. This indicates that *Homework Hints* are available for the exercise. These hints can be found on stewart-calculus.com as well as Enhanced WebAssign and CourseMate. The homework hints ask you questions that allow you to make progress toward a solution without actually giving you the answer. You need to pursue each hint in an active manner with pencil and paper to work out the details. If a particular hint doesn't enable you to solve the problem, you can click to reveal the next hint.

I recommend that you keep this book for reference purposes after you finish the course. Because you will likely forget some of the specific details of calculus, the book will serve as a useful reminder when you need to use calculus in subsequent courses. And, because this book contains more material than can be covered in any one course, it can also serve as a valuable resource for a working scientist or engineer.

Calculus is an exciting subject, justly considered to be one of the greatest achievements of the human intellect. I hope you will discover that it is not only useful but also intrinsically beautiful.

JAMES STEWART

10 Parametric Equations and Polar Coordinates

The photo shows Halley's comet as it passed Earth in 1986. Due to return in 2061, it was named after Edmond Halley (1656–1742), the English scientist who first recognized its periodicity. In Section 10.6 you will see how polar coordinates provide a convenient equation for the elliptical path of its orbit.

SO FAR WE HAVE DESCRIBED plane curves by giving y as a function of x $[y = f(x)]$ or x as a function of y $[x = g(y)]$ or by giving a relation between x and y that defines y implicitly as a function of x $[f(x, y) = 0]$. In this chapter we discuss two new methods for describing curves.

Some curves, such as the cycloid, are best handled when both x and y are given in terms of a third variable t called a parameter $[x = f(t), y = g(t)]$. Other curves, such as the cardioid, have their most convenient description when we use a new coordinate system, called the polar coordinate system.

10.1 Curves Defined by Parametric Equations

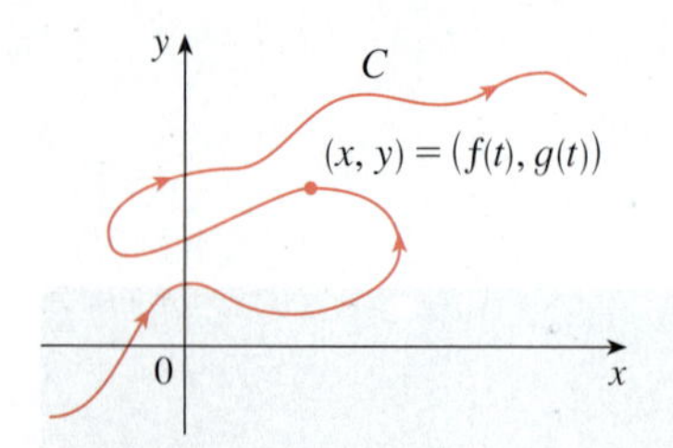

FIGURE 1

Imagine that a particle moves along the curve C shown in Figure 1. It is impossible to describe C by an equation of the form $y = f(x)$ because C fails the Vertical Line Test. But the x- and y-coordinates of the particle are functions of time and so we can write $x = f(t)$ and $y = g(t)$. Such a pair of equations is often a convenient way of describing a curve and gives rise to the following definition.

Suppose that x and y are both given as functions of a third variable t (called a **parameter**) by the equations

$$x = f(t) \qquad y = g(t)$$

(called **parametric equations**). Each value of t determines a point (x, y), which we can plot in a coordinate plane. As t varies, the point $(x, y) = (f(t), g(t))$ varies and traces out a curve C, which we call a **parametric curve**. The parameter t does not necessarily represent time and, in fact, we could use a letter other than t for the parameter. But in many applications of parametric curves, t does denote time and therefore we can interpret $(x, y) = (f(t), g(t))$ as the position of a particle at time t.

EXAMPLE 1 Sketch and identify the curve defined by the parametric equations

$$x = t^2 - 2t \qquad y = t + 1$$

SOLUTION Each value of t gives a point on the curve, as shown in the table. For instance, if $t = 0$, then $x = 0$, $y = 1$ and so the corresponding point is $(0, 1)$. In Figure 2 we plot the points (x, y) determined by several values of the parameter and we join them to produce a curve.

t	x	y
-2	8	-1
-1	3	0
0	0	1
1	-1	2
2	0	3
3	3	4
4	8	5

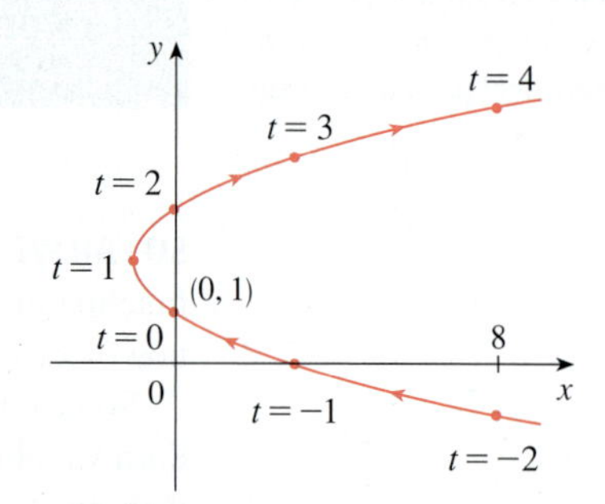

FIGURE 2

A particle whose position is given by the parametric equations moves along the curve in the direction of the arrows as t increases. Notice that the consecutive points marked on the curve appear at equal time intervals but not at equal distances. That is because the particle slows down and then speeds up as t increases.

It appears from Figure 2 that the curve traced out by the particle may be a parabola. This can be confirmed by eliminating the parameter t as follows. We obtain $t = y - 1$ from the second equation and substitute into the first equation. This gives

This equation in x and y describes *where* the particle has been, but it doesn't tell us *when* the particle was at a particular point. The parametric equations have an advantage—they tell us *when* the particle was at a point. They also indicate the *direction* of the motion.

$$x = t^2 - 2t = (y - 1)^2 - 2(y - 1) = y^2 - 4y + 3$$

and so the curve represented by the given parametric equations is the parabola $x = y^2 - 4y + 3$. ■

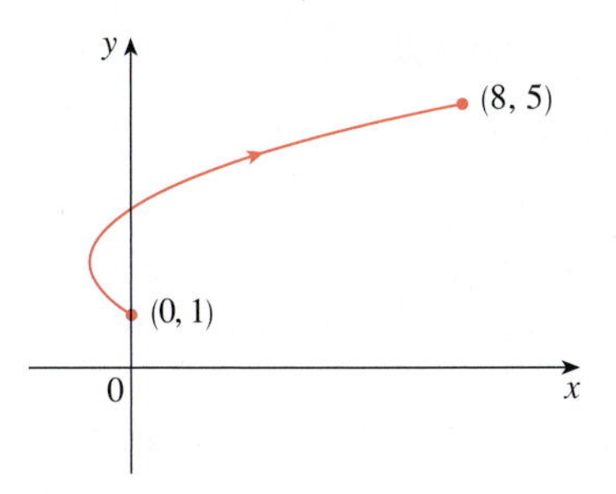

FIGURE 3

No restriction was placed on the parameter t in Example 1, so we assumed that t could be any real number. But sometimes we restrict t to lie in a finite interval. For instance, the parametric curve

$$x = t^2 - 2t \qquad y = t + 1 \qquad 0 \leqslant t \leqslant 4$$

shown in Figure 3 is the part of the parabola in Example 1 that starts at the point (0, 1) and ends at the point (8, 5). The arrowhead indicates the direction in which the curve is traced as t increases from 0 to 4.

In general, the curve with parametric equations

$$x = f(t) \qquad y = g(t) \qquad a \leqslant t \leqslant b$$

has **initial point** $(f(a), g(a))$ and **terminal point** $(f(b), g(b))$.

EXAMPLE 2 What curve is represented by the following parametric equations?

$$x = \cos t \qquad y = \sin t \qquad 0 \leqslant t \leqslant 2\pi$$

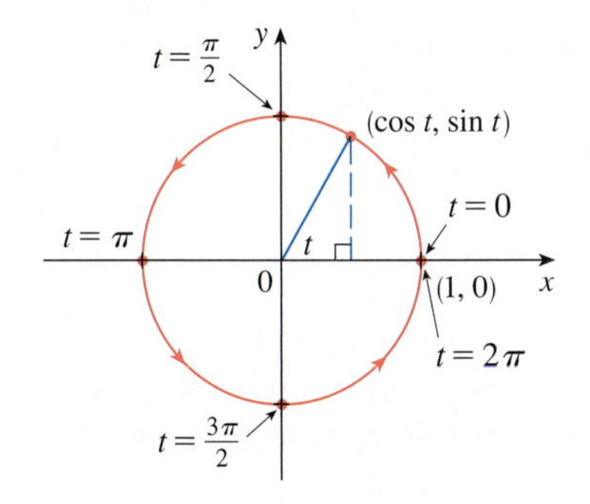

FIGURE 4

SOLUTION If we plot points, it appears that the curve is a circle. We can confirm this impression by eliminating t. Observe that

$$x^2 + y^2 = \cos^2 t + \sin^2 t = 1$$

Thus the point (x, y) moves on the unit circle $x^2 + y^2 = 1$. Notice that in this example the parameter t can be interpreted as the angle (in radians) shown in Figure 4. As t increases from 0 to 2π, the point $(x, y) = (\cos t, \sin t)$ moves once around the circle in the counterclockwise direction starting from the point (1, 0). ■

EXAMPLE 3 What curve is represented by the given parametric equations?

$$x = \sin 2t \qquad y = \cos 2t \qquad 0 \leqslant t \leqslant 2\pi$$

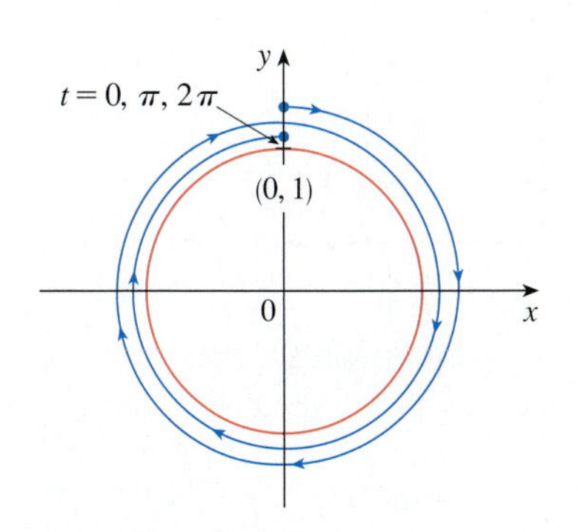

FIGURE 5

SOLUTION Again we have

$$x^2 + y^2 = \sin^2 2t + \cos^2 2t = 1$$

so the parametric equations again represent the unit circle $x^2 + y^2 = 1$. But as t increases from 0 to 2π, the point $(x, y) = (\sin 2t, \cos 2t)$ starts at (0, 1) and moves *twice* around the circle in the clockwise direction as indicated in Figure 5. ■

Examples 2 and 3 show that different sets of parametric equations can represent the same curve. Thus we distinguish between a *curve*, which is a set of points, and a *parametric curve*, in which the points are traced in a particular way.

EXAMPLE 4 Find parametric equations for the circle with center (h, k) and radius r.

SOLUTION If we take the equations of the unit circle in Example 2 and multiply the expressions for x and y by r, we get $x = r \cos t$, $y = r \sin t$. You can verify that these equations represent a circle with radius r and center the origin traced counterclockwise. We now shift h units in the x-direction and k units in the y-direction and obtain parametric equations of the circle (Figure 6) with center (h, k) and radius r:

$$x = h + r\cos t \qquad y = k + r\sin t \qquad 0 \leqslant t \leqslant 2\pi$$

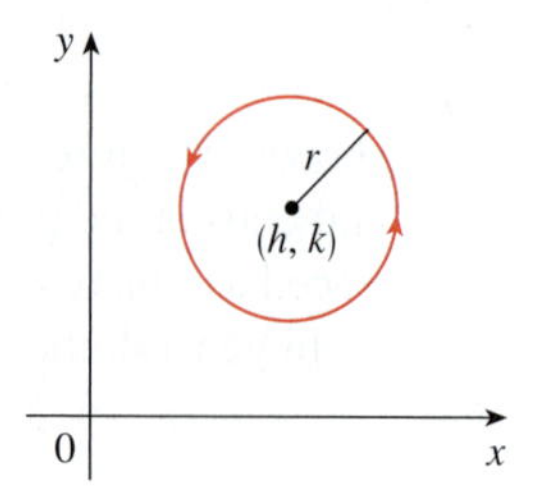

FIGURE 6
$x = h + r\cos t,\ y = k + r\sin t$

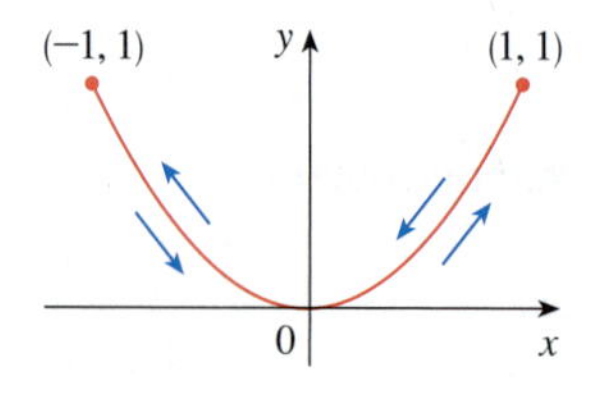

FIGURE 7

EXAMPLE 5 Sketch the curve with parametric equations $x = \sin t$, $y = \sin^2 t$.

SOLUTION Observe that $y = (\sin t)^2 = x^2$ and so the point (x, y) moves on the parabola $y = x^2$. But note also that, since $-1 \leqslant \sin t \leqslant 1$, we have $-1 \leqslant x \leqslant 1$, so the parametric equations represent only the part of the parabola for which $-1 \leqslant x \leqslant 1$. Since $\sin t$ is periodic, the point $(x, y) = (\sin t, \sin^2 t)$ moves back and forth infinitely often along the parabola from $(-1, 1)$ to $(1, 1)$. (See Figure 7.)

TEC Module 10.1A gives an animation of the relationship between motion along a parametric curve $x = f(t)$, $y = g(t)$ and motion along the graphs of f and g as functions of t. Clicking on TRIG gives you the family of parametric curves

$$x = a\cos bt \qquad y = c\sin dt$$

If you choose $a = b = c = d = 1$ and click on **animate**, you will see how the graphs of $x = \cos t$ and $y = \sin t$ relate to the circle in Example 2. If you choose $a = b = c = 1$, $d = 2$, you will see graphs as in Figure 8. By clicking on **animate** or moving the t-slider to the right, you can see from the color coding how motion along the graphs of $x = \cos t$ and $y = \sin 2t$ corresponds to motion along the parametric curve, which is called a **Lissajous figure**.

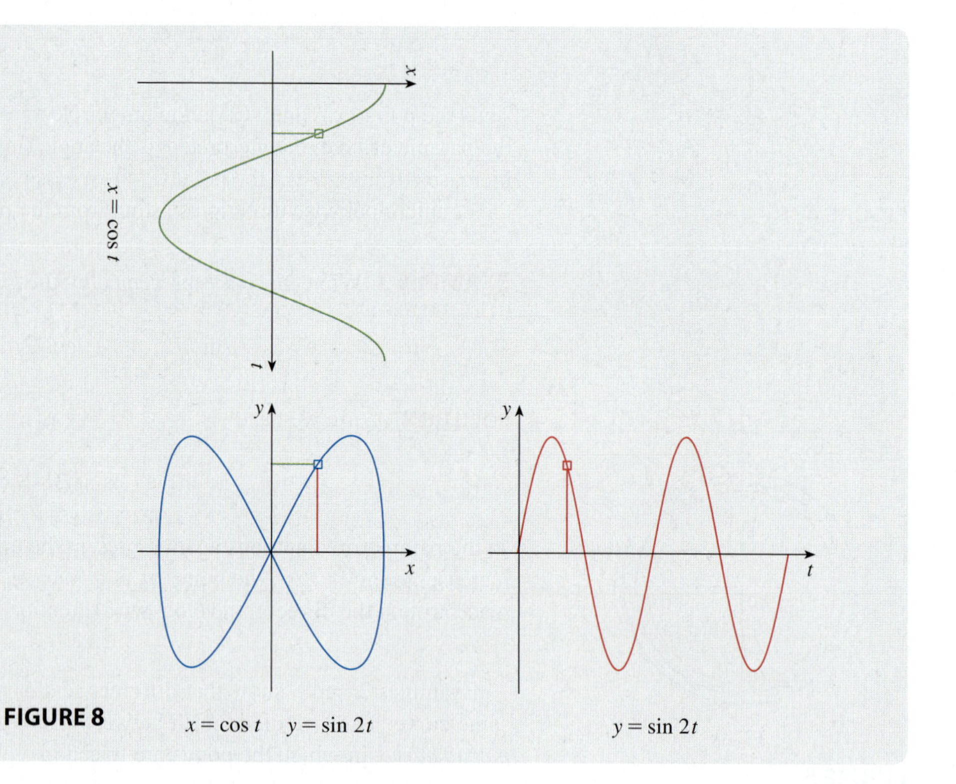

FIGURE 8 $x = \cos t \quad y = \sin 2t$ $\qquad$ $y = \sin 2t$

Graphing Devices

Most graphing calculators and other graphing devices can be used to graph curves defined by parametric equations. In fact, it's instructive to watch a parametric curve being drawn by a graphing calculator because the points are plotted in order as the corresponding parameter values increase.

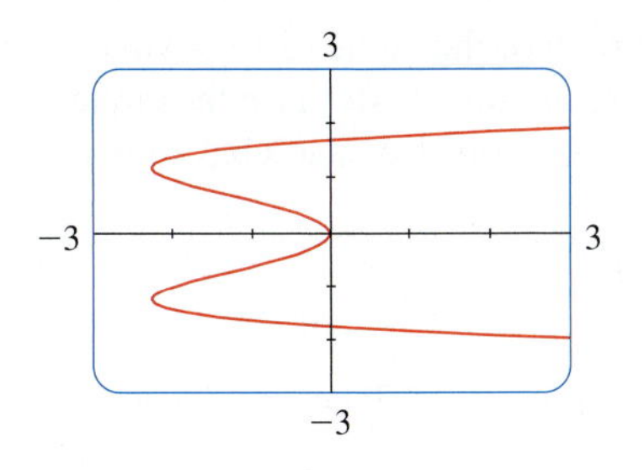

FIGURE 9

EXAMPLE 6 Use a graphing device to graph the curve $x = y^4 - 3y^2$.

SOLUTION If we let the parameter be $t = y$, then we have the equations

$$x = t^4 - 3t^2 \qquad y = t$$

Using these parametric equations to graph the curve, we obtain Figure 9. It would be possible to solve the given equation ($x = y^4 - 3y^2$) for y as four functions of x and graph them individually, but the parametric equations provide a much easier method. ■

In general, if we need to graph an equation of the form $x = g(y)$, we can use the parametric equations

$$x = g(t) \qquad y = t$$

Notice also that curves with equations $y = f(x)$ (the ones we are most familiar with—graphs of functions) can also be regarded as curves with parametric equations

$$x = t \qquad y = f(t)$$

Graphing devices are particularly useful for sketching complicated parametric curves. For instance, the curves shown in Figures 10, 11, and 12 would be virtually impossible to produce by hand.

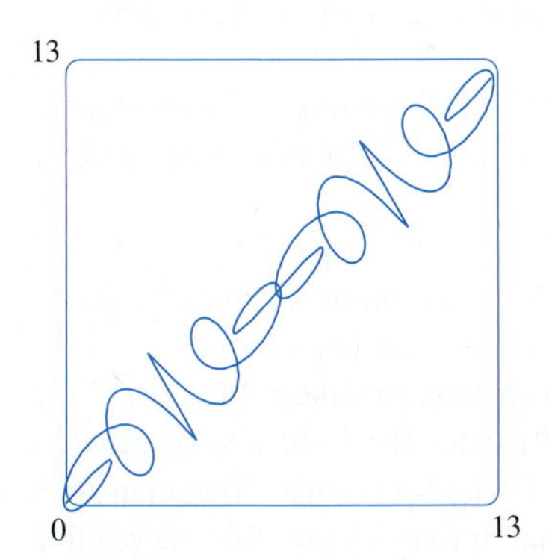

FIGURE 10
$x = t + \sin 5t$
$y = t + \sin 6t$

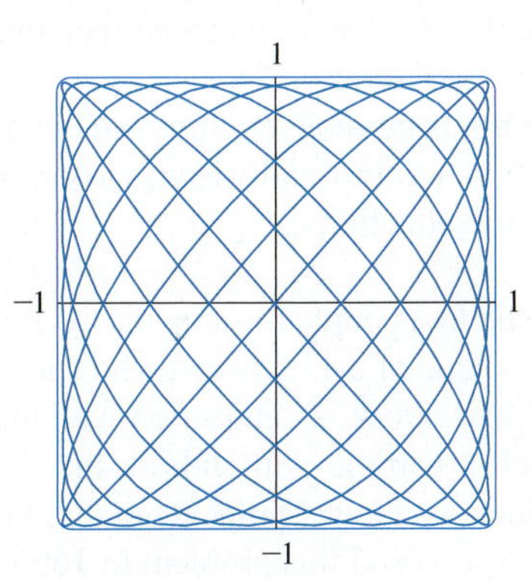

FIGURE 11
$x = \sin 9t$
$y = \sin 10t$

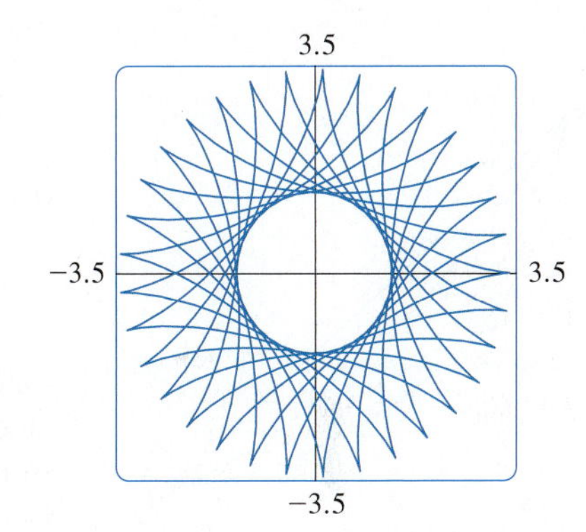

FIGURE 12
$x = 2.3 \cos 10t + \cos 23t$
$y = 2.3 \sin 10t - \sin 23t$

One of the most important uses of parametric curves is in computer-aided design (CAD). In the Laboratory Project after Section 10.2 we will investigate special parametric curves, called **Bézier curves**, that are used extensively in manufacturing, especially in the automotive industry. These curves are also employed in specifying the shapes of letters and other symbols in laser printers and in documents viewed electronically.

■ The Cycloid

TEC An animation in Module 10.1B shows how the cycloid is formed as the circle moves.

EXAMPLE 7 The curve traced out by a point P on the circumference of a circle as the circle rolls along a straight line is called a **cycloid** (see Figure 13). If the circle has radius r and rolls along the x-axis and if one position of P is the origin, find parametric equations for the cycloid.

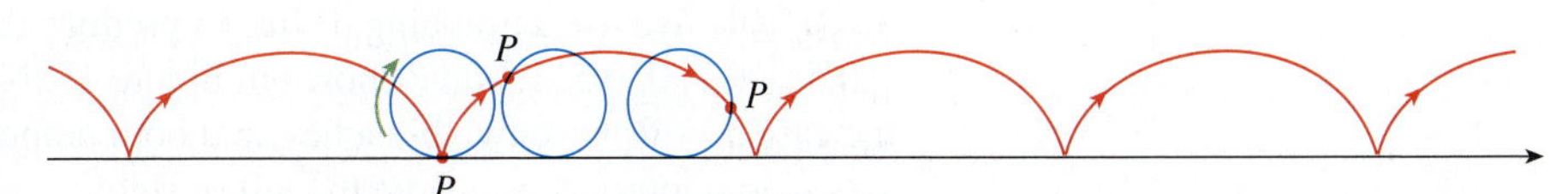

FIGURE 13

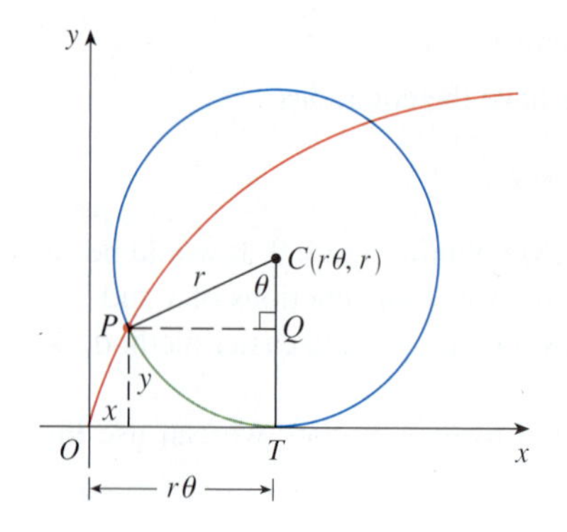

FIGURE 14

SOLUTION We choose as parameter the angle of rotation θ of the circle ($\theta = 0$ when P is at the origin). Suppose the circle has rotated through θ radians. Because the circle has been in contact with the line, we see from Figure 14 that the distance it has rolled from the origin is

$$|OT| = \text{arc } PT = r\theta$$

Therefore the center of the circle is $C(r\theta, r)$. Let the coordinates of P be (x, y). Then from Figure 14 we see that

$$x = |OT| - |PQ| = r\theta - r\sin\theta = r(\theta - \sin\theta)$$

$$y = |TC| - |QC| = r - r\cos\theta = r(1 - \cos\theta)$$

Therefore parametric equations of the cycloid are

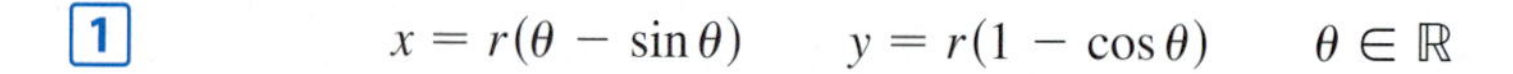

$$\boxed{1} \qquad x = r(\theta - \sin\theta) \qquad y = r(1 - \cos\theta) \qquad \theta \in \mathbb{R}$$

One arch of the cycloid comes from one rotation of the circle and so is described by $0 \leqslant \theta \leqslant 2\pi$. Although Equations 1 were derived from Figure 14, which illustrates the case where $0 < \theta < \pi/2$, it can be seen that these equations are still valid for other values of θ (see Exercise 39).

Although it is possible to eliminate the parameter θ from Equations 1, the resulting Cartesian equation in x and y is very complicated and not as convenient to work with as the parametric equations. ■

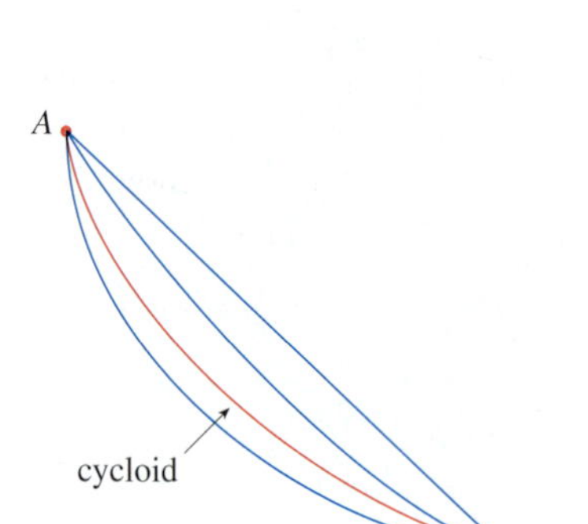

FIGURE 15

One of the first people to study the cycloid was Galileo, who proposed that bridges be built in the shape of cycloids and who tried to find the area under one arch of a cycloid. Later this curve arose in connection with the **brachistochrone problem**: Find the curve along which a particle will slide in the shortest time (under the influence of gravity) from a point A to a lower point B not directly beneath A. The Swiss mathematician John Bernoulli, who posed this problem in 1696, showed that among all possible curves that join A to B, as in Figure 15, the particle will take the least time sliding from A to B if the curve is part of an inverted arch of a cycloid.

The Dutch physicist Huygens had already shown that the cycloid is also the solution to the **tautochrone problem**; that is, no matter where a particle P is placed on an inverted cycloid, it takes the same time to slide to the bottom (see Figure 16). Huygens proposed that pendulum clocks (which he invented) should swing in cycloidal arcs because then the pendulum would take the same time to make a complete oscillation whether it swings through a wide or a small arc.

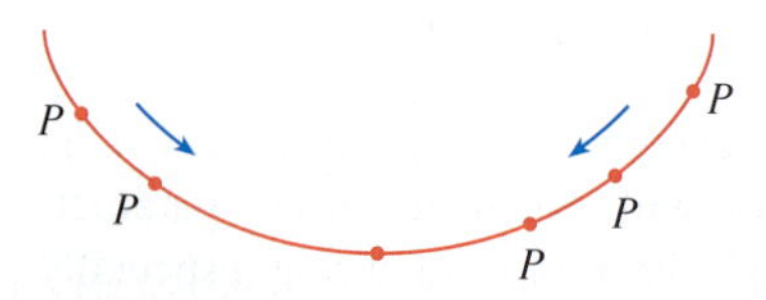

FIGURE 16

Families of Parametric Curves

EXAMPLE 8 Investigate the family of curves with parametric equations

$$x = a + \cos t \qquad y = a\tan t + \sin t$$

What do these curves have in common? How does the shape change as a increases?

SOLUTION We use a graphing device to produce the graphs for the cases $a = -2, -1, -0.5, -0.2, 0, 0.5, 1$, and 2 shown in Figure 17. Notice that all of these curves (except the case $a = 0$) have two branches, and both branches approach the vertical asymptote $x = a$ as x approaches a from the left or right.

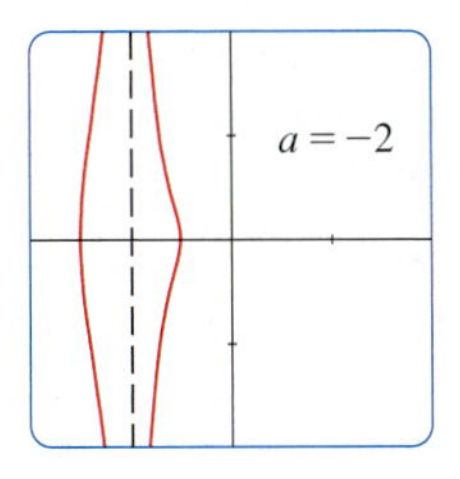

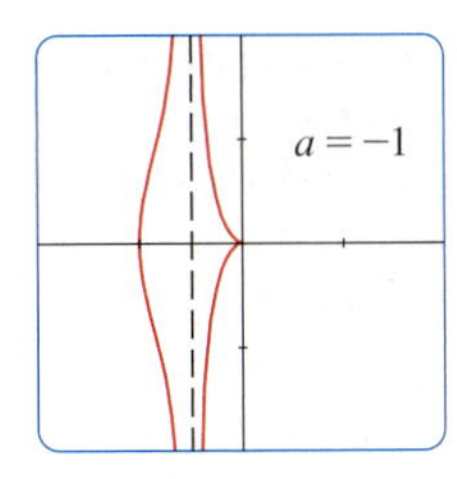

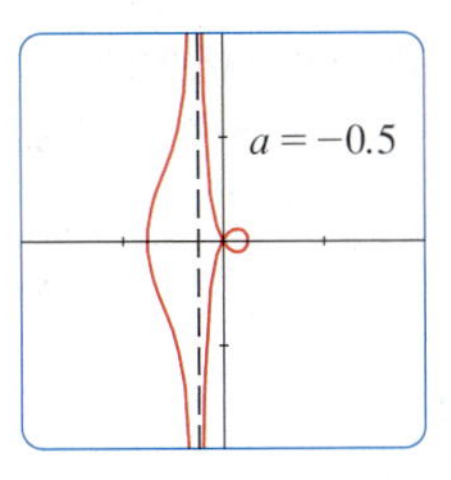

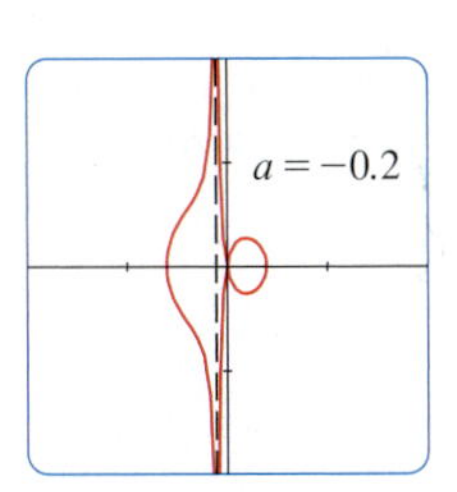

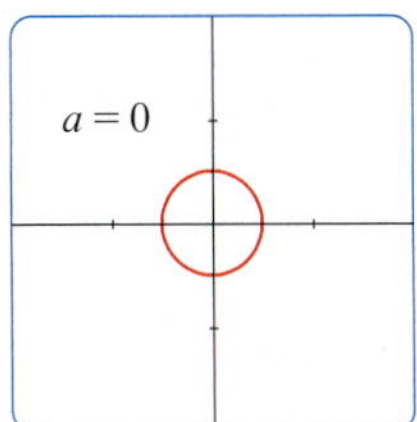

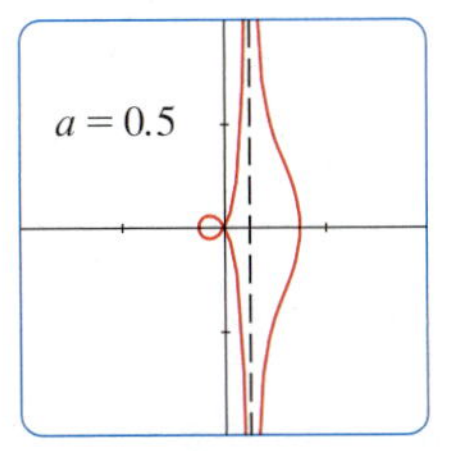

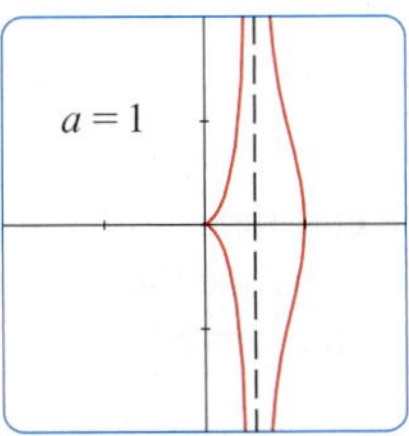

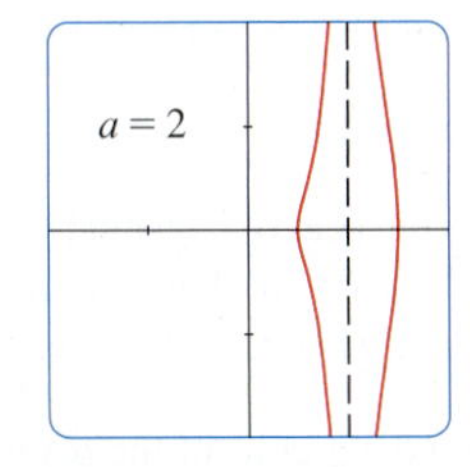

FIGURE 17
Members of the family $x = a + \cos t$, $y = a \tan t + \sin t$, all graphed in the viewing rectangle $[-4, 4]$ by $[-4, 4]$

When $a < -1$, both branches are smooth; but when a reaches -1, the right branch acquires a sharp point, called a *cusp*. For a between -1 and 0 the cusp turns into a loop, which becomes larger as a approaches 0. When $a = 0$, both branches come together and form a circle (see Example 2). For a between 0 and 1, the left branch has a loop, which shrinks to become a cusp when $a = 1$. For $a > 1$, the branches become smooth again, and as a increases further, they become less curved. Notice that the curves with a positive are reflections about the y-axis of the corresponding curves with a negative.

These curves are called **conchoids of Nicomedes** after the ancient Greek scholar Nicomedes. He called them conchoids because the shape of their outer branches resembles that of a conch shell or mussel shell. ■

10.1 EXERCISES

1–4 Sketch the curve by using the parametric equations to plot points. Indicate with an arrow the direction in which the curve is traced as t increases.

1. $x = 1 - t^2, \quad y = 2t - t^2, \quad -1 \leqslant t \leqslant 2$

2. $x = t^3 + t, \quad y = t^2 + 2, \quad -2 \leqslant t \leqslant 2$

3. $x = t + \sin t, \quad y = \cos t, \quad -\pi \leqslant t \leqslant \pi$

4. $x = e^{-t} + t, \quad y = e^t - t, \quad -2 \leqslant t \leqslant 2$

5–10
(a) Sketch the curve by using the parametric equations to plot points. Indicate with an arrow the direction in which the curve is traced as t increases.
(b) Eliminate the parameter to find a Cartesian equation of the curve.

5. $x = 2t - 1, \quad y = \frac{1}{2}t + 1$

6. $x = 3t + 2, \quad y = 2t + 3$

7. $x = t^2 - 3, \quad y = t + 2, \quad -3 \leqslant t \leqslant 3$

8. $x = \sin t, \quad y = 1 - \cos t, \quad 0 \leqslant t \leqslant 2\pi$

9. $x = \sqrt{t}, \quad y = 1 - t$

10. $x = t^2, \quad y = t^3$

11–18
(a) Eliminate the parameter to find a Cartesian equation of the curve.
(b) Sketch the curve and indicate with an arrow the direction in which the curve is traced as the parameter increases.

11. $x = \sin \frac{1}{2}\theta, \quad y = \cos \frac{1}{2}\theta, \quad -\pi \leqslant \theta \leqslant \pi$

12. $x = \frac{1}{2}\cos\theta, \quad y = 2\sin\theta, \quad 0 \leqslant \theta \leqslant \pi$

13. $x = \sin t, \quad y = \csc t, \quad 0 < t < \pi/2$

14. $x = e^t, \quad y = e^{-2t}$

15. $x = t^2, \quad y = \ln t$

16. $x = \sqrt{t + 1}, \quad y = \sqrt{t - 1}$

17. $x = \sinh t, \quad y = \cosh t$

18. $x = \tan^2\theta, \quad y = \sec\theta, \quad -\pi/2 < \theta < \pi/2$

19–22 Describe the motion of a particle with position (x, y) as t varies in the given interval.

19. $x = 5 + 2\cos \pi t, \quad y = 3 + 2\sin \pi t, \quad 1 \leqslant t \leqslant 2$

20. $x = 2 + \sin t, \quad y = 1 + 3\cos t, \quad \pi/2 \leqslant t \leqslant 2\pi$

21. $x = 5\sin t, \quad y = 2\cos t, \quad -\pi \leqslant t \leqslant 5\pi$

22. $x = \sin t, \quad y = \cos^2 t, \quad -2\pi \leqslant t \leqslant 2\pi$

---

23. Suppose a curve is given by the parametric equations $x = f(t)$, $y = g(t)$, where the range of f is $[1, 4]$ and the range of g is $[2, 3]$. What can you say about the curve?

24. Match the graphs of the parametric equations $x = f(t)$ and $y = g(t)$ in (a)–(d) with the parametric curves labeled I–IV. Give reasons for your choices.

(a)

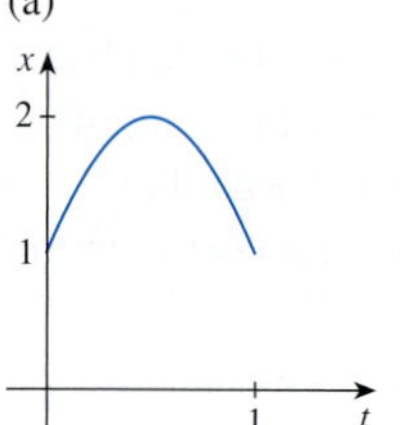

I

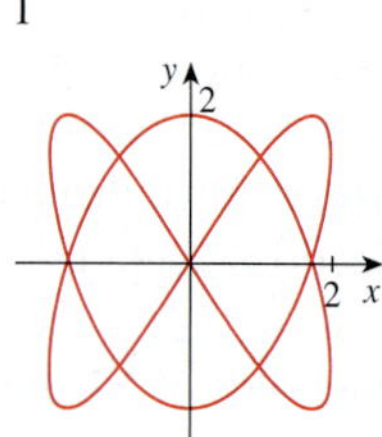

(b)

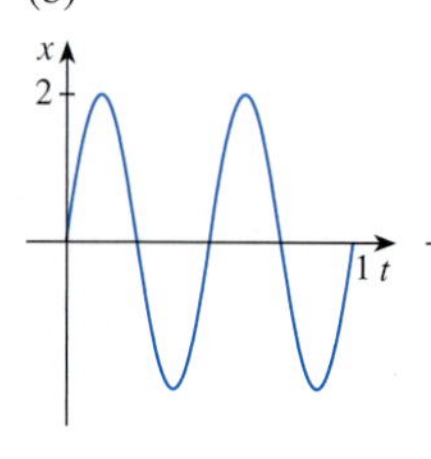

II

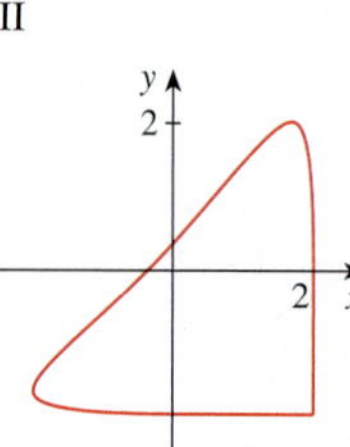

(c)

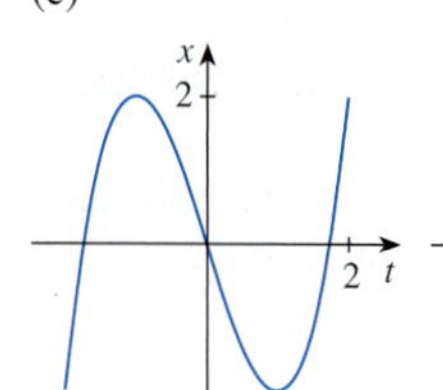

III

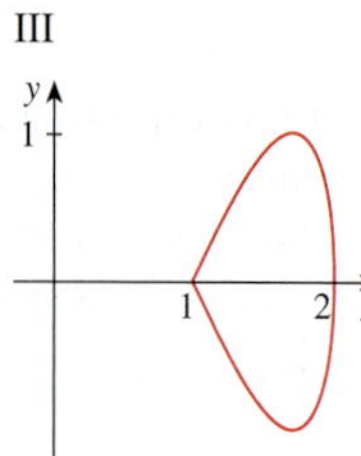

(d)

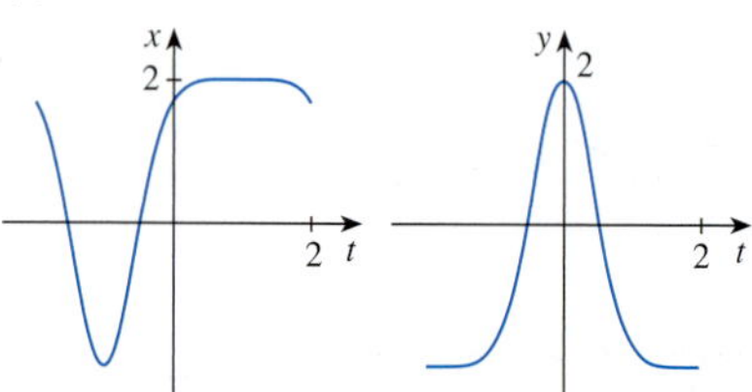

IV

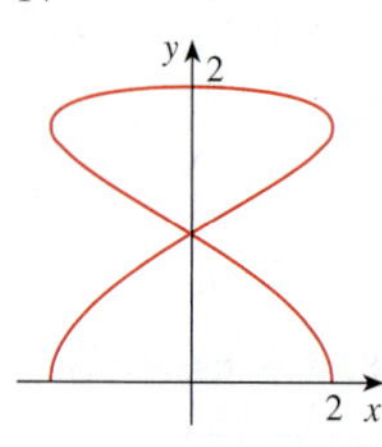

25–27 Use the graphs of $x = f(t)$ and $y = g(t)$ to sketch the parametric curve $x = f(t)$, $y = g(t)$. Indicate with arrows the direction in which the curve is traced as t increases.

25.

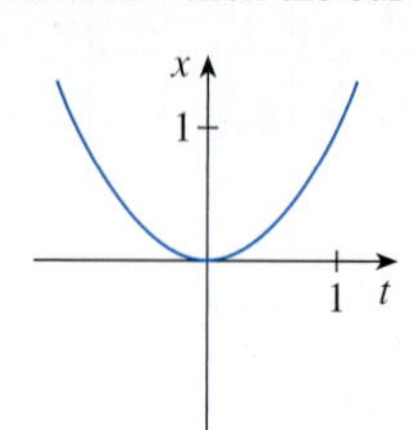

26.

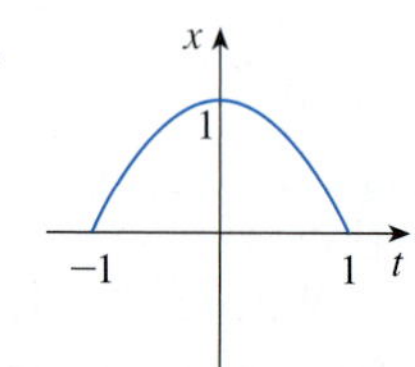

27.

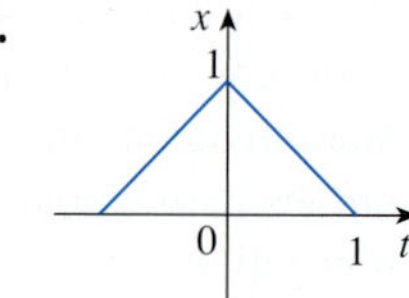

---

28. Match the parametric equations with the graphs labeled I–VI. Give reasons for your choices. (Do not use a graphing device.)

(a) $x = t^4 - t + 1, \quad y = t^2$

(b) $x = t^2 - 2t, \quad y = \sqrt{t}$

(c) $x = \sin 2t, \quad y = \sin(t + \sin 2t)$

(d) $x = \cos 5t, \quad y = \sin 2t$

(e) $x = t + \sin 4t, \quad y = t^2 + \cos 3t$

(f) $x = \dfrac{\sin 2t}{4 + t^2}, \quad y = \dfrac{\cos 2t}{4 + t^2}$

I

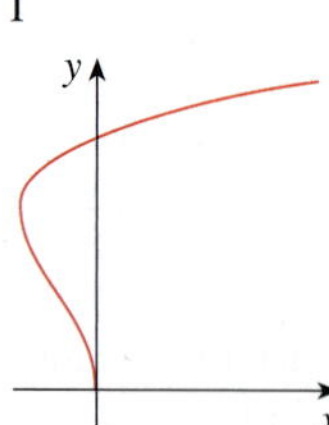

II

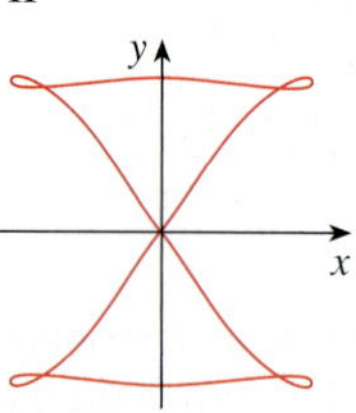

III

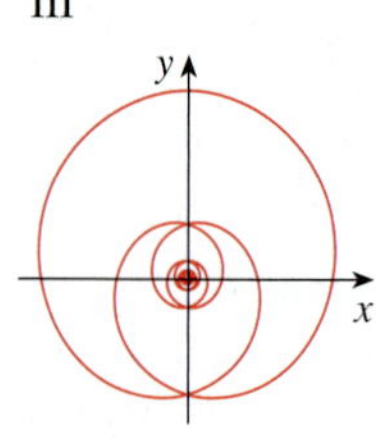

IV

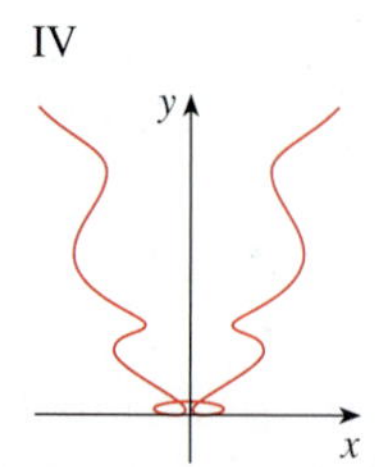

V

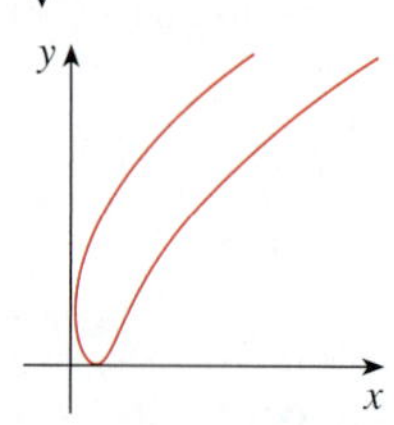

VI

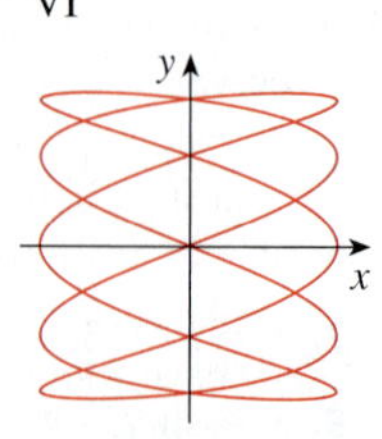

29. Graph the curve $x = y - 2 \sin \pi y$.

30. Graph the curves $y = x^3 - 4x$ and $x = y^3 - 4y$ and find their points of intersection correct to one decimal place.

31. (a) Show that the parametric equations

$$x = x_1 + (x_2 - x_1)t \qquad y = y_1 + (y_2 - y_1)t$$

where $0 \leqslant t \leqslant 1$, describe the line segment that joins the points $P_1(x_1, y_1)$ and $P_2(x_2, y_2)$.

(b) Find parametric equations to represent the line segment from $(-2, 7)$ to $(3, -1)$.

32. Use a graphing device and the result of Exercise 31(a) to draw the triangle with vertices $A(1, 1)$, $B(4, 2)$, and $C(1, 5)$.

33. Find parametric equations for the path of a particle that moves along the circle $x^2 + (y - 1)^2 = 4$ in the manner described.

(a) Once around clockwise, starting at $(2, 1)$

(b) Three times around counterclockwise, starting at $(2, 1)$

(c) Halfway around counterclockwise, starting at $(0, 3)$

34. (a) Find parametric equations for the ellipse $x^2/a^2 + y^2/b^2 = 1$. [*Hint:* Modify the equations of the circle in Example 2.]

(b) Use these parametric equations to graph the ellipse when $a = 3$ and $b = 1, 2, 4$, and 8.

(c) How does the shape of the ellipse change as b varies?

35–36 Use a graphing calculator or computer to reproduce the picture.

35.

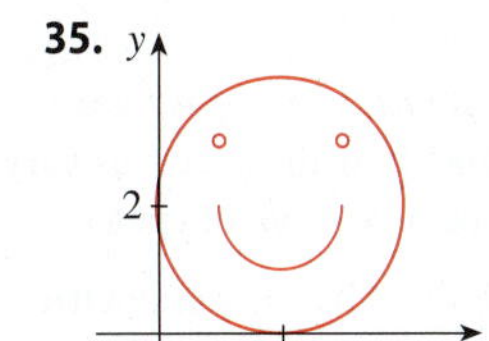

36.

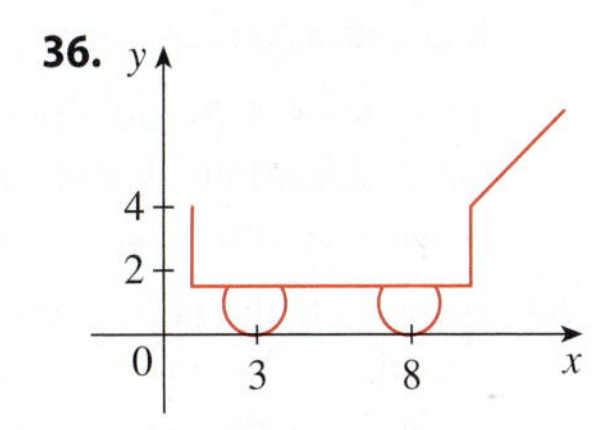

37–38 Compare the curves represented by the parametric equations. How do they differ?

37. (a) $x = t^3, \quad y = t^2$ (b) $x = t^6, \quad y = t^4$

(c) $x = e^{-3t}, \quad y = e^{-2t}$

38. (a) $x = t, \quad y = t^{-2}$ (b) $x = \cos t, \quad y = \sec^2 t$

(c) $x = e^t, \quad y = e^{-2t}$

39. Derive Equations 1 for the case $\pi/2 < \theta < \pi$.

40. Let P be a point at a distance d from the center of a circle of radius r. The curve traced out by P as the circle rolls along a straight line is called a **trochoid**. (Think of the motion of a point on a spoke of a bicycle wheel.) The cycloid is the special case of a trochoid with $d = r$. Using the same parameter θ as for the cycloid, and assuming the line is the x-axis and $\theta = 0$ when P is at one of its lowest points, show that parametric equations of the trochoid are

$$x = r\theta - d \sin\theta \qquad y = r - d \cos\theta$$

Sketch the trochoid for the cases $d < r$ and $d > r$.

41. If a and b are fixed numbers, find parametric equations for the curve that consists of all possible positions of the point P in the figure, using the angle θ as the parameter. Then eliminate the parameter and identify the curve.

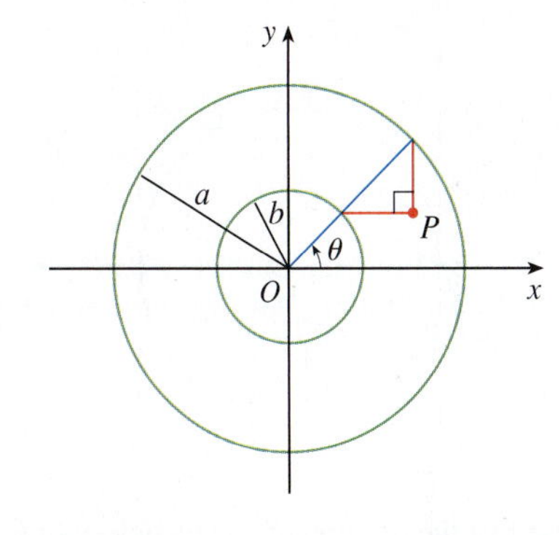

42. If a and b are fixed numbers, find parametric equations for the curve that consists of all possible positions of the point P in the figure, using the angle θ as the parameter. The line segment AB is tangent to the larger circle.

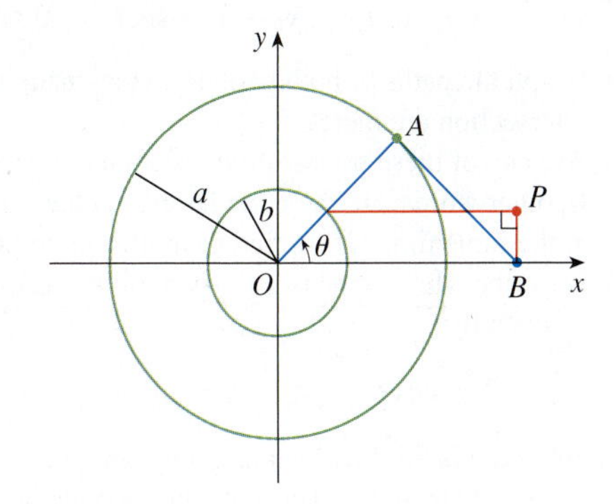

43. A curve, called a **witch of Maria Agnesi**, consists of all possible positions of the point P in the figure. Show that parametric equations for this curve can be written as

$$x = 2a \cot\theta \qquad y = 2a \sin^2\theta$$

Sketch the curve.

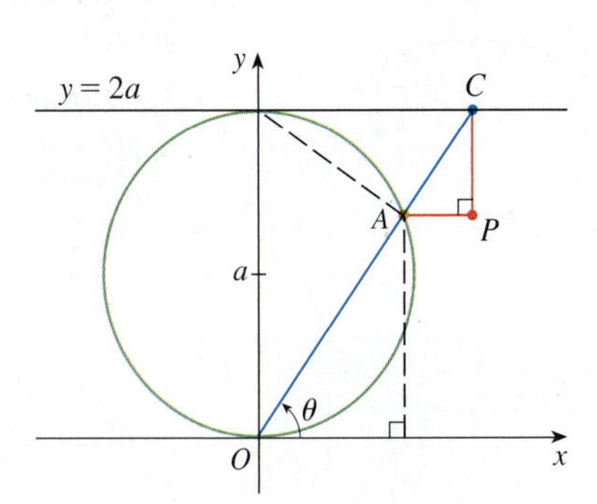

44. (a) Find parametric equations for the set of all points P as shown in the figure such that $|OP| = |AB|$. (This curve is called the **cissoid of Diocles** after the Greek scholar Diocles, who introduced the cissoid as a graphical method for constructing the edge of a cube whose volume is twice that of a given cube.)

(b) Use the geometric description of the curve to draw a rough sketch of the curve by hand. Check your work by using the parametric equations to graph the curve.

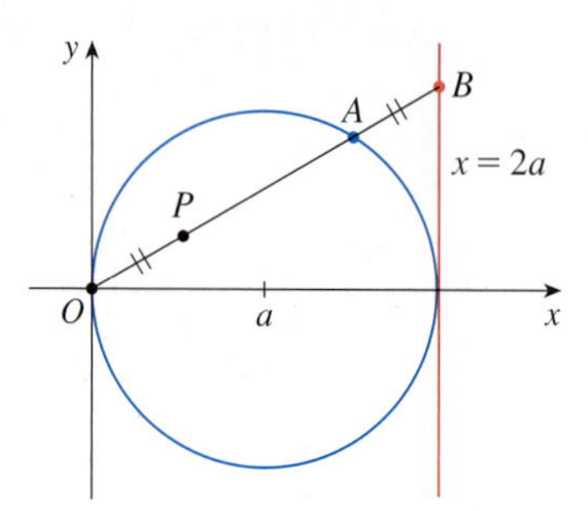

45. Suppose that the position of one particle at time t is given by

$$x_1 = 3 \sin t \qquad y_1 = 2 \cos t \qquad 0 \leqslant t \leqslant 2\pi$$

and the position of a second particle is given by

$$x_2 = -3 + \cos t \qquad y_2 = 1 + \sin t \qquad 0 \leqslant t \leqslant 2\pi$$

(a) Graph the paths of both particles. How many points of intersection are there?

(b) Are any of these points of intersection *collision points*? In other words, are the particles ever at the same place at the same time? If so, find the collision points.

(c) Describe what happens if the path of the second particle is given by

$$x_2 = 3 + \cos t \qquad y_2 = 1 + \sin t \qquad 0 \leqslant t \leqslant 2\pi$$

46. If a projectile is fired with an initial velocity of v_0 meters per second at an angle α above the horizontal and air resistance is assumed to be negligible, then its position after t seconds is given by the parametric equations

$$x = (v_0 \cos \alpha)t \qquad y = (v_0 \sin \alpha)t - \tfrac{1}{2}gt^2$$

where g is the acceleration due to gravity (9.8 m/s^2).

(a) If a gun is fired with $\alpha = 30°$ and $v_0 = 500$ m/s, when will the bullet hit the ground? How far from the gun will it hit the ground? What is the maximum height reached by the bullet?

(b) Use a graphing device to check your answers to part (a). Then graph the path of the projectile for several other values of the angle α to see where it hits the ground. Summarize your findings.

(c) Show that the path is parabolic by eliminating the parameter.

47. Investigate the family of curves defined by the parametric equations $x = t^2$, $y = t^3 - ct$. How does the shape change as c increases? Illustrate by graphing several members of the family.

48. The **swallowtail catastrophe curves** are defined by the parametric equations $x = 2ct - 4t^3$, $y = -ct^2 + 3t^4$. Graph several of these curves. What features do the curves have in common? How do they change when c increases?

49. Graph several members of the family of curves with parametric equations $x = t + a \cos t$, $y = t + a \sin t$, where $a > 0$. How does the shape change as a increases? For what values of a does the curve have a loop?

50. Graph several members of the family of curves $x = \sin t + \sin nt$, $y = \cos t + \cos nt$, where n is a positive integer. What features do the curves have in common? What happens as n increases?

51. The curves with equations $x = a \sin nt$, $y = b \cos t$ are called **Lissajous figures**. Investigate how these curves vary when a, b, and n vary. (Take n to be a positive integer.)

52. Investigate the family of curves defined by the parametric equations $x = \cos t$, $y = \sin t - \sin ct$, where $c > 0$. Start by letting c be a positive integer and see what happens to the shape as c increases. Then explore some of the possibilities that occur when c is a fraction.

LABORATORY PROJECT RUNNING CIRCLES AROUND CIRCLES

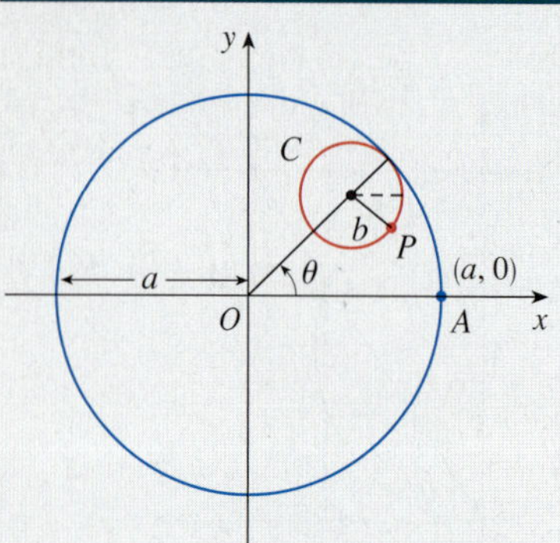

In this project we investigate families of curves, called *hypocycloids* and *epicycloids*, that are generated by the motion of a point on a circle that rolls inside or outside another circle.

1. A **hypocycloid** is a curve traced out by a fixed point P on a circle C of radius b as C rolls on the inside of a circle with center O and radius a. Show that if the initial position of P is $(a, 0)$ and the parameter θ is chosen as in the figure, then parametric equations of the hypocycloid are

$$x = (a - b)\cos\theta + b\cos\left(\frac{a-b}{b}\theta\right) \qquad y = (a - b)\sin\theta - b\sin\left(\frac{a-b}{b}\theta\right)$$

2. Use a graphing device (or the interactive graphic in TEC Module 10.1B) to draw the graphs of hypocycloids with a a positive integer and $b = 1$. How does the value of a affect the

TEC Look at Module 10.1B to see how hypocycloids and epicycloids are formed by the motion of rolling circles.

graph? Show that if we take $a = 4$, then the parametric equations of the hypocycloid reduce to

$$x = 4\cos^3\theta \qquad y = 4\sin^3\theta$$

This curve is called a **hypocycloid of four cusps**, or an **astroid**.

3. Now try $b = 1$ and $a = n/d$, a fraction where n and d have no common factor. First let $n = 1$ and try to determine graphically the effect of the denominator d on the shape of the graph. Then let n vary while keeping d constant. What happens when $n = d + 1$?

4. What happens if $b = 1$ and a is irrational? Experiment with an irrational number like $\sqrt{2}$ or $e - 2$. Take larger and larger values for θ and speculate on what would happen if we were to graph the hypocycloid for all real values of θ.

5. If the circle C rolls on the *outside* of the fixed circle, the curve traced out by P is called an **epicycloid**. Find parametric equations for the epicycloid.

6. Investigate the possible shapes for epicycloids. Use methods similar to Problems 2–4.

10.2 Calculus with Parametric Curves

Having seen how to represent curves by parametric equations, we now apply the methods of calculus to these parametric curves. In particular, we solve problems involving tangents, areas, arc length, and surface area.

Tangents

Suppose f and g are differentiable functions and we want to find the tangent line at a point on the parametric curve $x = f(t)$, $y = g(t)$, where y is also a differentiable function of x. Then the Chain Rule gives

$$\frac{dy}{dt} = \frac{dy}{dx} \cdot \frac{dx}{dt}$$

If $dx/dt \neq 0$, we can solve for dy/dx:

If we think of the curve as being traced out by a moving particle, then dy/dt and dx/dt are the vertical and horizontal velocities of the particle and Formula 1 says that the slope of the tangent is the ratio of these velocities.

1
$$\frac{dy}{dx} = \frac{\dfrac{dy}{dt}}{\dfrac{dx}{dt}} \qquad \text{if} \quad \frac{dx}{dt} \neq 0$$

Equation 1 (which you can remember by thinking of canceling the dt's) enables us to find the slope dy/dx of the tangent to a parametric curve without having to eliminate the parameter t. We see from (1) that the curve has a horizontal tangent when $dy/dt = 0$ (provided that $dx/dt \neq 0$) and it has a vertical tangent when $dx/dt = 0$ (provided that $dy/dt \neq 0$). This information is useful for sketching parametric curves.

As we know from Chapter 4, it is also useful to consider d^2y/dx^2. This can be found by replacing y by dy/dx in Equation 1:

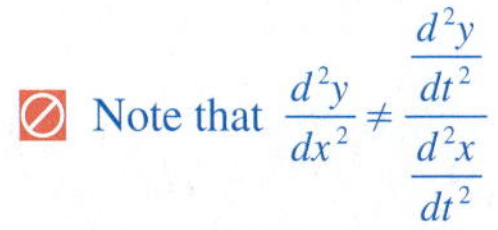
Note that $\dfrac{d^2y}{dx^2} \neq \dfrac{\dfrac{d^2y}{dt^2}}{\dfrac{d^2x}{dt^2}}$

$$\frac{d^2y}{dx^2} = \frac{d}{dx}\left(\frac{dy}{dx}\right) = \frac{\dfrac{d}{dt}\left(\dfrac{dy}{dx}\right)}{\dfrac{dx}{dt}}$$

EXAMPLE 1 A curve C is defined by the parametric equations $x = t^2$, $y = t^3 - 3t$.
(a) Show that C has two tangents at the point (3, 0) and find their equations.
(b) Find the points on C where the tangent is horizontal or vertical.
(c) Determine where the curve is concave upward or downward.
(d) Sketch the curve.

SOLUTION
(a) Notice that $y = t^3 - 3t = t(t^2 - 3) = 0$ when $t = 0$ or $t = \pm\sqrt{3}$. Therefore the point (3, 0) on C arises from two values of the parameter, $t = \sqrt{3}$ and $t = -\sqrt{3}$. This indicates that C crosses itself at (3, 0). Since

$$\frac{dy}{dx} = \frac{dy/dt}{dx/dt} = \frac{3t^2 - 3}{2t} = \frac{3}{2}\left(t - \frac{1}{t}\right)$$

the slope of the tangent when $t = \pm\sqrt{3}$ is $dy/dx = \pm 6/(2\sqrt{3}) = \pm\sqrt{3}$, so the equations of the tangents at (3, 0) are

$$y = \sqrt{3}\,(x - 3) \qquad \text{and} \qquad y = -\sqrt{3}\,(x - 3)$$

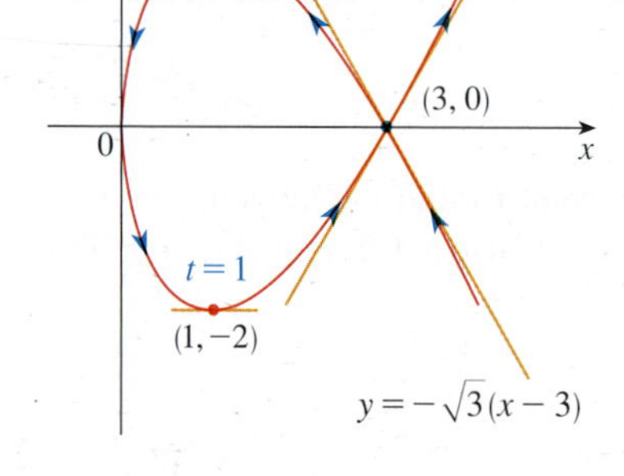

FIGURE 1

(b) C has a horizontal tangent when $dy/dx = 0$, that is, when $dy/dt = 0$ and $dx/dt \neq 0$. Since $dy/dt = 3t^2 - 3$, this happens when $t^2 = 1$, that is, $t = \pm 1$. The corresponding points on C are $(1, -2)$ and $(1, 2)$. C has a vertical tangent when $dx/dt = 2t = 0$, that is, $t = 0$. (Note that $dy/dt \neq 0$ there.) The corresponding point on C is (0, 0).

(c) To determine concavity we calculate the second derivative:

$$\frac{d^2y}{dx^2} = \frac{\dfrac{d}{dt}\left(\dfrac{dy}{dx}\right)}{\dfrac{dx}{dt}} = \frac{\dfrac{3}{2}\left(1 + \dfrac{1}{t^2}\right)}{2t} = \frac{3(t^2 + 1)}{4t^3}$$

Thus the curve is concave upward when $t > 0$ and concave downward when $t < 0$.

(d) Using the information from parts (b) and (c), we sketch C in Figure 1. ■

EXAMPLE 2
(a) Find the tangent to the cycloid $x = r(\theta - \sin\theta)$, $y = r(1 - \cos\theta)$ at the point where $\theta = \pi/3$. (See Example 10.1.7.)
(b) At what points is the tangent horizontal? When is it vertical?

SOLUTION
(a) The slope of the tangent line is

$$\frac{dy}{dx} = \frac{dy/d\theta}{dx/d\theta} = \frac{r\sin\theta}{r(1 - \cos\theta)} = \frac{\sin\theta}{1 - \cos\theta}$$

When $\theta = \pi/3$, we have

$$x = r\left(\frac{\pi}{3} - \sin\frac{\pi}{3}\right) = r\left(\frac{\pi}{3} - \frac{\sqrt{3}}{2}\right) \qquad y = r\left(1 - \cos\frac{\pi}{3}\right) = \frac{r}{2}$$

and

$$\frac{dy}{dx} = \frac{\sin(\pi/3)}{1 - \cos(\pi/3)} = \frac{\sqrt{3}/2}{1 - \frac{1}{2}} = \sqrt{3}$$

Therefore the slope of the tangent is $\sqrt{3}$ and its equation is

$$y - \frac{r}{2} = \sqrt{3}\left(x - \frac{r\pi}{3} + \frac{r\sqrt{3}}{2}\right) \qquad \text{or} \qquad \sqrt{3}\,x - y = r\left(\frac{\pi}{\sqrt{3}} - 2\right)$$

The tangent is sketched in Figure 2.

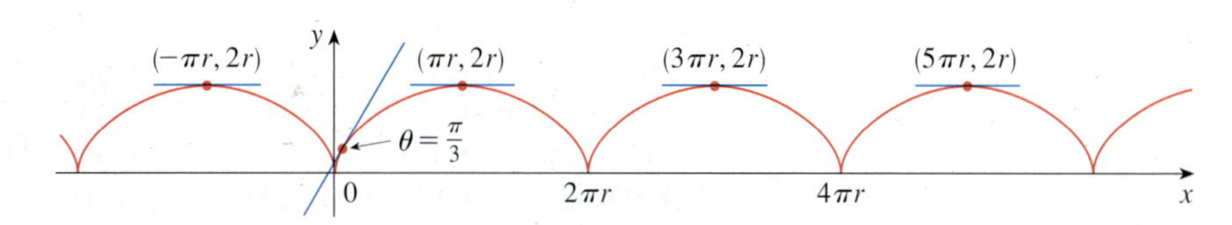

FIGURE 2

(b) The tangent is horizontal when $dy/dx = 0$, which occurs when $\sin\theta = 0$ and $1 - \cos\theta \neq 0$, that is, $\theta = (2n - 1)\pi$, n an integer. The corresponding point on the cycloid is $((2n - 1)\pi r, 2r)$.

When $\theta = 2n\pi$, both $dx/d\theta$ and $dy/d\theta$ are 0. It appears from the graph that there are vertical tangents at those points. We can verify this by using l'Hospital's Rule as follows:

$$\lim_{\theta \to 2n\pi^+} \frac{dy}{dx} = \lim_{\theta \to 2n\pi^+} \frac{\sin\theta}{1 - \cos\theta} = \lim_{\theta \to 2n\pi^+} \frac{\cos\theta}{\sin\theta} = \infty$$

A similar computation shows that $dy/dx \to -\infty$ as $\theta \to 2n\pi^-$, so indeed there are vertical tangents when $\theta = 2n\pi$, that is, when $x = 2n\pi r$. ■

Areas

We know that the area under a curve $y = F(x)$ from a to b is $A = \int_a^b F(x)\,dx$, where $F(x) \geqslant 0$. If the curve is traced out once by the parametric equations $x = f(t)$ and $y = g(t)$, $\alpha \leqslant t \leqslant \beta$, then we can calculate an area formula by using the Substitution Rule for Definite Integrals as follows:

The limits of integration for t are found as usual with the Substitution Rule. When $x = a$, t is either α or β. When $x = b$, t is the remaining value.

$$A = \int_a^b y\,dx = \int_\alpha^\beta g(t)f'(t)\,dt \qquad \left[\text{or } \int_\beta^\alpha g(t)f'(t)\,dt\right]$$

EXAMPLE 3 Find the area under one arch of the cycloid

$$x = r(\theta - \sin\theta) \qquad y = r(1 - \cos\theta)$$

(See Figure 3.)

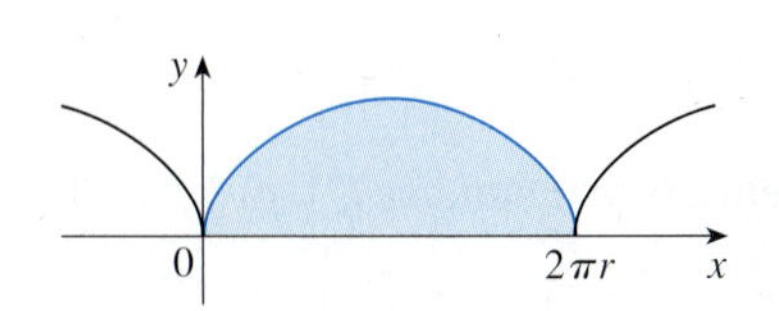

FIGURE 3

SOLUTION One arch of the cycloid is given by $0 \leqslant \theta \leqslant 2\pi$. Using the Substitution Rule with $y = r(1 - \cos\theta)$ and $dx = r(1 - \cos\theta)\,d\theta$, we have

$$\begin{aligned} A &= \int_0^{2\pi r} y\,dx = \int_0^{2\pi} r(1 - \cos\theta)\,r(1 - \cos\theta)\,d\theta \\ &= r^2 \int_0^{2\pi} (1 - \cos\theta)^2\,d\theta = r^2 \int_0^{2\pi} (1 - 2\cos\theta + \cos^2\theta)\,d\theta \\ &= r^2 \int_0^{2\pi} \left[1 - 2\cos\theta + \tfrac{1}{2}(1 + \cos 2\theta)\right] d\theta \\ &= r^2 \left[\tfrac{3}{2}\theta - 2\sin\theta + \tfrac{1}{4}\sin 2\theta\right]_0^{2\pi} \\ &= r^2\left(\tfrac{3}{2} \cdot 2\pi\right) = 3\pi r^2 \end{aligned}$$ ■

The result of Example 3 says that the area under one arch of the cycloid is three times the area of the rolling circle that generates the cycloid (see Example 10.1.7). Galileo guessed this result but it was first proved by the French mathematician Roberval and the Italian mathematician Torricelli.

Arc Length

We already know how to find the length L of a curve C given in the form $y = F(x)$, $a \leqslant x \leqslant b$. Formula 8.1.3 says that if F' is continuous, then

2 $$L = \int_a^b \sqrt{1 + \left(\frac{dy}{dx}\right)^2}\, dx$$

Suppose that C can also be described by the parametric equations $x = f(t)$ and $y = g(t)$, $\alpha \leqslant t \leqslant \beta$, where $dx/dt = f'(t) > 0$. This means that C is traversed once, from left to right, as t increases from α to β and $f(\alpha) = a$, $f(\beta) = b$. Putting Formula 1 into Formula 2 and using the Substitution Rule, we obtain

$$L = \int_a^b \sqrt{1 + \left(\frac{dy}{dx}\right)^2}\, dx = \int_\alpha^\beta \sqrt{1 + \left(\frac{dy/dt}{dx/dt}\right)^2}\, \frac{dx}{dt}\, dt$$

Since $dx/dt > 0$, we have

3 $$L = \int_\alpha^\beta \sqrt{\left(\frac{dx}{dt}\right)^2 + \left(\frac{dy}{dt}\right)^2}\, dt$$

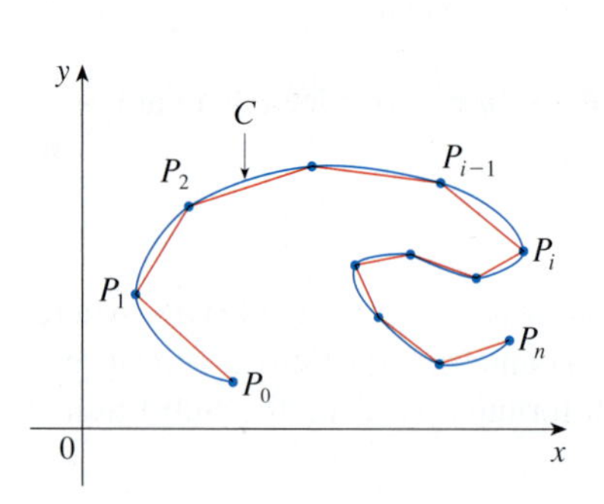

FIGURE 4

Even if C can't be expressed in the form $y = F(x)$, Formula 3 is still valid but we obtain it by polygonal approximations. We divide the parameter interval $[\alpha, \beta]$ into n subintervals of equal width Δt. If $t_0, t_1, t_2, \ldots, t_n$ are the endpoints of these subintervals, then $x_i = f(t_i)$ and $y_i = g(t_i)$ are the coordinates of points $P_i(x_i, y_i)$ that lie on C and the polygon with vertices $P_0, P_1, \ldots, P_n$ approximates C. (See Figure 4.)

As in Section 8.1, we define the length L of C to be the limit of the lengths of these approximating polygons as $n \to \infty$:

$$L = \lim_{n\to\infty} \sum_{i=1}^{n} |P_{i-1}P_i|$$

The Mean Value Theorem, when applied to f on the interval $[t_{i-1}, t_i]$, gives a number t_i^* in (t_{i-1}, t_i) such that

$$f(t_i) - f(t_{i-1}) = f'(t_i^*)(t_i - t_{i-1})$$

If we let $\Delta x_i = x_i - x_{i-1}$ and $\Delta y_i = y_i - y_{i-1}$, this equation becomes

$$\Delta x_i = f'(t_i^*)\, \Delta t$$

Similarly, when applied to g, the Mean Value Theorem gives a number t_i^{**} in (t_{i-1}, t_i) such that

$$\Delta y_i = g'(t_i^{**})\, \Delta t$$

Therefore

$$|P_{i-1}P_i| = \sqrt{(\Delta x_i)^2 + (\Delta y_i)^2} = \sqrt{[f'(t_i^*)\Delta t]^2 + [g'(t_i^{**})\Delta t]^2}$$
$$= \sqrt{[f'(t_i^*)]^2 + [g'(t_i^{**})]^2}\, \Delta t$$

and so

4 $$L = \lim_{n\to\infty} \sum_{i=1}^{n} \sqrt{[f'(t_i^*)]^2 + [g'(t_i^{**})]^2}\, \Delta t$$

The sum in (4) resembles a Riemann sum for the function $\sqrt{[f'(t)]^2 + [g'(t)]^2}$ but it is not exactly a Riemann sum because $t_i^* \neq t_i^{**}$ in general. Nevertheless, if f' and g' are continuous, it can be shown that the limit in (4) is the same as if t_i^* and t_i^{**} were equal, namely,

$$L = \int_\alpha^\beta \sqrt{[f'(t)]^2 + [g'(t)]^2}\, dt$$

Thus, using Leibniz notation, we have the following result, which has the same form as Formula 3.

5 Theorem If a curve C is described by the parametric equations $x = f(t)$, $y = g(t)$, $\alpha \leqslant t \leqslant \beta$, where f' and g' are continuous on $[\alpha, \beta]$ and C is traversed exactly once as t increases from α to β, then the length of C is

$$L = \int_\alpha^\beta \sqrt{\left(\frac{dx}{dt}\right)^2 + \left(\frac{dy}{dt}\right)^2}\, dt$$

Notice that the formula in Theorem 5 is consistent with the general formulas $L = \int ds$ and $(ds)^2 = (dx)^2 + (dy)^2$ of Section 8.1.

EXAMPLE 4 If we use the representation of the unit circle given in Example 10.1.2,

$$x = \cos t \qquad y = \sin t \qquad 0 \leqslant t \leqslant 2\pi$$

then $dx/dt = -\sin t$ and $dy/dt = \cos t$, so Theorem 5 gives

$$L = \int_0^{2\pi} \sqrt{\left(\frac{dx}{dt}\right)^2 + \left(\frac{dy}{dt}\right)^2}\, dt = \int_0^{2\pi} \sqrt{\sin^2 t + \cos^2 t}\, dt = \int_0^{2\pi} dt = 2\pi$$

as expected. If, on the other hand, we use the representation given in Example 10.1.3,

$$x = \sin 2t \qquad y = \cos 2t \qquad 0 \leqslant t \leqslant 2\pi$$

then $dx/dt = 2\cos 2t$, $dy/dt = -2\sin 2t$, and the integral in Theorem 5 gives

$$\int_0^{2\pi} \sqrt{\left(\frac{dx}{dt}\right)^2 + \left(\frac{dy}{dt}\right)^2}\, dt = \int_0^{2\pi} \sqrt{4\cos^2 2t + 4\sin^2 2t}\, dt = \int_0^{2\pi} 2\, dt = 4\pi$$

Notice that the integral gives twice the arc length of the circle because as t increases from 0 to 2π, the point $(\sin 2t, \cos 2t)$ traverses the circle twice. In general, when finding the length of a curve C from a parametric representation, we have to be careful to ensure that C is traversed only once as t increases from α to β. ■

EXAMPLE 5 Find the length of one arch of the cycloid $x = r(\theta - \sin\theta)$, $y = r(1 - \cos\theta)$.

SOLUTION From Example 3 we see that one arch is described by the parameter interval $0 \leqslant \theta \leqslant 2\pi$. Since

$$\frac{dx}{d\theta} = r(1 - \cos\theta) \qquad \text{and} \qquad \frac{dy}{d\theta} = r\sin\theta$$

we have

$$L = \int_0^{2\pi} \sqrt{\left(\frac{dx}{d\theta}\right)^2 + \left(\frac{dy}{d\theta}\right)^2}\, d\theta$$

$$= \int_0^{2\pi} \sqrt{r^2(1 - \cos\theta)^2 + r^2\sin^2\theta}\, d\theta$$

$$= \int_0^{2\pi} \sqrt{r^2(1 - 2\cos\theta + \cos^2\theta + \sin^2\theta)}\, d\theta$$

$$= r\int_0^{2\pi} \sqrt{2(1 - \cos\theta)}\, d\theta$$

The result of Example 5 says that the length of one arch of a cycloid is eight times the radius of the generating circle (see Figure 5). This was first proved in 1658 by Sir Christopher Wren, who later became the architect of St. Paul's Cathedral in London.

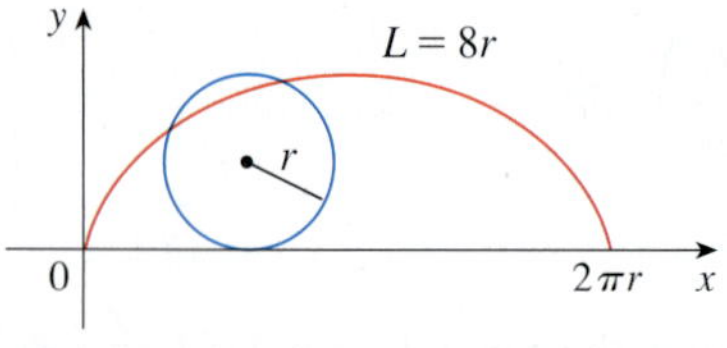

FIGURE 5

To evaluate this integral we use the identity $\sin^2 x = \frac{1}{2}(1 - \cos 2x)$ with $\theta = 2x$, which gives $1 - \cos\theta = 2\sin^2(\theta/2)$. Since $0 \leqslant \theta \leqslant 2\pi$, we have $0 \leqslant \theta/2 \leqslant \pi$ and so $\sin(\theta/2) \geqslant 0$. Therefore

$$\sqrt{2(1 - \cos\theta)} = \sqrt{4\sin^2(\theta/2)} = 2|\sin(\theta/2)| = 2\sin(\theta/2)$$

and so

$$L = 2r\int_0^{2\pi} \sin(\theta/2)\, d\theta = 2r\Big[-2\cos(\theta/2)\Big]_0^{2\pi}$$

$$= 2r[2 + 2] = 8r$$

Surface Area

In the same way as for arc length, we can adapt Formula 8.2.5 to obtain a formula for surface area. Suppose the curve c given by the parametric equations $x = f(t)$, $y = g(t)$, $\alpha \leqslant t \leqslant \beta$, where f', g' are continuous, $g(t) \geqslant 0$, is rotated about the x-axis. If C is traversed exactly once as t increases from α to β, then the area of the resulting surface is given by

6
$$S = \int_\alpha^\beta 2\pi y \sqrt{\left(\frac{dx}{dt}\right)^2 + \left(\frac{dy}{dt}\right)^2}\, dt$$

The general symbolic formulas $S = \int 2\pi y\, ds$ and $S = \int 2\pi x\, ds$ (Formulas 8.2.7 and 8.2.8) are still valid, but for parametric curves we use

$$ds = \sqrt{\left(\frac{dx}{dt}\right)^2 + \left(\frac{dy}{dt}\right)^2}\, dt$$

EXAMPLE 6 Show that the surface area of a sphere of radius r is $4\pi r^2$.

SOLUTION The sphere is obtained by rotating the semicircle

$$x = r\cos t \qquad y = r\sin t \qquad 0 \leqslant t \leqslant \pi$$

about the x-axis. Therefore, from Formula 6, we get

$$S = \int_0^\pi 2\pi r\sin t\,\sqrt{(-r\sin t)^2 + (r\cos t)^2}\, dt$$

$$= 2\pi\int_0^\pi r\sin t\,\sqrt{r^2(\sin^2 t + \cos^2 t)}\, dt = 2\pi\int_0^\pi r\sin t \cdot r\, dt$$

$$= 2\pi r^2\int_0^\pi \sin t\, dt = 2\pi r^2(-\cos t)\Big]_0^\pi = 4\pi r^2$$

10.2 EXERCISES

1–2 Find dy/dx.

1. $x = \dfrac{t}{1+t}, \quad y = \sqrt{1+t}$

2. $x = te^t, \quad y = t + \sin t$

3–6 Find an equation of the tangent to the curve at the point corresponding to the given value of the parameter.

3. $x = t^3 + 1, \quad y = t^4 + t; \quad t = -1$

4. $x = \sqrt{t}, \quad y = t^2 - 2t; \quad t = 4$

5. $x = t \cos t, \quad y = t \sin t; \quad t = \pi$

6. $x = e^t \sin \pi t, \quad y = e^{2t}; \quad t = 0$

7–8 Find an equation of the tangent to the curve at the given point by two methods: (a) without eliminating the parameter and (b) by first eliminating the parameter.

7. $x = 1 + \ln t, \quad y = t^2 + 2; \quad (1, 3)$

8. $x = 1 + \sqrt{t}, \quad y = e^{t^2}; \quad (2, e)$

9–10 Find an equation of the tangent to the curve at the given point. Then graph the curve and the tangent.

9. $x = t^2 - t, \quad y = t^2 + t + 1; \quad (0, 3)$

10. $x = \sin \pi t, \quad y = t^2 + t; \quad (0, 2)$

11–16 Find dy/dx and d^2y/dx^2. For which values of t is the curve concave upward?

11. $x = t^2 + 1, \quad y = t^2 + t$

12. $x = t^3 + 1, \quad y = t^2 - t$

13. $x = e^t, \quad y = te^{-t}$

14. $x = t^2 + 1, \quad y = e^t - 1$

15. $x = t - \ln t, \quad y = t + \ln t$

16. $x = \cos t, \quad y = \sin 2t, \quad 0 < t < \pi$

17–20 Find the points on the curve where the tangent is horizontal or vertical. If you have a graphing device, graph the curve to check your work.

17. $x = t^3 - 3t, \quad y = t^2 - 3$

18. $x = t^3 - 3t, \quad y = t^3 - 3t^2$

19. $x = \cos \theta, \quad y = \cos 3\theta$

20. $x = e^{\sin \theta}, \quad y = e^{\cos \theta}$

21. Use a graph to estimate the coordinates of the rightmost point on the curve $x = t - t^6$, $y = e^t$. Then use calculus to find the exact coordinates.

22. Use a graph to estimate the coordinates of the lowest point and the leftmost point on the curve $x = t^4 - 2t$, $y = t + t^4$. Then find the exact coordinates.

23–24 Graph the curve in a viewing rectangle that displays all the important aspects of the curve.

23. $x = t^4 - 2t^3 - 2t^2, \quad y = t^3 - t$

24. $x = t^4 + 4t^3 - 8t^2, \quad y = 2t^2 - t$

25. Show that the curve $x = \cos t$, $y = \sin t \cos t$ has two tangents at $(0, 0)$ and find their equations. Sketch the curve.

26. Graph the curve $x = -2 \cos t$, $y = \sin t + \sin 2t$ to discover where it crosses itself. Then find equations of both tangents at that point.

27. (a) Find the slope of the tangent line to the trochoid $x = r\theta - d \sin \theta$, $y = r - d \cos \theta$ in terms of θ. (See Exercise 10.1.40.)
(b) Show that if $d < r$, then the trochoid does not have a vertical tangent.

28. (a) Find the slope of the tangent to the astroid $x = a \cos^3\theta$, $y = a \sin^3\theta$ in terms of θ. (Astroids are explored in the Laboratory Project on page 689.)
(b) At what points is the tangent horizontal or vertical?
(c) At what points does the tangent have slope 1 or -1?

29. At what point(s) on the curve $x = 3t^2 + 1$, $y = t^3 - 1$ does the tangent line have slope $\frac{1}{2}$?

30. Find equations of the tangents to the curve $x = 3t^2 + 1$, $y = 2t^3 + 1$ that pass through the point $(4, 3)$.

31. Use the parametric equations of an ellipse, $x = a \cos \theta$, $y = b \sin \theta$, $0 \leqslant \theta \leqslant 2\pi$, to find the area that it encloses.

32. Find the area enclosed by the curve $x = t^2 - 2t$, $y = \sqrt{t}$ and the y-axis.

33. Find the area enclosed by the x-axis and the curve $x = t^3 + 1$, $y = 2t - t^2$.

34. Find the area of the region enclosed by the astroid $x = a \cos^3\theta$, $y = a \sin^3\theta$. (Astroids are explored in the Laboratory Project on page 689.)

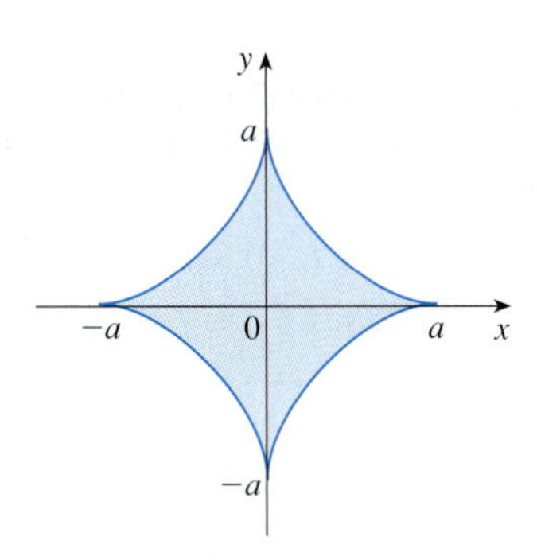

35. Find the area under one arch of the trochoid of Exercise 10.1.40 for the case $d < r$.

36. Let $\mathcal{R}$ be the region enclosed by the loop of the curve in Example 1.
(a) Find the area of $\mathcal{R}$.
(b) If $\mathcal{R}$ is rotated about the x-axis, find the volume of the resulting solid.
(c) Find the centroid of $\mathcal{R}$.

37–40 Set up an integral that represents the length of the curve. Then use your calculator to find the length correct to four decimal places.

37. $x = t + e^{-t}, \quad y = t - e^{-t}, \quad 0 \leqslant t \leqslant 2$

38. $x = t^2 - t, \quad y = t^4, \quad 1 \leqslant t \leqslant 4$

39. $x = t - 2 \sin t, \quad y = 1 - 2 \cos t, \quad 0 \leqslant t \leqslant 4\pi$

40. $x = t + \sqrt{t}, \quad y = t - \sqrt{t}, \quad 0 \leqslant t \leqslant 1$

41–44 Find the exact length of the curve.

41. $x = 1 + 3t^2, \quad y = 4 + 2t^3, \quad 0 \leqslant t \leqslant 1$

42. $x = e^t - t, \quad y = 4e^{t/2}, \quad 0 \leqslant t \leqslant 2$

43. $x = t \sin t, \quad y = t \cos t, \quad 0 \leqslant t \leqslant 1$

44. $x = 3 \cos t - \cos 3t, \quad y = 3 \sin t - \sin 3t, \quad 0 \leqslant t \leqslant \pi$

45–46 Graph the curve and find its exact length.

45. $x = e^t \cos t, \quad y = e^t \sin t, \quad 0 \leqslant t \leqslant \pi$

46. $x = \cos t + \ln(\tan \frac{1}{2}t), \quad y = \sin t, \quad \pi/4 \leqslant t \leqslant 3\pi/4$

47. Graph the curve $x = \sin t + \sin 1.5t$, $y = \cos t$ and find its length correct to four decimal places.

48. Find the length of the loop of the curve $x = 3t - t^3$, $y = 3t^2$.

49. Use Simpson's Rule with $n = 6$ to estimate the length of the curve $x = t - e^t$, $y = t + e^t$, $-6 \leqslant t \leqslant 6$.

50. In Exercise 10.1.43 you were asked to derive the parametric equations $x = 2a \cot \theta$, $y = 2a \sin^2\theta$ for the curve called the witch of Maria Agnesi. Use Simpson's Rule with $n = 4$ to estimate the length of the arc of this curve given by $\pi/4 \leqslant \theta \leqslant \pi/2$.

51–52 Find the distance traveled by a particle with position (x, y) as t varies in the given time interval. Compare with the length of the curve.

51. $x = \sin^2 t, \quad y = \cos^2 t, \quad 0 \leqslant t \leqslant 3\pi$

52. $x = \cos^2 t, \quad y = \cos t, \quad 0 \leqslant t \leqslant 4\pi$

53. Show that the total length of the ellipse $x = a \sin \theta$, $y = b \cos \theta$, $a > b > 0$, is

$$L = 4a \int_0^{\pi/2} \sqrt{1 - e^2 \sin^2\theta}\, d\theta$$

where e is the eccentricity of the ellipse $\left(e = c/a, \text{ where } c = \sqrt{a^2 - b^2}\,\right)$.

54. Find the total length of the astroid $x = a \cos^3\theta$, $y = a \sin^3\theta$, where $a > 0$.

CAS **55.** (a) Graph the **epitrochoid** with equations

$$x = 11 \cos t - 4 \cos(11t/2)$$
$$y = 11 \sin t - 4 \sin(11t/2)$$

What parameter interval gives the complete curve?
(b) Use your CAS to find the approximate length of this curve.

CAS **56.** A curve called **Cornu's spiral** is defined by the parametric equations

$$x = C(t) = \int_0^t \cos(\pi u^2/2)\, du$$
$$y = S(t) = \int_0^t \sin(\pi u^2/2)\, du$$

where C and S are the Fresnel functions that were introduced in Chapter 4.
(a) Graph this curve. What happens as $t \to \infty$ and as $t \to -\infty$?
(b) Find the length of Cornu's spiral from the origin to the point with parameter value t.

57–60 Set up an integral that represents the area of the surface obtained by rotating the given curve about the x-axis. Then use your calculator to find the surface area correct to four decimal places.

57. $x = t \sin t, \quad y = t \cos t, \quad 0 \leqslant t \leqslant \pi/2$

58. $x = \sin t, \quad y = \sin 2t, \quad 0 \leqslant t \leqslant \pi/2$

59. $x = t + e^t, \quad y = e^{-t}, \quad 0 \leqslant t \leqslant 1$

60. $x = t^2 - t^3, \quad y = t + t^4, \quad 0 \leqslant t \leqslant 1$

61–63 Find the exact area of the surface obtained by rotating the given curve about the x-axis.

61. $x = t^3, \quad y = t^2, \quad 0 \leqslant t \leqslant 1$

62. $x = 2t^2 + 1/t, \quad y = 8\sqrt{t}, \quad 1 \leqslant t \leqslant 3$

63. $x = a \cos^3\theta, \quad y = a \sin^3\theta, \quad 0 \leqslant \theta \leqslant \pi/2$

64. Graph the curve

$$x = 2 \cos \theta - \cos 2\theta \qquad y = 2 \sin \theta - \sin 2\theta$$

If this curve is rotated about the x-axis, find the exact area of the resulting surface. (Use your graph to help find the correct parameter interval.)

65–66 Find the surface area generated by rotating the given curve about the y-axis.

65. $x = 3t^2, \quad y = 2t^3, \quad 0 \leqslant t \leqslant 5$

66. $x = e^t - t, \quad y = 4e^{t/2}, \quad 0 \leqslant t \leqslant 1$

---

67. If f' is continuous and $f'(t) \neq 0$ for $a \leqslant t \leqslant b$, show that the parametric curve $x = f(t)$, $y = g(t)$, $a \leqslant t \leqslant b$, can be put in the form $y = F(x)$. [*Hint:* Show that f^{-1} exists.]

68. Use Formula 1 to derive Formula 6 from Formula 8.2.5 for the case in which the curve can be represented in the form $y = F(x)$, $a \leqslant x \leqslant b$.

69. The **curvature** at a point P of a curve is defined as

$$\kappa = \left| \frac{d\phi}{ds} \right|$$

where ϕ is the angle of inclination of the tangent line at P, as shown in the figure. Thus the curvature is the absolute value of the rate of change of ϕ with respect to arc length. It can be regarded as a measure of the rate of change of direction of the curve at P and will be studied in greater detail in Chapter 13.

(a) For a parametric curve $x = x(t)$, $y = y(t)$, derive the formula

$$\kappa = \frac{|\dot{x}\ddot{y} - \ddot{x}\dot{y}|}{[\dot{x}^2 + \dot{y}^2]^{3/2}}$$

where the dots indicate derivatives with respect to t, so $\dot{x} = dx/dt$. [*Hint:* Use $\phi = \tan^{-1}(dy/dx)$ and Formula 2 to find $d\phi/dt$. Then use the Chain Rule to find $d\phi/ds$.]

(b) By regarding a curve $y = f(x)$ as the parametric curve $x = x$, $y = f(x)$, with parameter x, show that the formula in part (a) becomes

$$\kappa = \frac{|d^2y/dx^2|}{[1 + (dy/dx)^2]^{3/2}}$$

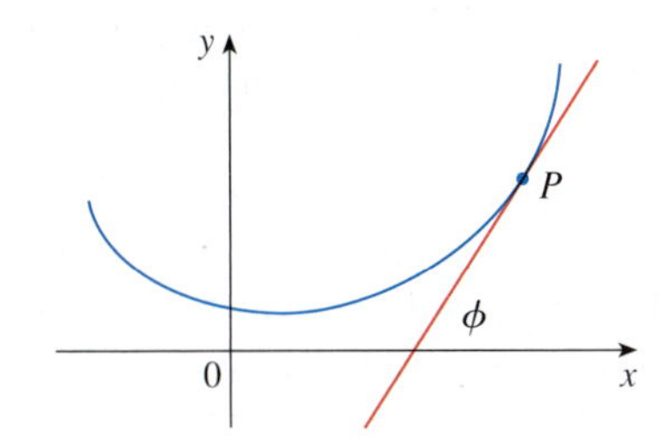

70. (a) Use the formula in Exercise 69(b) to find the curvature of the parabola $y = x^2$ at the point $(1, 1)$.

(b) At what point does this parabola have maximum curvature?

71. Use the formula in Exercise 69(a) to find the curvature of the cycloid $x = \theta - \sin\theta$, $y = 1 - \cos\theta$ at the top of one of its arches.

72. (a) Show that the curvature at each point of a straight line is $\kappa = 0$.

(b) Show that the curvature at each point of a circle of radius r is $\kappa = 1/r$.

73. A string is wound around a circle and then unwound while being held taut. The curve traced by the point P at the end of the string is called the **involute** of the circle. If the circle has radius r and center O and the initial position of P is $(r, 0)$, and if the parameter θ is chosen as in the figure, show that parametric equations of the involute are

$$x = r(\cos\theta + \theta\sin\theta) \qquad y = r(\sin\theta - \theta\cos\theta)$$

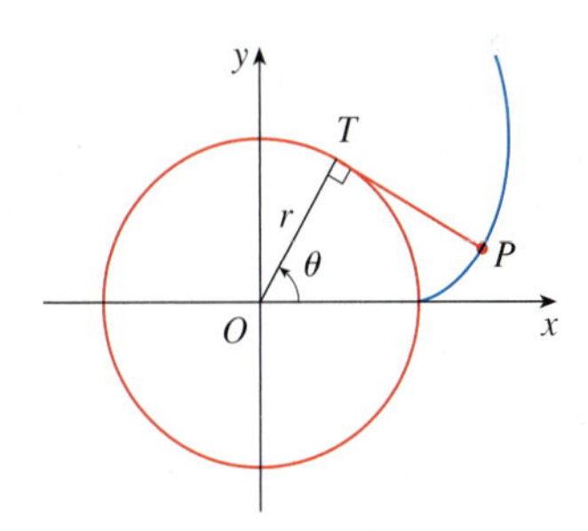

74. A cow is tied to a silo with radius r by a rope just long enough to reach the opposite side of the silo. Find the grazing area available for the cow.

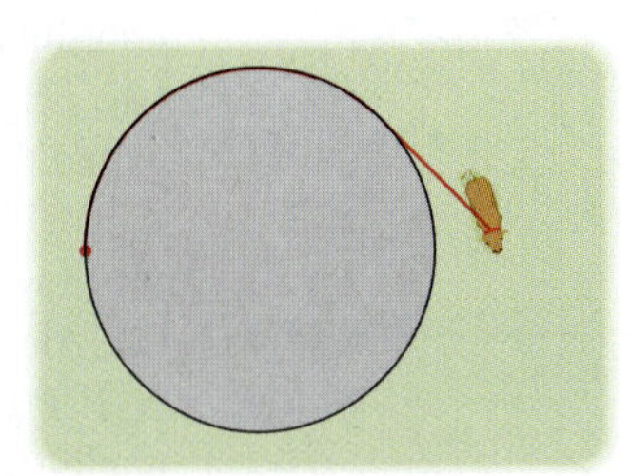

LABORATORY PROJECT BÉZIER CURVES

Bézier curves are used in computer-aided design and are named after the French mathematician Pierre Bézier (1910–1999), who worked in the automotive industry. A cubic Bézier curve is determined by four *control points*, $P_0(x_0, y_0)$, $P_1(x_1, y_1)$, $P_2(x_2, y_2)$, and $P_3(x_3, y_3)$, and is defined by the parametric equations

$$x = x_0(1 - t)^3 + 3x_1t(1 - t)^2 + 3x_2t^2(1 - t) + x_3t^3$$

$$y = y_0(1 - t)^3 + 3y_1t(1 - t)^2 + 3y_2t^2(1 - t) + y_3t^3$$

where $0 \leqslant t \leqslant 1$. Notice that when $t = 0$ we have $(x, y) = (x_0, y_0)$ and when $t = 1$ we have $(x, y) = (x_3, y_3)$, so the curve starts at P_0 and ends at P_3.

1. Graph the Bézier curve with control points $P_0(4, 1)$, $P_1(28, 48)$, $P_2(50, 42)$, and $P_3(40, 5)$. Then, on the same screen, graph the line segments P_0P_1, P_1P_2, and P_2P_3. (Exercise 10.1.31 shows how to do this.) Notice that the middle control points P_1 and P_2 don't lie on the curve; the curve starts at P_0, heads toward P_1 and P_2 without reaching them, and ends at P_3.

2. From the graph in Problem 1, it appears that the tangent at P_0 passes through P_1 and the tangent at P_3 passes through P_2. Prove it.

3. Try to produce a Bézier curve with a loop by changing the second control point in Problem 1.

4. Some laser printers use Bézier curves to represent letters and other symbols. Experiment with control points until you find a Bézier curve that gives a reasonable representation of the letter C.

5. More complicated shapes can be represented by piecing together two or more Bézier curves. Suppose the first Bézier curve has control points P_0, P_1, P_2, P_3 and the second one has control points P_3, P_4, P_5, P_6. If we want these two pieces to join together smoothly, then the tangents at P_3 should match and so the points P_2, P_3, and P_4 all have to lie on this common tangent line. Using this principle, find control points for a pair of Bézier curves that represent the letter S.

10.3 Polar Coordinates

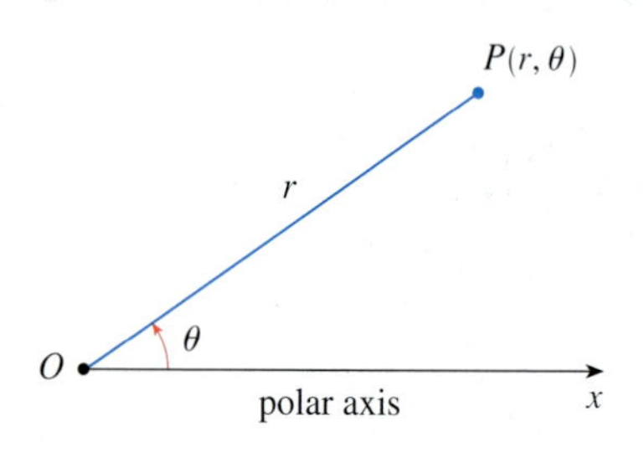

FIGURE 1

A coordinate system represents a point in the plane by an ordered pair of numbers called coordinates. Usually we use Cartesian coordinates, which are directed distances from two perpendicular axes. Here we describe a coordinate system introduced by Newton, called the **polar coordinate system**, which is more convenient for many purposes.

We choose a point in the plane that is called the **pole** (or origin) and is labeled O. Then we draw a ray (half-line) starting at O called the **polar axis**. This axis is usually drawn horizontally to the right and corresponds to the positive x-axis in Cartesian coordinates.

If P is any other point in the plane, let r be the distance from O to P and let θ be the angle (usually measured in radians) between the polar axis and the line OP as in Figure 1. Then the point P is represented by the ordered pair (r, θ) and r, θ are called **polar coordinates** of P. We use the convention that an angle is positive if measured in the counterclockwise direction from the polar axis and negative in the clockwise direction. If $P = O$, then $r = 0$ and we agree that $(0, \theta)$ represents the pole for any value of θ.

We extend the meaning of polar coordinates (r, θ) to the case in which r is negative by agreeing that, as in Figure 2, the points $(-r, \theta)$ and (r, θ) lie on the same line through O and at the same distance $|r|$ from O, but on opposite sides of O. If $r > 0$, the point (r, θ) lies in the same quadrant as θ; if $r < 0$, it lies in the quadrant on the opposite side of the pole. Notice that $(-r, \theta)$ represents the same point as $(r, \theta + \pi)$.

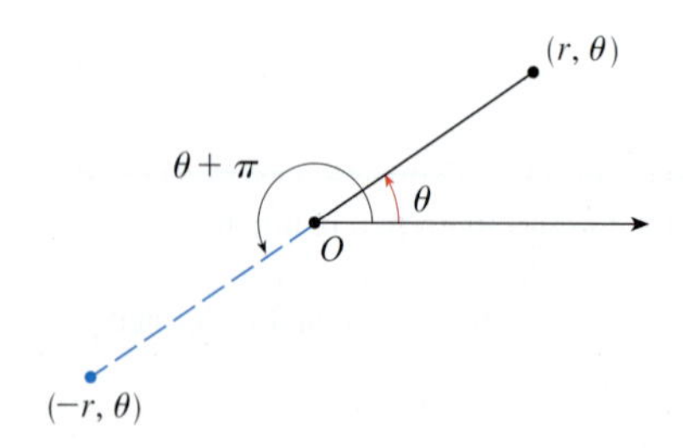

FIGURE 2

EXAMPLE 1 Plot the points whose polar coordinates are given.
(a) $(1, 5\pi/4)$ (b) $(2, 3\pi)$ (c) $(2, -2\pi/3)$ (d) $(-3, 3\pi/4)$

SOLUTION The points are plotted in Figure 3. In part (d) the point $(-3, 3\pi/4)$ is located three units from the pole in the fourth quadrant because the angle $3\pi/4$ is in the second quadrant and $r = -3$ is negative.

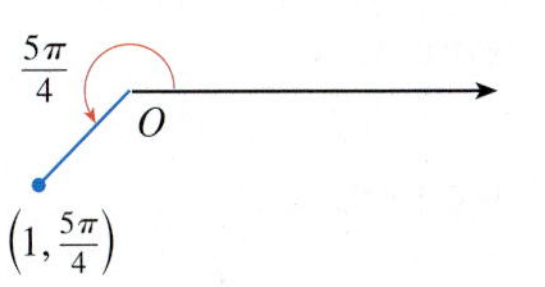

3π (2, 3π) O
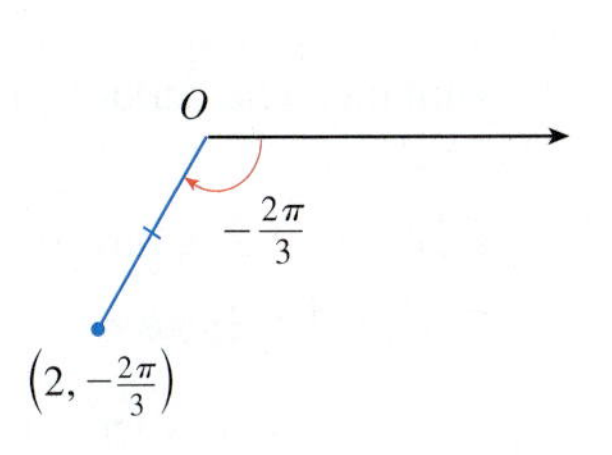

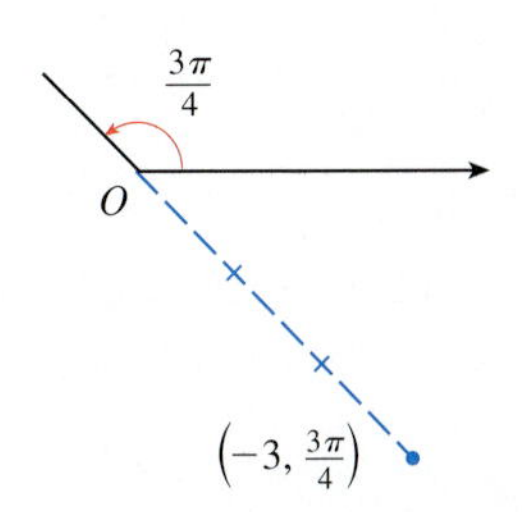

FIGURE 3

In the Cartesian coordinate system every point has only one representation, but in the polar coordinate system each point has many representations. For instance, the point $(1, 5\pi/4)$ in Example 1(a) could be written as $(1, -3\pi/4)$ or $(1, 13\pi/4)$ or $(-1, \pi/4)$. (See Figure 4.)

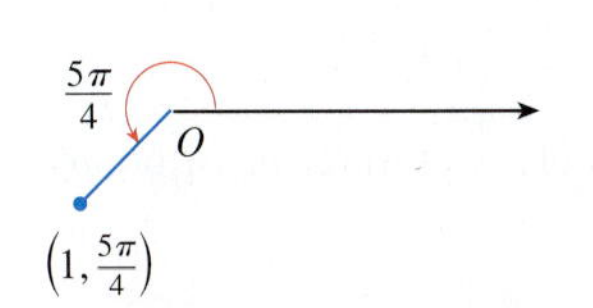

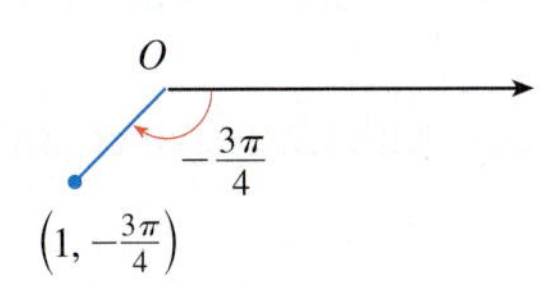

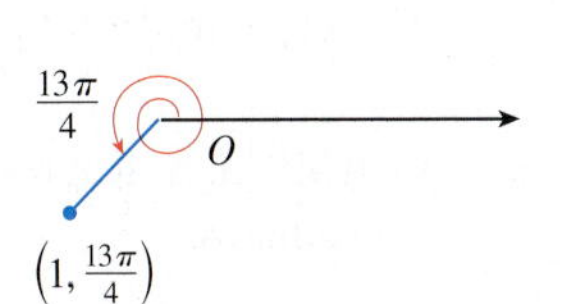

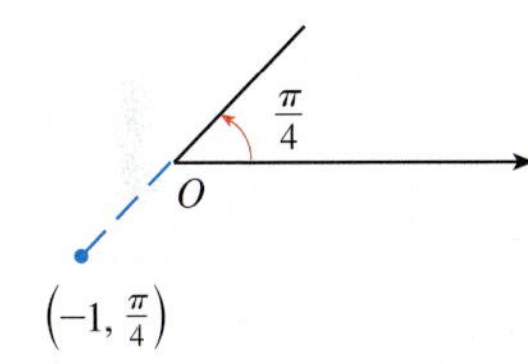

FIGURE 4

In fact, since a complete counterclockwise rotation is given by an angle 2π, the point represented by polar coordinates (r, θ) is also represented by

$$(r, \theta + 2n\pi) \qquad \text{and} \qquad (-r, \theta + (2n + 1)\pi)$$

where n is any integer.

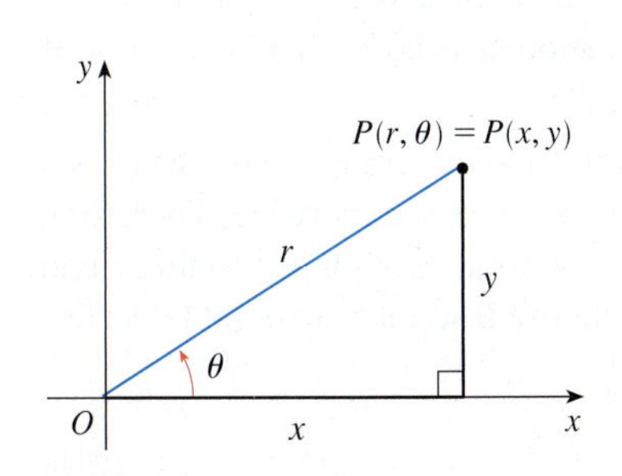

FIGURE 5

The connection between polar and Cartesian coordinates can be seen from Figure 5, in which the pole corresponds to the origin and the polar axis coincides with the positive x-axis. If the point P has Cartesian coordinates (x, y) and polar coordinates (r, θ), then, from the figure, we have

$$\cos\theta = \frac{x}{r} \qquad \sin\theta = \frac{y}{r}$$

and so

$$\boxed{1} \qquad x = r\cos\theta \qquad y = r\sin\theta$$

Although Equations 1 were deduced from Figure 5, which illustrates the case where $r > 0$ and $0 < \theta < \pi/2$, these equations are valid for all values of r and θ. (See the general definition of $\sin\theta$ and $\cos\theta$ in Appendix D.)

Equations 1 allow us to find the Cartesian coordinates of a point when the polar coordinates are known. To find r and θ when x and y are known, we use the equations

2
$$r^2 = x^2 + y^2 \qquad \tan\theta = \frac{y}{x}$$

which can be deduced from Equations 1 or simply read from Figure 5.

EXAMPLE 2 Convert the point $(2, \pi/3)$ from polar to Cartesian coordinates.

SOLUTION Since $r = 2$ and $\theta = \pi/3$, Equations 1 give

$$x = r\cos\theta = 2\cos\frac{\pi}{3} = 2 \cdot \frac{1}{2} = 1$$

$$y = r\sin\theta = 2\sin\frac{\pi}{3} = 2 \cdot \frac{\sqrt{3}}{2} = \sqrt{3}$$

Therefore the point is $(1, \sqrt{3})$ in Cartesian coordinates. ■

EXAMPLE 3 Represent the point with Cartesian coordinates $(1, -1)$ in terms of polar coordinates.

SOLUTION If we choose r to be positive, then Equations 2 give

$$r = \sqrt{x^2 + y^2} = \sqrt{1^2 + (-1)^2} = \sqrt{2}$$

$$\tan\theta = \frac{y}{x} = -1$$

Since the point $(1, -1)$ lies in the fourth quadrant, we can choose $\theta = -\pi/4$ or $\theta = 7\pi/4$. Thus one possible answer is $(\sqrt{2}, -\pi/4)$; another is $(\sqrt{2}, 7\pi/4)$. ■

NOTE Equations 2 do not uniquely determine θ when x and y are given because, as θ increases through the interval $0 \leqslant \theta < 2\pi$, each value of $\tan\theta$ occurs twice. Therefore, in converting from Cartesian to polar coordinates, it's not good enough just to find r and θ that satisfy Equations 2. As in Example 3, we must choose θ so that the point (r, θ) lies in the correct quadrant.

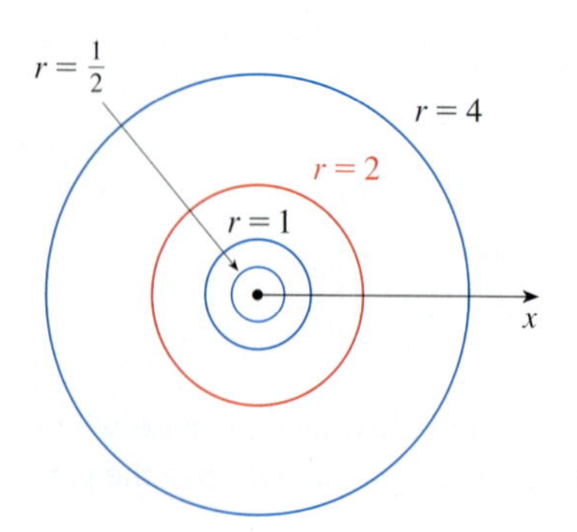

FIGURE 6

Polar Curves

The **graph of a polar equation** $r = f(\theta)$, or more generally $F(r, \theta) = 0$, consists of all points P that have at least one polar representation (r, θ) whose coordinates satisfy the equation.

EXAMPLE 4 What curve is represented by the polar equation $r = 2$?

SOLUTION The curve consists of all points (r, θ) with $r = 2$. Since r represents the distance from the point to the pole, the curve $r = 2$ represents the circle with center O and radius 2. In general, the equation $r = a$ represents a circle with center O and radius $|a|$. (See Figure 6.) ■

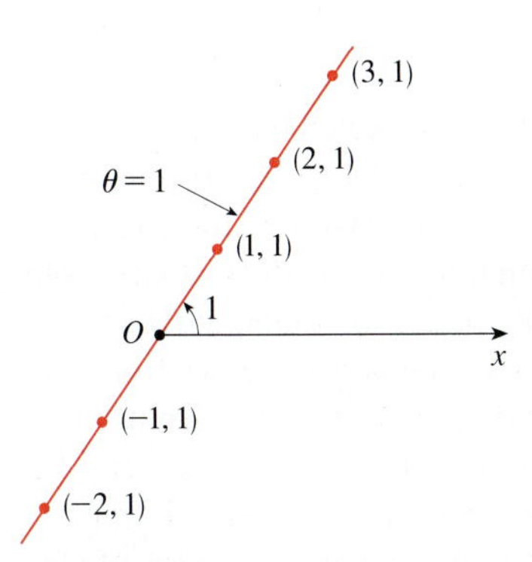

FIGURE 7

EXAMPLE 5 Sketch the polar curve $\theta = 1$.

SOLUTION This curve consists of all points (r, θ) such that the polar angle θ is 1 radian. It is the straight line that passes through O and makes an angle of 1 radian with the polar axis (see Figure 7). Notice that the points $(r, 1)$ on the line with $r > 0$ are in the first quadrant, whereas those with $r < 0$ are in the third quadrant. ■

EXAMPLE 6

(a) Sketch the curve with polar equation $r = 2\cos\theta$.

(b) Find a Cartesian equation for this curve.

SOLUTION

(a) In Figure 8 we find the values of r for some convenient values of θ and plot the corresponding points (r, θ). Then we join these points to sketch the curve, which appears to be a circle. We have used only values of θ between 0 and π, since if we let θ increase beyond π, we obtain the same points again.

θ	$r = 2\cos\theta$
0	2
$\pi/6$	$\sqrt{3}$
$\pi/4$	$\sqrt{2}$
$\pi/3$	1
$\pi/2$	0
$2\pi/3$	-1
$3\pi/4$	$-\sqrt{2}$
$5\pi/6$	$-\sqrt{3}$
π	-2

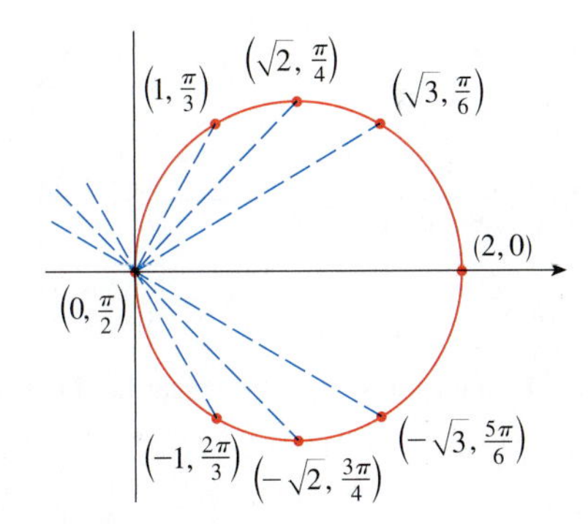

FIGURE 8
Table of values and graph of $r = 2\cos\theta$

(b) To convert the given equation to a Cartesian equation we use Equations 1 and 2. From $x = r\cos\theta$ we have $\cos\theta = x/r$, so the equation $r = 2\cos\theta$ becomes $r = 2x/r$, which gives

$$2x = r^2 = x^2 + y^2 \qquad \text{or} \qquad x^2 + y^2 - 2x = 0$$

Completing the square, we obtain

$$(x - 1)^2 + y^2 = 1$$

which is an equation of a circle with center $(1, 0)$ and radius 1. ■

Figure 9 shows a geometrical illustration that the circle in Example 6 has the equation $r = 2\cos\theta$. The angle OPQ is a right angle (Why?) and so $r/2 = \cos\theta$.

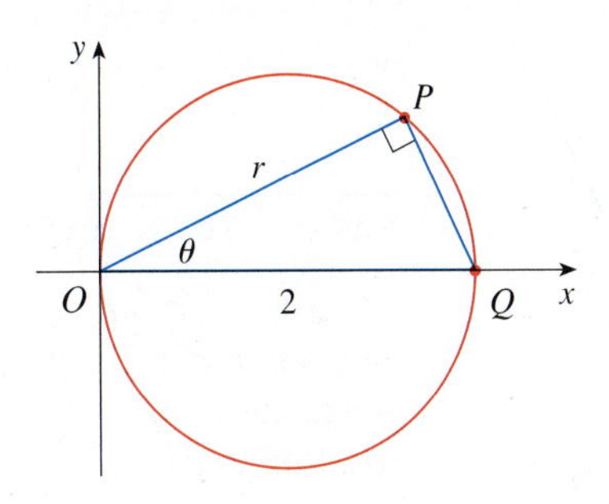

FIGURE 9

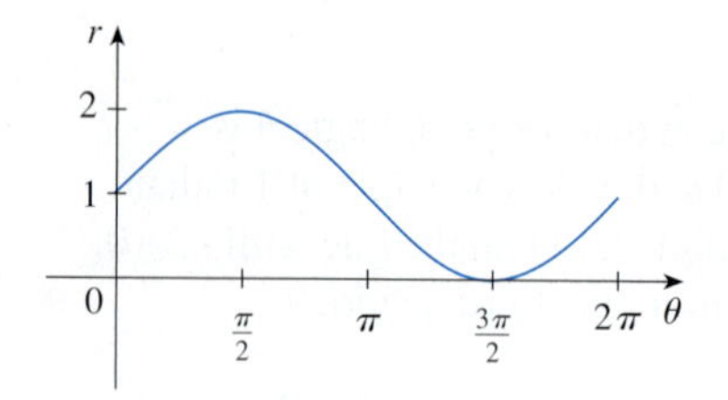

FIGURE 10
$r = 1 + \sin\theta$ in Cartesian coordinates, $0 \leqslant \theta \leqslant 2\pi$

EXAMPLE 7 Sketch the curve $r = 1 + \sin\theta$.

SOLUTION Instead of plotting points as in Example 6, we first sketch the graph of $r = 1 + \sin\theta$ in *Cartesian* coordinates in Figure 10 by shifting the sine curve up one unit. This enables us to read at a glance the values of r that correspond to increasing values of θ. For instance, we see that as θ increases from 0 to $\pi/2$, r (the distance from O) increases from 1 to 2, so we sketch the corresponding part of the polar curve in Figure 11(a). As θ increases from $\pi/2$ to π, Figure 10 shows that r decreases from 2 to 1, so we sketch the next part of the curve as in Figure 11(b). As θ increases from π to $3\pi/2$, r decreases from 1 to 0 as shown in part (c). Finally, as θ increases from $3\pi/2$ to 2π, r increases from 0 to 1 as shown in part (d). If we let θ increase beyond 2π or decrease beyond 0, we would simply retrace our path. Putting together the parts of the curve from Figure 11(a)–(d), we sketch the complete curve in part (e). It is called a **cardioid** because it's shaped like a heart.

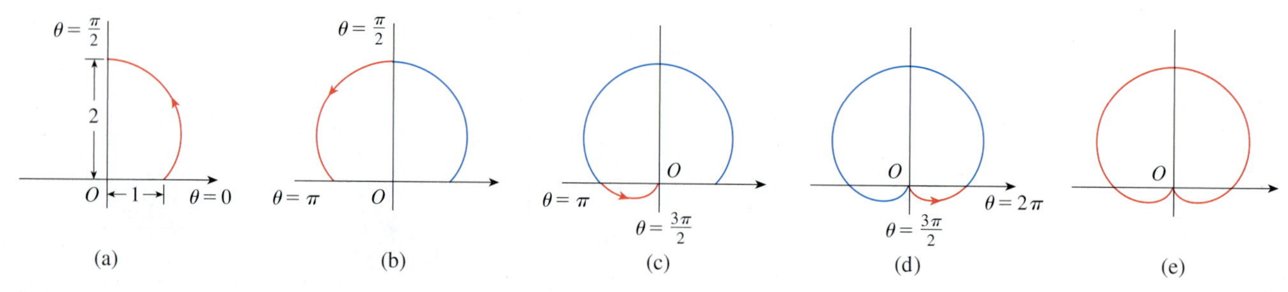

FIGURE 11 Stages in sketching the cardioid $r = 1 + \sin\theta$

EXAMPLE 8 Sketch the curve $r = \cos 2\theta$.

TEC Module 10.3 helps you see how polar curves are traced out by showing animations similar to Figures 10–13.

SOLUTION As in Example 7, we first sketch $r = \cos 2\theta$, $0 \leqslant \theta \leqslant 2\pi$, in Cartesian coordinates in Figure 12. As θ increases from 0 to $\pi/4$, Figure 12 shows that r decreases from 1 to 0 and so we draw the corresponding portion of the polar curve in Figure 13 (indicated by ①). As θ increases from $\pi/4$ to $\pi/2$, r goes from 0 to -1. This means that the distance from O increases from 0 to 1, but instead of being in the first quadrant this portion of the polar curve (indicated by ②) lies on the opposite side of the pole in the third quadrant. The remainder of the curve is drawn in a similar fashion, with the arrows and numbers indicating the order in which the portions are traced out. The resulting curve has four loops and is called a **four-leaved rose**.

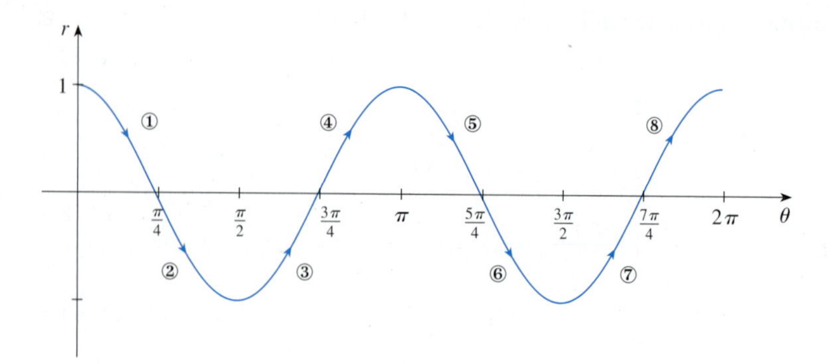

FIGURE 12
$r = \cos 2\theta$ in Cartesian coordinates

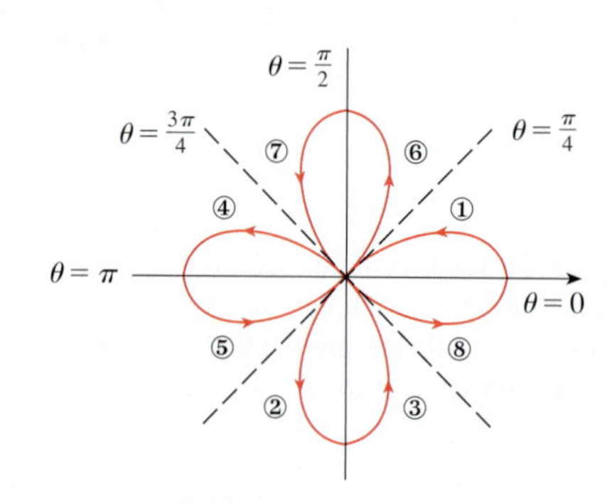

FIGURE 13
Four-leaved rose $r = \cos 2\theta$

Symmetry

When we sketch polar curves it is sometimes helpful to take advantage of symmetry. The following three rules are explained by Figure 14.

(a) If a polar equation is unchanged when θ is replaced by $-\theta$, the curve is symmetric about the polar axis.

(b) If the equation is unchanged when r is replaced by $-r$, or when θ is replaced by $\theta + \pi$, the curve is symmetric about the pole. (This means that the curve remains unchanged if we rotate it through 180° about the origin.)

(c) If the equation is unchanged when θ is replaced by $\pi - \theta$, the curve is symmetric about the vertical line $\theta = \pi/2$.

(r, θ)
θ
O
−θ
(r, −θ)

(a)

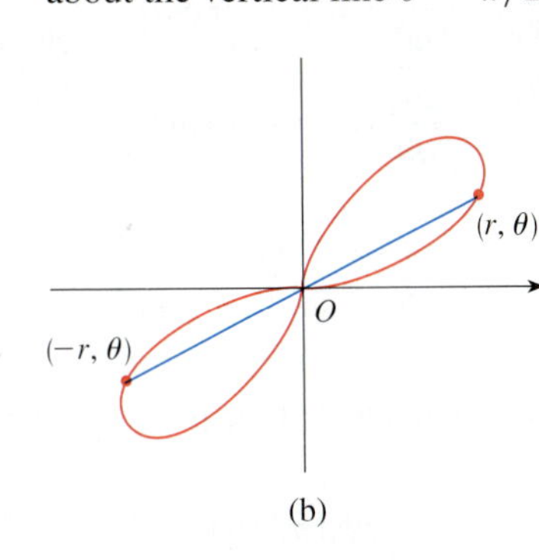

(b)

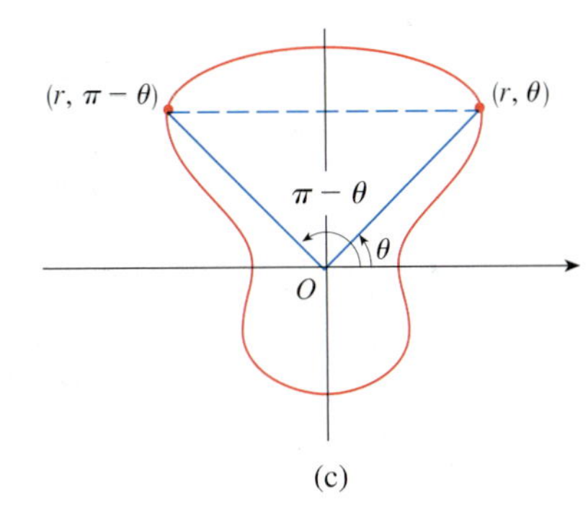

(c)

FIGURE 14

The curves sketched in Examples 6 and 8 are symmetric about the polar axis, since $\cos(-\theta) = \cos\theta$. The curves in Examples 7 and 8 are symmetric about $\theta = \pi/2$ because $\sin(\pi - \theta) = \sin\theta$ and $\cos 2(\pi - \theta) = \cos 2\theta$. The four-leaved rose is also symmetric about the pole. These symmetry properties could have been used in sketching the curves. For instance, in Example 6 we need only have plotted points for $0 \leqslant \theta \leqslant \pi/2$ and then reflected about the polar axis to obtain the complete circle.

Tangents to Polar Curves

To find a tangent line to a polar curve $r = f(\theta)$, we regard θ as a parameter and write its parametric equations as

$$x = r\cos\theta = f(\theta)\cos\theta \qquad y = r\sin\theta = f(\theta)\sin\theta$$

Then, using the method for finding slopes of parametric curves (Equation 10.2.1) and the Product Rule, we have

3
$$\frac{dy}{dx} = \frac{\dfrac{dy}{d\theta}}{\dfrac{dx}{d\theta}} = \frac{\dfrac{dr}{d\theta}\sin\theta + r\cos\theta}{\dfrac{dr}{d\theta}\cos\theta - r\sin\theta}$$

We locate horizontal tangents by finding the points where $dy/d\theta = 0$ (provided that $dx/d\theta \neq 0$). Likewise, we locate vertical tangents at the points where $dx/d\theta = 0$ (provided that $dy/d\theta \neq 0$).

Notice that if we are looking for tangent lines at the pole, then $r = 0$ and Equation 3 simplifies to

$$\frac{dy}{dx} = \tan\theta \qquad \text{if} \quad \frac{dr}{d\theta} \neq 0$$

For instance, in Example 8 we found that $r = \cos 2\theta = 0$ when $\theta = \pi/4$ or $3\pi/4$. This means that the lines $\theta = \pi/4$ and $\theta = 3\pi/4$ (or $y = x$ and $y = -x$) are tangent lines to $r = \cos 2\theta$ at the origin.

EXAMPLE 9

(a) For the cardioid $r = 1 + \sin\theta$ of Example 7, find the slope of the tangent line when $\theta = \pi/3$.

(b) Find the points on the cardioid where the tangent line is horizontal or vertical.

SOLUTION Using Equation 3 with $r = 1 + \sin\theta$, we have

$$\frac{dy}{dx} = \frac{\dfrac{dr}{d\theta}\sin\theta + r\cos\theta}{\dfrac{dr}{d\theta}\cos\theta - r\sin\theta} = \frac{\cos\theta\,\sin\theta + (1 + \sin\theta)\cos\theta}{\cos\theta\,\cos\theta - (1 + \sin\theta)\sin\theta}$$

$$= \frac{\cos\theta\,(1 + 2\sin\theta)}{1 - 2\sin^2\theta - \sin\theta} = \frac{\cos\theta\,(1 + 2\sin\theta)}{(1 + \sin\theta)(1 - 2\sin\theta)}$$

(a) The slope of the tangent at the point where $\theta = \pi/3$ is

$$\left.\frac{dy}{dx}\right|_{\theta=\pi/3} = \frac{\cos(\pi/3)(1 + 2\sin(\pi/3))}{(1 + \sin(\pi/3))(1 - 2\sin(\pi/3))} = \frac{\frac{1}{2}(1 + \sqrt{3})}{(1 + \sqrt{3}/2)(1 - \sqrt{3})}$$

$$= \frac{1 + \sqrt{3}}{(2 + \sqrt{3})(1 - \sqrt{3})} = \frac{1 + \sqrt{3}}{-1 - \sqrt{3}} = -1$$

(b) Observe that

$$\frac{dy}{d\theta} = \cos\theta\,(1 + 2\sin\theta) = 0 \qquad \text{when } \theta = \frac{\pi}{2}, \frac{3\pi}{2}, \frac{7\pi}{6}, \frac{11\pi}{6}$$

$$\frac{dx}{d\theta} = (1 + \sin\theta)(1 - 2\sin\theta) = 0 \qquad \text{when } \theta = \frac{3\pi}{2}, \frac{\pi}{6}, \frac{5\pi}{6}$$

Therefore there are horizontal tangents at the points $(2, \pi/2)$, $(\frac{1}{2}, 7\pi/6)$, $(\frac{1}{2}, 11\pi/6)$ and vertical tangents at $(\frac{3}{2}, \pi/6)$ and $(\frac{3}{2}, 5\pi/6)$. When $\theta = 3\pi/2$, both $dy/d\theta$ and $dx/d\theta$ are 0, so we must be careful. Using l'Hospital's Rule, we have

$$\lim_{\theta\to(3\pi/2)^-}\frac{dy}{dx} = \left(\lim_{\theta\to(3\pi/2)^-}\frac{1 + 2\sin\theta}{1 - 2\sin\theta}\right)\left(\lim_{\theta\to(3\pi/2)^-}\frac{\cos\theta}{1 + \sin\theta}\right)$$

$$= -\frac{1}{3}\lim_{\theta\to(3\pi/2)^-}\frac{\cos\theta}{1 + \sin\theta} = -\frac{1}{3}\lim_{\theta\to(3\pi/2)^-}\frac{-\sin\theta}{\cos\theta} = \infty$$

By symmetry,

$$\lim_{\theta\to(3\pi/2)^+}\frac{dy}{dx} = -\infty$$

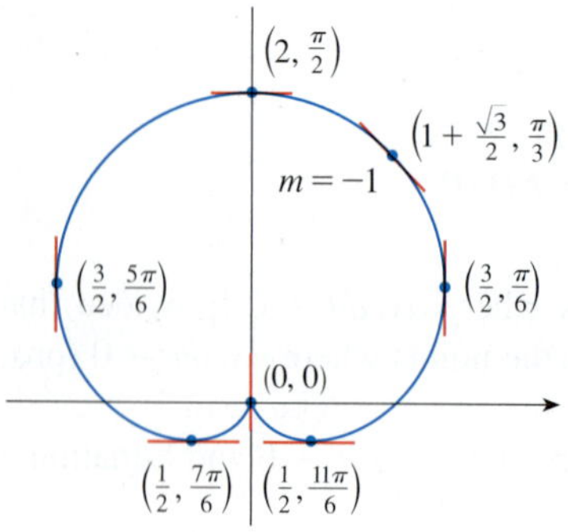

Thus there is a vertical tangent line at the pole (see Figure 15). ■

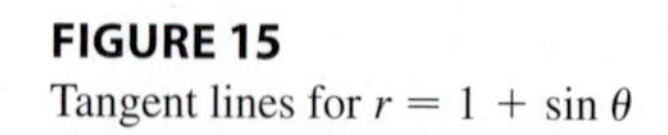

FIGURE 15
Tangent lines for $r = 1 + \sin\theta$

NOTE Instead of having to remember Equation 3, we could employ the method used to derive it. For instance, in Example 9 we could have written

$$x = r\cos\theta = (1 + \sin\theta)\cos\theta = \cos\theta + \tfrac{1}{2}\sin 2\theta$$

$$y = r\sin\theta = (1 + \sin\theta)\sin\theta = \sin\theta + \sin^2\theta$$

Then we have

$$\frac{dy}{dx} = \frac{dy/d\theta}{dx/d\theta} = \frac{\cos\theta + 2\sin\theta\ \cos\theta}{-\sin\theta + \cos 2\theta} = \frac{\cos\theta + \sin 2\theta}{-\sin\theta + \cos 2\theta}$$

which is equivalent to our previous expression.

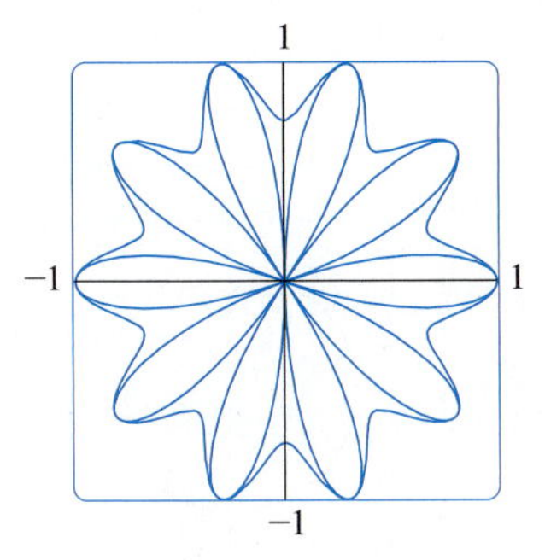

FIGURE 16
$r = \sin^3(2.5\theta) + \cos^3(2.5\theta)$

Graphing Polar Curves with Graphing Devices

Although it's useful to be able to sketch simple polar curves by hand, we need to use a graphing calculator or computer when we are faced with a curve as complicated as the ones shown in Figures 16 and 17.

Some graphing devices have commands that enable us to graph polar curves directly. With other machines we need to convert to parametric equations first. In this case we take the polar equation $r = f(\theta)$ and write its parametric equations as

$$x = r\cos\theta = f(\theta)\cos\theta \qquad y = r\sin\theta = f(\theta)\sin\theta$$

Some machines require that the parameter be called t rather than θ.

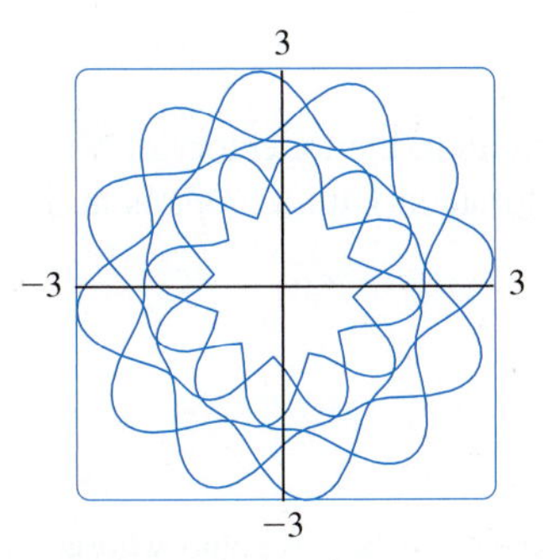

FIGURE 17
$r = 2 + \sin^3(2.4\theta)$

EXAMPLE 10 Graph the curve $r = \sin(8\theta/5)$.

SOLUTION Let's assume that our graphing device doesn't have a built-in polar graphing command. In this case we need to work with the corresponding parametric equations, which are

$$x = r\cos\theta = \sin(8\theta/5)\cos\theta \qquad y = r\sin\theta = \sin(8\theta/5)\sin\theta$$

In any case we need to determine the domain for θ. So we ask ourselves: How many complete rotations are required until the curve starts to repeat itself? If the answer is n, then

$$\sin\frac{8(\theta + 2n\pi)}{5} = \sin\left(\frac{8\theta}{5} + \frac{16n\pi}{5}\right) = \sin\frac{8\theta}{5}$$

and so we require that $16n\pi/5$ be an even multiple of π. This will first occur when $n = 5$. Therefore we will graph the entire curve if we specify that $0 \leqslant \theta \leqslant 10\pi$. Switching from θ to t, we have the equations

$$x = \sin(8t/5)\cos t \qquad y = \sin(8t/5)\sin t \qquad 0 \leqslant t \leqslant 10\pi$$

and Figure 18 shows the resulting curve. Notice that this rose has 16 loops. ■

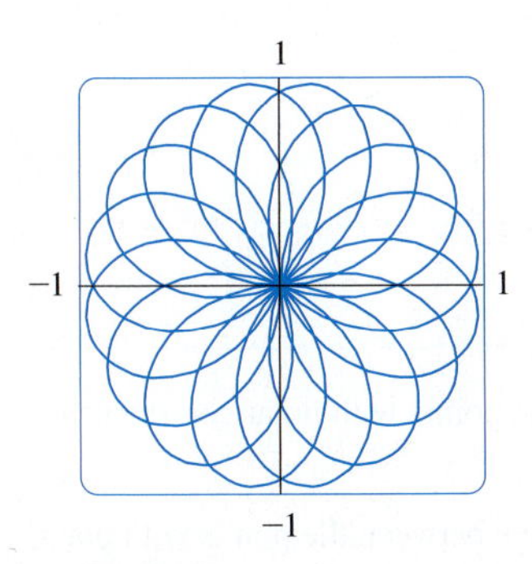

FIGURE 18
$r = \sin(8\theta/5)$

EXAMPLE 11 Investigate the family of polar curves given by $r = 1 + c\sin\theta$. How does the shape change as c changes? (These curves are called **limaçons**, after a French word for snail, because of the shape of the curves for certain values of c.)

SOLUTION Figure 19 on page 706 shows computer-drawn graphs for various values of c. For $c > 1$ there is a loop that decreases in size as c decreases. When $c = 1$ the loop disappears and the curve becomes the cardioid that we sketched in Example 7. For c between 1 and $\frac{1}{2}$ the cardioid's cusp is smoothed out and becomes a "dimple." When c

In Exercise 53 you are asked to prove analytically what we have discovered from the graphs in Figure 19.

decreases from $\frac{1}{2}$ to 0, the limaçon is shaped like an oval. This oval becomes more circular as $c \to 0$, and when $c = 0$ the curve is just the circle $r = 1$.

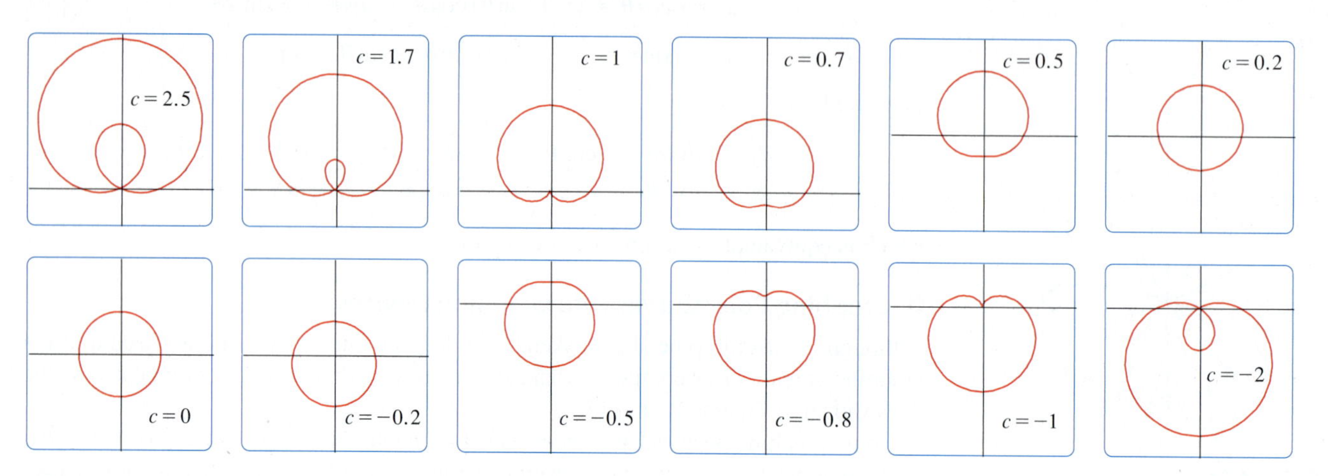

FIGURE 19
Members of the family of limaçons $r = 1 + c \sin\theta$

The remaining parts of Figure 19 show that as c becomes negative, the shapes change in reverse order. In fact, these curves are reflections about the horizontal axis of the corresponding curves with positive c. ■

Limaçons arise in the study of planetary motion. In particular, the trajectory of Mars, as viewed from the planet Earth, has been modeled by a limaçon with a loop, as in the parts of Figure 19 with $|c| > 1$.

10.3 EXERCISES

1–2 Plot the point whose polar coordinates are given. Then find two other pairs of polar coordinates of this point, one with $r > 0$ and one with $r < 0$.

1. (a) $(1, \pi/4)$ (b) $(-2, 3\pi/2)$ (c) $(3, -\pi/3)$

2. (a) $(2, 5\pi/6)$ (b) $(1, -2\pi/3)$ (c) $(-1, 5\pi/4)$

3–4 Plot the point whose polar coordinates are given. Then find the Cartesian coordinates of the point.

3. (a) $(2, 3\pi/2)$ (b) $(\sqrt{2}, \pi/4)$ (c) $(-1, -\pi/6)$

4. (a) $(4, 4\pi/3)$ (b) $(-2, 3\pi/4)$ (c) $(-3, -\pi/3)$

5–6 The Cartesian coordinates of a point are given.
(i) Find polar coordinates (r, θ) of the point, where $r > 0$ and $0 \leqslant \theta < 2\pi$.
(ii) Find polar coordinates (r, θ) of the point, where $r < 0$ and $0 \leqslant \theta < 2\pi$.

5. (a) $(-4, 4)$ (b) $(3, 3\sqrt{3})$

6. (a) $(\sqrt{3}, -1)$ (b) $(-6, 0)$

7–12 Sketch the region in the plane consisting of points whose polar coordinates satisfy the given conditions.

7. $r \geqslant 1$

8. $0 \leqslant r < 2, \quad \pi \leqslant \theta \leqslant 3\pi/2$

9. $r \geqslant 0, \quad \pi/4 \leqslant \theta \leqslant 3\pi/4$

10. $1 \leqslant r \leqslant 3, \quad \pi/6 < \theta < 5\pi/6$

11. $2 < r < 3, \quad 5\pi/3 \leqslant \theta \leqslant 7\pi/3$

12. $r \geqslant 1, \quad \pi \leqslant \theta \leqslant 2\pi$

13. Find the distance between the points with polar coordinates $(4, 4\pi/3)$ and $(6, 5\pi/3)$.

14. Find a formula for the distance between the points with polar coordinates (r_1, θ_1) and (r_2, θ_2).

15–20 Identify the curve by finding a Cartesian equation for the curve.

15. $r^2 = 5$

16. $r = 4 \sec\theta$

17. $r = 5 \cos\theta$

18. $\theta = \pi/3$

19. $r^2 \cos 2\theta = 1$

20. $r^2 \sin 2\theta = 1$

21–26 Find a polar equation for the curve represented by the given Cartesian equation.

21. $y = 2$

22. $y = x$

23. $y = 1 + 3x$

24. $4y^2 = x$

25. $x^2 + y^2 = 2cx$

26. $x^2 - y^2 = 4$

27–28 For each of the described curves, decide if the curve would be more easily given by a polar equation or a Cartesian equation. Then write an equation for the curve.

27. (a) A line through the origin that makes an angle of $\pi/6$ with the positive x-axis
(b) A vertical line through the point (3, 3)

28. (a) A circle with radius 5 and center (2, 3)
(b) A circle centered at the origin with radius 4

29–46 Sketch the curve with the given polar equation by first sketching the graph of r as a function of θ in Cartesian coordinates.

29. $r = -2 \sin\theta$

30. $r = 1 - \cos\theta$

31. $r = 2(1 + \cos\theta)$

32. $r = 1 + 2\cos\theta$

33. $r = \theta,\ \theta \geqslant 0$

34. $r = \theta^2,\ -2\pi \leqslant \theta \leqslant 2\pi$

35. $r = 3\cos 3\theta$

36. $r = -\sin 5\theta$

37. $r = 2\cos 4\theta$

38. $r = 2\sin 6\theta$

39. $r = 1 + 3\cos\theta$

40. $r = 1 + 5\sin\theta$

41. $r^2 = 9\sin 2\theta$

42. $r^2 = \cos 4\theta$

43. $r = 2 + \sin 3\theta$

44. $r^2\theta = 1$

45. $r = \sin(\theta/2)$

46. $r = \cos(\theta/3)$

47–48 The figure shows a graph of r as a function of θ in Cartesian coordinates. Use it to sketch the corresponding polar curve.

47.

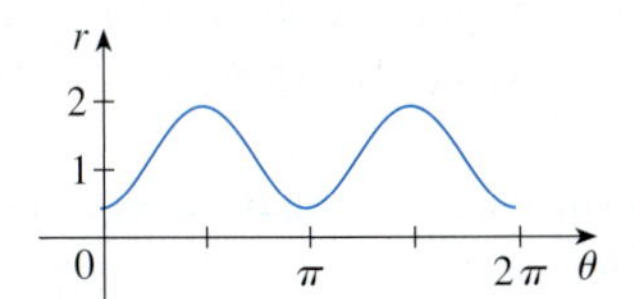

48.

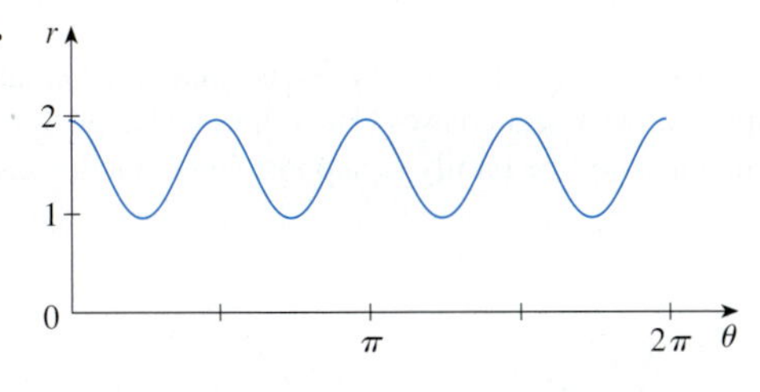

49. Show that the polar curve $r = 4 + 2\sec\theta$ (called a **conchoid**) has the line $x = 2$ as a vertical asymptote by showing that $\lim_{r\to\pm\infty} x = 2$. Use this fact to help sketch the conchoid.

50. Show that the curve $r = 2 - \csc\theta$ (also a conchoid) has the line $y = -1$ as a horizontal asymptote by showing that $\lim_{r\to\pm\infty} y = -1$. Use this fact to help sketch the conchoid.

51. Show that the curve $r = \sin\theta\ \tan\theta$ (called a **cissoid of Diocles**) has the line $x = 1$ as a vertical asymptote. Show also that the curve lies entirely within the vertical strip $0 \leqslant x < 1$. Use these facts to help sketch the cissoid.

52. Sketch the curve $(x^2 + y^2)^3 = 4x^2y^2$.

53. (a) In Example 11 the graphs suggest that the limaçon $r = 1 + c\sin\theta$ has an inner loop when $|c| > 1$. Prove that this is true, and find the values of θ that correspond to the inner loop.
(b) From Figure 19 it appears that the limaçon loses its dimple when $c = \frac{1}{2}$. Prove this.

54. Match the polar equations with the graphs labeled I–VI. Give reasons for your choices. (Don't use a graphing device.)

(a) $r = \ln\theta,\quad 1 \leqslant \theta \leqslant 6\pi$

(b) $r = \theta^2,\quad 0 \leqslant \theta \leqslant 8\pi$

(c) $r = \cos 3\theta$

(d) $r = 2 + \cos 3\theta$

(e) $r = \cos(\theta/2)$

(f) $r = 2 + \cos(3\theta/2)$

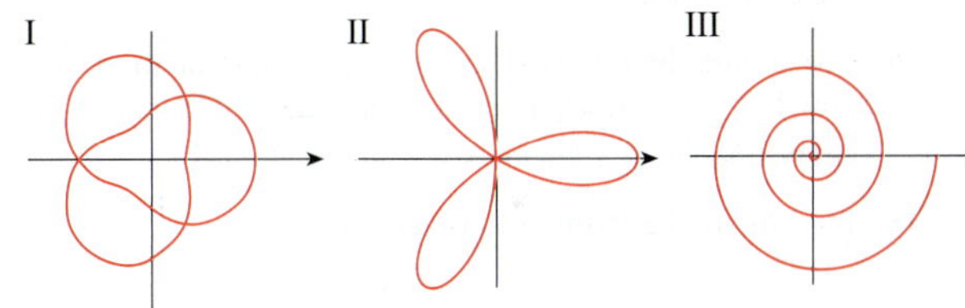

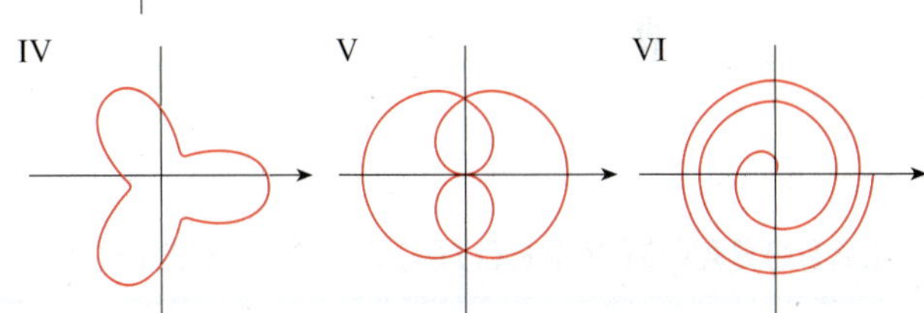

55–60 Find the slope of the tangent line to the given polar curve at the point specified by the value of θ.

55. $r = 2\cos\theta,\quad \theta = \pi/3$

56. $r = 2 + \sin 3\theta,\quad \theta = \pi/4$

57. $r = 1/\theta,\quad \theta = \pi$

58. $r = \cos(\theta/3),\quad \theta = \pi$

59. $r = \cos 2\theta,\quad \theta = \pi/4$

60. $r = 1 + 2\cos\theta,\quad \theta = \pi/3$

61–64 Find the points on the given curve where the tangent line is horizontal or vertical.

61. $r = 3\cos\theta$

62. $r = 1 - \sin\theta$

63. $r = 1 + \cos\theta$

64. $r = e^\theta$

65. Show that the polar equation $r = a \sin\theta + b \cos\theta$, where $ab \neq 0$, represents a circle, and find its center and radius.

66. Show that the curves $r = a \sin\theta$ and $r = a \cos\theta$ intersect at right angles.

67–72 Use a graphing device to graph the polar curve. Choose the parameter interval to make sure that you produce the entire curve.

67. $r = 1 + 2\sin(\theta/2)$ (nephroid of Freeth)

68. $r = \sqrt{1 - 0.8\sin^2\theta}$ (hippopede)

69. $r = e^{\sin\theta} - 2\cos(4\theta)$ (butterfly curve)

70. $r = |\tan\theta|^{|\cot\theta|}$ (valentine curve)

71. $r = 1 + \cos^{999}\theta$ (Pac-Man curve)

72. $r = 2 + \cos(9\theta/4)$

73. How are the graphs of $r = 1 + \sin(\theta - \pi/6)$ and $r = 1 + \sin(\theta - \pi/3)$ related to the graph of $r = 1 + \sin\theta$? In general, how is the graph of $r = f(\theta - \alpha)$ related to the graph of $r = f(\theta)$?

74. Use a graph to estimate the y-coordinate of the highest points on the curve $r = \sin 2\theta$. Then use calculus to find the exact value.

75. Investigate the family of curves with polar equations $r = 1 + c\cos\theta$, where c is a real number. How does the shape change as c changes?

76. Investigate the family of polar curves

$$r = 1 + \cos^n\theta$$

where n is a positive integer. How does the shape change as n increases? What happens as n becomes large? Explain the shape for large n by considering the graph of r as a function of θ in Cartesian coordinates.

77. Let P be any point (except the origin) on the curve $r = f(\theta)$. If ψ is the angle between the tangent line at P and the radial line OP, show that

$$\tan\psi = \frac{r}{dr/d\theta}$$

[*Hint:* Observe that $\psi = \phi - \theta$ in the figure.]

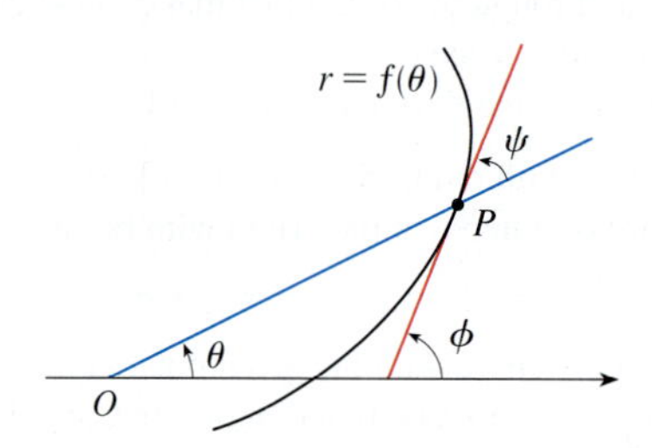

78. (a) Use Exercise 77 to show that the angle between the tangent line and the radial line is $\psi = \pi/4$ at every point on the curve $r = e^\theta$.

(b) Illustrate part (a) by graphing the curve and the tangent lines at the points where $\theta = 0$ and $\pi/2$.

(c) Prove that any polar curve $r = f(\theta)$ with the property that the angle ψ between the radial line and the tangent line is a constant must be of the form $r = Ce^{k\theta}$, where C and k are constants.

LABORATORY PROJECT FAMILIES OF POLAR CURVES

In this project you will discover the interesting and beautiful shapes that members of families of polar curves can take. You will also see how the shape of the curve changes when you vary the constants.

1. (a) Investigate the family of curves defined by the polar equations $r = \sin n\theta$, where n is a positive integer. How is the number of loops related to n?

(b) What happens if the equation in part (a) is replaced by $r = |\sin n\theta|$?

2. A family of curves is given by the equations $r = 1 + c\sin n\theta$, where c is a real number and n is a positive integer. How does the graph change as n increases? How does it change as c changes? Illustrate by graphing enough members of the family to support your conclusions.

3. A family of curves has polar equations

$$r = \frac{1 - a\cos\theta}{1 + a\cos\theta}$$

Investigate how the graph changes as the number a changes. In particular, you should identify the transitional values of a for which the basic shape of the curve changes.

4. The astronomer Giovanni Cassini (1625–1712) studied the family of curves with polar equations

$$r^4 - 2c^2r^2 \cos 2\theta + c^4 - a^4 = 0$$

where a and c are positive real numbers. These curves are called the **ovals of Cassini** even though they are oval shaped only for certain values of a and c. (Cassini thought that these curves might represent planetary orbits better than Kepler's ellipses.) Investigate the variety of shapes that these curves may have. In particular, how are a and c related to each other when the curve splits into two parts?

10.4 Areas and Lengths in Polar Coordinates

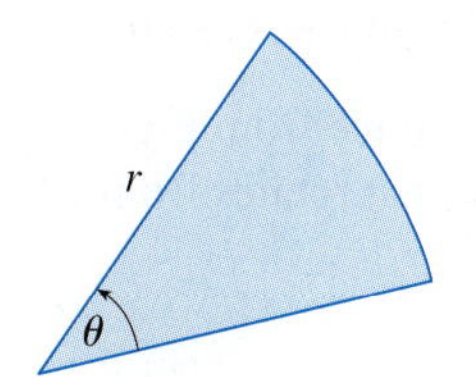

FIGURE 1

In this section we develop the formula for the area of a region whose boundary is given by a polar equation. We need to use the formula for the area of a sector of a circle:

1 $$A = \tfrac{1}{2}r^2\theta$$

where, as in Figure 1, r is the radius and θ is the radian measure of the central angle. Formula 1 follows from the fact that the area of a sector is proportional to its central angle: $A = (\theta/2\pi)\pi r^2 = \frac{1}{2}r^2\theta$. (See also Exercise 7.3.35.)

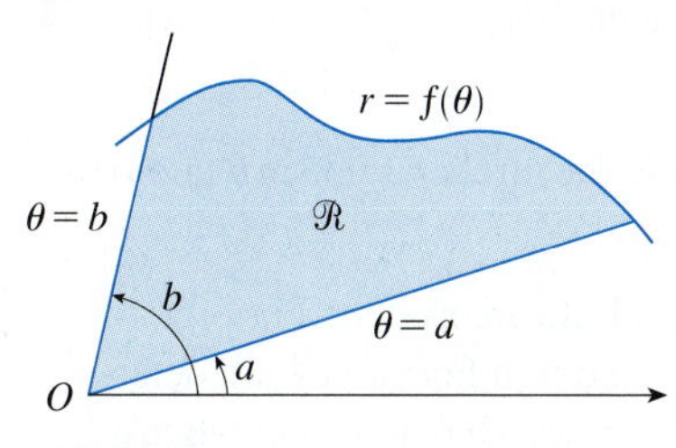

FIGURE 2

Let $\mathscr{R}$ be the region, illustrated in Figure 2, bounded by the polar curve $r = f(\theta)$ and by the rays $\theta = a$ and $\theta = b$, where f is a positive continuous function and where $0 < b - a \leqslant 2\pi$. We divide the interval $[a, b]$ into subintervals with endpoints $\theta_0, \theta_1, \theta_2, \ldots, \theta_n$ and equal width $\Delta\theta$. The rays $\theta = \theta_i$ then divide $\mathscr{R}$ into n smaller regions with central angle $\Delta\theta = \theta_i - \theta_{i-1}$. If we choose θ_i^* in the ith subinterval $[\theta_{i-1}, \theta_i]$, then the area ΔA_i of the ith region is approximated by the area of the sector of a circle with central angle $\Delta\theta$ and radius $f(\theta_i^*)$. (See Figure 3.)

Thus from Formula 1 we have

$$\Delta A_i \approx \tfrac{1}{2}[f(\theta_i^*)]^2 \,\Delta\theta$$

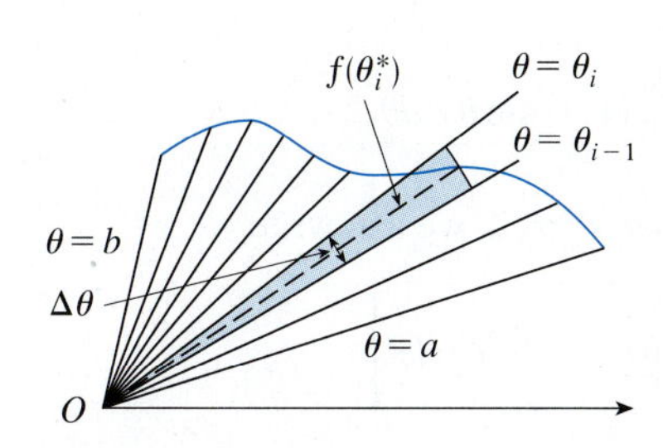

FIGURE 3

and so an approximation to the total area A of $\mathscr{R}$ is

2 $$A \approx \sum_{i=1}^{n} \tfrac{1}{2}[f(\theta_i^*)]^2 \,\Delta\theta$$

It appears from Figure 3 that the approximation in (2) improves as $n \to \infty$. But the sums in (2) are Riemann sums for the function $g(\theta) = \frac{1}{2}[f(\theta)]^2$, so

$$\lim_{n \to \infty} \sum_{i=1}^{n} \tfrac{1}{2}[f(\theta_i^*)]^2 \,\Delta\theta = \int_a^b \tfrac{1}{2}[f(\theta)]^2 \,d\theta$$

It therefore appears plausible (and can in fact be proved) that the formula for the area A of the polar region $\mathcal{R}$ is

3 $$A = \int_a^b \tfrac{1}{2}[f(\theta)]^2\, d\theta$$

Formula 3 is often written as

4 $$A = \int_a^b \tfrac{1}{2} r^2\, d\theta$$

with the understanding that $r = f(\theta)$. Note the similarity between Formulas 1 and 4.

When we apply Formula 3 or 4, it is helpful to think of the area as being swept out by a rotating ray through O that starts with angle a and ends with angle b.

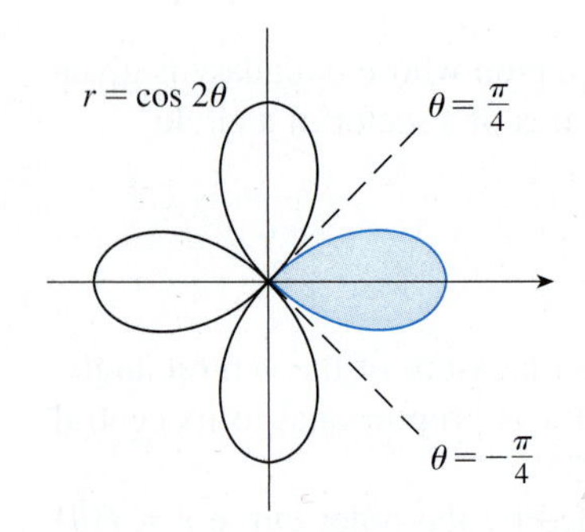

FIGURE 4

EXAMPLE 1 Find the area enclosed by one loop of the four-leaved rose $r = \cos 2\theta$.

SOLUTION The curve $r = \cos 2\theta$ was sketched in Example 10.3.8. Notice from Figure 4 that the region enclosed by the right loop is swept out by a ray that rotates from $\theta = -\pi/4$ to $\theta = \pi/4$. Therefore Formula 4 gives

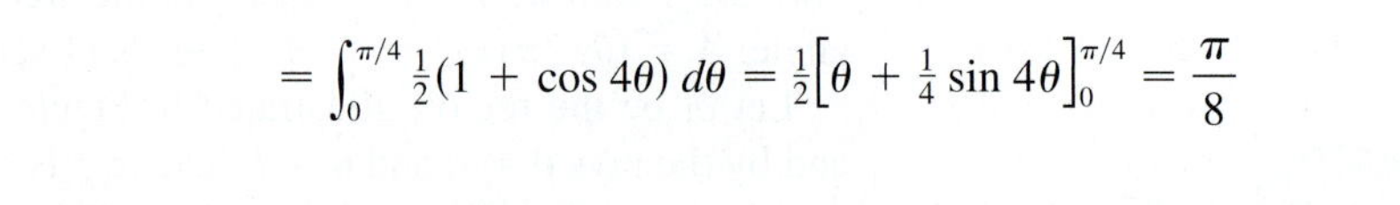

$$A = \int_{-\pi/4}^{\pi/4} \tfrac{1}{2} r^2\, d\theta = \tfrac{1}{2}\int_{-\pi/4}^{\pi/4} \cos^2 2\theta\, d\theta = \int_0^{\pi/4} \cos^2 2\theta\, d\theta$$

$$= \int_0^{\pi/4} \tfrac{1}{2}(1 + \cos 4\theta)\, d\theta = \tfrac{1}{2}\Big[\theta + \tfrac{1}{4}\sin 4\theta\Big]_0^{\pi/4} = \frac{\pi}{8}$$ ■

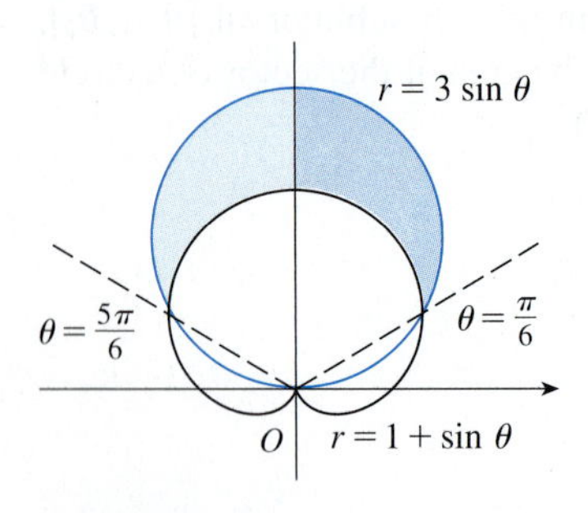

FIGURE 5

EXAMPLE 2 Find the area of the region that lies inside the circle $r = 3\sin\theta$ and outside the cardioid $r = 1 + \sin\theta$.

SOLUTION The cardioid (see Example 10.3.7) and the circle are sketched in Figure 5 and the desired region is shaded. The values of a and b in Formula 4 are determined by finding the points of intersection of the two curves. They intersect when $3\sin\theta = 1 + \sin\theta$, which gives $\sin\theta = \frac{1}{2}$, so $\theta = \pi/6, 5\pi/6$. The desired area can be found by subtracting the area inside the cardioid between $\theta = \pi/6$ and $\theta = 5\pi/6$ from the area inside the circle from $\pi/6$ to $5\pi/6$. Thus

$$A = \tfrac{1}{2}\int_{\pi/6}^{5\pi/6} (3\sin\theta)^2\, d\theta - \tfrac{1}{2}\int_{\pi/6}^{5\pi/6} (1 + \sin\theta)^2\, d\theta$$

Since the region is symmetric about the vertical axis $\theta = \pi/2$, we can write

$$A = 2\left[\tfrac{1}{2}\int_{\pi/6}^{\pi/2} 9\sin^2\theta\, d\theta - \tfrac{1}{2}\int_{\pi/6}^{\pi/2} (1 + 2\sin\theta + \sin^2\theta)\, d\theta\right]$$

$$= \int_{\pi/6}^{\pi/2} (8\sin^2\theta - 1 - 2\sin\theta)\, d\theta$$

$$= \int_{\pi/6}^{\pi/2} (3 - 4\cos 2\theta - 2\sin\theta)\, d\theta \qquad \left[\text{because } \sin^2\theta = \tfrac{1}{2}(1 - \cos 2\theta)\right]$$

$$= 3\theta - 2\sin 2\theta + 2\cos\theta\Big]_{\pi/6}^{\pi/2} = \pi$$ ■

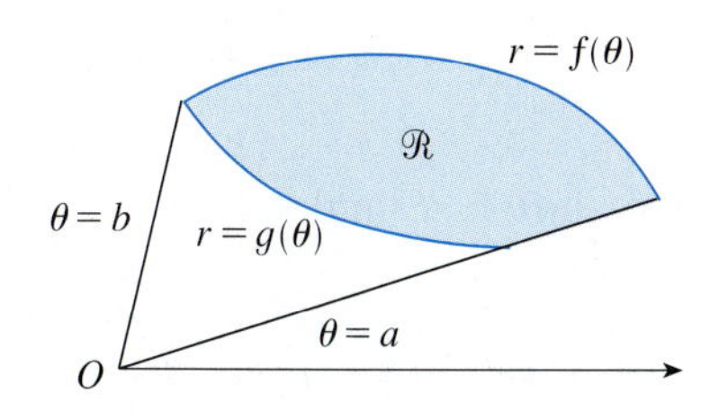

FIGURE 6

Example 2 illustrates the procedure for finding the area of the region bounded by two polar curves. In general, let $\mathcal{R}$ be a region, as illustrated in Figure 6, that is bounded by curves with polar equations $r = f(\theta)$, $r = g(\theta)$, $\theta = a$, and $\theta = b$, where $f(\theta) \geqslant g(\theta) \geqslant 0$ and $0 < b - a \leqslant 2\pi$. The area A of $\mathcal{R}$ is found by subtracting the area inside $r = g(\theta)$ from the area inside $r = f(\theta)$, so using Formula 3 we have

$$A = \int_a^b \tfrac{1}{2}[f(\theta)]^2\,d\theta - \int_a^b \tfrac{1}{2}[g(\theta)]^2\,d\theta$$

$$= \tfrac{1}{2}\int_a^b \left([f(\theta)]^2 - [g(\theta)]^2\right) d\theta$$

CAUTION The fact that a single point has many representations in polar coordinates sometimes makes it difficult to find all the points of intersection of two polar curves. For instance, it is obvious from Figure 5 that the circle and the cardioid have three points of intersection; however, in Example 2 we solved the equations $r = 3\sin\theta$ and $r = 1 + \sin\theta$ and found only two such points, $\left(\tfrac{3}{2}, \pi/6\right)$ and $\left(\tfrac{3}{2}, 5\pi/6\right)$. The origin is also a point of intersection, but we can't find it by solving the equations of the curves because the origin has no single representation in polar coordinates that satisfies both equations. Notice that, when represented as $(0, 0)$ or $(0, \pi)$, the origin satisfies $r = 3\sin\theta$ and so it lies on the circle; when represented as $(0, 3\pi/2)$, it satisfies $r = 1 + \sin\theta$ and so it lies on the cardioid. Think of two points moving along the curves as the parameter value θ increases from 0 to 2π. On one curve the origin is reached at $\theta = 0$ and $\theta = \pi$; on the other curve it is reached at $\theta = 3\pi/2$. The points don't collide at the origin because they reach the origin at different times, but the curves intersect there nonetheless.

Thus, to find *all* points of intersection of two polar curves, it is recommended that you draw the graphs of both curves. It is especially convenient to use a graphing calculator or computer to help with this task.

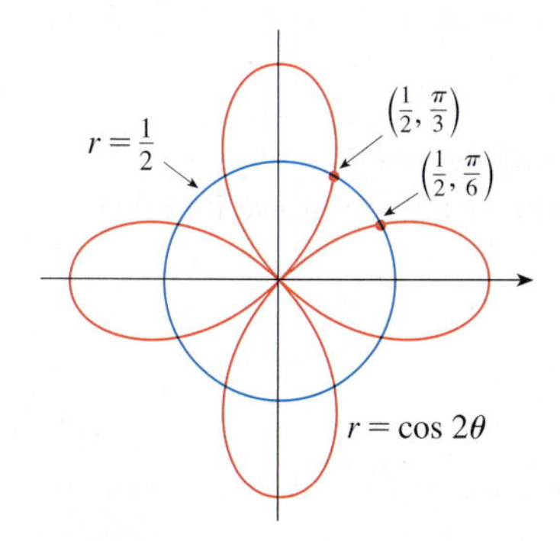

FIGURE 7

EXAMPLE 3 Find all points of intersection of the curves $r = \cos 2\theta$ and $r = \tfrac{1}{2}$.

SOLUTION If we solve the equations $r = \cos 2\theta$ and $r = \tfrac{1}{2}$, we get $\cos 2\theta = \tfrac{1}{2}$ and, therefore, $2\theta = \pi/3, 5\pi/3, 7\pi/3, 11\pi/3$. Thus the values of θ between 0 and 2π that satisfy both equations are $\theta = \pi/6, 5\pi/6, 7\pi/6, 11\pi/6$. We have found four points of intersection: $\left(\tfrac{1}{2}, \pi/6\right)$, $\left(\tfrac{1}{2}, 5\pi/6\right)$, $\left(\tfrac{1}{2}, 7\pi/6\right)$, and $\left(\tfrac{1}{2}, 11\pi/6\right)$.

However, you can see from Figure 7 that the curves have four other points of intersection—namely, $\left(\tfrac{1}{2}, \pi/3\right)$, $\left(\tfrac{1}{2}, 2\pi/3\right)$, $\left(\tfrac{1}{2}, 4\pi/3\right)$, and $\left(\tfrac{1}{2}, 5\pi/3\right)$. These can be found using symmetry or by noticing that another equation of the circle is $r = -\tfrac{1}{2}$ and then solving the equations $r = \cos 2\theta$ and $r = -\tfrac{1}{2}$. ■

Arc Length

To find the length of a polar curve $r = f(\theta)$, $a \leqslant \theta \leqslant b$, we regard θ as a parameter and write the parametric equations of the curve as

$$x = r\cos\theta = f(\theta)\cos\theta \qquad y = r\sin\theta = f(\theta)\sin\theta$$

Using the Product Rule and differentiating with respect to θ, we obtain

$$\frac{dx}{d\theta} = \frac{dr}{d\theta}\cos\theta - r\sin\theta \qquad \frac{dy}{d\theta} = \frac{dr}{d\theta}\sin\theta + r\cos\theta$$

so, using $\cos^2\theta + \sin^2\theta = 1$, we have

$$\left(\frac{dx}{d\theta}\right)^2 + \left(\frac{dy}{d\theta}\right)^2 = \left(\frac{dr}{d\theta}\right)^2 \cos^2\theta - 2r\frac{dr}{d\theta}\cos\theta\,\sin\theta + r^2\sin^2\theta$$
$$+ \left(\frac{dr}{d\theta}\right)^2 \sin^2\theta + 2r\frac{dr}{d\theta}\sin\theta\,\cos\theta + r^2\cos^2\theta$$
$$= \left(\frac{dr}{d\theta}\right)^2 + r^2$$

Assuming that f' is continuous, we can use Theorem 10.2.5 to write the arc length as

$$L = \int_a^b \sqrt{\left(\frac{dx}{d\theta}\right)^2 + \left(\frac{dy}{d\theta}\right)^2}\, d\theta$$

Therefore the length of a curve with polar equation $r = f(\theta)$, $a \leqslant \theta \leqslant b$, is

5
$$L = \int_a^b \sqrt{r^2 + \left(\frac{dr}{d\theta}\right)^2}\, d\theta$$

EXAMPLE 4 Find the length of the cardioid $r = 1 + \sin\theta$.

SOLUTION The cardioid is shown in Figure 8. (We sketched it in Example 10.3.7.) Its full length is given by the parameter interval $0 \leqslant \theta \leqslant 2\pi$, so Formula 5 gives

$$L = \int_0^{2\pi} \sqrt{r^2 + \left(\frac{dr}{d\theta}\right)^2}\, d\theta = \int_0^{2\pi} \sqrt{(1+\sin\theta)^2 + \cos^2\theta}\, d\theta = \int_0^{2\pi} \sqrt{2 + 2\sin\theta}\, d\theta$$

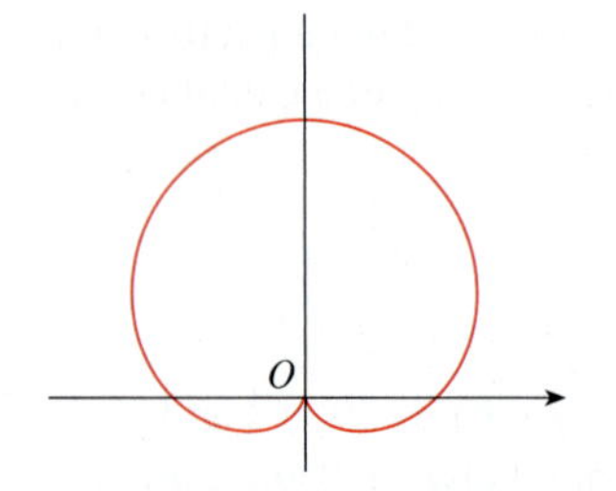

FIGURE 8
$r = 1 + \sin\theta$

We could evaluate this integral by multiplying and dividing the integrand by $\sqrt{2 - 2\sin\theta}$, or we could use a computer algebra system. In any event, we find that the length of the cardioid is $L = 8$. ■

10.4 EXERCISES

1–4 Find the area of the region that is bounded by the given curve and lies in the specified sector.

1. $r = e^{-\theta/4}$, $\pi/2 \leqslant \theta \leqslant \pi$

2. $r = \cos\theta$, $0 \leqslant \theta \leqslant \pi/6$

3. $r = \sin\theta + \cos\theta$, $0 \leqslant \theta \leqslant \pi$

4. $r = 1/\theta$, $\pi/2 \leqslant \theta \leqslant 2\pi$

5–8 Find the area of the shaded region.

5.

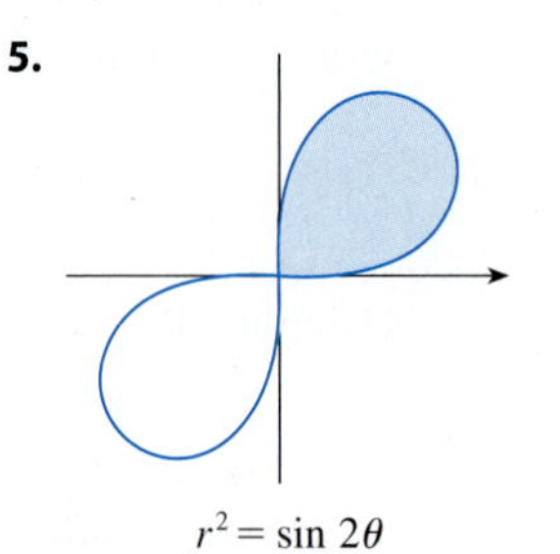

$r^2 = \sin 2\theta$

6.

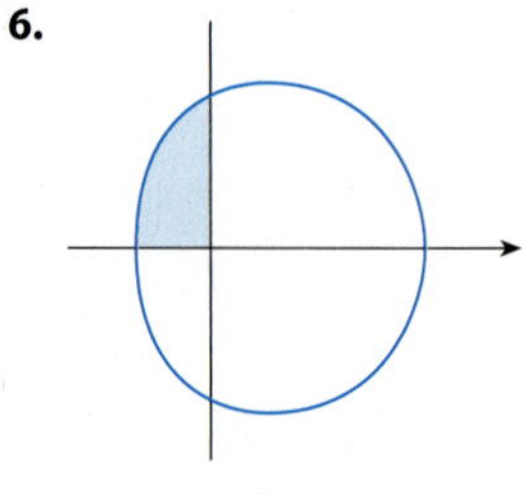

$r = 2 + \cos\theta$

7.

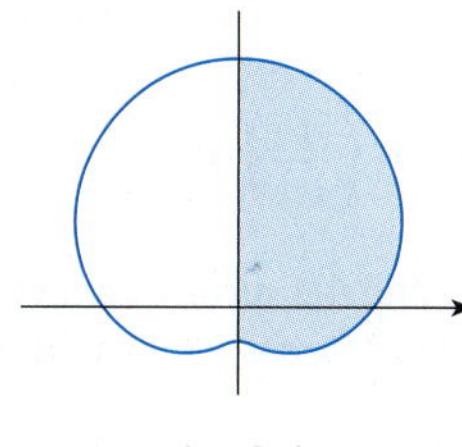

$r = 4 + 3\sin\theta$

8.

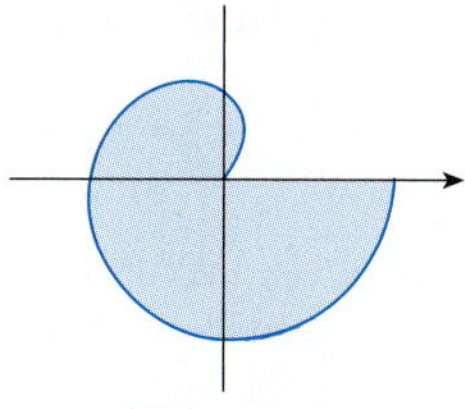

$r = \sqrt{\ln\theta},\ 1 \leqslant \theta \leqslant 2\pi$

9–12 Sketch the curve and find the area that it encloses.

9. $r = 2\sin\theta$

10. $r = 1 - \sin\theta$

11. $r = 3 + 2\cos\theta$

12. $r = 2 - \cos\theta$

13–16 Graph the curve and find the area that it encloses.

13. $r = 2 + \sin 4\theta$

14. $r = 3 - 2\cos 4\theta$

15. $r = \sqrt{1 + \cos^2(5\theta)}$

16. $r = 1 + 5\sin 6\theta$

17–21 Find the area of the region enclosed by one loop of the curve.

17. $r = 4\cos 3\theta$

18. $r^2 = 4\cos 2\theta$

19. $r = \sin 4\theta$

20. $r = 2\sin 5\theta$

21. $r = 1 + 2\sin\theta$ (inner loop)

22. Find the area enclosed by the loop of the **strophoid** $r = 2\cos\theta - \sec\theta$.

23–28 Find the area of the region that lies inside the first curve and outside the second curve.

23. $r = 4\sin\theta,\quad r = 2$

24. $r = 1 - \sin\theta,\quad r = 1$

25. $r^2 = 8\cos 2\theta,\quad r = 2$

26. $r = 1 + \cos\theta,\quad r = 2 - \cos\theta$

27. $r = 3\cos\theta,\quad r = 1 + \cos\theta$

28. $r = 3\sin\theta,\quad r = 2 - \sin\theta$

29–34 Find the area of the region that lies inside both curves.

29. $r = 3\sin\theta,\quad r = 3\cos\theta$

30. $r = 1 + \cos\theta,\quad r = 1 - \cos\theta$

31. $r = \sin 2\theta,\quad r = \cos 2\theta$

32. $r = 3 + 2\cos\theta,\quad r = 3 + 2\sin\theta$

33. $r^2 = 2\sin 2\theta,\quad r = 1$

34. $r = a\sin\theta,\quad r = b\cos\theta,\quad a > 0,\ b > 0$

35. Find the area inside the larger loop and outside the smaller loop of the limaçon $r = \frac{1}{2} + \cos\theta$.

36. Find the area between a large loop and the enclosed small loop of the curve $r = 1 + 2\cos 3\theta$.

37–42 Find all points of intersection of the given curves.

37. $r = \sin\theta,\quad r = 1 - \sin\theta$

38. $r = 1 + \cos\theta,\quad r = 1 - \sin\theta$

39. $r = 2\sin 2\theta,\quad r = 1$

40. $r = \cos 3\theta,\quad r = \sin 3\theta$

41. $r = \sin\theta,\quad r = \sin 2\theta$

42. $r^2 = \sin 2\theta,\quad r^2 = \cos 2\theta$

43. The points of intersection of the cardioid $r = 1 + \sin\theta$ and the spiral loop $r = 2\theta,\ -\pi/2 \leqslant \theta \leqslant \pi/2$, can't be found exactly. Use a graphing device to find the approximate values of θ at which they intersect. Then use these values to estimate the area that lies inside both curves.

44. When recording live performances, sound engineers often use a microphone with a cardioid pickup pattern because it suppresses noise from the audience. Suppose the microphone is placed 4 m from the front of the stage (as in the figure) and the boundary of the optimal pickup region is given by the cardioid $r = 8 + 8\sin\theta$, where r is measured in meters and the microphone is at the pole. The musicians want to know the area they will have on stage within the optimal pickup range of the microphone. Answer their question.

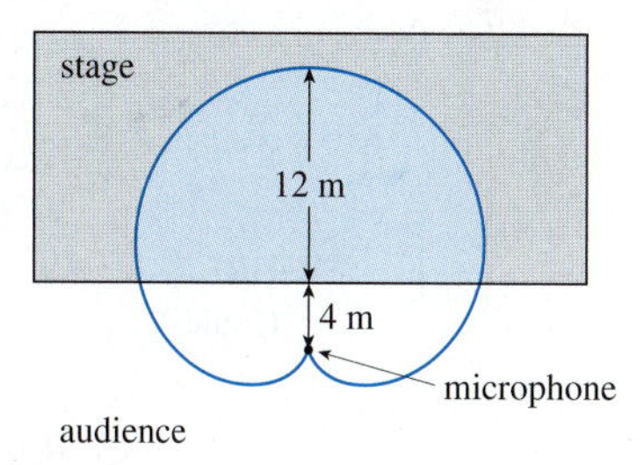

45–48 Find the exact length of the polar curve.

45. $r = 2\cos\theta,\quad 0 \leqslant \theta \leqslant \pi$

46. $r = 5^\theta,\quad 0 \leqslant \theta \leqslant 2\pi$

47. $r = \theta^2,\quad 0 \leqslant \theta \leqslant 2\pi$

48. $r = 2(1 + \cos\theta)$

49–50 Find the exact length of the curve. Use a graph to determine the parameter interval.

49. $r = \cos^4(\theta/4)$

50. $r = \cos^2(\theta/2)$

51–54 Use a calculator to find the length of the curve correct to four decimal places. If necessary, graph the curve to determine the parameter interval.

51. One loop of the curve $r = \cos 2\theta$

52. $r = \tan\theta, \quad \pi/6 \leqslant \theta \leqslant \pi/3$

53. $r = \sin(6 \sin\theta)$

54. $r = \sin(\theta/4)$

---

55. (a) Use Formula 10.2.6 to show that the area of the surface generated by rotating the polar curve

$$r = f(\theta) \qquad a \leqslant \theta \leqslant b$$

(where f' is continuous and $0 \leqslant a < b \leqslant \pi$) about the polar axis is

$$S = \int_a^b 2\pi r \sin\theta \sqrt{r^2 + \left(\frac{dr}{d\theta}\right)^2}\, d\theta$$

(b) Use the formula in part (a) to find the surface area generated by rotating the lemniscate $r^2 = \cos 2\theta$ about the polar axis.

56. (a) Find a formula for the area of the surface generated by rotating the polar curve $r = f(\theta)$, $a \leqslant \theta \leqslant b$ (where f' is continuous and $0 \leqslant a < b \leqslant \pi$), about the line $\theta = \pi/2$.

(b) Find the surface area generated by rotating the lemniscate $r^2 = \cos 2\theta$ about the line $\theta = \pi/2$.

10.5 Conic Sections

In this section we give geometric definitions of parabolas, ellipses, and hyperbolas and derive their standard equations. They are called **conic sections**, or **conics**, because they result from intersecting a cone with a plane as shown in Figure 1.

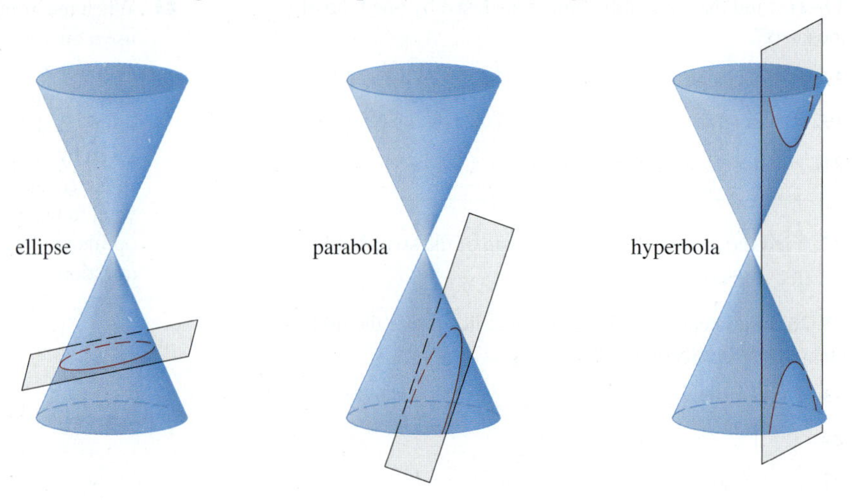

FIGURE 1
Conics

Parabolas

A **parabola** is the set of points in a plane that are equidistant from a fixed point F (called the **focus**) and a fixed line (called the **directrix**). This definition is illustrated by Figure 2. Notice that the point halfway between the focus and the directrix lies on the parabola; it is called the **vertex**. The line through the focus perpendicular to the directrix is called the **axis** of the parabola.

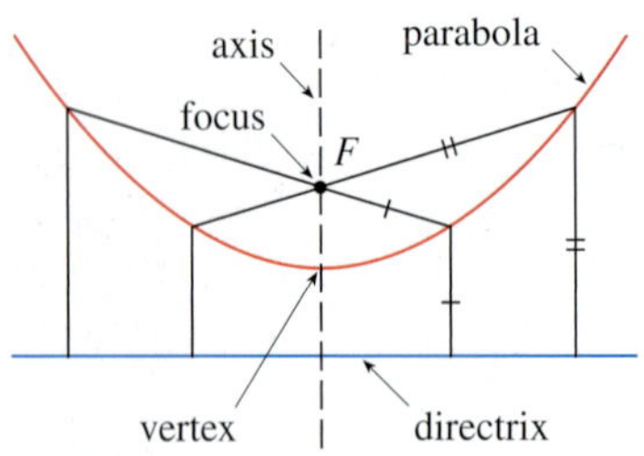

FIGURE 2

In the 16th century Galileo showed that the path of a projectile that is shot into the air at an angle to the ground is a parabola. Since then, parabolic shapes have been used in designing automobile headlights, reflecting telescopes, and suspension bridges. (See Problem 18 on page 202 for the reflection property of parabolas that makes them so useful.)

We obtain a particularly simple equation for a parabola if we place its vertex at the origin O and its directrix parallel to the x-axis as in Figure 3. If the focus is the point $(0, p)$, then the directrix has the equation $y = -p$. If $P(x, y)$ is any point on the parabola,

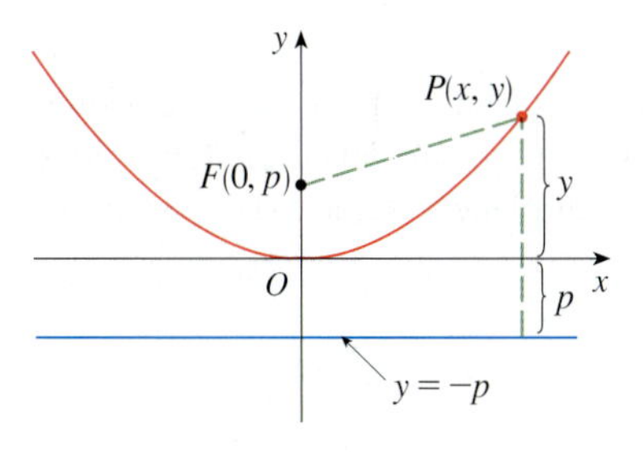

FIGURE 3

then the distance from P to the focus is

$$|PF| = \sqrt{x^2 + (y - p)^2}$$

and the distance from P to the directrix is $|y + p|$. (Figure 3 illustrates the case where $p > 0$.) The defining property of a parabola is that these distances are equal:

$$\sqrt{x^2 + (y - p)^2} = |y + p|$$

We get an equivalent equation by squaring and simplifying:

$$x^2 + (y - p)^2 = |y + p|^2 = (y + p)^2$$

$$x^2 + y^2 - 2py + p^2 = y^2 + 2py + p^2$$

$$x^2 = 4py$$

1 An equation of the parabola with focus $(0, p)$ and directrix $y = -p$ is

$$x^2 = 4py$$

If we write $a = 1/(4p)$, then the standard equation of a parabola (1) becomes $y = ax^2$. It opens upward if $p > 0$ and downward if $p < 0$ [see Figure 4, parts (a) and (b)]. The graph is symmetric with respect to the y-axis because (1) is unchanged when x is replaced by $-x$.

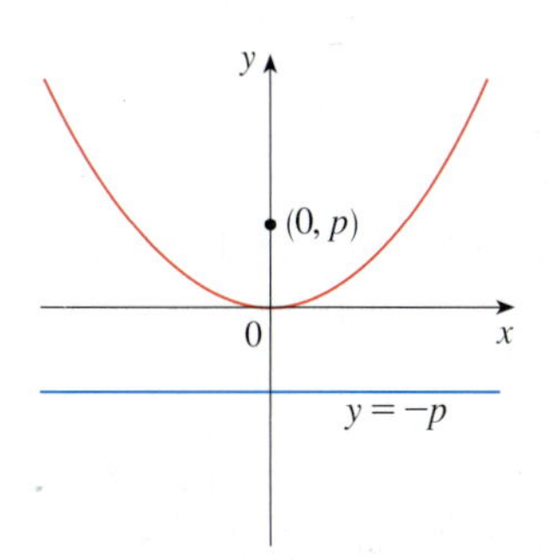

(a) $x^2 = 4py$, $p > 0$

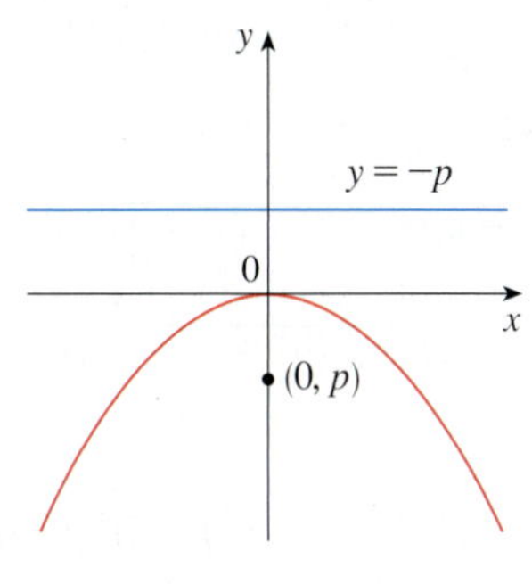

(b) $x^2 = 4py$, $p < 0$

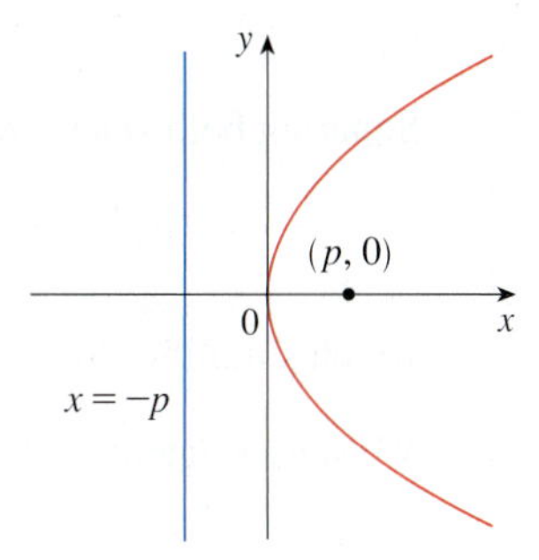

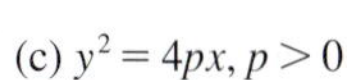

(c) $y^2 = 4px$, $p > 0$

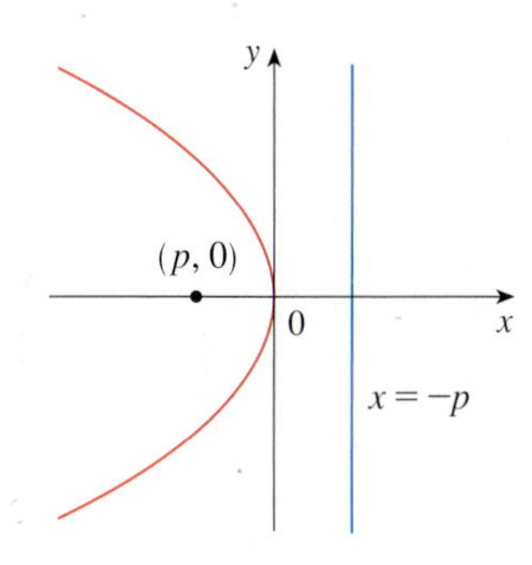

(d) $y^2 = 4px$, $p < 0$

FIGURE 4

If we interchange x and y in (1), we obtain

2
$$y^2 = 4px$$

which is an equation of the parabola with focus $(p, 0)$ and directrix $x = -p$. (Interchanging x and y amounts to reflecting about the diagonal line $y = x$.) The parabola opens to the right if $p > 0$ and to the left if $p < 0$ [see Figure 4, parts (c) and (d)]. In both cases the graph is symmetric with respect to the x-axis, which is the axis of the parabola.

EXAMPLE 1 Find the focus and directrix of the parabola $y^2 + 10x = 0$ and sketch the graph.

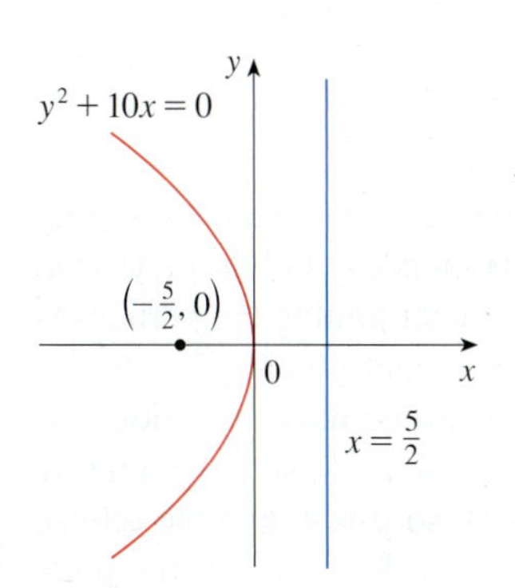

FIGURE 5

SOLUTION If we write the equation as $y^2 = -10x$ and compare it with Equation 2, we see that $4p = -10$, so $p = -\frac{5}{2}$. Thus the focus is $(p, 0) = \left(-\frac{5}{2}, 0\right)$ and the directrix is $x = \frac{5}{2}$. The sketch is shown in Figure 5. ■

Ellipses

An **ellipse** is the set of points in a plane the sum of whose distances from two fixed points F_1 and F_2 is a constant (see Figure 6). These two fixed points are called the **foci** (plural of **focus**). One of Kepler's laws is that the orbits of the planets in the solar system are ellipses with the sun at one focus.

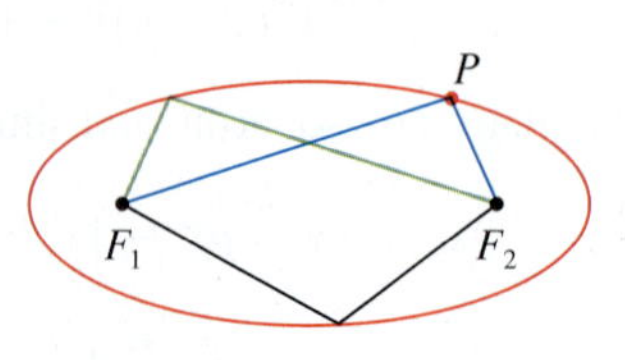

FIGURE 6

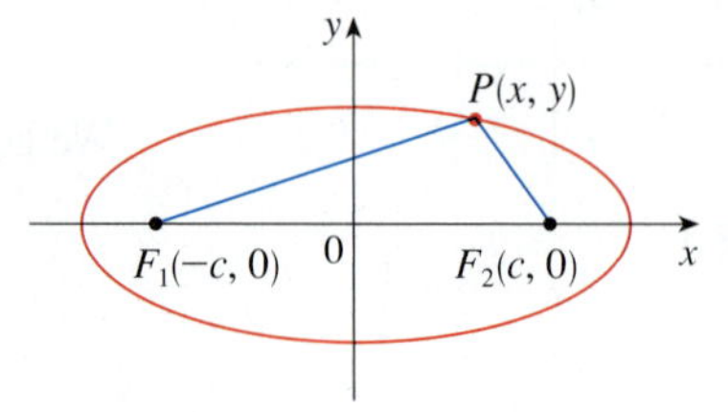

FIGURE 7

In order to obtain the simplest equation for an ellipse, we place the foci on the x-axis at the points $(-c, 0)$ and $(c, 0)$ as in Figure 7 so that the origin is halfway between the foci. Let the sum of the distances from a point on the ellipse to the foci be $2a > 0$. Then $P(x, y)$ is a point on the ellipse when

$$|PF_1| + |PF_2| = 2a$$

that is,

$$\sqrt{(x + c)^2 + y^2} + \sqrt{(x - c)^2 + y^2} = 2a$$

or

$$\sqrt{(x - c)^2 + y^2} = 2a - \sqrt{(x + c)^2 + y^2}$$

Squaring both sides, we have

$$x^2 - 2cx + c^2 + y^2 = 4a^2 - 4a\sqrt{(x + c)^2 + y^2} + x^2 + 2cx + c^2 + y^2$$

which simplifies to

$$a\sqrt{(x + c)^2 + y^2} = a^2 + cx$$

We square again:

$$a^2(x^2 + 2cx + c^2 + y^2) = a^4 + 2a^2cx + c^2x^2$$

which becomes

$$(a^2 - c^2)x^2 + a^2y^2 = a^2(a^2 - c^2)$$

From triangle F_1F_2P in Figure 7 we can see that $2c < 2a$, so $c < a$ and therefore $a^2 - c^2 > 0$. For convenience, let $b^2 = a^2 - c^2$. Then the equation of the ellipse becomes $b^2x^2 + a^2y^2 = a^2b^2$ or, if both sides are divided by a^2b^2,

3 $$\frac{x^2}{a^2} + \frac{y^2}{b^2} = 1$$

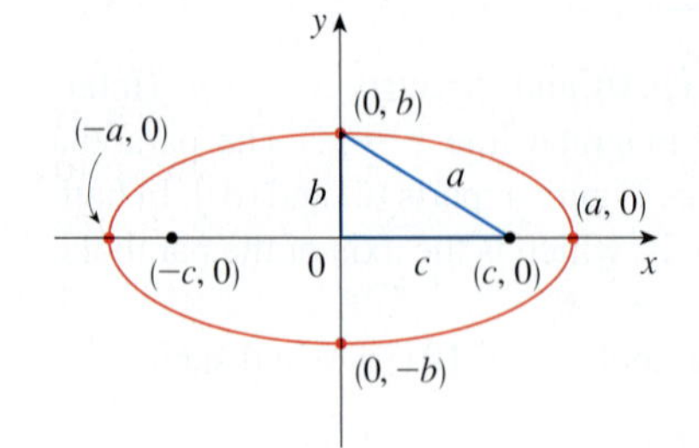

FIGURE 8
$\frac{x^2}{a^2} + \frac{y^2}{b^2} = 1, a \geqslant b$

Since $b^2 = a^2 - c^2 < a^2$, it follows that $b < a$. The x-intercepts are found by setting $y = 0$. Then $x^2/a^2 = 1$, or $x^2 = a^2$, so $x = \pm a$. The corresponding points $(a, 0)$ and $(-a, 0)$ are called the **vertices** of the ellipse and the line segment joining the vertices is called the **major axis**. To find the y-intercepts we set $x = 0$ and obtain $y^2 = b^2$, so $y = \pm b$. The line segment joining $(0, b)$ and $(0, -b)$ is the **minor axis**. Equation 3 is unchanged if x is replaced by $-x$ or y is replaced by $-y$, so the ellipse is symmetric about both axes. Notice that if the foci coincide, then $c = 0$, so $a = b$ and the ellipse becomes a circle with radius $r = a = b$.

We summarize this discussion as follows (see also Figure 8).

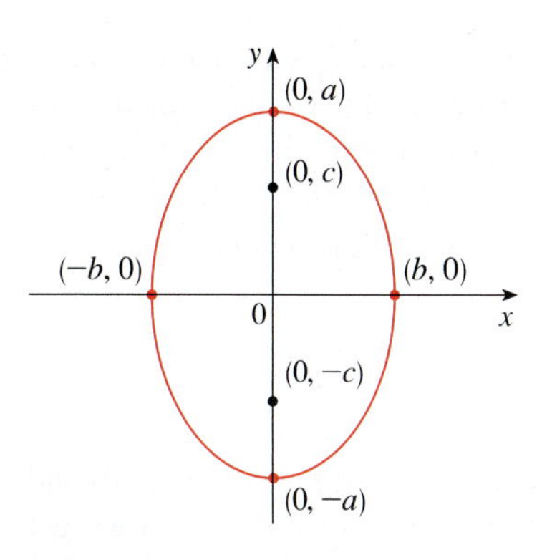

FIGURE 9
$\dfrac{x^2}{b^2} + \dfrac{y^2}{a^2} = 1, a \geqslant b$

4 The ellipse

$$\frac{x^2}{a^2} + \frac{y^2}{b^2} = 1 \qquad a \geqslant b > 0$$

has foci $(\pm c, 0)$, where $c^2 = a^2 - b^2$, and vertices $(\pm a, 0)$.

If the foci of an ellipse are located on the y-axis at $(0, \pm c)$, then we can find its equation by interchanging x and y in (4). (See Figure 9.)

5 The ellipse

$$\frac{x^2}{b^2} + \frac{y^2}{a^2} = 1 \qquad a \geqslant b > 0$$

has foci $(0, \pm c)$, where $c^2 = a^2 - b^2$, and vertices $(0, \pm a)$.

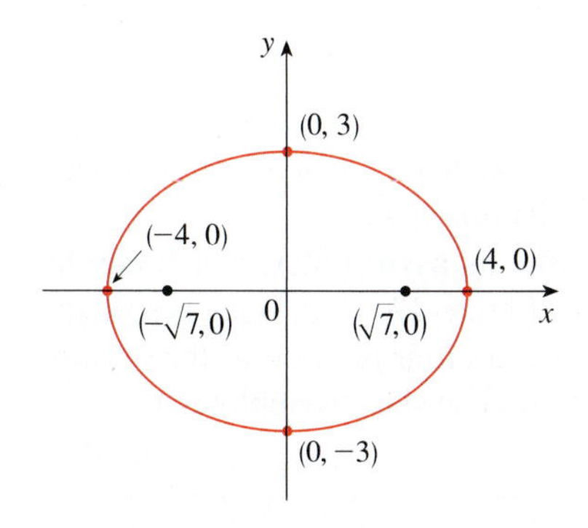

FIGURE 10
$9x^2 + 16y^2 = 144$

EXAMPLE 2 Sketch the graph of $9x^2 + 16y^2 = 144$ and locate the foci.

SOLUTION Divide both sides of the equation by 144:

$$\frac{x^2}{16} + \frac{y^2}{9} = 1$$

The equation is now in the standard form for an ellipse, so we have $a^2 = 16$, $b^2 = 9$, $a = 4$, and $b = 3$. The x-intercepts are ± 4 and the y-intercepts are ± 3. Also, $c^2 = a^2 - b^2 = 7$, so $c = \sqrt{7}$ and the foci are $(\pm\sqrt{7}, 0)$. The graph is sketched in Figure 10. ■

EXAMPLE 3 Find an equation of the ellipse with foci $(0, \pm 2)$ and vertices $(0, \pm 3)$.

SOLUTION Using the notation of (5), we have $c = 2$ and $a = 3$. Then we obtain $b^2 = a^2 - c^2 = 9 - 4 = 5$, so an equation of the ellipse is

$$\frac{x^2}{5} + \frac{y^2}{9} = 1$$

Another way of writing the equation is $9x^2 + 5y^2 = 45$. ■

Like parabolas, ellipses have an interesting reflection property that has practical consequences. If a source of light or sound is placed at one focus of a surface with elliptical cross-sections, then all the light or sound is reflected off the surface to the other focus (see Exercise 65). This principle is used in *lithotripsy,* a treatment for kidney stones. A reflector with elliptical cross-section is placed in such a way that the kidney stone is at one focus. High-intensity sound waves generated at the other focus are reflected to the stone and destroy it without damaging surrounding tissue. The patient is spared the trauma of surgery and recovers within a few days.

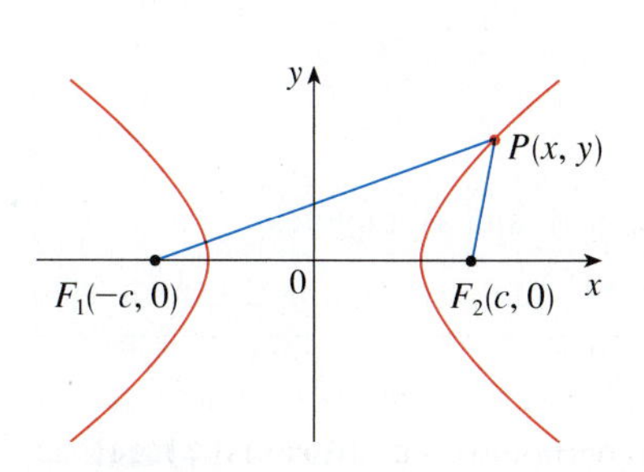

FIGURE 11
P is on the hyperbola when $|PF_1| - |PF_2| = \pm 2a$.

Hyperbolas

A **hyperbola** is the set of all points in a plane the difference of whose distances from two fixed points F_1 and F_2 (the **foci**) is a constant. This definition is illustrated in Figure 11.

Hyperbolas occur frequently as graphs of equations in chemistry, physics, biology, and economics (Boyle's Law, Ohm's Law, supply and demand curves). A particularly significant application of hyperbolas was found in the navigation systems developed in World Wars I and II (see Exercise 51).

Notice that the definition of a hyperbola is similar to that of an ellipse; the only change is that the sum of distances has become a difference of distances. In fact, the derivation of the equation of a hyperbola is also similar to the one given earlier for an ellipse. It is left as Exercise 52 to show that when the foci are on the x-axis at $(\pm c, 0)$ and the difference of distances is $|PF_1| - |PF_2| = \pm 2a$, then the equation of the hyperbola is

$$\boxed{6} \qquad \frac{x^2}{a^2} - \frac{y^2}{b^2} = 1$$

where $c^2 = a^2 + b^2$. Notice that the x-intercepts are again $\pm a$ and the points $(a, 0)$ and $(-a, 0)$ are the **vertices** of the hyperbola. But if we put $x = 0$ in Equation 6 we get $y^2 = -b^2$, which is impossible, so there is no y-intercept. The hyperbola is symmetric with respect to both axes.

To analyze the hyperbola further, we look at Equation 6 and obtain

$$\frac{x^2}{a^2} = 1 + \frac{y^2}{b^2} \geqslant 1$$

This shows that $x^2 \geqslant a^2$, so $|x| = \sqrt{x^2} \geqslant a$. Therefore we have $x \geqslant a$ or $x \leqslant -a$. This means that the hyperbola consists of two parts, called its *branches*.

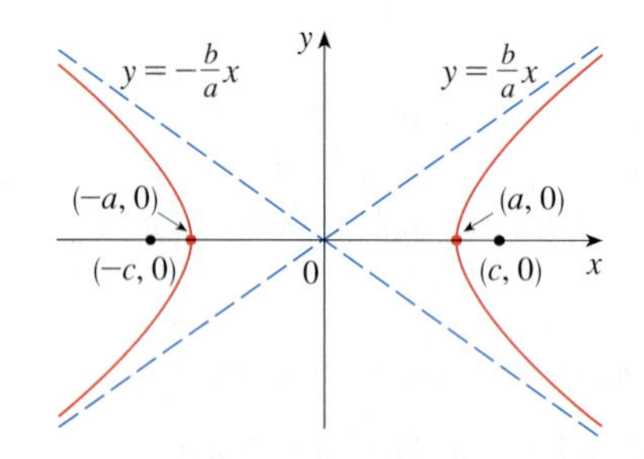

FIGURE 12
$\frac{x^2}{a^2} - \frac{y^2}{b^2} = 1$

When we draw a hyperbola it is useful to first draw its **asymptotes**, which are the dashed lines $y = (b/a)x$ and $y = -(b/a)x$ shown in Figure 12. Both branches of the hyperbola approach the asymptotes; that is, they come arbitrarily close to the asymptotes. (See Exercise 3.5.57, where these lines are shown to be slant asymptotes.)

> $\boxed{7}$ The hyperbola
> $$\frac{x^2}{a^2} - \frac{y^2}{b^2} = 1$$
> has foci $(\pm c, 0)$, where $c^2 = a^2 + b^2$, vertices $(\pm a, 0)$, and asymptotes $y = \pm(b/a)x$.

If the foci of a hyperbola are on the y-axis, then by reversing the roles of x and y we obtain the following information, which is illustrated in Figure 13.

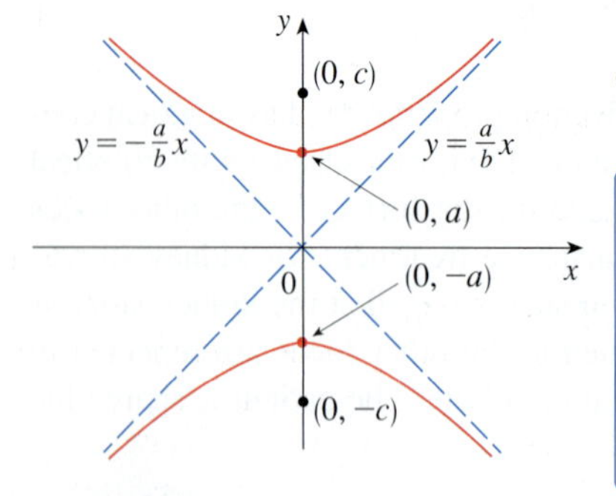

FIGURE 13
$\frac{y^2}{a^2} - \frac{x^2}{b^2} = 1$

> $\boxed{8}$ The hyperbola
> $$\frac{y^2}{a^2} - \frac{x^2}{b^2} = 1$$
> has foci $(0, \pm c)$, where $c^2 = a^2 + b^2$, vertices $(0, \pm a)$, and asymptotes $y = \pm(a/b)x$.

EXAMPLE 4 Find the foci and asymptotes of the hyperbola $9x^2 - 16y^2 = 144$ and sketch its graph.

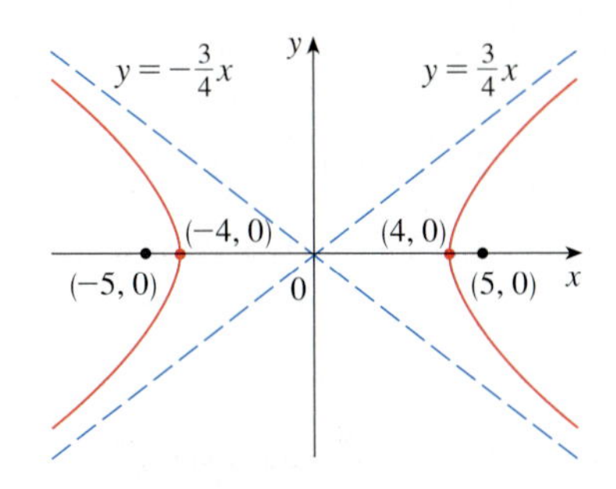

FIGURE 14
$9x^2 - 16y^2 = 144$

SOLUTION If we divide both sides of the equation by 144, it becomes

$$\frac{x^2}{16} - \frac{y^2}{9} = 1$$

which is of the form given in (7) with $a = 4$ and $b = 3$. Since $c^2 = 16 + 9 = 25$, the foci are $(\pm 5, 0)$. The asymptotes are the lines $y = \frac{3}{4}x$ and $y = -\frac{3}{4}x$. The graph is shown in Figure 14. ■

EXAMPLE 5 Find the foci and equation of the hyperbola with vertices $(0, \pm 1)$ and asymptote $y = 2x$.

SOLUTION From (8) and the given information, we see that $a = 1$ and $a/b = 2$. Thus $b = a/2 = \frac{1}{2}$ and $c^2 = a^2 + b^2 = \frac{5}{4}$. The foci are $\left(0, \pm\sqrt{5}/2\right)$ and the equation of the hyperbola is

$$y^2 - 4x^2 = 1$$ ■

Shifted Conics

As discussed in Appendix C, we shift conics by taking the standard equations (1), (2), (4), (5), (7), and (8) and replacing x and y by $x - h$ and $y - k$.

EXAMPLE 6 Find an equation of the ellipse with foci $(2, -2)$, $(4, -2)$ and vertices $(1, -2)$, $(5, -2)$.

SOLUTION The major axis is the line segment that joins the vertices $(1, -2)$, $(5, -2)$ and has length 4, so $a = 2$. The distance between the foci is 2, so $c = 1$. Thus $b^2 = a^2 - c^2 = 3$. Since the center of the ellipse is $(3, -2)$, we replace x and y in (4) by $x - 3$ and $y + 2$ to obtain

$$\frac{(x-3)^2}{4} + \frac{(y+2)^2}{3} = 1$$

as the equation of the ellipse. ■

EXAMPLE 7 Sketch the conic $9x^2 - 4y^2 - 72x + 8y + 176 = 0$ and find its foci.

SOLUTION We complete the squares as follows:

$$4(y^2 - 2y) - 9(x^2 - 8x) = 176$$

$$4(y^2 - 2y + 1) - 9(x^2 - 8x + 16) = 176 + 4 - 144$$

$$4(y-1)^2 - 9(x-4)^2 = 36$$

$$\frac{(y-1)^2}{9} - \frac{(x-4)^2}{4} = 1$$

This is in the form (8) except that x and y are replaced by $x - 4$ and $y - 1$. Thus $a^2 = 9$, $b^2 = 4$, and $c^2 = 13$. The hyperbola is shifted four units to the right and one unit upward. The foci are $\left(4, 1 + \sqrt{13}\right)$ and $\left(4, 1 - \sqrt{13}\right)$ and the vertices are $(4, 4)$ and $(4, -2)$. The asymptotes are $y - 1 = \pm\frac{3}{2}(x - 4)$. The hyperbola is sketched in Figure 15. ■

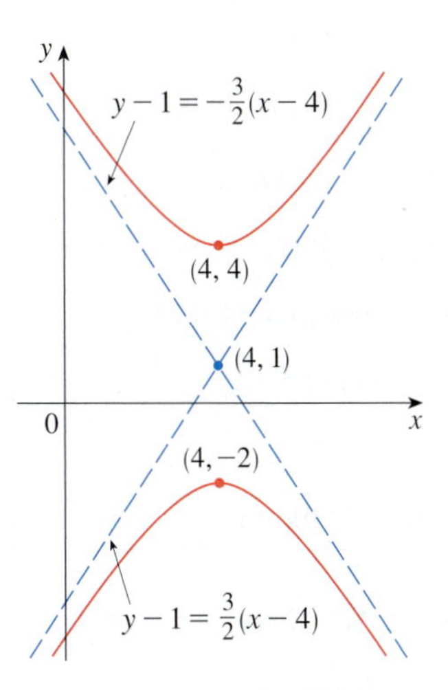

FIGURE 15
$9x^2 - 4y^2 - 72x + 8y + 176 = 0$

10.5 EXERCISES

1–8 Find the vertex, focus, and directrix of the parabola and sketch its graph.

1. $x^2 = 6y$

2. $2y^2 = 5x$

3. $2x = -y^2$

4. $3x^2 + 8y = 0$

5. $(x + 2)^2 = 8(y - 3)$

6. $(y - 2)^2 = 2x + 1$

7. $y^2 + 6y + 2x + 1 = 0$

8. $2x^2 - 16x - 3y + 38 = 0$

9–10 Find an equation of the parabola. Then find the focus and directrix.

9.

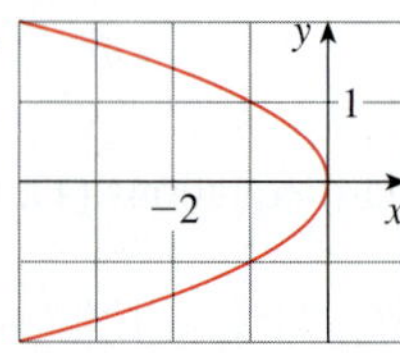

10.

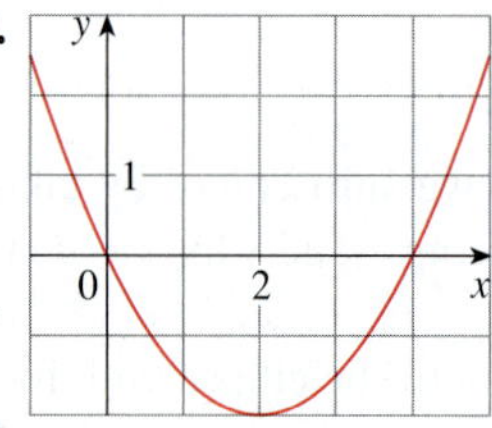

11–16 Find the vertices and foci of the ellipse and sketch its graph.

11. $\dfrac{x^2}{2} + \dfrac{y^2}{4} = 1$

12. $\dfrac{x^2}{36} + \dfrac{y^2}{8} = 1$

13. $x^2 + 9y^2 = 9$

14. $100x^2 + 36y^2 = 225$

15. $9x^2 - 18x + 4y^2 = 27$

16. $x^2 + 3y^2 + 2x - 12y + 10 = 0$

17–18 Find an equation of the ellipse. Then find its foci.

17.

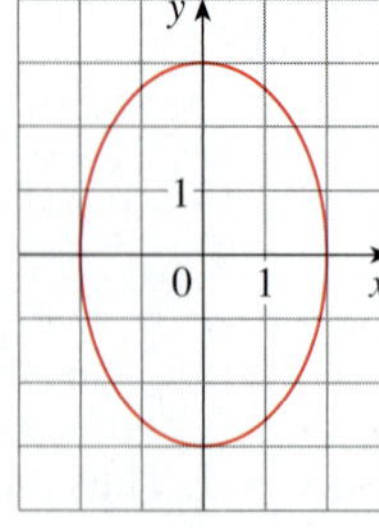

18.

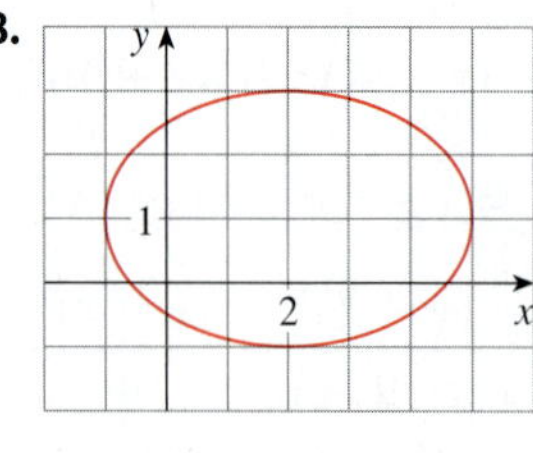

19–24 Find the vertices, foci, and asymptotes of the hyperbola and sketch its graph.

19. $\dfrac{y^2}{25} - \dfrac{x^2}{9} = 1$

20. $\dfrac{x^2}{36} - \dfrac{y^2}{64} = 1$

21. $x^2 - y^2 = 100$

22. $y^2 - 16x^2 = 16$

23. $x^2 - y^2 + 2y = 2$

24. $9y^2 - 4x^2 - 36y - 8x = 4$

25–30 Identify the type of conic section whose equation is given and find the vertices and foci.

25. $4x^2 = y^2 + 4$

26. $4x^2 = y + 4$

27. $x^2 = 4y - 2y^2$

28. $y^2 - 2 = x^2 - 2x$

29. $3x^2 - 6x - 2y = 1$

30. $x^2 - 2x + 2y^2 - 8y + 7 = 0$

31–48 Find an equation for the conic that satisfies the given conditions.

31. Parabola, vertex $(0, 0)$, focus $(1, 0)$

32. Parabola, focus $(0, 0)$, directrix $y = 6$

33. Parabola, focus $(-4, 0)$, directrix $x = 2$

34. Parabola, focus $(2, -1)$, vertex $(2, 3)$

35. Parabola, vertex $(3, -1)$, horizontal axis, passing through $(-15, 2)$

36. Parabola, vertical axis, passing through $(0, 4)$, $(1, 3)$, and $(-2, -6)$

37. Ellipse, foci $(\pm 2, 0)$, vertices $(\pm 5, 0)$

38. Ellipse, foci $(0, \pm\sqrt{2})$, vertices $(0, \pm 2)$

39. Ellipse, foci $(0, 2)$, $(0, 6)$, vertices $(0, 0)$, $(0, 8)$

40. Ellipse, foci $(0, -1)$, $(8, -1)$, vertex $(9, -1)$

41. Ellipse, center $(-1, 4)$, vertex $(-1, 0)$, focus $(-1, 6)$

42. Ellipse, foci $(\pm 4, 0)$, passing through $(-4, 1.8)$

43. Hyperbola, vertices $(\pm 3, 0)$, foci $(\pm 5, 0)$

44. Hyperbola, vertices $(0, \pm 2)$, foci $(0, \pm 5)$

45. Hyperbola, vertices $(-3, -4)$, $(-3, 6)$, foci $(-3, -7)$, $(-3, 9)$

46. Hyperbola, vertices $(-1, 2)$, $(7, 2)$, foci $(-2, 2)$, $(8, 2)$

47. Hyperbola, vertices $(\pm 3, 0)$, asymptotes $y = \pm 2x$

48. Hyperbola, foci $(2, 0)$, $(2, 8)$, asymptotes $y = 3 + \frac{1}{2}x$ and $y = 5 - \frac{1}{2}x$

49. The point in a lunar orbit nearest the surface of the moon is called *perilune* and the point farthest from the surface is called *apolune*. The *Apollo 11* spacecraft was placed in an elliptical lunar orbit with perilune altitude 110 km and apolune altitude 314 km (above the moon). Find an equation of this ellipse if the radius of the moon is 1728 km and the center of the moon is at one focus.

50. A cross-section of a parabolic reflector is shown in the figure. The bulb is located at the focus and the opening at the focus is 10 cm.

(a) Find an equation of the parabola.

(b) Find the diameter of the opening $|CD|$, 11 cm from the vertex.

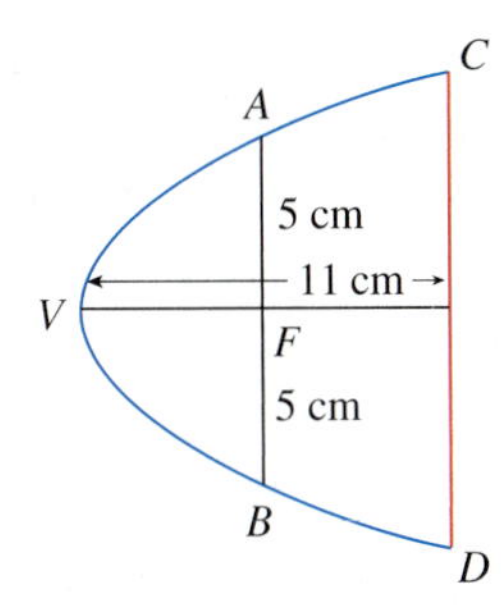

51. The LORAN (LOng RAnge Navigation) radio navigation system was widely used until the 1990s when it was superseded by the GPS system. In the LORAN system, two radio stations located at A and B transmit simultaneous signals to a ship or an aircraft located at P. The onboard computer converts the time difference in receiving these signals into a distance difference $|PA| - |PB|$, and this, according to the definition of a hyperbola, locates the ship or aircraft on one branch of a hyperbola (see the figure). Suppose that station B is located 400 mi due east of station A on a coastline. A ship received the signal from B 1200 microseconds (μs) before it received the signal from A.

(a) Assuming that radio signals travel at a speed of 980 ft/μs, find an equation of the hyperbola on which the ship lies.

(b) If the ship is due north of B, how far off the coastline is the ship?

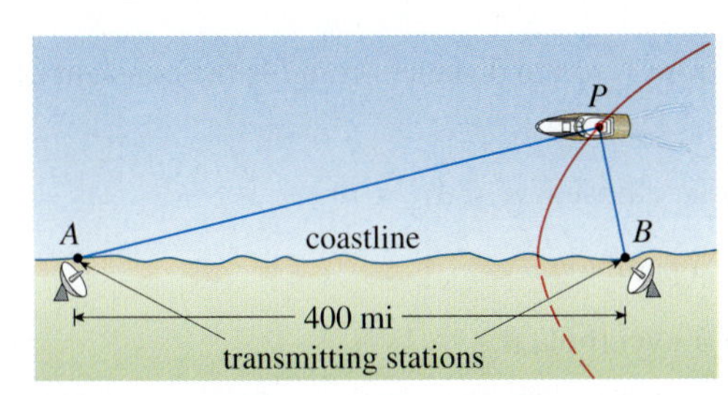

52. Use the definition of a hyperbola to derive Equation 6 for a hyperbola with foci $(\pm c, 0)$ and vertices $(\pm a, 0)$.

53. Show that the function defined by the upper branch of the hyperbola $y^2/a^2 - x^2/b^2 = 1$ is concave upward.

54. Find an equation for the ellipse with foci $(1, 1)$ and $(-1, -1)$ and major axis of length 4.

55. Determine the type of curve represented by the equation

$$\frac{x^2}{k} + \frac{y^2}{k - 16} = 1$$

in each of the following cases:

(a) $k > 16$ (b) $0 < k < 16$ (c) $k < 0$

(d) Show that all the curves in parts (a) and (b) have the same foci, no matter what the value of k is.

56. (a) Show that the equation of the tangent line to the parabola $y^2 = 4px$ at the point (x_0, y_0) can be written as

$$y_0 y = 2p(x + x_0)$$

(b) What is the x-intercept of this tangent line? Use this fact to draw the tangent line.

57. Show that the tangent lines to the parabola $x^2 = 4py$ drawn from any point on the directrix are perpendicular.

58. Show that if an ellipse and a hyperbola have the same foci, then their tangent lines at each point of intersection are perpendicular.

59. Use parametric equations and Simpson's Rule with $n = 8$ to estimate the circumference of the ellipse $9x^2 + 4y^2 = 36$.

60. The dwarf planet Pluto travels in an elliptical orbit around the sun (at one focus). The length of the major axis is 1.18×10^{10} km and the length of the minor axis is 1.14×10^{10} km. Use Simpson's Rule with $n = 10$ to estimate the distance traveled by the planet during one complete orbit around the sun.

61. Find the area of the region enclosed by the hyperbola $x^2/a^2 - y^2/b^2 = 1$ and the vertical line through a focus.

62. (a) If an ellipse is rotated about its major axis, find the volume of the resulting solid.

(b) If it is rotated about its minor axis, find the resulting volume.

63. Find the centroid of the region enclosed by the x-axis and the top half of the ellipse $9x^2 + 4y^2 = 36$.

64. (a) Calculate the surface area of the ellipsoid that is generated by rotating an ellipse about its major axis.

(b) What is the surface area if the ellipse is rotated about its minor axis?

65. Let $P(x_1, y_1)$ be a point on the ellipse $x^2/a^2 + y^2/b^2 = 1$ with foci F_1 and F_2 and let α and β be the angles between the lines

PF_1, PF_2 and the ellipse as shown in the figure. Prove that $\alpha = \beta$. This explains how whispering galleries and lithotripsy work. Sound coming from one focus is reflected and passes through the other focus. [*Hint:* Use the formula in Problem 17 on page 201 to show that $\tan \alpha = \tan \beta$.]

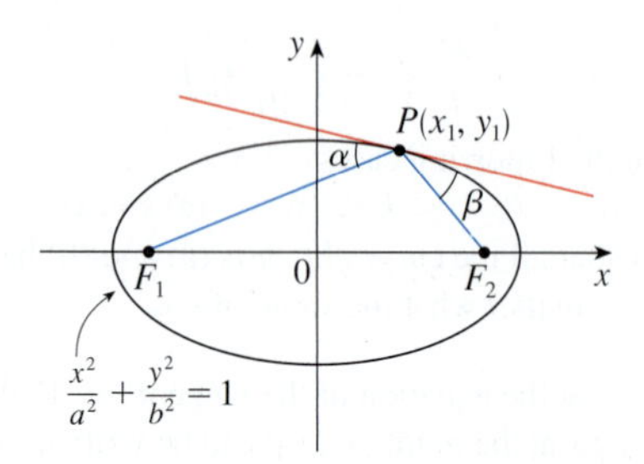

66. Let $P(x_1, y_1)$ be a point on the hyperbola $x^2/a^2 - y^2/b^2 = 1$ with foci F_1 and F_2 and let α and β be the angles between the lines PF_1, PF_2 and the hyperbola as shown in the figure. Prove that $\alpha = \beta$. (This is the reflection property of the hyperbola. It shows that light aimed at a focus F_2 of a hyperbolic mirror is reflected toward the other focus F_1.)

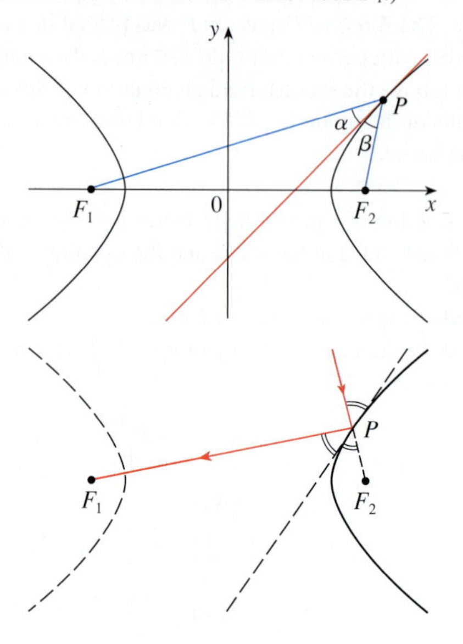

10.6 Conic Sections in Polar Coordinates

In the preceding section we defined the parabola in terms of a focus and directrix, but we defined the ellipse and hyperbola in terms of two foci. In this section we give a more unified treatment of all three types of conic sections in terms of a focus and directrix. Furthermore, if we place the focus at the origin, then a conic section has a simple polar equation, which provides a convenient description of the motion of planets, satellites, and comets.

1 Theorem Let F be a fixed point (called the **focus**) and l be a fixed line (called the **directrix**) in a plane. Let e be a fixed positive number (called the **eccentricity**). The set of all points P in the plane such that

$$\frac{|PF|}{|Pl|} = e$$

(that is, the ratio of the distance from F to the distance from l is the constant e) is a conic section. The conic is

(a) an ellipse if $e < 1$

(b) a parabola if $e = 1$

(c) a hyperbola if $e > 1$

PROOF Notice that if the eccentricity is $e = 1$, then $|PF| = |Pl|$ and so the given condition simply becomes the definition of a parabola as given in Section 10.5.

Let us place the focus F at the origin and the directrix parallel to the y-axis and d units to the right. Thus the directrix has equation $x = d$ and is perpendicular to the

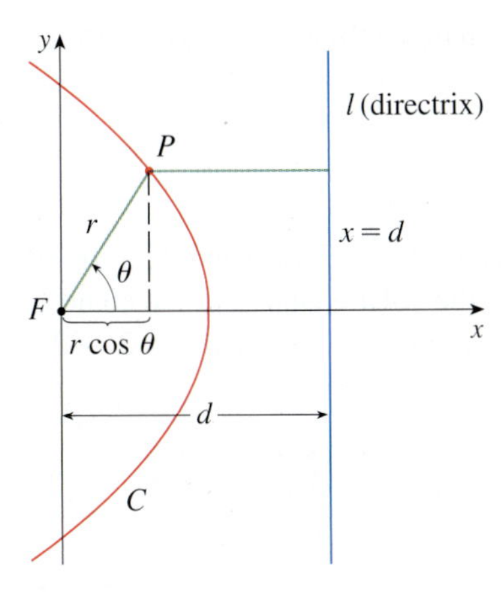

FIGURE 1

polar axis. If the point P has polar coordinates (r, θ), we see from Figure 1 that

$$|PF| = r \qquad |Pl| = d - r\cos\theta$$

Thus the condition $|PF|/|Pl| = e$, or $|PF| = e|Pl|$, becomes

2 $$r = e(d - r\cos\theta)$$

If we square both sides of this polar equation and convert to rectangular coordinates, we get

$$x^2 + y^2 = e^2(d - x)^2 = e^2(d^2 - 2dx + x^2)$$

or

$$(1 - e^2)x^2 + 2de^2x + y^2 = e^2d^2$$

After completing the square, we have

3 $$\left(x + \frac{e^2d}{1 - e^2}\right)^2 + \frac{y^2}{1 - e^2} = \frac{e^2d^2}{(1 - e^2)^2}$$

If $e < 1$, we recognize Equation 3 as the equation of an ellipse. In fact, it is of the form

$$\frac{(x - h)^2}{a^2} + \frac{y^2}{b^2} = 1$$

where

4 $$h = -\frac{e^2d}{1 - e^2} \qquad a^2 = \frac{e^2d^2}{(1 - e^2)^2} \qquad b^2 = \frac{e^2d^2}{1 - e^2}$$

In Section 10.5 we found that the foci of an ellipse are at a distance c from the center, where

5 $$c^2 = a^2 - b^2 = \frac{e^4d^2}{(1 - e^2)^2}$$

This shows that

$$c = \frac{e^2d}{1 - e^2} = -h$$

and confirms that the focus as defined in Theorem 1 means the same as the focus defined in Section 10.5. It also follows from Equations 4 and 5 that the eccentricity is given by

$$e = \frac{c}{a}$$

If $e > 1$, then $1 - e^2 < 0$ and we see that Equation 3 represents a hyperbola. Just as we did before, we could rewrite Equation 3 in the form

$$\frac{(x - h)^2}{a^2} - \frac{y^2}{b^2} = 1$$

and see that

$$e = \frac{c}{a} \qquad \text{where} \quad c^2 = a^2 + b^2$$

■

By solving Equation 2 for r, we see that the polar equation of the conic shown in Figure 1 can be written as

$$r = \frac{ed}{1 + e\cos\theta}$$

If the directrix is chosen to be to the left of the focus as $x = -d$, or if the directrix is chosen to be parallel to the polar axis as $y = \pm d$, then the polar equation of the conic is given by the following theorem, which is illustrated by Figure 2. (See Exercises 21–23.)

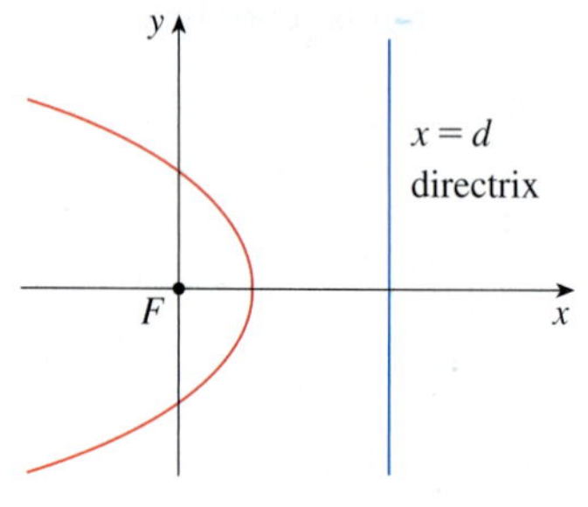

(a) $r = \dfrac{ed}{1 + e\cos\theta}$

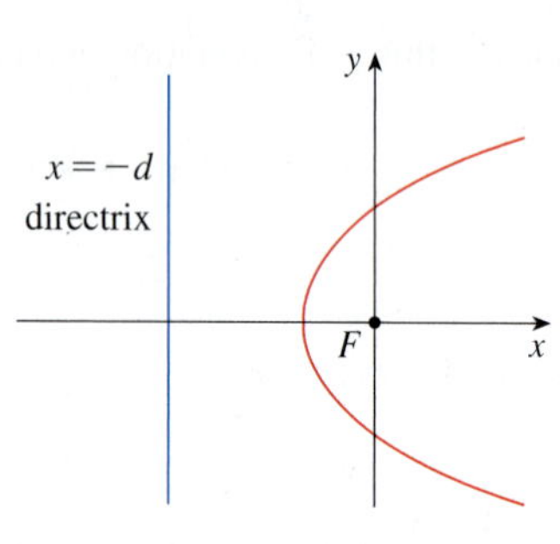

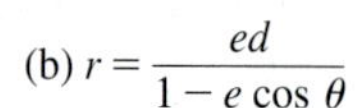
(b) $r = \dfrac{ed}{1 - e\cos\theta}$

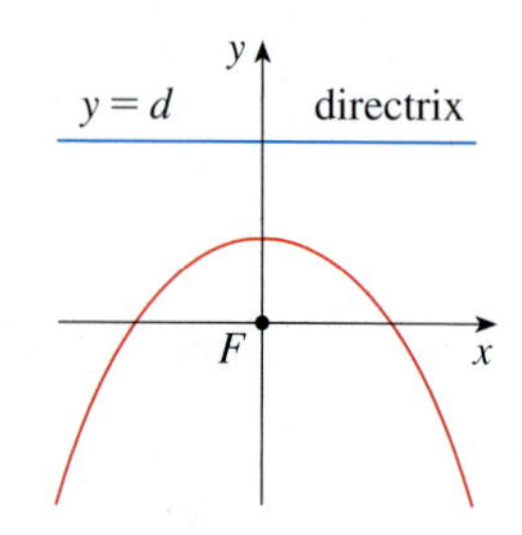

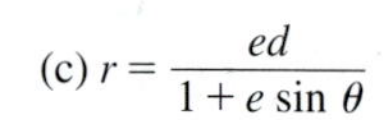
(c) $r = \dfrac{ed}{1 + e\sin\theta}$

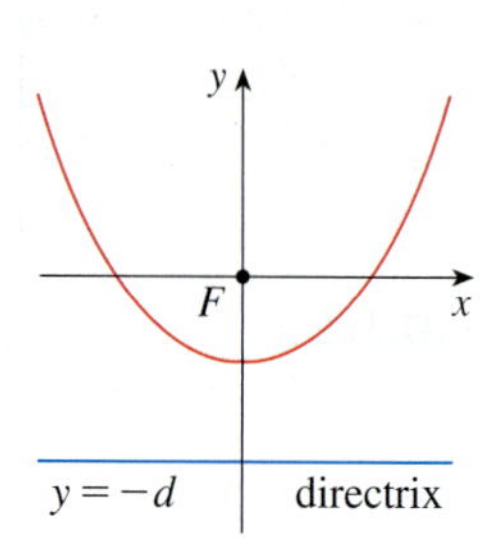

(d) $r = \dfrac{ed}{1 - e\sin\theta}$

FIGURE 2
Polar equations of conics

6 Theorem A polar equation of the form

$$r = \frac{ed}{1 \pm e\cos\theta} \qquad \text{or} \qquad r = \frac{ed}{1 \pm e\sin\theta}$$

represents a conic section with eccentricity e. The conic is an ellipse if $e < 1$, a parabola if $e = 1$, or a hyperbola if $e > 1$.

EXAMPLE 1 Find a polar equation for a parabola that has its focus at the origin and whose directrix is the line $y = -6$.

SOLUTION Using Theorem 6 with $e = 1$ and $d = 6$, and using part (d) of Figure 2, we see that the equation of the parabola is

$$r = \frac{6}{1 - \sin\theta}$$

■

EXAMPLE 2 A conic is given by the polar equation

$$r = \frac{10}{3 - 2\cos\theta}$$

Find the eccentricity, identify the conic, locate the directrix, and sketch the conic.

SOLUTION Dividing numerator and denominator by 3, we write the equation as

$$r = \frac{\frac{10}{3}}{1 - \frac{2}{3}\cos\theta}$$

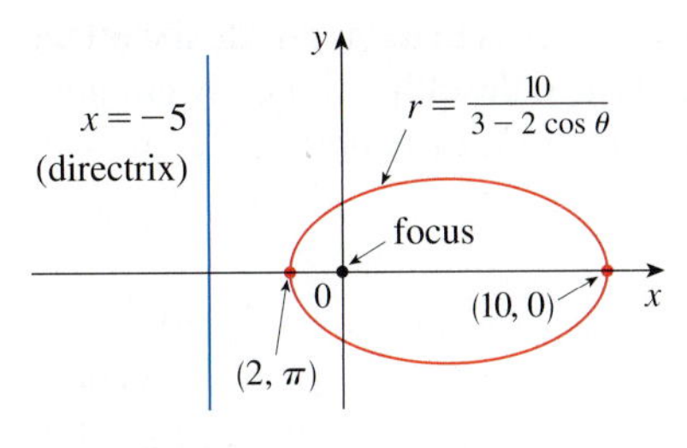

FIGURE 3

From Theorem 6 we see that this represents an ellipse with $e = \frac{2}{3}$. Since $ed = \frac{10}{3}$, we have

$$d = \frac{\frac{10}{3}}{e} = \frac{\frac{10}{3}}{\frac{2}{3}} = 5$$

so the directrix has Cartesian equation $x = -5$. When $\theta = 0$, $r = 10$; when $\theta = \pi$, $r = 2$. So the vertices have polar coordinates $(10, 0)$ and $(2, \pi)$. The ellipse is sketched in Figure 3. ■

EXAMPLE 3 Sketch the conic $r = \dfrac{12}{2 + 4\sin\theta}$.

SOLUTION Writing the equation in the form

$$r = \frac{6}{1 + 2\sin\theta}$$

we see that the eccentricity is $e = 2$ and the equation therefore represents a hyperbola. Since $ed = 6$, $d = 3$ and the directrix has equation $y = 3$. The vertices occur when $\theta = \pi/2$ and $3\pi/2$, so they are $(2, \pi/2)$ and $(-6, 3\pi/2) = (6, \pi/2)$. It is also useful to plot the x-intercepts. These occur when $\theta = 0, \pi$; in both cases $r = 6$. For additional accuracy we could draw the asymptotes. Note that $r \to \pm\infty$ when $1 + 2\sin\theta \to 0^+$ or 0^- and $1 + 2\sin\theta = 0$ when $\sin\theta = -\frac{1}{2}$. Thus the asymptotes are parallel to the rays $\theta = 7\pi/6$ and $\theta = 11\pi/6$. The hyperbola is sketched in Figure 4.

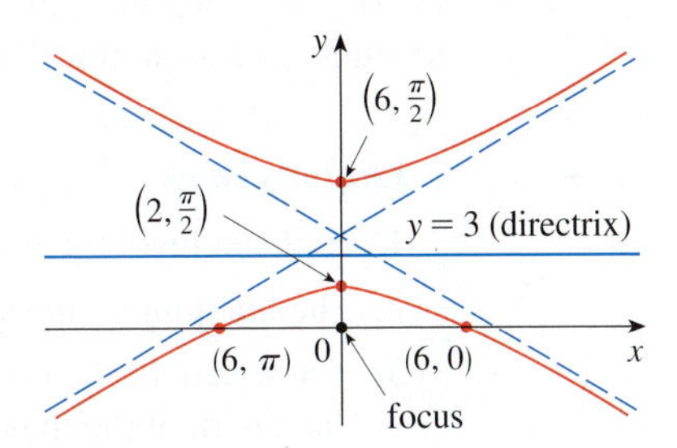

FIGURE 4

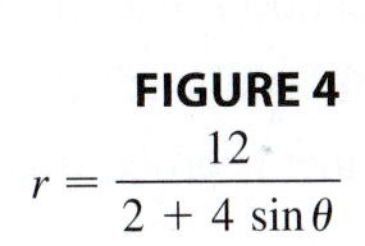

$r = \dfrac{12}{2 + 4\sin\theta}$ ■

When rotating conic sections, we find it much more convenient to use polar equations than Cartesian equations. We just use the fact (see Exercise 10.3.73) that the graph of $r = f(\theta - \alpha)$ is the graph of $r = f(\theta)$ rotated counterclockwise about the origin through an angle α.

EXAMPLE 4 If the ellipse of Example 2 is rotated through an angle $\pi/4$ about the origin, find a polar equation and graph the resulting ellipse.

SOLUTION We get the equation of the rotated ellipse by replacing θ with $\theta - \pi/4$ in the equation given in Example 2. So the new equation is

$$r = \frac{10}{3 - 2\cos(\theta - \pi/4)}$$

We use this equation to graph the rotated ellipse in Figure 5. Notice that the ellipse has been rotated about its left focus. ■

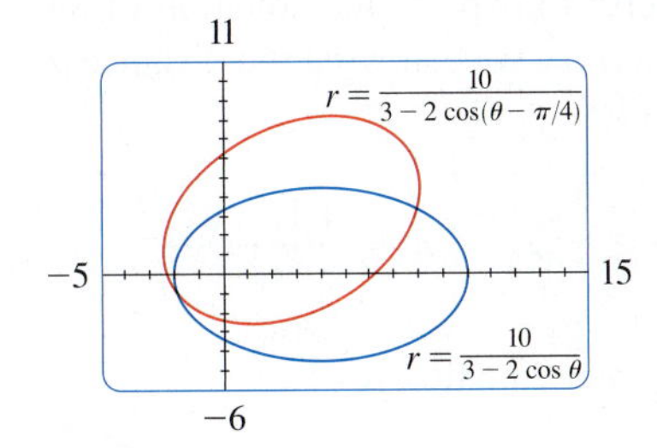

FIGURE 5

In Figure 6 we use a computer to sketch a number of conics to demonstrate the effect of varying the eccentricity e. Notice that when e is close to 0 the ellipse is nearly circular, whereas it becomes more elongated as $e \to 1^-$. When $e = 1$, of course, the conic is a parabola.

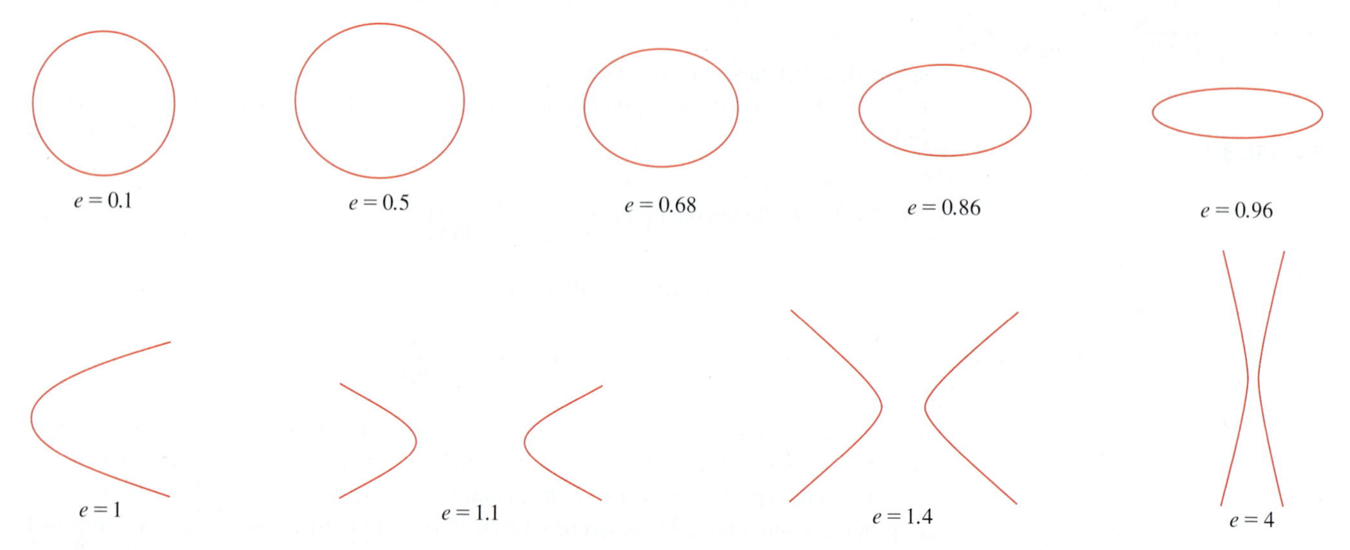

FIGURE 6

Kepler's Laws

In 1609 the German mathematician and astronomer Johannes Kepler, on the basis of huge amounts of astronomical data, published the following three laws of planetary motion.

> **Kepler's Laws**
>
> **1.** A planet revolves around the sun in an elliptical orbit with the sun at one focus.
>
> **2.** The line joining the sun to a planet sweeps out equal areas in equal times.
>
> **3.** The square of the period of revolution of a planet is proportional to the cube of the length of the major axis of its orbit.

Although Kepler formulated his laws in terms of the motion of planets around the sun, they apply equally well to the motion of moons, comets, satellites, and other bodies that orbit subject to a single gravitational force. In Section 13.4 we will show how to deduce Kepler's Laws from Newton's Laws. Here we use Kepler's First Law, together with the polar equation of an ellipse, to calculate quantities of interest in astronomy.

For purposes of astronomical calculations, it's useful to express the equation of an ellipse in terms of its eccentricity e and its semimajor axis a. We can write the distance d from the focus to the directrix in terms of a if we use (4):

$$a^2 = \frac{e^2 d^2}{(1 - e^2)^2} \quad \Rightarrow \quad d^2 = \frac{a^2(1 - e^2)^2}{e^2} \quad \Rightarrow \quad d = \frac{a(1 - e^2)}{e}$$

So $ed = a(1 - e^2)$. If the directrix is $x = d$, then the polar equation is

$$r = \frac{ed}{1 + e\cos\theta} = \frac{a(1 - e^2)}{1 + e\cos\theta}$$

7 The polar equation of an ellipse with focus at the origin, semimajor axis a, eccentricity e, and directrix $x = d$ can be written in the form

$$r = \frac{a(1 - e^2)}{1 + e\cos\theta}$$

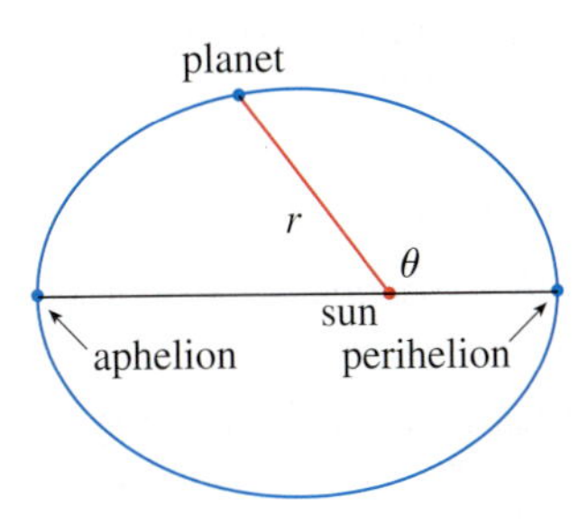

FIGURE 7

The positions of a planet that are closest to and farthest from the sun are called its **perihelion** and **aphelion**, respectively, and correspond to the vertices of the ellipse (see Figure 7). The distances from the sun to the perihelion and aphelion are called the **perihelion distance** and **aphelion distance**, respectively. In Figure 1 on page 723 the sun is at the focus F, so at perihelion we have $\theta = 0$ and, from Equation 7,

$$r = \frac{a(1 - e^2)}{1 + e\cos 0} = \frac{a(1 - e)(1 + e)}{1 + e} = a(1 - e)$$

Similarly, at aphelion $\theta = \pi$ and $r = a(1 + e)$.

8 The perihelion distance from a planet to the sun is $a(1 - e)$ and the aphelion distance is $a(1 + e)$.

EXAMPLE 5

(a) Find an approximate polar equation for the elliptical orbit of the earth around the sun (at one focus) given that the eccentricity is about 0.017 and the length of the major axis is about 2.99×10^8 km.
(b) Find the distance from the earth to the sun at perihelion and at aphelion.

SOLUTION
(a) The length of the major axis is $2a = 2.99 \times 10^8$, so $a = 1.495 \times 10^8$. We are given that $e = 0.017$ and so, from Equation 7, an equation of the earth's orbit around the sun is

$$r = \frac{a(1 - e^2)}{1 + e\cos\theta} = \frac{(1.495 \times 10^8)[1 - (0.017)^2]}{1 + 0.017\cos\theta}$$

or, approximately,

$$r = \frac{1.49 \times 10^8}{1 + 0.017\cos\theta}$$

(b) From (8), the perihelion distance from the earth to the sun is

$$a(1 - e) \approx (1.495 \times 10^8)(1 - 0.017) \approx 1.47 \times 10^8 \text{ km}$$

and the aphelion distance is

$$a(1 + e) \approx (1.495 \times 10^8)(1 + 0.017) \approx 1.52 \times 10^8 \text{ km}$$ ■

10.6 EXERCISES

1–8 Write a polar equation of a conic with the focus at the origin and the given data.

1. Ellipse, eccentricity $\frac{1}{2}$, directrix $x = 4$

2. Parabola, directrix $x = -3$

3. Hyperbola, eccentricity 1.5, directrix $y = 2$

4. Hyperbola, eccentricity 3, directrix $x = 3$

5. Ellipse, eccentricity $\frac{2}{3}$, vertex $(2, \pi)$

6. Ellipse, eccentricity 0.6, directrix $r = 4 \csc\theta$

7. Parabola, vertex $(3, \pi/2)$

8. Hyperbola, eccentricity 2, directrix $r = -2 \sec\theta$

9–16 (a) Find the eccentricity, (b) identify the conic, (c) give an equation of the directrix, and (d) sketch the conic.

9. $r = \dfrac{4}{5 - 4\sin\theta}$

10. $r = \dfrac{1}{2 + \sin\theta}$

11. $r = \dfrac{2}{3 + 3\sin\theta}$

12. $r = \dfrac{5}{2 - 4\cos\theta}$

13. $r = \dfrac{9}{6 + 2\cos\theta}$

14. $r = \dfrac{1}{3 - 3\sin\theta}$

15. $r = \dfrac{3}{4 - 8\cos\theta}$

16. $r = \dfrac{4}{2 + 3\cos\theta}$

17. (a) Find the eccentricity and directrix of the conic $r = 1/(1 - 2\sin\theta)$ and graph the conic and its directrix.
(b) If this conic is rotated counterclockwise about the origin through an angle $3\pi/4$, write the resulting equation and graph its curve.

18. Graph the conic $r = 4/(5 + 6\cos\theta)$ and its directrix. Also graph the conic obtained by rotating this curve about the origin through an angle $\pi/3$.

19. Graph the conics $r = e/(1 - e\cos\theta)$ with $e = 0.4, 0.6, 0.8$, and 1.0 on a common screen. How does the value of e affect the shape of the curve?

20. (a) Graph the conics $r = ed/(1 + e\sin\theta)$ for $e = 1$ and various values of d. How does the value of d affect the shape of the conic?
(b) Graph these conics for $d = 1$ and various values of e. How does the value of e affect the shape of the conic?

21. Show that a conic with focus at the origin, eccentricity e, and directrix $x = -d$ has polar equation

$$r = \frac{ed}{1 - e\cos\theta}$$

22. Show that a conic with focus at the origin, eccentricity e, and directrix $y = d$ has polar equation

$$r = \frac{ed}{1 + e\sin\theta}$$

23. Show that a conic with focus at the origin, eccentricity e, and directrix $y = -d$ has polar equation

$$r = \frac{ed}{1 - e\sin\theta}$$

24. Show that the parabolas $r = c/(1 + \cos\theta)$ and $r = d/(1 - \cos\theta)$ intersect at right angles.

25. The orbit of Mars around the sun is an ellipse with eccentricity 0.093 and semimajor axis 2.28×10^8 km. Find a polar equation for the orbit.

26. Jupiter's orbit has eccentricity 0.048 and the length of the major axis is 1.56×10^9 km. Find a polar equation for the orbit.

27. The orbit of Halley's comet, last seen in 1986 and due to return in 2061, is an ellipse with eccentricity 0.97 and one focus at the sun. The length of its major axis is 36.18 AU. [An astronomical unit (AU) is the mean distance between the earth and the sun, about 93 million miles.] Find a polar equation for the orbit of Halley's comet. What is the maximum distance from the comet to the sun?

28. Comet Hale-Bopp, discovered in 1995, has an elliptical orbit with eccentricity 0.9951. The length of the orbit's major axis is 356.5 AU. Find a polar equation for the orbit of this comet. How close to the sun does it come?

29. The planet Mercury travels in an elliptical orbit with eccentricity 0.206. Its minimum distance from the sun is 4.6×10^7 km. Find its maximum distance from the sun.

30. The distance from the dwarf planet Pluto to the sun is 4.43×10^9 km at perihelion and 7.37×10^9 km at aphelion. Find the eccentricity of Pluto's orbit.

31. Using the data from Exercise 29, find the distance traveled by the planet Mercury during one complete orbit around the sun. (If your calculator or computer algebra system evaluates definite integrals, use it. Otherwise, use Simpson's Rule.)

10 REVIEW

CONCEPT CHECK

Answers to the Concept Check can be found on the back endpapers.

1. (a) What is a parametric curve?
   (b) How do you sketch a parametric curve?
2. (a) How do you find the slope of a tangent to a parametric curve?
   (b) How do you find the area under a parametric curve?
3. Write an expression for each of the following:
   (a) The length of a parametric curve
   (b) The area of the surface obtained by rotating a parametric curve about the x-axis
4. (a) Use a diagram to explain the meaning of the polar coordinates (r, θ) of a point.
   (b) Write equations that express the Cartesian coordinates (x, y) of a point in terms of the polar coordinates.
   (c) What equations would you use to find the polar coordinates of a point if you knew the Cartesian coordinates?
5. (a) How do you find the slope of a tangent line to a polar curve?
   (b) How do you find the area of a region bounded by a polar curve?
   (c) How do you find the length of a polar curve?
6. (a) Give a geometric definition of a parabola.
   (b) Write an equation of a parabola with focus $(0, p)$ and directrix $y = -p$. What if the focus is $(p, 0)$ and the directrix is $x = -p$?
7. (a) Give a definition of an ellipse in terms of foci.
   (b) Write an equation for the ellipse with foci $(\pm c, 0)$ and vertices $(\pm a, 0)$.
8. (a) Give a definition of a hyperbola in terms of foci.
   (b) Write an equation for the hyperbola with foci $(\pm c, 0)$ and vertices $(\pm a, 0)$.
   (c) Write equations for the asymptotes of the hyperbola in part (b).
9. (a) What is the eccentricity of a conic section?
   (b) What can you say about the eccentricity if the conic section is an ellipse? A hyperbola? A parabola?
   (c) Write a polar equation for a conic section with eccentricity e and directrix $x = d$. What if the directrix is $x = -d$? $y = d$? $y = -d$?

TRUE-FALSE QUIZ

Determine whether the statement is true or false. If it is true, explain why. If it is false, explain why or give an example that disproves the statement.

1. If the parametric curve $x = f(t)$, $y = g(t)$ satisfies $g'(1) = 0$, then it has a horizontal tangent when $t = 1$.
2. If $x = f(t)$ and $y = g(t)$ are twice differentiable, then
$$\frac{d^2y}{dx^2} = \frac{d^2y/dt^2}{d^2x/dt^2}$$
3. The length of the curve $x = f(t)$, $y = g(t)$, $a \leqslant t \leqslant b$, is $\int_a^b \sqrt{[f'(t)]^2 + [g'(t)]^2}\, dt$.
4. If a point is represented by (x, y) in Cartesian coordinates (where $x \neq 0$) and (r, θ) in polar coordinates, then $\theta = \tan^{-1}(y/x)$.
5. The polar curves
$$r = 1 - \sin 2\theta \qquad r = \sin 2\theta - 1$$
have the same graph.
6. The equations $r = 2$, $x^2 + y^2 = 4$, and $x = 2 \sin 3t$, $y = 2 \cos 3t$ $(0 \leqslant t \leqslant 2\pi)$ all have the same graph.
7. The parametric equations $x = t^2$, $y = t^4$ have the same graph as $x = t^3$, $y = t^6$.
8. The graph of $y^2 = 2y + 3x$ is a parabola.
9. A tangent line to a parabola intersects the parabola only once.
10. A hyperbola never intersects its directrix.

EXERCISES

1–4 Sketch the parametric curve and eliminate the parameter to find the Cartesian equation of the curve.

1. $x = t^2 + 4t, \quad y = 2 - t, \quad -4 \leqslant t \leqslant 1$

2. $x = 1 + e^{2t}, \quad y = e^t$

3. $x = \cos\theta, \quad y = \sec\theta, \quad 0 \leqslant \theta < \pi/2$

4. $x = 2\cos\theta, \quad y = 1 + \sin\theta$

5. Write three different sets of parametric equations for the curve $y = \sqrt{x}$.

6. Use the graphs of $x = f(t)$ and $y = g(t)$ to sketch the parametric curve $x = f(t)$, $y = g(t)$. Indicate with arrows the direction in which the curve is traced as t increases.

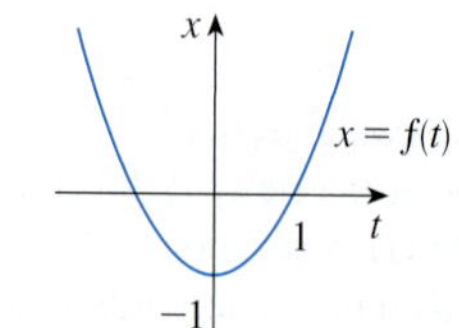

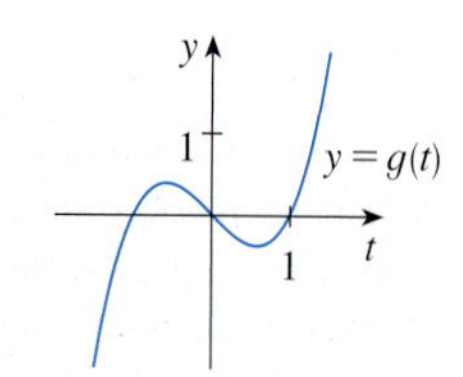

7. (a) Plot the point with polar coordinates $(4, 2\pi/3)$. Then find its Cartesian coordinates.
(b) The Cartesian coordinates of a point are $(-3, 3)$. Find two sets of polar coordinates for the point.

8. Sketch the region consisting of points whose polar coordinates satisfy $1 \leqslant r < 2$ and $\pi/6 \leqslant \theta \leqslant 5\pi/6$.

9–16 Sketch the polar curve.

9. $r = 1 + \sin\theta$

10. $r = \sin 4\theta$

11. $r = \cos 3\theta$

12. $r = 3 + \cos 3\theta$

13. $r = 1 + \cos 2\theta$

14. $r = 2\cos(\theta/2)$

15. $r = \dfrac{3}{1 + 2\sin\theta}$

16. $r = \dfrac{3}{2 - 2\cos\theta}$

17–18 Find a polar equation for the curve represented by the given Cartesian equation.

17. $x + y = 2$

18. $x^2 + y^2 = 2$

19. The curve with polar equation $r = (\sin\theta)/\theta$ is called a **cochleoid**. Use a graph of r as a function of θ in Cartesian coordinates to sketch the cochleoid by hand. Then graph it with a machine to check your sketch.

20. Graph the ellipse $r = 2/(4 - 3\cos\theta)$ and its directrix. Also graph the ellipse obtained by rotation about the origin through an angle $2\pi/3$.

21–24 Find the slope of the tangent line to the given curve at the point corresponding to the specified value of the parameter.

21. $x = \ln t, \quad y = 1 + t^2; \quad t = 1$

22. $x = t^3 + 6t + 1, \quad y = 2t - t^2; \quad t = -1$

23. $r = e^{-\theta}; \quad \theta = \pi$

24. $r = 3 + \cos 3\theta; \quad \theta = \pi/2$

25–26 Find dy/dx and d^2y/dx^2.

25. $x = t + \sin t, \quad y = t - \cos t$

26. $x = 1 + t^2, \quad y = t - t^3$

27. Use a graph to estimate the coordinates of the lowest point on the curve $x = t^3 - 3t$, $y = t^2 + t + 1$. Then use calculus to find the exact coordinates.

28. Find the area enclosed by the loop of the curve in Exercise 27.

29. At what points does the curve

$$x = 2a\cos t - a\cos 2t \qquad y = 2a\sin t - a\sin 2t$$

have vertical or horizontal tangents? Use this information to help sketch the curve.

30. Find the area enclosed by the curve in Exercise 29.

31. Find the area enclosed by the curve $r^2 = 9\cos 5\theta$.

32. Find the area enclosed by the inner loop of the curve $r = 1 - 3\sin\theta$.

33. Find the points of intersection of the curves $r = 2$ and $r = 4\cos\theta$.

34. Find the points of intersection of the curves $r = \cot\theta$ and $r = 2\cos\theta$.

35. Find the area of the region that lies inside both of the circles $r = 2\sin\theta$ and $r = \sin\theta + \cos\theta$.

36. Find the area of the region that lies inside the curve $r = 2 + \cos 2\theta$ but outside the curve $r = 2 + \sin\theta$.

37–40 Find the length of the curve.

37. $x = 3t^2, \quad y = 2t^3, \quad 0 \leqslant t \leqslant 2$

38. $x = 2 + 3t, \quad y = \cosh 3t, \quad 0 \leqslant t \leqslant 1$

39. $r = 1/\theta, \quad \pi \leqslant \theta \leqslant 2\pi$

40. $r = \sin^3(\theta/3), \quad 0 \leqslant \theta \leqslant \pi$

41–42 Find the area of the surface obtained by rotating the given curve about the x-axis.

41. $x = 4\sqrt{t}, \quad y = \dfrac{t^3}{3} + \dfrac{1}{2t^2}, \quad 1 \leqslant t \leqslant 4$

42. $x = 2 + 3t, \quad y = \cosh 3t, \quad 0 \leqslant t \leqslant 1$

43. The curves defined by the parametric equations

$$x = \frac{t^2 - c}{t^2 + 1} \qquad y = \frac{t(t^2 - c)}{t^2 + 1}$$

are called **strophoids** (from a Greek word meaning "to turn or twist"). Investigate how these curves vary as c varies.

44. A family of curves has polar equations $r^a = |\sin 2\theta|$ where a is a positive number. Investigate how the curves change as a changes.

45–48 Find the foci and vertices and sketch the graph.

45. $\dfrac{x^2}{9} + \dfrac{y^2}{8} = 1$

46. $4x^2 - y^2 = 16$

47. $6y^2 + x - 36y + 55 = 0$

48. $25x^2 + 4y^2 + 50x - 16y = 59$

49. Find an equation of the ellipse with foci $(\pm 4, 0)$ and vertices $(\pm 5, 0)$.

50. Find an equation of the parabola with focus $(2, 1)$ and directrix $x = -4$.

51. Find an equation of the hyperbola with foci $(0, \pm 4)$ and asymptotes $y = \pm 3x$.

52. Find an equation of the ellipse with foci $(3, \pm 2)$ and major axis with length 8.

53. Find an equation for the ellipse that shares a vertex and a focus with the parabola $x^2 + y = 100$ and that has its other focus at the origin.

54. Show that if m is any real number, then there are exactly two lines of slope m that are tangent to the ellipse $x^2/a^2 + y^2/b^2 = 1$ and their equations are

$$y = mx \pm \sqrt{a^2m^2 + b^2}$$

55. Find a polar equation for the ellipse with focus at the origin, eccentricity $\frac{1}{3}$, and directrix with equation $r = 4 \sec\theta$.

56. Show that the angles between the polar axis and the asymptotes of the hyperbola $r = ed/(1 - e\cos\theta)$, $e > 1$, are given by $\cos^{-1}(\pm 1/e)$.

57. In the figure the circle of radius a is stationary, and for every θ, the point P is the midpoint of the segment QR. The curve traced out by P for $0 < \theta < \pi$ is called the **long-bow curve**. Find parametric equations for this curve.

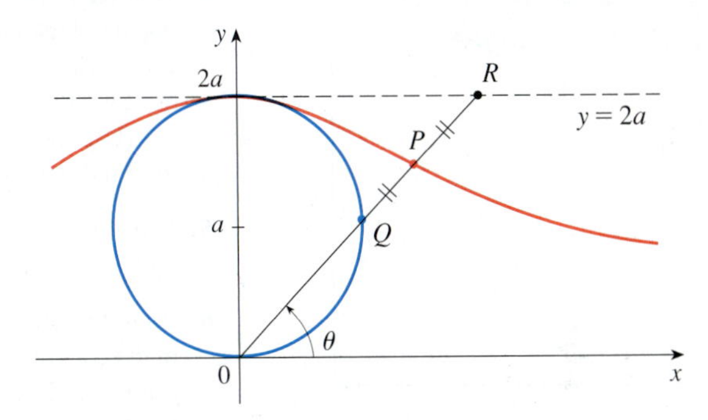

58. A curve called the **folium of Descartes** is defined by the parametric equations

$$x = \frac{3t}{1 + t^3} \qquad y = \frac{3t^2}{1 + t^3}$$

(a) Show that if (a, b) lies on the curve, then so does (b, a); that is, the curve is symmetric with respect to the line $y = x$. Where does the curve intersect this line?
(b) Find the points on the curve where the tangent lines are horizontal or vertical.
(c) Show that the line $y = -x - 1$ is a slant asymptote.
(d) Sketch the curve.
(e) Show that a Cartesian equation of this curve is $x^3 + y^3 = 3xy$.
(f) Show that the polar equation can be written in the form

$$r = \frac{3\sec\theta\tan\theta}{1 + \tan^3\theta}$$

(g) Find the area enclosed by the loop of this curve.
CAS (h) Show that the area of the loop is the same as the area that lies between the asymptote and the infinite branches of the curve. (Use a computer algebra system to evaluate the integral.)

Problems Plus

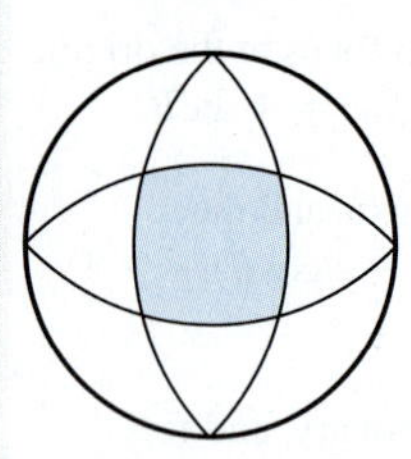

FIGURE FOR PROBLEM 1

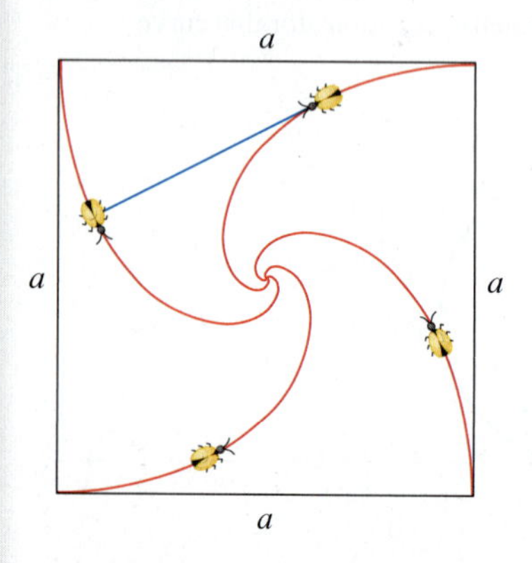

FIGURE FOR PROBLEM 4

1. The outer circle in the figure has radius 1 and the centers of the interior circular arcs lie on the outer circle. Find the area of the shaded region.

2. (a) Find the highest and lowest points on the curve $x^4 + y^4 = x^2 + y^2$.
   (b) Sketch the curve. (Notice that it is symmetric with respect to both axes and both of the lines $y = \pm x$, so it suffices to consider $y \geqslant x \geqslant 0$ initially.)
   CAS (c) Use polar coordinates and a computer algebra system to find the area enclosed by the curve.

3. What is the smallest viewing rectangle that contains every member of the family of polar curves $r = 1 + c \sin\theta$, where $0 \leqslant c \leqslant 1$? Illustrate your answer by graphing several members of the family in this viewing rectangle.

4. Four bugs are placed at the four corners of a square with side length a. The bugs crawl counterclockwise at the same speed and each bug crawls directly toward the next bug at all times. They approach the center of the square along spiral paths.
   (a) Find the polar equation of a bug's path assuming the pole is at the center of the square. (Use the fact that the line joining one bug to the next is tangent to the bug's path.)
   (b) Find the distance traveled by a bug by the time it meets the other bugs at the center.

5. Show that any tangent line to a hyperbola touches the hyperbola halfway between the points of intersection of the tangent and the asymptotes.

6. A circle C of radius $2r$ has its center at the origin. A circle of radius r rolls without slipping in the counterclockwise direction around C. A point P is located on a fixed radius of the rolling circle at a distance b from its center, $0 < b < r$. [See parts (i) and (ii) of the figure.] Let L be the line from the center of C to the center of the rolling circle and let θ be the angle that L makes with the positive x-axis.
   (a) Using θ as a parameter, show that parametric equations of the path traced out by P are

$$x = b\cos 3\theta + 3r\cos\theta \qquad y = b\sin 3\theta + 3r\sin\theta$$

   Note: If $b = 0$, the path is a circle of radius $3r$; if $b = r$, the path is an *epicycloid*. The path traced out by P for $0 < b < r$ is called an *epitrochoid*.
   (b) Graph the curve for various values of b between 0 and r.
   (c) Show that an equilateral triangle can be inscribed in the epitrochoid and that its centroid is on the circle of radius b centered at the origin.
   Note: This is the principle of the Wankel rotary engine. When the equilateral triangle rotates with its vertices on the epitrochoid, its centroid sweeps out a circle whose center is at the center of the curve.
   (d) In most rotary engines the sides of the equilateral triangles are replaced by arcs of circles centered at the opposite vertices as in part (iii) of the figure. (Then the diameter of the rotor is constant.) Show that the rotor will fit in the epitrochoid if $b \leqslant \frac{3}{2}(2 - \sqrt{3})r$.

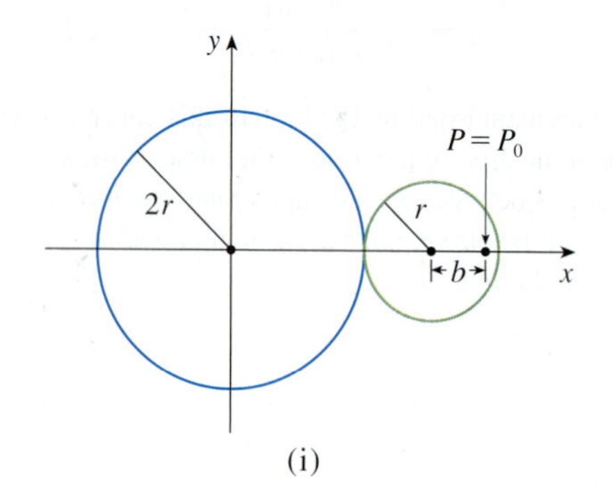

(i)

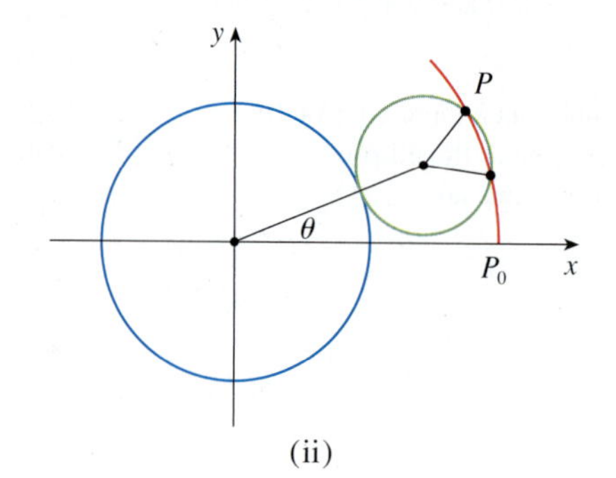

(ii)

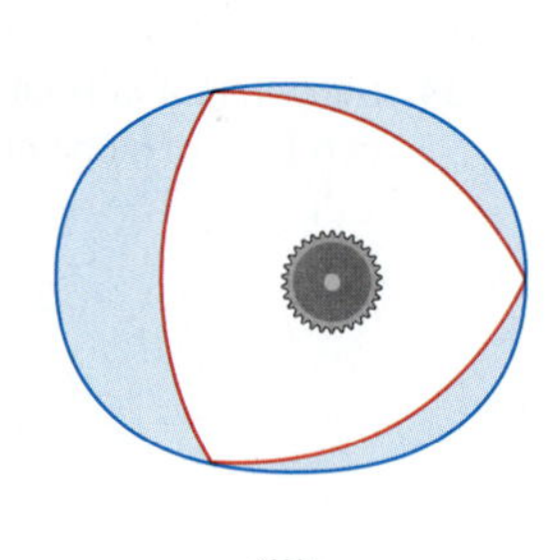

(iii)

11 Infinite Sequences and Series

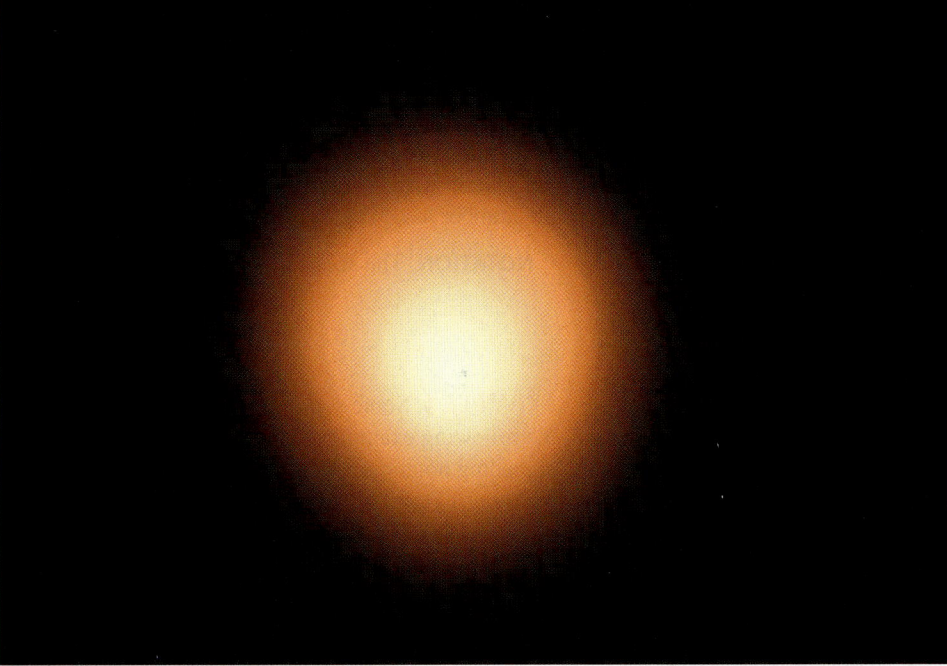

Betelgeuse is a red supergiant star, one of the largest and brightest of the observable stars. In the project on page 823 you are asked to compare the radiation emitted by Betelgeuse with that of other stars.

STScI / NASA / ESA / Galaxy / Galaxy Picture Library / Alamy

INFINITE SEQUENCES AND SERIES WERE introduced briefly in *A Preview of Calculus* in connection with Zeno's paradoxes and the decimal representation of numbers. Their importance in calculus stems from Newton's idea of representing functions as sums of infinite series. For instance, in finding areas he often integrated a function by first expressing it as a series and then integrating each term of the series. We will pursue his idea in Section 11.10 in order to integrate such functions as e^{-x^2}. (Recall that we have previously been unable to do this.) Many of the functions that arise in mathematical physics and chemistry, such as Bessel functions, are defined as sums of series, so it is important to be familiar with the basic concepts of convergence of infinite sequences and series.

Physicists also use series in another way, as we will see in Section 11.11. In studying fields as diverse as optics, special relativity, and electromagnetism, they analyze phenomena by replacing a function with the first few terms in the series that represents it.

11.1 Sequences

A **sequence** can be thought of as a list of numbers written in a definite order:

$$a_1,\ a_2,\ a_3,\ a_4,\ \ldots,\ a_n,\ \ldots$$

The number a_1 is called the *first term*, a_2 is the *second term*, and in general a_n is the *nth term*. We will deal exclusively with infinite sequences and so each term a_n will have a successor a_{n+1}.

Notice that for every positive integer n there is a corresponding number a_n and so a sequence can be defined as a function whose domain is the set of positive integers. But we usually write a_n instead of the function notation $f(n)$ for the value of the function at the number n.

NOTATION The sequence $\{a_1, a_2, a_3, \ldots\}$ is also denoted by

$$\{a_n\} \qquad \text{or} \qquad \{a_n\}_{n=1}^{\infty}$$

EXAMPLE 1 Some sequences can be defined by giving a formula for the nth term. In the following examples we give three descriptions of the sequence: one by using the preceding notation, another by using the defining formula, and a third by writing out the terms of the sequence. Notice that n doesn't have to start at 1.

(a) $\left\{\dfrac{n}{n+1}\right\}_{n=1}^{\infty}$ $\qquad a_n = \dfrac{n}{n+1}$ $\qquad \left\{\dfrac{1}{2}, \dfrac{2}{3}, \dfrac{3}{4}, \dfrac{4}{5}, \ldots, \dfrac{n}{n+1}, \ldots\right\}$

(b) $\left\{\dfrac{(-1)^n(n+1)}{3^n}\right\}$ $\qquad a_n = \dfrac{(-1)^n(n+1)}{3^n}$ $\qquad \left\{-\dfrac{2}{3}, \dfrac{3}{9}, -\dfrac{4}{27}, \dfrac{5}{81}, \ldots, \dfrac{(-1)^n(n+1)}{3^n}, \ldots\right\}$

(c) $\left\{\sqrt{n-3}\right\}_{n=3}^{\infty}$ $\qquad a_n = \sqrt{n-3},\ n \geqslant 3$ $\qquad \left\{0, 1, \sqrt{2}, \sqrt{3}, \ldots, \sqrt{n-3}, \ldots\right\}$

(d) $\left\{\cos\dfrac{n\pi}{6}\right\}_{n=0}^{\infty}$ $\qquad a_n = \cos\dfrac{n\pi}{6},\ n \geqslant 0$ $\qquad \left\{1, \dfrac{\sqrt{3}}{2}, \dfrac{1}{2}, 0, \ldots, \cos\dfrac{n\pi}{6}, \ldots\right\}$ ■

EXAMPLE 2 Find a formula for the general term a_n of the sequence

$$\left\{\frac{3}{5}, -\frac{4}{25}, \frac{5}{125}, -\frac{6}{625}, \frac{7}{3125}, \ldots\right\}$$

assuming that the pattern of the first few terms continues.

SOLUTION We are given that

$$a_1 = \frac{3}{5} \qquad a_2 = -\frac{4}{25} \qquad a_3 = \frac{5}{125} \qquad a_4 = -\frac{6}{625} \qquad a_5 = \frac{7}{3125}$$

Notice that the numerators of these fractions start with 3 and increase by 1 whenever we go to the next term. The second term has numerator 4, the third term has numerator 5; in general, the nth term will have numerator $n + 2$. The denominators are the powers of 5, so a_n has denominator 5^n. The signs of the terms are alternately positive and negative, so we need to multiply by a power of -1. In Example 1(b) the factor $(-1)^n$ meant we started with a negative term. Here we want to start with a positive term and so we use $(-1)^{n-1}$ or $(-1)^{n+1}$. Therefore

$$a_n = (-1)^{n-1}\frac{n+2}{5^n}$$ ■

EXAMPLE 3 Here are some sequences that don't have a simple defining equation.
(a) The sequence $\{p_n\}$, where p_n is the population of the world as of January 1 in the year n.

(b) If we let a_n be the digit in the nth decimal place of the number e, then $\{a_n\}$ is a well-defined sequence whose first few terms are

$$\{7, 1, 8, 2, 8, 1, 8, 2, 8, 4, 5, \ldots\}$$

(c) **The Fibonacci sequence** $\{f_n\}$ is defined recursively by the conditions

$$f_1 = 1 \qquad f_2 = 1 \qquad f_n = f_{n-1} + f_{n-2} \qquad n \geqslant 3$$

Each term is the sum of the two preceding terms. The first few terms are

$$\{1, 1, 2, 3, 5, 8, 13, 21, \ldots\}$$

This sequence arose when the 13th-century Italian mathematician known as Fibonacci solved a problem concerning the breeding of rabbits (see Exercise 83). ■

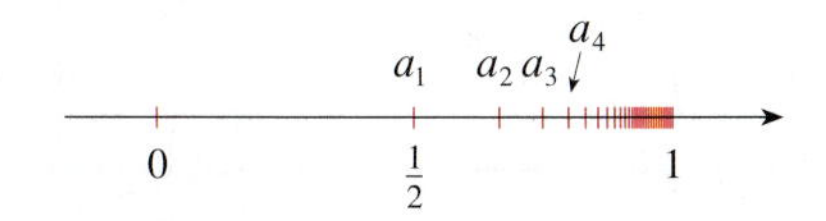

FIGURE 1

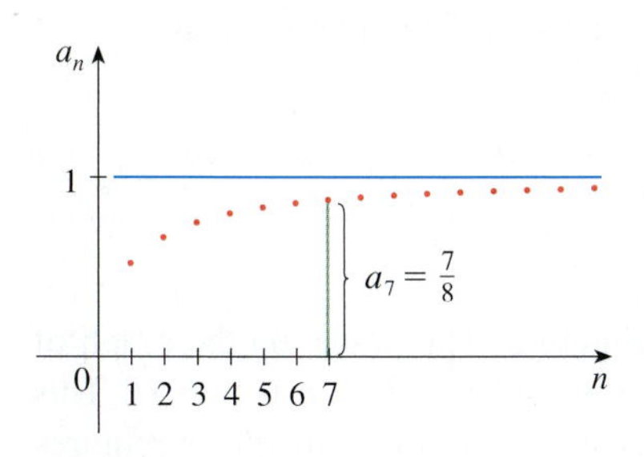

FIGURE 2

A sequence such as the one in Example 1(a), $a_n = n/(n+1)$, can be pictured either by plotting its terms on a number line, as in Figure 1, or by plotting its graph, as in Figure 2. Note that, since a sequence is a function whose domain is the set of positive integers, its graph consists of isolated points with coordinates

$$(1, a_1) \qquad (2, a_2) \qquad (3, a_3) \qquad \ldots \qquad (n, a_n) \qquad \ldots$$

From Figure 1 or Figure 2 it appears that the terms of the sequence $a_n = n/(n+1)$ are approaching 1 as n becomes large. In fact, the difference

$$1 - \frac{n}{n+1} = \frac{1}{n+1}$$

can be made as small as we like by taking n sufficiently large. We indicate this by writing

$$\lim_{n\to\infty} \frac{n}{n+1} = 1$$

In general, the notation

$$\lim_{n\to\infty} a_n = L$$

means that the terms of the sequence $\{a_n\}$ approach L as n becomes large. Notice that the following definition of the limit of a sequence is very similar to the definition of a limit of a function at infinity given in Section 3.4.

1 Definition A sequence $\{a_n\}$ has the **limit** L and we write

$$\lim_{n\to\infty} a_n = L \qquad \text{or} \qquad a_n \to L \text{ as } n \to \infty$$

if we can make the terms a_n as close to L as we like by taking n sufficiently large. If $\lim_{n\to\infty} a_n$ exists, we say the sequence **converges** (or is **convergent**). Otherwise, we say the sequence **diverges** (or is **divergent**).

Figure 3 illustrates Definition 1 by showing the graphs of two sequences that have the limit L.

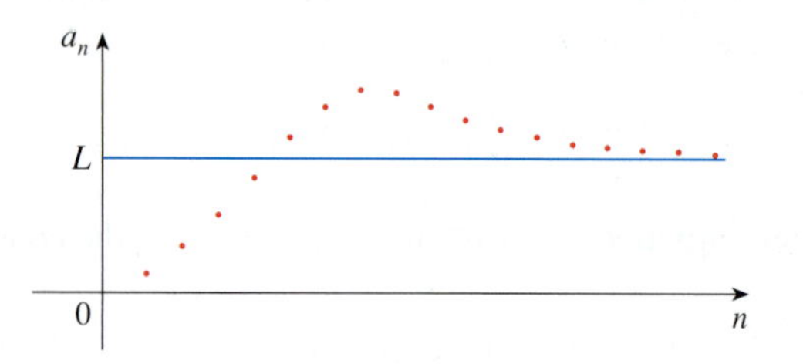

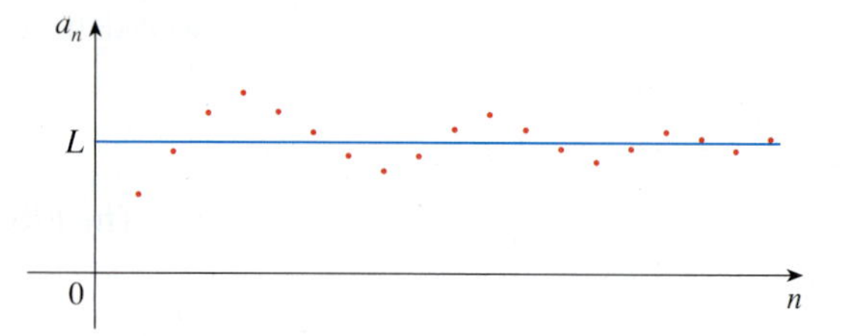

FIGURE 3
Graphs of two sequences with $\lim_{n\to\infty} a_n = L$

A more precise version of Definition 1 is as follows.

2 Definition A sequence $\{a_n\}$ has the **limit** L and we write

$$\lim_{n\to\infty} a_n = L \qquad \text{or} \qquad a_n \to L \text{ as } n \to \infty$$

if for every $\varepsilon > 0$ there is a corresponding integer N such that

$$\text{if} \qquad n > N \qquad \text{then} \qquad |a_n - L| < \varepsilon$$

Compare this definition with Definition 3.4.5.

Definition 2 is illustrated by Figure 4, in which the terms $a_1, a_2, a_3, \ldots$ are plotted on a number line. No matter how small an interval $(L - \varepsilon, L + \varepsilon)$ is chosen, there exists an N such that all terms of the sequence from a_{N+1} onward must lie in that interval.

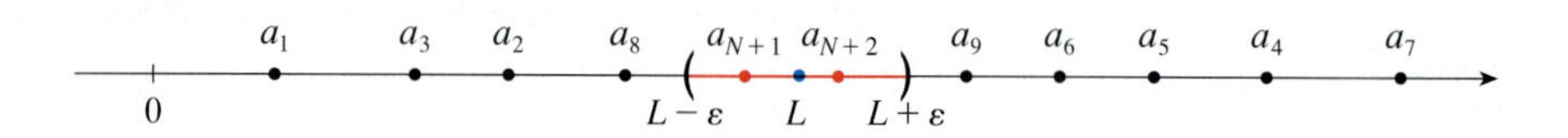

FIGURE 4

Another illustration of Definition 2 is given in Figure 5. The points on the graph of $\{a_n\}$ must lie between the horizontal lines $y = L + \varepsilon$ and $y = L - \varepsilon$ if $n > N$. This picture must be valid no matter how small ε is chosen, but usually a smaller ε requires a larger N.

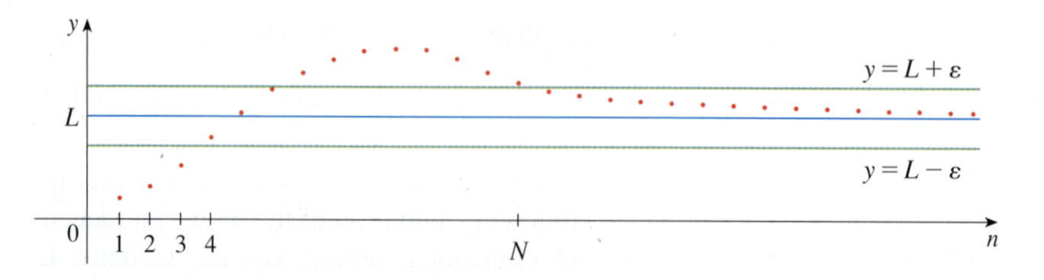

FIGURE 5

If you compare Definition 2 with Definition 3.4.5, you will see that the only difference between $\lim_{n\to\infty} a_n = L$ and $\lim_{x\to\infty} f(x) = L$ is that n is required to be an integer. Thus we have the following theorem, which is illustrated by Figure 6.

3 Theorem If $\lim_{x\to\infty} f(x) = L$ and $f(n) = a_n$ when n is an integer, then $\lim_{n\to\infty} a_n = L$.

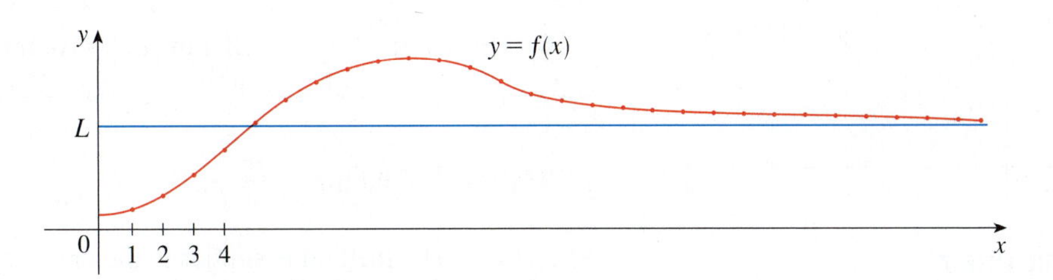

FIGURE 6

In particular, since we know that $\lim_{x\to\infty} (1/x^r) = 0$ when $r > 0$ (Theorem 3.4.4), we have

$$\boxed{4} \qquad \lim_{n\to\infty} \frac{1}{n^r} = 0 \qquad \text{if } r > 0$$

If a_n becomes large as n becomes large, we use the notation $\lim_{n\to\infty} a_n = \infty$. The following precise definition is similar to Definition 3.4.7.

5 Definition $\lim_{n\to\infty} a_n = \infty$ means that for every positive number M there is an integer N such that

$$\text{if} \quad n > N \quad \text{then} \quad a_n > M$$

If $\lim_{n\to\infty} a_n = \infty$, then the sequence $\{a_n\}$ is divergent but in a special way. We say that $\{a_n\}$ diverges to ∞.

The Limit Laws given in Section 1.6 also hold for the limits of sequences and their proofs are similar.

Limit Laws for Sequences

If $\{a_n\}$ and $\{b_n\}$ are convergent sequences and c is a constant, then

$$\lim_{n\to\infty} (a_n + b_n) = \lim_{n\to\infty} a_n + \lim_{n\to\infty} b_n$$

$$\lim_{n\to\infty} (a_n - b_n) = \lim_{n\to\infty} a_n - \lim_{n\to\infty} b_n$$

$$\lim_{n\to\infty} ca_n = c \lim_{n\to\infty} a_n \qquad \lim_{n\to\infty} c = c$$

$$\lim_{n\to\infty} (a_n b_n) = \lim_{n\to\infty} a_n \cdot \lim_{n\to\infty} b_n$$

$$\lim_{n\to\infty} \frac{a_n}{b_n} = \frac{\lim_{n\to\infty} a_n}{\lim_{n\to\infty} b_n} \quad \text{if } \lim_{n\to\infty} b_n \neq 0$$

$$\lim_{n\to\infty} a_n^p = \left[\lim_{n\to\infty} a_n\right]^p \quad \text{if } p > 0 \text{ and } a_n > 0$$

The Squeeze Theorem can also be adapted for sequences as follows (see Figure 7).

Squeeze Theorem for Sequences

If $a_n \leqslant b_n \leqslant c_n$ for $n \geqslant n_0$ and $\lim_{n\to\infty} a_n = \lim_{n\to\infty} c_n = L$, then $\lim_{n\to\infty} b_n = L$.

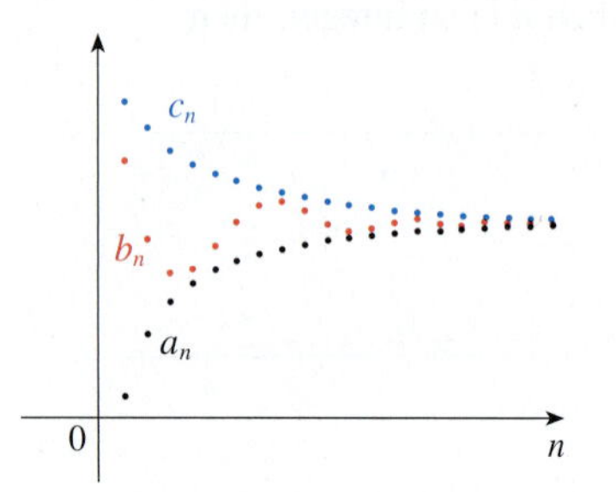

FIGURE 7
The sequence $\{b_n\}$ is squeezed between the sequences $\{a_n\}$ and $\{c_n\}$.

Another useful fact about limits of sequences is given by the following theorem, whose proof is left as Exercise 87.

6 Theorem If $\lim_{n\to\infty} |a_n| = 0$, then $\lim_{n\to\infty} a_n = 0$.

EXAMPLE 4 Find $\lim_{n\to\infty} \dfrac{n}{n+1}$.

SOLUTION The method is similar to the one we used in Section 3.4: Divide numerator and denominator by the highest power of n that occurs in the denominator and then use the Limit Laws.

$$\lim_{n\to\infty} \frac{n}{n+1} = \lim_{n\to\infty} \frac{1}{1+\dfrac{1}{n}} = \frac{\lim_{n\to\infty} 1}{\lim_{n\to\infty} 1 + \lim_{n\to\infty} \dfrac{1}{n}}$$

$$= \frac{1}{1+0} = 1$$

This shows that the guess we made earlier from Figures 1 and 2 was correct.

Here we used Equation 4 with $r = 1$. ∎

EXAMPLE 5 Is the sequence $a_n = \dfrac{n}{\sqrt{10+n}}$ convergent or divergent?

SOLUTION As in Example 4, we divide numerator and denominator by n:

$$\lim_{n\to\infty} \frac{n}{\sqrt{10+n}} = \lim_{n\to\infty} \frac{1}{\sqrt{\dfrac{10}{n^2} + \dfrac{1}{n}}} = \infty$$

because the numerator is constant and the denominator approaches 0. So $\{a_n\}$ is divergent. ∎

EXAMPLE 6 Calculate $\lim_{n\to\infty} \dfrac{\ln n}{n}$.

SOLUTION Notice that both numerator and denominator approach infinity as $n \to \infty$. We can't apply l'Hospital's Rule directly because it applies not to sequences but to functions of a real variable. However, we can apply l'Hospital's Rule to the related function $f(x) = (\ln x)/x$ and obtain

$$\lim_{x\to\infty} \frac{\ln x}{x} = \lim_{x\to\infty} \frac{1/x}{1} = 0$$

Therefore, by Theorem 3, we have

$$\lim_{n\to\infty} \frac{\ln n}{n} = 0$$ ∎

EXAMPLE 7 Determine whether the sequence $a_n = (-1)^n$ is convergent or divergent.

SOLUTION If we write out the terms of the sequence, we obtain

$$\{-1, 1, -1, 1, -1, 1, -1, \ldots\}$$

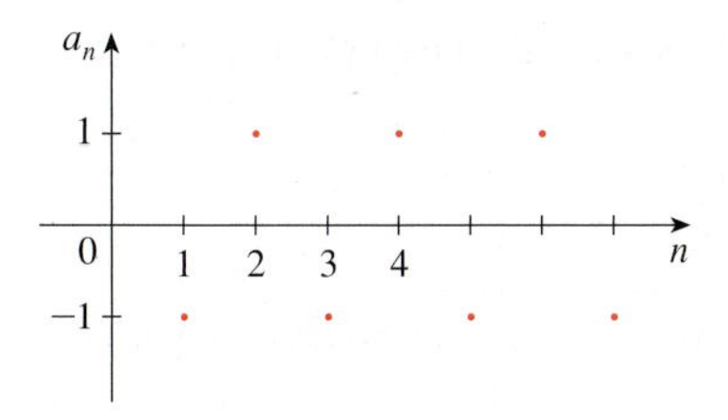

FIGURE 8

The graph of this sequence is shown in Figure 8. Since the terms oscillate between 1 and -1 infinitely often, a_n does not approach any number. Thus $\lim_{n\to\infty} (-1)^n$ does not exist; that is, the sequence $\{(-1)^n\}$ is divergent. ■

The graph of the sequence in Example 8 is shown in Figure 9 and supports our answer.

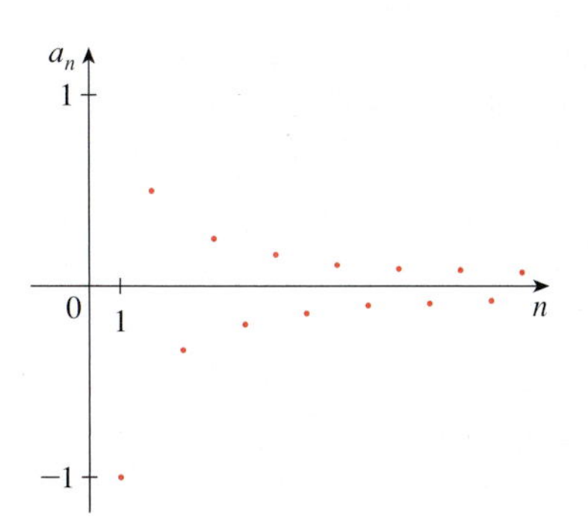

FIGURE 9

EXAMPLE 8 Evaluate $\displaystyle\lim_{n\to\infty} \frac{(-1)^n}{n}$ if it exists.

SOLUTION We first calculate the limit of the absolute value:

$$\lim_{n\to\infty} \left| \frac{(-1)^n}{n} \right| = \lim_{n\to\infty} \frac{1}{n} = 0$$

Therefore, by Theorem 6,

$$\lim_{n\to\infty} \frac{(-1)^n}{n} = 0$$ ■

The following theorem says that if we apply a continuous function to the terms of a convergent sequence, the result is also convergent. The proof is left as Exercise 88.

7 Theorem If $\lim_{n\to\infty} a_n = L$ and the function f is continuous at L, then

$$\lim_{n\to\infty} f(a_n) = f(L)$$

EXAMPLE 9 Find $\lim_{n\to\infty} \sin(\pi/n)$.

SOLUTION Because the sine function is continuous at 0, Theorem 7 enables us to write

$$\lim_{n\to\infty} \sin(\pi/n) = \sin\left(\lim_{n\to\infty} (\pi/n)\right) = \sin 0 = 0$$ ■

Creating Graphs of Sequences

Some computer algebra systems have special commands that enable us to create sequences and graph them directly. With most graphing calculators, however, sequences can be graphed by using parametric equations. For instance, the sequence in Example 10 can be graphed by entering the parametric equations

$$x = t \qquad y = t!/t^t$$

and graphing in dot mode, starting with $t = 1$ and setting the t-step equal to 1. The result is shown in Figure 10.

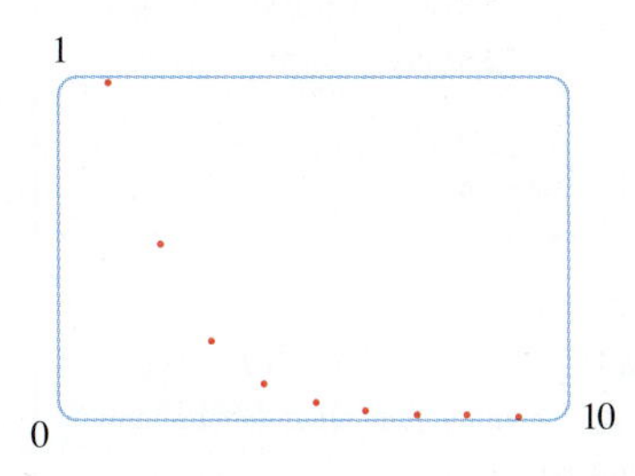

FIGURE 10

EXAMPLE 10 Discuss the convergence of the sequence $a_n = n!/n^n$, where $n! = 1 \cdot 2 \cdot 3 \cdot \cdots \cdot n$.

SOLUTION Both numerator and denominator approach infinity as $n \to \infty$ but here we have no corresponding function for use with l'Hospital's Rule ($x!$ is not defined when x is not an integer). Let's write out a few terms to get a feeling for what happens to a_n as n gets large:

$$a_1 = 1 \qquad a_2 = \frac{1 \cdot 2}{2 \cdot 2} \qquad a_3 = \frac{1 \cdot 2 \cdot 3}{3 \cdot 3 \cdot 3}$$

$$\boxed{8} \qquad a_n = \frac{1 \cdot 2 \cdot 3 \cdot \cdots \cdot n}{n \cdot n \cdot n \cdot \cdots \cdot n}$$

It appears from these expressions and the graph in Figure 10 that the terms are decreasing and perhaps approach 0. To confirm this, observe from Equation 8 that

$$a_n = \frac{1}{n}\left(\frac{2 \cdot 3 \cdot \cdots \cdot n}{n \cdot n \cdot \cdots \cdot n}\right)$$

Notice that the expression in parentheses is at most 1 because the numerator is less than (or equal to) the denominator. So

$$0 < a_n \leqslant \frac{1}{n}$$

We know that $1/n \to 0$ as $n \to \infty$. Therefore $a_n \to 0$ as $n \to \infty$ by the Squeeze Theorem. ■

EXAMPLE 11 For what values of r is the sequence $\{r^n\}$ convergent?

SOLUTION We know from Section 3.4 and the graphs of the exponential functions in Section 6.2 (or Section 6.4*) that $\lim_{x\to\infty} a^x = \infty$ for $a > 1$ and $\lim_{x\to\infty} a^x = 0$ for $0 < a < 1$. Therefore, putting $a = r$ and using Theorem 3, we have

$$\lim_{n\to\infty} r^n = \begin{cases} \infty & \text{if } r > 1 \\ 0 & \text{if } 0 < r < 1 \end{cases}$$

It is obvious that

$$\lim_{n\to\infty} 1^n = 1 \qquad \text{and} \qquad \lim_{n\to\infty} 0^n = 0$$

If $-1 < r < 0$, then $0 < |r| < 1$, so

$$\lim_{n\to\infty} |r^n| = \lim_{n\to\infty} |r|^n = 0$$

and therefore $\lim_{n\to\infty} r^n = 0$ by Theorem 6. If $r \leqslant -1$, then $\{r^n\}$ diverges as in Example 7. Figure 11 shows the graphs for various values of r. (The case $r = -1$ is shown in Figure 8.)

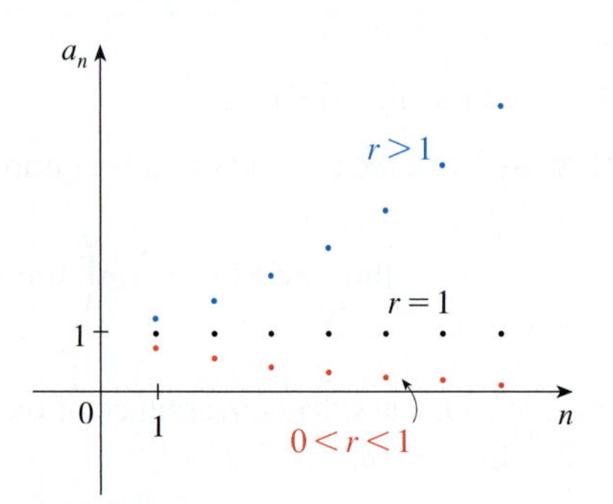

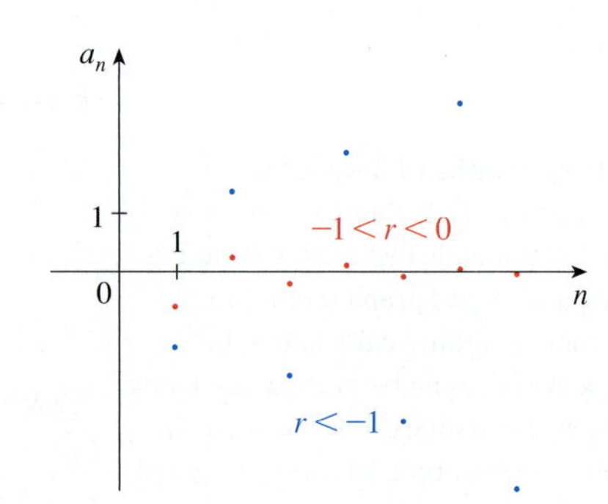

FIGURE 11
The sequence $a_n = r^n$

■

The results of Example 11 are summarized for future use as follows.

9 The sequence $\{r^n\}$ is convergent if $-1 < r \leqslant 1$ and divergent for all other values of r.

$$\lim_{n\to\infty} r^n = \begin{cases} 0 & \text{if } -1 < r < 1 \\ 1 & \text{if } r = 1 \end{cases}$$

10 Definition A sequence $\{a_n\}$ is called **increasing** if $a_n < a_{n+1}$ for all $n \geqslant 1$, that is, $a_1 < a_2 < a_3 < \cdots$. It is called **decreasing** if $a_n > a_{n+1}$ for all $n \geqslant 1$. A sequence is **monotonic** if it is either increasing or decreasing.

EXAMPLE 12 The sequence $\left\{\dfrac{3}{n+5}\right\}$ is decreasing because

The right side is smaller because it has a larger denominator.

$$\frac{3}{n+5} > \frac{3}{(n+1)+5} = \frac{3}{n+6}$$

and so $a_n > a_{n+1}$ for all $n \geqslant 1$. ■

EXAMPLE 13 Show that the sequence $a_n = \dfrac{n}{n^2+1}$ is decreasing.

SOLUTION 1 We must show that $a_{n+1} < a_n$, that is,

$$\frac{n+1}{(n+1)^2+1} < \frac{n}{n^2+1}$$

This inequality is equivalent to the one we get by cross-multiplication:

$$\frac{n+1}{(n+1)^2+1} < \frac{n}{n^2+1} \iff (n+1)(n^2+1) < n[(n+1)^2+1]$$

$$\iff n^3 + n^2 + n + 1 < n^3 + 2n^2 + 2n$$

$$\iff 1 < n^2 + n$$

Since $n \geqslant 1$, we know that the inequality $n^2 + n > 1$ is true. Therefore $a_{n+1} < a_n$ and so $\{a_n\}$ is decreasing.

SOLUTION 2 Consider the function $f(x) = \dfrac{x}{x^2+1}$:

$$f'(x) = \frac{x^2+1-2x^2}{(x^2+1)^2} = \frac{1-x^2}{(x^2+1)^2} < 0 \qquad \text{whenever } x^2 > 1$$

Thus f is decreasing on $(1, \infty)$ and so $f(n) > f(n+1)$. Therefore $\{a_n\}$ is decreasing. ■

11 Definition A sequence $\{a_n\}$ is **bounded above** if there is a number M such that

$$a_n \leqslant M \qquad \text{for all } n \geqslant 1$$

It is **bounded below** if there is a number m such that

$$m \leqslant a_n \qquad \text{for all } n \geqslant 1$$

If it is bounded above and below, then $\{a_n\}$ is a **bounded sequence**.

For instance, the sequence $a_n = n$ is bounded below $(a_n > 0)$ but not above. The sequence $a_n = n/(n+1)$ is bounded because $0 < a_n < 1$ for all n.

We know that not every bounded sequence is convergent [for instance, the sequence $a_n = (-1)^n$ satisfies $-1 \leqslant a_n \leqslant 1$ but is divergent from Example 7] and not every monotonic sequence is convergent ($a_n = n \to \infty$). But if a sequence is both bounded *and* monotonic, then it must be convergent. This fact is proved as Theorem 12, but intuitively you can understand why it is true by looking at Figure 12. If $\{a_n\}$ is increasing and $a_n \leqslant M$ for all n, then the terms are forced to crowd together and approach some number L.

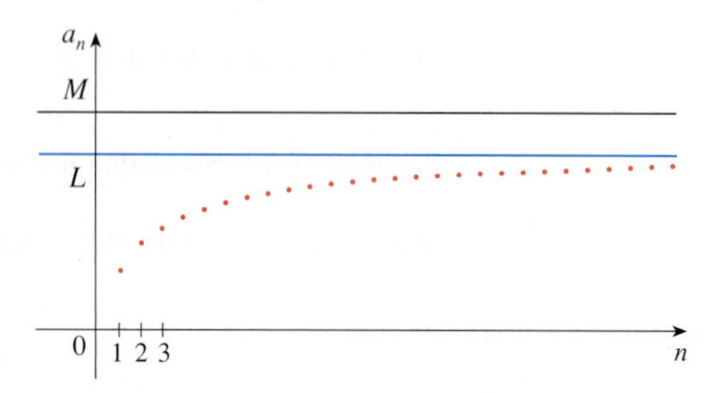

FIGURE 12

The proof of Theorem 12 is based on the **Completeness Axiom** for the set $\mathbb{R}$ of real numbers, which says that if S is a nonempty set of real numbers that has an upper bound M ($x \leqslant M$ for all x in S), then S has a **least upper bound** b. (This means that b is an upper bound for S, but if M is any other upper bound, then $b \leqslant M$.) The Completeness Axiom is an expression of the fact that there is no gap or hole in the real number line.

12 Monotonic Sequence Theorem Every bounded, monotonic sequence is convergent.

PROOF Suppose $\{a_n\}$ is an increasing sequence. Since $\{a_n\}$ is bounded, the set $S = \{a_n \mid n \geqslant 1\}$ has an upper bound. By the Completeness Axiom it has a least upper bound L. Given $\varepsilon > 0$, $L - \varepsilon$ is *not* an upper bound for S (since L is the *least* upper bound). Therefore

$$a_N > L - \varepsilon \qquad \text{for some integer } N$$

But the sequence is increasing so $a_n \geqslant a_N$ for every $n > N$. Thus if $n > N$, we have

$$a_n > L - \varepsilon$$

so

$$0 \leqslant L - a_n < \varepsilon$$

since $a_n \leqslant L$. Thus

$$|L - a_n| < \varepsilon \qquad \text{whenever } n > N$$

so $\lim_{n\to\infty} a_n = L$.

A similar proof (using the greatest lower bound) works if $\{a_n\}$ is decreasing. ■

The proof of Theorem 12 shows that a sequence that is increasing and bounded above is convergent. (Likewise, a decreasing sequence that is bounded below is convergent.) This fact is used many times in dealing with infinite series.

EXAMPLE 14 Investigate the sequence $\{a_n\}$ defined by the *recurrence relation*

$$a_1 = 2 \qquad a_{n+1} = \tfrac{1}{2}(a_n + 6) \qquad \text{for } n = 1, 2, 3, \ldots$$

SOLUTION We begin by computing the first several terms:

$$a_1 = 2 \qquad a_2 = \tfrac{1}{2}(2 + 6) = 4 \qquad a_3 = \tfrac{1}{2}(4 + 6) = 5$$

$$a_4 = \tfrac{1}{2}(5 + 6) = 5.5 \qquad a_5 = 5.75 \qquad a_6 = 5.875$$

$$a_7 = 5.9375 \qquad a_8 = 5.96875 \qquad a_9 = 5.984375$$

Mathematical induction is often used in dealing with recursive sequences. See page 99 for a discussion of the Principle of Mathematical Induction.

These initial terms suggest that the sequence is increasing and the terms are approaching 6. To confirm that the sequence is increasing, we use mathematical induction to show that $a_{n+1} > a_n$ for all $n \geqslant 1$. This is true for $n = 1$ because $a_2 = 4 > a_1$. If we assume that it is true for $n = k$, then we have

$$a_{k+1} > a_k$$

so

$$a_{k+1} + 6 > a_k + 6$$

and

$$\tfrac{1}{2}(a_{k+1} + 6) > \tfrac{1}{2}(a_k + 6)$$

Thus

$$a_{k+2} > a_{k+1}$$

We have deduced that $a_{n+1} > a_n$ is true for $n = k + 1$. Therefore the inequality is true for all n by induction.

Next we verify that $\{a_n\}$ is bounded by showing that $a_n < 6$ for all n. (Since the sequence is increasing, we already know that it has a lower bound: $a_n \geqslant a_1 = 2$ for all n.) We know that $a_1 < 6$, so the assertion is true for $n = 1$. Suppose it is true for $n = k$. Then

$$a_k < 6$$

so

$$a_k + 6 < 12$$

and

$$\tfrac{1}{2}(a_k + 6) < \tfrac{1}{2}(12) = 6$$

Thus

$$a_{k+1} < 6$$

This shows, by mathematical induction, that $a_n < 6$ for all n.

Since the sequence $\{a_n\}$ is increasing and bounded, Theorem 12 guarantees that it has a limit. The theorem doesn't tell us what the value of the limit is. But now that we know $L = \lim_{n \to \infty} a_n$ exists, we can use the given recurrence relation to write

$$\lim_{n \to \infty} a_{n+1} = \lim_{n \to \infty} \tfrac{1}{2}(a_n + 6) = \tfrac{1}{2}\left(\lim_{n \to \infty} a_n + 6\right) = \tfrac{1}{2}(L + 6)$$

A proof of this fact is requested in Exercise 70.

Since $a_n \to L$, it follows that $a_{n+1} \to L$ too (as $n \to \infty$, $n + 1 \to \infty$ also). So we have

$$L = \tfrac{1}{2}(L + 6)$$

Solving this equation for L, we get $L = 6$, as we predicted. ■

11.1 EXERCISES

1. (a) What is a sequence?
(b) What does it mean to say that $\lim_{n\to\infty} a_n = 8$?
(c) What does it mean to say that $\lim_{n\to\infty} a_n = \infty$?

2. (a) What is a convergent sequence? Give two examples.
(b) What is a divergent sequence? Give two examples.

3–12 List the first five terms of the sequence.

3. $a_n = \dfrac{2^n}{2n+1}$

4. $a_n = \dfrac{n^2-1}{n^2+1}$

5. $a_n = \dfrac{(-1)^{n-1}}{5^n}$

6. $a_n = \cos\dfrac{n\pi}{2}$

7. $a_n = \dfrac{1}{(n+1)!}$

8. $a_n = \dfrac{(-1)^n n}{n!+1}$

9. $a_1 = 1,\quad a_{n+1} = 5a_n - 3$

10. $a_1 = 6,\quad a_{n+1} = \dfrac{a_n}{n}$

11. $a_1 = 2,\quad a_{n+1} = \dfrac{a_n}{1+a_n}$

12. $a_1 = 2,\quad a_2 = 1,\quad a_{n+1} = a_n - a_{n-1}$

13–18 Find a formula for the general term a_n of the sequence, assuming that the pattern of the first few terms continues.

13. $\left\{\frac{1}{2}, \frac{1}{4}, \frac{1}{6}, \frac{1}{8}, \frac{1}{10}, \ldots\right\}$

14. $\left\{4, -1, \frac{1}{4}, -\frac{1}{16}, \frac{1}{64}, \ldots\right\}$

15. $\left\{-3, 2, -\frac{4}{3}, \frac{8}{9}, -\frac{16}{27}, \ldots\right\}$

16. $\{5, 8, 11, 14, 17, \ldots\}$

17. $\left\{\frac{1}{2}, -\frac{4}{3}, \frac{9}{4}, -\frac{16}{5}, \frac{25}{6}, \ldots\right\}$

18. $\{1, 0, -1, 0, 1, 0, -1, 0, \ldots\}$

19–22 Calculate, to four decimal places, the first ten terms of the sequence and use them to plot the graph of the sequence by hand. Does the sequence appear to have a limit? If so, calculate it. If not, explain why.

19. $a_n = \dfrac{3n}{1+6n}$

20. $a_n = 2 + \dfrac{(-1)^n}{n}$

21. $a_n = 1 + \left(-\frac{1}{2}\right)^n$

22. $a_n = 1 + \dfrac{10^n}{9^n}$

23–56 Determine whether the sequence converges or diverges. If it converges, find the limit.

23. $a_n = \dfrac{3+5n^2}{n+n^2}$

24. $a_n = \dfrac{3+5n^2}{1+n}$

25. $a_n = \dfrac{n^4}{n^3-2n}$

26. $a_n = 2 + (0.86)^n$

27. $a_n = 3^n 7^{-n}$

28. $a_n = \dfrac{3\sqrt{n}}{\sqrt{n}+2}$

29. $a_n = e^{-1/\sqrt{n}}$

30. $a_n = \dfrac{4^n}{1+9^n}$

31. $a_n = \sqrt{\dfrac{1+4n^2}{1+n^2}}$

32. $a_n = \cos\left(\dfrac{n\pi}{n+1}\right)$

33. $a_n = \dfrac{n^2}{\sqrt{n^3+4n}}$

34. $a_n = e^{2n/(n+2)}$

35. $a_n = \dfrac{(-1)^n}{2\sqrt{n}}$

36. $a_n = \dfrac{(-1)^{n+1} n}{n+\sqrt{n}}$

37. $\left\{\dfrac{(2n-1)!}{(2n+1)!}\right\}$

38. $\left\{\dfrac{\ln n}{\ln 2n}\right\}$

39. $\{\sin n\}$

40. $a_n = \dfrac{\tan^{-1} n}{n}$

41. $\{n^2 e^{-n}\}$

42. $a_n = \ln(n+1) - \ln n$

43. $a_n = \dfrac{\cos^2 n}{2^n}$

44. $a_n = \sqrt[n]{2^{1+3n}}$

45. $a_n = n\sin(1/n)$

46. $a_n = 2^{-n}\cos n\pi$

47. $a_n = \left(1 + \dfrac{2}{n}\right)^n$

48. $a_n = \sqrt[n]{n}$

49. $a_n = \ln(2n^2+1) - \ln(n^2+1)$

50. $a_n = \dfrac{(\ln n)^2}{n}$

51. $a_n = \arctan(\ln n)$

52. $a_n = n - \sqrt{n+1}\sqrt{n+3}$

53. $\{0, 1, 0, 0, 1, 0, 0, 0, 1, \ldots\}$

54. $\left\{\frac{1}{1}, \frac{1}{3}, \frac{1}{2}, \frac{1}{4}, \frac{1}{3}, \frac{1}{5}, \frac{1}{4}, \frac{1}{6}, \ldots\right\}$

55. $a_n = \dfrac{n!}{2^n}$

56. $a_n = \dfrac{(-3)^n}{n!}$

57–63 Use a graph of the sequence to decide whether the sequence is convergent or divergent. If the sequence is convergent, guess the value of the limit from the graph and then prove your guess. (See the margin note on page 739 for advice on graphing sequences.)

57. $a_n = (-1)^n \dfrac{n}{n+1}$

58. $a_n = \dfrac{\sin n}{n}$

59. $a_n = \arctan\left(\dfrac{n^2}{n^2+4}\right)$

60. $a_n = \sqrt[n]{3^n + 5^n}$

61. $a_n = \dfrac{n^2 \cos n}{1 + n^2}$

62. $a_n = \dfrac{1 \cdot 3 \cdot 5 \cdot \cdots \cdot (2n-1)}{n!}$

63. $a_n = \dfrac{1 \cdot 3 \cdot 5 \cdot \cdots \cdot (2n-1)}{(2n)^n}$

64. (a) Determine whether the sequence defined as follows is convergent or divergent:

$$a_1 = 1 \qquad a_{n+1} = 4 - a_n \qquad \text{for } n \geqslant 1$$

(b) What happens if the first term is $a_1 = 2$?

65. If \$1000 is invested at 6% interest, compounded annually, then after n years the investment is worth $a_n = 1000(1.06)^n$ dollars.
(a) Find the first five terms of the sequence $\{a_n\}$.
(b) Is the sequence convergent or divergent? Explain.

66. If you deposit \$100 at the end of every month into an account that pays 3% interest per year compounded monthly, the amount of interest accumulated after n months is given by the sequence

$$I_n = 100\left(\frac{1.0025^n - 1}{0.0025} - n\right)$$

(a) Find the first six terms of the sequence.
(b) How much interest will you have earned after two years?

67. A fish farmer has 5000 catfish in his pond. The number of catfish increases by 8% per month and the farmer harvests 300 catfish per month.
(a) Show that the catfish population P_n after n months is given recursively by

$$P_n = 1.08P_{n-1} - 300 \qquad P_0 = 5000$$

(b) How many catfish are in the pond after six months?

68. Find the first 40 terms of the sequence defined by

$$a_{n+1} = \begin{cases} \frac{1}{2}a_n & \text{if } a_n \text{ is an even number} \\ 3a_n + 1 & \text{if } a_n \text{ is an odd number} \end{cases}$$

and $a_1 = 11$. Do the same if $a_1 = 25$. Make a conjecture about this type of sequence.

69. For what values of r is the sequence $\{nr^n\}$ convergent?

70. (a) If $\{a_n\}$ is convergent, show that

$$\lim_{n\to\infty} a_{n+1} = \lim_{n\to\infty} a_n$$

(b) A sequence $\{a_n\}$ is defined by $a_1 = 1$ and $a_{n+1} = 1/(1 + a_n)$ for $n \geqslant 1$. Assuming that $\{a_n\}$ is convergent, find its limit.

71. Suppose you know that $\{a_n\}$ is a decreasing sequence and all its terms lie between the numbers 5 and 8. Explain why the sequence has a limit. What can you say about the value of the limit?

72–78 Determine whether the sequence is increasing, decreasing, or not monotonic. Is the sequence bounded?

72. $a_n = \cos n$

73. $a_n = \dfrac{1}{2n+3}$

74. $a_n = \dfrac{1-n}{2+n}$

75. $a_n = n(-1)^n$

76. $a_n = 2 + \dfrac{(-1)^n}{n}$

77. $a_n = 3 - 2ne^{-n}$

78. $a_n = n^3 - 3n + 3$

79. Find the limit of the sequence

$$\left\{\sqrt{2}, \sqrt{2\sqrt{2}}, \sqrt{2\sqrt{2\sqrt{2}}}, \ldots\right\}$$

80. A sequence $\{a_n\}$ is given by $a_1 = \sqrt{2}$, $a_{n+1} = \sqrt{2 + a_n}$.
(a) By induction or otherwise, show that $\{a_n\}$ is increasing and bounded above by 3. Apply the Monotonic Sequence Theorem to show that $\lim_{n\to\infty} a_n$ exists.
(b) Find $\lim_{n\to\infty} a_n$.

81. Show that the sequence defined by

$$a_1 = 1 \qquad a_{n+1} = 3 - \frac{1}{a_n}$$

is increasing and $a_n < 3$ for all n. Deduce that $\{a_n\}$ is convergent and find its limit.

82. Show that the sequence defined by

$$a_1 = 2 \qquad a_{n+1} = \frac{1}{3 - a_n}$$

satisfies $0 < a_n \leqslant 2$ and is decreasing. Deduce that the sequence is convergent and find its limit.

83. (a) Fibonacci posed the following problem: Suppose that rabbits live forever and that every month each pair produces a new pair which becomes productive at age 2 months. If we start with one newborn pair, how many pairs of rabbits will we have in the nth month? Show that the answer is f_n, where $\{f_n\}$ is the Fibonacci sequence defined in Example 3(c).
(b) Let $a_n = f_{n+1}/f_n$ and show that $a_{n-1} = 1 + 1/a_{n-2}$. Assuming that $\{a_n\}$ is convergent, find its limit.

84. (a) Let $a_1 = a$, $a_2 = f(a)$, $a_3 = f(a_2) = f(f(a))$, . . . , $a_{n+1} = f(a_n)$, where f is a continuous function. If $\lim_{n\to\infty} a_n = L$, show that $f(L) = L$.
(b) Illustrate part (a) by taking $f(x) = \cos x$, $a = 1$, and estimating the value of L to five decimal places.

85. (a) Use a graph to guess the value of the limit

$$\lim_{n\to\infty} \frac{n^5}{n!}$$

(b) Use a graph of the sequence in part (a) to find the smallest values of N that correspond to $\varepsilon = 0.1$ and $\varepsilon = 0.001$ in Definition 2.

86. Use Definition 2 directly to prove that $\lim_{n\to\infty} r^n = 0$ when $|r| < 1$.

87. Prove Theorem 6.
[*Hint:* Use either Definition 2 or the Squeeze Theorem.]

88. Prove Theorem 7.

89. Prove that if $\lim_{n\to\infty} a_n = 0$ and $\{b_n\}$ is bounded, then $\lim_{n\to\infty} (a_n b_n) = 0$.

90. Let $a_n = \left(1 + \dfrac{1}{n}\right)^n$.

(a) Show that if $0 \leqslant a < b$, then

$$\frac{b^{n+1} - a^{n+1}}{b - a} < (n + 1)b^n$$

(b) Deduce that $b^n[(n + 1)a - nb] < a^{n+1}$.
(c) Use $a = 1 + 1/(n + 1)$ and $b = 1 + 1/n$ in part (b) to show that $\{a_n\}$ is increasing.
(d) Use $a = 1$ and $b = 1 + 1/(2n)$ in part (b) to show that $a_{2n} < 4$.
(e) Use parts (c) and (d) to show that $a_n < 4$ for all n.
(f) Use Theorem 12 to show that $\lim_{n\to\infty} (1 + 1/n)^n$ exists. (The limit is e. See Equation 6.4.9 or 6.4*.9.)

91. Let a and b be positive numbers with $a > b$. Let a_1 be their arithmetic mean and b_1 their geometric mean:

$$a_1 = \frac{a + b}{2} \qquad b_1 = \sqrt{ab}$$

Repeat this process so that, in general,

$$a_{n+1} = \frac{a_n + b_n}{2} \qquad b_{n+1} = \sqrt{a_n b_n}$$

(a) Use mathematical induction to show that

$$a_n > a_{n+1} > b_{n+1} > b_n$$

(b) Deduce that both $\{a_n\}$ and $\{b_n\}$ are convergent.
(c) Show that $\lim_{n\to\infty} a_n = \lim_{n\to\infty} b_n$. Gauss called the common value of these limits the **arithmetic-geometric mean** of the numbers a and b.

92. (a) Show that if $\lim_{n\to\infty} a_{2n} = L$ and $\lim_{n\to\infty} a_{2n+1} = L$, then $\{a_n\}$ is convergent and $\lim_{n\to\infty} a_n = L$.
(b) If $a_1 = 1$ and

$$a_{n+1} = 1 + \frac{1}{1 + a_n}$$

find the first eight terms of the sequence $\{a_n\}$. Then use part (a) to show that $\lim_{n\to\infty} a_n = \sqrt{2}$. This gives the **continued fraction expansion**

$$\sqrt{2} = 1 + \cfrac{1}{2 + \cfrac{1}{2 + \cdots}}$$

93. The size of an undisturbed fish population has been modeled by the formula

$$p_{n+1} = \frac{bp_n}{a + p_n}$$

where p_n is the fish population after n years and a and b are positive constants that depend on the species and its environment. Suppose that the population in year 0 is $p_0 > 0$.
(a) Show that if $\{p_n\}$ is convergent, then the only possible values for its limit are 0 and $b - a$.
(b) Show that $p_{n+1} < (b/a)p_n$.
(c) Use part (b) to show that if $a > b$, then $\lim_{n\to\infty} p_n = 0$; in other words, the population dies out.
(d) Now assume that $a < b$. Show that if $p_0 < b - a$, then $\{p_n\}$ is increasing and $0 < p_n < b - a$. Show also that if $p_0 > b - a$, then $\{p_n\}$ is decreasing and $p_n > b - a$. Deduce that if $a < b$, then $\lim_{n\to\infty} p_n = b - a$.

LABORATORY PROJECT CAS LOGISTIC SEQUENCES

A sequence that arises in ecology as a model for population growth is defined by the **logistic difference equation**

$$p_{n+1} = kp_n(1 - p_n)$$

where p_n measures the size of the population of the nth generation of a single species. To keep the numbers manageable, p_n is a fraction of the maximal size of the population, so $0 \leqslant p_n \leqslant 1$. Notice that the form of this equation is similar to the logistic differential equation in Section 9.4. The discrete model—with sequences instead of continuous functions—is preferable for modeling insect populations, where mating and death occur in a periodic fashion.

An ecologist is interested in predicting the size of the population as time goes on, and asks these questions: Will it stabilize at a limiting value? Will it change in a cyclical fashion? Or will it exhibit random behavior?

Write a program to compute the first n terms of this sequence starting with an initial population p_0, where $0 < p_0 < 1$. Use this program to do the following.

1. Calculate 20 or 30 terms of the sequence for $p_0 = \frac{1}{2}$ and for two values of k such that $1 < k < 3$. Graph each sequence. Do the sequences appear to converge? Repeat for a different value of p_0 between 0 and 1. Does the limit depend on the choice of p_0? Does it depend on the choice of k?

2. Calculate terms of the sequence for a value of k between 3 and 3.4 and plot them. What do you notice about the behavior of the terms?

3. Experiment with values of k between 3.4 and 3.5. What happens to the terms?

4. For values of k between 3.6 and 4, compute and plot at least 100 terms and comment on the behavior of the sequence. What happens if you change p_0 by 0.001? This type of behavior is called *chaotic* and is exhibited by insect populations under certain conditions.

11.2 Series

The current record for computing a decimal approximation for π was obtained by Shigeru Kondo and Alexander Yee in 2011 and contains more than 10 trillion decimal places.

What do we mean when we express a number as an infinite decimal? For instance, what does it mean to write

$$\pi = 3.14159\ 26535\ 89793\ 23846\ 26433\ 83279\ 50288\ldots$$

The convention behind our decimal notation is that any number can be written as an infinite sum. Here it means that

$$\pi = 3 + \frac{1}{10} + \frac{4}{10^2} + \frac{1}{10^3} + \frac{5}{10^4} + \frac{9}{10^5} + \frac{2}{10^6} + \frac{6}{10^7} + \frac{5}{10^8} + \cdots$$

where the three dots $(\cdots)$ indicate that the sum continues forever, and the more terms we add, the closer we get to the actual value of π.

In general, if we try to add the terms of an infinite sequence $\{a_n\}_{n=1}^{\infty}$ we get an expression of the form

$$a_1 + a_2 + a_3 + \cdots + a_n + \cdots \tag{1}$$

which is called an **infinite series** (or just a **series**) and is denoted, for short, by the symbol

$$\sum_{n=1}^{\infty} a_n \qquad \text{or} \qquad \sum a_n$$

Does it make sense to talk about the sum of infinitely many terms?

It would be impossible to find a finite sum for the series

$$1 + 2 + 3 + 4 + 5 + \cdots + n + \cdots$$

because if we start adding the terms we get the cumulative sums 1, 3, 6, 10, 15, 21, . . . and, after the nth term, we get $n(n + 1)/2$, which becomes very large as n increases.

However, if we start to add the terms of the series

$$\frac{1}{2} + \frac{1}{4} + \frac{1}{8} + \frac{1}{16} + \frac{1}{32} + \frac{1}{64} + \cdots + \frac{1}{2^n} + \cdots$$

n	Sum of first n terms
1	0.50000000
2	0.75000000
3	0.87500000
4	0.93750000
5	0.96875000
6	0.98437500
7	0.99218750
10	0.99902344
15	0.99996948
20	0.99999905
25	0.99999997

we get $\frac{1}{2}, \frac{3}{4}, \frac{7}{8}, \frac{15}{16}, \frac{31}{32}, \frac{63}{64}, \ldots, 1 - 1/2^n, \ldots$. The table shows that as we add more and more terms, these *partial sums* become closer and closer to 1. (See also Figure 11 in *A Preview of Calculus*, page 6.) In fact, by adding sufficiently many terms of the series we can make the partial sums as close as we like to 1. So it seems reasonable to say that the sum of this infinite series is 1 and to write

$$\sum_{n=1}^{\infty} \frac{1}{2^n} = \frac{1}{2} + \frac{1}{4} + \frac{1}{8} + \frac{1}{16} + \cdots + \frac{1}{2^n} + \cdots = 1$$

We use a similar idea to determine whether or not a general series (1) has a sum. We consider the **partial sums**

$$s_1 = a_1$$

$$s_2 = a_1 + a_2$$

$$s_3 = a_1 + a_2 + a_3$$

$$s_4 = a_1 + a_2 + a_3 + a_4$$

and, in general,

$$s_n = a_1 + a_2 + a_3 + \cdots + a_n = \sum_{i=1}^{n} a_i$$

These partial sums form a new sequence $\{s_n\}$, which may or may not have a limit. If $\lim_{n\to\infty} s_n = s$ exists (as a finite number), then, as in the preceding example, we call it the sum of the infinite series $\Sigma\, a_n$.

2 Definition Given a series $\sum_{n=1}^{\infty} a_n = a_1 + a_2 + a_3 + \cdots$, let s_n denote its nth partial sum:

$$s_n = \sum_{i=1}^{n} a_i = a_1 + a_2 + \cdots + a_n$$

If the sequence $\{s_n\}$ is convergent and $\lim_{n\to\infty} s_n = s$ exists as a real number, then the series $\Sigma\, a_n$ is called **convergent** and we write

$$a_1 + a_2 + \cdots + a_n + \cdots = s \qquad \text{or} \qquad \sum_{n=1}^{\infty} a_n = s$$

The number s is called the **sum** of the series. If the sequence $\{s_n\}$ is divergent, then the series is called **divergent**.

Thus the sum of a series is the limit of the sequence of partial sums. So when we write $\sum_{n=1}^{\infty} a_n = s$, we mean that by adding sufficiently many terms of the series we can get as close as we like to the number s. Notice that

$$\sum_{n=1}^{\infty} a_n = \lim_{n \to \infty} \sum_{i=1}^{n} a_i$$

Compare with the improper integral

$$\int_1^{\infty} f(x)\,dx = \lim_{t \to \infty} \int_1^t f(x)\,dx$$

To find this integral we integrate from 1 to t and then let $t \to \infty$. For a series, we sum from 1 to n and then let $n \to \infty$.

EXAMPLE 1 Suppose we know that the sum of the first n terms of the series $\sum_{n=1}^{\infty} a_n$ is

$$s_n = a_1 + a_2 + \cdots + a_n = \frac{2n}{3n + 5}$$

Then the sum of the series is the limit of the sequence $\{s_n\}$:

$$\sum_{n=1}^{\infty} a_n = \lim_{n \to \infty} s_n = \lim_{n \to \infty} \frac{2n}{3n + 5} = \lim_{n \to \infty} \frac{2}{3 + \dfrac{5}{n}} = \frac{2}{3}$$ ■

In Example 1 we were *given* an expression for the sum of the first n terms, but it's usually not easy to *find* such an expression. In Example 2, however, we look at a famous series for which we *can* find an explicit formula for s_n.

EXAMPLE 2 An important example of an infinite series is the **geometric series**

$$a + ar + ar^2 + ar^3 + \cdots + ar^{n-1} + \cdots = \sum_{n=1}^{\infty} ar^{n-1} \qquad a \neq 0$$

Each term is obtained from the preceding one by multiplying it by the **common ratio** r. (We have already considered the special case where $a = \frac{1}{2}$ and $r = \frac{1}{2}$ on page 748.)

If $r = 1$, then $s_n = a + a + \cdots + a = na \to \pm\infty$. Since $\lim_{n \to \infty} s_n$ doesn't exist, the geometric series diverges in this case.

If $r \neq 1$, we have

$$s_n = a + ar + ar^2 + \cdots + ar^{n-1}$$

and

$$rs_n = \qquad ar + ar^2 + \cdots + ar^{n-1} + ar^n$$

Subtracting these equations, we get

$$s_n - rs_n = a - ar^n$$

3 $$s_n = \frac{a(1 - r^n)}{1 - r}$$

Figure 1 provides a geometric demonstration of the result in Example 2. If the triangles are constructed as shown and s is the sum of the series, then, by similar triangles,

$$\frac{s}{a} = \frac{a}{a - ar} \quad \text{so} \quad s = \frac{a}{1 - r}$$

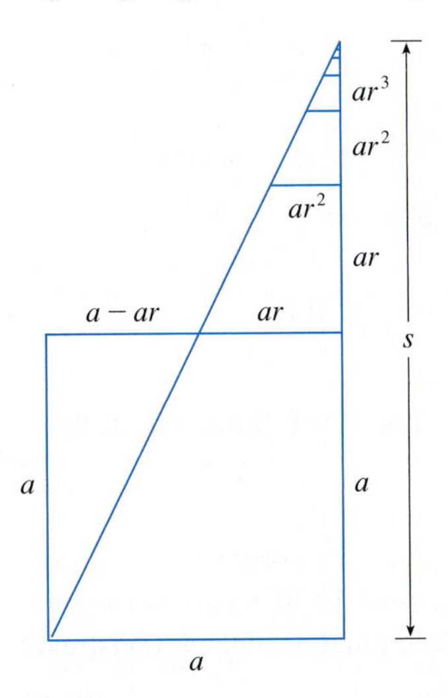

FIGURE 1

If $-1 < r < 1$, we know from (11.1.9) that $r^n \to 0$ as $n \to \infty$, so

$$\lim_{n \to \infty} s_n = \lim_{n \to \infty} \frac{a(1 - r^n)}{1 - r} = \frac{a}{1 - r} - \frac{a}{1 - r} \lim_{n \to \infty} r^n = \frac{a}{1 - r}$$

Thus when $|r| < 1$ the geometric series is convergent and its sum is $a/(1 - r)$.

If $r \leqslant -1$ or $r > 1$, the sequence $\{r^n\}$ is divergent by (11.1.9) and so, by Equation 3, $\lim_{n \to \infty} s_n$ does not exist. Therefore the geometric series diverges in those cases. ■

We summarize the results of Example 2 as follows.

4 The geometric series

$$\sum_{n=1}^{\infty} ar^{n-1} = a + ar + ar^2 + \cdots$$

is convergent if $|r| < 1$ and its sum is

$$\sum_{n=1}^{\infty} ar^{n-1} = \frac{a}{1-r} \qquad |r| < 1$$

If $|r| \geq 1$, the geometric series is divergent.

In words: The sum of a convergent geometric series is

$$\frac{\text{first term}}{1 - \text{common ratio}}$$

EXAMPLE 3 Find the sum of the geometric series

$$5 - \tfrac{10}{3} + \tfrac{20}{9} - \tfrac{40}{27} + \cdots$$

SOLUTION The first term is $a = 5$ and the common ratio is $r = -\frac{2}{3}$. Since $|r| = \frac{2}{3} < 1$, the series is convergent by (4) and its sum is

$$5 - \frac{10}{3} + \frac{20}{9} - \frac{40}{27} + \cdots = \frac{5}{1 - \left(-\frac{2}{3}\right)} = \frac{5}{\frac{5}{3}} = 3$$

What do we really mean when we say that the sum of the series in Example 3 is 3? Of course, we can't literally add an infinite number of terms, one by one. But, according to Definition 2, the total sum is the limit of the sequence of partial sums. So, by taking the sum of sufficiently many terms, we can get as close as we like to the number 3. The table shows the first ten partial sums s_n and the graph in Figure 2 shows how the sequence of partial sums approaches 3.

n	s_n
1	5.000000
2	1.666667
3	3.888889
4	2.407407
5	3.395062
6	2.736626
7	3.175583
8	2.882945
9	3.078037
10	2.947975

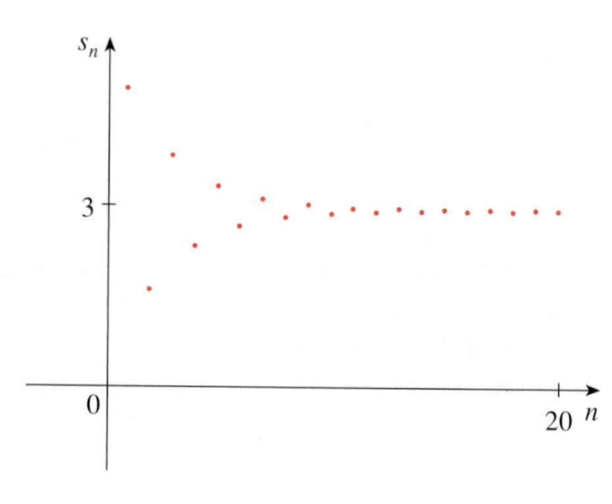

FIGURE 2

EXAMPLE 4 Is the series $\sum_{n=1}^{\infty} 2^{2n}3^{1-n}$ convergent or divergent?

SOLUTION Let's rewrite the nth term of the series in the form ar^{n-1}:

$$\sum_{n=1}^{\infty} 2^{2n}3^{1-n} = \sum_{n=1}^{\infty} (2^2)^n 3^{-(n-1)} = \sum_{n=1}^{\infty} \frac{4^n}{3^{n-1}} = \sum_{n=1}^{\infty} 4\left(\tfrac{4}{3}\right)^{n-1}$$

Another way to identify a and r is to write out the first few terms:

$$4 + \tfrac{16}{3} + \tfrac{64}{9} + \cdots$$

We recognize this series as a geometric series with $a = 4$ and $r = \frac{4}{3}$. Since $r > 1$, the series diverges by (4).

EXAMPLE 5 A drug is administered to a patient at the same time every day. Suppose the concentration of the drug is C_n (measured in mg/mL) after the injection on the nth day. Before the injection the next day, only 30% of the drug remains in the bloodstream and the daily dose raises the concentration by 0.2 mg/mL.

(a) Find the concentration after three days.

(b) What is the concentration after the nth dose?
(c) What is the limiting concentration?

SOLUTION
(a) Just before the daily dose of medication is administered, the concentration is reduced to 30% of the preceding day's concentration, that is, $0.3C_n$. With the new dose, the concentration is increased by 0.2 mg/mL and so

$$C_{n+1} = 0.2 + 0.3C_n$$

Starting with $C_0 = 0$ and putting $n = 0, 1, 2$ into this equation, we get

$$C_1 = 0.2 + 0.3C_0 = 0.2$$

$$C_2 = 0.2 + 0.3C_1 = 0.2 + 0.2(0.3) = 0.26$$

$$C_3 = 0.2 + 0.3C_2 = 0.2 + 0.2(0.3) + 0.2(0.3)^2 = 0.278$$

The concentration after three days is 0.278 mg/mL.
(b) After the nth dose the concentration is

$$C_n = 0.2 + 0.2(0.3) + 0.2(0.3)^2 + \cdots + 0.2(0.3)^{n-1}$$

This is a finite geometric series with $a = 0.2$ and $r = 0.3$, so by Formula 3 we have

$$C_n = \frac{0.2[1 - (0.3)^n]}{1 - 0.3} = \frac{2}{7}[1 - (0.3)^n] \text{ mg/mL}$$

(c) Because $0.3 < 1$, we know that $\lim_{n\to\infty} (0.3)^n = 0$. So the limiting concentration is

$$\lim_{n\to\infty} C_n = \lim_{n\to\infty} \frac{2}{7}[1 - (0.3)^n] = \frac{2}{7}(1 - 0) = \frac{2}{7} \text{ mg/mL}$$ ■

EXAMPLE 6 Write the number $2.3\overline{17} = 2.3171717\ldots$ as a ratio of integers.

SOLUTION

$$2.3171717\ldots = 2.3 + \frac{17}{10^3} + \frac{17}{10^5} + \frac{17}{10^7} + \cdots$$

After the first term we have a geometric series with $a = 17/10^3$ and $r = 1/10^2$. Therefore

$$2.3\overline{17} = 2.3 + \frac{\frac{17}{10^3}}{1 - \frac{1}{10^2}} = 2.3 + \frac{\frac{17}{1000}}{\frac{99}{100}}$$

$$= \frac{23}{10} + \frac{17}{990} = \frac{1147}{495}$$ ■

EXAMPLE 7 Find the sum of the series $\sum_{n=0}^{\infty} x^n$, where $|x| < 1$.

SOLUTION Notice that this series starts with $n = 0$ and so the first term is $x^0 = 1$. (With series, we adopt the convention that $x^0 = 1$ even when $x = 0$.)

TEC Module 11.2 explores a series that depends on an angle θ in a triangle and enables you to see how rapidly the series converges when θ varies.

Thus

$$\sum_{n=0}^{\infty} x^n = 1 + x + x^2 + x^3 + x^4 + \cdots$$

This is a geometric series with $a = 1$ and $r = x$. Since $|r| = |x| < 1$, it converges and (4) gives

$$\boxed{5} \qquad \sum_{n=0}^{\infty} x^n = \frac{1}{1 - x}$$

■

EXAMPLE 8 Show that the series $\displaystyle\sum_{n=1}^{\infty} \frac{1}{n(n+1)}$ is convergent, and find its sum.

SOLUTION This is not a geometric series, so we go back to the definition of a convergent series and compute the partial sums.

$$s_n = \sum_{i=1}^{n} \frac{1}{i(i+1)} = \frac{1}{1 \cdot 2} + \frac{1}{2 \cdot 3} + \frac{1}{3 \cdot 4} + \cdots + \frac{1}{n(n+1)}$$

We can simplify this expression if we use the partial fraction decomposition

$$\frac{1}{i(i+1)} = \frac{1}{i} - \frac{1}{i+1}$$

(see Section 7.4). Thus we have

$$s_n = \sum_{i=1}^{n} \frac{1}{i(i+1)} = \sum_{i=1}^{n} \left(\frac{1}{i} - \frac{1}{i+1} \right)$$

$$= \left(1 - \frac{1}{2}\right) + \left(\frac{1}{2} - \frac{1}{3}\right) + \left(\frac{1}{3} - \frac{1}{4}\right) + \cdots + \left(\frac{1}{n} - \frac{1}{n+1}\right)$$

$$= 1 - \frac{1}{n+1}$$

Notice that the terms cancel in pairs. This is an example of a **telescoping sum**: Because of all the cancellations, the sum collapses (like a pirate's collapsing telescope) into just two terms.

and so
$$\lim_{n\to\infty} s_n = \lim_{n\to\infty} \left(1 - \frac{1}{n+1}\right) = 1 - 0 = 1$$

Therefore the given series is convergent and

$$\sum_{n=1}^{\infty} \frac{1}{n(n+1)} = 1$$

■

Figure 3 illustrates Example 8 by showing the graphs of the sequence of terms $a_n = 1/[n(n+1)]$ and the sequence $\{s_n\}$ of partial sums. Notice that $a_n \to 0$ and $s_n \to 1$. See Exercises 78 and 79 for two geometric interpretations of Example 8.

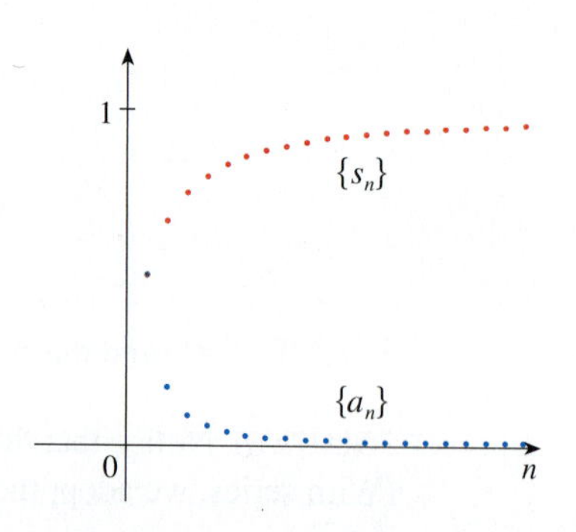

FIGURE 3

EXAMPLE 9 Show that the **harmonic series**

$$\sum_{n=1}^{\infty} \frac{1}{n} = 1 + \frac{1}{2} + \frac{1}{3} + \frac{1}{4} + \cdots$$

is divergent.

SOLUTION For this particular series it's convenient to consider the partial sums s_2, s_4, s_8, s_{16}, s_{32}, . . . and show that they become large.

$$s_2 = 1 + \tfrac{1}{2}$$

$$s_4 = 1 + \tfrac{1}{2} + \left(\tfrac{1}{3} + \tfrac{1}{4}\right) > 1 + \tfrac{1}{2} + \left(\tfrac{1}{4} + \tfrac{1}{4}\right) = 1 + \tfrac{2}{2}$$

$$\begin{aligned} s_8 &= 1 + \tfrac{1}{2} + \left(\tfrac{1}{3} + \tfrac{1}{4}\right) + \left(\tfrac{1}{5} + \tfrac{1}{6} + \tfrac{1}{7} + \tfrac{1}{8}\right) \\ &> 1 + \tfrac{1}{2} + \left(\tfrac{1}{4} + \tfrac{1}{4}\right) + \left(\tfrac{1}{8} + \tfrac{1}{8} + \tfrac{1}{8} + \tfrac{1}{8}\right) \\ &= 1 + \tfrac{1}{2} + \tfrac{1}{2} + \tfrac{1}{2} = 1 + \tfrac{3}{2} \end{aligned}$$

$$\begin{aligned} s_{16} &= 1 + \tfrac{1}{2} + \left(\tfrac{1}{3} + \tfrac{1}{4}\right) + \left(\tfrac{1}{5} + \cdots + \tfrac{1}{8}\right) + \left(\tfrac{1}{9} + \cdots + \tfrac{1}{16}\right) \\ &> 1 + \tfrac{1}{2} + \left(\tfrac{1}{4} + \tfrac{1}{4}\right) + \left(\tfrac{1}{8} + \cdots + \tfrac{1}{8}\right) + \left(\tfrac{1}{16} + \cdots + \tfrac{1}{16}\right) \\ &= 1 + \tfrac{1}{2} + \tfrac{1}{2} + \tfrac{1}{2} + \tfrac{1}{2} = 1 + \tfrac{4}{2} \end{aligned}$$

Similarly, $s_{32} > 1 + \frac{5}{2}$, $s_{64} > 1 + \frac{6}{2}$, and in general

$$s_{2^n} > 1 + \frac{n}{2}$$

The method used in Example 9 for showing that the harmonic series diverges is due to the French scholar Nicole Oresme (1323–1382).

This shows that $s_{2^n} \to \infty$ as $n \to \infty$ and so $\{s_n\}$ is divergent. Therefore the harmonic series diverges. ■

> **6 Theorem** If the series $\sum_{n=1}^{\infty} a_n$ is convergent, then $\lim_{n \to \infty} a_n = 0$.

PROOF Let $s_n = a_1 + a_2 + \cdots + a_n$. Then $a_n = s_n - s_{n-1}$. Since $\sum a_n$ is convergent, the sequence $\{s_n\}$ is convergent. Let $\lim_{n \to \infty} s_n = s$. Since $n - 1 \to \infty$ as $n \to \infty$, we also have $\lim_{n \to \infty} s_{n-1} = s$. Therefore

$$\lim_{n \to \infty} a_n = \lim_{n \to \infty} (s_n - s_{n-1}) = \lim_{n \to \infty} s_n - \lim_{n \to \infty} s_{n-1} = s - s = 0$$ ■

NOTE 1 With any *series* $\sum a_n$ we associate two *sequences:* the sequence $\{s_n\}$ of its partial sums and the sequence $\{a_n\}$ of its terms. If $\sum a_n$ is convergent, then the limit of the sequence $\{s_n\}$ is s (the sum of the series) and, as Theorem 6 asserts, the limit of the sequence $\{a_n\}$ is 0.

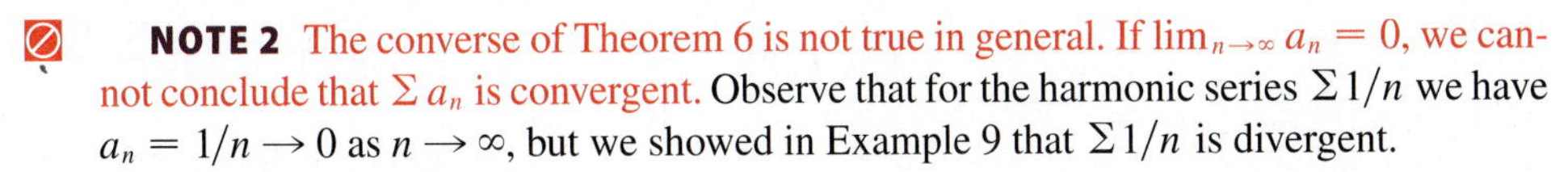

NOTE 2 The converse of Theorem 6 is not true in general. If $\lim_{n \to \infty} a_n = 0$, we cannot conclude that $\sum a_n$ is convergent. Observe that for the harmonic series $\sum 1/n$ we have $a_n = 1/n \to 0$ as $n \to \infty$, but we showed in Example 9 that $\sum 1/n$ is divergent.

> **7 Test for Divergence** If $\lim_{n \to \infty} a_n$ does not exist or if $\lim_{n \to \infty} a_n \neq 0$, then the series $\sum_{n=1}^{\infty} a_n$ is divergent.

The Test for Divergence follows from Theorem 6 because, if the series is not divergent, then it is convergent, and so $\lim_{n\to\infty} a_n = 0$.

EXAMPLE 10 Show that the series $\sum_{n=1}^{\infty} \dfrac{n^2}{5n^2+4}$ diverges.

SOLUTION

$$\lim_{n\to\infty} a_n = \lim_{n\to\infty} \frac{n^2}{5n^2+4} = \lim_{n\to\infty} \frac{1}{5+4/n^2} = \frac{1}{5} \neq 0$$

So the series diverges by the Test for Divergence. ■

NOTE 3 If we find that $\lim_{n\to\infty} a_n \neq 0$, we know that $\sum a_n$ is divergent. If we find that $\lim_{n\to\infty} a_n = 0$, we know *nothing* about the convergence or divergence of $\sum a_n$. Remember the warning in Note 2: if $\lim_{n\to\infty} a_n = 0$, the series $\sum a_n$ might converge or it might diverge.

8 Theorem If $\sum a_n$ and $\sum b_n$ are convergent series, then so are the series $\sum ca_n$ (where c is a constant), $\sum(a_n + b_n)$, and $\sum(a_n - b_n)$, and

(i) $\displaystyle\sum_{n=1}^{\infty} ca_n = c\sum_{n=1}^{\infty} a_n$

(ii) $\displaystyle\sum_{n=1}^{\infty} (a_n + b_n) = \sum_{n=1}^{\infty} a_n + \sum_{n=1}^{\infty} b_n$

(iii) $\displaystyle\sum_{n=1}^{\infty} (a_n - b_n) = \sum_{n=1}^{\infty} a_n - \sum_{n=1}^{\infty} b_n$

These properties of convergent series follow from the corresponding Limit Laws for Sequences in Section 11.1. For instance, here is how part (ii) of Theorem 8 is proved:

Let

$$s_n = \sum_{i=1}^{n} a_i \qquad s = \sum_{n=1}^{\infty} a_n \qquad t_n = \sum_{i=1}^{n} b_i \qquad t = \sum_{n=1}^{\infty} b_n$$

The nth partial sum for the series $\sum(a_n + b_n)$ is

$$u_n = \sum_{i=1}^{n} (a_i + b_i)$$

and, using Equation 4.2.10, we have

$$\begin{aligned}\lim_{n\to\infty} u_n &= \lim_{n\to\infty} \sum_{i=1}^{n} (a_i + b_i) = \lim_{n\to\infty} \left(\sum_{i=1}^{n} a_i + \sum_{i=1}^{n} b_i\right) \\ &= \lim_{n\to\infty} \sum_{i=1}^{n} a_i + \lim_{n\to\infty} \sum_{i=1}^{n} b_i \,, \\ &= \lim_{n\to\infty} s_n + \lim_{n\to\infty} t_n = s + t\end{aligned}$$

Therefore $\sum (a_n + b_n)$ is convergent and its sum is

$$\sum_{n=1}^{\infty} (a_n + b_n) = s + t = \sum_{n=1}^{\infty} a_n + \sum_{n=1}^{\infty} b_n$$ ■

EXAMPLE 11 Find the sum of the series $\sum_{n=1}^{\infty} \left(\frac{3}{n(n+1)} + \frac{1}{2^n} \right)$.

SOLUTION The series $\Sigma\, 1/2^n$ is a geometric series with $a = \frac{1}{2}$ and $r = \frac{1}{2}$, so

$$\sum_{n=1}^{\infty} \frac{1}{2^n} = \frac{\frac{1}{2}}{1 - \frac{1}{2}} = 1$$

In Example 8 we found that

$$\sum_{n=1}^{\infty} \frac{1}{n(n+1)} = 1$$

So, by Theorem 8, the given series is convergent and

$$\sum_{n=1}^{\infty} \left(\frac{3}{n(n+1)} + \frac{1}{2^n} \right) = 3 \sum_{n=1}^{\infty} \frac{1}{n(n+1)} + \sum_{n=1}^{\infty} \frac{1}{2^n}$$

$$= 3 \cdot 1 + 1 = 4$$

NOTE 4 A finite number of terms doesn't affect the convergence or divergence of a series. For instance, suppose that we were able to show that the series

$$\sum_{n=4}^{\infty} \frac{n}{n^3 + 1}$$

is convergent. Since

$$\sum_{n=1}^{\infty} \frac{n}{n^3 + 1} = \frac{1}{2} + \frac{2}{9} + \frac{3}{28} + \sum_{n=4}^{\infty} \frac{n}{n^3 + 1}$$

it follows that the entire series $\Sigma_{n=1}^{\infty}\, n/(n^3 + 1)$ is convergent. Similarly, if it is known that the series $\Sigma_{n=N+1}^{\infty}\, a_n$ converges, then the full series

$$\sum_{n=1}^{\infty} a_n = \sum_{n=1}^{N} a_n + \sum_{n=N+1}^{\infty} a_n$$

is also convergent.

11.2 EXERCISES

1. (a) What is the difference between a sequence and a series?
(b) What is a convergent series? What is a divergent series?

2. Explain what it means to say that $\Sigma_{n=1}^{\infty}\, a_n = 5$.

3–4 Calculate the sum of the series $\Sigma_{n=1}^{\infty}\, a_n$ whose partial sums are given.

3. $s_n = 2 - 3(0.8)^n$

4. $s_n = \frac{n^2 - 1}{4n^2 + 1}$

5–8 Calculate the first eight terms of the sequence of partial sums correct to four decimal places. Does it appear that the series is convergent or divergent?

5. $\sum_{n=1}^{\infty} \frac{1}{n^4 + n^2}$

6. $\sum_{n=1}^{\infty} \frac{1}{\sqrt[3]{n}}$

7. $\sum_{n=1}^{\infty} \sin n$

8. $\sum_{n=1}^{\infty} \frac{(-1)^{n-1}}{n!}$

9–14 Find at least 10 partial sums of the series. Graph both the sequence of terms and the sequence of partial sums on the same screen. Does it appear that the series is convergent or divergent? If it is convergent, find the sum. If it is divergent, explain why.

9. $\sum_{n=1}^{\infty} \frac{12}{(-5)^n}$

10. $\sum_{n=1}^{\infty} \cos n$

11. $\sum_{n=1}^{\infty} \frac{n}{\sqrt{n^2 + 4}}$

12. $\sum_{n=1}^{\infty} \frac{7^{n+1}}{10^n}$

13. $\displaystyle\sum_{n=1}^{\infty} \frac{1}{n^2+1}$

14. $\displaystyle\sum_{n=1}^{\infty} \left(\sin\frac{1}{n} - \sin\frac{1}{n+1} \right)$

15. Let $a_n = \dfrac{2n}{3n+1}$.

(a) Determine whether $\{a_n\}$ is convergent.

(b) Determine whether $\sum_{n=1}^{\infty} a_n$ is convergent.

16. (a) Explain the difference between

$$\sum_{i=1}^{n} a_i \quad \text{and} \quad \sum_{j=1}^{n} a_j$$

(b) Explain the difference between

$$\sum_{i=1}^{n} a_i \quad \text{and} \quad \sum_{i=1}^{n} a_j$$

17–26 Determine whether the geometric series is convergent or divergent. If it is convergent, find its sum.

17. $3 - 4 + \frac{16}{3} - \frac{64}{9} + \cdots$

18. $4 + 3 + \frac{9}{4} + \frac{27}{16} + \cdots$

19. $10 - 2 + 0.4 - 0.08 + \cdots$

20. $2 + 0.5 + 0.125 + 0.03125 + \cdots$

21. $\displaystyle\sum_{n=1}^{\infty} 12(0.73)^{n-1}$

22. $\displaystyle\sum_{n=1}^{\infty} \frac{5}{\pi^n}$

23. $\displaystyle\sum_{n=1}^{\infty} \frac{(-3)^{n-1}}{4^n}$

24. $\displaystyle\sum_{n=0}^{\infty} \frac{3^{n+1}}{(-2)^n}$

25. $\displaystyle\sum_{n=1}^{\infty} \frac{e^{2n}}{6^{n-1}}$

26. $\displaystyle\sum_{n=1}^{\infty} \frac{6 \cdot 2^{2n-1}}{3^n}$

27–42 Determine whether the series is convergent or divergent. If it is convergent, find its sum.

27. $\dfrac{1}{3} + \dfrac{1}{6} + \dfrac{1}{9} + \dfrac{1}{12} + \dfrac{1}{15} + \cdots$

28. $\dfrac{1}{3} + \dfrac{2}{9} + \dfrac{1}{27} + \dfrac{2}{81} + \dfrac{1}{243} + \dfrac{2}{729} + \cdots$

29. $\displaystyle\sum_{n=1}^{\infty} \frac{2+n}{1-2n}$

30. $\displaystyle\sum_{k=1}^{\infty} \frac{k^2}{k^2-2k+5}$

31. $\displaystyle\sum_{n=1}^{\infty} 3^{n+1}4^{-n}$

32. $\displaystyle\sum_{n=1}^{\infty} [(-0.2)^n + (0.6)^{n-1}]$

33. $\displaystyle\sum_{n=1}^{\infty} \frac{1}{4+e^{-n}}$

34. $\displaystyle\sum_{n=1}^{\infty} \frac{2^n+4^n}{e^n}$

35. $\displaystyle\sum_{k=1}^{\infty} (\sin 100)^k$

36. $\displaystyle\sum_{n=1}^{\infty} \frac{1}{1+\left(\frac{2}{3}\right)^n}$

37. $\displaystyle\sum_{n=1}^{\infty} \ln\left(\frac{n^2+1}{2n^2+1}\right)$

38. $\displaystyle\sum_{k=0}^{\infty} (\sqrt{2})^{-k}$

39. $\displaystyle\sum_{n=1}^{\infty} \arctan n$

40. $\displaystyle\sum_{n=1}^{\infty} \left(\frac{3}{5^n} + \frac{2}{n}\right)$

41. $\displaystyle\sum_{n=1}^{\infty} \left(\frac{1}{e^n} + \frac{1}{n(n+1)}\right)$

42. $\displaystyle\sum_{n=1}^{\infty} \frac{e^n}{n^2}$

43–48 Determine whether the series is convergent or divergent by expressing s_n as a telescoping sum (as in Example 8). If it is convergent, find its sum.

43. $\displaystyle\sum_{n=2}^{\infty} \frac{2}{n^2-1}$

44. $\displaystyle\sum_{n=1}^{\infty} \ln\frac{n}{n+1}$

45. $\displaystyle\sum_{n=1}^{\infty} \frac{3}{n(n+3)}$

46. $\displaystyle\sum_{n=4}^{\infty} \left(\frac{1}{\sqrt{n}} - \frac{1}{\sqrt{n+1}}\right)$

47. $\displaystyle\sum_{n=1}^{\infty} \left(e^{1/n} - e^{1/(n+1)}\right)$

48. $\displaystyle\sum_{n=2}^{\infty} \frac{1}{n^3-n}$

49. Let $x = 0.99999\ldots$.

(a) Do you think that $x < 1$ or $x = 1$?

(b) Sum a geometric series to find the value of x.

(c) How many decimal representations does the number 1 have?

(d) Which numbers have more than one decimal representation?

50. A sequence of terms is defined by

$$a_1 = 1 \qquad a_n = (5-n)a_{n-1}$$

Calculate $\sum_{n=1}^{\infty} a_n$.

51–56 Express the number as a ratio of integers.

51. $0.\overline{8} = 0.8888\ldots$

52. $0.\overline{46} = 0.46464646\ldots$

53. $2.\overline{516} = 2.516516516\ldots$

54. $10.1\overline{35} = 10.135353535\ldots$

55. $1.234\overline{567}$

56. $5.\overline{71358}$

57–63 Find the values of x for which the series converges. Find the sum of the series for those values of x.

57. $\displaystyle\sum_{n=1}^{\infty} (-5)^n x^n$

58. $\displaystyle\sum_{n=1}^{\infty} (x+2)^n$

59. $\displaystyle\sum_{n=0}^{\infty} \frac{(x-2)^n}{3^n}$

60. $\displaystyle\sum_{n=0}^{\infty} (-4)^n (x-5)^n$

61. $\displaystyle\sum_{n=0}^{\infty} \frac{2^n}{x^n}$

62. $\displaystyle\sum_{n=0}^{\infty} \frac{\sin^n x}{3^n}$

63. $\displaystyle\sum_{n=0}^{\infty} e^{nx}$

64. We have seen that the harmonic series is a divergent series whose terms approach 0. Show that

$$\sum_{n=1}^{\infty} \ln\left(1 + \frac{1}{n}\right)$$

is another series with this property.

CAS **65–66** Use the partial fraction command on your CAS to find a convenient expression for the partial sum, and then use this expression to find the sum of the series. Check your answer by using the CAS to sum the series directly.

65. $\displaystyle\sum_{n=1}^{\infty} \frac{3n^2 + 3n + 1}{(n^2 + n)^3}$

66. $\displaystyle\sum_{n=3}^{\infty} \frac{1}{n^5 - 5n^3 + 4n}$

---

67. If the nth partial sum of a series $\Sigma_{n=1}^{\infty} a_n$ is

$$s_n = \frac{n - 1}{n + 1}$$

find a_n and $\Sigma_{n=1}^{\infty} a_n$.

68. If the nth partial sum of a series $\Sigma_{n=1}^{\infty} a_n$ is $s_n = 3 - n2^{-n}$, find a_n and $\Sigma_{n=1}^{\infty} a_n$.

69. A doctor prescribes a 100-mg antibiotic tablet to be taken every eight hours. Just before each tablet is taken, 20% of the drug remains in the body.

(a) How much of the drug is in the body just after the second tablet is taken? After the third tablet?

(b) If Q_n is the quantity of the antibiotic in the body just after the nth tablet is taken, find an equation that expresses Q_{n+1} in terms of Q_n.

(c) What quantity of the antibiotic remains in the body in the long run?

70. A patient is injected with a drug every 12 hours. Immediately before each injection the concentration of the drug has been reduced by 90% and the new dose increases the concentration by 1.5 mg/L.

(a) What is the concentration after three doses?

(b) If C_n is the concentration after the nth dose, find a formula for C_n as a function of n.

(c) What is the limiting value of the concentration?

71. A patient takes 150 mg of a drug at the same time every day. Just before each tablet is taken, 5% of the drug remains in the body.

(a) What quantity of the drug is in the body after the third tablet? After the nth tablet?

(b) What quantity of the drug remains in the body in the long run?

72. After injection of a dose D of insulin, the concentration of insulin in a patient's system decays exponentially and so it can be written as De^{-at}, where t represents time in hours and a is a positive constant.

(a) If a dose D is injected every T hours, write an expression for the sum of the residual concentrations just before the $(n + 1)$st injection.

(b) Determine the limiting pre-injection concentration.

(c) If the concentration of insulin must always remain at or above a critical value C, determine a minimal dosage D in terms of C, a, and T.

73. When money is spent on goods and services, those who receive the money also spend some of it. The people receiving some of the twice-spent money will spend some of that, and so on. Economists call this chain reaction the *multiplier effect*. In a hypothetical isolated community, the local government begins the process by spending D dollars. Suppose that each recipient of spent money spends $100c\%$ and saves $100s\%$ of the money that he or she receives. The values c and s are called the *marginal propensity to consume* and the *marginal propensity to save* and, of course, $c + s = 1$.

(a) Let S_n be the total spending that has been generated after n transactions. Find an equation for S_n.

(b) Show that $\lim_{n\to\infty} S_n = kD$, where $k = 1/s$. The number k is called the *multiplier*. What is the multiplier if the marginal propensity to consume is 80%?

Note: The federal government uses this principle to justify deficit spending. Banks use this principle to justify lending a large percentage of the money that they receive in deposits.

74. A certain ball has the property that each time it falls from a height h onto a hard, level surface, it rebounds to a height rh, where $0 < r < 1$. Suppose that the ball is dropped from an initial height of H meters.

(a) Assuming that the ball continues to bounce indefinitely, find the total distance that it travels.

(b) Calculate the total time that the ball travels. (Use the fact that the ball falls $\frac{1}{2}gt^2$ meters in t seconds.)

(c) Suppose that each time the ball strikes the surface with velocity v it rebounds with velocity $-kv$, where $0 < k < 1$. How long will it take for the ball to come to rest?

75. Find the value of c if

$$\sum_{n=2}^{\infty} (1 + c)^{-n} = 2$$

76. Find the value of c such that

$$\sum_{n=0}^{\infty} e^{nc} = 10$$

77. In Example 9 we showed that the harmonic series is divergent. Here we outline another method, making use of the fact that $e^x > 1 + x$ for any $x > 0$. (See Exercise 6.2.109.)

If s_n is the nth partial sum of the harmonic series, show that $e^{s_n} > n + 1$. Why does this imply that the harmonic series is divergent?

78. Graph the curves $y = x^n$, $0 \leqslant x \leqslant 1$, for $n = 0, 1, 2, 3, 4, \ldots$ on a common screen. By finding the areas between successive curves, give a geometric demonstration of the fact, shown in Example 8, that

$$\sum_{n=1}^{\infty} \frac{1}{n(n + 1)} = 1$$

79. The figure shows two circles C and D of radius 1 that touch at P. The line T is a common tangent line; C_1 is the circle that touches C, D, and T; C_2 is the circle that touches C, D, and C_1; C_3 is the circle that touches C, D, and C_2. This procedure can be continued indefinitely and produces an infinite sequence of circles $\{C_n\}$. Find an expression for the diameter of C_n and thus provide another geometric demonstration of Example 8.

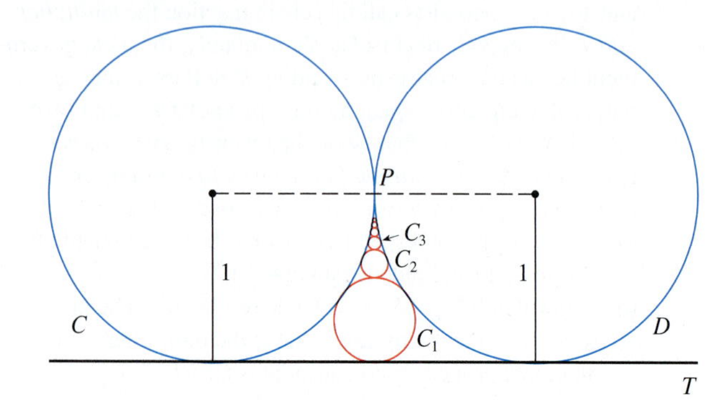

80. A right triangle ABC is given with $\angle A = \theta$ and $|AC| = b$. CD is drawn perpendicular to AB, DE is drawn perpendicular to BC, $EF \perp AB$, and this process is continued indefinitely, as shown in the figure. Find the total length of all the perpendiculars

$$|CD| + |DE| + |EF| + |FG| + \cdots$$

in terms of b and θ.

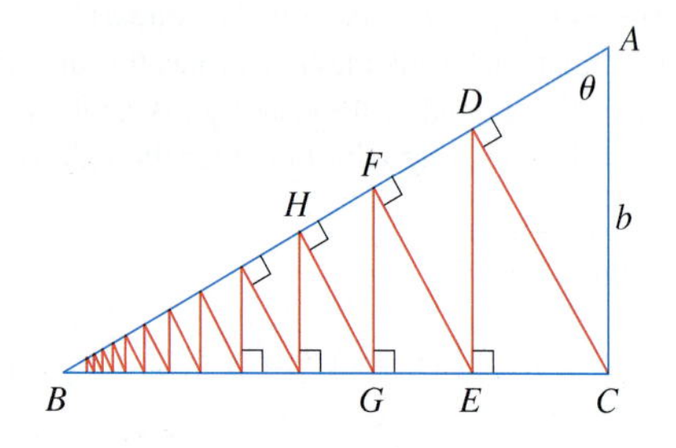

81. What is wrong with the following calculation?

$$\begin{aligned} 0 &= 0 + 0 + 0 + \cdots \\ &= (1 - 1) + (1 - 1) + (1 - 1) + \cdots \\ &= 1 - 1 + 1 - 1 + 1 - 1 + \cdots \\ &= 1 + (-1 + 1) + (-1 + 1) + (-1 + 1) + \cdots \\ &= 1 + 0 + 0 + 0 + \cdots = 1 \end{aligned}$$

(Guido Ubaldus thought that this proved the existence of God because "something has been created out of nothing.")

82. Suppose that $\sum_{n=1}^{\infty} a_n$ ($a_n \neq 0$) is known to be a convergent series. Prove that $\sum_{n=1}^{\infty} 1/a_n$ is a divergent series.

83. Prove part (i) of Theorem 8.

84. If $\sum a_n$ is divergent and $c \neq 0$, show that $\sum ca_n$ is divergent.

85. If $\sum a_n$ is convergent and $\sum b_n$ is divergent, show that the series $\sum (a_n + b_n)$ is divergent. [*Hint:* Argue by contradiction.]

86. If $\sum a_n$ and $\sum b_n$ are both divergent, is $\sum (a_n + b_n)$ necessarily divergent?

87. Suppose that a series $\sum a_n$ has positive terms and its partial sums s_n satisfy the inequality $s_n \leqslant 1000$ for all n. Explain why $\sum a_n$ must be convergent.

88. The Fibonacci sequence was defined in Section 11.1 by the equations

$$f_1 = 1, \quad f_2 = 1, \quad f_n = f_{n-1} + f_{n-2} \qquad n \geqslant 3$$

Show that each of the following statements is true.

(a) $\dfrac{1}{f_{n-1} f_{n+1}} = \dfrac{1}{f_{n-1} f_n} - \dfrac{1}{f_n f_{n+1}}$

(b) $\displaystyle\sum_{n=2}^{\infty} \frac{1}{f_{n-1} f_{n+1}} = 1$

(c) $\displaystyle\sum_{n=2}^{\infty} \frac{f_n}{f_{n-1} f_{n+1}} = 2$

89. The **Cantor set**, named after the German mathematician Georg Cantor (1845–1918), is constructed as follows. We start with the closed interval [0, 1] and remove the open interval $(\frac{1}{3}, \frac{2}{3})$. That leaves the two intervals $[0, \frac{1}{3}]$ and $[\frac{2}{3}, 1]$ and we remove the open middle third of each. Four intervals remain and again we remove the open middle third of each of them. We continue this procedure indefinitely, at each step removing the open middle third of every interval that remains from the preceding step. The Cantor set consists of the numbers that remain in [0, 1] after all those intervals have been removed.

(a) Show that the total length of all the intervals that are removed is 1. Despite that, the Cantor set contains infinitely many numbers. Give examples of some numbers in the Cantor set.

(b) The **Sierpinski carpet** is a two-dimensional counterpart of the Cantor set. It is constructed by removing the center one-ninth of a square of side 1, then removing the centers of the eight smaller remaining squares, and so on. (The figure shows the first three steps of the construction.) Show that the sum of the areas of the removed squares is 1. This implies that the Sierpinski carpet has area 0.

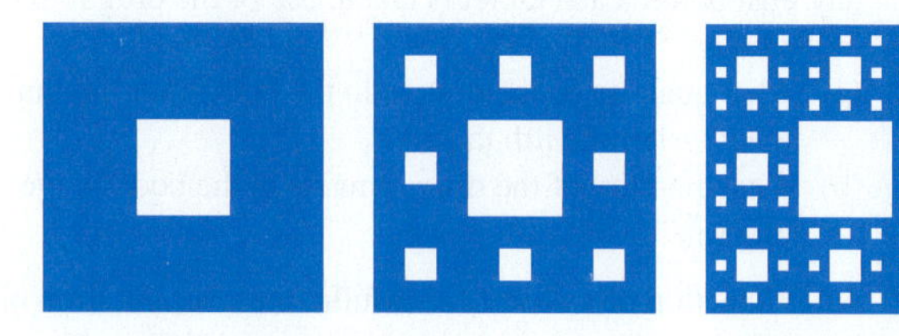

90. (a) A sequence $\{a_n\}$ is defined recursively by the equation $a_n = \frac{1}{2}(a_{n-1} + a_{n-2})$ for $n \geqslant 3$, where a_1 and a_2 can be any real numbers. Experiment with various values of a_1 and a_2 and use your calculator to guess the limit of the sequence.

(b) Find $\lim_{n \to \infty} a_n$ in terms of a_1 and a_2 by expressing $a_{n+1} - a_n$ in terms of $a_2 - a_1$ and summing a series.

91. Consider the series $\sum_{n=1}^{\infty} n/(n+1)!$.

(a) Find the partial sums s_1, s_2, s_3, and s_4. Do you recognize the denominators? Use the pattern to guess a formula for s_n.

(b) Use mathematical induction to prove your guess.

(c) Show that the given infinite series is convergent, and find its sum.

92. In the figure at the right there are infinitely many circles approaching the vertices of an equilateral triangle, each circle touching other circles and sides of the triangle. If the triangle has sides of length 1, find the total area occupied by the circles.

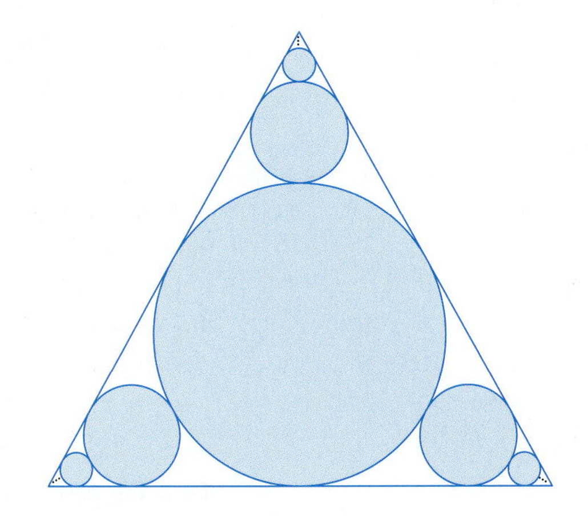

11.3 The Integral Test and Estimates of Sums

In general, it is difficult to find the exact sum of a series. We were able to accomplish this for geometric series and the series $\sum 1/[n(n+1)]$ because in each of those cases we could find a simple formula for the nth partial sum s_n. But usually it isn't easy to discover such a formula. Therefore, in the next few sections, we develop several tests that enable us to determine whether a series is convergent or divergent without explicitly finding its sum. (In some cases, however, our methods will enable us to find good estimates of the sum.) Our first test involves improper integrals.

We begin by investigating the series whose terms are the reciprocals of the squares of the positive integers:

$$\sum_{n=1}^{\infty} \frac{1}{n^2} = \frac{1}{1^2} + \frac{1}{2^2} + \frac{1}{3^2} + \frac{1}{4^2} + \frac{1}{5^2} + \cdots$$

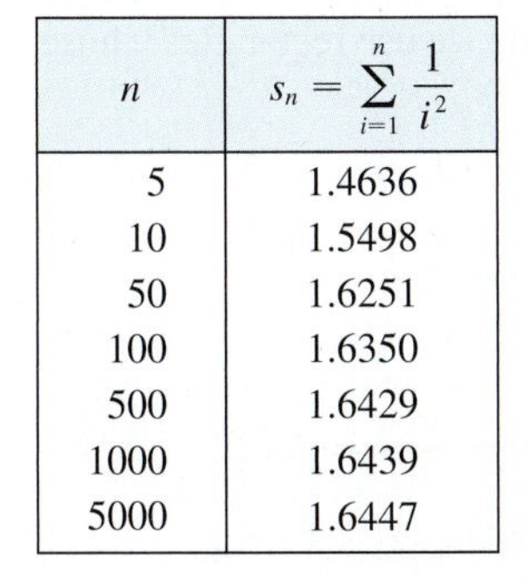

n	$s_n = \sum_{i=1}^{n} \frac{1}{i^2}$
5	1.4636
10	1.5498
50	1.6251
100	1.6350
500	1.6429
1000	1.6439
5000	1.6447

There's no simple formula for the sum s_n of the first n terms, but the computer-generated table of approximate values given in the margin suggests that the partial sums are approaching a number near 1.64 as $n \to \infty$ and so it looks as if the series is convergent.

We can confirm this impression with a geometric argument. Figure 1 shows the curve $y = 1/x^2$ and rectangles that lie below the curve. The base of each rectangle is an interval of length 1; the height is equal to the value of the function $y = 1/x^2$ at the right endpoint of the interval.

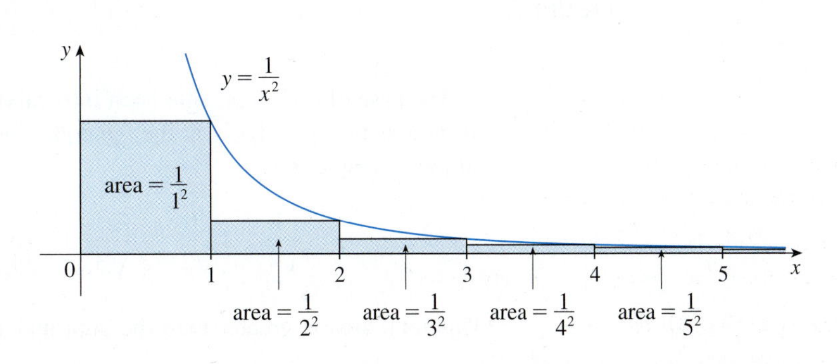

FIGURE 1

So the sum of the areas of the rectangles is

$$\frac{1}{1^2} + \frac{1}{2^2} + \frac{1}{3^2} + \frac{1}{4^2} + \frac{1}{5^2} + \cdots = \sum_{n=1}^{\infty} \frac{1}{n^2}$$

If we exclude the first rectangle, the total area of the remaining rectangles is smaller than the area under the curve $y = 1/x^2$ for $x \geqslant 1$, which is the value of the integral $\int_1^\infty (1/x^2)\,dx$. In Section 7.8 we discovered that this improper integral is convergent and has value 1. So the picture shows that all the partial sums are less than

$$\frac{1}{1^2} + \int_1^\infty \frac{1}{x^2}\,dx = 2$$

Thus the partial sums are bounded. We also know that the partial sums are increasing (because all the terms are positive). Therefore the partial sums converge (by the Monotonic Sequence Theorem) and so the series is convergent. The sum of the series (the limit of the partial sums) is also less than 2:

$$\sum_{n=1}^{\infty} \frac{1}{n^2} = \frac{1}{1^2} + \frac{1}{2^2} + \frac{1}{3^2} + \frac{1}{4^2} + \cdots < 2$$

[The exact sum of this series was found by the Swiss mathematician Leonhard Euler (1707–1783) to be $\pi^2/6$, but the proof of this fact is quite difficult. (See Problem 6 in the Problems Plus following Chapter 15.)]

Now let's look at the series

$$\sum_{n=1}^{\infty} \frac{1}{\sqrt{n}} = \frac{1}{\sqrt{1}} + \frac{1}{\sqrt{2}} + \frac{1}{\sqrt{3}} + \frac{1}{\sqrt{4}} + \frac{1}{\sqrt{5}} + \cdots$$

n	$s_n = \sum_{i=1}^{n} \frac{1}{\sqrt{i}}$
5	3.2317
10	5.0210
50	12.7524
100	18.5896
500	43.2834
1000	61.8010
5000	139.9681

The table of values of s_n suggests that the partial sums aren't approaching a finite number, so we suspect that the given series may be divergent. Again we use a picture for confirmation. Figure 2 shows the curve $y = 1/\sqrt{x}$, but this time we use rectangles whose tops lie *above* the curve.

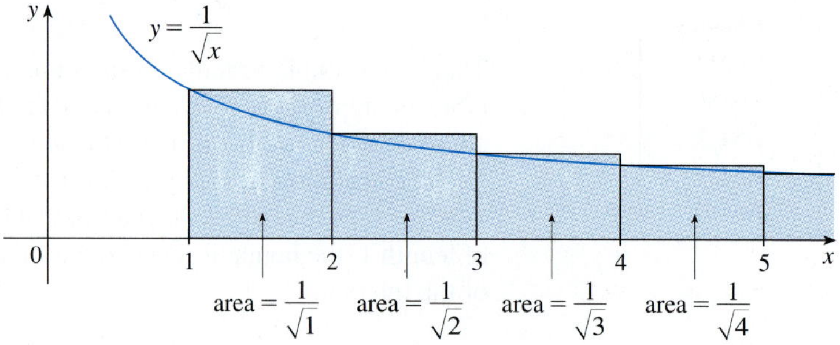

FIGURE 2

The base of each rectangle is an interval of length 1. The height is equal to the value of the function $y = 1/\sqrt{x}$ at the *left* endpoint of the interval. So the sum of the areas of all the rectangles is

$$\frac{1}{\sqrt{1}} + \frac{1}{\sqrt{2}} + \frac{1}{\sqrt{3}} + \frac{1}{\sqrt{4}} + \frac{1}{\sqrt{5}} + \cdots = \sum_{n=1}^{\infty} \frac{1}{\sqrt{n}}$$

This total area is greater than the area under the curve $y = 1/\sqrt{x}$ for $x \geqslant 1$, which is equal to the integral $\int_1^\infty (1/\sqrt{x})\,dx$. But we know from Section 7.8 that this improper integral is divergent. In other words, the area under the curve is infinite. So the sum of the series must be infinite; that is, the series is divergent.

The same sort of geometric reasoning that we used for these two series can be used to prove the following test. (The proof is given at the end of this section.)

The Integral Test Suppose f is a continuous, positive, decreasing function on $[1, \infty)$ and let $a_n = f(n)$. Then the series $\sum_{n=1}^{\infty} a_n$ is convergent if and only if the improper integral $\int_1^\infty f(x)\,dx$ is convergent. In other words:

(i) If $\displaystyle\int_1^\infty f(x)\,dx$ is convergent, then $\displaystyle\sum_{n=1}^{\infty} a_n$ is convergent.

(ii) If $\displaystyle\int_1^\infty f(x)\,dx$ is divergent, then $\displaystyle\sum_{n=1}^{\infty} a_n$ is divergent.

NOTE When we use the Integral Test, it is not necessary to start the series or the integral at $n = 1$. For instance, in testing the series

$$\sum_{n=4}^{\infty} \frac{1}{(n-3)^2} \qquad \text{we use} \qquad \int_4^\infty \frac{1}{(x-3)^2}\,dx$$

Also, it is not necessary that f be *always* decreasing. What is important is that f be *ultimately* decreasing, that is, decreasing for x larger than some number N. Then $\sum_{n=N}^{\infty} a_n$ is convergent, so $\sum_{n=1}^{\infty} a_n$ is convergent by Note 4 of Section 11.2.

EXAMPLE 1 Test the series $\displaystyle\sum_{n=1}^{\infty} \frac{1}{n^2+1}$ for convergence or divergence.

SOLUTION The function $f(x) = 1/(x^2+1)$ is continuous, positive, and decreasing on $[1, \infty)$ so we use the Integral Test:

$$\int_1^\infty \frac{1}{x^2+1}\,dx = \lim_{t\to\infty} \int_1^t \frac{1}{x^2+1}\,dx = \lim_{t\to\infty} \tan^{-1}x \Big]_1^t$$

$$= \lim_{t\to\infty} \left(\tan^{-1}t - \frac{\pi}{4}\right) = \frac{\pi}{2} - \frac{\pi}{4} = \frac{\pi}{4}$$

Thus $\int_1^\infty 1/(x^2+1)\,dx$ is a convergent integral and so, by the Integral Test, the series $\sum 1/(n^2+1)$ is convergent. ■

EXAMPLE 2 For what values of p is the series $\displaystyle\sum_{n=1}^{\infty} \frac{1}{n^p}$ convergent?

SOLUTION If $p < 0$, then $\lim_{n\to\infty} (1/n^p) = \infty$. If $p = 0$, then $\lim_{n\to\infty} (1/n^p) = 1$. In either case $\lim_{n\to\infty} (1/n^p) \neq 0$, so the given series diverges by the Test for Divergence (11.2.7).

If $p > 0$, then the function $f(x) = 1/x^p$ is clearly continuous, positive, and decreasing on $[1, \infty)$. We found in Chapter 7 [see (7.8.2)] that

$$\int_1^\infty \frac{1}{x^p}\,dx \quad \text{converges if } p > 1 \text{ and diverges if } p \leqslant 1$$

In order to use the Integral Test we need to be able to evaluate $\int_1^\infty f(x)\,dx$ and therefore we have to be able to find an antiderivative of f. Frequently this is difficult or impossible, so we need other tests for convergence too.

It follows from the Integral Test that the series $\Sigma\, 1/n^p$ converges if $p > 1$ and diverges if $0 < p \leqslant 1$. (For $p = 1$, this series is the harmonic series discussed in Example 11.2.9.) ■

The series in Example 2 is called the ***p*-series**. It is important in the rest of this chapter, so we summarize the results of Example 2 for future reference as follows.

1 The p-series $\displaystyle\sum_{n=1}^{\infty} \frac{1}{n^p}$ is convergent if $p > 1$ and divergent if $p \leqslant 1$.

EXAMPLE 3

(a) The series

$$\sum_{n=1}^{\infty} \frac{1}{n^3} = \frac{1}{1^3} + \frac{1}{2^3} + \frac{1}{3^3} + \frac{1}{4^3} + \cdots$$

is convergent because it is a p-series with $p = 3 > 1$.

(b) The series

$$\sum_{n=1}^{\infty} \frac{1}{n^{1/3}} = \sum_{n=1}^{\infty} \frac{1}{\sqrt[3]{n}} = 1 + \frac{1}{\sqrt[3]{2}} + \frac{1}{\sqrt[3]{3}} + \frac{1}{\sqrt[3]{4}} + \cdots$$

is divergent because it is a p-series with $p = \frac{1}{3} < 1$. ■

NOTE We should *not* infer from the Integral Test that the sum of the series is equal to the value of the integral. In fact,

$$\sum_{n=1}^{\infty} \frac{1}{n^2} = \frac{\pi^2}{6} \qquad \text{whereas} \qquad \int_1^{\infty} \frac{1}{x^2}\, dx = 1$$

Therefore, in general,

$$\sum_{n=1}^{\infty} a_n \neq \int_1^{\infty} f(x)\, dx$$

EXAMPLE 4 Determine whether the series $\displaystyle\sum_{n=1}^{\infty} \frac{\ln n}{n}$ converges or diverges.

SOLUTION The function $f(x) = (\ln x)/x$ is positive and continuous for $x > 1$ because the logarithm function is continuous. But it is not obvious whether or not f is decreasing, so we compute its derivative:

$$f'(x) = \frac{(1/x)x - \ln x}{x^2} = \frac{1 - \ln x}{x^2}$$

Thus $f'(x) < 0$ when $\ln x > 1$, that is, when $x > e$. It follows that f is decreasing when $x > e$ and so we can apply the Integral Test:

$$\int_1^{\infty} \frac{\ln x}{x}\, dx = \lim_{t \to \infty} \int_1^t \frac{\ln x}{x}\, dx = \lim_{t \to \infty} \frac{(\ln x)^2}{2} \bigg]_1^t$$

$$= \lim_{t \to \infty} \frac{(\ln t)^2}{2} = \infty$$

Since this improper integral is divergent, the series $\Sigma\, (\ln n)/n$ is also divergent by the Integral Test. ■

Estimating the Sum of a Series

Suppose we have been able to use the Integral Test to show that a series Σa_n is convergent and we now want to find an approximation to the sum s of the series. Of course, any partial sum s_n is an approximation to s because $\lim_{n \to \infty} s_n = s$. But how good is such an approximation? To find out, we need to estimate the size of the **remainder**

$$R_n = s - s_n = a_{n+1} + a_{n+2} + a_{n+3} + \cdots$$

The remainder R_n is the error made when s_n, the sum of the first n terms, is used as an approximation to the total sum.

We use the same notation and ideas as in the Integral Test, assuming that f is decreasing on $[n, \infty)$. Comparing the areas of the rectangles with the area under $y = f(x)$ for $x > n$ in Figure 3, we see that

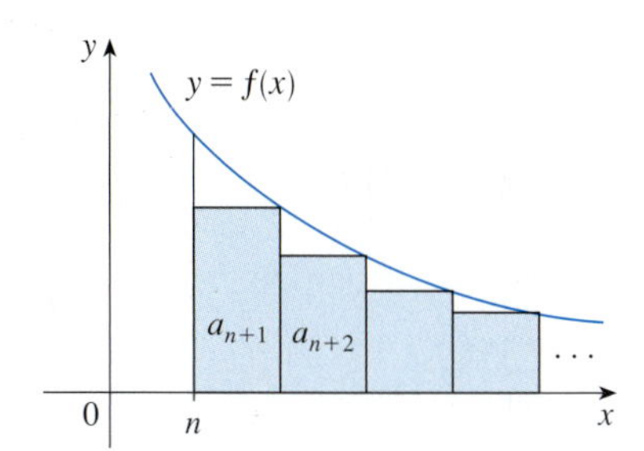

FIGURE 3

$$R_n = a_{n+1} + a_{n+2} + \cdots \leqslant \int_n^\infty f(x)\,dx$$

Similarly, we see from Figure 4 that

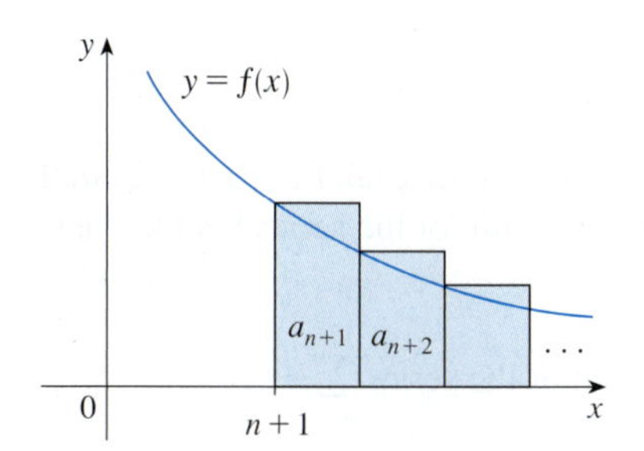

FIGURE 4

$$R_n = a_{n+1} + a_{n+2} + \cdots \geqslant \int_{n+1}^\infty f(x)\,dx$$

So we have proved the following error estimate.

> **2 Remainder Estimate for the Integral Test** Suppose $f(k) = a_k$, where f is a continuous, positive, decreasing function for $x \geqslant n$ and Σa_n is convergent. If $R_n = s - s_n$, then
>
> $$\int_{n+1}^\infty f(x)\,dx \leqslant R_n \leqslant \int_n^\infty f(x)\,dx$$

EXAMPLE 5

(a) Approximate the sum of the series $\Sigma 1/n^3$ by using the sum of the first 10 terms. Estimate the error involved in this approximation.

(b) How many terms are required to ensure that the sum is accurate to within 0.0005?

SOLUTION In both parts (a) and (b) we need to know $\int_n^\infty f(x)\,dx$. With $f(x) = 1/x^3$, which satisfies the conditions of the Integral Test, we have

$$\int_n^\infty \frac{1}{x^3}\,dx = \lim_{t \to \infty} \left[-\frac{1}{2x^2} \right]_n^t = \lim_{t \to \infty} \left(-\frac{1}{2t^2} + \frac{1}{2n^2} \right) = \frac{1}{2n^2}$$

(a) Approximating the sum of the series by the 10th partial sum, we have

$$\sum_{n=1}^\infty \frac{1}{n^3} \approx s_{10} = \frac{1}{1^3} + \frac{1}{2^3} + \frac{1}{3^3} + \cdots + \frac{1}{10^3} \approx 1.1975$$

According to the remainder estimate in (2), we have

$$R_{10} \leqslant \int_{10}^\infty \frac{1}{x^3}\,dx = \frac{1}{2(10)^2} = \frac{1}{200}$$

So the size of the error is at most 0.005.

(b) Accuracy to within 0.0005 means that we have to find a value of n such that $R_n \leqslant 0.0005$. Since

$$R_n \leqslant \int_n^\infty \frac{1}{x^3}\,dx = \frac{1}{2n^2}$$

we want

$$\frac{1}{2n^2} < 0.0005$$

Solving this inequality, we get

$$n^2 > \frac{1}{0.001} = 1000 \qquad \text{or} \qquad n > \sqrt{1000} \approx 31.6$$

We need 32 terms to ensure accuracy to within 0.0005. ■

If we add s_n to each side of the inequalities in (2), we get

3
$$s_n + \int_{n+1}^\infty f(x)\,dx \leqslant s \leqslant s_n + \int_n^\infty f(x)\,dx$$

because $s_n + R_n = s$. The inequalities in (3) give a lower bound and an upper bound for s. They provide a more accurate approximation to the sum of the series than the partial sum s_n does.

Although Euler was able to calculate the exact sum of the p-series for $p = 2$, nobody has been able to find the exact sum for $p = 3$. In Example 6, however, we show how to *estimate* this sum.

EXAMPLE 6 Use (3) with $n = 10$ to estimate the sum of the series $\sum_{n=1}^{\infty} \frac{1}{n^3}$.

SOLUTION The inequalities in (3) become

$$s_{10} + \int_{11}^\infty \frac{1}{x^3}\,dx \leqslant s \leqslant s_{10} + \int_{10}^\infty \frac{1}{x^3}\,dx$$

From Example 5 we know that

$$\int_n^\infty \frac{1}{x^3}\,dx = \frac{1}{2n^2}$$

so

$$s_{10} + \frac{1}{2(11)^2} \leqslant s \leqslant s_{10} + \frac{1}{2(10)^2}$$

Using $s_{10} \approx 1.197532$, we get

$$1.201664 \leqslant s \leqslant 1.202532$$

If we approximate s by the midpoint of this interval, then the error is at most half the length of the interval. So

$$\sum_{n=1}^{\infty} \frac{1}{n^3} \approx 1.2021 \qquad \text{with error} < 0.0005$$

■

If we compare Example 6 with Example 5, we see that the improved estimate in (3) can be much better than the estimate $s \approx s_n$. To make the error smaller than 0.0005 we had to use 32 terms in Example 5 but only 10 terms in Example 6.

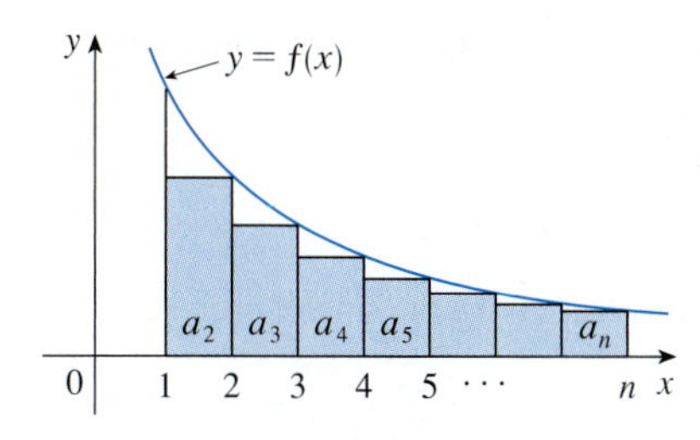

FIGURE 5

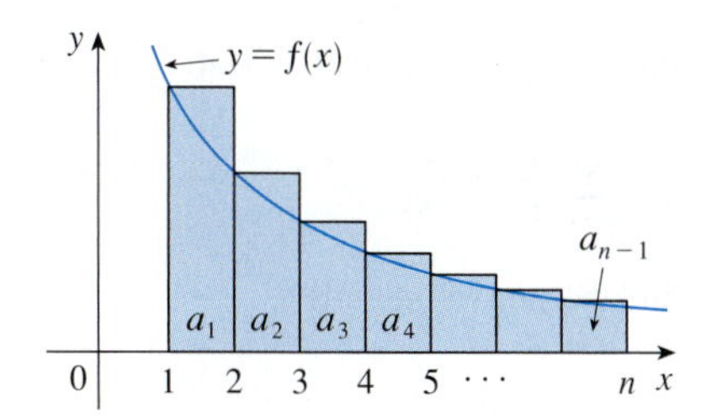

FIGURE 6

Proof of the Integral Test

We have already seen the basic idea behind the proof of the Integral Test in Figures 1 and 2 for the series $\Sigma\, 1/n^2$ and $\Sigma\, 1/\sqrt{n}$. For the general series $\Sigma\, a_n$, look at Figures 5 and 6. The area of the first shaded rectangle in Figure 5 is the value of f at the right endpoint of $[1, 2]$, that is, $f(2) = a_2$. So, comparing the areas of the shaded rectangles with the area under $y = f(x)$ from 1 to n, we see that

$$\boxed{4} \qquad a_2 + a_3 + \cdots + a_n \leqslant \int_1^n f(x)\, dx$$

(Notice that this inequality depends on the fact that f is decreasing.) Likewise, Figure 6 shows that

$$\boxed{5} \qquad \int_1^n f(x)\, dx \leqslant a_1 + a_2 + \cdots + a_{n-1}$$

(i) If $\int_1^\infty f(x)\, dx$ is convergent, then (4) gives

$$\sum_{i=2}^{n} a_i \leqslant \int_1^n f(x)\, dx \leqslant \int_1^\infty f(x)\, dx$$

since $f(x) \geqslant 0$. Therefore

$$s_n = a_1 + \sum_{i=2}^{n} a_i \leqslant a_1 + \int_1^\infty f(x)\, dx = M, \text{ say}$$

Since $s_n \leqslant M$ for all n, the sequence $\{s_n\}$ is bounded above. Also

$$s_{n+1} = s_n + a_{n+1} \geqslant s_n$$

since $a_{n+1} = f(n + 1) \geqslant 0$. Thus $\{s_n\}$ is an increasing bounded sequence and so it is convergent by the Monotonic Sequence Theorem (11.1.12). This means that $\Sigma\, a_n$ is convergent.

(ii) If $\int_1^\infty f(x)\, dx$ is divergent, then $\int_1^n f(x)\, dx \to \infty$ as $n \to \infty$ because $f(x) \geqslant 0$. But (5) gives

$$\int_1^n f(x)\, dx \leqslant \sum_{i=1}^{n-1} a_i = s_{n-1}$$

and so $s_{n-1} \to \infty$. This implies that $s_n \to \infty$ and so $\Sigma\, a_n$ diverges. ■

11.3 EXERCISES

1. Draw a picture to show that

$$\sum_{n=2}^{\infty} \frac{1}{n^{1.3}} < \int_1^\infty \frac{1}{x^{1.3}}\, dx$$

What can you conclude about the series?

2. Suppose f is a continuous positive decreasing function for $x \geqslant 1$ and $a_n = f(n)$. By drawing a picture, rank the following three quantities in increasing order:

$$\int_1^6 f(x)\, dx \qquad \sum_{i=1}^{5} a_i \qquad \sum_{i=2}^{6} a_i$$

3–8 Use the Integral Test to determine whether the series is convergent or divergent.

3. $\displaystyle\sum_{n=1}^{\infty} n^{-3}$

4. $\displaystyle\sum_{n=1}^{\infty} n^{-0.3}$

5. $\displaystyle\sum_{n=1}^{\infty} \frac{2}{5n - 1}$

6. $\displaystyle\sum_{n=1}^{\infty} \frac{1}{(3n - 1)^4}$

7. $\displaystyle\sum_{n=1}^{\infty} \frac{n}{n^2 + 1}$

8. $\displaystyle\sum_{n=1}^{\infty} n^2 e^{-n^3}$

9–26 Determine whether the series is convergent or divergent.

9. $\sum_{n=1}^{\infty} \frac{1}{n^{\sqrt{2}}}$

10. $\sum_{n=3}^{\infty} n^{-0.9999}$

11. $1 + \frac{1}{8} + \frac{1}{27} + \frac{1}{64} + \frac{1}{125} + \cdots$

12. $\frac{1}{5} + \frac{1}{7} + \frac{1}{9} + \frac{1}{11} + \frac{1}{13} + \cdots$

13. $\frac{1}{3} + \frac{1}{7} + \frac{1}{11} + \frac{1}{15} + \frac{1}{19} + \cdots$

14. $1 + \frac{1}{2\sqrt{2}} + \frac{1}{3\sqrt{3}} + \frac{1}{4\sqrt{4}} + \frac{1}{5\sqrt{5}} + \cdots$

15. $\sum_{n=1}^{\infty} \frac{\sqrt{n} + 4}{n^2}$

16. $\sum_{n=1}^{\infty} \frac{\sqrt{n}}{1 + n^{3/2}}$

17. $\sum_{n=1}^{\infty} \frac{1}{n^2 + 4}$

18. $\sum_{n=1}^{\infty} \frac{1}{n^2 + 2n + 2}$

19. $\sum_{n=1}^{\infty} \frac{n^3}{n^4 + 4}$

20. $\sum_{n=3}^{\infty} \frac{3n - 4}{n^2 - 2n}$

21. $\sum_{n=2}^{\infty} \frac{1}{n \ln n}$

22. $\sum_{n=2}^{\infty} \frac{\ln n}{n^2}$

23. $\sum_{k=1}^{\infty} ke^{-k}$

24. $\sum_{k=1}^{\infty} ke^{-k^2}$

25. $\sum_{n=1}^{\infty} \frac{1}{n^2 + n^3}$

26. $\sum_{n=1}^{\infty} \frac{n}{n^4 + 1}$

27–28 Explain why the Integral Test can't be used to determine whether the series is convergent.

27. $\sum_{n=1}^{\infty} \frac{\cos \pi n}{\sqrt{n}}$

28. $\sum_{n=1}^{\infty} \frac{\cos^2 n}{1 + n^2}$

29–32 Find the values of p for which the series is convergent.

29. $\sum_{n=2}^{\infty} \frac{1}{n(\ln n)^p}$

30. $\sum_{n=3}^{\infty} \frac{1}{n \ln n \, [\ln(\ln n)]^p}$

31. $\sum_{n=1}^{\infty} n(1 + n^2)^p$

32. $\sum_{n=1}^{\infty} \frac{\ln n}{n^p}$

---

33. The Riemann zeta-function ζ is defined by

$$\zeta(x) = \sum_{n=1}^{\infty} \frac{1}{n^x}$$

and is used in number theory to study the distribution of prime numbers. What is the domain of ζ?

34. Leonhard Euler was able to calculate the exact sum of the p-series with $p = 2$:

$$\zeta(2) = \sum_{n=1}^{\infty} \frac{1}{n^2} = \frac{\pi^2}{6}$$

(See page 760.) Use this fact to find the sum of each series.

(a) $\sum_{n=2}^{\infty} \frac{1}{n^2}$ (b) $\sum_{n=3}^{\infty} \frac{1}{(n + 1)^2}$

(c) $\sum_{n=1}^{\infty} \frac{1}{(2n)^2}$

35. Euler also found the sum of the p-series with $p = 4$:

$$\zeta(4) = \sum_{n=1}^{\infty} \frac{1}{n^4} = \frac{\pi^4}{90}$$

Use Euler's result to find the sum of the series.

(a) $\sum_{n=1}^{\infty} \left(\frac{3}{n}\right)^4$ (b) $\sum_{k=5}^{\infty} \frac{1}{(k - 2)^4}$

36. (a) Find the partial sum s_{10} of the series $\sum_{n=1}^{\infty} 1/n^4$. Estimate the error in using s_{10} as an approximation to the sum of the series.
(b) Use (3) with $n = 10$ to give an improved estimate of the sum.
(c) Compare your estimate in part (b) with the exact value given in Exercise 35.
(d) Find a value of n so that s_n is within 0.00001 of the sum.

37. (a) Use the sum of the first 10 terms to estimate the sum of the series $\sum_{n=1}^{\infty} 1/n^2$. How good is this estimate?
(b) Improve this estimate using (3) with $n = 10$.
(c) Compare your estimate in part (b) with the exact value given in Exercise 34.
(d) Find a value of n that will ensure that the error in the approximation $s \approx s_n$ is less than 0.001.

38. Find the sum of the series $\sum_{n=1}^{\infty} ne^{-2n}$ correct to four decimal places.

39. Estimate $\sum_{n=1}^{\infty} (2n + 1)^{-6}$ correct to five decimal places.

40. How many terms of the series $\sum_{n=2}^{\infty} 1/[n(\ln n)^2]$ would you need to add to find its sum to within 0.01?

41. Show that if we want to approximate the sum of the series $\sum_{n=1}^{\infty} n^{-1.001}$ so that the error is less than 5 in the ninth decimal place, then we need to add more than $10^{11,301}$ terms!

CAS **42.** (a) Show that the series $\sum_{n=1}^{\infty} (\ln n)^2/n^2$ is convergent.
(b) Find an upper bound for the error in the approximation $s \approx s_n$.
(c) What is the smallest value of n such that this upper bound is less than 0.05?
(d) Find s_n for this value of n.

43. (a) Use (4) to show that if s_n is the nth partial sum of the harmonic series, then

$$s_n \leqslant 1 + \ln n$$

(b) The harmonic series diverges, but very slowly. Use part (a) to show that the sum of the first million terms is less than 15 and the sum of the first billion terms is less than 22.

44. Use the following steps to show that the sequence

$$t_n = 1 + \frac{1}{2} + \frac{1}{3} + \cdots + \frac{1}{n} - \ln n$$

has a limit. (The value of the limit is denoted by γ and is called Euler's constant.)

(a) Draw a picture like Figure 6 with $f(x) = 1/x$ and interpret t_n as an area [or use (5)] to show that $t_n > 0$ for all n.

(b) Interpret

$$t_n - t_{n+1} = [\ln(n+1) - \ln n] - \frac{1}{n+1}$$

as a difference of areas to show that $t_n - t_{n+1} > 0$. Therefore $\{t_n\}$ is a decreasing sequence.

(c) Use the Monotonic Sequence Theorem to show that $\{t_n\}$ is convergent.

45. Find all positive values of b for which the series $\Sigma_{n=1}^{\infty} b^{\ln n}$ converges.

46. Find all values of c for which the following series converges.

$$\sum_{n=1}^{\infty} \left(\frac{c}{n} - \frac{1}{n+1} \right)$$

11.4 The Comparison Tests

In the comparison tests the idea is to compare a given series with a series that is known to be convergent or divergent. For instance, the series

$$\sum_{n=1}^{\infty} \frac{1}{2^n + 1} \tag{1}$$

reminds us of the series $\Sigma_{n=1}^{\infty} 1/2^n$, which is a geometric series with $a = \frac{1}{2}$ and $r = \frac{1}{2}$ and is therefore convergent. Because the series (1) is so similar to a convergent series, we have the feeling that it too must be convergent. Indeed, it is. The inequality

$$\frac{1}{2^n + 1} < \frac{1}{2^n}$$

shows that our given series (1) has smaller terms than those of the geometric series and therefore all its partial sums are also smaller than 1 (the sum of the geometric series). This means that its partial sums form a bounded increasing sequence, which is convergent. It also follows that the sum of the series is less than the sum of the geometric series:

$$\sum_{n=1}^{\infty} \frac{1}{2^n + 1} < 1$$

Similar reasoning can be used to prove the following test, which applies only to series whose terms are positive. The first part says that if we have a series whose terms are *smaller* than those of a known *convergent* series, then our series is also convergent. The second part says that if we start with a series whose terms are *larger* than those of a known *divergent* series, then it too is divergent.

The Comparison Test Suppose that Σa_n and Σb_n are series with positive terms.

(i) If Σb_n is convergent and $a_n \leqslant b_n$ for all n, then Σa_n is also convergent.

(ii) If Σb_n is divergent and $a_n \geqslant b_n$ for all n, then Σa_n is also divergent.

It is important to keep in mind the distinction between a sequence and a series. A sequence is a list of numbers, whereas a series is a sum. With every series Σa_n there are associated two sequences: the sequence $\{a_n\}$ of terms and the sequence $\{s_n\}$ of partial sums.

PROOF

(i) Let $$s_n = \sum_{i=1}^{n} a_i \qquad t_n = \sum_{i=1}^{n} b_i \qquad t = \sum_{n=1}^{\infty} b_n$$

Since both series have positive terms, the sequences $\{s_n\}$ and $\{t_n\}$ are increasing ($s_{n+1} = s_n + a_{n+1} \geqslant s_n$). Also $t_n \to t$, so $t_n \leqslant t$ for all n. Since $a_i \leqslant b_i$, we have $s_n \leqslant t_n$. Thus $s_n \leqslant t$ for all n. This means that $\{s_n\}$ is increasing and bounded above and therefore converges by the Monotonic Sequence Theorem. Thus Σa_n converges.

(ii) If Σb_n is divergent, then $t_n \to \infty$ (since $\{t_n\}$ is increasing). But $a_i \geqslant b_i$ so $s_n \geqslant t_n$. Thus $s_n \to \infty$. Therefore Σa_n diverges. ■

Standard Series for Use with the Comparison Test

In using the Comparison Test we must, of course, have some known series Σb_n for the purpose of comparison. Most of the time we use one of these series:

- A p-series $[\Sigma 1/n^p$ converges if $p > 1$ and diverges if $p \leqslant 1$; see (11.3.1)$]$
- A geometric series $[\Sigma ar^{n-1}$ converges if $|r| < 1$ and diverges if $|r| \geqslant 1$; see (11.2.4)$]$

EXAMPLE 1 Determine whether the series $\displaystyle\sum_{n=1}^{\infty} \frac{5}{2n^2 + 4n + 3}$ converges or diverges.

SOLUTION For large n the dominant term in the denominator is $2n^2$, so we compare the given series with the series $\Sigma 5/(2n^2)$. Observe that

$$\frac{5}{2n^2 + 4n + 3} < \frac{5}{2n^2}$$

because the left side has a bigger denominator. (In the notation of the Comparison Test, a_n is the left side and b_n is the right side.) We know that

$$\sum_{n=1}^{\infty} \frac{5}{2n^2} = \frac{5}{2} \sum_{n=1}^{\infty} \frac{1}{n^2}$$

is convergent because it's a constant times a p-series with $p = 2 > 1$. Therefore

$$\sum_{n=1}^{\infty} \frac{5}{2n^2 + 4n + 3}$$

is convergent by part (i) of the Comparison Test. ■

NOTE 1 Although the condition $a_n \leqslant b_n$ or $a_n \geqslant b_n$ in the Comparison Test is given for all n, we need verify only that it holds for $n \geqslant N$, where N is some fixed integer, because the convergence of a series is not affected by a finite number of terms. This is illustrated in the next example.

EXAMPLE 2 Test the series $\displaystyle\sum_{k=1}^{\infty} \frac{\ln k}{k}$ for convergence or divergence.

SOLUTION We used the Integral Test to test this series in Example 11.3.4, but we can also test it by comparing it with the harmonic series. Observe that $\ln k > 1$ for $k \geqslant 3$ and so

$$\frac{\ln k}{k} > \frac{1}{k} \qquad k \geqslant 3$$

We know that $\Sigma\, 1/k$ is divergent (p-series with $p = 1$). Thus the given series is divergent by the Comparison Test. ■

NOTE 2 The terms of the series being tested must be smaller than those of a convergent series or larger than those of a divergent series. If the terms are larger than the terms of a convergent series or smaller than those of a divergent series, then the Comparison Test doesn't apply. Consider, for instance, the series

$$\sum_{n=1}^{\infty} \frac{1}{2^n - 1}$$

The inequality

$$\frac{1}{2^n - 1} > \frac{1}{2^n}$$

is useless as far as the Comparison Test is concerned because $\Sigma\, b_n = \Sigma \left(\frac{1}{2}\right)^n$ is convergent and $a_n > b_n$. Nonetheless, we have the feeling that $\Sigma\, 1/(2^n - 1)$ ought to be convergent because it is very similar to the convergent geometric series $\Sigma \left(\frac{1}{2}\right)^n$. In such cases the following test can be used.

Exercises 40 and 41 deal with the cases $c = 0$ and $c = \infty$.

The Limit Comparison Test Suppose that $\Sigma\, a_n$ and $\Sigma\, b_n$ are series with positive terms. If

$$\lim_{n \to \infty} \frac{a_n}{b_n} = c$$

where c is a finite number and $c > 0$, then either both series converge or both diverge.

PROOF Let m and M be positive numbers such that $m < c < M$. Because a_n/b_n is close to c for large n, there is an integer N such that

$$m < \frac{a_n}{b_n} < M \qquad \text{when } n > N$$

and so

$$mb_n < a_n < Mb_n \qquad \text{when } n > N$$

If $\Sigma\, b_n$ converges, so does $\Sigma\, Mb_n$. Thus $\Sigma\, a_n$ converges by part (i) of the Comparison Test. If $\Sigma\, b_n$ diverges, so does $\Sigma\, mb_n$ and part (ii) of the Comparison Test shows that $\Sigma\, a_n$ diverges. ■

EXAMPLE 3 Test the series $\displaystyle\sum_{n=1}^{\infty} \frac{1}{2^n - 1}$ for convergence or divergence.

SOLUTION We use the Limit Comparison Test with

$$a_n = \frac{1}{2^n - 1} \qquad b_n = \frac{1}{2^n}$$

and obtain

$$\lim_{n \to \infty} \frac{a_n}{b_n} = \lim_{n \to \infty} \frac{1/(2^n - 1)}{1/2^n} = \lim_{n \to \infty} \frac{2^n}{2^n - 1} = \lim_{n \to \infty} \frac{1}{1 - 1/2^n} = 1 > 0$$

Since this limit exists and $\Sigma\, 1/2^n$ is a convergent geometric series, the given series converges by the Limit Comparison Test. ■

EXAMPLE 4 Determine whether the series $\displaystyle\sum_{n=1}^{\infty} \frac{2n^2 + 3n}{\sqrt{5 + n^5}}$ converges or diverges.

SOLUTION The dominant part of the numerator is $2n^2$ and the dominant part of the denominator is $\sqrt{n^5} = n^{5/2}$. This suggests taking

$$a_n = \frac{2n^2 + 3n}{\sqrt{5 + n^5}} \qquad b_n = \frac{2n^2}{n^{5/2}} = \frac{2}{n^{1/2}}$$

$$\lim_{n\to\infty} \frac{a_n}{b_n} = \lim_{n\to\infty} \frac{2n^2 + 3n}{\sqrt{5 + n^5}} \cdot \frac{n^{1/2}}{2} = \lim_{n\to\infty} \frac{2n^{5/2} + 3n^{3/2}}{2\sqrt{5 + n^5}}$$

$$= \lim_{n\to\infty} \frac{2 + \dfrac{3}{n}}{2\sqrt{\dfrac{5}{n^5} + 1}} = \frac{2 + 0}{2\sqrt{0 + 1}} = 1$$

Since $\Sigma\, b_n = 2\,\Sigma\, 1/n^{1/2}$ is divergent $\left(p\text{-series with } p = \frac{1}{2} < 1\right)$, the given series diverges by the Limit Comparison Test. ■

Notice that in testing many series we find a suitable comparison series $\Sigma\, b_n$ by keeping only the highest powers in the numerator and denominator.

Estimating Sums

If we have used the Comparison Test to show that a series $\Sigma\, a_n$ converges by comparison with a series $\Sigma\, b_n$, then we may be able to estimate the sum $\Sigma\, a_n$ by comparing remainders. As in Section 11.3, we consider the remainder

$$R_n = s - s_n = a_{n+1} + a_{n+2} + \cdots$$

For the comparison series $\Sigma\, b_n$ we consider the corresponding remainder

$$T_n = t - t_n = b_{n+1} + b_{n+2} + \cdots$$

Since $a_n \leqslant b_n$ for all n, we have $R_n \leqslant T_n$. If $\Sigma\, b_n$ is a p-series, we can estimate its remainder T_n as in Section 11.3. If $\Sigma\, b_n$ is a geometric series, then T_n is the sum of a geometric series and we can sum it exactly (see Exercises 35 and 36). In either case we know that R_n is smaller than T_n.

EXAMPLE 5 Use the sum of the first 100 terms to approximate the sum of the series $\Sigma\, 1/(n^3 + 1)$. Estimate the error involved in this approximation.

SOLUTION Since

$$\frac{1}{n^3 + 1} < \frac{1}{n^3}$$

the given series is convergent by the Comparison Test. The remainder T_n for the comparison series $\Sigma\, 1/n^3$ was estimated in Example 11.3.5 using the Remainder Estimate for the Integral Test. There we found that

$$T_n \leqslant \int_n^\infty \frac{1}{x^3}\, dx = \frac{1}{2n^2}$$

Therefore the remainder R_n for the given series satisfies

$$R_n \leqslant T_n \leqslant \frac{1}{2n^2}$$

With $n = 100$ we have

$$R_{100} \leqslant \frac{1}{2(100)^2} = 0.00005$$

Using a programmable calculator or a computer, we find that

$$\sum_{n=1}^{\infty} \frac{1}{n^3+1} \approx \sum_{n=1}^{100} \frac{1}{n^3+1} \approx 0.6864538$$

with error less than 0.00005. ■

11.4 EXERCISES

1. Suppose $\Sigma\, a_n$ and $\Sigma\, b_n$ are series with positive terms and $\Sigma\, b_n$ is known to be convergent.
   (a) If $a_n > b_n$ for all n, what can you say about $\Sigma\, a_n$? Why?
   (b) If $a_n < b_n$ for all n, what can you say about $\Sigma\, a_n$? Why?

2. Suppose $\Sigma\, a_n$ and $\Sigma\, b_n$ are series with positive terms and $\Sigma\, b_n$ is known to be divergent.
   (a) If $a_n > b_n$ for all n, what can you say about $\Sigma\, a_n$? Why?
   (b) If $a_n < b_n$ for all n, what can you say about $\Sigma\, a_n$? Why?

3–32 Determine whether the series converges or diverges.

3. $\displaystyle\sum_{n=1}^{\infty} \frac{1}{n^3+8}$

4. $\displaystyle\sum_{n=2}^{\infty} \frac{1}{\sqrt{n}-1}$

5. $\displaystyle\sum_{n=1}^{\infty} \frac{n+1}{n\sqrt{n}}$

6. $\displaystyle\sum_{n=1}^{\infty} \frac{n-1}{n^3+1}$

7. $\displaystyle\sum_{n=1}^{\infty} \frac{9^n}{3+10^n}$

8. $\displaystyle\sum_{n=1}^{\infty} \frac{6^n}{5^n-1}$

9. $\displaystyle\sum_{k=1}^{\infty} \frac{\ln k}{k}$

10. $\displaystyle\sum_{k=1}^{\infty} \frac{k\sin^2 k}{1+k^3}$

11. $\displaystyle\sum_{k=1}^{\infty} \frac{\sqrt[3]{k}}{\sqrt{k^3+4k+3}}$

12. $\displaystyle\sum_{k=1}^{\infty} \frac{(2k-1)(k^2-1)}{(k+1)(k^2+4)^2}$

13. $\displaystyle\sum_{n=1}^{\infty} \frac{1+\cos n}{e^n}$

14. $\displaystyle\sum_{n=1}^{\infty} \frac{1}{\sqrt[3]{3n^4+1}}$

15. $\displaystyle\sum_{n=1}^{\infty} \frac{4^{n+1}}{3^n-2}$

16. $\displaystyle\sum_{n=1}^{\infty} \frac{1}{n^n}$

17. $\displaystyle\sum_{n=1}^{\infty} \frac{1}{\sqrt{n^2+1}}$

18. $\displaystyle\sum_{n=1}^{\infty} \frac{2}{\sqrt{n}+2}$

19. $\displaystyle\sum_{n=1}^{\infty} \frac{n+1}{n^3+n}$

20. $\displaystyle\sum_{n=1}^{\infty} \frac{n^2+n+1}{n^4+n^2}$

21. $\displaystyle\sum_{n=1}^{\infty} \frac{\sqrt{1+n}}{2+n}$

22. $\displaystyle\sum_{n=3}^{\infty} \frac{n+2}{(n+1)^3}$

23. $\displaystyle\sum_{n=1}^{\infty} \frac{5+2n}{(1+n^2)^2}$

24. $\displaystyle\sum_{n=1}^{\infty} \frac{n+3^n}{n+2^n}$

25. $\displaystyle\sum_{n=1}^{\infty} \frac{e^n+1}{ne^n+1}$

26. $\displaystyle\sum_{n=2}^{\infty} \frac{1}{n\sqrt{n^2-1}}$

27. $\displaystyle\sum_{n=1}^{\infty} \left(1+\frac{1}{n}\right)^2 e^{-n}$

28. $\displaystyle\sum_{n=1}^{\infty} \frac{e^{1/n}}{n}$

29. $\displaystyle\sum_{n=1}^{\infty} \frac{1}{n!}$

30. $\displaystyle\sum_{n=1}^{\infty} \frac{n!}{n^n}$

31. $\displaystyle\sum_{n=1}^{\infty} \sin\left(\frac{1}{n}\right)$

32. $\displaystyle\sum_{n=1}^{\infty} \frac{1}{n^{1+1/n}}$

33–36 Use the sum of the first 10 terms to approximate the sum of the series. Estimate the error.

33. $\displaystyle\sum_{n=1}^{\infty} \frac{1}{5+n^5}$

34. $\displaystyle\sum_{n=1}^{\infty} \frac{e^{1/n}}{n^4}$

35. $\displaystyle\sum_{n=1}^{\infty} 5^{-n}\cos^2 n$

36. $\displaystyle\sum_{n=1}^{\infty} \frac{1}{3^n+4^n}$

37. The meaning of the decimal representation of a number $0.d_1d_2d_3\ldots$ (where the digit d_i is one of the numbers 0, 1, 2, . . . , 9) is that

$$0.d_1d_2d_3d_4\ldots = \frac{d_1}{10} + \frac{d_2}{10^2} + \frac{d_3}{10^3} + \frac{d_4}{10^4} + \cdots$$

Show that this series always converges.

38. For what values of p does the series $\sum_{n=2}^{\infty} 1/(n^p \ln n)$ converge?

39. Prove that if $a_n \geqslant 0$ and $\sum a_n$ converges, then $\sum a_n^2$ also converges.

40. (a) Suppose that $\sum a_n$ and $\sum b_n$ are series with positive terms and $\sum b_n$ is convergent. Prove that if

$$\lim_{n\to\infty} \frac{a_n}{b_n} = 0$$

then $\sum a_n$ is also convergent.

(b) Use part (a) to show that the series converges.

(i) $\displaystyle\sum_{n=1}^{\infty} \frac{\ln n}{n^3}$ (ii) $\displaystyle\sum_{n=1}^{\infty} \frac{\ln n}{\sqrt{n}\, e^n}$

41. (a) Suppose that $\sum a_n$ and $\sum b_n$ are series with positive terms and $\sum b_n$ is divergent. Prove that if

$$\lim_{n\to\infty} \frac{a_n}{b_n} = \infty$$

then $\sum a_n$ is also divergent.

(b) Use part (a) to show that the series diverges.

(i) $\displaystyle\sum_{n=2}^{\infty} \frac{1}{\ln n}$ (ii) $\displaystyle\sum_{n=1}^{\infty} \frac{\ln n}{n}$

42. Give an example of a pair of series $\sum a_n$ and $\sum b_n$ with positive terms where $\lim_{n\to\infty} (a_n/b_n) = 0$ and $\sum b_n$ diverges, but $\sum a_n$ converges. (Compare with Exercise 40.)

43. Show that if $a_n > 0$ and $\lim_{n\to\infty} na_n \neq 0$, then $\sum a_n$ is divergent.

44. Show that if $a_n > 0$ and $\sum a_n$ is convergent, then $\sum \ln(1 + a_n)$ is convergent.

45. If $\sum a_n$ is a convergent series with positive terms, is it true that $\sum \sin(a_n)$ is also convergent?

46. If $\sum a_n$ and $\sum b_n$ are both convergent series with positive terms, is it true that $\sum a_n b_n$ is also convergent?

11.5 Alternating Series

The convergence tests that we have looked at so far apply only to series with positive terms. In this section and the next we learn how to deal with series whose terms are not necessarily positive. Of particular importance are *alternating series,* whose terms alternate in sign.

An **alternating series** is a series whose terms are alternately positive and negative. Here are two examples:

$$1 - \frac{1}{2} + \frac{1}{3} - \frac{1}{4} + \frac{1}{5} - \frac{1}{6} + \cdots = \sum_{n=1}^{\infty} (-1)^{n-1} \frac{1}{n}$$

$$-\frac{1}{2} + \frac{2}{3} - \frac{3}{4} + \frac{4}{5} - \frac{5}{6} + \frac{6}{7} - \cdots = \sum_{n=1}^{\infty} (-1)^n \frac{n}{n+1}$$

We see from these examples that the nth term of an alternating series is of the form

$$a_n = (-1)^{n-1} b_n \qquad \text{or} \qquad a_n = (-1)^n b_n$$

where b_n is a positive number. (In fact, $b_n = |a_n|$.)

The following test says that if the terms of an alternating series decrease toward 0 in absolute value, then the series converges.

> **Alternating Series Test** If the alternating series
>
> $$\sum_{n=1}^{\infty} (-1)^{n-1} b_n = b_1 - b_2 + b_3 - b_4 + b_5 - b_6 + \cdots \qquad b_n > 0$$
>
> satisfies
>
> (i) $b_{n+1} \leqslant b_n$ for all n
>
> (ii) $\displaystyle\lim_{n\to\infty} b_n = 0$
>
> then the series is convergent.

Before giving the proof let's look at Figure 1, which gives a picture of the idea behind the proof. We first plot $s_1 = b_1$ on a number line. To find s_2 we subtract b_2, so s_2 is to the left of s_1. Then to find s_3 we add b_3, so s_3 is to the right of s_2. But, since $b_3 < b_2$, s_3 is to the left of s_1. Continuing in this manner, we see that the partial sums oscillate back and forth. Since $b_n \to 0$, the successive steps are becoming smaller and smaller. The even partial sums $s_2, s_4, s_6, \ldots$ are increasing and the odd partial sums $s_1, s_3, s_5, \ldots$ are decreasing. Thus it seems plausible that both are converging to some number s, which is the sum of the series. Therefore we consider the even and odd partial sums separately in the following proof.

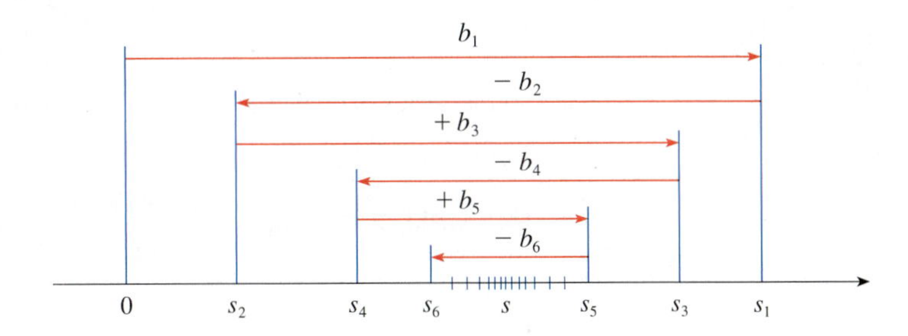

FIGURE 1

PROOF OF THE ALTERNATING SERIES TEST We first consider the even partial sums:

$$s_2 = b_1 - b_2 \geqslant 0 \qquad \text{since } b_2 \leqslant b_1$$

$$s_4 = s_2 + (b_3 - b_4) \geqslant s_2 \qquad \text{since } b_4 \leqslant b_3$$

In general $\quad s_{2n} = s_{2n-2} + (b_{2n-1} - b_{2n}) \geqslant s_{2n-2} \qquad \text{since } b_{2n} \leqslant b_{2n-1}$

Thus $\quad 0 \leqslant s_2 \leqslant s_4 \leqslant s_6 \leqslant \cdots \leqslant s_{2n} \leqslant \cdots$

But we can also write

$$s_{2n} = b_1 - (b_2 - b_3) - (b_4 - b_5) - \cdots - (b_{2n-2} - b_{2n-1}) - b_{2n}$$

Every term in parentheses is positive, so $s_{2n} \leqslant b_1$ for all n. Therefore the sequence $\{s_{2n}\}$ of even partial sums is increasing and bounded above. It is therefore convergent by the Monotonic Sequence Theorem. Let's call its limit s, that is,

$$\lim_{n \to \infty} s_{2n} = s$$

Now we compute the limit of the odd partial sums:

$$\begin{aligned} \lim_{n \to \infty} s_{2n+1} &= \lim_{n \to \infty} (s_{2n} + b_{2n+1}) \\ &= \lim_{n \to \infty} s_{2n} + \lim_{n \to \infty} b_{2n+1} \\ &= s + 0 \qquad \text{[by condition (ii)]} \\ &= s \end{aligned}$$

Since both the even and odd partial sums converge to s, we have $\lim_{n \to \infty} s_n = s$ [see Exercise 11.1.92(a)] and so the series is convergent. ■

Figure 2 illustrates Example 1 by showing the graphs of the terms $a_n = (-1)^{n-1}/n$ and the partial sums s_n. Notice how the values of s_n zigzag across the limiting value, which appears to be about 0.7. In fact, it can be proved that the exact sum of the series is $\ln 2 \approx 0.693$ (see Exercise 36).

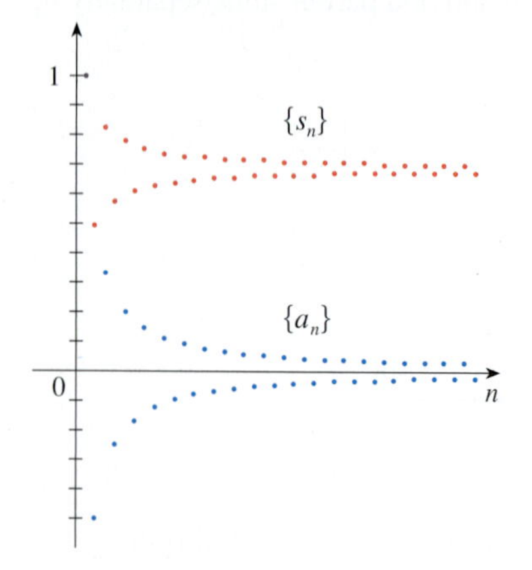

FIGURE 2

EXAMPLE 1 The alternating harmonic series

$$1 - \frac{1}{2} + \frac{1}{3} - \frac{1}{4} + \cdots = \sum_{n=1}^{\infty} \frac{(-1)^{n-1}}{n}$$

satisfies

$$\text{(i) } b_{n+1} < b_n \quad \text{because} \quad \frac{1}{n+1} < \frac{1}{n}$$

$$\text{(ii) } \lim_{n\to\infty} b_n = \lim_{n\to\infty} \frac{1}{n} = 0$$

so the series is convergent by the Alternating Series Test. ■

EXAMPLE 2 The series $\displaystyle\sum_{n=1}^{\infty} \frac{(-1)^n 3n}{4n-1}$ is alternating, but

$$\lim_{n\to\infty} b_n = \lim_{n\to\infty} \frac{3n}{4n-1} = \lim_{n\to\infty} \frac{3}{4 - \dfrac{1}{n}} = \frac{3}{4}$$

so condition (ii) is not satisfied. Instead, we look at the limit of the nth term of the series:

$$\lim_{n\to\infty} a_n = \lim_{n\to\infty} \frac{(-1)^n 3n}{4n-1}$$

This limit does not exist, so the series diverges by the Test for Divergence. ■

EXAMPLE 3 Test the series $\displaystyle\sum_{n=1}^{\infty} (-1)^{n+1} \frac{n^2}{n^3+1}$ for convergence or divergence.

SOLUTION The given series is alternating so we try to verify conditions (i) and (ii) of the Alternating Series Test.

Unlike the situation in Example 1, it is not obvious that the sequence given by $b_n = n^2/(n^3 + 1)$ is decreasing. However, if we consider the related function $f(x) = x^2/(x^3 + 1)$, we find that

$$f'(x) = \frac{x(2 - x^3)}{(x^3 + 1)^2}$$

Since we are considering only positive x, we see that $f'(x) < 0$ if $2 - x^3 < 0$, that is, $x > \sqrt[3]{2}$. Thus f is decreasing on the interval $(\sqrt[3]{2}, \infty)$. This means that $f(n + 1) < f(n)$ and therefore $b_{n+1} < b_n$ when $n \geqslant 2$. (The inequality $b_2 < b_1$ can be verified directly but all that really matters is that the sequence $\{b_n\}$ is eventually decreasing.)

Instead of verifying condition (i) of the Alternating Series Test by computing a derivative, we could verify that $b_{n+1} < b_n$ directly by using the technique of Solution 1 of Example 11.1.13.

Condition (ii) is readily verified:

$$\lim_{n\to\infty} b_n = \lim_{n\to\infty} \frac{n^2}{n^3+1} = \lim_{n\to\infty} \frac{\dfrac{1}{n}}{1 + \dfrac{1}{n^3}} = 0$$

Thus the given series is convergent by the Alternating Series Test. ■

Estimating Sums

A partial sum s_n of any convergent series can be used as an approximation to the total sum s, but this is not of much use unless we can estimate the accuracy of the approximation. The error involved in using $s \approx s_n$ is the remainder $R_n = s - s_n$. The next theorem says that for series that satisfy the conditions of the Alternating Series Test, the size of the error is smaller than b_{n+1}, which is the absolute value of the first neglected term.

You can see geometrically why the Alternating Series Estimation Theorem is true by looking at Figure 1 (on page 773). Notice that $s - s_4 < b_5$, $|s - s_5| < b_6$, and so on. Notice also that s lies between any two consecutive partial sums.

Alternating Series Estimation Theorem If $s = \Sigma(-1)^{n-1}b_n$, where $b_n > 0$, is the sum of an alternating series that satisfies

$$\text{(i)}\ b_{n+1} \leqslant b_n \quad \text{and} \quad \text{(ii)}\ \lim_{n\to\infty} b_n = 0$$

then

$$|R_n| = |s - s_n| \leqslant b_{n+1}$$

PROOF We know from the proof of the Alternating Series Test that s lies between any two consecutive partial sums s_n and s_{n+1}. (There we showed that s is larger than all the even partial sums. A similar argument shows that s is smaller than all the odd sums.) It follows that

$$|s - s_n| \leqslant |s_{n+1} - s_n| = b_{n+1}$$

By definition, $0! = 1$.

EXAMPLE 4 Find the sum of the series $\displaystyle\sum_{n=0}^{\infty} \frac{(-1)^n}{n!}$ correct to three decimal places.

SOLUTION We first observe that the series is convergent by the Alternating Series Test because

$$\text{(i)}\quad \frac{1}{(n+1)!} = \frac{1}{n!\,(n+1)} < \frac{1}{n!}$$

$$\text{(ii)}\quad 0 < \frac{1}{n!} < \frac{1}{n} \to 0 \quad \text{so} \quad \frac{1}{n!} \to 0 \text{ as } n \to \infty$$

To get a feel for how many terms we need to use in our approximation, let's write out the first few terms of the series:

$$s = \frac{1}{0!} - \frac{1}{1!} + \frac{1}{2!} - \frac{1}{3!} + \frac{1}{4!} - \frac{1}{5!} + \frac{1}{6!} - \frac{1}{7!} + \cdots$$

$$= 1 - 1 + \tfrac{1}{2} - \tfrac{1}{6} + \tfrac{1}{24} - \tfrac{1}{120} + \tfrac{1}{720} - \tfrac{1}{5040} + \cdots$$

Notice that $$b_7 = \tfrac{1}{5040} < \tfrac{1}{5000} = 0.0002$$

and $$s_6 = 1 - 1 + \tfrac{1}{2} - \tfrac{1}{6} + \tfrac{1}{24} - \tfrac{1}{120} + \tfrac{1}{720} \approx 0.368056$$

By the Alternating Series Estimation Theorem we know that

$$|s - s_6| \leqslant b_7 < 0.0002$$

This error of less than 0.0002 does not affect the third decimal place, so we have $s \approx 0.368$ correct to three decimal places.

In Section 11.10 we will prove that $e^x = \Sigma_{n=0}^{\infty} x^n/n!$ for all x, so what we have obtained in Example 4 is actually an approximation to the number e^{-1}.

NOTE The rule that the error (in using s_n to approximate s) is smaller than the first neglected term is, in general, valid only for alternating series that satisfy the conditions of the Alternating Series Estimation Theorem. The rule does not apply to other types of series.

11.5 EXERCISES

1. (a) What is an alternating series?
   (b) Under what conditions does an alternating series converge?
   (c) If these conditions are satisfied, what can you say about the remainder after n terms?

2–20 Test the series for convergence or divergence.

2. $\frac{2}{3} - \frac{2}{5} + \frac{2}{7} - \frac{2}{9} + \frac{2}{11} - \cdots$

3. $-\frac{2}{5} + \frac{4}{6} - \frac{6}{7} + \frac{8}{8} - \frac{10}{9} + \cdots$

4. $\dfrac{1}{\ln 3} - \dfrac{1}{\ln 4} + \dfrac{1}{\ln 5} - \dfrac{1}{\ln 6} + \dfrac{1}{\ln 7} - \cdots$

5. $\displaystyle\sum_{n=1}^{\infty} \frac{(-1)^{n-1}}{3 + 5n}$

6. $\displaystyle\sum_{n=0}^{\infty} \frac{(-1)^{n+1}}{\sqrt{n+1}}$

7. $\displaystyle\sum_{n=1}^{\infty} (-1)^n \frac{3n-1}{2n+1}$

8. $\displaystyle\sum_{n=1}^{\infty} (-1)^n \frac{n^2}{n^2+n+1}$

9. $\displaystyle\sum_{n=1}^{\infty} (-1)^n e^{-n}$

10. $\displaystyle\sum_{n=1}^{\infty} (-1)^n \frac{\sqrt{n}}{2n+3}$

11. $\displaystyle\sum_{n=1}^{\infty} (-1)^{n+1} \frac{n^2}{n^3+4}$

12. $\displaystyle\sum_{n=1}^{\infty} (-1)^{n+1} n e^{-n}$

13. $\displaystyle\sum_{n=1}^{\infty} (-1)^{n-1} e^{2/n}$

14. $\displaystyle\sum_{n=1}^{\infty} (-1)^{n-1} \arctan n$

15. $\displaystyle\sum_{n=0}^{\infty} \frac{\sin\left(n + \frac{1}{2}\right)\pi}{1 + \sqrt{n}}$

16. $\displaystyle\sum_{n=1}^{\infty} \frac{n \cos n\pi}{2^n}$

17. $\displaystyle\sum_{n=1}^{\infty} (-1)^n \sin\left(\frac{\pi}{n}\right)$

18. $\displaystyle\sum_{n=1}^{\infty} (-1)^n \cos\left(\frac{\pi}{n}\right)$

19. $\displaystyle\sum_{n=1}^{\infty} (-1)^n \frac{n^n}{n!}$

20. $\displaystyle\sum_{n=1}^{\infty} (-1)^n \left(\sqrt{n+1} - \sqrt{n}\right)$

21–22 Graph both the sequence of terms and the sequence of partial sums on the same screen. Use the graph to make a rough estimate of the sum of the series. Then use the Alternating Series Estimation Theorem to estimate the sum correct to four decimal places.

21. $\displaystyle\sum_{n=1}^{\infty} \frac{(-0.8)^n}{n!}$

22. $\displaystyle\sum_{n=1}^{\infty} (-1)^{n-1} \frac{n}{8^n}$

23–26 Show that the series is convergent. How many terms of the series do we need to add in order to find the sum to the indicated accuracy?

23. $\displaystyle\sum_{n=1}^{\infty} \frac{(-1)^{n+1}}{n^6}$ $\quad(|\text{error}| < 0.00005)$

24. $\displaystyle\sum_{n=1}^{\infty} \frac{\left(-\frac{1}{3}\right)^n}{n}$ $\quad(|\text{error}| < 0.0005)$

25. $\displaystyle\sum_{n=1}^{\infty} \frac{(-1)^{n-1}}{n^2 2^n}$ $\quad(|\text{error}| < 0.0005)$

26. $\displaystyle\sum_{n=1}^{\infty} \left(-\frac{1}{n}\right)^n$ $\quad(|\text{error}| < 0.00005)$

27–30 Approximate the sum of the series correct to four decimal places.

27. $\displaystyle\sum_{n=1}^{\infty} \frac{(-1)^n}{(2n)!}$

28. $\displaystyle\sum_{n=1}^{\infty} \frac{(-1)^{n+1}}{n^6}$

29. $\displaystyle\sum_{n=1}^{\infty} (-1)^n n e^{-2n}$

30. $\displaystyle\sum_{n=1}^{\infty} \frac{(-1)^{n-1}}{n\, 4^n}$

31. Is the 50th partial sum s_{50} of the alternating series $\sum_{n=1}^{\infty} (-1)^{n-1}/n$ an overestimate or an underestimate of the total sum? Explain.

32–34 For what values of p is each series convergent?

32. $\displaystyle\sum_{n=1}^{\infty} \frac{(-1)^{n-1}}{n^p}$

33. $\displaystyle\sum_{n=1}^{\infty} \frac{(-1)^n}{n + p}$

34. $\displaystyle\sum_{n=2}^{\infty} (-1)^{n-1} \frac{(\ln n)^p}{n}$

35. Show that the series $\Sigma\,(-1)^{n-1}b_n$, where $b_n = 1/n$ if n is odd and $b_n = 1/n^2$ if n is even, is divergent. Why does the Alternating Series Test not apply?

36. Use the following steps to show that

$$\sum_{n=1}^{\infty} \frac{(-1)^{n-1}}{n} = \ln 2$$

Let h_n and s_n be the partial sums of the harmonic and alternating harmonic series.

(a) Show that $s_{2n} = h_{2n} - h_n$.

(b) From Exercise 11.3.44 we have

$$h_n - \ln n \to \gamma \qquad \text{as } n \to \infty$$

and therefore

$$h_{2n} - \ln(2n) \to \gamma \qquad \text{as } n \to \infty$$

Use these facts together with part (a) to show that $s_{2n} \to \ln 2$ as $n \to \infty$.

11.6 Absolute Convergence and the Ratio and Root Tests

Given any series $\Sigma\, a_n$, we can consider the corresponding series

$$\sum_{n=1}^{\infty} |a_n| = |a_1| + |a_2| + |a_3| + \cdots$$

whose terms are the absolute values of the terms of the original series.

We have convergence tests for series with positive terms and for alternating series. But what if the signs of the terms switch back and forth irregularly? We will see in Example 3 that the idea of absolute convergence sometimes helps in such cases.

1 Definition A series $\Sigma\, a_n$ is called **absolutely convergent** if the series of absolute values $\Sigma\, |a_n|$ is convergent.

Notice that if $\Sigma\, a_n$ is a series with positive terms, then $|a_n| = a_n$ and so absolute convergence is the same as convergence in this case.

EXAMPLE 1 The series

$$\sum_{n=1}^{\infty} \frac{(-1)^{n-1}}{n^2} = 1 - \frac{1}{2^2} + \frac{1}{3^2} - \frac{1}{4^2} + \cdots$$

is absolutely convergent because

$$\sum_{n=1}^{\infty} \left| \frac{(-1)^{n-1}}{n^2} \right| = \sum_{n=1}^{\infty} \frac{1}{n^2} = 1 + \frac{1}{2^2} + \frac{1}{3^2} + \frac{1}{4^2} + \cdots$$

is a convergent p-series ($p = 2$). ■

EXAMPLE 2 We know that the alternating harmonic series

$$\sum_{n=1}^{\infty} \frac{(-1)^{n-1}}{n} = 1 - \frac{1}{2} + \frac{1}{3} - \frac{1}{4} + \cdots$$

is convergent (see Example 11.5.1), but it is not absolutely convergent because the corresponding series of absolute values is

$$\sum_{n=1}^{\infty} \left| \frac{(-1)^{n-1}}{n} \right| = \sum_{n=1}^{\infty} \frac{1}{n} = 1 + \frac{1}{2} + \frac{1}{3} + \frac{1}{4} + \cdots$$

which is the harmonic series (p-series with $p = 1$) and is therefore divergent. ■

2 Definition A series $\Sigma\, a_n$ is called **conditionally convergent** if it is convergent but not absolutely convergent.

Example 2 shows that the alternating harmonic series is conditionally convergent. Thus it is possible for a series to be convergent but not absolutely convergent. However, the next theorem shows that absolute convergence implies convergence.

3 Theorem If a series $\Sigma\, a_n$ is absolutely convergent, then it is convergent.

PROOF Observe that the inequality

$$0 \leqslant a_n + |a_n| \leqslant 2|a_n|$$

is true because $|a_n|$ is either a_n or $-a_n$. If $\Sigma\, a_n$ is absolutely convergent, then $\Sigma\, |a_n|$ is convergent, so $\Sigma\, 2|a_n|$ is convergent. Therefore, by the Comparison Test, $\Sigma\, (a_n + |a_n|)$ is convergent. Then

$$\sum a_n = \sum (a_n + |a_n|) - \sum |a_n|$$

is the difference of two convergent series and is therefore convergent. ■

EXAMPLE 3 Determine whether the series

$$\sum_{n=1}^{\infty} \frac{\cos n}{n^2} = \frac{\cos 1}{1^2} + \frac{\cos 2}{2^2} + \frac{\cos 3}{3^2} + \cdots$$

is convergent or divergent.

Figure 1 shows the graphs of the terms a_n and partial sums s_n of the series in Example 3. Notice that the series is not alternating but has positive and negative terms.

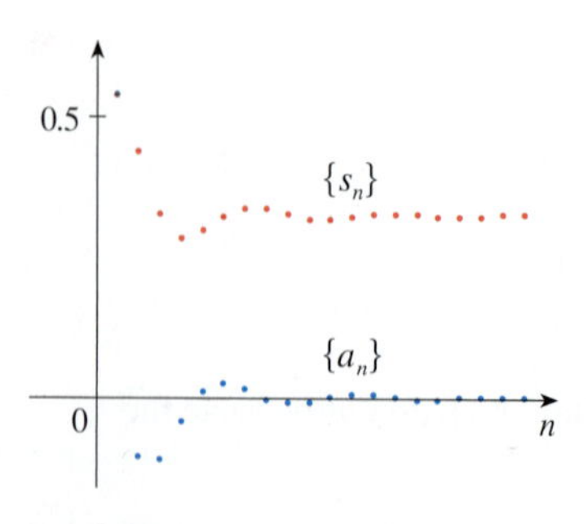

FIGURE 1

SOLUTION This series has both positive and negative terms, but it is not alternating. (The first term is positive, the next three are negative, and the following three are positive: the signs change irregularly.) We can apply the Comparison Test to the series of absolute values

$$\sum_{n=1}^{\infty} \left| \frac{\cos n}{n^2} \right| = \sum_{n=1}^{\infty} \frac{|\cos n|}{n^2}$$

Since $|\cos n| \leqslant 1$ for all n, we have

$$\frac{|\cos n|}{n^2} \leqslant \frac{1}{n^2}$$

We know that $\Sigma\, 1/n^2$ is convergent (p-series with $p = 2$) and therefore $\Sigma\, |\cos n|/n^2$ is convergent by the Comparison Test. Thus the given series $\Sigma\, (\cos n)/n^2$ is absolutely convergent and therefore convergent by Theorem 3. ■

The following test is very useful in determining whether a given series is absolutely convergent.

The Ratio Test

(i) If $\lim_{n\to\infty}\left|\frac{a_{n+1}}{a_n}\right| = L < 1$, then the series $\sum_{n=1}^{\infty} a_n$ is absolutely convergent (and therefore convergent).

(ii) If $\lim_{n\to\infty}\left|\frac{a_{n+1}}{a_n}\right| = L > 1$ or $\lim_{n\to\infty}\left|\frac{a_{n+1}}{a_n}\right| = \infty$, then the series $\sum_{n=1}^{\infty} a_n$ is divergent.

(iii) If $\lim_{n\to\infty}\left|\frac{a_{n+1}}{a_n}\right| = 1$, the Ratio Test is inconclusive; that is, no conclusion can be drawn about the convergence or divergence of $\Sigma\, a_n$.

PROOF

(i) The idea is to compare the given series with a convergent geometric series. Since $L < 1$, we can choose a number r such that $L < r < 1$. Since

$$\lim_{n\to\infty}\left|\frac{a_{n+1}}{a_n}\right| = L \qquad \text{and} \qquad L < r$$

the ratio $|a_{n+1}/a_n|$ will eventually be less than r; that is, there exists an integer N such that

$$\left|\frac{a_{n+1}}{a_n}\right| < r \qquad \text{whenever } n \geqslant N$$

or, equivalently,

4 $$|a_{n+1}| < |a_n|\, r \qquad \text{whenever } n \geqslant N$$

Putting n successively equal to $N, N + 1, N + 2, \ldots$ in (4), we obtain

$$|a_{N+1}| < |a_N|\, r$$
$$|a_{N+2}| < |a_{N+1}|\, r < |a_N|\, r^2$$
$$|a_{N+3}| < |a_{N+2}|\, r < |a_N|\, r^3$$

and, in general,

5 $$|a_{N+k}| < |a_N|\, r^k \qquad \text{for all } k \geqslant 1$$

Now the series

$$\sum_{k=1}^{\infty} |a_N|\, r^k = |a_N|\, r + |a_N|\, r^2 + |a_N|\, r^3 + \cdots$$

is convergent because it is a geometric series with $0 < r < 1$. So the inequality (5), together with the Comparison Test, shows that the series

$$\sum_{n=N+1}^{\infty} |a_n| = \sum_{k=1}^{\infty} |a_{N+k}| = |a_{N+1}| + |a_{N+2}| + |a_{N+3}| + \cdots$$

is also convergent. It follows that the series $\sum_{n=1}^{\infty} |a_n|$ is convergent. (Recall that a finite number of terms doesn't affect convergence.) Therefore $\sum a_n$ is absolutely convergent.

(ii) If $|a_{n+1}/a_n| \to L > 1$ or $|a_{n+1}/a_n| \to \infty$, then the ratio $|a_{n+1}/a_n|$ will eventually be greater than 1; that is, there exists an integer N such that

$$\left| \frac{a_{n+1}}{a_n} \right| > 1 \qquad \text{whenever } n \geqslant N$$

This means that $|a_{n+1}| > |a_n|$ whenever $n \geqslant N$ and so

$$\lim_{n \to \infty} a_n \neq 0$$

Therefore $\sum a_n$ diverges by the Test for Divergence. ■

NOTE Part (iii) of the Ratio Test says that if $\lim_{n \to \infty} |a_{n+1}/a_n| = 1$, the test gives no information. For instance, for the convergent series $\sum 1/n^2$ we have

$$\left| \frac{a_{n+1}}{a_n} \right| = \frac{\dfrac{1}{(n+1)^2}}{\dfrac{1}{n^2}} = \frac{n^2}{(n+1)^2} = \frac{1}{\left(1 + \dfrac{1}{n}\right)^2} \to 1 \qquad \text{as } n \to \infty$$

whereas for the divergent series $\sum 1/n$ we have

$$\left| \frac{a_{n+1}}{a_n} \right| = \frac{\dfrac{1}{n+1}}{\dfrac{1}{n}} = \frac{n}{n+1} = \frac{1}{1 + \dfrac{1}{n}} \to 1 \qquad \text{as } n \to \infty$$

The Ratio Test is usually conclusive if the nth term of the series contains an exponential or a factorial, as we will see in Examples 4 and 5.

Therefore, if $\lim_{n \to \infty} |a_{n+1}/a_n| = 1$, the series $\sum a_n$ might converge or it might diverge. In this case the Ratio Test fails and we must use some other test.

EXAMPLE 4 Test the series $\displaystyle\sum_{n=1}^{\infty} (-1)^n \frac{n^3}{3^n}$ for absolute convergence.

Estimating Sums

In the last three sections we used various methods for estimating the sum of a series—the method depended on which test was used to prove convergence. What about series for which the Ratio Test works? There are two possibilities: If the series happens to be an alternating series, as in Example 4, then it is best to use the methods of Section 11.5. If the terms are all positive, then use the special methods explained in Exercise 46.

SOLUTION We use the Ratio Test with $a_n = (-1)^n n^3/3^n$:

$$\left| \frac{a_{n+1}}{a_n} \right| = \left| \frac{\dfrac{(-1)^{n+1}(n+1)^3}{3^{n+1}}}{\dfrac{(-1)^n n^3}{3^n}} \right| = \frac{(n+1)^3}{3^{n+1}} \cdot \frac{3^n}{n^3}$$

$$= \frac{1}{3}\left(\frac{n+1}{n}\right)^3 = \frac{1}{3}\left(1 + \frac{1}{n}\right)^3 \to \frac{1}{3} < 1$$

Thus, by the Ratio Test, the given series is absolutely convergent. ■

EXAMPLE 5 Test the convergence of the series $\sum_{n=1}^{\infty} \frac{n^n}{n!}$.

SOLUTION Since the terms $a_n = n^n/n!$ are positive, we don't need the absolute value signs.

$$\frac{a_{n+1}}{a_n} = \frac{(n+1)^{n+1}}{(n+1)!} \cdot \frac{n!}{n^n} = \frac{(n+1)(n+1)^n}{(n+1)n!} \cdot \frac{n!}{n^n}$$

$$= \left(\frac{n+1}{n}\right)^n = \left(1 + \frac{1}{n}\right)^n \to e \quad \text{as } n \to \infty$$

(see Equation 6.4.9 or 6.4*.9). Since $e > 1$, the given series is divergent by the Ratio Test. ■

NOTE Although the Ratio Test works in Example 5, an easier method is to use the Test for Divergence. Since

$$a_n = \frac{n^n}{n!} = \frac{n \cdot n \cdot n \cdot \cdots \cdot n}{1 \cdot 2 \cdot 3 \cdot \cdots \cdot n} \geqslant n$$

it follows that a_n does not approach 0 as $n \to \infty$. Therefore the given series is divergent by the Test for Divergence.

The following test is convenient to apply when nth powers occur. Its proof is similar to the proof of the Ratio Test and is left as Exercise 49.

The Root Test

(i) If $\lim_{n \to \infty} \sqrt[n]{|a_n|} = L < 1$, then the series $\sum_{n=1}^{\infty} a_n$ is absolutely convergent (and therefore convergent).

(ii) If $\lim_{n \to \infty} \sqrt[n]{|a_n|} = L > 1$ or $\lim_{n \to \infty} \sqrt[n]{|a_n|} = \infty$, then the series $\sum_{n=1}^{\infty} a_n$ is divergent.

(iii) If $\lim_{n \to \infty} \sqrt[n]{|a_n|} = 1$, the Root Test is inconclusive.

If $\lim_{n \to \infty} \sqrt[n]{|a_n|} = 1$, then part (iii) of the Root Test says that the test gives no information. The series $\sum a_n$ could converge or diverge. (If $L = 1$ in the Ratio Test, don't try the Root Test because L will again be 1. And if $L = 1$ in the Root Test, don't try the Ratio Test because it will fail too.)

EXAMPLE 6 Test the convergence of the series $\sum_{n=1}^{\infty} \left(\frac{2n+3}{3n+2}\right)^n$.

SOLUTION

$$a_n = \left(\frac{2n+3}{3n+2}\right)^n$$

$$\sqrt[n]{|a_n|} = \frac{2n+3}{3n+2} = \frac{2 + \frac{3}{n}}{3 + \frac{2}{n}} \to \frac{2}{3} < 1$$

Thus the given series is absolutely convergent (and therefore convergent) by the Root Test. ■

Rearrangements

The question of whether a given convergent series is absolutely convergent or conditionally convergent has a bearing on the question of whether infinite sums behave like finite sums.

If we rearrange the order of the terms in a finite sum, then of course the value of the sum remains unchanged. But this is not always the case for an infinite series. By a **rearrangement** of an infinite series $\Sigma\, a_n$ we mean a series obtained by simply changing the order of the terms. For instance, a rearrangement of $\Sigma\, a_n$ could start as follows:

$$a_1 + a_2 + a_5 + a_3 + a_4 + a_{15} + a_6 + a_7 + a_{20} + \cdots$$

It turns out that

> if $\Sigma\, a_n$ is an absolutely convergent series with sum s, then any rearrangement of $\Sigma\, a_n$ has the same sum s.

However, any conditionally convergent series can be rearranged to give a different sum. To illustrate this fact let's consider the alternating harmonic series

6 $$1 - \tfrac{1}{2} + \tfrac{1}{3} - \tfrac{1}{4} + \tfrac{1}{5} - \tfrac{1}{6} + \tfrac{1}{7} - \tfrac{1}{8} + \cdots = \ln 2$$

(See Exercise 11.5.36.) If we multiply this series by $\frac{1}{2}$, we get

$$\tfrac{1}{2} - \tfrac{1}{4} + \tfrac{1}{6} - \tfrac{1}{8} + \cdots = \tfrac{1}{2}\ln 2$$

Inserting zeros between the terms of this series, we have

Adding these zeros does not affect the sum of the series; each term in the sequence of partial sums is repeated, but the limit is the same.

7 $$0 + \tfrac{1}{2} + 0 - \tfrac{1}{4} + 0 + \tfrac{1}{6} + 0 - \tfrac{1}{8} + \cdots = \tfrac{1}{2}\ln 2$$

Now we add the series in Equations 6 and 7 using Theorem 11.2.8:

8 $$1 + \tfrac{1}{3} - \tfrac{1}{2} + \tfrac{1}{5} + \tfrac{1}{7} - \tfrac{1}{4} + \cdots = \tfrac{3}{2}\ln 2$$

Notice that the series in (8) contains the same terms as in (6) but rearranged so that one negative term occurs after each pair of positive terms. The sums of these series, however, are different. In fact, Riemann proved that

> if $\Sigma\, a_n$ is a conditionally convergent series and r is any real number whatsoever, then there is a rearrangement of $\Sigma\, a_n$ that has a sum equal to r.

A proof of this fact is outlined in Exercise 52.

11.6 EXERCISES

1. What can you say about the series $\Sigma\, a_n$ in each of the following cases?

(a) $\displaystyle\lim_{n\to\infty}\left|\frac{a_{n+1}}{a_n}\right| = 8$ (b) $\displaystyle\lim_{n\to\infty}\left|\frac{a_{n+1}}{a_n}\right| = 0.8$

(c) $\displaystyle\lim_{n\to\infty}\left|\frac{a_{n+1}}{a_n}\right| = 1$

2–6 Determine whether the series is absolutely convergent or conditionally convergent.

2. $\displaystyle\sum_{n=1}^{\infty}\frac{(-1)^{n-1}}{\sqrt{n}}$

3. $\displaystyle\sum_{n=0}^{\infty}\frac{(-1)^n}{5n+1}$ **4.** $\displaystyle\sum_{n=1}^{\infty}\frac{(-1)^n}{n^3+1}$

5. $\sum_{n=1}^{\infty} \frac{\sin n}{2^n}$

6. $\sum_{n=1}^{\infty} (-1)^{n-1} \frac{n}{n^2+4}$

7–24 Use the Ratio Test to determine whether the series is convergent or divergent.

7. $\sum_{n=1}^{\infty} \frac{n}{5^n}$

8. $\sum_{n=1}^{\infty} \frac{(-2)^n}{n^2}$

9. $\sum_{n=1}^{\infty} (-1)^{n-1} \frac{3^n}{2^n n^3}$

10. $\sum_{n=0}^{\infty} \frac{(-3)^n}{(2n+1)!}$

11. $\sum_{k=1}^{\infty} \frac{1}{k!}$

12. $\sum_{k=1}^{\infty} ke^{-k}$

13. $\sum_{n=1}^{\infty} \frac{10^n}{(n+1)4^{2n+1}}$

14. $\sum_{n=1}^{\infty} \frac{n!}{100^n}$

15. $\sum_{n=1}^{\infty} \frac{n\pi^n}{(-3)^{n-1}}$

16. $\sum_{n=1}^{\infty} \frac{n^{10}}{(-10)^{n+1}}$

17. $\sum_{n=1}^{\infty} \frac{\cos(n\pi/3)}{n!}$

18. $\sum_{n=1}^{\infty} \frac{n!}{n^n}$

19. $\sum_{n=1}^{\infty} \frac{n^{100}100^n}{n!}$

20. $\sum_{n=1}^{\infty} \frac{(2n)!}{(n!)^2}$

21. $1 - \frac{2!}{1 \cdot 3} + \frac{3!}{1 \cdot 3 \cdot 5} - \frac{4!}{1 \cdot 3 \cdot 5 \cdot 7} + \cdots$

$$+ (-1)^{n-1}\frac{n!}{1 \cdot 3 \cdot 5 \cdot \cdots \cdot (2n-1)} + \cdots$$

22. $\frac{2}{3} + \frac{2 \cdot 5}{3 \cdot 5} + \frac{2 \cdot 5 \cdot 8}{3 \cdot 5 \cdot 7} + \frac{2 \cdot 5 \cdot 8 \cdot 11}{3 \cdot 5 \cdot 7 \cdot 9} + \cdots$

23. $\sum_{n=1}^{\infty} \frac{2 \cdot 4 \cdot 6 \cdot \cdots \cdot (2n)}{n!}$

24. $\sum_{n=1}^{\infty} (-1)^n \frac{2^n n!}{5 \cdot 8 \cdot 11 \cdot \cdots \cdot (3n+2)}$

25–30 Use the Root Test to determine whether the series is convergent or divergent.

25. $\sum_{n=1}^{\infty} \left(\frac{n^2+1}{2n^2+1}\right)^n$

26. $\sum_{n=1}^{\infty} \frac{(-2)^n}{n^n}$

27. $\sum_{n=2}^{\infty} \frac{(-1)^{n-1}}{(\ln n)^n}$

28. $\sum_{n=1}^{\infty} \left(\frac{-2n}{n+1}\right)^{5n}$

29. $\sum_{n=1}^{\infty} \left(1 + \frac{1}{n}\right)^{n^2}$

30. $\sum_{n=0}^{\infty} (\arctan n)^n$

31–38 Use any test to determine whether the series is absolutely convergent, conditionally convergent, or divergent.

31. $\sum_{n=2}^{\infty} \frac{(-1)^n}{\ln n}$

32. $\sum_{n=1}^{\infty} \left(\frac{1-n}{2+3n}\right)^n$

33. $\sum_{n=1}^{\infty} \frac{(-9)^n}{n10^{n+1}}$

34. $\sum_{n=1}^{\infty} \frac{n5^{2n}}{10^{n+1}}$

35. $\sum_{n=2}^{\infty} \left(\frac{n}{\ln n}\right)^n$

36. $\sum_{n=1}^{\infty} \frac{\sin(n\pi/6)}{1+n\sqrt{n}}$

37. $\sum_{n=1}^{\infty} \frac{(-1)^n \arctan n}{n^2}$

38. $\sum_{n=2}^{\infty} \frac{(-1)^n}{n \ln n}$

39. The terms of a series are defined recursively by the equations

$$a_1 = 2 \qquad a_{n+1} = \frac{5n+1}{4n+3}a_n$$

Determine whether $\Sigma\, a_n$ converges or diverges.

40. A series $\Sigma\, a_n$ is defined by the equations

$$a_1 = 1 \qquad a_{n+1} = \frac{2+\cos n}{\sqrt{n}}a_n$$

Determine whether $\Sigma\, a_n$ converges or diverges.

41–42 Let $\{b_n\}$ be a sequence of positive numbers that converges to $\frac{1}{2}$. Determine whether the given series is absolutely convergent.

41. $\sum_{n=1}^{\infty} \frac{b_n^n \cos n\pi}{n}$

42. $\sum_{n=1}^{\infty} \frac{(-1)^n n!}{n^n b_1 b_2 b_3 \cdots b_n}$

43. For which of the following series is the Ratio Test inconclusive (that is, it fails to give a definite answer)?

(a) $\sum_{n=1}^{\infty} \frac{1}{n^3}$ (b) $\sum_{n=1}^{\infty} \frac{n}{2^n}$

(c) $\sum_{n=1}^{\infty} \frac{(-3)^{n-1}}{\sqrt{n}}$ (d) $\sum_{n=1}^{\infty} \frac{\sqrt{n}}{1+n^2}$

44. For which positive integers k is the following series convergent?

$$\sum_{n=1}^{\infty} \frac{(n!)^2}{(kn)!}$$

45. (a) Show that $\Sigma_{n=0}^{\infty}\, x^n/n!$ converges for all x.
(b) Deduce that $\lim_{n\to\infty} x^n/n! = 0$ for all x.

46. Let $\Sigma\, a_n$ be a series with positive terms and let $r_n = a_{n+1}/a_n$. Suppose that $\lim_{n\to\infty} r_n = L < 1$, so $\Sigma\, a_n$ converges by the Ratio Test. As usual, we let R_n be the remainder after n terms, that is,

$$R_n = a_{n+1} + a_{n+2} + a_{n+3} + \cdots$$

(a) If $\{r_n\}$ is a decreasing sequence and $r_{n+1} < 1$, show, by summing a geometric series, that

$$R_n \leqslant \frac{a_{n+1}}{1 - r_{n+1}}$$

(b) If $\{r_n\}$ is an increasing sequence, show that

$$R_n \leqslant \frac{a_{n+1}}{1 - L}$$

47. (a) Find the partial sum s_5 of the series $\sum_{n=1}^{\infty} 1/(n2^n)$. Use Exercise 46 to estimate the error in using s_5 as an approximation to the sum of the series.
(b) Find a value of n so that s_n is within 0.00005 of the sum. Use this value of n to approximate the sum of the series.

48. Use the sum of the first 10 terms to approximate the sum of the series

$$\sum_{n=1}^{\infty} \frac{n}{2^n}$$

Use Exercise 46 to estimate the error.

49. Prove the Root Test. [*Hint for part (i):* Take any number r such that $L < r < 1$ and use the fact that there is an integer N such that $\sqrt[n]{|a_n|} < r$ whenever $n \geq N$.]

50. Around 1910, the Indian mathematician Srinivasa Ramanujan discovered the formula

$$\frac{1}{\pi} = \frac{2\sqrt{2}}{9801} \sum_{n=0}^{\infty} \frac{(4n)!(1103 + 26390n)}{(n!)^4 396^{4n}}$$

William Gosper used this series in 1985 to compute the first 17 million digits of π.
(a) Verify that the series is convergent.
(b) How many correct decimal places of π do you get if you use just the first term of the series? What if you use two terms?

51. Given any series $\sum a_n$, we define a series $\sum a_n^+$ whose terms are all the positive terms of $\sum a_n$ and a series $\sum a_n^-$ whose terms are all the negative terms of $\sum a_n$. To be specific, we let

$$a_n^+ = \frac{a_n + |a_n|}{2} \qquad a_n^- = \frac{a_n - |a_n|}{2}$$

Notice that if $a_n > 0$, then $a_n^+ = a_n$ and $a_n^- = 0$, whereas if $a_n < 0$, then $a_n^- = a_n$ and $a_n^+ = 0$.
(a) If $\sum a_n$ is absolutely convergent, show that both of the series $\sum a_n^+$ and $\sum a_n^-$ are convergent.
(b) If $\sum a_n$ is conditionally convergent, show that both of the series $\sum a_n^+$ and $\sum a_n^-$ are divergent.

52. Prove that if $\sum a_n$ is a conditionally convergent series and r is any real number, then there is a rearrangement of $\sum a_n$ whose sum is r. [*Hints:* Use the notation of Exercise 51. Take just enough positive terms a_n^+ so that their sum is greater than r. Then add just enough negative terms a_n^- so that the cumulative sum is less than r. Continue in this manner and use Theorem 11.2.6.]

53. Suppose the series $\sum a_n$ is conditionally convergent.
(a) Prove that the series $\sum n^2 a_n$ is divergent.
(b) Conditional convergence of $\sum a_n$ is not enough to determine whether $\sum n a_n$ is convergent. Show this by giving an example of a conditionally convergent series such that $\sum n a_n$ converges and an example where $\sum n a_n$ diverges.

11.7 Strategy for Testing Series

We now have several ways of testing a series for convergence or divergence; the problem is to decide which test to use on which series. In this respect, testing series is similar to integrating functions. Again there are no hard and fast rules about which test to apply to a given series, but you may find the following advice of some use.

It is not wise to apply a list of the tests in a specific order until one finally works. That would be a waste of time and effort. Instead, as with integration, the main strategy is to classify the series according to its *form*.

1. If the series is of the form $\sum 1/n^p$, it is a p-series, which we know to be convergent if $p > 1$ and divergent if $p \leq 1$.
2. If the series has the form $\sum ar^{n-1}$ or $\sum ar^n$, it is a geometric series, which converges if $|r| < 1$ and diverges if $|r| \geq 1$. Some preliminary algebraic manipulation may be required to bring the series into this form.
3. If the series has a form that is similar to a p-series or a geometric series, then one of the comparison tests should be considered. In particular, if a_n is a rational function or an algebraic function of n (involving roots of polynomials), then the series should be compared with a p-series. Notice that most of the series in Exercises 11.4 have this form. (The value of p should be chosen as in Section 11.4 by keeping only the highest powers of n in the numerator and denominator.) The comparison tests apply only to series with positive terms, but if $\sum a_n$ has some negative terms, then we can apply the Comparison Test to $\sum |a_n|$ and test for absolute convergence.

4. If you can see at a glance that $\lim_{n\to\infty} a_n \neq 0$, then the Test for Divergence should be used.
5. If the series is of the form $\Sigma\,(-1)^{n-1}b_n$ or $\Sigma\,(-1)^n b_n$, then the Alternating Series Test is an obvious possibility.
6. Series that involve factorials or other products (including a constant raised to the nth power) are often conveniently tested using the Ratio Test. Bear in mind that $|a_{n+1}/a_n| \to 1$ as $n \to \infty$ for all p-series and therefore all rational or algebraic functions of n. Thus the Ratio Test should not be used for such series.
7. If a_n is of the form $(b_n)^n$, then the Root Test may be useful.
8. If $a_n = f(n)$, where $\int_1^\infty f(x)\,dx$ is easily evaluated, then the Integral Test is effective (assuming the hypotheses of this test are satisfied).

In the following examples we don't work out all the details but simply indicate which tests should be used.

EXAMPLE 1 $\displaystyle\sum_{n=1}^{\infty} \frac{n-1}{2n+1}$

Since $a_n \to \frac{1}{2} \neq 0$ as $n \to \infty$, we should use the Test for Divergence. ■

EXAMPLE 2 $\displaystyle\sum_{n=1}^{\infty} \frac{\sqrt{n^3+1}}{3n^3+4n^2+2}$

Since a_n is an algebraic function of n, we compare the given series with a p-series. The comparison series for the Limit Comparison Test is $\Sigma\, b_n$, where

$$b_n = \frac{\sqrt{n^3}}{3n^3} = \frac{n^{3/2}}{3n^3} = \frac{1}{3n^{3/2}}$$ ■

EXAMPLE 3 $\displaystyle\sum_{n=1}^{\infty} ne^{-n^2}$

Since the integral $\int_1^\infty xe^{-x^2}\,dx$ is easily evaluated, we use the Integral Test. The Ratio Test also works. ■

EXAMPLE 4 $\displaystyle\sum_{n=1}^{\infty} (-1)^n \frac{n^3}{n^4+1}$

Since the series is alternating, we use the Alternating Series Test. ■

EXAMPLE 5 $\displaystyle\sum_{k=1}^{\infty} \frac{2^k}{k!}$

Since the series involves $k!$, we use the Ratio Test. ■

EXAMPLE 6 $\displaystyle\sum_{n=1}^{\infty} \frac{1}{2+3^n}$

Since the series is closely related to the geometric series $\Sigma\, 1/3^n$, we use the Comparison Test. ■

11.7 EXERCISES

1–38 Test the series for convergence or divergence.

1. $\displaystyle\sum_{n=1}^{\infty} \frac{n^2 - 1}{n^3 + 1}$

2. $\displaystyle\sum_{n=1}^{\infty} \frac{n - 1}{n^3 + 1}$

3. $\displaystyle\sum_{n=1}^{\infty} (-1)^n \frac{n^2 - 1}{n^3 + 1}$

4. $\displaystyle\sum_{n=1}^{\infty} (-1)^n \frac{n^2 - 1}{n^2 + 1}$

5. $\displaystyle\sum_{n=1}^{\infty} \frac{e^n}{n^2}$

6. $\displaystyle\sum_{n=1}^{\infty} \frac{n^{2n}}{(1 + n)^{3n}}$

7. $\displaystyle\sum_{n=2}^{\infty} \frac{1}{n\sqrt{\ln n}}$

8. $\displaystyle\sum_{n=1}^{\infty} (-1)^{n-1} \frac{n^4}{4^n}$

9. $\displaystyle\sum_{n=0}^{\infty} (-1)^n \frac{\pi^{2n}}{(2n)!}$

10. $\displaystyle\sum_{n=1}^{\infty} n^2 e^{-n^3}$

11. $\displaystyle\sum_{n=1}^{\infty} \left(\frac{1}{n^3} + \frac{1}{3^n}\right)$

12. $\displaystyle\sum_{k=1}^{\infty} \frac{1}{k\sqrt{k^2 + 1}}$

13. $\displaystyle\sum_{n=1}^{\infty} \frac{3^n n^2}{n!}$

14. $\displaystyle\sum_{n=1}^{\infty} \frac{\sin 2n}{1 + 2^n}$

15. $\displaystyle\sum_{k=1}^{\infty} \frac{2^{k-1} 3^{k+1}}{k^k}$

16. $\displaystyle\sum_{n=1}^{\infty} \frac{\sqrt{n^4 + 1}}{n^3 + n}$

17. $\displaystyle\sum_{n=1}^{\infty} \frac{1 \cdot 3 \cdot 5 \cdot \cdots \cdot (2n - 1)}{2 \cdot 5 \cdot 8 \cdot \cdots \cdot (3n - 1)}$

18. $\displaystyle\sum_{n=2}^{\infty} \frac{(-1)^{n-1}}{\sqrt{n} - 1}$

19. $\displaystyle\sum_{n=1}^{\infty} (-1)^n \frac{\ln n}{\sqrt{n}}$

20. $\displaystyle\sum_{k=1}^{\infty} \frac{\sqrt[3]{k} - 1}{k(\sqrt{k} + 1)}$

21. $\displaystyle\sum_{n=1}^{\infty} (-1)^n \cos(1/n^2)$

22. $\displaystyle\sum_{k=1}^{\infty} \frac{1}{2 + \sin k}$

23. $\displaystyle\sum_{n=1}^{\infty} \tan(1/n)$

24. $\displaystyle\sum_{n=1}^{\infty} n \sin(1/n)$

25. $\displaystyle\sum_{n=1}^{\infty} \frac{n!}{e^{n^2}}$

26. $\displaystyle\sum_{n=1}^{\infty} \frac{n^2 + 1}{5^n}$

27. $\displaystyle\sum_{k=1}^{\infty} \frac{k \ln k}{(k + 1)^3}$

28. $\displaystyle\sum_{n=1}^{\infty} \frac{e^{1/n}}{n^2}$

29. $\displaystyle\sum_{n=1}^{\infty} \frac{(-1)^n}{\cosh n}$

30. $\displaystyle\sum_{j=1}^{\infty} (-1)^j \frac{\sqrt{j}}{j + 5}$

31. $\displaystyle\sum_{k=1}^{\infty} \frac{5^k}{3^k + 4^k}$

32. $\displaystyle\sum_{n=1}^{\infty} \frac{(n!)^n}{n^{4n}}$

33. $\displaystyle\sum_{n=1}^{\infty} \left(\frac{n}{n + 1}\right)^{n^2}$

34. $\displaystyle\sum_{n=1}^{\infty} \frac{1}{n + n\cos^2 n}$

35. $\displaystyle\sum_{n=1}^{\infty} \frac{1}{n^{1+1/n}}$

36. $\displaystyle\sum_{n=2}^{\infty} \frac{1}{(\ln n)^{\ln n}}$

37. $\displaystyle\sum_{n=1}^{\infty} (\sqrt[n]{2} - 1)^n$

38. $\displaystyle\sum_{n=1}^{\infty} (\sqrt[n]{2} - 1)$

11.8 Power Series

A **power series** is a series of the form

$$\boxed{1} \qquad \sum_{n=0}^{\infty} c_n x^n = c_0 + c_1 x + c_2 x^2 + c_3 x^3 + \cdots$$

where x is a variable and the c_n's are constants called the **coefficients** of the series. For each fixed x, the series (1) is a series of constants that we can test for convergence or divergence. A power series may converge for some values of x and diverge for other values of x. The sum of the series is a function

$$f(x) = c_0 + c_1 x + c_2 x^2 + \cdots + c_n x^n + \cdots$$

whose domain is the set of all x for which the series converges. Notice that f resembles a polynomial. The only difference is that f has infinitely many terms.

For instance, if we take $c_n = 1$ for all n, the power series becomes the geometric series

$$\boxed{2} \qquad \sum_{n=0}^{\infty} x^n = 1 + x + x^2 + \cdots + x^n + \cdots$$

which converges when $-1 < x < 1$ and diverges when $|x| \geqslant 1$. (See Equation 11.2.5.)

Trigonometric Series
A power series is a series in which each term is a power function. A **trigonometric series**

$$\sum_{n=0}^{\infty} (a_n \cos nx + b_n \sin nx)$$

is a series whose terms are trigonometric functions. This type of series is discussed on the website

www.stewartcalculus.com

Click on *Additional Topics* and then on *Fourier Series.*

In fact if we put $x = \frac{1}{2}$ in the geometric series (2) we get the convergent series

$$\sum_{n=0}^{\infty} \left(\frac{1}{2}\right)^n = 1 + \frac{1}{2} + \frac{1}{4} + \frac{1}{8} + \frac{1}{16} + \cdots$$

but if we put $x = 2$ in (2) we get the divergent series

$$\sum_{n=0}^{\infty} 2^n = 1 + 2 + 4 + 8 + 16 + \cdots$$

More generally, a series of the form

$$\boxed{3} \qquad \sum_{n=0}^{\infty} c_n(x - a)^n = c_0 + c_1(x - a) + c_2(x - a)^2 + \cdots$$

is called a **power series in $(x - a)$** or a **power series centered at a** or a **power series about a**. Notice that in writing out the term corresponding to $n = 0$ in Equations 1 and 3 we have adopted the convention that $(x - a)^0 = 1$ even when $x = a$. Notice also that when $x = a$, all of the terms are 0 for $n \geqslant 1$ and so the power series (3) always converges when $x = a$.

EXAMPLE 1 For what values of x is the series $\sum_{n=0}^{\infty} n!x^n$ convergent?

SOLUTION We use the Ratio Test. If we let a_n, as usual, denote the nth term of the series, then $a_n = n!x^n$. If $x \neq 0$, we have

Notice that

$$(n + 1)! = (n + 1)n(n - 1) \cdot \dots \cdot 3 \cdot 2 \cdot 1$$
$$= (n + 1)n!$$

$$\lim_{n\to\infty} \left|\frac{a_{n+1}}{a_n}\right| = \lim_{n\to\infty} \left|\frac{(n + 1)!x^{n+1}}{n!x^n}\right| = \lim_{n\to\infty} (n + 1)|x| = \infty$$

By the Ratio Test, the series diverges when $x \neq 0$. Thus the given series converges only when $x = 0$. ■

EXAMPLE 2 For what values of x does the series $\sum_{n=1}^{\infty} \frac{(x - 3)^n}{n}$ converge?

SOLUTION Let $a_n = (x - 3)^n/n$. Then

$$\left|\frac{a_{n+1}}{a_n}\right| = \left|\frac{(x - 3)^{n+1}}{n + 1} \cdot \frac{n}{(x - 3)^n}\right|$$

$$= \frac{1}{1 + \frac{1}{n}}|x - 3| \to |x - 3| \qquad \text{as } n \to \infty$$

By the Ratio Test, the given series is absolutely convergent, and therefore convergent, when $|x - 3| < 1$ and divergent when $|x - 3| > 1$. Now

$$|x - 3| < 1 \iff -1 < x - 3 < 1 \iff 2 < x < 4$$

so the series converges when $2 < x < 4$ and diverges when $x < 2$ or $x > 4$.

The Ratio Test gives no information when $|x - 3| = 1$ so we must consider $x = 2$ and $x = 4$ separately. If we put $x = 4$ in the series, it becomes $\Sigma\ 1/n$, the harmonic series, which is divergent. If $x = 2$, the series is $\Sigma\ (-1)^n/n$, which converges by the Alternating Series Test. Thus the given power series converges for $2 \leqslant x < 4$. ■

Membrane courtesy of National Film Board of Canada

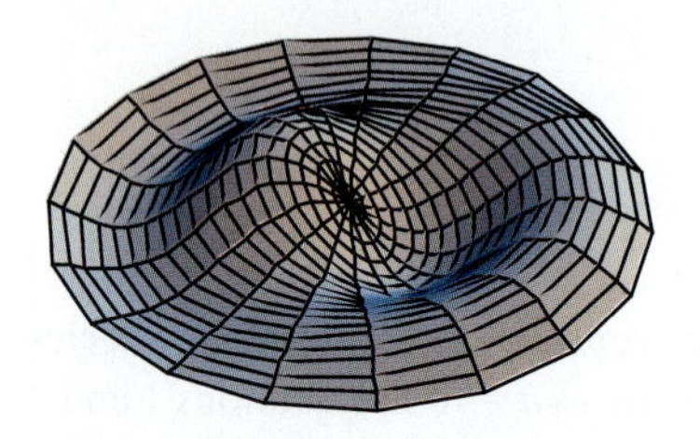

Notice how closely the computer-generated model (which involves Bessel functions and cosine functions) matches the photograph of a vibrating rubber membrane.

We will see that the main use of a power series is that it provides a way to represent some of the most important functions that arise in mathematics, physics, and chemistry. In particular, the sum of the power series in the next example is called a **Bessel function**, after the German astronomer Friedrich Bessel (1784–1846), and the function given in Exercise 35 is another example of a Bessel function. In fact, these functions first arose when Bessel solved Kepler's equation for describing planetary motion. Since that time, these functions have been applied in many different physical situations, including the temperature distribution in a circular plate and the shape of a vibrating drumhead.

EXAMPLE 3 Find the domain of the Bessel function of order 0 defined by

$$J_0(x) = \sum_{n=0}^{\infty} \frac{(-1)^n x^{2n}}{2^{2n}(n!)^2}$$

SOLUTION Let $a_n = (-1)^n x^{2n}/[2^{2n}(n!)^2]$. Then

$$\left|\frac{a_{n+1}}{a_n}\right| = \left|\frac{(-1)^{n+1}x^{2(n+1)}}{2^{2(n+1)}[(n+1)!]^2} \cdot \frac{2^{2n}(n!)^2}{(-1)^n x^{2n}}\right|$$

$$= \frac{x^{2n+2}}{2^{2n+2}(n+1)^2(n!)^2} \cdot \frac{2^{2n}(n!)^2}{x^{2n}}$$

$$= \frac{x^2}{4(n+1)^2} \to 0 < 1 \qquad \text{for all } x$$

Thus, by the Ratio Test, the given series converges for all values of x. In other words, the domain of the Bessel function J_0 is $(-\infty, \infty) = \mathbb{R}$. ■

Recall that the sum of a series is equal to the limit of the sequence of partial sums. So when we define the Bessel function in Example 3 as the sum of a series we mean that, for every real number x,

$$J_0(x) = \lim_{n\to\infty} s_n(x) \qquad \text{where} \qquad s_n(x) = \sum_{i=0}^{n} \frac{(-1)^i x^{2i}}{2^{2i}(i!)^2}$$

The first few partial sums are

$$s_0(x) = 1$$

$$s_1(x) = 1 - \frac{x^2}{4}$$

$$s_2(x) = 1 - \frac{x^2}{4} + \frac{x^4}{64}$$

$$s_3(x) = 1 - \frac{x^2}{4} + \frac{x^4}{64} - \frac{x^6}{2304}$$

$$s_4(x) = 1 - \frac{x^2}{4} + \frac{x^4}{64} - \frac{x^6}{2304} + \frac{x^8}{147{,}456}$$

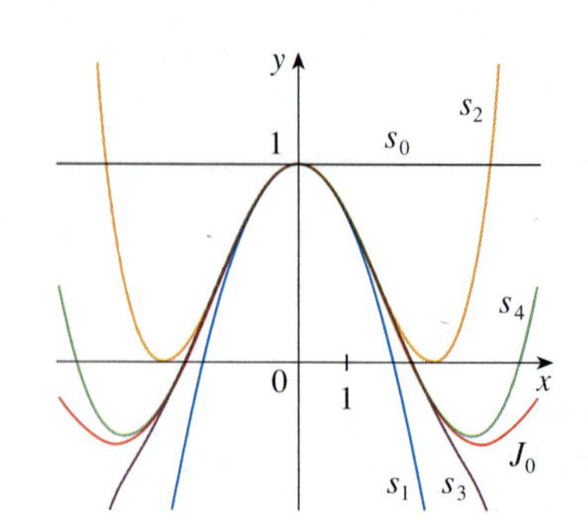

FIGURE 1
Partial sums of the Bessel function J_0

Figure 1 shows the graphs of these partial sums, which are polynomials. They are all approximations to the function J_0, but notice that the approximations become better when more terms are included. Figure 2 shows a more complete graph of the Bessel function.

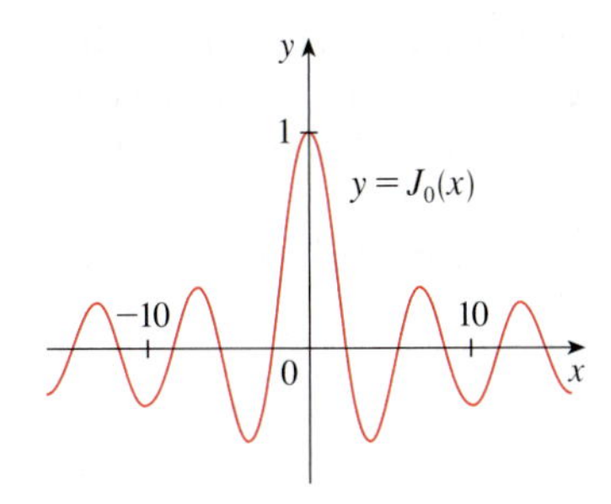

FIGURE 2

For the power series that we have looked at so far, the set of values of x for which the series is convergent has always turned out to be an interval [a finite interval for the geometric series and the series in Example 2, the infinite interval $(-\infty, \infty)$ in Example 3, and a collapsed interval $[0, 0] = \{0\}$ in Example 1]. The following theorem, proved in Appendix F, says that this is true in general.

> **4 Theorem** For a given power series $\sum_{n=0}^{\infty} c_n(x - a)^n$, there are only three possibilities:
>
> (i) The series converges only when $x = a$.
>
> (ii) The series converges for all x.
>
> (iii) There is a positive number R such that the series converges if $|x - a| < R$ and diverges if $|x - a| > R$.

The number R in case (iii) is called the **radius of convergence** of the power series. By convention, the radius of convergence is $R = 0$ in case (i) and $R = \infty$ in case (ii). The **interval of convergence** of a power series is the interval that consists of all values of x for which the series converges. In case (i) the interval consists of just a single point a. In case (ii) the interval is $(-\infty, \infty)$. In case (iii) note that the inequality $|x - a| < R$ can be rewritten as $a - R < x < a + R$. When x is an *endpoint* of the interval, that is, $x = a \pm R$, anything can happen—the series might converge at one or both endpoints or it might diverge at both endpoints. Thus in case (iii) there are four possibilities for the interval of convergence:

$$(a - R, a + R) \qquad (a - R, a + R] \qquad [a - R, a + R) \qquad [a - R, a + R]$$

The situation is illustrated in Figure 3.

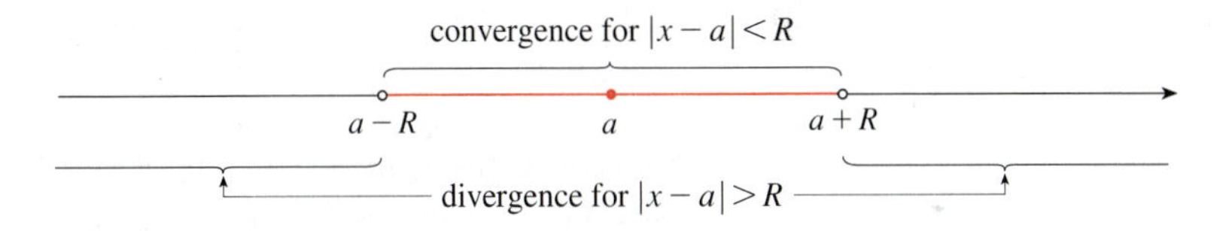

FIGURE 3

We summarize here the radius and interval of convergence for each of the examples already considered in this section.

	Series	Radius of convergence	Interval of convergence
Geometric series	$\sum_{n=0}^{\infty} x^n$	$R = 1$	$(-1, 1)$
Example 1	$\sum_{n=0}^{\infty} n!\, x^n$	$R = 0$	$\{0\}$
Example 2	$\sum_{n=1}^{\infty} \frac{(x - 3)^n}{n}$	$R = 1$	$[2, 4)$
Example 3	$\sum_{n=0}^{\infty} \frac{(-1)^n x^{2n}}{2^{2n}(n!)^2}$	$R = \infty$	$(-\infty, \infty)$

In general, the Ratio Test (or sometimes the Root Test) should be used to determine the radius of convergence R. The Ratio and Root Tests always fail when x is an endpoint of the interval of convergence, so the endpoints must be checked with some other test.

EXAMPLE 4 Find the radius of convergence and interval of convergence of the series

$$\sum_{n=0}^{\infty} \frac{(-3)^n x^n}{\sqrt{n+1}}$$

SOLUTION Let $a_n = (-3)^n x^n / \sqrt{n+1}$. Then

$$\left|\frac{a_{n+1}}{a_n}\right| = \left|\frac{(-3)^{n+1}x^{n+1}}{\sqrt{n+2}} \cdot \frac{\sqrt{n+1}}{(-3)^n x^n}\right| = \left|-3x\sqrt{\frac{n+1}{n+2}}\right|$$

$$= 3\sqrt{\frac{1+(1/n)}{1+(2/n)}}\,|x| \to 3|x| \qquad \text{as } n \to \infty$$

By the Ratio Test, the given series converges if $3|x| < 1$ and diverges if $3|x| > 1$. Thus it converges if $|x| < \frac{1}{3}$ and diverges if $|x| > \frac{1}{3}$. This means that the radius of convergence is $R = \frac{1}{3}$.

We know the series converges in the interval $\left(-\frac{1}{3}, \frac{1}{3}\right)$, but we must now test for convergence at the endpoints of this interval. If $x = -\frac{1}{3}$, the series becomes

$$\sum_{n=0}^{\infty} \frac{(-3)^n\left(-\frac{1}{3}\right)^n}{\sqrt{n+1}} = \sum_{n=0}^{\infty} \frac{1}{\sqrt{n+1}} = \frac{1}{\sqrt{1}} + \frac{1}{\sqrt{2}} + \frac{1}{\sqrt{3}} + \frac{1}{\sqrt{4}} + \cdots$$

which diverges. (Use the Integral Test or simply observe that it is a p-series with $p = \frac{1}{2} < 1$.) If $x = \frac{1}{3}$, the series is

$$\sum_{n=0}^{\infty} \frac{(-3)^n\left(\frac{1}{3}\right)^n}{\sqrt{n+1}} = \sum_{n=0}^{\infty} \frac{(-1)^n}{\sqrt{n+1}}$$

which converges by the Alternating Series Test. Therefore the given power series converges when $-\frac{1}{3} < x \leqslant \frac{1}{3}$, so the interval of convergence is $\left(-\frac{1}{3}, \frac{1}{3}\right]$. ■

EXAMPLE 5 Find the radius of convergence and interval of convergence of the series

$$\sum_{n=0}^{\infty} \frac{n(x+2)^n}{3^{n+1}}$$

SOLUTION If $a_n = n(x+2)^n / 3^{n+1}$, then

$$\left|\frac{a_{n+1}}{a_n}\right| = \left|\frac{(n+1)(x+2)^{n+1}}{3^{n+2}} \cdot \frac{3^{n+1}}{n(x+2)^n}\right|$$

$$= \left(1 + \frac{1}{n}\right)\frac{|x+2|}{3} \to \frac{|x+2|}{3} \qquad \text{as } n \to \infty$$

Using the Ratio Test, we see that the series converges if $|x+2|/3 < 1$ and it diverges if $|x+2|/3 > 1$. So it converges if $|x+2| < 3$ and diverges if $|x+2| > 3$. Thus the radius of convergence is $R = 3$.

The inequality $|x+2|<3$ can be written as $-5<x<1$, so we test the series at the endpoints -5 and 1. When $x=-5$, the series is

$$\sum_{n=0}^{\infty} \frac{n(-3)^n}{3^{n+1}} = \frac{1}{3}\sum_{n=0}^{\infty} (-1)^n n$$

which diverges by the Test for Divergence [$(-1)^n n$ doesn't converge to 0]. When $x=1$, the series is

$$\sum_{n=0}^{\infty} \frac{n(3)^n}{3^{n+1}} = \frac{1}{3}\sum_{n=0}^{\infty} n$$

which also diverges by the Test for Divergence. Thus the series converges only when $-5<x<1$, so the interval of convergence is $(-5, 1)$. ■

11.8 EXERCISES

1. What is a power series?

2. (a) What is the radius of convergence of a power series? How do you find it?
(b) What is the interval of convergence of a power series? How do you find it?

3–28 Find the radius of convergence and interval of convergence of the series.

3. $\sum_{n=1}^{\infty} (-1)^n n x^n$

4. $\sum_{n=1}^{\infty} \frac{(-1)^n x^n}{\sqrt[3]{n}}$

5. $\sum_{n=1}^{\infty} \frac{x^n}{2n-1}$

6. $\sum_{n=1}^{\infty} \frac{(-1)^n x^n}{n^2}$

7. $\sum_{n=0}^{\infty} \frac{x^n}{n!}$

8. $\sum_{n=1}^{\infty} n^n x^n$

9. $\sum_{n=1}^{\infty} \frac{x^n}{n^4 4^n}$

10. $\sum_{n=1}^{\infty} 2^n n^2 x^n$

11. $\sum_{n=1}^{\infty} \frac{(-1)^n 4^n}{\sqrt{n}} x^n$

12. $\sum_{n=1}^{\infty} \frac{(-1)^{n-1}}{n 5^n} x^n$

13. $\sum_{n=1}^{\infty} \frac{n}{2^n(n^2+1)} x^n$

14. $\sum_{n=1}^{\infty} \frac{x^{2n}}{n!}$

15. $\sum_{n=0}^{\infty} \frac{(x-2)^n}{n^2+1}$

16. $\sum_{n=1}^{\infty} \frac{(-1)^n}{(2n-1)2^n} (x-1)^n$

17. $\sum_{n=2}^{\infty} \frac{(x+2)^n}{2^n \ln n}$

18. $\sum_{n=1}^{\infty} \frac{\sqrt{n}}{8^n} (x+6)^n$

19. $\sum_{n=1}^{\infty} \frac{(x-2)^n}{n^n}$

20. $\sum_{n=1}^{\infty} \frac{(2x-1)^n}{5^n \sqrt{n}}$

21. $\sum_{n=1}^{\infty} \frac{n}{b^n} (x-a)^n, \quad b>0$

22. $\sum_{n=2}^{\infty} \frac{b^n}{\ln n} (x-a)^n, \quad b>0$

23. $\sum_{n=1}^{\infty} n!(2x-1)^n$

24. $\sum_{n=1}^{\infty} \frac{n^2 x^n}{2 \cdot 4 \cdot 6 \cdot \cdots \cdot (2n)}$

25. $\sum_{n=1}^{\infty} \frac{(5x-4)^n}{n^3}$

26. $\sum_{n=2}^{\infty} \frac{x^{2n}}{n(\ln n)^2}$

27. $\sum_{n=1}^{\infty} \frac{x^n}{1 \cdot 3 \cdot 5 \cdot \cdots \cdot (2n-1)}$

28. $\sum_{n=1}^{\infty} \frac{n! x^n}{1 \cdot 3 \cdot 5 \cdot \cdots \cdot (2n-1)}$

---

29. If $\sum_{n=0}^{\infty} c_n 4^n$ is convergent, can we conclude that each of the following series is convergent?

(a) $\sum_{n=0}^{\infty} c_n(-2)^n$ (b) $\sum_{n=0}^{\infty} c_n(-4)^n$

30. Suppose that $\sum_{n=0}^{\infty} c_n x^n$ converges when $x=-4$ and diverges when $x=6$. What can be said about the convergence or divergence of the following series?

(a) $\sum_{n=0}^{\infty} c_n$ (b) $\sum_{n=0}^{\infty} c_n 8^n$

(c) $\sum_{n=0}^{\infty} c_n(-3)^n$ (d) $\sum_{n=0}^{\infty} (-1)^n c_n 9^n$

31. If k is a positive integer, find the radius of convergence of the series

$$\sum_{n=0}^{\infty} \frac{(n!)^k}{(kn)!} x^n$$

32. Let p and q be real numbers with $p < q$. Find a power series whose interval of convergence is
(a) (p, q) (b) $(p, q]$ (c) $[p, q)$ (d) $[p, q]$

33. Is it possible to find a power series whose interval of convergence is $[0, \infty)$? Explain.

34. Graph the first several partial sums $s_n(x)$ of the series $\sum_{n=0}^{\infty} x^n$, together with the sum function $f(x) = 1/(1 - x)$, on a common screen. On what interval do these partial sums appear to be converging to $f(x)$?

35. The function J_1 defined by

$$J_1(x) = \sum_{n=0}^{\infty} \frac{(-1)^n x^{2n+1}}{n!(n+1)!2^{2n+1}}$$

is called the *Bessel function of order 1*.
(a) Find its domain.
(b) Graph the first several partial sums on a common screen.
CAS (c) If your CAS has built-in Bessel functions, graph J_1 on the same screen as the partial sums in part (b) and observe how the partial sums approximate J_1.

36. The function A defined by

$$A(x) = 1 + \frac{x^3}{2 \cdot 3} + \frac{x^6}{2 \cdot 3 \cdot 5 \cdot 6} + \frac{x^9}{2 \cdot 3 \cdot 5 \cdot 6 \cdot 8 \cdot 9} + \cdots$$

is called an *Airy function* after the English mathematician and astronomer Sir George Airy (1801–1892).
(a) Find the domain of the Airy function.
(b) Graph the first several partial sums on a common screen.
CAS (c) If your CAS has built-in Airy functions, graph A on the same screen as the partial sums in part (b) and observe how the partial sums approximate A.

37. A function f is defined by

$$f(x) = 1 + 2x + x^2 + 2x^3 + x^4 + \cdots$$

that is, its coefficients are $c_{2n} = 1$ and $c_{2n+1} = 2$ for all $n \geqslant 0$. Find the interval of convergence of the series and find an explicit formula for $f(x)$.

38. If $f(x) = \sum_{n=0}^{\infty} c_n x^n$, where $c_{n+4} = c_n$ for all $n \geqslant 0$, find the interval of convergence of the series and a formula for $f(x)$.

39. Show that if $\lim_{n \to \infty} \sqrt[n]{|c_n|} = c$, where $c \neq 0$, then the radius of convergence of the power series $\sum c_n x^n$ is $R = 1/c$.

40. Suppose that the power series $\sum c_n(x - a)^n$ satisfies $c_n \neq 0$ for all n. Show that if $\lim_{n \to \infty} |c_n/c_{n+1}|$ exists, then it is equal to the radius of convergence of the power series.

41. Suppose the series $\sum c_n x^n$ has radius of convergence 2 and the series $\sum d_n x^n$ has radius of convergence 3. What is the radius of convergence of the series $\sum (c_n + d_n)x^n$?

42. Suppose that the radius of convergence of the power series $\sum c_n x^n$ is R. What is the radius of convergence of the power series $\sum c_n x^{2n}$?

11.9 Representations of Functions as Power Series

In this section we learn how to represent certain types of functions as sums of power series by manipulating geometric series or by differentiating or integrating such a series. You might wonder why we would ever want to express a known function as a sum of infinitely many terms. We will see later that this strategy is useful for integrating functions that don't have elementary antiderivatives, for solving differential equations, and for approximating functions by polynomials. (Scientists do this to simplify the expressions they deal with; computer scientists do this to represent functions on calculators and computers.)

We start with an equation that we have seen before:

$$\boxed{1} \qquad \frac{1}{1-x} = 1 + x + x^2 + x^3 + \cdots = \sum_{n=0}^{\infty} x^n \qquad |x| < 1$$

We first encountered this equation in Example 11.2.7, where we obtained it by observing that the series is a geometric series with $a = 1$ and $r = x$. But here our point of view is different. We now regard Equation 1 as expressing the function $f(x) = 1/(1 - x)$ as a sum of a power series.

A geometric illustration of Equation 1 is shown in Figure 1. Because the sum of a series is the limit of the sequence of partial sums, we have

$$\frac{1}{1-x} = \lim_{n\to\infty} s_n(x)$$

where

$$s_n(x) = 1 + x + x^2 + \cdots + x^n$$

is the nth partial sum. Notice that as n increases, $s_n(x)$ becomes a better approximation to $f(x)$ for $-1 < x < 1$.

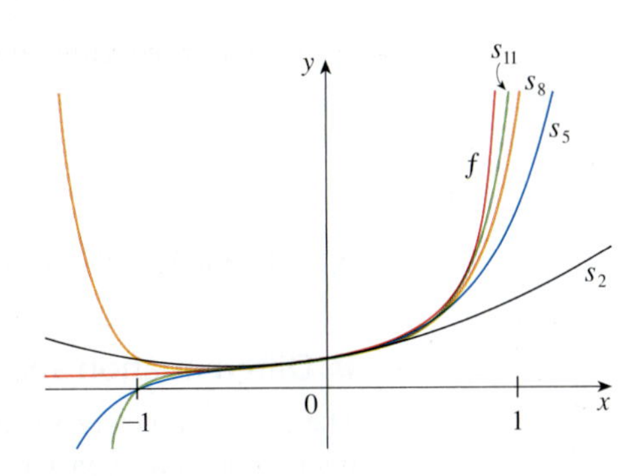

FIGURE 1 $f(x) = \dfrac{1}{1-x}$ and some partial sums

EXAMPLE 1 Express $1/(1+x^2)$ as the sum of a power series and find the interval of convergence.

SOLUTION Replacing x by $-x^2$ in Equation 1, we have

$$\frac{1}{1+x^2} = \frac{1}{1-(-x^2)} = \sum_{n=0}^{\infty} (-x^2)^n$$

$$= \sum_{n=0}^{\infty} (-1)^n x^{2n} = 1 - x^2 + x^4 - x^6 + x^8 - \cdots$$

Because this is a geometric series, it converges when $|-x^2| < 1$, that is, $x^2 < 1$, or $|x| < 1$. Therefore the interval of convergence is $(-1, 1)$. (Of course, we could have determined the radius of convergence by applying the Ratio Test, but that much work is unnecessary here.) ■

EXAMPLE 2 Find a power series representation for $1/(x+2)$.

SOLUTION In order to put this function in the form of the left side of Equation 1, we first factor a 2 from the denominator:

$$\frac{1}{2+x} = \frac{1}{2\left(1+\dfrac{x}{2}\right)} = \frac{1}{2\left[1-\left(-\dfrac{x}{2}\right)\right]}$$

$$= \frac{1}{2}\sum_{n=0}^{\infty}\left(-\frac{x}{2}\right)^n = \sum_{n=0}^{\infty}\frac{(-1)^n}{2^{n+1}}x^n$$

This series converges when $|-x/2| < 1$, that is, $|x| < 2$. So the interval of convergence is $(-2, 2)$. ■

EXAMPLE 3 Find a power series representation of $x^3/(x+2)$.

SOLUTION Since this function is just x^3 times the function in Example 2, all we have to do is to multiply that series by x^3:

It's legitimate to move x^3 across the sigma sign because it doesn't depend on n. [Use Theorem 11.2.8(i) with $c = x^3$.]

$$\frac{x^3}{x+2} = x^3 \cdot \frac{1}{x+2} = x^3 \sum_{n=0}^{\infty}\frac{(-1)^n}{2^{n+1}}x^n = \sum_{n=0}^{\infty}\frac{(-1)^n}{2^{n+1}}x^{n+3}$$

$$= \tfrac{1}{2}x^3 - \tfrac{1}{4}x^4 + \tfrac{1}{8}x^5 - \tfrac{1}{16}x^6 + \cdots$$

Another way of writing this series is as follows:

$$\frac{x^3}{x+2} = \sum_{n=3}^{\infty} \frac{(-1)^{n-1}}{2^{n-2}} x^n$$

As in Example 2, the interval of convergence is $(-2, 2)$. ■

Differentiation and Integration of Power Series

The sum of a power series is a function $f(x) = \sum_{n=0}^{\infty} c_n(x-a)^n$ whose domain is the interval of convergence of the series. We would like to be able to differentiate and integrate such functions, and the following theorem (which we won't prove) says that we can do so by differentiating or integrating each individual term in the series, just as we would for a polynomial. This is called **term-by-term differentiation and integration**.

2 Theorem If the power series $\sum c_n(x-a)^n$ has radius of convergence $R > 0$, then the function f defined by

$$f(x) = c_0 + c_1(x-a) + c_2(x-a)^2 + \cdots = \sum_{n=0}^{\infty} c_n(x-a)^n$$

is differentiable (and therefore continuous) on the interval $(a-R, a+R)$ and

(i) $f'(x) = c_1 + 2c_2(x-a) + 3c_3(x-a)^2 + \cdots = \sum_{n=1}^{\infty} nc_n(x-a)^{n-1}$

(ii) $\int f(x)\,dx = C + c_0(x-a) + c_1\frac{(x-a)^2}{2} + c_2\frac{(x-a)^3}{3} + \cdots$

$$= C + \sum_{n=0}^{\infty} c_n \frac{(x-a)^{n+1}}{n+1}$$

The radii of convergence of the power series in Equations (i) and (ii) are both R.

In part (ii), $\int c_0\,dx = c_0x + C_1$ is written as $c_0(x-a) + C$, where $C = C_1 + ac_0$, so all the terms of the series have the same form.

NOTE 1 Equations (i) and (ii) in Theorem 2 can be rewritten in the form

(iii) $\frac{d}{dx}\left[\sum_{n=0}^{\infty} c_n(x-a)^n\right] = \sum_{n=0}^{\infty} \frac{d}{dx}[c_n(x-a)^n]$

(iv) $\int\left[\sum_{n=0}^{\infty} c_n(x-a)^n\right]dx = \sum_{n=0}^{\infty} \int c_n(x-a)^n\,dx$

We know that, for finite sums, the derivative of a sum is the sum of the derivatives and the integral of a sum is the sum of the integrals. Equations (iii) and (iv) assert that the same is true for infinite sums, provided we are dealing with *power series.* (For other types of series of functions the situation is not as simple; see Exercise 38.)

NOTE 2 Although Theorem 2 says that the radius of convergence remains the same when a power series is differentiated or integrated, this does not mean that the *interval* of convergence remains the same. It may happen that the original series converges at an endpoint, whereas the differentiated series diverges there. (See Exercise 39.)

NOTE 3 The idea of differentiating a power series term by term is the basis for a powerful method for solving differential equations. We will discuss this method in Chapter 17.

EXAMPLE 4 In Example 11.8.3 we saw that the Bessel function

$$J_0(x) = \sum_{n=0}^{\infty} \frac{(-1)^n x^{2n}}{2^{2n}(n!)^2}$$

is defined for all x. Thus, by Theorem 2, J_0 is differentiable for all x and its derivative is found by term-by-term differentiation as follows:

$$J_0'(x) = \sum_{n=0}^{\infty} \frac{d}{dx} \frac{(-1)^n x^{2n}}{2^{2n}(n!)^2} = \sum_{n=1}^{\infty} \frac{(-1)^n 2n x^{2n-1}}{2^{2n}(n!)^2}$$

■

EXAMPLE 5 Express $1/(1 - x)^2$ as a power series by differentiating Equation 1. What is the radius of convergence?

SOLUTION Differentiating each side of the equation

$$\frac{1}{1 - x} = 1 + x + x^2 + x^3 + \cdots = \sum_{n=0}^{\infty} x^n$$

we get

$$\frac{1}{(1 - x)^2} = 1 + 2x + 3x^2 + \cdots = \sum_{n=1}^{\infty} n x^{n-1}$$

If we wish, we can replace n by $n + 1$ and write the answer as

$$\frac{1}{(1 - x)^2} = \sum_{n=0}^{\infty} (n + 1)x^n$$

According to Theorem 2, the radius of convergence of the differentiated series is the same as the radius of convergence of the original series, namely, $R = 1$. ■

EXAMPLE 6 Find a power series representation for $\ln(1 + x)$ and its radius of convergence.

SOLUTION We notice that the derivative of this function is $1/(1 + x)$. From Equation 1 we have

$$\frac{1}{1 + x} = \frac{1}{1 - (-x)} = 1 - x + x^2 - x^3 + \cdots \qquad |x| < 1$$

Integrating both sides of this equation, we get

$$\begin{aligned} \ln(1 + x) &= \int \frac{1}{1 + x}\,dx = \int (1 - x + x^2 - x^3 + \cdots)\,dx \\ &= x - \frac{x^2}{2} + \frac{x^3}{3} - \frac{x^4}{4} + \cdots + C \\ &= \sum_{n=1}^{\infty} (-1)^{n-1} \frac{x^n}{n} + C \qquad |x| < 1 \end{aligned}$$

To determine the value of C we put $x = 0$ in this equation and obtain $\ln(1 + 0) = C$.

Thus $C = 0$ and

$$\ln(1 + x) = x - \frac{x^2}{2} + \frac{x^3}{3} - \frac{x^4}{4} + \cdots = \sum_{n=1}^{\infty} (-1)^{n-1} \frac{x^n}{n} \qquad |x| < 1$$

The radius of convergence is the same as for the original series: $R = 1$. ■

EXAMPLE 7 Find a power series representation for $f(x) = \tan^{-1}x$.

SOLUTION We observe that $f'(x) = 1/(1 + x^2)$ and find the required series by integrating the power series for $1/(1 + x^2)$ found in Example 1.

$$\tan^{-1}x = \int \frac{1}{1 + x^2}\,dx = \int (1 - x^2 + x^4 - x^6 + \cdots)\,dx$$

$$= C + x - \frac{x^3}{3} + \frac{x^5}{5} - \frac{x^7}{7} + \cdots$$

The power series for $\tan^{-1}x$ obtained in Example 7 is called *Gregory's series* after the Scottish mathematician James Gregory (1638–1675), who had anticipated some of Newton's discoveries. We have shown that Gregory's series is valid when $-1 < x < 1$, but it turns out (although it isn't easy to prove) that it is also valid when $x = \pm 1$. Notice that when $x = 1$ the series becomes

$$\frac{\pi}{4} = 1 - \frac{1}{3} + \frac{1}{5} - \frac{1}{7} + \cdots$$

This beautiful result is known as the Leibniz formula for π.

To find C we put $x = 0$ and obtain $C = \tan^{-1}0 = 0$. Therefore

$$\tan^{-1}x = x - \frac{x^3}{3} + \frac{x^5}{5} - \frac{x^7}{7} + \cdots$$

$$= \sum_{n=0}^{\infty} (-1)^n \frac{x^{2n+1}}{2n + 1}$$

Since the radius of convergence of the series for $1/(1 + x^2)$ is 1, the radius of convergence of this series for $\tan^{-1}x$ is also 1. ■

EXAMPLE 8

(a) Evaluate $\int [1/(1 + x^7)]\,dx$ as a power series.

(b) Use part (a) to approximate $\int_0^{0.5} [1/(1 + x^7)]\,dx$ correct to within 10^{-7}.

SOLUTION

(a) The first step is to express the integrand, $1/(1 + x^7)$, as the sum of a power series. As in Example 1, we start with Equation 1 and replace x by $-x^7$:

$$\frac{1}{1 + x^7} = \frac{1}{1 - (-x^7)} = \sum_{n=0}^{\infty} (-x^7)^n$$

$$= \sum_{n=0}^{\infty} (-1)^n x^{7n} = 1 - x^7 + x^{14} - \cdots$$

This example demonstrates one way in which power series representations are useful. Integrating $1/(1 + x^7)$ by hand is incredibly difficult. Different computer algebra systems return different forms of the answer, but they are all extremely complicated. (If you have a CAS, try it yourself.) The infinite series answer that we obtain in Example 8(a) is actually much easier to deal with than the finite answer provided by a CAS.

Now we integrate term by term:

$$\int \frac{1}{1 + x^7}\,dx = \int \sum_{n=0}^{\infty} (-1)^n x^{7n}\,dx = C + \sum_{n=0}^{\infty} (-1)^n \frac{x^{7n+1}}{7n + 1}$$

$$= C + x - \frac{x^8}{8} + \frac{x^{15}}{15} - \frac{x^{22}}{22} + \cdots$$

This series converges for $|-x^7| < 1$, that is, for $|x| < 1$.

(b) In applying the Fundamental Theorem of Calculus, it doesn't matter which antiderivative we use, so let's use the antiderivative from part (a) with $C = 0$:

$$\int_0^{0.5} \frac{1}{1+x^7}\,dx = \left[x - \frac{x^8}{8} + \frac{x^{15}}{15} - \frac{x^{22}}{22} + \cdots\right]_0^{1/2}$$

$$= \frac{1}{2} - \frac{1}{8 \cdot 2^8} + \frac{1}{15 \cdot 2^{15}} - \frac{1}{22 \cdot 2^{22}} + \cdots + \frac{(-1)^n}{(7n+1)2^{7n+1}} + \cdots$$

This infinite series is the exact value of the definite integral, but since it is an alternating series, we can approximate the sum using the Alternating Series Estimation Theorem. If we stop adding after the term with $n = 3$, the error is smaller than the term with $n = 4$:

$$\frac{1}{29 \cdot 2^{29}} \approx 6.4 \times 10^{-11}$$

So we have

$$\int_0^{0.5} \frac{1}{1+x^7}\,dx \approx \frac{1}{2} - \frac{1}{8 \cdot 2^8} + \frac{1}{15 \cdot 2^{15}} - \frac{1}{22 \cdot 2^{22}} \approx 0.49951374$$ ■

11.9 EXERCISES

1. If the radius of convergence of the power series $\sum_{n=0}^{\infty} c_n x^n$ is 10, what is the radius of convergence of the series $\sum_{n=1}^{\infty} n c_n x^{n-1}$? Why?

2. Suppose you know that the series $\sum_{n=0}^{\infty} b_n x^n$ converges for $|x| < 2$. What can you say about the following series? Why?

$$\sum_{n=0}^{\infty} \frac{b_n}{n+1} x^{n+1}$$

3–10 Find a power series representation for the function and determine the interval of convergence.

3. $f(x) = \dfrac{1}{1+x}$

4. $f(x) = \dfrac{5}{1-4x^2}$

5. $f(x) = \dfrac{2}{3-x}$

6. $f(x) = \dfrac{4}{2x+3}$

7. $f(x) = \dfrac{x^2}{x^4+16}$

8. $f(x) = \dfrac{x}{2x^2+1}$

9. $f(x) = \dfrac{x-1}{x+2}$

10. $f(x) = \dfrac{x+a}{x^2+a^2}$, $a > 0$

11–12 Express the function as the sum of a power series by first using partial fractions. Find the interval of convergence.

11. $f(x) = \dfrac{2x-4}{x^2-4x+3}$

12. $f(x) = \dfrac{2x+3}{x^2+3x+2}$

13. (a) Use differentiation to find a power series representation for

$$f(x) = \frac{1}{(1+x)^2}$$

What is the radius of convergence?

(b) Use part (a) to find a power series for

$$f(x) = \frac{1}{(1+x)^3}$$

(c) Use part (b) to find a power series for

$$f(x) = \frac{x^2}{(1+x)^3}$$

14. (a) Use Equation 1 to find a power series representation for $f(x) = \ln(1-x)$. What is the radius of convergence?

(b) Use part (a) to find a power series for $f(x) = x\ln(1-x)$.

(c) By putting $x = \frac{1}{2}$ in your result from part (a), express ln 2 as the sum of an infinite series.

15–20 Find a power series representation for the function and determine the radius of convergence.

15. $f(x) = \ln(5-x)$

16. $f(x) = x^2 \tan^{-1}(x^3)$

17. $f(x) = \dfrac{x}{(1+4x)^2}$

18. $f(x) = \left(\dfrac{x}{2-x}\right)^3$

19. $f(x) = \dfrac{1+x}{(1-x)^2}$

20. $f(x) = \dfrac{x^2+x}{(1-x)^3}$

21–24 Find a power series representation for f, and graph f and several partial sums $s_n(x)$ on the same screen. What happens as n increases?

21. $f(x) = \dfrac{x^2}{x^2 + 1}$

22. $f(x) = \ln(1 + x^4)$

23. $f(x) = \ln\left(\dfrac{1 + x}{1 - x}\right)$

24. $f(x) = \tan^{-1}(2x)$

25–28 Evaluate the indefinite integral as a power series. What is the radius of convergence?

25. $\displaystyle\int \frac{t}{1 - t^8}\,dt$

26. $\displaystyle\int \frac{t}{1 + t^3}\,dt$

27. $\displaystyle\int x^2 \ln(1 + x)\,dx$

28. $\displaystyle\int \frac{\tan^{-1}x}{x}\,dx$

29–32 Use a power series to approximate the definite integral to six decimal places.

29. $\displaystyle\int_0^{0.3} \frac{x}{1 + x^3}\,dx$

30. $\displaystyle\int_0^{1/2} \arctan(x/2)\,dx$

31. $\displaystyle\int_0^{0.2} x\ln(1 + x^2)\,dx$

32. $\displaystyle\int_0^{0.3} \frac{x^2}{1 + x^4}\,dx$

33. Use the result of Example 7 to compute arctan 0.2 correct to five decimal places.

34. Show that the function

$$f(x) = \sum_{n=0}^{\infty} \frac{(-1)^n x^{2n}}{(2n)!}$$

is a solution of the differential equation

$$f''(x) + f(x) = 0$$

35. (a) Show that J_0 (the Bessel function of order 0 given in Example 4) satisfies the differential equation

$$x^2 J_0''(x) + x J_0'(x) + x^2 J_0(x) = 0$$

(b) Evaluate $\int_0^1 J_0(x)\,dx$ correct to three decimal places.

36. The Bessel function of order 1 is defined by

$$J_1(x) = \sum_{n=0}^{\infty} \frac{(-1)^n x^{2n+1}}{n!\,(n+1)!\,2^{2n+1}}$$

(a) Show that J_1 satisfies the differential equation

$$x^2 J_1''(x) + x J_1'(x) + (x^2 - 1)J_1(x) = 0$$

(b) Show that $J_0'(x) = -J_1(x)$.

37. (a) Show that the function

$$f(x) = \sum_{n=0}^{\infty} \frac{x^n}{n!}$$

is a solution of the differential equation

$$f'(x) = f(x)$$

(b) Show that $f(x) = e^x$.

38. Let $f_n(x) = (\sin nx)/n^2$. Show that the series $\Sigma f_n(x)$ converges for all values of x but the series of derivatives $\Sigma f_n'(x)$ diverges when $x = 2n\pi$, n an integer. For what values of x does the series $\Sigma f_n''(x)$ converge?

39. Let

$$f(x) = \sum_{n=1}^{\infty} \frac{x^n}{n^2}$$

Find the intervals of convergence for f, f', and f''.

40. (a) Starting with the geometric series $\Sigma_{n=0}^{\infty} x^n$, find the sum of the series

$$\sum_{n=1}^{\infty} nx^{n-1} \qquad |x| < 1$$

(b) Find the sum of each of the following series.

(i) $\displaystyle\sum_{n=1}^{\infty} nx^n, \quad |x| < 1$ (ii) $\displaystyle\sum_{n=1}^{\infty} \frac{n}{2^n}$

(c) Find the sum of each of the following series.

(i) $\displaystyle\sum_{n=2}^{\infty} n(n-1)x^n, \quad |x| < 1$

(ii) $\displaystyle\sum_{n=2}^{\infty} \frac{n^2 - n}{2^n}$ (iii) $\displaystyle\sum_{n=1}^{\infty} \frac{n^2}{2^n}$

41. Use the power series for $\tan^{-1}x$ to prove the following expression for π as the sum of an infinite series:

$$\pi = 2\sqrt{3} \sum_{n=0}^{\infty} \frac{(-1)^n}{(2n + 1)3^n}$$

42. (a) By completing the square, show that

$$\int_0^{1/2} \frac{dx}{x^2 - x + 1} = \frac{\pi}{3\sqrt{3}}$$

(b) By factoring $x^3 + 1$ as a sum of cubes, rewrite the integral in part (a). Then express $1/(x^3 + 1)$ as the sum of a power series and use it to prove the following formula for π:

$$\pi = \frac{3\sqrt{3}}{4} \sum_{n=0}^{\infty} \frac{(-1)^n}{8^n} \left(\frac{2}{3n + 1} + \frac{1}{3n + 2} \right)$$

11.10 Taylor and Maclaurin Series

In the preceding section we were able to find power series representations for a certain restricted class of functions. Here we investigate more general problems: Which functions have power series representations? How can we find such representations?

We start by supposing that f is any function that can be represented by a power series

$$\boxed{1} \quad f(x) = c_0 + c_1(x - a) + c_2(x - a)^2 + c_3(x - a)^3 + c_4(x - a)^4 + \cdots \qquad |x - a| < R$$

Let's try to determine what the coefficients c_n must be in terms of f. To begin, notice that if we put $x = a$ in Equation 1, then all terms after the first one are 0 and we get

$$f(a) = c_0$$

By Theorem 11.9.2, we can differentiate the series in Equation 1 term by term:

$$\boxed{2} \quad f'(x) = c_1 + 2c_2(x - a) + 3c_3(x - a)^2 + 4c_4(x - a)^3 + \cdots \qquad |x - a| < R$$

and substitution of $x = a$ in Equation 2 gives

$$f'(a) = c_1$$

Now we differentiate both sides of Equation 2 and obtain

$$\boxed{3} \quad f''(x) = 2c_2 + 2 \cdot 3c_3(x - a) + 3 \cdot 4c_4(x - a)^2 + \cdots \qquad |x - a| < R$$

Again we put $x = a$ in Equation 3. The result is

$$f''(a) = 2c_2$$

Let's apply the procedure one more time. Differentiation of the series in Equation 3 gives

$$\boxed{4} \quad f'''(x) = 2 \cdot 3c_3 + 2 \cdot 3 \cdot 4c_4(x - a) + 3 \cdot 4 \cdot 5c_5(x - a)^2 + \cdots \qquad |x - a| < R$$

and substitution of $x = a$ in Equation 4 gives

$$f'''(a) = 2 \cdot 3c_3 = 3!c_3$$

By now you can see the pattern. If we continue to differentiate and substitute $x = a$, we obtain

$$f^{(n)}(a) = 2 \cdot 3 \cdot 4 \cdot \cdots \cdot nc_n = n!c_n$$

Solving this equation for the nth coefficient c_n, we get

$$c_n = \frac{f^{(n)}(a)}{n!}$$

This formula remains valid even for $n = 0$ if we adopt the conventions that $0! = 1$ and $f^{(0)} = f$. Thus we have proved the following theorem.

5 Theorem If f has a power series representation (expansion) at a, that is, if

$$f(x) = \sum_{n=0}^{\infty} c_n(x-a)^n \qquad |x-a| < R$$

then its coefficients are given by the formula

$$c_n = \frac{f^{(n)}(a)}{n!}$$

Substituting this formula for c_n back into the series, we see that *if* f has a power series expansion at a, then it must be of the following form.

$$\boxed{6} \quad f(x) = \sum_{n=0}^{\infty} \frac{f^{(n)}(a)}{n!}(x-a)^n$$

$$= f(a) + \frac{f'(a)}{1!}(x-a) + \frac{f''(a)}{2!}(x-a)^2 + \frac{f'''(a)}{3!}(x-a)^3 + \cdots$$

Taylor and Maclaurin

The Taylor series is named after the English mathematician Brook Taylor (1685–1731) and the Maclaurin series is named in honor of the Scottish mathematician Colin Maclaurin (1698–1746) despite the fact that the Maclaurin series is really just a special case of the Taylor series. But the idea of representing particular functions as sums of power series goes back to Newton, and the general Taylor series was known to the Scottish mathematician James Gregory in 1668 and to the Swiss mathematician John Bernoulli in the 1690s. Taylor was apparently unaware of the work of Gregory and Bernoulli when he published his discoveries on series in 1715 in his book *Methodus incrementorum directa et inversa*. Maclaurin series are named after Colin Maclaurin because he popularized them in his calculus textbook *Treatise of Fluxions* published in 1742.

The series in Equation 6 is called the **Taylor series of the function *f* at *a*** (or **about *a*** or **centered at *a***). For the special case $a = 0$ the Taylor series becomes

$$\boxed{7} \quad f(x) = \sum_{n=0}^{\infty} \frac{f^{(n)}(0)}{n!}x^n = f(0) + \frac{f'(0)}{1!}x + \frac{f''(0)}{2!}x^2 + \cdots$$

This case arises frequently enough that it is given the special name **Maclaurin series**.

NOTE We have shown that *if* f can be represented as a power series about a, then f is equal to the sum of its Taylor series. But there exist functions that are not equal to the sum of their Taylor series. An example of such a function is given in Exercise 84.

EXAMPLE 1 Find the Maclaurin series of the function $f(x) = e^x$ and its radius of convergence.

SOLUTION If $f(x) = e^x$, then $f^{(n)}(x) = e^x$, so $f^{(n)}(0) = e^0 = 1$ for all n. Therefore the Taylor series for f at 0 (that is, the Maclaurin series) is

$$\sum_{n=0}^{\infty} \frac{f^{(n)}(0)}{n!}x^n = \sum_{n=0}^{\infty} \frac{x^n}{n!} = 1 + \frac{x}{1!} + \frac{x^2}{2!} + \frac{x^3}{3!} + \cdots$$

To find the radius of convergence we let $a_n = x^n/n!$. Then

$$\left|\frac{a_{n+1}}{a_n}\right| = \left|\frac{x^{n+1}}{(n+1)!} \cdot \frac{n!}{x^n}\right| = \frac{|x|}{n+1} \to 0 < 1$$

so, by the Ratio Test, the series converges for all x and the radius of convergence is $R = \infty$. ■

The conclusion we can draw from Theorem 5 and Example 1 is that *if* e^x has a power series expansion at 0, then

$$e^x = \sum_{n=0}^{\infty} \frac{x^n}{n!}$$

So how can we determine whether e^x *does* have a power series representation?

Let's investigate the more general question: under what circumstances is a function equal to the sum of its Taylor series? In other words, if f has derivatives of all orders, when is it true that

$$f(x) = \sum_{n=0}^{\infty} \frac{f^{(n)}(a)}{n!}(x-a)^n$$

As with any convergent series, this means that $f(x)$ is the limit of the sequence of partial sums. In the case of the Taylor series, the partial sums are

$$T_n(x) = \sum_{i=0}^{n} \frac{f^{(i)}(a)}{i!}(x-a)^i$$

$$= f(a) + \frac{f'(a)}{1!}(x-a) + \frac{f''(a)}{2!}(x-a)^2 + \cdots + \frac{f^{(n)}(a)}{n!}(x-a)^n$$

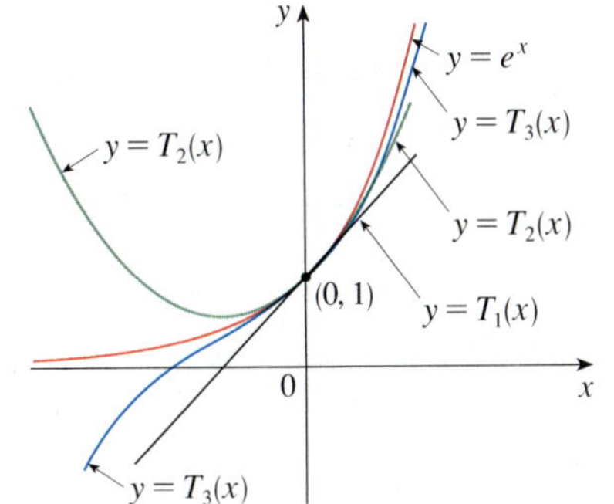

FIGURE 1

As n increases, $T_n(x)$ appears to approach e^x in Figure 1. This suggests that e^x is equal to the sum of its Taylor series.

Notice that T_n is a polynomial of degree n called the ***n*****th-degree Taylor polynomial of** ***f*** **at** ***a***. For instance, for the exponential function $f(x) = e^x$, the result of Example 1 shows that the Taylor polynomials at 0 (or Maclaurin polynomials) with $n = 1, 2$, and 3 are

$$T_1(x) = 1 + x \qquad T_2(x) = 1 + x + \frac{x^2}{2!} \qquad T_3(x) = 1 + x + \frac{x^2}{2!} + \frac{x^3}{3!}$$

The graphs of the exponential function and these three Taylor polynomials are drawn in Figure 1.

In general, $f(x)$ is the sum of its Taylor series if

$$f(x) = \lim_{n \to \infty} T_n(x)$$

If we let

$$R_n(x) = f(x) - T_n(x) \qquad \text{so that} \qquad f(x) = T_n(x) + R_n(x)$$

then $R_n(x)$ is called the **remainder** of the Taylor series. If we can somehow show that $\lim_{n \to \infty} R_n(x) = 0$, then it follows that

$$\lim_{n \to \infty} T_n(x) = \lim_{n \to \infty} [f(x) - R_n(x)] = f(x) - \lim_{n \to \infty} R_n(x) = f(x)$$

We have therefore proved the following theorem.

8 Theorem If $f(x) = T_n(x) + R_n(x)$, where T_n is the nth-degree Taylor polynomial of f at a and

$$\lim_{n \to \infty} R_n(x) = 0$$

for $|x - a| < R$, then f is equal to the sum of its Taylor series on the interval $|x - a| < R$.

In trying to show that $\lim_{n\to\infty} R_n(x) = 0$ for a specific function f, we usually use the following theorem.

9 **Taylor's Inequality** If $|f^{(n+1)}(x)| \leqslant M$ for $|x - a| \leqslant d$, then the remainder $R_n(x)$ of the Taylor series satisfies the inequality

$$|R_n(x)| \leqslant \frac{M}{(n+1)!}|x - a|^{n+1} \qquad \text{for } |x - a| \leqslant d$$

To see why this is true for $n = 1$, we assume that $|f''(x)| \leqslant M$. In particular, we have $f''(x) \leqslant M$, so for $a \leqslant x \leqslant a + d$ we have

$$\int_a^x f''(t)\,dt \leqslant \int_a^x M\,dt$$

An antiderivative of f'' is f', so by Part 2 of the Fundamental Theorem of Calculus, we have

$$f'(x) - f'(a) \leqslant M(x - a) \qquad \text{or} \qquad f'(x) \leqslant f'(a) + M(x - a)$$

Thus

$$\int_a^x f'(t)\,dt \leqslant \int_a^x [f'(a) + M(t - a)]\,dt$$

$$f(x) - f(a) \leqslant f'(a)(x - a) + M\frac{(x - a)^2}{2}$$

$$f(x) - f(a) - f'(a)(x - a) \leqslant \frac{M}{2}(x - a)^2$$

But $R_1(x) = f(x) - T_1(x) = f(x) - f(a) - f'(a)(x - a)$. So

$$R_1(x) \leqslant \frac{M}{2}(x - a)^2$$

A similar argument, using $f''(x) \geqslant -M$, shows that

$$R_1(x) \geqslant -\frac{M}{2}(x - a)^2$$

So

$$|R_1(x)| \leqslant \frac{M}{2}|x - a|^2$$

Although we have assumed that $x > a$, similar calculations show that this inequality is also true for $x < a$.

This proves Taylor's Inequality for the case where $n = 1$. The result for any n is proved in a similar way by integrating $n + 1$ times. (See Exercise 83 for the case $n = 2$.)

NOTE In Section 11.11 we will explore the use of Taylor's Inequality in approximating functions. Our immediate use of it is in conjunction with Theorem 8.

In applying Theorems 8 and 9 it is often helpful to make use of the following fact.

10
$$\lim_{n\to\infty} \frac{x^n}{n!} = 0 \qquad \text{for every real number } x$$

Formulas for the Taylor Remainder Term

As alternatives to Taylor's Inequality, we have the following formulas for the remainder term. If $f^{(n+1)}$ is continuous on an interval I and $x \in I$, then

$$R_n(x) = \frac{1}{n!}\int_a^x (x - t)^n f^{(n+1)}(t)\,dt$$

This is called the *integral form of the remainder term*. Another formula, called *Lagrange's form of the remainder term,* states that there is a number z between x and a such that

$$R_n(x) = \frac{f^{(n+1)}(z)}{(n+1)!}(x - a)^{n+1}$$

This version is an extension of the Mean Value Theorem (which is the case $n = 0$).

Proofs of these formulas, together with discussions of how to use them to solve the examples of Sections 11.10 and 11.11, are given on the website

www.stewartcalculus.com

Click on *Additional Topics* and then on *Formulas for the Remainder Term in Taylor series.*

This is true because we know from Example 1 that the series $\Sigma\, x^n/n!$ converges for all x and so its nth term approaches 0.

EXAMPLE 2 Prove that e^x is equal to the sum of its Maclaurin series.

SOLUTION If $f(x) = e^x$, then $f^{(n+1)}(x) = e^x$ for all n. If d is any positive number and $|x| \leqslant d$, then $|f^{(n+1)}(x)| = e^x \leqslant e^d$. So Taylor's Inequality, with $a = 0$ and $M = e^d$, says that

$$|R_n(x)| \leqslant \frac{e^d}{(n+1)!}|x|^{n+1} \qquad \text{for } |x| \leqslant d$$

Notice that the same constant $M = e^d$ works for every value of n. But, from Equation 10, we have

$$\lim_{n\to\infty} \frac{e^d}{(n+1)!}|x|^{n+1} = e^d \lim_{n\to\infty} \frac{|x|^{n+1}}{(n+1)!} = 0$$

It follows from the Squeeze Theorem that $\lim_{n\to\infty} |R_n(x)| = 0$ and therefore $\lim_{n\to\infty} R_n(x) = 0$ for all values of x. By Theorem 8, e^x is equal to the sum of its Maclaurin series, that is,

11
$$e^x = \sum_{n=0}^{\infty} \frac{x^n}{n!} \qquad \text{for all } x$$

■

In particular, if we put $x = 1$ in Equation 11, we obtain the following expression for the number e as a sum of an infinite series:

In 1748 Leonhard Euler used Equation 12 to find the value of e correct to 23 digits. In 2010 Shigeru Kondo, again using the series in (12), computed e to more than one trillion decimal places. The special techniques employed to speed up the computation are explained on the website

numbers.computation.free.fr

12
$$e = \sum_{n=0}^{\infty} \frac{1}{n!} = 1 + \frac{1}{1!} + \frac{1}{2!} + \frac{1}{3!} + \cdots$$

EXAMPLE 3 Find the Taylor series for $f(x) = e^x$ at $a = 2$.

SOLUTION We have $f^{(n)}(2) = e^2$ and so, putting $a = 2$ in the definition of a Taylor series (6), we get

$$\sum_{n=0}^{\infty} \frac{f^{(n)}(2)}{n!}(x-2)^n = \sum_{n=0}^{\infty} \frac{e^2}{n!}(x-2)^n$$

Again it can be verified, as in Example 1, that the radius of convergence is $R = \infty$. As in Example 2 we can verify that $\lim_{n\to\infty} R_n(x) = 0$, so

13
$$e^x = \sum_{n=0}^{\infty} \frac{e^2}{n!}(x-2)^n \qquad \text{for all } x$$

■

We have two power series expansions for e^x, the Maclaurin series in Equation 11 and the Taylor series in Equation 13. The first is better if we are interested in values of x near 0 and the second is better if x is near 2.

EXAMPLE 4 Find the Maclaurin series for $\sin x$ and prove that it represents $\sin x$ for all x.

SOLUTION We arrange our computation in two columns as follows:

$$\begin{aligned} f(x) &= \sin x & f(0) &= 0 \\ f'(x) &= \cos x & f'(0) &= 1 \\ f''(x) &= -\sin x & f''(0) &= 0 \\ f'''(x) &= -\cos x & f'''(0) &= -1 \\ f^{(4)}(x) &= \sin x & f^{(4)}(0) &= 0 \end{aligned}$$

Figure 2 shows the graph of $\sin x$ together with its Taylor (or Maclaurin) polynomials

$$T_1(x) = x$$

$$T_3(x) = x - \frac{x^3}{3!}$$

$$T_5(x) = x - \frac{x^3}{3!} + \frac{x^5}{5!}$$

Notice that, as n increases, $T_n(x)$ becomes a better approximation to $\sin x$.

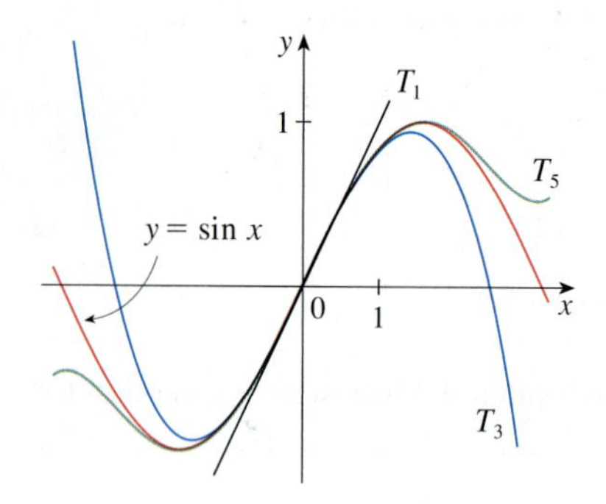

FIGURE 2

Since the derivatives repeat in a cycle of four, we can write the Maclaurin series as follows:

$$f(0) + \frac{f'(0)}{1!}x + \frac{f''(0)}{2!}x^2 + \frac{f'''(0)}{3!}x^3 + \cdots$$

$$= x - \frac{x^3}{3!} + \frac{x^5}{5!} - \frac{x^7}{7!} + \cdots = \sum_{n=0}^{\infty} (-1)^n \frac{x^{2n+1}}{(2n+1)!}$$

Since $f^{(n+1)}(x)$ is $\pm\sin x$ or $\pm\cos x$, we know that $|f^{(n+1)}(x)| \leqslant 1$ for all x. So we can take $M = 1$ in Taylor's Inequality:

14
$$|R_n(x)| \leqslant \frac{M}{(n+1)!}|x^{n+1}| = \frac{|x|^{n+1}}{(n+1)!}$$

By Equation 10 the right side of this inequality approaches 0 as $n \to \infty$, so $|R_n(x)| \to 0$ by the Squeeze Theorem. It follows that $R_n(x) \to 0$ as $n \to \infty$, so $\sin x$ is equal to the sum of its Maclaurin series by Theorem 8. ■

We state the result of Example 4 for future reference.

15
$$\sin x = x - \frac{x^3}{3!} + \frac{x^5}{5!} - \frac{x^7}{7!} + \cdots$$

$$= \sum_{n=0}^{\infty} (-1)^n \frac{x^{2n+1}}{(2n+1)!} \qquad \text{for all } x$$

EXAMPLE 5 Find the Maclaurin series for $\cos x$.

SOLUTION We could proceed directly as in Example 4, but it's easier to differentiate the Maclaurin series for $\sin x$ given by Equation 15:

$$\cos x = \frac{d}{dx}(\sin x) = \frac{d}{dx}\left(x - \frac{x^3}{3!} + \frac{x^5}{5!} - \frac{x^7}{7!} + \cdots\right)$$

$$= 1 - \frac{3x^2}{3!} + \frac{5x^4}{5!} - \frac{7x^6}{7!} + \cdots = 1 - \frac{x^2}{2!} + \frac{x^4}{4!} - \frac{x^6}{6!} + \cdots$$

The Maclaurin series for e^x, $\sin x$, and $\cos x$ that we found in Examples 2, 4, and 5 were discovered, using different methods, by Newton. These equations are remarkable because they say we know everything about each of these functions if we know all its derivatives at the single number 0.

Since the Maclaurin series for $\sin x$ converges for all x, Theorem 11.9.2 tells us that the differentiated series for $\cos x$ also converges for all x. Thus

16
$$\cos x = 1 - \frac{x^2}{2!} + \frac{x^4}{4!} - \frac{x^6}{6!} + \cdots$$
$$= \sum_{n=0}^{\infty} (-1)^n \frac{x^{2n}}{(2n)!} \qquad \text{for all } x$$

EXAMPLE 6 Find the Maclaurin series for the function $f(x) = x \cos x$.

SOLUTION Instead of computing derivatives and substituting in Equation 7, it's easier to multiply the series for $\cos x$ (Equation 16) by x:

$$x \cos x = x \sum_{n=0}^{\infty} (-1)^n \frac{x^{2n}}{(2n)!} = \sum_{n=0}^{\infty} (-1)^n \frac{x^{2n+1}}{(2n)!}$$

EXAMPLE 7 Represent $f(x) = \sin x$ as the sum of its Taylor series centered at $\pi/3$.

SOLUTION Arranging our work in columns, we have

$$f(x) = \sin x \qquad f\left(\frac{\pi}{3}\right) = \frac{\sqrt{3}}{2}$$

$$f'(x) = \cos x \qquad f'\left(\frac{\pi}{3}\right) = \frac{1}{2}$$

$$f''(x) = -\sin x \qquad f''\left(\frac{\pi}{3}\right) = -\frac{\sqrt{3}}{2}$$

$$f'''(x) = -\cos x \qquad f'''\left(\frac{\pi}{3}\right) = -\frac{1}{2}$$

We have obtained two different series representations for $\sin x$, the Maclaurin series in Example 4 and the Taylor series in Example 7. It is best to use the Maclaurin series for values of x near 0 and the Taylor series for x near $\pi/3$. Notice that the third Taylor polynomial T_3 in Figure 3 is a good approximation to $\sin x$ near $\pi/3$ but not as good near 0. Compare it with the third Maclaurin polynomial T_3 in Figure 2, where the opposite is true.

and this pattern repeats indefinitely. Therefore the Taylor series at $\pi/3$ is

$$f\left(\frac{\pi}{3}\right) + \frac{f'\left(\frac{\pi}{3}\right)}{1!}\left(x - \frac{\pi}{3}\right) + \frac{f''\left(\frac{\pi}{3}\right)}{2!}\left(x - \frac{\pi}{3}\right)^2 + \frac{f'''\left(\frac{\pi}{3}\right)}{3!}\left(x - \frac{\pi}{3}\right)^3 + \cdots$$

$$= \frac{\sqrt{3}}{2} + \frac{1}{2 \cdot 1!}\left(x - \frac{\pi}{3}\right) - \frac{\sqrt{3}}{2 \cdot 2!}\left(x - \frac{\pi}{3}\right)^2 - \frac{1}{2 \cdot 3!}\left(x - \frac{\pi}{3}\right)^3 + \cdots$$

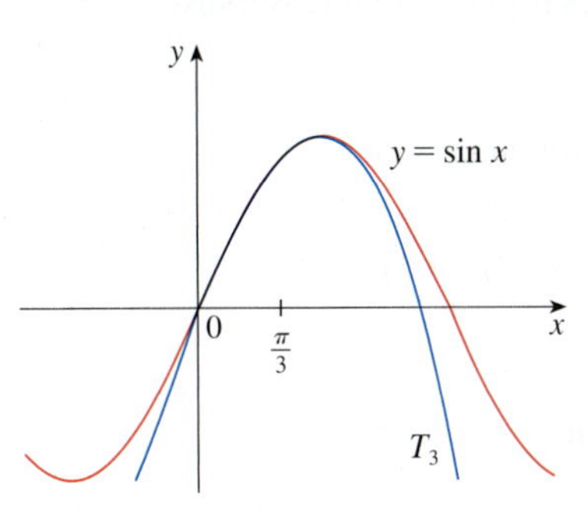

FIGURE 3

The proof that this series represents $\sin x$ for all x is very similar to that in Example 4. [Just replace x by $x - \pi/3$ in (14).] We can write the series in sigma notation if we separate the terms that contain $\sqrt{3}$:

$$\sin x = \sum_{n=0}^{\infty} \frac{(-1)^n \sqrt{3}}{2(2n)!}\left(x - \frac{\pi}{3}\right)^{2n} + \sum_{n=0}^{\infty} \frac{(-1)^n}{2(2n+1)!}\left(x - \frac{\pi}{3}\right)^{2n+1}$$

The power series that we obtained by indirect methods in Examples 5 and 6 and in Section 11.9 are indeed the Taylor or Maclaurin series of the given functions because Theorem 5 asserts that, no matter how a power series representation $f(x) = \sum c_n(x - a)^n$ is obtained, it is always true that $c_n = f^{(n)}(a)/n!$. In other words, the coefficients are uniquely determined.

EXAMPLE 8 Find the Maclaurin series for $f(x) = (1 + x)^k$, where k is any real number.

SOLUTION Arranging our work in columns, we have

$$\begin{aligned}
f(x) &= (1 + x)^k & f(0) &= 1 \\
f'(x) &= k(1 + x)^{k-1} & f'(0) &= k \\
f''(x) &= k(k - 1)(1 + x)^{k-2} & f''(0) &= k(k - 1) \\
f'''(x) &= k(k - 1)(k - 2)(1 + x)^{k-3} & f'''(0) &= k(k - 1)(k - 2) \\
&\vdots & &\vdots \\
f^{(n)}(x) &= k(k - 1) \cdots (k - n + 1)(1 + x)^{k-n} & f^{(n)}(0) &= k(k - 1) \cdots (k - n + 1)
\end{aligned}$$

Therefore the Maclaurin series of $f(x) = (1 + x)^k$ is

$$\sum_{n=0}^{\infty} \frac{f^{(n)}(0)}{n!} x^n = \sum_{n=0}^{\infty} \frac{k(k - 1) \cdots (k - n + 1)}{n!} x^n$$

This series is called the **binomial series**. Notice that if k is a nonnegative integer, then the terms are eventually 0 and so the series is finite. For other values of k none of the terms is 0 and so we can try the Ratio Test. If the nth term is a_n, then

$$\begin{aligned}
\left| \frac{a_{n+1}}{a_n} \right| &= \left| \frac{k(k - 1) \cdots (k - n + 1)(k - n)x^{n+1}}{(n + 1)!} \cdot \frac{n!}{k(k - 1) \cdots (k - n + 1)x^n} \right| \\
&= \frac{|k - n|}{n + 1} |x| = \frac{\left| 1 - \dfrac{k}{n} \right|}{1 + \dfrac{1}{n}} |x| \to |x| \quad \text{as } n \to \infty
\end{aligned}$$

Thus, by the Ratio Test, the binomial series converges if $|x| < 1$ and diverges if $|x| > 1$. ■

The traditional notation for the coefficients in the binomial series is

$$\binom{k}{n} = \frac{k(k - 1)(k - 2) \cdots (k - n + 1)}{n!}$$

and these numbers are called the **binomial coefficients**.

The following theorem states that $(1 + x)^k$ is equal to the sum of its Maclaurin series. It is possible to prove this by showing that the remainder term $R_n(x)$ approaches 0, but that turns out to be quite difficult. The proof outlined in Exercise 85 is much easier.

17 The Binomial Series If k is any real number and $|x| < 1$, then

$$(1 + x)^k = \sum_{n=0}^{\infty} \binom{k}{n} x^n = 1 + kx + \frac{k(k-1)}{2!}x^2 + \frac{k(k-1)(k-2)}{3!}x^3 + \cdots$$

Although the binomial series always converges when $|x| < 1$, the question of whether or not it converges at the endpoints, ± 1, depends on the value of k. It turns out that the series converges at 1 if $-1 < k \leqslant 0$ and at both endpoints if $k \geqslant 0$. Notice that if k is a positive integer and $n > k$, then the expression for $\binom{k}{n}$ contains a factor $(k - k)$, so $\binom{k}{n} = 0$ for $n > k$. This means that the series terminates and reduces to the ordinary Binomial Theorem when k is a positive integer. (See Reference Page 1.)

EXAMPLE 9 Find the Maclaurin series for the function $f(x) = \dfrac{1}{\sqrt{4 - x}}$ and its radius of convergence.

SOLUTION We rewrite $f(x)$ in a form where we can use the binomial series:

$$\frac{1}{\sqrt{4 - x}} = \frac{1}{\sqrt{4\left(1 - \dfrac{x}{4}\right)}} = \frac{1}{2\sqrt{1 - \dfrac{x}{4}}} = \frac{1}{2}\left(1 - \frac{x}{4}\right)^{-1/2}$$

Using the binomial series with $k = -\frac{1}{2}$ and with x replaced by $-x/4$, we have

$$\frac{1}{\sqrt{4 - x}} = \frac{1}{2}\left(1 - \frac{x}{4}\right)^{-1/2} = \frac{1}{2}\sum_{n=0}^{\infty}\binom{-\frac{1}{2}}{n}\left(-\frac{x}{4}\right)^n$$

$$= \frac{1}{2}\left[1 + \left(-\frac{1}{2}\right)\left(-\frac{x}{4}\right) + \frac{(-\frac{1}{2})(-\frac{3}{2})}{2!}\left(-\frac{x}{4}\right)^2 + \frac{(-\frac{1}{2})(-\frac{3}{2})(-\frac{5}{2})}{3!}\left(-\frac{x}{4}\right)^3\right.$$

$$\left. + \cdots + \frac{(-\frac{1}{2})(-\frac{3}{2})(-\frac{5}{2}) \cdots (-\frac{1}{2} - n + 1)}{n!}\left(-\frac{x}{4}\right)^n + \cdots\right]$$

$$= \frac{1}{2}\left[1 + \frac{1}{8}x + \frac{1 \cdot 3}{2!8^2}x^2 + \frac{1 \cdot 3 \cdot 5}{3!8^3}x^3 + \cdots + \frac{1 \cdot 3 \cdot 5 \cdot \cdots \cdot (2n-1)}{n!8^n}x^n + \cdots\right]$$

We know from (17) that this series converges when $|-x/4| < 1$, that is, $|x| < 4$, so the radius of convergence is $R = 4$. ■

We collect in the following table, for future reference, some important Maclaurin series that we have derived in this section and the preceding one.

Table 1

Important Maclaurin Series and Their Radii of Convergence

Series	Radius
$\dfrac{1}{1-x} = \displaystyle\sum_{n=0}^{\infty} x^n = 1 + x + x^2 + x^3 + \cdots$	$R = 1$
$e^x = \displaystyle\sum_{n=0}^{\infty} \frac{x^n}{n!} = 1 + \frac{x}{1!} + \frac{x^2}{2!} + \frac{x^3}{3!} + \cdots$	$R = \infty$
$\sin x = \displaystyle\sum_{n=0}^{\infty} (-1)^n \frac{x^{2n+1}}{(2n+1)!} = x - \frac{x^3}{3!} + \frac{x^5}{5!} - \frac{x^7}{7!} + \cdots$	$R = \infty$
$\cos x = \displaystyle\sum_{n=0}^{\infty} (-1)^n \frac{x^{2n}}{(2n)!} = 1 - \frac{x^2}{2!} + \frac{x^4}{4!} - \frac{x^6}{6!} + \cdots$	$R = \infty$
$\tan^{-1} x = \displaystyle\sum_{n=0}^{\infty} (-1)^n \frac{x^{2n+1}}{2n+1} = x - \frac{x^3}{3} + \frac{x^5}{5} - \frac{x^7}{7} + \cdots$	$R = 1$
$\ln(1+x) = \displaystyle\sum_{n=1}^{\infty} (-1)^{n-1} \frac{x^n}{n} = x - \frac{x^2}{2} + \frac{x^3}{3} - \frac{x^4}{4} + \cdots$	$R = 1$
$(1+x)^k = \displaystyle\sum_{n=0}^{\infty} \binom{k}{n} x^n = 1 + kx + \frac{k(k-1)}{2!}x^2 + \frac{k(k-1)(k-2)}{3!}x^3 + \cdots$	$R = 1$

EXAMPLE 10 Find the sum of the series $\dfrac{1}{1 \cdot 2} - \dfrac{1}{2 \cdot 2^2} + \dfrac{1}{3 \cdot 2^3} - \dfrac{1}{4 \cdot 2^4} + \cdots$.

SOLUTION With sigma notation we can write the given series as

$$\sum_{n=1}^{\infty} (-1)^{n-1} \frac{1}{n \cdot 2^n} = \sum_{n=1}^{\infty} (-1)^{n-1} \frac{\left(\frac{1}{2}\right)^n}{n}$$

Then from Table 1 we see that this series matches the entry for $\ln(1 + x)$ with $x = \frac{1}{2}$. So

$$\sum_{n=1}^{\infty} (-1)^{n-1} \frac{1}{n \cdot 2^n} = \ln\left(1 + \tfrac{1}{2}\right) = \ln \tfrac{3}{2}$$ ■

TEC Module 11.10/11.11 enables you to see how successive Taylor polynomials approach the original function.

One reason that Taylor series are important is that they enable us to integrate functions that we couldn't previously handle. In fact, in the introduction to this chapter we mentioned that Newton often integrated functions by first expressing them as power series and then integrating the series term by term. The function $f(x) = e^{-x^2}$ can't be integrated by techniques discussed so far because its antiderivative is not an elementary function (see Section 7.5). In the following example we use Newton's idea to integrate this function.

EXAMPLE 11

(a) Evaluate $\int e^{-x^2}\,dx$ as an infinite series.

(b) Evaluate $\int_0^1 e^{-x^2}\,dx$ correct to within an error of 0.001.

SOLUTION

(a) First we find the Maclaurin series for $f(x) = e^{-x^2}$. Although it's possible to use the direct method, let's find it simply by replacing x with $-x^2$ in the series for e^x given in

Table 1. Thus, for all values of x,

$$e^{-x^2} = \sum_{n=0}^{\infty} \frac{(-x^2)^n}{n!} = \sum_{n=0}^{\infty} (-1)^n \frac{x^{2n}}{n!} = 1 - \frac{x^2}{1!} + \frac{x^4}{2!} - \frac{x^6}{3!} + \cdots$$

Now we integrate term by term:

$$\int e^{-x^2}\,dx = \int \left(1 - \frac{x^2}{1!} + \frac{x^4}{2!} - \frac{x^6}{3!} + \cdots + (-1)^n \frac{x^{2n}}{n!} + \cdots\right) dx$$

$$= C + x - \frac{x^3}{3 \cdot 1!} + \frac{x^5}{5 \cdot 2!} - \frac{x^7}{7 \cdot 3!} + \cdots + (-1)^n \frac{x^{2n+1}}{(2n+1)n!} + \cdots$$

This series converges for all x because the original series for e^{-x^2} converges for all x.

(b) The Fundamental Theorem of Calculus gives

$$\int_0^1 e^{-x^2}\,dx = \left[x - \frac{x^3}{3 \cdot 1!} + \frac{x^5}{5 \cdot 2!} - \frac{x^7}{7 \cdot 3!} + \frac{x^9}{9 \cdot 4!} - \cdots\right]_0^1$$

We can take $C = 0$ in the antiderivative in part (a).

$$= 1 - \tfrac{1}{3} + \tfrac{1}{10} - \tfrac{1}{42} + \tfrac{1}{216} - \cdots$$

$$\approx 1 - \tfrac{1}{3} + \tfrac{1}{10} - \tfrac{1}{42} + \tfrac{1}{216} \approx 0.7475$$

The Alternating Series Estimation Theorem shows that the error involved in this approximation is less than

$$\frac{1}{11 \cdot 5!} = \frac{1}{1320} < 0.001$$

■

Another use of Taylor series is illustrated in the next example. The limit could be found with l'Hospital's Rule, but instead we use a series.

EXAMPLE 12 Evaluate $\displaystyle\lim_{x\to 0} \frac{e^x - 1 - x}{x^2}$.

SOLUTION Using the Maclaurin series for e^x, we have

$$\lim_{x\to 0} \frac{e^x - 1 - x}{x^2} = \lim_{x\to 0} \frac{\left(1 + \frac{x}{1!} + \frac{x^2}{2!} + \frac{x^3}{3!} + \cdots\right) - 1 - x}{x^2}$$

Some computer algebra systems compute limits in this way.

$$= \lim_{x\to 0} \frac{\frac{x^2}{2!} + \frac{x^3}{3!} + \frac{x^4}{4!} + \cdots}{x^2}$$

$$= \lim_{x\to 0} \left(\frac{1}{2} + \frac{x}{3!} + \frac{x^2}{4!} + \frac{x^3}{5!} + \cdots\right) = \frac{1}{2}$$

because power series are continuous functions. ■

Multiplication and Division of Power Series

If power series are added or subtracted, they behave like polynomials (Theorem 11.2.8 shows this). In fact, as the following example illustrates, they can also be multiplied and divided like polynomials. We find only the first few terms because the calculations for the later terms become tedious and the initial terms are the most important ones.

EXAMPLE 13 Find the first three nonzero terms in the Maclaurin series for (a) $e^x \sin x$ and (b) $\tan x$.

SOLUTION

(a) Using the Maclaurin series for e^x and $\sin x$ in Table 1, we have

$$e^x \sin x = \left(1 + \frac{x}{1!} + \frac{x^2}{2!} + \frac{x^3}{3!} + \cdots\right)\left(x - \frac{x^3}{3!} + \cdots\right)$$

We multiply these expressions, collecting like terms just as for polynomials:

$$\begin{array}{rl}
 & 1 + x + \tfrac{1}{2}x^2 + \tfrac{1}{6}x^3 + \cdots \\
\times & x \qquad\quad - \tfrac{1}{6}x^3 + \cdots \\
\hline
 & x + x^2 + \tfrac{1}{2}x^3 + \tfrac{1}{6}x^4 + \cdots \\
+ & \qquad - \tfrac{1}{6}x^3 - \tfrac{1}{6}x^4 - \cdots \\
\hline
 & x + x^2 + \tfrac{1}{3}x^3 + \cdots
\end{array}$$

Thus

$$e^x \sin x = x + x^2 + \tfrac{1}{3}x^3 + \cdots$$

(b) Using the Maclaurin series in Table 1, we have

$$\tan x = \frac{\sin x}{\cos x} = \frac{x - \dfrac{x^3}{3!} + \dfrac{x^5}{5!} - \cdots}{1 - \dfrac{x^2}{2!} + \dfrac{x^4}{4!} - \cdots}$$

We use a procedure like long division:

$$\begin{array}{r|l}
 & x + \tfrac{1}{3}x^3 + \tfrac{2}{15}x^5 + \cdots \\
\hline
1 - \tfrac{1}{2}x^2 + \tfrac{1}{24}x^4 - \cdots & x - \tfrac{1}{6}x^3 + \tfrac{1}{120}x^5 - \cdots \\
 & x - \tfrac{1}{2}x^3 + \tfrac{1}{24}x^5 - \cdots \\
\hline
 & \quad \tfrac{1}{3}x^3 - \tfrac{1}{30}x^5 + \cdots \\
 & \quad \tfrac{1}{3}x^3 - \tfrac{1}{6}x^5 + \cdots \\
\hline
 & \qquad \tfrac{2}{15}x^5 + \cdots
\end{array}$$

Thus

$$\tan x = x + \tfrac{1}{3}x^3 + \tfrac{2}{15}x^5 + \cdots$$

Although we have not attempted to justify the formal manipulations used in Example 13, they are legitimate. There is a theorem which states that if both $f(x) = \Sigma\, c_n x^n$ and $g(x) = \Sigma\, b_n x^n$ converge for $|x| < R$ and the series are multiplied as if they were polynomials, then the resulting series also converges for $|x| < R$ and represents $f(x)g(x)$. For division we require $b_0 \neq 0$; the resulting series converges for sufficiently small $|x|$.

11.10 EXERCISES

1. If $f(x) = \sum_{n=0}^{\infty} b_n(x-5)^n$ for all x, write a formula for b_8.

2. The graph of f is shown.

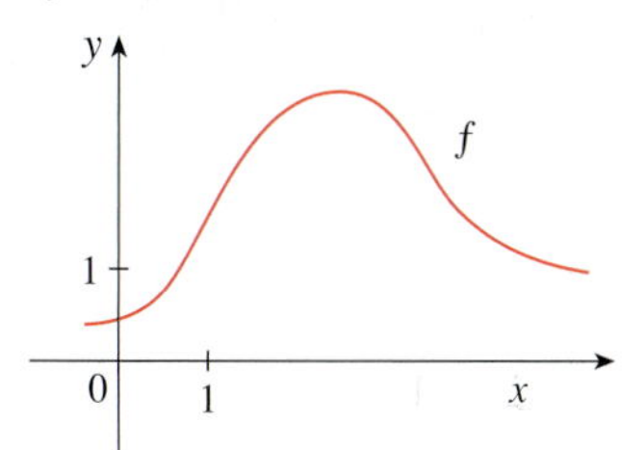

(a) Explain why the series

$$1.6 - 0.8(x-1) + 0.4(x-1)^2 - 0.1(x-1)^3 + \cdots$$

is *not* the Taylor series of f centered at 1.

(b) Explain why the series

$$2.8 + 0.5(x-2) + 1.5(x-2)^2 - 0.1(x-2)^3 + \cdots$$

is *not* the Taylor series of f centered at 2.

3. If $f^{(n)}(0) = (n+1)!$ for $n = 0, 1, 2, \ldots$, find the Maclaurin series for f and its radius of convergence.

4. Find the Taylor series for f centered at 4 if

$$f^{(n)}(4) = \frac{(-1)^n n!}{3^n(n+1)}$$

What is the radius of convergence of the Taylor series?

5–10 Use the definition of a Taylor series to find the first four nonzero terms of the series for $f(x)$ centered at the given value of a.

5. $f(x) = xe^x, \quad a = 0$

6. $f(x) = \dfrac{1}{1+x}, \quad a = 2$

7. $f(x) = \sqrt[3]{x}, \quad a = 8$

8. $f(x) = \ln x, \quad a = 1$

9. $f(x) = \sin x, \quad a = \pi/6$

10. $f(x) = \cos^2 x, \quad a = 0$

11–18 Find the Maclaurin series for $f(x)$ using the definition of a Maclaurin series. [Assume that f has a power series expansion. Do not show that $R_n(x) \to 0$.] Also find the associated radius of convergence.

11. $f(x) = (1-x)^{-2}$

12. $f(x) = \ln(1+x)$

13. $f(x) = \cos x$

14. $f(x) = e^{-2x}$

15. $f(x) = 2^x$

16. $f(x) = x \cos x$

17. $f(x) = \sinh x$

18. $f(x) = \cosh x$

19–26 Find the Taylor series for $f(x)$ centered at the given value of a. [Assume that f has a power series expansion. Do not show that $R_n(x) \to 0$.] Also find the associated radius of convergence.

19. $f(x) = x^5 + 2x^3 + x, \quad a = 2$

20. $f(x) = x^6 - x^4 + 2, \quad a = -2$

21. $f(x) = \ln x, \quad a = 2$

22. $f(x) = 1/x, \quad a = -3$

23. $f(x) = e^{2x}, \quad a = 3$

24. $f(x) = \cos x, \quad a = \pi/2$

25. $f(x) = \sin x, \quad a = \pi$

26. $f(x) = \sqrt{x}, \quad a = 16$

27. Prove that the series obtained in Exercise 13 represents $\cos x$ for all x.

28. Prove that the series obtained in Exercise 25 represents $\sin x$ for all x.

29. Prove that the series obtained in Exercise 17 represents $\sinh x$ for all x.

30. Prove that the series obtained in Exercise 18 represents $\cosh x$ for all x.

31–34 Use the binomial series to expand the function as a power series. State the radius of convergence.

31. $\sqrt[4]{1-x}$

32. $\sqrt[3]{8+x}$

33. $\dfrac{1}{(2+x)^3}$

34. $(1-x)^{3/4}$

35–44 Use a Maclaurin series in Table 1 to obtain the Maclaurin series for the given function.

35. $f(x) = \arctan(x^2)$

36. $f(x) = \sin(\pi x/4)$

37. $f(x) = x \cos 2x$

38. $f(x) = e^{3x} - e^{2x}$

39. $f(x) = x \cos\left(\frac{1}{2}x^2\right)$

40. $f(x) = x^2 \ln(1 + x^3)$

41. $f(x) = \dfrac{x}{\sqrt{4 + x^2}}$

42. $f(x) = \dfrac{x^2}{\sqrt{2 + x}}$

43. $f(x) = \sin^2 x \quad \left[\textit{Hint: } \text{Use } \sin^2 x = \frac{1}{2}(1 - \cos 2x).\right]$

44. $f(x) = \begin{cases} \dfrac{x - \sin x}{x^3} & \text{if } x \neq 0 \\ \frac{1}{6} & \text{if } x = 0 \end{cases}$

45–48 Find the Maclaurin series of f (by any method) and its radius of convergence. Graph f and its first few Taylor polynomials on the same screen. What do you notice about the relationship between these polynomials and f?

45. $f(x) = \cos(x^2)$

46. $f(x) = \ln(1 + x^2)$

47. $f(x) = xe^{-x}$

48. $f(x) = \tan^{-1}(x^3)$

49. Use the Maclaurin series for $\cos x$ to compute $\cos 5°$ correct to five decimal places.

50. Use the Maclaurin series for e^x to calculate $1/\sqrt[10]{e}$ correct to five decimal places.

51. (a) Use the binomial series to expand $1/\sqrt{1 - x^2}$.
(b) Use part (a) to find the Maclaurin series for $\sin^{-1}x$.

52. (a) Expand $1/\sqrt[4]{1 + x}$ as a power series.
(b) Use part (a) to estimate $1/\sqrt[4]{1.1}$ correct to three decimal places.

53–56 Evaluate the indefinite integral as an infinite series.

53. $\int \sqrt{1 + x^3}\, dx$

54. $\int x^2 \sin(x^2)\, dx$

55. $\int \dfrac{\cos x - 1}{x}\, dx$

56. $\int \arctan(x^2)\, dx$

57–60 Use series to approximate the definite integral to within the indicated accuracy.

57. $\int_0^{1/2} x^3 \arctan x\, dx$ (four decimal places)

58. $\int_0^1 \sin(x^4)\, dx$ (four decimal places)

59. $\int_0^{0.4} \sqrt{1 + x^4}\, dx$ $\left(|\text{error}| < 5 \times 10^{-6}\right)$

60. $\int_0^{0.5} x^2 e^{-x^2}\, dx$ $\left(|\text{error}| < 0.001\right)$

61–65 Use series to evaluate the limit.

61. $\lim\limits_{x \to 0} \dfrac{x - \ln(1 + x)}{x^2}$

62. $\lim\limits_{x \to 0} \dfrac{1 - \cos x}{1 + x - e^x}$

63. $\lim\limits_{x \to 0} \dfrac{\sin x - x + \frac{1}{6}x^3}{x^5}$

64. $\lim\limits_{x \to 0} \dfrac{\sqrt{1 + x} - 1 - \frac{1}{2}x}{x^2}$

65. $\lim\limits_{x \to 0} \dfrac{x^3 - 3x + 3\tan^{-1}x}{x^5}$

66. Use the series in Example 13(b) to evaluate

$$\lim_{x \to 0} \frac{\tan x - x}{x^3}$$

We found this limit in Example 6.8.4 using l'Hospital's Rule three times. Which method do you prefer?

67–72 Use multiplication or division of power series to find the first three nonzero terms in the Maclaurin series for each function.

67. $y = e^{-x^2} \cos x$

68. $y = \sec x$

69. $y = \dfrac{x}{\sin x}$

70. $y = e^x \ln(1 + x)$

71. $y = (\arctan x)^2$

72. $y = e^x \sin^2 x$

73–80 Find the sum of the series.

73. $\displaystyle\sum_{n=0}^{\infty} (-1)^n \frac{x^{4n}}{n!}$

74. $\displaystyle\sum_{n=0}^{\infty} \frac{(-1)^n \pi^{2n}}{6^{2n}(2n)!}$

75. $\displaystyle\sum_{n=1}^{\infty} (-1)^{n-1} \frac{3^n}{n\,5^n}$

76. $\displaystyle\sum_{n=0}^{\infty} \frac{3^n}{5^n n!}$

77. $\displaystyle\sum_{n=0}^{\infty} \frac{(-1)^n \pi^{2n+1}}{4^{2n+1}(2n + 1)!}$

78. $1 - \ln 2 + \dfrac{(\ln 2)^2}{2!} - \dfrac{(\ln 2)^3}{3!} + \cdots$

79. $3 + \dfrac{9}{2!} + \dfrac{27}{3!} + \dfrac{81}{4!} + \cdots$

80. $\dfrac{1}{1 \cdot 2} - \dfrac{1}{3 \cdot 2^3} + \dfrac{1}{5 \cdot 2^5} - \dfrac{1}{7 \cdot 2^7} + \cdots$

81. Show that if p is an nth-degree polynomial, then

$$p(x + 1) = \sum_{i=0}^{n} \frac{p^{(i)}(x)}{i!}$$

82. If $f(x) = (1 + x^3)^{30}$, what is $f^{(58)}(0)$?

83. Prove Taylor's Inequality for $n = 2$, that is, prove that if $|f'''(x)| \leqslant M$ for $|x - a| \leqslant d$, then

$$|R_2(x)| \leqslant \frac{M}{6}|x - a|^3 \qquad \text{for } |x - a| \leqslant d$$

84. (a) Show that the function defined by

$$f(x) = \begin{cases} e^{-1/x^2} & \text{if } x \neq 0 \\ 0 & \text{if } x = 0 \end{cases}$$

is not equal to its Maclaurin series.

(b) Graph the function in part (a) and comment on its behavior near the origin.

85. Use the following steps to prove (17).

(a) Let $g(x) = \sum_{n=0}^{\infty} \binom{k}{n} x^n$. Differentiate this series to show that

$$g'(x) = \frac{kg(x)}{1 + x} \qquad -1 < x < 1$$

(b) Let $h(x) = (1 + x)^{-k} g(x)$ and show that $h'(x) = 0$.
(c) Deduce that $g(x) = (1 + x)^k$.

86. In Exercise 10.2.53 it was shown that the length of the ellipse $x = a \sin\theta$, $y = b \cos\theta$, where $a > b > 0$, is

$$L = 4a \int_0^{\pi/2} \sqrt{1 - e^2 \sin^2\theta}\, d\theta$$

where $e = \sqrt{a^2 - b^2}/a$ is the eccentricity of the ellipse.

Expand the integrand as a binomial series and use the result of Exercise 7.1.50 to express L as a series in powers of the eccentricity up to the term in e^6.

LABORATORY PROJECT CAS AN ELUSIVE LIMIT

This project deals with the function

$$f(x) = \frac{\sin(\tan x) - \tan(\sin x)}{\arcsin(\arctan x) - \arctan(\arcsin x)}$$

1. Use your computer algebra system to evaluate $f(x)$ for $x = 1, 0.1, 0.01, 0.001$, and 0.0001. Does it appear that f has a limit as $x \to 0$?
2. Use the CAS to graph f near $x = 0$. Does it appear that f has a limit as $x \to 0$?
3. Try to evaluate $\lim_{x \to 0} f(x)$ with l'Hospital's Rule, using the CAS to find derivatives of the numerator and denominator. What do you discover? How many applications of l'Hospital's Rule are required?
4. Evaluate $\lim_{x \to 0} f(x)$ by using the CAS to find sufficiently many terms in the Taylor series of the numerator and denominator. (Use the command `taylor` in Maple or `Series` in Mathematica.)
5. Use the limit command on your CAS to find $\lim_{x \to 0} f(x)$ directly. (Most computer algebra systems use the method of Problem 4 to compute limits.)
6. In view of the answers to Problems 4 and 5, how do you explain the results of Problems 1 and 2?

WRITING PROJECT HOW NEWTON DISCOVERED THE BINOMIAL SERIES

The Binomial Theorem, which gives the expansion of $(a + b)^k$, was known to Chinese mathematicians many centuries before the time of Newton for the case where the exponent k is a positive integer. In 1665, when he was 22, Newton was the first to discover the infinite series expansion of $(a + b)^k$ when k is a fractional exponent (positive or negative). He didn't publish his discovery, but he stated it and gave examples of how to use it in a letter (now called the *epistola prior*) dated June 13, 1676, that he sent to Henry Oldenburg, secretary of the Royal Society of London, to transmit to Leibniz. When Leibniz replied, he asked how Newton had discovered the binomial series. Newton wrote a second letter, the *epistola posterior* of October 24, 1676, in which he explained in great detail how he arrived at his discovery by a very indirect route. He was investigating the areas under the curves $y = (1 - x^2)^{n/2}$ from 0 to x for $n = 0, 1, 2, 3, 4, \ldots$. These are easy to calculate if n is even. By observing patterns and interpolating, Newton was able to guess the answers for odd values of n. Then he realized he could get the same answers by expressing $(1 - x^2)^{n/2}$ as an infinite series.

Write a report on Newton's discovery of the binomial series. Start by giving the statement of the binomial series in Newton's notation (see the *epistola prior* on page 285 of [4] or page 402 of [2]). Explain why Newton's version is equivalent to Theorem 17 on page 807. Then read Newton's *epistola posterior* (page 287 in [4] or page 404 in [2]) and explain the patterns that Newton discovered in the areas under the curves $y = (1 - x^2)^{n/2}$. Show how he was able to guess the areas under the remaining curves and how he verified his answers. Finally, explain how these discoveries led to the binomial series. The books by Edwards [1] and Katz [3] contain commentaries on Newton's letters.

1. C. H. Edwards, *The Historical Development of the Calculus* (New York: Springer-Verlag, 1979), pp. 178–187.
2. John Fauvel and Jeremy Gray, eds., *The History of Mathematics: A Reader* (London: MacMillan Press, 1987).
3. Victor Katz, *A History of Mathematics: An Introduction* (New York: HarperCollins, 1993), pp. 463–466.
4. D. J. Struik, ed., *A Sourcebook in Mathematics, 1200–1800* (Princeton, NJ: Princeton University Press, 1969).

11.11 Applications of Taylor Polynomials

In this section we explore two types of applications of Taylor polynomials. First we look at how they are used to approximate functions—computer scientists like them because polynomials are the simplest of functions. Then we investigate how physicists and engineers use them in such fields as relativity, optics, blackbody radiation, electric dipoles, the velocity of water waves, and building highways across a desert.

Approximating Functions by Polynomials

Suppose that $f(x)$ is equal to the sum of its Taylor series at a:

$$f(x) = \sum_{n=0}^{\infty} \frac{f^{(n)}(a)}{n!}(x - a)^n$$

In Section 11.10 we introduced the notation $T_n(x)$ for the nth partial sum of this series and called it the nth-degree Taylor polynomial of f at a. Thus

$$T_n(x) = \sum_{i=0}^{n} \frac{f^{(i)}(a)}{i!}(x - a)^i$$

$$= f(a) + \frac{f'(a)}{1!}(x - a) + \frac{f''(a)}{2!}(x - a)^2 + \cdots + \frac{f^{(n)}(a)}{n!}(x - a)^n$$

Since f is the sum of its Taylor series, we know that $T_n(x) \to f(x)$ as $n \to \infty$ and so T_n can be used as an approximation to f: $f(x) \approx T_n(x)$.

Notice that the first-degree Taylor polynomial

$$T_1(x) = f(a) + f'(a)(x - a)$$

is the same as the linearization of f at a that we discussed in Section 2.9. Notice also that T_1 and its derivative have the same values at a that f and f' have. In general, it can be shown that the derivatives of T_n at a agree with those of f up to and including derivatives of order n.

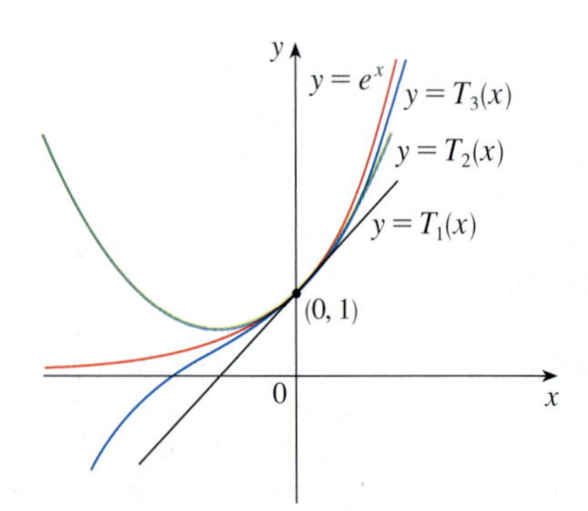

FIGURE 1

To illustrate these ideas let's take another look at the graphs of $y = e^x$ and its first few Taylor polynomials, as shown in Figure 1. The graph of T_1 is the tangent line to $y = e^x$ at $(0, 1)$; this tangent line is the best linear approximation to e^x near $(0, 1)$. The graph of T_2 is the parabola $y = 1 + x + x^2/2$, and the graph of T_3 is the cubic curve $y = 1 + x + x^2/2 + x^3/6$, which is a closer fit to the exponential curve $y = e^x$ than T_2. The next Taylor polynomial T_4 would be an even better approximation, and so on.

	$x = 0.2$	$x = 3.0$
$T_2(x)$	1.220000	8.500000
$T_4(x)$	1.221400	16.375000
$T_6(x)$	1.221403	19.412500
$T_8(x)$	1.221403	20.009152
$T_{10}(x)$	1.221403	20.079665
e^x	1.221403	20.085537

The values in the table give a numerical demonstration of the convergence of the Taylor polynomials $T_n(x)$ to the function $y = e^x$. We see that when $x = 0.2$ the convergence is very rapid, but when $x = 3$ it is somewhat slower. In fact, the farther x is from 0, the more slowly $T_n(x)$ converges to e^x.

When using a Taylor polynomial T_n to approximate a function f, we have to ask the questions: How good an approximation is it? How large should we take n to be in order to achieve a desired accuracy? To answer these questions we need to look at the absolute value of the remainder:

$$|R_n(x)| = |f(x) - T_n(x)|$$

There are three possible methods for estimating the size of the error:

1. If a graphing device is available, we can use it to graph $|R_n(x)|$ and thereby estimate the error.
2. If the series happens to be an alternating series, we can use the Alternating Series Estimation Theorem.
3. In all cases we can use Taylor's Inequality (Theorem 11.10.9), which says that if $|f^{(n+1)}(x)| \leqslant M$, then

$$|R_n(x)| \leqslant \frac{M}{(n+1)!}|x-a|^{n+1}$$

EXAMPLE 1

(a) Approximate the function $f(x) = \sqrt[3]{x}$ by a Taylor polynomial of degree 2 at $a = 8$.
(b) How accurate is this approximation when $7 \leqslant x \leqslant 9$?

SOLUTION

(a)

$$f(x) = \sqrt[3]{x} = x^{1/3} \qquad f(8) = 2$$

$$f'(x) = \tfrac{1}{3}x^{-2/3} \qquad f'(8) = \tfrac{1}{12}$$

$$f''(x) = -\tfrac{2}{9}x^{-5/3} \qquad f''(8) = -\tfrac{1}{144}$$

$$f'''(x) = \tfrac{10}{27}x^{-8/3}$$

Thus the second-degree Taylor polynomial is

$$\begin{aligned} T_2(x) &= f(8) + \frac{f'(8)}{1!}(x-8) + \frac{f''(8)}{2!}(x-8)^2 \\ &= 2 + \tfrac{1}{12}(x-8) - \tfrac{1}{288}(x-8)^2 \end{aligned}$$

The desired approximation is

$$\sqrt[3]{x} \approx T_2(x) = 2 + \tfrac{1}{12}(x-8) - \tfrac{1}{288}(x-8)^2$$

(b) The Taylor series is not alternating when $x < 8$, so we can't use the Alternating Series Estimation Theorem in this example. But we can use Taylor's Inequality with $n = 2$ and $a = 8$:

$$|R_2(x)| \leqslant \frac{M}{3!}|x-8|^3$$

where $|f'''(x)| \leqslant M$. Because $x \geqslant 7$, we have $x^{8/3} \geqslant 7^{8/3}$ and so

$$f'''(x) = \frac{10}{27} \cdot \frac{1}{x^{8/3}} \leqslant \frac{10}{27} \cdot \frac{1}{7^{8/3}} < 0.0021$$

Therefore we can take $M = 0.0021$. Also $7 \leqslant x \leqslant 9$, so $-1 \leqslant x - 8 \leqslant 1$ and $|x - 8| \leqslant 1$. Then Taylor's Inequality gives

$$|R_2(x)| \leqslant \frac{0.0021}{3!} \cdot 1^3 = \frac{0.0021}{6} < 0.0004$$

Thus, if $7 \leqslant x \leqslant 9$, the approximation in part (a) is accurate to within 0.0004. ■

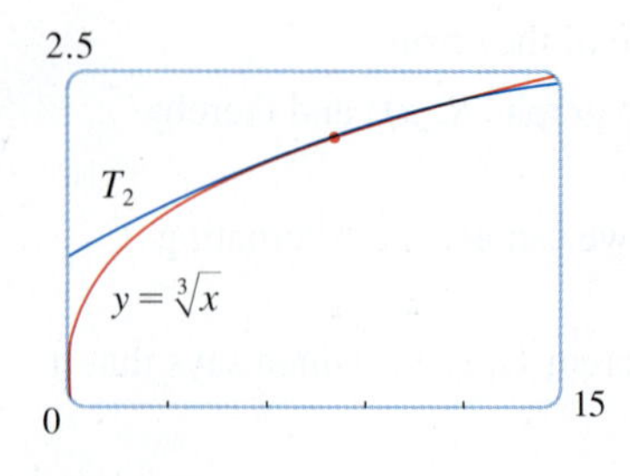

FIGURE 2

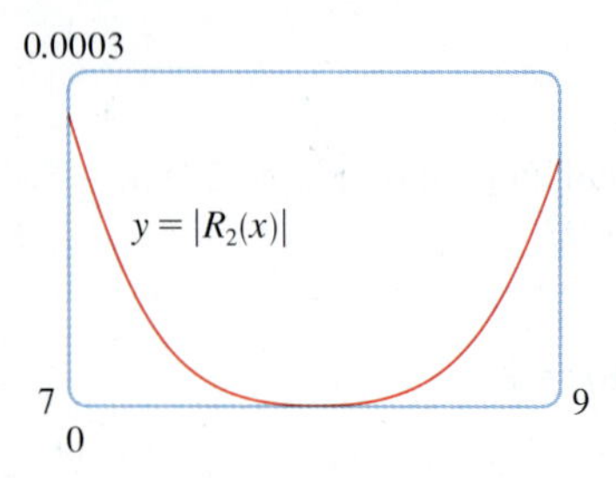

FIGURE 3

Let's use a graphing device to check the calculation in Example 1. Figure 2 shows that the graphs of $y = \sqrt[3]{x}$ and $y = T_2(x)$ are very close to each other when x is near 8. Figure 3 shows the graph of $|R_2(x)|$ computed from the expression

$$|R_2(x)| = |\sqrt[3]{x} - T_2(x)|$$

We see from the graph that

$$|R_2(x)| < 0.0003$$

when $7 \leqslant x \leqslant 9$. Thus the error estimate from graphical methods is slightly better than the error estimate from Taylor's Inequality in this case.

EXAMPLE 2

(a) What is the maximum error possible in using the approximation

$$\sin x \approx x - \frac{x^3}{3!} + \frac{x^5}{5!}$$

when $-0.3 \leqslant x \leqslant 0.3$? Use this approximation to find sin 12° correct to six decimal places.

(b) For what values of x is this approximation accurate to within 0.00005?

SOLUTION

(a) Notice that the Maclaurin series

$$\sin x = x - \frac{x^3}{3!} + \frac{x^5}{5!} - \frac{x^7}{7!} + \cdots$$

is alternating for all nonzero values of x, and the successive terms decrease in size because $|x| < 1$, so we can use the Alternating Series Estimation Theorem. The error in approximating sin x by the first three terms of its Maclaurin series is at most

$$\left|\frac{x^7}{7!}\right| = \frac{|x|^7}{5040}$$

If $-0.3 \leqslant x \leqslant 0.3$, then $|x| \leqslant 0.3$, so the error is smaller than

$$\frac{(0.3)^7}{5040} \approx 4.3 \times 10^{-8}$$

To find sin 12° we first convert to radian measure:

$$\sin 12° = \sin\left(\frac{12\pi}{180}\right) = \sin\left(\frac{\pi}{15}\right)$$

$$\approx \frac{\pi}{15} - \left(\frac{\pi}{15}\right)^3 \frac{1}{3!} + \left(\frac{\pi}{15}\right)^5 \frac{1}{5!} \approx 0.20791169$$

Thus, correct to six decimal places, $\sin 12° \approx 0.207912$.

(b) The error will be smaller than 0.00005 if

$$\frac{|x|^7}{5040} < 0.00005$$

Solving this inequality for x, we get

$$|x|^7 < 0.252 \quad \text{or} \quad |x| < (0.252)^{1/7} \approx 0.821$$

So the given approximation is accurate to within 0.00005 when $|x| < 0.82$. ■

TEC Module 11.10/11.11 graphically shows the remainders in Taylor polynomial approximations.

What if we use Taylor's Inequality to solve Example 2? Since $f^{(7)}(x) = -\cos x$, we have $|f^{(7)}(x)| \leq 1$ and so

$$|R_6(x)| \leq \frac{1}{7!}|x|^7$$

So we get the same estimates as with the Alternating Series Estimation Theorem.

What about graphical methods? Figure 4 shows the graph of

$$|R_6(x)| = \left|\sin x - \left(x - \tfrac{1}{6}x^3 + \tfrac{1}{120}x^5\right)\right|$$

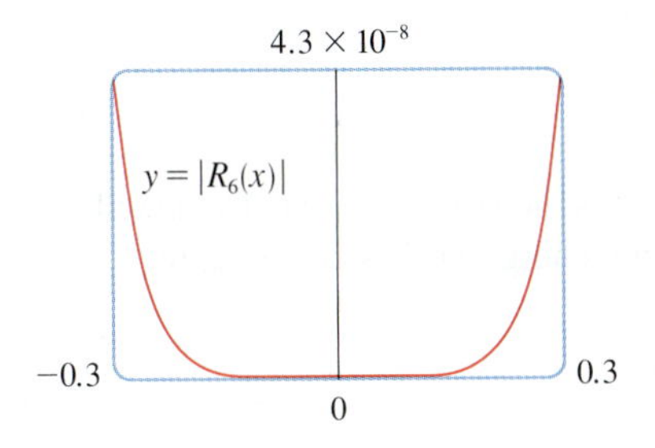

FIGURE 4

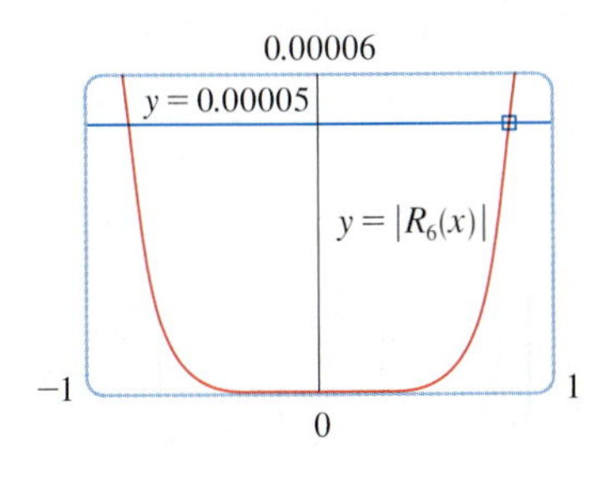

FIGURE 5

and we see from it that $|R_6(x)| < 4.3 \times 10^{-8}$ when $|x| \leq 0.3$. This is the same estimate that we obtained in Example 2. For part (b) we want $|R_6(x)| < 0.00005$, so we graph both $y = |R_6(x)|$ and $y = 0.00005$ in Figure 5. By placing the cursor on the right intersection point we find that the inequality is satisfied when $|x| < 0.82$. Again this is the same estimate that we obtained in the solution to Example 2.

If we had been asked to approximate $\sin 72°$ instead of $\sin 12°$ in Example 2, it would have been wise to use the Taylor polynomials at $a = \pi/3$ (instead of $a = 0$) because they are better approximations to $\sin x$ for values of x close to $\pi/3$. Notice that $72°$ is close to $60°$ (or $\pi/3$ radians) and the derivatives of $\sin x$ are easy to compute at $\pi/3$.

Figure 6 shows the graphs of the Maclaurin polynomial approximations

$$T_1(x) = x \qquad T_3(x) = x - \frac{x^3}{3!}$$

$$T_5(x) = x - \frac{x^3}{3!} + \frac{x^5}{5!} \qquad T_7(x) = x - \frac{x^3}{3!} + \frac{x^5}{5!} - \frac{x^7}{7!}$$

to the sine curve. You can see that as n increases, $T_n(x)$ is a good approximation to $\sin x$ on a larger and larger interval.

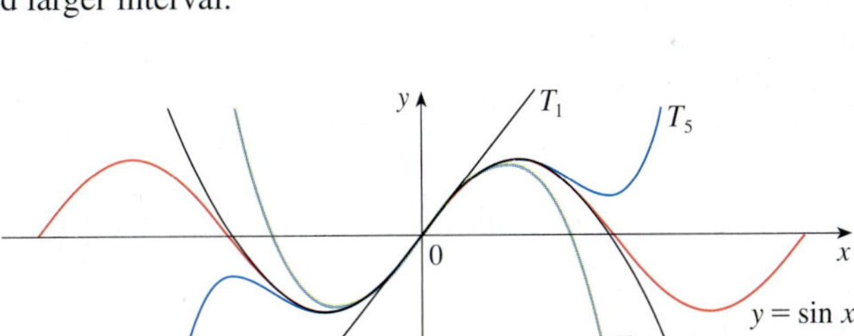

FIGURE 6

One use of the type of calculation done in Examples 1 and 2 occurs in calculators and computers. For instance, when you press the sin or e^x key on your calculator, or when a computer programmer uses a subroutine for a trigonometric or exponential or Bessel function, in many machines a polynomial approximation is calculated. The polynomial is often a Taylor polynomial that has been modified so that the error is spread more evenly throughout an interval.

Applications to Physics

Taylor polynomials are also used frequently in physics. In order to gain insight into an equation, a physicist often simplifies a function by considering only the first two or three terms in its Taylor series. In other words, the physicist uses a Taylor polynomial as an approximation to the function. Taylor's Inequality can then be used to gauge the accuracy of the approximation. The following example shows one way in which this idea is used in special relativity.

EXAMPLE 3 In Einstein's theory of special relativity the mass of an object moving with velocity v is

$$m = \frac{m_0}{\sqrt{1 - v^2/c^2}}$$

where m_0 is the mass of the object when at rest and c is the speed of light. The kinetic energy of the object is the difference between its total energy and its energy at rest:

$$K = mc^2 - m_0c^2$$

(a) Show that when v is very small compared with c, this expression for K agrees with classical Newtonian physics: $K = \frac{1}{2}m_0v^2$.
(b) Use Taylor's Inequality to estimate the difference in these expressions for K when $|v| \leqslant 100$ m/s.

SOLUTION
(a) Using the expressions given for K and m, we get

$$K = mc^2 - m_0c^2 = \frac{m_0c^2}{\sqrt{1 - v^2/c^2}} - m_0c^2 = m_0c^2\left[\left(1 - \frac{v^2}{c^2}\right)^{-1/2} - 1\right]$$

With $x = -v^2/c^2$, the Maclaurin series for $(1 + x)^{-1/2}$ is most easily computed as a binomial series with $k = -\frac{1}{2}$. (Notice that $|x| < 1$ because $v < c$.) Therefore we have

The upper curve in Figure 7 is the graph of the expression for the kinetic energy K of an object with velocity v in special relativity. The lower curve shows the function used for K in classical Newtonian physics. When v is much smaller than the speed of light, the curves are practically identical.

$$(1 + x)^{-1/2} = 1 - \tfrac{1}{2}x + \frac{(-\frac{1}{2})(-\frac{3}{2})}{2!}x^2 + \frac{(-\frac{1}{2})(-\frac{3}{2})(-\frac{5}{2})}{3!}x^3 + \cdots$$

$$= 1 - \tfrac{1}{2}x + \tfrac{3}{8}x^2 - \tfrac{5}{16}x^3 + \cdots$$

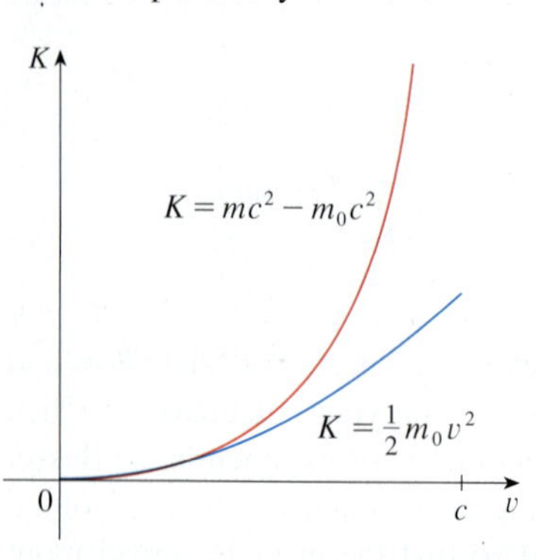

FIGURE 7

and

$$K = m_0c^2\left[\left(1 + \frac{1}{2}\frac{v^2}{c^2} + \frac{3}{8}\frac{v^4}{c^4} + \frac{5}{16}\frac{v^6}{c^6} + \cdots\right) - 1\right]$$

$$= m_0c^2\left(\frac{1}{2}\frac{v^2}{c^2} + \frac{3}{8}\frac{v^4}{c^4} + \frac{5}{16}\frac{v^6}{c^6} + \cdots\right)$$

If v is much smaller than c, then all terms after the first are very small when compared with the first term. If we omit them, we get

$$K \approx m_0c^2\left(\frac{1}{2}\frac{v^2}{c^2}\right) = \tfrac{1}{2}m_0v^2$$

(b) If $x = -v^2/c^2$, $f(x) = m_0c^2[(1 + x)^{-1/2} - 1]$, and M is a number such that $|f''(x)| \leq M$, then we can use Taylor's Inequality to write

$$|R_1(x)| \leq \frac{M}{2!}x^2$$

We have $f''(x) = \frac{3}{4}m_0c^2(1 + x)^{-5/2}$ and we are given that $|v| \leq 100$ m/s, so

$$|f''(x)| = \frac{3m_0c^2}{4(1 - v^2/c^2)^{5/2}} \leq \frac{3m_0c^2}{4(1 - 100^2/c^2)^{5/2}} \quad (=M)$$

Thus, with $c = 3 \times 10^8$ m/s,

$$|R_1(x)| \leq \frac{1}{2} \cdot \frac{3m_0c^2}{4(1 - 100^2/c^2)^{5/2}} \cdot \frac{100^4}{c^4} < (4.17 \times 10^{-10})m_0$$

So when $|v| \leq 100$ m/s, the magnitude of the error in using the Newtonian expression for kinetic energy is at most $(4.2 \times 10^{-10})m_0$. ■

Another application to physics occurs in optics. Figure 8 is adapted from *Optics*, 4th ed., by Eugene Hecht (San Francisco, 2002), page 153. It depicts a wave from the point source S meeting a spherical interface of radius R centered at C. The ray SA is refracted toward P.

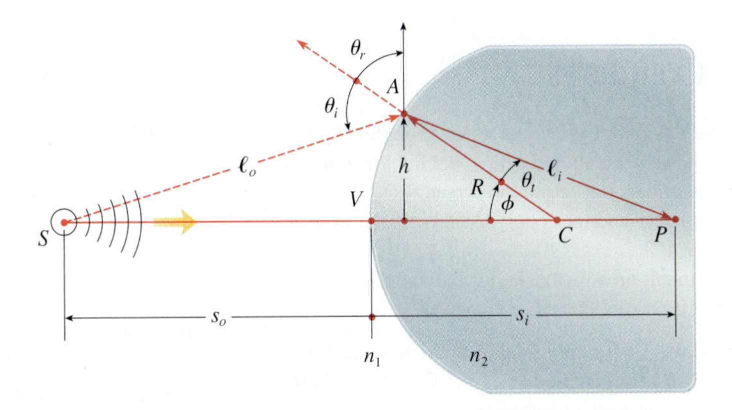

FIGURE 8
Refraction at a spherical interface

Source: Adapted from E. Hecht, *Optics*, 4e (Upper Saddle River, NJ: Pearson Education, 2002).

Using Fermat's principle that light travels so as to minimize the time taken, Hecht derives the equation

1 $$\frac{n_1}{\ell_o} + \frac{n_2}{\ell_i} = \frac{1}{R}\left(\frac{n_2 s_i}{\ell_i} - \frac{n_1 s_o}{\ell_o}\right)$$

where n_1 and n_2 are indexes of refraction and ℓ_o, ℓ_i, s_o, and s_i are the distances indicated in Figure 8. By the Law of Cosines, applied to triangles ACS and ACP, we have

2 $$\ell_o = \sqrt{R^2 + (s_o + R)^2 - 2R(s_o + R)\cos\phi}$$

$$\ell_i = \sqrt{R^2 + (s_i - R)^2 + 2R(s_i - R)\cos\phi}$$

Here we use the identity

$$\cos(\pi - \phi) = -\cos\phi$$

Because Equation 1 is cumbersome to work with, Gauss, in 1841, simplified it by using the linear approximation $\cos \phi \approx 1$ for small values of ϕ. (This amounts to using the Taylor polynomial of degree 1.) Then Equation 1 becomes the following simpler equation [as you are asked to show in Exercise 34(a)]:

$$\boxed{3} \qquad \frac{n_1}{s_o} + \frac{n_2}{s_i} = \frac{n_2 - n_1}{R}$$

The resulting optical theory is known as *Gaussian optics*, or *first-order optics*, and has become the basic theoretical tool used to design lenses.

A more accurate theory is obtained by approximating $\cos \phi$ by its Taylor polynomial of degree 3 (which is the same as the Taylor polynomial of degree 2). This takes into account rays for which ϕ is not so small, that is, rays that strike the surface at greater distances h above the axis. In Exercise 34(b) you are asked to use this approximation to derive the more accurate equation

$$\boxed{4} \qquad \frac{n_1}{s_o} + \frac{n_2}{s_i} = \frac{n_2 - n_1}{R} + h^2\left[\frac{n_1}{2s_o}\left(\frac{1}{s_o} + \frac{1}{R}\right)^2 + \frac{n_2}{2s_i}\left(\frac{1}{R} - \frac{1}{s_i}\right)^2\right]$$

The resulting optical theory is known as *third-order optics*.

Other applications of Taylor polynomials to physics and engineering are explored in Exercises 32, 33, 35, 36, 37, and 38, and in the Applied Project on page 823.

11.11 EXERCISES

1. (a) Find the Taylor polynomials up to degree 5 for $f(x) = \sin x$ centered at $a = 0$. Graph f and these polynomials on a common screen.
(b) Evaluate f and these polynomials at $x = \pi/4$, $\pi/2$, and π.
(c) Comment on how the Taylor polynomials converge to $f(x)$.

2. (a) Find the Taylor polynomials up to degree 3 for $f(x) = \tan x$ centered at $a = 0$. Graph f and these polynomials on a common screen.
(b) Evaluate f and these polynomials at $x = \pi/6$, $\pi/4$, and $\pi/3$.
(c) Comment on how the Taylor polynomials converge to $f(x)$.

3–10 Find the Taylor polynomial $T_3(x)$ for the function f centered at the number a. Graph f and T_3 on the same screen.

3. $f(x) = e^x$, $a = 1$

4. $f(x) = \sin x$, $a = \pi/6$

5. $f(x) = \cos x$, $a = \pi/2$

6. $f(x) = e^{-x} \sin x$, $a = 0$

7. $f(x) = \ln x$, $a = 1$

8. $f(x) = x \cos x$, $a = 0$

9. $f(x) = xe^{-2x}$, $a = 0$

10. $f(x) = \tan^{-1}x$, $a = 1$

CAS **11–12** Use a computer algebra system to find the Taylor polynomials T_n centered at a for $n = 2, 3, 4, 5$. Then graph these polynomials and f on the same screen.

11. $f(x) = \cot x$, $a = \pi/4$

12. $f(x) = \sqrt[3]{1 + x^2}$, $a = 0$

13–22
(a) Approximate f by a Taylor polynomial with degree n at the number a.
(b) Use Taylor's Inequality to estimate the accuracy of the approximation $f(x) \approx T_n(x)$ when x lies in the given interval.
(c) Check your result in part (b) by graphing $|R_n(x)|$.

13. $f(x) = 1/x$, $a = 1$, $n = 2$, $0.7 \leqslant x \leqslant 1.3$

14. $f(x) = x^{-1/2}$, $a = 4$, $n = 2$, $3.5 \leqslant x \leqslant 4.5$

15. $f(x) = x^{2/3}, \quad a = 1, \quad n = 3, \quad 0.8 \leqslant x \leqslant 1.2$

16. $f(x) = \sin x, \quad a = \pi/6, \quad n = 4, \quad 0 \leqslant x \leqslant \pi/3$

17. $f(x) = \sec x, \quad a = 0, \quad n = 2, \quad -0.2 \leqslant x \leqslant 0.2$

18. $f(x) = \ln(1 + 2x), \quad a = 1, \quad n = 3, \quad 0.5 \leqslant x \leqslant 1.5$

19. $f(x) = e^{x^2}, \quad a = 0, \quad n = 3, \quad 0 \leqslant x \leqslant 0.1$

20. $f(x) = x \ln x, \quad a = 1, \quad n = 3, \quad 0.5 \leqslant x \leqslant 1.5$

21. $f(x) = x \sin x, \quad a = 0, \quad n = 4, \quad -1 \leqslant x \leqslant 1$

22. $f(x) = \sinh 2x, \quad a = 0, \quad n = 5, \quad -1 \leqslant x \leqslant 1$

---

23. Use the information from Exercise 5 to estimate cos 80° correct to five decimal places.

24. Use the information from Exercise 16 to estimate sin 38° correct to five decimal places.

25. Use Taylor's Inequality to determine the number of terms of the Maclaurin series for e^x that should be used to estimate $e^{0.1}$ to within 0.00001.

26. How many terms of the Maclaurin series for $\ln(1 + x)$ do you need to use to estimate ln 1.4 to within 0.001?

27–29 Use the Alternating Series Estimation Theorem or Taylor's Inequality to estimate the range of values of x for which the given approximation is accurate to within the stated error. Check your answer graphically.

27. $\sin x \approx x - \dfrac{x^3}{6} \quad (|\text{error}| < 0.01)$

28. $\cos x \approx 1 - \dfrac{x^2}{2} + \dfrac{x^4}{24} \quad (|\text{error}| < 0.005)$

29. $\arctan x \approx x - \dfrac{x^3}{3} + \dfrac{x^5}{5} \quad (|\text{error}| < 0.05)$

---

30. Suppose you know that

$$f^{(n)}(4) = \frac{(-1)^n n!}{3^n(n + 1)}$$

and the Taylor series of f centered at 4 converges to $f(x)$ for all x in the interval of convergence. Show that the fifth-degree Taylor polynomial approximates $f(5)$ with error less than 0.0002.

31. A car is moving with speed 20 m/s and acceleration 2 m/s² at a given instant. Using a second-degree Taylor polynomial, estimate how far the car moves in the next second. Would it be reasonable to use this polynomial to estimate the distance traveled during the next minute?

32. The resistivity ρ of a conducting wire is the reciprocal of the conductivity and is measured in units of ohm-meters (Ω-m). The resistivity of a given metal depends on the temperature according to the equation

$$\rho(t) = \rho_{20}e^{\alpha(t-20)}$$

where t is the temperature in °C. There are tables that list the values of α (called the temperature coefficient) and ρ_{20} (the resistivity at 20°C) for various metals. Except at very low temperatures, the resistivity varies almost linearly with temperature and so it is common to approximate the expression for $\rho(t)$ by its first- or second-degree Taylor polynomial at $t = 20$.

(a) Find expressions for these linear and quadratic approximations.

(b) For copper, the tables give $\alpha = 0.0039/°\text{C}$ and $\rho_{20} = 1.7 \times 10^{-8}\ \Omega$-m. Graph the resistivity of copper and the linear and quadratic approximations for $-250°\text{C} \leqslant t \leqslant 1000°\text{C}$.

(c) For what values of t does the linear approximation agree with the exponential expression to within one percent?

33. An electric dipole consists of two electric charges of equal magnitude and opposite sign. If the charges are q and $-q$ and are located at a distance d from each other, then the electric field E at the point P in the figure is

$$E = \frac{q}{D^2} - \frac{q}{(D + d)^2}$$

By expanding this expression for E as a series in powers of d/D, show that E is approximately proportional to $1/D^3$ when P is far away from the dipole.

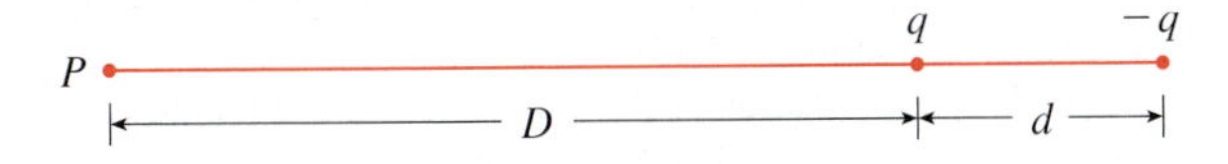

34. (a) Derive Equation 3 for Gaussian optics from Equation 1 by approximating $\cos \phi$ in Equation 2 by its first-degree Taylor polynomial.

(b) Show that if $\cos \phi$ is replaced by its third-degree Taylor polynomial in Equation 2, then Equation 1 becomes Equation 4 for third-order optics. [*Hint:* Use the first two terms in the binomial series for ℓ_o^{-1} and ℓ_i^{-1}. Also, use $\phi \approx \sin \phi$.]

35. If a water wave with length L moves with velocity v across a body of water with depth d, as in the figure on page 822, then

$$v^2 = \frac{gL}{2\pi} \tanh \frac{2\pi d}{L}$$

(a) If the water is deep, show that $v \approx \sqrt{gL/(2\pi)}$.

(b) If the water is shallow, use the Maclaurin series for tanh to show that $v \approx \sqrt{gd}$. (Thus in shallow water the velocity of a wave tends to be independent of the length of the wave.)

(c) Use the Alternating Series Estimation Theorem to show that if $L > 10d$, then the estimate $v^2 \approx gd$ is accurate to within $0.014gL$.

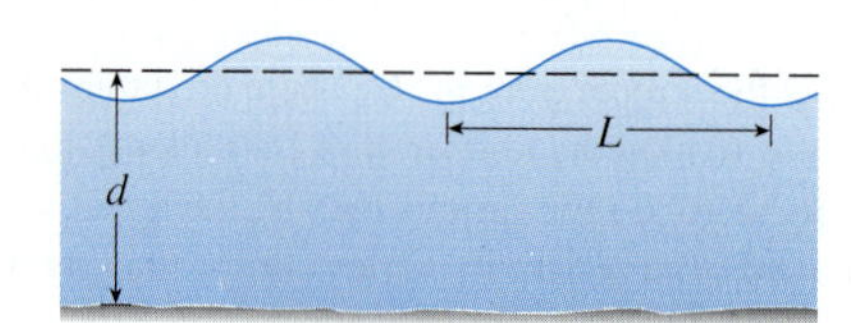

36. A uniformly charged disk has radius R and surface charge density σ as in the figure. The electric potential V at a point P at a distance d along the perpendicular central axis of the disk is

$$V = 2\pi k_e \sigma\left(\sqrt{d^2 + R^2} - d\right)$$

where k_e is a constant (called Coulomb's constant). Show that

$$V \approx \frac{\pi k_e R^2 \sigma}{d} \qquad \text{for large } d$$

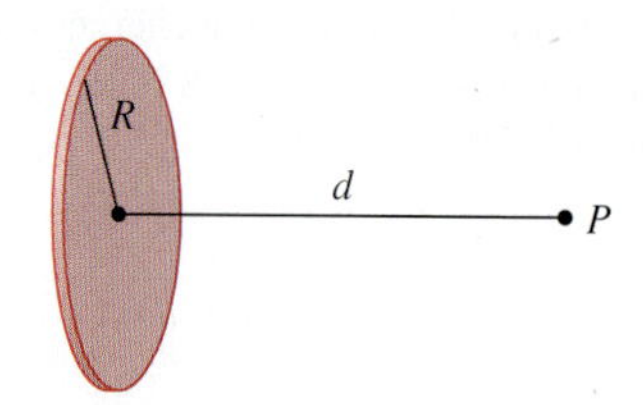

37. If a surveyor measures differences in elevation when making plans for a highway across a desert, corrections must be made for the curvature of the earth.

(a) If R is the radius of the earth and L is the length of the highway, show that the correction is

$$C = R\sec(L/R) - R$$

(b) Use a Taylor polynomial to show that

$$C \approx \frac{L^2}{2R} + \frac{5L^4}{24R^3}$$

(c) Compare the corrections given by the formulas in parts (a) and (b) for a highway that is 100 km long. (Take the radius of the earth to be 6370 km.)

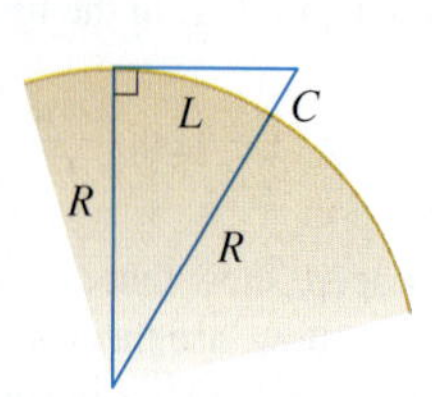

38. The period of a pendulum with length L that makes a maximum angle θ_0 with the vertical is

$$T = 4\sqrt{\frac{L}{g}}\int_0^{\pi/2} \frac{dx}{\sqrt{1 - k^2\sin^2 x}}$$

where $k = \sin\left(\frac{1}{2}\theta_0\right)$ and g is the acceleration due to gravity. (In Exercise 7.7.42 we approximated this integral using Simpson's Rule.)

(a) Expand the integrand as a binomial series and use the result of Exercise 7.1.50 to show that

$$T = 2\pi\sqrt{\frac{L}{g}}\left[1 + \frac{1^2}{2^2}k^2 + \frac{1^2 3^2}{2^2 4^2}k^4 + \frac{1^2 3^2 5^2}{2^2 4^2 6^2}k^6 + \cdots\right]$$

If θ_0 is not too large, the approximation $T \approx 2\pi\sqrt{L/g}$, obtained by using only the first term in the series, is often used. A better approximation is obtained by using two terms:

$$T \approx 2\pi\sqrt{\frac{L}{g}}\left(1 + \tfrac{1}{4}k^2\right)$$

(b) Notice that all the terms in the series after the first one have coefficients that are at most $\frac{1}{4}$. Use this fact to compare this series with a geometric series and show that

$$2\pi\sqrt{\frac{L}{g}}\left(1 + \tfrac{1}{4}k^2\right) \leqslant T \leqslant 2\pi\sqrt{\frac{L}{g}}\,\frac{4 - 3k^2}{4 - 4k^2}$$

(c) Use the inequalities in part (b) to estimate the period of a pendulum with $L = 1$ meter and $\theta_0 = 10°$. How does it compare with the estimate $T \approx 2\pi\sqrt{L/g}$? What if $\theta_0 = 42°$?

39. In Section 3.8 we considered Newton's method for approximating a root r of the equation $f(x) = 0$, and from an initial approximation x_1 we obtained successive approximations $x_2, x_3, \ldots,$ where

$$x_{n+1} = x_n - \frac{f(x_n)}{f'(x_n)}$$

Use Taylor's Inequality with $n = 1$, $a = x_n$, and $x = r$ to show that if $f''(x)$ exists on an interval I containing r, x_n, and x_{n+1}, and $|f''(x)| \leqslant M$, $|f'(x)| \geqslant K$ for all $x \in I$, then

$$|x_{n+1} - r| \leqslant \frac{M}{2K}|x_n - r|^2$$

[This means that if x_n is accurate to d decimal places, then x_{n+1} is accurate to about $2d$ decimal places. More precisely, if the error at stage n is at most 10^{-m}, then the error at stage $n + 1$ is at most $(M/2K)10^{-2m}$.]

APPLIED PROJECT RADIATION FROM THE STARS

Luke Dodd / Science Source

Any object emits radiation when heated. A *blackbody* is a system that absorbs all the radiation that falls on it. For instance, a matte black surface or a large cavity with a small hole in its wall (like a blast furnace) is a blackbody and emits blackbody radiation. Even the radiation from the sun is close to being blackbody radiation.

Proposed in the late 19th century, the Rayleigh-Jeans Law expresses the energy density of blackbody radiation of wavelength λ as

$$f(\lambda) = \frac{8\pi kT}{\lambda^4}$$

where λ is measured in meters, T is the temperature in kelvins (K), and k is Boltzmann's constant. The Rayleigh-Jeans Law agrees with experimental measurements for long wavelengths but disagrees drastically for short wavelengths. [The law predicts that $f(\lambda) \to \infty$ as $\lambda \to 0^+$ but experiments have shown that $f(\lambda) \to 0$.] This fact is known as the *ultraviolet catastrophe.*

In 1900 Max Planck found a better model (known now as Planck's Law) for blackbody radiation:

$$f(\lambda) = \frac{8\pi hc\lambda^{-5}}{e^{hc/(\lambda kT)} - 1}$$

where λ is measured in meters, T is the temperature (in kelvins), and

$$h = \text{Planck's constant} = 6.6262 \times 10^{-34}\ \text{J}\cdot\text{s}$$

$$c = \text{speed of light} = 2.997925 \times 10^{8}\ \text{m/s}$$

$$k = \text{Boltzmann's constant} = 1.3807 \times 10^{-23}\ \text{J/K}$$

1. Use l'Hospital's Rule to show that

$$\lim_{\lambda \to 0^+} f(\lambda) = 0 \qquad \text{and} \qquad \lim_{\lambda \to \infty} f(\lambda) = 0$$

for Planck's Law. So this law models blackbody radiation better than the Rayleigh-Jeans Law for short wavelengths.

2. Use a Taylor polynomial to show that, for large wavelengths, Planck's Law gives approximately the same values as the Rayleigh-Jeans Law.

3. Graph f as given by both laws on the same screen and comment on the similarities and differences. Use $T = 5700$ K (the temperature of the sun). (You may want to change from meters to the more convenient unit of micrometers: 1 mm $= 10^{-6}$ m.)

4. Use your graph in Problem 3 to estimate the value of λ for which $f(\lambda)$ is a maximum under Planck's Law.

5. Investigate how the graph of f changes as T varies. (Use Planck's Law.) In particular, graph f for the stars Betelgeuse ($T = 3400$ K), Procyon ($T = 6400$ K), and Sirius ($T = 9200$ K), as well as the sun. How does the total radiation emitted (the area under the curve) vary with T? Use the graph to comment on why Sirius is known as a blue star and Betelgeuse as a red star.

11 REVIEW

CONCEPT CHECK

Answers to the Concept Check can be found on the back endpapers.

1. (a) What is a convergent sequence?
(b) What is a convergent series?
(c) What does $\lim_{n\to\infty} a_n = 3$ mean?
(d) What does $\sum_{n=1}^{\infty} a_n = 3$ mean?

2. (a) What is a bounded sequence?
(b) What is a monotonic sequence?
(c) What can you say about a bounded monotonic sequence?

3. (a) What is a geometric series? Under what circumstances is it convergent? What is its sum?
(b) What is a p-series? Under what circumstances is it convergent?

4. Suppose $\sum a_n = 3$ and s_n is the nth partial sum of the series. What is $\lim_{n\to\infty} a_n$? What is $\lim_{n\to\infty} s_n$?

5. State the following.
(a) The Test for Divergence
(b) The Integral Test
(c) The Comparison Test
(d) The Limit Comparison Test
(e) The Alternating Series Test
(f) The Ratio Test
(g) The Root Test

6. (a) What is an absolutely convergent series?
(b) What can you say about such a series?
(c) What is a conditionally convergent series?

7. (a) If a series is convergent by the Integral Test, how do you estimate its sum?
(b) If a series is convergent by the Comparison Test, how do you estimate its sum?
(c) If a series is convergent by the Alternating Series Test, how do you estimate its sum?

8. (a) Write the general form of a power series.
(b) What is the radius of convergence of a power series?
(c) What is the interval of convergence of a power series?

9. Suppose $f(x)$ is the sum of a power series with radius of convergence R.
(a) How do you differentiate f? What is the radius of convergence of the series for f'?
(b) How do you integrate f? What is the radius of convergence of the series for $\int f(x)\,dx$?

10. (a) Write an expression for the nth-degree Taylor polynomial of f centered at a.
(b) Write an expression for the Taylor series of f centered at a.
(c) Write an expression for the Maclaurin series of f.
(d) How do you show that $f(x)$ is equal to the sum of its Taylor series?
(e) State Taylor's Inequality.

11. Write the Maclaurin series and the interval of convergence for each of the following functions.
(a) $1/(1-x)$ (b) e^x (c) $\sin x$
(d) $\cos x$ (e) $\tan^{-1}x$ (f) $\ln(1+x)$

12. Write the binomial series expansion of $(1+x)^k$. What is the radius of convergence of this series?

TRUE-FALSE QUIZ

Determine whether the statement is true or false. If it is true, explain why. If it is false, explain why or give an example that disproves the statement.

1. If $\lim_{n\to\infty} a_n = 0$, then $\sum a_n$ is convergent.

2. The series $\sum_{n=1}^{\infty} n^{-\sin 1}$ is convergent.

3. If $\lim_{n\to\infty} a_n = L$, then $\lim_{n\to\infty} a_{2n+1} = L$.

4. If $\sum c_n 6^n$ is convergent, then $\sum c_n(-2)^n$ is convergent.

5. If $\sum c_n 6^n$ is convergent, then $\sum c_n(-6)^n$ is convergent.

6. If $\sum c_n x^n$ diverges when $x = 6$, then it diverges when $x = 10$.

7. The Ratio Test can be used to determine whether $\sum 1/n^3$ converges.

8. The Ratio Test can be used to determine whether $\sum 1/n!$ converges.

9. If $0 \leqslant a_n \leqslant b_n$ and $\sum b_n$ diverges, then $\sum a_n$ diverges.

10. $\displaystyle\sum_{n=0}^{\infty} \frac{(-1)^n}{n!} = \frac{1}{e}$

11. If $-1 < \alpha < 1$, then $\lim_{n\to\infty} \alpha^n = 0$.

12. If $\sum a_n$ is divergent, then $\sum |a_n|$ is divergent.

13. If $f(x) = 2x - x^2 + \frac{1}{3}x^3 - \cdots$ converges for all x, then $f'''(0) = 2$.

14. If $\{a_n\}$ and $\{b_n\}$ are divergent, then $\{a_n + b_n\}$ is divergent.

15. If $\{a_n\}$ and $\{b_n\}$ are divergent, then $\{a_n b_n\}$ is divergent.

16. If $\{a_n\}$ is decreasing and $a_n > 0$ for all n, then $\{a_n\}$ is convergent.

17. If $a_n > 0$ and $\sum a_n$ converges, then $\sum (-1)^n a_n$ converges.

18. If $a_n > 0$ and $\lim_{n\to\infty} (a_{n+1}/a_n) < 1$, then $\lim_{n\to\infty} a_n = 0$.

19. $0.99999\ldots = 1$

20. If $\lim_{n\to\infty} a_n = 2$, then $\lim_{n\to\infty} (a_{n+3} - a_n) = 0$.

21. If a finite number of terms are added to a convergent series, then the new series is still convergent.

22. If $\sum_{n-1}^{\infty} a_n = A$ and $\sum_{n=1}^{\infty} b_n = B$, then $\sum_{n=1}^{\infty} a_n b_n = AB$.

EXERCISES

1–8 Determine whether the sequence is convergent or divergent. If it is convergent, find its limit.

1. $a_n = \dfrac{2 + n^3}{1 + 2n^3}$

2. $a_n = \dfrac{9^{n+1}}{10^n}$

3. $a_n = \dfrac{n^3}{1 + n^2}$

4. $a_n = \cos(n\pi/2)$

5. $a_n = \dfrac{n \sin n}{n^2 + 1}$

6. $a_n = \dfrac{\ln n}{\sqrt{n}}$

7. $\{(1 + 3/n)^{4n}\}$

8. $\{(-10)^n/n!\}$

9. A sequence is defined recursively by the equations $a_1 = 1$, $a_{n+1} = \frac{1}{3}(a_n + 4)$. Show that $\{a_n\}$ is increasing and $a_n < 2$ for all n. Deduce that $\{a_n\}$ is convergent and find its limit.

10. Show that $\lim_{n\to\infty} n^4 e^{-n} = 0$ and use a graph to find the smallest value of N that corresponds to $\varepsilon = 0.1$ in the precise definition of a limit.

11–22 Determine whether the series is convergent or divergent.

11. $\displaystyle\sum_{n=1}^{\infty} \frac{n}{n^3 + 1}$

12. $\displaystyle\sum_{n=1}^{\infty} \frac{n^2 + 1}{n^3 + 1}$

13. $\displaystyle\sum_{n=1}^{\infty} \frac{n^3}{5^n}$

14. $\displaystyle\sum_{n=1}^{\infty} \frac{(-1)^n}{\sqrt{n + 1}}$

15. $\displaystyle\sum_{n=2}^{\infty} \frac{1}{n\sqrt{\ln n}}$

16. $\displaystyle\sum_{n=1}^{\infty} \ln\left(\frac{n}{3n + 1}\right)$

17. $\displaystyle\sum_{n=1}^{\infty} \frac{\cos 3n}{1 + (1.2)^n}$

18. $\displaystyle\sum_{n=1}^{\infty} \frac{n^{2n}}{(1 + 2n^2)^n}$

19. $\displaystyle\sum_{n=1}^{\infty} \frac{1 \cdot 3 \cdot 5 \cdot \cdots \cdot (2n - 1)}{5^n n!}$

20. $\displaystyle\sum_{n=1}^{\infty} \frac{(-5)^{2n}}{n^2 9^n}$

21. $\displaystyle\sum_{n=1}^{\infty} (-1)^{n-1} \frac{\sqrt{n}}{n + 1}$

22. $\displaystyle\sum_{n=1}^{\infty} \frac{\sqrt{n + 1} - \sqrt{n - 1}}{n}$

23–26 Determine whether the series is conditionally convergent, absolutely convergent, or divergent.

23. $\displaystyle\sum_{n=1}^{\infty} (-1)^{n-1} n^{-1/3}$

24. $\displaystyle\sum_{n=1}^{\infty} (-1)^{n-1} n^{-3}$

25. $\displaystyle\sum_{n=1}^{\infty} \frac{(-1)^n (n + 1)3^n}{2^{2n+1}}$

26. $\displaystyle\sum_{n=2}^{\infty} \frac{(-1)^n \sqrt{n}}{\ln n}$

27–31 Find the sum of the series.

27. $\displaystyle\sum_{n=1}^{\infty} \frac{(-3)^{n-1}}{2^{3n}}$

28. $\displaystyle\sum_{n=1}^{\infty} \frac{1}{n(n + 3)}$

29. $\displaystyle\sum_{n=1}^{\infty} [\tan^{-1}(n + 1) - \tan^{-1} n]$

30. $\displaystyle\sum_{n=0}^{\infty} \frac{(-1)^n \pi^n}{3^{2n}(2n)!}$

31. $1 - e + \dfrac{e^2}{2!} - \dfrac{e^3}{3!} + \dfrac{e^4}{4!} - \cdots$

32. Express the repeating decimal $4.17326326326\ldots$ as a fraction.

33. Show that $\cosh x \geqslant 1 + \frac{1}{2}x^2$ for all x.

34. For what values of x does the series $\sum_{n=1}^{\infty} (\ln x)^n$ converge?

35. Find the sum of the series $\displaystyle\sum_{n=1}^{\infty} \frac{(-1)^{n+1}}{n^5}$ correct to four decimal places.

36. (a) Find the partial sum s_5 of the series $\sum_{n=1}^{\infty} 1/n^6$ and estimate the error in using it as an approximation to the sum of the series.

(b) Find the sum of this series correct to five decimal places.

37. Use the sum of the first eight terms to approximate the sum of the series $\sum_{n=1}^{\infty} (2 + 5^n)^{-1}$. Estimate the error involved in this approximation.

38. (a) Show that the series $\displaystyle\sum_{n=1}^{\infty} \frac{n^n}{(2n)!}$ is convergent.

(b) Deduce that $\displaystyle\lim_{n\to\infty} \frac{n^n}{(2n)!} = 0$.

39. Prove that if the series $\sum_{n=1}^{\infty} a_n$ is absolutely convergent, then the series

$$\sum_{n=1}^{\infty} \left(\frac{n + 1}{n}\right) a_n$$

is also absolutely convergent.

40–43 Find the radius of convergence and interval of convergence of the series.

40. $\sum_{n=1}^{\infty} (-1)^n \frac{x^n}{n^2 5^n}$

41. $\sum_{n=1}^{\infty} \frac{(x+2)^n}{n 4^n}$

42. $\sum_{n=1}^{\infty} \frac{2^n (x-2)^n}{(n+2)!}$

43. $\sum_{n=0}^{\infty} \frac{2^n (x-3)^n}{\sqrt{n+3}}$

44. Find the radius of convergence of the series

$$\sum_{n=1}^{\infty} \frac{(2n)!}{(n!)^2} x^n$$

45. Find the Taylor series of $f(x) = \sin x$ at $a = \pi/6$.

46. Find the Taylor series of $f(x) = \cos x$ at $a = \pi/3$.

47–54 Find the Maclaurin series for f and its radius of convergence. You may use either the direct method (definition of a Maclaurin series) or known series such as geometric series, binomial series, or the Maclaurin series for e^x, $\sin x$, $\tan^{-1}x$, and $\ln(1 + x)$.

47. $f(x) = \frac{x^2}{1+x}$

48. $f(x) = \tan^{-1}(x^2)$

49. $f(x) = \ln(4 - x)$

50. $f(x) = xe^{2x}$

51. $f(x) = \sin(x^4)$

52. $f(x) = 10^x$

53. $f(x) = 1/\sqrt[4]{16 - x}$

54. $f(x) = (1 - 3x)^{-5}$

55. Evaluate $\int \frac{e^x}{x}\,dx$ as an infinite series.

56. Use series to approximate $\int_0^1 \sqrt{1 + x^4}\,dx$ correct to two decimal places.

57–58

(a) Approximate f by a Taylor polynomial with degree n at the number a.

(b) Graph f and T_n on a common screen.

(c) Use Taylor's Inequality to estimate the accuracy of the approximation $f(x) \approx T_n(x)$ when x lies in the given interval.

(d) Check your result in part (c) by graphing $|R_n(x)|$.

57. $f(x) = \sqrt{x}, \quad a = 1, \quad n = 3, \quad 0.9 \leqslant x \leqslant 1.1$

58. $f(x) = \sec x, \quad a = 0, \quad n = 2, \quad 0 \leqslant x \leqslant \pi/6$

59. Use series to evaluate the following limit.

$$\lim_{x \to 0} \frac{\sin x - x}{x^3}$$

60. The force due to gravity on an object with mass m at a height h above the surface of the earth is

$$F = \frac{mgR^2}{(R + h)^2}$$

where R is the radius of the earth and g is the acceleration due to gravity for an object on the surface of the earth.

(a) Express F as a series in powers of h/R.

(b) Observe that if we approximate F by the first term in the series, we get the expression $F \approx mg$ that is usually used when h is much smaller than R. Use the Alternating Series Estimation Theorem to estimate the range of values of h for which the approximation $F \approx mg$ is accurate to within one percent. (Use $R = 6400$ km.)

61. Suppose that $f(x) = \sum_{n=0}^{\infty} c_n x^n$ for all x.

(a) If f is an odd function, show that

$$c_0 = c_2 = c_4 = \cdots = 0$$

(b) If f is an even function, show that

$$c_1 = c_3 = c_5 = \cdots = 0$$

62. If $f(x) = e^{x^2}$, show that $f^{(2n)}(0) = \frac{(2n)!}{n!}$.

Problems Plus

Before you look at the solution of the example, cover it up and first try to solve the problem yourself.

EXAMPLE Find the sum of the series $\sum_{n=0}^{\infty} \frac{(x+2)^n}{(n+3)!}$.

SOLUTION The problem-solving principle that is relevant here is *recognizing something familiar*. Does the given series look anything like a series that we already know? Well, it does have some ingredients in common with the Maclaurin series for the exponential function:

$$e^x = \sum_{n=0}^{\infty} \frac{x^n}{n!} = 1 + x + \frac{x^2}{2!} + \frac{x^3}{3!} + \cdots$$

We can make this series look more like our given series by replacing x by $x + 2$:

$$e^{x+2} = \sum_{n=0}^{\infty} \frac{(x+2)^n}{n!} = 1 + (x+2) + \frac{(x+2)^2}{2!} + \frac{(x+2)^3}{3!} + \cdots$$

But here the exponent in the numerator matches the number in the denominator whose factorial is taken. To make that happen in the given series, let's multiply and divide by $(x+2)^3$:

$$\sum_{n=0}^{\infty} \frac{(x+2)^n}{(n+3)!} = \frac{1}{(x+2)^3} \sum_{n=0}^{\infty} \frac{(x+2)^{n+3}}{(n+3)!}$$

$$= (x+2)^{-3}\left[\frac{(x+2)^3}{3!} + \frac{(x+2)^4}{4!} + \cdots\right]$$

We see that the series between brackets is just the series for e^{x+2} with the first three terms missing. So

$$\sum_{n=0}^{\infty} \frac{(x+2)^n}{(n+3)!} = (x+2)^{-3}\left[e^{x+2} - 1 - (x+2) - \frac{(x+2)^2}{2!}\right]$$

■

Problems

1. If $f(x) = \sin(x^3)$, find $f^{(15)}(0)$.

2. A function f is defined by

$$f(x) = \lim_{n \to \infty} \frac{x^{2n} - 1}{x^{2n} + 1}$$

Where is f continuous?

3. (a) Show that $\tan \frac{1}{2}x = \cot \frac{1}{2}x - 2 \cot x$.

(b) Find the sum of the series

$$\sum_{n=1}^{\infty} \frac{1}{2^n} \tan \frac{x}{2^n}$$

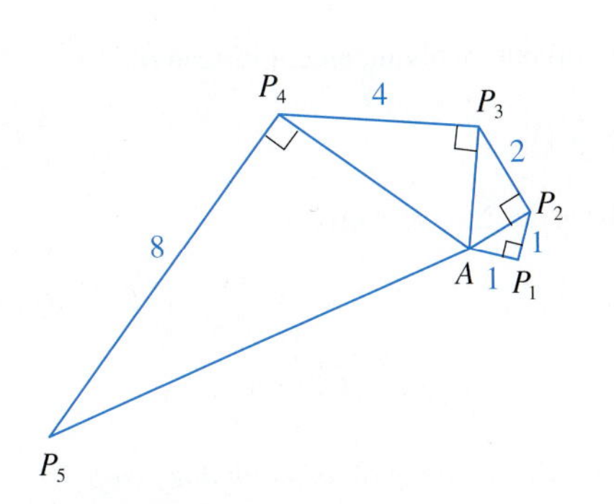

FIGURE FOR PROBLEM 4

4. Let $\{P_n\}$ be a sequence of points determined as in the figure. Thus $|AP_1| = 1$, $|P_n P_{n+1}| = 2^{n-1}$, and angle $AP_n P_{n+1}$ is a right angle. Find $\lim_{n \to \infty} \angle P_n A P_{n+1}$.

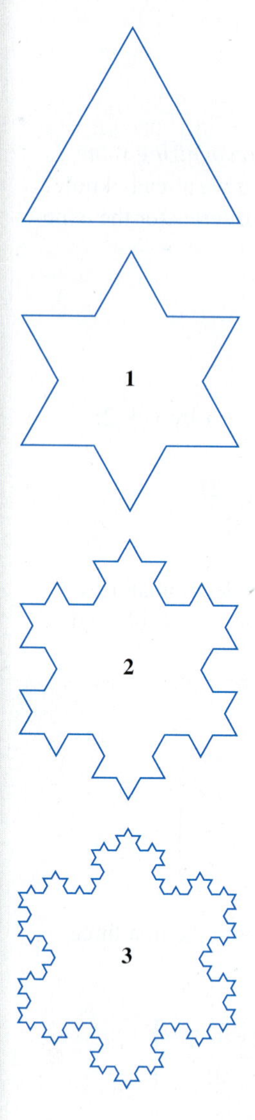

FIGURE FOR PROBLEM 5

5. To construct the **snowflake curve**, start with an equilateral triangle with sides of length 1. Step 1 in the construction is to divide each side into three equal parts, construct an equilateral triangle on the middle part, and then delete the middle part (see the figure). Step 2 is to repeat step 1 for each side of the resulting polygon. This process is repeated at each succeeding step. The snowflake curve is the curve that results from repeating this process indefinitely.

(a) Let s_n, l_n, and p_n represent the number of sides, the length of a side, and the total length of the nth approximating curve (the curve obtained after step n of the construction), respectively. Find formulas for s_n, l_n, and p_n.

(b) Show that $p_n \to \infty$ as $n \to \infty$.

(c) Sum an infinite series to find the area enclosed by the snowflake curve.

Note: Parts (b) and (c) show that the snowflake curve is infinitely long but encloses only a finite area.

6. Find the sum of the series

$$1 + \frac{1}{2} + \frac{1}{3} + \frac{1}{4} + \frac{1}{6} + \frac{1}{8} + \frac{1}{9} + \frac{1}{12} + \cdots$$

where the terms are the reciprocals of the positive integers whose only prime factors are 2s and 3s.

7. (a) Show that for $xy \neq -1$,

$$\arctan x - \arctan y = \arctan \frac{x - y}{1 + xy}$$

if the left side lies between $-\pi/2$ and $\pi/2$.

(b) Show that $\arctan \frac{120}{119} - \arctan \frac{1}{239} = \pi/4$.

(c) Deduce the following formula of John Machin (1680–1751):

$$4 \arctan \tfrac{1}{5} - \arctan \tfrac{1}{239} = \frac{\pi}{4}$$

(d) Use the Maclaurin series for arctan to show that

$$0.1973955597 < \arctan \tfrac{1}{5} < 0.1973955616$$

(e) Show that

$$0.004184075 < \arctan \tfrac{1}{239} < 0.004184077$$

(f) Deduce that, correct to seven decimal places, $\pi \approx 3.1415927$.

Machin used this method in 1706 to find π correct to 100 decimal places. Recently, with the aid of computers, the value of π has been computed to increasingly greater accuracy. In 2013 Shigeru Kondo and Alexander Yee computed the value of π to more than 12 trillion decimal places!

8. (a) Prove a formula similar to the one in Problem 7(a) but involving arccot instead of arctan.

(b) Find the sum of the series $\Sigma_{n=0}^{\infty} \operatorname{arccot}(n^2 + n + 1)$.

9. Use the result of Problem 7(a) to find the sum of the series $\Sigma_{n=1}^{\infty} \arctan(2/n^2)$.

10. If $a_0 + a_1 + a_2 + \cdots + a_k = 0$, show that

$$\lim_{n \to \infty} \left(a_0\sqrt{n} + a_1\sqrt{n + 1} + a_2\sqrt{n + 2} + \cdots + a_k\sqrt{n + k} \right) = 0$$

If you don't see how to prove this, try the problem-solving strategy of *using analogy* (see page 98). Try the special cases $k = 1$ and $k = 2$ first. If you can see how to prove the assertion for these cases, then you will probably see how to prove it in general.

11. Find the interval of convergence of $\sum_{n=1}^{\infty} n^3x^n$ and find its sum.

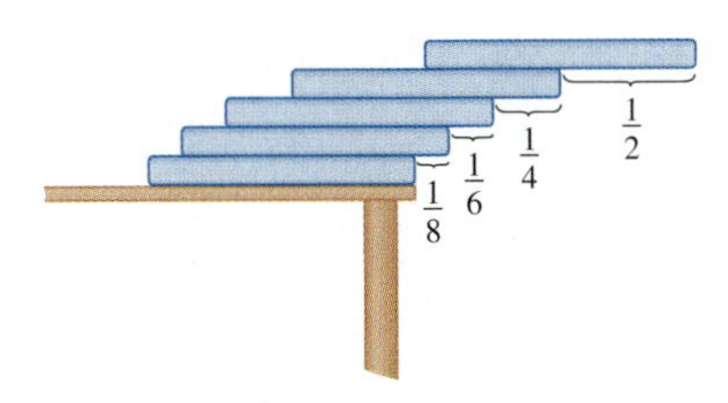

FIGURE FOR PROBLEM 12

12. Suppose you have a large supply of books, all the same size, and you stack them at the edge of a table, with each book extending farther beyond the edge of the table than the one beneath it. Show that it is possible to do this so that the top book extends entirely beyond the table. In fact, show that the top book can extend any distance at all beyond the edge of the table if the stack is high enough. Use the following method of stacking: The top book extends half its length beyond the second book. The second book extends a quarter of its length beyond the third. The third extends one-sixth of its length beyond the fourth, and so on. (Try it yourself with a deck of cards.) Consider centers of mass.

13. Find the sum of the series $\displaystyle\sum_{n=2}^{\infty} \ln\left(1 - \frac{1}{n^2}\right)$.

14. If $p > 1$, evaluate the expression

$$\frac{1 + \dfrac{1}{2^p} + \dfrac{1}{3^p} + \dfrac{1}{4^p} + \cdots}{1 - \dfrac{1}{2^p} + \dfrac{1}{3^p} - \dfrac{1}{4^p} + \cdots}$$

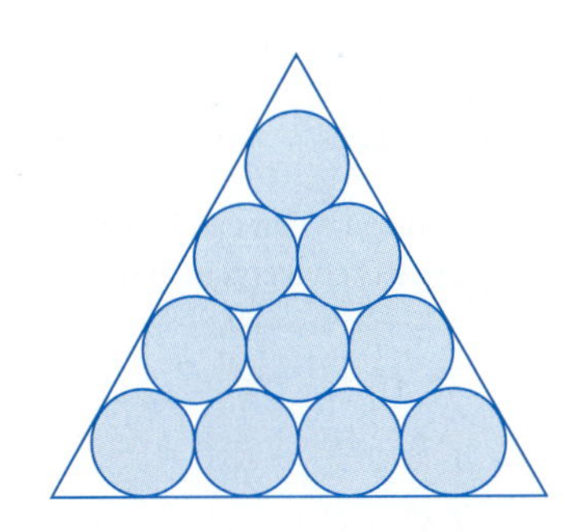

FIGURE FOR PROBLEM 15

15. Suppose that circles of equal diameter are packed tightly in n rows inside an equilateral triangle. (The figure illustrates the case $n = 4$.) If A is the area of the triangle and A_n is the total area occupied by the n rows of circles, show that

$$\lim_{n\to\infty} \frac{A_n}{A} = \frac{\pi}{2\sqrt{3}}$$

16. A sequence $\{a_n\}$ is defined recursively by the equations

$$a_0 = a_1 = 1 \qquad n(n-1)a_n = (n-1)(n-2)a_{n-1} - (n-3)a_{n-2}$$

Find the sum of the series $\sum_{n=0}^{\infty} a_n$.

17. If the curve $y = e^{-x/10}\sin x$, $x \geqslant 0$, is rotated about the x-axis, the resulting solid looks like an infinite decreasing string of beads.

(a) Find the exact volume of the nth bead. (Use either a table of integrals or a computer algebra system.)

(b) Find the total volume of the beads.

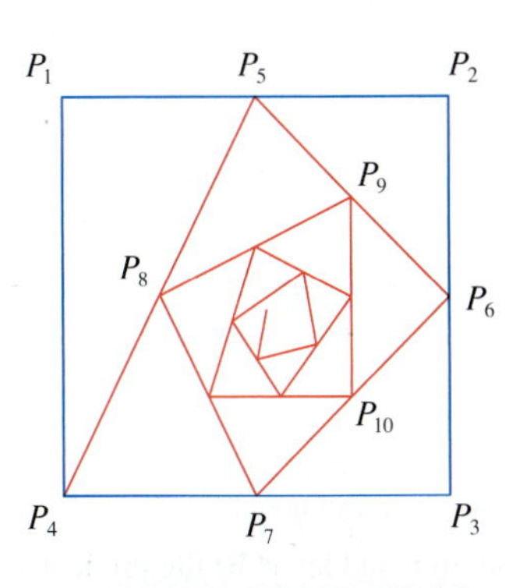

FIGURE FOR PROBLEM 18

18. Starting with the vertices $P_1(0, 1)$, $P_2(1, 1)$, $P_3(1, 0)$, $P_4(0, 0)$ of a square, we construct further points as shown in the figure: P_5 is the midpoint of P_1P_2, P_6 is the midpoint of P_2P_3, P_7 is the midpoint of P_3P_4, and so on. The polygonal spiral path $P_1P_2P_3P_4P_5P_6P_7\ldots$ approaches a point P inside the square.

(a) If the coordinates of P_n are (x_n, y_n), show that $\frac{1}{2}x_n + x_{n+1} + x_{n+2} + x_{n+3} = 2$ and find a similar equation for the y-coordinates.

(b) Find the coordinates of P.

19. Find the sum of the series $\displaystyle\sum_{n=1}^{\infty} \frac{(-1)^n}{(2n+1)3^n}$.

20. Carry out the following steps to show that

$$\frac{1}{1\cdot 2} + \frac{1}{3\cdot 4} + \frac{1}{5\cdot 6} + \frac{1}{7\cdot 8} + \cdots = \ln 2$$

(a) Use the formula for the sum of a finite geometric series (11.2.3) to get an expression for

$$1 - x + x^2 - x^3 + \cdots + x^{2n-2} - x^{2n-1}$$

(b) Integrate the result of part (a) from 0 to 1 to get an expression for

$$1 - \frac{1}{2} + \frac{1}{3} - \frac{1}{4} + \cdots + \frac{1}{2n-1} - \frac{1}{2n}$$

as an integral.

(c) Deduce from part (b) that

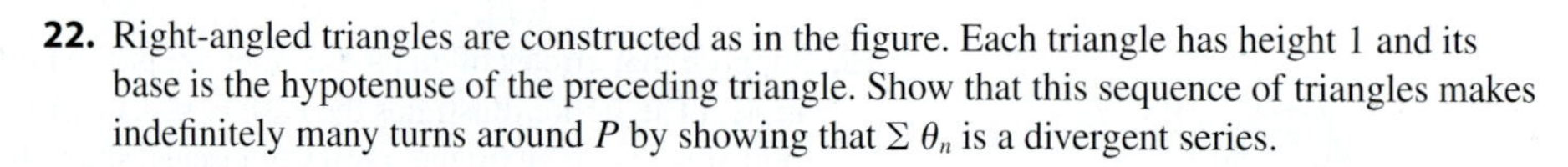

$$\left| \frac{1}{1 \cdot 2} + \frac{1}{3 \cdot 4} + \frac{1}{5 \cdot 6} + \cdots + \frac{1}{(2n-1)(2n)} - \int_0^1 \frac{dx}{1+x} \right| < \int_0^1 x^{2n}\, dx$$

(d) Use part (c) to show that the sum of the given series is ln 2.

21. Find all the solutions of the equation

$$1 + \frac{x}{2!} + \frac{x^2}{4!} + \frac{x^3}{6!} + \frac{x^4}{8!} + \cdots = 0$$

[*Hint:* Consider the cases $x \geqslant 0$ and $x < 0$ separately.]

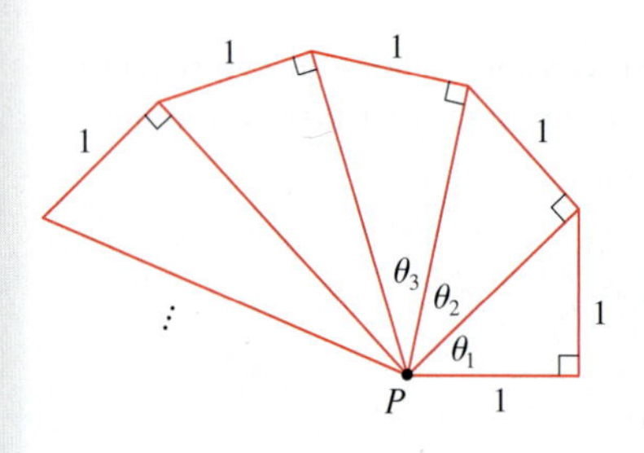

FIGURE FOR PROBLEM 22

22. Right-angled triangles are constructed as in the figure. Each triangle has height 1 and its base is the hypotenuse of the preceding triangle. Show that this sequence of triangles makes indefinitely many turns around P by showing that $\Sigma\, \theta_n$ is a divergent series.

23. Consider the series whose terms are the reciprocals of the positive integers that can be written in base 10 notation without using the digit 0. Show that this series is convergent and the sum is less than 90.

24. (a) Show that the Maclaurin series of the function

$$f(x) = \frac{x}{1 - x - x^2} \qquad \text{is} \qquad \sum_{n=1}^{\infty} f_n x^n$$

where f_n is the nth Fibonacci number, that is, $f_1 = 1$, $f_2 = 1$, and $f_n = f_{n-1} + f_{n-2}$ for $n \geqslant 3$. [*Hint:* Write $x/(1 - x - x^2) = c_0 + c_1 x + c_2 x^2 + \cdots$ and multiply both sides of this equation by $1 - x - x^2$.]

(b) By writing $f(x)$ as a sum of partial fractions and thereby obtaining the Maclaurin series in a different way, find an explicit formula for the nth Fibonacci number.

25. Let

$$u = 1 + \frac{x^3}{3!} + \frac{x^6}{6!} + \frac{x^9}{9!} + \cdots$$

$$v = x + \frac{x^4}{4!} + \frac{x^7}{7!} + \frac{x^{10}}{10!} + \cdots$$

$$w = \frac{x^2}{2!} + \frac{x^5}{5!} + \frac{x^8}{8!} + \cdots$$

Show that $u^3 + v^3 + w^3 - 3uvw = 1$.

26. Prove that if $n > 1$, the nth partial sum of the harmonic series is not an integer.

Hint: Let 2^k be the largest power of 2 that is less than or equal to n and let M be the product of all odd integers that are less than or equal to n. Suppose that $s_n = m$, an integer. Then $M2^k s_n = M2^k m$. The right side of this equation is even. Prove that the left side is odd by showing that each of its terms is an even integer, except for the last one.

12 Vectors and the Geometry of Space

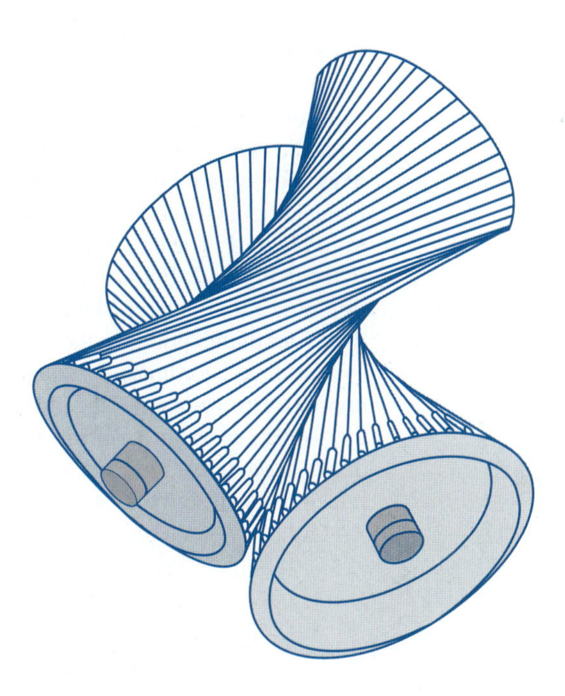

Each of these gears has the shape of a hyperboloid, a type of surface we will study in Section 12.6. The shape allows the gears to transmit motion between skew (neither parallel nor intersecting) axes.

IN THIS CHAPTER WE INTRODUCE vectors and coordinate systems for three-dimensional space. This will be the setting for our study of the calculus of functions of two variables in Chapter 14 because the graph of such a function is a surface in space. In this chapter we will see that vectors provide particularly simple descriptions of lines and planes in space.

12.1 Three-Dimensional Coordinate Systems

3D Space

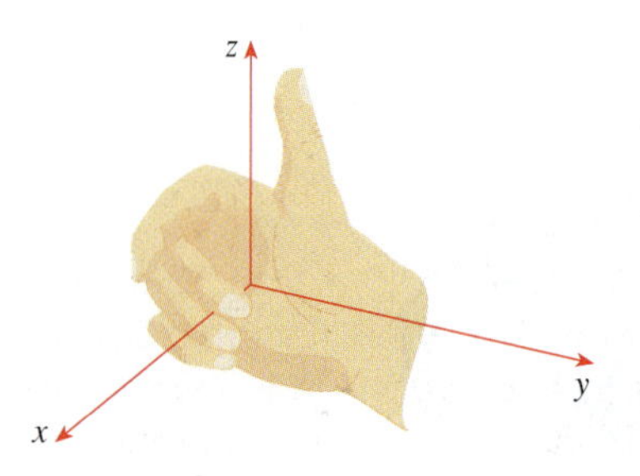

FIGURE 1
Coordinate axes

To locate a point in a plane, we need two numbers. We know that any point in the plane can be represented as an ordered pair (a, b) of real numbers, where a is the x-coordinate and b is the y-coordinate. For this reason, a plane is called two-dimensional. To locate a point in space, three numbers are required. We represent any point in space by an ordered triple (a, b, c) of real numbers.

In order to represent points in space, we first choose a fixed point O (the origin) and three directed lines through O that are perpendicular to each other, called the **coordinate axes** and labeled the x-axis, y-axis, and z-axis. Usually we think of the x- and y-axes as being horizontal and the z-axis as being vertical, and we draw the orientation of the axes as in Figure 1. The direction of the z-axis is determined by the **right-hand rule** as illustrated in Figure 2: If you curl the fingers of your right hand around the z-axis in the direction of a 90° counterclockwise rotation from the positive x-axis to the positive y-axis, then your thumb points in the positive direction of the z-axis.

FIGURE 2
Right-hand rule

The three coordinate axes determine the three **coordinate planes** illustrated in Figure 3(a). The xy-plane is the plane that contains the x- and y-axes; the yz-plane contains the y- and z-axes; the xz-plane contains the x- and z-axes. These three coordinate planes divide space into eight parts, called **octants**. The **first octant**, in the foreground, is determined by the positive axes.

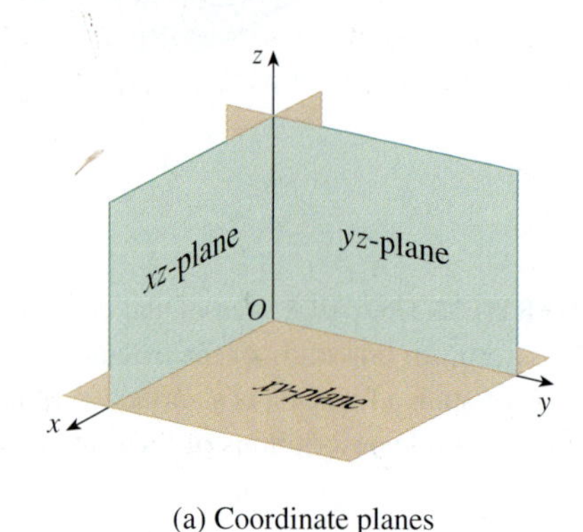

(a) Coordinate planes

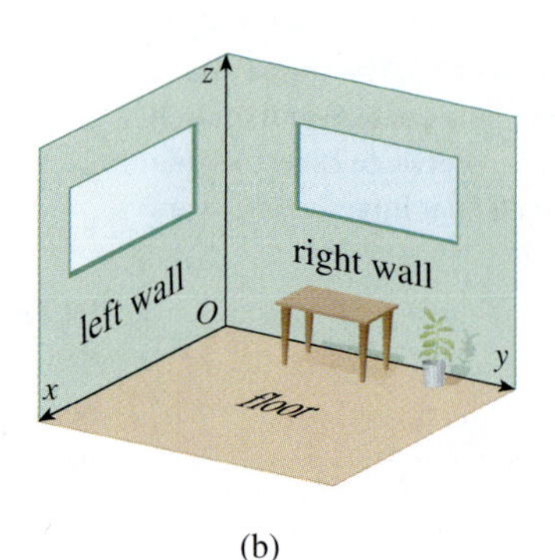

(b)

FIGURE 3

Because many people have some difficulty visualizing diagrams of three-dimensional figures, you may find it helpful to do the following [see Figure 3(b)]. Look at any bottom corner of a room and call the corner the origin. The wall on your left is in the xz-plane, the wall on your right is in the yz-plane, and the floor is in the xy-plane. The x-axis runs along the intersection of the floor and the left wall. The y-axis runs along the intersection of the floor and the right wall. The z-axis runs up from the floor toward the ceiling along the intersection of the two walls. You are situated in the first octant, and you can now imagine seven other rooms situated in the other seven octants (three on the same floor and four on the floor below), all connected by the common corner point O.

Now if P is any point in space, let a be the (directed) distance from the yz-plane to P, let b be the distance from the xz-plane to P, and let c be the distance from the xy-plane to P. We represent the point P by the ordered triple (a, b, c) of real numbers and we call a, b, and c the **coordinates** of P; a is the x-coordinate, b is the y-coordinate, and c is the z-coordinate. Thus, to locate the point (a, b, c), we can start at the origin O and move a units along the x-axis, then b units parallel to the y-axis, and then c units parallel to the z-axis as in Figure 4.

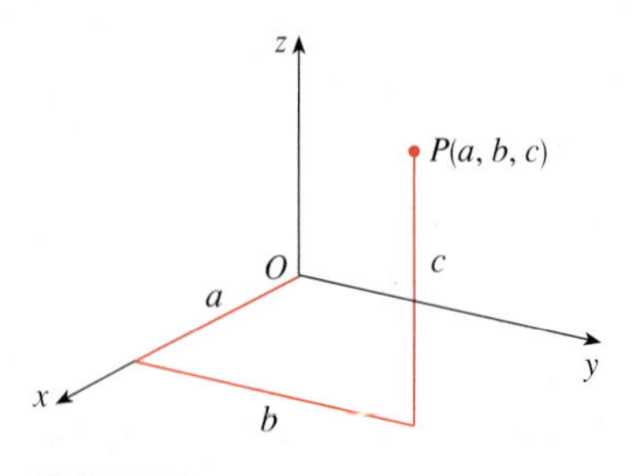

FIGURE 4

The point $P(a, b, c)$ determines a rectangular box as in Figure 5. If we drop a perpendicular from P to the xy-plane, we get a point Q with coordinates $(a, b, 0)$ called the **projection** of P onto the xy-plane. Similarly, $R(0, b, c)$ and $S(a, 0, c)$ are the projections of P onto the yz-plane and xz-plane, respectively.

As numerical illustrations, the points $(-4, 3, -5)$ and $(3, -2, -6)$ are plotted in Figure 6.

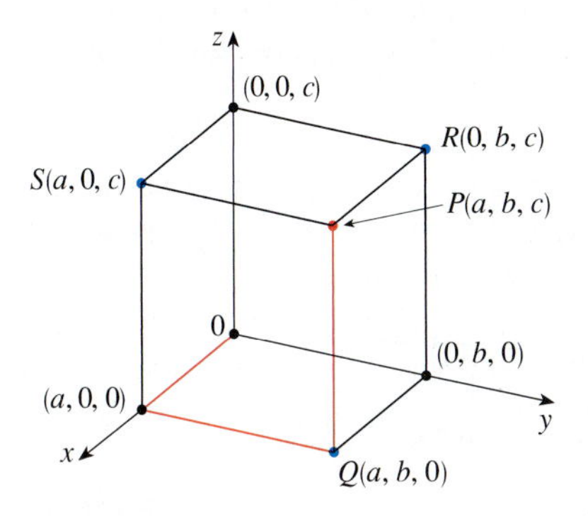

FIGURE 5

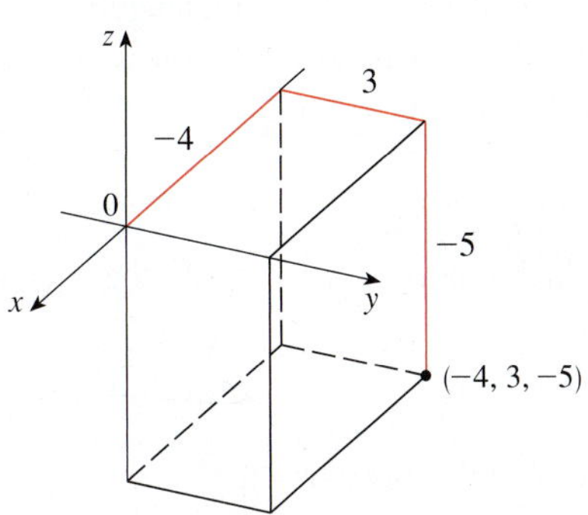

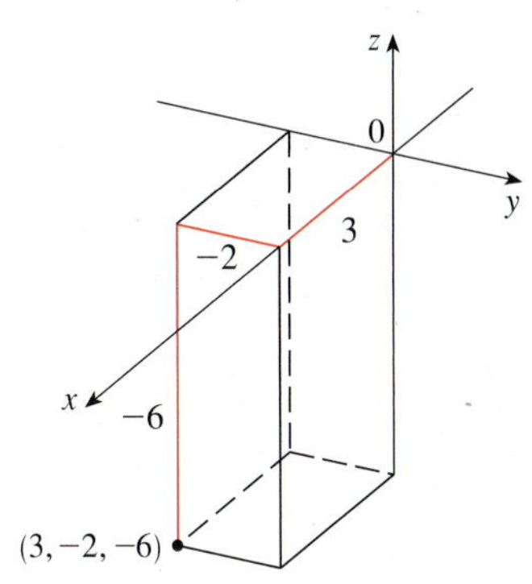

FIGURE 6

The Cartesian product $\mathbb{R} \times \mathbb{R} \times \mathbb{R} = \{(x, y, z) \mid x, y, z \in \mathbb{R}\}$ is the set of all ordered triples of real numbers and is denoted by $\mathbb{R}^3$. We have given a one-to-one correspondence between points P in space and ordered triples (a, b, c) in $\mathbb{R}^3$. It is called a **three-dimensional rectangular coordinate system**. Notice that, in terms of coordinates, the first octant can be described as the set of points whose coordinates are all positive.

Surfaces

In two-dimensional analytic geometry, the graph of an equation involving x and y is a curve in $\mathbb{R}^2$. In three-dimensional analytic geometry, an equation in x, y, and z represents a *surface* in $\mathbb{R}^3$.

EXAMPLE 1 What surfaces in $\mathbb{R}^3$ are represented by the following equations?

(a) $z = 3$ (b) $y = 5$

SOLUTION

(a) The equation $z = 3$ represents the set $\{(x, y, z) \mid z = 3\}$, which is the set of all points in $\mathbb{R}^3$ whose z-coordinate is 3 (x and y can each be any value). This is the horizontal plane that is parallel to the xy-plane and three units above it as in Figure 7(a).

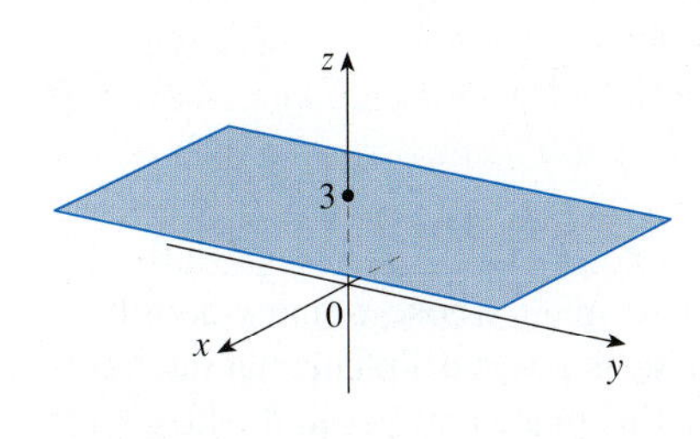

(a) $z = 3$, a plane in $\mathbb{R}^3$

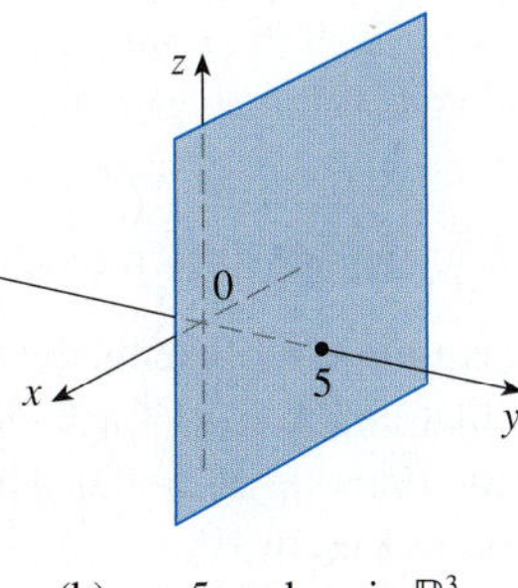

(b) $y = 5$, a plane in $\mathbb{R}^3$

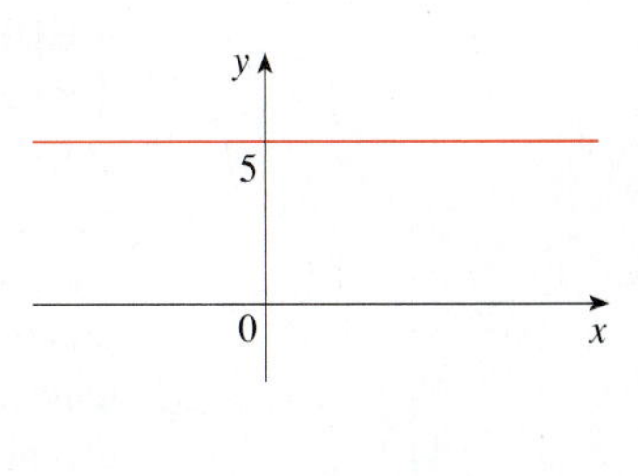

(c) $y = 5$, a line in $\mathbb{R}^2$

FIGURE 7

(b) The equation $y = 5$ represents the set of all points in $\mathbb{R}^3$ whose y-coordinate is 5. This is the vertical plane that is parallel to the xz-plane and five units to the right of it as in Figure 7(b). ■

NOTE When an equation is given, we must understand from the context whether it represents a curve in $\mathbb{R}^2$ or a surface in $\mathbb{R}^3$. In Example 1, $y = 5$ represents a plane in $\mathbb{R}^3$, but of course $y = 5$ can also represent a line in $\mathbb{R}^2$ if we are dealing with two-dimensional analytic geometry. See Figure 7(b) and (c).

In general, if k is a constant, then $x = k$ represents a plane parallel to the yz-plane, $y = k$ is a plane parallel to the xz-plane, and $z = k$ is a plane parallel to the xy-plane. In Figure 5, the faces of the rectangular box are formed by the three coordinate planes $x = 0$ (the yz-plane), $y = 0$ (the xz-plane), and $z = 0$ (the xy-plane), and the planes $x = a$, $y = b$, and $z = c$.

EXAMPLE 2

(a) Which points (x, y, z) satisfy the equations

$$x^2 + y^2 = 1 \qquad \text{and} \qquad z = 3$$

(b) What does the equation $x^2 + y^2 = 1$ represent as a surface in $\mathbb{R}^3$?

SOLUTION

(a) Because $z = 3$, the points lie in the horizontal plane $z = 3$ from Example 1(a). Because $x^2 + y^2 = 1$, the points lie on the circle with radius 1 and center on the z-axis. See Figure 8.

(b) Given that $x^2 + y^2 = 1$, with no restrictions on z, we see that the point (x, y, z) could lie on a circle in any horizontal plane $z = k$. So the surface $x^2 + y^2 = 1$ in $\mathbb{R}^3$ consists of all possible horizontal circles $x^2 + y^2 = 1$, $z = k$, and is therefore the circular cylinder with radius 1 whose axis is the z-axis. See Figure 9.

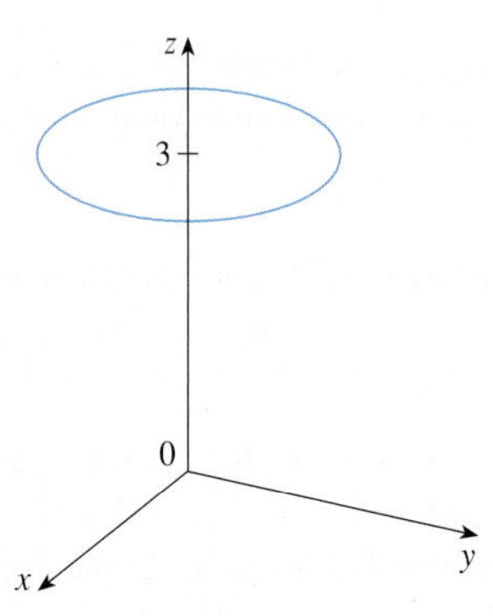

FIGURE 8
The circle $x^2 + y^2 = 1$, $z = 3$

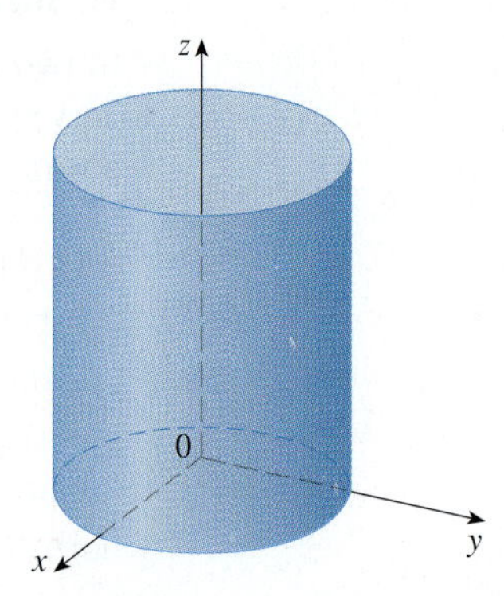

FIGURE 9
The cylinder $x^2 + y^2 = 1$ ■

EXAMPLE 3 Describe and sketch the surface in $\mathbb{R}^3$ represented by the equation $y = x$.

SOLUTION The equation represents the set of all points in $\mathbb{R}^3$ whose x- and y-coordinates are equal, that is, $\{(x, x, z) \mid x \in \mathbb{R}, z \in \mathbb{R}\}$. This is a vertical plane that intersects the xy-plane in the line $y = x$, $z = 0$. The portion of this plane that lies in the first octant is sketched in Figure 10. ■

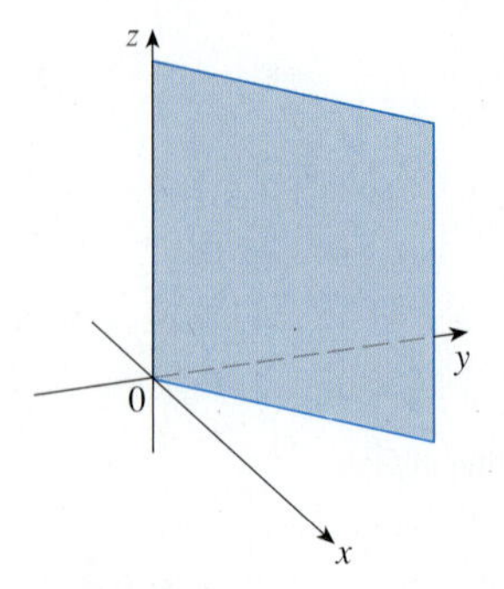

FIGURE 10
The plane $y = x$

■ Distance and Spheres

The familiar formula for the distance between two points in a plane is easily extended to the following three-dimensional formula.

> **Distance Formula in Three Dimensions** The distance $|P_1P_2|$ between the points $P_1(x_1, y_1, z_1)$ and $P_2(x_2, y_2, z_2)$ is
>
> $$|P_1P_2| = \sqrt{(x_2 - x_1)^2 + (y_2 - y_1)^2 + (z_2 - z_1)^2}$$

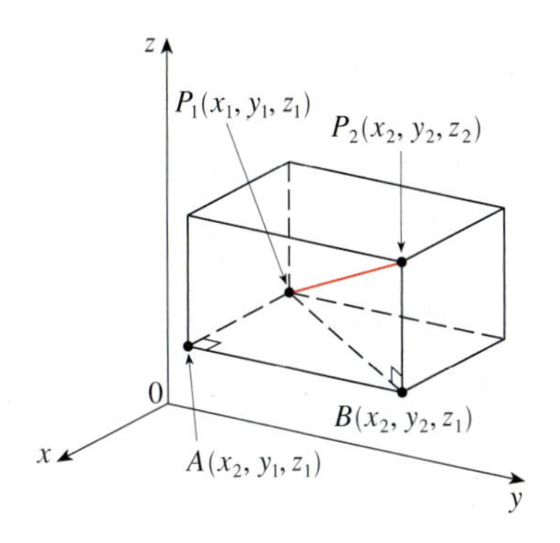

FIGURE 11

To see why this formula is true, we construct a rectangular box as in Figure 11, where P_1 and P_2 are opposite vertices and the faces of the box are parallel to the coordinate planes. If $A(x_2, y_1, z_1)$ and $B(x_2, y_2, z_1)$ are the vertices of the box indicated in the figure, then

$$|P_1A| = |x_2 - x_1| \qquad |AB| = |y_2 - y_1| \qquad |BP_2| = |z_2 - z_1|$$

Because triangles P_1BP_2 and P_1AB are both right-angled, two applications of the Pythagorean Theorem give

$$|P_1P_2|^2 = |P_1B|^2 + |BP_2|^2$$

and

$$|P_1B|^2 = |P_1A|^2 + |AB|^2$$

Combining these equations, we get

$$\begin{aligned} |P_1P_2|^2 &= |P_1A|^2 + |AB|^2 + |BP_2|^2 \\ &= |x_2 - x_1|^2 + |y_2 - y_1|^2 + |z_2 - z_1|^2 \\ &= (x_2 - x_1)^2 + (y_2 - y_1)^2 + (z_2 - z_1)^2 \end{aligned}$$

Therefore $$|P_1P_2| = \sqrt{(x_2 - x_1)^2 + (y_2 - y_1)^2 + (z_2 - z_1)^2}$$

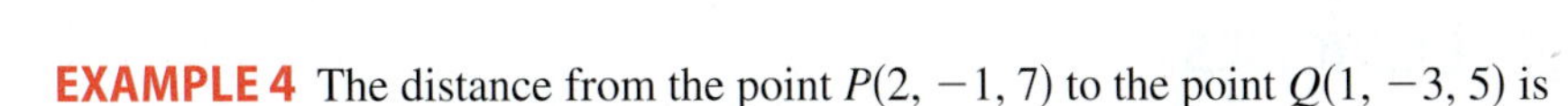

EXAMPLE 4 The distance from the point $P(2, -1, 7)$ to the point $Q(1, -3, 5)$ is

$$|PQ| = \sqrt{(1 - 2)^2 + (-3 + 1)^2 + (5 - 7)^2} = \sqrt{1 + 4 + 4} = 3$$ ■

EXAMPLE 5 Find an equation of a sphere with radius r and center $C(h, k, l)$.

SOLUTION By definition, a sphere is the set of all points $P(x, y, z)$ whose distance from C is r. (See Figure 12.) Thus P is on the sphere if and only if $|PC| = r$. Squaring both sides, we have $|PC|^2 = r^2$ or

$$(x - h)^2 + (y - k)^2 + (z - l)^2 = r^2$$ ■

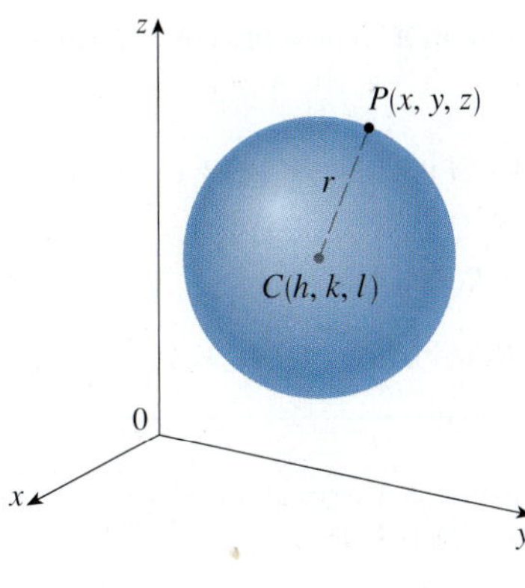

FIGURE 12

The result of Example 5 is worth remembering.

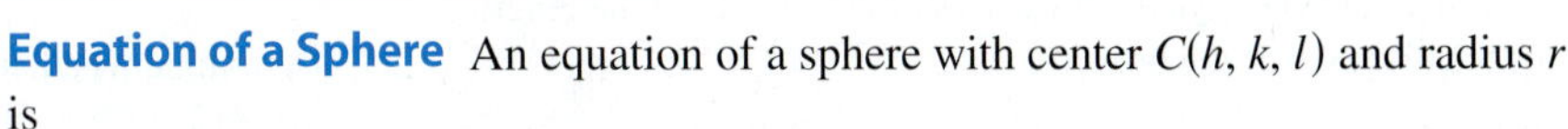

> **Equation of a Sphere** An equation of a sphere with center $C(h, k, l)$ and radius r is
>
> $$(x - h)^2 + (y - k)^2 + (z - l)^2 = r^2$$
>
> In particular, if the center is the origin O, then an equation of the sphere is
>
> $$x^2 + y^2 + z^2 = r^2$$

EXAMPLE 6 Show that $x^2 + y^2 + z^2 + 4x - 6y + 2z + 6 = 0$ is the equation of a sphere, and find its center and radius.

SOLUTION We can rewrite the given equation in the form of an equation of a sphere if we complete squares:

$$(x^2 + 4x + 4) + (y^2 - 6y + 9) + (z^2 + 2z + 1) = -6 + 4 + 9 + 1$$

$$(x + 2)^2 + (y - 3)^2 + (z + 1)^2 = 8$$

Comparing this equation with the standard form, we see that it is the equation of a sphere with center $(-2, 3, -1)$ and radius $\sqrt{8} = 2\sqrt{2}$. ■

EXAMPLE 7 What region in $\mathbb{R}^3$ is represented by the following inequalities?

$$1 \leqslant x^2 + y^2 + z^2 \leqslant 4 \qquad z \leqslant 0$$

SOLUTION The inequalities

$$1 \leqslant x^2 + y^2 + z^2 \leqslant 4$$

can be rewritten as

$$1 \leqslant \sqrt{x^2 + y^2 + z^2} \leqslant 2$$

so they represent the points (x, y, z) whose distance from the origin is at least 1 and at most 2. But we are also given that $z \leqslant 0$, so the points lie on or below the xy-plane. Thus the given inequalities represent the region that lies between (or on) the spheres $x^2 + y^2 + z^2 = 1$ and $x^2 + y^2 + z^2 = 4$ and beneath (or on) the xy-plane. It is sketched in Figure 13. ■

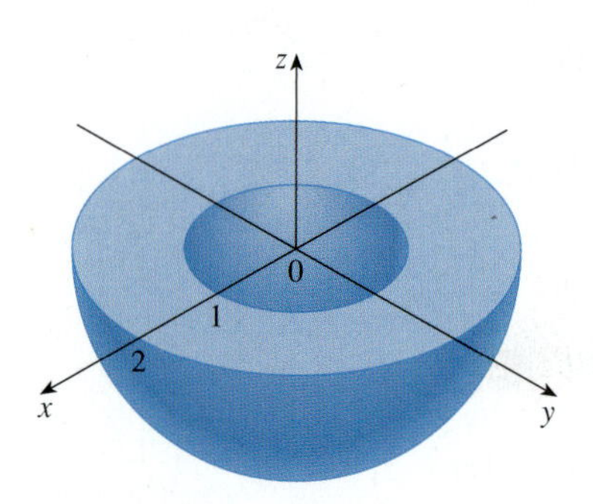

FIGURE 13

12.1 EXERCISES

1. Suppose you start at the origin, move along the x-axis a distance of 4 units in the positive direction, and then move downward a distance of 3 units. What are the coordinates of your position?

2. Sketch the points $(1, 5, 3)$, $(0, 2, -3)$, $(-3, 0, 2)$, and $(2, -2, -1)$ on a single set of coordinate axes.

3. Which of the points $A(-4, 0, -1)$, $B(3, 1, -5)$, and $C(2, 4, 6)$ is closest to the yz-plane? Which point lies in the xz-plane?

4. What are the projections of the point $(2, 3, 5)$ on the xy-, yz-, and xz-planes? Draw a rectangular box with the origin and $(2, 3, 5)$ as opposite vertices and with its faces parallel to the coordinate planes. Label all vertices of the box. Find the length of the diagonal of the box.

5. What does the equation $x = 4$ represent in $\mathbb{R}^2$? What does it represent in $\mathbb{R}^3$? Illustrate with sketches.

6. What does the equation $y = 3$ represent in $\mathbb{R}^3$? What does $z = 5$ represent? What does the pair of equations $y = 3$, $z = 5$ represent? In other words, describe the set of points (x, y, z) such that $y = 3$ and $z = 5$. Illustrate with a sketch.

7. Describe and sketch the surface in $\mathbb{R}^3$ represented by the equation $x + y = 2$.

8. Describe and sketch the surface in $\mathbb{R}^3$ represented by the equation $x^2 + z^2 = 9$.

9–10 Find the lengths of the sides of the triangle PQR. Is it a right triangle? Is it an isosceles triangle?

9. $P(3, -2, -3)$, $Q(7, 0, 1)$, $R(1, 2, 1)$

10. $P(2, -1, 0)$, $Q(4, 1, 1)$, $R(4, -5, 4)$

11. Determine whether the points lie on a straight line.
(a) $A(2, 4, 2)$, $B(3, 7, -2)$, $C(1, 3, 3)$
(b) $D(0, -5, 5)$, $E(1, -2, 4)$, $F(3, 4, 2)$

12. Find the distance from $(4, -2, 6)$ to each of the following.
(a) The xy-plane
(b) The yz-plane
(c) The xz-plane
(d) The x-axis
(e) The y-axis
(f) The z-axis

13. Find an equation of the sphere with center $(-3, 2, 5)$ and radius 4. What is the intersection of this sphere with the yz-plane?

14. Find an equation of the sphere with center $(2, -6, 4)$ and radius 5. Describe its intersection with each of the coordinate planes.

15. Find an equation of the sphere that passes through the point $(4, 3, -1)$ and has center $(3, 8, 1)$.

16. Find an equation of the sphere that passes through the origin and whose center is $(1, 2, 3)$.

17–20 Show that the equation represents a sphere, and find its center and radius.

17. $x^2 + y^2 + z^2 - 2x - 4y + 8z = 15$

18. $x^2 + y^2 + z^2 + 8x - 6y + 2z + 17 = 0$

19. $2x^2 + 2y^2 + 2z^2 = 8x - 24z + 1$

20. $3x^2 + 3y^2 + 3z^2 = 10 + 6y + 12z$

21. (a) Prove that the midpoint of the line segment from $P_1(x_1, y_1, z_1)$ to $P_2(x_2, y_2, z_2)$ is

$$\left(\frac{x_1 + x_2}{2}, \frac{y_1 + y_2}{2}, \frac{z_1 + z_2}{2}\right)$$

(b) Find the lengths of the medians of the triangle with vertices $A(1, 2, 3)$, $B(-2, 0, 5)$, and $C(4, 1, 5)$. (A *median* of a triangle is a line segment that joins a vertex to the midpoint of the opposite side.)

22. Find an equation of a sphere if one of its diameters has endpoints $(5, 4, 3)$ and $(1, 6, -9)$.

23. Find equations of the spheres with center $(2, -3, 6)$ that touch (a) the xy-plane, (b) the yz-plane, (c) the xz-plane.

24. Find an equation of the largest sphere with center $(5, 4, 9)$ that is contained in the first octant.

25–38 Describe in words the region of $\mathbb{R}^3$ represented by the equation(s) or inequality.

25. $x = 5$

26. $y = -2$

27. $y < 8$

28. $z \geqslant -1$

29. $0 \leqslant z \leqslant 6$

30. $y^2 = 4$

31. $x^2 + y^2 = 4, \quad z = -1$

32. $x^2 + y^2 = 4$

33. $x^2 + y^2 + z^2 = 4$

34. $x^2 + y^2 + z^2 \leqslant 4$

35. $1 \leqslant x^2 + y^2 + z^2 \leqslant 5$

36. $x = z$

37. $x^2 + z^2 \leqslant 9$

38. $x^2 + y^2 + z^2 > 2z$

39–42 Write inequalities to describe the region.

39. The region between the yz-plane and the vertical plane $x = 5$

40. The solid cylinder that lies on or below the plane $z = 8$ and on or above the disk in the xy-plane with center the origin and radius 2

41. The region consisting of all points between (but not on) the spheres of radius r and R centered at the origin, where $r < R$

42. The solid upper hemisphere of the sphere of radius 2 centered at the origin

43. The figure shows a line L_1 in space and a second line L_2, which is the projection of L_1 onto the xy-plane. (In other words, the points on L_2 are directly beneath, or above, the points on L_1.)

(a) Find the coordinates of the point P on the line L_1.

(b) Locate on the diagram the points A, B, and C, where the line L_1 intersects the xy-plane, the yz-plane, and the xz-plane, respectively.

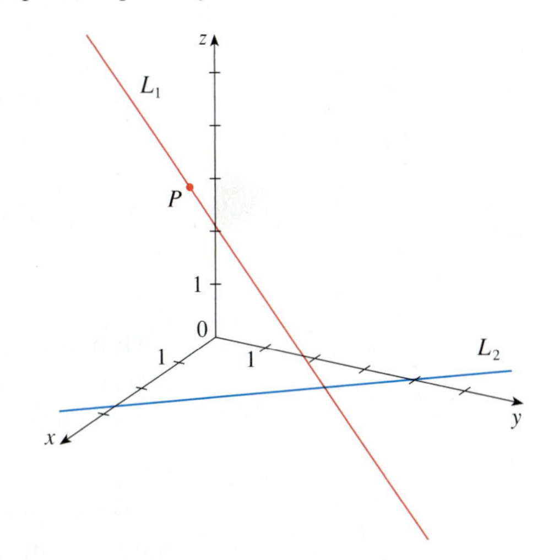

44. Consider the points P such that the distance from P to $A(-1, 5, 3)$ is twice the distance from P to $B(6, 2, -2)$. Show that the set of all such points is a sphere, and find its center and radius.

45. Find an equation of the set of all points equidistant from the points $A(-1, 5, 3)$ and $B(6, 2, -2)$. Describe the set.

46. Find the volume of the solid that lies inside both of the spheres

$$x^2 + y^2 + z^2 + 4x - 2y + 4z + 5 = 0$$

and

$$x^2 + y^2 + z^2 = 4$$

47. Find the distance between the spheres $x^2 + y^2 + z^2 = 4$ and $x^2 + y^2 + z^2 = 4x + 4y + 4z - 11$.

48. Describe and sketch a solid with the following properties. When illuminated by rays parallel to the z-axis, its shadow is a circular disk. If the rays are parallel to the y-axis, its shadow is a square. If the rays are parallel to the x-axis, its shadow is an isosceles triangle.

12.2 Vectors

The term **vector** is used by scientists to indicate a quantity (such as displacement or velocity or force) that has both magnitude and direction. A vector is often represented by an arrow or a directed line segment. The length of the arrow represents the magnitude of the vector and the arrow points in the direction of the vector. We denote a vector by printing a letter in boldface ($\mathbf{v}$) or by putting an arrow above the letter ($\vec{v}$).

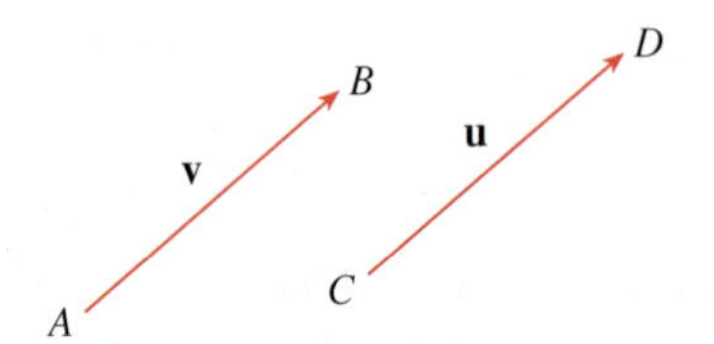

FIGURE 1
Equivalent vectors

For instance, suppose a particle moves along a line segment from point A to point B. The corresponding **displacement vector** $\mathbf{v}$, shown in Figure 1, has **initial point** A (the tail) and **terminal point** B (the tip) and we indicate this by writing $\mathbf{v} = \overrightarrow{AB}$. Notice that the vector $\mathbf{u} = \overrightarrow{CD}$ has the same length and the same direction as $\mathbf{v}$ even though it is in a different position. We say that $\mathbf{u}$ and $\mathbf{v}$ are **equivalent** (or **equal**) and we write $\mathbf{u} = \mathbf{v}$. The **zero vector**, denoted by $\mathbf{0}$, has length 0. It is the only vector with no specific direction.

Combining Vectors

Suppose a particle moves from A to B, so its displacement vector is $\overrightarrow{AB}$. Then the particle changes direction and moves from B to C, with displacement vector $\overrightarrow{BC}$ as in Figure 2. The combined effect of these displacements is that the particle has moved from A to C. The resulting displacement vector $\overrightarrow{AC}$ is called the *sum* of $\overrightarrow{AB}$ and $\overrightarrow{BC}$ and we write

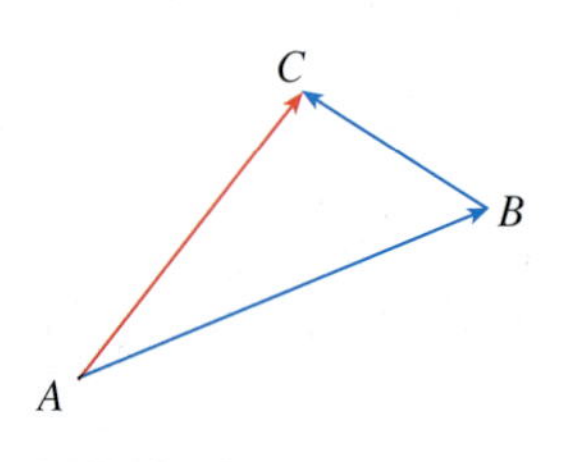

FIGURE 2

$$\overrightarrow{AC} = \overrightarrow{AB} + \overrightarrow{BC}$$

In general, if we start with vectors $\mathbf{u}$ and $\mathbf{v}$, we first move $\mathbf{v}$ so that its tail coincides with the tip of $\mathbf{u}$ and define the sum of $\mathbf{u}$ and $\mathbf{v}$ as follows.

Definition of Vector Addition If $\mathbf{u}$ and $\mathbf{v}$ are vectors positioned so the initial point of $\mathbf{v}$ is at the terminal point of $\mathbf{u}$, then the **sum** $\mathbf{u} + \mathbf{v}$ is the vector from the initial point of $\mathbf{u}$ to the terminal point of $\mathbf{v}$.

The definition of vector addition is illustrated in Figure 3. You can see why this definition is sometimes called the **Triangle Law**.

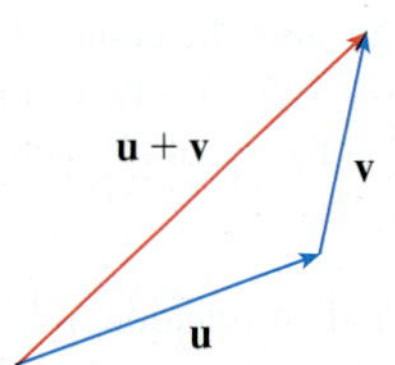

FIGURE 3
The Triangle Law

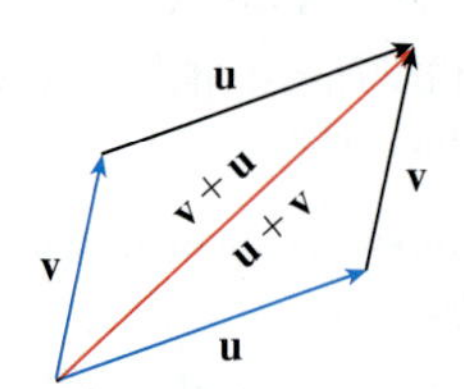

FIGURE 4
The Parallelogram Law

In Figure 4 we start with the same vectors $\mathbf{u}$ and $\mathbf{v}$ as in Figure 3 and draw another copy of $\mathbf{v}$ with the same initial point as $\mathbf{u}$. Completing the parallelogram, we see that $\mathbf{u} + \mathbf{v} = \mathbf{v} + \mathbf{u}$. This also gives another way to construct the sum: if we place $\mathbf{u}$ and $\mathbf{v}$ so they start at the same point, then $\mathbf{u} + \mathbf{v}$ lies along the diagonal of the parallelogram with $\mathbf{u}$ and $\mathbf{v}$ as sides. (This is called the **Parallelogram Law**.)

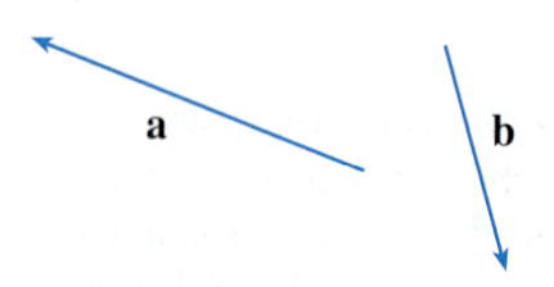

FIGURE 5

EXAMPLE 1 Draw the sum of the vectors $\mathbf{a}$ and $\mathbf{b}$ shown in Figure 5.

SOLUTION First we move $\mathbf{b}$ and place its tail at the tip of $\mathbf{a}$, being careful to draw a copy of $\mathbf{b}$ that has the same length and direction. Then we draw the vector $\mathbf{a} + \mathbf{b}$ [see

Figure 6(a)] starting at the initial point of **a** and ending at the terminal point of the copy of **b**.

Alternatively, we could place **b** so it starts where **a** starts and construct **a** + **b** by the Parallelogram Law as in Figure 6(b).

TEC Visual 12.2 shows how the Triangle and Parallelogram Laws work for various vectors **a** and **b**.

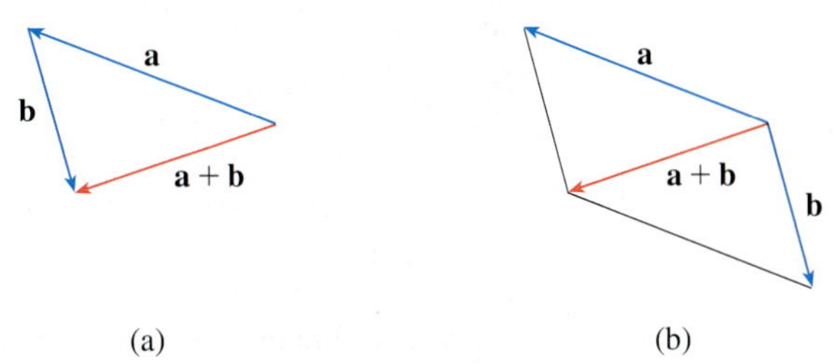

FIGURE 6 (a) (b)

It is possible to multiply a vector by a real number c. (In this context we call the real number c a **scalar** to distinguish it from a vector.) For instance, we want $2\mathbf{v}$ to be the same vector as $\mathbf{v} + \mathbf{v}$, which has the same direction as $\mathbf{v}$ but is twice as long. In general, we multiply a vector by a scalar as follows.

> **Definition of Scalar Multiplication** If c is a scalar and $\mathbf{v}$ is a vector, then the **scalar multiple** $c\mathbf{v}$ is the vector whose length is $|c|$ times the length of $\mathbf{v}$ and whose direction is the same as $\mathbf{v}$ if $c > 0$ and is opposite to $\mathbf{v}$ if $c < 0$. If $c = 0$ or $\mathbf{v} = \mathbf{0}$, then $c\mathbf{v} = \mathbf{0}$.

This definition is illustrated in Figure 7. We see that real numbers work like scaling factors here; that's why we call them scalars. Notice that two nonzero vectors are **parallel** if they are scalar multiples of one another. In particular, the vector $-\mathbf{v} = (-1)\mathbf{v}$ has the same length as $\mathbf{v}$ but points in the opposite direction. We call it the **negative** of $\mathbf{v}$.

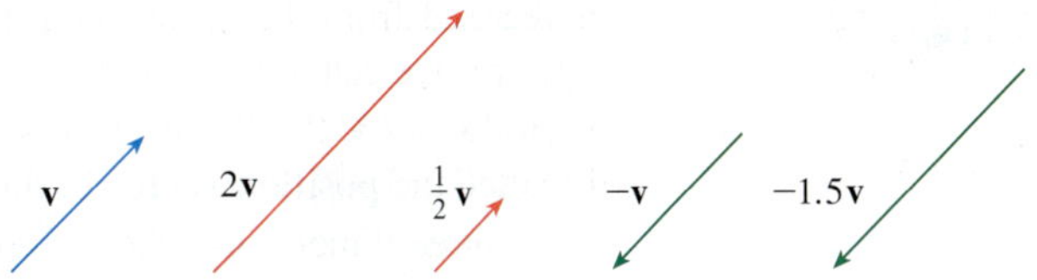

$\mathbf{v}$ $\quad 2\mathbf{v}$ $\quad \frac{1}{2}\mathbf{v}$ $\quad -\mathbf{v}$ $\quad -1.5\mathbf{v}$

FIGURE 7
Scalar multiples of **v**

By the **difference u − v** of two vectors we mean

$$\mathbf{u} - \mathbf{v} = \mathbf{u} + (-\mathbf{v})$$

So we can construct $\mathbf{u} - \mathbf{v}$ by first drawing the negative of $\mathbf{v}$, $-\mathbf{v}$, and then adding it to $\mathbf{u}$ by the Parallelogram Law as in Figure 8(a). Alternatively, since $\mathbf{v} + (\mathbf{u} - \mathbf{v}) = \mathbf{u}$, the vector $\mathbf{u} - \mathbf{v}$, when added to $\mathbf{v}$, gives $\mathbf{u}$. So we could construct $\mathbf{u} - \mathbf{v}$ as in Figure 8(b) by means of the Triangle Law. Notice that if $\mathbf{u}$ and $\mathbf{v}$ both start from the same initial point, then $\mathbf{u} - \mathbf{v}$ connects the tip of $\mathbf{v}$ to the tip of $\mathbf{u}$.

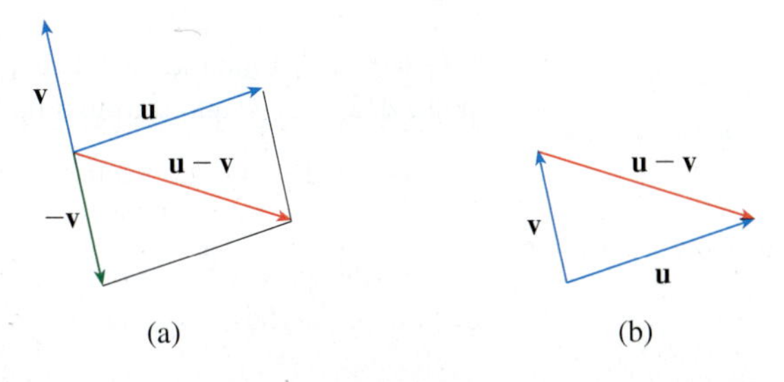

FIGURE 8
Drawing **u − v** (a) (b)

FIGURE 9

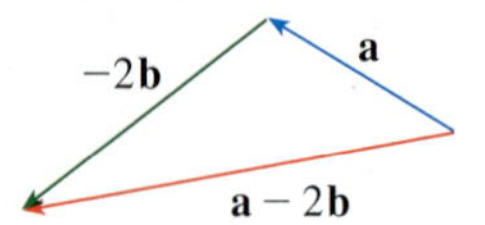

FIGURE 10

EXAMPLE 2 If **a** and **b** are the vectors shown in Figure 9, draw **a** − 2**b**.

SOLUTION We first draw the vector −2**b** pointing in the direction opposite to **b** and twice as long. We place it with its tail at the tip of **a** and then use the Triangle Law to draw **a** + (−2**b**) as in Figure 10. ■

Components

For some purposes it's best to introduce a coordinate system and treat vectors algebraically. If we place the initial point of a vector **a** at the origin of a rectangular coordinate system, then the terminal point of **a** has coordinates of the form (a_1, a_2) or (a_1, a_2, a_3), depending on whether our coordinate system is two- or three-dimensional (see Figure 11). These coordinates are called the **components** of **a** and we write

$$\mathbf{a} = \langle a_1, a_2 \rangle \qquad \text{or} \qquad \mathbf{a} = \langle a_1, a_2, a_3 \rangle$$

We use the notation $\langle a_1, a_2 \rangle$ for the ordered pair that refers to a vector so as not to confuse it with the ordered pair (a_1, a_2) that refers to a point in the plane.

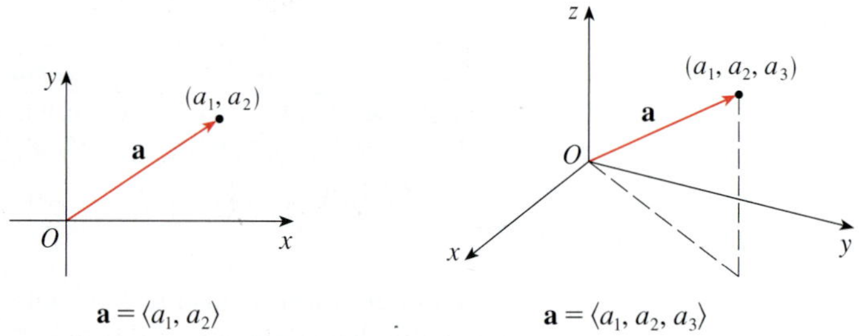

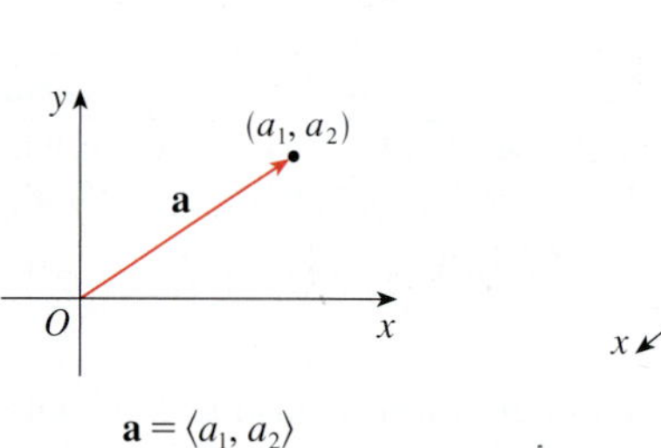

z
(a_1, a_2, a_3)
a
O
x
y
$\mathbf{a} = \langle a_1, a_2, a_3 \rangle$

FIGURE 11

For instance, the vectors shown in Figure 12 are all equivalent to the vector $\overrightarrow{OP} = \langle 3, 2 \rangle$ whose terminal point is $P(3, 2)$. What they have in common is that the terminal point is reached from the initial point by a displacement of three units to the right and two upward. We can think of all these geometric vectors as **representations** of the algebraic vector $\mathbf{a} = \langle 3, 2 \rangle$. The particular representation $\overrightarrow{OP}$ from the origin to the point $P(3, 2)$ is called the **position vector** of the point P.

In three dimensions, the vector $\mathbf{a} = \overrightarrow{OP} = \langle a_1, a_2, a_3 \rangle$ is the **position vector** of the point $P(a_1, a_2, a_3)$. (See Figure 13.) Let's consider any other representation $\overrightarrow{AB}$ of **a**, where the initial point is $A(x_1, y_1, z_1)$ and the terminal point is $B(x_2, y_2, z_2)$. Then we must have $x_1 + a_1 = x_2$, $y_1 + a_2 = y_2$, and $z_1 + a_3 = z_2$ and so $a_1 = x_2 - x_1$, $a_2 = y_2 - y_1$, and $a_3 = z_2 - z_1$. Thus we have the following result.

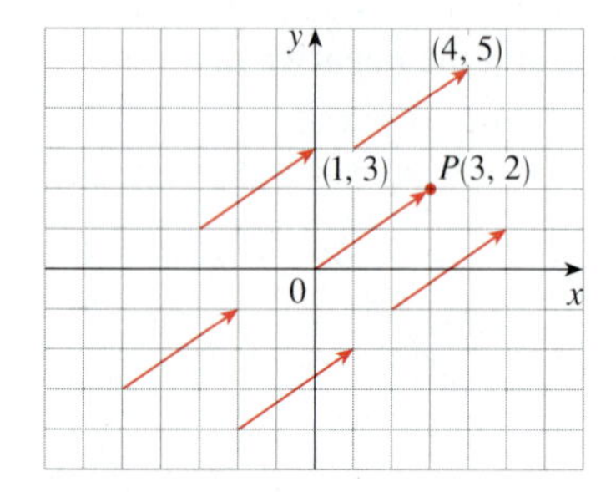

FIGURE 12
Representations of $\mathbf{a} = \langle 3, 2 \rangle$

> **1** Given the points $A(x_1, y_1, z_1)$ and $B(x_2, y_2, z_2)$, the vector **a** with representation $\overrightarrow{AB}$ is
>
> $$\mathbf{a} = \langle x_2 - x_1, y_2 - y_1, z_2 - z_1 \rangle$$

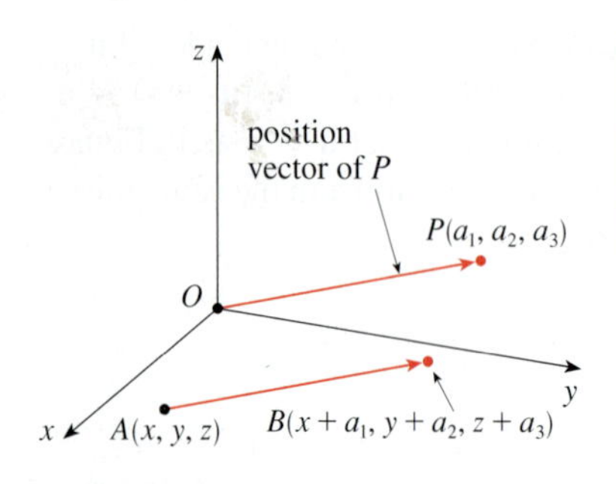

FIGURE 13
Representations of $\mathbf{a} = \langle a_1, a_2, a_3 \rangle$

EXAMPLE 3 Find the vector represented by the directed line segment with initial point $A(2, -3, 4)$ and terminal point $B(-2, 1, 1)$.

SOLUTION By (1), the vector corresponding to $\overrightarrow{AB}$ is

$$\mathbf{a} = \langle -2 - 2, 1 - (-3), 1 - 4 \rangle = \langle -4, 4, -3 \rangle$$ ■

The **magnitude** or **length** of the vector **v** is the length of any of its representations and is denoted by the symbol $|\mathbf{v}|$ or $\|\mathbf{v}\|$. By using the distance formula to compute the length of a segment OP, we obtain the following formulas.

> The length of the two-dimensional vector $\mathbf{a} = \langle a_1, a_2 \rangle$ is
>
> $$|\mathbf{a}| = \sqrt{a_1^2 + a_2^2}$$
>
> The length of the three-dimensional vector $\mathbf{a} = \langle a_1, a_2, a_3 \rangle$ is
>
> $$|\mathbf{a}| = \sqrt{a_1^2 + a_2^2 + a_3^2}$$

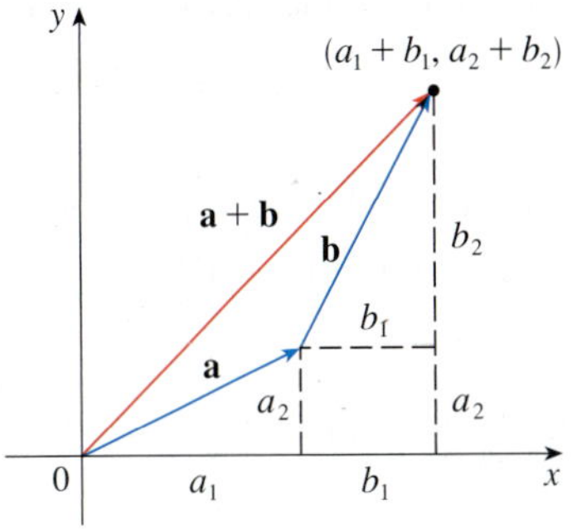

FIGURE 14

How do we add vectors algebraically? Figure 14 shows that if $\mathbf{a} = \langle a_1, a_2 \rangle$ and $\mathbf{b} = \langle b_1, b_2 \rangle$, then the sum is $\mathbf{a} + \mathbf{b} = \langle a_1 + b_1, a_2 + b_2 \rangle$, at least for the case where the components are positive. In other words, *to add algebraic vectors we add corresponding components.* Similarly, *to subtract vectors we subtract corresponding components.* From the similar triangles in Figure 15 we see that the components of $c\mathbf{a}$ are ca_1 and ca_2. So *to multiply a vector by a scalar we multiply each component by that scalar.*

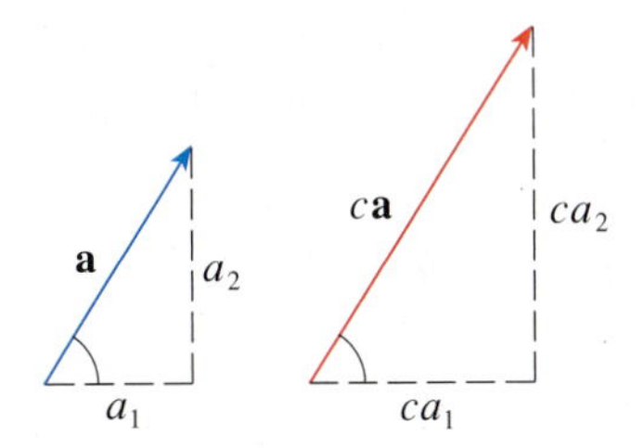

FIGURE 15

> If $\mathbf{a} = \langle a_1, a_2 \rangle$ and $\mathbf{b} = \langle b_1, b_2 \rangle$, then
>
> $$\mathbf{a} + \mathbf{b} = \langle a_1 + b_1, a_2 + b_2 \rangle \qquad \mathbf{a} - \mathbf{b} = \langle a_1 - b_1, a_2 - b_2 \rangle$$
>
> $$c\mathbf{a} = \langle ca_1, ca_2 \rangle$$
>
> Similarly, for three-dimensional vectors,
>
> $$\langle a_1, a_2, a_3 \rangle + \langle b_1, b_2, b_3 \rangle = \langle a_1 + b_1, a_2 + b_2, a_3 + b_3 \rangle$$
>
> $$\langle a_1, a_2, a_3 \rangle - \langle b_1, b_2, b_3 \rangle = \langle a_1 - b_1, a_2 - b_2, a_3 - b_3 \rangle$$
>
> $$c\langle a_1, a_2, a_3 \rangle = \langle ca_1, ca_2, ca_3 \rangle$$

EXAMPLE 4 If $\mathbf{a} = \langle 4, 0, 3 \rangle$ and $\mathbf{b} = \langle -2, 1, 5 \rangle$, find $|\mathbf{a}|$ and the vectors $\mathbf{a} + \mathbf{b}$, $\mathbf{a} - \mathbf{b}$, $3\mathbf{b}$, and $2\mathbf{a} + 5\mathbf{b}$.

SOLUTION

$$|\mathbf{a}| = \sqrt{4^2 + 0^2 + 3^2} = \sqrt{25} = 5$$

$$\begin{aligned} \mathbf{a} + \mathbf{b} &= \langle 4, 0, 3 \rangle + \langle -2, 1, 5 \rangle \\ &= \langle 4 + (-2), 0 + 1, 3 + 5 \rangle = \langle 2, 1, 8 \rangle \end{aligned}$$

$$\begin{aligned} \mathbf{a} - \mathbf{b} &= \langle 4, 0, 3 \rangle - \langle -2, 1, 5 \rangle \\ &= \langle 4 - (-2), 0 - 1, 3 - 5 \rangle = \langle 6, -1, -2 \rangle \end{aligned}$$

$$3\mathbf{b} = 3\langle -2, 1, 5 \rangle = \langle 3(-2), 3(1), 3(5) \rangle = \langle -6, 3, 15 \rangle$$

$$\begin{aligned} 2\mathbf{a} + 5\mathbf{b} &= 2\langle 4, 0, 3 \rangle + 5\langle -2, 1, 5 \rangle \\ &= \langle 8, 0, 6 \rangle + \langle -10, 5, 25 \rangle = \langle -2, 5, 31 \rangle \end{aligned}$$

■

Vectors in n dimensions are used to list various quantities in an organized way. For instance, the components of a six-dimensional vector

$$\mathbf{p} = \langle p_1, p_2, p_3, p_4, p_5, p_6 \rangle$$

might represent the prices of six different ingredients required to make a particular product. Four-dimensional vectors $\langle x, y, z, t \rangle$ are used in relativity theory, where the first three components specify a position in space and the fourth represents time.

We denote by V_2 the set of all two-dimensional vectors and by V_3 the set of all three-dimensional vectors. More generally, we will later need to consider the set V_n of all n-dimensional vectors. An n-dimensional vector is an ordered n-tuple:

$$\mathbf{a} = \langle a_1, a_2, \ldots, a_n \rangle$$

where $a_1, a_2, \ldots, a_n$ are real numbers that are called the components of $\mathbf{a}$. Addition and scalar multiplication are defined in terms of components just as for the cases $n = 2$ and $n = 3$.

Properties of Vectors If $\mathbf{a}$, $\mathbf{b}$, and $\mathbf{c}$ are vectors in V_n and c and d are scalars, then

1. $\mathbf{a} + \mathbf{b} = \mathbf{b} + \mathbf{a}$
2. $\mathbf{a} + (\mathbf{b} + \mathbf{c}) = (\mathbf{a} + \mathbf{b}) + \mathbf{c}$
3. $\mathbf{a} + \mathbf{0} = \mathbf{a}$
4. $\mathbf{a} + (-\mathbf{a}) = \mathbf{0}$
5. $c(\mathbf{a} + \mathbf{b}) = c\mathbf{a} + c\mathbf{b}$
6. $(c + d)\mathbf{a} = c\mathbf{a} + d\mathbf{a}$
7. $(cd)\mathbf{a} = c(d\mathbf{a})$
8. $1\mathbf{a} = \mathbf{a}$

These eight properties of vectors can be readily verified either geometrically or algebraically. For instance, Property 1 can be seen from Figure 4 (it's equivalent to the Parallelogram Law) or as follows for the case $n = 2$:

$$\begin{aligned} \mathbf{a} + \mathbf{b} &= \langle a_1, a_2 \rangle + \langle b_1, b_2 \rangle = \langle a_1 + b_1, a_2 + b_2 \rangle \\ &= \langle b_1 + a_1, b_2 + a_2 \rangle = \langle b_1, b_2 \rangle + \langle a_1, a_2 \rangle \\ &= \mathbf{b} + \mathbf{a} \end{aligned}$$

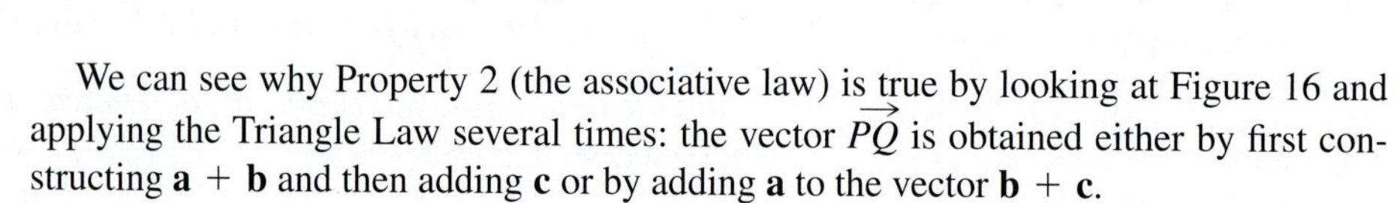

FIGURE 16

We can see why Property 2 (the associative law) is true by looking at Figure 16 and applying the Triangle Law several times: the vector $\overrightarrow{PQ}$ is obtained either by first constructing $\mathbf{a} + \mathbf{b}$ and then adding $\mathbf{c}$ or by adding $\mathbf{a}$ to the vector $\mathbf{b} + \mathbf{c}$.

Three vectors in V_3 play a special role. Let

$$\mathbf{i} = \langle 1, 0, 0 \rangle \qquad \mathbf{j} = \langle 0, 1, 0 \rangle \qquad \mathbf{k} = \langle 0, 0, 1 \rangle$$

These vectors $\mathbf{i}$, $\mathbf{j}$, and $\mathbf{k}$ are called the **standard basis vectors**. They have length 1 and point in the directions of the positive x-, y-, and z-axes. Similarly, in two dimensions we define $\mathbf{i} = \langle 1, 0 \rangle$ and $\mathbf{j} = \langle 0, 1 \rangle$. (See Figure 17.)

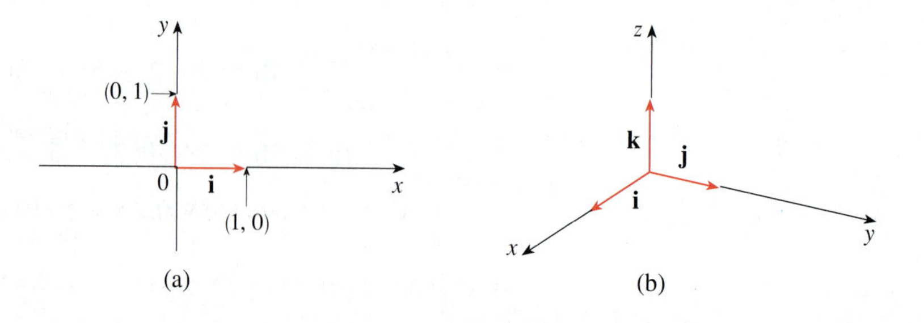

FIGURE 17
Standard basis vectors in V_2 and V_3

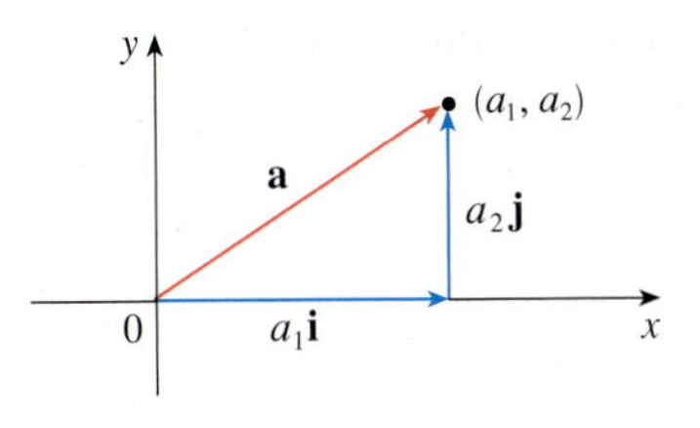

(a) $\mathbf{a} = a_1\mathbf{i} + a_2\mathbf{j}$

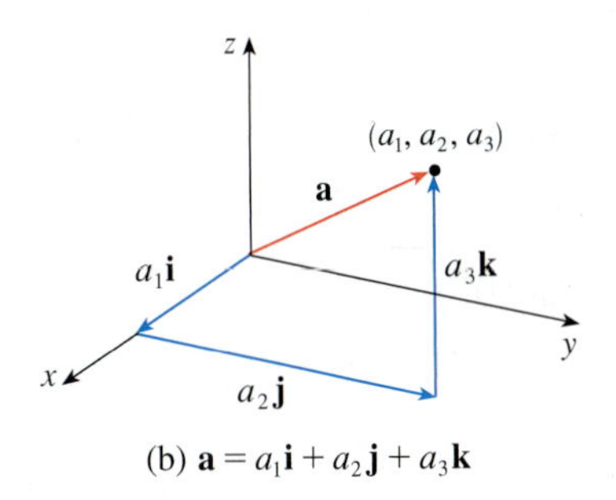

(b) $\mathbf{a} = a_1\mathbf{i} + a_2\mathbf{j} + a_3\mathbf{k}$

FIGURE 18

If $\mathbf{a} = \langle a_1, a_2, a_3 \rangle$, then we can write

$$\mathbf{a} = \langle a_1, a_2, a_3 \rangle = \langle a_1, 0, 0 \rangle + \langle 0, a_2, 0 \rangle + \langle 0, 0, a_3 \rangle$$

$$= a_1\langle 1, 0, 0 \rangle + a_2\langle 0, 1, 0 \rangle + a_3\langle 0, 0, 1 \rangle$$

2 $$\mathbf{a} = a_1\,\mathbf{i} + a_2\,\mathbf{j} + a_3\,\mathbf{k}$$

Thus any vector in V_3 can be expressed in terms of **i**, **j**, and **k**. For instance,

$$\langle 1, -2, 6 \rangle = \mathbf{i} - 2\mathbf{j} + 6\mathbf{k}$$

Similarly, in two dimensions, we can write

3 $$\mathbf{a} = \langle a_1, a_2 \rangle = a_1\,\mathbf{i} + a_2\,\mathbf{j}$$

See Figure 18 for the geometric interpretation of Equations 3 and 2 and compare with Figure 17.

EXAMPLE 5 If $\mathbf{a} = \mathbf{i} + 2\mathbf{j} - 3\mathbf{k}$ and $\mathbf{b} = 4\mathbf{i} + 7\mathbf{k}$, express the vector $2\mathbf{a} + 3\mathbf{b}$ in terms of **i**, **j**, and **k**.

SOLUTION Using Properties 1, 2, 5, 6, and 7 of vectors, we have

$$2\mathbf{a} + 3\mathbf{b} = 2(\mathbf{i} + 2\mathbf{j} - 3\mathbf{k}) + 3(4\mathbf{i} + 7\mathbf{k})$$

$$= 2\mathbf{i} + 4\mathbf{j} - 6\mathbf{k} + 12\mathbf{i} + 21\mathbf{k} = 14\mathbf{i} + 4\mathbf{j} + 15\mathbf{k}$$ ■

> **Gibbs**
> Josiah Willard Gibbs (1839–1903), a professor of mathematical physics at Yale College, published the first book on vectors, *Vector Analysis*, in 1881. More complicated objects, called quaternions, had earlier been invented by Hamilton as mathematical tools for describing space, but they weren't easy for scientists to use. Quaternions have a scalar part and a vector part. Gibb's idea was to use the vector part separately. Maxwell and Heaviside had similar ideas, but Gibb's approach has proved to be the most convenient way to study space.

A **unit vector** is a vector whose length is 1. For instance, **i**, **j**, and **k** are all unit vectors. In general, if $\mathbf{a} \neq \mathbf{0}$, then the unit vector that has the same direction as **a** is

4 $$\mathbf{u} = \frac{1}{|\mathbf{a}|}\mathbf{a} = \frac{\mathbf{a}}{|\mathbf{a}|}$$

In order to verify this, we let $c = 1/|\mathbf{a}|$. Then $\mathbf{u} = c\mathbf{a}$ and c is a positive scalar, so **u** has the same direction as **a**. Also

$$|\mathbf{u}| = |c\mathbf{a}| = |c||\mathbf{a}| = \frac{1}{|\mathbf{a}|}|\mathbf{a}| = 1$$

EXAMPLE 6 Find the unit vector in the direction of the vector $2\mathbf{i} - \mathbf{j} - 2\mathbf{k}$.

SOLUTION The given vector has length

$$|2\mathbf{i} - \mathbf{j} - 2\mathbf{k}| = \sqrt{2^2 + (-1)^2 + (-2)^2} = \sqrt{9} = 3$$

so, by Equation 4, the unit vector with the same direction is

$$\tfrac{1}{3}(2\mathbf{i} - \mathbf{j} - 2\mathbf{k}) = \tfrac{2}{3}\mathbf{i} - \tfrac{1}{3}\mathbf{j} - \tfrac{2}{3}\mathbf{k}$$ ■

Applications

Vectors are useful in many aspects of physics and engineering. In Chapter 13 we will see how they describe the velocity and acceleration of objects moving in space. Here we look at forces.

A force is represented by a vector because it has both a magnitude (measured in pounds or newtons) and a direction. If several forces are acting on an object, the **resultant force** experienced by the object is the vector sum of these forces.

EXAMPLE 7 A 100-lb weight hangs from two wires as shown in Figure 19. Find the tensions (forces) $\mathbf{T}_1$ and $\mathbf{T}_2$ in both wires and the magnitudes of the tensions.

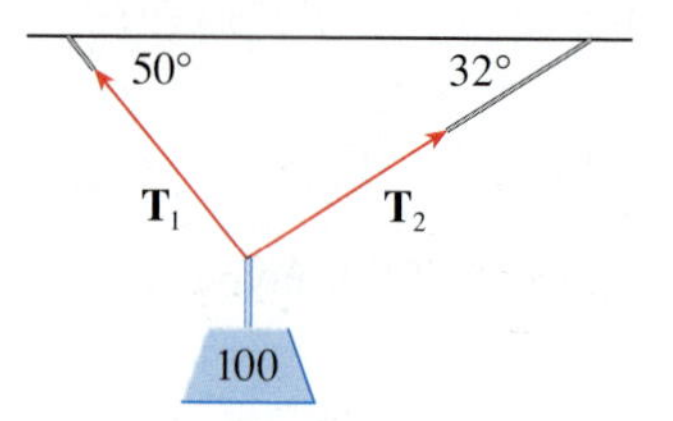

FIGURE 19

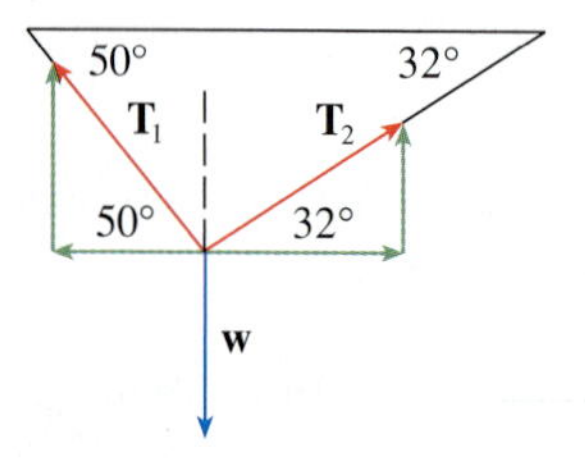

FIGURE 20

SOLUTION We first express $\mathbf{T}_1$ and $\mathbf{T}_2$ in terms of their horizontal and vertical components. From Figure 20 we see that

$$\boxed{5} \qquad \mathbf{T}_1 = -|\mathbf{T}_1|\cos 50^\circ\,\mathbf{i} + |\mathbf{T}_1|\sin 50^\circ\,\mathbf{j}$$

$$\boxed{6} \qquad \mathbf{T}_2 = |\mathbf{T}_2|\cos 32^\circ\,\mathbf{i} + |\mathbf{T}_2|\sin 32^\circ\,\mathbf{j}$$

The resultant $\mathbf{T}_1 + \mathbf{T}_2$ of the tensions counterbalances the weight $\mathbf{w} = -100\,\mathbf{j}$ and so we must have

$$\mathbf{T}_1 + \mathbf{T}_2 = -\mathbf{w} = 100\,\mathbf{j}$$

Thus

$$\left(-|\mathbf{T}_1|\cos 50^\circ + |\mathbf{T}_2|\cos 32^\circ\right)\mathbf{i} + \left(|\mathbf{T}_1|\sin 50^\circ + |\mathbf{T}_2|\sin 32^\circ\right)\mathbf{j} = 100\,\mathbf{j}$$

Equating components, we get

$$-|\mathbf{T}_1|\cos 50^\circ + |\mathbf{T}_2|\cos 32^\circ = 0$$

$$|\mathbf{T}_1|\sin 50^\circ + |\mathbf{T}_2|\sin 32^\circ = 100$$

Solving the first of these equations for $|\mathbf{T}_2|$ and substituting into the second, we get

$$|\mathbf{T}_1|\sin 50^\circ + \frac{|\mathbf{T}_1|\cos 50^\circ}{\cos 32^\circ}\sin 32^\circ = 100$$

$$|\mathbf{T}_1|\left(\sin 50^\circ + \cos 50^\circ\,\frac{\sin 32^\circ}{\cos 32^\circ}\right) = 100$$

So the magnitudes of the tensions are

$$|\mathbf{T}_1| = \frac{100}{\sin 50^\circ + \tan 32^\circ\cos 50^\circ} \approx 85.64 \text{ lb}$$

and $$|\mathbf{T}_2| = \frac{|\mathbf{T}_1|\cos 50^\circ}{\cos 32^\circ} \approx 64.91 \text{ lb}$$

Substituting these values in (5) and (6), we obtain the tension vectors

$$\mathbf{T}_1 \approx -55.05\,\mathbf{i} + 65.60\,\mathbf{j}$$

$$\mathbf{T}_2 \approx 55.05\,\mathbf{i} + 34.40\,\mathbf{j}$$ ■

12.2 EXERCISES

1. Are the following quantities vectors or scalars? Explain.
(a) The cost of a theater ticket
(b) The current in a river
(c) The initial flight path from Houston to Dallas
(d) The population of the world

2. What is the relationship between the point (4, 7) and the vector $\langle 4, 7 \rangle$? Illustrate with a sketch.

3. Name all the equal vectors in the parallelogram shown.

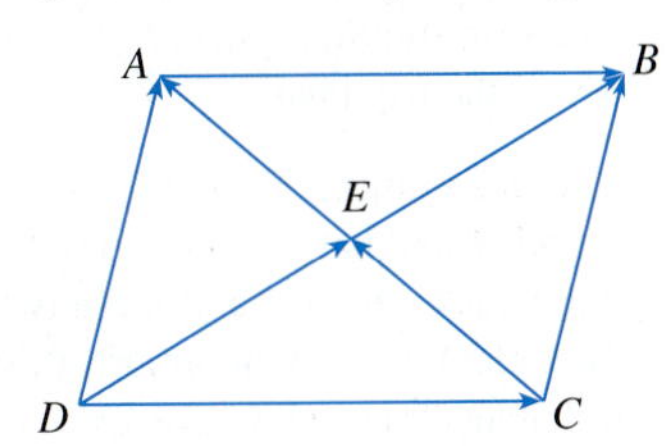

4. Write each combination of vectors as a single vector.

(a) $\overrightarrow{AB} + \overrightarrow{BC}$ (b) $\overrightarrow{CD} + \overrightarrow{DB}$
(c) $\overrightarrow{DB} - \overrightarrow{AB}$ (d) $\overrightarrow{DC} + \overrightarrow{CA} + \overrightarrow{AB}$

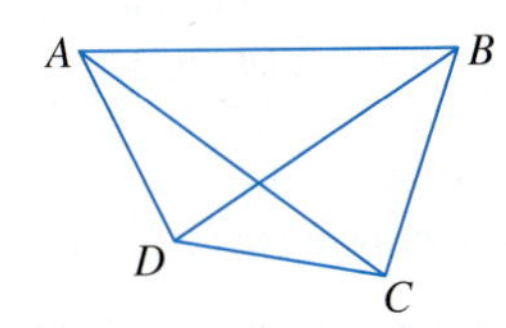

5. Copy the vectors in the figure and use them to draw the following vectors.

(a) $\mathbf{u} + \mathbf{v}$ (b) $\mathbf{u} + \mathbf{w}$
(c) $\mathbf{v} + \mathbf{w}$ (d) $\mathbf{u} - \mathbf{v}$
(e) $\mathbf{v} + \mathbf{u} + \mathbf{w}$ (f) $\mathbf{u} - \mathbf{w} - \mathbf{v}$

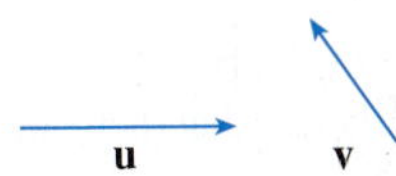

6. Copy the vectors in the figure and use them to draw the following vectors.

(a) $\mathbf{a} + \mathbf{b}$ (b) $\mathbf{a} - \mathbf{b}$
(c) $\frac{1}{2}\mathbf{a}$ (d) $-3\mathbf{b}$
(e) $\mathbf{a} + 2\mathbf{b}$ (f) $2\mathbf{b} - \mathbf{a}$

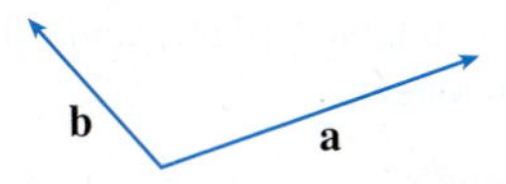

7. In the figure, the tip of $\mathbf{c}$ and the tail of $\mathbf{d}$ are both the midpoint of QR. Express $\mathbf{c}$ and $\mathbf{d}$ in terms of $\mathbf{a}$ and $\mathbf{b}$.

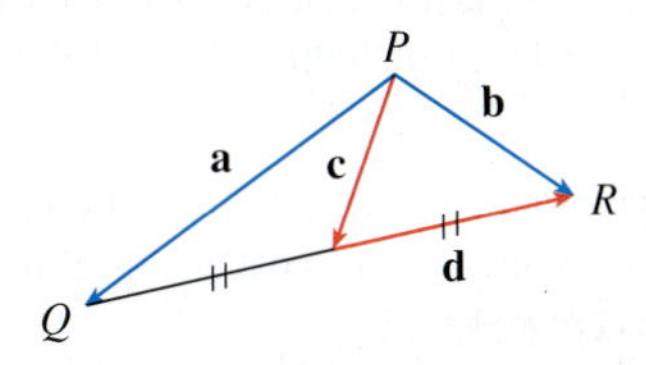

8. If the vectors in the figure satisfy $|\mathbf{u}| = |\mathbf{v}| = 1$ and $\mathbf{u} + \mathbf{v} + \mathbf{w} = \mathbf{0}$, what is $|\mathbf{w}|$?

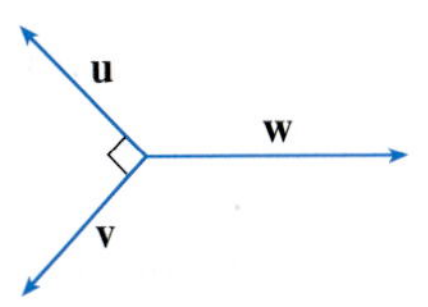

9–14 Find a vector $\mathbf{a}$ with representation given by the directed line segment $\overrightarrow{AB}$. Draw $\overrightarrow{AB}$ and the equivalent representation starting at the origin.

9. $A(-2, 1)$, $B(1, 2)$
10. $A(-5, -1)$, $B(-3, 3)$
11. $A(3, -1)$, $B(2, 3)$
12. $A(3, 2)$, $B(1, 0)$
13. $A(0, 3, 1)$, $B(2, 3, -1)$
14. $A(0, 6, -1)$, $B(3, 4, 4)$

15–18 Find the sum of the given vectors and illustrate geometrically.

15. $\langle -1, 4 \rangle$, $\langle 6, -2 \rangle$
16. $\langle 3, -1 \rangle$, $\langle -1, 5 \rangle$
17. $\langle 3, 0, 1 \rangle$, $\langle 0, 8, 0 \rangle$
18. $\langle 1, 3, -2 \rangle$, $\langle 0, 0, 6 \rangle$

19–22 Find $\mathbf{a} + \mathbf{b}$, $4\mathbf{a} + 2\mathbf{b}$, $|\mathbf{a}|$, and $|\mathbf{a} - \mathbf{b}|$.

19. $\mathbf{a} = \langle -3, 4 \rangle$, $\mathbf{b} = \langle 9, -1 \rangle$
20. $\mathbf{a} = 5\mathbf{i} + 3\mathbf{j}$, $\mathbf{b} = -\mathbf{i} - 2\mathbf{j}$
21. $\mathbf{a} = 4\mathbf{i} - 3\mathbf{j} + 2\mathbf{k}$, $\mathbf{b} = 2\mathbf{i} - 4\mathbf{k}$
22. $\mathbf{a} = \langle 8, 1, -4 \rangle$, $\mathbf{b} = \langle 5, -2, 1 \rangle$

23–25 Find a unit vector that has the same direction as the given vector.

23. $\langle 6, -2 \rangle$
24. $-5\mathbf{i} + 3\mathbf{j} - \mathbf{k}$
25. $8\mathbf{i} - \mathbf{j} + 4\mathbf{k}$

26. Find the vector that has the same direction as $\langle 6, 2, -3 \rangle$ but has length 4.

27–28 What is the angle between the given vector and the positive direction of the x-axis?

27. $\mathbf{i} + \sqrt{3}\,\mathbf{j}$
28. $8\mathbf{i} + 6\mathbf{j}$

29. If $\mathbf{v}$ lies in the first quadrant and makes an angle $\pi/3$ with the positive x-axis and $|\mathbf{v}| = 4$, find $\mathbf{v}$ in component form.

30. If a child pulls a sled through the snow on a level path with a force of 50 N exerted at an angle of 38° above the horizontal, find the horizontal and vertical components of the force.

31. A quarterback throws a football with angle of elevation 40° and speed 60 ft/s. Find the horizontal and vertical components of the velocity vector.

32–33 Find the magnitude of the resultant force and the angle it makes with the positive x-axis.

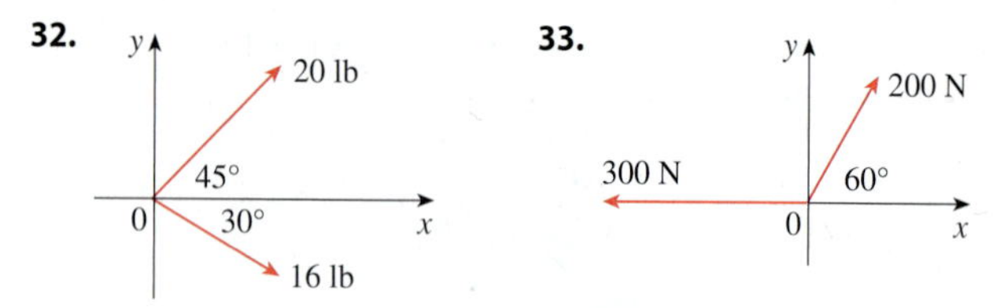

34. The magnitude of a velocity vector is called *speed*. Suppose that a wind is blowing *from* the direction N45°W at a speed of 50 km/h. (This means that the direction from which the wind blows is 45° west of the northerly direction.) A pilot is steering a plane in the direction N60°E at an airspeed (speed in still air) of 250 km/h. The *true course*, or *track*, of the plane is the direction of the resultant of the velocity vectors of the plane and the wind. The *ground speed* of the plane is the magnitude of the resultant. Find the true course and the ground speed of the plane.

35. A woman walks due west on the deck of a ship at 3 mi/h. The ship is moving north at a speed of 22 mi/h. Find the speed and direction of the woman relative to the surface of the water.

36. A crane suspends a 500-lb steel beam horizontally by support cables (with negligible weight) attached from a hook to each end of the beam. The support cables each make an angle of 60° with the beam. Find the tension vector in each support cable and the magnitude of each tension.

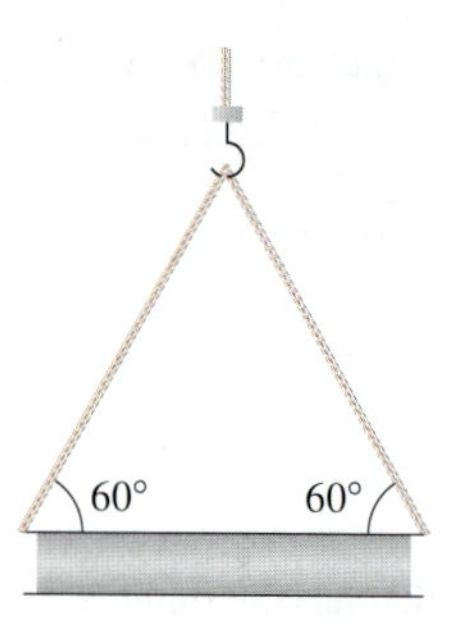

37. A block-and-tackle pulley hoist is suspended in a warehouse by ropes of lengths 2 m and 3 m. The hoist weighs 350 N. The ropes, fastened at different heights, make angles of 50° and 38° with the horizontal. Find the tension in each rope and the magnitude of each tension.

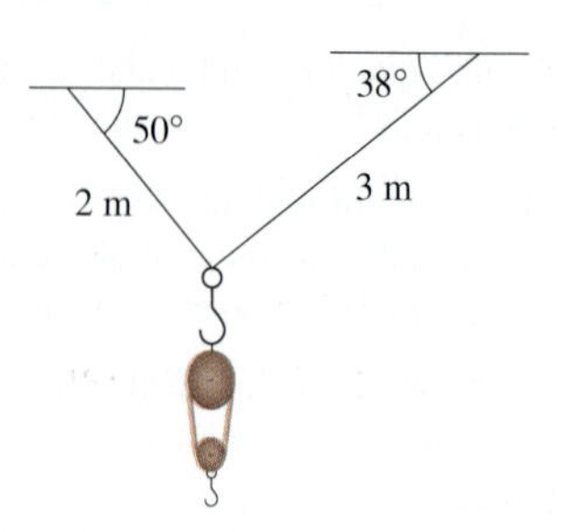

38. The tension $\mathbf{T}$ at each end of a chain has magnitude 25 N (see the figure). What is the weight of the chain?

39. A boatman wants to cross a canal that is 3 km wide and wants to land at a point 2 km upstream from his starting point. The current in the canal flows at 3.5 km/h and the speed of his boat is 13 km/h.
(a) In what direction should he steer?
(b) How long will the trip take?

40. Three forces act on an object. Two of the forces are at an angle of 100° to each other and have magnitudes 25 N and 12 N. The third is perpendicular to the plane of these two forces and has magnitude 4 N. Calculate the magnitude of the force that would exactly counterbalance these three forces.

41. Find the unit vectors that are parallel to the tangent line to the parabola $y = x^2$ at the point (2, 4).

42. (a) Find the unit vectors that are parallel to the tangent line to the curve $y = 2 \sin x$ at the point $(\pi/6, 1)$.
(b) Find the unit vectors that are perpendicular to the tangent line.
(c) Sketch the curve $y = 2 \sin x$ and the vectors in parts (a) and (b), all starting at $(\pi/6, 1)$.

43. If A, B, and C are the vertices of a triangle, find

$$\overrightarrow{AB} + \overrightarrow{BC} + \overrightarrow{CA}$$

44. Let C be the point on the line segment AB that is twice as far from B as it is from A. If $\mathbf{a} = \overrightarrow{OA}$, $\mathbf{b} = \overrightarrow{OB}$, and $\mathbf{c} = \overrightarrow{OC}$, show that $\mathbf{c} = \frac{2}{3}\mathbf{a} + \frac{1}{3}\mathbf{b}$.

45. (a) Draw the vectors $\mathbf{a} = \langle 3, 2 \rangle$, $\mathbf{b} = \langle 2, -1 \rangle$, and $\mathbf{c} = \langle 7, 1 \rangle$.
(b) Show, by means of a sketch, that there are scalars s and t such that $\mathbf{c} = s\mathbf{a} + t\mathbf{b}$.
(c) Use the sketch to estimate the values of s and t.
(d) Find the exact values of s and t.

46. Suppose that $\mathbf{a}$ and $\mathbf{b}$ are nonzero vectors that are not parallel and $\mathbf{c}$ is any vector in the plane determined by $\mathbf{a}$ and $\mathbf{b}$. Give a geometric argument to show that $\mathbf{c}$ can be written as $\mathbf{c} = s\mathbf{a} + t\mathbf{b}$ for suitable scalars s and t. Then give an argument using components.

47. If $\mathbf{r} = \langle x, y, z \rangle$ and $\mathbf{r}_0 = \langle x_0, y_0, z_0 \rangle$, describe the set of all points (x, y, z) such that $|\mathbf{r} - \mathbf{r}_0| = 1$.

48. If $\mathbf{r} = \langle x, y \rangle$, $\mathbf{r}_1 = \langle x_1, y_1 \rangle$, and $\mathbf{r}_2 = \langle x_2, y_2 \rangle$, describe the set of all points (x, y) such that $|\mathbf{r} - \mathbf{r}_1| + |\mathbf{r} - \mathbf{r}_2| = k$, where $k > |\mathbf{r}_1 - \mathbf{r}_2|$.

49. Figure 16 gives a geometric demonstration of Property 2 of vectors. Use components to give an algebraic proof of this fact for the case $n = 2$.

50. Prove Property 5 of vectors algebraically for the case $n = 3$. Then use similar triangles to give a geometric proof.

51. Use vectors to prove that the line joining the midpoints of two sides of a triangle is parallel to the third side and half its length.

52. Suppose the three coordinate planes are all mirrored and a light ray given by the vector $\mathbf{a} = \langle a_1, a_2, a_3 \rangle$ first strikes the xz-plane, as shown in the figure. Use the fact that the angle of incidence equals the angle of reflection to show that the direction of the reflected ray is given by $\mathbf{b} = \langle a_1, -a_2, a_3 \rangle$. Deduce that, after being reflected by all three mutually perpendicular mirrors, the resulting ray is parallel to the initial ray. (American space scientists used this principle, together with laser beams and an array of corner mirrors on the moon, to calculate very precisely the distance from the earth to the moon.)

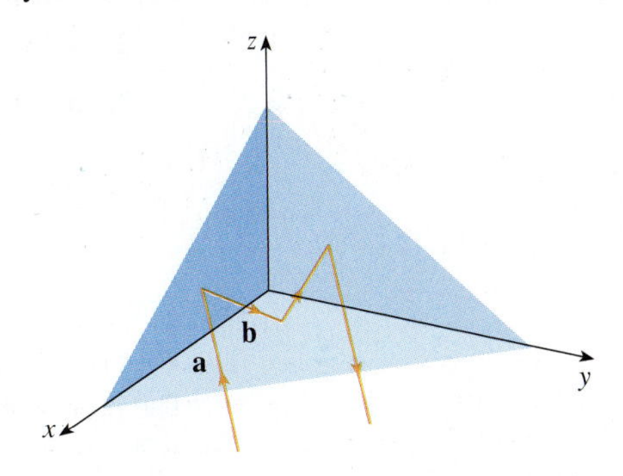

12.3 The Dot Product

So far we have added two vectors and multiplied a vector by a scalar. The question arises: is it possible to multiply two vectors so that their product is a useful quantity? One such product is the dot product, whose definition follows. Another is the cross product, which is discussed in the next section.

> **1 Definition** If $\mathbf{a} = \langle a_1, a_2, a_3 \rangle$ and $\mathbf{b} = \langle b_1, b_2, b_3 \rangle$, then the **dot product** of **a** and **b** is the number $\mathbf{a} \cdot \mathbf{b}$ given by
>
> $$\mathbf{a} \cdot \mathbf{b} = a_1 b_1 + a_2 b_2 + a_3 b_3$$

Thus, to find the dot product of **a** and **b**, we multiply corresponding components and add. The result is not a vector. It is a real number, that is, a scalar. For this reason, the dot product is sometimes called the **scalar product** (or **inner product**). Although Definition 1 is given for three-dimensional vectors, the dot product of two-dimensional vectors is defined in a similar fashion:

$$\langle a_1, a_2 \rangle \cdot \langle b_1, b_2 \rangle = a_1 b_1 + a_2 b_2$$

EXAMPLE 1

$$\langle 2, 4 \rangle \cdot \langle 3, -1 \rangle = 2(3) + 4(-1) = 2$$

$$\langle -1, 7, 4 \rangle \cdot \left\langle 6, 2, -\tfrac{1}{2} \right\rangle = (-1)(6) + 7(2) + 4\left(-\tfrac{1}{2}\right) = 6$$

$$(\mathbf{i} + 2\mathbf{j} - 3\mathbf{k}) \cdot (2\mathbf{j} - \mathbf{k}) = 1(0) + 2(2) + (-3)(-1) = 7$$ ■

The dot product obeys many of the laws that hold for ordinary products of real numbers. These are stated in the following theorem.

> **2 Properties of the Dot Product** If **a**, **b**, and **c** are vectors in V_3 and c is a scalar, then
>
> **1.** $\mathbf{a} \cdot \mathbf{a} = |\mathbf{a}|^2$ **2.** $\mathbf{a} \cdot \mathbf{b} = \mathbf{b} \cdot \mathbf{a}$
>
> **3.** $\mathbf{a} \cdot (\mathbf{b} + \mathbf{c}) = \mathbf{a} \cdot \mathbf{b} + \mathbf{a} \cdot \mathbf{c}$ **4.** $(c\mathbf{a}) \cdot \mathbf{b} = c(\mathbf{a} \cdot \mathbf{b}) = \mathbf{a} \cdot (c\mathbf{b})$
>
> **5.** $\mathbf{0} \cdot \mathbf{a} = 0$

These properties are easily proved using Definition 1. For instance, here are the proofs of Properties 1 and 3:

1. $\mathbf{a} \cdot \mathbf{a} = a_1^2 + a_2^2 + a_3^2 = |\mathbf{a}|^2$

3. $$\begin{aligned}\mathbf{a} \cdot (\mathbf{b} + \mathbf{c}) &= \langle a_1, a_2, a_3 \rangle \cdot \langle b_1 + c_1, b_2 + c_2, b_3 + c_3 \rangle \\ &= a_1(b_1 + c_1) + a_2(b_2 + c_2) + a_3(b_3 + c_3) \\ &= a_1b_1 + a_1c_1 + a_2b_2 + a_2c_2 + a_3b_3 + a_3c_3 \\ &= (a_1b_1 + a_2b_2 + a_3b_3) + (a_1c_1 + a_2c_2 + a_3c_3) \\ &= \mathbf{a} \cdot \mathbf{b} + \mathbf{a} \cdot \mathbf{c}\end{aligned}$$

The proofs of the remaining properties are left as exercises. ■

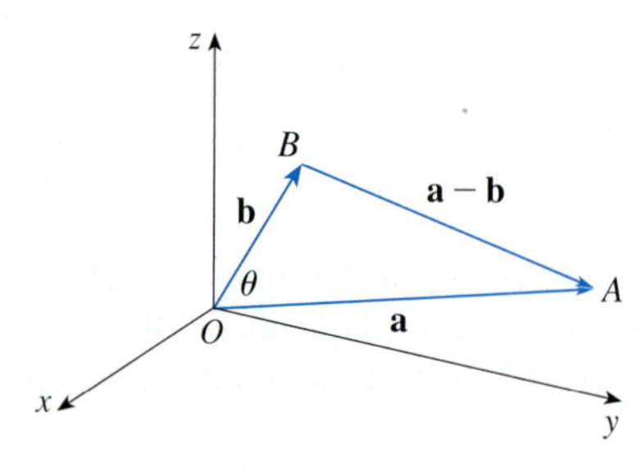

FIGURE 1

The dot product $\mathbf{a} \cdot \mathbf{b}$ can be given a geometric interpretation in terms of the **angle θ between a and b**, which is defined to be the angle between the representations of **a** and **b** that start at the origin, where $0 \leqslant \theta \leqslant \pi$. In other words, θ is the angle between the line segments $\overrightarrow{OA}$ and $\overrightarrow{OB}$ in Figure 1. Note that if **a** and **b** are parallel vectors, then $\theta = 0$ or $\theta = \pi$.

The formula in the following theorem is used by physicists as the *definition* of the dot product.

3 Theorem If θ is the angle between the vectors **a** and **b**, then

$$\mathbf{a} \cdot \mathbf{b} = |\mathbf{a}||\mathbf{b}| \cos \theta$$

PROOF If we apply the Law of Cosines to triangle OAB in Figure 1, we get

$$|AB|^2 = |OA|^2 + |OB|^2 - 2|OA||OB| \cos \theta \tag{4}$$

(Observe that the Law of Cosines still applies in the limiting cases when $\theta = 0$ or π, or $\mathbf{a} = \mathbf{0}$ or $\mathbf{b} = \mathbf{0}$.) But $|OA| = |\mathbf{a}|$, $|OB| = |\mathbf{b}|$, and $|AB| = |\mathbf{a} - \mathbf{b}|$, so Equation 4 becomes

$$|\mathbf{a} - \mathbf{b}|^2 = |\mathbf{a}|^2 + |\mathbf{b}|^2 - 2|\mathbf{a}||\mathbf{b}| \cos \theta \tag{5}$$

Using Properties 1, 2, and 3 of the dot product, we can rewrite the left side of this equation as follows:

$$\begin{aligned}|\mathbf{a} - \mathbf{b}|^2 &= (\mathbf{a} - \mathbf{b}) \cdot (\mathbf{a} - \mathbf{b}) \\ &= \mathbf{a} \cdot \mathbf{a} - \mathbf{a} \cdot \mathbf{b} - \mathbf{b} \cdot \mathbf{a} + \mathbf{b} \cdot \mathbf{b} \\ &= |\mathbf{a}|^2 - 2\mathbf{a} \cdot \mathbf{b} + |\mathbf{b}|^2\end{aligned}$$

Therefore Equation 5 gives

$$|\mathbf{a}|^2 - 2\mathbf{a} \cdot \mathbf{b} + |\mathbf{b}|^2 = |\mathbf{a}|^2 + |\mathbf{b}|^2 - 2|\mathbf{a}||\mathbf{b}| \cos \theta$$

Thus $$-2\mathbf{a} \cdot \mathbf{b} = -2|\mathbf{a}||\mathbf{b}| \cos \theta$$

or $$\mathbf{a} \cdot \mathbf{b} = |\mathbf{a}||\mathbf{b}| \cos \theta$$ ■

EXAMPLE 2 If the vectors $\mathbf{a}$ and $\mathbf{b}$ have lengths 4 and 6, and the angle between them is $\pi/3$, find $\mathbf{a} \cdot \mathbf{b}$.

SOLUTION Using Theorem 3, we have

$$\mathbf{a} \cdot \mathbf{b} = |\mathbf{a}||\mathbf{b}| \cos(\pi/3) = 4 \cdot 6 \cdot \tfrac{1}{2} = 12$$ ■

The formula in Theorem 3 also enables us to find the angle between two vectors.

6 Corollary If θ is the angle between the nonzero vectors $\mathbf{a}$ and $\mathbf{b}$, then

$$\cos\theta = \frac{\mathbf{a} \cdot \mathbf{b}}{|\mathbf{a}||\mathbf{b}|}$$

EXAMPLE 3 Find the angle between the vectors $\mathbf{a} = \langle 2, 2, -1 \rangle$ and $\mathbf{b} = \langle 5, -3, 2 \rangle$.

SOLUTION Since

$$|\mathbf{a}| = \sqrt{2^2 + 2^2 + (-1)^2} = 3 \qquad \text{and} \qquad |\mathbf{b}| = \sqrt{5^2 + (-3)^2 + 2^2} = \sqrt{38}$$

and since

$$\mathbf{a} \cdot \mathbf{b} = 2(5) + 2(-3) + (-1)(2) = 2$$

we have, from Corollary 6,

$$\cos\theta = \frac{\mathbf{a} \cdot \mathbf{b}}{|\mathbf{a}||\mathbf{b}|} = \frac{2}{3\sqrt{38}}$$

So the angle between $\mathbf{a}$ and $\mathbf{b}$ is

$$\theta = \cos^{-1}\left(\frac{2}{3\sqrt{38}}\right) \approx 1.46 \quad (\text{or } 84°)$$ ■

Two nonzero vectors $\mathbf{a}$ and $\mathbf{b}$ are called **perpendicular** or **orthogonal** if the angle between them is $\theta = \pi/2$. Then Theorem 3 gives

$$\mathbf{a} \cdot \mathbf{b} = |\mathbf{a}||\mathbf{b}| \cos(\pi/2) = 0$$

and conversely if $\mathbf{a} \cdot \mathbf{b} = 0$, then $\cos\theta = 0$, so $\theta = \pi/2$. The zero vector $\mathbf{0}$ is considered to be perpendicular to all vectors. Therefore we have the following method for determining whether two vectors are orthogonal.

7 Two vectors $\mathbf{a}$ and $\mathbf{b}$ are orthogonal if and only if $\mathbf{a} \cdot \mathbf{b} = 0$.

EXAMPLE 4 Show that $2\mathbf{i} + 2\mathbf{j} - \mathbf{k}$ is perpendicular to $5\mathbf{i} - 4\mathbf{j} + 2\mathbf{k}$.

SOLUTION Since

$$(2\mathbf{i} + 2\mathbf{j} - \mathbf{k}) \cdot (5\mathbf{i} - 4\mathbf{j} + 2\mathbf{k}) = 2(5) + 2(-4) + (-1)(2) = 0$$

these vectors are perpendicular by (7). ■

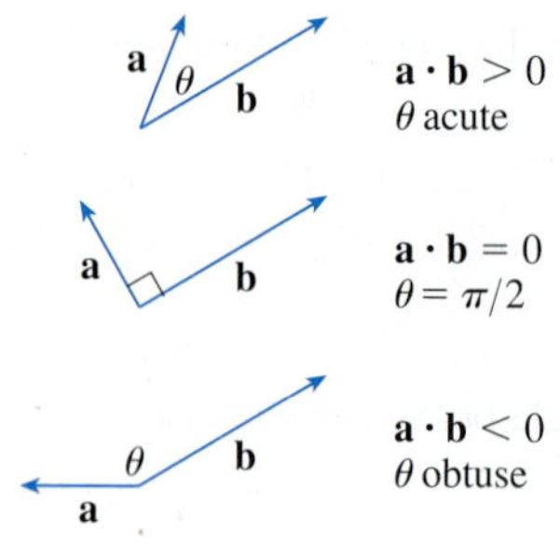

FIGURE 2

TEC Visual 12.3A shows an animation of Figure 2.

Because $\cos\theta > 0$ if $0 \leqslant \theta < \pi/2$ and $\cos\theta < 0$ if $\pi/2 < \theta \leqslant \pi$, we see that $\mathbf{a}\cdot\mathbf{b}$ is positive for $\theta < \pi/2$ and negative for $\theta > \pi/2$. We can think of $\mathbf{a}\cdot\mathbf{b}$ as measuring the extent to which $\mathbf{a}$ and $\mathbf{b}$ point in the same direction. The dot product $\mathbf{a}\cdot\mathbf{b}$ is positive if $\mathbf{a}$ and $\mathbf{b}$ point in the same general direction, 0 if they are perpendicular, and negative if they point in generally opposite directions (see Figure 2). In the extreme case where $\mathbf{a}$ and $\mathbf{b}$ point in exactly the same direction, we have $\theta = 0$, so $\cos\theta = 1$ and

$$\mathbf{a}\cdot\mathbf{b} = |\mathbf{a}|\,|\mathbf{b}|$$

If $\mathbf{a}$ and $\mathbf{b}$ point in exactly opposite directions, then we have $\theta = \pi$ and so $\cos\theta = -1$ and $\mathbf{a}\cdot\mathbf{b} = -|\mathbf{a}|\,|\mathbf{b}|$.

Direction Angles and Direction Cosines

The **direction angles** of a nonzero vector $\mathbf{a}$ are the angles α, β, and γ (in the interval $[0, \pi]$) that $\mathbf{a}$ makes with the positive x-, y-, and z-axes, respectively. (See Figure 3.)

The cosines of these direction angles, $\cos\alpha$, $\cos\beta$, and $\cos\gamma$, are called the **direction cosines** of the vector $\mathbf{a}$. Using Corollary 6 with $\mathbf{b}$ replaced by $\mathbf{i}$, we obtain

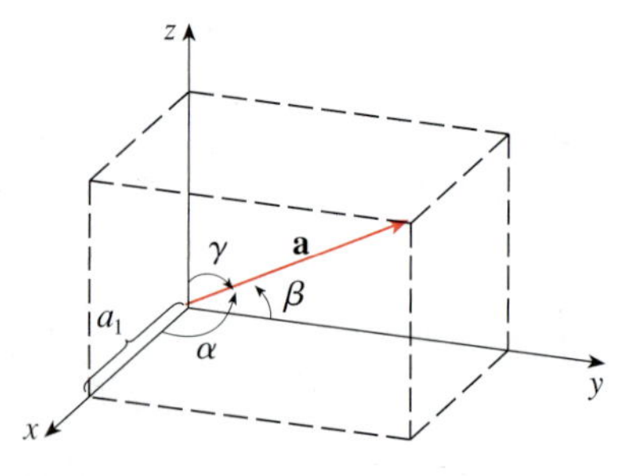

FIGURE 3

8 $$\cos\alpha = \frac{\mathbf{a}\cdot\mathbf{i}}{|\mathbf{a}|\,|\mathbf{i}|} = \frac{a_1}{|\mathbf{a}|}$$

(This can also be seen directly from Figure 3.)

Similarly, we also have

9 $$\cos\beta = \frac{a_2}{|\mathbf{a}|} \qquad \cos\gamma = \frac{a_3}{|\mathbf{a}|}$$

By squaring the expressions in Equations 8 and 9 and adding, we see that

10 $$\cos^2\alpha + \cos^2\beta + \cos^2\gamma = 1$$

We can also use Equations 8 and 9 to write

$$\mathbf{a} = \langle a_1, a_2, a_3 \rangle = \langle |\mathbf{a}|\cos\alpha, |\mathbf{a}|\cos\beta, |\mathbf{a}|\cos\gamma \rangle$$

$$= |\mathbf{a}|\langle \cos\alpha, \cos\beta, \cos\gamma \rangle$$

Therefore

11 $$\frac{1}{|\mathbf{a}|}\mathbf{a} = \langle \cos\alpha, \cos\beta, \cos\gamma \rangle$$

which says that the direction cosines of $\mathbf{a}$ are the components of the unit vector in the direction of $\mathbf{a}$.

EXAMPLE 5 Find the direction angles of the vector $\mathbf{a} = \langle 1, 2, 3 \rangle$.

SOLUTION Since $|\mathbf{a}| = \sqrt{1^2 + 2^2 + 3^2} = \sqrt{14}$, Equations 8 and 9 give

$$\cos\alpha = \frac{1}{\sqrt{14}} \qquad \cos\beta = \frac{2}{\sqrt{14}} \qquad \cos\gamma = \frac{3}{\sqrt{14}}$$

and so

$$\alpha = \cos^{-1}\left(\frac{1}{\sqrt{14}}\right) \approx 74° \qquad \beta = \cos^{-1}\left(\frac{2}{\sqrt{14}}\right) \approx 58° \qquad \gamma = \cos^{-1}\left(\frac{3}{\sqrt{14}}\right) \approx 37°$$

■

Projections

Figure 4 shows representations $\overrightarrow{PQ}$ and $\overrightarrow{PR}$ of two vectors **a** and **b** with the same initial point P. If S is the foot of the perpendicular from R to the line containing $\overrightarrow{PQ}$, then the vector with representation $\overrightarrow{PS}$ is called the **vector projection** of **b** onto **a** and is denoted by $\text{proj}_{\mathbf{a}}\,\mathbf{b}$. (You can think of it as a shadow of **b**).

TEC Visual 12.3B shows how Figure 4 changes when we vary **a** and **b**.

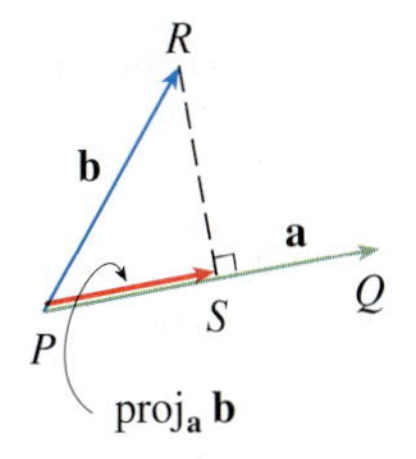

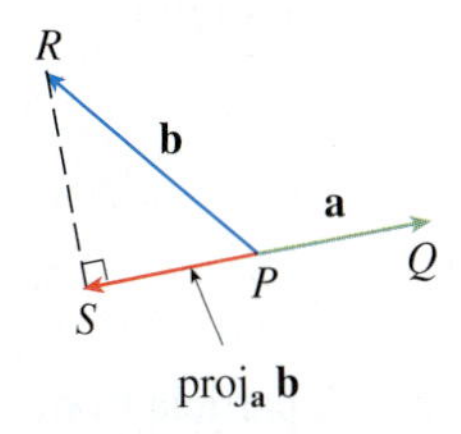

FIGURE 4
Vector projections

The **scalar projection** of **b** onto **a** (also called the **component of b along a**) is defined to be the signed magnitude of the vector projection, which is the number $|\mathbf{b}|\cos\theta$, where θ is the angle between **a** and **b**. (See Figure 5.) This is denoted by $\text{comp}_{\mathbf{a}}\,\mathbf{b}$. Observe that it is negative if $\pi/2 < \theta \leqslant \pi$. The equation

$$\mathbf{a}\cdot\mathbf{b} = |\mathbf{a}||\mathbf{b}|\cos\theta = |\mathbf{a}|\big(|\mathbf{b}|\cos\theta\big)$$

shows that the dot product of **a** and **b** can be interpreted as the length of **a** times the scalar projection of **b** onto **a**. Since

$$|\mathbf{b}|\cos\theta = \frac{\mathbf{a}\cdot\mathbf{b}}{|\mathbf{a}|} = \frac{\mathbf{a}}{|\mathbf{a}|}\cdot\mathbf{b}$$

the component of **b** along **a** can be computed by taking the dot product of **b** with the unit vector in the direction of **a**. We summarize these ideas as follows.

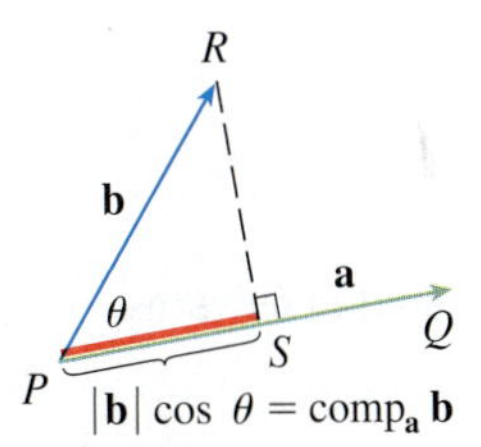

FIGURE 5
Scalar projection

Scalar projection of **b** onto **a**: $\text{comp}_{\mathbf{a}}\,\mathbf{b} = \dfrac{\mathbf{a}\cdot\mathbf{b}}{|\mathbf{a}|}$

Vector projection of **b** onto **a**: $\text{proj}_{\mathbf{a}}\,\mathbf{b} = \left(\dfrac{\mathbf{a}\cdot\mathbf{b}}{|\mathbf{a}|}\right)\dfrac{\mathbf{a}}{|\mathbf{a}|} = \dfrac{\mathbf{a}\cdot\mathbf{b}}{|\mathbf{a}|^2}\,\mathbf{a}$

Notice that the vector projection is the scalar projection times the unit vector in the direction of **a**.

EXAMPLE 6 Find the scalar projection and vector projection of $\mathbf{b} = \langle 1, 1, 2\rangle$ onto $\mathbf{a} = \langle -2, 3, 1\rangle$.

SOLUTION Since $|\mathbf{a}| = \sqrt{(-2)^2 + 3^2 + 1^2} = \sqrt{14}$, the scalar projection of **b** onto **a** is

$$\text{comp}_{\mathbf{a}}\,\mathbf{b} = \frac{\mathbf{a}\cdot\mathbf{b}}{|\mathbf{a}|} = \frac{(-2)(1) + 3(1) + 1(2)}{\sqrt{14}} = \frac{3}{\sqrt{14}}$$

The vector projection is this scalar projection times the unit vector in the direction of **a**:

$$\text{proj}_{\mathbf{a}}\,\mathbf{b} = \frac{3}{\sqrt{14}}\frac{\mathbf{a}}{|\mathbf{a}|} = \frac{3}{14}\mathbf{a} = \left\langle -\frac{3}{7}, \frac{9}{14}, \frac{3}{14}\right\rangle$$

One use of projections occurs in physics in calculating work. In Section 5.4 we defined the work done by a constant force F in moving an object through a distance d as $W = Fd$, but this applies only when the force is directed along the line of motion of the object. Suppose, however, that the constant force is a vector $\mathbf{F} = \overrightarrow{PR}$ pointing in some other direction, as in Figure 6. If the force moves the object from P to Q, then the **displacement vector** is $\mathbf{D} = \overrightarrow{PQ}$. The **work** done by this force is defined to be the product of the component of the force along **D** and the distance moved:

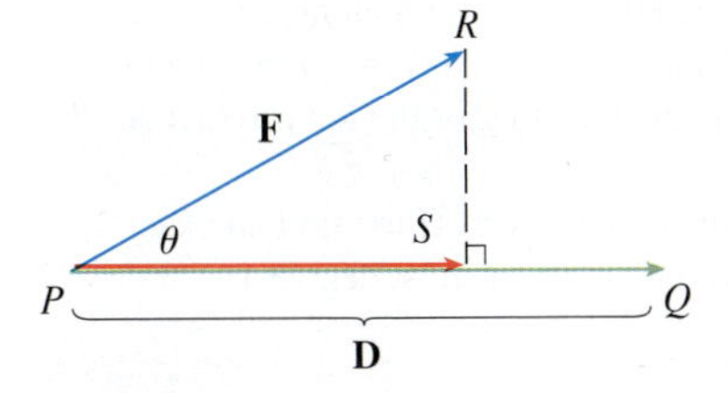

FIGURE 6

$$W = \big(|\mathbf{F}|\cos\theta\big)|\mathbf{D}|$$

But then, from Theorem 3, we have

$$W = |\mathbf{F}||\mathbf{D}| \cos\theta = \mathbf{F} \cdot \mathbf{D} \tag{12}$$

Thus the work done by a constant force **F** is the dot product $\mathbf{F} \cdot \mathbf{D}$, where **D** is the displacement vector.

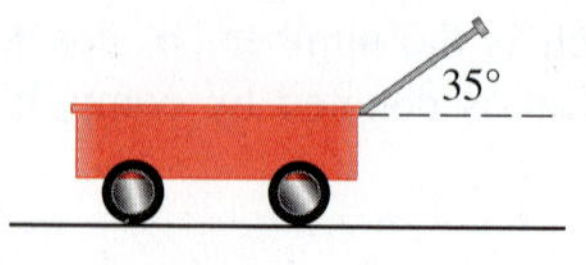

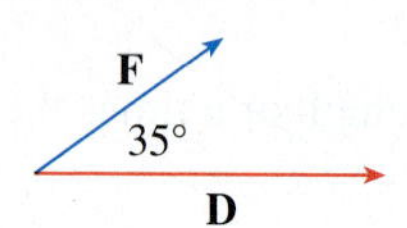

FIGURE 7

EXAMPLE 7 A wagon is pulled a distance of 100 m along a horizontal path by a constant force of 70 N. The handle of the wagon is held at an angle of 35° above the horizontal. Find the work done by the force.

SOLUTION If **F** and **D** are the force and displacement vectors, as pictured in Figure 7, then the work done is

$$W = \mathbf{F} \cdot \mathbf{D} = |\mathbf{F}||\mathbf{D}| \cos 35°$$

$$= (70)(100) \cos 35° \approx 5734 \text{ N}\cdot\text{m} = 5734 \text{ J}$$

EXAMPLE 8 A force is given by a vector $\mathbf{F} = 3\mathbf{i} + 4\mathbf{j} + 5\mathbf{k}$ and moves a particle from the point $P(2, 1, 0)$ to the point $Q(4, 6, 2)$. Find the work done.

SOLUTION The displacement vector is $\mathbf{D} = \overrightarrow{PQ} = \langle 2, 5, 2 \rangle$, so by Equation 12, the work done is

$$W = \mathbf{F} \cdot \mathbf{D} = \langle 3, 4, 5 \rangle \cdot \langle 2, 5, 2 \rangle$$

$$= 6 + 20 + 10 = 36$$

If the unit of length is meters and the magnitude of the force is measured in newtons, then the work done is 36 J.

12.3 EXERCISES

1. Which of the following expressions are meaningful? Which are meaningless? Explain.

(a) $(\mathbf{a} \cdot \mathbf{b}) \cdot \mathbf{c}$ (b) $(\mathbf{a} \cdot \mathbf{b})\mathbf{c}$
(c) $|\mathbf{a}|(\mathbf{b} \cdot \mathbf{c})$ (d) $\mathbf{a} \cdot (\mathbf{b} + \mathbf{c})$
(e) $\mathbf{a} \cdot \mathbf{b} + \mathbf{c}$ (f) $|\mathbf{a}| \cdot (\mathbf{b} + \mathbf{c})$

2–10 Find $\mathbf{a} \cdot \mathbf{b}$.

2. $\mathbf{a} = \langle 5, -2 \rangle$, $\mathbf{b} = \langle 3, 4 \rangle$

3. $\mathbf{a} = \langle 1.5, 0.4 \rangle$, $\mathbf{b} = \langle -4, 6 \rangle$

4. $\mathbf{a} = \langle 6, -2, 3 \rangle$, $\mathbf{b} = \langle 2, 5, -1 \rangle$

5. $\mathbf{a} = \langle 4, 1, \frac{1}{4} \rangle$, $\mathbf{b} = \langle 6, -3, -8 \rangle$

6. $\mathbf{a} = \langle p, -p, 2p \rangle$, $\mathbf{b} = \langle 2q, q, -q \rangle$

7. $\mathbf{a} = 2\mathbf{i} + \mathbf{j}$, $\mathbf{b} = \mathbf{i} - \mathbf{j} + \mathbf{k}$

8. $\mathbf{a} = 3\mathbf{i} + 2\mathbf{j} - \mathbf{k}$, $\mathbf{b} = 4\mathbf{i} + 5\mathbf{k}$

9. $|\mathbf{a}| = 7$, $|\mathbf{b}| = 4$, the angle between **a** and **b** is 30°

10. $|\mathbf{a}| = 80$, $|\mathbf{b}| = 50$, the angle between **a** and **b** is $3\pi/4$

11–12 If **u** is a unit vector, find $\mathbf{u} \cdot \mathbf{v}$ and $\mathbf{u} \cdot \mathbf{w}$.

11.

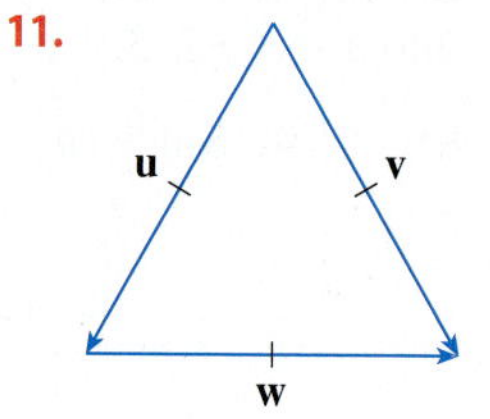

12.

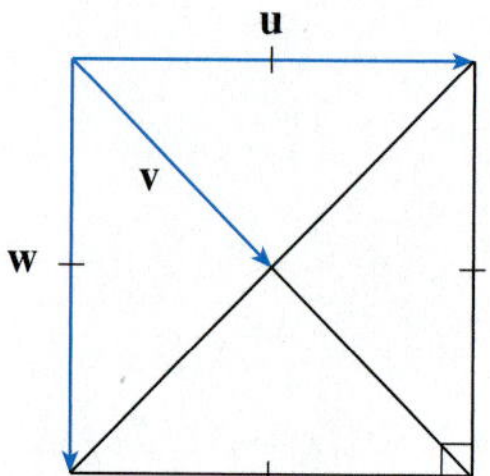

13. (a) Show that $\mathbf{i} \cdot \mathbf{j} = \mathbf{j} \cdot \mathbf{k} = \mathbf{k} \cdot \mathbf{i} = 0$.
(b) Show that $\mathbf{i} \cdot \mathbf{i} = \mathbf{j} \cdot \mathbf{j} = \mathbf{k} \cdot \mathbf{k} = 1$.

14. A street vendor sells a hamburgers, b hot dogs, and c soft drinks on a given day. He charges \$4 for a hamburger, \$2.50 for a hot dog, and \$1 for a soft drink. If $\mathbf{A} = \langle a, b, c \rangle$ and $\mathbf{P} = \langle 4, 2.5, 1 \rangle$, what is the meaning of the dot product $\mathbf{A} \cdot \mathbf{P}$?

15–20 Find the angle between the vectors. (First find an exact expression and then approximate to the nearest degree.)

15. $\mathbf{a} = \langle 4, 3 \rangle$, $\mathbf{b} = \langle 2, -1 \rangle$

16. $\mathbf{a} = \langle -2, 5 \rangle$, $\mathbf{b} = \langle 5, 12 \rangle$

17. $\mathbf{a} = \langle 1, -4, 1 \rangle$, $\mathbf{b} = \langle 0, 2, -2 \rangle$

18. $\mathbf{a} = \langle -1, 3, 4 \rangle$, $\mathbf{b} = \langle 5, 2, 1 \rangle$

19. $\mathbf{a} = 4\mathbf{i} - 3\mathbf{j} + \mathbf{k}$, $\mathbf{b} = 2\mathbf{i} - \mathbf{k}$

20. $\mathbf{a} = 8\mathbf{i} - \mathbf{j} + 4\mathbf{k}$, $\mathbf{b} = 4\mathbf{j} + 2\mathbf{k}$

21–22 Find, correct to the nearest degree, the three angles of the triangle with the given vertices.

21. $P(2, 0)$, $Q(0, 3)$, $R(3, 4)$

22. $A(1, 0, -1)$, $B(3, -2, 0)$, $C(1, 3, 3)$

23–24 Determine whether the given vectors are orthogonal, parallel, or neither.

23. (a) $\mathbf{a} = \langle 9, 3 \rangle$, $\mathbf{b} = \langle -2, 6 \rangle$
(b) $\mathbf{a} = \langle 4, 5, -2 \rangle$, $\mathbf{b} = \langle 3, -1, 5 \rangle$
(c) $\mathbf{a} = -8\mathbf{i} + 12\mathbf{j} + 4\mathbf{k}$, $\mathbf{b} = 6\mathbf{i} - 9\mathbf{j} - 3\mathbf{k}$
(d) $\mathbf{a} = 3\mathbf{i} - \mathbf{j} + 3\mathbf{k}$, $\mathbf{b} = 5\mathbf{i} + 9\mathbf{j} - 2\mathbf{k}$

24. (a) $\mathbf{u} = \langle -5, 4, -2 \rangle$, $\mathbf{v} = \langle 3, 4, -1 \rangle$
(b) $\mathbf{u} = 9\mathbf{i} - 6\mathbf{j} + 3\mathbf{k}$, $\mathbf{v} = -6\mathbf{i} + 4\mathbf{j} - 2\mathbf{k}$
(c) $\mathbf{u} = \langle c, c, c \rangle$, $\mathbf{v} = \langle c, 0, -c \rangle$

25. Use vectors to decide whether the triangle with vertices $P(1, -3, -2)$, $Q(2, 0, -4)$, and $R(6, -2, -5)$ is right-angled.

26. Find the values of x such that the angle between the vectors $\langle 2, 1, -1 \rangle$, and $\langle 1, x, 0 \rangle$ is 45°.

27. Find a unit vector that is orthogonal to both $\mathbf{i} + \mathbf{j}$ and $\mathbf{i} + \mathbf{k}$.

28. Find two unit vectors that make an angle of 60° with $\mathbf{v} = \langle 3, 4 \rangle$.

29–30 Find the acute angle between the lines.

29. $2x - y = 3$, $3x + y = 7$

30. $x + 2y = 7$, $5x - y = 2$

31–32 Find the acute angles between the curves at their points of intersection. (The angle between two curves is the angle between their tangent lines at the point of intersection.)

31. $y = x^2$, $y = x^3$

32. $y = \sin x$, $y = \cos x$, $0 \leqslant x \leqslant \pi/2$

33–37 Find the direction cosines and direction angles of the vector. (Give the direction angles correct to the nearest degree.)

33. $\langle 2, 1, 2 \rangle$

34. $\langle 6, 3, -2 \rangle$

35. $\mathbf{i} - 2\mathbf{j} - 3\mathbf{k}$

36. $\frac{1}{2}\mathbf{i} + \mathbf{j} + \mathbf{k}$

37. $\langle c, c, c \rangle$, where $c > 0$

38. If a vector has direction angles $\alpha = \pi/4$ and $\beta = \pi/3$, find the third direction angle γ.

39–44 Find the scalar and vector projections of $\mathbf{b}$ onto $\mathbf{a}$.

39. $\mathbf{a} = \langle -5, 12 \rangle$, $\mathbf{b} = \langle 4, 6 \rangle$

40. $\mathbf{a} = \langle 1, 4 \rangle$, $\mathbf{b} = \langle 2, 3 \rangle$

41. $\mathbf{a} = \langle 4, 7, -4 \rangle$, $\mathbf{b} = \langle 3, -1, 1 \rangle$

42. $\mathbf{a} = \langle -1, 4, 8 \rangle$, $\mathbf{b} = \langle 12, 1, 2 \rangle$

43. $\mathbf{a} = 3\mathbf{i} - 3\mathbf{j} + \mathbf{k}$, $\mathbf{b} = 2\mathbf{i} + 4\mathbf{j} - \mathbf{k}$

44. $\mathbf{a} = \mathbf{i} + 2\mathbf{j} + 3\mathbf{k}$, $\mathbf{b} = 5\mathbf{i} - \mathbf{k}$

45. Show that the vector $\text{orth}_{\mathbf{a}}\,\mathbf{b} = \mathbf{b} - \text{proj}_{\mathbf{a}}\,\mathbf{b}$ is orthogonal to $\mathbf{a}$. (It is called an **orthogonal projection** of $\mathbf{b}$.)

46. For the vectors in Exercise 40, find $\text{orth}_{\mathbf{a}}\,\mathbf{b}$ and illustrate by drawing the vectors $\mathbf{a}$, $\mathbf{b}$, $\text{proj}_{\mathbf{a}}\,\mathbf{b}$, and $\text{orth}_{\mathbf{a}}\,\mathbf{b}$.

47. If $\mathbf{a} = \langle 3, 0, -1 \rangle$, find a vector $\mathbf{b}$ such that $\text{comp}_{\mathbf{a}}\,\mathbf{b} = 2$.

48. Suppose that $\mathbf{a}$ and $\mathbf{b}$ are nonzero vectors.
(a) Under what circumstances is $\text{comp}_{\mathbf{a}}\,\mathbf{b} = \text{comp}_{\mathbf{b}}\,\mathbf{a}$?
(b) Under what circumstances is $\text{proj}_{\mathbf{a}}\,\mathbf{b} = \text{proj}_{\mathbf{b}}\,\mathbf{a}$?

49. Find the work done by a force $\mathbf{F} = 8\mathbf{i} - 6\mathbf{j} + 9\mathbf{k}$ that moves an object from the point $(0, 10, 8)$ to the point $(6, 12, 20)$ along a straight line. The distance is measured in meters and the force in newtons.

50. A tow truck drags a stalled car along a road. The chain makes an angle of 30° with the road and the tension in the chain is 1500 N. How much work is done by the truck in pulling the car 1 km?

51. A sled is pulled along a level path through snow by a rope. A 30-lb force acting at an angle of 40° above the horizontal moves the sled 80 ft. Find the work done by the force.

52. A boat sails south with the help of a wind blowing in the direction S36°E with magnitude 400 lb. Find the work done by the wind as the boat moves 120 ft.

53. Use a scalar projection to show that the distance from a point $P_1(x_1, y_1)$ to the line $ax + by + c = 0$ is

$$\frac{|ax_1 + by_1 + c|}{\sqrt{a^2 + b^2}}$$

Use this formula to find the distance from the point $(-2, 3)$ to the line $3x - 4y + 5 = 0$.

54. If $\mathbf{r} = \langle x, y, z \rangle$, $\mathbf{a} = \langle a_1, a_2, a_3 \rangle$, and $\mathbf{b} = \langle b_1, b_2, b_3 \rangle$, show that the vector equation $(\mathbf{r} - \mathbf{a}) \cdot (\mathbf{r} - \mathbf{b}) = 0$ represents a sphere, and find its center and radius.

55. Find the angle between a diagonal of a cube and one of its edges.

56. Find the angle between a diagonal of a cube and a diagonal of one of its faces.

57. A molecule of methane, CH_4, is structured with the four hydrogen atoms at the vertices of a regular tetrahedron and the carbon atom at the centroid. The *bond angle* is the angle formed by the H—C—H combination; it is the angle between the lines that join the carbon atom to two of the hydrogen atoms. Show that the bond angle is about 109.5°. [*Hint:* Take the vertices of the tetrahedron to be the points (1, 0, 0), (0, 1, 0), (0, 0, 1), and (1, 1, 1), as shown in the figure. Then the centroid is $\left(\frac{1}{2}, \frac{1}{2}, \frac{1}{2}\right)$.]

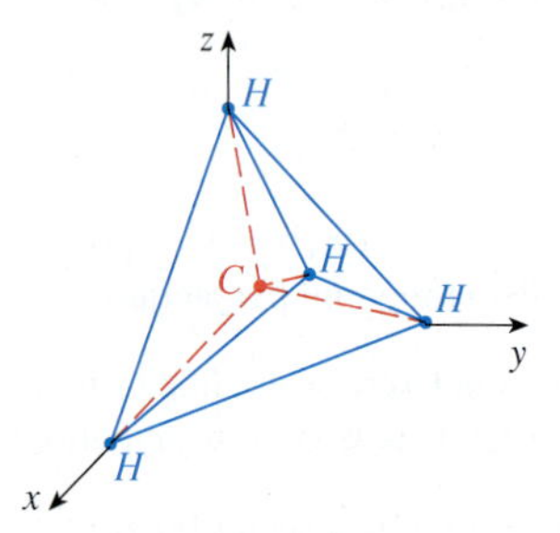

58. If $\mathbf{c} = |\mathbf{a}|\mathbf{b} + |\mathbf{b}|\mathbf{a}$, where **a**, **b**, and **c** are all nonzero vectors, show that **c** bisects the angle between **a** and **b**.

59. Prove Properties 2, 4, and 5 of the dot product (Theorem 2).

60. Suppose that all sides of a quadrilateral are equal in length and opposite sides are parallel. Use vector methods to show that the diagonals are perpendicular.

61. Use Theorem 3 to prove the Cauchy-Schwarz Inequality:

$$|\mathbf{a} \cdot \mathbf{b}| \leq |\mathbf{a}||\mathbf{b}|$$

62. The Triangle Inequality for vectors is

$$|\mathbf{a} + \mathbf{b}| \leq |\mathbf{a}| + |\mathbf{b}|$$

(a) Give a geometric interpretation of the Triangle Inequality.
(b) Use the Cauchy-Schwarz Inequality from Exercise 61 to prove the Triangle Inequality. [*Hint:* Use the fact that $|\mathbf{a} + \mathbf{b}|^2 = (\mathbf{a} + \mathbf{b}) \cdot (\mathbf{a} + \mathbf{b})$ and use Property 3 of the dot product.]

63. The Parallelogram Law states that

$$|\mathbf{a} + \mathbf{b}|^2 + |\mathbf{a} - \mathbf{b}|^2 = 2|\mathbf{a}|^2 + 2|\mathbf{b}|^2$$

(a) Give a geometric interpretation of the Parallelogram Law.
(b) Prove the Parallelogram Law. (See the hint in Exercise 62.)

64. Show that if $\mathbf{u} + \mathbf{v}$ and $\mathbf{u} - \mathbf{v}$ are orthogonal, then the vectors **u** and **v** must have the same length.

65. If θ is the angle between vectors **a** and **b**, show that

$$\text{proj}_{\mathbf{a}} \mathbf{b} \cdot \text{proj}_{\mathbf{b}} \mathbf{a} = (\mathbf{a} \cdot \mathbf{b}) \cos^2 \theta$$

12.4 The Cross Product

Given two nonzero vectors $\mathbf{a} = \langle a_1, a_2, a_3 \rangle$ and $\mathbf{b} = \langle b_1, b_2, b_3 \rangle$, it is very useful to be able to find a nonzero vector **c** that is perpendicular to both **a** and **b**, as we will see in the next section and in Chapters 13 and 14. If $\mathbf{c} = \langle c_1, c_2, c_3 \rangle$ is such a vector, then $\mathbf{a} \cdot \mathbf{c} = 0$ and $\mathbf{b} \cdot \mathbf{c} = 0$ and so

1 $$a_1c_1 + a_2c_2 + a_3c_3 = 0$$

2 $$b_1c_1 + b_2c_2 + b_3c_3 = 0$$

To eliminate c_3 we multiply (1) by b_3 and (2) by a_3 and subtract:

3 $$(a_1b_3 - a_3b_1)c_1 + (a_2b_3 - a_3b_2)c_2 = 0$$

Equation 3 has the form $pc_1 + qc_2 = 0$, for which an obvious solution is $c_1 = q$ and $c_2 = -p$. So a solution of (3) is

$$c_1 = a_2b_3 - a_3b_2 \qquad c_2 = a_3b_1 - a_1b_3$$

Substituting these values into (1) and (2), we then get

$$c_3 = a_1b_2 - a_2b_1$$

This means that a vector perpendicular to both **a** and **b** is

$$\langle c_1, c_2, c_3 \rangle = \langle a_2b_3 - a_3b_2, a_3b_1 - a_1b_3, a_1b_2 - a_2b_1 \rangle$$

The resulting vector is called the *cross product* of **a** and **b** and is denoted by $\mathbf{a} \times \mathbf{b}$.

Hamilton

The cross product was invented by the Irish mathematician Sir William Rowan Hamilton (1805–1865), who had created a precursor of vectors, called quaternions. When he was five years old Hamilton could read Latin, Greek, and Hebrew. At age eight he added French and Italian and when ten he could read Arabic and Sanskrit. At the age of 21, while still an undergraduate at Trinity College in Dublin, Hamilton was appointed Professor of Astronomy at the university and Royal Astronomer of Ireland!

4 Definition If $\mathbf{a} = \langle a_1, a_2, a_3 \rangle$ and $\mathbf{b} = \langle b_1, b_2, b_3 \rangle$, then the **cross product** of $\mathbf{a}$ and $\mathbf{b}$ is the vector

$$\mathbf{a} \times \mathbf{b} = \langle a_2 b_3 - a_3 b_2,\ a_3 b_1 - a_1 b_3,\ a_1 b_2 - a_2 b_1 \rangle$$

Notice that the **cross product a × b** of two vectors **a** and **b**, unlike the dot product, is a vector. For this reason it is also called the **vector product**. Note that $\mathbf{a} \times \mathbf{b}$ is defined only when **a** and **b** are *three-dimensional* vectors.

In order to make Definition 4 easier to remember, we use the notation of determinants. A **determinant of order 2** is defined by

$$\begin{vmatrix} a & b \\ c & d \end{vmatrix} = ad - bc$$

(Multiply across the diagonals and subtract.) For example,

$$\begin{vmatrix} 2 & 1 \\ -6 & 4 \end{vmatrix} = 2(4) - 1(-6) = 14$$

A **determinant of order 3** can be defined in terms of second-order determinants as follows:

$$\boxed{5} \qquad \begin{vmatrix} a_1 & a_2 & a_3 \\ b_1 & b_2 & b_3 \\ c_1 & c_2 & c_3 \end{vmatrix} = a_1 \begin{vmatrix} b_2 & b_3 \\ c_2 & c_3 \end{vmatrix} - a_2 \begin{vmatrix} b_1 & b_3 \\ c_1 & c_3 \end{vmatrix} + a_3 \begin{vmatrix} b_1 & b_2 \\ c_1 & c_2 \end{vmatrix}$$

Observe that each term on the right side of Equation 5 involves a number a_i in the first row of the determinant, and a_i is multiplied by the second-order determinant obtained from the left side by deleting the row and column in which a_i appears. Notice also the minus sign in the second term. For example,

$$\begin{vmatrix} 1 & 2 & -1 \\ 3 & 0 & 1 \\ -5 & 4 & 2 \end{vmatrix} = 1 \begin{vmatrix} 0 & 1 \\ 4 & 2 \end{vmatrix} - 2 \begin{vmatrix} 3 & 1 \\ -5 & 2 \end{vmatrix} + (-1) \begin{vmatrix} 3 & 0 \\ -5 & 4 \end{vmatrix}$$

$$= 1(0 - 4) - 2(6 + 5) + (-1)(12 - 0) = -38$$

If we now rewrite Definition 4 using second-order determinants and the standard basis vectors **i**, **j**, and **k**, we see that the cross product of the vectors $\mathbf{a} = a_1\mathbf{i} + a_2\mathbf{j} + a_3\mathbf{k}$ and $\mathbf{b} = b_1\mathbf{i} + b_2\mathbf{j} + b_3\mathbf{k}$ is

$$\boxed{6} \qquad \mathbf{a} \times \mathbf{b} = \begin{vmatrix} a_2 & a_3 \\ b_2 & b_3 \end{vmatrix} \mathbf{i} - \begin{vmatrix} a_1 & a_3 \\ b_1 & b_3 \end{vmatrix} \mathbf{j} + \begin{vmatrix} a_1 & a_2 \\ b_1 & b_2 \end{vmatrix} \mathbf{k}$$

In view of the similarity between Equations 5 and 6, we often write

$$\boxed{7} \qquad \mathbf{a} \times \mathbf{b} = \begin{vmatrix} \mathbf{i} & \mathbf{j} & \mathbf{k} \\ a_1 & a_2 & a_3 \\ b_1 & b_2 & b_3 \end{vmatrix}$$

Although the first row of the symbolic determinant in Equation 7 consists of vectors, if we expand it as if it were an ordinary determinant using the rule in Equation 5, we obtain

Equation 6. The symbolic formula in Equation 7 is probably the easiest way of remembering and computing cross products.

EXAMPLE 1 If $\mathbf{a} = \langle 1, 3, 4 \rangle$ and $\mathbf{b} = \langle 2, 7, -5 \rangle$, then

$$\mathbf{a} \times \mathbf{b} = \begin{vmatrix} \mathbf{i} & \mathbf{j} & \mathbf{k} \\ 1 & 3 & 4 \\ 2 & 7 & -5 \end{vmatrix}$$

$$= \begin{vmatrix} 3 & 4 \\ 7 & -5 \end{vmatrix} \mathbf{i} - \begin{vmatrix} 1 & 4 \\ 2 & -5 \end{vmatrix} \mathbf{j} + \begin{vmatrix} 1 & 3 \\ 2 & 7 \end{vmatrix} \mathbf{k}$$

$$= (-15 - 28)\,\mathbf{i} - (-5 - 8)\,\mathbf{j} + (7 - 6)\,\mathbf{k} = -43\mathbf{i} + 13\mathbf{j} + \mathbf{k}$$ ■

EXAMPLE 2 Show that $\mathbf{a} \times \mathbf{a} = \mathbf{0}$ for any vector $\mathbf{a}$ in V_3.

SOLUTION If $\mathbf{a} = \langle a_1, a_2, a_3 \rangle$, then

$$\mathbf{a} \times \mathbf{a} = \begin{vmatrix} \mathbf{i} & \mathbf{j} & \mathbf{k} \\ a_1 & a_2 & a_3 \\ a_1 & a_2 & a_3 \end{vmatrix}$$

$$= (a_2a_3 - a_3a_2)\,\mathbf{i} - (a_1a_3 - a_3a_1)\,\mathbf{j} + (a_1a_2 - a_2a_1)\,\mathbf{k}$$

$$= 0\mathbf{i} - 0\mathbf{j} + 0\mathbf{k} = \mathbf{0}$$ ■

We constructed the cross product $\mathbf{a} \times \mathbf{b}$ so that it would be perpendicular to both $\mathbf{a}$ and $\mathbf{b}$. This is one of the most important properties of a cross product, so let's emphasize and verify it in the following theorem and give a formal proof.

8 Theorem The vector $\mathbf{a} \times \mathbf{b}$ is orthogonal to both $\mathbf{a}$ and $\mathbf{b}$.

PROOF In order to show that $\mathbf{a} \times \mathbf{b}$ is orthogonal to $\mathbf{a}$, we compute their dot product as follows:

$$(\mathbf{a} \times \mathbf{b}) \cdot \mathbf{a} = \begin{vmatrix} a_2 & a_3 \\ b_2 & b_3 \end{vmatrix} a_1 - \begin{vmatrix} a_1 & a_3 \\ b_1 & b_3 \end{vmatrix} a_2 + \begin{vmatrix} a_1 & a_2 \\ b_1 & b_2 \end{vmatrix} a_3$$

$$= a_1(a_2b_3 - a_3b_2) - a_2(a_1b_3 - a_3b_1) + a_3(a_1b_2 - a_2b_1)$$

$$= a_1a_2b_3 - a_1b_2a_3 - a_1a_2b_3 + b_1a_2a_3 + a_1b_2a_3 - b_1a_2a_3$$

$$= 0$$

A similar computation shows that $(\mathbf{a} \times \mathbf{b}) \cdot \mathbf{b} = 0$. Therefore $\mathbf{a} \times \mathbf{b}$ is orthogonal to both $\mathbf{a}$ and $\mathbf{b}$. ■

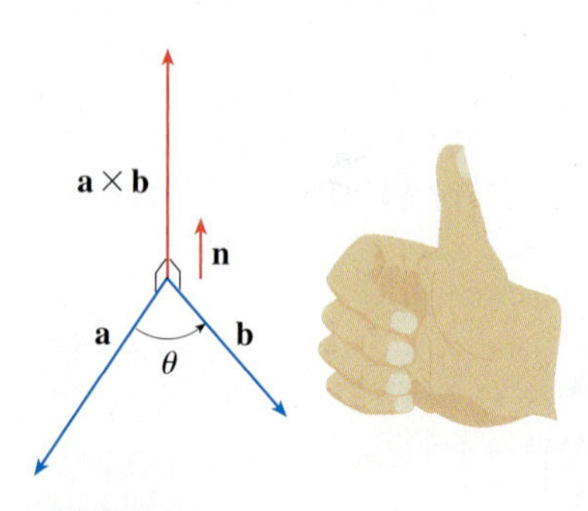

FIGURE 1
The right-hand rule gives the direction of $\mathbf{a} \times \mathbf{b}$.

If $\mathbf{a}$ and $\mathbf{b}$ are represented by directed line segments with the same initial point (as in Figure 1), then Theorem 8 says that the cross product $\mathbf{a} \times \mathbf{b}$ points in a direction perpendicular to the plane through $\mathbf{a}$ and $\mathbf{b}$. It turns out that the direction of $\mathbf{a} \times \mathbf{b}$ is given by the *right-hand rule:* if the fingers of your right hand curl in the direction of a rotation (through an angle less than 180°) from **a** to **b**, then your thumb points in the direction of $\mathbf{a} \times \mathbf{b}$.

Now that we know the direction of the vector $\mathbf{a} \times \mathbf{b}$, the remaining thing we need to complete its geometric description is its length $|\mathbf{a} \times \mathbf{b}|$. This is given by the following theorem.

9 Theorem If θ is the angle between **a** and **b** (so $0 \leqslant \theta \leqslant \pi$), then

$$|\mathbf{a} \times \mathbf{b}| = |\mathbf{a}||\mathbf{b}| \sin \theta$$

TEC Visual 12.4 shows how $\mathbf{a} \times \mathbf{b}$ changes as **b** changes.

PROOF From the definitions of the cross product and length of a vector, we have

$$\begin{aligned}
|\mathbf{a} \times \mathbf{b}|^2 &= (a_2b_3 - a_3b_2)^2 + (a_3b_1 - a_1b_3)^2 + (a_1b_2 - a_2b_1)^2 \\
&= a_2^2b_3^2 - 2a_2a_3b_2b_3 + a_3^2b_2^2 + a_3^2b_1^2 - 2a_1a_3b_1b_3 + a_1^2b_3^2 \\
&\qquad + a_1^2b_2^2 - 2a_1a_2b_1b_2 + a_2^2b_1^2 \\
&= (a_1^2 + a_2^2 + a_3^2)(b_1^2 + b_2^2 + b_3^2) - (a_1b_1 + a_2b_2 + a_3b_3)^2 \\
&= |\mathbf{a}|^2|\mathbf{b}|^2 - (\mathbf{a} \cdot \mathbf{b})^2 \\
&= |\mathbf{a}|^2|\mathbf{b}|^2 - |\mathbf{a}|^2|\mathbf{b}|^2\cos^2\theta \qquad \text{(by Theorem 12.3.3)} \\
&= |\mathbf{a}|^2|\mathbf{b}|^2(1 - \cos^2\theta) \\
&= |\mathbf{a}|^2|\mathbf{b}|^2\sin^2\theta
\end{aligned}$$

Taking square roots and observing that $\sqrt{\sin^2\theta} = \sin\theta$ because $\sin\theta \geqslant 0$ when $0 \leqslant \theta \leqslant \pi$, we have

$$|\mathbf{a} \times \mathbf{b}| = |\mathbf{a}||\mathbf{b}| \sin \theta$$ ■

Geometric characterization of $\mathbf{a} \times \mathbf{b}$

Since a vector is completely determined by its magnitude and direction, we can now say that $\mathbf{a} \times \mathbf{b}$ is the vector that is perpendicular to both **a** and **b**, whose orientation is determined by the right-hand rule, and whose length is $|\mathbf{a}||\mathbf{b}| \sin \theta$. In fact, that is exactly how physicists *define* $\mathbf{a} \times \mathbf{b}$.

10 Corollary Two nonzero vectors **a** and **b** are parallel if and only if

$$\mathbf{a} \times \mathbf{b} = \mathbf{0}$$

PROOF Two nonzero vectors **a** and **b** are parallel if and only if $\theta = 0$ or π. In either case $\sin\theta = 0$, so $|\mathbf{a} \times \mathbf{b}| = 0$ and therefore $\mathbf{a} \times \mathbf{b} = \mathbf{0}$. ■

The geometric interpretation of Theorem 9 can be seen by looking at Figure 2. If **a** and **b** are represented by directed line segments with the same initial point, then they determine a parallelogram with base $|\mathbf{a}|$, altitude $|\mathbf{b}| \sin\theta$, and area

$$A = |\mathbf{a}|\big(|\mathbf{b}| \sin\theta\big) = |\mathbf{a} \times \mathbf{b}|$$

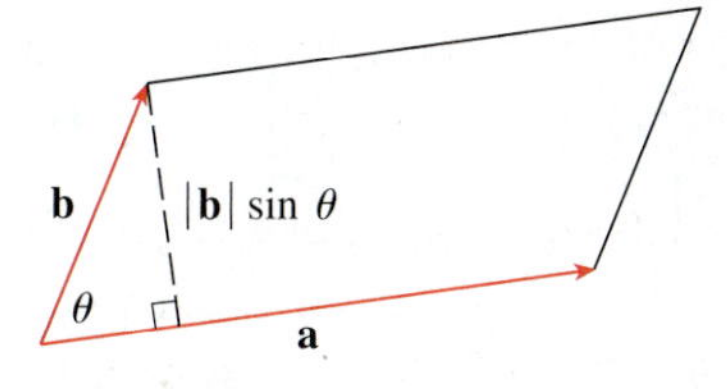

FIGURE 2

Thus we have the following way of interpreting the magnitude of a cross product.

> The length of the cross product $\mathbf{a} \times \mathbf{b}$ is equal to the area of the parallelogram determined by $\mathbf{a}$ and $\mathbf{b}$.

EXAMPLE 3 Find a vector perpendicular to the plane that passes through the points $P(1, 4, 6)$, $Q(-2, 5, -1)$, and $R(1, -1, 1)$.

SOLUTION The vector $\overrightarrow{PQ} \times \overrightarrow{PR}$ is perpendicular to both $\overrightarrow{PQ}$ and $\overrightarrow{PR}$ and is therefore perpendicular to the plane through P, Q, and R. We know from (12.2.1) that

$$\overrightarrow{PQ} = (-2 - 1)\,\mathbf{i} + (5 - 4)\,\mathbf{j} + (-1 - 6)\,\mathbf{k} = -3\mathbf{i} + \mathbf{j} - 7\mathbf{k}$$

$$\overrightarrow{PR} = (1 - 1)\,\mathbf{i} + (-1 - 4)\,\mathbf{j} + (1 - 6)\,\mathbf{k} = -5\mathbf{j} - 5\mathbf{k}$$

We compute the cross product of these vectors:

$$\overrightarrow{PQ} \times \overrightarrow{PR} = \begin{vmatrix} \mathbf{i} & \mathbf{j} & \mathbf{k} \\ -3 & 1 & -7 \\ 0 & -5 & -5 \end{vmatrix}$$

$$= (-5 - 35)\,\mathbf{i} - (15 - 0)\,\mathbf{j} + (15 - 0)\,\mathbf{k} = -40\mathbf{i} - 15\mathbf{j} + 15\mathbf{k}$$

So the vector $\langle -40, -15, 15 \rangle$ is perpendicular to the given plane. Any nonzero scalar multiple of this vector, such as $\langle -8, -3, 3 \rangle$, is also perpendicular to the plane. ■

EXAMPLE 4 Find the area of the triangle with vertices $P(1, 4, 6)$, $Q(-2, 5, -1)$, and $R(1, -1, 1)$.

SOLUTION In Example 3 we computed that $\overrightarrow{PQ} \times \overrightarrow{PR} = \langle -40, -15, 15 \rangle$. The area of the parallelogram with adjacent sides PQ and PR is the length of this cross product:

$$\left| \overrightarrow{PQ} \times \overrightarrow{PR} \right| = \sqrt{(-40)^2 + (-15)^2 + 15^2} = 5\sqrt{82}$$

The area A of the triangle PQR is half the area of this parallelogram, that is, $\frac{5}{2}\sqrt{82}$. ■

If we apply Theorems 8 and 9 to the standard basis vectors $\mathbf{i}$, $\mathbf{j}$, and $\mathbf{k}$ using $\theta = \pi/2$, we obtain

$\mathbf{i} \times \mathbf{j} = \mathbf{k}$	$\mathbf{j} \times \mathbf{k} = \mathbf{i}$	$\mathbf{k} \times \mathbf{i} = \mathbf{j}$
$\mathbf{j} \times \mathbf{i} = -\mathbf{k}$	$\mathbf{k} \times \mathbf{j} = -\mathbf{i}$	$\mathbf{i} \times \mathbf{k} = -\mathbf{j}$

Observe that

$$\mathbf{i} \times \mathbf{j} \neq \mathbf{j} \times \mathbf{i}$$

Thus the cross product is not commutative. Also

$$\mathbf{i} \times (\mathbf{i} \times \mathbf{j}) = \mathbf{i} \times \mathbf{k} = -\mathbf{j}$$

whereas

$$(\mathbf{i} \times \mathbf{i}) \times \mathbf{j} = \mathbf{0} \times \mathbf{j} = \mathbf{0}$$

So the associative law for multiplication does not usually hold; that is, in general,

$$(\mathbf{a} \times \mathbf{b}) \times \mathbf{c} \neq \mathbf{a} \times (\mathbf{b} \times \mathbf{c})$$

However, some of the usual laws of algebra *do* hold for cross products. The following theorem summarizes the properties of vector products.

11 Properties of the Cross Product If $\mathbf{a}$, $\mathbf{b}$, and $\mathbf{c}$ are vectors and c is a scalar, then

1. $\mathbf{a} \times \mathbf{b} = -\mathbf{b} \times \mathbf{a}$
2. $(c\mathbf{a}) \times \mathbf{b} = c(\mathbf{a} \times \mathbf{b}) = \mathbf{a} \times (c\mathbf{b})$
3. $\mathbf{a} \times (\mathbf{b} + \mathbf{c}) = \mathbf{a} \times \mathbf{b} + \mathbf{a} \times \mathbf{c}$
4. $(\mathbf{a} + \mathbf{b}) \times \mathbf{c} = \mathbf{a} \times \mathbf{c} + \mathbf{b} \times \mathbf{c}$
5. $\mathbf{a} \cdot (\mathbf{b} \times \mathbf{c}) = (\mathbf{a} \times \mathbf{b}) \cdot \mathbf{c}$
6. $\mathbf{a} \times (\mathbf{b} \times \mathbf{c}) = (\mathbf{a} \cdot \mathbf{c})\mathbf{b} - (\mathbf{a} \cdot \mathbf{b})\mathbf{c}$

These properties can be proved by writing the vectors in terms of their components and using the definition of a cross product. We give the proof of Property 5 and leave the remaining proofs as exercises.

PROOF OF PROPERTY 5 If $\mathbf{a} = \langle a_1, a_2, a_3 \rangle$, $\mathbf{b} = \langle b_1, b_2, b_3 \rangle$, and $\mathbf{c} = \langle c_1, c_2, c_3 \rangle$, then

$$\boxed{12} \quad \begin{aligned} \mathbf{a} \cdot (\mathbf{b} \times \mathbf{c}) &= a_1(b_2c_3 - b_3c_2) + a_2(b_3c_1 - b_1c_3) + a_3(b_1c_2 - b_2c_1) \\ &= a_1b_2c_3 - a_1b_3c_2 + a_2b_3c_1 - a_2b_1c_3 + a_3b_1c_2 - a_3b_2c_1 \\ &= (a_2b_3 - a_3b_2)c_1 + (a_3b_1 - a_1b_3)c_2 + (a_1b_2 - a_2b_1)c_3 \\ &= (\mathbf{a} \times \mathbf{b}) \cdot \mathbf{c} \end{aligned}$$

■

Triple Products

The product $\mathbf{a} \cdot (\mathbf{b} \times \mathbf{c})$ that occurs in Property 5 is called the **scalar triple product** of the vectors $\mathbf{a}$, $\mathbf{b}$, and $\mathbf{c}$. Notice from Equation 12 that we can write the scalar triple product as a determinant:

$$\boxed{13} \qquad \mathbf{a} \cdot (\mathbf{b} \times \mathbf{c}) = \begin{vmatrix} a_1 & a_2 & a_3 \\ b_1 & b_2 & b_3 \\ c_1 & c_2 & c_3 \end{vmatrix}$$

The geometric significance of the scalar triple product can be seen by considering the parallelepiped determined by the vectors $\mathbf{a}$, $\mathbf{b}$, and $\mathbf{c}$. (See Figure 3.) The area of the base parallelogram is $A = |\mathbf{b} \times \mathbf{c}|$. If θ is the angle between $\mathbf{a}$ and $\mathbf{b} \times \mathbf{c}$, then the height h of the parallelepiped is $h = |\mathbf{a}||\cos\theta|$. (We must use $|\cos\theta|$ instead of $\cos\theta$ in case $\theta > \pi/2$.) Therefore the volume of the parallelepiped is

$$V = Ah = |\mathbf{b} \times \mathbf{c}||\mathbf{a}||\cos\theta| = |\mathbf{a} \cdot (\mathbf{b} \times \mathbf{c})|$$

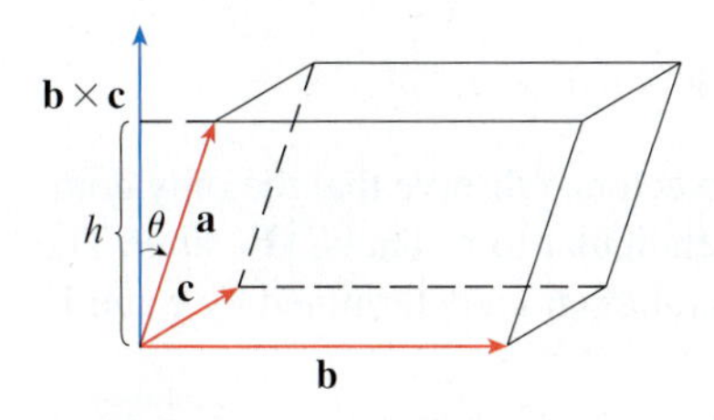

FIGURE 3

Thus we have proved the following formula.

14 The volume of the parallelepiped determined by the vectors **a**, **b**, and **c** is the magnitude of their scalar triple product:

$$V = |\mathbf{a} \cdot (\mathbf{b} \times \mathbf{c})|$$

If we use the formula in (14) and discover that the volume of the parallelepiped determined by **a**, **b**, and **c** is 0, then the vectors must lie in the same plane; that is, they are **coplanar**.

EXAMPLE 5 Use the scalar triple product to show that the vectors $\mathbf{a} = \langle 1, 4, -7 \rangle$, $\mathbf{b} = \langle 2, -1, 4 \rangle$, and $\mathbf{c} = \langle 0, -9, 18 \rangle$ are coplanar.

SOLUTION We use Equation 13 to compute their scalar triple product:

$$\mathbf{a} \cdot (\mathbf{b} \times \mathbf{c}) = \begin{vmatrix} 1 & 4 & -7 \\ 2 & -1 & 4 \\ 0 & -9 & 18 \end{vmatrix}$$

$$= 1\begin{vmatrix} -1 & 4 \\ -9 & 18 \end{vmatrix} - 4\begin{vmatrix} 2 & 4 \\ 0 & 18 \end{vmatrix} - 7\begin{vmatrix} 2 & -1 \\ 0 & -9 \end{vmatrix}$$

$$= 1(18) - 4(36) - 7(-18) = 0$$

Therefore, by (14), the volume of the parallelepiped determined by **a**, **b**, and **c** is 0. This means that **a**, **b**, and **c** are coplanar. ■

The product $\mathbf{a} \times (\mathbf{b} \times \mathbf{c})$ that occurs in Property 6 is called the **vector triple product** of **a**, **b**, and **c**. Property 6 will be used to derive Kepler's First Law of planetary motion in Chapter 13. Its proof is left as Exercise 50.

Torque

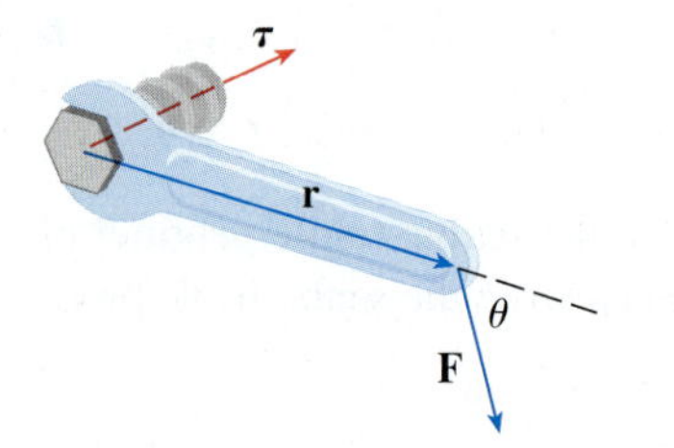

FIGURE 4

The idea of a cross product occurs often in physics. In particular, we consider a force **F** acting on a rigid body at a point given by a position vector **r**. (For instance, if we tighten a bolt by applying a force to a wrench as in Figure 4, we produce a turning effect.) The **torque** $\boldsymbol{\tau}$ (relative to the origin) is defined to be the cross product of the position and force vectors

$$\boldsymbol{\tau} = \mathbf{r} \times \mathbf{F}$$

and measures the tendency of the body to rotate about the origin. The direction of the torque vector indicates the axis of rotation. According to Theorem 9, the magnitude of the torque vector is

$$|\boldsymbol{\tau}| = |\mathbf{r} \times \mathbf{F}| = |\mathbf{r}||\mathbf{F}| \sin\theta$$

where θ is the angle between the position and force vectors. Observe that the only component of **F** that can cause a rotation is the one perpendicular to **r**, that is, $|\mathbf{F}| \sin\theta$. The magnitude of the torque is equal to the area of the parallelogram determined by **r** and **F**.

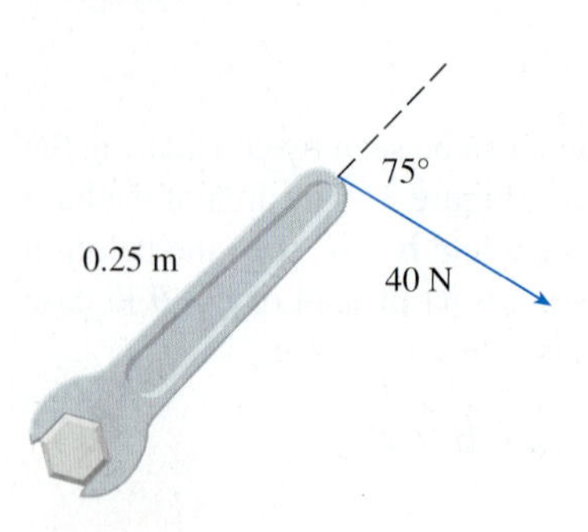

FIGURE 5

EXAMPLE 6 A bolt is tightened by applying a 40-N force to a 0.25-m wrench as shown in Figure 5. Find the magnitude of the torque about the center of the bolt.

SOLUTION The magnitude of the torque vector is

$$|\boldsymbol{\tau}| = |\mathbf{r} \times \mathbf{F}| = |\mathbf{r}||\mathbf{F}| \sin 75° = (0.25)(40) \sin 75°$$
$$= 10 \sin 75° \approx 9.66 \text{ N}\cdot\text{m}$$

If the bolt is right-threaded, then the torque vector itself is

$$\boldsymbol{\tau} = |\boldsymbol{\tau}|\,\mathbf{n} \approx 9.66\,\mathbf{n}$$

where **n** is a unit vector directed down into the page (by the right-hand rule). ■

12.4 EXERCISES

1–7 Find the cross product $\mathbf{a} \times \mathbf{b}$ and verify that it is orthogonal to both **a** and **b**.

1. $\mathbf{a} = \langle 2, 3, 0 \rangle$, $\mathbf{b} = \langle 1, 0, 5 \rangle$
2. $\mathbf{a} = \langle 4, 3, -2 \rangle$, $\mathbf{b} = \langle 2, -1, 1 \rangle$
3. $\mathbf{a} = 2\mathbf{j} - 4\mathbf{k}$, $\mathbf{b} = -\mathbf{i} + 3\mathbf{j} + \mathbf{k}$
4. $\mathbf{a} = 3\mathbf{i} + 3\mathbf{j} - 3\mathbf{k}$, $\mathbf{b} = 3\mathbf{i} - 3\mathbf{j} + 3\mathbf{k}$
5. $\mathbf{a} = \frac{1}{2}\mathbf{i} + \frac{1}{3}\mathbf{j} + \frac{1}{4}\mathbf{k}$, $\mathbf{b} = \mathbf{i} + 2\mathbf{j} - 3\mathbf{k}$
6. $\mathbf{a} = t\mathbf{i} + \cos t\mathbf{j} + \sin t\mathbf{k}$, $\mathbf{b} = \mathbf{i} - \sin t\mathbf{j} + \cos t\mathbf{k}$
7. $\mathbf{a} = \langle t, 1, 1/t \rangle$, $\mathbf{b} = \langle t^2, t^2, 1 \rangle$

8. If $\mathbf{a} = \mathbf{i} - 2\mathbf{k}$ and $\mathbf{b} = \mathbf{j} + \mathbf{k}$, find $\mathbf{a} \times \mathbf{b}$. Sketch **a**, **b**, and $\mathbf{a} \times \mathbf{b}$ as vectors starting at the origin.

9–12 Find the vector, not with determinants, but by using properties of cross products.

9. $(\mathbf{i} \times \mathbf{j}) \times \mathbf{k}$
10. $\mathbf{k} \times (\mathbf{i} - 2\mathbf{j})$
11. $(\mathbf{j} - \mathbf{k}) \times (\mathbf{k} - \mathbf{i})$
12. $(\mathbf{i} + \mathbf{j}) \times (\mathbf{i} - \mathbf{j})$

13. State whether each expression is meaningful. If not, explain why. If so, state whether it is a vector or a scalar.
(a) $\mathbf{a} \cdot (\mathbf{b} \times \mathbf{c})$ (b) $\mathbf{a} \times (\mathbf{b} \cdot \mathbf{c})$
(c) $\mathbf{a} \times (\mathbf{b} \times \mathbf{c})$ (d) $\mathbf{a} \cdot (\mathbf{b} \cdot \mathbf{c})$
(e) $(\mathbf{a} \cdot \mathbf{b}) \times (\mathbf{c} \cdot \mathbf{d})$ (f) $(\mathbf{a} \times \mathbf{b}) \cdot (\mathbf{c} \times \mathbf{d})$

14–15 Find $|\mathbf{u} \times \mathbf{v}|$ and determine whether $\mathbf{u} \times \mathbf{v}$ is directed into the page or out of the page.

14.
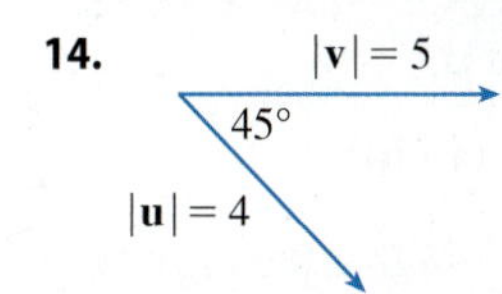

15.
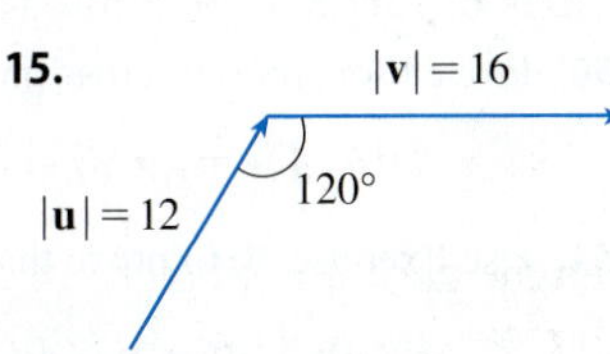

16. The figure shows a vector **a** in the xy-plane and a vector **b** in the direction of **k**. Their lengths are $|\mathbf{a}| = 3$ and $|\mathbf{b}| = 2$.
(a) Find $|\mathbf{a} \times \mathbf{b}|$.
(b) Use the right-hand rule to decide whether the components of $\mathbf{a} \times \mathbf{b}$ are positive, negative, or 0.

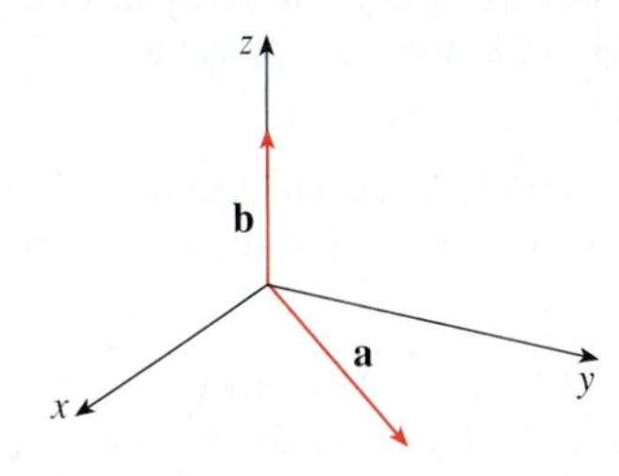

17. If $\mathbf{a} = \langle 2, -1, 3 \rangle$ and $\mathbf{b} = \langle 4, 2, 1 \rangle$, find $\mathbf{a} \times \mathbf{b}$ and $\mathbf{b} \times \mathbf{a}$.
18. If $\mathbf{a} = \langle 1, 0, 1 \rangle$, $\mathbf{b} = \langle 2, 1, -1 \rangle$, and $\mathbf{c} = \langle 0, 1, 3 \rangle$, show that $\mathbf{a} \times (\mathbf{b} \times \mathbf{c}) \neq (\mathbf{a} \times \mathbf{b}) \times \mathbf{c}$.
19. Find two unit vectors orthogonal to both $\langle 3, 2, 1 \rangle$ and $\langle -1, 1, 0 \rangle$.
20. Find two unit vectors orthogonal to both $\mathbf{j} - \mathbf{k}$ and $\mathbf{i} + \mathbf{j}$.
21. Show that $\mathbf{0} \times \mathbf{a} = \mathbf{0} = \mathbf{a} \times \mathbf{0}$ for any vector **a** in V_3.
22. Show that $(\mathbf{a} \times \mathbf{b}) \cdot \mathbf{b} = 0$ for all vectors **a** and **b** in V_3.

23–26 Prove the property of cross products (Theorem 11).

23. Property 1: $\mathbf{a} \times \mathbf{b} = -\mathbf{b} \times \mathbf{a}$
24. Property 2: $(c\mathbf{a}) \times \mathbf{b} = c(\mathbf{a} \times \mathbf{b}) = \mathbf{a} \times (c\mathbf{b})$
25. Property 3: $\mathbf{a} \times (\mathbf{b} + \mathbf{c}) = \mathbf{a} \times \mathbf{b} + \mathbf{a} \times \mathbf{c}$
26. Property 4: $(\mathbf{a} + \mathbf{b}) \times \mathbf{c} = \mathbf{a} \times \mathbf{c} + \mathbf{b} \times \mathbf{c}$

27. Find the area of the parallelogram with vertices $A(-3, 0)$, $B(-1, 3)$, $C(5, 2)$, and $D(3, -1)$.
28. Find the area of the parallelogram with vertices $P(1, 0, 2)$, $Q(3, 3, 3)$, $R(7, 5, 8)$, and $S(5, 2, 7)$.

29–32 (a) Find a nonzero vector orthogonal to the plane through the points P, Q, and R, and (b) find the area of triangle PQR.

29. $P(1, 0, 1)$, $Q(-2, 1, 3)$, $R(4, 2, 5)$
30. $P(0, 0, -3)$, $Q(4, 2, 0)$, $R(3, 3, 1)$
31. $P(0, -2, 0)$, $Q(4, 1, -2)$, $R(5, 3, 1)$

32. $P(2, -3, 4)$, $Q(-1, -2, 2)$, $R(3, 1, -3)$

33–34 Find the volume of the parallelepiped determined by the vectors **a**, **b**, and **c**.

33. $\mathbf{a} = \langle 1, 2, 3 \rangle$, $\mathbf{b} = \langle -1, 1, 2 \rangle$, $\mathbf{c} = \langle 2, 1, 4 \rangle$

34. $\mathbf{a} = \mathbf{i} + \mathbf{j}$, $\mathbf{b} = \mathbf{j} + \mathbf{k}$, $\mathbf{c} = \mathbf{i} + \mathbf{j} + \mathbf{k}$

35–36 Find the volume of the parallelepiped with adjacent edges PQ, PR, and PS.

35. $P(-2, 1, 0)$, $Q(2, 3, 2)$, $R(1, 4, -1)$, $S(3, 6, 1)$

36. $P(3, 0, 1)$, $Q(-1, 2, 5)$, $R(5, 1, -1)$, $S(0, 4, 2)$

37. Use the scalar triple product to verify that the vectors $\mathbf{u} = \mathbf{i} + 5\mathbf{j} - 2\mathbf{k}$, $\mathbf{v} = 3\mathbf{i} - \mathbf{j}$, and $\mathbf{w} = 5\mathbf{i} + 9\mathbf{j} - 4\mathbf{k}$ are coplanar.

38. Use the scalar triple product to determine whether the points $A(1, 3, 2)$, $B(3, -1, 6)$, $C(5, 2, 0)$, and $D(3, 6, -4)$ lie in the same plane.

39. A bicycle pedal is pushed by a foot with a 60-N force as shown. The shaft of the pedal is 18 cm long. Find the magnitude of the torque about P.

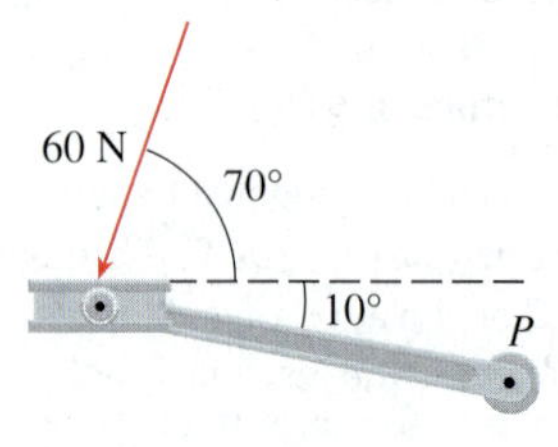

40. (a) A horizontal force of 20 lb is applied to the handle of a gearshift lever as shown. Find the magnitude of the torque about the pivot point P.
(b) Find the magnitude of the torque about P if the same force is applied at the elbow Q of the lever.

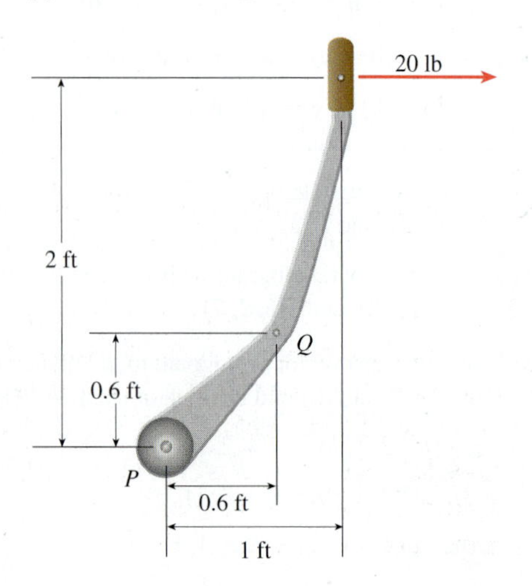

41. A wrench 30 cm long lies along the positive y-axis and grips a bolt at the origin. A force is applied in the direction $\langle 0, 3, -4 \rangle$ at the end of the wrench. Find the magnitude of the force needed to supply 100 N·m of torque to the bolt.

42. Let $\mathbf{v} = 5\mathbf{j}$ and let $\mathbf{u}$ be a vector with length 3 that starts at the origin and rotates in the xy-plane. Find the maximum and minimum values of the length of the vector $\mathbf{u} \times \mathbf{v}$. In what direction does $\mathbf{u} \times \mathbf{v}$ point?

43. If $\mathbf{a} \cdot \mathbf{b} = \sqrt{3}$ and $\mathbf{a} \times \mathbf{b} = \langle 1, 2, 2 \rangle$, find the angle between **a** and **b**.

44. (a) Find all vectors **v** such that

$$\langle 1, 2, 1 \rangle \times \mathbf{v} = \langle 3, 1, -5 \rangle$$

(b) Explain why there is no vector **v** such that

$$\langle 1, 2, 1 \rangle \times \mathbf{v} = \langle 3, 1, 5 \rangle$$

45. (a) Let P be a point not on the line L that passes through the points Q and R. Show that the distance d from the point P to the line L is

$$d = \frac{|\mathbf{a} \times \mathbf{b}|}{|\mathbf{a}|}$$

where $\mathbf{a} = \overrightarrow{QR}$ and $\mathbf{b} = \overrightarrow{QP}$.
(b) Use the formula in part (a) to find the distance from the point $P(1, 1, 1)$ to the line through $Q(0, 6, 8)$ and $R(-1, 4, 7)$.

46. (a) Let P be a point not on the plane that passes through the points Q, R, and S. Show that the distance d from P to the plane is

$$d = \frac{|\mathbf{a} \cdot (\mathbf{b} \times \mathbf{c})|}{|\mathbf{a} \times \mathbf{b}|}$$

where $\mathbf{a} = \overrightarrow{QR}$, $\mathbf{b} = \overrightarrow{QS}$, and $\mathbf{c} = \overrightarrow{QP}$.
(b) Use the formula in part (a) to find the distance from the point $P(2, 1, 4)$ to the plane through the points $Q(1, 0, 0)$, $R(0, 2, 0)$, and $S(0, 0, 3)$.

47. Show that $|\mathbf{a} \times \mathbf{b}|^2 = |\mathbf{a}|^2|\mathbf{b}|^2 - (\mathbf{a} \cdot \mathbf{b})^2$.

48. If $\mathbf{a} + \mathbf{b} + \mathbf{c} = \mathbf{0}$, show that

$$\mathbf{a} \times \mathbf{b} = \mathbf{b} \times \mathbf{c} = \mathbf{c} \times \mathbf{a}$$

49. Prove that $(\mathbf{a} - \mathbf{b}) \times (\mathbf{a} + \mathbf{b}) = 2(\mathbf{a} \times \mathbf{b})$.

50. Prove Property 6 of cross products, that is,

$$\mathbf{a} \times (\mathbf{b} \times \mathbf{c}) = (\mathbf{a} \cdot \mathbf{c})\mathbf{b} - (\mathbf{a} \cdot \mathbf{b})\mathbf{c}$$

51. Use Exercise 50 to prove that

$$\mathbf{a} \times (\mathbf{b} \times \mathbf{c}) + \mathbf{b} \times (\mathbf{c} \times \mathbf{a}) + \mathbf{c} \times (\mathbf{a} \times \mathbf{b}) = \mathbf{0}$$

52. Prove that

$$(\mathbf{a} \times \mathbf{b}) \cdot (\mathbf{c} \times \mathbf{d}) = \begin{vmatrix} \mathbf{a} \cdot \mathbf{c} & \mathbf{b} \cdot \mathbf{c} \\ \mathbf{a} \cdot \mathbf{d} & \mathbf{b} \cdot \mathbf{d} \end{vmatrix}$$

53. Suppose that $\mathbf{a} \neq \mathbf{0}$.
(a) If $\mathbf{a} \cdot \mathbf{b} = \mathbf{a} \cdot \mathbf{c}$, does it follow that $\mathbf{b} = \mathbf{c}$?
(b) If $\mathbf{a} \times \mathbf{b} = \mathbf{a} \times \mathbf{c}$, does it follow that $\mathbf{b} = \mathbf{c}$?
(c) If $\mathbf{a} \cdot \mathbf{b} = \mathbf{a} \cdot \mathbf{c}$ and $\mathbf{a} \times \mathbf{b} = \mathbf{a} \times \mathbf{c}$, does it follow that $\mathbf{b} = \mathbf{c}$?

54. If $\mathbf{v}_1$, $\mathbf{v}_2$, and $\mathbf{v}_3$ are noncoplanar vectors, let

$$\mathbf{k}_1 = \frac{\mathbf{v}_2 \times \mathbf{v}_3}{\mathbf{v}_1 \cdot (\mathbf{v}_2 \times \mathbf{v}_3)} \qquad \mathbf{k}_2 = \frac{\mathbf{v}_3 \times \mathbf{v}_1}{\mathbf{v}_1 \cdot (\mathbf{v}_2 \times \mathbf{v}_3)}$$

$$\mathbf{k}_3 = \frac{\mathbf{v}_1 \times \mathbf{v}_2}{\mathbf{v}_1 \cdot (\mathbf{v}_2 \times \mathbf{v}_3)}$$

(These vectors occur in the study of crystallography. Vectors of the form $n_1\mathbf{v}_1 + n_2\mathbf{v}_2 + n_3\mathbf{v}_3$, where each n_i is an integer, form a *lattice* for a crystal. Vectors written similarly in terms of $\mathbf{k}_1$, $\mathbf{k}_2$, and $\mathbf{k}_3$ form the *reciprocal lattice*.)
(a) Show that $\mathbf{k}_i$ is perpendicular to $\mathbf{v}_j$ if $i \neq j$.
(b) Show that $\mathbf{k}_i \cdot \mathbf{v}_i = 1$ for $i = 1, 2, 3$.
(c) Show that $\mathbf{k}_1 \cdot (\mathbf{k}_2 \times \mathbf{k}_3) = \dfrac{1}{\mathbf{v}_1 \cdot (\mathbf{v}_2 \times \mathbf{v}_3)}$.

DISCOVERY PROJECT THE GEOMETRY OF A TETRAHEDRON

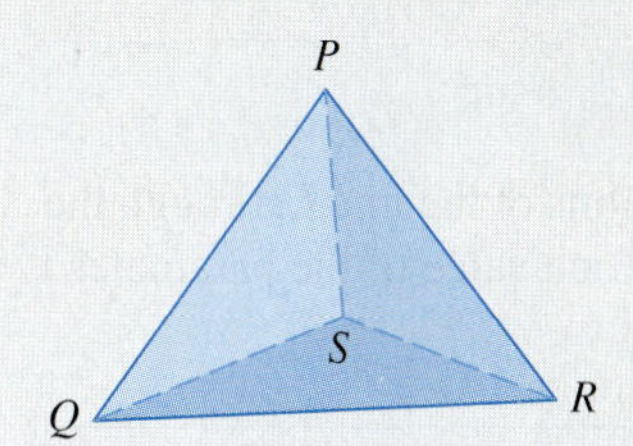

A tetrahedron is a solid with four vertices, P, Q, R, and S, and four triangular faces, as shown in the figure.

1. Let $\mathbf{v}_1$, $\mathbf{v}_2$, $\mathbf{v}_3$, and $\mathbf{v}_4$ be vectors with lengths equal to the areas of the faces opposite the vertices P, Q, R, and S, respectively, and directions perpendicular to the respective faces and pointing outward. Show that

$$\mathbf{v}_1 + \mathbf{v}_2 + \mathbf{v}_3 + \mathbf{v}_4 = \mathbf{0}$$

2. The volume V of a tetrahedron is one-third the distance from a vertex to the opposite face, times the area of that face.
(a) Find a formula for the volume of a tetrahedron in terms of the coordinates of its vertices P, Q, R, and S.
(b) Find the volume of the tetrahedron whose vertices are $P(1, 1, 1)$, $Q(1, 2, 3)$, $R(1, 1, 2)$, and $S(3, -1, 2)$.

3. Suppose the tetrahedron in the figure has a trirectangular vertex S. (This means that the three angles at S are all right angles.) Let A, B, and C be the areas of the three faces that meet at S, and let D be the area of the opposite face PQR. Using the result of Problem 1, or otherwise, show that

$$D^2 = A^2 + B^2 + C^2$$

(This is a three-dimensional version of the Pythagorean Theorem.)

12.5 Equations of Lines and Planes

Lines

A line in the xy-plane is determined when a point on the line and the direction of the line (its slope or angle of inclination) are given. The equation of the line can then be written using the point-slope form.

Likewise, a line L in three-dimensional space is determined when we know a point $P_0(x_0, y_0, z_0)$ on L and the direction of L. In three dimensions the direction of a line is conveniently described by a vector, so we let $\mathbf{v}$ be a vector parallel to L. Let $P(x, y, z)$ be an arbitrary point on L and let $\mathbf{r}_0$ and $\mathbf{r}$ be the position vectors of P_0 and P (that is, they have representations $\overrightarrow{OP_0}$ and $\overrightarrow{OP}$). If $\mathbf{a}$ is the vector with representation $\overrightarrow{P_0P}$, as in Figure 1, then the Triangle Law for vector addition gives $\mathbf{r} = \mathbf{r}_0 + \mathbf{a}$. But, since $\mathbf{a}$ and $\mathbf{v}$ are parallel vectors, there is a scalar t such that $\mathbf{a} = t\mathbf{v}$. Thus

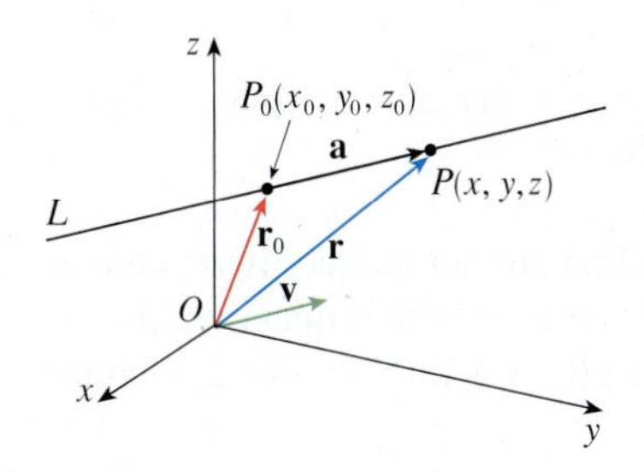

FIGURE 1

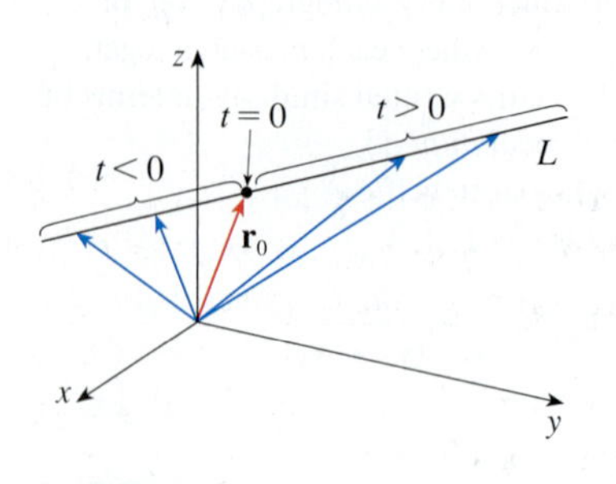

FIGURE 2

1
$$\mathbf{r} = \mathbf{r}_0 + t\mathbf{v}$$

which is a **vector equation** of L. Each value of the **parameter** t gives the position vector $\mathbf{r}$ of a point on L. In other words, as t varies, the line is traced out by the tip of the vector $\mathbf{r}$. As Figure 2 indicates, positive values of t correspond to points on L that lie on one side of P_0, whereas negative values of t correspond to points that lie on the other side of P_0.

If the vector $\mathbf{v}$ that gives the direction of the line L is written in component form as $\mathbf{v} = \langle a, b, c\rangle$, then we have $t\mathbf{v} = \langle ta, tb, tc\rangle$. We can also write $\mathbf{r} = \langle x, y, z\rangle$ and $\mathbf{r}_0 = \langle x_0, y_0, z_0\rangle$, so the vector equation (1) becomes

$$\langle x, y, z\rangle = \langle x_0 + ta, y_0 + tb, z_0 + tc\rangle$$

Two vectors are equal if and only if corresponding components are equal. Therefore we have the three scalar equations:

$$x = x_0 + at \qquad y = y_0 + bt \qquad z = z_0 + ct$$

where $t \in \mathbb{R}$. These equations are called **parametric equations** of the line L through the point $P_0(x_0, y_0, z_0)$ and parallel to the vector $\mathbf{v} = \langle a, b, c\rangle$. Each value of the parameter t gives a point (x, y, z) on L.

2 Parametric equations for a line through the point (x_0, y_0, z_0) and parallel to the direction vector $\langle a, b, c\rangle$ are

$$x = x_0 + at \qquad y = y_0 + bt \qquad z = z_0 + ct$$

Figure 3 shows the line L in Example 1 and its relation to the given point and to the vector that gives its direction.

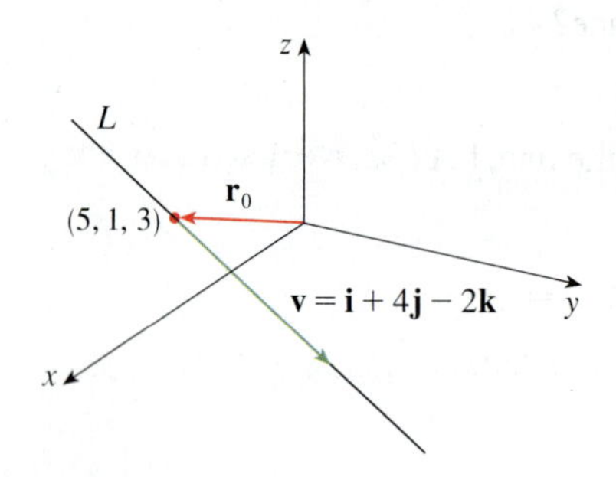

FIGURE 3

EXAMPLE 1

(a) Find a vector equation and parametric equations for the line that passes through the point (5, 1, 3) and is parallel to the vector $\mathbf{i} + 4\mathbf{j} - 2\mathbf{k}$.
(b) Find two other points on the line.

SOLUTION

(a) Here $\mathbf{r}_0 = \langle 5, 1, 3\rangle = 5\mathbf{i} + \mathbf{j} + 3\mathbf{k}$ and $\mathbf{v} = \mathbf{i} + 4\mathbf{j} - 2\mathbf{k}$, so the vector equation (1) becomes

$$\mathbf{r} = (5\mathbf{i} + \mathbf{j} + 3\mathbf{k}) + t(\mathbf{i} + 4\mathbf{j} - 2\mathbf{k})$$

or

$$\mathbf{r} = (5 + t)\,\mathbf{i} + (1 + 4t)\,\mathbf{j} + (3 - 2t)\,\mathbf{k}$$

Parametric equations are

$$x = 5 + t \qquad y = 1 + 4t \qquad z = 3 - 2t$$

(b) Choosing the parameter value $t = 1$ gives $x = 6$, $y = 5$, and $z = 1$, so (6, 5, 1) is a point on the line. Similarly, $t = -1$ gives the point $(4, -3, 5)$. ■

The vector equation and parametric equations of a line are not unique. If we change the point or the parameter or choose a different parallel vector, then the equations change. For instance, if, instead of (5, 1, 3), we choose the point (6, 5, 1) in Example 1, then the parametric equations of the line become

$$x = 6 + t \qquad y = 5 + 4t \qquad z = 1 - 2t$$

Or, if we stay with the point (5, 1, 3) but choose the parallel vector $2\mathbf{i} + 8\mathbf{j} - 4\mathbf{k}$, we arrive at the equations

$$x = 5 + 2t \qquad y = 1 + 8t \qquad z = 3 - 4t$$

In general, if a vector $\mathbf{v} = \langle a, b, c \rangle$ is used to describe the direction of a line L, then the numbers a, b, and c are called **direction numbers** of L. Since any vector parallel to $\mathbf{v}$ could also be used, we see that any three numbers proportional to a, b, and c could also be used as a set of direction numbers for L.

Another way of describing a line L is to eliminate the parameter t from Equations 2. If none of a, b, or c is 0, we can solve each of these equations for t:

$$t = \frac{x - x_0}{a} \qquad t = \frac{y - y_0}{b} \qquad t = \frac{z - z_0}{c}$$

Equating the results, we obtain

3
$$\frac{x - x_0}{a} = \frac{y - y_0}{b} = \frac{z - z_0}{c}$$

These equations are called **symmetric equations** of L. Notice that the numbers a, b, and c that appear in the denominators of Equations 3 are direction numbers of L, that is, components of a vector parallel to L. If one of a, b, or c is 0, we can still eliminate t. For instance, if $a = 0$, we could write the equations of L as

$$x = x_0 \qquad \frac{y - y_0}{b} = \frac{z - z_0}{c}$$

This means that L lies in the vertical plane $x = x_0$.

Figure 4 shows the line L in Example 2 and the point P where it intersects the xy-plane.

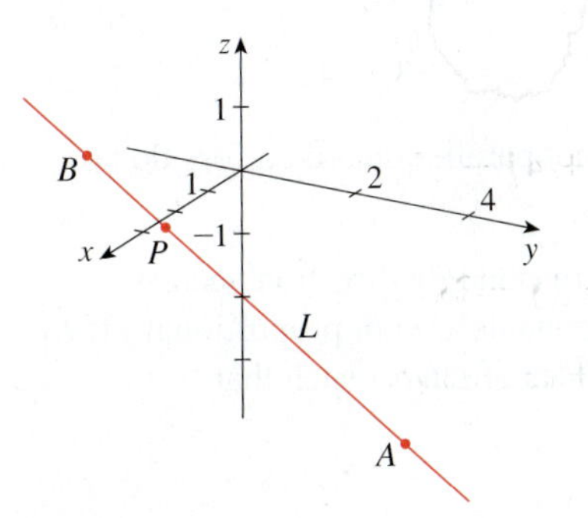

FIGURE 4

EXAMPLE 2

(a) Find parametric equations and symmetric equations of the line that passes through the points $A(2, 4, -3)$ and $B(3, -1, 1)$.
(b) At what point does this line intersect the xy-plane?

SOLUTION

(a) We are not explicitly given a vector parallel to the line, but observe that the vector $\mathbf{v}$ with representation $\overrightarrow{AB}$ is parallel to the line and

$$\mathbf{v} = \langle 3 - 2, -1 - 4, 1 - (-3) \rangle = \langle 1, -5, 4 \rangle$$

Thus direction numbers are $a = 1$, $b = -5$, and $c = 4$. Taking the point $(2, 4, -3)$ as P_0, we see that parametric equations (2) are

$$x = 2 + t \qquad y = 4 - 5t \qquad z = -3 + 4t$$

and symmetric equations (3) are

$$\frac{x - 2}{1} = \frac{y - 4}{-5} = \frac{z + 3}{4}$$

(b) The line intersects the xy-plane when $z = 0$, so we put $z = 0$ in the symmetric equations and obtain

$$\frac{x - 2}{1} = \frac{y - 4}{-5} = \frac{3}{4}$$

This gives $x = \frac{11}{4}$ and $y = \frac{1}{4}$, so the line intersects the xy-plane at the point $\left(\frac{11}{4}, \frac{1}{4}, 0\right)$. ■

In general, the procedure of Example 2 shows that direction numbers of the line L through the points $P_0(x_0, y_0, z_0)$ and $P_1(x_1, y_1, z_1)$ are $x_1 - x_0$, $y_1 - y_0$, and $z_1 - z_0$ and so symmetric equations of L are

$$\frac{x - x_0}{x_1 - x_0} = \frac{y - y_0}{y_1 - y_0} = \frac{z - z_0}{z_1 - z_0}$$

Often, we need a description, not of an entire line, but of just a line segment. How, for instance, could we describe the line segment AB in Example 2? If we put $t = 0$ in the parametric equations in Example 2(a), we get the point $(2, 4, -3)$ and if we put $t = 1$ we get $(3, -1, 1)$. So the line segment AB is described by the parametric equations

$$x = 2 + t \qquad y = 4 - 5t \qquad z = -3 + 4t \qquad 0 \leqslant t \leqslant 1$$

or by the corresponding vector equation

$$\mathbf{r}(t) = \langle 2 + t, 4 - 5t, -3 + 4t \rangle \qquad 0 \leqslant t \leqslant 1$$

In general, we know from Equation 1 that the vector equation of a line through the (tip of the) vector $\mathbf{r}_0$ in the direction of a vector $\mathbf{v}$ is $\mathbf{r} = \mathbf{r}_0 + t\mathbf{v}$. If the line also passes through (the tip of) $\mathbf{r}_1$, then we can take $\mathbf{v} = \mathbf{r}_1 - \mathbf{r}_0$ and so its vector equation is

$$\mathbf{r} = \mathbf{r}_0 + t(\mathbf{r}_1 - \mathbf{r}_0) = (1 - t)\mathbf{r}_0 + t\mathbf{r}_1$$

The line segment from $\mathbf{r}_0$ to $\mathbf{r}_1$ is given by the parameter interval $0 \leqslant t \leqslant 1$.

4 The line segment from $\mathbf{r}_0$ to $\mathbf{r}_1$ is given by the vector equation

$$\mathbf{r}(t) = (1 - t)\mathbf{r}_0 + t\mathbf{r}_1 \qquad 0 \leqslant t \leqslant 1$$

The lines L_1 and L_2 in Example 3, shown in Figure 5, are skew lines.

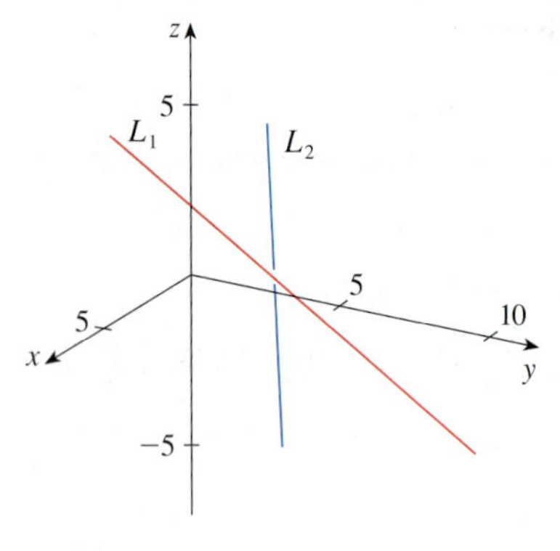

FIGURE 5

EXAMPLE 3 Show that the lines L_1 and L_2 with parametric equations

$$L_1\colon \quad x = 1 + t \qquad y = -2 + 3t \qquad z = 4 - t$$

$$L_2\colon \quad x = 2s \qquad y = 3 + s \qquad z = -3 + 4s$$

are **skew lines**; that is, they do not intersect and are not parallel (and therefore do not lie in the same plane).

SOLUTION The lines are not parallel because the corresponding direction vectors $\langle 1, 3, -1 \rangle$ and $\langle 2, 1, 4 \rangle$ are not parallel. (Their components are not proportional.) If L_1 and L_2 had a point of intersection, there would be values of t and s such that

$$1 + t = 2s$$

$$-2 + 3t = 3 + s$$

$$4 - t = -3 + 4s$$

But if we solve the first two equations, we get $t = \frac{11}{5}$ and $s = \frac{8}{5}$, and these values don't satisfy the third equation. Therefore there are no values of t and s that satisfy the three equations, so L_1 and L_2 do not intersect. Thus L_1 and L_2 are skew lines. ■

Planes

Although a line in space is determined by a point and a direction, a plane in space is more difficult to describe. A single vector parallel to a plane is not enough to convey the "direction" of the plane, but a vector perpendicular to the plane does completely specify its direction. Thus a plane in space is determined by a point $P_0(x_0, y_0, z_0)$ in the plane and a vector $\mathbf{n}$ that is orthogonal to the plane. This orthogonal vector $\mathbf{n}$ is called a **normal vector**. Let $P(x, y, z)$ be an arbitrary point in the plane, and let $\mathbf{r}_0$ and $\mathbf{r}$ be the position vectors of P_0 and P. Then the vector $\mathbf{r} - \mathbf{r}_0$ is represented by $\overrightarrow{P_0P}$. (See Figure 6.) The normal vector $\mathbf{n}$ is orthogonal to every vector in the given plane. In particular, $\mathbf{n}$ is orthogonal to $\mathbf{r} - \mathbf{r}_0$ and so we have

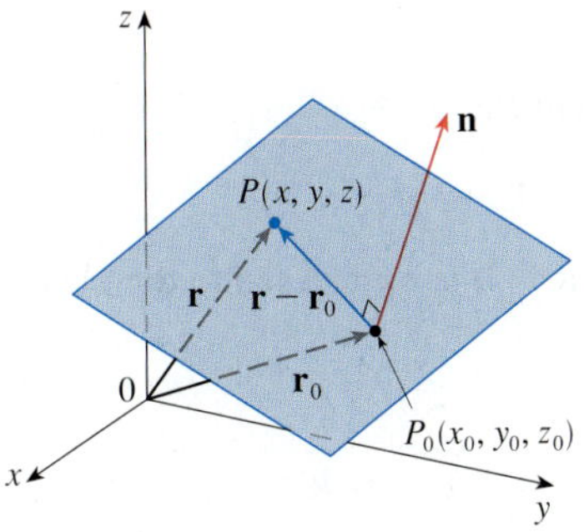

FIGURE 6

5
$$\mathbf{n} \cdot (\mathbf{r} - \mathbf{r}_0) = 0$$

which can be rewritten as

6
$$\mathbf{n} \cdot \mathbf{r} = \mathbf{n} \cdot \mathbf{r}_0$$

Either Equation 5 or Equation 6 is called a **vector equation of the plane**.

To obtain a scalar equation for the plane, we write $\mathbf{n} = \langle a, b, c \rangle$, $\mathbf{r} = \langle x, y, z \rangle$, and $\mathbf{r}_0 = \langle x_0, y_0, z_0 \rangle$. Then the vector equation (5) becomes

$$\langle a, b, c \rangle \cdot \langle x - x_0, y - y_0, z - z_0 \rangle = 0$$

or

$$a(x - x_0) + b(y - y_0) + c(z - z_0) = 0$$

7 A **scalar equation of the plane** through point $P_0(x_0, y_0, z_0)$ with normal vector $\mathbf{n} = \langle a, b, c \rangle$ is

$$a(x - x_0) + b(y - y_0) + c(z - z_0) = 0$$

EXAMPLE 4 Find an equation of the plane through the point $(2, 4, -1)$ with normal vector $\mathbf{n} = \langle 2, 3, 4 \rangle$. Find the intercepts and sketch the plane.

SOLUTION Putting $a = 2$, $b = 3$, $c = 4$, $x_0 = 2$, $y_0 = 4$, and $z_0 = -1$ in Equation 7, we see that an equation of the plane is

$$2(x - 2) + 3(y - 4) + 4(z + 1) = 0$$

or

$$2x + 3y + 4z = 12$$

To find the x-intercept we set $y = z = 0$ in this equation and obtain $x = 6$. Similarly, the y-intercept is 4 and the z-intercept is 3. This enables us to sketch the portion of the plane that lies in the first octant (see Figure 7). ■

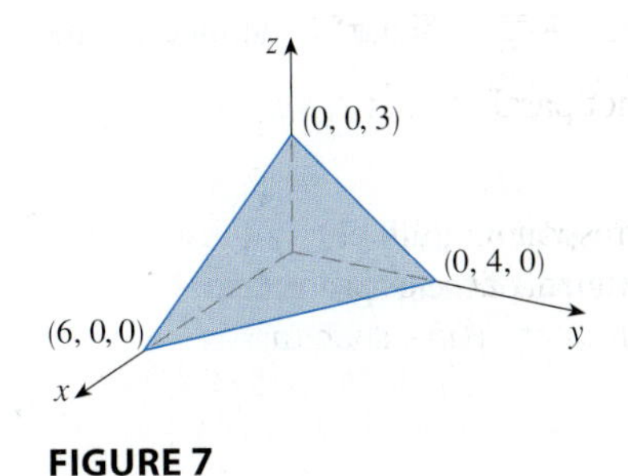

FIGURE 7

By collecting terms in Equation 7 as we did in Example 4, we can rewrite the equation of a plane as

8
$$ax + by + cz + d = 0$$

where $d = -(ax_0 + by_0 + cz_0)$. Equation 8 is called a **linear equation** in x, y, and z. Conversely, it can be shown that if a, b, and c are not all 0, then the linear equation (8) represents a plane with normal vector $\langle a, b, c \rangle$. (See Exercise 83.)

Figure 8 shows the portion of the plane in Example 5 that is enclosed by triangle PQR.

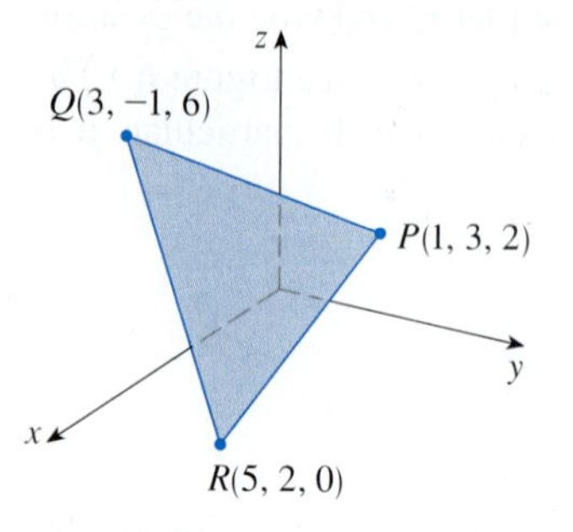

FIGURE 8

EXAMPLE 5 Find an equation of the plane that passes through the points $P(1, 3, 2)$, $Q(3, -1, 6)$, and $R(5, 2, 0)$.

SOLUTION The vectors $\mathbf{a}$ and $\mathbf{b}$ corresponding to $\overrightarrow{PQ}$ and $\overrightarrow{PR}$ are

$$\mathbf{a} = \langle 2, -4, 4 \rangle \qquad \mathbf{b} = \langle 4, -1, -2 \rangle$$

Since both $\mathbf{a}$ and $\mathbf{b}$ lie in the plane, their cross product $\mathbf{a} \times \mathbf{b}$ is orthogonal to the plane and can be taken as the normal vector. Thus

$$\mathbf{n} = \mathbf{a} \times \mathbf{b} = \begin{vmatrix} \mathbf{i} & \mathbf{j} & \mathbf{k} \\ 2 & -4 & 4 \\ 4 & -1 & -2 \end{vmatrix} = 12\mathbf{i} + 20\mathbf{j} + 14\mathbf{k}$$

With the point $P(1, 3, 2)$ and the normal vector $\mathbf{n}$, an equation of the plane is

$$12(x - 1) + 20(y - 3) + 14(z - 2) = 0$$

or

$$6x + 10y + 7z = 50$$

■

EXAMPLE 6 Find the point at which the line with parametric equations $x = 2 + 3t$, $y = -4t$, $z = 5 + t$ intersects the plane $4x + 5y - 2z = 18$.

SOLUTION We substitute the expressions for x, y, and z from the parametric equations into the equation of the plane:

$$4(2 + 3t) + 5(-4t) - 2(5 + t) = 18$$

This simplifies to $-10t = 20$, so $t = -2$. Therefore the point of intersection occurs when the parameter value is $t = -2$. Then $x = 2 + 3(-2) = -4$, $y = -4(-2) = 8$, $z = 5 - 2 = 3$ and so the point of intersection is $(-4, 8, 3)$. ■

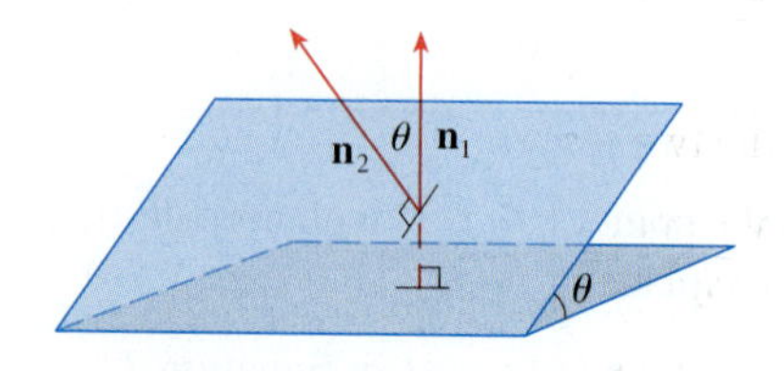

FIGURE 9

Two planes are **parallel** if their normal vectors are parallel. For instance, the planes $x + 2y - 3z = 4$ and $2x + 4y - 6z = 3$ are parallel because their normal vectors are $\mathbf{n}_1 = \langle 1, 2, -3 \rangle$ and $\mathbf{n}_2 = \langle 2, 4, -6 \rangle$ and $\mathbf{n}_2 = 2\mathbf{n}_1$. If two planes are not parallel, then they intersect in a straight line and the angle between the two planes is defined as the acute angle between their normal vectors (see angle θ in Figure 9).

Figure 10 shows the planes in Example 7 and their line of intersection L.

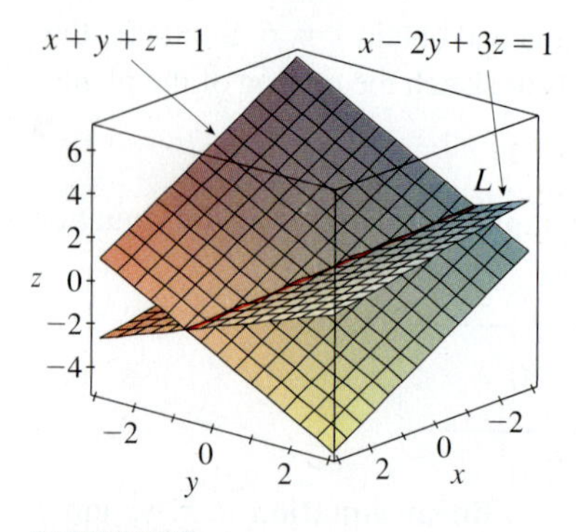

FIGURE 10

EXAMPLE 7

(a) Find the angle between the planes $x + y + z = 1$ and $x - 2y + 3z = 1$.

(b) Find symmetric equations for the line of intersection L of these two planes.

SOLUTION

(a) The normal vectors of these planes are

$$\mathbf{n}_1 = \langle 1, 1, 1 \rangle \qquad \mathbf{n}_2 = \langle 1, -2, 3 \rangle$$

and so, if θ is the angle between the planes, Corollary 12.3.6 gives

$$\cos\theta = \frac{\mathbf{n}_1 \cdot \mathbf{n}_2}{|\mathbf{n}_1||\mathbf{n}_2|} = \frac{1(1) + 1(-2) + 1(3)}{\sqrt{1 + 1 + 1}\,\sqrt{1 + 4 + 9}} = \frac{2}{\sqrt{42}}$$

$$\theta = \cos^{-1}\left(\frac{2}{\sqrt{42}}\right) \approx 72^\circ$$

(b) We first need to find a point on L. For instance, we can find the point where the line intersects the xy-plane by setting $z = 0$ in the equations of both planes. This gives the equations $x + y = 1$ and $x - 2y = 1$, whose solution is $x = 1$, $y = 0$. So the point $(1, 0, 0)$ lies on L.

Now we observe that, since L lies in both planes, it is perpendicular to both of the normal vectors. Thus a vector $\mathbf{v}$ parallel to L is given by the cross product

Another way to find the line of intersection is to solve the equations of the planes for two of the variables in terms of the third, which can be taken as the parameter.

$$\mathbf{v} = \mathbf{n}_1 \times \mathbf{n}_2 = \begin{vmatrix} \mathbf{i} & \mathbf{j} & \mathbf{k} \\ 1 & 1 & 1 \\ 1 & -2 & 3 \end{vmatrix} = 5\mathbf{i} - 2\mathbf{j} - 3\mathbf{k}$$

and so the symmetric equations of L can be written as

$$\frac{x-1}{5} = \frac{y}{-2} = \frac{z}{-3}$$

NOTE Since a linear equation in x, y, and z represents a plane and two nonparallel planes intersect in a line, it follows that two linear equations can represent a line. The points (x, y, z) that satisfy both $a_1x + b_1y + c_1z + d_1 = 0$ and $a_2x + b_2y + c_2z + d_2 = 0$ lie on both of these planes, and so the pair of linear equations represents the line of intersection of the planes (if they are not parallel). For instance, in Example 7 the line L was given as the line of intersection of the planes $x + y + z = 1$ and $x - 2y + 3z = 1$. The symmetric equations that we found for L could be written as

$$\frac{x-1}{5} = \frac{y}{-2} \quad \text{and} \quad \frac{y}{-2} = \frac{z}{-3}$$

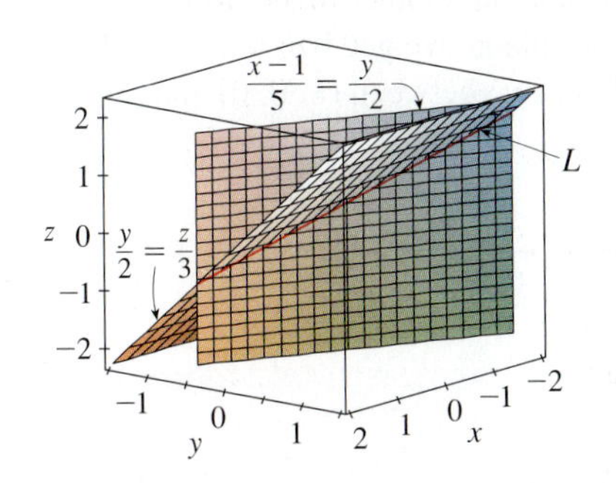

FIGURE 11

Figure 11 shows how the line L in Example 7 can also be regarded as the line of intersection of planes derived from its symmetric equations.

which is again a pair of linear equations. They exhibit L as the line of intersection of the planes $(x - 1)/5 = y/(-2)$ and $y/(-2) = z/(-3)$. (See Figure 11.)

In general, when we write the equations of a line in the symmetric form

$$\frac{x - x_0}{a} = \frac{y - y_0}{b} = \frac{z - z_0}{c}$$

we can regard the line as the line of intersection of the two planes

$$\frac{x - x_0}{a} = \frac{y - y_0}{b} \quad \text{and} \quad \frac{y - y_0}{b} = \frac{z - z_0}{c}$$

Distances

EXAMPLE 8 Find a formula for the distance D from a point $P_1(x_1, y_1, z_1)$ to the plane $ax + by + cz + d = 0$.

SOLUTION Let $P_0(x_0, y_0, z_0)$ be any point in the given plane and let $\mathbf{b}$ be the vector corresponding to $\overrightarrow{P_0P_1}$. Then

$$\mathbf{b} = \langle x_1 - x_0, y_1 - y_0, z_1 - z_0 \rangle$$

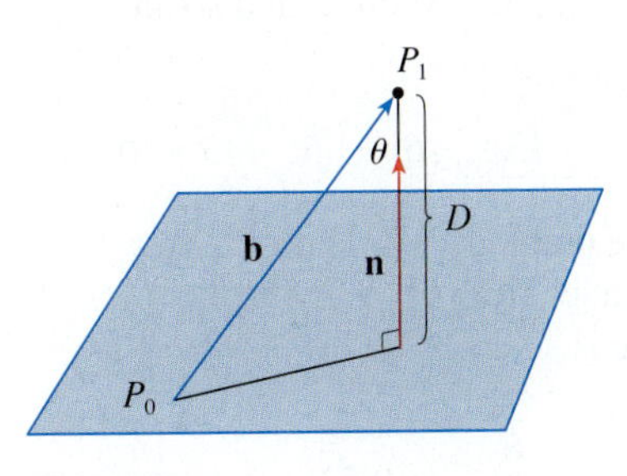

FIGURE 12

From Figure 12 you can see that the distance D from P_1 to the plane is equal to the absolute value of the scalar projection of $\mathbf{b}$ onto the normal vector $\mathbf{n} = \langle a, b, c \rangle$. (See Section 12.3.) Thus

$$D = |\text{comp}_{\mathbf{n}} \mathbf{b}| = \frac{|\mathbf{n} \cdot \mathbf{b}|}{|\mathbf{n}|}$$

$$= \frac{|a(x_1 - x_0) + b(y_1 - y_0) + c(z_1 - z_0)|}{\sqrt{a^2 + b^2 + c^2}}$$

$$= \frac{|(ax_1 + by_1 + cz_1) - (ax_0 + by_0 + cz_0)|}{\sqrt{a^2 + b^2 + c^2}}$$

Since P_0 lies in the plane, its coordinates satisfy the equation of the plane and so we have $ax_0 + by_0 + cz_0 + d = 0$. Thus the formula for D can be written as

9
$$D = \frac{|ax_1 + by_1 + cz_1 + d|}{\sqrt{a^2 + b^2 + c^2}}$$

■

EXAMPLE 9 Find the distance between the parallel planes $10x + 2y - 2z = 5$ and $5x + y - z = 1$.

SOLUTION First we note that the planes are parallel because their normal vectors $\langle 10, 2, -2 \rangle$ and $\langle 5, 1, -1 \rangle$ are parallel. To find the distance D between the planes, we choose any point on one plane and calculate its distance to the other plane. In particular, if we put $y = z = 0$ in the equation of the first plane, we get $10x = 5$ and so $(\frac{1}{2}, 0, 0)$ is a point in this plane. By Formula 9, the distance between $(\frac{1}{2}, 0, 0)$ and the plane $5x + y - z - 1 = 0$ is

$$D = \frac{|5(\frac{1}{2}) + 1(0) - 1(0) - 1|}{\sqrt{5^2 + 1^2 + (-1)^2}} = \frac{\frac{3}{2}}{3\sqrt{3}} = \frac{\sqrt{3}}{6}$$

So the distance between the planes is $\sqrt{3}/6$. ■

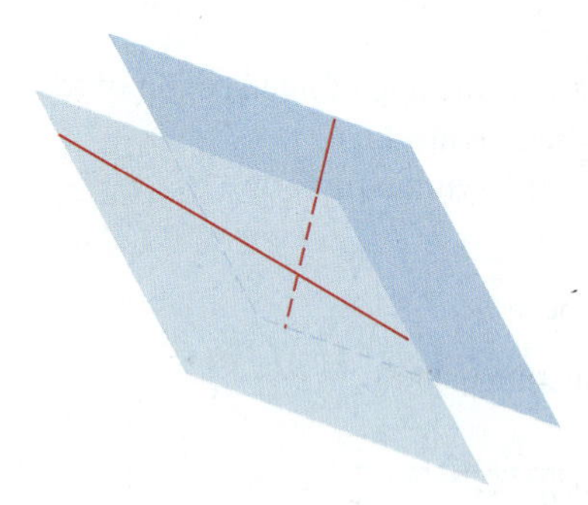

FIGURE 13
Skew lines, like those in Example 10, always lie on (nonidentical) parallel planes.

EXAMPLE 10 In Example 3 we showed that the lines

$$L_1: \quad x = 1 + t \qquad y = -2 + 3t \qquad z = 4 - t$$

$$L_2: \quad x = 2s \qquad y = 3 + s \qquad z = -3 + 4s$$

are skew. Find the distance between them.

SOLUTION Since the two lines L_1 and L_2 are skew, they can be viewed as lying on two parallel planes P_1 and P_2. The distance between L_1 and L_2 is the same as the distance between P_1 and P_2, which can be computed as in Example 9. The common normal vector to both planes must be orthogonal to both $\mathbf{v}_1 = \langle 1, 3, -1 \rangle$ (the direction of L_1) and $\mathbf{v}_2 = \langle 2, 1, 4 \rangle$ (the direction of L_2). So a normal vector is

$$\mathbf{n} = \mathbf{v}_1 \times \mathbf{v}_2 = \begin{vmatrix} \mathbf{i} & \mathbf{j} & \mathbf{k} \\ 1 & 3 & -1 \\ 2 & 1 & 4 \end{vmatrix} = 13\mathbf{i} - 6\mathbf{j} - 5\mathbf{k}$$

If we put $s = 0$ in the equations of L_2, we get the point $(0, 3, -3)$ on L_2 and so an equation for P_2 is

$$13(x - 0) - 6(y - 3) - 5(z + 3) = 0 \qquad \text{or} \qquad 13x - 6y - 5z + 3 = 0$$

If we now set $t = 0$ in the equations for L_1, we get the point $(1, -2, 4)$ on P_1. So the distance between L_1 and L_2 is the same as the distance from $(1, -2, 4)$ to $13x - 6y - 5z + 3 = 0$. By Formula 9, this distance is

$$D = \frac{|13(1) - 6(-2) - 5(4) + 3|}{\sqrt{13^2 + (-6)^2 + (-5)^2}} = \frac{8}{\sqrt{230}} \approx 0.53$$

■

12.5 EXERCISES

1. Determine whether each statement is true or false in $\mathbb{R}^3$.
(a) Two lines parallel to a third line are parallel.
(b) Two lines perpendicular to a third line are parallel.
(c) Two planes parallel to a third plane are parallel.
(d) Two planes perpendicular to a third plane are parallel.
(e) Two lines parallel to a plane are parallel.
(f) Two lines perpendicular to a plane are parallel.
(g) Two planes parallel to a line are parallel.
(h) Two planes perpendicular to a line are parallel.
(i) Two planes either intersect or are parallel.
(j) Two lines either intersect or are parallel.
(k) A plane and a line either intersect or are parallel.

2–5 Find a vector equation and parametric equations for the line.

2. The line through the point $(6, -5, 2)$ and parallel to the vector $\langle 1, 3, -\frac{2}{3} \rangle$

3. The line through the point $(2, 2.4, 3.5)$ and parallel to the vector $3\mathbf{i} + 2\mathbf{j} - \mathbf{k}$

4. The line through the point $(0, 14, -10)$ and parallel to the line $x = -1 + 2t$, $y = 6 - 3t$, $z = 3 + 9t$

5. The line through the point $(1, 0, 6)$ and perpendicular to the plane $x + 3y + z = 5$

6–12 Find parametric equations and symmetric equations for the line.

6. The line through the origin and the point $(4, 3, -1)$

7. The line through the points $(0, \frac{1}{2}, 1)$ and $(2, 1, -3)$

8. The line through the points $(1, 2.4, 4.6)$ and $(2.6, 1.2, 0.3)$

9. The line through the points $(-8, 1, 4)$ and $(3, -2, 4)$

10. The line through $(2, 1, 0)$ and perpendicular to both $\mathbf{i} + \mathbf{j}$ and $\mathbf{j} + \mathbf{k}$

11. The line through $(-6, 2, 3)$ and parallel to the line $\frac{1}{2}x = \frac{1}{3}y = z + 1$

12. The line of intersection of the planes $x + 2y + 3z = 1$ and $x - y + z = 1$

13. Is the line through $(-4, -6, 1)$ and $(-2, 0, -3)$ parallel to the line through $(10, 18, 4)$ and $(5, 3, 14)$?

14. Is the line through $(-2, 4, 0)$ and $(1, 1, 1)$ perpendicular to the line through $(2, 3, 4)$ and $(3, -1, -8)$?

15. (a) Find symmetric equations for the line that passes through the point $(1, -5, 6)$ and is parallel to the vector $\langle -1, 2, -3 \rangle$.
(b) Find the points in which the required line in part (a) intersects the coordinate planes.

16. (a) Find parametric equations for the line through $(2, 4, 6)$ that is perpendicular to the plane $x - y + 3z = 7$.
(b) In what points does this line intersect the coordinate planes?

17. Find a vector equation for the line segment from $(6, -1, 9)$ to $(7, 6, 0)$.

18. Find parametric equations for the line segment from $(-2, 18, 31)$ to $(11, -4, 48)$.

19–22 Determine whether the lines L_1 and L_2 are parallel, skew, or intersecting. If they intersect, find the point of intersection.

19. L_1: $x = 3 + 2t, \quad y = 4 - t, \quad z = 1 + 3t$

L_2: $x = 1 + 4s, \quad y = 3 - 2s, \quad z = 4 + 5s$

20. L_1: $x = 5 - 12t, \quad y = 3 + 9t, \quad z = 1 - 3t$

L_2: $x = 3 + 8s, \quad y = -6s, \quad z = 7 + 2s$

21. L_1: $\dfrac{x-2}{1} = \dfrac{y-3}{-2} = \dfrac{z-1}{-3}$

L_2: $\dfrac{x-3}{1} = \dfrac{y+4}{3} = \dfrac{z-2}{-7}$

22. L_1: $\dfrac{x}{1} = \dfrac{y-1}{-1} = \dfrac{z-2}{3}$

L_2: $\dfrac{x-2}{2} = \dfrac{y-3}{-2} = \dfrac{z}{7}$

23–40 Find an equation of the plane.

23. The plane through the origin and perpendicular to the vector $\langle 1, -2, 5 \rangle$

24. The plane through the point $(5, 3, 5)$ and with normal vector $2\mathbf{i} + \mathbf{j} - \mathbf{k}$

25. The plane through the point $(-1, \frac{1}{2}, 3)$ and with normal vector $\mathbf{i} + 4\mathbf{j} + \mathbf{k}$

26. The plane through the point $(2, 0, 1)$ and perpendicular to the line $x = 3t$, $y = 2 - t$, $z = 3 + 4t$

27. The plane through the point $(1, -1, -1)$ and parallel to the plane $5x - y - z = 6$

28. The plane through the point $(3, -2, 8)$ and parallel to the plane $z = x + y$

29. The plane through the point $(1, \frac{1}{2}, \frac{1}{3})$ and parallel to the plane $x + y + z = 0$

30. The plane that contains the line $x = 1 + t$, $y = 2 - t$, $z = 4 - 3t$ and is parallel to the plane $5x + 2y + z = 1$

31. The plane through the points $(0, 1, 1)$, $(1, 0, 1)$, and $(1, 1, 0)$

32. The plane through the origin and the points $(3, -2, 1)$ and $(1, 1, 1)$

33. The plane through the points $(2, 1, 2)$, $(3, -8, 6)$, and $(-2, -3, 1)$

34. The plane through the points $(3, 0, -1)$, $(-2, -2, 3)$, and $(7, 1, -4)$

35. The plane that passes through the point $(3, 5, -1)$ and contains the line $x = 4 - t$, $y = 2t - 1$, $z = -3t$

36. The plane that passes through the point $(6, -1, 3)$ and contains the line with symmetric equations $x/3 = y + 4 = z/2$

37. The plane that passes through the point $(3, 1, 4)$ and contains the line of intersection of the planes $x + 2y + 3z = 1$ and $2x - y + z = -3$

38. The plane that passes through the points $(0, -2, 5)$ and $(-1, 3, 1)$ and is perpendicular to the plane $2z = 5x + 4y$

39. The plane that passes through the point $(1, 5, 1)$ and is perpendicular to the planes $2x + y - 2z = 2$ and $x + 3z = 4$

40. The plane that passes through the line of intersection of the planes $x - z = 1$ and $y + 2z = 3$ and is perpendicular to the plane $x + y - 2z = 1$

41–44 Use intercepts to help sketch the plane.

41. $2x + 5y + z = 10$

42. $3x + y + 2z = 6$

43. $6x - 3y + 4z = 6$

44. $6x + 5y - 3z = 15$

45–47 Find the point at which the line intersects the given plane.

45. $x = 2 - 2t$, $y = 3t$, $z = 1 + t$; $x + 2y - z = 7$

46. $x = t - 1$, $y = 1 + 2t$, $z = 3 - t$; $3x - y + 2z = 5$

47. $5x = y/2 = z + 2$; $10x - 7y + 3z + 24 = 0$

48. Where does the line through $(-3, 1, 0)$ and $(-1, 5, 6)$ intersect the plane $2x + y - z = -2$?

49. Find direction numbers for the line of intersection of the planes $x + y + z = 1$ and $x + z = 0$.

50. Find the cosine of the angle between the planes $x + y + z = 0$ and $x + 2y + 3z = 1$.

51–56 Determine whether the planes are parallel, perpendicular, or neither. If neither, find the angle between them. (Round to one decimal place.)

51. $x + 4y - 3z = 1$, $-3x + 6y + 7z = 0$

52. $9x - 3y + 6z = 2$, $2y = 6x + 4z$

53. $x + 2y - z = 2$, $2x - 2y + z = 1$

54. $x - y + 3z = 1$, $3x + y - z = 2$

55. $2x - 3y = z$, $4x = 3 + 6y + 2z$

56. $5x + 2y + 3z = 2$, $y = 4x - 6z$

57–58 (a) Find parametric equations for the line of intersection of the planes and (b) find the angle between the planes.

57. $x + y + z = 1$, $x + 2y + 2z = 1$

58. $3x - 2y + z = 1$, $2x + y - 3z = 3$

59–60 Find symmetric equations for the line of intersection of the planes.

59. $5x - 2y - 2z = 1$, $4x + y + z = 6$

60. $z = 2x - y - 5$, $z = 4x + 3y - 5$

61. Find an equation for the plane consisting of all points that are equidistant from the points $(1, 0, -2)$ and $(3, 4, 0)$.

62. Find an equation for the plane consisting of all points that are equidistant from the points $(2, 5, 5)$ and $(-6, 3, 1)$.

63. Find an equation of the plane with x-intercept a, y-intercept b, and z-intercept c.

64. (a) Find the point at which the given lines intersect:

$$\mathbf{r} = \langle 1, 1, 0 \rangle + t\langle 1, -1, 2 \rangle$$
$$\mathbf{r} = \langle 2, 0, 2 \rangle + s\langle -1, 1, 0 \rangle$$

(b) Find an equation of the plane that contains these lines.

65. Find parametric equations for the line through the point $(0, 1, 2)$ that is parallel to the plane $x + y + z = 2$ and perpendicular to the line $x = 1 + t$, $y = 1 - t$, $z = 2t$.

66. Find parametric equations for the line through the point $(0, 1, 2)$ that is perpendicular to the line $x = 1 + t$, $y = 1 - t$, $z = 2t$ and intersects this line.

67. Which of the following four planes are parallel? Are any of them identical?

P_1: $3x + 6y - 3z = 6$ $\quad$ P_2: $4x - 12y + 8z = 5$

P_3: $9y = 1 + 3x + 6z$ $\quad$ P_4: $z = x + 2y - 2$

68. Which of the following four lines are parallel? Are any of them identical?

L_1: $x = 1 + 6t$, $y = 1 - 3t$, $z = 12t + 5$

L_2: $x = 1 + 2t$, $y = t$, $z = 1 + 4t$

L_3: $2x - 2 = 4 - 4y = z + 1$

L_4: $\mathbf{r} = \langle 3, 1, 5 \rangle + t\langle 4, 2, 8 \rangle$

69–70 Use the formula in Exercise 12.4.45 to find the distance from the point to the given line.

69. $(4, 1, -2)$; $x = 1 + t$, $y = 3 - 2t$, $z = 4 - 3t$

70. $(0, 1, 3)$; $x = 2t$, $y = 6 - 2t$, $z = 3 + t$

71–72 Find the distance from the point to the given plane.

71. $(1, -2, 4)$, $3x + 2y + 6z = 5$

72. $(-6, 3, 5)$, $x - 2y - 4z = 8$

73–74 Find the distance between the given parallel planes.

73. $2x - 3y + z = 4$, $4x - 6y + 2z = 3$

74. $6z = 4y - 2x$, $9z = 1 - 3x + 6y$

75. Show that the distance between the parallel planes $ax + by + cz + d_1 = 0$ and $ax + by + cz + d_2 = 0$ is

$$D = \frac{|d_1 - d_2|}{\sqrt{a^2 + b^2 + c^2}}$$

76. Find equations of the planes that are parallel to the plane $x + 2y - 2z = 1$ and two units away from it.

77. Show that the lines with symmetric equations $x = y = z$ and $x + 1 = y/2 = z/3$ are skew, and find the distance between these lines.

78. Find the distance between the skew lines with parametric equations $x = 1 + t$, $y = 1 + 6t$, $z = 2t$, and $x = 1 + 2s$, $y = 5 + 15s$, $z = -2 + 6s$.

79. Let L_1 be the line through the origin and the point $(2, 0, -1)$. Let L_2 be the line through the points $(1, -1, 1)$ and $(4, 1, 3)$. Find the distance between L_1 and L_2.

80. Let L_1 be the line through the points $(1, 2, 6)$ and $(2, 4, 8)$. Let L_2 be the line of intersection of the planes P_1 and P_2, where P_1 is the plane $x - y + 2z + 1 = 0$ and P_2 is the plane through the points $(3, 2, -1)$, $(0, 0, 1)$, and $(1, 2, 1)$. Calculate the distance between L_1 and L_2.

81. Two tanks are participating in a battle simulation. Tank A is at point $(325, 810, 561)$ and tank B is positioned at point $(765, 675, 599)$.
(a) Find parametric equations for the line of sight between the tanks.
(b) If we divide the line of sight into 5 equal segments, the elevations of the terrain at the four intermediate points from tank A to tank B are 549, 566, 586, and 589. Can the tanks see each other?

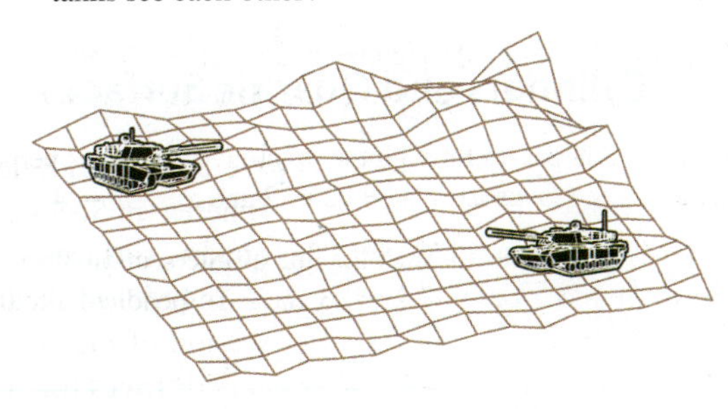

82. Give a geometric description of each family of planes.
(a) $x + y + z = c$ (b) $x + y + cz = 1$
(c) $y \cos\theta + z \sin\theta = 1$

83. If a, b, and c are not all 0, show that the equation $ax + by + cz + d = 0$ represents a plane and $\langle a, b, c \rangle$ is a normal vector to the plane.
Hint: Suppose $a \neq 0$ and rewrite the equation in the form

$$a\left(x + \frac{d}{a}\right) + b(y - 0) + c(z - 0) = 0$$

LABORATORY PROJECT PUTTING 3D IN PERSPECTIVE

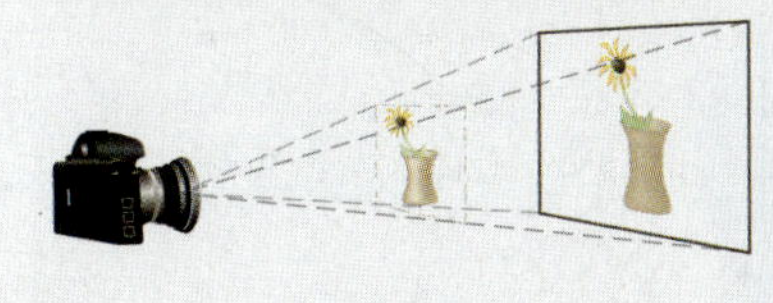

Computer graphics programmers face the same challenge as the great painters of the past: how to represent a three-dimensional scene as a flat image on a two-dimensional plane (a screen or a canvas). To create the illusion of perspective, in which closer objects appear larger than those farther away, three-dimensional objects in the computer's memory are projected onto a rectangular screen window from a viewpoint where the eye, or camera, is located. The viewing volume—the portion of space that will be visible—is the region contained by the four planes that pass through the viewpoint and an edge of the screen window. If objects in the scene extend beyond these four planes, they must be truncated before pixel data are sent to the screen. These planes are therefore called *clipping planes*.

1. Suppose the screen is represented by a rectangle in the yz-plane with vertices $(0, \pm 400, 0)$ and $(0, \pm 400, 600)$, and the camera is placed at $(1000, 0, 0)$. A line L in the scene passes through the points $(230, -285, 102)$ and $(860, 105, 264)$. At what points should L be clipped by the clipping planes?

2. If the clipped line segment is projected onto the screen window, identify the resulting line segment.

3. Use parametric equations to plot the edges of the screen window, the clipped line segment, and its projection onto the screen window. Then add sight lines connecting the viewpoint to each end of the clipped segments to verify that the projection is correct.

4. A rectangle with vertices $(621, -147, 206)$, $(563, 31, 242)$, $(657, -111, 86)$, and $(599, 67, 122)$ is added to the scene. The line L intersects this rectangle. To make the rectangle appear opaque, a programmer can use *hidden line rendering,* which removes portions of objects that are behind other objects. Identify the portion of L that should be removed.

12.6 Cylinders and Quadric Surfaces

We have already looked at two special types of surfaces: planes (in Section 12.5) and spheres (in Section 12.1). Here we investigate two other types of surfaces: cylinders and quadric surfaces.

In order to sketch the graph of a surface, it is useful to determine the curves of intersection of the surface with planes parallel to the coordinate planes. These curves are called **traces** (or cross-sections) of the surface.

Cylinders

A **cylinder** is a surface that consists of all lines (called **rulings**) that are parallel to a given line and pass through a given plane curve.

EXAMPLE 1 Sketch the graph of the surface $z = x^2$.

SOLUTION Notice that the equation of the graph, $z = x^2$, doesn't involve y. This means that any vertical plane with equation $y = k$ (parallel to the xz-plane) intersects the graph in a curve with equation $z = x^2$. So these vertical traces are parabolas. Figure 1 shows how the graph is formed by taking the parabola $z = x^2$ in the xz-plane and moving it in the direction of the y-axis. The graph is a surface, called a **parabolic cylinder**, made up of infinitely many shifted copies of the same parabola. Here the rulings of the cylinder are parallel to the y-axis.

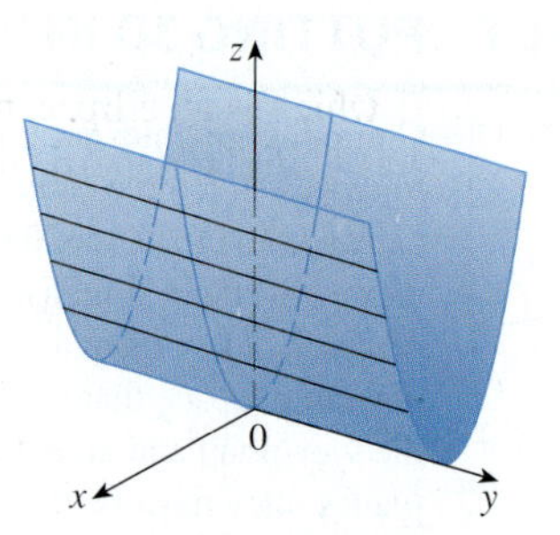

FIGURE 1
The surface $z = x^2$ is a parabolic cylinder.

We noticed that the variable y is missing from the equation of the cylinder in Example 1. This is typical of a surface whose rulings are parallel to one of the coordinate axes. If one of the variables x, y, or z is missing from the equation of a surface, then the surface is a cylinder.

EXAMPLE 2 Identify and sketch the surfaces.

(a) $x^2 + y^2 = 1$ (b) $y^2 + z^2 = 1$

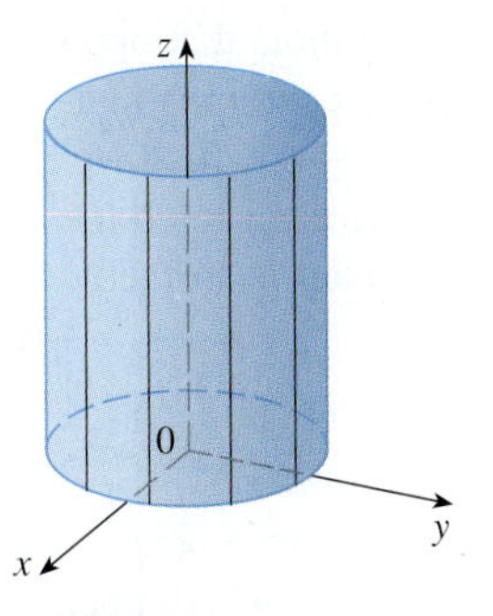

FIGURE 2
$x^2 + y^2 = 1$

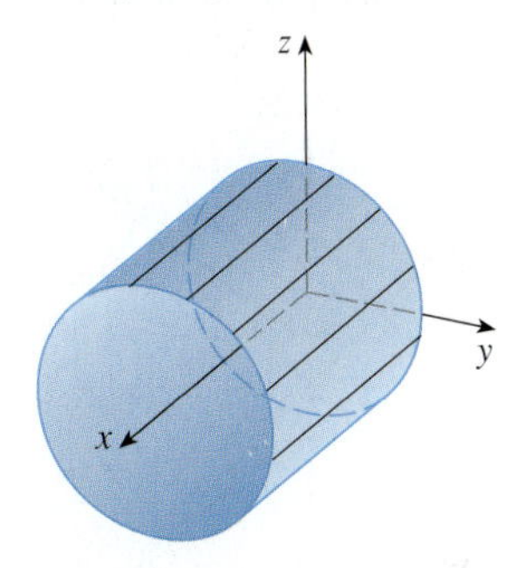

FIGURE 3
$y^2 + z^2 = 1$

SOLUTION

(a) Since z is missing and the equations $x^2 + y^2 = 1$, $z = k$ represent a circle with radius 1 in the plane $z = k$, the surface $x^2 + y^2 = 1$ is a circular cylinder whose axis is the z-axis. (See Figure 2.) Here the rulings are vertical lines.

(b) In this case x is missing and the surface is a circular cylinder whose axis is the x-axis. (See Figure 3.) It is obtained by taking the circle $y^2 + z^2 = 1$, $x = 0$ in the yz-plane and moving it parallel to the x-axis. ■

NOTE When you are dealing with surfaces, it is important to recognize that an equation like $x^2 + y^2 = 1$ represents a cylinder and not a circle. The trace of the cylinder $x^2 + y^2 = 1$ in the xy-plane is the circle with equations $x^2 + y^2 = 1$, $z = 0$.

Quadric Surfaces

A **quadric surface** is the graph of a second-degree equation in three variables x, y, and z. The most general such equation is

$$Ax^2 + By^2 + Cz^2 + Dxy + Eyz + Fxz + Gx + Hy + Iz + J = 0$$

where $A, B, C, \ldots, J$ are constants, but by translation and rotation it can be brought into one of the two standard forms

$$Ax^2 + By^2 + Cz^2 + J = 0 \qquad \text{or} \qquad Ax^2 + By^2 + Iz = 0$$

Quadric surfaces are the counterparts in three dimensions of the conic sections in the plane. (See Section 10.5 for a review of conic sections.)

EXAMPLE 3 Use traces to sketch the quadric surface with equation

$$x^2 + \frac{y^2}{9} + \frac{z^2}{4} = 1$$

SOLUTION By substituting $z = 0$, we find that the trace in the xy-plane is $x^2 + y^2/9 = 1$, which we recognize as an equation of an ellipse. In general, the horizontal trace in the plane $z = k$ is

$$x^2 + \frac{y^2}{9} = 1 - \frac{k^2}{4} \qquad z = k$$

which is an ellipse, provided that $k^2 < 4$, that is, $-2 < k < 2$.

Similarly, vertical traces parallel to the yz- and xz-planes are also ellipses:

$$\frac{y^2}{9} + \frac{z^2}{4} = 1 - k^2 \qquad x = k \qquad (\text{if } -1 < k < 1)$$

$$x^2 + \frac{z^2}{4} = 1 - \frac{k^2}{9} \qquad y = k \qquad (\text{if } -3 < k < 3)$$

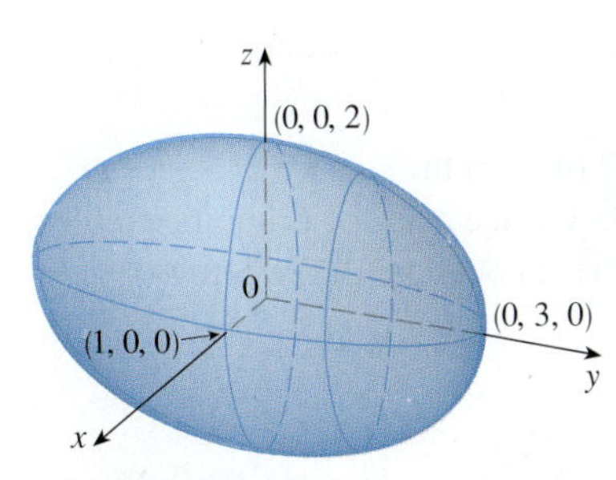

FIGURE 4
The ellipsoid $x^2 + \frac{y^2}{9} + \frac{z^2}{4} = 1$

Figure 4 shows how drawing some traces indicates the shape of the surface. It's called an **ellipsoid** because all of its traces are ellipses. Notice that it is symmetric with respect to each coordinate plane; this is a reflection of the fact that its equation involves only even powers of x, y, and z. ■

EXAMPLE 4 Use traces to sketch the surface $z = 4x^2 + y^2$.

SOLUTION If we put $x = 0$, we get $z = y^2$, so the yz-plane intersects the surface in a parabola. If we put $x = k$ (a constant), we get $z = y^2 + 4k^2$. This means that if we

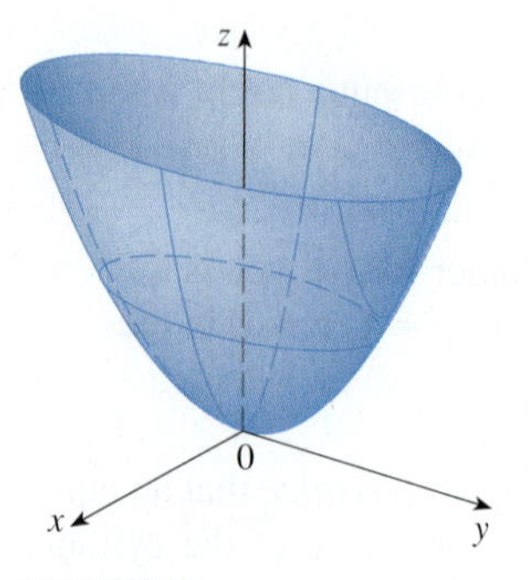

FIGURE 5
The surface $z = 4x^2 + y^2$ is an elliptic paraboloid. Horizontal traces are ellipses; vertical traces are parabolas.

slice the graph with any plane parallel to the yz-plane, we obtain a parabola that opens upward. Similarly, if $y = k$, the trace is $z = 4x^2 + k^2$, which is again a parabola that opens upward. If we put $z = k$, we get the horizontal traces $4x^2 + y^2 = k$, which we recognize as a family of ellipses. Knowing the shapes of the traces, we can sketch the graph in Figure 5. Because of the elliptical and parabolic traces, the quadric surface $z = 4x^2 + y^2$ is called an **elliptic paraboloid**. ■

EXAMPLE 5 Sketch the surface $z = y^2 - x^2$.

SOLUTION The traces in the vertical planes $x = k$ are the parabolas $z = y^2 - k^2$, which open upward. The traces in $y = k$ are the parabolas $z = -x^2 + k^2$, which open downward. The horizontal traces are $y^2 - x^2 = k$, a family of hyperbolas. We draw the families of traces in Figure 6, and we show how the traces appear when placed in their correct planes in Figure 7.

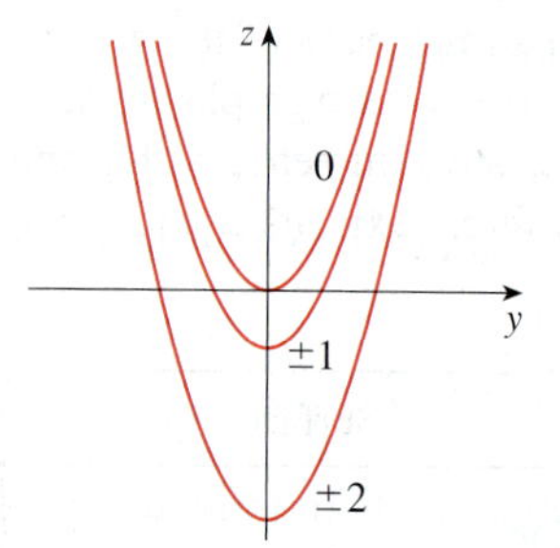

Traces in $x = k$ are $z = y^2 - k^2$.

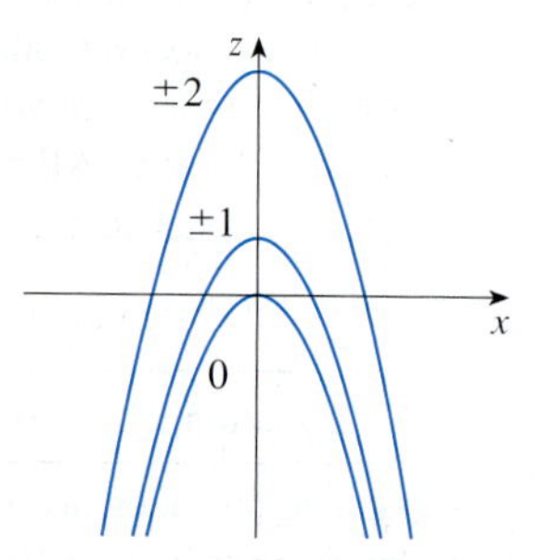

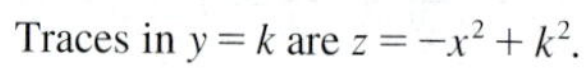

Traces in $y = k$ are $z = -x^2 + k^2$.

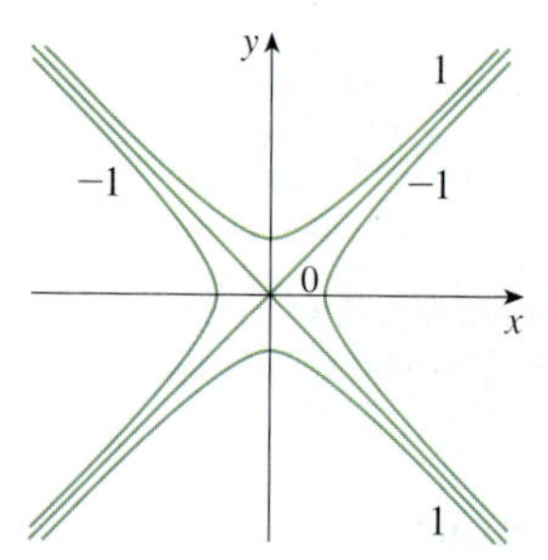

Traces in $z = k$ are $y^2 - x^2 = k$.

FIGURE 6
Vertical traces are parabolas; horizontal traces are hyperbolas. All traces are labeled with the value of k.

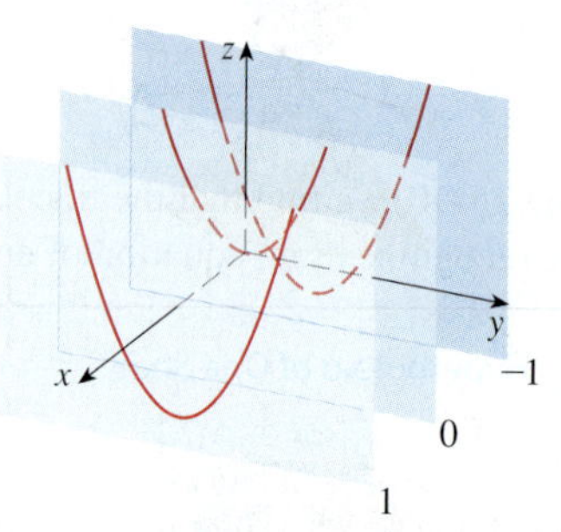

Traces in $x = k$

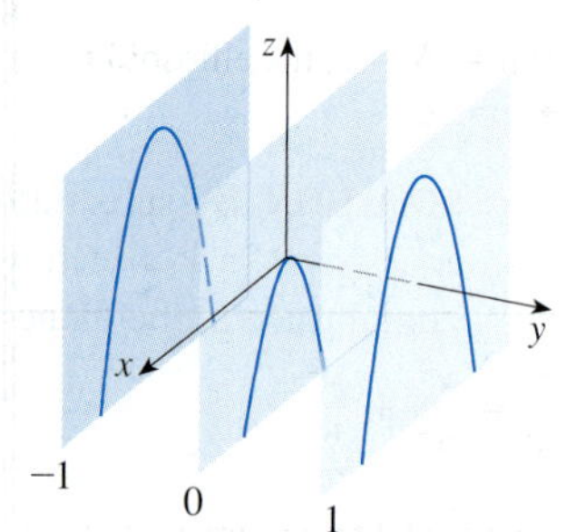

Traces in $y = k$

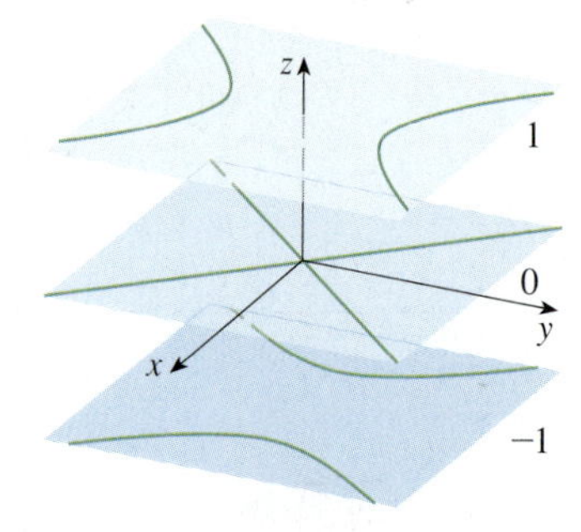

Traces in $z = k$

FIGURE 7
Traces moved to their correct planes

TEC In Module 12.6A you can investigate how traces determine the shape of a surface.

In Figure 8 we fit together the traces from Figure 7 to form the surface $z = y^2 - x^2$, a **hyperbolic paraboloid**. Notice that the shape of the surface near the origin resembles that of a saddle. This surface will be investigated further in Section 14.7 when we discuss saddle points.

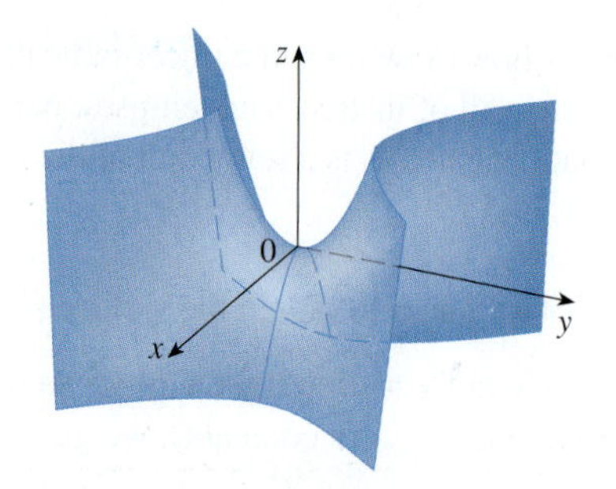

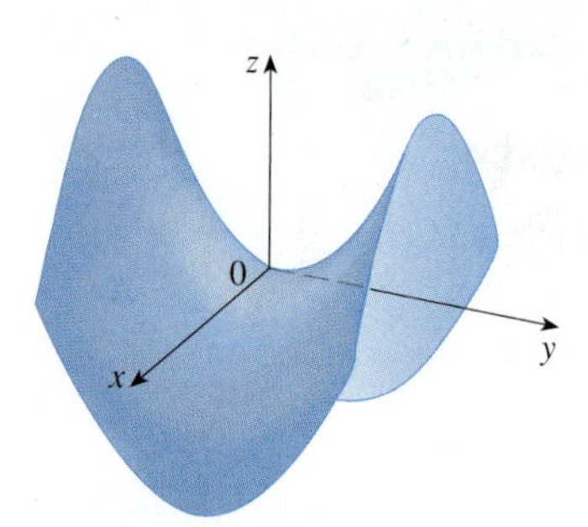

FIGURE 8
Two views of the surface $z = y^2 - x^2$, a hyperbolic paraboloid

■

EXAMPLE 6 Sketch the surface $\frac{x^2}{4} + y^2 - \frac{z^2}{4} = 1$.

SOLUTION The trace in any horizontal plane $z = k$ is the ellipse

$$\frac{x^2}{4} + y^2 = 1 + \frac{k^2}{4} \qquad z = k$$

but the traces in the xz- and yz-planes are the hyperbolas

$$\frac{x^2}{4} - \frac{z^2}{4} = 1 \quad y = 0 \qquad \text{and} \qquad y^2 - \frac{z^2}{4} = 1 \quad x = 0$$

This surface is called a **hyperboloid of one sheet** and is sketched in Figure 9. ■

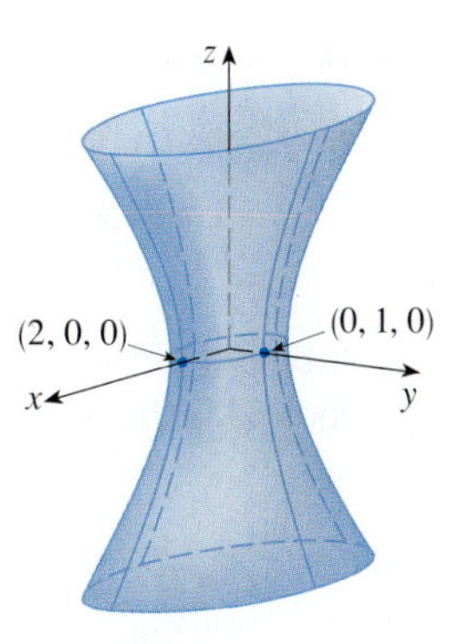

FIGURE 9

The idea of using traces to draw a surface is employed in three-dimensional graphing software. In most such software, traces in the vertical planes $x = k$ and $y = k$ are drawn for equally spaced values of k, and parts of the graph are eliminated using hidden line removal. Table 1 shows computer-drawn graphs of the six basic types of quadric surfaces in standard form. All surfaces are symmetric with respect to the z-axis. If a quadric surface is symmetric about a different axis, its equation changes accordingly.

Table 1 Graphs of Quadric Surfaces

Surface	Equation	Surface	Equation
Ellipsoid	$\frac{x^2}{a^2} + \frac{y^2}{b^2} + \frac{z^2}{c^2} = 1$ All traces are ellipses. If $a = b = c$, the ellipsoid is a sphere.	Cone	$\frac{z^2}{c^2} = \frac{x^2}{a^2} + \frac{y^2}{b^2}$ Horizontal traces are ellipses. Vertical traces in the planes $x = k$ and $y = k$ are hyperbolas if $k \neq 0$ but are pairs of lines if $k = 0$.
Elliptic Paraboloid	$\frac{z}{c} = \frac{x^2}{a^2} + \frac{y^2}{b^2}$ Horizontal traces are ellipses. Vertical traces are parabolas. The variable raised to the first power indicates the axis of the paraboloid.	Hyperboloid of One Sheet	$\frac{x^2}{a^2} + \frac{y^2}{b^2} - \frac{z^2}{c^2} = 1$ Horizontal traces are ellipses. Vertical traces are hyperbolas. The axis of symmetry corresponds to the variable whose coefficient is negative.
Hyperbolic Paraboloid	$\frac{z}{c} = \frac{x^2}{a^2} - \frac{y^2}{b^2}$ Horizontal traces are hyperbolas. Vertical traces are parabolas. The case where $c < 0$ is illustrated.	Hyperboloid of Two Sheets	$-\frac{x^2}{a^2} - \frac{y^2}{b^2} + \frac{z^2}{c^2} = 1$ Horizontal traces in $z = k$ are ellipses if $k > c$ or $k < -c$. Vertical traces are hyperbolas. The two minus signs indicate two sheets.

TEC In Module 12.6B you can see how changing a, b, and c in Table 1 affects the shape of the quadric surface.

EXAMPLE 7 Identify and sketch the surface $4x^2 - y^2 + 2z^2 + 4 = 0$.

SOLUTION Dividing by -4, we first put the equation in standard form:

$$-x^2 + \frac{y^2}{4} - \frac{z^2}{2} = 1$$

Comparing this equation with Table 1, we see that it represents a hyperboloid of two sheets, the only difference being that in this case the axis of the hyperboloid is the y-axis. The traces in the xy- and yz-planes are the hyperbolas

$$-x^2 + \frac{y^2}{4} = 1 \qquad z = 0 \qquad \text{and} \qquad \frac{y^2}{4} - \frac{z^2}{2} = 1 \qquad x = 0$$

The surface has no trace in the xz-plane, but traces in the vertical planes $y = k$ for $|k| > 2$ are the ellipses

$$x^2 + \frac{z^2}{2} = \frac{k^2}{4} - 1 \qquad y = k$$

which can be written as

$$\frac{x^2}{\frac{k^2}{4} - 1} + \frac{z^2}{2\left(\frac{k^2}{4} - 1\right)} = 1 \qquad y = k$$

These traces are used to make the sketch in Figure 10. ■

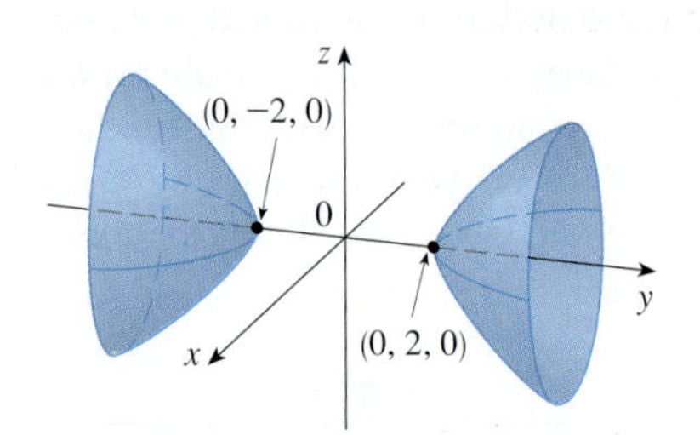

FIGURE 10
$4x^2 - y^2 + 2z^2 + 4 = 0$

EXAMPLE 8 Classify the quadric surface $x^2 + 2z^2 - 6x - y + 10 = 0$.

SOLUTION By completing the square we rewrite the equation as

$$y - 1 = (x - 3)^2 + 2z^2$$

Comparing this equation with Table 1, we see that it represents an elliptic paraboloid. Here, however, the axis of the paraboloid is parallel to the y-axis, and it has been shifted so that its vertex is the point $(3, 1, 0)$. The traces in the plane $y = k$ $(k > 1)$ are the ellipses

$$(x - 3)^2 + 2z^2 = k - 1 \qquad y = k$$

The trace in the xy-plane is the parabola with equation $y = 1 + (x - 3)^2$, $z = 0$. The paraboloid is sketched in Figure 11.

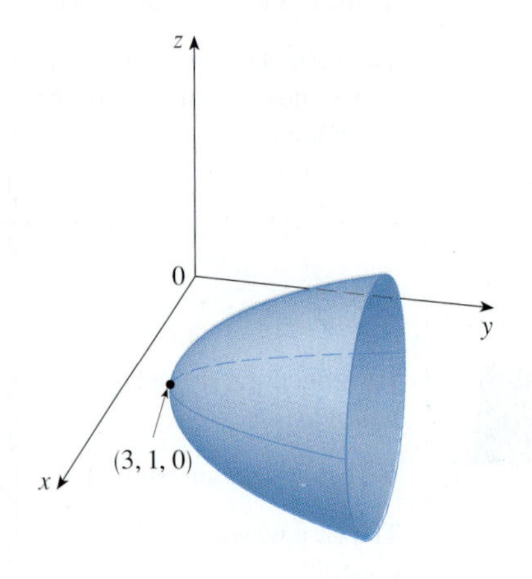

FIGURE 11
$x^2 + 2z^2 - 6x - y + 10 = 0$

■

Applications of Quadric Surfaces

Examples of quadric surfaces can be found in the world around us. In fact, the world itself is a good example. Although the earth is commonly modeled as a sphere, a more accurate model is an ellipsoid because the earth's rotation has caused a flattening at the poles. (See Exercise 49.)

Circular paraboloids, obtained by rotating a parabola about its axis, are used to collect and reflect light, sound, and radio and television signals. In a radio telescope, for instance, signals from distant stars that strike the bowl are all reflected to the receiver at the focus and are therefore amplified. (The idea is explained in Problem 18 on page 202.) The same principle applies to microphones and satellite dishes in the shape of paraboloids.

Cooling towers for nuclear reactors are usually designed in the shape of hyperboloids of one sheet for reasons of structural stability. Pairs of hyperboloids are used to transmit rotational motion between skew axes. (The cogs of the gears are the generating lines of the hyperboloids. See Exercise 51.)

David Frazier / Spirit / Corbis

A satellite dish reflects signals to the focus of a paraboloid.

Mark C. Burnett / Science Source

Nuclear reactors have cooling towers in the shape of hyperboloids.

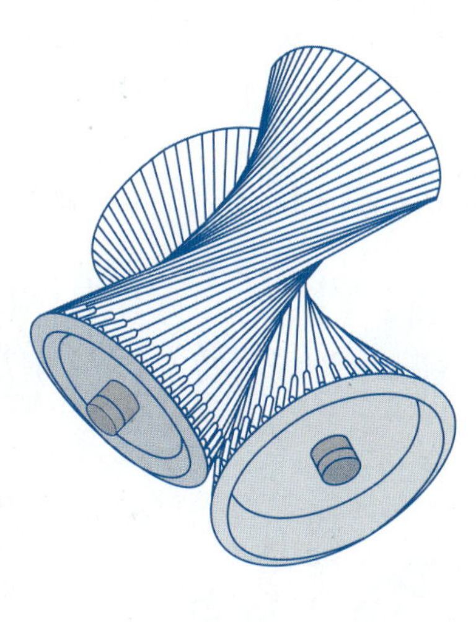

Hperboloids produce gear transmission.

12.6 EXERCISES

1. (a) What does the equation $y = x^2$ represent as a curve in $\mathbb{R}^2$?
(b) What does it represent as a surface in $\mathbb{R}^3$?
(c) What does the equation $z = y^2$ represent?

2. (a) Sketch the graph of $y = e^x$ as a curve in $\mathbb{R}^2$.
(b) Sketch the graph of $y = e^x$ as a surface in $\mathbb{R}^3$.
(c) Describe and sketch the surface $z = e^y$.

3–8 Describe and sketch the surface.

3. $x^2 + z^2 = 1$

4. $4x^2 + y^2 = 4$

5. $z = 1 - y^2$

6. $y = z^2$

7. $xy = 1$

8. $z = \sin y$

9. (a) Find and identify the traces of the quadric surface $x^2 + y^2 - z^2 = 1$ and explain why the graph looks like the graph of the hyperboloid of one sheet in Table 1.
(b) If we change the equation in part (a) to $x^2 - y^2 + z^2 = 1$, how is the graph affected?
(c) What if we change the equation in part (a) to $x^2 + y^2 + 2y - z^2 = 0$?

10. (a) Find and identify the traces of the quadric surface $-x^2 - y^2 + z^2 = 1$ and explain why the graph looks like the graph of the hyperboloid of two sheets in Table 1.
(b) If the equation in part (a) is changed to $x^2 - y^2 - z^2 = 1$, what happens to the graph? Sketch the new graph.

11–20 Use traces to sketch and identify the surface.

11. $x = y^2 + 4z^2$

12. $4x^2 + 9y^2 + 9z^2 = 36$

13. $x^2 = 4y^2 + z^2$

14. $z^2 - 4x^2 - y^2 = 4$

15. $9y^2 + 4z^2 = x^2 + 36$

16. $3x^2 + y + 3z^2 = 0$

17. $\dfrac{x^2}{9} + \dfrac{y^2}{25} + \dfrac{z^2}{4} = 1$

18. $3x^2 - y^2 + 3z^2 = 0$

19. $y = z^2 - x^2$

20. $x = y^2 - z^2$

21–28 Match the equation with its graph (labeled I–VIII). Give reasons for your choice.

21. $x^2 + 4y^2 + 9z^2 = 1$

22. $9x^2 + 4y^2 + z^2 = 1$

23. $x^2 - y^2 + z^2 = 1$

24. $-x^2 + y^2 - z^2 = 1$

25. $y = 2x^2 + z^2$

26. $y^2 = x^2 + 2z^2$

27. $x^2 + 2z^2 = 1$

28. $y = x^2 - z^2$

I

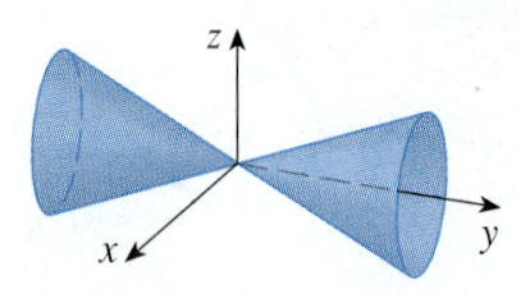

II

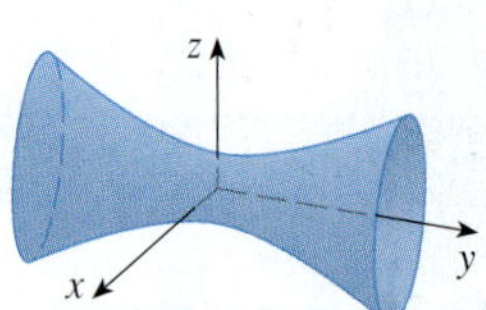

III

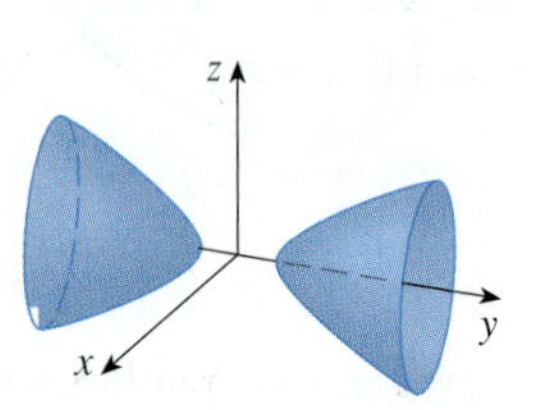

IV

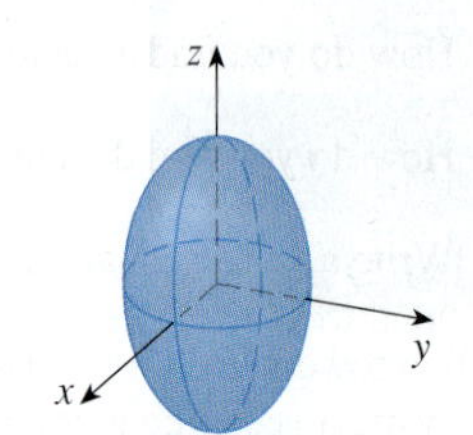

V

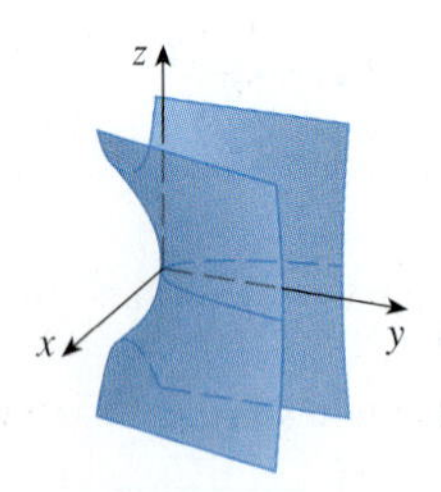

VI

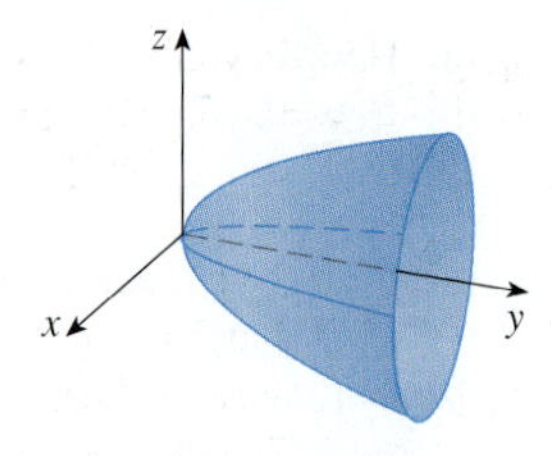

VII

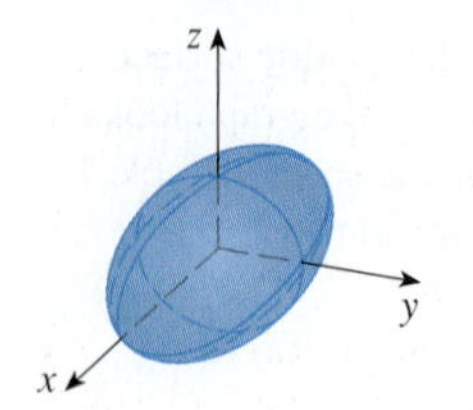

VIII

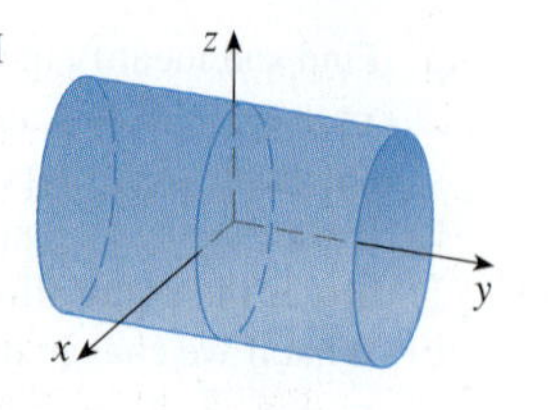

29–30 Sketch and identify a quadric surface that could have the traces shown.

29. Traces in $x = k$ Traces in $y = k$

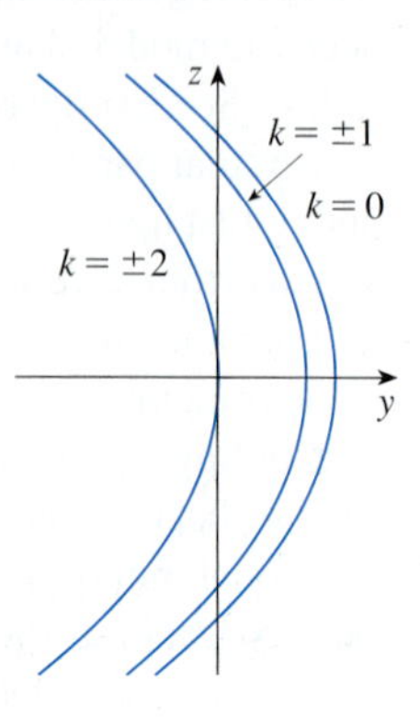

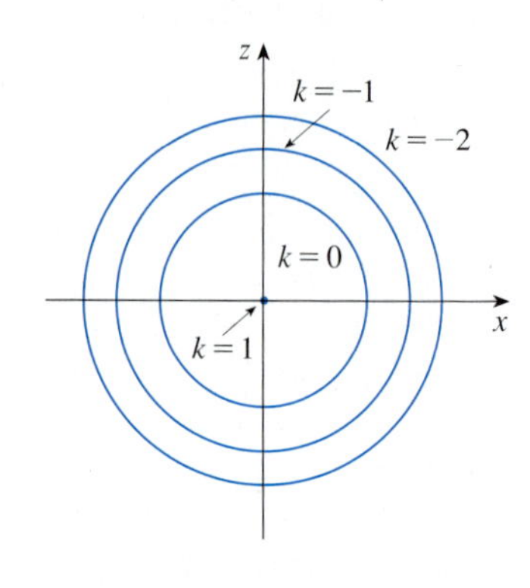

30. Traces in $x = k$ Traces in $z = k$

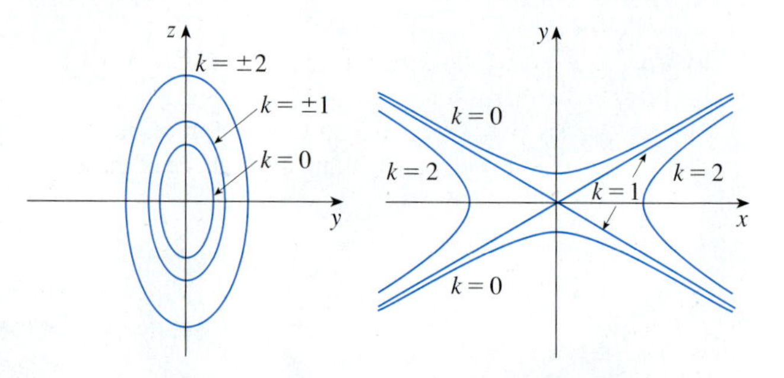

31–38 Reduce the equation to one of the standard forms, classify the surface, and sketch it.

31. $y^2 = x^2 + \frac{1}{9}z^2$

32. $4x^2 - y + 2z^2 = 0$

33. $x^2 + 2y - 2z^2 = 0$

34. $y^2 = x^2 + 4z^2 + 4$

35. $x^2 + y^2 - 2x - 6y - z + 10 = 0$

36. $x^2 - y^2 - z^2 - 4x - 2z + 3 = 0$

37. $x^2 - y^2 + z^2 - 4x - 2z = 0$

38. $4x^2 + y^2 + z^2 - 24x - 8y + 4z + 55 = 0$

39–42 Use a computer with three-dimensional graphing software to graph the surface. Experiment with viewpoints and with domains for the variables until you get a good view of the surface.

39. $-4x^2 - y^2 + z^2 = 1$

40. $x^2 - y^2 - z = 0$

41. $-4x^2 - y^2 + z^2 = 0$

42. $x^2 - 6x + 4y^2 - z = 0$

43. Sketch the region bounded by the surfaces $z = \sqrt{x^2 + y^2}$ and $x^2 + y^2 = 1$ for $1 \leqslant z \leqslant 2$.

44. Sketch the region bounded by the paraboloids $z = x^2 + y^2$ and $z = 2 - x^2 - y^2$.

45. Find an equation for the surface obtained by rotating the curve $y = \sqrt{x}$ about the x-axis.

46. Find an equation for the surface obtained by rotating the line $z = 2y$ about the z-axis.

47. Find an equation for the surface consisting of all points that are equidistant from the point $(-1, 0, 0)$ and the plane $x = 1$. Identify the surface.

48. Find an equation for the surface consisting of all points P for which the distance from P to the x-axis is twice the distance from P to the yz-plane. Identify the surface.

49. Traditionally, the earth's surface has been modeled as a sphere, but the World Geodetic System of 1984 (WGS-84) uses an ellipsoid as a more accurate model. It places the center of the earth at the origin and the north pole on the positive z-axis. The distance from the center to the poles is 6356.523 km and the distance to a point on the equator is 6378.137 km.
(a) Find an equation of the earth's surface as used by WGS-84.
(b) Curves of equal latitude are traces in the planes $z = k$. What is the shape of these curves?
(c) Meridians (curves of equal longitude) are traces in planes of the form $y = mx$. What is the shape of these meridians?

50. A cooling tower for a nuclear reactor is to be constructed in the shape of a hyperboloid of one sheet (see the photo on page 879). The diameter at the base is 280 m and the minimum diameter, 500 m above the base, is 200 m. Find an equation for the tower.

51. Show that if the point (a, b, c) lies on the hyperbolic paraboloid $z = y^2 - x^2$, then the lines with parametric equations $x = a + t$, $y = b + t$, $z = c + 2(b - a)t$ and $x = a + t$, $y = b - t$, $z = c - 2(b + a)t$ both lie entirely on this paraboloid. (This shows that the hyperbolic paraboloid is what is called a **ruled surface**; that is, it can be generated by the motion of a straight line. In fact, this exercise shows that through each point on the hyperbolic paraboloid there are two generating lines. The only other quadric surfaces that are ruled surfaces are cylinders, cones, and hyperboloids of one sheet.)

52. Show that the curve of intersection of the surfaces $x^2 + 2y^2 - z^2 + 3x = 1$ and $2x^2 + 4y^2 - 2z^2 - 5y = 0$ lies in a plane.

53. Graph the surfaces $z = x^2 + y^2$ and $z = 1 - y^2$ on a common screen using the domain $|x| \leq 1.2$, $|y| \leq 1.2$ and observe the curve of intersection of these surfaces. Show that the projection of this curve onto the xy-plane is an ellipse.

12 REVIEW

CONCEPT CHECK

Answers to the Concept Check can be found on the back endpapers.

1. What is the difference between a vector and a scalar?

2. How do you add two vectors geometrically? How do you add them algebraically?

3. If **a** is a vector and c is a scalar, how is c**a** related to **a** geometrically? How do you find c**a** algebraically?

4. How do you find the vector from one point to another?

5. How do you find the dot product **a** · **b** of two vectors if you know their lengths and the angle between them? What if you know their components?

6. How are dot products useful?

7. Write expressions for the scalar and vector projections of **b** onto **a**. Illustrate with diagrams.

8. How do you find the cross product **a** × **b** of two vectors if you know their lengths and the angle between them? What if you know their components?

9. How are cross products useful?

10. (a) How do you find the area of the parallelogram determined by **a** and **b**?
(b) How do you find the volume of the parallelepiped determined by **a**, **b**, and **c**?

11. How do you find a vector perpendicular to a plane?

12. How do you find the angle between two intersecting planes?

13. Write a vector equation, parametric equations, and symmetric equations for a line.

14. Write a vector equation and a scalar equation for a plane.

15. (a) How do you tell if two vectors are parallel?
(b) How do you tell if two vectors are perpendicular?
(c) How do you tell if two planes are parallel?

16. (a) Describe a method for determining whether three points P, Q, and R lie on the same line.
(b) Describe a method for determining whether four points P, Q, R, and S lie in the same plane.

17. (a) How do you find the distance from a point to a line?
(b) How do you find the distance from a point to a plane?
(c) How do you find the distance between two lines?

18. What are the traces of a surface? How do you find them?

19. Write equations in standard form of the six types of quadric surfaces.

TRUE-FALSE QUIZ

Determine whether the statement is true or false. If it is true, explain why. If it is false, explain why or give an example that disproves the statement.

1. If $\mathbf{u} = \langle u_1, u_2 \rangle$ and $\mathbf{v} = \langle v_1, v_2 \rangle$, then $\mathbf{u} \cdot \mathbf{v} = \langle u_1 v_1, u_2 v_2 \rangle$.
2. For any vectors $\mathbf{u}$ and $\mathbf{v}$ in V_3, $|\mathbf{u} + \mathbf{v}| = |\mathbf{u}| + |\mathbf{v}|$.
3. For any vectors $\mathbf{u}$ and $\mathbf{v}$ in V_3, $|\mathbf{u} \cdot \mathbf{v}| = |\mathbf{u}||\mathbf{v}|$.
4. For any vectors $\mathbf{u}$ and $\mathbf{v}$ in V_3, $|\mathbf{u} \times \mathbf{v}| = |\mathbf{u}||\mathbf{v}|$.
5. For any vectors $\mathbf{u}$ and $\mathbf{v}$ in V_3, $\mathbf{u} \cdot \mathbf{v} = \mathbf{v} \cdot \mathbf{u}$.
6. For any vectors $\mathbf{u}$ and $\mathbf{v}$ in V_3, $\mathbf{u} \times \mathbf{v} = \mathbf{v} \times \mathbf{u}$.
7. For any vectors $\mathbf{u}$ and $\mathbf{v}$ in V_3, $|\mathbf{u} \times \mathbf{v}| = |\mathbf{v} \times \mathbf{u}|$.
8. For any vectors $\mathbf{u}$ and $\mathbf{v}$ in V_3 and any scalar k,
$$k(\mathbf{u} \cdot \mathbf{v}) = (k\mathbf{u}) \cdot \mathbf{v}$$
9. For any vectors $\mathbf{u}$ and $\mathbf{v}$ in V_3 and any scalar k,
$$k(\mathbf{u} \times \mathbf{v}) = (k\mathbf{u}) \times \mathbf{v}$$
10. For any vectors $\mathbf{u}$, $\mathbf{v}$, and $\mathbf{w}$ in V_3,
$$(\mathbf{u} + \mathbf{v}) \times \mathbf{w} = \mathbf{u} \times \mathbf{w} + \mathbf{v} \times \mathbf{w}$$
11. For any vectors $\mathbf{u}$, $\mathbf{v}$, and $\mathbf{w}$ in V_3,
$$\mathbf{u} \cdot (\mathbf{v} \times \mathbf{w}) = (\mathbf{u} \times \mathbf{v}) \cdot \mathbf{w}$$
12. For any vectors $\mathbf{u}$, $\mathbf{v}$, and $\mathbf{w}$ in V_3,
$$\mathbf{u} \times (\mathbf{v} \times \mathbf{w}) = (\mathbf{u} \times \mathbf{v}) \times \mathbf{w}$$
13. For any vectors $\mathbf{u}$ and $\mathbf{v}$ in V_3, $(\mathbf{u} \times \mathbf{v}) \cdot \mathbf{u} = 0$.
14. For any vectors $\mathbf{u}$ and $\mathbf{v}$ in V_3, $(\mathbf{u} + \mathbf{v}) \times \mathbf{v} = \mathbf{u} \times \mathbf{v}$.
15. The vector $\langle 3, -1, 2 \rangle$ is parallel to the plane
$$6x - 2y + 4z = 1$$
16. A linear equation $Ax + By + Cz + D = 0$ represents a line in space.
17. The set of points $\{(x, y, z) \mid x^2 + y^2 = 1\}$ is a circle.
18. In $\mathbb{R}^3$ the graph of $y = x^2$ is a paraboloid.
19. If $\mathbf{u} \cdot \mathbf{v} = 0$, then $\mathbf{u} = \mathbf{0}$ or $\mathbf{v} = \mathbf{0}$.
20. If $\mathbf{u} \times \mathbf{v} = \mathbf{0}$, then $\mathbf{u} = \mathbf{0}$ or $\mathbf{v} = \mathbf{0}$.
21. If $\mathbf{u} \cdot \mathbf{v} = 0$ and $\mathbf{u} \times \mathbf{v} = \mathbf{0}$, then $\mathbf{u} = \mathbf{0}$ or $\mathbf{v} = \mathbf{0}$.
22. If $\mathbf{u}$ and $\mathbf{v}$ are in V_3, then $|\mathbf{u} \cdot \mathbf{v}| \leqslant |\mathbf{u}||\mathbf{v}|$.

EXERCISES

1. (a) Find an equation of the sphere that passes through the point $(6, -2, 3)$ and has center $(-1, 2, 1)$.
   (b) Find the curve in which this sphere intersects the yz-plane.
   (c) Find the center and radius of the sphere
$$x^2 + y^2 + z^2 - 8x + 2y + 6z + 1 = 0$$
2. Copy the vectors in the figure and use them to draw each of the following vectors.
   (a) $\mathbf{a} + \mathbf{b}$ (b) $\mathbf{a} - \mathbf{b}$ (c) $-\frac{1}{2}\mathbf{a}$ (d) $2\mathbf{a} + \mathbf{b}$

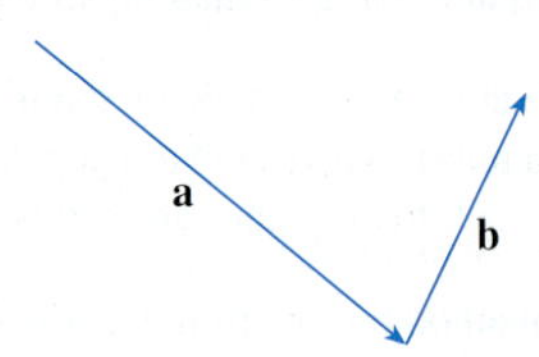

3. If $\mathbf{u}$ and $\mathbf{v}$ are the vectors shown in the figure, find $\mathbf{u} \cdot \mathbf{v}$ and $|\mathbf{u} \times \mathbf{v}|$. Is $\mathbf{u} \times \mathbf{v}$ directed into the page or out of it?

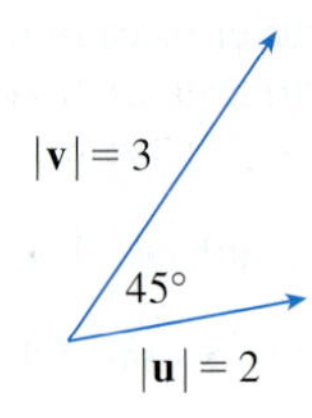

4. Calculate the given quantity if
$$\mathbf{a} = \mathbf{i} + \mathbf{j} - 2\mathbf{k}$$
$$\mathbf{b} = 3\mathbf{i} - 2\mathbf{j} + \mathbf{k}$$
$$\mathbf{c} = \mathbf{j} - 5\mathbf{k}$$
   (a) $2\mathbf{a} + 3\mathbf{b}$ (b) $|\mathbf{b}|$
   (c) $\mathbf{a} \cdot \mathbf{b}$ (d) $\mathbf{a} \times \mathbf{b}$
   (e) $|\mathbf{b} \times \mathbf{c}|$ (f) $\mathbf{a} \cdot (\mathbf{b} \times \mathbf{c})$
   (g) $\mathbf{c} \times \mathbf{c}$ (h) $\mathbf{a} \times (\mathbf{b} \times \mathbf{c})$
   (i) $\text{comp}_{\mathbf{a}} \mathbf{b}$ (j) $\text{proj}_{\mathbf{a}} \mathbf{b}$
   (k) The angle between $\mathbf{a}$ and $\mathbf{b}$ (correct to the nearest degree)
5. Find the values of x such that the vectors $\langle 3, 2, x \rangle$ and $\langle 2x, 4, x \rangle$ are orthogonal.
6. Find two unit vectors that are orthogonal to both $\mathbf{j} + 2\mathbf{k}$ and $\mathbf{i} - 2\mathbf{j} + 3\mathbf{k}$.
7. Suppose that $\mathbf{u} \cdot (\mathbf{v} \times \mathbf{w}) = 2$. Find
   (a) $(\mathbf{u} \times \mathbf{v}) \cdot \mathbf{w}$ (b) $\mathbf{u} \cdot (\mathbf{w} \times \mathbf{v})$
   (c) $\mathbf{v} \cdot (\mathbf{u} \times \mathbf{w})$ (d) $(\mathbf{u} \times \mathbf{v}) \cdot \mathbf{v}$
8. Show that if $\mathbf{a}$, $\mathbf{b}$, and $\mathbf{c}$ are in V_3, then
$$(\mathbf{a} \times \mathbf{b}) \cdot [(\mathbf{b} \times \mathbf{c}) \times (\mathbf{c} \times \mathbf{a})] = [\mathbf{a} \cdot (\mathbf{b} \times \mathbf{c})]^2$$
9. Find the acute angle between two diagonals of a cube.

10. Given the points $A(1, 0, 1)$, $B(2, 3, 0)$, $C(-1, 1, 4)$, and $D(0, 3, 2)$, find the volume of the parallelepiped with adjacent edges AB, AC, and AD.

11. (a) Find a vector perpendicular to the plane through the points $A(1, 0, 0)$, $B(2, 0, -1)$, and $C(1, 4, 3)$.
(b) Find the area of triangle ABC.

12. A constant force $\mathbf{F} = 3\mathbf{i} + 5\mathbf{j} + 10\mathbf{k}$ moves an object along the line segment from $(1, 0, 2)$ to $(5, 3, 8)$. Find the work done if the distance is measured in meters and the force in newtons.

13. A boat is pulled onto shore using two ropes, as shown in the diagram. If a force of 255 N is needed, find the magnitude of the force in each rope.

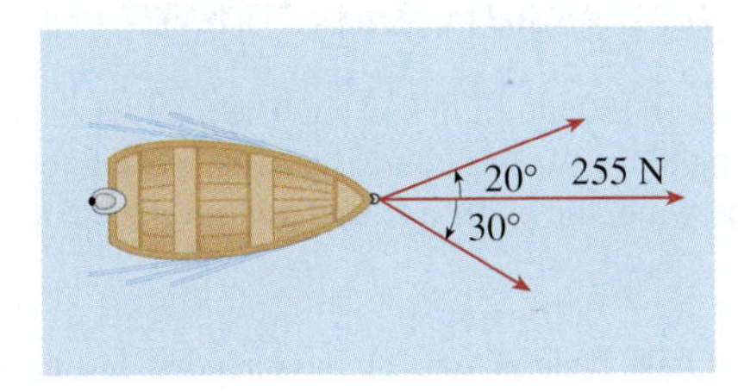

14. Find the magnitude of the torque about P if a 50-N force is applied as shown.

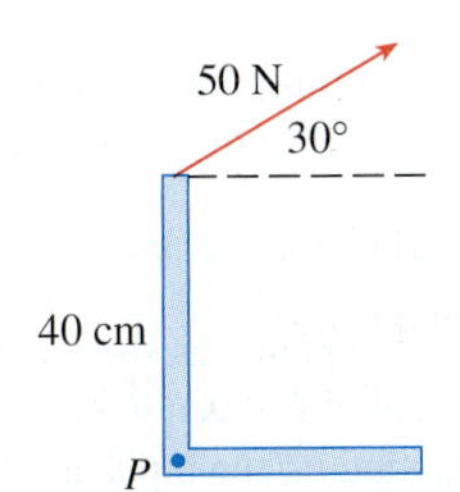

15–17 Find parametric equations for the line.

15. The line through $(4, -1, 2)$ and $(1, 1, 5)$

16. The line through $(1, 0, -1)$ and parallel to the line $\frac{1}{3}(x - 4) = \frac{1}{2}y = z + 2$

17. The line through $(-2, 2, 4)$ and perpendicular to the plane $2x - y + 5z = 12$

18–20 Find an equation of the plane.

18. The plane through $(2, 1, 0)$ and parallel to $x + 4y - 3z = 1$

19. The plane through $(3, -1, 1)$, $(4, 0, 2)$, and $(6, 3, 1)$

20. The plane through $(1, 2, -2)$ that contains the line $x = 2t$, $y = 3 - t$, $z = 1 + 3t$

21. Find the point in which the line with parametric equations $x = 2 - t$, $y = 1 + 3t$, $z = 4t$ intersects the plane $2x - y + z = 2$.

22. Find the distance from the origin to the line $x = 1 + t$, $y = 2 - t$, $z = -1 + 2t$.

23. Determine whether the lines given by the symmetric equations

$$\frac{x - 1}{2} = \frac{y - 2}{3} = \frac{z - 3}{4}$$

and

$$\frac{x + 1}{6} = \frac{y - 3}{-1} = \frac{z + 5}{2}$$

are parallel, skew, or intersecting.

24. (a) Show that the planes $x + y - z = 1$ and $2x - 3y + 4z = 5$ are neither parallel nor perpendicular.
(b) Find, correct to the nearest degree, the angle between these planes.

25. Find an equation of the plane through the line of intersection of the planes $x - z = 1$ and $y + 2z = 3$ and perpendicular to the plane $x + y - 2z = 1$.

26. (a) Find an equation of the plane that passes through the points $A(2, 1, 1)$, $B(-1, -1, 10)$, and $C(1, 3, -4)$.
(b) Find symmetric equations for the line through B that is perpendicular to the plane in part (a).
(c) A second plane passes through $(2, 0, 4)$ and has normal vector $\langle 2, -4, -3 \rangle$. Show that the acute angle between the planes is approximately 43°.
(d) Find parametric equations for the line of intersection of the two planes.

27. Find the distance between the planes $3x + y - 4z = 2$ and $3x + y - 4z = 24$.

28–36 Identify and sketch the graph of each surface.

28. $x = 3$

29. $x = z$

30. $y = z^2$

31. $x^2 = y^2 + 4z^2$

32. $4x - y + 2z = 4$

33. $-4x^2 + y^2 - 4z^2 = 4$

34. $y^2 + z^2 = 1 + x^2$

35. $4x^2 + 4y^2 - 8y + z^2 = 0$

36. $x = y^2 + z^2 - 2y - 4z + 5$

37. An ellipsoid is created by rotating the ellipse $4x^2 + y^2 = 16$ about the x-axis. Find an equation of the ellipsoid.

38. A surface consists of all points P such that the distance from P to the plane $y = 1$ is twice the distance from P to the point $(0, -1, 0)$. Find an equation for this surface and identify it.

Problems Plus

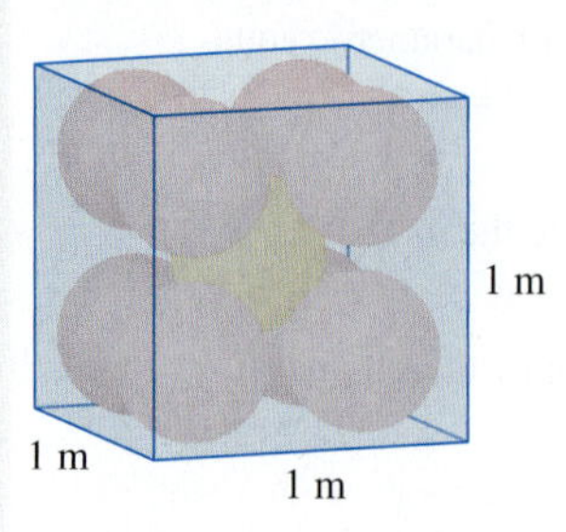

FIGURE FOR PROBLEM 1

1. Each edge of a cubical box has length 1 m. The box contains nine spherical balls with the same radius r. The center of one ball is at the center of the cube and it touches the other eight balls. Each of the other eight balls touches three sides of the box. Thus the balls are tightly packed in the box (see the figure). Find r. (If you have trouble with this problem, read about the problem-solving strategy entitled *Use Analogy* on page 98.)

2. Let B be a solid box with length L, width W, and height H. Let S be the set of all points that are a distance at most 1 from some point of B. Express the volume of S in terms of L, W, and H.

3. Let L be the line of intersection of the planes $cx + y + z = c$ and $x - cy + cz = -1$, where c is a real number.

(a) Find symmetric equations for L.

(b) As the number c varies, the line L sweeps out a surface S. Find an equation for the curve of intersection of S with the horizontal plane $z = t$ (the trace of S in the plane $z = t$).

(c) Find the volume of the solid bounded by S and the planes $z = 0$ and $z = 1$.

4. A plane is capable of flying at a speed of 180 km/h in still air. The pilot takes off from an airfield and heads due north according to the plane's compass. After 30 minutes of flight time, the pilot notices that, due to the wind, the plane has actually traveled 80 km at an angle 5° east of north.

(a) What is the wind velocity?

(b) In what direction should the pilot have headed to reach the intended destination?

5. Suppose $\mathbf{v}_1$ and $\mathbf{v}_2$ are vectors with $|\mathbf{v}_1| = 2$, $|\mathbf{v}_2| = 3$, and $\mathbf{v}_1 \cdot \mathbf{v}_2 = 5$. Let $\mathbf{v}_3 = \text{proj}_{\mathbf{v}_1}\mathbf{v}_2$, $\mathbf{v}_4 = \text{proj}_{\mathbf{v}_2}\mathbf{v}_3$, $\mathbf{v}_5 = \text{proj}_{\mathbf{v}_3}\mathbf{v}_4$, and so on. Compute $\sum_{n=1}^{\infty} |\mathbf{v}_n|$.

6. Find an equation of the largest sphere that passes through the point $(-1, 1, 4)$ and is such that each of the points (x, y, z) inside the sphere satisfies the condition

$$x^2 + y^2 + z^2 < 136 + 2(x + 2y + 3z)$$

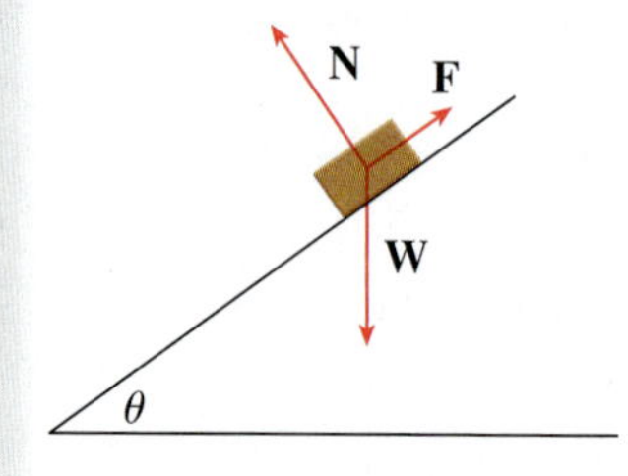

FIGURE FOR PROBLEM 7

7. Suppose a block of mass m is placed on an inclined plane, as shown in the figure. The block's descent down the plane is slowed by friction; if θ is not too large, friction will prevent the block from moving at all. The forces acting on the block are the weight $\mathbf{W}$, where $|\mathbf{W}| = mg$ (g is the acceleration due to gravity); the normal force $\mathbf{N}$ (the normal component of the reactionary force of the plane on the block), where $|\mathbf{N}| = n$; and the force $\mathbf{F}$ due to friction, which acts parallel to the inclined plane, opposing the direction of motion. If the block is at rest and θ is increased, $|\mathbf{F}|$ must also increase until ultimately $|\mathbf{F}|$ reaches its maximum, beyond which the block begins to slide. At this angle θ_s, it has been observed that $|\mathbf{F}|$ is proportional to n. Thus, when $|\mathbf{F}|$ is maximal, we can say that $|\mathbf{F}| = \mu_s n$, where μ_s is called the *coefficient of static friction* and depends on the materials that are in contact.

(a) Observe that $\mathbf{N} + \mathbf{F} + \mathbf{W} = \mathbf{0}$ and deduce that $\mu_s = \tan(\theta_s)$.

(b) Suppose that, for $\theta > \theta_s$, an additional outside force $\mathbf{H}$ is applied to the block, horizontally from the left, and let $|\mathbf{H}| = h$. If h is small, the block may still slide down the plane; if h is large enough, the block will move up the plane. Let $h_{\min}$ be the smallest value of h that allows the block to remain motionless (so that $|\mathbf{F}|$ is maximal).

By choosing the coordinate axes so that $\mathbf{F}$ lies along the x-axis, resolve each force into components parallel and perpendicular to the inclined plane and show that

$$h_{\min} \sin\theta + mg\cos\theta = n \qquad \text{and} \qquad h_{\min}\cos\theta + \mu_s n = mg\sin\theta$$

(c) Show that

$$h_{\min} = mg\tan(\theta - \theta_s)$$

Does this equation seem reasonable? Does it make sense for $\theta = \theta_s$? Does it make sense as $\theta \to 90°$? Explain.

(d) Let $h_{\max}$ be the largest value of h that allows the block to remain motionless. (In which direction is **F** heading?) Show that

$$h_{\max} = mg\tan(\theta + \theta_s)$$

Does this equation seem reasonable? Explain.

8. A solid has the following properties. When illuminated by rays parallel to the z-axis, its shadow is a circular disk. If the rays are parallel to the y-axis, its shadow is a square. If the rays are parallel to the x-axis, its shadow is an isosceles triangle. (In Exercise 12.1.48 you were asked to describe and sketch an example of such a solid, but there are many such solids.) Assume that the projection onto the xz-plane is a square whose sides have length 1.

(a) What is the volume of the largest such solid?

(b) Is there a smallest volume?

13 Vector Functions

The paths of objects moving through space like the jet planes pictured here can be described by vector functions. In Section 13.1 we will see how to use these vector functions to determine whether or not two such objects will collide.

THE FUNCTIONS THAT WE HAVE been using so far have been real-valued functions. We now study functions whose values are vectors because such functions are needed to describe curves and surfaces in space. We will also use vector-valued functions to describe the motion of objects through space. In particular, we will use them to derive Kepler's laws of planetary motion.

13.1 Vector Functions and Space Curves

In general, a function is a rule that assigns to each element in the domain an element in the range. A **vector-valued function**, or **vector function**, is simply a function whose domain is a set of real numbers and whose range is a set of vectors. We are most interested in vector functions $\mathbf{r}$ whose values are three-dimensional vectors. This means that for every number t in the domain of $\mathbf{r}$ there is a unique vector in V_3 denoted by $\mathbf{r}(t)$. If $f(t)$, $g(t)$, and $h(t)$ are the components of the vector $\mathbf{r}(t)$, then f, g, and h are real-valued functions called the **component functions** of $\mathbf{r}$ and we can write

$$\mathbf{r}(t) = \langle f(t), g(t), h(t)\rangle = f(t)\,\mathbf{i} + g(t)\,\mathbf{j} + h(t)\,\mathbf{k}$$

We use the letter t to denote the independent variable because it represents time in most applications of vector functions.

EXAMPLE 1 If

$$\mathbf{r}(t) = \left\langle t^3, \ln(3 - t), \sqrt{t}\right\rangle$$

then the component functions are

$$f(t) = t^3 \qquad g(t) = \ln(3 - t) \qquad h(t) = \sqrt{t}$$

By our usual convention, the domain of $\mathbf{r}$ consists of all values of t for which the expression for $\mathbf{r}(t)$ is defined. The expressions t^3, $\ln(3 - t)$, and $\sqrt{t}$ are all defined when $3 - t > 0$ and $t \geqslant 0$. Therefore the domain of $\mathbf{r}$ is the interval $[0, 3)$. ■

■ Limits and Continuity

The **limit** of a vector function $\mathbf{r}$ is defined by taking the limits of its component functions as follows.

If $\lim_{t\to a}\mathbf{r}(t) = \mathbf{L}$, this definition is equivalent to saying that the length and direction of the vector $\mathbf{r}(t)$ approach the length and direction of the vector $\mathbf{L}$.

1 If $\mathbf{r}(t) = \langle f(t), g(t), h(t)\rangle$, then

$$\lim_{t\to a}\mathbf{r}(t) = \left\langle \lim_{t\to a} f(t), \lim_{t\to a} g(t), \lim_{t\to a} h(t)\right\rangle$$

provided the limits of the component functions exist.

Equivalently, we could have used an ε-δ definition (see Exercise 54). Limits of vector functions obey the same rules as limits of real-valued functions (see Exercise 53).

EXAMPLE 2 Find $\lim_{t\to 0}\mathbf{r}(t)$, where $\mathbf{r}(t) = (1 + t^3)\,\mathbf{i} + te^{-t}\,\mathbf{j} + \dfrac{\sin t}{t}\,\mathbf{k}$.

SOLUTION According to Definition 1, the limit of $\mathbf{r}$ is the vector whose components are the limits of the component functions of $\mathbf{r}$:

$$\lim_{t\to 0}\mathbf{r}(t) = \left[\lim_{t\to 0}(1 + t^3)\right]\mathbf{i} + \left[\lim_{t\to 0} te^{-t}\right]\mathbf{j} + \left[\lim_{t\to 0}\frac{\sin t}{t}\right]\mathbf{k}$$

$$= \mathbf{i} + \mathbf{k} \qquad \text{(by Equation 2.4.2)}$$ ■

A vector function $\mathbf{r}$ is **continuous at *a*** if

$$\lim_{t \to a} \mathbf{r}(t) = \mathbf{r}(a)$$

In view of Definition 1, we see that $\mathbf{r}$ is continuous at a if and only if its component functions f, g, and h are continuous at a.

Space Curves

There is a close connection between continuous vector functions and space curves. Suppose that f, g, and h are continuous real-valued functions on an interval I. Then the set C of all points (x, y, z) in space, where

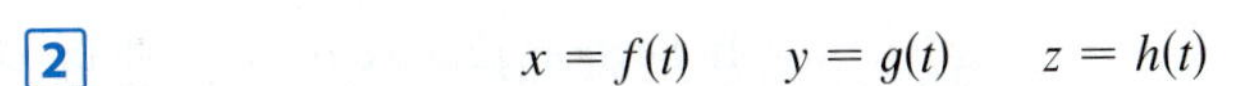

2 $$x = f(t) \qquad y = g(t) \qquad z = h(t)$$

and t varies throughout the interval I, is called a **space curve**. The equations in (2) are called **parametric equations of *C*** and t is called a **parameter**. We can think of C as being traced out by a moving particle whose position at time t is $(f(t), g(t), h(t))$. If we now consider the vector function $\mathbf{r}(t) = \langle f(t), g(t), h(t) \rangle$, then $\mathbf{r}(t)$ is the position vector of the point $P(f(t), g(t), h(t))$ on C. Thus any continuous vector function $\mathbf{r}$ defines a space curve C that is traced out by the tip of the moving vector $\mathbf{r}(t)$, as shown in Figure 1.

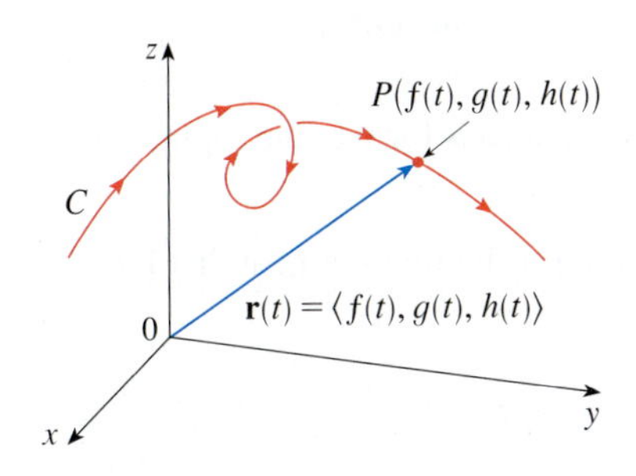

FIGURE 1
C is traced out by the tip of a moving position vector $\mathbf{r}(t)$.

EXAMPLE 3 Describe the curve defined by the vector function

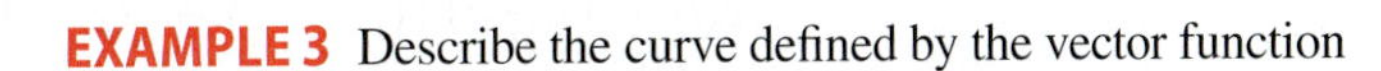

$$\mathbf{r}(t) = \langle 1 + t, 2 + 5t, -1 + 6t \rangle$$

SOLUTION The corresponding parametric equations are

$$x = 1 + t \qquad y = 2 + 5t \qquad z = -1 + 6t$$

which we recognize from Equations 12.5.2 as parametric equations of a line passing through the point $(1, 2, -1)$ and parallel to the vector $\langle 1, 5, 6 \rangle$. Alternatively, we could observe that the function can be written as $\mathbf{r} = \mathbf{r}_0 + t\mathbf{v}$, where $\mathbf{r}_0 = \langle 1, 2, -1 \rangle$ and $\mathbf{v} = \langle 1, 5, 6 \rangle$, and this is the vector equation of a line as given by Equation 12.5.1. ■

TEC Visual 13.1A shows several curves being traced out by position vectors, including those in Figures 1 and 2.

Plane curves can also be represented in vector notation. For instance, the curve given by the parametric equations $x = t^2 - 2t$ and $y = t + 1$ (see Example 10.1.1) could also be described by the vector equation

$$\mathbf{r}(t) = \langle t^2 - 2t, t + 1 \rangle = (t^2 - 2t)\,\mathbf{i} + (t + 1)\,\mathbf{j}$$

where $\mathbf{i} = \langle 1, 0 \rangle$ and $\mathbf{j} = \langle 0, 1 \rangle$.

EXAMPLE 4 Sketch the curve whose vector equation is

$$\mathbf{r}(t) = \cos t\,\mathbf{i} + \sin t\,\mathbf{j} + t\,\mathbf{k}$$

SOLUTION The parametric equations for this curve are

$$x = \cos t \qquad y = \sin t \qquad z = t$$

Since $x^2 + y^2 = \cos^2 t + \sin^2 t = 1$ for all values of t, the curve must lie on the circular cylinder $x^2 + y^2 = 1$. The point (x, y, z) lies directly above the point $(x, y, 0)$, which moves counterclockwise around the circle $x^2 + y^2 = 1$ in the xy-plane. (The projection of the curve onto the xy-plane has vector equation $\mathbf{r}(t) = \langle \cos t, \sin t, 0 \rangle$. See Example 10.1.2.) Since $z = t$, the curve spirals upward around the cylinder as t increases. The curve, shown in Figure 2, is called a **helix**. ■

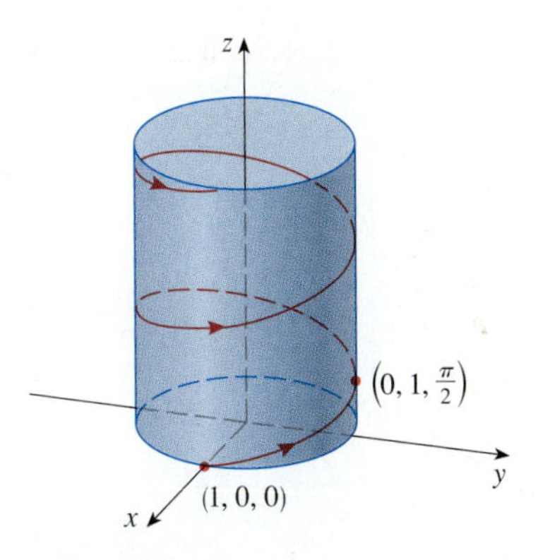

FIGURE 2

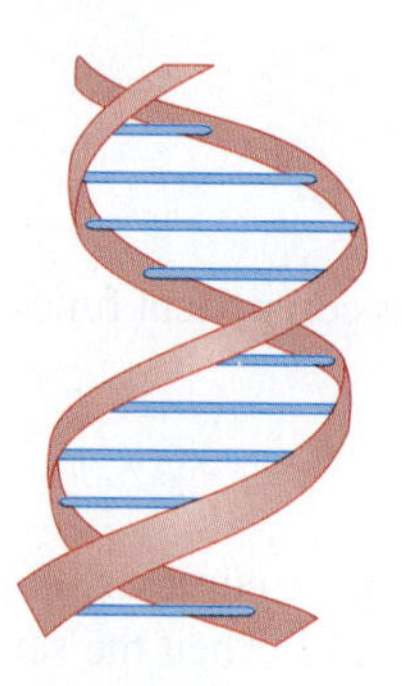

FIGURE 3
A double helix

The corkscrew shape of the helix in Example 4 is familiar from its occurrence in coiled springs. It also occurs in the model of DNA (deoxyribonucleic acid, the genetic material of living cells). In 1953 James Watson and Francis Crick showed that the structure of the DNA molecule is that of two linked, parallel helixes that are intertwined as in Figure 3.

In Examples 3 and 4 we were given vector equations of curves and asked for a geometric description or sketch. In the next two examples we are given a geometric description of a curve and are asked to find parametric equations for the curve.

EXAMPLE 5 Find a vector equation and parametric equations for the line segment that joins the point $P(1, 3, -2)$ to the point $Q(2, -1, 3)$.

Figure 4 shows the line segment PQ in Example 5.

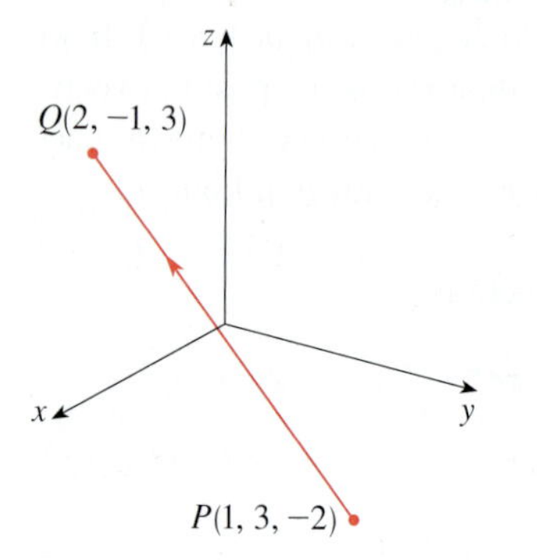

FIGURE 4

SOLUTION In Section 12.5 we found a vector equation for the line segment that joins the tip of the vector $\mathbf{r}_0$ to the tip of the vector $\mathbf{r}_1$:

$$\mathbf{r}(t) = (1 - t)\mathbf{r}_0 + t\mathbf{r}_1 \qquad 0 \leqslant t \leqslant 1$$

(See Equation 12.5.4.) Here we take $\mathbf{r}_0 = \langle 1, 3, -2 \rangle$ and $\mathbf{r}_1 = \langle 2, -1, 3 \rangle$ to obtain a vector equation of the line segment from P to Q:

$$\mathbf{r}(t) = (1 - t)\langle 1, 3, -2 \rangle + t\langle 2, -1, 3 \rangle \qquad 0 \leqslant t \leqslant 1$$

or

$$\mathbf{r}(t) = \langle 1 + t, 3 - 4t, -2 + 5t \rangle \qquad 0 \leqslant t \leqslant 1$$

The corresponding parametric equations are

$$x = 1 + t \qquad y = 3 - 4t \qquad z = -2 + 5t \qquad 0 \leqslant t \leqslant 1$$ ■

EXAMPLE 6 Find a vector function that represents the curve of intersection of the cylinder $x^2 + y^2 = 1$ and the plane $y + z = 2$.

SOLUTION Figure 5 shows how the plane and the cylinder intersect, and Figure 6 shows the curve of intersection C, which is an ellipse.

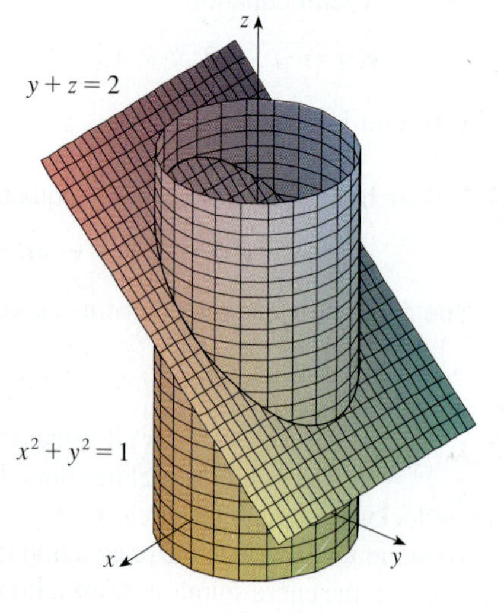

FIGURE 5

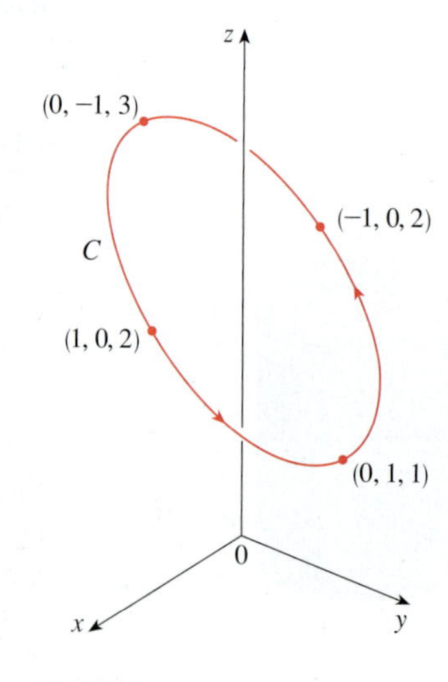

FIGURE 6

The projection of C onto the xy-plane is the circle $x^2 + y^2 = 1$, $z = 0$. So we know from Example 10.1.2 that we can write

$$x = \cos t \qquad y = \sin t \qquad 0 \leqslant t \leqslant 2\pi$$

From the equation of the plane, we have

$$z = 2 - y = 2 - \sin t$$

So we can write parametric equations for C as

$$x = \cos t \qquad y = \sin t \qquad z = 2 - \sin t \qquad 0 \leqslant t \leqslant 2\pi$$

The corresponding vector equation is

$$\mathbf{r}(t) = \cos t\,\mathbf{i} + \sin t\,\mathbf{j} + (2 - \sin t)\,\mathbf{k} \qquad 0 \leqslant t \leqslant 2\pi$$

This equation is called a *parametrization* of the curve C. The arrows in Figure 6 indicate the direction in which C is traced as the parameter t increases. ■

Using Computers to Draw Space Curves

Space curves are inherently more difficult to draw by hand than plane curves; for an accurate representation we need to use technology. For instance, Figure 7 shows a computer-generated graph of the curve with parametric equations

$$x = (4 + \sin 20t)\cos t \qquad y = (4 + \sin 20t)\sin t \qquad z = \cos 20t$$

It's called a **toroidal spiral** because it lies on a torus. Another interesting curve, the **trefoil knot**, with equations

$$x = (2 + \cos 1.5t)\cos t \qquad y = (2 + \cos 1.5t)\sin t \qquad z = \sin 1.5t$$

is graphed in Figure 8. It wouldn't be easy to plot either of these curves by hand.

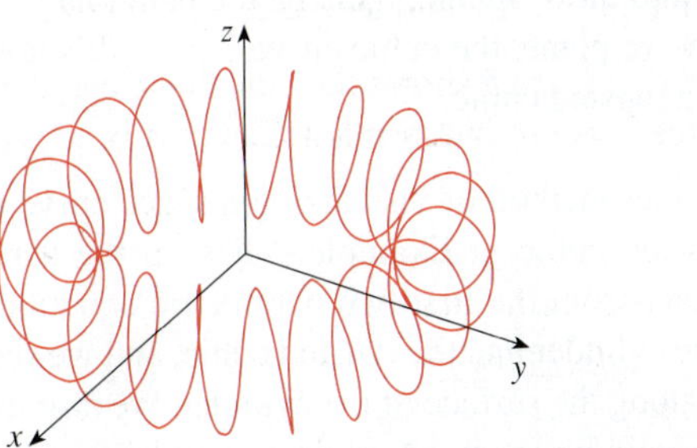

FIGURE 7
A toroidal spiral

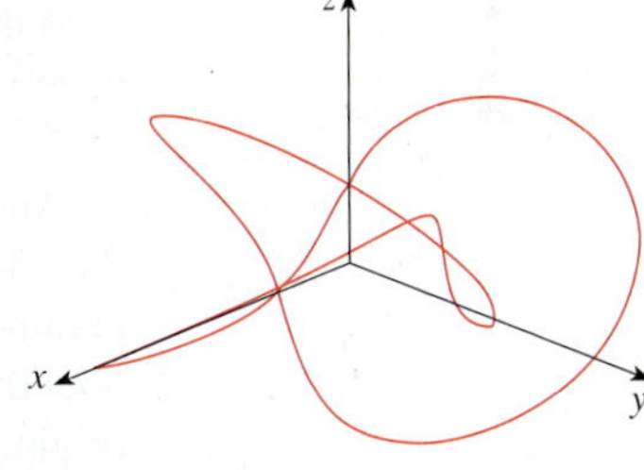

FIGURE 8
A trefoil knot

Even when a computer is used to draw a space curve, optical illusions make it difficult to get a good impression of what the curve really looks like. (This is especially true in Figure 8. See Exercise 52.) The next example shows how to cope with this problem.

EXAMPLE 7 Use a computer to draw the curve with vector equation $\mathbf{r}(t) = \langle t, t^2, t^3 \rangle$. This curve is called a **twisted cubic**.

SOLUTION We start by using the computer to plot the curve with parametric equations $x = t$, $y = t^2$, $z = t^3$ for $-2 \leqslant t \leqslant 2$. The result is shown in Figure 9(a), but it's hard to see the true nature of the curve from that graph alone. Most three-dimensional computer graphing programs allow the user to enclose a curve or surface in a box instead of displaying the coordinate axes. When we look at the same curve in a box in Figure 9(b), we have a much clearer picture of the curve. We can see that it climbs from a lower corner of the box to the upper corner nearest us, and it twists as it climbs.

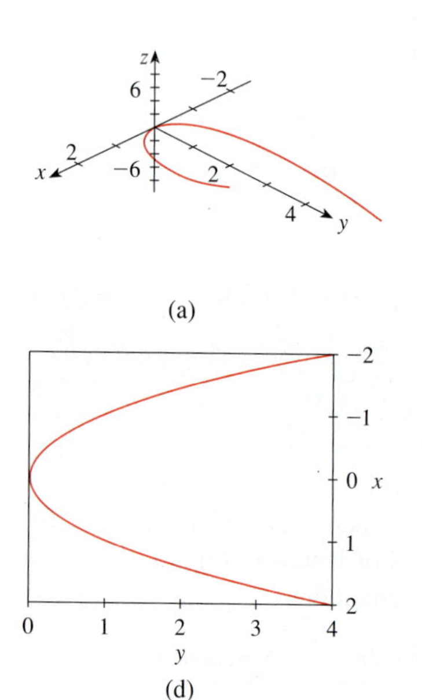

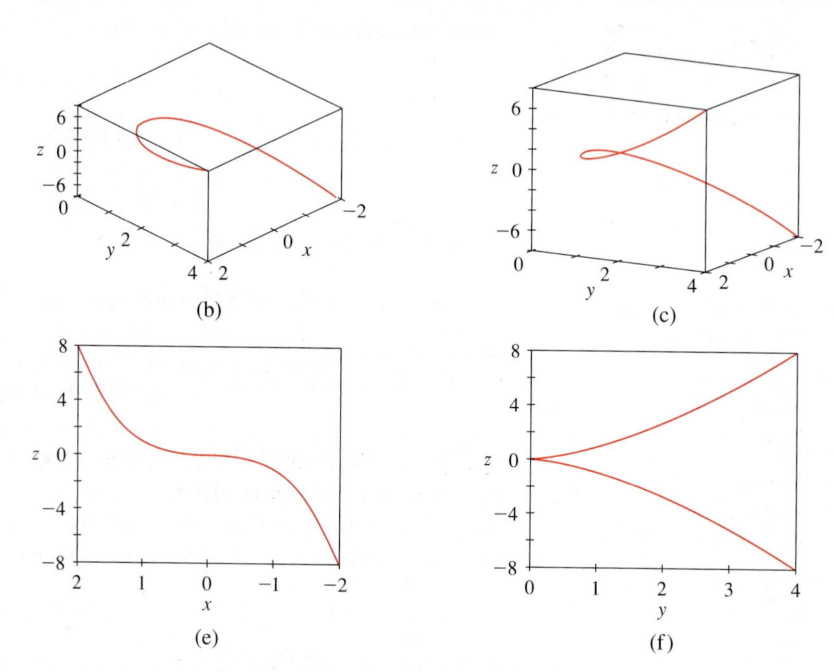

FIGURE 9 Views of the twisted cubic

TEC In Visual 13.1B you can rotate the box in Figure 9 to see the curve from any viewpoint.

We get an even better idea of the curve when we view it from different vantage points. Part (c) shows the result of rotating the box to give another viewpoint. Parts (d), (e), and (f) show the views we get when we look directly at a face of the box. In particular, part (d) shows the view from directly above the box. It is the projection of the curve onto the xy-plane, namely, the parabola $y = x^2$. Part (e) shows the projection onto the xz-plane, the cubic curve $z = x^3$. It's now obvious why the given curve is called a twisted cubic. ■

Another method of visualizing a space curve is to draw it on a surface. For instance, the twisted cubic in Example 7 lies on the parabolic cylinder $y = x^2$. (Eliminate the parameter from the first two parametric equations, $x = t$ and $y = t^2$.) Figure 10 shows both the cylinder and the twisted cubic, and we see that the curve moves upward from the origin along the surface of the cylinder. We also used this method in Example 4 to visualize the helix lying on the circular cylinder (see Figure 2).

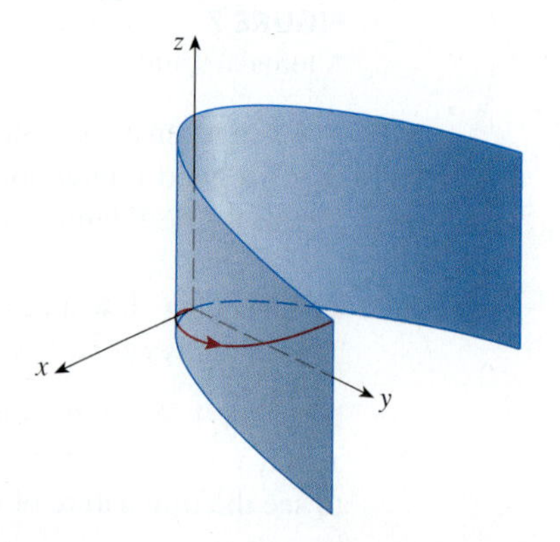

FIGURE 10

A third method for visualizing the twisted cubic is to realize that it also lies on the cylinder $z = x^3$. So it can be viewed as the curve of intersection of the cylinders $y = x^2$ and $z = x^3$. (See Figure 11.)

TEC Visual 13.1C shows how curves arise as intersections of surfaces.

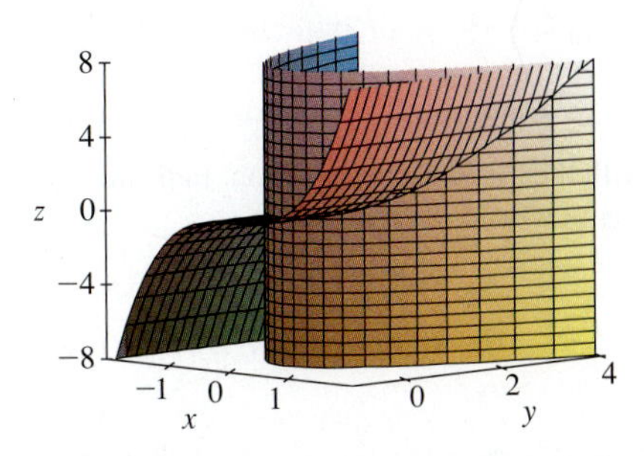

FIGURE 11

Some computer algebra systems provide us with a clearer picture of a space curve by enclosing it in a tube. Such a plot enables us to see whether one part of a curve passes in front of or behind another part of the curve. For example, Figure 13 shows the curve of Figure 12(b) as rendered by the `tubeplot` command in Maple.

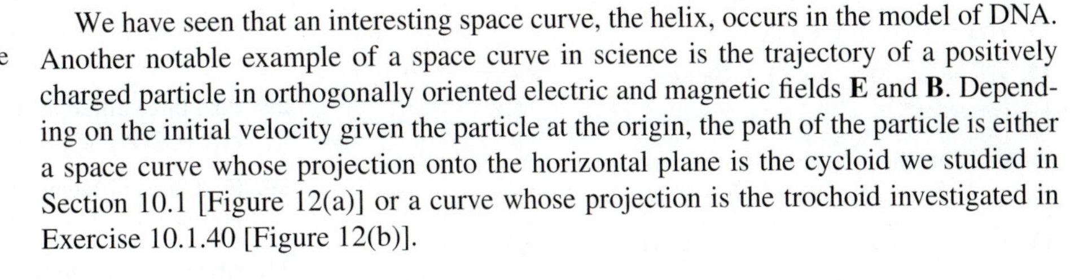

We have seen that an interesting space curve, the helix, occurs in the model of DNA. Another notable example of a space curve in science is the trajectory of a positively charged particle in orthogonally oriented electric and magnetic fields **E** and **B**. Depending on the initial velocity given the particle at the origin, the path of the particle is either a space curve whose projection onto the horizontal plane is the cycloid we studied in Section 10.1 [Figure 12(a)] or a curve whose projection is the trochoid investigated in Exercise 10.1.40 [Figure 12(b)].

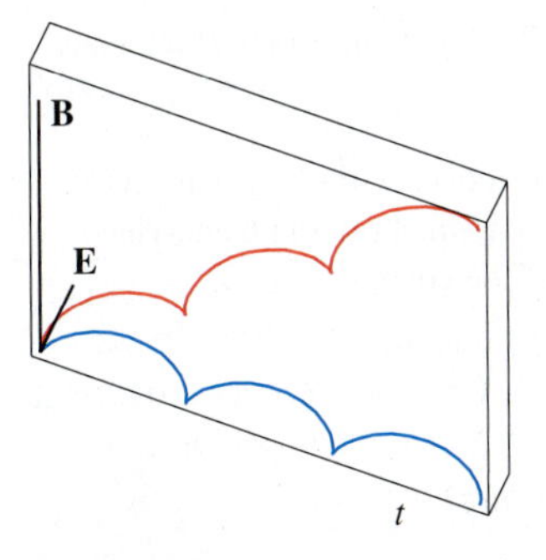

(a) $\mathbf{r}(t) = \langle t - \sin t, 1 - \cos t, t \rangle$

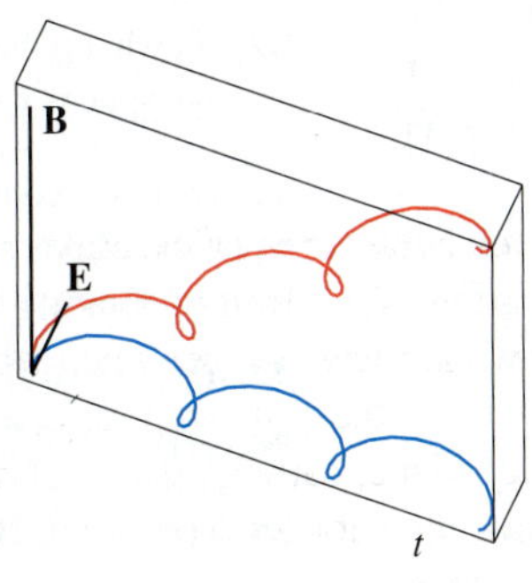

(b) $\mathbf{r}(t) = \left\langle t - \frac{3}{2}\sin t, 1 - \frac{3}{2}\cos t, t \right\rangle$

FIGURE 12
Motion of a charged particle in orthogonally oriented electric and magnetic fields

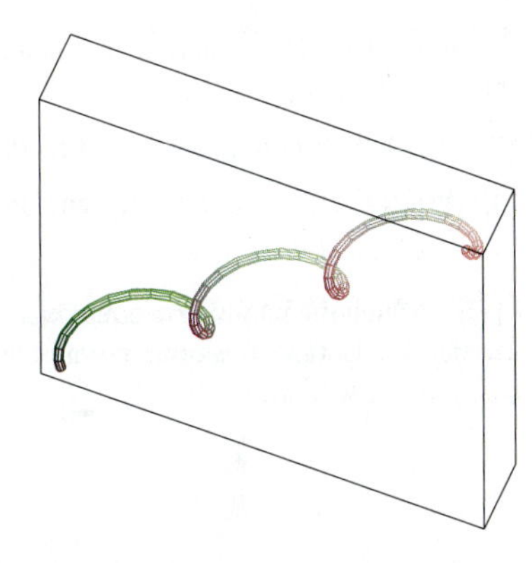

FIGURE 13

For further details concerning the physics involved and animations of the trajectories of the particles, see the following websites:

- www.physics.ucla.edu/plasma-exp/Beam/
- www.phy.ntnu.edu.tw/ntnujava/index.php?topic=36

13.1 EXERCISES

1–2 Find the domain of the vector function.

1. $\mathbf{r}(t) = \left\langle \ln(t + 1), \dfrac{t}{\sqrt{9 - t^2}}, 2^t \right\rangle$

2. $\mathbf{r}(t) = \cos t\,\mathbf{i} + \ln t\,\mathbf{j} + \dfrac{1}{t - 2}\,\mathbf{k}$

3–6 Find the limit.

3. $\displaystyle\lim_{t \to 0} \left(e^{-3t}\,\mathbf{i} + \frac{t^2}{\sin^2 t}\,\mathbf{j} + \cos 2t\,\mathbf{k} \right)$

4. $\displaystyle\lim_{t \to 1} \left(\frac{t^2 - t}{t - 1}\,\mathbf{i} + \sqrt{t + 8}\,\mathbf{j} + \frac{\sin \pi t}{\ln t}\,\mathbf{k} \right)$

5. $\lim_{t \to \infty} \left\langle \frac{1+t^2}{1-t^2}, \tan^{-1}t, \frac{1-e^{-2t}}{t} \right\rangle$

6. $\lim_{t \to \infty} \left\langle te^{-t}, \frac{t^3+t}{2t^3-1}, t \sin \frac{1}{t} \right\rangle$

7–14 Sketch the curve with the given vector equation. Indicate with an arrow the direction in which t increases.

7. $\mathbf{r}(t) = \langle \sin t, t \rangle$

8. $\mathbf{r}(t) = \langle t^2 - 1, t \rangle$

9. $\mathbf{r}(t) = \langle t, 2 - t, 2t \rangle$

10. $\mathbf{r}(t) = \langle \sin \pi t, t, \cos \pi t \rangle$

11. $\mathbf{r}(t) = \langle 3, t, 2 - t^2 \rangle$

12. $\mathbf{r}(t) = 2 \cos t\, \mathbf{i} + 2 \sin t\, \mathbf{j} + \mathbf{k}$

13. $\mathbf{r}(t) = t^2\, \mathbf{i} + t^4\, \mathbf{j} + t^6\, \mathbf{k}$

14. $\mathbf{r}(t) = \cos t\, \mathbf{i} - \cos t\, \mathbf{j} + \sin t\, \mathbf{k}$

15–16 Draw the projections of the curve on the three coordinate planes. Use these projections to help sketch the curve.

15. $\mathbf{r}(t) = \langle t, \sin t, 2 \cos t \rangle$

16. $\mathbf{r}(t) = \langle t, t, t^2 \rangle$

17–20 Find a vector equation and parametric equations for the line segment that joins P to Q.

17. $P(2, 0, 0)$, $Q(6, 2, -2)$

18. $P(-1, 2, -2)$, $Q(-3, 5, 1)$

19. $P(0, -1, 1)$, $Q(\frac{1}{2}, \frac{1}{3}, \frac{1}{4})$

20. $P(a, b, c)$, $Q(u, v, w)$

21–26 Match the parametric equations with the graphs (labeled I–VI). Give reasons for your choices.

I

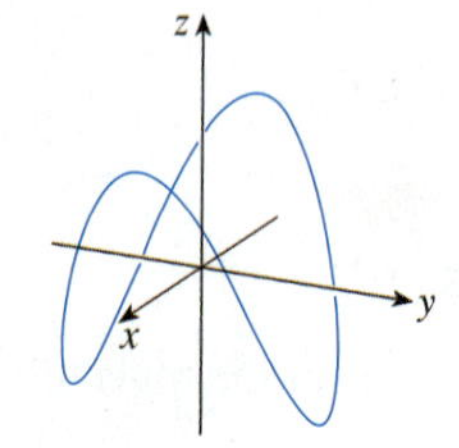

II

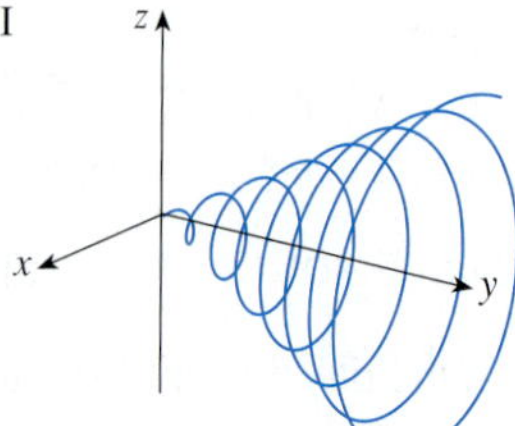

III

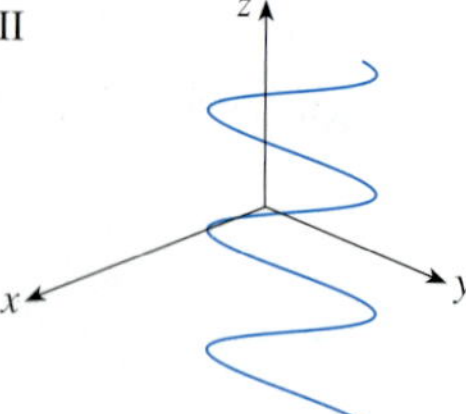

IV

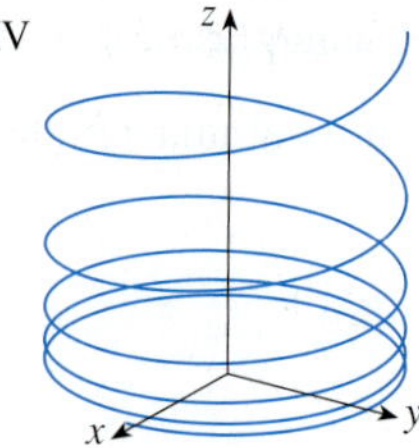

V

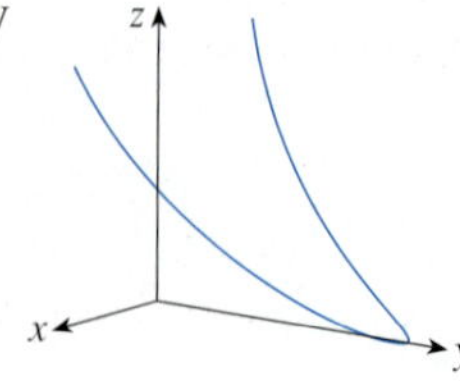

VI

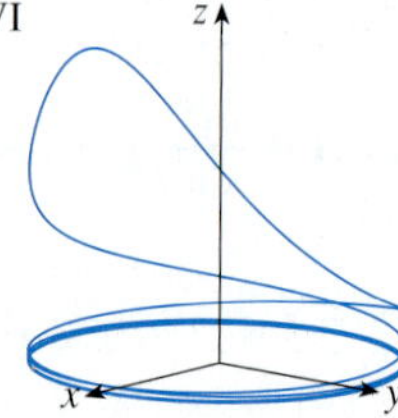

21. $x = t \cos t$, $y = t$, $z = t \sin t$, $t \geqslant 0$

22. $x = \cos t$, $y = \sin t$, $z = 1/(1 + t^2)$

23. $x = t$, $y = 1/(1 + t^2)$, $z = t^2$

24. $x = \cos t$, $y = \sin t$, $z = \cos 2t$

25. $x = \cos 8t$, $y = \sin 8t$, $z = e^{0.8t}$, $t \geqslant 0$

26. $x = \cos^2 t$, $y = \sin^2 t$, $z = t$

27. Show that the curve with parametric equations $x = t \cos t$, $y = t \sin t$, $z = t$ lies on the cone $z^2 = x^2 + y^2$, and use this fact to help sketch the curve.

28. Show that the curve with parametric equations $x = \sin t$, $y = \cos t$, $z = \sin^2 t$ is the curve of intersection of the surfaces $z = x^2$ and $x^2 + y^2 = 1$. Use this fact to help sketch the curve.

29. Find three different surfaces that contain the curve $\mathbf{r}(t) = 2t\, \mathbf{i} + e^t\, \mathbf{j} + e^{2t}\, \mathbf{k}$.

30. Find three different surfaces that contain the curve $\mathbf{r}(t) = t^2\, \mathbf{i} + \ln t\, \mathbf{j} + (1/t)\, \mathbf{k}$.

31. At what points does the curve $\mathbf{r}(t) = t\, \mathbf{i} + (2t - t^2)\, \mathbf{k}$ intersect the paraboloid $z = x^2 + y^2$?

32. At what points does the helix $\mathbf{r}(t) = \langle \sin t, \cos t, t \rangle$ intersect the sphere $x^2 + y^2 + z^2 = 5$?

33–37 Use a computer to graph the curve with the given vector equation. Make sure you choose a parameter domain and viewpoints that reveal the true nature of the curve.

33. $\mathbf{r}(t) = \langle \cos t \sin 2t, \sin t \sin 2t, \cos 2t \rangle$

34. $\mathbf{r}(t) = \langle te^t, e^{-t}, t \rangle$

35. $\mathbf{r}(t) = \left\langle \sin 3t \cos t, \frac{1}{4}t, \sin 3t \sin t \right\rangle$

36. $\mathbf{r}(t) = \langle \cos(8 \cos t) \sin t, \sin(8 \cos t) \sin t, \cos t \rangle$

37. $\mathbf{r}(t) = \langle \cos 2t, \cos 3t, \cos 4t \rangle$

38. Graph the curve with parametric equations $x = \sin t$, $y = \sin 2t$, $z = \cos 4t$. Explain its shape by graphing its projections onto the three coordinate planes.

39. Graph the curve with parametric equations

$$x = (1 + \cos 16t) \cos t$$

$$y = (1 + \cos 16t) \sin t$$

$$z = 1 + \cos 16t$$

Explain the appearance of the graph by showing that it lies on a cone.

40. Graph the curve with parametric equations

$$x = \sqrt{1 - 0.25 \cos^2 10t}\, \cos t$$

$$y = \sqrt{1 - 0.25 \cos^2 10t}\, \sin t$$

$$z = 0.5 \cos 10t$$

Explain the appearance of the graph by showing that it lies on a sphere.

41. Show that the curve with parametric equations $x = t^2$, $y = 1 - 3t$, $z = 1 + t^3$ passes through the points $(1, 4, 0)$ and $(9, -8, 28)$ but not through the point $(4, 7, -6)$.

42–46 Find a vector function that represents the curve of intersection of the two surfaces.

42. The cylinder $x^2 + y^2 = 4$ and the surface $z = xy$

43. The cone $z = \sqrt{x^2 + y^2}$ and the plane $z = 1 + y$

44. The paraboloid $z = 4x^2 + y^2$ and the parabolic cylinder $y = x^2$

45. The hyperboloid $z = x^2 - y^2$ and the cylinder $x^2 + y^2 = 1$

46. The semiellipsoid $x^2 + y^2 + 4z^2 = 4$, $y \geqslant 0$, and the cylinder $x^2 + z^2 = 1$

---

47. Try to sketch by hand the curve of intersection of the circular cylinder $x^2 + y^2 = 4$ and the parabolic cylinder $z = x^2$. Then find parametric equations for this curve and use these equations and a computer to graph the curve.

48. Try to sketch by hand the curve of intersection of the parabolic cylinder $y = x^2$ and the top half of the ellipsoid $x^2 + 4y^2 + 4z^2 = 16$. Then find parametric equations for this curve and use these equations and a computer to graph the curve.

49. If two objects travel through space along two different curves, it's often important to know whether they will collide. (Will a missile hit its moving target? Will two aircraft collide?) The curves might intersect, but we need to know whether the objects are in the same position *at the same time.* Suppose the trajectories of two particles are given by the vector functions

$$\mathbf{r}_1(t) = \langle t^2, 7t - 12, t^2 \rangle \qquad \mathbf{r}_2(t) = \langle 4t - 3, t^2, 5t - 6 \rangle$$

for $t \geqslant 0$. Do the particles collide?

50. Two particles travel along the space curves

$$\mathbf{r}_1(t) = \langle t, t^2, t^3 \rangle \qquad \mathbf{r}_2(t) = \langle 1 + 2t, 1 + 6t, 1 + 14t \rangle$$

Do the particles collide? Do their paths intersect?

51. (a) Graph the curve with parametric equations

$$x = \tfrac{27}{26} \sin 8t - \tfrac{8}{39} \sin 18t$$

$$y = -\tfrac{27}{26} \cos 8t + \tfrac{8}{39} \cos 18t$$

$$z = \tfrac{144}{65} \sin 5t$$

(b) Show that the curve lies on the hyperboloid of one sheet $144x^2 + 144y^2 - 25z^2 = 100$.

52. The view of the trefoil knot shown in Figure 8 is accurate, but it doesn't reveal the whole story. Use the parametric equations

$$x = (2 + \cos 1.5t) \cos t$$

$$y = (2 + \cos 1.5t) \sin t$$

$$z = \sin 1.5t$$

to sketch the curve by hand as viewed from above, with gaps indicating where the curve passes over itself. Start by showing that the projection of the curve onto the xy-plane has polar coordinates $r = 2 + \cos 1.5t$ and $\theta = t$, so r varies between 1 and 3. Then show that z has maximum and minimum values when the projection is halfway between $r = 1$ and $r = 3$.

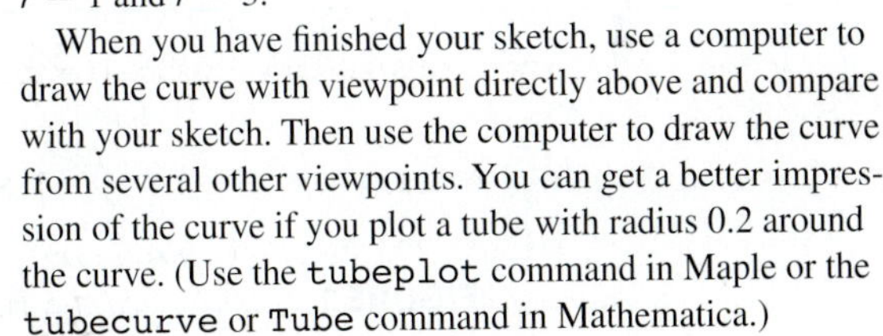

When you have finished your sketch, use a computer to draw the curve with viewpoint directly above and compare with your sketch. Then use the computer to draw the curve from several other viewpoints. You can get a better impression of the curve if you plot a tube with radius 0.2 around the curve. (Use the `tubeplot` command in Maple or the `tubecurve` or `Tube` command in Mathematica.)

53. Suppose $\mathbf{u}$ and $\mathbf{v}$ are vector functions that possess limits as $t \to a$ and let c be a constant. Prove the following properties of limits.

(a) $\lim_{t \to a} [\mathbf{u}(t) + \mathbf{v}(t)] = \lim_{t \to a} \mathbf{u}(t) + \lim_{t \to a} \mathbf{v}(t)$

(b) $\lim_{t \to a} c\mathbf{u}(t) = c \lim_{t \to a} \mathbf{u}(t)$

(c) $\lim_{t \to a} [\mathbf{u}(t) \cdot \mathbf{v}(t)] = \lim_{t \to a} \mathbf{u}(t) \cdot \lim_{t \to a} \mathbf{v}(t)$

(d) $\lim_{t \to a} [\mathbf{u}(t) \times \mathbf{v}(t)] = \lim_{t \to a} \mathbf{u}(t) \times \lim_{t \to a} \mathbf{v}(t)$

54. Show that $\lim_{t \to a} \mathbf{r}(t) = \mathbf{b}$ if and only if for every $\varepsilon > 0$ there is a number $\delta > 0$ such that

$$\text{if } 0 < |t - a| < \delta \quad \text{then} \quad |\mathbf{r}(t) - \mathbf{b}| < \varepsilon$$

13.2 Derivatives and Integrals of Vector Functions

Later in this chapter we are going to use vector functions to describe the motion of planets and other objects through space. Here we prepare the way by developing the calculus of vector functions.

Derivatives

The derivative $\mathbf{r}'$ of a vector function $\mathbf{r}$ is defined in much the same way as for real-valued functions:

1
$$\frac{d\mathbf{r}}{dt} = \mathbf{r}'(t) = \lim_{h \to 0} \frac{\mathbf{r}(t+h) - \mathbf{r}(t)}{h}$$

if this limit exists. The geometric significance of this definition is shown in Figure 1. If the points P and Q have position vectors $\mathbf{r}(t)$ and $\mathbf{r}(t+h)$, then $\overrightarrow{PQ}$ represents the vector $\mathbf{r}(t+h) - \mathbf{r}(t)$, which can therefore be regarded as a secant vector. If $h > 0$, the scalar multiple $(1/h)(\mathbf{r}(t+h) - \mathbf{r}(t))$ has the same direction as $\mathbf{r}(t+h) - \mathbf{r}(t)$. As $h \to 0$, it appears that this vector approaches a vector that lies on the tangent line. For this reason, the vector $\mathbf{r}'(t)$ is called the **tangent vector** to the curve defined by $\mathbf{r}$ at the point P, provided that $\mathbf{r}'(t)$ exists and $\mathbf{r}'(t) \neq \mathbf{0}$. The **tangent line** to C at P is defined to be the line through P parallel to the tangent vector $\mathbf{r}'(t)$. We will also have occasion to consider the **unit tangent vector**, which is

Notice that when $0 < h < 1$, multiplying the secant vector by $1/h$ *stretches* the vector, as shown in Figure 1(b).

$$\mathbf{T}(t) = \frac{\mathbf{r}'(t)}{|\mathbf{r}'(t)|}$$

TEC Visual 13.2 shows an animation of Figure 1.

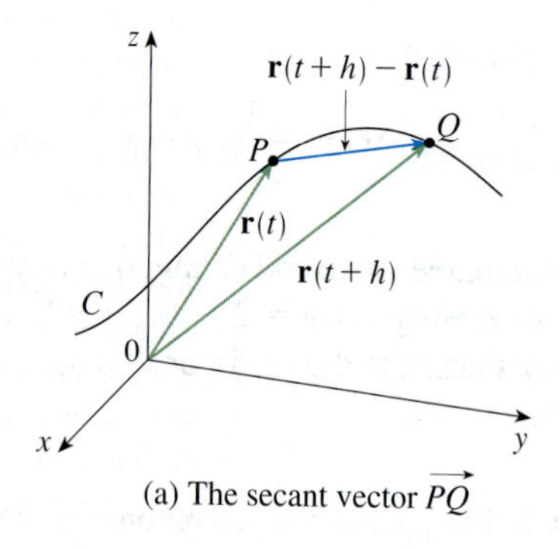

(a) The secant vector $\overrightarrow{PQ}$

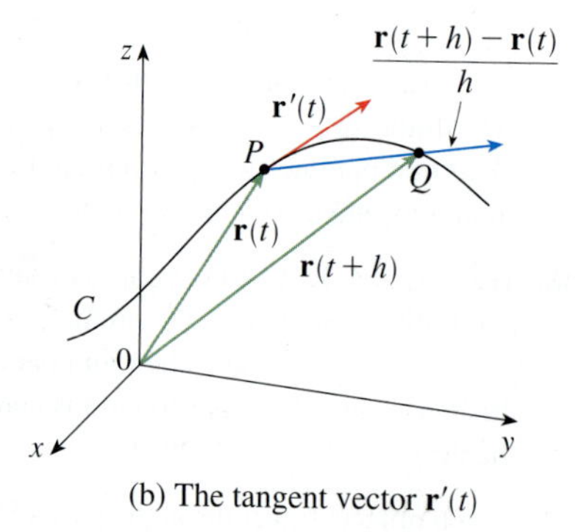

(b) The tangent vector $\mathbf{r}'(t)$

FIGURE 1

The following theorem gives us a convenient method for computing the derivative of a vector function $\mathbf{r}$: just differentiate each component of $\mathbf{r}$.

2 Theorem If $\mathbf{r}(t) = \langle f(t), g(t), h(t) \rangle = f(t)\,\mathbf{i} + g(t)\,\mathbf{j} + h(t)\,\mathbf{k}$, where f, g, and h are differentiable functions, then

$$\mathbf{r}'(t) = \langle f'(t), g'(t), h'(t) \rangle = f'(t)\,\mathbf{i} + g'(t)\,\mathbf{j} + h'(t)\,\mathbf{k}$$

PROOF

$$\mathbf{r}'(t) = \lim_{\Delta t \to 0} \frac{1}{\Delta t}[\mathbf{r}(t + \Delta t) - \mathbf{r}(t)]$$

$$= \lim_{\Delta t \to 0} \frac{1}{\Delta t}[\langle f(t + \Delta t), g(t + \Delta t), h(t + \Delta t) \rangle - \langle f(t), g(t), h(t) \rangle]$$

$$= \lim_{\Delta t \to 0} \left\langle \frac{f(t + \Delta t) - f(t)}{\Delta t}, \frac{g(t + \Delta t) - g(t)}{\Delta t}, \frac{h(t + \Delta t) - h(t)}{\Delta t} \right\rangle$$

$$= \left\langle \lim_{\Delta t \to 0} \frac{f(t + \Delta t) - f(t)}{\Delta t}, \lim_{\Delta t \to 0} \frac{g(t + \Delta t) - g(t)}{\Delta t}, \lim_{\Delta t \to 0} \frac{h(t + \Delta t) - h(t)}{\Delta t} \right\rangle$$

$$= \langle f'(t), g'(t), h'(t) \rangle$$

■

EXAMPLE 1
(a) Find the derivative of $\mathbf{r}(t) = (1 + t^3)\mathbf{i} + te^{-t}\mathbf{j} + \sin 2t\,\mathbf{k}$.
(b) Find the unit tangent vector at the point where $t = 0$.

SOLUTION
(a) According to Theorem 2, we differentiate each component of $\mathbf{r}$:

$$\mathbf{r}'(t) = 3t^2\,\mathbf{i} + (1 - t)e^{-t}\mathbf{j} + 2\cos 2t\,\mathbf{k}$$

(b) Since $\mathbf{r}(0) = \mathbf{i}$ and $\mathbf{r}'(0) = \mathbf{j} + 2\mathbf{k}$, the unit tangent vector at the point $(1, 0, 0)$ is

$$\mathbf{T}(0) = \frac{\mathbf{r}'(0)}{|\mathbf{r}'(0)|} = \frac{\mathbf{j} + 2\mathbf{k}}{\sqrt{1 + 4}} = \frac{1}{\sqrt{5}}\mathbf{j} + \frac{2}{\sqrt{5}}\mathbf{k}$$ ■

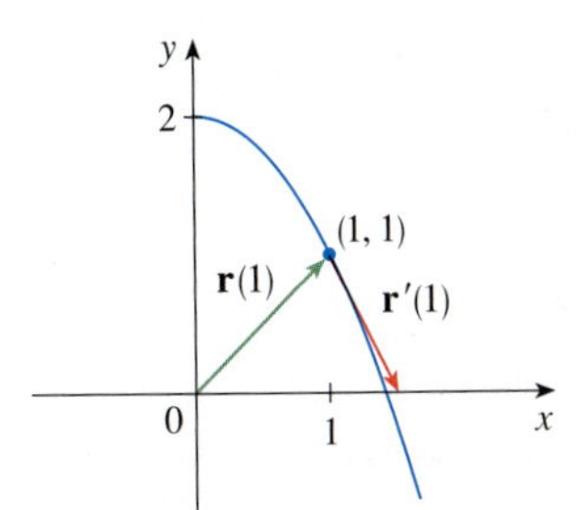

FIGURE 2

Notice from Figure 2 that the tangent vector points in the direction of increasing t. (See Exercise 58.)

EXAMPLE 2 For the curve $\mathbf{r}(t) = \sqrt{t}\,\mathbf{i} + (2 - t)\,\mathbf{j}$, find $\mathbf{r}'(t)$ and sketch the position vector $\mathbf{r}(1)$ and the tangent vector $\mathbf{r}'(1)$.

SOLUTION We have

$$\mathbf{r}'(t) = \frac{1}{2\sqrt{t}}\mathbf{i} - \mathbf{j} \qquad \text{and} \qquad \mathbf{r}'(1) = \frac{1}{2}\mathbf{i} - \mathbf{j}$$

The curve is a plane curve and elimination of the parameter from the equations $x = \sqrt{t}$, $y = 2 - t$ gives $y = 2 - x^2$, $x \geqslant 0$. In Figure 2 we draw the position vector $\mathbf{r}(1) = \mathbf{i} + \mathbf{j}$ starting at the origin and the tangent vector $\mathbf{r}'(1)$ starting at the corresponding point $(1, 1)$. ■

EXAMPLE 3 Find parametric equations for the tangent line to the helix with parametric equations

$$x = 2\cos t \qquad y = \sin t \qquad z = t$$

at the point $(0, 1, \pi/2)$.

SOLUTION The vector equation of the helix is $\mathbf{r}(t) = \langle 2\cos t, \sin t, t\rangle$, so

$$\mathbf{r}'(t) = \langle -2\sin t, \cos t, 1\rangle$$

The parameter value corresponding to the point $(0, 1, \pi/2)$ is $t = \pi/2$, so the tangent vector there is $\mathbf{r}'(\pi/2) = \langle -2, 0, 1\rangle$. The tangent line is the line through $(0, 1, \pi/2)$ parallel to the vector $\langle -2, 0, 1\rangle$, so by Equations 12.5.2 its parametric equations are

$$x = -2t \qquad y = 1 \qquad z = \frac{\pi}{2} + t$$ ■

The helix and the tangent line in Example 3 are shown in Figure 3.

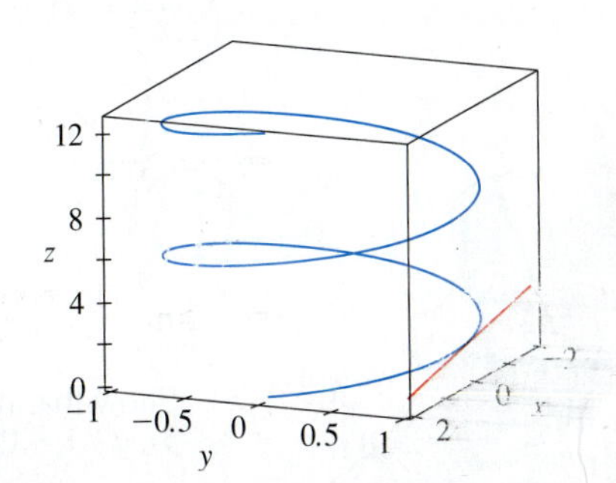

FIGURE 3

In Section 13.4 we will see how $\mathbf{r}'(t)$ and $\mathbf{r}''(t)$ can be interpreted as the velocity and acceleration vectors of a particle moving through space with position vector $\mathbf{r}(t)$ at time t.

Just as for real-valued functions, the **second derivative** of a vector function $\mathbf{r}$ is the derivative of $\mathbf{r}'$, that is, $\mathbf{r}'' = (\mathbf{r}')'$. For instance, the second derivative of the function in Example 3 is

$$\mathbf{r}''(t) = \langle -2\cos t, -\sin t, 0\rangle$$

Differentiation Rules

The next theorem shows that the differentiation formulas for real-valued functions have their counterparts for vector-valued functions.

3 Theorem Suppose $\mathbf{u}$ and $\mathbf{v}$ are differentiable vector functions, c is a scalar, and f is a real-valued function. Then

1. $\dfrac{d}{dt}[\mathbf{u}(t) + \mathbf{v}(t)] = \mathbf{u}'(t) + \mathbf{v}'(t)$
2. $\dfrac{d}{dt}[c\mathbf{u}(t)] = c\mathbf{u}'(t)$
3. $\dfrac{d}{dt}[f(t)\,\mathbf{u}(t)] = f'(t)\,\mathbf{u}(t) + f(t)\,\mathbf{u}'(t)$
4. $\dfrac{d}{dt}[\mathbf{u}(t) \cdot \mathbf{v}(t)] = \mathbf{u}'(t) \cdot \mathbf{v}(t) + \mathbf{u}(t) \cdot \mathbf{v}'(t)$
5. $\dfrac{d}{dt}[\mathbf{u}(t) \times \mathbf{v}(t)] = \mathbf{u}'(t) \times \mathbf{v}(t) + \mathbf{u}(t) \times \mathbf{v}'(t)$
6. $\dfrac{d}{dt}[\mathbf{u}(f(t))] = f'(t)\,\mathbf{u}'(f(t))$ (Chain Rule)

This theorem can be proved either directly from Definition 1 or by using Theorem 2 and the corresponding differentiation formulas for real-valued functions. The proof of Formula 4 follows; the remaining formulas are left as exercises.

PROOF OF FORMULA 4 Let

$$\mathbf{u}(t) = \langle f_1(t), f_2(t), f_3(t)\rangle \qquad \mathbf{v}(t) = \langle g_1(t), g_2(t), g_3(t)\rangle$$

Then

$$\mathbf{u}(t) \cdot \mathbf{v}(t) = f_1(t)\,g_1(t) + f_2(t)\,g_2(t) + f_3(t)\,g_3(t) = \sum_{i=1}^{3} f_i(t)\,g_i(t)$$

so the ordinary Product Rule gives

$$\begin{aligned}
\frac{d}{dt}[\mathbf{u}(t) \cdot \mathbf{v}(t)] &= \frac{d}{dt}\sum_{i=1}^{3} f_i(t)\,g_i(t) = \sum_{i=1}^{3} \frac{d}{dt}[f_i(t)\,g_i(t)] \\
&= \sum_{i=1}^{3} [f_i'(t)\,g_i(t) + f_i(t)\,g_i'(t)] \\
&= \sum_{i=1}^{3} f_i'(t)\,g_i(t) + \sum_{i=1}^{3} f_i(t)\,g_i'(t) \\
&= \mathbf{u}'(t) \cdot \mathbf{v}(t) + \mathbf{u}(t) \cdot \mathbf{v}'(t)
\end{aligned}$$

■

EXAMPLE 4 Show that if $|\mathbf{r}(t)| = c$ (a constant), then $\mathbf{r}'(t)$ is orthogonal to $\mathbf{r}(t)$ for all t.

SOLUTION Since

$$\mathbf{r}(t) \cdot \mathbf{r}(t) = |\mathbf{r}(t)|^2 = c^2$$

and c^2 is a constant, Formula 4 of Theorem 3 gives

$$0 = \frac{d}{dt}[\mathbf{r}(t) \cdot \mathbf{r}(t)] = \mathbf{r}'(t) \cdot \mathbf{r}(t) + \mathbf{r}(t) \cdot \mathbf{r}'(t) = 2\mathbf{r}'(t) \cdot \mathbf{r}(t)$$

Thus $\mathbf{r}'(t) \cdot \mathbf{r}(t) = 0$, which says that $\mathbf{r}'(t)$ is orthogonal to $\mathbf{r}(t)$.

Geometrically, this result says that if a curve lies on a sphere with center the origin, then the tangent vector $\mathbf{r}'(t)$ is always perpendicular to the position vector $\mathbf{r}(t)$. (See Figure 4.) ■

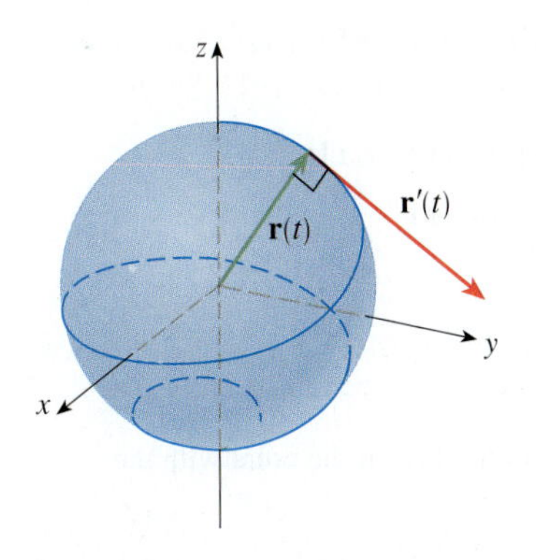

FIGURE 4

Integrals

The **definite integral** of a continuous vector function $\mathbf{r}(t)$ can be defined in much the same way as for real-valued functions except that the integral is a vector. But then we can express the integral of $\mathbf{r}$ in terms of the integrals of its component functions f, g, and h as follows. (We use the notation of Chapter 4.)

$$\int_a^b \mathbf{r}(t)\, dt = \lim_{n\to\infty} \sum_{i=1}^{n} \mathbf{r}(t_i^*)\, \Delta t$$

$$= \lim_{n\to\infty} \left[\left(\sum_{i=1}^{n} f(t_i^*)\, \Delta t \right) \mathbf{i} + \left(\sum_{i=1}^{n} g(t_i^*)\, \Delta t \right) \mathbf{j} + \left(\sum_{i=1}^{n} h(t_i^*)\, \Delta t \right) \mathbf{k} \right]$$

and so

$$\int_a^b \mathbf{r}(t)\, dt = \left(\int_a^b f(t)\, dt \right) \mathbf{i} + \left(\int_a^b g(t)\, dt \right) \mathbf{j} + \left(\int_a^b h(t)\, dt \right) \mathbf{k}$$

This means that we can evaluate an integral of a vector function by integrating each component function.

We can extend the Fundamental Theorem of Calculus to continuous vector functions as follows:

$$\int_a^b \mathbf{r}(t)\, dt = \mathbf{R}(t)\Big]_a^b = \mathbf{R}(b) - \mathbf{R}(a)$$

where $\mathbf{R}$ is an antiderivative of $\mathbf{r}$, that is, $\mathbf{R}'(t) = \mathbf{r}(t)$. We use the notation $\int \mathbf{r}(t)\, dt$ for indefinite integrals (antiderivatives).

EXAMPLE 5 If $\mathbf{r}(t) = 2\cos t\, \mathbf{i} + \sin t\, \mathbf{j} + 2t\, \mathbf{k}$, then

$$\int \mathbf{r}(t)\, dt = \left(\int 2\cos t\, dt \right) \mathbf{i} + \left(\int \sin t\, dt \right) \mathbf{j} + \left(\int 2t\, dt \right) \mathbf{k}$$

$$= 2\sin t\, \mathbf{i} - \cos t\, \mathbf{j} + t^2\, \mathbf{k} + \mathbf{C}$$

where $\mathbf{C}$ is a vector constant of integration, and

$$\int_0^{\pi/2} \mathbf{r}(t)\, dt = \Big[2\sin t\, \mathbf{i} - \cos t\, \mathbf{j} + t^2\, \mathbf{k} \Big]_0^{\pi/2} = 2\,\mathbf{i} + \mathbf{j} + \frac{\pi^2}{4}\,\mathbf{k}$$ ■

13.2 EXERCISES

1. The figure shows a curve C given by a vector function $\mathbf{r}(t)$.
(a) Draw the vectors $\mathbf{r}(4.5) - \mathbf{r}(4)$ and $\mathbf{r}(4.2) - \mathbf{r}(4)$.
(b) Draw the vectors

$$\frac{\mathbf{r}(4.5) - \mathbf{r}(4)}{0.5} \quad \text{and} \quad \frac{\mathbf{r}(4.2) - \mathbf{r}(4)}{0.2}$$

(c) Write expressions for $\mathbf{r}'(4)$ and the unit tangent vector $\mathbf{T}(4)$.
(d) Draw the vector $\mathbf{T}(4)$.

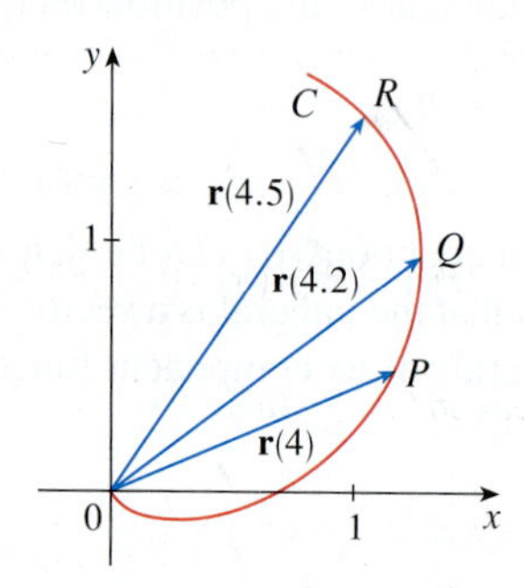

2. (a) Make a large sketch of the curve described by the vector function $\mathbf{r}(t) = \langle t^2, t \rangle$, $0 \leqslant t \leqslant 2$, and draw the vectors $\mathbf{r}(1)$, $\mathbf{r}(1.1)$, and $\mathbf{r}(1.1) - \mathbf{r}(1)$.
(b) Draw the vector $\mathbf{r}'(1)$ starting at $(1, 1)$, and compare it with the vector

$$\frac{\mathbf{r}(1.1) - \mathbf{r}(1)}{0.1}$$

Explain why these vectors are so close to each other in length and direction.

3–8
(a) Sketch the plane curve with the given vector equation.
(b) Find $\mathbf{r}'(t)$.
(c) Sketch the position vector $\mathbf{r}(t)$ and the tangent vector $\mathbf{r}'(t)$ for the given value of t.

3. $\mathbf{r}(t) = \langle t - 2, t^2 + 1 \rangle$, $t = -1$

4. $\mathbf{r}(t) = \langle t^2, t^3 \rangle$, $t = 1$

5. $\mathbf{r}(t) = e^{2t}\,\mathbf{i} + e^t\,\mathbf{j}$, $t = 0$

6. $\mathbf{r}(t) = e^t\mathbf{i} + 2t\,\mathbf{j}$, $t = 0$

7. $\mathbf{r}(t) = 4 \sin t\,\mathbf{i} - 2 \cos t\,\mathbf{j}$, $t = 3\pi/4$

8. $\mathbf{r}(t) = (\cos t + 1)\,\mathbf{i} + (\sin t - 1)\,\mathbf{j}$, $t = -\pi/3$

9–16 Find the derivative of the vector function.

9. $\mathbf{r}(t) = \left\langle \sqrt{t - 2}, 3, 1/t^2 \right\rangle$

10. $\mathbf{r}(t) = \langle e^{-t}, t - t^3, \ln t \rangle$

11. $\mathbf{r}(t) = t^2\,\mathbf{i} + \cos(t^2)\,\mathbf{j} + \sin^2 t\,\mathbf{k}$

12. $\mathbf{r}(t) = \dfrac{1}{1 + t}\mathbf{i} + \dfrac{t}{1 + t}\mathbf{j} + \dfrac{t^2}{1 + t}\mathbf{k}$

13. $\mathbf{r}(t) = t \sin t\,\mathbf{i} + e^t \cos t\,\mathbf{j} + \sin t \cos t\,\mathbf{k}$

14. $\mathbf{r}(t) = \sin^2 at\,\mathbf{i} + te^{bt}\,\mathbf{j} + \cos^2 ct\,\mathbf{k}$

15. $\mathbf{r}(t) = \mathbf{a} + t\,\mathbf{b} + t^2\mathbf{c}$

16. $\mathbf{r}(t) = t\,\mathbf{a} \times (\mathbf{b} + t\,\mathbf{c})$

17–20 Find the unit tangent vector $\mathbf{T}(t)$ at the point with the given value of the parameter t.

17. $\mathbf{r}(t) = \left\langle t^2 - 2t, 1 + 3t, \frac{1}{3}t^3 + \frac{1}{2}t^2 \right\rangle$, $t = 2$

18. $\mathbf{r}(t) = \langle \tan^{-1}t, 2e^{2t}, 8te^t \rangle$, $t = 0$

19. $\mathbf{r}(t) = \cos t\,\mathbf{i} + 3t\,\mathbf{j} + 2 \sin 2t\,\mathbf{k}$, $t = 0$

20. $\mathbf{r}(t) = \sin^2 t\,\mathbf{i} + \cos^2 t\,\mathbf{j} + \tan^2 t\,\mathbf{k}$, $t = \pi/4$

21. If $\mathbf{r}(t) = \langle t, t^2, t^3 \rangle$, find $\mathbf{r}'(t)$, $\mathbf{T}(1)$, $\mathbf{r}''(t)$, and $\mathbf{r}'(t) \times \mathbf{r}''(t)$.

22. If $\mathbf{r}(t) = \langle e^{2t}, e^{-2t}, te^{2t} \rangle$, find $\mathbf{T}(0)$, $\mathbf{r}''(0)$, and $\mathbf{r}'(t) \cdot \mathbf{r}''(t)$.

23–26 Find parametric equations for the tangent line to the curve with the given parametric equations at the specified point.

23. $x = t^2 + 1$, $y = 4\sqrt{t}$, $z = e^{t^2 - t}$; $(2, 4, 1)$

24. $x = \ln(t + 1)$, $y = t \cos 2t$, $z = 2^t$; $(0, 0, 1)$

25. $x = e^{-t} \cos t$, $y = e^{-t} \sin t$, $z = e^{-t}$; $(1, 0, 1)$

26. $x = \sqrt{t^2 + 3}$, $y = \ln(t^2 + 3)$, $z = t$; $(2, \ln 4, 1)$

27. Find a vector equation for the tangent line to the curve of intersection of the cylinders $x^2 + y^2 = 25$ and $y^2 + z^2 = 20$ at the point $(3, 4, 2)$.

28. Find the point on the curve $\mathbf{r}(t) = \langle 2 \cos t, 2 \sin t, e^t \rangle$, $0 \leqslant t \leqslant \pi$, where the tangent line is parallel to the plane $\sqrt{3}x + y = 1$.

CAS **29–31** Find parametric equations for the tangent line to the curve with the given parametric equations at the specified point. Illustrate by graphing both the curve and the tangent line on a common screen.

29. $x = t$, $y = e^{-t}$, $z = 2t - t^2$; $(0, 1, 0)$

30. $x = 2 \cos t$, $y = 2 \sin t$, $z = 4 \cos 2t$; $\left(\sqrt{3}, 1, 2\right)$

31. $x = t \cos t$, $y = t$, $z = t \sin t$; $(-\pi, \pi, 0)$

32. (a) Find the point of intersection of the tangent lines to the curve $\mathbf{r}(t) = \langle \sin \pi t, 2 \sin \pi t, \cos \pi t \rangle$ at the points where $t = 0$ and $t = 0.5$.
(b) Illustrate by graphing the curve and both tangent lines.

33. The curves $\mathbf{r}_1(t) = \langle t, t^2, t^3 \rangle$ and $\mathbf{r}_2(t) = \langle \sin t, \sin 2t, t \rangle$ intersect at the origin. Find their angle of intersection correct to the nearest degree.

34. At what point do the curves $\mathbf{r}_1(t) = \langle t, 1 - t, 3 + t^2 \rangle$ and $\mathbf{r}_2(s) = \langle 3 - s, s - 2, s^2 \rangle$ intersect? Find their angle of intersection correct to the nearest degree.

35–40 Evaluate the integral.

35. $\int_0^2 (t\,\mathbf{i} - t^3\,\mathbf{j} + 3t^5\,\mathbf{k})\,dt$

36. $\int_1^4 (2t^{3/2}\,\mathbf{i} + (t + 1)\sqrt{t}\,\mathbf{k})\,dt$

37. $\int_0^1 \left(\frac{1}{t+1}\,\mathbf{i} + \frac{1}{t^2+1}\,\mathbf{j} + \frac{t}{t^2+1}\,\mathbf{k} \right) dt$

38. $\int_0^{\pi/4} (\sec t \tan t\,\mathbf{i} + t \cos 2t\,\mathbf{j} + \sin^2 2t \cos 2t\,\mathbf{k})\,dt$

39. $\int (\sec^2 t\,\mathbf{i} + t(t^2 + 1)^3\,\mathbf{j} + t^2 \ln t\,\mathbf{k})\,dt$

40. $\int \left(te^{2t}\,\mathbf{i} + \frac{t}{1-t}\,\mathbf{j} + \frac{1}{\sqrt{1-t^2}}\,\mathbf{k} \right) dt$

---

41. Find $\mathbf{r}(t)$ if $\mathbf{r}'(t) = 2t\,\mathbf{i} + 3t^2\,\mathbf{j} + \sqrt{t}\,\mathbf{k}$ and $\mathbf{r}(1) = \mathbf{i} + \mathbf{j}$.

42. Find $\mathbf{r}(t)$ if $\mathbf{r}'(t) = t\,\mathbf{i} + e^t\,\mathbf{j} + te^t\,\mathbf{k}$ and $\mathbf{r}(0) = \mathbf{i} + \mathbf{j} + \mathbf{k}$.

43. Prove Formula 1 of Theorem 3.

44. Prove Formula 3 of Theorem 3.

45. Prove Formula 5 of Theorem 3.

46. Prove Formula 6 of Theorem 3.

47. If $\mathbf{u}(t) = \langle \sin t, \cos t, t \rangle$ and $\mathbf{v}(t) = \langle t, \cos t, \sin t \rangle$, use Formula 4 of Theorem 3 to find

$$\frac{d}{dt}[\mathbf{u}(t) \cdot \mathbf{v}(t)]$$

48. If $\mathbf{u}$ and $\mathbf{v}$ are the vector functions in Exercise 47, use Formula 5 of Theorem 3 to find

$$\frac{d}{dt}[\mathbf{u}(t) \times \mathbf{v}(t)]$$

49. Find $f'(2)$, where $f(t) = \mathbf{u}(t) \cdot \mathbf{v}(t)$, $\mathbf{u}(2) = \langle 1, 2, -1 \rangle$, $\mathbf{u}'(2) = \langle 3, 0, 4 \rangle$, and $\mathbf{v}(t) = \langle t, t^2, t^3 \rangle$.

50. If $\mathbf{r}(t) = \mathbf{u}(t) \times \mathbf{v}(t)$, where $\mathbf{u}$ and $\mathbf{v}$ are the vector functions in Exercise 49, find $\mathbf{r}'(2)$.

51. If $\mathbf{r}(t) = \mathbf{a} \cos \omega t + \mathbf{b} \sin \omega t$, where $\mathbf{a}$ and $\mathbf{b}$ are constant vectors, show that $\mathbf{r}(t) \times \mathbf{r}'(t) = \omega\mathbf{a} \times \mathbf{b}$.

52. If $\mathbf{r}$ is the vector function in Exercise 51, show that $\mathbf{r}''(t) + \omega^2\mathbf{r}(t) = \mathbf{0}$.

53. Show that if $\mathbf{r}$ is a vector function such that $\mathbf{r}''$ exists, then

$$\frac{d}{dt}[\mathbf{r}(t) \times \mathbf{r}'(t)] = \mathbf{r}(t) \times \mathbf{r}''(t)$$

54. Find an expression for $\frac{d}{dt}[\mathbf{u}(t) \cdot (\mathbf{v}(t) \times \mathbf{w}(t))]$.

55. If $\mathbf{r}(t) \neq \mathbf{0}$, show that $\frac{d}{dt}|\mathbf{r}(t)| = \frac{1}{|\mathbf{r}(t)|}\mathbf{r}(t) \cdot \mathbf{r}'(t)$.

[*Hint:* $|\mathbf{r}(t)|^2 = \mathbf{r}(t) \cdot \mathbf{r}(t)$]

56. If a curve has the property that the position vector $\mathbf{r}(t)$ is always perpendicular to the tangent vector $\mathbf{r}'(t)$, show that the curve lies on a sphere with center the origin.

57. If $\mathbf{u}(t) = \mathbf{r}(t) \cdot [\mathbf{r}'(t) \times \mathbf{r}''(t)]$, show that

$$\mathbf{u}'(t) = \mathbf{r}(t) \cdot [\mathbf{r}'(t) \times \mathbf{r}'''(t)]$$

58. Show that the tangent vector to a curve defined by a vector function $\mathbf{r}(t)$ points in the direction of increasing t. [*Hint:* Refer to Figure 1 and consider the cases $h > 0$ and $h < 0$ separately.]

13.3 Arc Length and Curvature

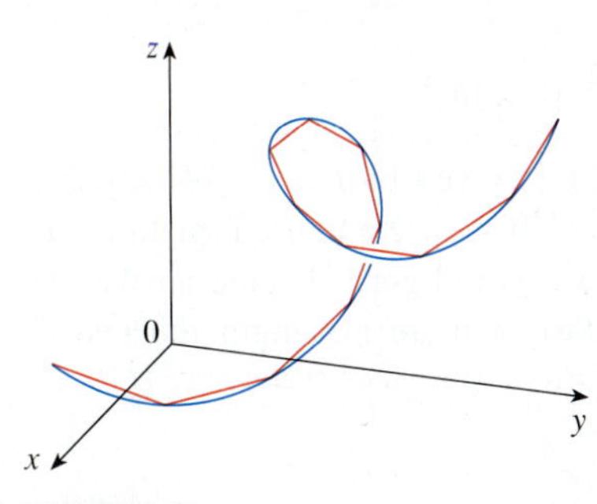

FIGURE 1
The length of a space curve is the limit of lengths of inscribed polygons.

Length of a Curve

In Section 10.2 we defined the length of a plane curve with parametric equations $x = f(t)$, $y = g(t)$, $a \leqslant t \leqslant b$, as the limit of lengths of inscribed polygons and, for the case where f' and g' are continuous, we arrived at the formula

$$\boxed{1} \qquad L = \int_a^b \sqrt{[f'(t)]^2 + [g'(t)]^2}\,dt = \int_a^b \sqrt{\left(\frac{dx}{dt}\right)^2 + \left(\frac{dy}{dt}\right)^2}\,dt$$

The length of a space curve is defined in exactly the same way (see Figure 1). Suppose that the curve has the vector equation $\mathbf{r}(t) = \langle f(t), g(t), h(t) \rangle$, $a \leqslant t \leqslant b$, or, equivalently, the parametric equations $x = f(t)$, $y = g(t)$, $z = h(t)$, where f', g', and h' are continuous. If the curve is traversed exactly once as t increases from a to b, then it can be shown that its length is

2
$$L = \int_a^b \sqrt{[f'(t)]^2 + [g'(t)]^2 + [h'(t)]^2}\, dt$$
$$= \int_a^b \sqrt{\left(\frac{dx}{dt}\right)^2 + \left(\frac{dy}{dt}\right)^2 + \left(\frac{dz}{dt}\right)^2}\, dt$$

Notice that both of the arc length formulas (1) and (2) can be put into the more compact form

3
$$L = \int_a^b |\mathbf{r}'(t)|\, dt$$

In the next section we will see that if $\mathbf{r}(t)$ is the position vector of a moving object at time t, then $\mathbf{r}'(t)$ is the velocity vector and $|\mathbf{r}'(t)|$ is the speed. Thus Equation 3 says that to compute distance traveled, we integrate speed.

because, for plane curves $\mathbf{r}(t) = f(t)\,\mathbf{i} + g(t)\,\mathbf{j}$,

$$|\mathbf{r}'(t)| = |f'(t)\,\mathbf{i} + g'(t)\,\mathbf{j}| = \sqrt{[f'(t)]^2 + [g'(t)]^2}$$

and for space curves $\mathbf{r}(t) = f(t)\,\mathbf{i} + g(t)\,\mathbf{j} + h(t)\,\mathbf{k}$,

$$|\mathbf{r}'(t)| = |f'(t)\,\mathbf{i} + g'(t)\,\mathbf{j} + h'(t)\,\mathbf{k}| = \sqrt{[f'(t)]^2 + [g'(t)]^2 + [h'(t)]^2}$$

Figure 2 shows the arc of the helix whose length is computed in Example 1.

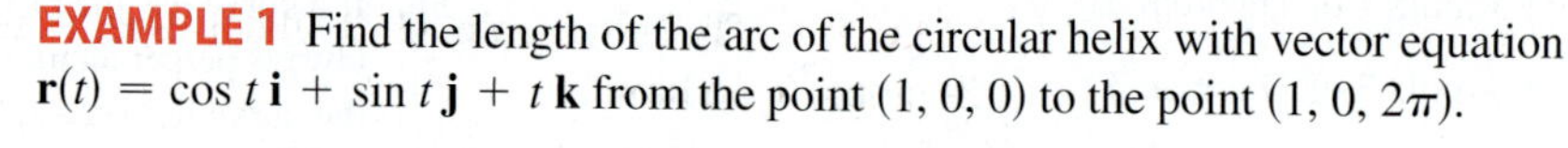

EXAMPLE 1 Find the length of the arc of the circular helix with vector equation $\mathbf{r}(t) = \cos t\,\mathbf{i} + \sin t\,\mathbf{j} + t\,\mathbf{k}$ from the point $(1, 0, 0)$ to the point $(1, 0, 2\pi)$.

SOLUTION Since $\mathbf{r}'(t) = -\sin t\,\mathbf{i} + \cos t\,\mathbf{j} + \mathbf{k}$, we have

$$|\mathbf{r}'(t)| = \sqrt{(-\sin t)^2 + \cos^2 t + 1} = \sqrt{2}$$

The arc from $(1, 0, 0)$ to $(1, 0, 2\pi)$ is described by the parameter interval $0 \leqslant t \leqslant 2\pi$ and so, from Formula 3, we have

$$L = \int_0^{2\pi} |\mathbf{r}'(t)|\, dt = \int_0^{2\pi} \sqrt{2}\, dt = 2\sqrt{2}\pi$$ ■

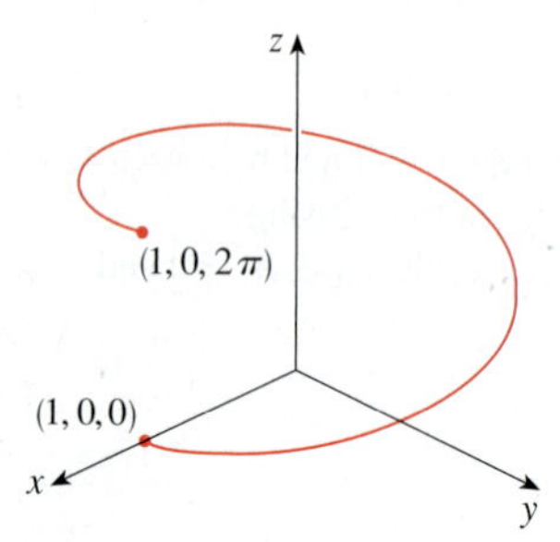

FIGURE 2

A single curve C can be represented by more than one vector function. For instance, the twisted cubic

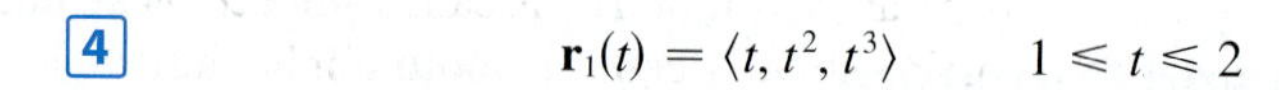

4
$$\mathbf{r}_1(t) = \langle t, t^2, t^3 \rangle \qquad 1 \leqslant t \leqslant 2$$

could also be represented by the function

5
$$\mathbf{r}_2(u) = \langle e^u, e^{2u}, e^{3u} \rangle \qquad 0 \leqslant u \leqslant \ln 2$$

where the connection between the parameters t and u is given by $t = e^u$. We say that Equations 4 and 5 are **parametrizations** of the curve C. If we were to use Equation 3 to compute the length of C using Equations 4 and 5, we would get the same answer. In general, it can be shown that when Equation 3 is used to compute arc length, the answer is independent of the parametrization that is used.

The Arc Length Function

Now we suppose that C is a curve given by a vector function

$$\mathbf{r}(t) = f(t)\,\mathbf{i} + g(t)\,\mathbf{j} + h(t)\,\mathbf{k} \qquad a \leqslant t \leqslant b$$

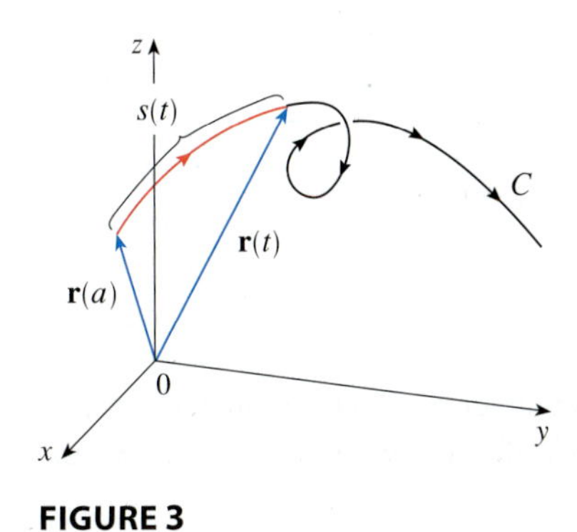

FIGURE 3

where $\mathbf{r}'$ is continuous and C is traversed exactly once as t increases from a to b. We define its **arc length function** s by

$$\boxed{6} \qquad s(t) = \int_a^t |\mathbf{r}'(u)|\,du = \int_a^t \sqrt{\left(\frac{dx}{du}\right)^2 + \left(\frac{dy}{du}\right)^2 + \left(\frac{dz}{du}\right)^2}\,du$$

Thus $s(t)$ is the length of the part of C between $\mathbf{r}(a)$ and $\mathbf{r}(t)$. (See Figure 3.) If we differentiate both sides of Equation 6 using Part 1 of the Fundamental Theorem of Calculus, we obtain

$$\boxed{7} \qquad \frac{ds}{dt} = |\mathbf{r}'(t)|$$

It is often useful to **parametrize a curve with respect to arc length** because arc length arises naturally from the shape of the curve and does not depend on a particular coordinate system. If a curve $\mathbf{r}(t)$ is already given in terms of a parameter t and $s(t)$ is the arc length function given by Equation 6, then we may be able to solve for t as a function of s: $t = t(s)$. Then the curve can be reparametrized in terms of s by substituting for t: $\mathbf{r} = \mathbf{r}(t(s))$. Thus, if $s = 3$ for instance, $\mathbf{r}(t(3))$ is the position vector of the point 3 units of length along the curve from its starting point.

EXAMPLE 2 Reparametrize the helix $\mathbf{r}(t) = \cos t\,\mathbf{i} + \sin t\,\mathbf{j} + t\,\mathbf{k}$ with respect to arc length measured from $(1, 0, 0)$ in the direction of increasing t.

SOLUTION The initial point $(1, 0, 0)$ corresponds to the parameter value $t = 0$. From Example 1 we have

$$\frac{ds}{dt} = |\mathbf{r}'(t)| = \sqrt{2}$$

and so
$$s = s(t) = \int_0^t |\mathbf{r}'(u)|\,du = \int_0^t \sqrt{2}\,du = \sqrt{2}\,t$$

Therefore $t = s/\sqrt{2}$ and the required reparametrization is obtained by substituting for t:

$$\mathbf{r}(t(s)) = \cos\!\left(s/\sqrt{2}\right)\mathbf{i} + \sin\!\left(s/\sqrt{2}\right)\mathbf{j} + \left(s/\sqrt{2}\right)\mathbf{k}$$

■

Curvature

TEC Visual 13.3A shows animated unit tangent vectors, like those in Figure 4, for a variety of plane curves and space curves.

A parametrization $\mathbf{r}(t)$ is called **smooth** on an interval I if $\mathbf{r}'$ is continuous and $\mathbf{r}'(t) \neq \mathbf{0}$ on I. A curve is called **smooth** if it has a smooth parametrization. A smooth curve has no sharp corners or cusps; when the tangent vector turns, it does so continuously.

If C is a smooth curve defined by the vector function $\mathbf{r}$, recall that the unit tangent vector $\mathbf{T}(t)$ is given by

$$\mathbf{T}(t) = \frac{\mathbf{r}'(t)}{|\mathbf{r}'(t)|}$$

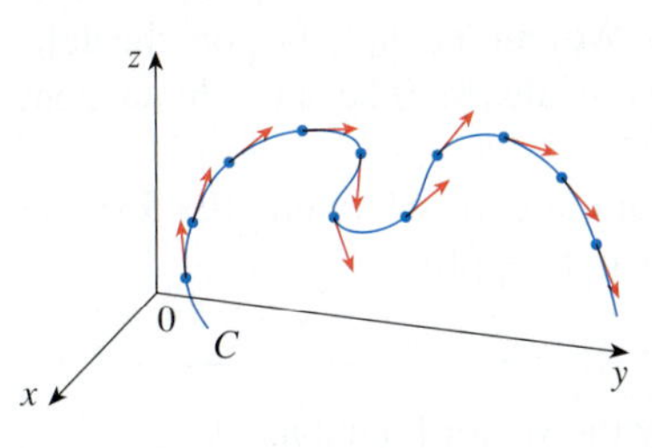

FIGURE 4
Unit tangent vectors at equally spaced points on C

and indicates the direction of the curve. From Figure 4 you can see that $\mathbf{T}(t)$ changes direction very slowly when C is fairly straight, but it changes direction more quickly when C bends or twists more sharply.

The curvature of C at a given point is a measure of how quickly the curve changes direction at that point. Specifically, we define it to be the magnitude of the rate of change of the unit tangent vector with respect to arc length. (We use arc length so that the curvature will be independent of the parametrization.) Because the unit tangent vector has constant length, only changes in direction contribute to the rate of change of $\mathbf{T}$.

8 Definition The **curvature** of a curve is

$$\kappa = \left| \frac{d\mathbf{T}}{ds} \right|$$

where $\mathbf{T}$ is the unit tangent vector.

The curvature is easier to compute if it is expressed in terms of the parameter t instead of s, so we use the Chain Rule (Theorem 13.2.3, Formula 6) to write

$$\frac{d\mathbf{T}}{dt} = \frac{d\mathbf{T}}{ds}\frac{ds}{dt} \qquad \text{and} \qquad \kappa = \left| \frac{d\mathbf{T}}{ds} \right| = \left| \frac{d\mathbf{T}/dt}{ds/dt} \right|$$

But $ds/dt = |\mathbf{r}'(t)|$ from Equation 7, so

9
$$\kappa(t) = \frac{|\mathbf{T}'(t)|}{|\mathbf{r}'(t)|}$$

EXAMPLE 3 Show that the curvature of a circle of radius a is $1/a$.

SOLUTION We can take the circle to have center the origin, and then a parametrization is

$$\mathbf{r}(t) = a \cos t\, \mathbf{i} + a \sin t\, \mathbf{j}$$

Therefore $\quad \mathbf{r}'(t) = -a \sin t\, \mathbf{i} + a \cos t\, \mathbf{j} \quad$ and $\quad |\mathbf{r}'(t)| = a$

so
$$\mathbf{T}(t) = \frac{\mathbf{r}'(t)}{|\mathbf{r}'(t)|} = -\sin t\, \mathbf{i} + \cos t\, \mathbf{j}$$

and
$$\mathbf{T}'(t) = -\cos t\, \mathbf{i} - \sin t\, \mathbf{j}$$

This gives $|\mathbf{T}'(t)| = 1$, so using Formula 9, we have

$$\kappa(t) = \frac{|\mathbf{T}'(t)|}{|\mathbf{r}'(t)|} = \frac{1}{a}$$ ■

The result of Example 3 shows that small circles have large curvature and large circles have small curvature, in accordance with our intuition. We can see directly from the definition of curvature that the curvature of a straight line is always 0 because the tangent vector is constant.

Although Formula 9 can be used in all cases to compute the curvature, the formula given by the following theorem is often more convenient to apply.

10 Theorem The curvature of the curve given by the vector function $\mathbf{r}$ is

$$\kappa(t) = \frac{|\mathbf{r}'(t) \times \mathbf{r}''(t)|}{|\mathbf{r}'(t)|^3}$$

PROOF Since $\mathbf{T} = \mathbf{r}'/|\mathbf{r}'|$ and $|\mathbf{r}'| = ds/dt$, we have

$$\mathbf{r}' = |\mathbf{r}'|\mathbf{T} = \frac{ds}{dt}\mathbf{T}$$

so the Product Rule (Theorem 13.2.3, Formula 3) gives

$$\mathbf{r}'' = \frac{d^2s}{dt^2}\mathbf{T} + \frac{ds}{dt}\mathbf{T}'$$

Using the fact that $\mathbf{T} \times \mathbf{T} = \mathbf{0}$ (see Example 12.4.2), we have

$$\mathbf{r}' \times \mathbf{r}'' = \left(\frac{ds}{dt}\right)^2 (\mathbf{T} \times \mathbf{T}')$$

Now $|\mathbf{T}(t)| = 1$ for all t, so $\mathbf{T}$ and $\mathbf{T}'$ are orthogonal by Example 13.2.4. Therefore, by Theorem 12.4.9,

$$|\mathbf{r}' \times \mathbf{r}''| = \left(\frac{ds}{dt}\right)^2 |\mathbf{T} \times \mathbf{T}'| = \left(\frac{ds}{dt}\right)^2 |\mathbf{T}||\mathbf{T}'| = \left(\frac{ds}{dt}\right)^2 |\mathbf{T}'|$$

Thus
$$|\mathbf{T}'| = \frac{|\mathbf{r}' \times \mathbf{r}''|}{(ds/dt)^2} = \frac{|\mathbf{r}' \times \mathbf{r}''|}{|\mathbf{r}'|^2}$$

and
$$\kappa = \frac{|\mathbf{T}'|}{|\mathbf{r}'|} = \frac{|\mathbf{r}' \times \mathbf{r}''|}{|\mathbf{r}'|^3}$$ ■

EXAMPLE 4 Find the curvature of the twisted cubic $\mathbf{r}(t) = \langle t, t^2, t^3 \rangle$ at a general point and at (0, 0, 0).

SOLUTION We first compute the required ingredients:

$$\mathbf{r}'(t) = \langle 1, 2t, 3t^2 \rangle \qquad \mathbf{r}''(t) = \langle 0, 2, 6t \rangle$$

$$|\mathbf{r}'(t)| = \sqrt{1 + 4t^2 + 9t^4}$$

$$\mathbf{r}'(t) \times \mathbf{r}''(t) = \begin{vmatrix} \mathbf{i} & \mathbf{j} & \mathbf{k} \\ 1 & 2t & 3t^2 \\ 0 & 2 & 6t \end{vmatrix} = 6t^2\,\mathbf{i} - 6t\,\mathbf{j} + 2\,\mathbf{k}$$

$$|\mathbf{r}'(t) \times \mathbf{r}''(t)| = \sqrt{36t^4 + 36t^2 + 4} = 2\sqrt{9t^4 + 9t^2 + 1}$$

Theorem 10 then gives

$$\kappa(t) = \frac{|\mathbf{r}'(t) \times \mathbf{r}''(t)|}{|\mathbf{r}'(t)|^3} = \frac{2\sqrt{1 + 9t^2 + 9t^4}}{(1 + 4t^2 + 9t^4)^{3/2}}$$

At the origin, where $t = 0$, the curvature is $\kappa(0) = 2$. ■

For the special case of a plane curve with equation $y = f(x)$, we choose x as the parameter and write $\mathbf{r}(x) = x\,\mathbf{i} + f(x)\,\mathbf{j}$. Then $\mathbf{r}'(x) = \mathbf{i} + f'(x)\,\mathbf{j}$ and $\mathbf{r}''(x) = f''(x)\,\mathbf{j}$. Since $\mathbf{i} \times \mathbf{j} = \mathbf{k}$ and $\mathbf{j} \times \mathbf{j} = \mathbf{0}$, it follows that $\mathbf{r}'(x) \times \mathbf{r}''(x) = f''(x)\,\mathbf{k}$. We also have $|\mathbf{r}'(x)| = \sqrt{1 + [f'(x)]^2}$ and so, by Theorem 10,

11
$$\kappa(x) = \frac{|f''(x)|}{[1 + (f'(x))^2]^{3/2}}$$

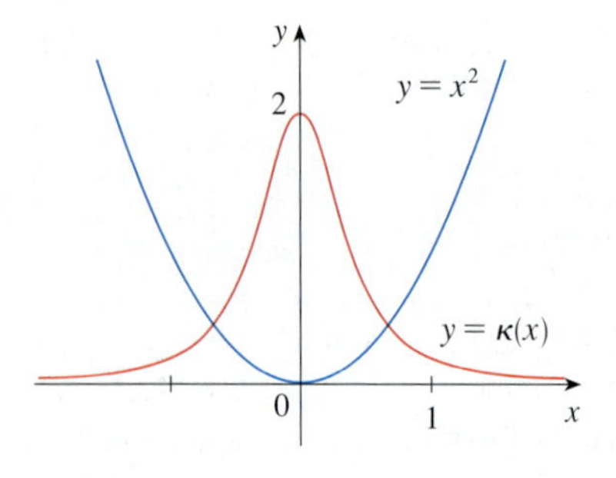

FIGURE 5
The parabola $y = x^2$ and its curvature function

EXAMPLE 5 Find the curvature of the parabola $y = x^2$ at the points (0, 0), (1, 1), and (2, 4).

SOLUTION Since $y' = 2x$ and $y'' = 2$, Formula 11 gives

$$\kappa(x) = \frac{|y''|}{[1 + (y')^2]^{3/2}} = \frac{2}{(1 + 4x^2)^{3/2}}$$

The curvature at (0, 0) is $\kappa(0) = 2$. At (1, 1) it is $\kappa(1) = 2/5^{3/2} \approx 0.18$. At (2, 4) it is $\kappa(2) = 2/17^{3/2} \approx 0.03$. Observe from the expression for $\kappa(x)$ or the graph of κ in Figure 5 that $\kappa(x) \to 0$ as $x \to \pm\infty$. This corresponds to the fact that the parabola appears to become flatter as $x \to \pm\infty$. ■

The Normal and Binormal Vectors

At a given point on a smooth space curve $\mathbf{r}(t)$, there are many vectors that are orthogonal to the unit tangent vector $\mathbf{T}(t)$. We single out one by observing that, because $|\mathbf{T}(t)| = 1$ for all t, we have $\mathbf{T}(t) \cdot \mathbf{T}'(t) = 0$ by Example 13.2.4, so $\mathbf{T}'(t)$ is orthogonal to $\mathbf{T}(t)$. Note that, typically, $\mathbf{T}'(t)$ is itself not a unit vector. But at any point where $\kappa \neq 0$ we can define the **principal unit normal vector** $\mathbf{N}(t)$ (or simply **unit normal**) as

$$\mathbf{N}(t) = \frac{\mathbf{T}'(t)}{|\mathbf{T}'(t)|}$$

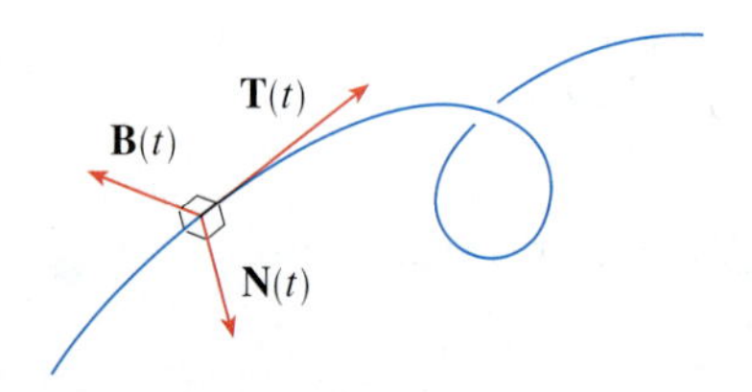

FIGURE 6

We can think of the unit normal vector as indicating the direction in which the curve is turning at each point. The vector $\mathbf{B}(t) = \mathbf{T}(t) \times \mathbf{N}(t)$ is called the **binormal vector**. It is perpendicular to both $\mathbf{T}$ and $\mathbf{N}$ and is also a unit vector. (See Figure 6.)

Figure 7 illustrates Example 6 by showing the vectors **T**, **N**, and **B** at two locations on the helix. In general, the vectors **T**, **N**, and **B**, starting at the various points on a curve, form a set of orthogonal vectors, called the **TNB** frame, that moves along the curve as t varies. This **TNB** frame plays an important role in the branch of mathematics known as differential geometry and in its applications to the motion of spacecraft.

EXAMPLE 6 Find the unit normal and binormal vectors for the circular helix

$$\mathbf{r}(t) = \cos t\,\mathbf{i} + \sin t\,\mathbf{j} + t\,\mathbf{k}$$

SOLUTION We first compute the ingredients needed for the unit normal vector:

$$\mathbf{r}'(t) = -\sin t\,\mathbf{i} + \cos t\,\mathbf{j} + \mathbf{k} \qquad |\mathbf{r}'(t)| = \sqrt{2}$$

$$\mathbf{T}(t) = \frac{\mathbf{r}'(t)}{|\mathbf{r}'(t)|} = \frac{1}{\sqrt{2}}(-\sin t\,\mathbf{i} + \cos t\,\mathbf{j} + \mathbf{k})$$

$$\mathbf{T}'(t) = \frac{1}{\sqrt{2}}(-\cos t\,\mathbf{i} - \sin t\,\mathbf{j}) \qquad |\mathbf{T}'(t)| = \frac{1}{\sqrt{2}}$$

$$\mathbf{N}(t) = \frac{\mathbf{T}'(t)}{|\mathbf{T}'(t)|} = -\cos t\,\mathbf{i} - \sin t\,\mathbf{j} = \langle -\cos t, -\sin t, 0 \rangle$$

This shows that the normal vector at any point on the helix is horizontal and points toward the z-axis. The binormal vector is

$$\mathbf{B}(t) = \mathbf{T}(t) \times \mathbf{N}(t) = \frac{1}{\sqrt{2}} \begin{bmatrix} \mathbf{i} & \mathbf{j} & \mathbf{k} \\ -\sin t & \cos t & 1 \\ -\cos t & -\sin t & 0 \end{bmatrix}$$

$$= \frac{1}{\sqrt{2}} \langle \sin t, -\cos t, 1 \rangle$$ ■

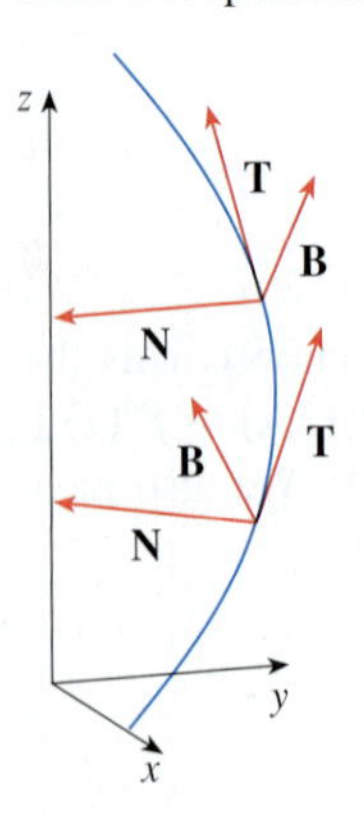

FIGURE 7

TEC Visual 13.3B shows how the **TNB** frame moves along several curves.

The plane determined by the normal and binormal vectors **N** and **B** at a point P on a curve C is called the **normal plane** of C at P. It consists of all lines that are orthogonal to the tangent vector **T**. The plane determined by the vectors **T** and **N** is called the **osculating plane** of C at P. The name comes from the Latin *osculum*, meaning "kiss." It is the plane that comes closest to containing the part of the curve near P. (For a plane curve, the osculating plane is simply the plane that contains the curve.)

The circle that lies in the osculating plane of C at P, has the same tangent as C at P, lies on the concave side of C (toward which **N** points), and has radius $\rho = 1/\kappa$ (the reciprocal of the curvature) is called the **osculating circle** (or the **circle of curvature**) of C at P. It is the circle that best describes how C behaves near P; it shares the same tangent, normal, and curvature at P.

EXAMPLE 7 Find equations of the normal plane and osculating plane of the helix in Example 6 at the point $P(0, 1, \pi/2)$.

Figure 8 shows the helix and the osculating plane in Example 7.

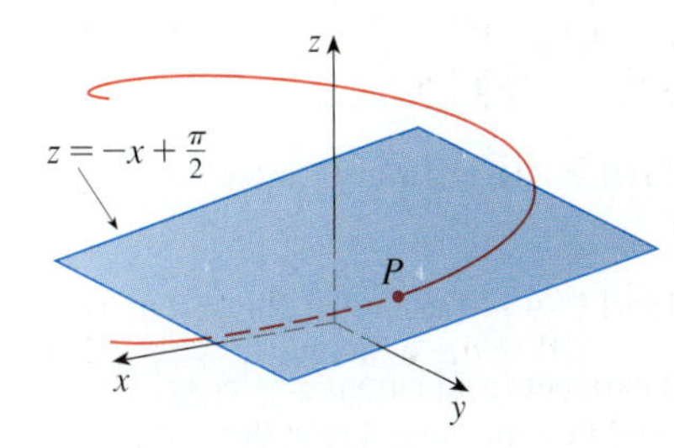

FIGURE 8

SOLUTION The point P corresponds to $t = \pi/2$ and the normal plane there has normal vector $\mathbf{r}'(\pi/2) = \langle -1, 0, 1 \rangle$, so an equation is

$$-1(x - 0) + 0(y - 1) + 1\left(z - \frac{\pi}{2}\right) = 0 \qquad \text{or} \qquad z = x + \frac{\pi}{2}$$

The osculating plane at P contains the vectors **T** and **N**, so its normal vector is $\mathbf{T} \times \mathbf{N} = \mathbf{B}$. From Example 6 we have

$$\mathbf{B}(t) = \frac{1}{\sqrt{2}} \langle \sin t, -\cos t, 1 \rangle \qquad \mathbf{B}\left(\frac{\pi}{2}\right) = \left\langle \frac{1}{\sqrt{2}}, 0, \frac{1}{\sqrt{2}} \right\rangle$$

A simpler normal vector is $\langle 1, 0, 1 \rangle$, so an equation of the osculating plane is

$$1(x - 0) + 0(y - 1) + 1\left(z - \frac{\pi}{2}\right) = 0 \qquad \text{or} \qquad z = -x + \frac{\pi}{2}$$

■

EXAMPLE 8 Find and graph the osculating circle of the parabola $y = x^2$ at the origin.

SOLUTION From Example 5, the curvature of the parabola at the origin is $\kappa(0) = 2$. So the radius of the osculating circle at the origin is $1/\kappa = \frac{1}{2}$ and its center is $\left(0, \frac{1}{2}\right)$. Its equation is therefore

$$x^2 + \left(y - \tfrac{1}{2}\right)^2 = \tfrac{1}{4}$$

For the graph in Figure 9 we use parametric equations of this circle:

$$x = \tfrac{1}{2}\cos t \qquad y = \tfrac{1}{2} + \tfrac{1}{2}\sin t$$

■

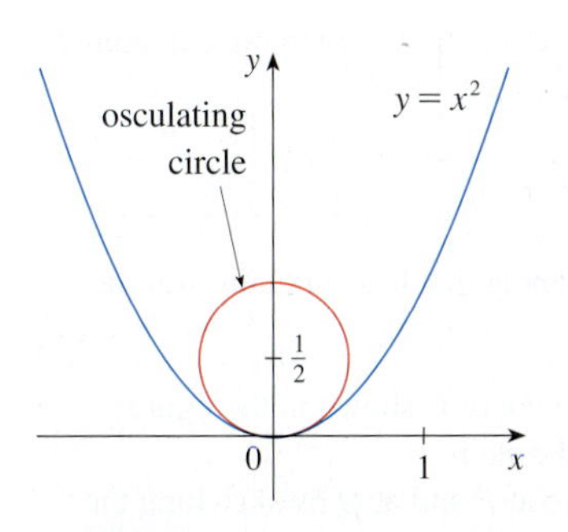

FIGURE 9
Notice that the circle and the parabola appear to bend similarly at the origin.

We summarize here the formulas for unit tangent, unit normal and binormal vectors, and curvature.

TEC Visual 13.3C shows how the osculating circle changes as a point moves along a curve.

$$\mathbf{T}(t) = \frac{\mathbf{r}'(t)}{|\mathbf{r}'(t)|} \qquad \mathbf{N}(t) = \frac{\mathbf{T}'(t)}{|\mathbf{T}'(t)|} \qquad \mathbf{B}(t) = \mathbf{T}(t) \times \mathbf{N}(t)$$

$$\kappa = \left|\frac{d\mathbf{T}}{ds}\right| = \frac{|\mathbf{T}'(t)|}{|\mathbf{r}'(t)|} = \frac{|\mathbf{r}'(t) \times \mathbf{r}''(t)|}{|\mathbf{r}'(t)|^3}$$

13.3 EXERCISES

1–6 Find the length of the curve.

1. $\mathbf{r}(t) = \langle t, 3\cos t, 3\sin t \rangle, \quad -5 \leqslant t \leqslant 5$

2. $\mathbf{r}(t) = \langle 2t, t^2, \frac{1}{3}t^3 \rangle, \quad 0 \leqslant t \leqslant 1$

3. $\mathbf{r}(t) = \sqrt{2}\, t\,\mathbf{i} + e^t\,\mathbf{j} + e^{-t}\,\mathbf{k}, \quad 0 \leqslant t \leqslant 1$

4. $\mathbf{r}(t) = \cos t\,\mathbf{i} + \sin t\,\mathbf{j} + \ln \cos t\,\mathbf{k}, \quad 0 \leqslant t \leqslant \pi/4$

5. $\mathbf{r}(t) = \mathbf{i} + t^2\,\mathbf{j} + t^3\,\mathbf{k}, \quad 0 \leqslant t \leqslant 1$

6. $\mathbf{r}(t) = t^2\,\mathbf{i} + 9t\,\mathbf{j} + 4t^{3/2}\,\mathbf{k}, \quad 1 \leqslant t \leqslant 4$

7–9 Find the length of the curve correct to four decimal places. (Use a calculator to approximate the integral.)

7. $\mathbf{r}(t) = \langle t^2, t^3, t^4 \rangle, \quad 0 \leqslant t \leqslant 2$

8. $\mathbf{r}(t) = \langle t, e^{-t}, te^{-t} \rangle, \quad 1 \leqslant t \leqslant 3$

9. $\mathbf{r}(t) = \langle \cos \pi t, 2t, \sin 2\pi t \rangle$, from $(1, 0, 0)$ to $(1, 4, 0)$

10. Graph the curve with parametric equations $x = \sin t$, $y = \sin 2t$, $z = \sin 3t$. Find the total length of this curve correct to four decimal places.

11. Let C be the curve of intersection of the parabolic cylinder $x^2 = 2y$ and the surface $3z = xy$. Find the exact length of C from the origin to the point $(6, 18, 36)$.

12. Find, correct to four decimal places, the length of the curve of intersection of the cylinder $4x^2 + y^2 = 4$ and the plane $x + y + z = 2$.

13–14 (a) Find the arc length function for the curve measured from the point P in the direction of increasing t and then reparametrize the curve with respect to arc length starting from P, and (b) find the point 4 units along the curve (in the direction of increasing t) from P.

13. $\mathbf{r}(t) = (5 - t)\,\mathbf{i} + (4t - 3)\,\mathbf{j} + 3t\,\mathbf{k}, \quad P(4, 1, 3)$

14. $\mathbf{r}(t) = e^t \sin t\,\mathbf{i} + e^t \cos t\,\mathbf{j} + \sqrt{2}\,e^t\,\mathbf{k}, \quad P(0, 1, \sqrt{2})$

15. Suppose you start at the point $(0, 0, 3)$ and move 5 units along the curve $x = 3\sin t$, $y = 4t$, $z = 3\cos t$ in the positive direction. Where are you now?

16. Reparametrize the curve

$$\mathbf{r}(t) = \left(\frac{2}{t^2 + 1} - 1\right)\mathbf{i} + \frac{2t}{t^2 + 1}\,\mathbf{j}$$

with respect to arc length measured from the point $(1, 0)$ in the direction of increasing t. Express the reparametrization in its simplest form. What can you conclude about the curve?

17–20
(a) Find the unit tangent and unit normal vectors $\mathbf{T}(t)$ and $\mathbf{N}(t)$.
(b) Use Formula 9 to find the curvature.

17. $\mathbf{r}(t) = \langle t, 3\cos t, 3\sin t \rangle$

18. $\mathbf{r}(t) = \langle t^2, \sin t - t\cos t, \cos t + t\sin t \rangle, \quad t > 0$

19. $\mathbf{r}(t) = \langle \sqrt{2}\,t, e^t, e^{-t} \rangle$

20. $\mathbf{r}(t) = \langle t, \frac{1}{2}t^2, t^2 \rangle$

21–23 Use Theorem 10 to find the curvature.

21. $\mathbf{r}(t) = t^3\,\mathbf{j} + t^2\,\mathbf{k}$

22. $\mathbf{r}(t) = t\,\mathbf{i} + t^2\,\mathbf{j} + e^t\,\mathbf{k}$

23. $\mathbf{r}(t) = \sqrt{6}\,t^2\,\mathbf{i} + 2t\,\mathbf{j} + 2t^3\,\mathbf{k}$

24. Find the curvature of $\mathbf{r}(t) = \langle t^2, \ln t, t\ln t \rangle$ at the point $(1, 0, 0)$.

25. Find the curvature of $\mathbf{r}(t) = \langle t, t^2, t^3 \rangle$ at the point $(1, 1, 1)$.

26. Graph the curve with parametric equations $x = \cos t$, $y = \sin t$, $z = \sin 5t$ and find the curvature at the point $(1, 0, 0)$.

27–29 Use Formula 11 to find the curvature.

27. $y = x^4$ **28.** $y = \tan x$ **29.** $y = xe^x$

30–31 At what point does the curve have maximum curvature? What happens to the curvature as $x \to \infty$?

30. $y = \ln x$ **31.** $y = e^x$

32. Find an equation of a parabola that has curvature 4 at the origin.

33. (a) Is the curvature of the curve C shown in the figure greater at P or at Q? Explain.
(b) Estimate the curvature at P and at Q by sketching the osculating circles at those points.

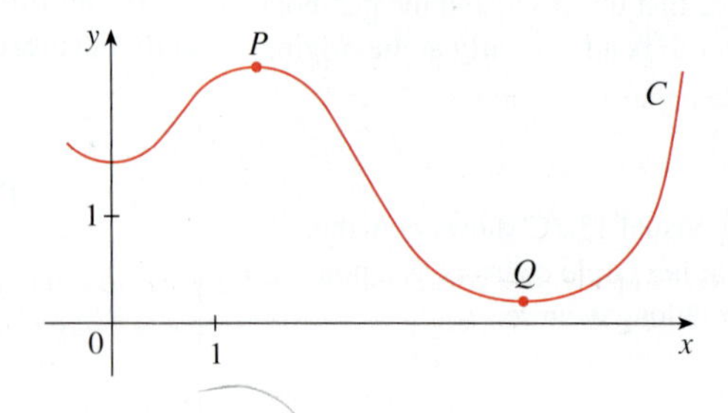

34–35 Use a graphing calculator or computer to graph both the curve and its curvature function $\kappa(x)$ on the same screen. Is the graph of κ what you would expect?

34. $y = x^4 - 2x^2$ **35.** $y = x^{-2}$

36–37 Plot the space curve and its curvature function $\kappa(t)$. Comment on how the curvature reflects the shape of the curve.

36. $\mathbf{r}(t) = \langle t - \sin t,\ 1 - \cos t,\ 4\cos(t/2)\rangle, \quad 0 \leqslant t \leqslant 8\pi$

37. $\mathbf{r}(t) = \langle te^t,\ e^{-t},\ \sqrt{2}\,t\rangle, \quad -5 \leqslant t \leqslant 5$

38–39 Two graphs, a and b, are shown. One is a curve $y = f(x)$ and the other is the graph of its curvature function $y = \kappa(x)$. Identify each curve and explain your choices.

38.

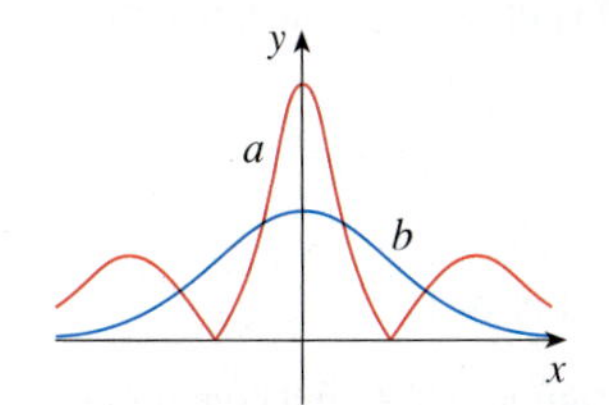

39.

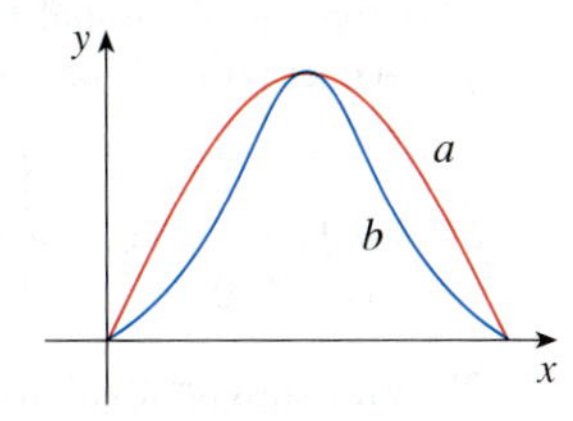

40. (a) Graph the curve $\mathbf{r}(t) = \langle \sin 3t, \sin 2t, \sin 3t\rangle$. At how many points on the curve does it appear that the curvature has a local or absolute maximum?
(b) Use a CAS to find and graph the curvature function. Does this graph confirm your conclusion from part (a)?

41. The graph of $\mathbf{r}(t) = \langle t - \frac{3}{2}\sin t,\ 1 - \frac{3}{2}\cos t,\ t\rangle$ is shown in Figure 13.1.12(b). Where do you think the curvature is largest? Use a CAS to find and graph the curvature function. For which values of t is the curvature largest?

42. Use Theorem 10 to show that the curvature of a plane parametric curve $x = f(t)$, $y = g(t)$ is

$$\kappa = \frac{|\dot{x}\ddot{y} - \dot{y}\ddot{x}|}{[\dot{x}^2 + \dot{y}^2]^{3/2}}$$

where the dots indicate derivatives with respect to t.

43–45 Use the formula in Exercise 42 to find the curvature.

43. $x = t^2, \quad y = t^3$

44. $x = a\cos\omega t, \quad y = b\sin\omega t$

45. $x = e^t\cos t, \quad y = e^t\sin t$

46. Consider the curvature at $x = 0$ for each member of the family of functions $f(x) = e^{cx}$. For which members is $\kappa(0)$ largest?

47–48 Find the vectors $\mathbf{T}$, $\mathbf{N}$, and $\mathbf{B}$ at the given point.

47. $\mathbf{r}(t) = \langle t^2, \frac{2}{3}t^3, t\rangle, \quad (1, \frac{2}{3}, 1)$

48. $\mathbf{r}(t) = \langle \cos t, \sin t, \ln\cos t\rangle, \quad (1, 0, 0)$

49–50 Find equations of the normal plane and osculating plane of the curve at the given point.

49. $x = \sin 2t,\ y = -\cos 2t,\ z = 4t; \quad (0, 1, 2\pi)$

50. $x = \ln t,\ y = 2t,\ z = t^2; \quad (0, 2, 1)$

51. Find equations of the osculating circles of the ellipse $9x^2 + 4y^2 = 36$ at the points $(2, 0)$ and $(0, 3)$. Use a graphing calculator or computer to graph the ellipse and both osculating circles on the same screen.

52. Find equations of the osculating circles of the parabola $y = \frac{1}{2}x^2$ at the points $(0, 0)$ and $(1, \frac{1}{2})$. Graph both osculating circles and the parabola on the same screen.

53. At what point on the curve $x = t^3$, $y = 3t$, $z = t^4$ is the normal plane parallel to the plane $6x + 6y - 8z = 1$?

54. Is there a point on the curve in Exercise 53 where the osculating plane is parallel to the plane $x + y + z = 1$? [*Note:* You will need a CAS for differentiating, for simplifying, and for computing a cross product.]

55. Find equations of the normal and osculating planes of the curve of intersection of the parabolic cylinders $x = y^2$ and $z = x^2$ at the point $(1, 1, 1)$.

56. Show that the osculating plane at every point on the curve $\mathbf{r}(t) = \langle t + 2,\ 1 - t,\ \frac{1}{2}t^2\rangle$ is the same plane. What can you conclude about the curve?

57. Show that at every point on the curve

$$\mathbf{r}(t) = \langle e^t\cos t,\ e^t\sin t,\ e^t\rangle$$

the angle between the unit tangent vector and the z-axis is the same. Then show that the same result holds true for the unit normal and binormal vectors.

58. The *rectifying plane* of a curve at a point is the plane that contains the vectors $\mathbf{T}$ and $\mathbf{B}$ at that point. Find the rectifying plane of the curve $\mathbf{r}(t) = \sin t\,\mathbf{i} + \cos t\,\mathbf{j} + \tan t\,\mathbf{k}$ at the point $(\sqrt{2}/2, \sqrt{2}/2, 1)$.

59. Show that the curvature κ is related to the tangent and normal vectors by the equation

$$\frac{d\mathbf{T}}{ds} = \kappa\mathbf{N}$$

60. Show that the curvature of a plane curve is $\kappa = |d\phi/ds|$, where ϕ is the angle between $\mathbf{T}$ and $\mathbf{i}$; that is, ϕ is the angle of inclination of the tangent line. (This shows that the definition of curvature is consistent with the definition for plane curves given in Exercise 10.2.69.)

61. (a) Show that $d\mathbf{B}/ds$ is perpendicular to $\mathbf{B}$.
(b) Show that $d\mathbf{B}/ds$ is perpendicular to $\mathbf{T}$.
(c) Deduce from parts (a) and (b) that $d\mathbf{B}/ds = -\tau(s)\mathbf{N}$ for some number $\tau(s)$ called the **torsion** of the curve. (The torsion measures the degree of twisting of a curve.)
(d) Show that for a plane curve the torsion is $\tau(s) = 0$.

62. The following formulas, called the **Frenet-Serret formulas**, are of fundamental importance in differential geometry:

1. $d\mathbf{T}/ds = \kappa\mathbf{N}$

2. $d\mathbf{N}/ds = -\kappa\mathbf{T} + \tau\mathbf{B}$

3. $d\mathbf{B}/ds = -\tau\mathbf{N}$

(Formula 1 comes from Exercise 59 and Formula 3 comes from Exercise 61.) Use the fact that $\mathbf{N} = \mathbf{B} \times \mathbf{T}$ to deduce Formula 2 from Formulas 1 and 3.

63. Use the Frenet-Serret formulas to prove each of the following. (Primes denote derivatives with respect to t. Start as in the proof of Theorem 10.)
(a) $\mathbf{r}'' = s''\mathbf{T} + \kappa(s')^2\mathbf{N}$
(b) $\mathbf{r}' \times \mathbf{r}'' = \kappa(s')^3\mathbf{B}$
(c) $\mathbf{r}''' = [s''' - \kappa^2(s')^3]\mathbf{T} + [3\kappa s's'' + \kappa'(s')^2]\mathbf{N} + \kappa\tau(s')^3\mathbf{B}$
(d) $\tau = \dfrac{(\mathbf{r}' \times \mathbf{r}'') \cdot \mathbf{r}'''}{|\mathbf{r}' \times \mathbf{r}''|^2}$

64. Show that the circular helix $\mathbf{r}(t) = \langle a\cos t, a\sin t, bt\rangle$, where a and b are positive constants, has constant curvature and constant torsion. [Use the result of Exercise 63(d).]

65. Use the formula in Exercise 63(d) to find the torsion of the curve $\mathbf{r}(t) = \langle t, \frac{1}{2}t^2, \frac{1}{3}t^3\rangle$.

66. Find the curvature and torsion of the curve $x = \sinh t$, $y = \cosh t$, $z = t$ at the point $(0, 1, 0)$.

67. The DNA molecule has the shape of a double helix (see Figure 3 on page 890). The radius of each helix is about 10 angstroms (1 Å $= 10^{-8}$ cm). Each helix rises about 34 Å during each complete turn, and there are about 2.9×10^8 complete turns. Estimate the length of each helix.

68. Let's consider the problem of designing a railroad track to make a smooth transition between sections of straight track. Existing track along the negative x-axis is to be joined smoothly to a track along the line $y = 1$ for $x \geqslant 1$.
(a) Find a polynomial $P = P(x)$ of degree 5 such that the function F defined by

$$F(x) = \begin{cases} 0 & \text{if } x \leqslant 0 \\ P(x) & \text{if } 0 < x < 1 \\ 1 & \text{if } x \geqslant 1 \end{cases}$$

is continuous and has continuous slope and continuous curvature.
 (b) Graph F.

13.4 Motion in Space: Velocity and Acceleration

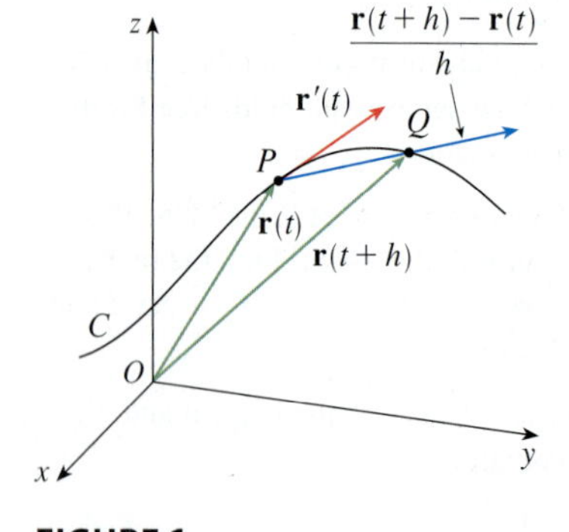

FIGURE 1

In this section we show how the ideas of tangent and normal vectors and curvature can be used in physics to study the motion of an object, including its velocity and acceleration, along a space curve. In particular, we follow in the footsteps of Newton by using these methods to derive Kepler's First Law of planetary motion.

Suppose a particle moves through space so that its position vector at time t is $\mathbf{r}(t)$. Notice from Figure 1 that, for small values of h, the vector

$$\boxed{1} \qquad \frac{\mathbf{r}(t+h) - \mathbf{r}(t)}{h}$$

approximates the direction of the particle moving along the curve $\mathbf{r}(t)$. Its magnitude measures the size of the displacement vector per unit time. The vector (1) gives the average velocity over a time interval of length h and its limit is the **velocity vector** $\mathbf{v}(t)$ at time t:

$$\boxed{2} \qquad \mathbf{v}(t) = \lim_{h \to 0} \frac{\mathbf{r}(t+h) - \mathbf{r}(t)}{h} = \mathbf{r}'(t)$$

Thus the velocity vector is also the tangent vector and points in the direction of the tangent line.

The **speed** of the particle at time t is the magnitude of the velocity vector, that is, $|\mathbf{v}(t)|$. This is appropriate because, from (2) and from Equation 13.3.7, we have

$$|\mathbf{v}(t)| = |\mathbf{r}'(t)| = \frac{ds}{dt} = \text{rate of change of distance with respect to time}$$

As in the case of one-dimensional motion, the **acceleration** of the particle is defined as the derivative of the velocity:

$$\mathbf{a}(t) = \mathbf{v}'(t) = \mathbf{r}''(t)$$

EXAMPLE 1 The position vector of an object moving in a plane is given by $\mathbf{r}(t) = t^3\,\mathbf{i} + t^2\,\mathbf{j}$. Find its velocity, speed, and acceleration when $t = 1$ and illustrate geometrically.

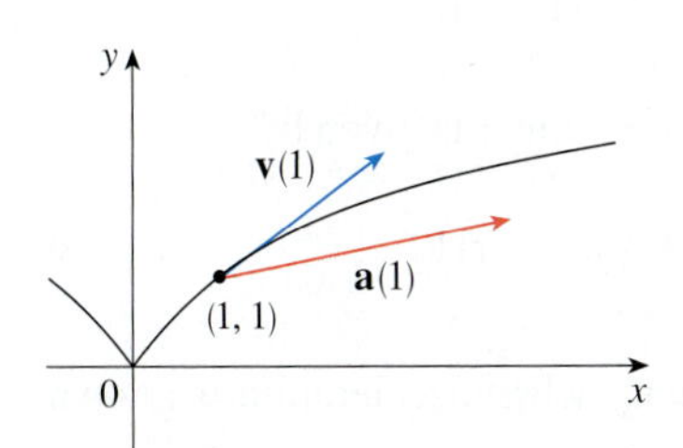

FIGURE 2

SOLUTION The velocity and acceleration at time t are

$$\mathbf{v}(t) = \mathbf{r}'(t) = 3t^2\,\mathbf{i} + 2t\,\mathbf{j}$$

$$\mathbf{a}(t) = \mathbf{r}''(t) = 6t\,\mathbf{i} + 2\,\mathbf{j}$$

and the speed is

$$|\mathbf{v}(t)| = \sqrt{(3t^2)^2 + (2t)^2} = \sqrt{9t^4 + 4t^2}$$

TEC Visual 13.4 shows animated velocity and acceleration vectors for objects moving along various curves.

When $t = 1$, we have

$$\mathbf{v}(1) = 3\,\mathbf{i} + 2\,\mathbf{j} \qquad \mathbf{a}(1) = 6\,\mathbf{i} + 2\,\mathbf{j} \qquad |\mathbf{v}(1)| = \sqrt{13}$$

These velocity and acceleration vectors are shown in Figure 2. ■

Figure 3 shows the path of the particle in Example 2 with the velocity and acceleration vectors when $t = 1$.

EXAMPLE 2 Find the velocity, acceleration, and speed of a particle with position vector $\mathbf{r}(t) = \langle t^2, e^t, te^t \rangle$.

SOLUTION

$$\mathbf{v}(t) = \mathbf{r}'(t) = \langle 2t, e^t, (1 + t)e^t \rangle$$

$$\mathbf{a}(t) = \mathbf{v}'(t) = \langle 2, e^t, (2 + t)e^t \rangle$$

$$|\mathbf{v}(t)| = \sqrt{4t^2 + e^{2t} + (1 + t)^2 e^{2t}}$$ ■

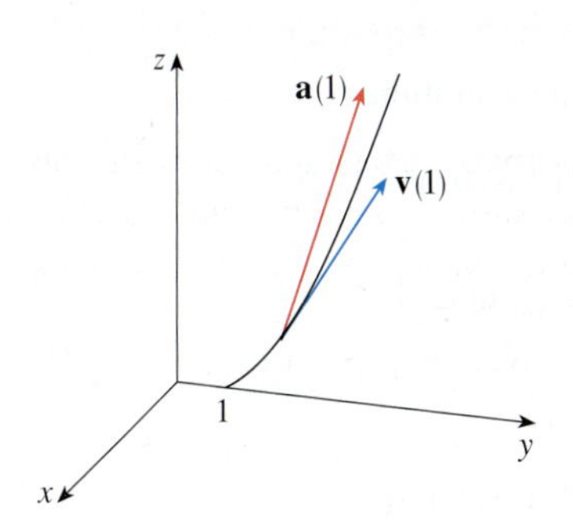

FIGURE 3

The vector integrals that were introduced in Section 13.2 can be used to find position vectors when velocity or acceleration vectors are known, as in the next example.

EXAMPLE 3 A moving particle starts at an initial position $\mathbf{r}(0) = \langle 1, 0, 0 \rangle$ with initial velocity $\mathbf{v}(0) = \mathbf{i} - \mathbf{j} + \mathbf{k}$. Its acceleration is $\mathbf{a}(t) = 4t\,\mathbf{i} + 6t\,\mathbf{j} + \mathbf{k}$. Find its velocity and position at time t.

SOLUTION Since $\mathbf{a}(t) = \mathbf{v}'(t)$, we have

$$\mathbf{v}(t) = \int \mathbf{a}(t)\,dt = \int (4t\,\mathbf{i} + 6t\,\mathbf{j} + \mathbf{k})\,dt$$

$$= 2t^2\,\mathbf{i} + 3t^2\,\mathbf{j} + t\,\mathbf{k} + \mathbf{C}$$

To determine the value of the constant vector $\mathbf{C}$, we use the fact that $\mathbf{v}(0) = \mathbf{i} - \mathbf{j} + \mathbf{k}$. The preceding equation gives $\mathbf{v}(0) = \mathbf{C}$, so $\mathbf{C} = \mathbf{i} - \mathbf{j} + \mathbf{k}$ and

$$\mathbf{v}(t) = 2t^2\,\mathbf{i} + 3t^2\,\mathbf{j} + t\,\mathbf{k} + \mathbf{i} - \mathbf{j} + \mathbf{k}$$

$$= (2t^2 + 1)\,\mathbf{i} + (3t^2 - 1)\,\mathbf{j} + (t + 1)\,\mathbf{k}$$

The expression for $\mathbf{r}(t)$ that we obtained in Example 3 was used to plot the path of the particle in Figure 4 for $0 \leqslant t \leqslant 3$.

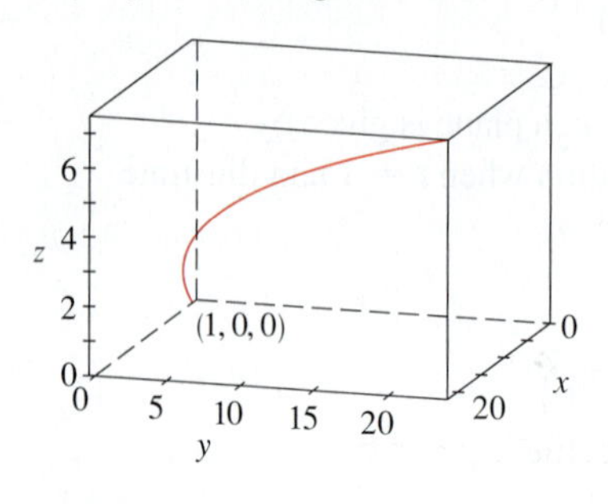

FIGURE 4

Since $\mathbf{v}(t) = \mathbf{r}'(t)$, we have

$$\mathbf{r}(t) = \int \mathbf{v}(t)\,dt$$

$$= \int [(2t^2 + 1)\,\mathbf{i} + (3t^2 - 1)\,\mathbf{j} + (t + 1)\,\mathbf{k}]\,dt$$

$$= \left(\tfrac{2}{3}t^3 + t\right)\mathbf{i} + (t^3 - t)\,\mathbf{j} + \left(\tfrac{1}{2}t^2 + t\right)\mathbf{k} + \mathbf{D}$$

Putting $t = 0$, we find that $\mathbf{D} = \mathbf{r}(0) = \mathbf{i}$, so the position at time t is given by

$$\mathbf{r}(t) = \left(\tfrac{2}{3}t^3 + t + 1\right)\mathbf{i} + (t^3 - t)\,\mathbf{j} + \left(\tfrac{1}{2}t^2 + t\right)\mathbf{k}$$ ■

In general, vector integrals allow us to recover velocity when acceleration is known and position when velocity is known:

$$\mathbf{v}(t) = \mathbf{v}(t_0) + \int_{t_0}^{t} \mathbf{a}(u)\,du \qquad \mathbf{r}(t) = \mathbf{r}(t_0) + \int_{t_0}^{t} \mathbf{v}(u)\,du$$

If the force that acts on a particle is known, then the acceleration can be found from **Newton's Second Law of Motion**. The vector version of this law states that if, at any time t, a force $\mathbf{F}(t)$ acts on an object of mass m producing an acceleration $\mathbf{a}(t)$, then

$$\mathbf{F}(t) = m\mathbf{a}(t)$$

EXAMPLE 4 An object with mass m that moves in a circular path with constant angular speed ω has position vector $\mathbf{r}(t) = a\cos\omega t\,\mathbf{i} + a\sin\omega t\,\mathbf{j}$. Find the force acting on the object and show that it is directed toward the origin.

The object moving with position P has angular speed $\omega = d\theta/dt$, where θ is the angle shown in Figure 5.

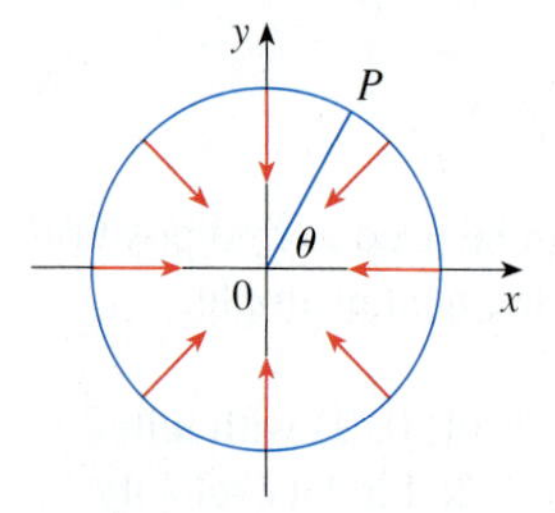

FIGURE 5

SOLUTION To find the force, we first need to know the acceleration:

$$\mathbf{v}(t) = \mathbf{r}'(t) = -a\omega\sin\omega t\,\mathbf{i} + a\omega\cos\omega t\,\mathbf{j}$$

$$\mathbf{a}(t) = \mathbf{v}'(t) = -a\omega^2\cos\omega t\,\mathbf{i} - a\omega^2\sin\omega t\,\mathbf{j}$$

Therefore Newton's Second Law gives the force as

$$\mathbf{F}(t) = m\mathbf{a}(t) = -m\omega^2(a\cos\omega t\,\mathbf{i} + a\sin\omega t\,\mathbf{j})$$

Notice that $\mathbf{F}(t) = -m\omega^2\mathbf{r}(t)$. This shows that the force acts in the direction opposite to the radius vector $\mathbf{r}(t)$ and therefore points toward the origin (see Figure 5). Such a force is called a *centripetal* (center-seeking) force. ■

Projectile Motion

EXAMPLE 5 A projectile is fired with angle of elevation α and initial velocity $\mathbf{v}_0$. (See Figure 6.) Assuming that air resistance is negligible and the only external force is due to gravity, find the position function $\mathbf{r}(t)$ of the projectile. What value of α maximizes the range (the horizontal distance traveled)?

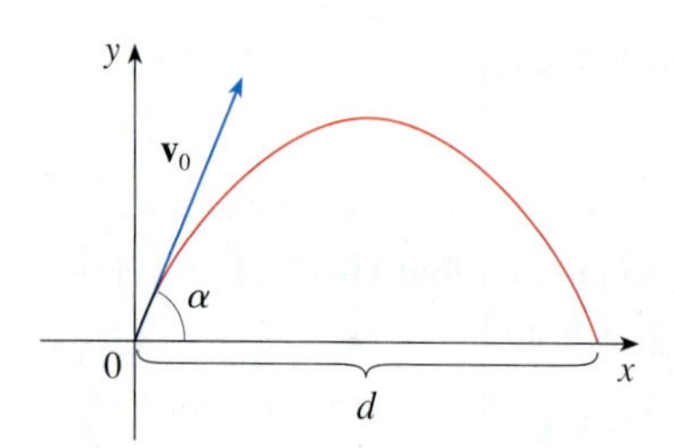

FIGURE 6

SOLUTION We set up the axes so that the projectile starts at the origin. Since the force due to gravity acts downward, we have

$$\mathbf{F} = m\mathbf{a} = -mg\,\mathbf{j}$$

where $g = |\mathbf{a}| \approx 9.8 \text{ m/s}^2$. Thus

$$\mathbf{a} = -g\,\mathbf{j}$$

Since $\mathbf{v}'(t) = \mathbf{a}$, we have

$$\mathbf{v}(t) = -gt\,\mathbf{j} + \mathbf{C}$$

where $\mathbf{C} = \mathbf{v}(0) = \mathbf{v}_0$. Therefore

$$\mathbf{r}'(t) = \mathbf{v}(t) = -gt\,\mathbf{j} + \mathbf{v}_0$$

Integrating again, we obtain

$$\mathbf{r}(t) = -\tfrac{1}{2}gt^2\,\mathbf{j} + t\,\mathbf{v}_0 + \mathbf{D}$$

But $\mathbf{D} = \mathbf{r}(0) = \mathbf{0}$, so the position vector of the projectile is given by

3 $$\mathbf{r}(t) = -\tfrac{1}{2}gt^2\,\mathbf{j} + t\,\mathbf{v}_0$$

If we write $|\mathbf{v}_0| = v_0$ (the initial speed of the projectile), then

$$\mathbf{v}_0 = v_0 \cos\alpha\,\mathbf{i} + v_0 \sin\alpha\,\mathbf{j}$$

and Equation 3 becomes

$$\mathbf{r}(t) = (v_0 \cos\alpha)t\,\mathbf{i} + \left[(v_0 \sin\alpha)t - \tfrac{1}{2}gt^2\right]\mathbf{j}$$

The parametric equations of the trajectory are therefore

If you eliminate t from Equations 4, you will see that y is a quadratic function of x. So the path of the projectile is part of a parabola.

4 $$x = (v_0 \cos\alpha)t \qquad y = (v_0 \sin\alpha)t - \tfrac{1}{2}gt^2$$

The horizontal distance d is the value of x when $y = 0$. Setting $y = 0$, we obtain $t = 0$ or $t = (2v_0 \sin\alpha)/g$. This second value of t then gives

$$d = x = (v_0 \cos\alpha)\frac{2v_0 \sin\alpha}{g} = \frac{v_0^2(2 \sin\alpha \cos\alpha)}{g} = \frac{v_0^2 \sin 2\alpha}{g}$$

Clearly, d has its maximum value when $\sin 2\alpha = 1$, that is, $\alpha = 45°$. ■

EXAMPLE 6 A projectile is fired with muzzle speed 150 m/s and angle of elevation 45° from a position 10 m above ground level. Where does the projectile hit the ground, and with what speed?

SOLUTION If we place the origin at ground level, then the initial position of the projectile is (0, 10) and so we need to adjust Equations 4 by adding 10 to the expression for y. With $v_0 = 150$ m/s, $\alpha = 45°$, and $g = 9.8 \text{ m/s}^2$, we have

$$x = 150 \cos(45°)t = 75\sqrt{2}\,t$$

$$y = 10 + 150 \sin(45°)t - \tfrac{1}{2}(9.8)t^2 = 10 + 75\sqrt{2}\,t - 4.9t^2$$

Impact occurs when $y = 0$, that is, $4.9t^2 - 75\sqrt{2}\,t - 10 = 0$. Using the quadratic formula to solve this equation (and taking only the positive value of t), we get

$$t = \frac{75\sqrt{2} + \sqrt{11{,}250 + 196}}{9.8} \approx 21.74$$

Then $x \approx 75\sqrt{2}\,(21.74) \approx 2306$, so the projectile hits the ground about 2306 m away.

The velocity of the projectile is

$$\mathbf{v}(t) = \mathbf{r}'(t) = 75\sqrt{2}\,\mathbf{i} + \left(75\sqrt{2} - 9.8t\right)\mathbf{j}$$

So its speed at impact is

$$|\mathbf{v}(21.74)| = \sqrt{(75\sqrt{2}\,)^2 + (75\sqrt{2} - 9.8 \cdot 21.74)^2} \approx 151 \text{ m/s}$$

Tangential and Normal Components of Acceleration

When we study the motion of a particle, it is often useful to resolve the acceleration into two components, one in the direction of the tangent and the other in the direction of the normal. If we write $v = |\mathbf{v}|$ for the speed of the particle, then

$$\mathbf{T}(t) = \frac{\mathbf{r}'(t)}{|\mathbf{r}'(t)|} = \frac{\mathbf{v}(t)}{|\mathbf{v}(t)|} = \frac{\mathbf{v}}{v}$$

and so

$$\mathbf{v} = v\mathbf{T}$$

If we differentiate both sides of this equation with respect to t, we get

$$\boxed{5} \qquad \mathbf{a} = \mathbf{v}' = v'\mathbf{T} + v\mathbf{T}'$$

If we use the expression for the curvature given by Equation 13.3.9, then we have

$$\boxed{6} \qquad \kappa = \frac{|\mathbf{T}'|}{|\mathbf{r}'|} = \frac{|\mathbf{T}'|}{v} \qquad \text{so} \qquad |\mathbf{T}'| = \kappa v$$

The unit normal vector was defined in the preceding section as $\mathbf{N} = \mathbf{T}'/|\mathbf{T}'|$, so (6) gives

$$\mathbf{T}' = |\mathbf{T}'|\mathbf{N} = \kappa v \mathbf{N}$$

and Equation 5 becomes

$$\boxed{7} \qquad \boxed{\mathbf{a} = v'\mathbf{T} + \kappa v^2 \mathbf{N}}$$

Writing a_T and a_N for the tangential and normal components of acceleration, we have

$$\mathbf{a} = a_T\mathbf{T} + a_N\mathbf{N}$$

where

$$\boxed{8} \qquad a_T = v' \qquad \text{and} \qquad a_N = \kappa v^2$$

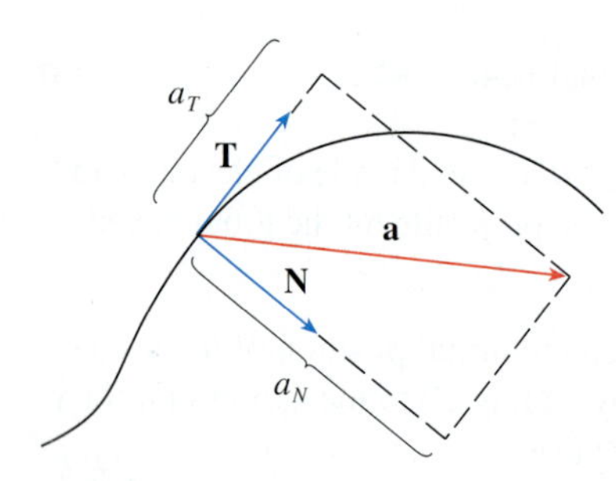

FIGURE 7

This resolution is illustrated in Figure 7.

Let's look at what Formula 7 says. The first thing to notice is that the binormal vector **B** is absent. No matter how an object moves through space, its acceleration always lies in the plane of **T** and **N** (the osculating plane). (Recall that **T** gives the direction of motion and **N** points in the direction the curve is turning.) Next we notice that the tangential component of acceleration is v', the rate of change of speed, and the normal component of acceleration is κv^2, the curvature times the square of the speed. This makes sense if we think of a passenger in a car—a sharp turn in a road means a large value of the curvature κ, so the component of the acceleration perpendicular to the motion is large and the passenger is thrown against a car door. High speed around the turn has the same effect; in fact, if you double your speed, a_N is increased by a factor of 4.

Although we have expressions for the tangential and normal components of acceleration in Equations 8, it's desirable to have expressions that depend only on **r**, **r**′, and **r**″. To this end we take the dot product of $\mathbf{v} = v\mathbf{T}$ with **a** as given by Equation 7:

$$\begin{aligned}\mathbf{v}\cdot\mathbf{a} &= v\mathbf{T}\cdot(v'\mathbf{T} + \kappa v^2\mathbf{N}) \\ &= vv'\mathbf{T}\cdot\mathbf{T} + \kappa v^3\mathbf{T}\cdot\mathbf{N} \\ &= vv' \qquad (\text{since } \mathbf{T}\cdot\mathbf{T} = 1 \text{ and } \mathbf{T}\cdot\mathbf{N} = 0)\end{aligned}$$

Therefore

9 $$a_T = v' = \frac{\mathbf{v}\cdot\mathbf{a}}{v} = \frac{\mathbf{r}'(t)\cdot\mathbf{r}''(t)}{|\mathbf{r}'(t)|}$$

Using the formula for curvature given by Theorem 13.3.10, we have

10 $$a_N = \kappa v^2 = \frac{|\mathbf{r}'(t)\times\mathbf{r}''(t)|}{|\mathbf{r}'(t)|^3}|\mathbf{r}'(t)|^2 = \frac{|\mathbf{r}'(t)\times\mathbf{r}''(t)|}{|\mathbf{r}'(t)|}$$

EXAMPLE 7 A particle moves with position function $\mathbf{r}(t) = \langle t^2, t^2, t^3\rangle$. Find the tangential and normal components of acceleration.

SOLUTION

$$\begin{aligned}\mathbf{r}(t) &= t^2\,\mathbf{i} + t^2\,\mathbf{j} + t^3\,\mathbf{k} \\ \mathbf{r}'(t) &= 2t\,\mathbf{i} + 2t\,\mathbf{j} + 3t^2\,\mathbf{k} \\ \mathbf{r}''(t) &= 2\,\mathbf{i} + 2\,\mathbf{j} + 6t\,\mathbf{k} \\ |\mathbf{r}'(t)| &= \sqrt{8t^2 + 9t^4}\end{aligned}$$

Therefore Equation 9 gives the tangential component as

$$a_T = \frac{\mathbf{r}'(t)\cdot\mathbf{r}''(t)}{|\mathbf{r}'(t)|} = \frac{8t + 18t^3}{\sqrt{8t^2 + 9t^4}}$$

Since $$\mathbf{r}'(t)\times\mathbf{r}''(t) = \begin{vmatrix}\mathbf{i} & \mathbf{j} & \mathbf{k} \\ 2t & 2t & 3t^2 \\ 2 & 2 & 6t\end{vmatrix} = 6t^2\,\mathbf{i} - 6t^2\,\mathbf{j}$$

Equation 10 gives the normal component as

$$a_N = \frac{|\mathbf{r}'(t)\times\mathbf{r}''(t)|}{|\mathbf{r}'(t)|} = \frac{6\sqrt{2}\,t^2}{\sqrt{8t^2 + 9t^4}}$$ ■

Kepler's Laws of Planetary Motion

We now describe one of the great accomplishments of calculus by showing how the material of this chapter can be used to prove Kepler's laws of planetary motion. After 20 years of studying the astronomical observations of the Danish astronomer Tycho Brahe, the German mathematician and astronomer Johannes Kepler (1571–1630) formulated the following three laws.

Kepler's Laws

1. A planet revolves around the sun in an elliptical orbit with the sun at one focus.
2. The line joining the sun to a planet sweeps out equal areas in equal times.
3. The square of the period of revolution of a planet is proportional to the cube of the length of the major axis of its orbit.

In his book *Principia Mathematica* of 1687, Sir Isaac Newton was able to show that these three laws are consequences of two of his own laws, the Second Law of Motion and the Law of Universal Gravitation. In what follows we prove Kepler's First Law. The remaining laws are left as exercises (with hints).

Since the gravitational force of the sun on a planet is so much larger than the forces exerted by other celestial bodies, we can safely ignore all bodies in the universe except the sun and one planet revolving about it. We use a coordinate system with the sun at the origin and we let $\mathbf{r} = \mathbf{r}(t)$ be the position vector of the planet. (Equally well, $\mathbf{r}$ could be the position vector of the moon or a satellite moving around the earth or a comet moving around a star.) The velocity vector is $\mathbf{v} = \mathbf{r}'$ and the acceleration vector is $\mathbf{a} = \mathbf{r}''$. We use the following laws of Newton:

Second Law of Motion: $\mathbf{F} = m\mathbf{a}$

Law of Gravitation: $\mathbf{F} = -\dfrac{GMm}{r^3}\mathbf{r} = -\dfrac{GMm}{r^2}\mathbf{u}$

where $\mathbf{F}$ is the gravitational force on the planet, m and M are the masses of the planet and the sun, G is the gravitational constant, $r = |\mathbf{r}|$, and $\mathbf{u} = (1/r)\mathbf{r}$ is the unit vector in the direction of $\mathbf{r}$.

We first show that the planet moves in one plane. By equating the expressions for $\mathbf{F}$ in Newton's two laws, we find that

$$\mathbf{a} = -\frac{GM}{r^3}\mathbf{r}$$

and so $\mathbf{a}$ is parallel to $\mathbf{r}$. It follows that $\mathbf{r} \times \mathbf{a} = \mathbf{0}$. We use Formula 5 in Theorem 13.2.3 to write

$$\frac{d}{dt}(\mathbf{r} \times \mathbf{v}) = \mathbf{r}' \times \mathbf{v} + \mathbf{r} \times \mathbf{v}'$$

$$= \mathbf{v} \times \mathbf{v} + \mathbf{r} \times \mathbf{a} = \mathbf{0} + \mathbf{0} = \mathbf{0}$$

Therefore $$\mathbf{r} \times \mathbf{v} = \mathbf{h}$$

where $\mathbf{h}$ is a constant vector. (We may assume that $\mathbf{h} \neq \mathbf{0}$; that is, $\mathbf{r}$ and $\mathbf{v}$ are not parallel.) This means that the vector $\mathbf{r} = \mathbf{r}(t)$ is perpendicular to $\mathbf{h}$ for all values of t, so the planet always lies in the plane through the origin perpendicular to $\mathbf{h}$. Thus the orbit of the planet is a plane curve.

To prove Kepler's First Law we rewrite the vector $\mathbf{h}$ as follows:

$$\mathbf{h} = \mathbf{r} \times \mathbf{v} = \mathbf{r} \times \mathbf{r}' = r\mathbf{u} \times (r\mathbf{u})'$$

$$= r\mathbf{u} \times (r\mathbf{u}' + r'\mathbf{u}) = r^2(\mathbf{u} \times \mathbf{u}') + rr'(\mathbf{u} \times \mathbf{u})$$

$$= r^2(\mathbf{u} \times \mathbf{u}')$$

Then

$$\mathbf{a} \times \mathbf{h} = \frac{-GM}{r^2}\mathbf{u} \times (r^2\mathbf{u} \times \mathbf{u}') = -GM\,\mathbf{u} \times (\mathbf{u} \times \mathbf{u}')$$

$$= -GM[(\mathbf{u} \cdot \mathbf{u}')\mathbf{u} - (\mathbf{u} \cdot \mathbf{u})\mathbf{u}'] \qquad \text{(by Theorem 12.4.11, Property 6)}$$

But $\mathbf{u} \cdot \mathbf{u} = |\mathbf{u}|^2 = 1$ and, since $|\mathbf{u}(t)| = 1$, it follows from Example 13.2.4 that

$$\mathbf{u} \cdot \mathbf{u}' = 0$$

Therefore

$$\mathbf{a} \times \mathbf{h} = GM\,\mathbf{u}'$$

and so

$$(\mathbf{v} \times \mathbf{h})' = \mathbf{v}' \times \mathbf{h} = \mathbf{a} \times \mathbf{h} = GM\,\mathbf{u}'$$

Integrating both sides of this equation, we get

11
$$\mathbf{v} \times \mathbf{h} = GM\,\mathbf{u} + \mathbf{c}$$

where **c** is a constant vector.

At this point it is convenient to choose the coordinate axes so that the standard basis vector **k** points in the direction of the vector **h**. Then the planet moves in the xy-plane. Since both $\mathbf{v} \times \mathbf{h}$ and **u** are perpendicular to **h**, Equation 11 shows that **c** lies in the xy-plane. This means that we can choose the x- and y-axes so that the vector **i** lies in the direction of **c**, as shown in Figure 8.

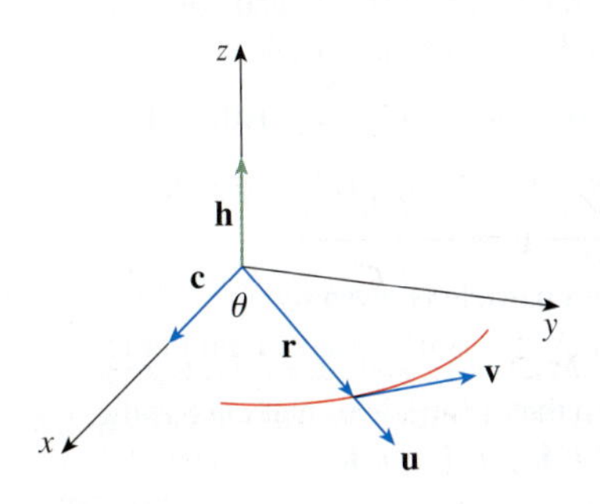

FIGURE 8

If θ is the angle between **c** and **r**, then (r, θ) are polar coordinates of the planet. From Equation 11 we have

$$\mathbf{r} \cdot (\mathbf{v} \times \mathbf{h}) = \mathbf{r} \cdot (GM\,\mathbf{u} + \mathbf{c}) = GM\,\mathbf{r} \cdot \mathbf{u} + \mathbf{r} \cdot \mathbf{c}$$

$$= GMr\,\mathbf{u} \cdot \mathbf{u} + |\mathbf{r}||\mathbf{c}|\cos\theta = GMr + rc\cos\theta$$

where $c = |\mathbf{c}|$. Then

$$r = \frac{\mathbf{r} \cdot (\mathbf{v} \times \mathbf{h})}{GM + c\cos\theta} = \frac{1}{GM}\frac{\mathbf{r} \cdot (\mathbf{v} \times \mathbf{h})}{1 + e\cos\theta}$$

where $e = c/(GM)$. But

$$\mathbf{r} \cdot (\mathbf{v} \times \mathbf{h}) = (\mathbf{r} \times \mathbf{v}) \cdot \mathbf{h} = \mathbf{h} \cdot \mathbf{h} = |\mathbf{h}|^2 = h^2$$

where $h = |\mathbf{h}|$. So

$$r = \frac{h^2/(GM)}{1 + e\cos\theta} = \frac{eh^2/c}{1 + e\cos\theta}$$

Writing $d = h^2/c$, we obtain the equation

12
$$r = \frac{ed}{1 + e\cos\theta}$$

Comparing with Theorem 10.6.6, we see that Equation 12 is the polar equation of a conic section with focus at the origin and eccentricity e. We know that the orbit of a planet is a closed curve and so the conic must be an ellipse.

This completes the derivation of Kepler's First Law. We will guide you through the derivation of the Second and Third Laws in the Applied Project on page 920. The proofs of these three laws show that the methods of this chapter provide a powerful tool for describing some of the laws of nature.

13.4 EXERCISES

1. The table gives coordinates of a particle moving through space along a smooth curve.
(a) Find the average velocities over the time intervals [0, 1], [0.5, 1], [1, 2], and [1, 1.5].
(b) Estimate the velocity and speed of the particle at $t = 1$.

t	x	y	z
0	2.7	9.8	3.7
0.5	3.5	7.2	3.3
1.0	4.5	6.0	3.0
1.5	5.9	6.4	2.8
2.0	7.3	7.8	2.7

2. The figure shows the path of a particle that moves with position vector $\mathbf{r}(t)$ at time t.
(a) Draw a vector that represents the average velocity of the particle over the time interval $2 \leqslant t \leqslant 2.4$.
(b) Draw a vector that represents the average velocity over the time interval $1.5 \leqslant t \leqslant 2$.
(c) Write an expression for the velocity vector $\mathbf{v}(2)$.
(d) Draw an approximation to the vector $\mathbf{v}(2)$ and estimate the speed of the particle at $t = 2$.

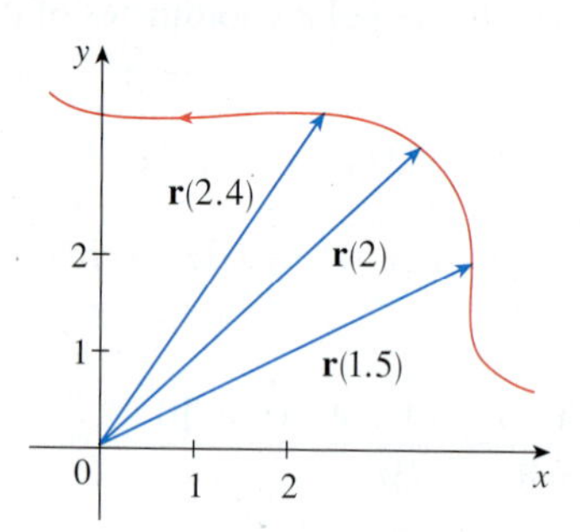

3–8 Find the velocity, acceleration, and speed of a particle with the given position function. Sketch the path of the particle and draw the velocity and acceleration vectors for the specified value of t.

3. $\mathbf{r}(t) = \langle -\frac{1}{2}t^2, t \rangle, \quad t = 2$

4. $\mathbf{r}(t) = \langle t^2, 1/t^2 \rangle, \quad t = 1$

5. $\mathbf{r}(t) = 3 \cos t\, \mathbf{i} + 2 \sin t\, \mathbf{j}, \quad t = \pi/3$

6. $\mathbf{r}(t) = e^t\, \mathbf{i} + e^{2t}\, \mathbf{j}, \quad t = 0$

7. $\mathbf{r}(t) = t\, \mathbf{i} + t^2 \mathbf{j} + 2\, \mathbf{k}, \quad t = 1$

8. $\mathbf{r}(t) = t\, \mathbf{i} + 2 \cos t\, \mathbf{j} + \sin t\, \mathbf{k}, \quad t = 0$

9–14 Find the velocity, acceleration, and speed of a particle with the given position function.

9. $\mathbf{r}(t) = \langle t^2 + t, t^2 - t, t^3 \rangle$

10. $\mathbf{r}(t) = \langle 2 \cos t, 3t, 2 \sin t \rangle$

11. $\mathbf{r}(t) = \sqrt{2}\, t\, \mathbf{i} + e^t\, \mathbf{j} + e^{-t}\, \mathbf{k}$

12. $\mathbf{r}(t) = t^2\, \mathbf{i} + 2t\, \mathbf{j} + \ln t\, \mathbf{k}$

13. $\mathbf{r}(t) = e^t(\cos t\, \mathbf{i} + \sin t\, \mathbf{j} + t\, \mathbf{k})$

14. $\mathbf{r}(t) = \langle t^2, \sin t - t \cos t, \cos t + t \sin t \rangle, \quad t \geqslant 0$

15–16 Find the velocity and position vectors of a particle that has the given acceleration and the given initial velocity and position.

15. $\mathbf{a}(t) = 2\, \mathbf{i} + 2t\, \mathbf{k}, \quad \mathbf{v}(0) = 3\, \mathbf{i} - \mathbf{j}, \quad \mathbf{r}(0) = \mathbf{j} + \mathbf{k}$

16. $\mathbf{a}(t) = \sin t\, \mathbf{i} + 2 \cos t\, \mathbf{j} + 6t\, \mathbf{k},$
$\mathbf{v}(0) = -\mathbf{k}, \quad \mathbf{r}(0) = \mathbf{j} - 4\, \mathbf{k}$

17–18
(a) Find the position vector of a particle that has the given acceleration and the specified initial velocity and position.
(b) Use a computer to graph the path of the particle.

17. $\mathbf{a}(t) = 2t\, \mathbf{i} + \sin t\, \mathbf{j} + \cos 2t\, \mathbf{k}, \quad \mathbf{v}(0) = \mathbf{i}, \quad \mathbf{r}(0) = \mathbf{j}$

18. $\mathbf{a}(t) = t\, \mathbf{i} + e^t\, \mathbf{j} + e^{-t}\, \mathbf{k}, \quad \mathbf{v}(0) = \mathbf{k}, \quad \mathbf{r}(0) = \mathbf{j} + \mathbf{k}$

19. The position function of a particle is given by $\mathbf{r}(t) = \langle t^2, 5t, t^2 - 16t \rangle$. When is the speed a minimum?

20. What force is required so that a particle of mass m has the position function $\mathbf{r}(t) = t^3\, \mathbf{i} + t^2\, \mathbf{j} + t^3\, \mathbf{k}$?

21. A force with magnitude 20 N acts directly upward from the xy-plane on an object with mass 4 kg. The object starts at the origin with initial velocity $\mathbf{v}(0) = \mathbf{i} - \mathbf{j}$. Find its position function and its speed at time t.

22. Show that if a particle moves with constant speed, then the velocity and acceleration vectors are orthogonal.

23. A projectile is fired with an initial speed of 200 m/s and angle of elevation 60°. Find (a) the range of the projectile, (b) the maximum height reached, and (c) the speed at impact.

24. Rework Exercise 23 if the projectile is fired from a position 100 m above the ground.

25. A ball is thrown at an angle of 45° to the ground. If the ball lands 90 m away, what was the initial speed of the ball?

26. A projectile is fired from a tank with initial speed 400 m/s. Find two angles of elevation that can be used to hit a target 3000 m away.

27. A rifle is fired with angle of elevation 36°. What is the muzzle speed if the maximum height of the bullet is 1600 ft?

28. A batter hits a baseball 3 ft above the ground toward the center field fence, which is 10 ft high and 400 ft from home plate. The ball leaves the bat with speed 115 ft/s at an angle 50° above the horizontal. Is it a home run? (In other words, does the ball clear the fence?)

29. A medieval city has the shape of a square and is protected by walls with length 500 m and height 15 m. You are the commander of an attacking army and the closest you can get to the wall is 100 m. Your plan is to set fire to the city by catapulting heated rocks over the wall (with an initial speed of 80 m/s). At what range of angles should you tell your men to set the catapult? (Assume the path of the rocks is perpendicular to the wall.)

30. Show that a projectile reaches three-quarters of its maximum height in half the time needed to reach its maximum height.

31. A ball is thrown eastward into the air from the origin (in the direction of the positive x-axis). The initial velocity is $50\,\mathbf{i} + 80\,\mathbf{k}$, with speed measured in feet per second. The spin of the ball results in a southward acceleration of 4 ft/s^2, so the acceleration vector is $\mathbf{a} = -4\mathbf{j} - 32\mathbf{k}$. Where does the ball land and with what speed?

32. A ball with mass 0.8 kg is thrown southward into the air with a speed of 30 m/s at an angle of 30° to the ground. A west wind applies a steady force of 4 N to the ball in an easterly direction. Where does the ball land and with what speed?

33. Water traveling along a straight portion of a river normally flows fastest in the middle, and the speed slows to almost zero at the banks. Consider a long straight stretch of river flowing north, with parallel banks 40 m apart. If the maximum water speed is 3 m/s, we can use a quadratic function as a basic model for the rate of water flow x units from the west bank: $f(x) = \frac{3}{400}x(40 - x)$.

(a) A boat proceeds at a constant speed of 5 m/s from a point A on the west bank while maintaining a heading perpendicular to the bank. How far down the river on the opposite bank will the boat touch shore? Graph the path of the boat.

(b) Suppose we would like to pilot the boat to land at the point B on the east bank directly opposite A. If we maintain a constant speed of 5 m/s and a constant heading, find the angle at which the boat should head. Then graph the actual path the boat follows. Does the path seem realistic?

34. Another reasonable model for the water speed of the river in Exercise 33 is a sine function: $f(x) = 3\sin(\pi x/40)$. If a boater would like to cross the river from A to B with constant heading and a constant speed of 5 m/s, determine the angle at which the boat should head.

35. A particle has position function $\mathbf{r}(t)$. If $\mathbf{r}'(t) = \mathbf{c} \times \mathbf{r}(t)$, where $\mathbf{c}$ is a constant vector, describe the path of the particle.

36. (a) If a particle moves along a straight line, what can you say about its acceleration vector?

(b) If a particle moves with constant speed along a curve, what can you say about its acceleration vector?

37–40 Find the tangential and normal components of the acceleration vector.

37. $\mathbf{r}(t) = (t^2 + 1)\,\mathbf{i} + t^3\,\mathbf{j}, \quad t \geqslant 0$

38. $\mathbf{r}(t) = 2t^2\,\mathbf{i} + \left(\frac{2}{3}t^3 - 2t\right)\mathbf{j}$

39. $\mathbf{r}(t) = \cos t\,\mathbf{i} + \sin t\,\mathbf{j} + t\,\mathbf{k}$

40. $\mathbf{r}(t) = t\,\mathbf{i} + 2e^t\,\mathbf{j} + e^{2t}\,\mathbf{k}$

41–42 Find the tangential and normal components of the acceleration vector at the given point.

41. $\mathbf{r}(t) = \ln t\,\mathbf{i} + (t^2 + 3t)\,\mathbf{j} + 4\sqrt{t}\,\mathbf{k}, \quad (0, 4, 4)$

42. $\mathbf{r}(t) = \dfrac{1}{t}\,\mathbf{i} + \dfrac{1}{t^2}\,\mathbf{j} + \dfrac{1}{t^3}\,\mathbf{k}, \quad (1, 1, 1)$

43. The magnitude of the acceleration vector $\mathbf{a}$ is 10 cm/s^2. Use the figure to estimate the tangential and normal components of $\mathbf{a}$.

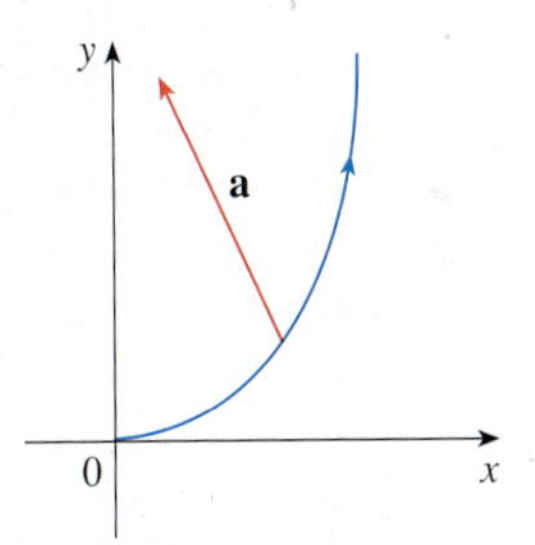

44. If a particle with mass m moves with position vector $\mathbf{r}(t)$, then its **angular momentum** is defined as $\mathbf{L}(t) = m\mathbf{r}(t) \times \mathbf{v}(t)$ and its **torque** as $\boldsymbol{\tau}(t) = m\mathbf{r}(t) \times \mathbf{a}(t)$. Show that $\mathbf{L}'(t) = \boldsymbol{\tau}(t)$. Deduce that if $\boldsymbol{\tau}(t) = \mathbf{0}$ for all t, then $\mathbf{L}(t)$ is constant. (This is the *law of conservation of angular momentum.*)

45. The position function of a spaceship is

$$\mathbf{r}(t) = (3 + t)\,\mathbf{i} + (2 + \ln t)\,\mathbf{j} + \left(7 - \frac{4}{t^2 + 1}\right)\mathbf{k}$$

and the coordinates of a space station are (6, 4, 9). The captain wants the spaceship to coast into the space station. When should the engines be turned off?

46. A rocket burning its onboard fuel while moving through space has velocity $\mathbf{v}(t)$ and mass $m(t)$ at time t. If the exhaust gases escape with velocity $\mathbf{v}_e$ relative to the rocket, it can be deduced from Newton's Second Law of Motion that

$$m\frac{d\mathbf{v}}{dt} = \frac{dm}{dt}\mathbf{v}_e$$

(a) Show that $\mathbf{v}(t) = \mathbf{v}(0) - \ln\dfrac{m(0)}{m(t)}\mathbf{v}_e$.

(b) For the rocket to accelerate in a straight line from rest to twice the speed of its own exhaust gases, what fraction of its initial mass would the rocket have to burn as fuel?

APPLIED PROJECT KEPLER'S LAWS

Johannes Kepler stated the following three laws of planetary motion on the basis of massive amounts of data on the positions of the planets at various times.

Kepler's Laws

1. A planet revolves around the sun in an elliptical orbit with the sun at one focus.
2. The line joining the sun to a planet sweeps out equal areas in equal times.
3. The square of the period of revolution of a planet is proportional to the cube of the length of the major axis of its orbit.

Kepler formulated these laws because they fitted the astronomical data. He wasn't able to see why they were true or how they related to each other. But Sir Isaac Newton, in his *Principia Mathematica* of 1687, showed how to deduce Kepler's three laws from two of Newton's own laws, the Second Law of Motion and the Law of Universal Gravitation. In Section 13.4 we proved Kepler's First Law using the calculus of vector functions. In this project we guide you through the proofs of Kepler's Second and Third Laws and explore some of their consequences.

1. Use the following steps to prove Kepler's Second Law. The notation is the same as in the proof of the First Law in Section 13.4. In particular, use polar coordinates so that $\mathbf{r} = (r\cos\theta)\,\mathbf{i} + (r\sin\theta)\,\mathbf{j}$.

(a) Show that $\mathbf{h} = r^2 \dfrac{d\theta}{dt}\,\mathbf{k}$.

(b) Deduce that $r^2 \dfrac{d\theta}{dt} = h$.

(c) If $A = A(t)$ is the area swept out by the radius vector $\mathbf{r} = \mathbf{r}(t)$ in the time interval $[t_0, t]$ as in the figure, show that

$$\frac{dA}{dt} = \tfrac{1}{2} r^2 \frac{d\theta}{dt}$$

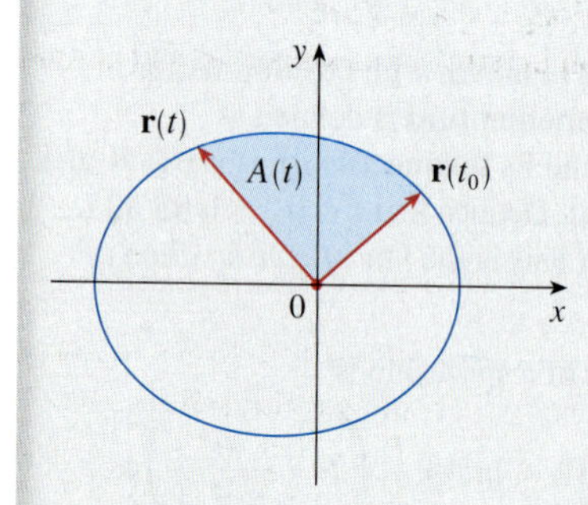

(d) Deduce that

$$\frac{dA}{dt} = \tfrac{1}{2} h = \text{constant}$$

This says that the rate at which A is swept out is constant and proves Kepler's Second Law.

2. Let T be the period of a planet about the sun; that is, T is the time required for it to travel once around its elliptical orbit. Suppose that the lengths of the major and minor axes of the ellipse are $2a$ and $2b$.

(a) Use part (d) of Problem 1 to show that $T = 2\pi ab/h$.

(b) Show that $\dfrac{h^2}{GM} = ed = \dfrac{b^2}{a}$.

(c) Use parts (a) and (b) to show that $T^2 = \dfrac{4\pi^2}{GM} a^3$.

This proves Kepler's Third Law. [Notice that the proportionality constant $4\pi^2/(GM)$ is independent of the planet.]

3. The period of the earth's orbit is approximately 365.25 days. Use this fact and Kepler's Third Law to find the length of the major axis of the earth's orbit. You will need the mass of the sun, $M = 1.99 \times 10^{30}$ kg, and the gravitational constant, $G = 6.67 \times 10^{-11}\ \text{N} \cdot \text{m}^2/\text{kg}^2$.

4. It's possible to place a satellite into orbit about the earth so that it remains fixed above a given location on the equator. Compute the altitude that is needed for such a satellite. The earth's mass is 5.98×10^{24} kg; its radius is 6.37×10^6 m. (This orbit is called the Clarke Geosynchronous Orbit after Arthur C. Clarke, who first proposed the idea in 1945. The first such satellite, *Syncom II,* was launched in July 1963.)

13 REVIEW

CONCEPT CHECK

Answers to the Concept Check can be found on the back endpapers.

1. What is a vector function? How do you find its derivative and its integral?
2. What is the connection between vector functions and space curves?
3. How do you find the tangent vector to a smooth curve at a point? How do you find the tangent line? The unit tangent vector?
4. If $\mathbf{u}$ and $\mathbf{v}$ are differentiable vector functions, c is a scalar, and f is a real-valued function, write the rules for differentiating the following vector functions.
   (a) $\mathbf{u}(t) + \mathbf{v}(t)$ (b) $c\mathbf{u}(t)$ (c) $f(t)\,\mathbf{u}(t)$
   (d) $\mathbf{u}(t) \cdot \mathbf{v}(t)$ (e) $\mathbf{u}(t) \times \mathbf{v}(t)$ (f) $\mathbf{u}(f(t))$
5. How do you find the length of a space curve given by a vector function $\mathbf{r}(t)$?
6. (a) What is the definition of curvature?
   (b) Write a formula for curvature in terms of $\mathbf{r}'(t)$ and $\mathbf{T}'(t)$.
   (c) Write a formula for curvature in terms of $\mathbf{r}'(t)$ and $\mathbf{r}''(t)$.
   (d) Write a formula for the curvature of a plane curve with equation $y = f(x)$.
7. (a) Write formulas for the unit normal and binormal vectors of a smooth space curve $\mathbf{r}(t)$.
   (b) What is the normal plane of a curve at a point? What is the osculating plane? What is the osculating circle?
8. (a) How do you find the velocity, speed, and acceleration of a particle that moves along a space curve?
   (b) Write the acceleration in terms of its tangential and normal components.
9. State Kepler's Laws.

TRUE-FALSE QUIZ

Determine whether the statement is true or false. If it is true, explain why. If it is false, explain why or give an example that disproves the statement.

1. The curve with vector equation $\mathbf{r}(t) = t^3\,\mathbf{i} + 2t^3\,\mathbf{j} + 3t^3\,\mathbf{k}$ is a line.
2. The curve $\mathbf{r}(t) = \langle 0, t^2, 4t \rangle$ is a parabola.
3. The curve $\mathbf{r}(t) = \langle 2t, 3 - t, 0 \rangle$ is a line that passes through the origin.
4. The derivative of a vector function is obtained by differentiating each component function.
5. If $\mathbf{u}(t)$ and $\mathbf{v}(t)$ are differentiable vector functions, then
$$\frac{d}{dt}[\mathbf{u}(t) \times \mathbf{v}(t)] = \mathbf{u}'(t) \times \mathbf{v}'(t)$$
6. If $\mathbf{r}(t)$ is a differentiable vector function, then
$$\frac{d}{dt}|\mathbf{r}(t)| = |\mathbf{r}'(t)|$$
7. If $\mathbf{T}(t)$ is the unit tangent vector of a smooth curve, then the curvature is $\kappa = |d\mathbf{T}/dt|$.
8. The binormal vector is $\mathbf{B}(t) = \mathbf{N}(t) \times \mathbf{T}(t)$.
9. Suppose f is twice continuously differentiable. At an inflection point of the curve $y = f(x)$, the curvature is 0.
10. If $\kappa(t) = 0$ for all t, the curve is a straight line.
11. If $|\mathbf{r}(t)| = 1$ for all t, then $|\mathbf{r}'(t)|$ is a constant.
12. If $|\mathbf{r}(t)| = 1$ for all t, then $\mathbf{r}'(t)$ is orthogonal to $\mathbf{r}(t)$ for all t.
13. The osculating circle of a curve C at a point has the same tangent vector, normal vector, and curvature as C at that point.
14. Different parametrizations of the same curve result in identical tangent vectors at a given point on the curve.

EXERCISES

1. (a) Sketch the curve with vector function

$$\mathbf{r}(t) = t\,\mathbf{i} + \cos \pi t\,\mathbf{j} + \sin \pi t\,\mathbf{k} \qquad t \geqslant 0$$

(b) Find $\mathbf{r}'(t)$ and $\mathbf{r}''(t)$.

2. Let $\mathbf{r}(t) = \langle \sqrt{2 - t}, (e^t - 1)/t, \ln(t + 1) \rangle$.
(a) Find the domain of $\mathbf{r}$.
(b) Find $\lim_{t \to 0} \mathbf{r}(t)$.
(c) Find $\mathbf{r}'(t)$.

3. Find a vector function that represents the curve of intersection of the cylinder $x^2 + y^2 = 16$ and the plane $x + z = 5$.

4. Find parametric equations for the tangent line to the curve $x = 2 \sin t$, $y = 2 \sin 2t$, $z = 2 \sin 3t$ at the point $(1, \sqrt{3}, 2)$. Graph the curve and the tangent line on a common screen.

5. If $\mathbf{r}(t) = t^2\,\mathbf{i} + t \cos \pi t\,\mathbf{j} + \sin \pi t\,\mathbf{k}$, evaluate $\int_0^1 \mathbf{r}(t)\,dt$.

6. Let C be the curve with equations $x = 2 - t^3$, $y = 2t - 1$, $z = \ln t$. Find (a) the point where C intersects the xz-plane, (b) parametric equations of the tangent line at $(1, 1, 0)$, and (c) an equation of the normal plane to C at $(1, 1, 0)$.

7. Use Simpson's Rule with $n = 6$ to estimate the length of the arc of the curve with equations $x = t^2$, $y = t^3$, $z = t^4$, $0 \leqslant t \leqslant 3$.

8. Find the length of the curve $\mathbf{r}(t) = \langle 2t^{3/2}, \cos 2t, \sin 2t \rangle$, $0 \leqslant t \leqslant 1$.

9. The helix $\mathbf{r}_1(t) = \cos t\,\mathbf{i} + \sin t\,\mathbf{j} + t\,\mathbf{k}$ intersects the curve $\mathbf{r}_2(t) = (1 + t)\mathbf{i} + t^2\,\mathbf{j} + t^3\,\mathbf{k}$ at the point $(1, 0, 0)$. Find the angle of intersection of these curves.

10. Reparametrize the curve $\mathbf{r}(t) = e^t\,\mathbf{i} + e^t \sin t\,\mathbf{j} + e^t \cos t\,\mathbf{k}$ with respect to arc length measured from the point $(1, 0, 1)$ in the direction of increasing t.

11. For the curve given by $\mathbf{r}(t) = \langle \sin^3 t, \cos^3 t, \sin^2 t \rangle$, $0 \leqslant t \leqslant \pi/2$, find
(a) the unit tangent vector,
(b) the unit normal vector,
(c) the unit binormal vector, and
(d) the curvature.

12. Find the curvature of the ellipse $x = 3 \cos t$, $y = 4 \sin t$ at the points $(3, 0)$ and $(0, 4)$.

13. Find the curvature of the curve $y = x^4$ at the point $(1, 1)$.

14. Find an equation of the osculating circle of the curve $y = x^4 - x^2$ at the origin. Graph both the curve and its osculating circle.

15. Find an equation of the osculating plane of the curve $x = \sin 2t$, $y = t$, $z = \cos 2t$ at the point $(0, \pi, 1)$.

16. The figure shows the curve C traced by a particle with position vector $\mathbf{r}(t)$ at time t.
(a) Draw a vector that represents the average velocity of the particle over the time interval $3 \leqslant t \leqslant 3.2$.
(b) Write an expression for the velocity $\mathbf{v}(3)$.
(c) Write an expression for the unit tangent vector $\mathbf{T}(3)$ and draw it.

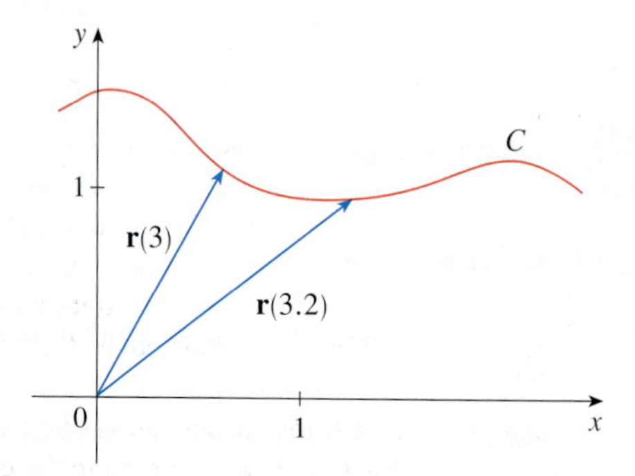

17. A particle moves with position function $\mathbf{r}(t) = t \ln t\,\mathbf{i} + t\,\mathbf{j} + e^{-t}\,\mathbf{k}$. Find the velocity, speed, and acceleration of the particle.

18. Find the velocity, speed, and acceleration of a particle moving with position function $\mathbf{r}(t) = (2t^2 - 3)\,\mathbf{i} + 2t\,\mathbf{j}$. Sketch the path of the particle and draw the position, velocity, and acceleration vectors for $t = 1$.

19. A particle starts at the origin with initial velocity $\mathbf{i} - \mathbf{j} + 3\mathbf{k}$. Its acceleration is $\mathbf{a}(t) = 6t\,\mathbf{i} + 12t^2\,\mathbf{j} - 6t\,\mathbf{k}$. Find its position function.

20. An athlete throws a shot at an angle of 45° to the horizontal at an initial speed of 43 ft/s. It leaves his hand 7 ft above the ground.
(a) Where is the shot 2 seconds later?
(b) How high does the shot go?
(c) Where does the shot land?

21. A projectile is launched with an initial speed of 40 m/s from the floor of a tunnel whose height is 30 m. What angle of elevation should be used to achieve the maximum possible horizontal range of the projectile? What is the maximum range?

22. Find the tangential and normal components of the acceleration vector of a particle with position function

$$\mathbf{r}(t) = t\,\mathbf{i} + 2t\,\mathbf{j} + t^2\,\mathbf{k}$$

23. A disk of radius 1 is rotating in the counterclockwise direction at a constant angular speed ω. A particle starts at the center of the disk and moves toward the edge along a fixed radius so that its position at time t, $t \geqslant 0$, is given by

$\mathbf{r}(t) = t\mathbf{R}(t)$, where

$$\mathbf{R}(t) = \cos \omega t \, \mathbf{i} + \sin \omega t \, \mathbf{j}$$

(a) Show that the velocity $\mathbf{v}$ of the particle is

$$\mathbf{v} = \cos \omega t \, \mathbf{i} + \sin \omega t \, \mathbf{j} + t\mathbf{v}_d$$

where $\mathbf{v}_d = \mathbf{R}'(t)$ is the velocity of a point on the edge of the disk.

(b) Show that the acceleration $\mathbf{a}$ of the particle is

$$\mathbf{a} = 2\mathbf{v}_d + t\mathbf{a}_d$$

where $\mathbf{a}_d = \mathbf{R}''(t)$ is the acceleration of a point on the edge of the disk. The extra term $2\mathbf{v}_d$ is called the *Coriolis acceleration*; it is the result of the interaction of the rotation of the disk and the motion of the particle. One can obtain a physical demonstration of this acceleration by walking toward the edge of a moving merry-go-round.

(c) Determine the Coriolis acceleration of a particle that moves on a rotating disk according to the equation

$$\mathbf{r}(t) = e^{-t}\cos \omega t \, \mathbf{i} + e^{-t}\sin \omega t \, \mathbf{j}$$

24. In designing *transfer curves* to connect sections of straight railroad tracks, it's important to realize that the acceleration of the train should be continuous so that the reactive force exerted by the train on the track is also continuous. Because of the formulas for the components of acceleration in Section 13.4, this will be the case if the curvature varies continuously.

(a) A logical candidate for a transfer curve to join existing tracks given by $y = 1$ for $x \leqslant 0$ and $y = \sqrt{2} - x$ for $x \geqslant 1/\sqrt{2}$ might be the function $f(x) = \sqrt{1 - x^2}$, $0 < x < 1/\sqrt{2}$, whose graph is the arc of the circle shown in the figure. It looks reasonable at first glance. Show that the function

$$F(x) = \begin{cases} 1 & \text{if } x \leqslant 0 \\ \sqrt{1 - x^2} & \text{if } 0 < x < 1/\sqrt{2} \\ \sqrt{2} - x & \text{if } x \geqslant 1/\sqrt{2} \end{cases}$$

is continuous and has continuous slope, but does not have continuous curvature. Therefore f is not an appropriate transfer curve.

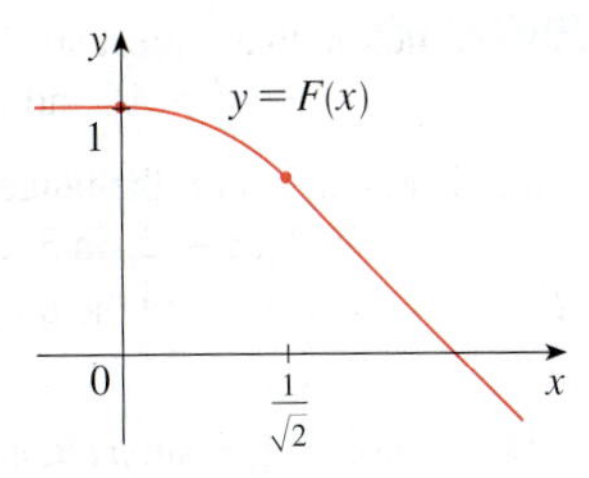

(b) Find a fifth-degree polynomial to serve as a transfer curve between the following straight line segments: $y = 0$ for $x \leqslant 0$ and $y = x$ for $x \geqslant 1$. Could this be done with a fourth-degree polynomial? Use a graphing calculator or computer to sketch the graph of the "connected" function and check to see that it looks like the one in the figure.

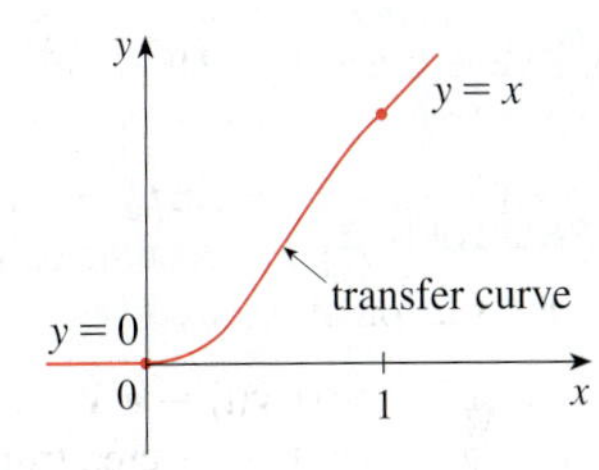

Problems Plus

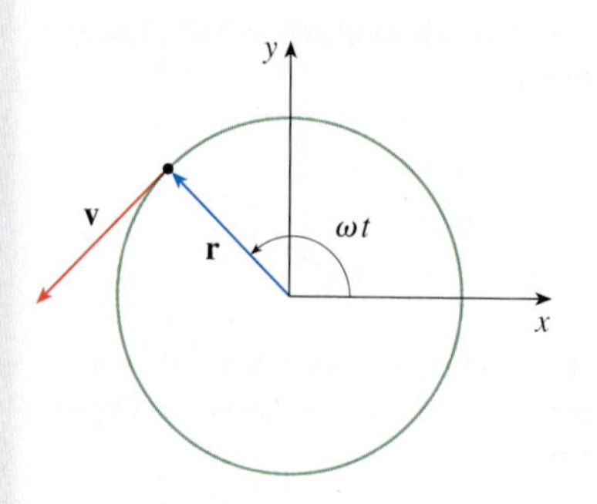

FIGURE FOR PROBLEM 1

1. A particle P moves with constant angular speed ω around a circle whose center is at the origin and whose radius is R. The particle is said to be in *uniform circular motion.* Assume that the motion is counterclockwise and that the particle is at the point $(R, 0)$ when $t = 0$. The position vector at time $t \geq 0$ is $\mathbf{r}(t) = R\cos\omega t\,\mathbf{i} + R\sin\omega t\,\mathbf{j}$.

(a) Find the velocity vector $\mathbf{v}$ and show that $\mathbf{v}\cdot\mathbf{r} = 0$. Conclude that $\mathbf{v}$ is tangent to the circle and points in the direction of the motion.

(b) Show that the speed $|\mathbf{v}|$ of the particle is the constant ωR. The *period* T of the particle is the time required for one complete revolution. Conclude that

$$T = \frac{2\pi R}{|\mathbf{v}|} = \frac{2\pi}{\omega}$$

(c) Find the acceleration vector $\mathbf{a}$. Show that it is proportional to $\mathbf{r}$ and that it points toward the origin. An acceleration with this property is called a *centripetal acceleration.* Show that the magnitude of the acceleration vector is $|\mathbf{a}| = R\omega^2$.

(d) Suppose that the particle has mass m. Show that the magnitude of the force $\mathbf{F}$ that is required to produce this motion, called a *centripetal force,* is

$$|\mathbf{F}| = \frac{m|\mathbf{v}|^2}{R}$$

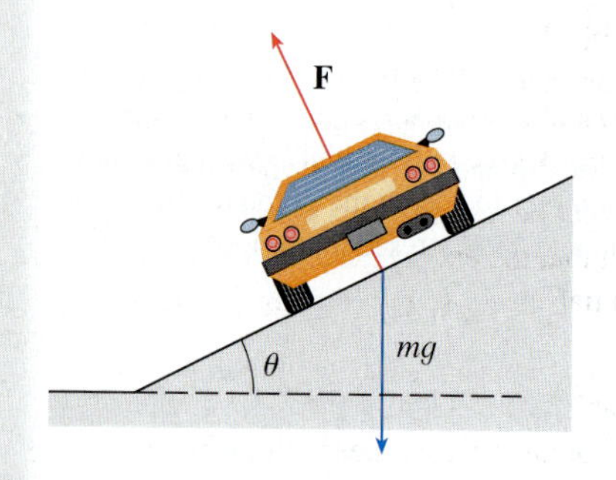

FIGURE FOR PROBLEM 2

2. A circular curve of radius R on a highway is banked at an angle θ so that a car can safely traverse the curve without skidding when there is no friction between the road and the tires. The loss of friction could occur, for example, if the road is covered with a film of water or ice. The rated speed v_R of the curve is the maximum speed that a car can attain without skidding. Suppose a car of mass m is traversing the curve at the rated speed v_R. Two forces are acting on the car: the vertical force, mg, due to the weight of the car, and a force $\mathbf{F}$ exerted by, and normal to, the road (see the figure).

The vertical component of $\mathbf{F}$ balances the weight of the car, so that $|\mathbf{F}|\cos\theta = mg$. The horizontal component of $\mathbf{F}$ produces a centripetal force on the car so that, by Newton's Second Law and part (d) of Problem 1,

$$|\mathbf{F}|\sin\theta = \frac{mv_R^2}{R}$$

(a) Show that $v_R^2 = Rg\tan\theta$.

(b) Find the rated speed of a circular curve with radius 400 ft that is banked at an angle of 12°.

(c) Suppose the design engineers want to keep the banking at 12°, but wish to increase the rated speed by 50%. What should the radius of the curve be?

3. A projectile is fired from the origin with angle of elevation α and initial speed v_0. Assuming that air resistance is negligible and that the only force acting on the projectile is gravity, g, we showed in Example 13.4.5 that the position vector of the projectile is

$$\mathbf{r}(t) = (v_0\cos\alpha)t\,\mathbf{i} + \left[(v_0\sin\alpha)t - \tfrac{1}{2}gt^2\right]\mathbf{j}$$

We also showed that the maximum horizontal distance of the projectile is achieved when $\alpha = 45°$ and in this case the range is $R = v_0^2/g$.

(a) At what angle should the projectile be fired to achieve maximum height and what is the maximum height?

(b) Fix the initial speed v_0 and consider the parabola $x^2 + 2Ry - R^2 = 0$, whose graph is shown in the figure at the left. Show that the projectile can hit any target inside or on the boundary of the region bounded by the parabola and the x-axis, and that it can't hit any target outside this region.

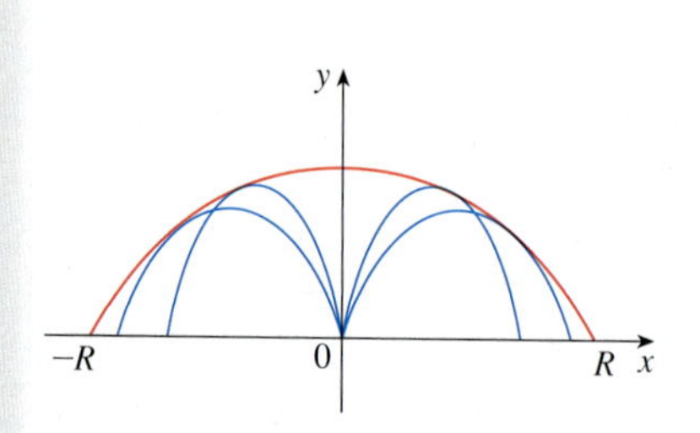

FIGURE FOR PROBLEM 3

(c) Suppose that the gun is elevated to an angle of inclination α in order to aim at a target that is suspended at a height h directly over a point D units downrange (see the figure below). The target is released at the instant the gun is fired. Show that the projectile always hits the target, regardless of the value v_0, provided the projectile does not hit the ground "before" D.

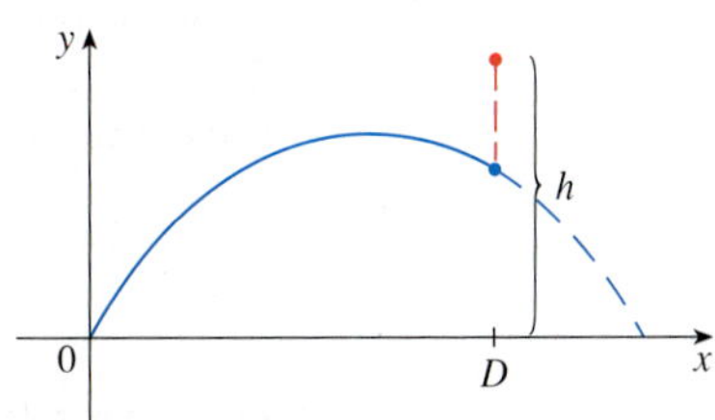

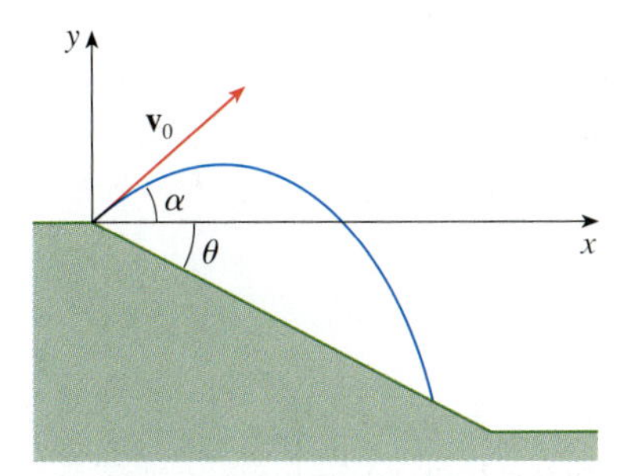

FIGURE FOR PROBLEM 4

4. (a) A projectile is fired from the origin down an inclined plane that makes an angle θ with the horizontal. The angle of elevation of the gun and the initial speed of the projectile are α and v_0, respectively. Find the position vector of the projectile and the parametric equations of the path of the projectile as functions of the time t. (Ignore air resistance.)

(b) Show that the angle of elevation α that will maximize the downhill range is the angle halfway between the plane and the vertical.

(c) Suppose the projectile is fired up an inclined plane whose angle of inclination is θ. Show that, in order to maximize the (uphill) range, the projectile should be fired in the direction halfway between the plane and the vertical.

(d) In a paper presented in 1686, Edmond Halley summarized the laws of gravity and projectile motion and applied them to gunnery. One problem he posed involved firing a projectile to hit a target a distance R up an inclined plane. Show that the angle at which the projectile should be fired to hit the target but use the least amount of energy is the same as the angle in part (c). (Use the fact that the energy needed to fire the projectile is proportional to the square of the initial speed, so minimizing the energy is equivalent to minimizing the initial speed.)

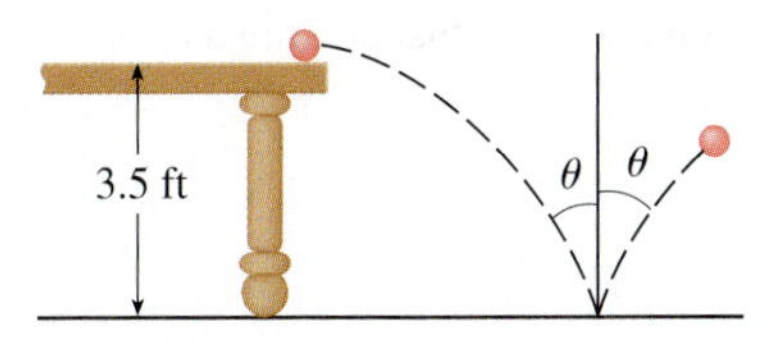

FIGURE FOR PROBLEM 5

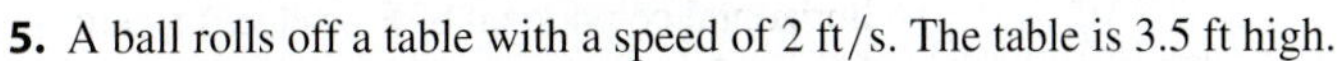

5. A ball rolls off a table with a speed of 2 ft/s. The table is 3.5 ft high.

(a) Determine the point at which the ball hits the floor and find its speed at the instant of impact.

(b) Find the angle θ between the path of the ball and the vertical line drawn through the point of impact (see the figure).

(c) Suppose the ball rebounds from the floor at the same angle with which it hits the floor, but loses 20% of its speed due to energy absorbed by the ball on impact. Where does the ball strike the floor on the second bounce?

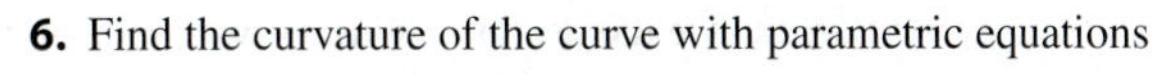

6. Find the curvature of the curve with parametric equations

$$x = \int_0^t \sin\left(\tfrac{1}{2}\pi\theta^2\right) d\theta \qquad y = \int_0^t \cos\left(\tfrac{1}{2}\pi\theta^2\right) d\theta$$

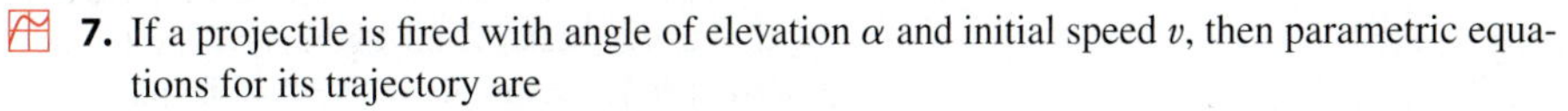

7. If a projectile is fired with angle of elevation α and initial speed v, then parametric equations for its trajectory are

$$x = (v\cos\alpha)t \qquad y = (v\sin\alpha)t - \tfrac{1}{2}gt^2$$

(See Example 13.4.5.) We know that the range (horizontal distance traveled) is maximized when $\alpha = 45°$. What value of α maximizes the total distance traveled by the projectile? (State your answer correct to the nearest degree.)

8. A cable has radius r and length L and is wound around a spool with radius R without overlapping. What is the shortest length along the spool that is covered by the cable?

9. Show that the curve with vector equation

$$\mathbf{r}(t) = \langle a_1t^2 + b_1t + c_1,\ a_2t^2 + b_2t + c_2,\ a_3t^2 + b_3t + c_3 \rangle$$

lies in a plane and find an equation of the plane.

14 Partial Derivatives

In 2008 Speedo introduced the LZR Racer and, because it reduced drag in the water, many swimming records were broken. In the project on page 976 you are asked to use partial derivatives to explain why a small decrease in drag can have a big effect on performance.

Courtesy Speedo and ANSYS, Inc.

SO FAR WE HAVE DEALT with the calculus of functions of a single variable. But, in the real world, physical quantities often depend on two or more variables, so in this chapter we turn our attention to functions of several variables and extend the basic ideas of differential calculus to such functions.

14.1 Functions of Several Variables

In this section we study functions of two or more variables from four points of view:

- verbally (by a description in words)
- numerically (by a table of values)
- algebraically (by an explicit formula)
- visually (by a graph or level curves)

Functions of Two Variables

The temperature T at a point on the surface of the earth at any given time depends on the longitude x and latitude y of the point. We can think of T as being a function of the two variables x and y, or as a function of the pair (x, y). We indicate this functional dependence by writing $T = f(x, y)$.

The volume V of a circular cylinder depends on its radius r and its height h. In fact, we know that $V = \pi r^2 h$. We say that V is a function of r and h, and we write $V(r, h) = \pi r^2 h$.

Definition A **function f of two variables** is a rule that assigns to each ordered pair of real numbers (x, y) in a set D a unique real number denoted by $f(x, y)$. The set D is the **domain** of f and its **range** is the set of values that f takes on, that is, $\{f(x, y) \mid (x, y) \in D\}$.

We often write $z = f(x, y)$ to make explicit the value taken on by f at the general point (x, y). The variables x and y are **independent variables** and z is the **dependent variable**. [Compare this with the notation $y = f(x)$ for functions of a single variable.]

A function of two variables is just a function whose domain is a subset of $\mathbb{R}^2$ and whose range is a subset of $\mathbb{R}$. One way of visualizing such a function is by means of an arrow diagram (see Figure 1), where the domain D is represented as a subset of the xy-plane and the range is a set of numbers on a real line, shown as a z-axis. For instance, if $f(x, y)$ represents the temperature at a point (x, y) in a flat metal plate with the shape of D, we can think of the z-axis as a thermometer displaying the recorded temperatures.

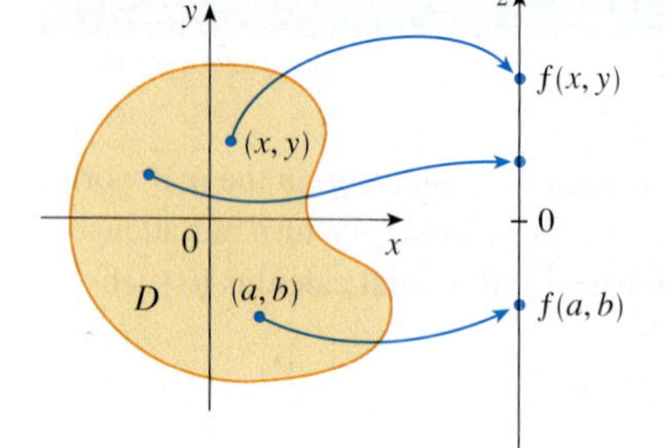

FIGURE 1

If a function f is given by a formula and no domain is specified, then the domain of f is understood to be the set of all pairs (x, y) for which the given expression is a well-defined real number.

EXAMPLE 1 For each of the following functions, evaluate $f(3, 2)$ and find and sketch the domain.

(a) $f(x, y) = \dfrac{\sqrt{x + y + 1}}{x - 1}$ (b) $f(x, y) = x \ln(y^2 - x)$

SOLUTION

(a) $f(3, 2) = \dfrac{\sqrt{3 + 2 + 1}}{3 - 1} = \dfrac{\sqrt{6}}{2}$

The expression for f makes sense if the denominator is not 0 and the quantity under the square root sign is nonnegative. So the domain of f is

$$D = \{(x, y) \mid x + y + 1 \geqslant 0,\ x \neq 1\}$$

The inequality $x + y + 1 \geqslant 0$, or $y \geqslant -x - 1$, describes the points that lie on or

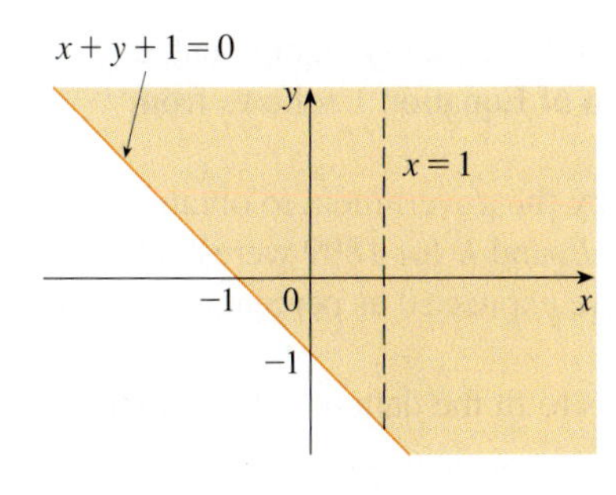

FIGURE 2

Domain of $f(x, y) = \dfrac{\sqrt{x + y + 1}}{x - 1}$

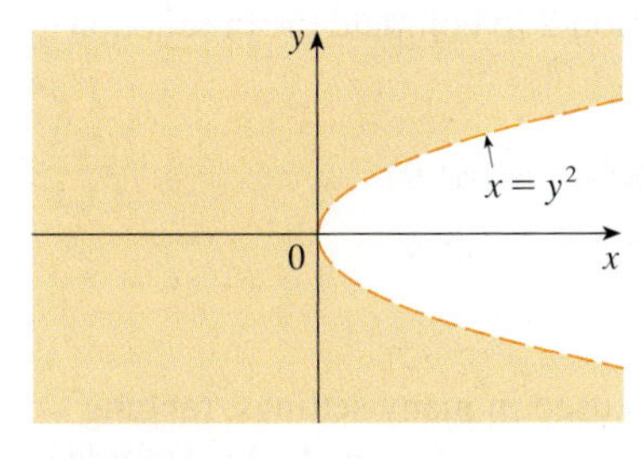

FIGURE 3

Domain of $f(x, y) = x \ln(y^2 - x)$

above the line $y = -x - 1$, while $x \neq 1$ means that the points on the line $x = 1$ must be excluded from the domain. (See Figure 2.)

(b)
$$f(3, 2) = 3 \ln(2^2 - 3) = 3 \ln 1 = 0$$

Since $\ln(y^2 - x)$ is defined only when $y^2 - x > 0$, that is, $x < y^2$, the domain of f is $D = \{(x, y) \mid x < y^2\}$. This is the set of points to the left of the parabola $x = y^2$. (See Figure 3.) ■

Not all functions can be represented by explicit formulas. The function in the next example is described verbally and by numerical estimates of its values.

EXAMPLE 2 In regions with severe winter weather, the *wind-chill index* is often used to describe the apparent severity of the cold. This index W is a subjective temperature that depends on the actual temperature T and the wind speed v. So W is a function of T and v, and we can write $W = f(T, v)$. Table 1 records values of W compiled by the US National Weather Service and the Meteorological Service of Canada.

Table 1 Wind-chill index as a function of air temperature and wind speed

Wind speed (km/h)

Actual temperature (°C) T \ v	5	10	15	20	25	30	40	50	60	70	80
5	4	3	2	1	1	0	−1	−1	−2	−2	−3
0	−2	−3	−4	−5	−6	−6	−7	−8	−9	−9	−10
−5	−7	−9	−11	−12	−12	−13	−14	−15	−16	−16	−17
−10	−13	−15	−17	−18	−19	−20	−21	−22	−23	−23	−24
−15	−19	−21	−23	−24	−25	−26	−27	−29	−30	−30	−31
−20	−24	−27	−29	−30	−32	−33	−34	−35	−36	−37	−38
−25	−30	−33	−35	−37	−38	−39	−41	−42	−43	−44	−45
−30	−36	−39	−41	−43	−44	−46	−48	−49	−50	−51	−52
−35	−41	−45	−48	−49	−51	−52	−54	−56	−57	−58	−60
−40	−47	−51	−54	−56	−57	−59	−61	−63	−64	−65	−67

The Wind-Chill Index

The wind-chill index measures how cold it feels when it's windy. It is based on a model of how fast a human face loses heat. It was developed through clinical trials in which volunteers were exposed to a variety of temperatures and wind speeds in a refrigerated wind tunnel.

For instance, the table shows that if the temperature is −5°C and the wind speed is 50 km/h, then subjectively it would feel as cold as a temperature of about −15°C with no wind. So

$$f(-5, 50) = -15$$ ■

EXAMPLE 3 In 1928 Charles Cobb and Paul Douglas published a study in which they modeled the growth of the American economy during the period 1899–1922. They considered a simplified view of the economy in which production output is determined by the amount of labor involved and the amount of capital invested. While there are many other factors affecting economic performance, their model proved to be remarkably accurate. The function they used to model production was of the form

$$P(L, K) = bL^{\alpha}K^{1-\alpha} \tag{1}$$

where P is the total production (the monetary value of all goods produced in a year), L is the amount of labor (the total number of person-hours worked in a year), and K is the amount of capital invested (the monetary worth of all machinery, equipment, and buildings). In Section 14.3 we will show how the form of Equation 1 follows from certain economic assumptions.

Cobb and Douglas used economic data published by the government to obtain Table 2. They took the year 1899 as a baseline and P, L, and K for 1899 were each assigned the value 100. The values for other years were expressed as percentages of the 1899 figures.

Table 2

Year	P	L	K
1899	100	100	100
1900	101	105	107
1901	112	110	114
1902	122	117	122
1903	124	122	131
1904	122	121	138
1905	143	125	149
1906	152	134	163
1907	151	140	176
1908	126	123	185
1909	155	143	198
1910	159	147	208
1911	153	148	216
1912	177	155	226
1913	184	156	236
1914	169	152	244
1915	189	156	266
1916	225	183	298
1917	227	198	335
1918	223	201	366
1919	218	196	387
1920	231	194	407
1921	179	146	417
1922	240	161	431

Cobb and Douglas used the method of least squares to fit the data of Table 2 to the function

$$P(L, K) = 1.01L^{0.75}K^{0.25} \tag{2}$$

(See Exercise 81 for the details.)

If we use the model given by the function in Equation 2 to compute the production in the years 1910 and 1920, we get the values

$$P(147, 208) = 1.01(147)^{0.75}(208)^{0.25} \approx 161.9$$

$$P(194, 407) = 1.01(194)^{0.75}(407)^{0.25} \approx 235.8$$

which are quite close to the actual values, 159 and 231.

The production function (1) has subsequently been used in many settings, ranging from individual firms to global economics. It has become known as the **Cobb-Douglas production function**. Its domain is $\{(L, K) \mid L \geqslant 0, K \geqslant 0\}$ because L and K represent labor and capital and are therefore never negative. ■

EXAMPLE 4 Find the domain and range of $g(x, y) = \sqrt{9 - x^2 - y^2}$.

SOLUTION The domain of g is

$$D = \{(x, y) \mid 9 - x^2 - y^2 \geqslant 0\} = \{(x, y) \mid x^2 + y^2 \leqslant 9\}$$

which is the disk with center (0, 0) and radius 3. (See Figure 4.) The range of g is

$$\left\{z \mid z = \sqrt{9 - x^2 - y^2}, (x, y) \in D\right\}$$

Since z is a positive square root, $z \geqslant 0$. Also, because $9 - x^2 - y^2 \leqslant 9$, we have

$$\sqrt{9 - x^2 - y^2} \leqslant 3$$

So the range is

$$\{z \mid 0 \leqslant z \leqslant 3\} = [0, 3]$$ ■

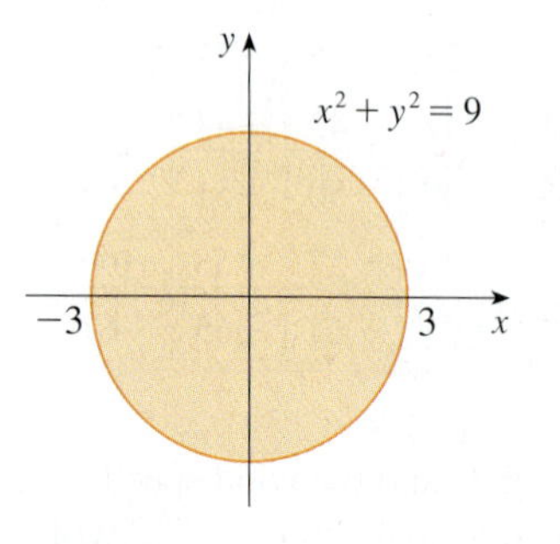

FIGURE 4
Domain of $g(x, y) = \sqrt{9 - x^2 - y^2}$

Graphs

Another way of visualizing the behavior of a function of two variables is to consider its graph.

> **Definition** If f is a function of two variables with domain D, then the **graph** of f is the set of all points (x, y, z) in $\mathbb{R}^3$ such that $z = f(x, y)$ and (x, y) is in D.

Just as the graph of a function f of one variable is a curve C with equation $y = f(x)$, so the graph of a function f of two variables is a surface S with equation $z = f(x, y)$. We can visualize the graph S of f as lying directly above or below its domain D in the xy-plane (see Figure 5).

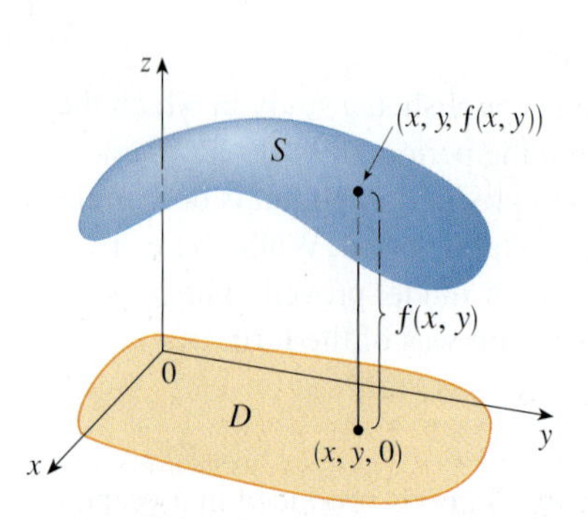

FIGURE 5

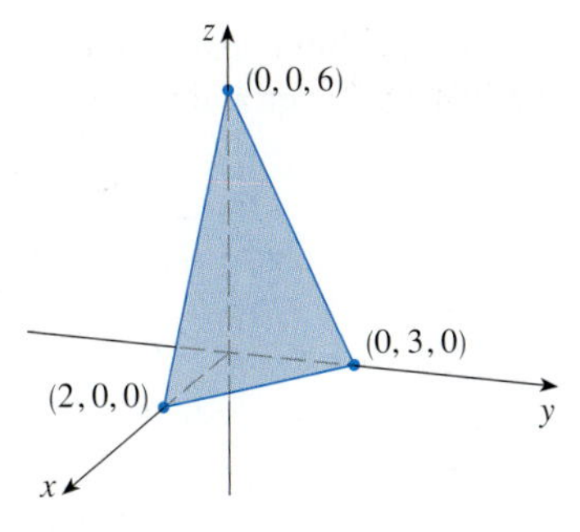

FIGURE 6

EXAMPLE 5 Sketch the graph of the function $f(x, y) = 6 - 3x - 2y$.

SOLUTION The graph of f has the equation $z = 6 - 3x - 2y$, or $3x + 2y + z = 6$, which represents a plane. To graph the plane we first find the intercepts. Putting $y = z = 0$ in the equation, we get $x = 2$ as the x-intercept. Similarly, the y-intercept is 3 and the z-intercept is 6. This helps us sketch the portion of the graph that lies in the first octant in Figure 6. ■

The function in Example 5 is a special case of the function

$$f(x, y) = ax + by + c$$

which is called a **linear function**. The graph of such a function has the equation

$$z = ax + by + c \qquad \text{or} \qquad ax + by - z + c = 0$$

so it is a plane. In much the same way that linear functions of one variable are important in single-variable calculus, we will see that linear functions of two variables play a central role in multivariable calculus.

EXAMPLE 6 Sketch the graph of $g(x, y) = \sqrt{9 - x^2 - y^2}$.

SOLUTION The graph has equation $z = \sqrt{9 - x^2 - y^2}$. We square both sides of this equation to obtain $z^2 = 9 - x^2 - y^2$, or $x^2 + y^2 + z^2 = 9$, which we recognize as an equation of the sphere with center the origin and radius 3. But, since $z \geqslant 0$, the graph of g is just the top half of this sphere (see Figure 7). ■

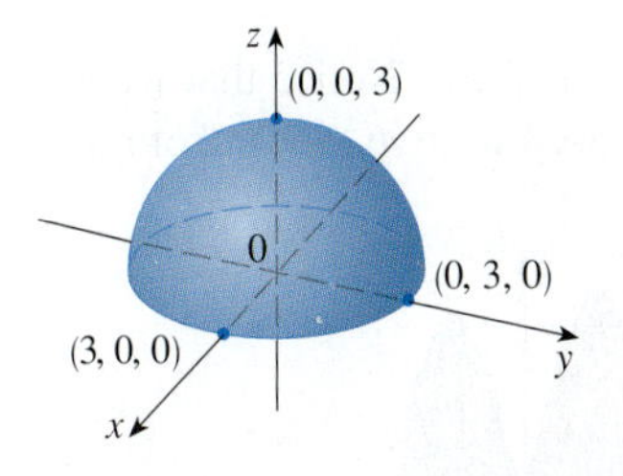

FIGURE 7
Graph of $g(x, y) = \sqrt{9 - x^2 - y^2}$

NOTE An entire sphere can't be represented by a single function of x and y. As we saw in Example 6, the upper hemisphere of the sphere $x^2 + y^2 + z^2 = 9$ is represented by the function $g(x, y) = \sqrt{9 - x^2 - y^2}$. The lower hemisphere is represented by the function $h(x, y) = -\sqrt{9 - x^2 - y^2}$.

EXAMPLE 7 Use a computer to draw the graph of the Cobb-Douglas production function $P(L, K) = 1.01L^{0.75}K^{0.25}$.

SOLUTION Figure 8 shows the graph of P for values of the labor L and capital K that lie between 0 and 300. The computer has drawn the surface by plotting vertical traces. We see from these traces that the value of the production P increases as either L or K increases, as is to be expected.

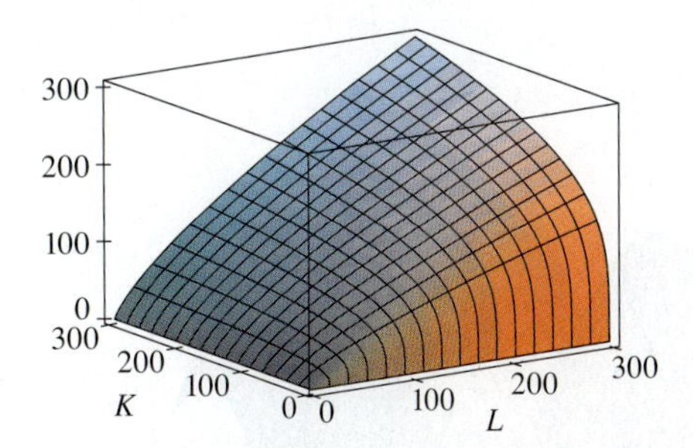

FIGURE 8 ■

EXAMPLE 8 Find the domain and range and sketch the graph of $h(x, y) = 4x^2 + y^2$.

SOLUTION Notice that $h(x, y)$ is defined for all possible ordered pairs of real numbers (x, y), so the domain is $\mathbb{R}^2$, the entire xy-plane. The range of h is the set $[0, \infty)$ of all nonnegative real numbers. [Notice that $x^2 \geqslant 0$ and $y^2 \geqslant 0$, so $h(x, y) \geqslant 0$ for all x

and y.] The graph of h has the equation $z = 4x^2 + y^2$, which is the elliptic paraboloid that we sketched in Example 12.6.4. Horizontal traces are ellipses and vertical traces are parabolas (see Figure 9).

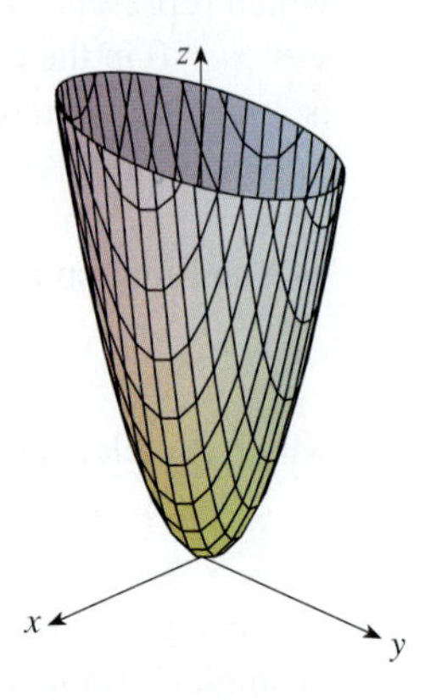

FIGURE 9
Graph of $h(x, y) = 4x^2 + y^2$

Computer programs are readily available for graphing functions of two variables. In most such programs, traces in the vertical planes $x = k$ and $y = k$ are drawn for equally spaced values of k and parts of the graph are eliminated using hidden line removal.

Figure 10 shows computer-generated graphs of several functions. Notice that we get an especially good picture of a function when rotation is used to give views from dif-

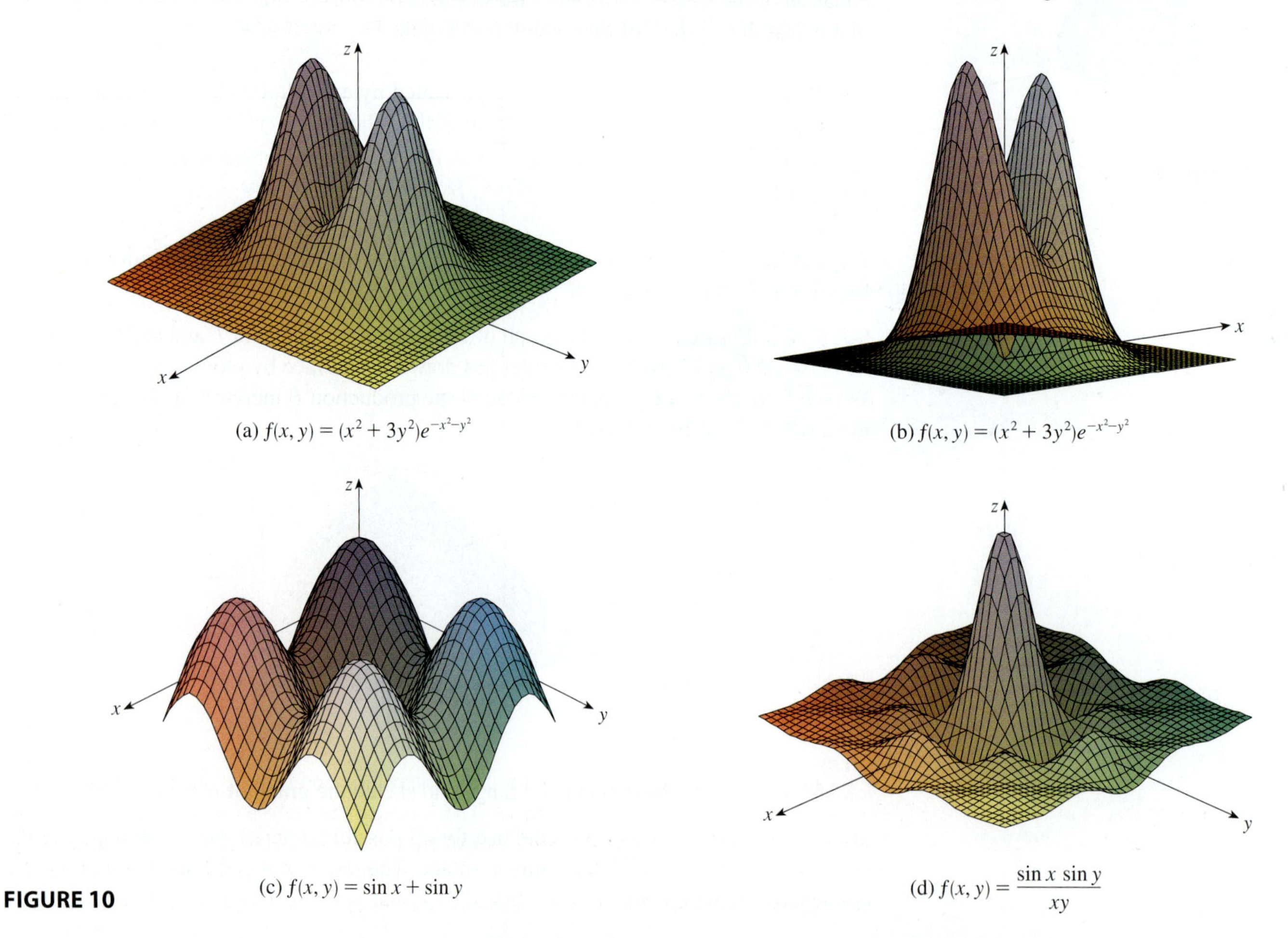

(a) $f(x, y) = (x^2 + 3y^2)e^{-x^2-y^2}$

(b) $f(x, y) = (x^2 + 3y^2)e^{-x^2-y^2}$

(c) $f(x, y) = \sin x + \sin y$

(d) $f(x, y) = \dfrac{\sin x \sin y}{xy}$

FIGURE 10

ferent vantage points. In parts (a) and (b) the graph of f is very flat and close to the xy-plane except near the origin; this is because $e^{-x^2-y^2}$ is very small when x or y is large.

Level Curves

So far we have two methods for visualizing functions: arrow diagrams and graphs. A third method, borrowed from mapmakers, is a contour map on which points of constant elevation are joined to form *contour curves*, or *level curves*.

Definition The **level curves** of a function f of two variables are the curves with equations $f(x, y) = k$, where k is a constant (in the range of f).

A level curve $f(x, y) = k$ is the set of all points in the domain of f at which f takes on a given value k. In other words, it shows where the graph of f has height k.

You can see from Figure 11 the relation between level curves and horizontal traces. The level curves $f(x, y) = k$ are just the traces of the graph of f in the horizontal plane $z = k$ projected down to the xy-plane. So if you draw the level curves of a function and visualize them being lifted up to the surface at the indicated height, then you can mentally piece together a picture of the graph. The surface is steep where the level curves are close together. It is somewhat flatter where they are farther apart.

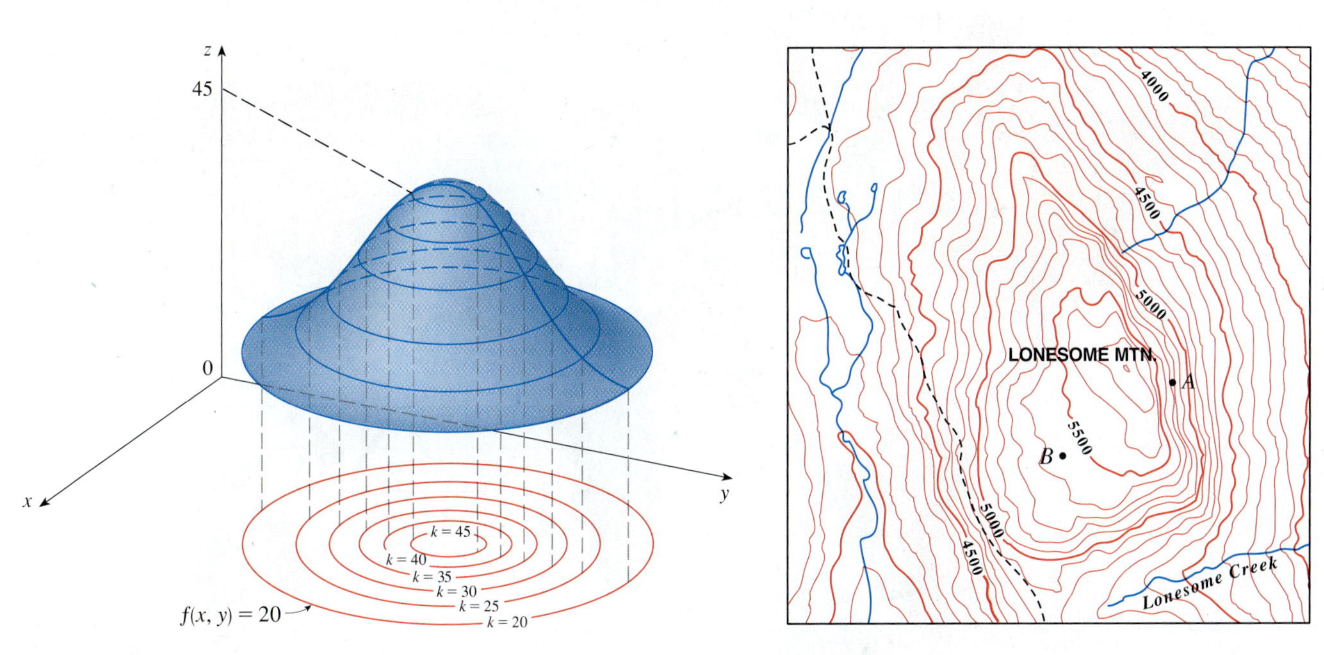

FIGURE 11

FIGURE 12

TEC Visual 14.1A animates Figure 11 by showing level curves being lifted up to graphs of functions.

One common example of level curves occurs in topographic maps of mountainous regions, such as the map in Figure 12. The level curves are curves of constant elevation above sea level. If you walk along one of these contour lines, you neither ascend nor descend. Another common example is the temperature function introduced in the opening paragraph of this section. Here the level curves are called **isothermals** and join loca-

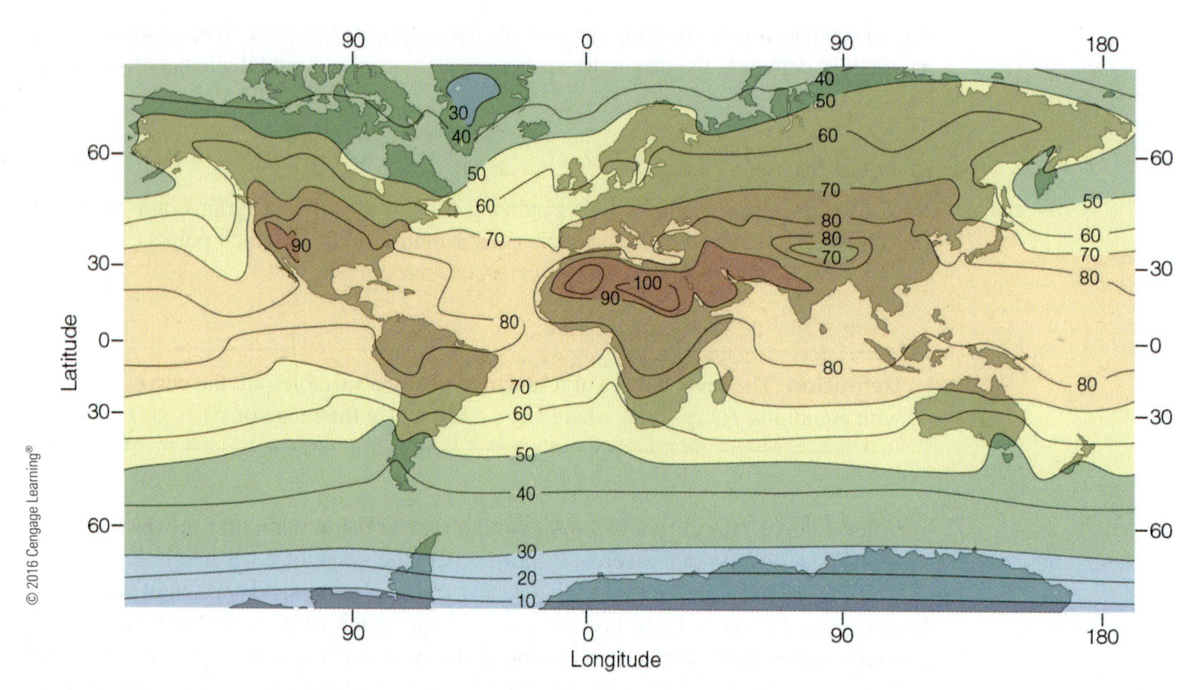

FIGURE 13 Average air temperature near sea level in July (°F)

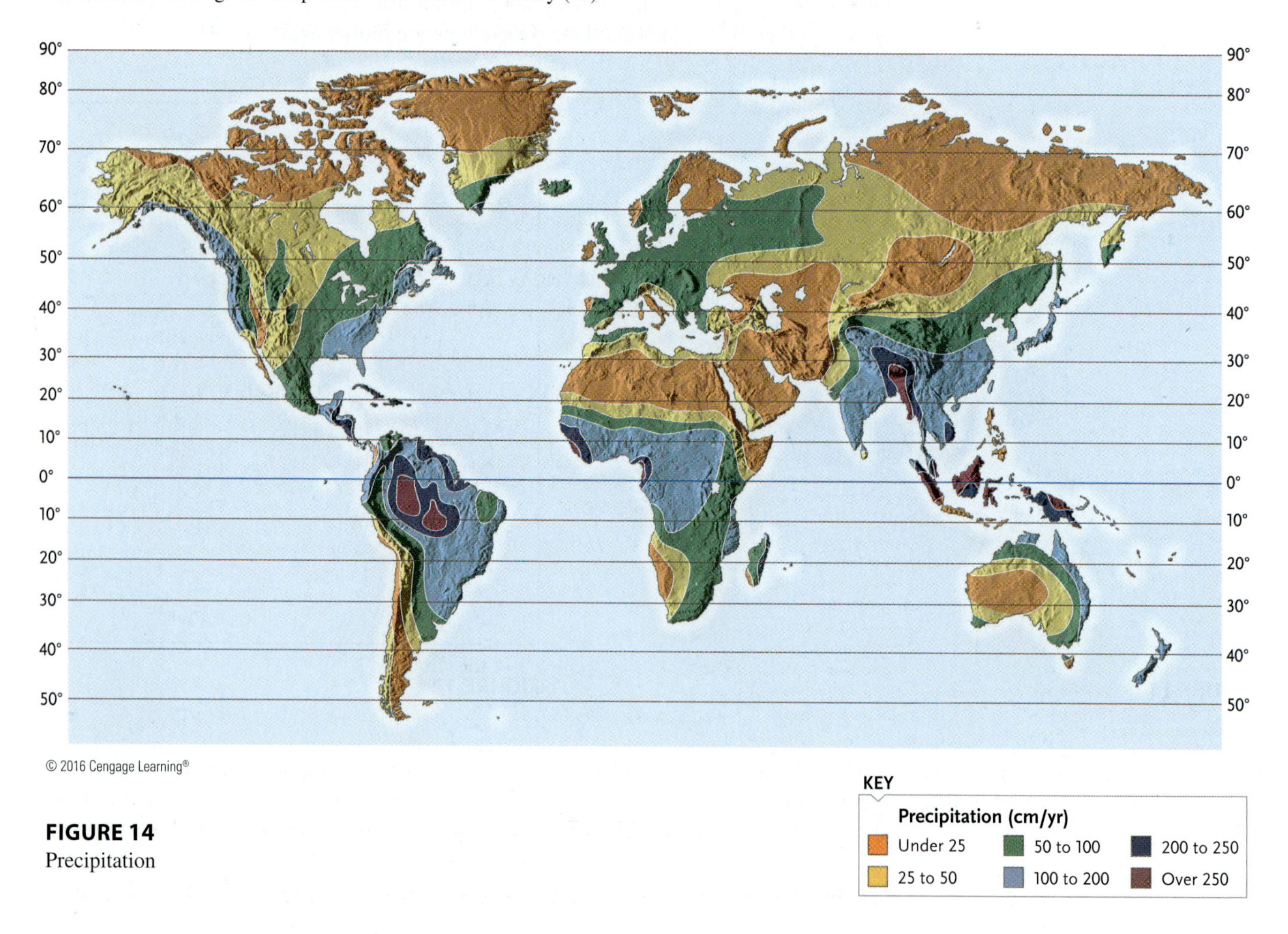

FIGURE 14
Precipitation

tions with the same temperature. Figure 13 shows a weather map of the world indicating the average July temperatures. The isothermals are the curves that separate the colored bands.

In weather maps of atmospheric pressure at a given time as a function of longitude and latitude, the level curves are called **isobars** and join locations with the same pressure. (See Exercise 34.) Surface winds tend to flow from areas of high pressure across the isobars toward areas of low pressure, and are strongest where the isobars are tightly packed.

A contour map of world-wide precipitation is shown in Figure 14. Here the level curves are not labeled but they separate the colored regions and the amount of precipitation in each region is indicated in the color key.

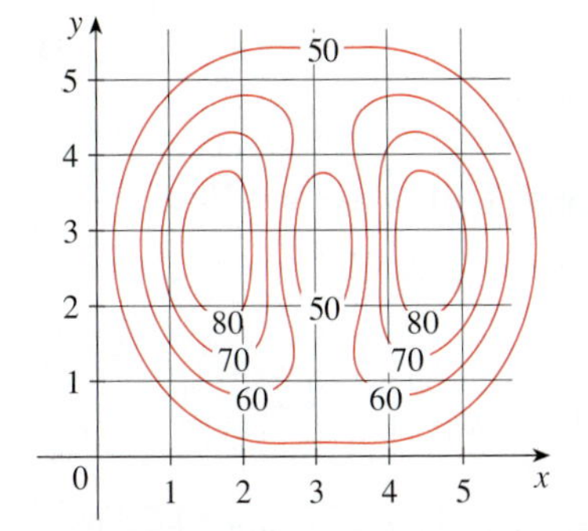

FIGURE 15

EXAMPLE 9 A contour map for a function f is shown in Figure 15. Use it to estimate the values of $f(1, 3)$ and $f(4, 5)$.

SOLUTION The point $(1, 3)$ lies partway between the level curves with z-values 70 and 80. We estimate that

$$f(1, 3) \approx 73$$

Similarly, we estimate that $\qquad f(4, 5) \approx 56$ ■

EXAMPLE 10 Sketch the level curves of the function $f(x, y) = 6 - 3x - 2y$ for the values $k = -6, 0, 6, 12$.

SOLUTION The level curves are

$$6 - 3x - 2y = k \qquad \text{or} \qquad 3x + 2y + (k - 6) = 0$$

This is a family of lines with slope $-\frac{3}{2}$. The four particular level curves with $k = -6, 0, 6$, and 12 are $3x + 2y - 12 = 0$, $3x + 2y - 6 = 0$, $3x + 2y = 0$, and $3x + 2y + 6 = 0$. They are sketched in Figure 16. The level curves are equally spaced parallel lines because the graph of f is a plane (see Figure 6).

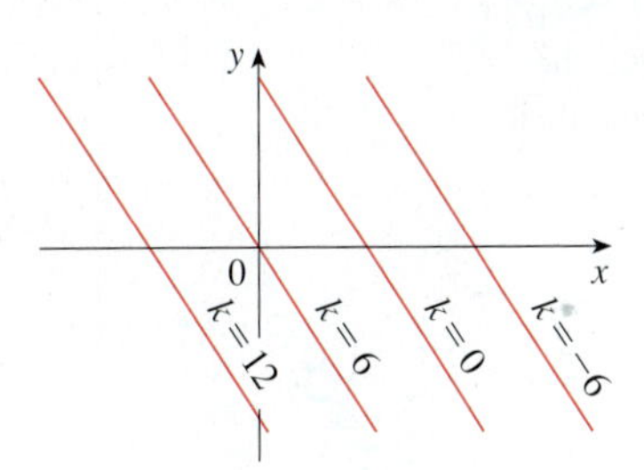

FIGURE 16
Contour map of
$f(x, y) = 6 - 3x - 2y$

■

EXAMPLE 11 Sketch the level curves of the function

$$g(x, y) = \sqrt{9 - x^2 - y^2} \qquad \text{for} \quad k = 0, 1, 2, 3$$

SOLUTION The level curves are

$$\sqrt{9 - x^2 - y^2} = k \qquad \text{or} \qquad x^2 + y^2 = 9 - k^2$$

This is a family of concentric circles with center $(0, 0)$ and radius $\sqrt{9 - k^2}$. The cases

$k = 0, 1, 2, 3$ are shown in Figure 17. Try to visualize these level curves lifted up to form a surface and compare with the graph of g (a hemisphere) in Figure 7. (See TEC Visual 14.1A.)

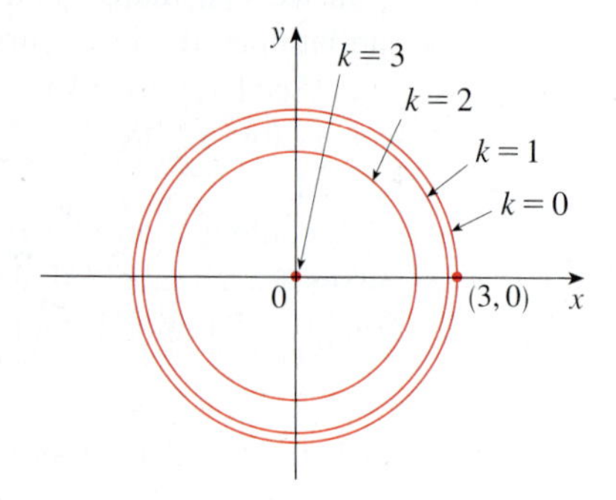

FIGURE 17
Contour map of
$g(x, y) = \sqrt{9 - x^2 - y^2}$

■

EXAMPLE 12 Sketch some level curves of the function $h(x, y) = 4x^2 + y^2 + 1$.

SOLUTION The level curves are

$$4x^2 + y^2 + 1 = k \qquad \text{or} \qquad \frac{x^2}{\frac{1}{4}(k-1)} + \frac{y^2}{k-1} = 1$$

which, for $k > 1$, describes a family of ellipses with semiaxes $\frac{1}{2}\sqrt{k-1}$ and $\sqrt{k-1}$. Figure 18(a) shows a contour map of h drawn by a computer. Figure 18(b) shows these level curves lifted up to the graph of h (an elliptic paraboloid) where they become horizontal traces. We see from Figure 18 how the graph of h is put together from the level curves.

TEC Visual 14.1B demonstrates the connection between surfaces and their contour maps.

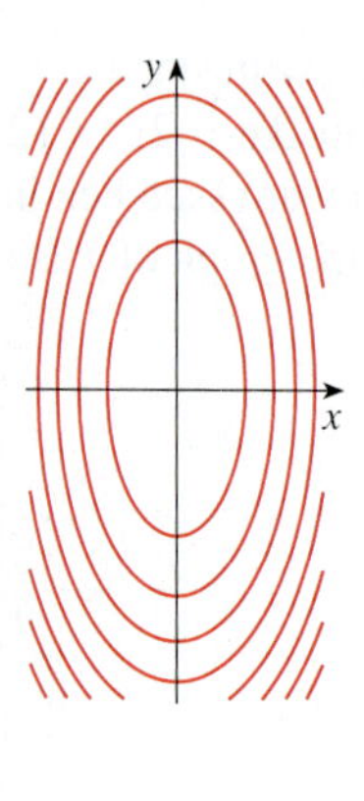

(a) Contour map

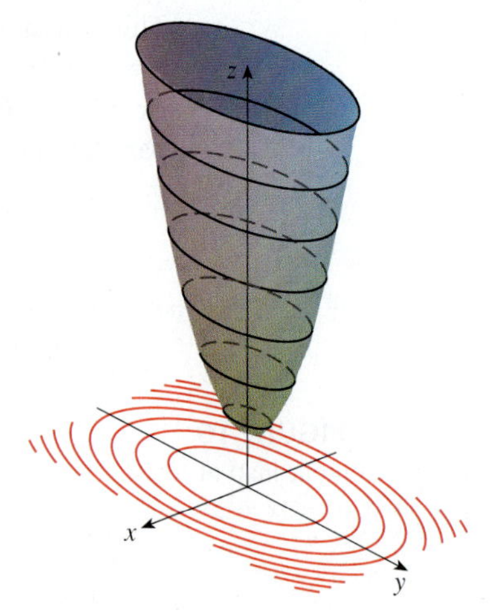

(b) Horizontal traces are raised level curves

FIGURE 18
The graph of $h(x, y) = 4x^2 + y^2 + 1$ is formed by lifting the level curves.

■

EXAMPLE 13 Plot level curves for the Cobb-Douglas production function of Example 3.

SOLUTION In Figure 19 we use a computer to draw a contour plot for the Cobb-Douglas production function

$$P(L, K) = 1.01L^{0.75}K^{0.25}$$

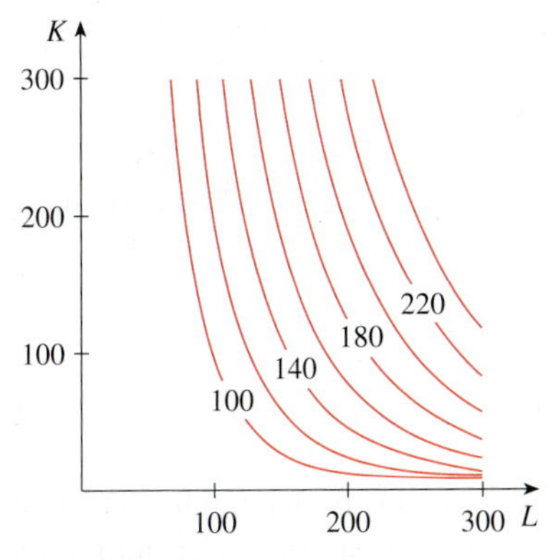

FIGURE 19

Level curves are labeled with the value of the production P. For instance, the level curve labeled 140 shows all values of the labor L and capital investment K that result in a production of $P = 140$. We see that, for a fixed value of P, as L increases K decreases, and vice versa. ■

For some purposes, a contour map is more useful than a graph. That is certainly true in Example 13. (Compare Figure 19 with Figure 8.) It is also true in estimating function values, as in Example 9.

Figure 20 shows some computer-generated level curves together with the corresponding computer-generated graphs. Notice that the level curves in part (c) crowd together near the origin. That corresponds to the fact that the graph in part (d) is very steep near the origin.

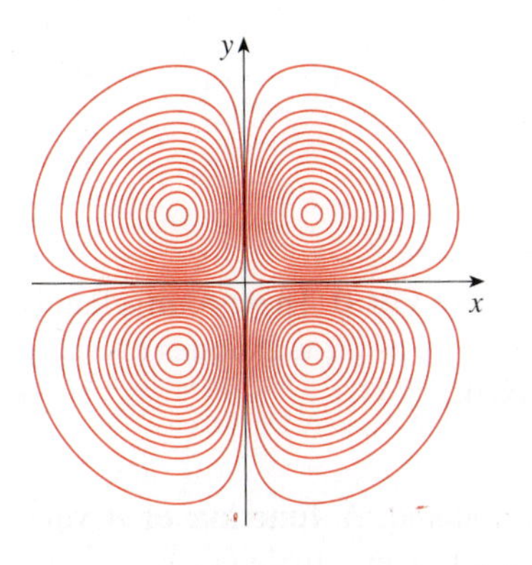

(a) Level curves of $f(x, y) = -xye^{-x^2-y^2}$

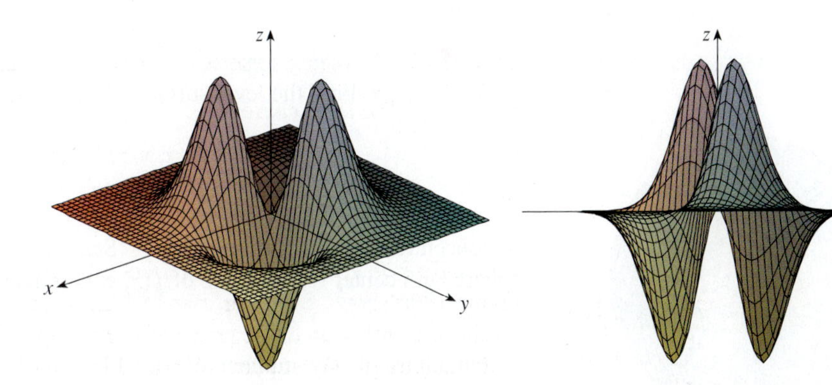

(b) Two views of $f(x, y) = -xye^{-x^2-y^2}$

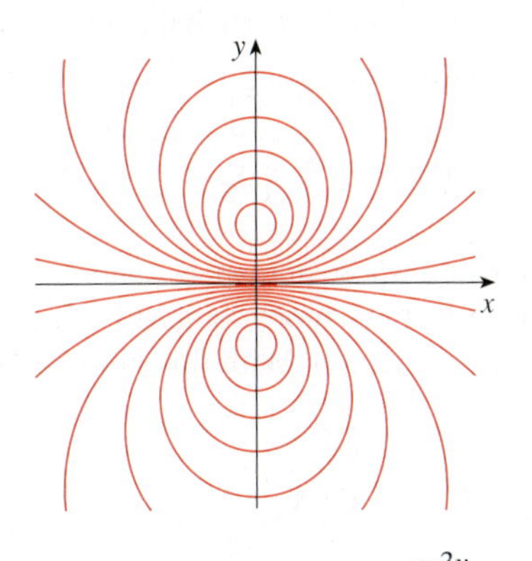

(c) Level curves of $f(x, y) = \dfrac{-3y}{x^2 + y^2 + 1}$

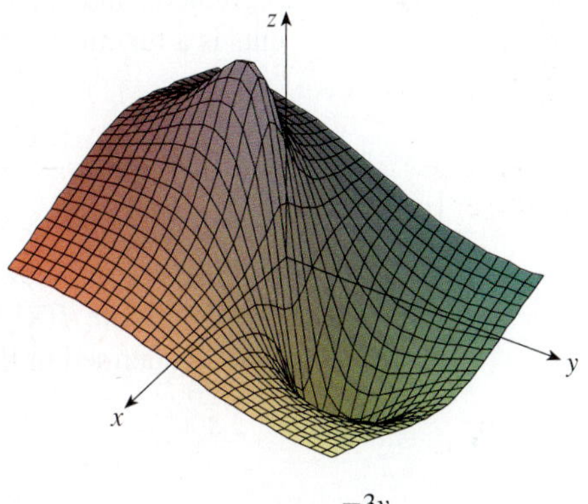

(d) $f(x, y) = \dfrac{-3y}{x^2 + y^2 + 1}$

FIGURE 20

Functions of Three or More Variables

A **function of three variables**, f, is a rule that assigns to each ordered triple (x, y, z) in a domain $D \subset \mathbb{R}^3$ a unique real number denoted by $f(x, y, z)$. For instance, the temperature T at a point on the surface of the earth depends on the longitude x and latitude y of the point and on the time t, so we could write $T = f(x, y, t)$.

EXAMPLE 14 Find the domain of f if

$$f(x, y, z) = \ln(z - y) + xy \sin z$$

SOLUTION The expression for $f(x, y, z)$ is defined as long as $z - y > 0$, so the domain of f is

$$D = \{(x, y, z) \in \mathbb{R}^3 \mid z > y\}$$

This is a **half-space** consisting of all points that lie above the plane $z = y$. ■

It's very difficult to visualize a function f of three variables by its graph, since that would lie in a four-dimensional space. However, we do gain some insight into f by examining its **level surfaces**, which are the surfaces with equations $f(x, y, z) = k$, where k is a constant. If the point (x, y, z) moves along a level surface, the value of $f(x, y, z)$ remains fixed.

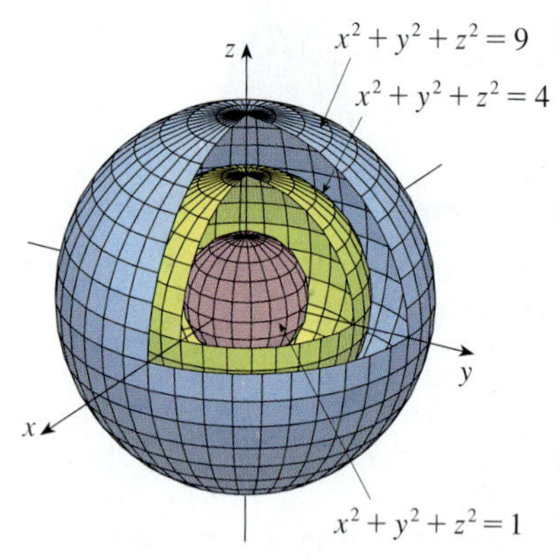

FIGURE 21

EXAMPLE 15 Find the level surfaces of the function

$$f(x, y, z) = x^2 + y^2 + z^2$$

SOLUTION The level surfaces are $x^2 + y^2 + z^2 = k$, where $k \geqslant 0$. These form a family of concentric spheres with radius $\sqrt{k}$. (See Figure 21.) Thus, as (x, y, z) varies over any sphere with center O, the value of $f(x, y, z)$ remains fixed. ■

Functions of any number of variables can be considered. A **function of *n* variables** is a rule that assigns a number $z = f(x_1, x_2, \ldots, x_n)$ to an n-tuple $(x_1, x_2, \ldots, x_n)$ of real numbers. We denote by $\mathbb{R}^n$ the set of all such n-tuples. For example, if a company uses n different ingredients in making a food product, c_i is the cost per unit of the ith ingredient, and x_i units of the ith ingredient are used, then the total cost C of the ingredients is a function of the n variables $x_1, x_2, \ldots, x_n$:

3 $$C = f(x_1, x_2, \ldots, x_n) = c_1x_1 + c_2x_2 + \cdots + c_nx_n$$

The function f is a real-valued function whose domain is a subset of $\mathbb{R}^n$. Sometimes we will use vector notation to write such functions more compactly: If $\mathbf{x} = \langle x_1, x_2, \ldots, x_n \rangle$, we often write $f(\mathbf{x})$ in place of $f(x_1, x_2, \ldots, x_n)$. With this notation we can rewrite the function defined in Equation 3 as

$$f(\mathbf{x}) = \mathbf{c} \cdot \mathbf{x}$$

where $\mathbf{c} = \langle c_1, c_2, \ldots, c_n \rangle$ and $\mathbf{c} \cdot \mathbf{x}$ denotes the dot product of the vectors $\mathbf{c}$ and $\mathbf{x}$ in V_n.

In view of the one-to-one correspondence between points $(x_1, x_2, \ldots, x_n)$ in $\mathbb{R}^n$ and their position vectors $\mathbf{x} = \langle x_1, x_2, \ldots, x_n \rangle$ in V_n, we have three ways of looking at a function f defined on a subset of $\mathbb{R}^n$:

1. As a function of n real variables $x_1, x_2, \ldots, x_n$
2. As a function of a single point variable $(x_1, x_2, \ldots, x_n)$
3. As a function of a single vector variable $\mathbf{x} = \langle x_1, x_2, \ldots, x_n \rangle$

We will see that all three points of view are useful.

14.1 EXERCISES

1. In Example 2 we considered the function $W = f(T, v)$, where W is the wind-chill index, T is the actual temperature, and v is the wind speed. A numerical representation is given in Table 1 on page 929.
(a) What is the value of $f(-15, 40)$? What is its meaning?
(b) Describe in words the meaning of the question "For what value of v is $f(-20, v) = -30$?" Then answer the question.
(c) Describe in words the meaning of the question "For what value of T is $f(T, 20) = -49$?" Then answer the question.
(d) What is the meaning of the function $W = f(-5, v)$? Describe the behavior of this function.
(e) What is the meaning of the function $W = f(T, 50)$? Describe the behavior of this function.

2. The *temperature-humidity index* I (or humidex, for short) is the perceived air temperature when the actual temperature is T and the relative humidity is h, so we can write $I = f(T, h)$. The following table of values of I is an excerpt from a table compiled by the National Oceanic & Atmospheric Administration.

Table 3 Apparent temperature as a function of temperature and humidity

Relative humidity (%)

Actual temperature (°F) T \ h	20	30	40	50	60	70
80	77	78	79	81	82	83
85	82	84	86	88	90	93
90	87	90	93	96	100	106
95	93	96	101	107	114	124
100	99	104	110	120	132	144

(a) What is the value of $f(95, 70)$? What is its meaning?
(b) For what value of h is $f(90, h) = 100$?
(c) For what value of T is $f(T, 50) = 88$?
(d) What are the meanings of the functions $I = f(80, h)$ and $I = f(100, h)$? Compare the behavior of these two functions of h.

3. A manufacturer has modeled its yearly production function P (the monetary value of its entire production in millions of dollars) as a Cobb-Douglas function

$$P(L, K) = 1.47L^{0.65}K^{0.35}$$

where L is the number of labor hours (in thousands) and K is the invested capital (in millions of dollars). Find $P(120, 20)$ and interpret it.

4. Verify for the Cobb-Douglas production function

$$P(L, K) = 1.01L^{0.75}K^{0.25}$$

discussed in Example 3 that the production will be doubled if both the amount of labor and the amount of capital are doubled. Determine whether this is also true for the general production function

$$P(L, K) = bL^{\alpha}K^{1-\alpha}$$

5. A model for the surface area of a human body is given by the function

$$S = f(w, h) = 0.1091w^{0.425}h^{0.725}$$

where w is the weight (in pounds), h is the height (in inches), and S is measured in square feet.
(a) Find $f(160, 70)$ and interpret it.
(b) What is your own surface area?

6. The wind-chill index W discussed in Example 2 has been modeled by the following function:

$$W(T, v) = 13.12 + 0.6215T - 11.37v^{0.16} + 0.3965Tv^{0.16}$$

Check to see how closely this model agrees with the values in Table 1 for a few values of T and v.

7. The wave heights h in the open sea depend on the speed v of the wind and the length of time t that the wind has been blowing at that speed. Values of the function $h = f(v, t)$ are recorded in feet in Table 4.
(a) What is the value of $f(40, 15)$? What is its meaning?
(b) What is the meaning of the function $h = f(30, t)$? Describe the behavior of this function.
(c) What is the meaning of the function $h = f(v, 30)$? Describe the behavior of this function.

Table 4

Duration (hours)

Wind speed (knots) v \ t	5	10	15	20	30	40	50
10	2	2	2	2	2	2	2
15	4	4	5	5	5	5	5
20	5	7	8	8	9	9	9
30	9	13	16	17	18	19	19
40	14	21	25	28	31	33	33
50	19	29	36	40	45	48	50
60	24	37	47	54	62	67	69

8. A company makes three sizes of cardboard boxes: small, medium, and large. It costs \$2.50 to make a small box,

\$4.00 for a medium box, and \$4.50 for a large box. Fixed costs are \$8000.

(a) Express the cost of making x small boxes, y medium boxes, and z large boxes as a function of three variables: $C = f(x, y, z)$.

(b) Find $f(3000, 5000, 4000)$ and interpret it.

(c) What is the domain of f?

9. Let $g(x, y) = \cos(x + 2y)$.

(a) Evaluate $g(2, -1)$.

(b) Find the domain of g.

(c) Find the range of g.

10. Let $F(x, y) = 1 + \sqrt{4 - y^2}$.

(a) Evaluate $F(3, 1)$.

(b) Find and sketch the domain of F.

(c) Find the range of F.

11. Let $f(x, y, z) = \sqrt{x} + \sqrt{y} + \sqrt{z} + \ln(4 - x^2 - y^2 - z^2)$.

(a) Evaluate $f(1, 1, 1)$.

(b) Find and describe the domain of f.

12. Let $g(x, y, z) = x^3y^2z\sqrt{10 - x - y - z}$.

(a) Evaluate $g(1, 2, 3)$.

(b) Find and describe the domain of g.

13–22 Find and sketch the domain of the function.

13. $f(x, y) = \sqrt{x - 2} + \sqrt{y - 1}$

14. $f(x, y) = \sqrt[4]{x - 3y}$

15. $f(x, y) = \ln(9 - x^2 - 9y^2)$

16. $f(x, y) = \sqrt{x^2 + y^2 - 4}$

17. $g(x, y) = \dfrac{x - y}{x + y}$

18. $g(x, y) = \dfrac{\ln(2 - x)}{1 - x^2 - y^2}$

19. $f(x, y) = \dfrac{\sqrt{y - x^2}}{1 - x^2}$

20. $f(x, y) = \sin^{-1}(x + y)$

21. $f(x, y, z) = \sqrt{4 - x^2} + \sqrt{9 - y^2} + \sqrt{1 - z^2}$

22. $f(x, y, z) = \ln(16 - 4x^2 - 4y^2 - z^2)$

23–31 Sketch the graph of the function.

23. $f(x, y) = y$

24. $f(x, y) = x^2$

25. $f(x, y) = 10 - 4x - 5y$

26. $f(x, y) = \cos y$

27. $f(x, y) = \sin x$

28. $f(x, y) = 2 - x^2 - y^2$

29. $f(x, y) = x^2 + 4y^2 + 1$

30. $f(x, y) = \sqrt{4x^2 + y^2}$

31. $f(x, y) = \sqrt{4 - 4x^2 - y^2}$

32. Match the function with its graph (labeled I–VI). Give reasons for your choices.

(a) $f(x, y) = \dfrac{1}{1 + x^2 + y^2}$

(b) $f(x, y) = \dfrac{1}{1 + x^2y^2}$

(c) $f(x, y) = \ln(x^2 + y^2)$

(d) $f(x, y) = \cos\sqrt{x^2 + y^2}$

(e) $f(x, y) = |xy|$

(f) $f(x, y) = \cos(xy)$

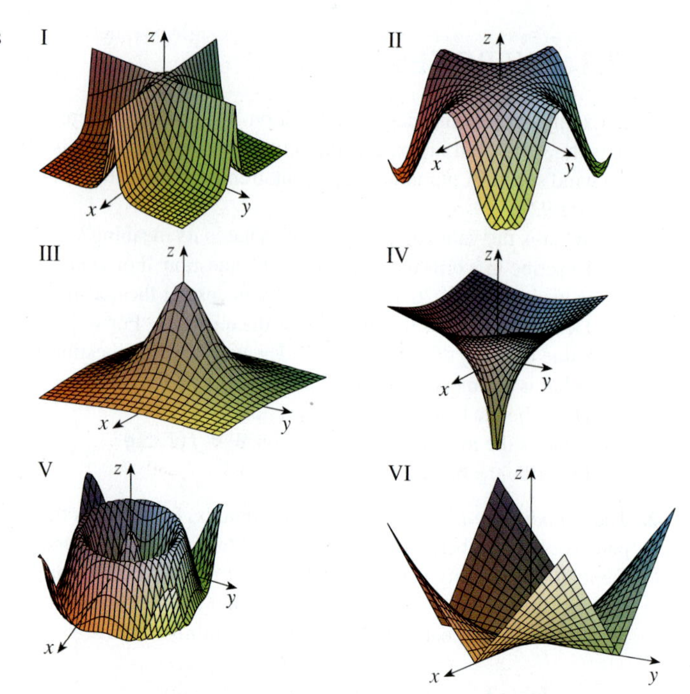

33. A contour map for a function f is shown. Use it to estimate the values of $f(-3, 3)$ and $f(3, -2)$. What can you say about the shape of the graph?

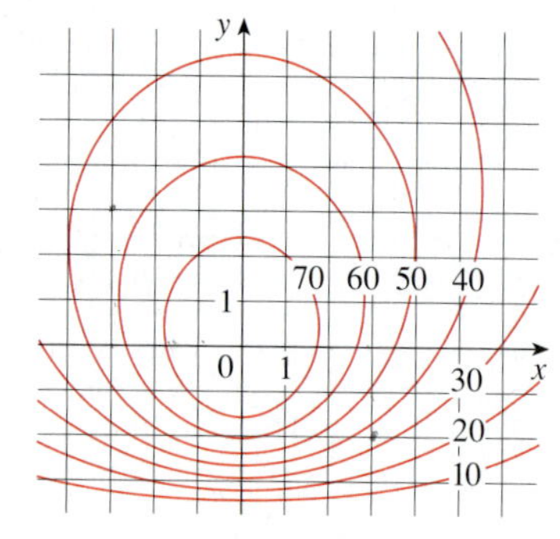

34. Shown is a contour map of atmospheric pressure in North America on August 12, 2008. On the level curves (called isobars) the pressure is indicated in millibars (mb).

(a) Estimate the pressure at C (Chicago), N (Nashville), S (San Francisco), and V (Vancouver).

(b) At which of these locations were the winds strongest?

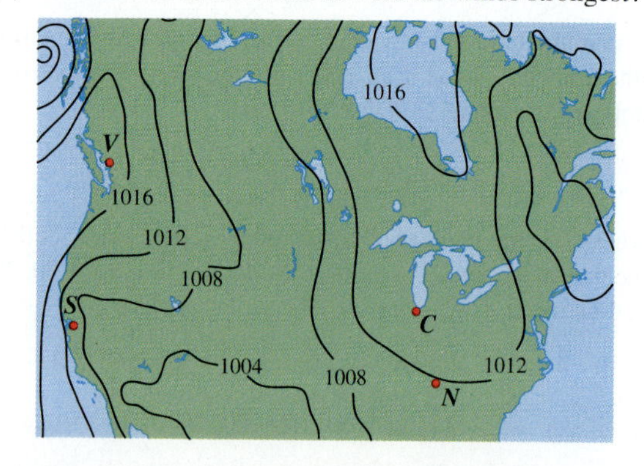

35. Level curves (isothermals) are shown for the typical water temperature (in °C) in Long Lake (Minnesota) as a function of depth and time of year. Estimate the temperature in the lake on June 9 (day 160) at a depth of 10 m and on June 29 (day 180) at a depth of 5 m.

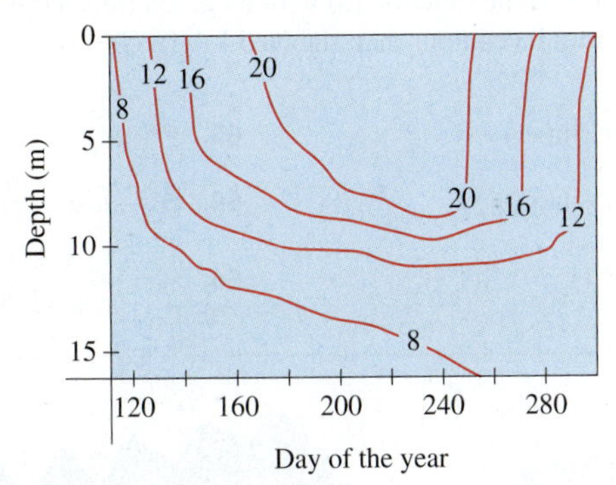

36. Two contour maps are shown. One is for a function f whose graph is a cone. The other is for a function g whose graph is a paraboloid. Which is which, and why?

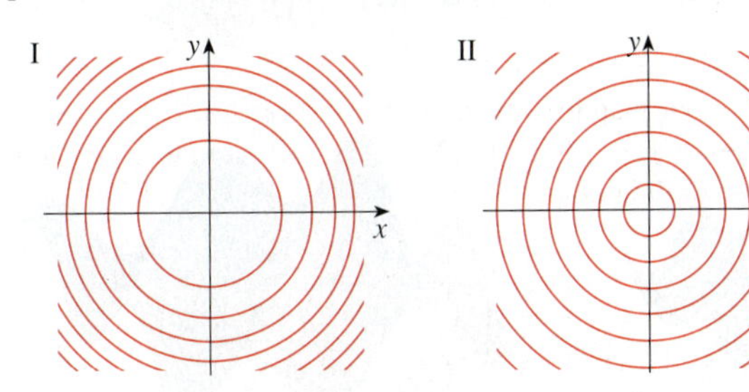

37. Locate the points A and B on the map of Lonesome Mountain (Figure 12). How would you describe the terrain near A? Near B?

38. Make a rough sketch of a contour map for the function whose graph is shown.

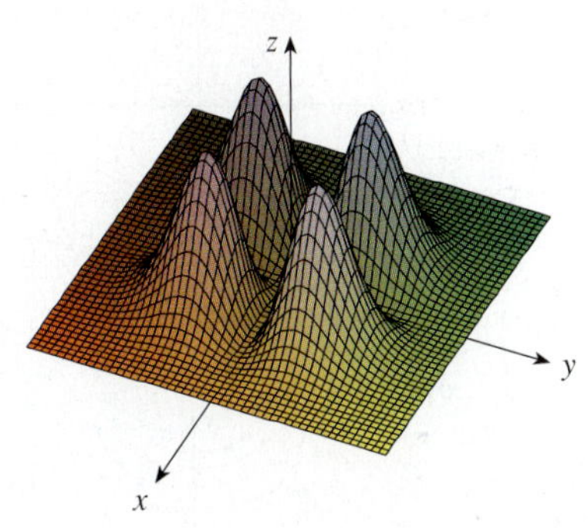

39. The *body mass index* (BMI) of a person is defined by

$$B(m, h) = \frac{m}{h^2}$$

where m is the person's mass (in kilograms) and h is the height (in meters). Draw the level curves $B(m, h) = 18.5$, $B(m, h) = 25$, $B(m, h) = 30$, and $B(m, h) = 40$. A rough guideline is that a person is underweight if the BMI is less than 18.5; optimal if the BMI lies between 18.5 and 25; overweight if the BMI lies between 25 and 30; and obese if the BMI exceeds 30. Shade the region corresponding to optimal BMI. Does someone who weighs 62 kg and is 152 cm tall fall into this category?

40. The body mass index is defined in Exercise 39. Draw the level curve of this function corresponding to someone who is 200 cm tall and weighs 80 kg. Find the weights and heights of two other people with that same level curve.

41–44 A contour map of a function is shown. Use it to make a rough sketch of the graph of f.

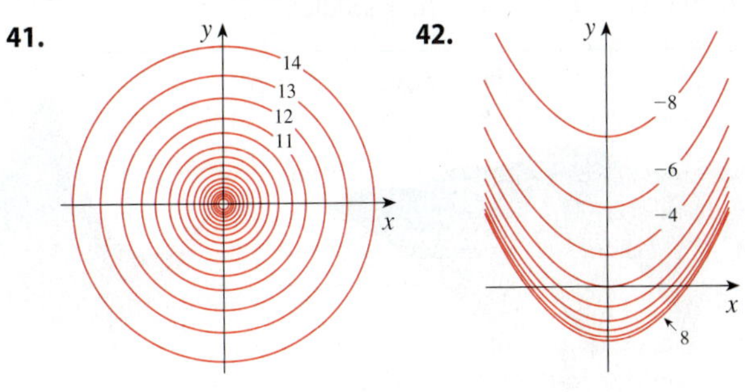

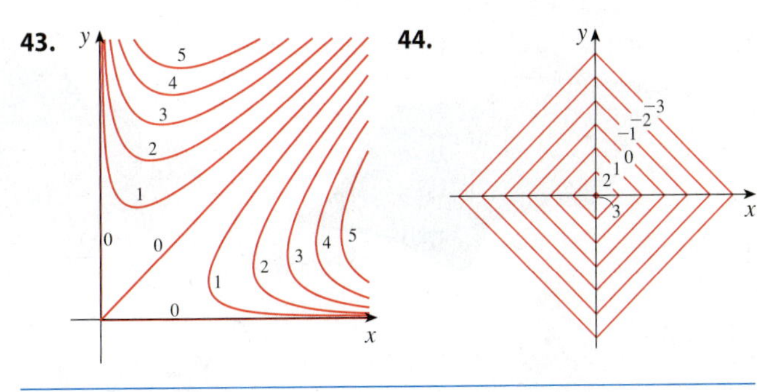

45–52 Draw a contour map of the function showing several level curves.

45. $f(x, y) = x^2 - y^2$

46. $f(x, y) = xy$

47. $f(x, y) = \sqrt{x} + y$

48. $f(x, y) = \ln(x^2 + 4y^2)$

49. $f(x, y) = ye^x$

50. $f(x, y) = y - \arctan x$

51. $f(x, y) = \sqrt[3]{x^2 + y^2}$

52. $f(x, y) = y/(x^2 + y^2)$

53–54 Sketch both a contour map and a graph of the function and compare them.

53. $f(x, y) = x^2 + 9y^2$

54. $f(x, y) = \sqrt{36 - 9x^2 - 4y^2}$

55. A thin metal plate, located in the xy-plane, has temperature $T(x, y)$ at the point (x, y). Sketch some level curves (isothermals) if the temperature function is given by

$$T(x, y) = \frac{100}{1 + x^2 + 2y^2}$$

56. If $V(x, y)$ is the electric potential at a point (x, y) in the xy-plane, then the level curves of V are called *equipotential curves* because at all points on such a curve the electric potential is the same. Sketch some equipotential curves if $V(x, y) = c/\sqrt{r^2 - x^2 - y^2}$, where c is a positive constant.

57–60 Use a computer to graph the function using various domains and viewpoints. Get a printout of one that, in your opinion, gives a good view. If your software also produces level curves, then plot some contour lines of the same function and compare with the graph.

57. $f(x, y) = xy^2 - x^3$ (monkey saddle)

58. $f(x, y) = xy^3 - yx^3$ (dog saddle)

59. $f(x, y) = e^{-(x^2+y^2)/3}(\sin(x^2) + \cos(y^2))$

60. $f(x, y) = \cos x \cos y$

61–66 Match the function (a) with its graph (labeled A–F below) and (b) with its contour map (labeled I–VI). Give reasons for your choices.

61. $z = \sin(xy)$

62. $z = e^x \cos y$

63. $z = \sin(x - y)$

64. $z = \sin x - \sin y$

65. $z = (1 - x^2)(1 - y^2)$

66. $z = \dfrac{x - y}{1 + x^2 + y^2}$

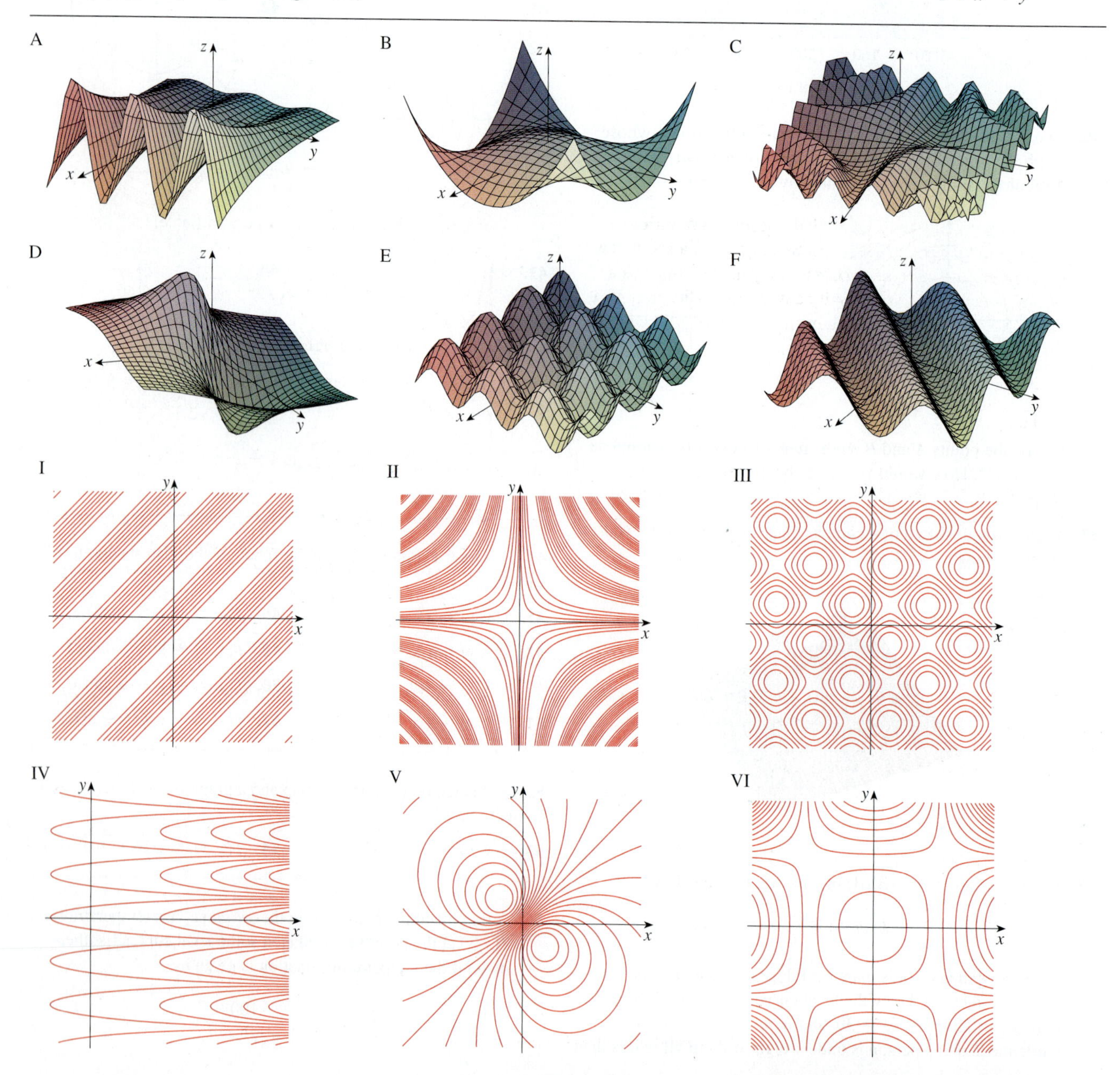

67–70 Describe the level surfaces of the function.

67. $f(x, y, z) = x + 3y + 5z$

68. $f(x, y, z) = x^2 + 3y^2 + 5z^2$

69. $f(x, y, z) = y^2 + z^2$

70. $f(x, y, z) = x^2 - y^2 - z^2$

71–72 Describe how the graph of g is obtained from the graph of f.

71. (a) $g(x, y) = f(x, y) + 2$
(b) $g(x, y) = 2f(x, y)$
(c) $g(x, y) = -f(x, y)$
(d) $g(x, y) = 2 - f(x, y)$

72. (a) $g(x, y) = f(x - 2, y)$
(b) $g(x, y) = f(x, y + 2)$
(c) $g(x, y) = f(x + 3, y - 4)$

73–74 Use a computer to graph the function using various domains and viewpoints. Get a printout that gives a good view of the "peaks and valleys." Would you say the function has a maximum value? Can you identify any points on the graph that you might consider to be "local maximum points"? What about "local minimum points"?

73. $f(x, y) = 3x - x^4 - 4y^2 - 10xy$

74. $f(x, y) = xye^{-x^2-y^2}$

75–76 Graph the function using various domains and viewpoints. Comment on the limiting behavior of the function. What happens as both x and y become large? What happens as (x, y) approaches the origin?

75. $f(x, y) = \dfrac{x + y}{x^2 + y^2}$ **76.** $f(x, y) = \dfrac{xy}{x^2 + y^2}$

---

77. Investigate the family of functions $f(x, y) = e^{cx^2+y^2}$. How does the shape of the graph depend on c?

78. Use a computer to investigate the family of surfaces

$$z = (ax^2 + by^2)e^{-x^2-y^2}$$

How does the shape of the graph depend on the numbers a and b?

79. Use a computer to investigate the family of surfaces $z = x^2 + y^2 + cxy$. In particular, you should determine the transitional values of c for which the surface changes from one type of quadric surface to another.

80. Graph the functions

$$f(x, y) = \sqrt{x^2 + y^2}$$

$$f(x, y) = e^{\sqrt{x^2+y^2}}$$

$$f(x, y) = \ln\sqrt{x^2 + y^2}$$

$$f(x, y) = \sin\left(\sqrt{x^2 + y^2}\right)$$

and $$f(x, y) = \frac{1}{\sqrt{x^2 + y^2}}$$

In general, if g is a function of one variable, how is the graph of

$$f(x, y) = g\left(\sqrt{x^2 + y^2}\right)$$

obtained from the graph of g?

81. (a) Show that, by taking logarithms, the general Cobb-Douglas function $P = bL^{\alpha}K^{1-\alpha}$ can be expressed as

$$\ln\frac{P}{K} = \ln b + \alpha \ln\frac{L}{K}$$

(b) If we let $x = \ln(L/K)$ and $y = \ln(P/K)$, the equation in part (a) becomes the linear equation $y = \alpha x + \ln b$. Use Table 2 (in Example 3) to make a table of values of $\ln(L/K)$ and $\ln(P/K)$ for the years 1899–1922. Then use a graphing calculator or computer to find the least squares regression line through the points $(\ln(L/K), \ln(P/K))$.

(c) Deduce that the Cobb-Douglas production function is $P = 1.01L^{0.75}K^{0.25}$.

14.2 Limits and Continuity

Let's compare the behavior of the functions

$$f(x, y) = \frac{\sin(x^2 + y^2)}{x^2 + y^2} \quad \text{and} \quad g(x, y) = \frac{x^2 - y^2}{x^2 + y^2}$$

as x and y both approach 0 [and therefore the point (x, y) approaches the origin].

Tables 1 and 2 show values of $f(x, y)$ and $g(x, y)$, correct to three decimal places, for points (x, y) near the origin. (Notice that neither function is defined at the origin.)

Table 1 Values of $f(x, y)$

x \ y	−1.0	−0.5	−0.2	0	0.2	0.5	1.0
−1.0	0.455	0.759	0.829	0.841	0.829	0.759	0.455
−0.5	0.759	0.959	0.986	0.990	0.986	0.959	0.759
−0.2	0.829	0.986	0.999	1.000	0.999	0.986	0.829
0	0.841	0.990	1.000		1.000	0.990	0.841
0.2	0.829	0.986	0.999	1.000	0.999	0.986	0.829
0.5	0.759	0.959	0.986	0.990	0.986	0.959	0.759
1.0	0.455	0.759	0.829	0.841	0.829	0.759	0.455

Table 2 Values of $g(x, y)$

x \ y	−1.0	−0.5	−0.2	0	0.2	0.5	1.0
−1.0	0.000	0.600	0.923	1.000	0.923	0.600	0.000
−0.5	−0.600	0.000	0.724	1.000	0.724	0.000	−0.600
−0.2	−0.923	−0.724	0.000	1.000	0.000	−0.724	−0.923
0	−1.000	−1.000	−1.000		−1.000	−1.000	−1.000
0.2	−0.923	−0.724	0.000	1.000	0.000	−0.724	−0.923
0.5	−0.600	0.000	0.724	1.000	0.724	0.000	−0.600
1.0	0.000	0.600	0.923	1.000	0.923	0.600	0.000

It appears that as (x, y) approaches $(0, 0)$, the values of $f(x, y)$ are approaching 1 whereas the values of $g(x, y)$ aren't approaching any number. It turns out that these guesses based on numerical evidence are correct, and we write

$$\lim_{(x, y)\to(0, 0)} \frac{\sin(x^2 + y^2)}{x^2 + y^2} = 1 \qquad \text{and} \qquad \lim_{(x, y)\to(0, 0)} \frac{x^2 - y^2}{x^2 + y^2} \quad \text{does not exist}$$

In general, we use the notation

$$\lim_{(x, y)\to(a, b)} f(x, y) = L$$

to indicate that the values of $f(x, y)$ approach the number L as the point (x, y) approaches the point (a, b) along any path that stays within the domain of f. In other words, we can make the values of $f(x, y)$ as close to L as we like by taking the point (x, y) sufficiently close to the point (a, b), but not equal to (a, b). A more precise definition follows.

1 Definition Let f be a function of two variables whose domain D includes points arbitrarily close to (a, b). Then we say that the **limit of $f(x, y)$ as (x, y) approaches (a, b)** is L and we write

$$\lim_{(x, y)\to(a, b)} f(x, y) = L$$

if for every number $\varepsilon > 0$ there is a corresponding number $\delta > 0$ such that

$$\text{if} \quad (x, y) \in D \quad \text{and} \quad 0 < \sqrt{(x - a)^2 + (y - b)^2} < \delta \quad \text{then} \quad |f(x, y) - L| < \varepsilon$$

Other notations for the limit in Definition 1 are

$$\lim_{\substack{x\to a\\ y\to b}} f(x, y) = L \qquad \text{and} \qquad f(x, y) \to L \text{ as } (x, y) \to (a, b)$$

Notice that $|f(x, y) - L|$ is the distance between the numbers $f(x, y)$ and L, and $\sqrt{(x - a)^2 + (y - b)^2}$ is the distance between the point (x, y) and the point (a, b). Thus Definition 1 says that the distance between $f(x, y)$ and L can be made arbitrarily small by making the distance from (x, y) to (a, b) sufficiently small (but not 0). Figure 1 illustrates Definition 1 by means of an arrow diagram. If any small interval $(L - \varepsilon, L + \varepsilon)$ is given around L, then we can find a disk D_δ with center (a, b) and radius $\delta > 0$ such that f maps all the points in D_δ [except possibly (a, b)] into the interval $(L - \varepsilon, L + \varepsilon)$.

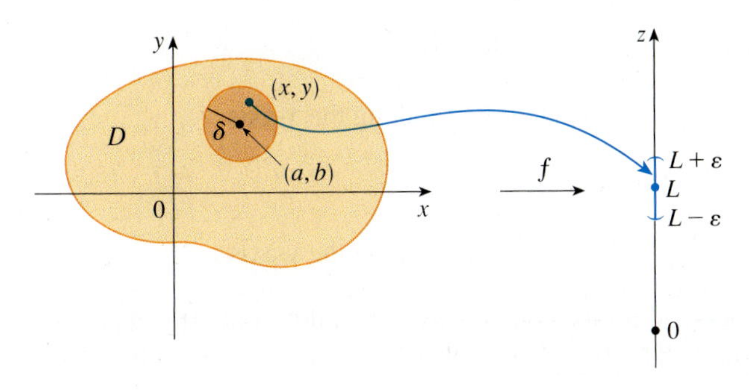

FIGURE 1

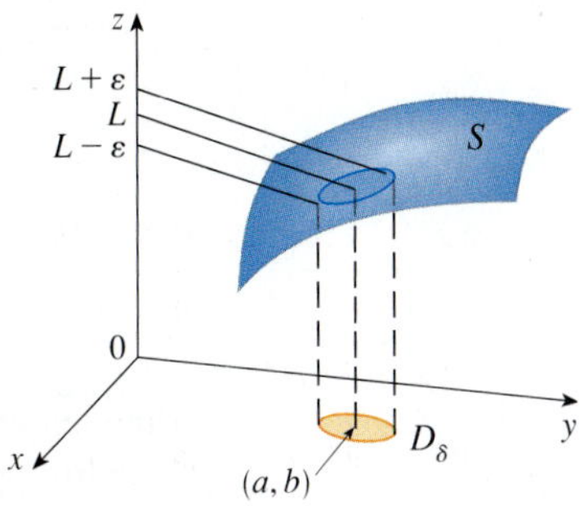

FIGURE 2

Another illustration of Definition 1 is given in Figure 2 where the surface S is the graph of f. If $\varepsilon > 0$ is given, we can find $\delta > 0$ such that if (x, y) is restricted to lie in the disk D_δ and $(x, y) \neq (a, b)$, then the corresponding part of S lies between the horizontal planes $z = L - \varepsilon$ and $z = L + \varepsilon$.

For functions of a single variable, when we let x approach a, there are only two possible directions of approach, from the left or from the right. We recall from Chapter 1 that if $\lim_{x \to a^-} f(x) \neq \lim_{x \to a^+} f(x)$, then $\lim_{x \to a} f(x)$ does not exist.

For functions of two variables the situation is not as simple because we can let (x, y) approach (a, b) from an infinite number of directions in any manner whatsoever (see Figure 3) as long as (x, y) stays within the domain of f.

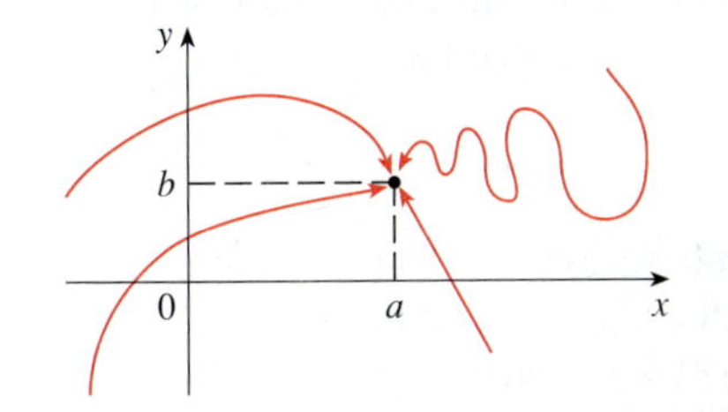

FIGURE 3

Definition 1 says that the distance between $f(x, y)$ and L can be made arbitrarily small by making the distance from (x, y) to (a, b) sufficiently small (but not 0). The definition refers only to the *distance* between (x, y) and (a, b). It does not refer to the direction of approach. Therefore, if the limit exists, then $f(x, y)$ must approach the same limit no matter how (x, y) approaches (a, b). Thus, if we can find two different paths of approach along which the function $f(x, y)$ has different limits, then it follows that $\lim_{(x, y) \to (a, b)} f(x, y)$ does not exist.

> If $f(x, y) \to L_1$ as $(x, y) \to (a, b)$ along a path C_1 and $f(x, y) \to L_2$ as $(x, y) \to (a, b)$ along a path C_2, where $L_1 \neq L_2$, then $\lim_{(x, y) \to (a, b)} f(x, y)$ does not exist.

EXAMPLE 1 Show that $\displaystyle\lim_{(x, y) \to (0, 0)} \frac{x^2 - y^2}{x^2 + y^2}$ does not exist.

SOLUTION Let $f(x, y) = (x^2 - y^2)/(x^2 + y^2)$. First let's approach $(0, 0)$ along the x-axis. Then $y = 0$ gives $f(x, 0) = x^2/x^2 = 1$ for all $x \neq 0$, so

$$f(x, y) \to 1 \qquad \text{as} \qquad (x, y) \to (0, 0) \text{ along the } x\text{-axis}$$

We now approach along the y-axis by putting $x = 0$. Then $f(0, y) = \dfrac{-y^2}{y^2} = -1$ for all $y \neq 0$, so

$$f(x, y) \to -1 \qquad \text{as} \qquad (x, y) \to (0, 0) \text{ along the } y\text{-axis}$$

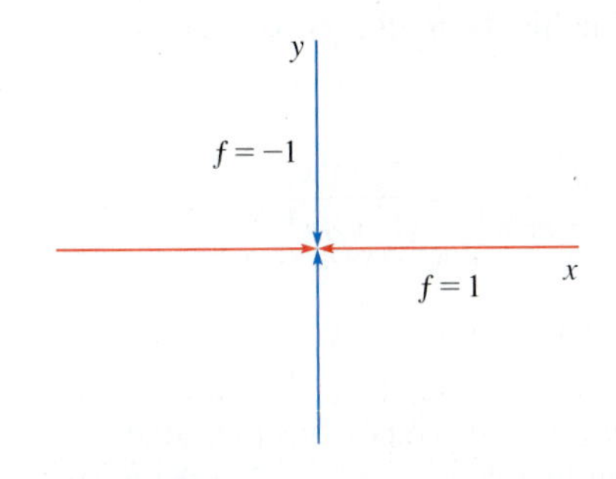

FIGURE 4

(See Figure 4.) Since f has two different limits along two different lines, the given limit does not exist. (This confirms the conjecture we made on the basis of numerical evidence at the beginning of this section.) ■

EXAMPLE 2 If $f(x, y) = xy/(x^2 + y^2)$, does $\lim_{(x, y)\to(0, 0)} f(x, y)$ exist?

SOLUTION If $y = 0$, then $f(x, 0) = 0/x^2 = 0$. Therefore

$$f(x, y) \to 0 \qquad \text{as} \qquad (x, y) \to (0, 0) \text{ along the } x\text{-axis}$$

If $x = 0$, then $f(0, y) = 0/y^2 = 0$, so

$$f(x, y) \to 0 \qquad \text{as} \qquad (x, y) \to (0, 0) \text{ along the } y\text{-axis}$$

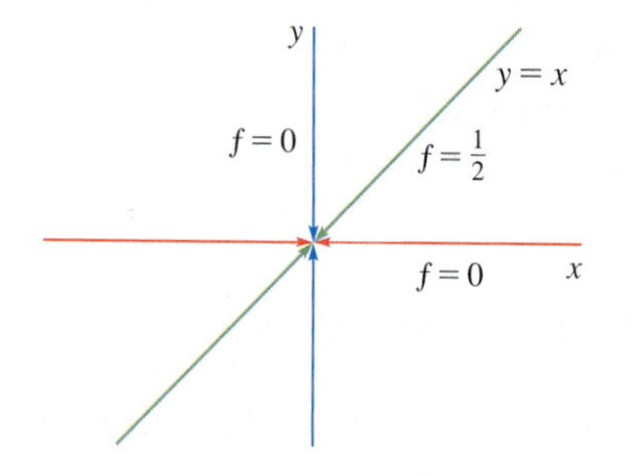

FIGURE 5

Although we have obtained identical limits along the axes, that does not show that the given limit is 0. Let's now approach (0, 0) along another line, say $y = x$. For all $x \neq 0$,

$$f(x, x) = \frac{x^2}{x^2 + x^2} = \frac{1}{2}$$

Therefore $\qquad f(x, y) \to \frac{1}{2} \qquad \text{as} \qquad (x, y) \to (0, 0) \text{ along } y = x$

(See Figure 5.) Since we have obtained different limits along different paths, the given limit does not exist. ■

Figure 6 sheds some light on Example 2. The ridge that occurs above the line $y = x$ corresponds to the fact that $f(x, y) = \frac{1}{2}$ for all points (x, y) on that line except the origin.

TEC In Visual 14.2 a rotating line on the surface in Figure 6 shows different limits at the origin from different directions.

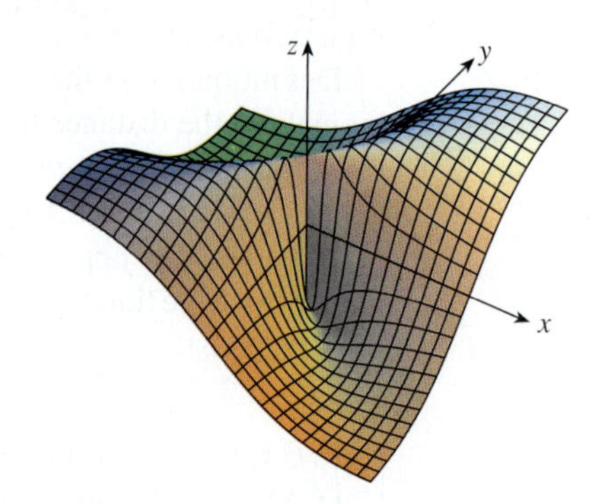

FIGURE 6

$$f(x, y) = \frac{xy}{x^2 + y^2}$$

EXAMPLE 3 If $f(x, y) = \dfrac{xy^2}{x^2 + y^4}$, does $\lim_{(x, y)\to(0, 0)} f(x, y)$ exist?

SOLUTION With the solution of Example 2 in mind, let's try to save time by letting $(x, y) \to (0, 0)$ along any line through the origin. If the line is not the y-axis, then $y = mx$, where m is the slope, and

$$f(x, y) = f(x, mx) = \frac{x(mx)^2}{x^2 + (mx)^4} = \frac{m^2x^3}{x^2 + m^4x^4} = \frac{m^2x}{1 + m^4x^2}$$

So $\qquad f(x, y) \to 0 \qquad \text{as} \qquad (x, y) \to (0, 0) \text{ along } y = mx$

We get the same result as $(x, y) \to (0, 0)$ along the line $x = 0$. Thus f has the same limiting value along every line through the origin. But that does not show that the given

Figure 7 shows the graph of the function in Example 3. Notice the ridge above the parabola $x = y^2$.

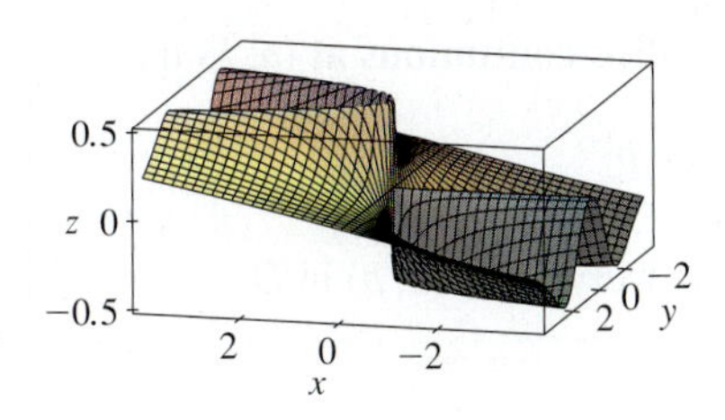

FIGURE 7

limit is 0, for if we now let $(x, y) \to (0, 0)$ along the parabola $x = y^2$, we have

$$f(x, y) = f(y^2, y) = \frac{y^2 \cdot y^2}{(y^2)^2 + y^4} = \frac{y^4}{2y^4} = \frac{1}{2}$$

so $\quad f(x, y) \to \frac{1}{2} \quad$ as $\quad (x, y) \to (0, 0)$ along $x = y^2$

Since different paths lead to different limiting values, the given limit does not exist. ■

Now let's look at limits that *do* exist. Just as for functions of one variable, the calculation of limits for functions of two variables can be greatly simplified by the use of properties of limits. The Limit Laws listed in Section 1.6 can be extended to functions of two variables: the limit of a sum is the sum of the limits, the limit of a product is the product of the limits, and so on. In particular, the following equations are true.

2 $$\lim_{(x, y) \to (a, b)} x = a \qquad \lim_{(x, y) \to (a, b)} y = b \qquad \lim_{(x, y) \to (a, b)} c = c$$

The Squeeze Theorem also holds.

EXAMPLE 4 Find $\displaystyle\lim_{(x, y) \to (0, 0)} \frac{3x^2y}{x^2 + y^2}$ if it exists.

SOLUTION As in Example 3, we could show that the limit along any line through the origin is 0. This doesn't prove that the given limit is 0, but the limits along the parabolas $y = x^2$ and $x = y^2$ also turn out to be 0, so we begin to suspect that the limit does exist and is equal to 0.

Let $\varepsilon > 0$. We want to find $\delta > 0$ such that

$$\text{if} \quad 0 < \sqrt{x^2 + y^2} < \delta \quad \text{then} \quad \left| \frac{3x^2y}{x^2 + y^2} - 0 \right| < \varepsilon$$

that is,
$$\text{if} \quad 0 < \sqrt{x^2 + y^2} < \delta \quad \text{then} \quad \frac{3x^2|y|}{x^2 + y^2} < \varepsilon$$

But $x^2 \leqslant x^2 + y^2$ since $y^2 \geqslant 0$, so $x^2/(x^2 + y^2) \leqslant 1$ and therefore

3 $$\frac{3x^2|y|}{x^2 + y^2} \leqslant 3|y| = 3\sqrt{y^2} \leqslant 3\sqrt{x^2 + y^2}$$

Thus if we choose $\delta = \varepsilon/3$ and let $0 < \sqrt{x^2 + y^2} < \delta$, then

$$\left| \frac{3x^2y}{x^2 + y^2} - 0 \right| \leqslant 3\sqrt{x^2 + y^2} < 3\delta = 3\left(\frac{\varepsilon}{3}\right) = \varepsilon$$

Hence, by Definition 1,

$$\lim_{(x, y) \to (0, 0)} \frac{3x^2y}{x^2 + y^2} = 0$$ ■

Another way to do Example 4 is to use the Squeeze Theorem instead of Definition 1. From (2) it follows that

$$\lim_{(x, y) \to (0, 0)} 3|y| = 0$$

and so the first inequality in (3) shows that the given limit is 0.

Continuity

Recall that evaluating limits of *continuous* functions of a single variable is easy. It can be accomplished by direct substitution because the defining property of a continuous function is $\lim_{x \to a} f(x) = f(a)$. Continuous functions of two variables are also defined by the direct substitution property.

4 Definition A function f of two variables is called **continuous at** (a, b) if

$$\lim_{(x, y) \to (a, b)} f(x, y) = f(a, b)$$

We say f is **continuous on** D if f is continuous at every point (a, b) in D.

The intuitive meaning of continuity is that if the point (x, y) changes by a small amount, then the value of $f(x, y)$ changes by a small amount. This means that a surface that is the graph of a continuous function has no hole or break.

Using the properties of limits, you can see that sums, differences, products, and quotients of continuous functions are continuous on their domains. Let's use this fact to give examples of continuous functions.

A **polynomial function of two variables** (or polynomial, for short) is a sum of terms of the form $cx^m y^n$, where c is a constant and m and n are nonnegative integers. A **rational function** is a ratio of polynomials. For instance,

$$f(x, y) = x^4 + 5x^3y^2 + 6xy^4 - 7y + 6$$

is a polynomial, whereas

$$g(x, y) = \frac{2xy + 1}{x^2 + y^2}$$

is a rational function.

The limits in (2) show that the functions $f(x, y) = x$, $g(x, y) = y$, and $h(x, y) = c$ are continuous. Since any polynomial can be built up out of the simple functions f, g, and h by multiplication and addition, it follows that *all polynomials are continuous on* $\mathbb{R}^2$. Likewise, any rational function is continuous on its domain because it is a quotient of continuous functions.

EXAMPLE 5 Evaluate $\displaystyle\lim_{(x, y) \to (1, 2)} (x^2y^3 - x^3y^2 + 3x + 2y)$.

SOLUTION Since $f(x, y) = x^2y^3 - x^3y^2 + 3x + 2y$ is a polynomial, it is continuous everywhere, so we can find the limit by direct substitution:

$$\lim_{(x, y) \to (1, 2)} (x^2y^3 - x^3y^2 + 3x + 2y) = 1^2 \cdot 2^3 - 1^3 \cdot 2^2 + 3 \cdot 1 + 2 \cdot 2 = 11 \quad \blacksquare$$

EXAMPLE 6 Where is the function $f(x, y) = \dfrac{x^2 - y^2}{x^2 + y^2}$ continuous?

SOLUTION The function f is discontinuous at $(0, 0)$ because it is not defined there. Since f is a rational function, it is continuous on its domain, which is the set $D = \{(x, y) \mid (x, y) \neq (0, 0)\}$. $\blacksquare$

EXAMPLE 7 Let

$$g(x, y) = \begin{cases} \dfrac{x^2 - y^2}{x^2 + y^2} & \text{if } (x, y) \neq (0, 0) \\ 0 & \text{if } (x, y) = (0, 0) \end{cases}$$

Here g is defined at $(0, 0)$ but g is still discontinuous there because $\lim_{(x, y)\to(0, 0)} g(x, y)$ does not exist (see Example 1). ■

Figure 8 shows the graph of the continuous function in Example 8.

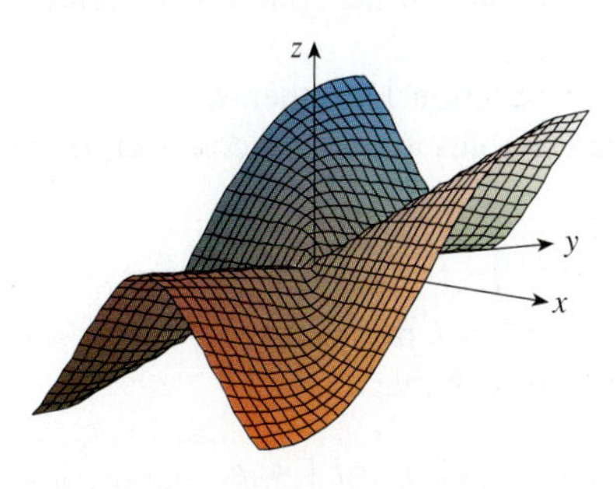

FIGURE 8

EXAMPLE 8 Let

$$f(x, y) = \begin{cases} \dfrac{3x^2y}{x^2 + y^2} & \text{if } (x, y) \neq (0, 0) \\ 0 & \text{if } (x, y) = (0, 0) \end{cases}$$

We know f is continuous for $(x, y) \neq (0, 0)$ since it is equal to a rational function there. Also, from Example 4, we have

$$\lim_{(x, y)\to(0, 0)} f(x, y) = \lim_{(x, y)\to(0, 0)} \frac{3x^2y}{x^2 + y^2} = 0 = f(0, 0)$$

Therefore f is continuous at $(0, 0)$, and so it is continuous on $\mathbb{R}^2$. ■

Just as for functions of one variable, composition is another way of combining two continuous functions to get a third. In fact, it can be shown that if f is a continuous function of two variables and g is a continuous function of a single variable that is defined on the range of f, then the composite function $h = g \circ f$ defined by $h(x, y) = g(f(x, y))$ is also a continuous function.

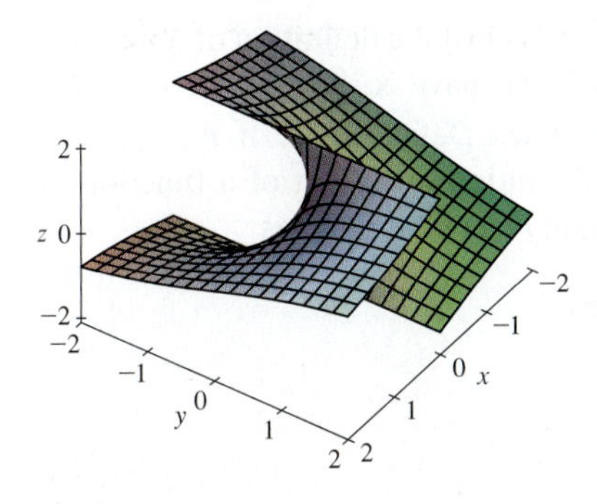

FIGURE 9
The function $h(x, y) = \arctan(y/x)$ is discontinuous where $x = 0$.

EXAMPLE 9 Where is the function $h(x, y) = \arctan(y/x)$ continuous?

SOLUTION The function $f(x, y) = y/x$ is a rational function and therefore continuous except on the line $x = 0$. The function $g(t) = \arctan t$ is continuous everywhere. So the composite function

$$g(f(x, y)) = \arctan(y/x) = h(x, y)$$

is continuous except where $x = 0$. The graph in Figure 9 shows the break in the graph of h above the y-axis. ■

■ Functions of Three or More Variables

Everything that we have done in this section can be extended to functions of three or more variables. The notation

$$\lim_{(x, y, z)\to(a, b, c)} f(x, y, z) = L$$

means that the values of $f(x, y, z)$ approach the number L as the point (x, y, z) approaches the point (a, b, c) along any path in the domain of f. Because the distance between two points (x, y, z) and (a, b, c) in $\mathbb{R}^3$ is given by $\sqrt{(x - a)^2 + (y - b)^2 + (z - c)^2}$, we can write the precise definition as follows: for every number $\varepsilon > 0$ there is a corresponding number $\delta > 0$ such that

$$\text{if } (x, y, z) \text{ is in the domain of } f \text{ and } 0 < \sqrt{(x - a)^2 + (y - b)^2 + (z - c)^2} < \delta$$

$$\text{then } |f(x, y, z) - L| < \varepsilon$$

The function f is **continuous** at (a, b, c) if

$$\lim_{(x, y, z)\to(a, b, c)} f(x, y, z) = f(a, b, c)$$

For instance, the function

$$f(x, y, z) = \frac{1}{x^2 + y^2 + z^2 - 1}$$

is a rational function of three variables and so is continuous at every point in $\mathbb{R}^3$ except where $x^2 + y^2 + z^2 = 1$. In other words, it is discontinuous on the sphere with center the origin and radius 1.

If we use the vector notation introduced at the end of Section 14.1, then we can write the definitions of a limit for functions of two or three variables in a single compact form as follows.

5 If f is defined on a subset D of $\mathbb{R}^n$, then $\lim_{\mathbf{x} \to \mathbf{a}} f(\mathbf{x}) = L$ means that for every number $\varepsilon > 0$ there is a corresponding number $\delta > 0$ such that

$$\text{if} \quad \mathbf{x} \in D \quad \text{and} \quad 0 < |\mathbf{x} - \mathbf{a}| < \delta \quad \text{then} \quad |f(\mathbf{x}) - L| < \varepsilon$$

Notice that if $n = 1$, then $\mathbf{x} = x$ and $\mathbf{a} = a$, and (5) is just the definition of a limit for functions of a single variable. For the case $n = 2$, we have $\mathbf{x} = \langle x, y \rangle$, $\mathbf{a} = \langle a, b \rangle$, and $|\mathbf{x} - \mathbf{a}| = \sqrt{(x - a)^2 + (y - b)^2}$, so (5) becomes Definition 1. If $n = 3$, then $\mathbf{x} = \langle x, y, z \rangle$, $\mathbf{a} = \langle a, b, c \rangle$, and (5) becomes the definition of a limit of a function of three variables. In each case the definition of continuity can be written as

$$\lim_{\mathbf{x} \to \mathbf{a}} f(\mathbf{x}) = f(\mathbf{a})$$

14.2 EXERCISES

1. Suppose that $\lim_{(x, y) \to (3, 1)} f(x, y) = 6$. What can you say about the value of $f(3, 1)$? What if f is continuous?

2. Explain why each function is continuous or discontinuous.
(a) The outdoor temperature as a function of longitude, latitude, and time
(b) Elevation (height above sea level) as a function of longitude, latitude, and time
(c) The cost of a taxi ride as a function of distance traveled and time

3–4 Use a table of numerical values of $f(x, y)$ for (x, y) near the origin to make a conjecture about the value of the limit of $f(x, y)$ as $(x, y) \to (0, 0)$. Then explain why your guess is correct.

3. $f(x, y) = \dfrac{x^2y^3 + x^3y^2 - 5}{2 - xy}$

4. $f(x, y) = \dfrac{2xy}{x^2 + 2y^2}$

5–22 Find the limit, if it exists, or show that the limit does not exist.

5. $\displaystyle\lim_{(x, y) \to (3, 2)} (x^2y^3 - 4y^2)$

6. $\displaystyle\lim_{(x, y) \to (2, -1)} \frac{x^2y + xy^2}{x^2 - y^2}$

7. $\displaystyle\lim_{(x, y) \to (\pi, \pi/2)} y \sin(x - y)$

8. $\displaystyle\lim_{(x, y) \to (3, 2)} e^{\sqrt{2x - y}}$

9. $\displaystyle\lim_{(x, y) \to (0, 0)} \frac{x^4 - 4y^2}{x^2 + 2y^2}$

10. $\displaystyle\lim_{(x, y) \to (0, 0)} \frac{5y^4\cos^2 x}{x^4 + y^4}$

11. $\displaystyle\lim_{(x, y) \to (0, 0)} \frac{y^2 \sin^2 x}{x^4 + y^4}$

12. $\displaystyle\lim_{(x, y) \to (1, 0)} \frac{xy - y}{(x - 1)^2 + y^2}$

13. $\displaystyle\lim_{(x, y) \to (0, 0)} \frac{xy}{\sqrt{x^2 + y^2}}$

14. $\displaystyle\lim_{(x, y) \to (0, 0)} \frac{x^3 - y^3}{x^2 + xy + y^2}$

15. $\displaystyle\lim_{(x, y) \to (0, 0)} \frac{xy^2 \cos y}{x^2 + y^4}$

16. $\displaystyle\lim_{(x, y) \to (0, 0)} \frac{xy^4}{x^4 + y^4}$

17. $\displaystyle\lim_{(x, y) \to (0, 0)} \frac{x^2 + y^2}{\sqrt{x^2 + y^2 + 1} - 1}$

18. $\displaystyle\lim_{(x, y) \to (0, 0)} \frac{xy^4}{x^2 + y^8}$

19. $\displaystyle\lim_{(x, y, z) \to (\pi, 0, 1/3)} e^{y^2} \tan(xz)$

20. $\displaystyle\lim_{(x, y, z) \to (0, 0, 0)} \frac{xy + yz}{x^2 + y^2 + z^2}$

21. $\displaystyle \lim_{(x, y, z) \to (0, 0, 0)} \frac{xy + yz^2 + xz^2}{x^2 + y^2 + z^4}$

22. $\displaystyle \lim_{(x, y, z) \to (0, 0, 0)} \frac{x^2y^2z^2}{x^2 + y^2 + z^2}$

23–24 Use a computer graph of the function to explain why the limit does not exist.

23. $\displaystyle \lim_{(x, y) \to (0, 0)} \frac{2x^2 + 3xy + 4y^2}{3x^2 + 5y^2}$ **24.** $\displaystyle \lim_{(x, y) \to (0, 0)} \frac{xy^3}{x^2 + y^6}$

25–26 Find $h(x, y) = g(f(x, y))$ and the set of points at which h is continuous.

25. $g(t) = t^2 + \sqrt{t}, \quad f(x, y) = 2x + 3y - 6$

26. $g(t) = t + \ln t, \quad f(x, y) = \dfrac{1 - xy}{1 + x^2y^2}$

27–28 Graph the function and observe where it is discontinuous. Then use the formula to explain what you have observed.

27. $f(x, y) = e^{1/(x-y)}$ **28.** $f(x, y) = \dfrac{1}{1 - x^2 - y^2}$

29–38 Determine the set of points at which the function is continuous.

29. $F(x, y) = \dfrac{xy}{1 + e^{x-y}}$ **30.** $F(x, y) = \cos\sqrt{1 + x - y}$

31. $F(x, y) = \dfrac{1 + x^2 + y^2}{1 - x^2 - y^2}$ **32.** $H(x, y) = \dfrac{e^x + e^y}{e^{xy} - 1}$

33. $G(x, y) = \sqrt{x} + \sqrt{1 - x^2 - y^2}$

34. $G(x, y) = \ln(1 + x - y)$

35. $f(x, y, z) = \arcsin(x^2 + y^2 + z^2)$

36. $f(x, y, z) = \sqrt{y - x^2}\,\ln z$

37. $f(x, y) = \begin{cases} \dfrac{x^2y^3}{2x^2 + y^2} & \text{if } (x, y) \neq (0, 0) \\ 1 & \text{if } (x, y) = (0, 0) \end{cases}$

38. $f(x, y) = \begin{cases} \dfrac{xy}{x^2 + xy + y^2} & \text{if } (x, y) \neq (0, 0) \\ 0 & \text{if } (x, y) = (0, 0) \end{cases}$

39–41 Use polar coordinates to find the limit. [If (r, θ) are polar coordinates of the point (x, y) with $r \geqslant 0$, note that $r \to 0^+$ as $(x, y) \to (0, 0)$.]

39. $\displaystyle \lim_{(x, y) \to (0, 0)} \frac{x^3 + y^3}{x^2 + y^2}$

40. $\displaystyle \lim_{(x, y) \to (0, 0)} (x^2 + y^2)\ln(x^2 + y^2)$

41. $\displaystyle \lim_{(x, y) \to (0, 0)} \frac{e^{-x^2-y^2} - 1}{x^2 + y^2}$

42. At the beginning of this section we considered the function

$$f(x, y) = \frac{\sin(x^2 + y^2)}{x^2 + y^2}$$

and guessed on the basis of numerical evidence that $f(x, y) \to 1$ as $(x, y) \to (0, 0)$. Use polar coordinates to confirm the value of the limit. Then graph the function.

43. Graph and discuss the continuity of the function

$$f(x, y) = \begin{cases} \dfrac{\sin xy}{xy} & \text{if } xy \neq 0 \\ 1 & \text{if } xy = 0 \end{cases}$$

44. Let

$$f(x, y) = \begin{cases} 0 & \text{if } y \leqslant 0 \quad \text{or} \quad y \geqslant x^4 \\ 1 & \text{if } 0 < y < x^4 \end{cases}$$

(a) Show that $f(x, y) \to 0$ as $(x, y) \to (0, 0)$ along any path through $(0, 0)$ of the form $y = mx^a$ with $0 < a < 4$.
(b) Despite part (a), show that f is discontinuous at $(0, 0)$.
(c) Show that f is discontinuous on two entire curves.

45. Show that the function f given by $f(\mathbf{x}) = |\mathbf{x}|$ is continuous on $\mathbb{R}^n$. [*Hint:* Consider $|\mathbf{x} - \mathbf{a}|^2 = (\mathbf{x} - \mathbf{a}) \cdot (\mathbf{x} - \mathbf{a})$.]

46. If $\mathbf{c} \in V_n$, show that the function f given by $f(\mathbf{x}) = \mathbf{c} \cdot \mathbf{x}$ is continuous on $\mathbb{R}^n$.

14.3 Partial Derivatives

On a hot day, extreme humidity makes us think the temperature is higher than it really is, whereas in very dry air we perceive the temperature to be lower than the thermometer indicates. The National Weather Service has devised the *heat index* (also called the temperature-humidity index, or humidex, in some countries) to describe the combined effects of temperature and humidity. The heat index I is the perceived air temperature when the actual temperature is T and the relative humidity is H. So I is a function of T and H and we can write $I = f(T, H)$. The following table of values of I is an excerpt from a table compiled by the National Weather Service.

Table 1 Heat index I as a function of temperature and humidity

Actual temperature (°F)	Relative humidity (%)								
T \ H	50	55	60	65	70	75	80	85	90
90	96	98	100	103	106	109	112	115	119
92	100	103	105	108	112	115	119	123	128
94	104	107	111	114	118	122	127	132	137
96	109	113	116	121	125	130	135	141	146
98	114	118	123	127	133	138	144	150	157
100	119	124	129	135	141	147	154	161	168

If we concentrate on the highlighted column of the table, which corresponds to a relative humidity of $H = 70\%$, we are considering the heat index as a function of the single variable T for a fixed value of H. Let's write $g(T) = f(T, 70)$. Then $g(T)$ describes how the heat index I increases as the actual temperature T increases when the relative humidity is 70%. The derivative of g when $T = 96°\text{F}$ is the rate of change of I with respect to T when $T = 96°\text{F}$:

$$g'(96) = \lim_{h \to 0} \frac{g(96 + h) - g(96)}{h} = \lim_{h \to 0} \frac{f(96 + h, 70) - f(96, 70)}{h}$$

We can approximate $g'(96)$ using the values in Table 1 by taking $h = 2$ and -2:

$$g'(96) \approx \frac{g(98) - g(96)}{2} = \frac{f(98, 70) - f(96, 70)}{2} = \frac{133 - 125}{2} = 4$$

$$g'(96) \approx \frac{g(94) - g(96)}{-2} = \frac{f(94, 70) - f(96, 70)}{-2} = \frac{118 - 125}{-2} = 3.5$$

Averaging these values, we can say that the derivative $g'(96)$ is approximately 3.75. This means that, when the actual temperature is 96°F and the relative humidity is 70%, the apparent temperature (heat index) rises by about 3.75°F for every degree that the actual temperature rises!

Now let's look at the highlighted row in Table 1, which corresponds to a fixed temperature of $T = 96°\text{F}$. The numbers in this row are values of the function $G(H) = f(96, H)$, which describes how the heat index increases as the relative humidity H increases when the actual temperature is $T = 96°\text{F}$. The derivative of this function when $H = 70\%$ is the rate of change of I with respect to H when $H = 70\%$:

$$G'(70) = \lim_{h \to 0} \frac{G(70 + h) - G(70)}{h} = \lim_{h \to 0} \frac{f(96, 70 + h) - f(96, 70)}{h}$$

By taking $h = 5$ and -5, we approximate $G'(70)$ using the tabular values:

$$G'(70) \approx \frac{G(75) - G(70)}{5} = \frac{f(96, 75) - f(96, 70)}{5} = \frac{130 - 125}{5} = 1$$

$$G'(70) \approx \frac{G(65) - G(70)}{-5} = \frac{f(96, 65) - f(96, 70)}{-5} = \frac{121 - 125}{-5} = 0.8$$

By averaging these values we get the estimate $G'(70) \approx 0.9$. This says that, when the temperature is 96°F and the relative humidity is 70%, the heat index rises about 0.9°F for every percent that the relative humidity rises.

In general, if f is a function of two variables x and y, suppose we let only x vary while keeping y fixed, say $y = b$, where b is a constant. Then we are really considering a function of a single variable x, namely, $g(x) = f(x, b)$. If g has a derivative at a, then we call it the **partial derivative of f with respect to x at (a, b)** and denote it by $f_x(a, b)$. Thus

1
$$f_x(a, b) = g'(a) \qquad \text{where} \qquad g(x) = f(x, b)$$

By the definition of a derivative, we have

$$g'(a) = \lim_{h \to 0} \frac{g(a + h) - g(a)}{h}$$

and so Equation 1 becomes

2
$$f_x(a, b) = \lim_{h \to 0} \frac{f(a + h, b) - f(a, b)}{h}$$

Similarly, the **partial derivative of f with respect to y at (a, b)**, denoted by $f_y(a, b)$, is obtained by keeping x fixed ($x = a$) and finding the ordinary derivative at b of the function $G(y) = f(a, y)$:

3
$$f_y(a, b) = \lim_{h \to 0} \frac{f(a, b + h) - f(a, b)}{h}$$

With this notation for partial derivatives, we can write the rates of change of the heat index I with respect to the actual temperature T and relative humidity H when $T = 96$°F and $H = 70$% as follows:

$$f_T(96, 70) \approx 3.75 \qquad f_H(96, 70) \approx 0.9$$

If we now let the point (a, b) vary in Equations 2 and 3, f_x and f_y become functions of two variables.

4 If f is a function of two variables, its **partial derivatives** are the functions f_x and f_y defined by

$$f_x(x, y) = \lim_{h \to 0} \frac{f(x + h, y) - f(x, y)}{h}$$

$$f_y(x, y) = \lim_{h \to 0} \frac{f(x, y + h) - f(x, y)}{h}$$

There are many alternative notations for partial derivatives. For instance, instead of f_x we can write f_1 or $D_1 f$ (to indicate differentiation with respect to the *first* variable) or $\partial f/\partial x$. But here $\partial f/\partial x$ can't be interpreted as a ratio of differentials.

Notations for Partial Derivatives If $z = f(x, y)$, we write

$$f_x(x, y) = f_x = \frac{\partial f}{\partial x} = \frac{\partial}{\partial x} f(x, y) = \frac{\partial z}{\partial x} = f_1 = D_1 f = D_x f$$

$$f_y(x, y) = f_y = \frac{\partial f}{\partial y} = \frac{\partial}{\partial y} f(x, y) = \frac{\partial z}{\partial y} = f_2 = D_2 f = D_y f$$

To compute partial derivatives, all we have to do is remember from Equation 1 that the partial derivative with respect to x is just the *ordinary* derivative of the function g of a single variable that we get by keeping y fixed. Thus we have the following rule.

Rule for Finding Partial Derivatives of $z = f(x, y)$

1. To find f_x, regard y as a constant and differentiate $f(x, y)$ with respect to x.
2. To find f_y, regard x as a constant and differentiate $f(x, y)$ with respect to y.

EXAMPLE 1 If $f(x, y) = x^3 + x^2y^3 - 2y^2$, find $f_x(2, 1)$ and $f_y(2, 1)$.

SOLUTION Holding y constant and differentiating with respect to x, we get

$$f_x(x, y) = 3x^2 + 2xy^3$$

and so

$$f_x(2, 1) = 3 \cdot 2^2 + 2 \cdot 2 \cdot 1^3 = 16$$

Holding x constant and differentiating with respect to y, we get

$$f_y(x, y) = 3x^2y^2 - 4y$$

$$f_y(2, 1) = 3 \cdot 2^2 \cdot 1^2 - 4 \cdot 1 = 8$$

■

Interpretations of Partial Derivatives

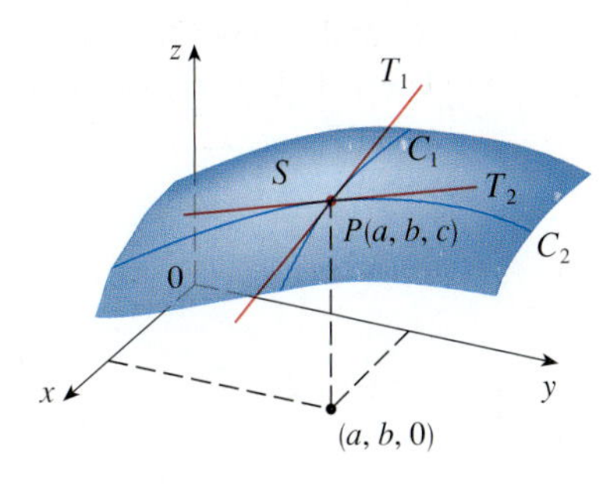

FIGURE 1
The partial derivatives of f at (a, b) are the slopes of the tangents to C_1 and C_2.

To give a geometric interpretation of partial derivatives, we recall that the equation $z = f(x, y)$ represents a surface S (the graph of f). If $f(a, b) = c$, then the point $P(a, b, c)$ lies on S. By fixing $y = b$, we are restricting our attention to the curve C_1 in which the vertical plane $y = b$ intersects S. (In other words, C_1 is the trace of S in the plane $y = b$.) Likewise, the vertical plane $x = a$ intersects S in a curve C_2. Both of the curves C_1 and C_2 pass through the point P. (See Figure 1.)

Note that the curve C_1 is the graph of the function $g(x) = f(x, b)$, so the slope of its tangent T_1 at P is $g'(a) = f_x(a, b)$. The curve C_2 is the graph of the function $G(y) = f(a, y)$, so the slope of its tangent T_2 at P is $G'(b) = f_y(a, b)$.

Thus the partial derivatives $f_x(a, b)$ and $f_y(a, b)$ can be interpreted geometrically as the slopes of the tangent lines at $P(a, b, c)$ to the traces C_1 and C_2 of S in the planes $y = b$ and $x = a$.

As we have seen in the case of the heat index function, partial derivatives can also be interpreted as *rates of change*. If $z = f(x, y)$, then $\partial z/\partial x$ represents the rate of change of z with respect to x when y is fixed. Similarly, $\partial z/\partial y$ represents the rate of change of z with respect to y when x is fixed.

EXAMPLE 2 If $f(x, y) = 4 - x^2 - 2y^2$, find $f_x(1, 1)$ and $f_y(1, 1)$ and interpret these numbers as slopes.

SOLUTION We have

$$f_x(x, y) = -2x \qquad f_y(x, y) = -4y$$

$$f_x(1, 1) = -2 \qquad f_y(1, 1) = -4$$

The graph of f is the paraboloid $z = 4 - x^2 - 2y^2$ and the vertical plane $y = 1$ intersects it in the parabola $z = 2 - x^2$, $y = 1$. (As in the preceding discussion, we label it C_1 in Figure 2.) The slope of the tangent line to this parabola at the point $(1, 1, 1)$ is $f_x(1, 1) = -2$. Similarly, the curve C_2 in which the plane $x = 1$ intersects the paraboloid is the parabola $z = 3 - 2y^2$, $x = 1$, and the slope of the tangent line at $(1, 1, 1)$ is $f_y(1, 1) = -4$. (See Figure 3.)

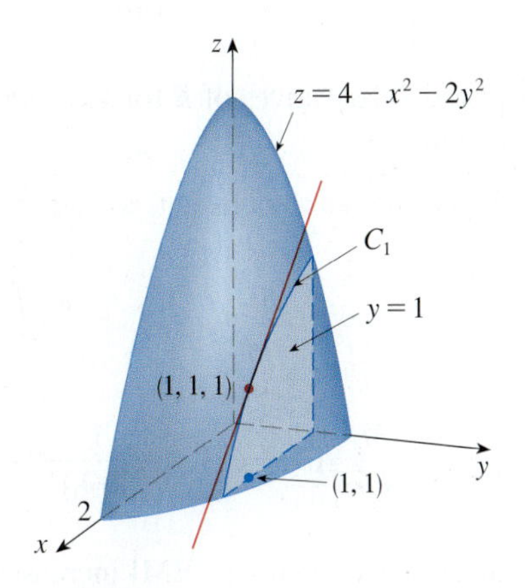

FIGURE 2

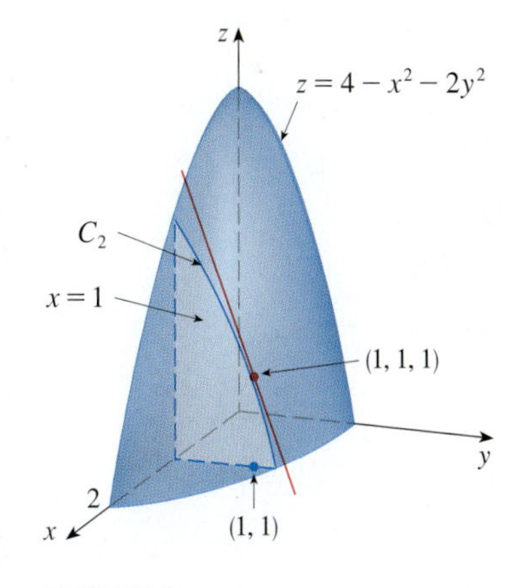

FIGURE 3

Figure 4 is a computer-drawn counterpart to Figure 2. Part (a) shows the plane $y = 1$ intersecting the surface to form the curve C_1 and part (b) shows C_1 and T_1. [We have used the vector equations $\mathbf{r}(t) = \langle t, 1, 2 - t^2 \rangle$ for C_1 and $\mathbf{r}(t) = \langle 1 + t, 1, 1 - 2t \rangle$ for T_1.] Similarly, Figure 5 corresponds to Figure 3.

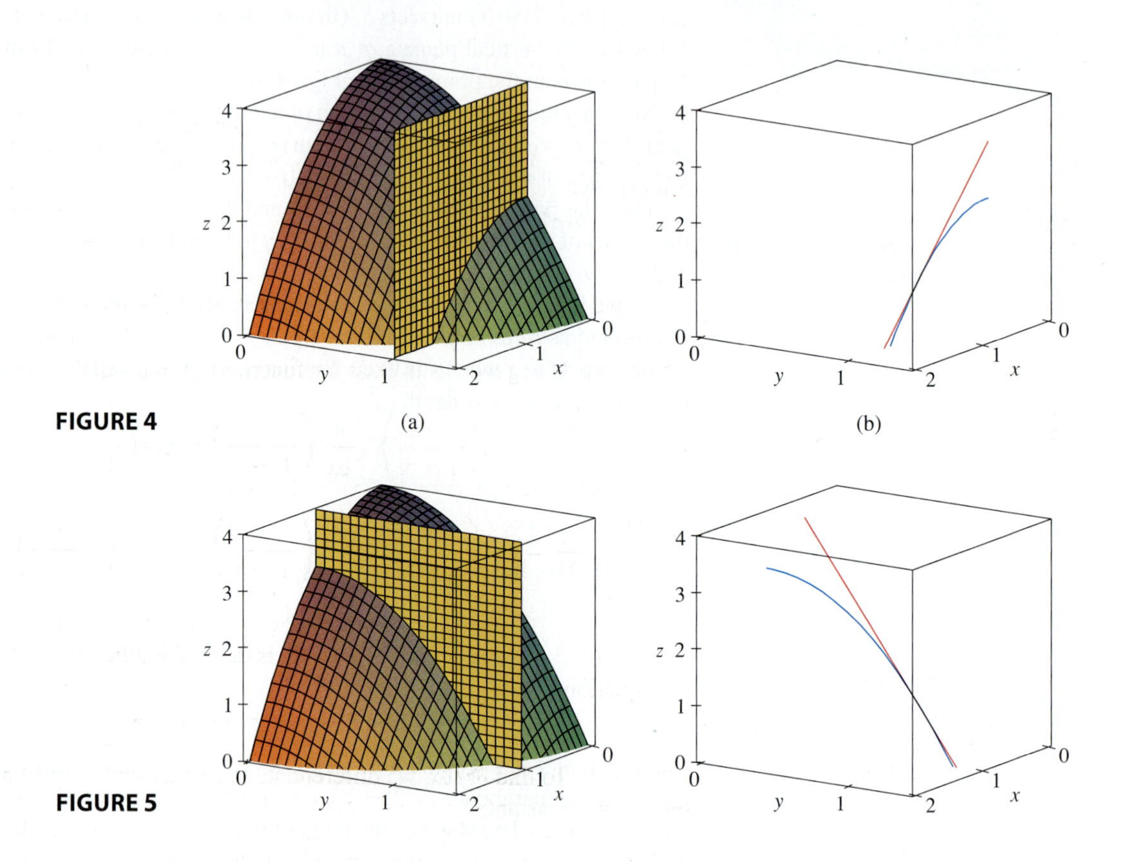

FIGURE 4 (a) (b)

FIGURE 5

EXAMPLE 3 In Exercise 14.1.39 we defined the body mass index of a person as

$$B(m, h) = \frac{m}{h^2}$$

Calculate the partial derivatives of B for a young man with $m = 64$ kg and $h = 1.68$ m and interpret them.

SOLUTION Regarding h as a constant, we see that the partial derivative with respect to m is

$$\frac{\partial B}{\partial m}(m, h) = \frac{\partial}{\partial m}\left(\frac{m}{h^2}\right) = \frac{1}{h^2}$$

so

$$\frac{\partial B}{\partial m}(64, 1.68) = \frac{1}{(1.68)^2} \approx 0.35 \text{ (kg/m}^2\text{)/kg}$$

This is the rate at which the man's BMI increases with respect to his weight when he weighs 64 kg and his height is 1.68 m. So if his weight increases by a small amount, one kilogram for instance, and his height remains unchanged, then his BMI will increase by about 0.35.

Now we regard m as a constant. The partial derivative with respect to h is

$$\frac{\partial B}{\partial h}(m, h) = \frac{\partial}{\partial h}\left(\frac{m}{h^2}\right) = m\left(-\frac{2}{h^3}\right) = -\frac{2m}{h^3}$$

so
$$\frac{\partial B}{\partial h}(64, 1.68) = -\frac{2 \cdot 64}{(1.68)^3} \approx -27\ (\text{kg/m}^2)/\text{m}$$

This is the rate at which the man's BMI increases with respect to his height when he weighs 64 kg and his height is 1.68 m. So if the man is still growing and his weight stays unchanged while his height increases by a small amount, say 1 cm, then his BMI will *decrease* by about $27(0.01) = 0.27$. ■

EXAMPLE 4 If $f(x, y) = \sin\left(\dfrac{x}{1 + y}\right)$, calculate $\dfrac{\partial f}{\partial x}$ and $\dfrac{\partial f}{\partial y}$.

SOLUTION Using the Chain Rule for functions of one variable, we have

$$\frac{\partial f}{\partial x} = \cos\left(\frac{x}{1 + y}\right) \cdot \frac{\partial}{\partial x}\left(\frac{x}{1 + y}\right) = \cos\left(\frac{x}{1 + y}\right) \cdot \frac{1}{1 + y}$$

$$\frac{\partial f}{\partial y} = \cos\left(\frac{x}{1 + y}\right) \cdot \frac{\partial}{\partial y}\left(\frac{x}{1 + y}\right) = -\cos\left(\frac{x}{1 + y}\right) \cdot \frac{x}{(1 + y)^2}$$ ■

Some computer software can plot surfaces defined by implicit equations in three variables. Figure 6 shows such a plot of the surface defined by the equation in Example 5.

FIGURE 6

EXAMPLE 5 Find $\partial z/\partial x$ and $\partial z/\partial y$ if z is defined implicitly as a function of x and y by the equation

$$x^3 + y^3 + z^3 + 6xyz = 1$$

SOLUTION To find $\partial z/\partial x$, we differentiate implicitly with respect to x, being careful to treat y as a constant:

$$3x^2 + 3z^2 \frac{\partial z}{\partial x} + 6yz + 6xy \frac{\partial z}{\partial x} = 0$$

Solving this equation for $\partial z/\partial x$, we obtain

$$\frac{\partial z}{\partial x} = -\frac{x^2 + 2yz}{z^2 + 2xy}$$

Similarly, implicit differentiation with respect to y gives

$$\frac{\partial z}{\partial y} = -\frac{y^2 + 2xz}{z^2 + 2xy}$$ ■

Functions of More Than Two Variables

Partial derivatives can also be defined for functions of three or more variables. For example, if f is a function of three variables x, y, and z, then its partial derivative with respect to x is defined as

$$f_x(x, y, z) = \lim_{h \to 0} \frac{f(x + h, y, z) - f(x, y, z)}{h}$$

and it is found by regarding y and z as constants and differentiating $f(x, y, z)$ with respect to x. If $w = f(x, y, z)$, then $f_x = \partial w/\partial x$ can be interpreted as the rate of change of w with respect to x when y and z are held fixed. But we can't interpret it geometrically because the graph of f lies in four-dimensional space.

In general, if u is a function of n variables, $u = f(x_1, x_2, \ldots, x_n)$, its partial derivative with respect to the ith variable x_i is

$$\frac{\partial u}{\partial x_i} = \lim_{h \to 0} \frac{f(x_1, \ldots, x_{i-1}, x_i + h, x_{i+1}, \ldots, x_n) - f(x_1, \ldots, x_i, \ldots, x_n)}{h}$$

and we also write

$$\frac{\partial u}{\partial x_i} = \frac{\partial f}{\partial x_i} = f_{x_i} = f_i = D_i f$$

EXAMPLE 6 Find f_x, f_y, and f_z if $f(x, y, z) = e^{xy} \ln z$.

SOLUTION Holding y and z constant and differentiating with respect to x, we have

$$f_x = ye^{xy} \ln z$$

Similarly, $$f_y = xe^{xy} \ln z \qquad \text{and} \qquad f_z = \frac{e^{xy}}{z}$$ ■

Higher Derivatives

If f is a function of two variables, then its partial derivatives f_x and f_y are also functions of two variables, so we can consider their partial derivatives $(f_x)_x$, $(f_x)_y$, $(f_y)_x$, and $(f_y)_y$, which are called the **second partial derivatives** of f. If $z = f(x, y)$, we use the following notation:

$$(f_x)_x = f_{xx} = f_{11} = \frac{\partial}{\partial x}\left(\frac{\partial f}{\partial x}\right) = \frac{\partial^2 f}{\partial x^2} = \frac{\partial^2 z}{\partial x^2}$$

$$(f_x)_y = f_{xy} = f_{12} = \frac{\partial}{\partial y}\left(\frac{\partial f}{\partial x}\right) = \frac{\partial^2 f}{\partial y\, \partial x} = \frac{\partial^2 z}{\partial y\, \partial x}$$

$$(f_y)_x = f_{yx} = f_{21} = \frac{\partial}{\partial x}\left(\frac{\partial f}{\partial y}\right) = \frac{\partial^2 f}{\partial x\, \partial y} = \frac{\partial^2 z}{\partial x\, \partial y}$$

$$(f_y)_y = f_{yy} = f_{22} = \frac{\partial}{\partial y}\left(\frac{\partial f}{\partial y}\right) = \frac{\partial^2 f}{\partial y^2} = \frac{\partial^2 z}{\partial y^2}$$

Thus the notation f_{xy} (or $\partial^2 f/\partial y\, \partial x$) means that we first differentiate with respect to x and then with respect to y, whereas in computing f_{yx} the order is reversed.

EXAMPLE 7 Find the second partial derivatives of

$$f(x, y) = x^3 + x^2y^3 - 2y^2$$

SOLUTION In Example 1 we found that

$$f_x(x, y) = 3x^2 + 2xy^3 \qquad f_y(x, y) = 3x^2y^2 - 4y$$

Therefore

$$f_{xx} = \frac{\partial}{\partial x}(3x^2 + 2xy^3) = 6x + 2y^3 \qquad f_{xy} = \frac{\partial}{\partial y}(3x^2 + 2xy^3) = 6xy^2$$

$$f_{yx} = \frac{\partial}{\partial x}(3x^2y^2 - 4y) = 6xy^2 \qquad f_{yy} = \frac{\partial}{\partial y}(3x^2y^2 - 4y) = 6x^2y - 4$$ ■

Figure 7 shows the graph of the function f in Example 7 and the graphs of its first- and second-order partial derivatives for $-2 \leq x \leq 2$, $-2 \leq y \leq 2$. Notice that these graphs are consistent with our interpretations of f_x and f_y as slopes of tangent lines to traces of the graph of f. For instance, the graph of f decreases if we start at $(0, -2)$ and move in the positive x-direction. This is reflected in the negative values of f_x. You should compare the graphs of f_{yx} and f_{yy} with the graph of f_y to see the relationships.

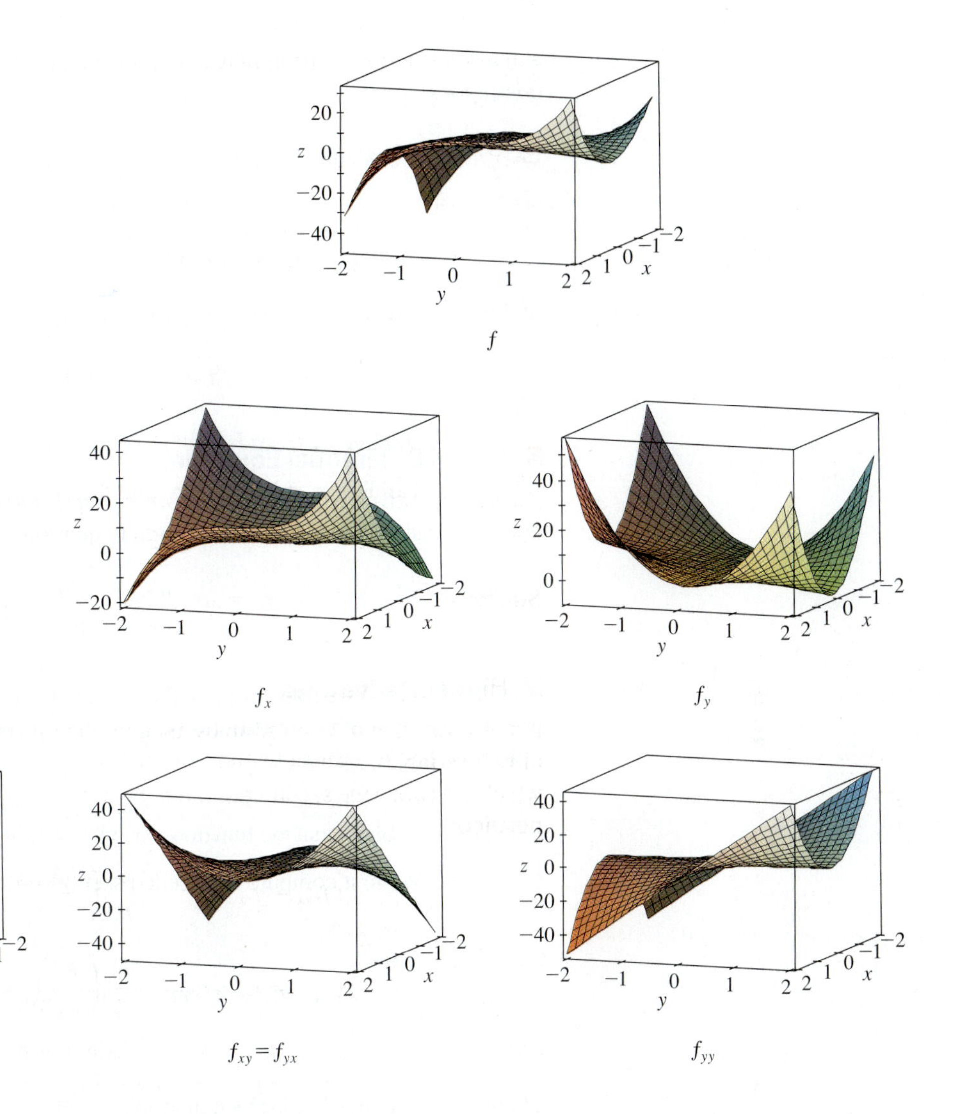

FIGURE 7

Notice that $f_{xy} = f_{yx}$ in Example 7. This is not just a coincidence. It turns out that the mixed partial derivatives f_{xy} and f_{yx} are equal for most functions that one meets in practice. The following theorem, which was discovered by the French mathematician Alexis Clairaut (1713–1765), gives conditions under which we can assert that $f_{xy} = f_{yx}$. The proof is given in Appendix F.

Clairaut

Alexis Clairaut was a child prodigy in mathematics: he read l'Hospital's textbook on calculus when he was ten and presented a paper on geometry to the French Academy of Sciences when he was 13. At the age of 18, Clairaut published *Recherches sur les courbes à double courbure,* which was the first systematic treatise on three-dimensional analytic geometry and included the calculus of space curves.

Clairaut's Theorem Suppose f is defined on a disk D that contains the point (a, b). If the functions f_{xy} and f_{yx} are both continuous on D, then

$$f_{xy}(a, b) = f_{yx}(a, b)$$

Partial derivatives of order 3 or higher can also be defined. For instance,

$$f_{xyy} = (f_{xy})_y = \frac{\partial}{\partial y}\left(\frac{\partial^2 f}{\partial y\,\partial x}\right) = \frac{\partial^3 f}{\partial y^2\,\partial x}$$

and using Clairaut's Theorem it can be shown that $f_{xyy} = f_{yxy} = f_{yyx}$ if these functions are continuous.

EXAMPLE 8 Calculate f_{xxyz} if $f(x, y, z) = \sin(3x + yz)$.

SOLUTION

$$f_x = 3\cos(3x + yz)$$

$$f_{xx} = -9\sin(3x + yz)$$

$$f_{xxy} = -9z\cos(3x + yz)$$

$$f_{xxyz} = -9\cos(3x + yz) + 9yz\sin(3x + yz)$$

■

Partial Differential Equations

Partial derivatives occur in *partial differential equations* that express certain physical laws. For instance, the partial differential equation

$$\frac{\partial^2 u}{\partial x^2} + \frac{\partial^2 u}{\partial y^2} = 0$$

is called **Laplace's equation** after Pierre Laplace (1749–1827). Solutions of this equation are called **harmonic functions**; they play a role in problems of heat conduction, fluid flow, and electric potential.

EXAMPLE 9 Show that the function $u(x, y) = e^x \sin y$ is a solution of Laplace's equation.

SOLUTION We first compute the needed second-order partial derivatives:

$$u_x = e^x \sin y \qquad u_y = e^x \cos y$$

$$u_{xx} = e^x \sin y \qquad u_{yy} = -e^x \sin y$$

So

$$u_{xx} + u_{yy} = e^x \sin y - e^x \sin y = 0$$

Therefore u satisfies Laplace's equation. ■

The **wave equation**

$$\frac{\partial^2 u}{\partial t^2} = a^2 \frac{\partial^2 u}{\partial x^2}$$

describes the motion of a waveform, which could be an ocean wave, a sound wave, a light wave, or a wave traveling along a vibrating string. For instance, if $u(x, t)$ represents the displacement of a vibrating violin string at time t and at a distance x from one end of the string (as in Figure 8), then $u(x, t)$ satisfies the wave equation. Here the constant a depends on the density of the string and on the tension in the string.

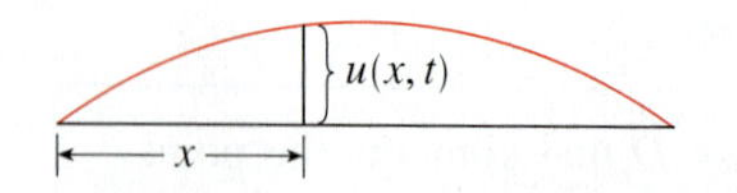

FIGURE 8

EXAMPLE 10 Verify that the function $u(x, t) = \sin(x - at)$ satisfies the wave equation.

SOLUTION

$$u_x = \cos(x - at) \qquad u_t = -a\cos(x - at)$$

$$u_{xx} = -\sin(x - at) \qquad u_{tt} = -a^2\sin(x - at) = a^2 u_{xx}$$

So u satisfies the wave equation. ■

Partial differential equations involving functions of three variables are also very important in science and engineering. The three-dimensional Laplace equation is

5
$$\frac{\partial^2 u}{\partial x^2} + \frac{\partial^2 u}{\partial y^2} + \frac{\partial^2 u}{\partial z^2} = 0$$

and one place it occurs is in geophysics. If $u(x, y, z)$ represents magnetic field strength at position (x, y, z), then it satisfies Equation 5. The strength of the magnetic field indicates the distribution of iron-rich minerals and reflects different rock types and the location of faults. Figure 9 shows a contour map of the earth's magnetic field as recorded from an aircraft carrying a magnetometer and flying 200 m above the surface of the ground. The contour map is enhanced by color-coding of the regions between the level curves.

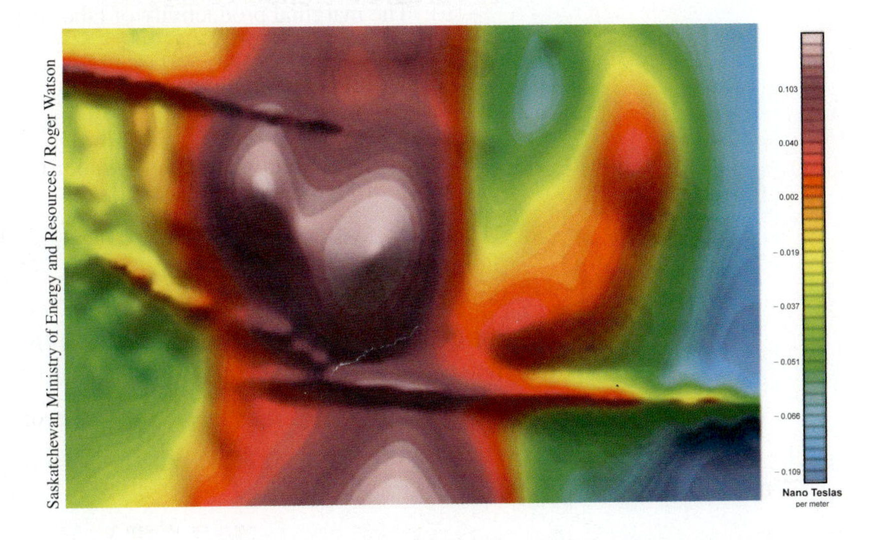

FIGURE 9
Magnetic field strength of the earth

Figure 10 shows a contour map for the second-order partial derivative of u in the vertical direction, that is, u_{zz}. It turns out that the values of the partial derivatives u_{xx} and u_{yy} are relatively easily measured from a map of the magnetic field. Then values of u_{zz} can be calculated from Laplace's equation (5).

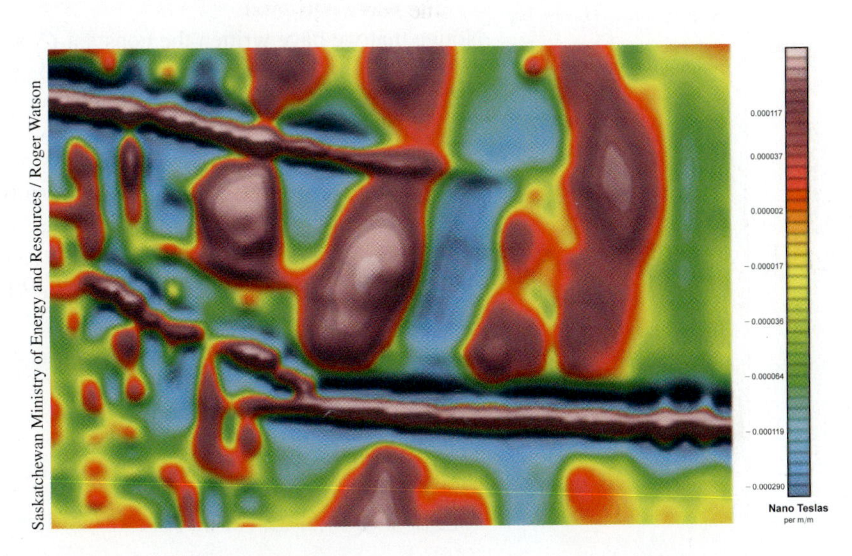

FIGURE 10
Second vertical derivative of the magnetic field

The Cobb-Douglas Production Function

In Example 14.1.3 we described the work of Cobb and Douglas in modeling the total production P of an economic system as a function of the amount of labor L and the capital investment K. Here we use partial derivatives to show how the particular form of their model follows from certain assumptions they made about the economy.

If the production function is denoted by $P = P(L, K)$, then the partial derivative $\partial P/\partial L$ is the rate at which production changes with respect to the amount of labor. Economists call it the marginal production with respect to labor or the **marginal productivity of labor**. Likewise, the partial derivative $\partial P/\partial K$ is the rate of change of production with respect to capital and is called the **marginal productivity of capital**. In these terms, the assumptions made by Cobb and Douglas can be stated as follows.

(i) If either labor or capital vanishes, then so will production.

(ii) The marginal productivity of labor is proportional to the amount of production per unit of labor.

(iii) The marginal productivity of capital is proportional to the amount of production per unit of capital.

Because the production per unit of labor is P/L, assumption (ii) says that

$$\frac{\partial P}{\partial L} = \alpha \frac{P}{L}$$

for some constant α. If we keep K constant ($K = K_0$), then this partial differential equation becomes an ordinary differential equation:

$$\frac{dP}{dL} = \alpha \frac{P}{L} \tag{6}$$

If we solve this separable differential equation by the methods of Section 9.3 (see also Exercise 85), we get

$$P(L, K_0) = C_1(K_0)L^{\alpha} \tag{7}$$

Notice that we have written the constant C_1 as a function of K_0 because it could depend on the value of K_0.

Similarly, assumption (iii) says that

$$\frac{\partial P}{\partial K} = \beta \frac{P}{K}$$

and we can solve this differential equation to get

$$P(L_0, K) = C_2(L_0)K^{\beta} \tag{8}$$

Comparing Equations 7 and 8, we have

$$P(L, K) = bL^{\alpha}K^{\beta} \tag{9}$$

where b is a constant that is independent of both L and K. Assumption (i) shows that $\alpha > 0$ and $\beta > 0$.

Notice from Equation 9 that if labor and capital are both increased by a factor m, then

$$P(mL, mK) = b(mL)^{\alpha}(mK)^{\beta} = m^{\alpha+\beta}bL^{\alpha}K^{\beta} = m^{\alpha+\beta}P(L, K)$$

If $\alpha + \beta = 1$, then $P(mL, mK) = mP(L, K)$, which means that production is also increased by a factor of m. That is why Cobb and Douglas assumed that $\alpha + \beta = 1$ and therefore

$$P(L, K) = bL^{\alpha}K^{1-\alpha}$$

This is the Cobb-Douglas production function that we discussed in Section 14.1.

14.3 EXERCISES

1. The temperature T (in °C) at a location in the Northern Hemisphere depends on the longitude x, latitude y, and time t, so we can write $T = f(x, y, t)$. Let's measure time in hours from the beginning of January.

(a) What are the meanings of the partial derivatives $\partial T/\partial x$, $\partial T/\partial y$, and $\partial T/\partial t$?

(b) Honolulu has longitude 158°W and latitude 21°N. Suppose that at 9:00 AM on January 1 the wind is blowing hot air to the northeast, so the air to the west and south is warm and the air to the north and east is cooler. Would you expect $f_x(158, 21, 9)$, $f_y(158, 21, 9)$, and $f_t(158, 21, 9)$ to be positive or negative? Explain.

2. At the beginning of this section we discussed the function $I = f(T, H)$, where I is the heat index, T is the temperature, and H is the relative humidity. Use Table 1 to estimate $f_T(92, 60)$ and $f_H(92, 60)$. What are the practical interpretations of these values?

3. The wind-chill index W is the perceived temperature when the actual temperature is T and the wind speed is v, so we can write $W = f(T, v)$. The following table of values is an excerpt from Table 1 in Section 14.1.

Wind speed (km/h)

Actual temperature (°C): T \ v	20	30	40	50	60	70
−10	−18	−20	−21	−22	−23	−23
−15	−24	−26	−27	−29	−30	−30
−20	−30	−33	−34	−35	−36	−37
−25	−37	−39	−41	−42	−43	−44

(a) Estimate the values of $f_T(-15, 30)$ and $f_v(-15, 30)$. What are the practical interpretations of these values?

(b) In general, what can you say about the signs of $\partial W/\partial T$ and $\partial W/\partial v$?

(c) What appears to be the value of the following limit?

$$\lim_{v\to\infty} \frac{\partial W}{\partial v}$$

4. The wave heights h in the open sea depend on the speed v of the wind and the length of time t that the wind has been blowing at that speed. Values of the function $h = f(v, t)$ are recorded in feet in the following table.

Duration (hours)

Wind speed (knots): v \ t	5	10	15	20	30	40	50
10	2	2	2	2	2	2	2
15	4	4	5	5	5	5	5
20	5	7	8	8	9	9	9
30	9	13	16	17	18	19	19
40	14	21	25	28	31	33	33
50	19	29	36	40	45	48	50
60	24	37	47	54	62	67	69

(a) What are the meanings of the partial derivatives $\partial h/\partial v$ and $\partial h/\partial t$?

(b) Estimate the values of $f_v(40, 15)$ and $f_t(40, 15)$. What are the practical interpretations of these values?

(c) What appears to be the value of the following limit?

$$\lim_{t\to\infty} \frac{\partial h}{\partial t}$$

5–8 Determine the signs of the partial derivatives for the function f whose graph is shown.

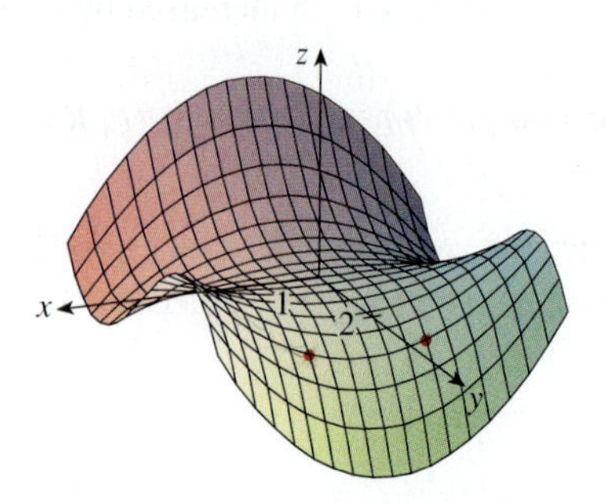

5. (a) $f_x(1, 2)$ (b) $f_y(1, 2)$

6. (a) $f_x(-1, 2)$ (b) $f_y(-1, 2)$

7. (a) $f_{xx}(-1, 2)$ (b) $f_{yy}(-1, 2)$

8. (a) $f_{xy}(1, 2)$ (b) $f_{xy}(-1, 2)$

---

9. The following surfaces, labeled a, b, and c, are graphs of a function f and its partial derivatives f_x and f_y. Identify each surface and give reasons for your choices.

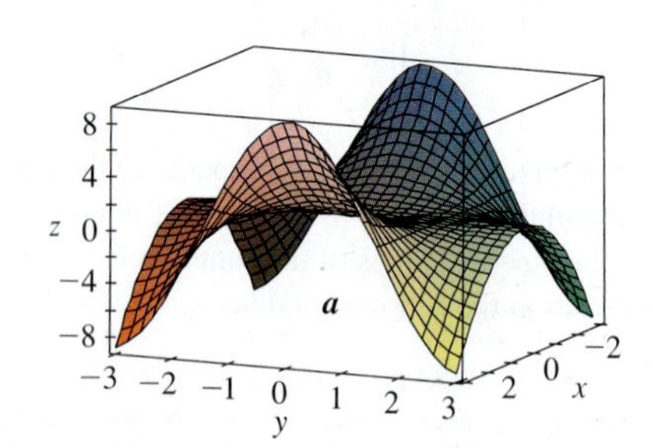

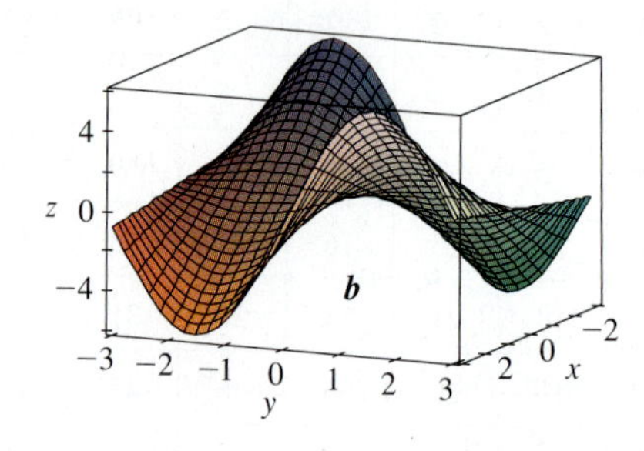

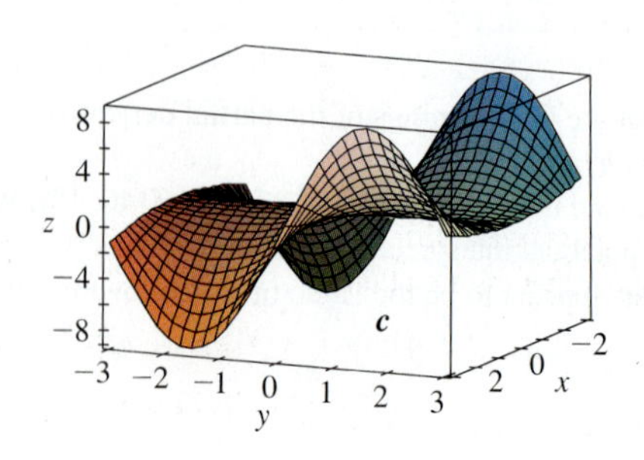

10. A contour map is given for a function f. Use it to estimate $f_x(2, 1)$ and $f_y(2, 1)$.

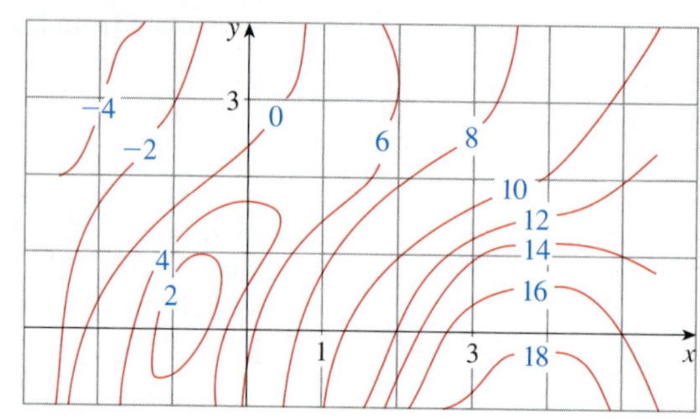

11. If $f(x, y) = 16 - 4x^2 - y^2$, find $f_x(1, 2)$ and $f_y(1, 2)$ and interpret these numbers as slopes. Illustrate with either hand-drawn sketches or computer plots.

12. If $f(x, y) = \sqrt{4 - x^2 - 4y^2}$, find $f_x(1, 0)$ and $f_y(1, 0)$ and interpret these numbers as slopes. Illustrate with either hand-drawn sketches or computer plots.

13–14 Find f_x and f_y and graph f, f_x, and f_y with domains and viewpoints that enable you to see the relationships between them.

13. $f(x, y) = x^2y^3$

14. $f(x, y) = \dfrac{y}{1 + x^2y^2}$

15–40 Find the first partial derivatives of the function.

15. $f(x, y) = x^4 + 5xy^3$

16. $f(x, y) = x^2y - 3y^4$

17. $f(x, t) = t^2e^{-x}$

18. $f(x, t) = \sqrt{3x + 4t}$

19. $z = \ln(x + t^2)$

20. $z = x\sin(xy)$

21. $f(x, y) = \dfrac{x}{y}$

22. $f(x, y) = \dfrac{x}{(x + y)^2}$

23. $f(x, y) = \dfrac{ax + by}{cx + dy}$

24. $w = \dfrac{e^v}{u + v^2}$

25. $g(u, v) = (u^2v - v^3)^5$

26. $u(r, \theta) = \sin(r\cos\theta)$

27. $R(p, q) = \tan^{-1}(pq^2)$

28. $f(x, y) = x^y$

29. $F(x, y) = \displaystyle\int_y^x \cos(e^t)\,dt$

30. $F(\alpha, \beta) = \displaystyle\int_\alpha^\beta \sqrt{t^3 + 1}\,dt$

31. $f(x, y, z) = x^3yz^2 + 2yz$

32. $f(x, y, z) = xy^2e^{-xz}$

33. $w = \ln(x + 2y + 3z)$

34. $w = y\tan(x + 2z)$

35. $p = \sqrt{t^4 + u^2\cos v}$

36. $u = x^{y/z}$

37. $h(x, y, z, t) = x^2y\cos(z/t)$

38. $\phi(x, y, z, t) = \dfrac{\alpha x + \beta y^2}{\gamma z + \delta t^2}$

39. $u = \sqrt{x_1^2 + x_2^2 + \cdots + x_n^2}$

40. $u = \sin(x_1 + 2x_2 + \cdots + nx_n)$

41–44 Find the indicated partial derivative.

41. $R(s, t) = te^{s/t}$; $R_t(0, 1)$

42. $f(x, y) = y\sin^{-1}(xy)$; $f_y(1, \frac{1}{2})$

43. $f(x, y, z) = \ln\dfrac{1 - \sqrt{x^2 + y^2 + z^2}}{1 + \sqrt{x^2 + y^2 + z^2}}$; $f_y(1, 2, 2)$

44. $f(x, y, z) = x^{yz}$; $f_z(e, 1, 0)$

45–46 Use the definition of partial derivatives as limits (4) to find $f_x(x, y)$ and $f_y(x, y)$.

45. $f(x, y) = xy^2 - x^3y$

46. $f(x, y) = \dfrac{x}{x + y^2}$

47–50 Use implicit differentiation to find $\partial z/\partial x$ and $\partial z/\partial y$.

47. $x^2 + 2y^2 + 3z^2 = 1$

48. $x^2 - y^2 + z^2 - 2z = 4$

49. $e^z = xyz$

50. $yz + x\ln y = z^2$

51–52 Find $\partial z/\partial x$ and $\partial z/\partial y$.

51. (a) $z = f(x) + g(y)$ (b) $z = f(x + y)$

52. (a) $z = f(x)g(y)$ (b) $z = f(xy)$
(c) $z = f(x/y)$

53–58 Find all the second partial derivatives.

53. $f(x, y) = x^4y - 2x^3y^2$

54. $f(x, y) = \ln(ax + by)$

55. $z = \dfrac{y}{2x + 3y}$

56. $T = e^{-2r}\cos\theta$

57. $v = \sin(s^2 - t^2)$

58. $w = \sqrt{1 + uv^2}$

59–62 Verify that the conclusion of Clairaut's Theorem holds, that is, $u_{xy} = u_{yx}$.

59. $u = x^4y^3 - y^4$

60. $u = e^{xy}\sin y$

61. $u = \cos(x^2y)$

62. $u = \ln(x + 2y)$

63–70 Find the indicated partial derivative(s).

63. $f(x, y) = x^4y^2 - x^3y$; f_{xxx}, f_{xyx}

64. $f(x, y) = \sin(2x + 5y)$; f_{yxy}

65. $f(x, y, z) = e^{xyz^2}$; f_{xyz}

66. $g(r, s, t) = e^r\sin(st)$; g_{rst}

67. $W = \sqrt{u + v^2}$; $\dfrac{\partial^3 W}{\partial u^2\,\partial v}$

68. $V = \ln(r + s^2 + t^3)$; $\dfrac{\partial^3 V}{\partial r\,\partial s\,\partial t}$

69. $w = \dfrac{x}{y + 2z}$; $\dfrac{\partial^3 w}{\partial z\,\partial y\,\partial x}$, $\dfrac{\partial^3 w}{\partial x^2\,\partial y}$

70. $u = x^a y^b z^c$; $\dfrac{\partial^6 u}{\partial x\,\partial y^2\,\partial z^3}$

71. If $f(x, y, z) = xy^2z^3 + \arcsin(x\sqrt{z})$, find f_{xzy}. [*Hint:* Which order of differentiation is easiest?]

72. If $g(x, y, z) = \sqrt{1 + xz} + \sqrt{1 - xy}$, find g_{xyz}. [*Hint:* Use a different order of differentiation for each term.]

73. Use the table of values of $f(x, y)$ to estimate the values of $f_x(3, 2)$, $f_x(3, 2.2)$, and $f_{xy}(3, 2)$.

x \ y	1.8	2.0	2.2
2.5	12.5	10.2	9.3
3.0	18.1	17.5	15.9
3.5	20.0	22.4	26.1

74. Level curves are shown for a function f. Determine whether the following partial derivatives are positive or negative at the point P.
(a) f_x (b) f_y (c) f_{xx}
(d) f_{xy} (e) f_{yy}

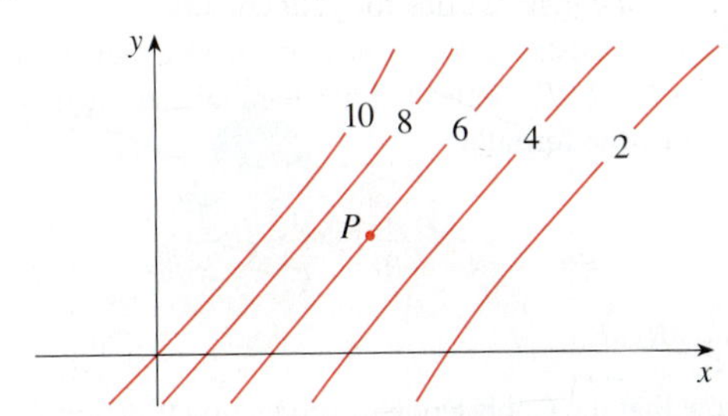

75. Verify that the function $u = e^{-\alpha^2k^2t}\sin kx$ is a solution of the heat conduction equation $u_t = \alpha^2 u_{xx}$.

76. Determine whether each of the following functions is a solution of Laplace's equation $u_{xx} + u_{yy} = 0$.
(a) $u = x^2 + y^2$ (b) $u = x^2 - y^2$
(c) $u = x^3 + 3xy^2$ (d) $u = \ln\sqrt{x^2 + y^2}$
(e) $u = \sin x\cosh y + \cos x\sinh y$
(f) $u = e^{-x}\cos y - e^{-y}\cos x$

77. Verify that the function $u = 1/\sqrt{x^2 + y^2 + z^2}$ is a solution of the three-dimensional Laplace equation $u_{xx} + u_{yy} + u_{zz} = 0$.

78. Show that each of the following functions is a solution of the wave equation $u_{tt} = a^2u_{xx}$.
(a) $u = \sin(kx)\sin(akt)$ (b) $u = t/(a^2t^2 - x^2)$
(c) $u = (x - at)^6 + (x + at)^6$
(d) $u = \sin(x - at) + \ln(x + at)$

79. If f and g are twice differentiable functions of a single variable, show that the function

$$u(x, t) = f(x + at) + g(x - at)$$

is a solution of the wave equation given in Exercise 78.

80. If $u = e^{a_1x_1+a_2x_2+\cdots+a_nx_n}$, where $a_1^2 + a_2^2 + \cdots + a_n^2 = 1$, show that

$$\frac{\partial^2 u}{\partial x_1^2} + \frac{\partial^2 u}{\partial x_2^2} + \cdots + \frac{\partial^2 u}{\partial x_n^2} = u$$

81. The *diffusion equation*

$$\frac{\partial c}{\partial t} = D\frac{\partial^2 c}{\partial x^2}$$

where D is a positive constant, describes the diffusion of heat through a solid, or the concentration of a pollutant at time t at a distance x from the source of the pollution, or the invasion of alien species into a new habitat. Verify that the function

$$c(x, t) = \frac{1}{\sqrt{4\pi Dt}}e^{-x^2/(4Dt)}$$

is a solution of the diffusion equation.

82. The temperature at a point (x, y) on a flat metal plate is given by $T(x, y) = 60/(1 + x^2 + y^2)$, where T is measured in °C and x, y in meters. Find the rate of change of temperature with respect to distance at the point $(2, 1)$ in (a) the x-direction and (b) the y-direction.

83. The total resistance R produced by three conductors with resistances R_1, R_2, R_3 connected in a parallel electrical circuit is given by the formula

$$\frac{1}{R} = \frac{1}{R_1} + \frac{1}{R_2} + \frac{1}{R_3}$$

Find $\partial R/\partial R_1$.

84. Show that the Cobb-Douglas production function $P = bL^{\alpha}K^{\beta}$ satisfies the equation

$$L\frac{\partial P}{\partial L} + K\frac{\partial P}{\partial K} = (\alpha + \beta)P$$

85. Show that the Cobb-Douglas production function satisfies $P(L, K_0) = C_1(K_0)L^{\alpha}$ by solving the differential equation

$$\frac{dP}{dL} = \alpha\frac{P}{L}$$

(See Equation 6.)

86. Cobb and Douglas used the equation $P(L, K) = 1.01L^{0.75}K^{0.25}$ to model the American economy from 1899 to 1922, where L is the amount of labor and K is the amount of capital. (See Example 14.1.3.)

(a) Calculate P_L and P_K.

(b) Find the marginal productivity of labor and the marginal productivity of capital in the year 1920, when $L = 194$ and $K = 407$ (compared with the assigned values $L = 100$ and $K = 100$ in 1899). Interpret the results.

(c) In the year 1920, which would have benefited production more, an increase in capital investment or an increase in spending on labor?

87. The *van der Waals equation* for n moles of a gas is

$$\left(P + \frac{n^2 a}{V^2}\right)(V - nb) = nRT$$

where P is the pressure, V is the volume, and T is the temperature of the gas. The constant R is the universal gas constant and a and b are positive constants that are characteristic of a particular gas. Calculate $\partial T/\partial P$ and $\partial P/\partial V$.

88. The gas law for a fixed mass m of an ideal gas at absolute temperature T, pressure P, and volume V is $PV = mRT$, where R is the gas constant. Show that

$$\frac{\partial P}{\partial V}\frac{\partial V}{\partial T}\frac{\partial T}{\partial P} = -1$$

89. For the ideal gas of Exercise 88, show that

$$T\frac{\partial P}{\partial T}\frac{\partial V}{\partial T} = mR$$

90. The wind-chill index is modeled by the function

$$W = 13.12 + 0.6215T - 11.37v^{0.16} + 0.3965Tv^{0.16}$$

where T is the temperature (°C) and v is the wind speed (km/h). When $T = -15$°C and $v = 30$ km/h, by how much would you expect the apparent temperature W to drop if the actual temperature decreases by 1°C? What if the wind speed increases by 1 km/h?

91. A model for the surface area of a human body is given by the function

$$S = f(w, h) = 0.1091w^{0.425}h^{0.725}$$

where w is the weight (in pounds), h is the height (in inches), and S is measured in square feet. Calculate and interpret the partial derivatives.

(a) $\dfrac{\partial S}{\partial w}(160, 70)$ (b) $\dfrac{\partial S}{\partial h}(160, 70)$

92. One of Poiseuille's laws states that the resistance of blood flowing through an artery is

$$R = C\frac{L}{r^4}$$

where L and r are the length and radius of the artery and C is a positive constant determined by the viscosity of the blood. Calculate $\partial R/\partial L$ and $\partial R/\partial r$ and interpret them.

93. In the project on page 271 we expressed the power needed by a bird during its flapping mode as

$$P(v, x, m) = Av^3 + \frac{B(mg/x)^2}{v}$$

where A and B are constants specific to a species of bird, v is the velocity of the bird, m is the mass of the bird, and x is the fraction of the flying time spent in flapping mode. Calculate $\partial P/\partial v$, $\partial P/\partial x$, and $\partial P/\partial m$ and interpret them.

94. The average energy E (in kcal) needed for a lizard to walk or run a distance of 1 km has been modeled by the equation

$$E(m, v) = 2.65m^{0.66} + \frac{3.5m^{0.75}}{v}$$

where m is the body mass of the lizard (in grams) and v is its speed (in km/h). Calculate $E_m(400, 8)$ and $E_v(400, 8)$ and interpret your answers.

Source: C. Robbins, *Wildlife Feeding and Nutrition*, 2d ed. (San Diego: Academic Press, 1993).

95. The kinetic energy of a body with mass m and velocity v is $K = \frac{1}{2}mv^2$. Show that

$$\frac{\partial K}{\partial m}\frac{\partial^2 K}{\partial v^2} = K$$

96. If a, b, c are the sides of a triangle and A, B, C are the opposite angles, find $\partial A/\partial a$, $\partial A/\partial b$, $\partial A/\partial c$ by implicit differentiation of the Law of Cosines.

97. You are told that there is a function f whose partial derivatives are $f_x(x, y) = x + 4y$ and $f_y(x, y) = 3x - y$. Should you believe it?

98. The paraboloid $z = 6 - x - x^2 - 2y^2$ intersects the plane $x = 1$ in a parabola. Find parametric equations for the tangent line to this parabola at the point $(1, 2, -4)$. Use a computer to graph the paraboloid, the parabola, and the tangent line on the same screen.

99. The ellipsoid $4x^2 + 2y^2 + z^2 = 16$ intersects the plane $y = 2$ in an ellipse. Find parametric equations for the tangent line to this ellipse at the point $(1, 2, 2)$.

100. In a study of frost penetration it was found that the temperature T at time t (measured in days) at a depth x (measured in feet) can be modeled by the function

$$T(x, t) = T_0 + T_1 e^{-\lambda x} \sin(\omega t - \lambda x)$$

where $\omega = 2\pi/365$ and λ is a positive constant.

(a) Find $\partial T/\partial x$. What is its physical significance?

(b) Find $\partial T/\partial t$. What is its physical significance?

(c) Show that T satisfies the heat equation $T_t = kT_{xx}$ for a certain constant k.

(d) If $\lambda = 0.2$, $T_0 = 0$, and $T_1 = 10$, use a computer to graph $T(x, t)$.

(e) What is the physical significance of the term $-\lambda x$ in the expression $\sin(\omega t - \lambda x)$?

101. Use Clairaut's Theorem to show that if the third-order partial derivatives of f are continuous, then

$$f_{xyy} = f_{yxy} = f_{yyx}$$

102. (a) How many nth-order partial derivatives does a function of two variables have?

(b) If these partial derivatives are all continuous, how many of them can be distinct?

(c) Answer the question in part (a) for a function of three variables.

103. If

$$f(x, y) = x(x^2 + y^2)^{-3/2} e^{\sin(x^2 y)}$$

find $f_x(1, 0)$. [*Hint:* Instead of finding $f_x(x, y)$ first, note that it's easier to use Equation 1 or Equation 2.]

104. If $f(x, y) = \sqrt[3]{x^3 + y^3}$, find $f_x(0, 0)$.

105. Let

$$f(x, y) = \begin{cases} \dfrac{x^3 y - xy^3}{x^2 + y^2} & \text{if } (x, y) \neq (0, 0) \\ 0 & \text{if } (x, y) = (0, 0) \end{cases}$$

(a) Use a computer to graph f.

(b) Find $f_x(x, y)$ and $f_y(x, y)$ when $(x, y) \neq (0, 0)$.

(c) Find $f_x(0, 0)$ and $f_y(0, 0)$ using Equations 2 and 3.

(d) Show that $f_{xy}(0, 0) = -1$ and $f_{yx}(0, 0) = 1$.

CAS (e) Does the result of part (d) contradict Clairaut's Theorem? Use graphs of f_{xy} and f_{yx} to illustrate your answer.

14.4 Tangent Planes and Linear Approximations

One of the most important ideas in single-variable calculus is that as we zoom in toward a point on the graph of a differentiable function, the graph becomes indistinguishable from its tangent line and we can approximate the function by a linear function. (See Section 2.9.) Here we develop similar ideas in three dimensions. As we zoom in toward a point on a surface that is the graph of a differentiable function of two variables, the surface looks more and more like a plane (its tangent plane) and we can approximate the function by a linear function of two variables. We also extend the idea of a differential to functions of two or more variables.

Tangent Planes

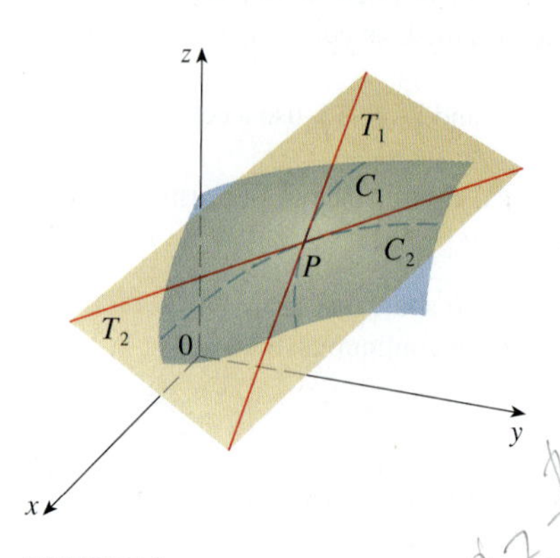

FIGURE 1
The tangent plane contains the tangent lines T_1 and T_2.

Suppose a surface S has equation $z = f(x, y)$, where f has continuous first partial derivatives, and let $P(x_0, y_0, z_0)$ be a point on S. As in the preceding section, let C_1 and C_2 be the curves obtained by intersecting the vertical planes $y = y_0$ and $x = x_0$ with the surface S. Then the point P lies on both C_1 and C_2. Let T_1 and T_2 be the tangent lines to the curves C_1 and C_2 at the point P. Then the **tangent plane** to the surface S at the point P is defined to be the plane that contains both tangent lines T_1 and T_2. (See Figure 1.)

We will see in Section 14.6 that if C is any other curve that lies on the surface S and passes through P, then its tangent line at P also lies in the tangent plane. Therefore you can think of the tangent plane to S at P as consisting of all possible tangent lines at P to curves that lie on S and pass through P. The tangent plane at P is the plane that most closely approximates the surface S near the point P.

We know from Equation 12.5.7 that any plane passing through the point $P(x_0, y_0, z_0)$ has an equation of the form

$$A(x - x_0) + B(y - y_0) + C(z - z_0) = 0$$

By dividing this equation by C and letting $a = -A/C$ and $b = -B/C$, we can write it in the form

$$\boxed{1} \qquad z - z_0 = a(x - x_0) + b(y - y_0)$$

If Equation 1 represents the tangent plane at P, then its intersection with the plane $y = y_0$ must be the tangent line T_1. Setting $y = y_0$ in Equation 1 gives

$$z - z_0 = a(x - x_0) \qquad \text{where } y = y_0$$

and we recognize this as the equation (in point-slope form) of a line with slope a. But from Section 14.3 we know that the slope of the tangent T_1 is $f_x(x_0, y_0)$. Therefore $a = f_x(x_0, y_0)$.

Similarly, putting $x = x_0$ in Equation 1, we get $z - z_0 = b(y - y_0)$, which must represent the tangent line T_2, so $b = f_y(x_0, y_0)$.

Note the similarity between the equation of a tangent plane and the equation of a tangent line:

$$y - y_0 = f'(x_0)(x - x_0)$$

2 Suppose f has continuous partial derivatives. An equation of the tangent plane to the surface $z = f(x, y)$ at the point $P(x_0, y_0, z_0)$ is

$$z - z_0 = f_x(x_0, y_0)(x - x_0) + f_y(x_0, y_0)(y - y_0)$$

EXAMPLE 1 Find the tangent plane to the elliptic paraboloid $z = 2x^2 + y^2$ at the point $(1, 1, 3)$.

SOLUTION Let $f(x, y) = 2x^2 + y^2$. Then

$$f_x(x, y) = 4x \qquad f_y(x, y) = 2y$$

$$f_x(1, 1) = 4 \qquad f_y(1, 1) = 2$$

Then (2) gives the equation of the tangent plane at $(1, 1, 3)$ as

$$z - 3 = 4(x - 1) + 2(y - 1)$$

or

$$z = 4x + 2y - 3$$

■

Figure 2(a) shows the elliptic paraboloid and its tangent plane at $(1, 1, 3)$ that we found in Example 1. In parts (b) and (c) we zoom in toward the point $(1, 1, 3)$ by restrict-

TEC Visual 14.4 shows an animation of Figures 2 and 3.

ing the domain of the function $f(x, y) = 2x^2 + y^2$. Notice that the more we zoom in, the flatter the graph appears and the more it resembles its tangent plane.

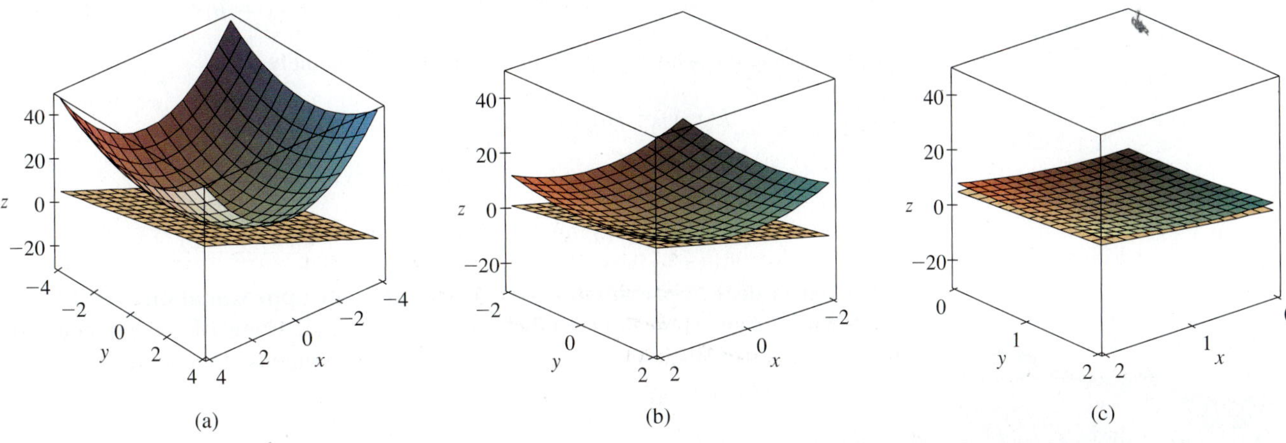

FIGURE 2 The elliptic paraboloid $z = 2x^2 + y^2$ appears to coincide with its tangent plane as we zoom in toward (1, 1, 3).

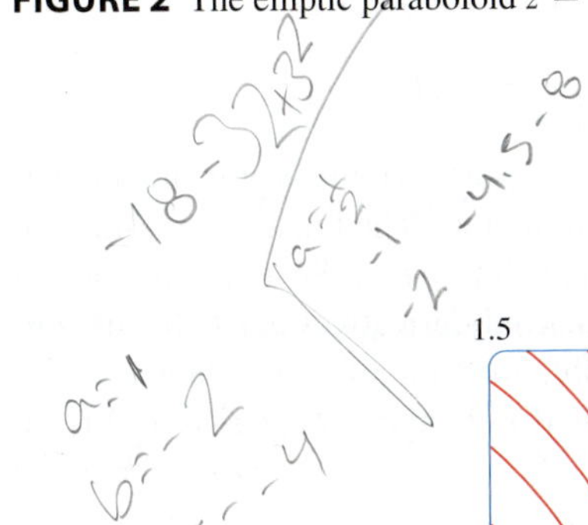

In Figure 3 we corroborate this impression by zooming in toward the point (1, 1) on a contour map of the function $f(x, y) = 2x^2 + y^2$. Notice that the more we zoom in, the more the level curves look like equally spaced parallel lines, which is characteristic of a plane.

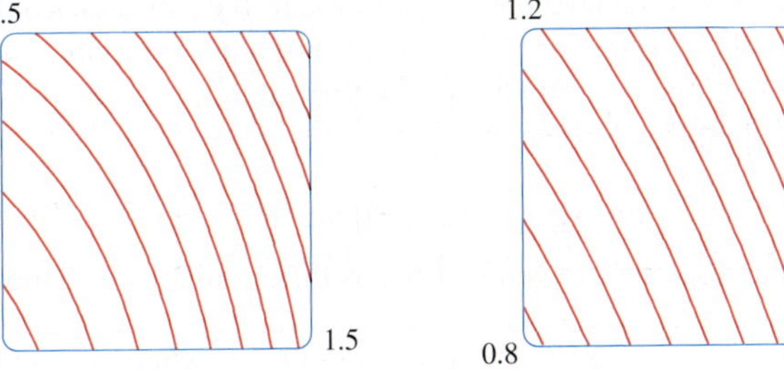

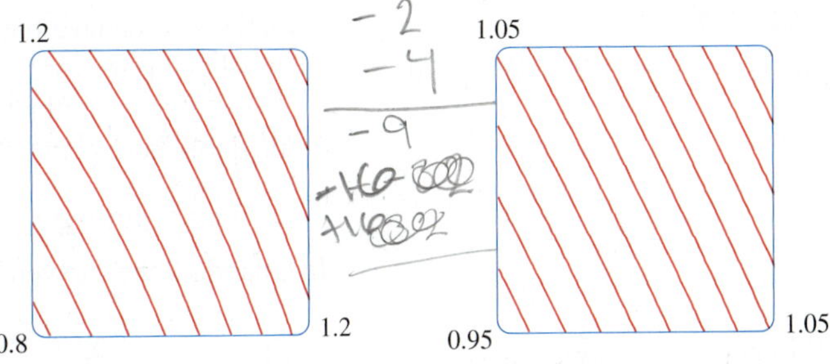

FIGURE 3
Zooming in toward (1, 1) on a contour map of $f(x, y) = 2x^2 + y^2$

Linear Approximations

In Example 1 we found that an equation of the tangent plane to the graph of the function $f(x, y) = 2x^2 + y^2$ at the point (1, 1, 3) is $z = 4x + 2y - 3$. Therefore, in view of the visual evidence in Figures 2 and 3, the linear function of two variables

$$L(x, y) = 4x + 2y - 3$$

is a good approximation to $f(x, y)$ when (x, y) is near (1, 1). The function L is called the *linearization* of f at (1, 1) and the approximation

$$f(x, y) \approx 4x + 2y - 3$$

is called the *linear approximation* or *tangent plane approximation* of f at (1, 1).

For instance, at the point (1.1, 0.95) the linear approximation gives

$$f(1.1, 0.95) \approx 4(1.1) + 2(0.95) - 3 = 3.3$$

which is quite close to the true value of $f(1.1, 0.95) = 2(1.1)^2 + (0.95)^2 = 3.3225$. But if we take a point farther away from (1, 1), such as (2, 3), we no longer get a good approximation. In fact, $L(2, 3) = 11$ whereas $f(2, 3) = 17$.

In general, we know from (2) that an equation of the tangent plane to the graph of a function f of two variables at the point $(a, b, f(a, b))$ is

$$z = f(a, b) + f_x(a, b)(x - a) + f_y(a, b)(y - b)$$

The linear function whose graph is this tangent plane, namely

$$\boxed{3} \qquad L(x, y) = f(a, b) + f_x(a, b)(x - a) + f_y(a, b)(y - b)$$

is called the **linearization** of f at (a, b) and the approximation

$$\boxed{4} \qquad f(x, y) \approx f(a, b) + f_x(a, b)(x - a) + f_y(a, b)(y - b)$$

is called the **linear approximation** or the **tangent plane approximation** of f at (a, b).

We have defined tangent planes for surfaces $z = f(x, y)$, where f has continuous first partial derivatives. What happens if f_x and f_y are not continuous? Figure 4 pictures such a function; its equation is

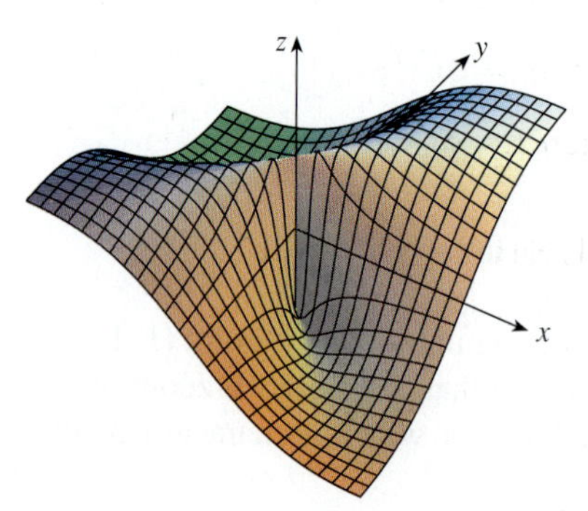

FIGURE 4
$f(x, y) = \dfrac{xy}{x^2 + y^2}$ if $(x, y) \neq (0, 0)$,
$f(0, 0) = 0$

$$f(x, y) = \begin{cases} \dfrac{xy}{x^2 + y^2} & \text{if } (x, y) \neq (0, 0) \\ 0 & \text{if } (x, y) = (0, 0) \end{cases}$$

You can verify (see Exercise 46) that its partial derivatives exist at the origin and, in fact, $f_x(0, 0) = 0$ and $f_y(0, 0) = 0$, but f_x and f_y are not continuous. The linear approximation would be $f(x, y) \approx 0$, but $f(x, y) = \frac{1}{2}$ at all points on the line $y = x$. So a function of two variables can behave badly even though both of its partial derivatives exist. To rule out such behavior, we formulate the idea of a differentiable function of two variables.

Recall that for a function of one variable, $y = f(x)$, if x changes from a to $a + \Delta x$, we defined the increment of y as

$$\Delta y = f(a + \Delta x) - f(a)$$

In Chapter 2 we showed that if f is differentiable at a, then

This is Equation 2.5.5.

$$\boxed{5} \qquad \Delta y = f'(a)\,\Delta x + \varepsilon\,\Delta x \qquad \text{where } \varepsilon \to 0 \text{ as } \Delta x \to 0$$

Now consider a function of two variables, $z = f(x, y)$, and suppose x changes from a to $a + \Delta x$ and y changes from b to $b + \Delta y$. Then the corresponding **increment** of z is

$$\boxed{6} \qquad \Delta z = f(a + \Delta x, b + \Delta y) - f(a, b)$$

Thus the increment Δz represents the change in the value of f when (x, y) changes from (a, b) to $(a + \Delta x, b + \Delta y)$. By analogy with (5) we define the differentiability of a function of two variables as follows.

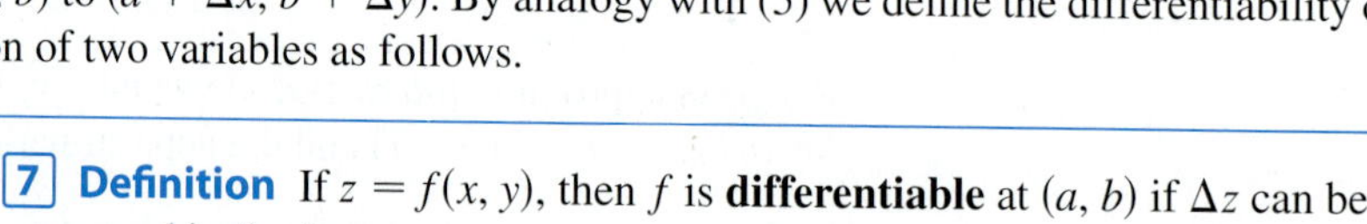

7 Definition If $z = f(x, y)$, then f is **differentiable** at (a, b) if Δz can be expressed in the form

$$\Delta z = f_x(a, b)\,\Delta x + f_y(a, b)\,\Delta y + \varepsilon_1\,\Delta x + \varepsilon_2\,\Delta y$$

where ε_1 and $\varepsilon_2 \to 0$ as $(\Delta x, \Delta y) \to (0, 0)$.

Definition 7 says that a differentiable function is one for which the linear approximation (4) is a good approximation when (x, y) is near (a, b). In other words, the tangent plane approximates the graph of f well near the point of tangency.

It's sometimes hard to use Definition 7 directly to check the differentiability of a function, but the next theorem provides a convenient sufficient condition for differentiability.

Theorem 8 is proved in Appendix F.

8 Theorem If the partial derivatives f_x and f_y exist near (a, b) and are continuous at (a, b), then f is differentiable at (a, b).

EXAMPLE 2 Show that $f(x, y) = xe^{xy}$ is differentiable at $(1, 0)$ and find its linearization there. Then use it to approximate $f(1.1, -0.1)$.

Figure 5 shows the graphs of the function f and its linearization L in Example 2.

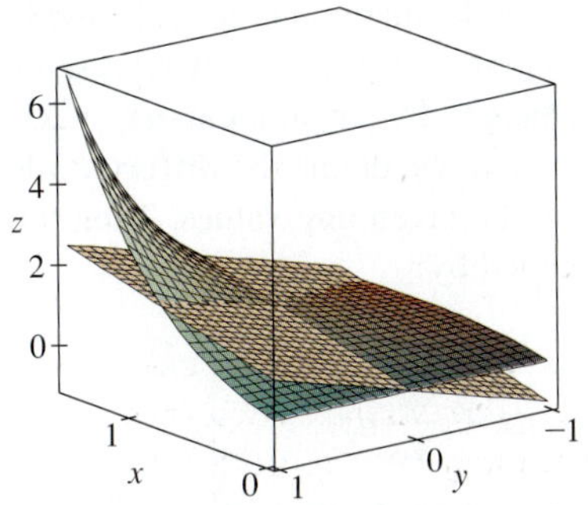

SOLUTION The partial derivatives are

$$f_x(x, y) = e^{xy} + xye^{xy} \qquad f_y(x, y) = x^2e^{xy}$$

$$f_x(1, 0) = 1 \qquad f_y(1, 0) = 1$$

Both f_x and f_y are continuous functions, so f is differentiable by Theorem 8. The linearization is

$$\begin{aligned} L(x, y) &= f(1, 0) + f_x(1, 0)(x - 1) + f_y(1, 0)(y - 0) \\ &= 1 + 1(x - 1) + 1 \cdot y = x + y \end{aligned}$$

The corresponding linear approximation is

$$xe^{xy} \approx x + y$$

so

$$f(1.1, -0.1) \approx 1.1 - 0.1 = 1$$

FIGURE 5

Compare this with the actual value of $f(1.1, -0.1) = 1.1e^{-0.11} \approx 0.98542$. ■

EXAMPLE 3 At the beginning of Section 14.3 we discussed the heat index (perceived temperature) I as a function of the actual temperature T and the relative humidity H and gave the following table of values from the National Weather Service.

Relative humidity (%)

Actual temperature (°F) T \ H	50	55	60	65	70	75	80	85	90
90	96	98	100	103	106	109	112	115	119
92	100	103	105	108	112	115	119	123	128
94	104	107	111	114	118	122	127	132	137
96	109	113	116	121	125	130	135	141	146
98	114	118	123	127	133	138	144	150	157
100	119	124	129	135	141	147	154	161	168

Find a linear approximation for the heat index $I = f(T, H)$ when T is near 96°F and H is near 70%. Use it to estimate the heat index when the temperature is 97°F and the relative humidity is 72%.

SOLUTION We read from the table that $f(96, 70) = 125$. In Section 14.3 we used the tabular values to estimate that $f_T(96, 70) \approx 3.75$ and $f_H(96, 70) \approx 0.9$. (See pages 952–53.) So the linear approximation is

$$\begin{aligned} f(T, H) &\approx f(96, 70) + f_T(96, 70)(T - 96) + f_H(96, 70)(H - 70) \\ &\approx 125 + 3.75(T - 96) + 0.9(H - 70) \end{aligned}$$

In particular,

$$f(97, 72) \approx 125 + 3.75(1) + 0.9(2) = 130.55$$

Therefore, when $T = 97°\text{F}$ and $H = 72\%$, the heat index is

$$I \approx 131°\text{F}$$

Differentials

For a differentiable function of one variable, $y = f(x)$, we define the differential dx to be an independent variable; that is, dx can be given the value of any real number. The differential of y is then defined as

$$dy = f'(x)\, dx \tag{9}$$

(See Section 2.9.) Figure 6 shows the relationship between the increment Δy and the differential dy: Δy represents the change in height of the curve $y = f(x)$ and dy represents the change in height of the tangent line when x changes by an amount $dx = \Delta x$.

FIGURE 6

For a differentiable function of two variables, $z = f(x, y)$, we define the **differentials** dx and dy to be independent variables; that is, they can be given any values. Then the **differential** dz, also called the **total differential**, is defined by

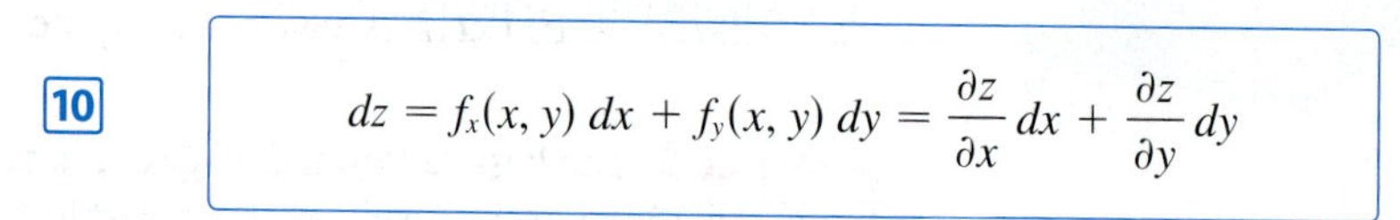

$$dz = f_x(x, y)\, dx + f_y(x, y)\, dy = \frac{\partial z}{\partial x}\, dx + \frac{\partial z}{\partial y}\, dy \tag{10}$$

(Compare with Equation 9.) Sometimes the notation df is used in place of dz.

If we take $dx = \Delta x = x - a$ and $dy = \Delta y = y - b$ in Equation 10, then the differential of z is

$$dz = f_x(a, b)(x - a) + f_y(a, b)(y - b)$$

So, in the notation of differentials, the linear approximation (4) can be written as

$$f(x, y) \approx f(a, b) + dz$$

Figure 7 is the three-dimensional counterpart of Figure 6 and shows the geometric interpretation of the differential dz and the increment Δz: dz represents the change in height of the tangent plane, whereas Δz represents the change in height of the surface $z = f(x, y)$ when (x, y) changes from (a, b) to $(a + \Delta x, b + \Delta y)$.

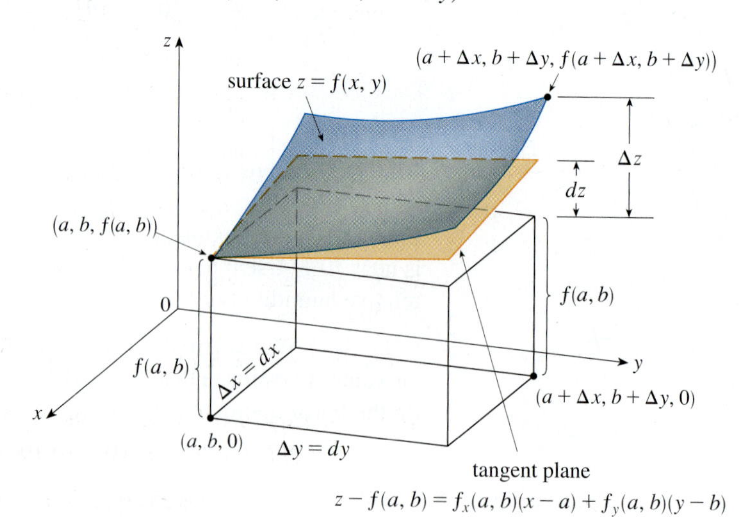

FIGURE 7

EXAMPLE 4
(a) If $z = f(x, y) = x^2 + 3xy - y^2$, find the differential dz.
(b) If x changes from 2 to 2.05 and y changes from 3 to 2.96, compare the values of Δz and dz.

In Example 4, dz is close to Δz because the tangent plane is a good approximation to the surface $z = x^2 + 3xy - y^2$ near (2, 3, 13). (See Figure 8.)

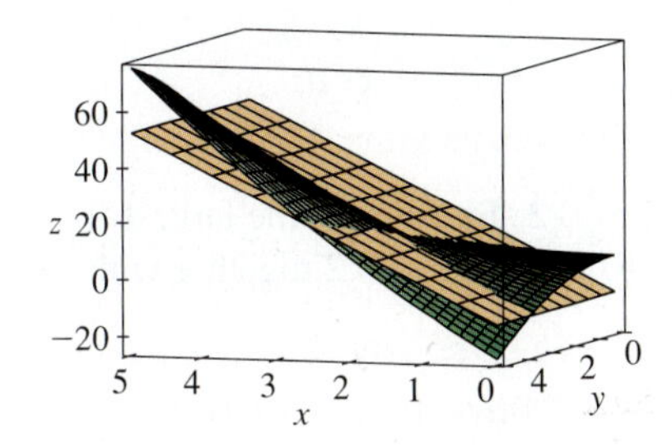

FIGURE 8

SOLUTION
(a) Definition 10 gives

$$dz = \frac{\partial z}{\partial x}\,dx + \frac{\partial z}{\partial y}\,dy = (2x + 3y)\,dx + (3x - 2y)\,dy$$

(b) Putting $x = 2$, $dx = \Delta x = 0.05$, $y = 3$, and $dy = \Delta y = -0.04$, we get

$$dz = [2(2) + 3(3)]0.05 + [3(2) - 2(3)](-0.04) = 0.65$$

The increment of z is

$$\begin{aligned}\Delta z &= f(2.05, 2.96) - f(2, 3)\\ &= [(2.05)^2 + 3(2.05)(2.96) - (2.96)^2] - [2^2 + 3(2)(3) - 3^2]\\ &= 0.6449\end{aligned}$$

Notice that $\Delta z \approx dz$ but dz is easier to compute. ■

EXAMPLE 5 The base radius and height of a right circular cone are measured as 10 cm and 25 cm, respectively, with a possible error in measurement of as much as 0.1 cm in each. Use differentials to estimate the maximum error in the calculated volume of the cone.

SOLUTION The volume V of a cone with base radius r and height h is $V = \pi r^2 h/3$. So the differential of V is

$$dV = \frac{\partial V}{\partial r}\,dr + \frac{\partial V}{\partial h}\,dh = \frac{2\pi rh}{3}\,dr + \frac{\pi r^2}{3}\,dh$$

Since each error is at most 0.1 cm, we have $|\Delta r| \leqslant 0.1$, $|\Delta h| \leqslant 0.1$. To estimate the largest error in the volume we take the largest error in the measurement of r and of h. Therefore we take $dr = 0.1$ and $dh = 0.1$ along with $r = 10$, $h = 25$. This gives

$$dV = \frac{500\pi}{3}(0.1) + \frac{100\pi}{3}(0.1) = 20\pi$$

Thus the maximum error in the calculated volume is about $20\pi\ \text{cm}^3 \approx 63\ \text{cm}^3$. ■

Functions of Three or More Variables

Linear approximations, differentiability, and differentials can be defined in a similar manner for functions of more than two variables. A differentiable function is defined by an expression similar to the one in Definition 7. For such functions the **linear approximation** is

$$f(x, y, z) \approx f(a, b, c) + f_x(a, b, c)(x - a) + f_y(a, b, c)(y - b) + f_z(a, b, c)(z - c)$$

and the linearization $L(x, y, z)$ is the right side of this expression.

If $w = f(x, y, z)$, then the **increment** of w is

$$\Delta w = f(x + \Delta x, y + \Delta y, z + \Delta z) - f(x, y, z)$$

The **differential** dw is defined in terms of the differentials dx, dy, and dz of the independent variables by

$$dw = \frac{\partial w}{\partial x}\,dx + \frac{\partial w}{\partial y}\,dy + \frac{\partial w}{\partial z}\,dz$$

EXAMPLE 6 The dimensions of a rectangular box are measured to be 75 cm, 60 cm, and 40 cm, and each measurement is correct to within 0.2 cm. Use differentials to estimate the largest possible error when the volume of the box is calculated from these measurements.

SOLUTION If the dimensions of the box are x, y, and z, its volume is $V = xyz$ and so

$$dV = \frac{\partial V}{\partial x}\,dx + \frac{\partial V}{\partial y}\,dy + \frac{\partial V}{\partial z}\,dz = yz\,dx + xz\,dy + xy\,dz$$

We are given that $|\Delta x| \leqslant 0.2$, $|\Delta y| \leqslant 0.2$, and $|\Delta z| \leqslant 0.2$. To estimate the largest error in the volume, we therefore use $dx = 0.2$, $dy = 0.2$, and $dz = 0.2$ together with $x = 75$, $y = 60$, and $z = 40$:

$$\Delta V \approx dV = (60)(40)(0.2) + (75)(40)(0.2) + (75)(60)(0.2) = 1980$$

Thus an error of only 0.2 cm in measuring each dimension could lead to an error of approximately 1980 cm^3 in the calculated volume! This may seem like a large error, but it's only about 1% of the volume of the box. ■

14.4 EXERCISES

1–6 Find an equation of the tangent plane to the given surface at the specified point.

1. $z = 2x^2 + y^2 - 5y$, $(1, 2, -4)$

2. $z = (x + 2)^2 - 2(y - 1)^2 - 5$, $(2, 3, 3)$

3. $z = e^{x-y}$, $(2, 2, 1)$

4. $z = x/y^2$, $(-4, 2, -1)$

5. $z = x\sin(x + y)$, $(-1, 1, 0)$

6. $z = \ln(x - 2y)$, $(3, 1, 0)$

7–8 Graph the surface and the tangent plane at the given point. (Choose the domain and viewpoint so that you get a good view of both the surface and the tangent plane.) Then zoom in until the surface and the tangent plane become indistinguishable.

7. $z = x^2 + xy + 3y^2$, $(1, 1, 5)$

8. $z = \sqrt{9 + x^2y^2}$, $(2, 2, 5)$

CAS **9–10** Draw the graph of f and its tangent plane at the given point. (Use your computer algebra system both to compute the partial derivatives and to graph the surface and its tangent plane.) Then zoom in until the surface and the tangent plane become indistinguishable.

9. $f(x, y) = \dfrac{1 + \cos^2(x - y)}{1 + \cos^2(x + y)}$, $\left(\dfrac{\pi}{3}, \dfrac{\pi}{6}, \dfrac{7}{4}\right)$

10. $f(x, y) = e^{-xy/10}\left(\sqrt{x} + \sqrt{y} + \sqrt{xy}\right)$, $(1, 1, 3e^{-0.1})$

11–16 Explain why the function is differentiable at the given point. Then find the linearization $L(x, y)$ of the function at that point.

11. $f(x, y) = 1 + x\ln(xy - 5)$, $(2, 3)$

12. $f(x, y) = \sqrt{xy}$, $(1, 4)$

13. $f(x, y) = x^2e^y$, $(1, 0)$

14. $f(x, y) = \dfrac{1 + y}{1 + x}$, $(1, 3)$

15. $f(x, y) = 4\arctan(xy)$, $(1, 1)$

16. $f(x, y) = y + \sin(x/y)$, $(0, 3)$

17–18 Verify the linear approximation at $(0, 0)$.

17. $e^x\cos(xy) \approx x + 1$

18. $\dfrac{y - 1}{x + 1} \approx x + y - 1$

19. Given that f is a differentiable function with $f(2, 5) = 6$, $f_x(2, 5) = 1$, and $f_y(2, 5) = -1$, use a linear approximation to estimate $f(2.2, 4.9)$.

20. Find the linear approximation of the function $f(x, y) = 1 - xy \cos \pi y$ at $(1, 1)$ and use it to approximate $f(1.02, 0.97)$. Illustrate by graphing f and the tangent plane.

21. Find the linear approximation of the function $f(x, y, z) = \sqrt{x^2 + y^2 + z^2}$ at $(3, 2, 6)$ and use it to approximate the number $\sqrt{(3.02)^2 + (1.97)^2 + (5.99)^2}$.

22. The wave heights h in the open sea depend on the speed v of the wind and the length of time t that the wind has been blowing at that speed. Values of the function $h = f(v, t)$ are recorded in feet in the following table. Use the table to find a linear approximation to the wave height function when v is near 40 knots and t is near 20 hours. Then estimate the wave heights when the wind has been blowing for 24 hours at 43 knots.

Duration (hours)

Wind speed (knots) v \ t	5	10	15	20	30	40	50
20	5	7	8	8	9	9	9
30	9	13	16	17	18	19	19
40	14	21	25	28	31	33	33
50	19	29	36	40	45	48	50
60	24	37	47	54	62	67	69

23. Use the table in Example 3 to find a linear approximation to the heat index function when the temperature is near 94°F and the relative humidity is near 80%. Then estimate the heat index when the temperature is 95°F and the relative humidity is 78%.

24. The wind-chill index W is the perceived temperature when the actual temperature is T and the wind speed is v, so we can write $W = f(T, v)$. The following table of values is an excerpt from Table 1 in Section 14.1. Use the table to find a linear approximation to the wind-chill index function when T is near -15°C and v is near 50 km/h. Then estimate the wind-chill index when the temperature is -17°C and the wind speed is 55 km/h.

Wind speed (km/h)

Actual temperature (°C) T \ v	20	30	40	50	60	70
−10	−18	−20	−21	−22	−23	−23
−15	−24	−26	−27	−29	−30	−30
−20	−30	−33	−34	−35	−36	−37
−25	−37	−39	−41	−42	−43	−44

25–30 Find the differential of the function.

25. $z = e^{-2x} \cos 2\pi t$

26. $u = \sqrt{x^2 + 3y^2}$

27. $m = p^5 q^3$

28. $T = \dfrac{v}{1 + uvw}$

29. $R = \alpha\beta^2 \cos \gamma$

30. $L = xze^{-y^2-z^2}$

31. If $z = 5x^2 + y^2$ and (x, y) changes from $(1, 2)$ to $(1.05, 2.1)$, compare the values of Δz and dz.

32. If $z = x^2 - xy + 3y^2$ and (x, y) changes from $(3, -1)$ to $(2.96, -0.95)$, compare the values of Δz and dz.

33. The length and width of a rectangle are measured as 30 cm and 24 cm, respectively, with an error in measurement of at most 0.1 cm in each. Use differentials to estimate the maximum error in the calculated area of the rectangle.

34. Use differentials to estimate the amount of metal in a closed cylindrical can that is 10 cm high and 4 cm in diameter if the metal in the top and bottom is 0.1 cm thick and the metal in the sides is 0.05 cm thick.

35. Use differentials to estimate the amount of tin in a closed tin can with diameter 8 cm and height 12 cm if the tin is 0.04 cm thick.

36. The wind-chill index is modeled by the function

$$W = 13.12 + 0.6215T - 11.37v^{0.16} + 0.3965Tv^{0.16}$$

where T is the temperature (in °C) and v is the wind speed (in km/h). The wind speed is measured as 26 km/h, with a possible error of ± 2 km/h, and the temperature is measured as -11°C, with a possible error of ± 1°C. Use differentials to estimate the maximum error in the calculated value of W due to the measurement errors in T and v.

37. The tension T in the string of the yo-yo in the figure is

$$T = \frac{mgR}{2r^2 + R^2}$$

where m is the mass of the yo-yo and g is acceleration due to gravity. Use differentials to estimate the change in the tension if R is increased from 3 cm to 3.1 cm and r is increased from 0.7 cm to 0.8 cm. Does the tension increase or decrease?

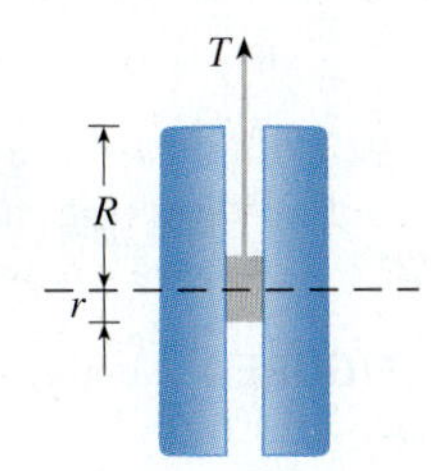

38. The pressure, volume, and temperature of a mole of an ideal gas are related by the equation $PV = 8.31T$, where P is measured in kilopascals, V in liters, and T in kelvins. Use differentials to find the approximate change in the pressure if the volume increases from 12 L to 12.3 L and the temperature decreases from 310 K to 305 K.

39. If R is the total resistance of three resistors, connected in parallel, with resistances R_1, R_2, R_3, then

$$\frac{1}{R} = \frac{1}{R_1} + \frac{1}{R_2} + \frac{1}{R_3}$$

If the resistances are measured in ohms as $R_1 = 25\ \Omega$, $R_2 = 40\ \Omega$, and $R_3 = 50\ \Omega$, with a possible error of 0.5% in each case, estimate the maximum error in the calculated value of R.

40. A model for the surface area of a human body is given by $S = 0.1091 w^{0.425} h^{0.725}$, where w is the weight (in pounds), h is the height (in inches), and S is measured in square feet. If the errors in measurement of w and h are at most 2%, use differentials to estimate the maximum percentage error in the calculated surface area.

41. In Exercise 14.1.39 and Example 14.3.3, the body mass index of a person was defined as $B(m, h) = m/h^2$, where m is the mass in kilograms and h is the height in meters.

(a) What is the linear approximation of $B(m, h)$ for a child with mass 23 kg and height 1.10 m?

(b) If the child's mass increases by 1 kg and height by 3 cm, use the linear approximation to estimate the new BMI. Compare with the actual new BMI.

42. Suppose you need to know an equation of the tangent plane to a surface S at the point $P(2, 1, 3)$. You don't have an equation for S but you know that the curves

$$\mathbf{r}_1(t) = \langle 2 + 3t, 1 - t^2, 3 - 4t + t^2 \rangle$$
$$\mathbf{r}_2(u) = \langle 1 + u^2, 2u^3 - 1, 2u + 1 \rangle$$

both lie on S. Find an equation of the tangent plane at P.

43–44 Show that the function is differentiable by finding values of ε_1 and ε_2 that satisfy Definition 7.

43. $f(x, y) = x^2 + y^2$

44. $f(x, y) = xy - 5y^2$

45. Prove that if f is a function of two variables that is differentiable at (a, b), then f is continuous at (a, b).
Hint: Show that

$$\lim_{(\Delta x, \Delta y) \to (0,0)} f(a + \Delta x, b + \Delta y) = f(a, b)$$

46. (a) The function

$$f(x, y) = \begin{cases} \dfrac{xy}{x^2 + y^2} & \text{if } (x, y) \neq (0, 0) \\ 0 & \text{if } (x, y) = (0, 0) \end{cases}$$

was graphed in Figure 4. Show that $f_x(0, 0)$ and $f_y(0, 0)$ both exist but f is not differentiable at $(0, 0)$. [*Hint:* Use the result of Exercise 45.]

(b) Explain why f_x and f_y are not continuous at $(0, 0)$.

APPLIED PROJECT THE SPEEDO LZR RACER

Many technological advances have occurred in sports that have contributed to increased athletic performance. One of the best known is the introduction, in 2008, of the Speedo LZR racer. It was claimed that this full-body swimsuit reduced a swimmer's drag in the water. Figure 1 shows the number of world records broken in men's and women's long-course freestyle swimming events from 1990 to 2011.[1] The dramatic increase in 2008 when the suit was introduced led people to claim that such suits are a form of technological doping. As a result all full-body suits were banned from competition starting in 2010.

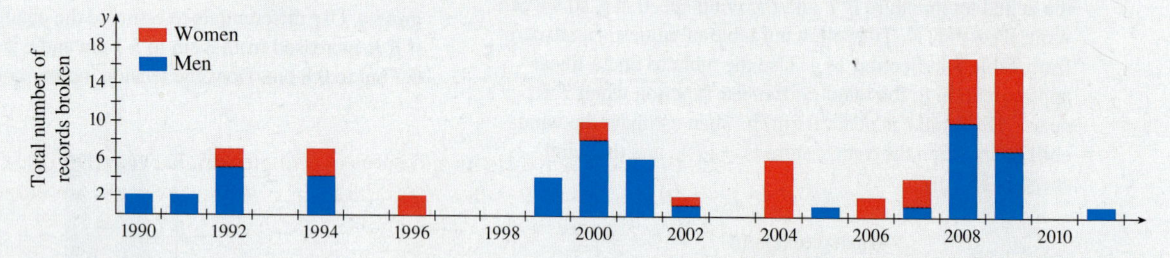

FIGURE 1 Number of world records set in long-course men's and women's freestyle swimming event 1990–2011

It might be surprising that a simple reduction in drag could have such a big effect on performance. We can gain some insight into this using a simple mathematical model.[2]

1. L. Foster et al., "Influence of Full Body Swimsuits on Competitive Performance," *Procedia Engineering* 34 (2012): 712–17.
2. Adapted from http://plus.maths.org/content/swimming.

Courtesy Speedo and ANSYS, Inc.

The speed v of an object being propelled through water is given by

$$v(P, C) = \left(\frac{2P}{kC}\right)^{1/3}$$

where P is the power being used to propel the object, C is the drag coefficient, and k is a positive constant. Athletes can therefore increase their swimming speeds by increasing their power or reducing their drag coefficients. But how effective is each of these?

To compare the effect of increasing power versus reducing drag, we need to somehow compare the two in common units. The most common approach is to determine the percentage change in speed that results from a given percentage change in power and in drag.

If we work with percentages as fractions, then when power is changed by a fraction x (with x corresponding to $100x$ percent), P changes from P to $P + xP$. Likewise, if the drag coefficient is changed by a fraction y, this means that it has changed from C to $C + yC$. Finally, the fractional change in speed resulting from both effects is

$$\boxed{1} \qquad \frac{v(P + xP, C + yC) - v(P, C)}{v(P, C)}$$

1. Expression 1 gives the fractional change in speed that results from a change x in power and a change y in drag. Show that this reduces to the function

$$f(x, y) = \left(\frac{1 + x}{1 + y}\right)^{1/3} - 1$$

Given the context, what is the domain of f?

2. Suppose that the possible changes in power x and drag y are small. Find the linear approximation to the function $f(x, y)$. What does this approximation tell you about the effect of a small increase in power versus a small decrease in drag?

3. Calculate $f_{xx}(x, y)$ and $f_{yy}(x, y)$. Based on the signs of these derivatives, does the linear approximation in Problem 2 result in an overestimate or an underestimate for an increase in power? What about for a decrease in drag? Use your answer to explain why, for changes in power or drag that are not very small, a decrease in drag is more effective.

4. Graph the level curves of $f(x, y)$. Explain how the shapes of these curves relate to your answers to Problems 2 and 3.

14.5 The Chain Rule

Recall that the Chain Rule for functions of a single variable gives the rule for differentiating a composite function: If $y = f(x)$ and $x = g(t)$, where f and g are differentiable functions, then y is indirectly a differentiable function of t and

$$\boxed{1} \qquad \frac{dy}{dt} = \frac{dy}{dx}\frac{dx}{dt}$$

For functions of more than one variable, the Chain Rule has several versions, each of them giving a rule for differentiating a composite function. The first version (Theorem 2) deals with the case where $z = f(x, y)$ and each of the variables x and y is, in turn, a function of a variable t. This means that z is indirectly a function of t, $z = f(g(t), h(t))$, and the Chain Rule gives a formula for differentiating z as a function of t. We assume that f

is differentiable (Definition 14.4.7). Recall that this is the case when f_x and f_y are continuous (Theorem 14.4.8).

2 The Chain Rule (Case 1) Suppose that $z = f(x, y)$ is a differentiable function of x and y, where $x = g(t)$ and $y = h(t)$ are both differentiable functions of t. Then z is a differentiable function of t and

$$\frac{dz}{dt} = \frac{\partial f}{\partial x}\frac{dx}{dt} + \frac{\partial f}{\partial y}\frac{dy}{dt}$$

PROOF A change of Δt in t produces changes of Δx in x and Δy in y. These, in turn, produce a change of Δz in z, and from Definition 14.4.7 we have

$$\Delta z = \frac{\partial f}{\partial x}\Delta x + \frac{\partial f}{\partial y}\Delta y + \varepsilon_1 \Delta x + \varepsilon_2 \Delta y$$

where $\varepsilon_1 \to 0$ and $\varepsilon_2 \to 0$ as $(\Delta x, \Delta y) \to (0, 0)$. [If the functions ε_1 and ε_2 are not defined at $(0, 0)$, we can define them to be 0 there.] Dividing both sides of this equation by Δt, we have

$$\frac{\Delta z}{\Delta t} = \frac{\partial f}{\partial x}\frac{\Delta x}{\Delta t} + \frac{\partial f}{\partial y}\frac{\Delta y}{\Delta t} + \varepsilon_1 \frac{\Delta x}{\Delta t} + \varepsilon_2 \frac{\Delta y}{\Delta t}$$

If we now let $\Delta t \to 0$, then $\Delta x = g(t + \Delta t) - g(t) \to 0$ because g is differentiable and therefore continuous. Similarly, $\Delta y \to 0$. This, in turn, means that $\varepsilon_1 \to 0$ and $\varepsilon_2 \to 0$, so

$$\begin{aligned}\frac{dz}{dt} &= \lim_{\Delta t \to 0} \frac{\Delta z}{\Delta t}\\ &= \frac{\partial f}{\partial x}\lim_{\Delta t \to 0}\frac{\Delta x}{\Delta t} + \frac{\partial f}{\partial y}\lim_{\Delta t \to 0}\frac{\Delta y}{\Delta t} + \left(\lim_{\Delta t \to 0}\varepsilon_1\right)\lim_{\Delta t \to 0}\frac{\Delta x}{\Delta t} + \left(\lim_{\Delta t \to 0}\varepsilon_2\right)\lim_{\Delta t \to 0}\frac{\Delta y}{\Delta t}\\ &= \frac{\partial f}{\partial x}\frac{dx}{dt} + \frac{\partial f}{\partial y}\frac{dy}{dt} + 0 \cdot \frac{dx}{dt} + 0 \cdot \frac{dy}{dt}\\ &= \frac{\partial f}{\partial x}\frac{dx}{dt} + \frac{\partial f}{\partial y}\frac{dy}{dt}\end{aligned}$$

■

Since we often write $\partial z/\partial x$ in place of $\partial f/\partial x$, we can rewrite the Chain Rule in the form

Notice the similarity to the definition of the differential:

$$dz = \frac{\partial z}{\partial x}dx + \frac{\partial z}{\partial y}dy$$

$$\frac{dz}{dt} = \frac{\partial z}{\partial x}\frac{dx}{dt} + \frac{\partial z}{\partial y}\frac{dy}{dt}$$

EXAMPLE 1 If $z = x^2y + 3xy^4$, where $x = \sin 2t$ and $y = \cos t$, find dz/dt when $t = 0$.

SOLUTION The Chain Rule gives

$$\begin{aligned}\frac{dz}{dt} &= \frac{\partial z}{\partial x}\frac{dx}{dt} + \frac{\partial z}{\partial y}\frac{dy}{dt}\\ &= (2xy + 3y^4)(2\cos 2t) + (x^2 + 12xy^3)(-\sin t)\end{aligned}$$

It's not necessary to substitute the expressions for x and y in terms of t. We simply observe that when $t = 0$, we have $x = \sin 0 = 0$ and $y = \cos 0 = 1$. Therefore

$$\left.\frac{dz}{dt}\right|_{t=0} = (0 + 3)(2 \cos 0) + (0 + 0)(-\sin 0) = 6$$

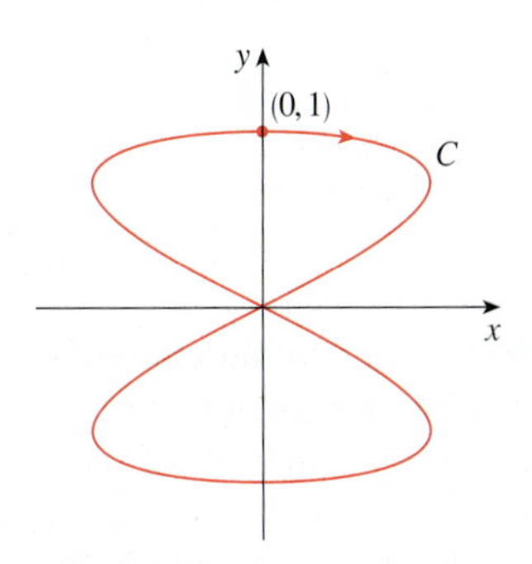

FIGURE 1
The curve $x = \sin 2t$, $y = \cos t$

The derivative in Example 1 can be interpreted as the rate of change of z with respect to t as the point (x, y) moves along the curve C with parametric equations $x = \sin 2t$, $y = \cos t$. (See Figure 1.) In particular, when $t = 0$, the point (x, y) is $(0, 1)$ and $dz/dt = 6$ is the rate of increase as we move along the curve C through $(0, 1)$. If, for instance, $z = T(x, y) = x^2y + 3xy^4$ represents the temperature at the point (x, y), then the composite function $z = T(\sin 2t, \cos t)$ represents the temperature at points on C and the derivative dz/dt represents the rate at which the temperature changes along C.

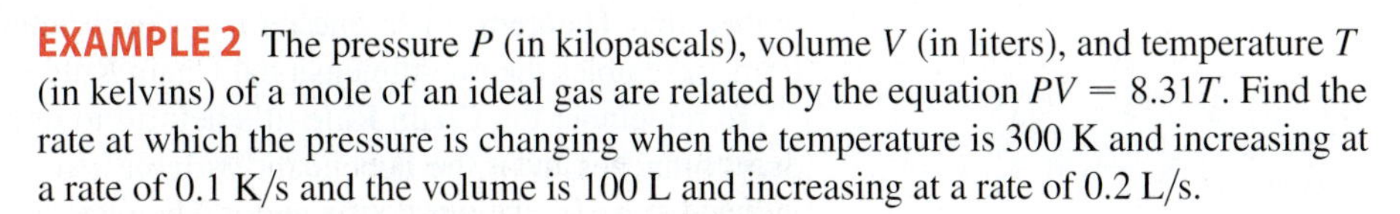

EXAMPLE 2 The pressure P (in kilopascals), volume V (in liters), and temperature T (in kelvins) of a mole of an ideal gas are related by the equation $PV = 8.31T$. Find the rate at which the pressure is changing when the temperature is 300 K and increasing at a rate of 0.1 K/s and the volume is 100 L and increasing at a rate of 0.2 L/s.

SOLUTION If t represents the time elapsed in seconds, then at the given instant we have $T = 300$, $dT/dt = 0.1$, $V = 100$, $dV/dt = 0.2$. Since

$$P = 8.31\frac{T}{V}$$

the Chain Rule gives

$$\frac{dP}{dt} = \frac{\partial P}{\partial T}\frac{dT}{dt} + \frac{\partial P}{\partial V}\frac{dV}{dt} = \frac{8.31}{V}\frac{dT}{dt} - \frac{8.31T}{V^2}\frac{dV}{dt}$$

$$= \frac{8.31}{100}(0.1) - \frac{8.31(300)}{100^2}(0.2) = -0.04155$$

The pressure is decreasing at a rate of about 0.042 kPa/s.

We now consider the situation where $z = f(x, y)$ but each of x and y is a function of two variables s and t: $x = g(s, t)$, $y = h(s, t)$. Then z is indirectly a function of s and t and we wish to find $\partial z/\partial s$ and $\partial z/\partial t$. Recall that in computing $\partial z/\partial t$ we hold s fixed and compute the ordinary derivative of z with respect to t. Therefore we can apply Theorem 2 to obtain

$$\frac{\partial z}{\partial t} = \frac{\partial z}{\partial x}\frac{\partial x}{\partial t} + \frac{\partial z}{\partial y}\frac{\partial y}{\partial t}$$

A similar argument holds for $\partial z/\partial s$ and so we have proved the following version of the Chain Rule.

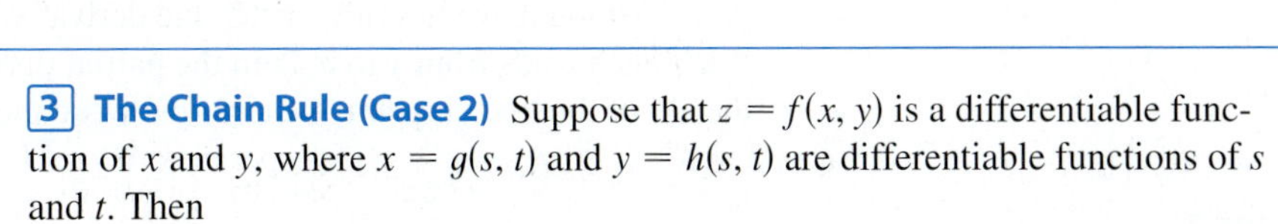

3 The Chain Rule (Case 2) Suppose that $z = f(x, y)$ is a differentiable function of x and y, where $x = g(s, t)$ and $y = h(s, t)$ are differentiable functions of s and t. Then

$$\frac{\partial z}{\partial s} = \frac{\partial z}{\partial x}\frac{\partial x}{\partial s} + \frac{\partial z}{\partial y}\frac{\partial y}{\partial s} \qquad \frac{\partial z}{\partial t} = \frac{\partial z}{\partial x}\frac{\partial x}{\partial t} + \frac{\partial z}{\partial y}\frac{\partial y}{\partial t}$$

EXAMPLE 3 If $z = e^x \sin y$, where $x = st^2$ and $y = s^2t$, find $\partial z/\partial s$ and $\partial z/\partial t$.

SOLUTION Applying Case 2 of the Chain Rule, we get

$$\frac{\partial z}{\partial s} = \frac{\partial z}{\partial x}\frac{\partial x}{\partial s} + \frac{\partial z}{\partial y}\frac{\partial y}{\partial s} = (e^x \sin y)(t^2) + (e^x \cos y)(2st)$$
$$= t^2 e^{st^2} \sin(s^2t) + 2ste^{st^2} \cos(s^2t)$$

$$\frac{\partial z}{\partial t} = \frac{\partial z}{\partial x}\frac{\partial x}{\partial t} + \frac{\partial z}{\partial y}\frac{\partial y}{\partial t} = (e^x \sin y)(2st) + (e^x \cos y)(s^2)$$
$$= 2ste^{st^2} \sin(s^2t) + s^2 e^{st^2} \cos(s^2t)$$

Case 2 of the Chain Rule contains three types of variables: s and t are **independent** variables, x and y are called **intermediate** variables, and z is the **dependent** variable. Notice that Theorem 3 has one term for each intermediate variable and each of these terms resembles the one-dimensional Chain Rule in Equation 1.

To remember the Chain Rule, it's helpful to draw the **tree diagram** in Figure 2. We draw branches from the dependent variable z to the intermediate variables x and y to indicate that z is a function of x and y. Then we draw branches from x and y to the independent variables s and t. On each branch we write the corresponding partial derivative. To find $\partial z/\partial s$, we find the product of the partial derivatives along each path from z to s and then add these products:

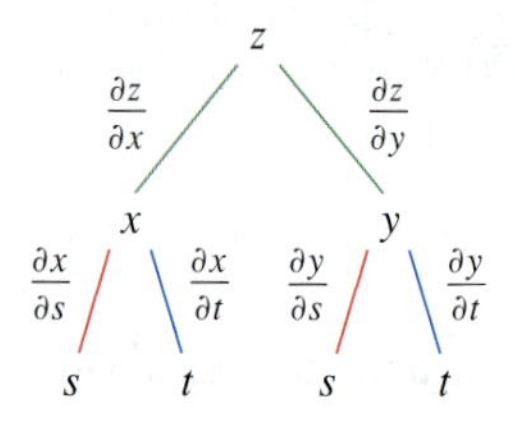

FIGURE 2

$$\frac{\partial z}{\partial s} = \frac{\partial z}{\partial x}\frac{\partial x}{\partial s} + \frac{\partial z}{\partial y}\frac{\partial y}{\partial s}$$

Similarly, we find $\partial z/\partial t$ by using the paths from z to t.

Now we consider the general situation in which a dependent variable u is a function of n intermediate variables $x_1, \ldots, x_n$, each of which is, in turn, a function of m independent variables $t_1, \ldots, t_m$. Notice that there are n terms, one for each intermediate variable. The proof is similar to that of Case 1.

4 The Chain Rule (General Version) Suppose that u is a differentiable function of the n variables $x_1, x_2, \ldots, x_n$ and each x_j is a differentiable function of the m variables $t_1, t_2, \ldots, t_m$. Then u is a function of $t_1, t_2, \ldots, t_m$ and

$$\frac{\partial u}{\partial t_i} = \frac{\partial u}{\partial x_1}\frac{\partial x_1}{\partial t_i} + \frac{\partial u}{\partial x_2}\frac{\partial x_2}{\partial t_i} + \cdots + \frac{\partial u}{\partial x_n}\frac{\partial x_n}{\partial t_i}$$

for each $i = 1, 2, \ldots, m$.

EXAMPLE 4 Write out the Chain Rule for the case where $w = f(x, y, z, t)$ and $x = x(u, v)$, $y = y(u, v)$, $z = z(u, v)$, and $t = t(u, v)$.

SOLUTION We apply Theorem 4 with $n = 4$ and $m = 2$. Figure 3 shows the tree diagram. Although we haven't written the derivatives on the branches, it's understood that if a branch leads from y to u, then the partial derivative for that branch is $\partial y/\partial u$. With the aid of the tree diagram, we can now write the required expressions:

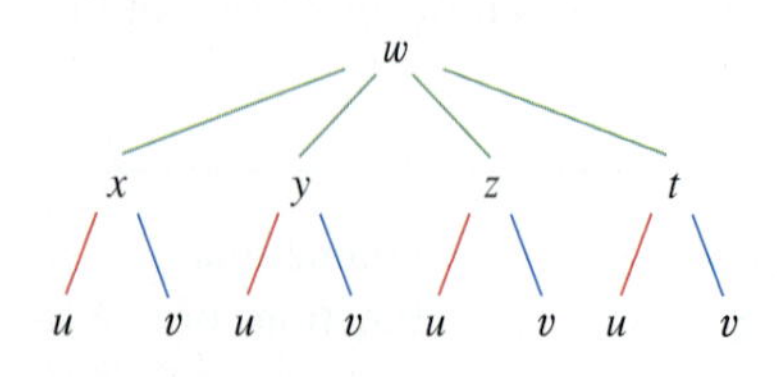

FIGURE 3

$$\frac{\partial w}{\partial u} = \frac{\partial w}{\partial x}\frac{\partial x}{\partial u} + \frac{\partial w}{\partial y}\frac{\partial y}{\partial u} + \frac{\partial w}{\partial z}\frac{\partial z}{\partial u} + \frac{\partial w}{\partial t}\frac{\partial t}{\partial u}$$

$$\frac{\partial w}{\partial v} = \frac{\partial w}{\partial x}\frac{\partial x}{\partial v} + \frac{\partial w}{\partial y}\frac{\partial y}{\partial v} + \frac{\partial w}{\partial z}\frac{\partial z}{\partial v} + \frac{\partial w}{\partial t}\frac{\partial t}{\partial v}$$

EXAMPLE 5 If $u = x^4y + y^2z^3$, where $x = rse^t$, $y = rs^2e^{-t}$, and $z = r^2s\sin t$, find the value of $\partial u/\partial s$ when $r = 2$, $s = 1$, $t = 0$.

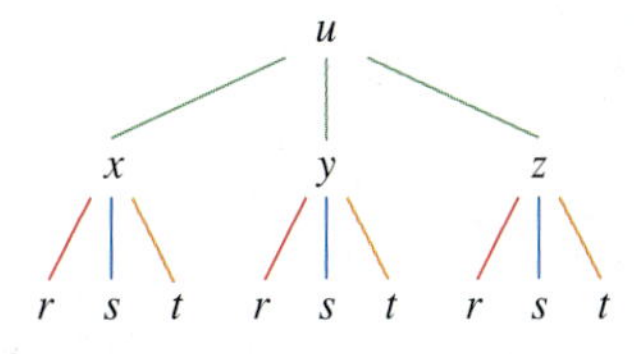

FIGURE 4

SOLUTION With the help of the tree diagram in Figure 4, we have

$$\begin{aligned}\frac{\partial u}{\partial s} &= \frac{\partial u}{\partial x}\frac{\partial x}{\partial s} + \frac{\partial u}{\partial y}\frac{\partial y}{\partial s} + \frac{\partial u}{\partial z}\frac{\partial z}{\partial s}\\ &= (4x^3y)(re^t) + (x^4 + 2yz^3)(2rse^{-t}) + (3y^2z^2)(r^2\sin t)\end{aligned}$$

When $r = 2$, $s = 1$, and $t = 0$, we have $x = 2$, $y = 2$, and $z = 0$, so

$$\frac{\partial u}{\partial s} = (64)(2) + (16)(4) + (0)(0) = 192$$ ■

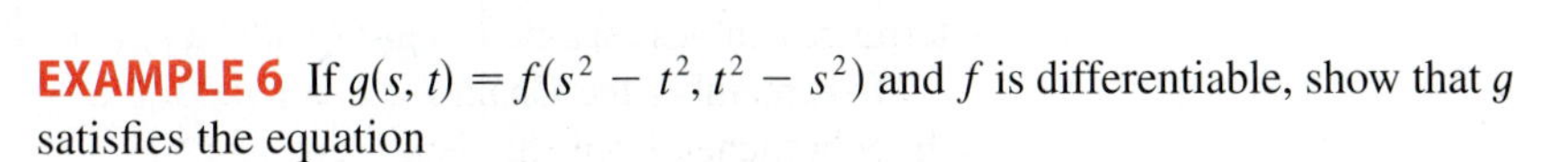

EXAMPLE 6 If $g(s, t) = f(s^2 - t^2, t^2 - s^2)$ and f is differentiable, show that g satisfies the equation

$$t\frac{\partial g}{\partial s} + s\frac{\partial g}{\partial t} = 0$$

SOLUTION Let $x = s^2 - t^2$ and $y = t^2 - s^2$. Then $g(s, t) = f(x, y)$ and the Chain Rule gives

$$\frac{\partial g}{\partial s} = \frac{\partial f}{\partial x}\frac{\partial x}{\partial s} + \frac{\partial f}{\partial y}\frac{\partial y}{\partial s} = \frac{\partial f}{\partial x}(2s) + \frac{\partial f}{\partial y}(-2s)$$

$$\frac{\partial g}{\partial t} = \frac{\partial f}{\partial x}\frac{\partial x}{\partial t} + \frac{\partial f}{\partial y}\frac{\partial y}{\partial t} = \frac{\partial f}{\partial x}(-2t) + \frac{\partial f}{\partial y}(2t)$$

Therefore

$$t\frac{\partial g}{\partial s} + s\frac{\partial g}{\partial t} = \left(2st\frac{\partial f}{\partial x} - 2st\frac{\partial f}{\partial y}\right) + \left(-2st\frac{\partial f}{\partial x} + 2st\frac{\partial f}{\partial y}\right) = 0$$ ■

EXAMPLE 7 If $z = f(x, y)$ has continuous second-order partial derivatives and $x = r^2 + s^2$ and $y = 2rs$, find (a) $\partial z/\partial r$ and (b) $\partial^2 z/\partial r^2$.

SOLUTION

(a) The Chain Rule gives

$$\frac{\partial z}{\partial r} = \frac{\partial z}{\partial x}\frac{\partial x}{\partial r} + \frac{\partial z}{\partial y}\frac{\partial y}{\partial r} = \frac{\partial z}{\partial x}(2r) + \frac{\partial z}{\partial y}(2s)$$

(b) Applying the Product Rule to the expression in part (a), we get

$$\begin{aligned}\frac{\partial^2 z}{\partial r^2} &= \frac{\partial}{\partial r}\left(2r\frac{\partial z}{\partial x} + 2s\frac{\partial z}{\partial y}\right)\\ &= 2\frac{\partial z}{\partial x} + 2r\frac{\partial}{\partial r}\left(\frac{\partial z}{\partial x}\right) + 2s\frac{\partial}{\partial r}\left(\frac{\partial z}{\partial y}\right)\end{aligned} \qquad \boxed{5}$$

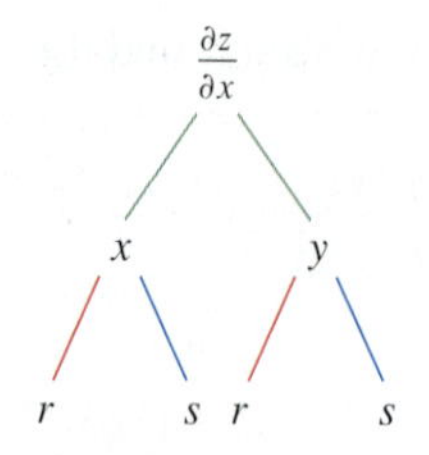

FIGURE 5

But, using the Chain Rule again (see Figure 5), we have

$$\frac{\partial}{\partial r}\left(\frac{\partial z}{\partial x}\right) = \frac{\partial}{\partial x}\left(\frac{\partial z}{\partial x}\right)\frac{\partial x}{\partial r} + \frac{\partial}{\partial y}\left(\frac{\partial z}{\partial x}\right)\frac{\partial y}{\partial r} = \frac{\partial^2 z}{\partial x^2}(2r) + \frac{\partial^2 z}{\partial y\,\partial x}(2s)$$

$$\frac{\partial}{\partial r}\left(\frac{\partial z}{\partial y}\right) = \frac{\partial}{\partial x}\left(\frac{\partial z}{\partial y}\right)\frac{\partial x}{\partial r} + \frac{\partial}{\partial y}\left(\frac{\partial z}{\partial y}\right)\frac{\partial y}{\partial r} = \frac{\partial^2 z}{\partial x\,\partial y}(2r) + \frac{\partial^2 z}{\partial y^2}(2s)$$

Putting these expressions into Equation 5 and using the equality of the mixed second-order derivatives, we obtain

$$\begin{aligned}\frac{\partial^2 z}{\partial r^2} &= 2\frac{\partial z}{\partial x} + 2r\left(2r\frac{\partial^2 z}{\partial x^2} + 2s\frac{\partial^2 z}{\partial y\,\partial x}\right) + 2s\left(2r\frac{\partial^2 z}{\partial x\,\partial y} + 2s\frac{\partial^2 z}{\partial y^2}\right)\\ &= 2\frac{\partial z}{\partial x} + 4r^2\frac{\partial^2 z}{\partial x^2} + 8rs\frac{\partial^2 z}{\partial x\,\partial y} + 4s^2\frac{\partial^2 z}{\partial y^2}\end{aligned}$$

Implicit Differentiation

The Chain Rule can be used to give a more complete description of the process of implicit differentiation that was introduced in Sections 2.6 and 14.3. We suppose that an equation of the form $F(x, y) = 0$ defines y implicitly as a differentiable function of x, that is, $y = f(x)$, where $F(x, f(x)) = 0$ for all x in the domain of f. If F is differentiable, we can apply Case 1 of the Chain Rule to differentiate both sides of the equation $F(x, y) = 0$ with respect to x. Since both x and y are functions of x, we obtain

$$\frac{\partial F}{\partial x}\frac{dx}{dx} + \frac{\partial F}{\partial y}\frac{dy}{dx} = 0$$

But $dx/dx = 1$, so if $\partial F/\partial y \neq 0$ we solve for dy/dx and obtain

$$\boxed{6} \qquad \frac{dy}{dx} = -\frac{\dfrac{\partial F}{\partial x}}{\dfrac{\partial F}{\partial y}} = -\frac{F_x}{F_y}$$

To derive this equation we assumed that $F(x, y) = 0$ defines y implicitly as a function of x. The **Implicit Function Theorem**, proved in advanced calculus, gives conditions under which this assumption is valid: it states that if F is defined on a disk containing (a, b), where $F(a, b) = 0$, $F_y(a, b) \neq 0$, and F_x and F_y are continuous on the disk, then the equation $F(x, y) = 0$ defines y as a function of x near the point (a, b) and the derivative of this function is given by Equation 6.

EXAMPLE 8 Find y' if $x^3 + y^3 = 6xy$.

SOLUTION The given equation can be written as

$$F(x, y) = x^3 + y^3 - 6xy = 0$$

so Equation 6 gives

$$\frac{dy}{dx} = -\frac{F_x}{F_y} = -\frac{3x^2 - 6y}{3y^2 - 6x} = -\frac{x^2 - 2y}{y^2 - 2x}$$

The solution to Example 8 should be compared to the one in Example 2.6.2.

Now we suppose that z is given implicitly as a function $z = f(x, y)$ by an equation of the form $F(x, y, z) = 0$. This means that $F(x, y, f(x, y)) = 0$ for all (x, y) in the domain of f. If F and f are differentiable, then we can use the Chain Rule to differentiate the equation $F(x, y, z) = 0$ as follows:

$$\frac{\partial F}{\partial x}\frac{\partial x}{\partial x} + \frac{\partial F}{\partial y}\frac{\partial y}{\partial x} + \frac{\partial F}{\partial z}\frac{\partial z}{\partial x} = 0$$

But

$$\frac{\partial}{\partial x}(x) = 1 \quad \text{and} \quad \frac{\partial}{\partial x}(y) = 0$$

so this equation becomes

$$\frac{\partial F}{\partial x} + \frac{\partial F}{\partial z}\frac{\partial z}{\partial x} = 0$$

If $\partial F/\partial z \neq 0$, we solve for $\partial z/\partial x$ and obtain the first formula in Equations 7. The formula for $\partial z/\partial y$ is obtained in a similar manner.

7

$$\frac{\partial z}{\partial x} = -\frac{\dfrac{\partial F}{\partial x}}{\dfrac{\partial F}{\partial z}} \qquad \frac{\partial z}{\partial y} = -\frac{\dfrac{\partial F}{\partial y}}{\dfrac{\partial F}{\partial z}}$$

Again, a version of the **Implicit Function Theorem** stipulates conditions under which our assumption is valid: if F is defined within a sphere containing (a, b, c), where $F(a, b, c) = 0$, $F_z(a, b, c) \neq 0$, and F_x, F_y, and F_z are continuous inside the sphere, then the equation $F(x, y, z) = 0$ defines z as a function of x and y near the point (a, b, c) and this function is differentiable, with partial derivatives given by (7).

EXAMPLE 9 Find $\dfrac{\partial z}{\partial x}$ and $\dfrac{\partial z}{\partial y}$ if $x^3 + y^3 + z^3 + 6xyz = 1$.

SOLUTION Let $F(x, y, z) = x^3 + y^3 + z^3 + 6xyz - 1$. Then, from Equations 7, we have

$$\frac{\partial z}{\partial x} = -\frac{F_x}{F_z} = -\frac{3x^2 + 6yz}{3z^2 + 6xy} = -\frac{x^2 + 2yz}{z^2 + 2xy}$$

$$\frac{\partial z}{\partial y} = -\frac{F_y}{F_z} = -\frac{3y^2 + 6xz}{3z^2 + 6xy} = -\frac{y^2 + 2xz}{z^2 + 2xy}$$ ■

The solution to Example 9 should be compared to the one in Example 14.3.5.

14.5 EXERCISES

1–6 Use the Chain Rule to find dz/dt or dw/dt.

1. $z = xy^3 - x^2y$, $x = t^2 + 1$, $y = t^2 - 1$

2. $z = \dfrac{x - y}{x + 2y}$, $x = e^{\pi t}$, $y = e^{-\pi t}$

3. $z = \sin x \cos y$, $x = \sqrt{t}$, $y = 1/t$

4. $z = \sqrt{1 + xy}$, $x = \tan t$, $y = \arctan t$

5. $w = xe^{y/z}$, $x = t^2$, $y = 1 - t$, $z = 1 + 2t$

6. $w = \ln\sqrt{x^2 + y^2 + z^2}$, $x = \sin t$, $y = \cos t$, $z = \tan t$

7–12 Use the Chain Rule to find $\partial z/\partial s$ and $\partial z/\partial t$.

7. $z = (x - y)^5$, $x = s^2t$, $y = st^2$

8. $z = \tan^{-1}(x^2 + y^2)$, $x = s \ln t$, $y = te^s$

9. $z = \ln(3x + 2y)$, $x = s \sin t$, $y = t \cos s$

10. $z = \sqrt{x}\, e^{xy}$, $x = 1 + st$, $y = s^2 - t^2$

11. $z = e^r \cos \theta$, $r = st$, $\theta = \sqrt{s^2 + t^2}$

12. $z = \tan(u/v)$, $u = 2s + 3t$, $v = 3s - 2t$

---

13. Let $p(t) = f(g(t), h(t))$, where f is differentiable, $g(2) = 4$, $g'(2) = -3$, $h(2) = 5$, $h'(2) = 6$, $f_x(4, 5) = 2$, $f_y(4, 5) = 8$. Find $p'(2)$.

14. Let $R(s, t) = G(u(s, t), v(s, t))$, where G, u, and v are differentiable, $u(1, 2) = 5$, $u_s(1, 2) = 4$, $u_t(1, 2) = -3$, $v(1, 2) = 7$, $v_s(1, 2) = 2$, $v_t(1, 2) = 6$, $G_u(5, 7) = 9$, $G_v(5, 7) = -2$. Find $R_s(1, 2)$ and $R_t(1, 2)$.

15. Suppose f is a differentiable function of x and y, and $g(u, v) = f(e^u + \sin v, e^u + \cos v)$. Use the table of values to calculate $g_u(0, 0)$ and $g_v(0, 0)$.

	f	g	f_x	f_y
(0, 0)	3	6	4	8
(1, 2)	6	3	2	5

16. Suppose f is a differentiable function of x and y, and $g(r, s) = f(2r - s, s^2 - 4r)$. Use the table of values in Exercise 15 to calculate $g_r(1, 2)$ and $g_s(1, 2)$.

17–20 Use a tree diagram to write out the Chain Rule for the given case. Assume all functions are differentiable.

17. $u = f(x, y)$, where $x = x(r, s, t)$, $y = y(r, s, t)$

18. $w = f(x, y, z)$, where $x = x(u, v)$, $y = y(u, v)$, $z = z(u, v)$

19. $T = F(p, q, r)$, where $p = p(x, y, z)$, $q = q(x, y, z)$, $r = r(x, y, z)$

20. $R = F(t, u)$ where $t = t(w, x, y, z)$, $u = u(w, x, y, z)$

21–26 Use the Chain Rule to find the indicated partial derivatives.

21. $z = x^4 + x^2 y$, $x = s + 2t - u$, $y = stu^2$;

$\dfrac{\partial z}{\partial s}, \dfrac{\partial z}{\partial t}, \dfrac{\partial z}{\partial u}$ when $s = 4$, $t = 2$, $u = 1$

22. $T = \dfrac{v}{2u + v}$, $u = pq\sqrt{r}$, $v = p\sqrt{q}\, r$;

$\dfrac{\partial T}{\partial p}, \dfrac{\partial T}{\partial q}, \dfrac{\partial T}{\partial r}$ when $p = 2$, $q = 1$, $r = 4$

23. $w = xy + yz + zx$, $x = r \cos \theta$, $y = r \sin \theta$, $z = r\theta$;

$\dfrac{\partial w}{\partial r}, \dfrac{\partial w}{\partial \theta}$ when $r = 2$, $\theta = \pi/2$

24. $P = \sqrt{u^2 + v^2 + w^2}$, $u = xe^y$, $v = ye^x$, $w = e^{xy}$;

$\dfrac{\partial P}{\partial x}, \dfrac{\partial P}{\partial y}$ when $x = 0$, $y = 2$

25. $N = \dfrac{p + q}{p + r}$, $p = u + vw$, $q = v + uw$, $r = w + uv$;

$\dfrac{\partial N}{\partial u}, \dfrac{\partial N}{\partial v}, \dfrac{\partial N}{\partial w}$ when $u = 2$, $v = 3$, $w = 4$

26. $u = xe^{ty}$, $x = \alpha^2\beta$, $y = \beta^2\gamma$, $t = \gamma^2\alpha$;

$\dfrac{\partial u}{\partial \alpha}, \dfrac{\partial u}{\partial \beta}, \dfrac{\partial u}{\partial \gamma}$ when $\alpha = -1$, $\beta = 2$, $\gamma = 1$

27–30 Use Equation 6 to find dy/dx.

27. $y \cos x = x^2 + y^2$

28. $\cos(xy) = 1 + \sin y$

29. $\tan^{-1}(x^2 y) = x + xy^2$

30. $e^y \sin x = x + xy$

31–34 Use Equations 7 to find $\partial z/\partial x$ and $\partial z/\partial y$.

31. $x^2 + 2y^2 + 3z^2 = 1$

32. $x^2 - y^2 + z^2 - 2z = 4$

33. $e^z = xyz$

34. $yz + x \ln y = z^2$

---

35. The temperature at a point (x, y) is $T(x, y)$, measured in degrees Celsius. A bug crawls so that its position after t seconds is given by $x = \sqrt{1 + t}$, $y = 2 + \frac{1}{3}t$, where x and y are measured in centimeters. The temperature function satisfies $T_x(2, 3) = 4$ and $T_y(2, 3) = 3$. How fast is the temperature rising on the bug's path after 3 seconds?

36. Wheat production W in a given year depends on the average temperature T and the annual rainfall R. Scientists estimate that the average temperature is rising at a rate of 0.15°C/year and rainfall is decreasing at a rate of 0.1 cm/year. They also estimate that at current production levels, $\partial W/\partial T = -2$ and $\partial W/\partial R = 8$.

(a) What is the significance of the signs of these partial derivatives?

(b) Estimate the current rate of change of wheat production, dW/dt.

37. The speed of sound traveling through ocean water with salinity 35 parts per thousand has been modeled by the equation

$$C = 1449.2 + 4.6T - 0.055T^2 + 0.00029T^3 + 0.016D$$

where C is the speed of sound (in meters per second), T is the temperature (in degrees Celsius), and D is the depth below the ocean surface (in meters). A scuba diver began a leisurely dive into the ocean water; the diver's depth and the surrounding water temperature over time are recorded in the following graphs. Estimate the rate of change (with respect to time) of the speed of sound through the ocean water experienced by the diver 20 minutes into the dive. What are the units?

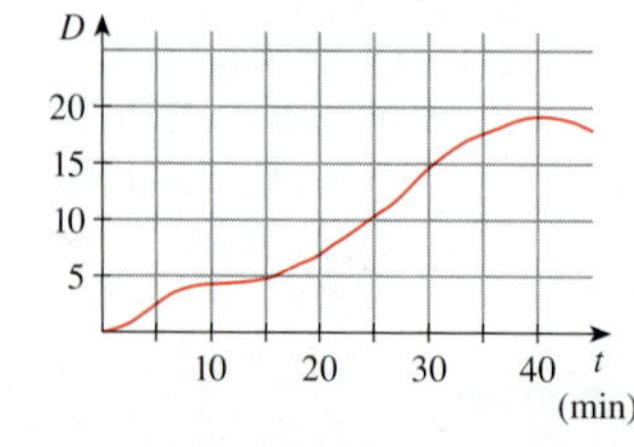

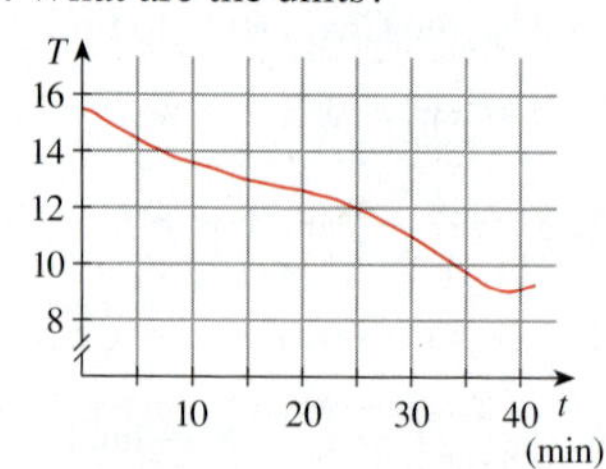

38. The radius of a right circular cone is increasing at a rate of 1.8 in/s while its height is decreasing at a rate of 2.5 in/s. At what rate is the volume of the cone changing when the radius is 120 in. and the height is 140 in.?

39. The length ℓ, width w, and height h of a box change with time. At a certain instant the dimensions are $\ell = 1$ m and $w = h = 2$ m, and ℓ and w are increasing at a rate of 2 m/s while h is decreasing at a rate of 3 m/s. At that instant find the rates at which the following quantities are changing.
(a) The volume
(b) The surface area
(c) The length of a diagonal

40. The voltage V in a simple electrical circuit is slowly decreasing as the battery wears out. The resistance R is slowly increasing as the resistor heats up. Use Ohm's Law, $V = IR$, to find how the current I is changing at the moment when $R = 400\ \Omega$, $I = 0.08$ A, $dV/dt = -0.01$ V/s, and $dR/dt = 0.03\ \Omega$/s.

41. The pressure of 1 mole of an ideal gas is increasing at a rate of 0.05 kPa/s and the temperature is increasing at a rate of 0.15 K/s. Use the equation $PV = 8.31T$ in Example 2 to find the rate of change of the volume when the pressure is 20 kPa and the temperature is 320 K.

42. A manufacturer has modeled its yearly production function P (the value of its entire production, in millions of dollars) as a Cobb-Douglas function

$$P(L, K) = 1.47L^{0.65}K^{0.35}$$

where L is the number of labor hours (in thousands) and K is the invested capital (in millions of dollars). Suppose that when $L = 30$ and $K = 8$, the labor force is decreasing at a rate of 2000 labor hours per year and capital is increasing at a rate of \$500,000 per year. Find the rate of change of production.

43. One side of a triangle is increasing at a rate of 3 cm/s and a second side is decreasing at a rate of 2 cm/s. If the area of the triangle remains constant, at what rate does the angle between the sides change when the first side is 20 cm long, the second side is 30 cm, and the angle is $\pi/6$?

44. A sound with frequency f_s is produced by a source traveling along a line with speed v_s. If an observer is traveling with speed v_o along the same line from the opposite direction toward the source, then the frequency of the sound heard by the observer is

$$f_o = \left(\frac{c + v_o}{c - v_s}\right) f_s$$

where c is the speed of sound, about 332 m/s. (This is the **Doppler effect**.) Suppose that, at a particular moment, you are in a train traveling at 34 m/s and accelerating at 1.2 m/s^2. A train is approaching you from the opposite direction on the other track at 40 m/s, accelerating at 1.4 m/s^2, and sounds its whistle, which has a frequency of 460 Hz. At that instant, what is the perceived frequency that you hear and how fast is it changing?

45–48 Assume that all the given functions are differentiable.

45. If $z = f(x, y)$, where $x = r\cos\theta$ and $y = r\sin\theta$, (a) find $\partial z/\partial r$ and $\partial z/\partial\theta$ and (b) show that

$$\left(\frac{\partial z}{\partial x}\right)^2 + \left(\frac{\partial z}{\partial y}\right)^2 = \left(\frac{\partial z}{\partial r}\right)^2 + \frac{1}{r^2}\left(\frac{\partial z}{\partial \theta}\right)^2$$

46. If $u = f(x, y)$, where $x = e^s\cos t$ and $y = e^s\sin t$, show that

$$\left(\frac{\partial u}{\partial x}\right)^2 + \left(\frac{\partial u}{\partial y}\right)^2 = e^{-2s}\left[\left(\frac{\partial u}{\partial s}\right)^2 + \left(\frac{\partial u}{\partial t}\right)^2\right]$$

47. If $z = \dfrac{1}{x}[f(x - y) + g(x + y)]$, show that

$$\frac{\partial}{\partial x}\left(x^2\frac{\partial z}{\partial x}\right) = x^2\frac{\partial^2 z}{\partial y^2}$$

48. If $z = \dfrac{1}{y}[f(ax + y) + g(ax - y)]$, show that

$$\frac{\partial^2 z}{\partial x^2} = \frac{a^2}{y^2}\frac{\partial}{\partial y}\left(y^2\frac{\partial z}{\partial y}\right)$$

49–54 Assume that all the given functions have continuous second-order partial derivatives.

49. Show that any function of the form

$$z = f(x + at) + g(x - at)$$

is a solution of the wave equation

$$\frac{\partial^2 z}{\partial t^2} = a^2\frac{\partial^2 z}{\partial x^2}$$

[*Hint:* Let $u = x + at$, $v = x - at$.]

50. If $u = f(x, y)$, where $x = e^s\cos t$ and $y = e^s\sin t$, show that

$$\frac{\partial^2 u}{\partial x^2} + \frac{\partial^2 u}{\partial y^2} = e^{-2s}\left[\frac{\partial^2 u}{\partial s^2} + \frac{\partial^2 u}{\partial t^2}\right]$$

51. If $z = f(x, y)$, where $x = r^2 + s^2$ and $y = 2rs$, find $\partial^2 z/\partial r\,\partial s$. (Compare with Example 7.)

52. If $z = f(x, y)$, where $x = r\cos\theta$ and $y = r\sin\theta$, find (a) $\partial z/\partial r$, (b) $\partial z/\partial\theta$, and (c) $\partial^2 z/\partial r\,\partial\theta$.

53. If $z = f(x, y)$, where $x = r\cos\theta$ and $y = r\sin\theta$, show that

$$\frac{\partial^2 z}{\partial x^2} + \frac{\partial^2 z}{\partial y^2} = \frac{\partial^2 z}{\partial r^2} + \frac{1}{r^2}\frac{\partial^2 z}{\partial \theta^2} + \frac{1}{r}\frac{\partial z}{\partial r}$$

54. Suppose $z = f(x, y)$, where $x = g(s, t)$ and $y = h(s, t)$.
(a) Show that

$$\frac{\partial^2 z}{\partial t^2} = \frac{\partial^2 z}{\partial x^2}\left(\frac{\partial x}{\partial t}\right)^2 + 2\frac{\partial^2 z}{\partial x\,\partial y}\frac{\partial x}{\partial t}\frac{\partial y}{\partial t} + \frac{\partial^2 z}{\partial y^2}\left(\frac{\partial y}{\partial t}\right)^2 + \frac{\partial z}{\partial x}\frac{\partial^2 x}{\partial t^2} + \frac{\partial z}{\partial y}\frac{\partial^2 y}{\partial t^2}$$

(b) Find a similar formula for $\partial^2 z/\partial s\,\partial t$.

55. A function f is called **homogeneous of degree n** if it satisfies the equation

$$f(tx, ty) = t^n f(x, y)$$

for all t, where n is a positive integer and f has continuous second-order partial derivatives.

(a) Verify that $f(x, y) = x^2y + 2xy^2 + 5y^3$ is homogeneous of degree 3.

(b) Show that if f is homogeneous of degree n, then

$$x\frac{\partial f}{\partial x} + y\frac{\partial f}{\partial y} = nf(x, y)$$

[*Hint:* Use the Chain Rule to differentiate $f(tx, ty)$ with respect to t.]

56. If f is homogeneous of degree n, show that

$$x^2\frac{\partial^2 f}{\partial x^2} + 2xy\frac{\partial^2 f}{\partial x\,\partial y} + y^2\frac{\partial^2 f}{\partial y^2} = n(n-1)f(x, y)$$

57. If f is homogeneous of degree n, show that

$$f_x(tx, ty) = t^{n-1}f_x(x, y)$$

58. Suppose that the equation $F(x, y, z) = 0$ implicitly defines each of the three variables x, y, and z as functions of the other two: $z = f(x, y)$, $y = g(x, z)$, $x = h(y, z)$. If F is differentiable and F_x, F_y, and F_z are all nonzero, show that

$$\frac{\partial z}{\partial x}\frac{\partial x}{\partial y}\frac{\partial y}{\partial z} = -1$$

59. Equation 6 is a formula for the derivative dy/dx of a function defined implicitly by an equation $F(x, y) = 0$, provided that F is differentiable and $F_y \neq 0$. Prove that if F has continuous second derivatives, then a formula for the second derivative of y is

$$\frac{d^2y}{dx^2} = -\frac{F_{xx}F_y^2 - 2F_{xy}F_xF_y + F_{yy}F_x^2}{F_y^3}$$

14.6 Directional Derivatives and the Gradient Vector

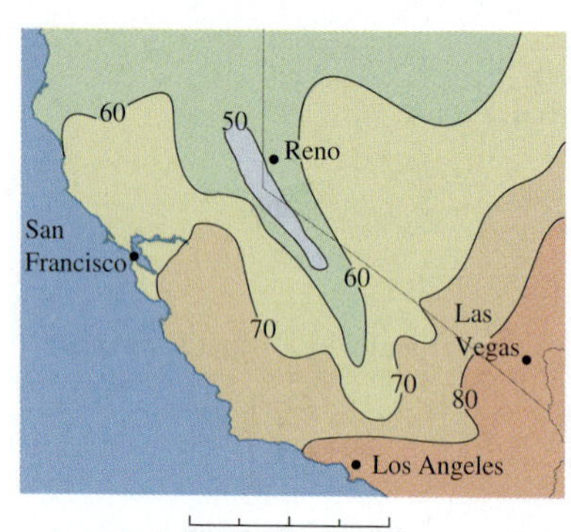

0 50 100 150 200
(Distance in miles)

FIGURE 1

The weather map in Figure 1 shows a contour map of the temperature function $T(x, y)$ for the states of California and Nevada at 3:00 PM on a day in October. The level curves, or isothermals, join locations with the same temperature. The partial derivative T_x at a location such as Reno is the rate of change of temperature with respect to distance if we travel east from Reno; T_y is the rate of change of temperature if we travel north. But what if we want to know the rate of change of temperature when we travel southeast (toward Las Vegas), or in some other direction? In this section we introduce a type of derivative, called a *directional derivative,* that enables us to find the rate of change of a function of two or more variables in any direction.

Directional Derivatives

Recall that if $z = f(x, y)$, then the partial derivatives f_x and f_y are defined as

1
$$f_x(x_0, y_0) = \lim_{h \to 0} \frac{f(x_0 + h, y_0) - f(x_0, y_0)}{h}$$

$$f_y(x_0, y_0) = \lim_{h \to 0} \frac{f(x_0, y_0 + h) - f(x_0, y_0)}{h}$$

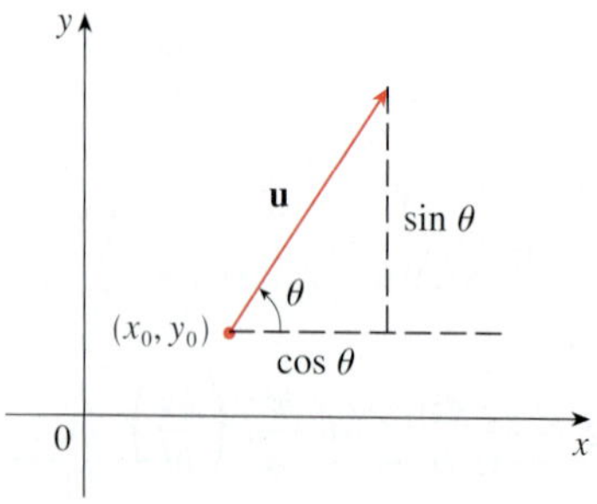

and represent the rates of change of z in the x- and y-directions, that is, in the directions of the unit vectors **i** and **j**.

Suppose that we now wish to find the rate of change of z at (x_0, y_0) in the direction of an arbitrary unit vector $\mathbf{u} = \langle a, b \rangle$. (See Figure 2.) To do this we consider the surface S with the equation $z = f(x, y)$ (the graph of f) and we let $z_0 = f(x_0, y_0)$. Then the point $P(x_0, y_0, z_0)$ lies on S. The vertical plane that passes through P in the direction of **u** inter-

FIGURE 2
A unit vector
$\mathbf{u} = \langle a, b \rangle = \langle \cos u, \sin u \rangle$

sects S in a curve C. (See Figure 3.) The slope of the tangent line T to C at the point P is the rate of change of z in the direction of $\mathbf{u}$.

TEC Visual 14.6A animates Figure 3 by rotating $\mathbf{u}$ and therefore T.

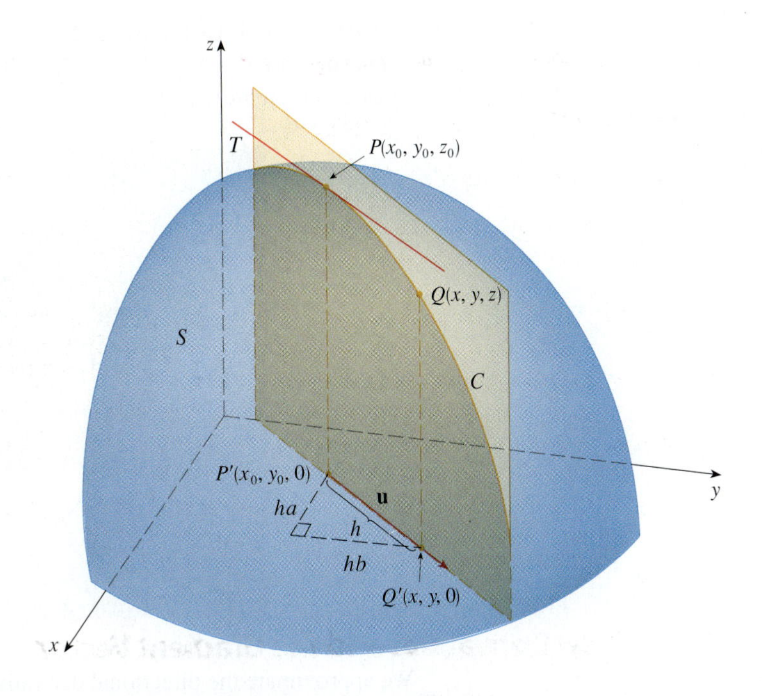

FIGURE 3

If $Q(x, y, z)$ is another point on C and P', Q' are the projections of P, Q onto the xy-plane, then the vector $\overrightarrow{P'Q'}$ is parallel to $\mathbf{u}$ and so

$$\overrightarrow{P'Q'} = h\mathbf{u} = \langle ha, hb \rangle$$

for some scalar h. Therefore $x - x_0 = ha$, $y - y_0 = hb$, so $x = x_0 + ha$, $y = y_0 + hb$, and

$$\frac{\Delta z}{h} = \frac{z - z_0}{h} = \frac{f(x_0 + ha, y_0 + hb) - f(x_0, y_0)}{h}$$

If we take the limit as $h \to 0$, we obtain the rate of change of z (with respect to distance) in the direction of $\mathbf{u}$, which is called the directional derivative of f in the direction of $\mathbf{u}$.

2 Definition The **directional derivative** of f at (x_0, y_0) in the direction of a unit vector $\mathbf{u} = \langle a, b \rangle$ is

$$D_{\mathbf{u}} f(x_0, y_0) = \lim_{h \to 0} \frac{f(x_0 + ha, y_0 + hb) - f(x_0, y_0)}{h}$$

if this limit exists.

By comparing Definition 2 with Equations 1, we see that if $\mathbf{u} = \mathbf{i} = \langle 1, 0 \rangle$, then $D_{\mathbf{i}} f = f_x$ and if $\mathbf{u} = \mathbf{j} = \langle 0, 1 \rangle$, then $D_{\mathbf{j}} f = f_y$. In other words, the partial derivatives of f with respect to x and y are just special cases of the directional derivative.

EXAMPLE 1 Use the weather map in Figure 1 to estimate the value of the directional derivative of the temperature function at Reno in the southeasterly direction.

SOLUTION The unit vector directed toward the southeast is $\mathbf{u} = (\mathbf{i} - \mathbf{j})/\sqrt{2}$, but we won't need to use this expression. We start by drawing a line through Reno toward the southeast (see Figure 4).

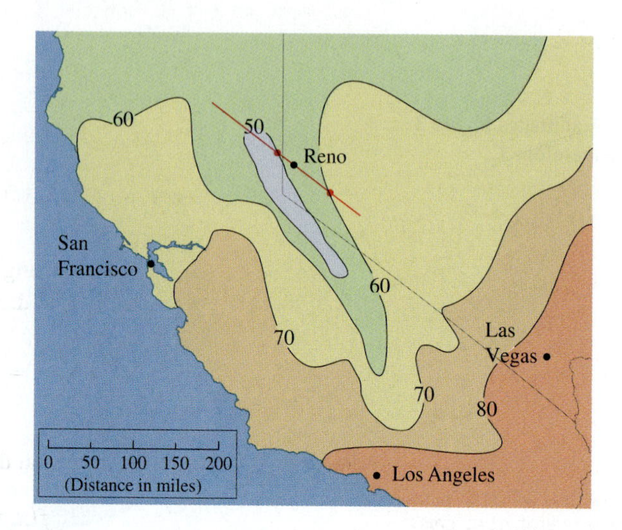

FIGURE 4

We approximate the directional derivative $D_{\mathbf{u}} T$ by the average rate of change of the temperature between the points where this line intersects the isothermals $T = 50$ and $T = 60$. The temperature at the point southeast of Reno is $T = 60°\text{F}$ and the temperature at the point northwest of Reno is $T = 50°\text{F}$. The distance between these points looks to be about 75 miles. So the rate of change of the temperature in the southeasterly direction is

$$D_{\mathbf{u}} T \approx \frac{60 - 50}{75} = \frac{10}{75} \approx 0.13°\text{F/mi}$$

When we compute the directional derivative of a function defined by a formula, we generally use the following theorem.

> **3 Theorem** If f is a differentiable function of x and y, then f has a directional derivative in the direction of any unit vector $\mathbf{u} = \langle a, b \rangle$ and
>
> $$D_{\mathbf{u}} f(x, y) = f_x(x, y)\, a + f_y(x, y)\, b$$

PROOF If we define a function g of the single variable h by

$$g(h) = f(x_0 + ha, y_0 + hb)$$

then, by the definition of a derivative, we have

$$\boxed{4} \qquad g'(0) = \lim_{h \to 0} \frac{g(h) - g(0)}{h} = \lim_{h \to 0} \frac{f(x_0 + ha, y_0 + hb) - f(x_0, y_0)}{h}$$

$$= D_{\mathbf{u}} f(x_0, y_0)$$

On the other hand, we can write $g(h) = f(x, y)$, where $x = x_0 + ha$, $y = y_0 + hb$, so the Chain Rule (Theorem 14.5.2) gives

$$g'(h) = \frac{\partial f}{\partial x}\frac{dx}{dh} + \frac{\partial f}{\partial y}\frac{dy}{dh} = f_x(x, y)\,a + f_y(x, y)\,b$$

If we now put $h = 0$, then $x = x_0$, $y = y_0$, and

5 $$g'(0) = f_x(x_0, y_0)\,a + f_y(x_0, y_0)\,b$$

Comparing Equations 4 and 5, we see that

$$D_{\mathbf{u}} f(x_0, y_0) = f_x(x_0, y_0)\,a + f_y(x_0, y_0)\,b$$ ■

If the unit vector **u** makes an angle θ with the positive x-axis (as in Figure 2), then we can write $\mathbf{u} = \langle \cos\theta, \sin\theta \rangle$ and the formula in Theorem 3 becomes

6 $$D_{\mathbf{u}} f(x, y) = f_x(x, y)\cos\theta + f_y(x, y)\sin\theta$$

The directional derivative $D_{\mathbf{u}} f(1, 2)$ in Example 2 represents the rate of change of z in the direction of **u**. This is the slope of the tangent line to the curve of intersection of the surface $z = x^3 - 3xy + 4y^2$ and the vertical plane through $(1, 2, 0)$ in the direction of **u** shown in Figure 5.

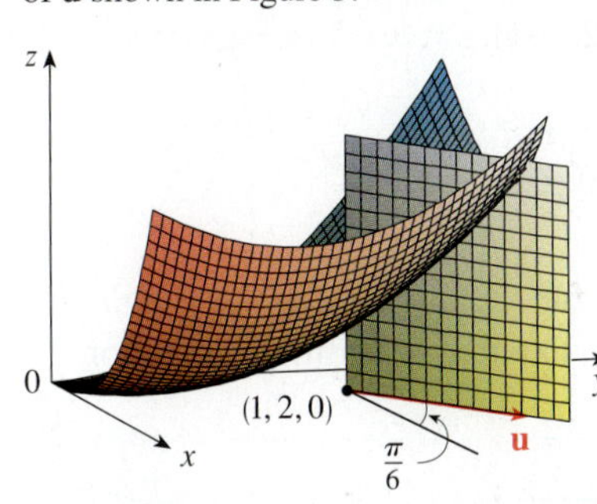

FIGURE 5

EXAMPLE 2 Find the directional derivative $D_{\mathbf{u}} f(x, y)$ if

$$f(x, y) = x^3 - 3xy + 4y^2$$

and **u** is the unit vector given by angle $\theta = \pi/6$. What is $D_{\mathbf{u}} f(1, 2)$?

SOLUTION Formula 6 gives

$$\begin{aligned} D_{\mathbf{u}} f(x, y) &= f_x(x, y)\cos\frac{\pi}{6} + f_y(x, y)\sin\frac{\pi}{6} \\ &= (3x^2 - 3y)\frac{\sqrt{3}}{2} + (-3x + 8y)\tfrac{1}{2} \\ &= \tfrac{1}{2}\left[3\sqrt{3}\,x^2 - 3x + \left(8 - 3\sqrt{3}\right)y\right] \end{aligned}$$

Therefore

$$D_{\mathbf{u}} f(1, 2) = \tfrac{1}{2}\left[3\sqrt{3}(1)^2 - 3(1) + \left(8 - 3\sqrt{3}\right)(2)\right] = \frac{13 - 3\sqrt{3}}{2}$$ ■

The Gradient Vector

Notice from Theorem 3 that the directional derivative of a differentiable function can be written as the dot product of two vectors:

7 $$\begin{aligned} D_{\mathbf{u}} f(x, y) &= f_x(x, y)\,a + f_y(x, y)\,b \\ &= \langle f_x(x, y), f_y(x, y) \rangle \cdot \langle a, b \rangle \\ &= \langle f_x(x, y), f_y(x, y) \rangle \cdot \mathbf{u} \end{aligned}$$

The first vector in this dot product occurs not only in computing directional derivatives but in many other contexts as well. So we give it a special name (the *gradient* of f) and a special notation (**grad** f or ∇f, which is read "del f").

8 Definition If f is a function of two variables x and y, then the **gradient** of f is the vector function ∇f defined by

$$\nabla f(x, y) = \langle f_x(x, y), f_y(x, y) \rangle = \frac{\partial f}{\partial x}\mathbf{i} + \frac{\partial f}{\partial y}\mathbf{j}$$

EXAMPLE 3 If $f(x, y) = \sin x + e^{xy}$, then

$$\nabla f(x, y) = \langle f_x, f_y \rangle = \langle \cos x + ye^{xy}, xe^{xy} \rangle$$

and $$\nabla f(0, 1) = \langle 2, 0 \rangle$$ ■

With this notation for the gradient vector, we can rewrite Equation 7 for the directional derivative of a differentiable function as

9 $$D_{\mathbf{u}} f(x, y) = \nabla f(x, y) \cdot \mathbf{u}$$

This expresses the directional derivative in the direction of a unit vector $\mathbf{u}$ as the scalar projection of the gradient vector onto $\mathbf{u}$.

The gradient vector $\nabla f(2, -1)$ in Example 4 is shown in Figure 6 with initial point $(2, -1)$. Also shown is the vector $\mathbf{v}$ that gives the direction of the directional derivative. Both of these vectors are superimposed on a contour plot of the graph of f.

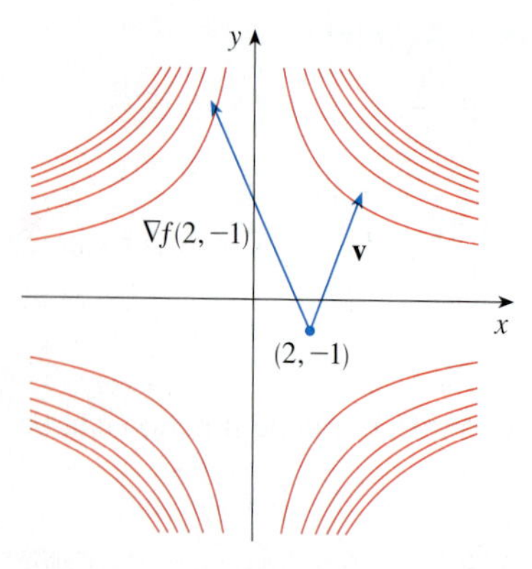

FIGURE 6

EXAMPLE 4 Find the directional derivative of the function $f(x, y) = x^2y^3 - 4y$ at the point $(2, -1)$ in the direction of the vector $\mathbf{v} = 2\mathbf{i} + 5\mathbf{j}$.

SOLUTION We first compute the gradient vector at $(2, -1)$:

$$\nabla f(x, y) = 2xy^3\mathbf{i} + (3x^2y^2 - 4)\mathbf{j}$$

$$\nabla f(2, -1) = -4\mathbf{i} + 8\mathbf{j}$$

Note that $\mathbf{v}$ is not a unit vector, but since $|\mathbf{v}| = \sqrt{29}$, the unit vector in the direction of $\mathbf{v}$ is

$$\mathbf{u} = \frac{\mathbf{v}}{|\mathbf{v}|} = \frac{2}{\sqrt{29}}\mathbf{i} + \frac{5}{\sqrt{29}}\mathbf{j}$$

Therefore, by Equation 9, we have

$$D_{\mathbf{u}} f(2, -1) = \nabla f(2, -1) \cdot \mathbf{u} = (-4\mathbf{i} + 8\mathbf{j}) \cdot \left(\frac{2}{\sqrt{29}}\mathbf{i} + \frac{5}{\sqrt{29}}\mathbf{j}\right)$$

$$= \frac{-4 \cdot 2 + 8 \cdot 5}{\sqrt{29}} = \frac{32}{\sqrt{29}}$$ ■

Functions of Three Variables

For functions of three variables we can define directional derivatives in a similar manner. Again $D_{\mathbf{u}} f(x, y, z)$ can be interpreted as the rate of change of the function in the direction of a unit vector $\mathbf{u}$.

10 Definition The **directional derivative** of f at (x_0, y_0, z_0) in the direction of a unit vector $\mathbf{u} = \langle a, b, c \rangle$ is

$$D_{\mathbf{u}} f(x_0, y_0, z_0) = \lim_{h \to 0} \frac{f(x_0 + ha, y_0 + hb, z_0 + hc) - f(x_0, y_0, z_0)}{h}$$

if this limit exists.

If we use vector notation, then we can write both definitions (2 and 10) of the directional derivative in the compact form

11
$$D_{\mathbf{u}} f(\mathbf{x}_0) = \lim_{h \to 0} \frac{f(\mathbf{x}_0 + h\mathbf{u}) - f(\mathbf{x}_0)}{h}$$

where $\mathbf{x}_0 = \langle x_0, y_0 \rangle$ if $n = 2$ and $\mathbf{x}_0 = \langle x_0, y_0, z_0 \rangle$ if $n = 3$. This is reasonable because the vector equation of the line through $\mathbf{x}_0$ in the direction of the vector $\mathbf{u}$ is given by $\mathbf{x} = \mathbf{x}_0 + t\mathbf{u}$ (Equation 12.5.1) and so $f(\mathbf{x}_0 + h\mathbf{u})$ represents the value of f at a point on this line.

If $f(x, y, z)$ is differentiable and $\mathbf{u} = \langle a, b, c \rangle$, then the same method that was used to prove Theorem 3 can be used to show that

12
$$D_{\mathbf{u}} f(x, y, z) = f_x(x, y, z)\, a + f_y(x, y, z)\, b + f_z(x, y, z)\, c$$

For a function f of three variables, the **gradient vector**, denoted by ∇f or **grad** f, is

$$\nabla f(x, y, z) = \langle f_x(x, y, z), f_y(x, y, z), f_z(x, y, z) \rangle$$

or, for short,

13
$$\nabla f = \langle f_x, f_y, f_z \rangle = \frac{\partial f}{\partial x}\mathbf{i} + \frac{\partial f}{\partial y}\mathbf{j} + \frac{\partial f}{\partial z}\mathbf{k}$$

Then, just as with functions of two variables, Formula 12 for the directional derivative can be rewritten as

14
$$D_{\mathbf{u}} f(x, y, z) = \nabla f(x, y, z) \cdot \mathbf{u}$$

EXAMPLE 5 If $f(x, y, z) = x \sin yz$, (a) find the gradient of f and (b) find the directional derivative of f at $(1, 3, 0)$ in the direction of $\mathbf{v} = \mathbf{i} + 2\mathbf{j} - \mathbf{k}$.

SOLUTION

(a) The gradient of f is

$$\begin{aligned} \nabla f(x, y, z) &= \langle f_x(x, y, z), f_y(x, y, z), f_z(x, y, z) \rangle \\ &= \langle \sin yz,\ xz \cos yz,\ xy \cos yz \rangle \end{aligned}$$

(b) At $(1, 3, 0)$ we have $\nabla f(1, 3, 0) = \langle 0, 0, 3 \rangle$. The unit vector in the direction of $\mathbf{v} = \mathbf{i} + 2\mathbf{j} - \mathbf{k}$ is

$$\mathbf{u} = \frac{1}{\sqrt{6}}\mathbf{i} + \frac{2}{\sqrt{6}}\mathbf{j} - \frac{1}{\sqrt{6}}\mathbf{k}$$

Therefore Equation 14 gives

$$\begin{aligned} D_{\mathbf{u}} f(1, 3, 0) &= \nabla f(1, 3, 0) \cdot \mathbf{u} \\ &= 3\mathbf{k} \cdot \left(\frac{1}{\sqrt{6}}\mathbf{i} + \frac{2}{\sqrt{6}}\mathbf{j} - \frac{1}{\sqrt{6}}\mathbf{k} \right) \\ &= 3\left(-\frac{1}{\sqrt{6}} \right) = -\sqrt{\frac{3}{2}} \end{aligned}$$

Maximizing the Directional Derivative

Suppose we have a function f of two or three variables and we consider all possible directional derivatives of f at a given point. These give the rates of change of f in all possible directions. We can then ask the questions: in which of these directions does f change fastest and what is the maximum rate of change? The answers are provided by the following theorem.

TEC Visual 14.6B provides visual confirmation of Theorem 15.

15 Theorem Suppose f is a differentiable function of two or three variables. The maximum value of the directional derivative $D_{\mathbf{u}} f(\mathbf{x})$ is $|\nabla f(\mathbf{x})|$ and it occurs when $\mathbf{u}$ has the same direction as the gradient vector $\nabla f(\mathbf{x})$.

PROOF From Equation 9 or 14 we have

$$D_{\mathbf{u}} f = \nabla f \cdot \mathbf{u} = |\nabla f||\mathbf{u}| \cos\theta = |\nabla f| \cos\theta$$

where θ is the angle between ∇f and $\mathbf{u}$. The maximum value of $\cos\theta$ is 1 and this occurs when $\theta = 0$. Therefore the maximum value of $D_{\mathbf{u}} f$ is $|\nabla f|$ and it occurs when $\theta = 0$, that is, when $\mathbf{u}$ has the same direction as ∇f.

EXAMPLE 6

(a) If $f(x, y) = xe^y$, find the rate of change of f at the point $P(2, 0)$ in the direction from P to $Q(\frac{1}{2}, 2)$.

(b) In what direction does f have the maximum rate of change? What is this maximum rate of change?

SOLUTION

(a) We first compute the gradient vector:

$$\nabla f(x, y) = \langle f_x, f_y \rangle = \langle e^y, xe^y \rangle$$

$$\nabla f(2, 0) = \langle 1, 2 \rangle$$

The unit vector in the direction of $\overrightarrow{PQ} = \langle -\frac{3}{2}, 2 \rangle$ is $\mathbf{u} = \langle -\frac{3}{5}, \frac{4}{5} \rangle$, so the rate of change of f in the direction from P to Q is

$$D_{\mathbf{u}} f(2, 0) = \nabla f(2, 0) \cdot \mathbf{u} = \langle 1, 2 \rangle \cdot \langle -\tfrac{3}{5}, \tfrac{4}{5} \rangle$$
$$= 1\left(-\tfrac{3}{5}\right) + 2\left(\tfrac{4}{5}\right) = 1$$

(b) According to Theorem 15, f increases fastest in the direction of the gradient vector $\nabla f(2, 0) = \langle 1, 2 \rangle$. The maximum rate of change is

$$|\nabla f(2, 0)| = |\langle 1, 2 \rangle| = \sqrt{5}$$

At (2, 0) the function in Example 6 increases fastest in the direction of the gradient vector $\nabla f(2, 0) = \langle 1, 2 \rangle$. Notice from Figure 7 that this vector appears to be perpendicular to the level curve through (2, 0). Figure 8 shows the graph of f and the gradient vector.

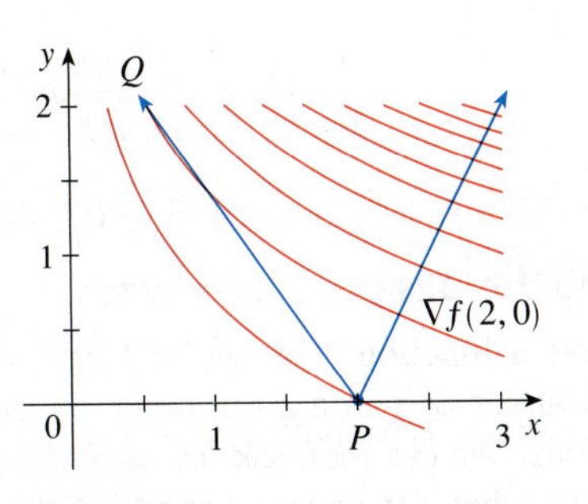

FIGURE 7

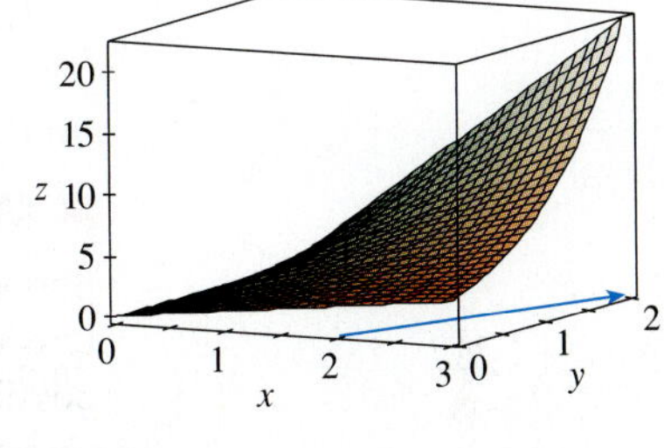

FIGURE 8

EXAMPLE 7 Suppose that the temperature at a point (x, y, z) in space is given by $T(x, y, z) = 80/(1 + x^2 + 2y^2 + 3z^2)$, where T is measured in degrees Celsius and x, y, z in meters. In which direction does the temperature increase fastest at the point $(1, 1, -2)$? What is the maximum rate of increase?

SOLUTION The gradient of T is

$$\nabla T = \frac{\partial T}{\partial x}\mathbf{i} + \frac{\partial T}{\partial y}\mathbf{j} + \frac{\partial T}{\partial z}\mathbf{k}$$
$$= -\frac{160x}{(1 + x^2 + 2y^2 + 3z^2)^2}\mathbf{i} - \frac{320y}{(1 + x^2 + 2y^2 + 3z^2)^2}\mathbf{j} - \frac{480z}{(1 + x^2 + 2y^2 + 3z^2)^2}\mathbf{k}$$
$$= \frac{160}{(1 + x^2 + 2y^2 + 3z^2)^2}(-x\,\mathbf{i} - 2y\,\mathbf{j} - 3z\,\mathbf{k})$$

At the point $(1, 1, -2)$ the gradient vector is

$$\nabla T(1, 1, -2) = \tfrac{160}{256}(-\mathbf{i} - 2\mathbf{j} + 6\mathbf{k}) = \tfrac{5}{8}(-\mathbf{i} - 2\mathbf{j} + 6\mathbf{k})$$

By Theorem 15 the temperature increases fastest in the direction of the gradient vector $\nabla T(1, 1, -2) = \frac{5}{8}(-\mathbf{i} - 2\mathbf{j} + 6\mathbf{k})$ or, equivalently, in the direction of $-\mathbf{i} - 2\mathbf{j} + 6\mathbf{k}$ or the unit vector $(-\mathbf{i} - 2\mathbf{j} + 6\mathbf{k})/\sqrt{41}$. The maximum rate of increase is the length of the gradient vector:

$$|\nabla T(1, 1, -2)| = \tfrac{5}{8}|-\mathbf{i} - 2\mathbf{j} + 6\mathbf{k}| = \tfrac{5}{8}\sqrt{41}$$

Therefore the maximum rate of increase of temperature is $\frac{5}{8}\sqrt{41} \approx 4°\text{C/m}$.

Tangent Planes to Level Surfaces

Suppose S is a surface with equation $F(x, y, z) = k$, that is, it is a level surface of a function F of three variables, and let $P(x_0, y_0, z_0)$ be a point on S. Let C be any curve that lies on the surface S and passes through the point P. Recall from Section 13.1 that the curve C is described by a continuous vector function $\mathbf{r}(t) = \langle x(t), y(t), z(t) \rangle$. Let t_0 be the parameter value corresponding to P; that is, $\mathbf{r}(t_0) = \langle x_0, y_0, z_0 \rangle$. Since C lies on S, any point $(x(t), y(t), z(t))$ must satisfy the equation of S, that is,

$$F(x(t), y(t), z(t)) = k \tag{16}$$

If x, y, and z are differentiable functions of t and F is also differentiable, then we can use the Chain Rule to differentiate both sides of Equation 16 as follows:

$$\frac{\partial F}{\partial x}\frac{dx}{dt} + \frac{\partial F}{\partial y}\frac{dy}{dt} + \frac{\partial F}{\partial z}\frac{dz}{dt} = 0 \tag{17}$$

But, since $\nabla F = \langle F_x, F_y, F_z \rangle$ and $\mathbf{r}'(t) = \langle x'(t), y'(t), z'(t) \rangle$, Equation 17 can be written in terms of a dot product as

$$\nabla F \cdot \mathbf{r}'(t) = 0$$

In particular, when $t = t_0$ we have $\mathbf{r}(t_0) = \langle x_0, y_0, z_0 \rangle$, so

$$\nabla F(x_0, y_0, z_0) \cdot \mathbf{r}'(t_0) = 0 \tag{18}$$

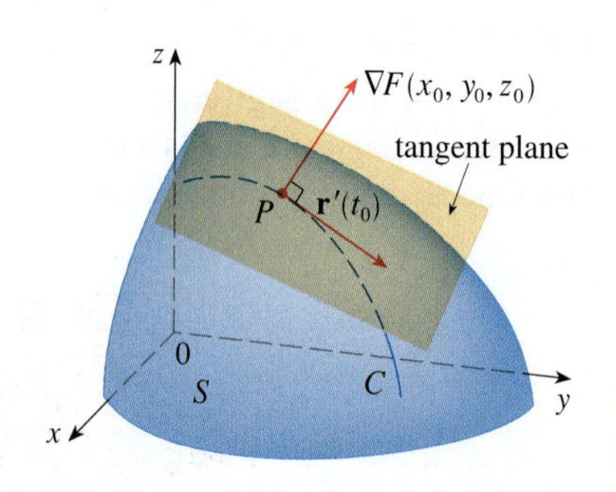

FIGURE 9

Equation 18 says that *the gradient vector at P, $\nabla F(x_0, y_0, z_0)$, is perpendicular to the tangent vector $\mathbf{r}'(t_0)$ to any curve C on S that passes through P.* (See Figure 9.) If $\nabla F(x_0, y_0, z_0) \neq \mathbf{0}$, it is therefore natural to define the **tangent plane to the level surface $F(x, y, z) = k$ at $P(x_0, y_0, z_0)$** as the plane that passes through P and has normal vector $\nabla F(x_0, y_0, z_0)$. Using the standard equation of a plane (Equation 12.5.7), we can write the equation of this tangent plane as

$$F_x(x_0, y_0, z_0)(x - x_0) + F_y(x_0, y_0, z_0)(y - y_0) + F_z(x_0, y_0, z_0)(z - z_0) = 0 \tag{19}$$

The **normal line** to S at P is the line passing through P and perpendicular to the tangent plane. The direction of the normal line is therefore given by the gradient vector $\nabla F(x_0, y_0, z_0)$ and so, by Equation 12.5.3, its symmetric equations are

$$\frac{x - x_0}{F_x(x_0, y_0, z_0)} = \frac{y - y_0}{F_y(x_0, y_0, z_0)} = \frac{z - z_0}{F_z(x_0, y_0, z_0)} \tag{20}$$

In the special case in which the equation of a surface S is of the form $z = f(x, y)$ (that is, S is the graph of a function f of two variables), we can rewrite the equation as

$$F(x, y, z) = f(x, y) - z = 0$$

and regard S as a level surface (with $k = 0$) of F. Then

$$F_x(x_0, y_0, z_0) = f_x(x_0, y_0)$$
$$F_y(x_0, y_0, z_0) = f_y(x_0, y_0)$$
$$F_z(x_0, y_0, z_0) = -1$$

so Equation 19 becomes

$$f_x(x_0, y_0)(x - x_0) + f_y(x_0, y_0)(y - y_0) - (z - z_0) = 0$$

which is equivalent to Equation 14.4.2. Thus our new, more general, definition of a tangent plane is consistent with the definition that was given for the special case of Section 14.4.

EXAMPLE 8 Find the equations of the tangent plane and normal line at the point $(-2, 1, -3)$ to the ellipsoid

$$\frac{x^2}{4} + y^2 + \frac{z^2}{9} = 3$$

SOLUTION The ellipsoid is the level surface (with $k = 3$) of the function

Figure 10 shows the ellipsoid, tangent plane, and normal line in Example 8.

$$F(x, y, z) = \frac{x^2}{4} + y^2 + \frac{z^2}{9}$$

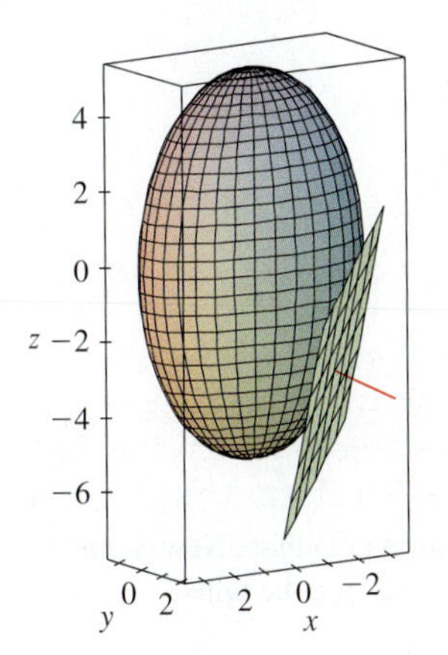

FIGURE 10

Therefore we have

$$F_x(x, y, z) = \frac{x}{2} \qquad F_y(x, y, z) = 2y \qquad F_z(x, y, z) = \frac{2z}{9}$$

$$F_x(-2, 1, -3) = -1 \qquad F_y(-2, 1, -3) = 2 \qquad F_z(-2, 1, -3) = -\tfrac{2}{3}$$

Then Equation 19 gives the equation of the tangent plane at $(-2, 1, -3)$ as

$$-1(x + 2) + 2(y - 1) - \tfrac{2}{3}(z + 3) = 0$$

which simplifies to $3x - 6y + 2z + 18 = 0$.

By Equation 20, symmetric equations of the normal line are

$$\frac{x + 2}{-1} = \frac{y - 1}{2} = \frac{z + 3}{-\frac{2}{3}}$$

Significance of the Gradient Vector

We now summarize the ways in which the gradient vector is significant. We first consider a function f of three variables and a point $P(x_0, y_0, z_0)$ in its domain. On the one hand, we know from Theorem 15 that the gradient vector $\nabla f(x_0, y_0, z_0)$ gives the direction of fastest increase of f. On the other hand, we know that $\nabla f(x_0, y_0, z_0)$ is orthogonal to the level surface S of f through P. (Refer to Figure 9.) These two properties are quite compatible intuitively because as we move away from P on the level surface S, the value of f does not change at all. So it seems reasonable that if we move in the perpendicular direction, we get the maximum increase.

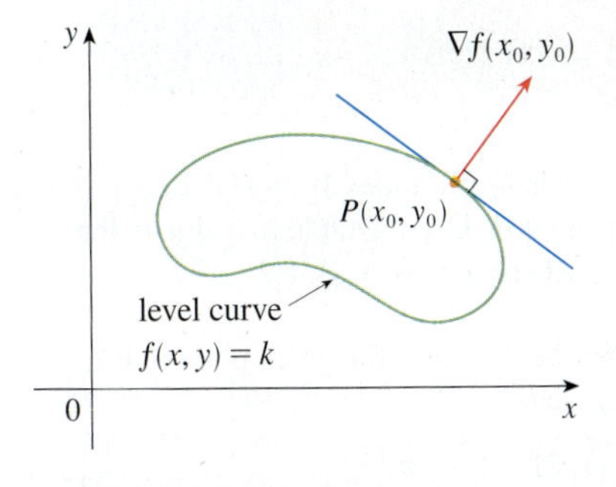

FIGURE 11

In like manner we consider a function f of two variables and a point $P(x_0, y_0)$ in its domain. Again the gradient vector $\nabla f(x_0, y_0)$ gives the direction of fastest increase of f. Also, by considerations similar to our discussion of tangent planes, it can be shown that $\nabla f(x_0, y_0)$ is perpendicular to the level curve $f(x, y) = k$ that passes through P. Again this is intuitively plausible because the values of f remain constant as we move along the curve. (See Figure 11.)

If we consider a topographical map of a hill and let $f(x, y)$ represent the height above sea level at a point with coordinates (x, y), then a curve of steepest ascent can be drawn

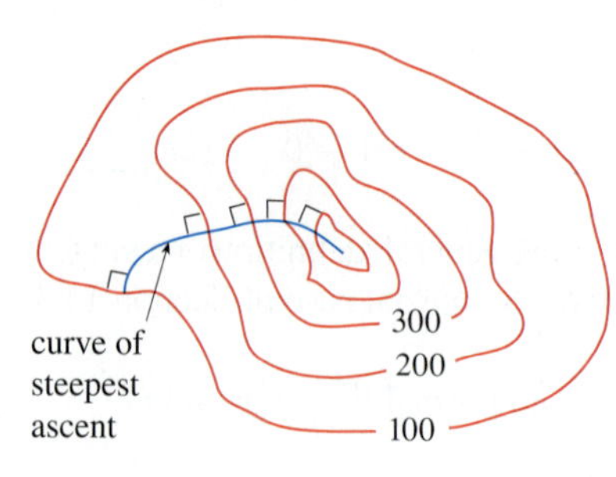

FIGURE 12

as in Figure 12 by making it perpendicular to all of the contour lines. This phenomenon can also be noticed in Figure 14.1.12, where Lonesome Creek follows a curve of steepest descent.

Computer algebra systems have commands that plot sample gradient vectors. Each gradient vector $\nabla f(a, b)$ is plotted starting at the point (a, b). Figure 13 shows such a plot (called a *gradient vector field*) for the function $f(x, y) = x^2 - y^2$ superimposed on a contour map of f. As expected, the gradient vectors point "uphill" and are perpendicular to the level curves.

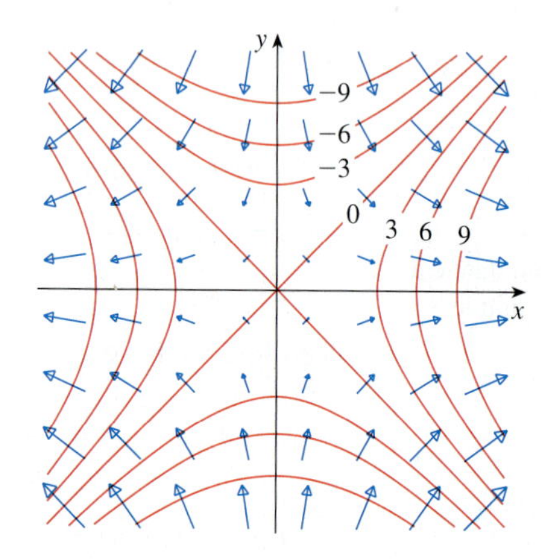

FIGURE 13

14.6 EXERCISES

1. Level curves for barometric pressure (in millibars) are shown for 6:00 AM on a day in November. A deep low with pressure 972 mb is moving over northeast Iowa. The distance along the red line from K (Kearney, Nebraska) to S (Sioux City, Iowa) is 300 km. Estimate the value of the directional derivative of the pressure function at Kearney in the direction of Sioux City. What are the units of the directional derivative?

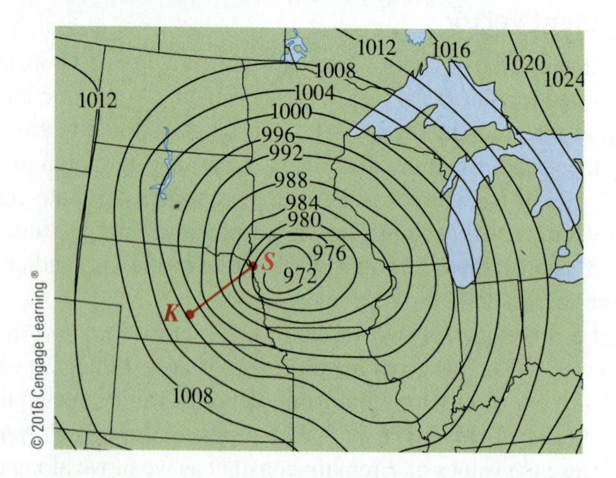

2. The contour map shows the average maximum temperature for November 2004 (in °C). Estimate the value of the directional derivative of this temperature function at Dubbo, New South Wales, in the direction of Sydney. What are the units?

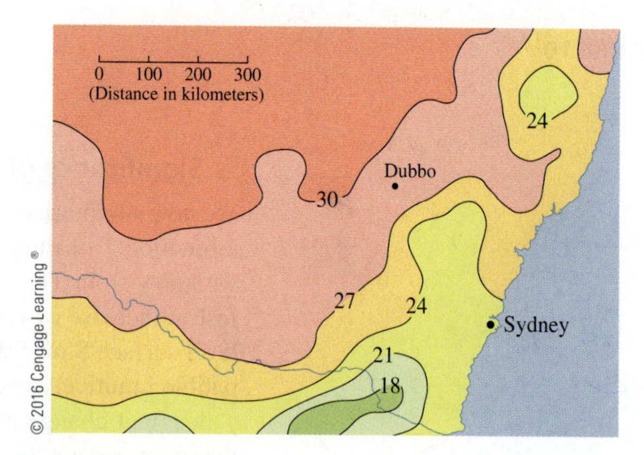

3. A table of values for the wind-chill index $W = f(T, v)$ is given in Exercise 14.3.3 on page 963. Use the table to estimate the value of $D_{\mathbf{u}} f(-20, 30)$, where $\mathbf{u} = (\mathbf{i} + \mathbf{j})/\sqrt{2}$.

4–6 Find the directional derivative of f at the given point in the direction indicated by the angle θ.

4. $f(x, y) = xy^3 - x^2$, $(1, 2)$, $\theta = \pi/3$

5. $f(x, y) = y\cos(xy)$, $(0, 1)$, $\theta = \pi/4$

6. $f(x, y) = \sqrt{2x + 3y}$, $(3, 1)$, $\theta = -\pi/6$

7–10
(a) Find the gradient of f.
(b) Evaluate the gradient at the point P.
(c) Find the rate of change of f at P in the direction of the vector $\mathbf{u}$.

7. $f(x, y) = x/y$, $P(2, 1)$, $\mathbf{u} = \frac{3}{5}\mathbf{i} + \frac{4}{5}\mathbf{j}$

8. $f(x, y) = x^2 \ln y$, $P(3, 1)$, $\mathbf{u} = -\frac{5}{13}\mathbf{i} + \frac{12}{13}\mathbf{j}$

9. $f(x, y, z) = x^2yz - xyz^3$, $P(2, -1, 1)$, $\mathbf{u} = \langle 0, \frac{4}{5}, -\frac{3}{5} \rangle$

10. $f(x, y, z) = y^2e^{xyz}$, $P(0, 1, -1)$, $\mathbf{u} = \langle \frac{3}{13}, \frac{4}{13}, \frac{12}{13} \rangle$

11–17 Find the directional derivative of the function at the given point in the direction of the vector $\mathbf{v}$.

11. $f(x, y) = e^x \sin y$, $(0, \pi/3)$, $\mathbf{v} = \langle -6, 8 \rangle$

12. $f(x, y) = \dfrac{x}{x^2 + y^2}$, $(1, 2)$, $\mathbf{v} = \langle 3, 5 \rangle$

13. $g(s, t) = s\sqrt{t}$, $(2, 4)$, $\mathbf{v} = 2\mathbf{i} - \mathbf{j}$

14. $g(u, v) = u^2e^{-v}$, $(3, 0)$, $\mathbf{v} = 3\mathbf{i} + 4\mathbf{j}$

15. $f(x, y, z) = x^2y + y^2z$, $(1, 2, 3)$, $\mathbf{v} = \langle 2, -1, 2 \rangle$

16. $f(x, y, z) = xy^2 \tan^{-1}z$, $(2, 1, 1)$, $\mathbf{v} = \langle 1, 1, 1 \rangle$

17. $h(r, s, t) = \ln(3r + 6s + 9t)$, $(1, 1, 1)$, $\mathbf{v} = 4\mathbf{i} + 12\mathbf{j} + 6\mathbf{k}$

18. Use the figure to estimate $D_{\mathbf{u}} f(2, 2)$.

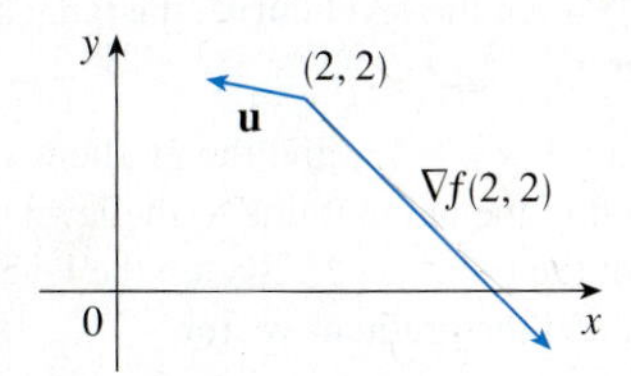

19. Find the directional derivative of $f(x, y) = \sqrt{xy}$ at $P(2, 8)$ in the direction of $Q(5, 4)$.

20. Find the directional derivative of $f(x, y, z) = xy^2z^3$ at $P(2, 1, 1)$ in the direction of $Q(0, -3, 5)$.

21–26 Find the maximum rate of change of f at the given point and the direction in which it occurs.

21. $f(x, y) = 4y\sqrt{x}$, $(4, 1)$

22. $f(s, t) = te^{st}$, $(0, 2)$

23. $f(x, y) = \sin(xy)$, $(1, 0)$

24. $f(x, y, z) = x\ln(yz)$, $(1, 2, \frac{1}{2})$

25. $f(x, y, z) = x/(y + z)$, $(8, 1, 3)$

26. $f(p, q, r) = \arctan(pqr)$, $(1, 2, 1)$

27. (a) Show that a differentiable function f decreases most rapidly at $\mathbf{x}$ in the direction opposite to the gradient vector, that is, in the direction of $-\nabla f(\mathbf{x})$.
(b) Use the result of part (a) to find the direction in which the function $f(x, y) = x^4y - x^2y^3$ decreases fastest at the point $(2, -3)$.

28. Find the directions in which the directional derivative of $f(x, y) = x^2 + xy^3$ at the point $(2, 1)$ has the value 2.

29. Find all points at which the direction of fastest change of the function $f(x, y) = x^2 + y^2 - 2x - 4y$ is $\mathbf{i} + \mathbf{j}$.

30. Near a buoy, the depth of a lake at the point with coordinates (x, y) is $z = 200 + 0.02x^2 - 0.001y^3$, where x, y, and z are measured in meters. A fisherman in a small boat starts at the point $(80, 60)$ and moves toward the buoy, which is located at $(0, 0)$. Is the water under the boat getting deeper or shallower when he departs? Explain.

31. The temperature T in a metal ball is inversely proportional to the distance from the center of the ball, which we take to be the origin. The temperature at the point $(1, 2, 2)$ is $120°$.
(a) Find the rate of change of T at $(1, 2, 2)$ in the direction toward the point $(2, 1, 3)$.
(b) Show that at any point in the ball the direction of greatest increase in temperature is given by a vector that points toward the origin.

32. The temperature at a point (x, y, z) is given by

$$T(x, y, z) = 200e^{-x^2-3y^2-9z^2}$$

where T is measured in °C and x, y, z in meters.
(a) Find the rate of change of temperature at the point $P(2, -1, 2)$ in the direction toward the point $(3, -3, 3)$.
(b) In which direction does the temperature increase fastest at P?
(c) Find the maximum rate of increase at P.

33. Suppose that over a certain region of space the electrical potential V is given by $V(x, y, z) = 5x^2 - 3xy + xyz$.
(a) Find the rate of change of the potential at $P(3, 4, 5)$ in the direction of the vector $\mathbf{v} = \mathbf{i} + \mathbf{j} - \mathbf{k}$.
(b) In which direction does V change most rapidly at P?
(c) What is the maximum rate of change at P?

34. Suppose you are climbing a hill whose shape is given by the equation $z = 1000 - 0.005x^2 - 0.01y^2$, where x, y, and z are measured in meters, and you are standing at a point with coordinates $(60, 40, 966)$. The positive x-axis points east and the positive y-axis points north.
(a) If you walk due south, will you start to ascend or descend? At what rate?

(b) If you walk northwest, will you start to ascend or descend? At what rate?

(c) In which direction is the slope largest? What is the rate of ascent in that direction? At what angle above the horizontal does the path in that direction begin?

35. Let f be a function of two variables that has continuous partial derivatives and consider the points $A(1, 3)$, $B(3, 3)$, $C(1, 7)$, and $D(6, 15)$. The directional derivative of f at A in the direction of the vector $\overrightarrow{AB}$ is 3 and the directional derivative at A in the direction of $\overrightarrow{AC}$ is 26. Find the directional derivative of f at A in the direction of the vector $\overrightarrow{AD}$.

36. Shown is a topographic map of Blue River Pine Provincial Park in British Columbia. Draw curves of steepest descent from point A (descending to Mud Lake) and from point B.

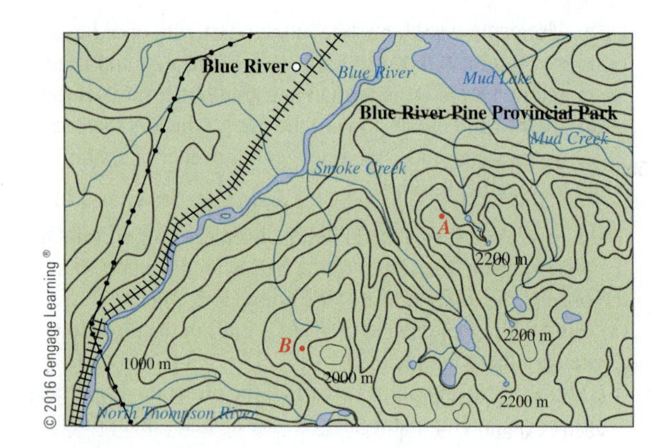

37. Show that the operation of taking the gradient of a function has the given property. Assume that u and v are differentiable functions of x and y and that a, b are constants.

(a) $\nabla(au + bv) = a\,\nabla u + b\,\nabla v$

(b) $\nabla(uv) = u\,\nabla v + v\,\nabla u$

(c) $\nabla\left(\dfrac{u}{v}\right) = \dfrac{v\,\nabla u - u\,\nabla v}{v^2}$

(d) $\nabla u^n = nu^{n-1}\,\nabla u$

38. Sketch the gradient vector $\nabla f(4, 6)$ for the function f whose level curves are shown. Explain how you chose the direction and length of this vector.

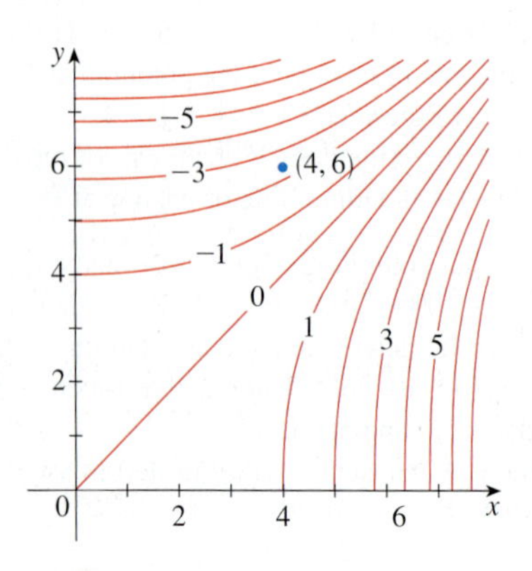

39. The **second directional derivative** of $f(x, y)$ is

$$D_{\mathbf{u}}^2 f(x, y) = D_{\mathbf{u}}[D_{\mathbf{u}} f(x, y)]$$

If $f(x, y) = x^3 + 5x^2y + y^3$ and $\mathbf{u} = \left\langle \frac{3}{5}, \frac{4}{5} \right\rangle$, calculate $D_{\mathbf{u}}^2 f(2, 1)$.

40. (a) If $\mathbf{u} = \langle a, b \rangle$ is a unit vector and f has continuous second partial derivatives, show that

$$D_{\mathbf{u}}^2 f = f_{xx}a^2 + 2f_{xy}ab + f_{yy}b^2$$

(b) Find the second directional derivative of $f(x, y) = xe^{2y}$ in the direction of $\mathbf{v} = \langle 4, 6 \rangle$.

41–46 Find equations of (a) the tangent plane and (b) the normal line to the given surface at the specified point.

41. $2(x - 2)^2 + (y - 1)^2 + (z - 3)^2 = 10, \quad (3, 3, 5)$

42. $x = y^2 + z^2 + 1, \quad (3, 1, -1)$

43. $xy^2z^3 = 8, \quad (2, 2, 1)$

44. $xy + yz + zx = 5, \quad (1, 2, 1)$

45. $x + y + z = e^{xyz}, \quad (0, 0, 1)$

46. $x^4 + y^4 + z^4 = 3x^2y^2z^2, \quad (1, 1, 1)$

47–48 Use a computer to graph the surface, the tangent plane, and the normal line on the same screen. Choose the domain carefully so that you avoid extraneous vertical planes. Choose the viewpoint so that you get a good view of all three objects.

47. $xy + yz + zx = 3, \quad (1, 1, 1)$

48. $xyz = 6, \quad (1, 2, 3)$

49. If $f(x, y) = xy$, find the gradient vector $\nabla f(3, 2)$ and use it to find the tangent line to the level curve $f(x, y) = 6$ at the point $(3, 2)$. Sketch the level curve, the tangent line, and the gradient vector.

50. If $g(x, y) = x^2 + y^2 - 4x$, find the gradient vector $\nabla g(1, 2)$ and use it to find the tangent line to the level curve $g(x, y) = 1$ at the point $(1, 2)$. Sketch the level curve, the tangent line, and the gradient vector.

51. Show that the equation of the tangent plane to the ellipsoid $x^2/a^2 + y^2/b^2 + z^2/c^2 = 1$ at the point (x_0, y_0, z_0) can be written as

$$\frac{xx_0}{a^2} + \frac{yy_0}{b^2} + \frac{zz_0}{c^2} = 1$$

52. Find the equation of the tangent plane to the hyperboloid $x^2/a^2 + y^2/b^2 - z^2/c^2 = 1$ at (x_0, y_0, z_0) and express it in a form similar to the one in Exercise 51.

53. Show that the equation of the tangent plane to the elliptic paraboloid $z/c = x^2/a^2 + y^2/b^2$ at the point (x_0, y_0, z_0) can be written as

$$\frac{2xx_0}{a^2} + \frac{2yy_0}{b^2} = \frac{z + z_0}{c}$$

54. At what point on the ellipsoid $x^2 + y^2 + 2z^2 = 1$ is the tangent plane parallel to the plane $x + 2y + z = 1$?

55. Are there any points on the hyperboloid $x^2 - y^2 - z^2 = 1$ where the tangent plane is parallel to the plane $z = x + y$?

56. Show that the ellipsoid $3x^2 + 2y^2 + z^2 = 9$ and the sphere $x^2 + y^2 + z^2 - 8x - 6y - 8z + 24 = 0$ are tangent to each other at the point $(1, 1, 2)$. (This means that they have a common tangent plane at the point.)

57. Show that every plane that is tangent to the cone $x^2 + y^2 = z^2$ passes through the origin.

58. Show that every normal line to the sphere $x^2 + y^2 + z^2 = r^2$ passes through the center of the sphere.

59. Where does the normal line to the paraboloid $z = x^2 + y^2$ at the point $(1, 1, 2)$ intersect the paraboloid a second time?

60. At what points does the normal line through the point $(1, 2, 1)$ on the ellipsoid $4x^2 + y^2 + 4z^2 = 12$ intersect the sphere $x^2 + y^2 + z^2 = 102$?

61. Show that the sum of the x-, y-, and z-intercepts of any tangent plane to the surface $\sqrt{x} + \sqrt{y} + \sqrt{z} = \sqrt{c}$ is a constant.

62. Show that the pyramids cut off from the first octant by any tangent planes to the surface $xyz = 1$ at points in the first octant must all have the same volume.

63. Find parametric equations for the tangent line to the curve of intersection of the paraboloid $z = x^2 + y^2$ and the ellipsoid $4x^2 + y^2 + z^2 = 9$ at the point $(-1, 1, 2)$.

64. (a) The plane $y + z = 3$ intersects the cylinder $x^2 + y^2 = 5$ in an ellipse. Find parametric equations for the tangent line to this ellipse at the point $(1, 2, 1)$.

(b) Graph the cylinder, the plane, and the tangent line on the same screen.

65. Where does the helix $\mathbf{r}(t) = \langle \cos \pi t, \sin \pi t, t \rangle$ intersect the paraboloid $z = x^2 + y^2$? What is the angle of intersection between the helix and the paraboloid? (This is the angle between the tangent vector to the curve and the tangent plane to the paraboloid.)

66. The helix $\mathbf{r}(t) = \langle \cos(\pi t/2), \sin(\pi t/2), t \rangle$ intersects the sphere $x^2 + y^2 + z^2 = 2$ in two points. Find the angle of intersection at each point.

67. (a) Two surfaces are called **orthogonal** at a point of intersection if their normal lines are perpendicular at that point. Show that surfaces with equations $F(x, y, z) = 0$ and $G(x, y, z) = 0$ are orthogonal at a point P where $\nabla F \neq \mathbf{0}$ and $\nabla G \neq \mathbf{0}$ if and only if

$$F_x G_x + F_y G_y + F_z G_z = 0 \quad \text{at } P$$

(b) Use part (a) to show that the surfaces $z^2 = x^2 + y^2$ and $x^2 + y^2 + z^2 = r^2$ are orthogonal at every point of intersection. Can you see why this is true without using calculus?

68. (a) Show that the function $f(x, y) = \sqrt[3]{xy}$ is continuous and the partial derivatives f_x and f_y exist at the origin but the directional derivatives in all other directions do not exist.

(b) Graph f near the origin and comment on how the graph confirms part (a).

69. Suppose that the directional derivatives of $f(x, y)$ are known at a given point in two nonparallel directions given by unit vectors $\mathbf{u}$ and $\mathbf{v}$. Is it possible to find ∇f at this point? If so, how would you do it?

70. Show that if $z = f(x, y)$ is differentiable at $\mathbf{x}_0 = \langle x_0, y_0 \rangle$, then

$$\lim_{\mathbf{x} \to \mathbf{x}_0} \frac{f(\mathbf{x}) - f(\mathbf{x}_0) - \nabla f(\mathbf{x}_0) \cdot (\mathbf{x} - \mathbf{x}_0)}{|\mathbf{x} - \mathbf{x}_0|} = 0$$

[*Hint:* Use Definition 14.4.7 directly.]

14.7 Maximum and Minimum Values

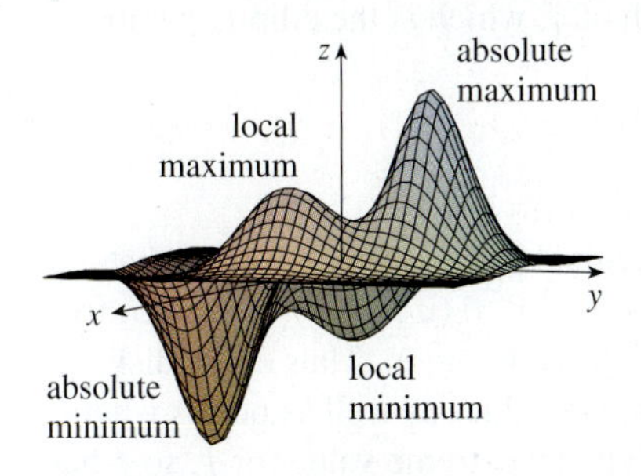

FIGURE 1

As we saw in Chapter 3, one of the main uses of ordinary derivatives is in finding maximum and minimum values (extreme values). In this section we see how to use partial derivatives to locate maxima and minima of functions of two variables. In particular, in Example 6 we will see how to maximize the volume of a box without a lid if we have a fixed amount of cardboard to work with.

Look at the hills and valleys in the graph of f shown in Figure 1. There are two points (a, b) where f has a *local maximum*, that is, where $f(a, b)$ is larger than nearby values of $f(x, y)$. The larger of these two values is the *absolute maximum*. Likewise, f has two *local minima*, where $f(a, b)$ is smaller than nearby values. The smaller of these two values is the *absolute minimum*.

1 Definition A function of two variables has a **local maximum** at (a, b) if $f(x, y) \leqslant f(a, b)$ when (x, y) is near (a, b). [This means that $f(x, y) \leqslant f(a, b)$ for all points (x, y) in some disk with center (a, b).] The number $f(a, b)$ is called a **local maximum value**. If $f(x, y) \geqslant f(a, b)$ when (x, y) is near (a, b), then f has a **local minimum** at (a, b) and $f(a, b)$ is a **local minimum value**.

If the inequalities in Definition 1 hold for *all* points (x, y) in the domain of f, then f has an **absolute maximum** (or **absolute minimum**) at (a, b).

Notice that the conclusion of Theorem 2 can be stated in the notation of gradient vectors as $\nabla f(a, b) = \mathbf{0}$.

2 Theorem If f has a local maximum or minimum at (a, b) and the first-order partial derivatives of f exist there, then $f_x(a, b) = 0$ and $f_y(a, b) = 0$.

PROOF Let $g(x) = f(x, b)$. If f has a local maximum (or minimum) at (a, b), then g has a local maximum (or minimum) at a, so $g'(a) = 0$ by Fermat's Theorem (see Theorem 3.1.4). But $g'(a) = f_x(a, b)$ (see Equation 14.3.1) and so $f_x(a, b) = 0$. Similarly, by applying Fermat's Theorem to the function $G(y) = f(a, y)$, we obtain $f_y(a, b) = 0$. ■

If we put $f_x(a, b) = 0$ and $f_y(a, b) = 0$ in the equation of a tangent plane (Equation 14.4.2), we get $z = z_0$. Thus the geometric interpretation of Theorem 2 is that if the graph of f has a tangent plane at a local maximum or minimum, then the tangent plane must be horizontal.

A point (a, b) is called a **critical point** (or *stationary point*) of f if $f_x(a, b) = 0$ and $f_y(a, b) = 0$, or if one of these partial derivatives does not exist. Theorem 2 says that if f has a local maximum or minimum at (a, b), then (a, b) is a critical point of f. However, as in single-variable calculus, not all critical points give rise to maxima or minima. At a critical point, a function could have a local maximum or a local minimum or neither.

EXAMPLE 1 Let $f(x, y) = x^2 + y^2 - 2x - 6y + 14$. Then

$$f_x(x, y) = 2x - 2 \qquad f_y(x, y) = 2y - 6$$

These partial derivatives are equal to 0 when $x = 1$ and $y = 3$, so the only critical point is $(1, 3)$. By completing the square, we find that

$$f(x, y) = 4 + (x - 1)^2 + (y - 3)^2$$

Since $(x - 1)^2 \geqslant 0$ and $(y - 3)^2 \geqslant 0$, we have $f(x, y) \geqslant 4$ for all values of x and y. Therefore $f(1, 3) = 4$ is a local minimum, and in fact it is the absolute minimum of f. This can be confirmed geometrically from the graph of f, which is the elliptic paraboloid with vertex $(1, 3, 4)$ shown in Figure 2. ■

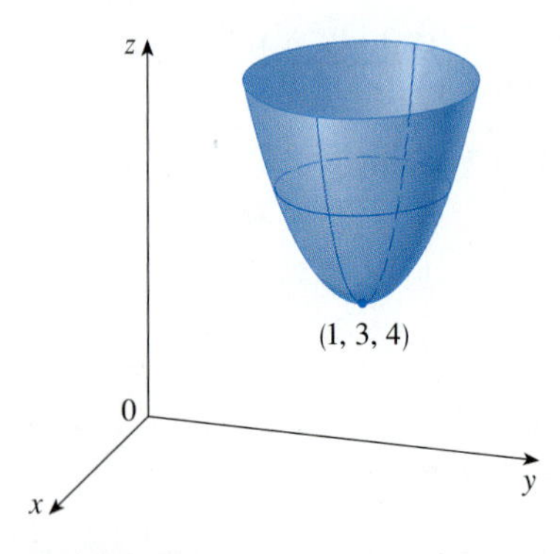

FIGURE 2
$z = x^2 + y^2 - 2x - 6y + 14$

EXAMPLE 2 Find the extreme values of $f(x, y) = y^2 - x^2$.

SOLUTION Since $f_x = -2x$ and $f_y = 2y$, the only critical point is $(0, 0)$. Notice that for points on the x-axis we have $y = 0$, so $f(x, y) = -x^2 < 0$ (if $x \neq 0$). However, for points on the y-axis we have $x = 0$, so $f(x, y) = y^2 > 0$ (if $y \neq 0$). Thus every disk with center $(0, 0)$ contains points where f takes positive values as well as points where f takes negative values. Therefore $f(0, 0) = 0$ can't be an extreme value for f, so f has no extreme value. ■

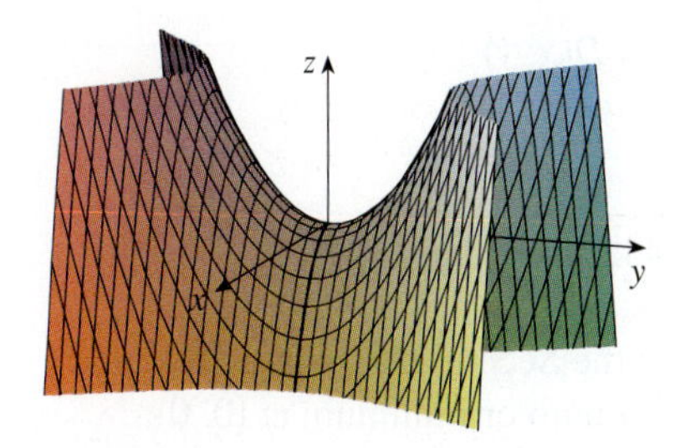

FIGURE 3
$z = y^2 - x^2$

Example 2 illustrates the fact that a function need not have a maximum or minimum value at a critical point. Figure 3 shows how this is possible. The graph of f is the hyperbolic paraboloid $z = y^2 - x^2$, which has a horizontal tangent plane ($z = 0$) at the origin. You can see that $f(0, 0) = 0$ is a maximum in the direction of the x-axis but a minimum in the direction of the y-axis. Near the origin the graph has the shape of a saddle and so $(0, 0)$ is called a *saddle point* of f.

A mountain pass also has the shape of a saddle. As the photograph of the geological formation illustrates, for people hiking in one direction the saddle point is the lowest point on their route, while for those traveling in a different direction the saddle point is the highest point.

We need to be able to determine whether or not a function has an extreme value at a critical point. The following test, which is proved at the end of this section, is analogous to the Second Derivative Test for functions of one variable.

3 Second Derivatives Test Suppose the second partial derivatives of f are continuous on a disk with center (a, b), and suppose that $f_x(a, b) = 0$ and $f_y(a, b) = 0$ [that is, (a, b) is a critical point of f]. Let

$$D = D(a, b) = f_{xx}(a, b)f_{yy}(a, b) - [f_{xy}(a, b)]^2$$

(a) If $D > 0$ and $f_{xx}(a, b) > 0$, then $f(a, b)$ is a local minimum.

(b) If $D > 0$ and $f_{xx}(a, b) < 0$, then $f(a, b)$ is a local maximum.

(c) If $D < 0$, then $f(a, b)$ is not a local maximum or minimum.

NOTE 1 In case (c) the point (a, b) is called a **saddle point** of f and the graph of f crosses its tangent plane at (a, b).

NOTE 2 If $D = 0$, the test gives no information: f could have a local maximum or local minimum at (a, b), or (a, b) could be a saddle point of f.

NOTE 3 To remember the formula for D, it's helpful to write it as a determinant:

$$D = \begin{vmatrix} f_{xx} & f_{xy} \\ f_{yx} & f_{yy} \end{vmatrix} = f_{xx}f_{yy} - (f_{xy})^2$$

EXAMPLE 3 Find the local maximum and minimum values and saddle points of $f(x, y) = x^4 + y^4 - 4xy + 1$.

SOLUTION We first locate the critical points:

$$f_x = 4x^3 - 4y \qquad f_y = 4y^3 - 4x$$

Setting these partial derivatives equal to 0, we obtain the equations

$$x^3 - y = 0 \quad \text{and} \quad y^3 - x = 0$$

To solve these equations we substitute $y = x^3$ from the first equation into the second one. This gives

$$0 = x^9 - x = x(x^8 - 1) = x(x^4 - 1)(x^4 + 1) = x(x^2 - 1)(x^2 + 1)(x^4 + 1)$$

so there are three real roots: $x = 0, 1, -1$. The three critical points are $(0, 0)$, $(1, 1)$, and $(-1, -1)$.

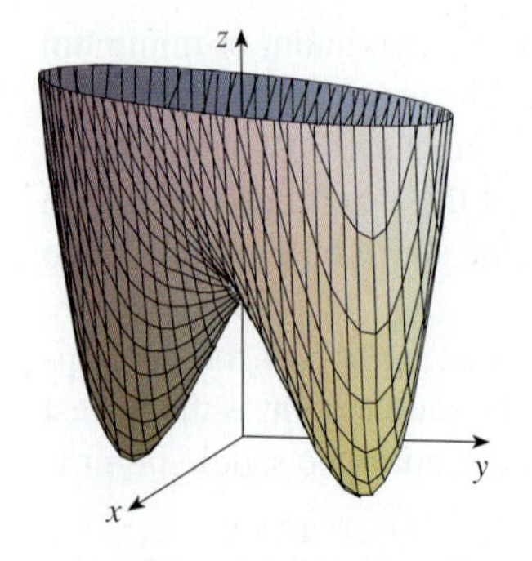

FIGURE 4
$z = x^4 + y^4 - 4xy + 1$

Next we calculate the second partial derivatives and $D(x, y)$:

$$f_{xx} = 12x^2 \qquad f_{xy} = -4 \qquad f_{yy} = 12y^2$$

$$D(x, y) = f_{xx} f_{yy} - (f_{xy})^2 = 144x^2y^2 - 16$$

Since $D(0, 0) = -16 < 0$, it follows from case (c) of the Second Derivatives Test that the origin is a saddle point; that is, f has no local maximum or minimum at $(0, 0)$. Since $D(1, 1) = 128 > 0$ and $f_{xx}(1, 1) = 12 > 0$, we see from case (a) of the test that $f(1, 1) = -1$ is a local minimum. Similarly, we have $D(-1, -1) = 128 > 0$ and $f_{xx}(-1, -1) = 12 > 0$, so $f(-1, -1) = -1$ is also a local minimum.

The graph of f is shown in Figure 4. ■

A contour map of the function f in Example 3 is shown in Figure 5. The level curves near $(1, 1)$ and $(-1, -1)$ are oval in shape and indicate that as we move away from $(1, 1)$ or $(-1, -1)$ in any direction the values of f are increasing. The level curves near $(0, 0)$, on the other hand, resemble hyperbolas. They reveal that as we move away from the origin (where the value of f is 1), the values of f decrease in some directions but increase in other directions. Thus the contour map suggests the presence of the minima and saddle point that we found in Example 3.

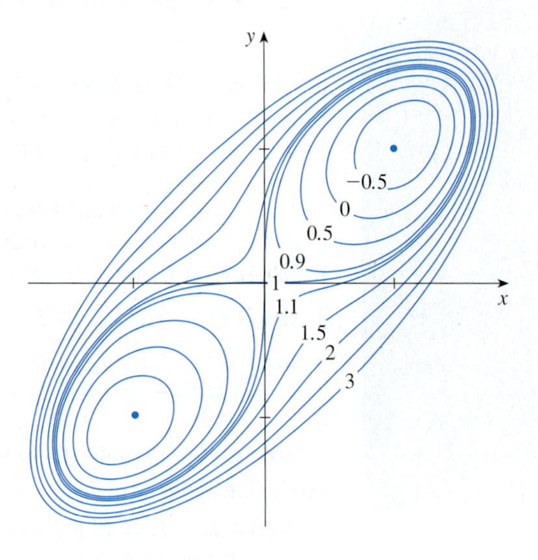

FIGURE 5

TEC In Module 14.7 you can use contour maps to estimate the locations of critical points.

EXAMPLE 4 Find and classify the critical points of the function

$$f(x, y) = 10x^2y - 5x^2 - 4y^2 - x^4 - 2y^4$$

Also find the highest point on the graph of f.

SOLUTION The first-order partial derivatives are

$$f_x = 20xy - 10x - 4x^3 \qquad f_y = 10x^2 - 8y - 8y^3$$

So to find the critical points we need to solve the equations

$$\boxed{4} \qquad 2x(10y - 5 - 2x^2) = 0$$

$$\boxed{5} \qquad 5x^2 - 4y - 4y^3 = 0$$

From Equation 4 we see that either

$$x = 0 \qquad \text{or} \qquad 10y - 5 - 2x^2 = 0$$

In the first case ($x = 0$), Equation 5 becomes $-4y(1 + y^2) = 0$, so $y = 0$ and we have the critical point $(0, 0)$.

In the second case ($10y - 5 - 2x^2 = 0$), we get

$$\boxed{6} \qquad x^2 = 5y - 2.5$$

and, putting this in Equation 5, we have $25y - 12.5 - 4y - 4y^3 = 0$. So we have to solve the cubic equation

$$\boxed{7} \qquad 4y^3 - 21y + 12.5 = 0$$

Using a graphing calculator or computer to graph the function

$$g(y) = 4y^3 - 21y + 12.5$$

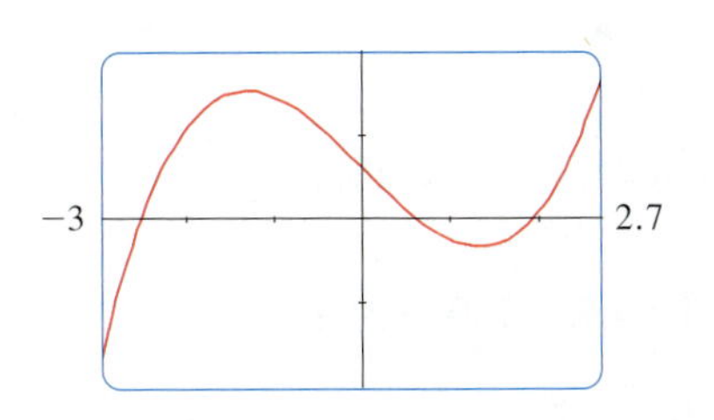

FIGURE 6

as in Figure 6, we see that Equation 7 has three real roots. By zooming in, we can find the roots to four decimal places:

$$y \approx -2.5452 \qquad y \approx 0.6468 \qquad y \approx 1.8984$$

(Alternatively, we could have used Newton's method or solved numerically using a calculator or computer to locate these roots.) From Equation 6, the corresponding x-values are given by

$$x = \pm\sqrt{5y - 2.5}$$

If $y \approx -2.5452$, then x has no corresponding real values. If $y \approx 0.6468$, then $x \approx \pm 0.8567$. If $y \approx 1.8984$, then $x \approx \pm 2.6442$. So we have a total of five critical points, which are analyzed in the following chart. All quantities are rounded to two decimal places.

Critical point	Value of f	f_{xx}	D	Conclusion
$(0, 0)$	0.00	−10.00	80.00	local maximum
$(\pm 2.64, 1.90)$	8.50	−55.93	2488.72	local maximum
$(\pm 0.86, 0.65)$	−1.48	−5.87	−187.64	saddle point

Figures 7 and 8 give two views of the graph of f and we see that the surface opens downward. [This can also be seen from the expression for $f(x, y)$: the dominant terms are $-x^4 - 2y^4$ when $|x|$ and $|y|$ are large.] Comparing the values of f at its local maximum points, we see that the absolute maximum value of f is $f(\pm 2.64, 1.90) \approx 8.50$. In other words, the highest points on the graph of f are $(\pm 2.64, 1.90, 8.50)$.

TEC Visual 14.7 shows several families of surfaces. The surface in Figures 7 and 8 is a member of one of these families.

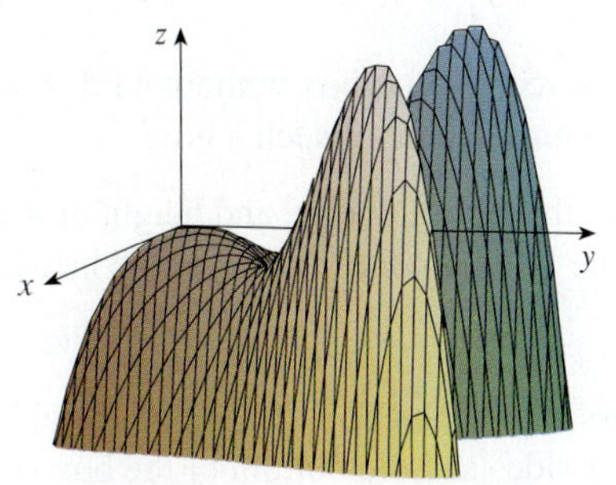

FIGURE 7

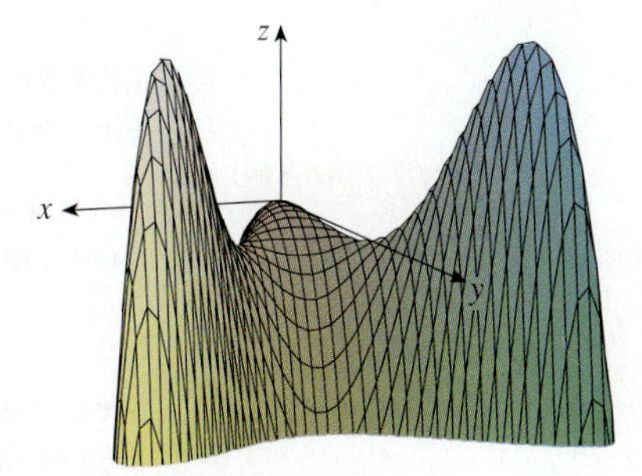

FIGURE 8

The five critical points of the function f in Example 4 are shown in red in the contour map of f in Figure 9.

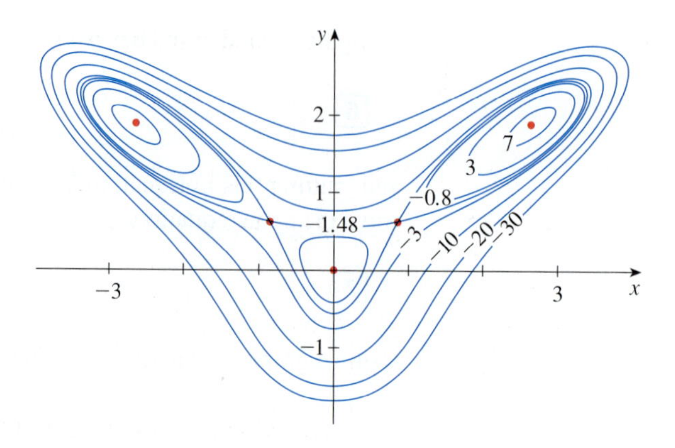

FIGURE 9

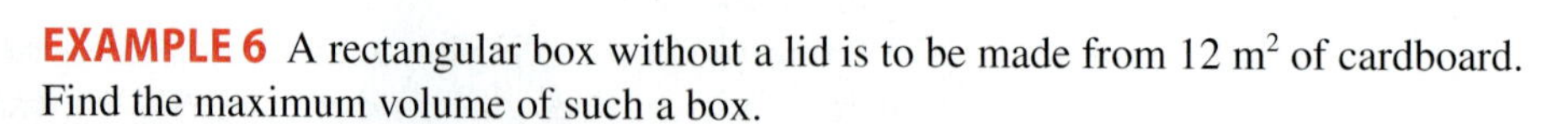

EXAMPLE 5 Find the shortest distance from the point $(1, 0, -2)$ to the plane $x + 2y + z = 4$.

SOLUTION The distance from any point (x, y, z) to the point $(1, 0, -2)$ is

$$d = \sqrt{(x-1)^2 + y^2 + (z+2)^2}$$

but if (x, y, z) lies on the plane $x + 2y + z = 4$, then $z = 4 - x - 2y$ and so we have $d = \sqrt{(x-1)^2 + y^2 + (6-x-2y)^2}$. We can minimize d by minimizing the simpler expression

$$d^2 = f(x, y) = (x-1)^2 + y^2 + (6-x-2y)^2$$

By solving the equations

$$f_x = 2(x-1) - 2(6-x-2y) = 4x + 4y - 14 = 0$$

$$f_y = 2y - 4(6-x-2y) = 4x + 10y - 24 = 0$$

we find that the only critical point is $\left(\frac{11}{6}, \frac{5}{3}\right)$. Since $f_{xx} = 4$, $f_{xy} = 4$, and $f_{yy} = 10$, we have $D(x, y) = f_{xx}f_{yy} - (f_{xy})^2 = 24 > 0$ and $f_{xx} > 0$, so by the Second Derivatives Test f has a local minimum at $\left(\frac{11}{6}, \frac{5}{3}\right)$. Intuitively, we can see that this local minimum is actually an absolute minimum because there must be a point on the given plane that is closest to $(1, 0, -2)$. If $x = \frac{11}{6}$ and $y = \frac{5}{3}$, then

Example 5 could also be solved using vectors. Compare with the methods of Section 12.5.

$$d = \sqrt{(x-1)^2 + y^2 + (6-x-2y)^2} = \sqrt{\left(\tfrac{5}{6}\right)^2 + \left(\tfrac{5}{3}\right)^2 + \left(\tfrac{5}{6}\right)^2} = \tfrac{5}{6}\sqrt{6}$$

The shortest distance from $(1, 0, -2)$ to the plane $x + 2y + z = 4$ is $\frac{5}{6}\sqrt{6}$. ■

EXAMPLE 6 A rectangular box without a lid is to be made from 12 m² of cardboard. Find the maximum volume of such a box.

SOLUTION Let the length, width, and height of the box (in meters) be x, y, and z, as shown in Figure 10. Then the volume of the box is

$$V = xyz$$

We can express V as a function of just two variables x and y by using the fact that the area of the four sides and the bottom of the box is

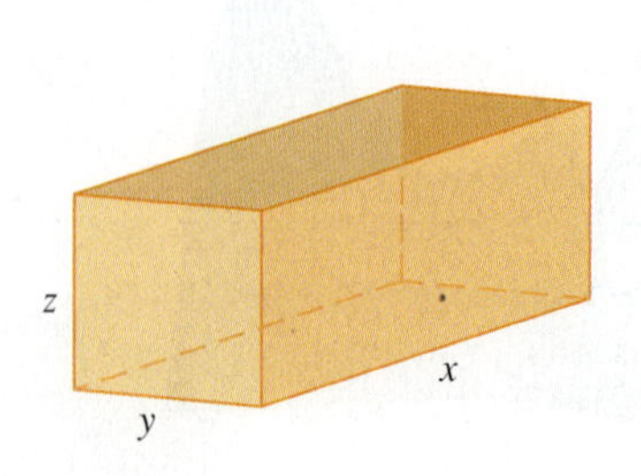

FIGURE 10

$$2xz + 2yz + xy = 12$$

Solving this equation for z, we get $z = (12 - xy)/[2(x + y)]$, so the expression for V becomes

$$V = xy\,\frac{12 - xy}{2(x + y)} = \frac{12xy - x^2y^2}{2(x + y)}$$

We compute the partial derivatives:

$$\frac{\partial V}{\partial x} = \frac{y^2(12 - 2xy - x^2)}{2(x + y)^2} \qquad \frac{\partial V}{\partial y} = \frac{x^2(12 - 2xy - y^2)}{2(x + y)^2}$$

If V is a maximum, then $\partial V/\partial x = \partial V/\partial y = 0$, but $x = 0$ or $y = 0$ gives $V = 0$, so we must solve the equations

$$12 - 2xy - x^2 = 0 \qquad 12 - 2xy - y^2 = 0$$

These imply that $x^2 = y^2$ and so $x = y$. (Note that x and y must both be positive in this problem.) If we put $x = y$ in either equation we get $12 - 3x^2 = 0$, which gives $x = 2$, $y = 2$, and $z = (12 - 2 \cdot 2)/[2(2 + 2)] = 1$.

We could use the Second Derivatives Test to show that this gives a local maximum of V, or we could simply argue from the physical nature of this problem that there must be an absolute maximum volume, which has to occur at a critical point of V, so it must occur when $x = 2$, $y = 2$, $z = 1$. Then $V = 2 \cdot 2 \cdot 1 = 4$, so the maximum volume of the box is 4 m^3. ■

Absolute Maximum and Minimum Values

For a function f of one variable, the Extreme Value Theorem says that if f is continuous on a closed interval $[a, b]$, then f has an absolute minimum value and an absolute maximum value. According to the Closed Interval Method in Section 3.1, we found these by evaluating f not only at the critical numbers but also at the endpoints a and b.

There is a similar situation for functions of two variables. Just as a closed interval contains its endpoints, a **closed set** in $\mathbb{R}^2$ is one that contains all its boundary points. [A boundary point of D is a point (a, b) such that every disk with center (a, b) contains points in D and also points not in D.] For instance, the disk

$$D = \{(x, y) \mid x^2 + y^2 \leqslant 1\}$$

which consists of all points on or inside the circle $x^2 + y^2 = 1$, is a closed set because it contains all of its boundary points (which are the points on the circle $x^2 + y^2 = 1$). But if even one point on the boundary curve were omitted, the set would not be closed. (See Figure 11.)

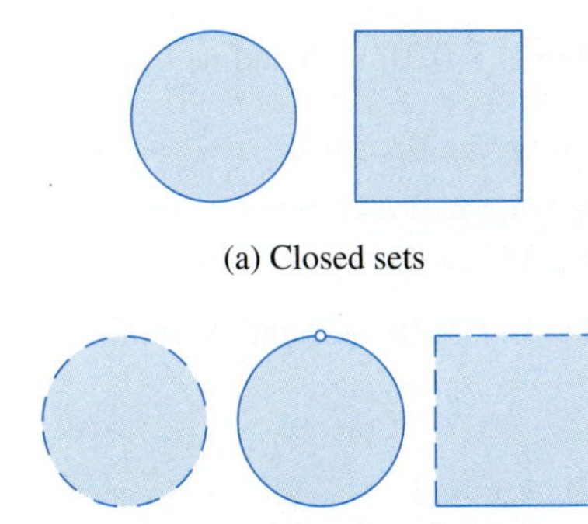

FIGURE 11

A **bounded set** in $\mathbb{R}^2$ is one that is contained within some disk. In other words, it is finite in extent. Then, in terms of closed and bounded sets, we can state the following counterpart of the Extreme Value Theorem in two dimensions.

> **8 Extreme Value Theorem for Functions of Two Variables** If f is continuous on a closed, bounded set D in $\mathbb{R}^2$, then f attains an absolute maximum value $f(x_1, y_1)$ and an absolute minimum value $f(x_2, y_2)$ at some points (x_1, y_1) and (x_2, y_2) in D.

To find the extreme values guaranteed by Theorem 8, we note that, by Theorem 2, if f has an extreme value at (x_1, y_1), then (x_1, y_1) is either a critical point of f or a boundary point of D. Thus we have the following extension of the Closed Interval Method.

9 To find the absolute maximum and minimum values of a continuous function f on a closed, bounded set D:

1. Find the values of f at the critical points of f in D.
2. Find the extreme values of f on the boundary of D.
3. The largest of the values from steps 1 and 2 is the absolute maximum value; the smallest of these values is the absolute minimum value.

EXAMPLE 7 Find the absolute maximum and minimum values of the function $f(x, y) = x^2 - 2xy + 2y$ on the rectangle $D = \{(x, y) \mid 0 \leqslant x \leqslant 3, 0 \leqslant y \leqslant 2\}$.

SOLUTION Since f is a polynomial, it is continuous on the closed, bounded rectangle D, so Theorem 8 tells us there is both an absolute maximum and an absolute minimum. According to step 1 in (9), we first find the critical points. These occur when

$$f_x = 2x - 2y = 0 \qquad f_y = -2x + 2 = 0$$

so the only critical point is $(1, 1)$, and the value of f there is $f(1, 1) = 1$.

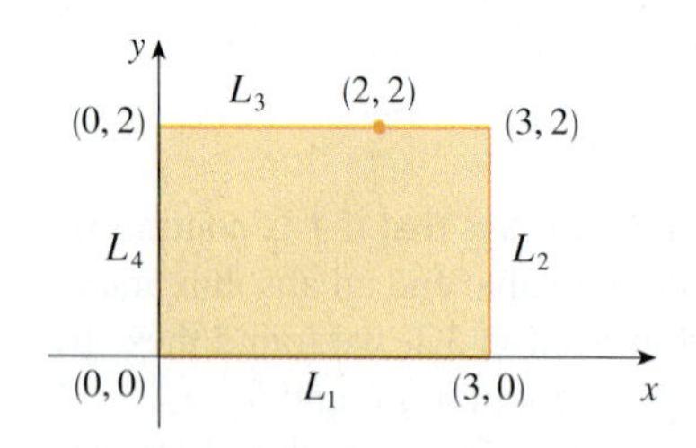

FIGURE 12

In step 2 we look at the values of f on the boundary of D, which consists of the four line segments L_1, L_2, L_3, L_4 shown in Figure 12. On L_1 we have $y = 0$ and

$$f(x, 0) = x^2 \qquad 0 \leqslant x \leqslant 3$$

This is an increasing function of x, so its minimum value is $f(0, 0) = 0$ and its maximum value is $f(3, 0) = 9$. On L_2 we have $x = 3$ and

$$f(3, y) = 9 - 4y \qquad 0 \leqslant y \leqslant 2$$

This is a decreasing function of y, so its maximum value is $f(3, 0) = 9$ and its minimum value is $f(3, 2) = 1$. On L_3 we have $y = 2$ and

$$f(x, 2) = x^2 - 4x + 4 \qquad 0 \leqslant x \leqslant 3$$

By the methods of Chapter 3, or simply by observing that $f(x, 2) = (x - 2)^2$, we see that the minimum value of this function is $f(2, 2) = 0$ and the maximum value is $f(0, 2) = 4$. Finally, on L_4 we have $x = 0$ and

$$f(0, y) = 2y \qquad 0 \leqslant y \leqslant 2$$

with maximum value $f(0, 2) = 4$ and minimum value $f(0, 0) = 0$. Thus, on the boundary, the minimum value of f is 0 and the maximum is 9.

In step 3 we compare these values with the value $f(1, 1) = 1$ at the critical point and conclude that the absolute maximum value of f on D is $f(3, 0) = 9$ and the absolute minimum value is $f(0, 0) = f(2, 2) = 0$. Figure 13 shows the graph of f. ■

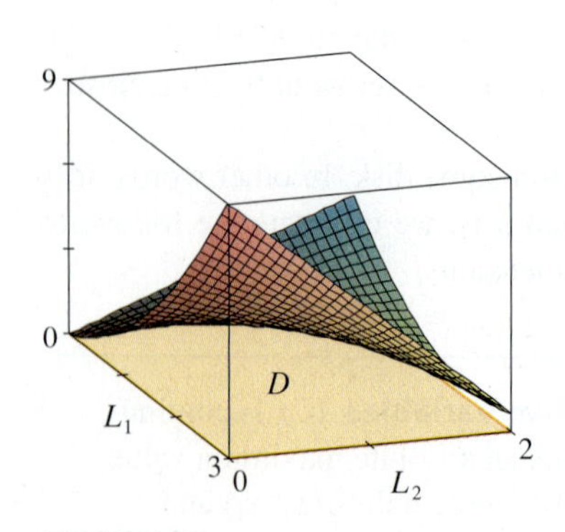

FIGURE 13
$f(x, y) = x^2 - 2xy + 2y$

We close this section by giving a proof of the first part of the Second Derivatives Test. Part (b) has a similar proof.

PROOF OF THEOREM 3, PART (a) We compute the second-order directional derivative of f in the direction of $\mathbf{u} = \langle h, k \rangle$. The first-order derivative is given by Theorem 14.6.3:

$$D_{\mathbf{u}} f = f_x h + f_y k$$

Applying this theorem a second time, we have

$$\begin{aligned} D_{\mathbf{u}}^2 f = D_{\mathbf{u}}(D_{\mathbf{u}} f) &= \frac{\partial}{\partial x}(D_{\mathbf{u}} f)h + \frac{\partial}{\partial y}(D_{\mathbf{u}} f)k \\ &= (f_{xx}h + f_{yx}k)h + (f_{xy}h + f_{yy}k)k \\ &= f_{xx}h^2 + 2f_{xy}hk + f_{yy}k^2 \qquad \text{(by Clairaut's Theorem)} \end{aligned}$$

If we complete the square in this expression, we obtain

$$\boxed{10} \qquad D_{\mathbf{u}}^2 f = f_{xx}\left(h + \frac{f_{xy}}{f_{xx}}k\right)^2 + \frac{k^2}{f_{xx}}(f_{xx}f_{yy} - f_{xy}^2)$$

We are given that $f_{xx}(a, b) > 0$ and $D(a, b) > 0$. But f_{xx} and $D = f_{xx}f_{yy} - f_{xy}^2$ are continuous functions, so there is a disk B with center (a, b) and radius $\delta > 0$ such that $f_{xx}(x, y) > 0$ and $D(x, y) > 0$ whenever (x, y) is in B. Therefore, by looking at Equation 10, we see that $D_{\mathbf{u}}^2 f(x, y) > 0$ whenever (x, y) is in B. This means that if C is the curve obtained by intersecting the graph of f with the vertical plane through $P(a, b, f(a, b))$ in the direction of $\mathbf{u}$, then C is concave upward on an interval of length 2δ. This is true in the direction of every vector $\mathbf{u}$, so if we restrict (x, y) to lie in B, the graph of f lies above its horizontal tangent plane at P. Thus $f(x, y) \geqslant f(a, b)$ whenever (x, y) is in B. This shows that $f(a, b)$ is a local minimum. ■

14.7 EXERCISES

1. Suppose $(1, 1)$ is a critical point of a function f with continuous second derivatives. In each case, what can you say about f?

(a) $f_{xx}(1, 1) = 4, \quad f_{xy}(1, 1) = 1, \quad f_{yy}(1, 1) = 2$

(b) $f_{xx}(1, 1) = 4, \quad f_{xy}(1, 1) = 3, \quad f_{yy}(1, 1) = 2$

2. Suppose $(0, 2)$ is a critical point of a function g with continuous second derivatives. In each case, what can you say about g?

(a) $g_{xx}(0, 2) = -1, \quad g_{xy}(0, 2) = 6, \quad g_{yy}(0, 2) = 1$

(b) $g_{xx}(0, 2) = -1, \quad g_{xy}(0, 2) = 2, \quad g_{yy}(0, 2) = -8$

(c) $g_{xx}(0, 2) = 4, \quad g_{xy}(0, 2) = 6, \quad g_{yy}(0, 2) = 9$

3–4 Use the level curves in the figure to predict the location of the critical points of f and whether f has a saddle point or a local maximum or minimum at each critical point. Explain your reasoning. Then use the Second Derivatives Test to confirm your predictions.

3. $f(x, y) = 4 + x^3 + y^3 - 3xy$

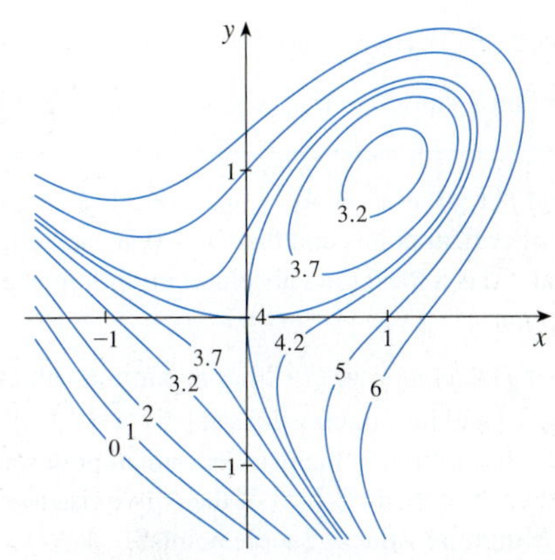

4. $f(x, y) = 3x - x^3 - 2y^2 + y^4$

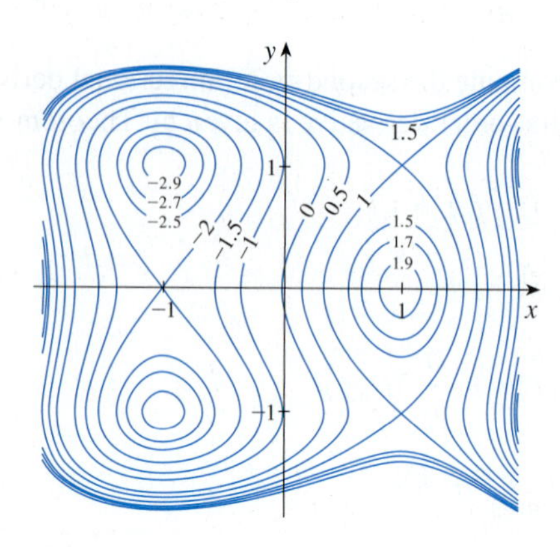

5–20 Find the local maximum and minimum values and saddle point(s) of the function. If you have three-dimensional graphing software, graph the function with a domain and viewpoint that reveal all the important aspects of the function.

5. $f(x, y) = x^2 + xy + y^2 + y$

6. $f(x, y) = xy - 2x - 2y - x^2 - y^2$

7. $f(x, y) = (x - y)(1 - xy)$

8. $f(x, y) = y(e^x - 1)$

9. $f(x, y) = x^2 + y^4 + 2xy$

10. $f(x, y) = 2 - x^4 + 2x^2 - y^2$

11. $f(x, y) = x^3 - 3x + 3xy^2$

12. $f(x, y) = x^3 + y^3 - 3x^2 - 3y^2 - 9x$

13. $f(x, y) = x^4 - 2x^2 + y^3 - 3y$

14. $f(x, y) = y \cos x$

15. $f(x, y) = e^x \cos y$

16. $f(x, y) = xye^{-(x^2+y^2)/2}$

17. $f(x, y) = xy + e^{-xy}$

18. $f(x, y) = (x^2 + y^2)e^{-x}$

19. $f(x, y) = y^2 - 2y \cos x, \quad -1 \leqslant x \leqslant 7$

20. $f(x, y) = \sin x \sin y, \quad -\pi < x < \pi, \quad -\pi < y < \pi$

21. Show that $f(x, y) = x^2 + 4y^2 - 4xy + 2$ has an infinite number of critical points and that $D = 0$ at each one. Then show that f has a local (and absolute) minimum at each critical point.

22. Show that $f(x, y) = x^2ye^{-x^2-y^2}$ has maximum values at $(\pm 1, 1/\sqrt{2})$ and minimum values at $(\pm 1, -1/\sqrt{2})$. Show also that f has infinitely many other critical points and $D = 0$ at each of them. Which of them give rise to maximum values? Minimum values? Saddle points?

23–26 Use a graph or level curves or both to estimate the local maximum and minimum values and saddle point(s) of the function. Then use calculus to find these values precisely.

23. $f(x, y) = x^2 + y^2 + x^{-2}y^{-2}$

24. $f(x, y) = (x - y)e^{-x^2-y^2}$

25. $f(x, y) = \sin x + \sin y + \sin(x + y)$,
$0 \leqslant x \leqslant 2\pi, \ 0 \leqslant y \leqslant 2\pi$

26. $f(x, y) = \sin x + \sin y + \cos(x + y)$,
$0 \leqslant x \leqslant \pi/4, \ 0 \leqslant y \leqslant \pi/4$

27–30 Use a graphing device as in Example 4 (or Newton's method or solve numerically using a calculator or computer) to find the critical points of f correct to three decimal places. Then classify the critical points and find the highest or lowest points on the graph, if any.

27. $f(x, y) = x^4 + y^4 - 4x^2y + 2y$

28. $f(x, y) = y^6 - 2y^4 + x^2 - y^2 + y$

29. $f(x, y) = x^4 + y^3 - 3x^2 + y^2 + x - 2y + 1$

30. $f(x, y) = 20e^{-x^2-y^2} \sin 3x \cos 3y, \quad |x| \leqslant 1, \quad |y| \leqslant 1$

31–38 Find the absolute maximum and minimum values of f on the set D.

31. $f(x, y) = x^2 + y^2 - 2x$, D is the closed triangular region with vertices $(2, 0)$, $(0, 2)$, and $(0, -2)$

32. $f(x, y) = x + y - xy$, D is the closed triangular region with vertices $(0, 0)$, $(0, 2)$, and $(4, 0)$

33. $f(x, y) = x^2 + y^2 + x^2y + 4$,
$D = \{(x, y) \mid |x| \leqslant 1, |y| \leqslant 1\}$

34. $f(x, y) = x^2 + xy + y^2 - 6y$,
$D = \{(x, y) \mid -3 \leqslant x \leqslant 3, 0 \leqslant y \leqslant 5\}$

35. $f(x, y) = x^2 + 2y^2 - 2x - 4y + 1$,
$D = \{(x, y) \mid 0 \leqslant x \leqslant 2, 0 \leqslant y \leqslant 3\}$

36. $f(x, y) = xy^2, \quad D = \{(x, y) \mid x \geqslant 0, y \geqslant 0, x^2 + y^2 \leqslant 3\}$

37. $f(x, y) = 2x^3 + y^4, \quad D = \{(x, y) \mid x^2 + y^2 \leqslant 1\}$

38. $f(x, y) = x^3 - 3x - y^3 + 12y$, D is the quadrilateral whose vertices are $(-2, 3)$, $(2, 3)$, $(2, 2)$, and $(-2, -2)$

39. For functions of one variable it is impossible for a continuous function to have two local maxima and no local minimum. But for functions of two variables such functions exist. Show that the function

$$f(x, y) = -(x^2 - 1)^2 - (x^2y - x - 1)^2$$

has only two critical points, but has local maxima at both of them. Then use a computer to produce a graph with a carefully chosen domain and viewpoint to see how this is possible.

40. If a function of one variable is continuous on an interval and has only one critical number, then a local maximum has to be an absolute maximum. But this is not true for functions of two variables. Show that the function

$$f(x, y) = 3xe^y - x^3 - e^{3y}$$

has exactly one critical point, and that f has a local maximum there that is not an absolute maximum. Then use a computer to produce a graph with a carefully chosen domain and viewpoint to see how this is possible.

41. Find the shortest distance from the point $(2, 0, -3)$ to the plane $x + y + z = 1$.

42. Find the point on the plane $x - 2y + 3z = 6$ that is closest to the point $(0, 1, 1)$.

43. Find the points on the cone $z^2 = x^2 + y^2$ that are closest to the point $(4, 2, 0)$.

44. Find the points on the surface $y^2 = 9 + xz$ that are closest to the origin.

45. Find three positive numbers whose sum is 100 and whose product is a maximum.

46. Find three positive numbers whose sum is 12 and the sum of whose squares is as small as possible.

47. Find the maximum volume of a rectangular box that is inscribed in a sphere of radius r.

48. Find the dimensions of the box with volume 1000 cm^3 that has minimal surface area.

49. Find the volume of the largest rectangular box in the first octant with three faces in the coordinate planes and one vertex in the plane $x + 2y + 3z = 6$.

50. Find the dimensions of the rectangular box with largest volume if the total surface area is given as 64 cm^2.

51. Find the dimensions of a rectangular box of maximum volume such that the sum of the lengths of its 12 edges is a constant c.

52. The base of an aquarium with given volume V is made of slate and the sides are made of glass. If slate costs five times as much (per unit area) as glass, find the dimensions of the aquarium that minimize the cost of the materials.

53. A cardboard box without a lid is to have a volume of 32,000 cm^3. Find the dimensions that minimize the amount of cardboard used.

54. A rectangular building is being designed to minimize heat loss. The east and west walls lose heat at a rate of 10 units/m^2 per day, the north and south walls at a rate of 8 units/m^2 per day, the floor at a rate of 1 unit/m^2 per day, and the roof at a rate of 5 units/m^2 per day. Each wall must be at least 30 m long, the height must be at least 4 m, and the volume must be exactly 4000 m^3.

(a) Find and sketch the domain of the heat loss as a function of the lengths of the sides.

(b) Find the dimensions that minimize heat loss. (Check both the critical points and the points on the boundary of the domain.)

(c) Could you design a building with even less heat loss if the restrictions on the lengths of the walls were removed?

55. If the length of the diagonal of a rectangular box must be L, what is the largest possible volume?

56. A model for the yield Y of an agricultural crop as a function of the nitrogen level N and phosphorus level P in the soil (measured in appropriate units) is

$$Y(N, P) = kNPe^{-N-P}$$

where k is a positive constant. What levels of nitrogen and phosphorus result in the best yield?

57. The Shannon index (sometimes called the Shannon-Wiener index or Shannon-Weaver index) is a measure of diversity in an ecosystem. For the case of three species, it is defined as

$$H = -p_1 \ln p_1 - p_2 \ln p_2 - p_3 \ln p_3$$

where p_i is the proportion of species i in the ecosystem.

(a) Express H as a function of two variables using the fact that $p_1 + p_2 + p_3 = 1$.

(b) What is the domain of H?

(c) Find the maximum value of H. For what values of p_1, p_2, p_3 does it occur?

58. Three alleles (alternative versions of a gene) A, B, and O determine the four blood types A (AA or AO), B (BB or BO), O (OO), and AB. The Hardy-Weinberg Law states that the proportion of individuals in a population who carry two different alleles is

$$P = 2pq + 2pr + 2rq$$

where p, q, and r are the proportions of A, B, and O in the population. Use the fact that $p + q + r = 1$ to show that P is at most $\frac{2}{3}$.

59. Suppose that a scientist has reason to believe that two quantities x and y are related linearly, that is, $y = mx + b$, at least approximately, for some values of m and b. The scientist performs an experiment and collects data in the form of points $(x_1, y_1), (x_2, y_2), \ldots, (x_n, y_n)$, and then plots these points. The points don't lie exactly on a straight line, so the scientist wants to find constants m and b so that the line $y = mx + b$ "fits" the points as well as possible (see the figure).

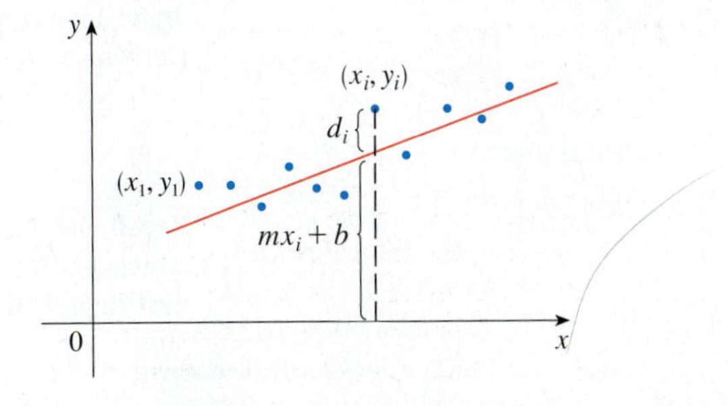

Let $d_i = y_i - (mx_i + b)$ be the vertical deviation of the point (x_i, y_i) from the line. The **method of least squares** determines m and b so as to minimize $\sum_{i=1}^{n} d_i^2$, the sum of the squares of these deviations. Show that, according to this method, the line of best fit is obtained when

$$m \sum_{i=1}^{n} x_i + bn = \sum_{i=1}^{n} y_i$$

and

$$m \sum_{i=1}^{n} x_i^2 + b \sum_{i=1}^{n} x_i = \sum_{i=1}^{n} x_i y_i$$

Thus the line is found by solving these two equations in the two unknowns m and b. (See Section 1.2 for a further discussion and applications of the method of least squares.)

60. Find an equation of the plane that passes through the point (1, 2, 3) and cuts off the smallest volume in the first octant.

APPLIED PROJECT DESIGNING A DUMPSTER

For this project we locate a rectangular trash Dumpster in order to study its shape and construction. We then attempt to determine the dimensions of a container of similar design that minimize construction cost.

1. First locate a trash Dumpster in your area. Carefully study and describe all details of its construction, and determine its volume. Include a sketch of the container.

2. While maintaining the general shape and method of construction, determine the dimensions such a container of the same volume should have in order to minimize the cost of construction. Use the following assumptions in your analysis:

- The sides, back, and front are to be made from 12-gauge (0.1046 inch thick) steel sheets, which cost \$0.70 per square foot (including any required cuts or bends).
- The base is to be made from a 10-gauge (0.1345 inch thick) steel sheet, which costs \$0.90 per square foot.
- Lids cost approximately \$50.00 each, regardless of dimensions.
- Welding costs approximately \$0.18 per foot for material and labor combined.

Give justification of any further assumptions or simplifications made of the details of construction.

3. Describe how any of your assumptions or simplifications may affect the final result.

4. If you were hired as a consultant on this investigation, what would your conclusions be? Would you recommend altering the design of the Dumpster? If so, describe the savings that would result.

DISCOVERY PROJECT QUADRATIC APPROXIMATIONS AND CRITICAL POINTS

The Taylor polynomial approximation to functions of one variable that we discussed in Chapter 11 can be extended to functions of two or more variables. Here we investigate quadratic approximations to functions of two variables and use them to give insight into the Second Derivatives Test for classifying critical points.

In Section 14.4 we discussed the linearization of a function f of two variables at a point (a, b):

$$L(x, y) = f(a, b) + f_x(a, b)(x - a) + f_y(a, b)(y - b)$$

Recall that the graph of L is the tangent plane to the surface $z = f(x, y)$ at $(a, b, f(a, b))$ and the corresponding linear approximation is $f(x, y) \approx L(x, y)$. The linearization L is also called the **first-degree Taylor polynomial** of f at (a, b).

1. If f has continuous second-order partial derivatives at (a, b), then the **second-degree Taylor polynomial** of f at (a, b) is

$$Q(x, y) = f(a, b) + f_x(a, b)(x - a) + f_y(a, b)(y - b)$$
$$+ \tfrac{1}{2}f_{xx}(a, b)(x - a)^2 + f_{xy}(a, b)(x - a)(y - b) + \tfrac{1}{2}f_{yy}(a, b)(y - b)^2$$

and the approximation $f(x, y) \approx Q(x, y)$ is called the **quadratic approximation** to f at (a, b). Verify that Q has the same first- and second-order partial derivatives as f at (a, b).

2. (a) Find the first- and second-degree Taylor polynomials L and Q of $f(x, y) = e^{-x^2-y^2}$ at $(0, 0)$.
(b) Graph f, L, and Q. Comment on how well L and Q approximate f.

3. (a) Find the first- and second-degree Taylor polynomials L and Q for $f(x, y) = xe^y$ at $(1, 0)$.
(b) Compare the values of L, Q, and f at $(0.9, 0.1)$.
(c) Graph f, L, and Q. Comment on how well L and Q approximate f.

4. In this problem we analyze the behavior of the polynomial $f(x, y) = ax^2 + bxy + cy^2$ (without using the Second Derivatives Test) by identifying the graph as a paraboloid.
(a) By completing the square, show that if $a \neq 0$, then

$$f(x, y) = ax^2 + bxy + cy^2 = a\left[\left(x + \frac{b}{2a}y\right)^2 + \left(\frac{4ac - b^2}{4a^2}\right)y^2\right]$$

(b) Let $D = 4ac - b^2$. Show that if $D > 0$ and $a > 0$, then f has a local minimum at $(0, 0)$.
(c) Show that if $D > 0$ and $a < 0$, then f has a local maximum at $(0, 0)$.
(d) Show that if $D < 0$, then $(0, 0)$ is a saddle point.

5. (a) Suppose f is any function with continuous second-order partial derivatives such that $f(0, 0) = 0$ and $(0, 0)$ is a critical point of f. Write an expression for the second-degree Taylor polynomial, Q, of f at $(0, 0)$.
(b) What can you conclude about Q from Problem 4?
(c) In view of the quadratic approximation $f(x, y) \approx Q(x, y)$, what does part (b) suggest about f?

14.8 Lagrange Multipliers

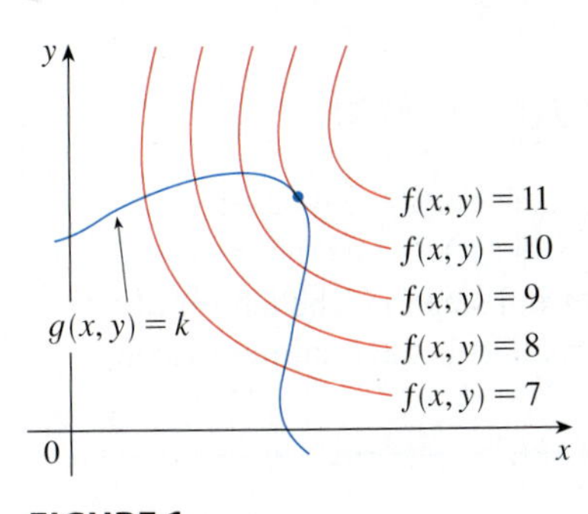

FIGURE 1

TEC Visual 14.8 animates Figure 1 for both level curves and level surfaces.

In Example 14.7.6 we maximized a volume function $V = xyz$ subject to the constraint $2xz + 2yz + xy = 12$, which expressed the side condition that the surface area was 12 m^2. In this section we present Lagrange's method for maximizing or minimizing a general function $f(x, y, z)$ subject to a constraint (or side condition) of the form $g(x, y, z) = k$.

It's easier to explain the geometric basis of Lagrange's method for functions of two variables. So we start by trying to find the extreme values of $f(x, y)$ subject to a constraint of the form $g(x, y) = k$. In other words, we seek the extreme values of $f(x, y)$ when the point (x, y) is restricted to lie on the level curve $g(x, y) = k$. Figure 1 shows this curve together with several level curves of f. These have the equations $f(x, y) = c$, where $c = 7, 8, 9, 10, 11$. To maximize $f(x, y)$ subject to $g(x, y) = k$ is to find the largest value of c such that the level curve $f(x, y) = c$ intersects $g(x, y) = k$. It appears from Figure 1 that this happens when these curves just touch each other, that is, when they have a common tangent line. (Otherwise, the value of c could be increased further.) This means that the normal lines at the point (x_0, y_0) where they touch are identical. So the gradient vectors are parallel; that is, $\nabla f(x_0, y_0) = \lambda \nabla g(x_0, y_0)$ for some scalar λ.

This kind of argument also applies to the problem of finding the extreme values of $f(x, y, z)$ subject to the constraint $g(x, y, z) = k$. Thus the point (x, y, z) is restricted to lie on the level surface S with equation $g(x, y, z) = k$. Instead of the level curves in Figure 1, we consider the level surfaces $f(x, y, z) = c$ and argue that if the maximum value of f is $f(x_0, y_0, z_0) = c$, then the level surface $f(x, y, z) = c$ is tangent to the level surface $g(x, y, z) = k$ and so the corresponding gradient vectors are parallel.

This intuitive argument can be made precise as follows. Suppose that a function f has an extreme value at a point $P(x_0, y_0, z_0)$ on the surface S and let C be a curve with vector equation $\mathbf{r}(t) = \langle x(t), y(t), z(t)\rangle$ that lies on S and passes through P. If t_0 is the parameter value corresponding to the point P, then $\mathbf{r}(t_0) = \langle x_0, y_0, z_0\rangle$. The composite function $h(t) = f(x(t), y(t), z(t))$ represents the values that f takes on the curve C. Since f has an extreme value at (x_0, y_0, z_0), it follows that h has an extreme value at t_0, so $h'(t_0) = 0$. But if f is differentiable, we can use the Chain Rule to write

$$\begin{aligned} 0 &= h'(t_0) \\ &= f_x(x_0, y_0, z_0)x'(t_0) + f_y(x_0, y_0, z_0)y'(t_0) + f_z(x_0, y_0, z_0)z'(t_0) \\ &= \nabla f(x_0, y_0, z_0) \cdot \mathbf{r}'(t_0) \end{aligned}$$

This shows that the gradient vector $\nabla f(x_0, y_0, z_0)$ is orthogonal to the tangent vector $\mathbf{r}'(t_0)$ to every such curve C. But we already know from Section 14.6 that the gradient vector of g, $\nabla g(x_0, y_0, z_0)$, is also orthogonal to $\mathbf{r}'(t_0)$ for every such curve. (See Equation 14.6.18.) This means that the gradient vectors $\nabla f(x_0, y_0, z_0)$ and $\nabla g(x_0, y_0, z_0)$ must be parallel. Therefore, if $\nabla g(x_0, y_0, z_0) \neq \mathbf{0}$, there is a number λ such that

$$\boxed{1} \qquad \nabla f(x_0, y_0, z_0) = \lambda \nabla g(x_0, y_0, z_0)$$

Lagrange multipliers are named after the French-Italian mathematician Joseph-Louis Lagrange (1736–1813). See page 217 for a biographical sketch of Lagrange.

The number λ in Equation 1 is called a **Lagrange multiplier**. The procedure based on Equation 1 is as follows.

Method of Lagrange Multipliers To find the maximum and minimum values of $f(x, y, z)$ subject to the constraint $g(x, y, z) = k$ [assuming that these extreme values exist and $\nabla g \neq \mathbf{0}$ on the surface $g(x, y, z) = k$]:

(a) Find all values of x, y, z, and λ such that

$$\nabla f(x, y, z) = \lambda \nabla g(x, y, z)$$

and

$$g(x, y, z) = k$$

(b) Evaluate f at all the points (x, y, z) that result from step (a). The largest of these values is the maximum value of f; the smallest is the minimum value of f.

In deriving Lagrange's method we assumed that $\nabla g \neq \mathbf{0}$. In each of our examples you can check that $\nabla g \neq \mathbf{0}$ at all points where $g(x, y, z) = k$. See Exercise 25 for what can go wrong if $\nabla g = \mathbf{0}$.

If we write the vector equation $\nabla f = \lambda \nabla g$ in terms of components, then the equations in step (a) become

$$f_x = \lambda g_x \qquad f_y = \lambda g_y \qquad f_z = \lambda g_z \qquad g(x, y, z) = k$$

This is a system of four equations in the four unknowns x, y, z, and λ, but it is not necessary to find explicit values for λ.

For functions of two variables the method of Lagrange multipliers is similar to the method just described. To find the extreme values of $f(x, y)$ subject to the constraint $g(x, y) = k$, we look for values of x, y, and λ such that

$$\nabla f(x, y) = \lambda \nabla g(x, y) \qquad \text{and} \qquad g(x, y) = k$$

This amounts to solving three equations in three unknowns:

$$f_x = \lambda g_x \qquad f_y = \lambda g_y \qquad g(x, y) = k$$

Our first illustration of Lagrange's method is to reconsider the problem given in Example 14.7.6.

EXAMPLE 1 A rectangular box without a lid is to be made from 12 m^2 of cardboard. Find the maximum volume of such a box.

SOLUTION As in Example 14.7.6, we let x, y, and z be the length, width, and height, respectively, of the box in meters. Then we wish to maximize

$$V = xyz$$

subject to the constraint

$$g(x, y, z) = 2xz + 2yz + xy = 12$$

Using the method of Lagrange multipliers, we look for values of x, y, z, and λ such that $\nabla V = \lambda \nabla g$ and $g(x, y, z) = 12$. This gives the equations

$$V_x = \lambda g_x$$
$$V_y = \lambda g_y$$
$$V_z = \lambda g_z$$
$$2xz + 2yz + xy = 12$$

which become

$$\boxed{2} \qquad yz = \lambda(2z + y)$$

$$\boxed{3} \qquad xz = \lambda(2z + x)$$

$$\boxed{4} \qquad xy = \lambda(2x + 2y)$$

$$\boxed{5} \qquad 2xz + 2yz + xy = 12$$

There are no general rules for solving systems of equations. Sometimes some ingenuity is required. In the present example you might notice that if we multiply (2) by x, (3) by y, and (4) by z, then the left sides of these equations will be identical. Doing this, we have

$$\boxed{6} \qquad xyz = \lambda(2xz + xy)$$

$$\boxed{7} \qquad xyz = \lambda(2yz + xy)$$

$$\boxed{8} \qquad xyz = \lambda(2xz + 2yz)$$

Another method for solving the system of equations (2–5) is to solve each of Equations 2, 3, and 4 for λ and then to equate the resulting expressions.

We observe that $\lambda \neq 0$ because $\lambda = 0$ would imply $yz = xz = xy = 0$ from (2), (3), and (4) and this would contradict (5). Therefore, from (6) and (7), we have

$$2xz + xy = 2yz + xy$$

which gives $xz = yz$. But $z \neq 0$ (since $z = 0$ would give $V = 0$), so $x = y$. From (7) and (8) we have

$$2yz + xy = 2xz + 2yz$$

which gives $2xz = xy$ and so (since $x \neq 0$) $y = 2z$. If we now put $x = y = 2z$ in (5), we get

$$4z^2 + 4z^2 + 4z^2 = 12$$

Since x, y, and z are all positive, we therefore have $z = 1$ and so $x = 2$ and $y = 2$. This agrees with our answer in Section 14.7. ■

EXAMPLE 2 Find the extreme values of the function $f(x, y) = x^2 + 2y^2$ on the circle $x^2 + y^2 = 1$.

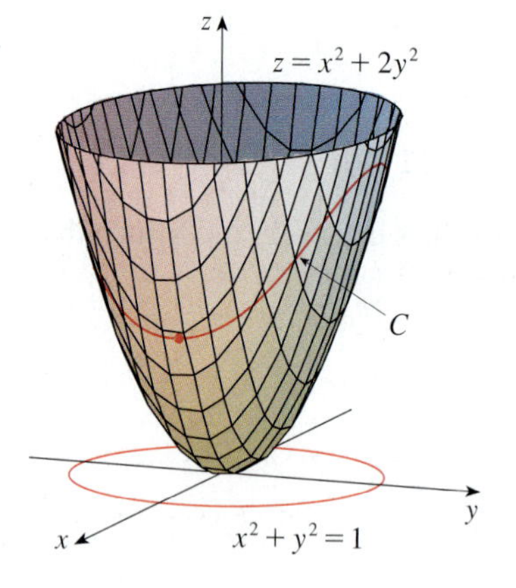

FIGURE 2

SOLUTION We are asked for the extreme values of f subject to the constraint $g(x, y) = x^2 + y^2 = 1$. Using Lagrange multipliers, we solve the equations $\nabla f = \lambda \nabla g$ and $g(x, y) = 1$, which can be written as

$$f_x = \lambda g_x \qquad f_y = \lambda g_y \qquad g(x, y) = 1$$

or as

9 $$2x = 2x\lambda$$

10 $$4y = 2y\lambda$$

11 $$x^2 + y^2 = 1$$

From (9) we have $x = 0$ or $\lambda = 1$. If $x = 0$, then (11) gives $y = \pm 1$. If $\lambda = 1$, then $y = 0$ from (10), so then (11) gives $x = \pm 1$. Therefore f has possible extreme values at the points $(0, 1)$, $(0, -1)$, $(1, 0)$, and $(-1, 0)$. Evaluating f at these four points, we find that

$$f(0, 1) = 2 \qquad f(0, -1) = 2 \qquad f(1, 0) = 1 \qquad f(-1, 0) = 1$$

Therefore the maximum value of f on the circle $x^2 + y^2 = 1$ is $f(0, \pm 1) = 2$ and the minimum value is $f(\pm 1, 0) = 1$. In geometric terms, these correspond to the highest and lowest points on the curve C in Figure 2, where C consists of those points on the paraboloid $z = x^2 + 2y^2$ that are directly above the constraint circle $x^2 + y^2 = 1$. ■

The geometry behind the use of Lagrange multipliers in Example 2 is shown in Figure 3. The extreme values of $f(x, y) = x^2 + 2y^2$ correspond to the level curves that touch the circle $x^2 + y^2 = 1$.

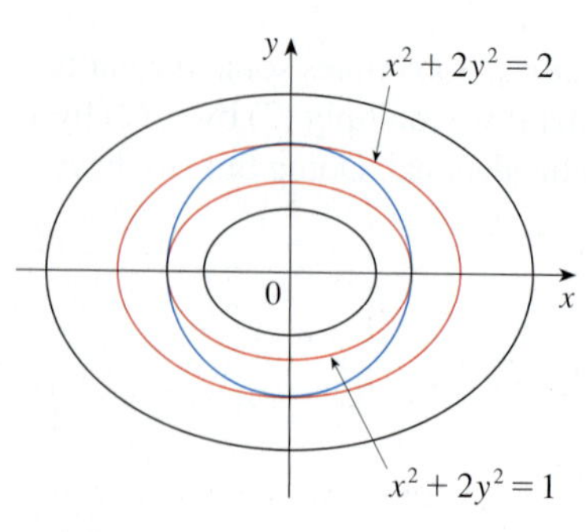

FIGURE 3

EXAMPLE 3 Find the extreme values of $f(x, y) = x^2 + 2y^2$ on the disk $x^2 + y^2 \leq 1$.

SOLUTION According to the procedure in (14.7.9), we compare the values of f at the critical points with values at the points on the boundary. Since $f_x = 2x$ and $f_y = 4y$, the only critical point is $(0, 0)$. We compare the value of f at that point with the extreme values on the boundary from Example 2:

$$f(0, 0) = 0 \qquad f(\pm 1, 0) = 1 \qquad f(0, \pm 1) = 2$$

Therefore the maximum value of f on the disk $x^2 + y^2 \leq 1$ is $f(0, \pm 1) = 2$ and the minimum value is $f(0, 0) = 0$. ■

EXAMPLE 4 Find the points on the sphere $x^2 + y^2 + z^2 = 4$ that are closest to and farthest from the point $(3, 1, -1)$.

SOLUTION The distance from a point (x, y, z) to the point $(3, 1, -1)$ is

$$d = \sqrt{(x - 3)^2 + (y - 1)^2 + (z + 1)^2}$$

but the algebra is simpler if we instead maximize and minimize the square of the distance:

$$d^2 = f(x, y, z) = (x - 3)^2 + (y - 1)^2 + (z + 1)^2$$

The constraint is that the point (x, y, z) lies on the sphere, that is,

$$g(x, y, z) = x^2 + y^2 + z^2 = 4$$

According to the method of Lagrange multipliers, we solve $\nabla f = \lambda \nabla g$, $g = 4$. This gives

$$2(x - 3) = 2x\lambda \tag{12}$$

$$2(y - 1) = 2y\lambda \tag{13}$$

$$2(z + 1) = 2z\lambda \tag{14}$$

$$x^2 + y^2 + z^2 = 4 \tag{15}$$

The simplest way to solve these equations is to solve for x, y, and z in terms of λ from (12), (13), and (14), and then substitute these values into (15). From (12) we have

$$x - 3 = x\lambda \qquad \text{or} \qquad x(1 - \lambda) = 3 \qquad \text{or} \qquad x = \frac{3}{1 - \lambda}$$

[Note that $1 - \lambda \neq 0$ because $\lambda = 1$ is impossible from (12).] Similarly, (13) and (14) give

$$y = \frac{1}{1 - \lambda} \qquad z = -\frac{1}{1 - \lambda}$$

Figure 4 shows the sphere and the nearest point P in Example 4. Can you see how to find the coordinates of P without using calculus?

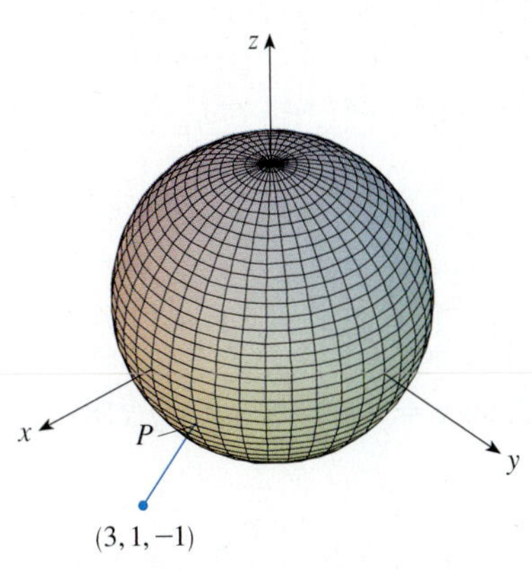

FIGURE 4

Therefore, from (15), we have

$$\frac{3^2}{(1 - \lambda)^2} + \frac{1^2}{(1 - \lambda)^2} + \frac{(-1)^2}{(1 - \lambda)^2} = 4$$

which gives $(1 - \lambda)^2 = \frac{11}{4}$, $1 - \lambda = \pm\sqrt{11}/2$, so

$$\lambda = 1 \pm \frac{\sqrt{11}}{2}$$

These values of λ then give the corresponding points (x, y, z):

$$\left(\frac{6}{\sqrt{11}}, \frac{2}{\sqrt{11}}, -\frac{2}{\sqrt{11}}\right) \qquad \text{and} \qquad \left(-\frac{6}{\sqrt{11}}, -\frac{2}{\sqrt{11}}, \frac{2}{\sqrt{11}}\right)$$

It's easy to see that f has a smaller value at the first of these points, so the closest point is $\left(6/\sqrt{11}, 2/\sqrt{11}, -2/\sqrt{11}\right)$ and the farthest is $\left(-6/\sqrt{11}, -2/\sqrt{11}, 2/\sqrt{11}\right)$. ■

Two Constraints

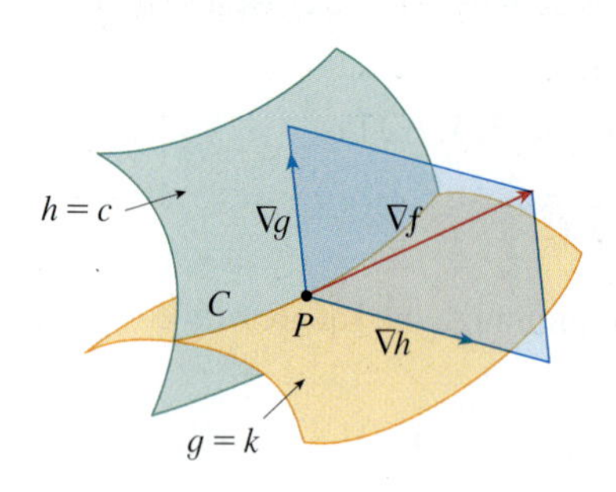

FIGURE 5

Suppose now that we want to find the maximum and minimum values of a function $f(x, y, z)$ subject to two constraints (side conditions) of the form $g(x, y, z) = k$ and $h(x, y, z) = c$. Geometrically, this means that we are looking for the extreme values of f when (x, y, z) is restricted to lie on the curve of intersection C of the level surfaces $g(x, y, z) = k$ and $h(x, y, z) = c$. (See Figure 5.) Suppose f has such an extreme value at a point $P(x_0, y_0, z_0)$. We know from the beginning of this section that ∇f is orthogonal to C at P. But we also know that ∇g is orthogonal to $g(x, y, z) = k$ and ∇h is orthogonal to $h(x, y, z) = c$, so ∇g and ∇h are both orthogonal to C. This means that the gradient vector $\nabla f(x_0, y_0, z_0)$ is in the plane determined by $\nabla g(x_0, y_0, z_0)$ and $\nabla h(x_0, y_0, z_0)$. (We assume that these gradient vectors are not zero and not parallel.) So there are numbers λ and μ (called Lagrange multipliers) such that

16 $$\nabla f(x_0, y_0, z_0) = \lambda \, \nabla g(x_0, y_0, z_0) + \mu \, \nabla h(x_0, y_0, z_0)$$

In this case Lagrange's method is to look for extreme values by solving five equations in the five unknowns x, y, z, λ, and μ. These equations are obtained by writing Equation 16 in terms of its components and using the constraint equations:

$$f_x = \lambda g_x + \mu h_x$$

$$f_y = \lambda g_y + \mu h_y$$

$$f_z = \lambda g_z + \mu h_z$$

$$g(x, y, z) = k$$

$$h(x, y, z) = c$$

The cylinder $x^2 + y^2 = 1$ intersects the plane $x - y + z = 1$ in an ellipse (Figure 6). Example 5 asks for the maximum value of f when (x, y, z) is restricted to lie on the ellipse.

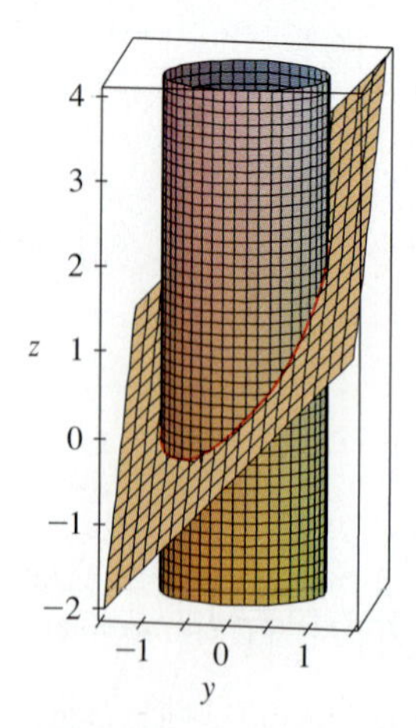

FIGURE 6

EXAMPLE 5 Find the maximum value of the function $f(x, y, z) = x + 2y + 3z$ on the curve of intersection of the plane $x - y + z = 1$ and the cylinder $x^2 + y^2 = 1$.

SOLUTION We maximize the function $f(x, y, z) = x + 2y + 3z$ subject to the constraints $g(x, y, z) = x - y + z = 1$ and $h(x, y, z) = x^2 + y^2 = 1$. The Lagrange condition is $\nabla f = \lambda \, \nabla g + \mu \, \nabla h$, so we solve the equations

17 $$1 = \lambda + 2x\mu$$

18 $$2 = -\lambda + 2y\mu$$

19 $$3 = \lambda$$

20 $$x - y + z = 1$$

21 $$x^2 + y^2 = 1$$

Putting $\lambda = 3$ [from (19)] in (17), we get $2x\mu = -2$, so $x = -1/\mu$. Similarly, (18) gives $y = 5/(2\mu)$. Substitution in (21) then gives

$$\frac{1}{\mu^2} + \frac{25}{4\mu^2} = 1$$

and so $\mu^2 = \frac{29}{4}$, $\mu = \pm\sqrt{29}/2$. Then $x = \mp 2/\sqrt{29}$, $y = \pm 5/\sqrt{29}$, and, from (20), $z = 1 - x + y = 1 \pm 7/\sqrt{29}$. The corresponding values of f are

$$\mp \frac{2}{\sqrt{29}} + 2\left(\pm \frac{5}{\sqrt{29}}\right) + 3\left(1 \pm \frac{7}{\sqrt{29}}\right) = 3 \pm \sqrt{29}$$

Therefore the maximum value of f on the given curve is $3 + \sqrt{29}$. ■

14.8 EXERCISES

1. Pictured are a contour map of f and a curve with equation $g(x, y) = 8$. Estimate the maximum and minimum values of f subject to the constraint that $g(x, y) = 8$. Explain your reasoning.

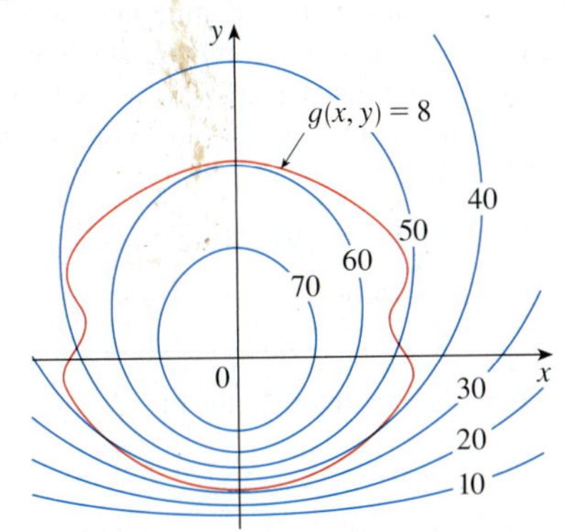

2. (a) Use a graphing calculator or computer to graph the circle $x^2 + y^2 = 1$. On the same screen, graph several curves of the form $x^2 + y = c$ until you find two that just touch the circle. What is the significance of the values of c for these two curves?

(b) Use Lagrange multipliers to find the extreme values of $f(x, y) = x^2 + y$ subject to the constraint $x^2 + y^2 = 1$. Compare your answers with those in part (a).

3–14 Each of these extreme value problems has a solution with both a maximum value and a minimum value. Use Lagrange multipliers to find the extreme values of the function subject to the given constraint.

3. $f(x, y) = x^2 - y^2; \quad x^2 + y^2 = 1$

4. $f(x, y) = 3x + y; \quad x^2 + y^2 = 10$

5. $f(x, y) = xy; \quad 4x^2 + y^2 = 8$

6. $f(x, y) = xe^y; \quad x^2 + y^2 = 2$

7. $f(x, y, z) = 2x + 2y + z; \quad x^2 + y^2 + z^2 = 9$

8. $f(x, y, z) = e^{xyz}; \quad 2x^2 + y^2 + z^2 = 24$

9. $f(x, y, z) = xy^2z; \quad x^2 + y^2 + z^2 = 4$

10. $f(x, y, z) = \ln(x^2 + 1) + \ln(y^2 + 1) + \ln(z^2 + 1);$
$x^2 + y^2 + z^2 = 12$

11. $f(x, y, z) = x^2 + y^2 + z^2; \quad x^4 + y^4 + z^4 = 1$

12. $f(x, y, z) = x^4 + y^4 + z^4; \quad x^2 + y^2 + z^2 = 1$

13. $f(x, y, z, t) = x + y + z + t; \quad x^2 + y^2 + z^2 + t^2 = 1$

14. $f(x_1, x_2, \ldots, x_n) = x_1 + x_2 + \cdots + x_n;$
$x_1^2 + x_2^2 + \cdots + x_n^2 = 1$

15. The method of Lagrange multipliers assumes that the extreme values exist, but that is not always the case. Show that the problem of finding the minimum value of $f(x, y) = x^2 + y^2$ subject to the constraint $xy = 1$ can be solved using Lagrange multipliers, but f does not have a maximum value with that constraint.

16. Find the minimum value of $f(x, y, z) = x^2 + 2y^2 + 3z^2$ subject to the constraint $x + 2y + 3z = 10$. Show that f has no maximum value with this constraint.

17–20 Find the extreme values of f subject to both constraints.

17. $f(x, y, z) = x + y + z; \quad x^2 + z^2 = 2, \quad x + y = 1$

18. $f(x, y, z) = z; \quad x^2 + y^2 = z^2, \quad x + y + z = 24$

19. $f(x, y, z) = yz + xy; \quad xy = 1, \quad y^2 + z^2 = 1$

20. $f(x, y, z) = x^2 + y^2 + z^2; \quad x - y = 1, \quad y^2 - z^2 = 1$

21–23 Find the extreme values of f on the region described by the inequality.

21. $f(x, y) = x^2 + y^2 + 4x - 4y, \quad x^2 + y^2 \leqslant 9$

22. $f(x, y) = 2x^2 + 3y^2 - 4x - 5, \quad x^2 + y^2 \leqslant 16$

23. $f(x, y) = e^{-xy}, \quad x^2 + 4y^2 \leqslant 1$

24. Consider the problem of maximizing the function $f(x, y) = 2x + 3y$ subject to the constraint $\sqrt{x} + \sqrt{y} = 5$.

(a) Try using Lagrange multipliers to solve the problem.

(b) Does $f(25, 0)$ give a larger value than the one in part (a)?

(c) Solve the problem by graphing the constraint equation and several level curves of f.

(d) Explain why the method of Lagrange multipliers fails to solve the problem.

(e) What is the significance of $f(9, 4)$?

25. Consider the problem of minimizing the function $f(x, y) = x$ on the curve $y^2 + x^4 - x^3 = 0$ (a piriform).
(a) Try using Lagrange multipliers to solve the problem.
(b) Show that the minimum value is $f(0, 0) = 0$ but the Lagrange condition $\nabla f(0, 0) = \lambda \nabla g(0, 0)$ is not satisfied for any value of λ.
(c) Explain why Lagrange multipliers fail to find the minimum value in this case.

CAS **26.** (a) If your computer algebra system plots implicitly defined curves, use it to estimate the minimum and maximum values of $f(x, y) = x^3 + y^3 + 3xy$ subject to the constraint $(x - 3)^2 + (y - 3)^2 = 9$ by graphical methods.
(b) Solve the problem in part (a) with the aid of Lagrange multipliers. Use your CAS to solve the equations numerically. Compare your answers with those in part (a).

27. The total production P of a certain product depends on the amount L of labor used and the amount K of capital investment. In Sections 14.1 and 14.3 we discussed how the Cobb-Douglas model $P = bL^\alpha K^{1-\alpha}$ follows from certain economic assumptions, where b and α are positive constants and $\alpha < 1$. If the cost of a unit of labor is m and the cost of a unit of capital is n, and the company can spend only p dollars as its total budget, then maximizing the production P is subject to the constraint $mL + nK = p$. Show that the maximum production occurs when

$$L = \frac{\alpha p}{m} \quad \text{and} \quad K = \frac{(1 - \alpha)p}{n}$$

28. Referring to Exercise 27, we now suppose that the production is fixed at $bL^\alpha K^{1-\alpha} = Q$, where Q is a constant. What values of L and K minimize the cost function $C(L, K) = mL + nK$?

29. Use Lagrange multipliers to prove that the rectangle with maximum area that has a given perimeter p is a square.

30. Use Lagrange multipliers to prove that the triangle with maximum area that has a given perimeter p is equilateral.
Hint: Use Heron's formula for the area:

$$A = \sqrt{s(s - x)(s - y)(s - z)}$$

where $s = p/2$ and x, y, z are the lengths of the sides.

31–43 Use Lagrange multipliers to give an alternate solution to the indicated exercise in Section 14.7.

31. Exercise 41 **32.** Exercise 42
33. Exercise 43 **34.** Exercise 44
35. Exercise 45 **36.** Exercise 46
37. Exercise 47 **38.** Exercise 48
39. Exercise 49 **40.** Exercise 50
41. Exercise 51 **42.** Exercise 52
43. Exercise 55

---

44. Find the maximum and minimum volumes of a rectangular box whose surface area is 1500 cm^2 and whose total edge length is 200 cm.

45. The plane $x + y + 2z = 2$ intersects the paraboloid $z = x^2 + y^2$ in an ellipse. Find the points on this ellipse that are nearest to and farthest from the origin.

46. The plane $4x - 3y + 8z = 5$ intersects the cone $z^2 = x^2 + y^2$ in an ellipse.
(a) Graph the cone and the plane, and observe the elliptical intersection.
(b) Use Lagrange multipliers to find the highest and lowest points on the ellipse.

CAS **47–48** Find the maximum and minimum values of f subject to the given constraints. Use a computer algebra system to solve the system of equations that arises in using Lagrange multipliers. (If your CAS finds only one solution, you may need to use additional commands.)

47. $f(x, y, z) = ye^{x-z}$; $9x^2 + 4y^2 + 36z^2 = 36$, $xy + yz = 1$

48. $f(x, y, z) = x + y + z$; $x^2 - y^2 = z$, $x^2 + z^2 = 4$

---

49. (a) Find the maximum value of

$$f(x_1, x_2, \ldots, x_n) = \sqrt[n]{x_1 x_2 \cdots x_n}$$

given that $x_1, x_2, \ldots, x_n$ are positive numbers and $x_1 + x_2 + \cdots + x_n = c$, where c is a constant.
(b) Deduce from part (a) that if $x_1, x_2, \ldots, x_n$ are positive numbers, then

$$\sqrt[n]{x_1 x_2 \cdots x_n} \leqslant \frac{x_1 + x_2 + \cdots + x_n}{n}$$

This inequality says that the geometric mean of n numbers is no larger than the arithmetic mean of the numbers. Under what circumstances are these two means equal?

50. (a) Maximize $\sum_{i=1}^{n} x_i y_i$ subject to the constraints $\sum_{i=1}^{n} x_i^2 = 1$ and $\sum_{i=1}^{n} y_i^2 = 1$.
(b) Put

$$x_i = \frac{a_i}{\sqrt{\sum a_j^2}} \quad \text{and} \quad y_i = \frac{b_i}{\sqrt{\sum b_j^2}}$$

to show that

$$\sum a_i b_i \leqslant \sqrt{\sum a_j^2}\,\sqrt{\sum b_j^2}$$

for any numbers $a_1, \ldots, a_n, b_1, \ldots, b_n$. This inequality is known as the Cauchy-Schwarz Inequality.

APPLIED PROJECT ROCKET SCIENCE

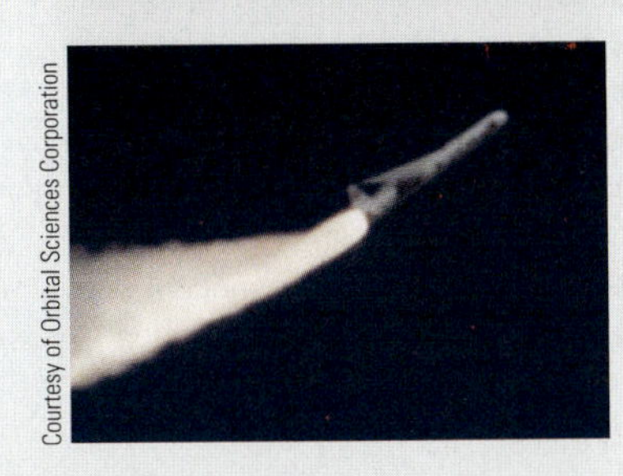

Courtesy of Orbital Sciences Corporation

Many rockets, such as the *Pegasus XL* currently used to launch satellites and the *Saturn V* that first put men on the moon, are designed to use three stages in their ascent into space. A large first stage initially propels the rocket until its fuel is consumed, at which point the stage is jettisoned to reduce the mass of the rocket. The smaller second and third stages function similarly in order to place the rocket's payload into orbit about the earth. (With this design, at least two stages are required in order to reach the necessary velocities, and using three stages has proven to be a good compromise between cost and performance.) Our goal here is to determine the individual masses of the three stages, which are to be designed to minimize the total mass of the rocket while enabling it to reach a desired velocity.

For a single-stage rocket consuming fuel at a constant rate, the change in velocity resulting from the acceleration of the rocket vehicle has been modeled by

$$\Delta V = -c \ln\left(1 - \frac{(1 - S)M_r}{P + M_r}\right)$$

where M_r is the mass of the rocket engine including initial fuel, P is the mass of the payload, S is a *structural factor* determined by the design of the rocket (specifically, it is the ratio of the mass of the rocket vehicle without fuel to the total mass of the rocket with payload), and c is the (constant) speed of exhaust relative to the rocket.

Now consider a rocket with three stages and a payload of mass A. Assume that outside forces are negligible and that c and S remain constant for each stage. If M_i is the mass of the ith stage, we can initially consider the rocket engine to have mass M_1 and its payload to have mass $M_2 + M_3 + A$; the second and third stages can be handled similarly.

1. Show that the velocity attained after all three stages have been jettisoned is given by

$$v_f = c\left[\ln\left(\frac{M_1 + M_2 + M_3 + A}{SM_1 + M_2 + M_3 + A}\right) + \ln\left(\frac{M_2 + M_3 + A}{SM_2 + M_3 + A}\right) + \ln\left(\frac{M_3 + A}{SM_3 + A}\right)\right]$$

2. We wish to minimize the total mass $M = M_1 + M_2 + M_3$ of the rocket engine subject to the constraint that the desired velocity v_f from Problem 1 is attained. The method of Lagrange multipliers is appropriate here, but difficult to implement using the current expressions. To simplify, we define variables N_i so that the constraint equation may be expressed as $v_f = c(\ln N_1 + \ln N_2 + \ln N_3)$. Since M is now difficult to express in terms of the N_i's, we wish to use a simpler function that will be minimized at the same place as M. Show that

$$\frac{M_1 + M_2 + M_3 + A}{M_2 + M_3 + A} = \frac{(1 - S)N_1}{1 - SN_1}$$

$$\frac{M_2 + M_3 + A}{M_3 + A} = \frac{(1 - S)N_2}{1 - SN_2}$$

$$\frac{M_3 + A}{A} = \frac{(1 - S)N_3}{1 - SN_3}$$

and conclude that

$$\frac{M + A}{A} = \frac{(1 - S)^3 N_1 N_2 N_3}{(1 - SN_1)(1 - SN_2)(1 - SN_3)}$$

3. Verify that $\ln((M + A)/A)$ is minimized at the same location as M; use Lagrange multipliers and the results of Problem 2 to find expressions for the values of N_i where the minimum occurs subject to the constraint $v_f = c(\ln N_1 + \ln N_2 + \ln N_3)$. [*Hint:* Use properties of logarithms to help simplify the expressions.]

4. Find an expression for the minimum value of M as a function of v_f.

5. If we want to put a three-stage rocket into orbit 100 miles above the earth's surface, a final velocity of approximately 17,500 mi/h is required. Suppose that each stage is built with a structural factor $S = 0.2$ and an exhaust speed of $c = 6000$ mi/h.
   (a) Find the minimum total mass M of the rocket engines as a function of A.
   (b) Find the mass of each individual stage as a function of A. (They are not equally sized!)

6. The same rocket would require a final velocity of approximately 24,700 mi/h in order to escape earth's gravity. Find the mass of each individual stage that would minimize the total mass of the rocket engines and allow the rocket to propel a 500-pound probe into deep space.

APPLIED PROJECT HYDRO-TURBINE OPTIMIZATION

At a hydroelectric generating station (once operated by the Katahdin Paper Company) in Millinocket, Maine, water is piped from a dam to the power station. The rate at which the water flows through the pipe varies, depending on external conditions.

The power station has three different hydroelectric turbines, each with a known (and unique) power function that gives the amount of electric power generated as a function of the water flow arriving at the turbine. The incoming water can be apportioned in different volumes to each turbine, so the goal is to determine how to distribute water among the turbines to give the maximum total energy production for any rate of flow.

Using experimental evidence and *Bernoulli's equation*, the following quadratic models were determined for the power output of each turbine, along with the allowable flows of operation:

$$KW_1 = (-18.89 + 0.1277Q_1 - 4.08 \cdot 10^{-5}Q_1^2)(170 - 1.6 \cdot 10^{-6}Q_T^2)$$

$$KW_2 = (-24.51 + 0.1358Q_2 - 4.69 \cdot 10^{-5}Q_2^2)(170 - 1.6 \cdot 10^{-6}Q_T^2)$$

$$KW_3 = (-27.02 + 0.1380Q_3 - 3.84 \cdot 10^{-5}Q_3^2)(170 - 1.6 \cdot 10^{-6}Q_T^2)$$

$$250 \leqslant Q_1 \leqslant 1110, \quad 250 \leqslant Q_2 \leqslant 1110, \quad 250 \leqslant Q_3 \leqslant 1225$$

where

Q_i = flow through turbine i in cubic feet per second

KW_i = power generated by turbine i in kilowatts

Q_T = total flow through the station in cubic feet per second

1. If all three turbines are being used, we wish to determine the flow Q_i to each turbine that will give the maximum total energy production. Our limitations are that the flows must sum to the total incoming flow and the given domain restrictions must be observed. Consequently, use Lagrange multipliers to find the values for the individual flows (as functions of Q_T) that maximize the total energy production $KW_1 + KW_2 + KW_3$ subject to the constraints $Q_1 + Q_2 + Q_3 = Q_T$ and the domain restrictions on each Q_i.

2. For which values of Q_T is your result valid?

3. For an incoming flow of 2500 ft^3/s, determine the distribution to the turbines and verify (by trying some nearby distributions) that your result is indeed a maximum.

4. Until now we have assumed that all three turbines are operating; is it possible in some situations that more power could be produced by using only one turbine? Make a graph of the three power functions and use it to help decide if an incoming flow of 1000 ft^3/s should be

distributed to all three turbines or routed to just one. (If you determine that only one turbine should be used, which one would it be?) What if the flow is only 600 ft^3/s?

5. Perhaps for some flow levels it would be advantageous to use two turbines. If the incoming flow is 1500 ft^3/s, which two turbines would you recommend using? Use Lagrange multipliers to determine how the flow should be distributed between the two turbines to maximize the energy produced. For this flow, is using two turbines more efficient than using all three?

6. If the incoming flow is 3400 ft^3/s, what would you recommend to the station management?

14 REVIEW

CONCEPT CHECK

Answers to the Concept Check can be found on the back endpapers.

1. (a) What is a function of two variables?
(b) Describe three methods for visualizing a function of two variables.

2. What is a function of three variables? How can you visualize such a function?

3. What does

$$\lim_{(x, y) \to (a, b)} f(x, y) = L$$

mean? How can you show that such a limit does not exist?

4. (a) What does it mean to say that f is continuous at (a, b)?
(b) If f is continuous on $\mathbb{R}^2$, what can you say about its graph?

5. (a) Write expressions for the partial derivatives $f_x(a, b)$ and $f_y(a, b)$ as limits.
(b) How do you interpret $f_x(a, b)$ and $f_y(a, b)$ geometrically? How do you interpret them as rates of change?
(c) If $f(x, y)$ is given by a formula, how do you calculate f_x and f_y?

6. What does Clairaut's Theorem say?

7. How do you find a tangent plane to each of the following types of surfaces?
(a) A graph of a function of two variables, $z = f(x, y)$
(b) A level surface of a function of three variables, $F(x, y, z) = k$

8. Define the linearization of f at (a, b). What is the corresponding linear approximation? What is the geometric interpretation of the linear approximation?

9. (a) What does it mean to say that f is differentiable at (a, b)?
(b) How do you usually verify that f is differentiable?

10. If $z = f(x, y)$, what are the differentials dx, dy, and dz?

11. State the Chain Rule for the case where $z = f(x, y)$ and x and y are functions of one variable. What if x and y are functions of two variables?

12. If z is defined implicitly as a function of x and y by an equation of the form $F(x, y, z) = 0$, how do you find $\partial z/\partial x$ and $\partial z/\partial y$?

13. (a) Write an expression as a limit for the directional derivative of f at (x_0, y_0) in the direction of a unit vector $\mathbf{u} = \langle a, b \rangle$. How do you interpret it as a rate? How do you interpret it geometrically?
(b) If f is differentiable, write an expression for $D_{\mathbf{u}} f(x_0, y_0)$ in terms of f_x and f_y.

14. (a) Define the gradient vector ∇f for a function f of two or three variables.
(b) Express $D_{\mathbf{u}} f$ in terms of ∇f.
(c) Explain the geometric significance of the gradient.

15. What do the following statements mean?
(a) f has a local maximum at (a, b).
(b) f has an absolute maximum at (a, b).
(c) f has a local minimum at (a, b).
(d) f has an absolute minimum at (a, b).
(e) f has a saddle point at (a, b).

16. (a) If f has a local maximum at (a, b), what can you say about its partial derivatives at (a, b)?
(b) What is a critical point of f?

17. State the Second Derivatives Test.

18. (a) What is a closed set in $\mathbb{R}^2$? What is a bounded set?
(b) State the Extreme Value Theorem for functions of two variables.
(c) How do you find the values that the Extreme Value Theorem guarantees?

19. Explain how the method of Lagrange multipliers works in finding the extreme values of $f(x, y, z)$ subject to the constraint $g(x, y, z) = k$. What if there is a second constraint $h(x, y, z) = c$?

TRUE-FALSE QUIZ

Determine whether the statement is true or false. If it is true, explain why. If it is false, explain why or give an example that disproves the statement.

1. $f_y(a, b) = \lim_{y \to b} \dfrac{f(a, y) - f(a, b)}{y - b}$

2. There exists a function f with continuous second-order partial derivatives such that $f_x(x, y) = x + y^2$ and $f_y(x, y) = x - y^2$.

3. $f_{xy} = \dfrac{\partial^2 f}{\partial x\, \partial y}$

4. $D_{\mathbf{k}} f(x, y, z) = f_z(x, y, z)$

5. If $f(x, y) \to L$ as $(x, y) \to (a, b)$ along every straight line through (a, b), then $\lim_{(x, y) \to (a, b)} f(x, y) = L$.

6. If $f_x(a, b)$ and $f_y(a, b)$ both exist, then f is differentiable at (a, b).

7. If f has a local minimum at (a, b) and f is differentiable at (a, b), then $\nabla f(a, b) = \mathbf{0}$.

8. If f is a function, then

$$\lim_{(x, y) \to (2, 5)} f(x, y) = f(2, 5)$$

9. If $f(x, y) = \ln y$, then $\nabla f(x, y) = 1/y$.

10. If $(2, 1)$ is a critical point of f and

$$f_{xx}(2, 1)\, f_{yy}(2, 1) < [f_{xy}(2, 1)]^2$$

then f has a saddle point at $(2, 1)$.

11. If $f(x, y) = \sin x + \sin y$, then $-\sqrt{2} \leqslant D_{\mathbf{u}} f(x, y) \leqslant \sqrt{2}$.

12. If $f(x, y)$ has two local maxima, then f must have a local minimum.

EXERCISES

1–2 Find and sketch the domain of the function.

1. $f(x, y) = \ln(x + y + 1)$

2. $f(x, y) = \sqrt{4 - x^2 - y^2} + \sqrt{1 - x^2}$

3–4 Sketch the graph of the function.

3. $f(x, y) = 1 - y^2$

4. $f(x, y) = x^2 + (y - 2)^2$

5–6 Sketch several level curves of the function.

5. $f(x, y) = \sqrt{4x^2 + y^2}$

6. $f(x, y) = e^x + y$

7. Make a rough sketch of a contour map for the function whose graph is shown.

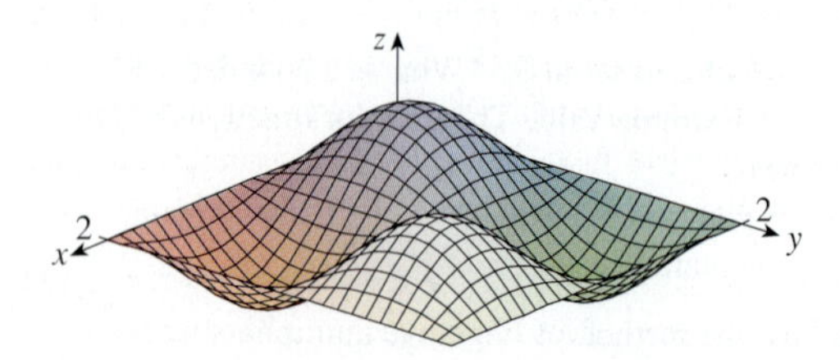

8. The contour map of a function f is shown.
   (a) Estimate the value of $f(3, 2)$.
   (b) Is $f_x(3, 2)$ positive or negative? Explain.
   (c) Which is greater, $f_y(2, 1)$ or $f_y(2, 2)$? Explain.

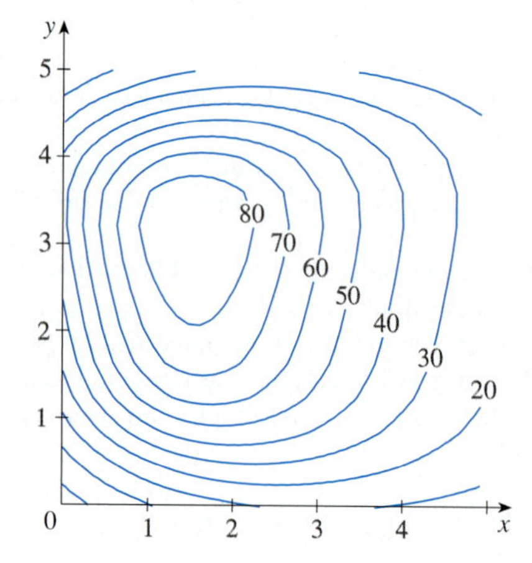

9–10 Evaluate the limit or show that it does not exist.

9. $\lim_{(x, y) \to (1, 1)} \dfrac{2xy}{x^2 + 2y^2}$

10. $\lim_{(x, y) \to (0, 0)} \dfrac{2xy}{x^2 + 2y^2}$

11. A metal plate is situated in the xy-plane and occupies the rectangle $0 \leqslant x \leqslant 10$, $0 \leqslant y \leqslant 8$, where x and y are measured in meters. The temperature at the point (x, y) in the plate is $T(x, y)$, where T is measured in degrees Celsius. Temperatures at equally spaced points were measured and recorded in the table.
   (a) Estimate the values of the partial derivatives $T_x(6, 4)$ and $T_y(6, 4)$. What are the units?

(b) Estimate the value of $D_{\mathbf{u}}T(6, 4)$, where $\mathbf{u} = (\mathbf{i} + \mathbf{j})/\sqrt{2}$. Interpret your result.
(c) Estimate the value of $T_{xy}(6, 4)$.

x \ y	0	2	4	6	8
0	30	38	45	51	55
2	52	56	60	62	61
4	78	74	72	68	66
6	98	87	80	75	71
8	96	90	86	80	75
10	92	92	91	87	78

12. Find a linear approximation to the temperature function $T(x, y)$ in Exercise 11 near the point (6, 4). Then use it to estimate the temperature at the point (5, 3.8).

13–17 Find the first partial derivatives.

13. $f(x, y) = (5y^3 + 2x^2y)^8$

14. $g(u, v) = \dfrac{u + 2v}{u^2 + v^2}$

15. $F(\alpha, \beta) = \alpha^2 \ln(\alpha^2 + \beta^2)$

16. $G(x, y, z) = e^{xz} \sin(y/z)$

17. $S(u, v, w) = u \arctan(v\sqrt{w})$

18. The speed of sound traveling through ocean water is a function of temperature, salinity, and pressure. It has been modeled by the function

$$C = 1449.2 + 4.6T - 0.055T^2 + 0.00029T^3 + (1.34 - 0.01T)(S - 35) + 0.016D$$

where C is the speed of sound (in meters per second), T is the temperature (in degrees Celsius), S is the salinity (the concentration of salts in parts per thousand, which means the number of grams of dissolved solids per 1000 g of water), and D is the depth below the ocean surface (in meters). Compute $\partial C/\partial T$, $\partial C/\partial S$, and $\partial C/\partial D$ when $T = 10°C$, $S = 35$ parts per thousand, and $D = 100$ m. Explain the physical significance of these partial derivatives.

19–22 Find all second partial derivatives of f.

19. $f(x, y) = 4x^3 - xy^2$

20. $z = xe^{-2y}$

21. $f(x, y, z) = x^k y^l z^m$

22. $v = r\cos(s + 2t)$

23. If $z = xy + xe^{y/x}$, show that $x\dfrac{\partial z}{\partial x} + y\dfrac{\partial z}{\partial y} = xy + z$.

24. If $z = \sin(x + \sin t)$, show that

$$\frac{\partial z}{\partial x}\frac{\partial^2 z}{\partial x\,\partial t} = \frac{\partial z}{\partial t}\frac{\partial^2 z}{\partial x^2}$$

25–29 Find equations of (a) the tangent plane and (b) the normal line to the given surface at the specified point.

25. $z = 3x^2 - y^2 + 2x$, $(1, -2, 1)$

26. $z = e^x \cos y$, $(0, 0, 1)$

27. $x^2 + 2y^2 - 3z^2 = 3$, $(2, -1, 1)$

28. $xy + yz + zx = 3$, $(1, 1, 1)$

29. $\sin(xyz) = x + 2y + 3z$, $(2, -1, 0)$

30. Use a computer to graph the surface $z = x^2 + y^4$ and its tangent plane and normal line at $(1, 1, 2)$ on the same screen. Choose the domain and viewpoint so that you get a good view of all three objects.

31. Find the points on the hyperboloid $x^2 + 4y^2 - z^2 = 4$ where the tangent plane is parallel to the plane $2x + 2y + z = 5$.

32. Find du if $u = \ln(1 + se^{2t})$.

33. Find the linear approximation of the function $f(x, y, z) = x^3\sqrt{y^2 + z^2}$ at the point (2, 3, 4) and use it to estimate the number $(1.98)^3\sqrt{(3.01)^2 + (3.97)^2}$.

34. The two legs of a right triangle are measured as 5 m and 12 m with a possible error in measurement of at most 0.2 cm in each. Use differentials to estimate the maximum error in the calculated value of (a) the area of the triangle and (b) the length of the hypotenuse.

35. If $u = x^2y^3 + z^4$, where $x = p + 3p^2$, $y = pe^p$, and $z = p\sin p$, use the Chain Rule to find du/dp.

36. If $v = x^2 \sin y + ye^{xy}$, where $x = s + 2t$ and $y = st$, use the Chain Rule to find $\partial v/\partial s$ and $\partial v/\partial t$ when $s = 0$ and $t = 1$.

37. Suppose $z = f(x, y)$, where $x = g(s, t)$, $y = h(s, t)$, $g(1, 2) = 3$, $g_s(1, 2) = -1$, $g_t(1, 2) = 4$, $h(1, 2) = 6$, $h_s(1, 2) = -5$, $h_t(1, 2) = 10$, $f_x(3, 6) = 7$, and $f_y(3, 6) = 8$. Find $\partial z/\partial s$ and $\partial z/\partial t$ when $s = 1$ and $t = 2$.

38. Use a tree diagram to write out the Chain Rule for the case where $w = f(t, u, v)$, $t = t(p, q, r, s)$, $u = u(p, q, r, s)$, and $v = v(p, q, r, s)$ are all differentiable functions.

39. If $z = y + f(x^2 - y^2)$, where f is differentiable, show that

$$y\frac{\partial z}{\partial x} + x\frac{\partial z}{\partial y} = x$$

40. The length x of a side of a triangle is increasing at a rate of 3 in/s, the length y of another side is decreasing at a rate of 2 in/s, and the contained angle θ is increasing at a rate of 0.05 radian/s. How fast is the area of the triangle changing when $x = 40$ in, $y = 50$ in, and $\theta = \pi/6$?

41. If $z = f(u, v)$, where $u = xy$, $v = y/x$, and f has continuous second partial derivatives, show that

$$x^2\frac{\partial^2 z}{\partial x^2} - y^2\frac{\partial^2 z}{\partial y^2} = -4uv\frac{\partial^2 z}{\partial u\,\partial v} + 2v\frac{\partial z}{\partial v}$$

42. If $\cos(xyz) = 1 + x^2y^2 + z^2$, find $\dfrac{\partial z}{\partial x}$ and $\dfrac{\partial z}{\partial y}$.

43. Find the gradient of the function $f(x, y, z) = x^2e^{yz^2}$.

44. (a) When is the directional derivative of f a maximum?
(b) When is it a minimum?
(c) When is it 0?
(d) When is it half of its maximum value?

45–46 Find the directional derivative of f at the given point in the indicated direction.

45. $f(x, y) = x^2e^{-y}$, $(-2, 0)$,
in the direction toward the point $(2, -3)$

46. $f(x, y, z) = x^2y + x\sqrt{1 + z}$, $(1, 2, 3)$,
in the direction of $\mathbf{v} = 2\mathbf{i} + \mathbf{j} - 2\mathbf{k}$

47. Find the maximum rate of change of $f(x, y) = x^2y + \sqrt{y}$ at the point $(2, 1)$. In which direction does it occur?

48. Find the direction in which $f(x, y, z) = ze^{xy}$ increases most rapidly at the point $(0, 1, 2)$. What is the maximum rate of increase?

49. The contour map shows wind speed in knots during Hurricane Andrew on August 24, 1992. Use it to estimate the value of the directional derivative of the wind speed at Homestead, Florida, in the direction of the eye of the hurricane.

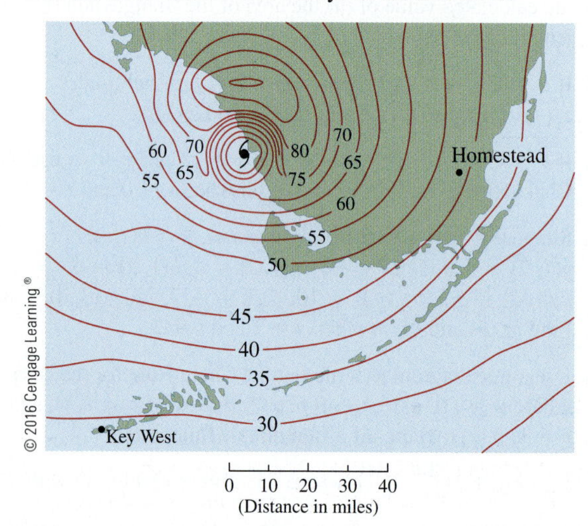

50. Find parametric equations of the tangent line at the point $(-2, 2, 4)$ to the curve of intersection of the surface $z = 2x^2 - y^2$ and the plane $z = 4$.

51–54 Find the local maximum and minimum values and saddle points of the function. If you have three-dimensional graphing software, graph the function with a domain and viewpoint that reveal all the important aspects of the function.

51. $f(x, y) = x^2 - xy + y^2 + 9x - 6y + 10$

52. $f(x, y) = x^3 - 6xy + 8y^3$

53. $f(x, y) = 3xy - x^2y - xy^2$

54. $f(x, y) = (x^2 + y)e^{y/2}$

55–56 Find the absolute maximum and minimum values of f on the set D.

55. $f(x, y) = 4xy^2 - x^2y^2 - xy^3$; D is the closed triangular region in the xy-plane with vertices $(0, 0)$, $(0, 6)$, and $(6, 0)$

56. $f(x, y) = e^{-x^2-y^2}(x^2 + 2y^2)$; D is the disk $x^2 + y^2 \leq 4$

57. Use a graph or level curves or both to estimate the local maximum and minimum values and saddle points of $f(x, y) = x^3 - 3x + y^4 - 2y^2$. Then use calculus to find these values precisely.

58. Use a graphing calculator or computer (or Newton's method or a computer algebra system) to find the critical points of $f(x, y) = 12 + 10y - 2x^2 - 8xy - y^4$ correct to three decimal places. Then classify the critical points and find the highest point on the graph.

59–62 Use Lagrange multipliers to find the maximum and minimum values of f subject to the given constraint(s).

59. $f(x, y) = x^2y$; $x^2 + y^2 = 1$

60. $f(x, y) = \dfrac{1}{x} + \dfrac{1}{y}$; $\dfrac{1}{x^2} + \dfrac{1}{y^2} = 1$

61. $f(x, y, z) = xyz$; $x^2 + y^2 + z^2 = 3$

62. $f(x, y, z) = x^2 + 2y^2 + 3z^2$;
$x + y + z = 1$, $x - y + 2z = 2$

63. Find the points on the surface $xy^2z^3 = 2$ that are closest to the origin.

64. A package in the shape of a rectangular box can be mailed by the US Postal Service if the sum of its length and girth (the perimeter of a cross-section perpendicular to the length) is at most 108 in. Find the dimensions of the package with largest volume that can be mailed.

65. A pentagon is formed by placing an isosceles triangle on a rectangle, as shown in the figure. If the pentagon has fixed perimeter P, find the lengths of the sides of the pentagon that maximize the area of the pentagon.

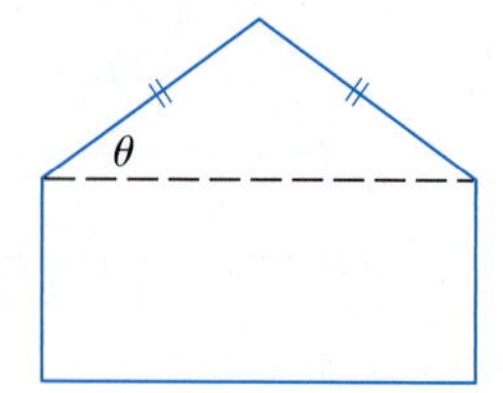

Problems Plus

1. A rectangle with length L and width W is cut into four smaller rectangles by two lines parallel to the sides. Find the maximum and minimum values of the sum of the squares of the areas of the smaller rectangles.

2. Marine biologists have determined that when a shark detects the presence of blood in the water, it will swim in the direction in which the concentration of the blood increases most rapidly. Based on certain tests, the concentration of blood (in parts per million) at a point $P(x, y)$ on the surface of seawater is approximated by

$$C(x, y) = e^{-(x^2+2y^2)/10^4}$$

where x and y are measured in meters in a rectangular coordinate system with the blood source at the origin.

(a) Identify the level curves of the concentration function and sketch several members of this family together with a path that a shark will follow to the source.

(b) Suppose a shark is at the point (x_0, y_0) when it first detects the presence of blood in the water. Find an equation of the shark's path by setting up and solving a differential equation.

3. A long piece of galvanized sheet metal with width w is to be bent into a symmetric form with three straight sides to make a rain gutter. A cross-section is shown in the figure.

(a) Determine the dimensions that allow the maximum possible flow; that is, find the dimensions that give the maximum possible cross-sectional area.

(b) Would it be better to bend the metal into a gutter with a semicircular cross-section?

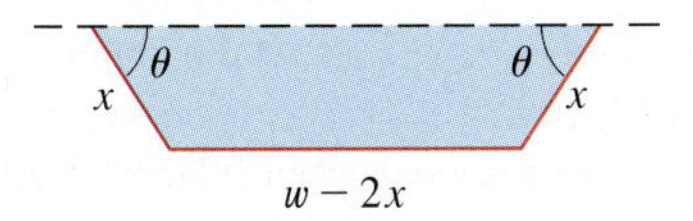

4. For what values of the number r is the function

$$f(x, y, z) = \begin{cases} \dfrac{(x + y + z)^r}{x^2 + y^2 + z^2} & \text{if } (x, y, z) \neq (0, 0, 0) \\ 0 & \text{if } (x, y, z) = (0, 0, 0) \end{cases}$$

continuous on $\mathbb{R}^3$?

5. Suppose f is a differentiable function of one variable. Show that all tangent planes to the surface $z = xf(y/x)$ intersect in a common point.

6. (a) Newton's method for approximating a root of an equation $f(x) = 0$ (see Section 3.8) can be adapted to approximating a solution of a system of equations $f(x, y) = 0$ and $g(x, y) = 0$. The surfaces $z = f(x, y)$ and $z = g(x, y)$ intersect in a curve that intersects the xy-plane at the point (r, s), which is the solution of the system. If an initial approximation (x_1, y_1) is close to this point, then the tangent planes to the surfaces at (x_1, y_1) intersect in a straight line that intersects the xy-plane in a point (x_2, y_2), which should be closer to (r, s). (Compare with Figure 3.8.2.) Show that

$$x_2 = x_1 - \frac{fg_y - f_y g}{f_x g_y - f_y g_x} \qquad \text{and} \qquad y_2 = y_1 - \frac{f_x g - fg_x}{f_x g_y - f_y g_x}$$

where f, g, and their partial derivatives are evaluated at (x_1, y_1). If we continue this procedure, we obtain successive approximations (x_n, y_n).

(b) It was Thomas Simpson (1710–1761) who formulated Newton's method as we know it today and who extended it to functions of two variables as in part (a). (See the biography of Simpson on page 560.) The example that he gave to illustrate the method was to solve the system of equations

$$x^x + y^y = 1000 \qquad x^y + y^x = 100$$

In other words, he found the points of intersection of the curves in the figure. Use the method of part (a) to find the coordinates of the points of intersection correct to six decimal places.

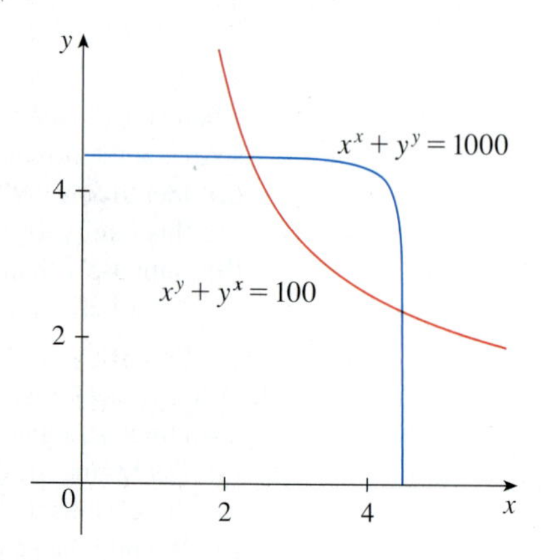

7. If the ellipse $x^2/a^2 + y^2/b^2 = 1$ is to enclose the circle $x^2 + y^2 = 2y$, what values of a and b minimize the area of the ellipse?

8. Show that the maximum value of the function

$$f(x, y) = \frac{(ax + by + c)^2}{x^2 + y^2 + 1}$$

is $a^2 + b^2 + c^2$.
Hint: One method for attacking this problem is to use the Cauchy-Schwarz Inequality:

$$|\mathbf{a} \cdot \mathbf{b}| \leqslant |\mathbf{a}||\mathbf{b}|$$

(See Exercise 12.3.61.)

15 Multiple Integrals

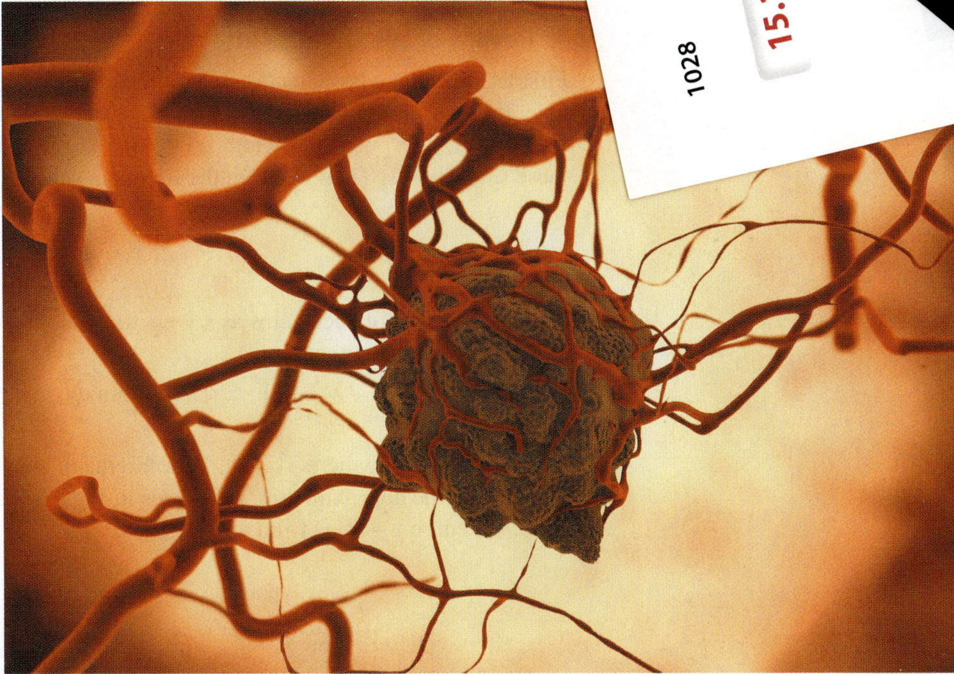

© Juan Gaertner / Shutterstock.com

Tumors, like the one shown, have been modeled as "bumpy spheres." In Exercise 47 in Section 15.8 you are asked to compute the volume enclosed by such a surface.

IN THIS CHAPTER WE EXTEND the idea of a definite integral to double and triple integrals of functions of two or three variables. These ideas are then used to compute volumes, masses, and centroids of more general regions than we were able to consider in Chapters 5 and 8. We also use double integrals to calculate probabilities when two random variables are involved.

We will see that polar coordinates are useful in computing double integrals over some types of regions. In a similar way, we will introduce two new coordinate systems in three-dimensional space—cylindrical coordinates and spherical coordinates—that greatly simplify the computation of triple integrals over certain commonly occurring solid regions.

Integrals over Rectangles

In much the same way that our attempt to solve the area problem led to the definition of a definite integral, we now seek to find the volume of a solid and in the process we arrive at the definition of a double integral.

Review of the Definite Integral

First let's recall the basic facts concerning definite integrals of functions of a single variable. If $f(x)$ is defined for $a \leqslant x \leqslant b$, we start by dividing the interval $[a, b]$ into n subintervals $[x_{i-1}, x_i]$ of equal width $\Delta x = (b - a)/n$ and we choose sample points x_i^* in these subintervals. Then we form the Riemann sum

1
$$\sum_{i=1}^{n} f(x_i^*)\,\Delta x$$

and take the limit of such sums as $n \to \infty$ to obtain the definite integral of f from a to b:

2
$$\int_a^b f(x)\,dx = \lim_{n\to\infty} \sum_{i=1}^{n} f(x_i^*)\,\Delta x$$

In the special case where $f(x) \geqslant 0$, the Riemann sum can be interpreted as the sum of the areas of the approximating rectangles in Figure 1, and $\int_a^b f(x)\,dx$ represents the area under the curve $y = f(x)$ from a to b.

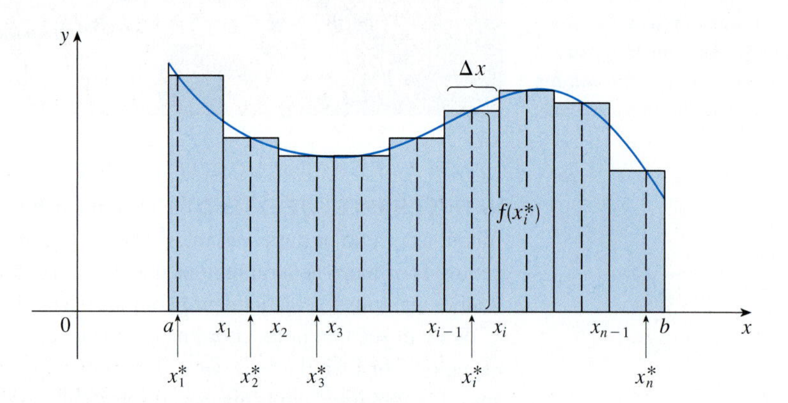

FIGURE 1

Volumes and Double Integrals

In a similar manner we consider a function f of two variables defined on a closed rectangle

$$R = [a, b] \times [c, d] = \{(x, y) \in \mathbb{R}^2 \mid a \leqslant x \leqslant b,\ c \leqslant y \leqslant d\}$$

and we first suppose that $f(x, y) \geqslant 0$. The graph of f is a surface with equation $z = f(x, y)$. Let S be the solid that lies above R and under the graph of f, that is,

$$S = \{(x, y, z) \in \mathbb{R}^3 \mid 0 \leqslant z \leqslant f(x, y),\ (x, y) \in R\}$$

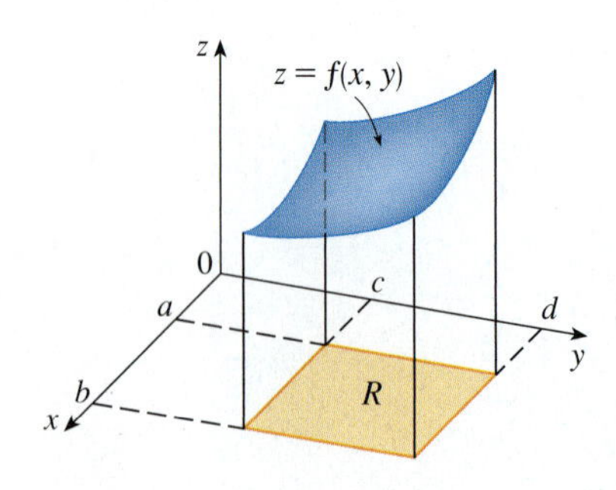

FIGURE 2

(See Figure 2.) Our goal is to find the volume of S.

The first step is to divide the rectangle R into subrectangles. We accomplish this by dividing the interval $[a, b]$ into m subintervals $[x_{i-1}, x_i]$ of equal width $\Delta x = (b - a)/m$ and dividing $[c, d]$ into n subintervals $[y_{j-1}, y_j]$ of equal width $\Delta y = (d - c)/n$. By drawing lines parallel to the coordinate axes through the endpoints of these subintervals, as in Figure 3, we form the subrectangles

$$R_{ij} = [x_{i-1}, x_i] \times [y_{j-1}, y_j] = \{(x, y) \mid x_{i-1} \leq x \leq x_i,\ y_{j-1} \leq y \leq y_j\}$$

each with area $\Delta A = \Delta x \, \Delta y$.

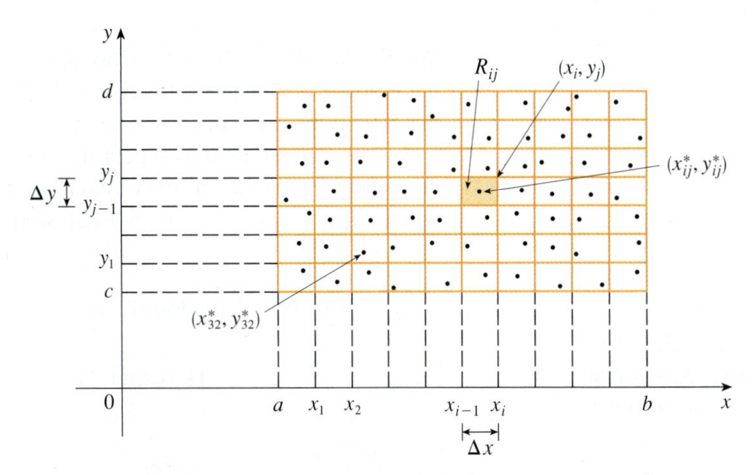

FIGURE 3
Dividing R into subrectangles

If we choose a **sample point** (x_{ij}^*, y_{ij}^*) in each R_{ij}, then we can approximate the part of S that lies above each R_{ij} by a thin rectangular box (or "column") with base R_{ij} and height $f(x_{ij}^*, y_{ij}^*)$ as shown in Figure 4. (Compare with Figure 1.) The volume of this box is the height of the box times the area of the base rectangle:

$$f(x_{ij}^*, y_{ij}^*) \, \Delta A$$

If we follow this procedure for all the rectangles and add the volumes of the corresponding boxes, we get an approximation to the total volume of S:

3
$$V \approx \sum_{i=1}^{m} \sum_{j=1}^{n} f(x_{ij}^*, y_{ij}^*) \, \Delta A$$

(See Figure 5.) This double sum means that for each subrectangle we evaluate f at the chosen point and multiply by the area of the subrectangle, and then we add the results.

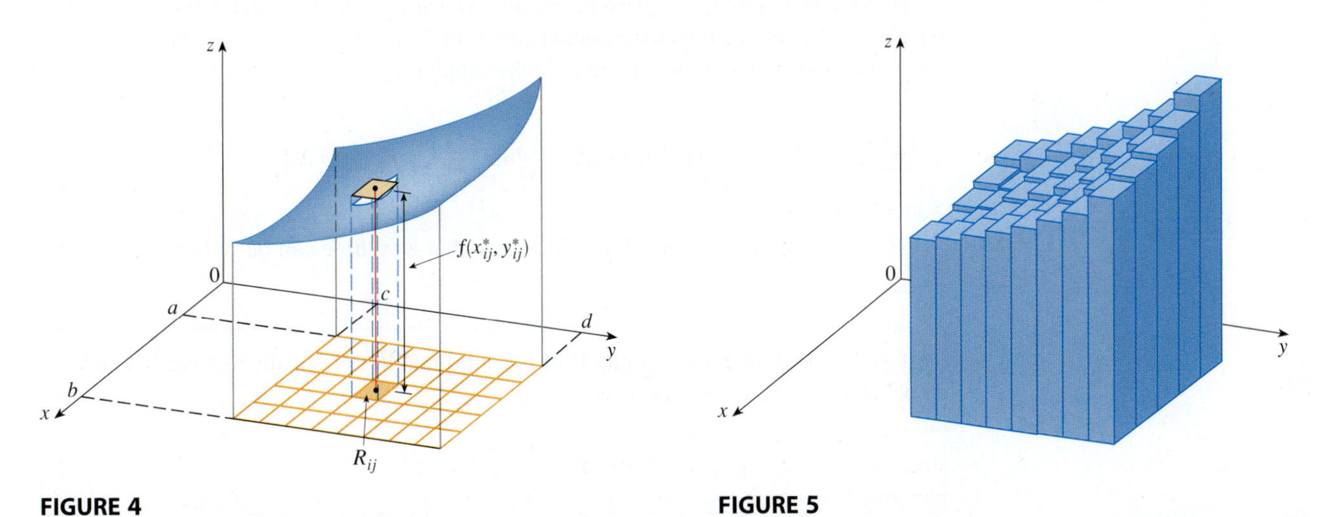

FIGURE 4

FIGURE 5

Our intuition tells us that the approximation given in (3) becomes better as m and n become larger and so we would expect that

The meaning of the double limit in Equation 4 is that we can make the double sum as close as we like to the number V [for any choice of (x_{ij}^*, y_{ij}^*) in R_{ij}] by taking m and n sufficiently large.

4
$$V = \lim_{m, n \to \infty} \sum_{i=1}^{m} \sum_{j=1}^{n} f(x_{ij}^*, y_{ij}^*) \, \Delta A$$

We use the expression in Equation 4 to define the **volume** of the solid S that lies under the graph of f and above the rectangle R. (It can be shown that this definition is consistent with our formula for volume in Section 5.2.)

Limits of the type that appear in Equation 4 occur frequently, not just in finding volumes but in a variety of other situations as well—as we will see in Section 15.4—even when f is not a positive function. So we make the following definition.

Notice the similarity between Definition 5 and the definition of a single integral in Equation 2.

5 Definition The **double integral** of f over the rectangle R is

$$\iint_R f(x, y) \, dA = \lim_{m, n \to \infty} \sum_{i=1}^{m} \sum_{j=1}^{n} f(x_{ij}^*, y_{ij}^*) \, \Delta A$$

if this limit exists.

The precise meaning of the limit in Definition 5 is that for every number $\varepsilon > 0$ there is an integer N such that

Although we have defined the double integral by dividing R into equal-sized subrectangles, we could have used subrectangles R_{ij} of unequal size. But then we would have to ensure that all of their dimensions approach 0 in the limiting process.

$$\left| \iint_R f(x, y) \, dA - \sum_{i=1}^{m} \sum_{j=1}^{n} f(x_{ij}^*, y_{ij}^*) \, \Delta A \right| < \varepsilon$$

for all integers m and n greater than N and for any choice of sample points (x_{ij}^*, y_{ij}^*) in R_{ij}.

A function f is called **integrable** if the limit in Definition 5 exists. It is shown in courses on advanced calculus that all continuous functions are integrable. In fact, the double integral of f exists provided that f is "not too discontinuous." In particular, if f is bounded on R, [that is, there is a constant M such that $|f(x, y)| \leqslant M$ for all (x, y) in R], and f is continuous there, except on a finite number of smooth curves, then f is integrable over R.

The sample point (x_{ij}^*, y_{ij}^*) can be chosen to be any point in the subrectangle R_{ij}, but if we choose it to be the upper right-hand corner of R_{ij} [namely (x_i, y_j), see Figure 3], then the expression for the double integral looks simpler:

6
$$\iint_R f(x, y) \, dA = \lim_{m, n \to \infty} \sum_{i=1}^{m} \sum_{j=1}^{n} f(x_i, y_j) \, \Delta A$$

By comparing Definitions 4 and 5, we see that a volume can be written as a double integral:

If $f(x, y) \geqslant 0$, then the volume V of the solid that lies above the rectangle R and below the surface $z = f(x, y)$ is

$$V = \iint_R f(x, y) \, dA$$

The sum in Definition 5,

$$\sum_{i=1}^{m} \sum_{j=1}^{n} f(x_{ij}^*, y_{ij}^*) \Delta A$$

is called a **double Riemann sum** and is used as an approximation to the value of the double integral. [Notice how similar it is to the Riemann sum in (1) for a function of a single variable.] If f happens to be a *positive* function, then the double Riemann sum represents the sum of volumes of columns, as in Figure 5, and is an approximation to the volume under the graph of f.

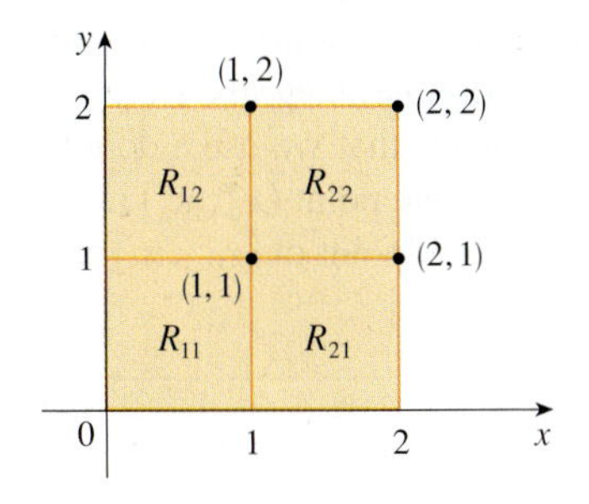

FIGURE 6

EXAMPLE 1 Estimate the volume of the solid that lies above the square $R = [0, 2] \times [0, 2]$ and below the elliptic paraboloid $z = 16 - x^2 - 2y^2$. Divide R into four equal squares and choose the sample point to be the upper right corner of each square R_{ij}. Sketch the solid and the approximating rectangular boxes.

SOLUTION The squares are shown in Figure 6. The paraboloid is the graph of $f(x, y) = 16 - x^2 - 2y^2$ and the area of each square is $\Delta A = 1$. Approximating the volume by the Riemann sum with $m = n = 2$, we have

$$V \approx \sum_{i=1}^{2} \sum_{j=1}^{2} f(x_i, y_j) \Delta A$$

$$= f(1, 1) \Delta A + f(1, 2) \Delta A + f(2, 1) \Delta A + f(2, 2) \Delta A$$

$$= 13(1) + 7(1) + 10(1) + 4(1) = 34$$

This is the volume of the approximating rectangular boxes shown in Figure 7. ■

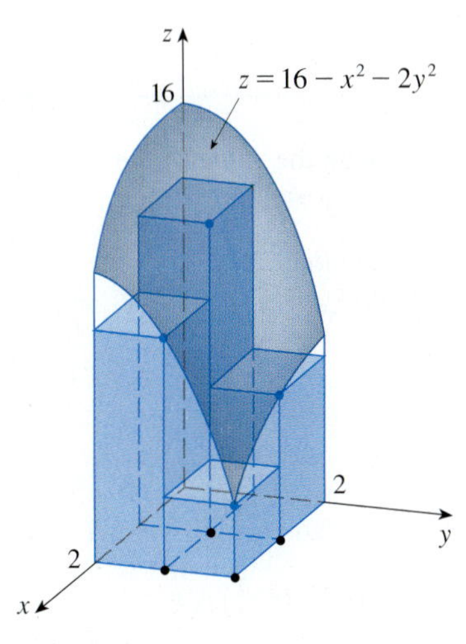

FIGURE 7

We get better approximations to the volume in Example 1 if we increase the number of squares. Figure 8 shows how the columns start to look more like the actual solid and the corresponding approximations become more accurate when we use 16, 64, and 256 squares. In Example 7 we will be able to show that the exact volume is 48.

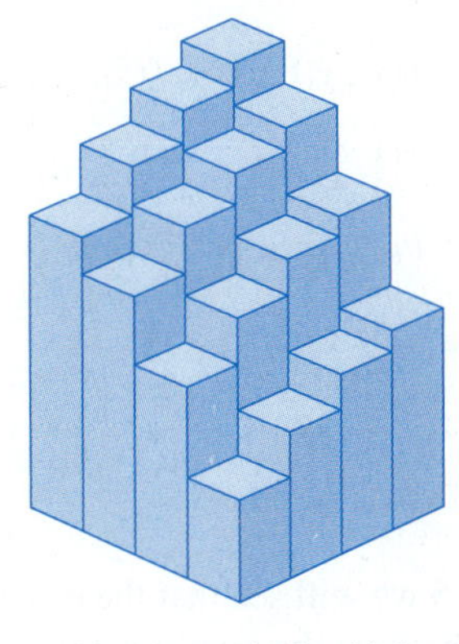

(a) $m = n = 4$, $V \approx 41.5$

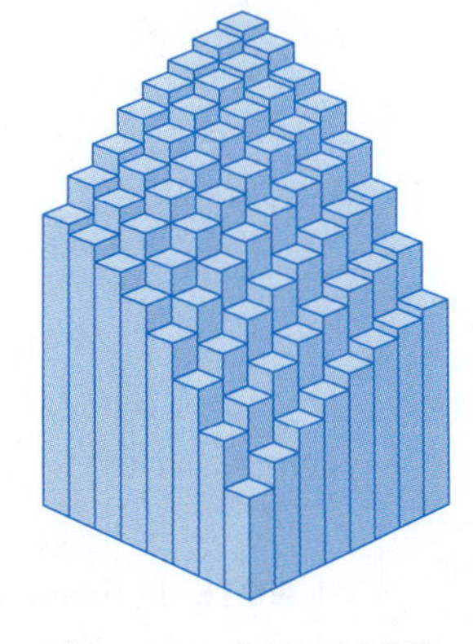

(b) $m = n = 8$, $V \approx 44.875$

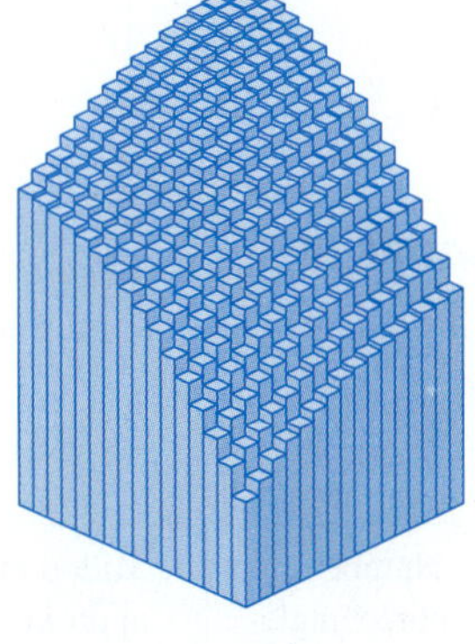

(c) $m = n = 16$, $V \approx 46.46875$

FIGURE 8
The Riemann sum approximations to the volume under $z = 16 - x^2 - 2y^2$ become more accurate as m and n increase.

EXAMPLE 2 If $R = \{(x, y) \mid -1 \leq x \leq 1, -2 \leq y \leq 2\}$, evaluate the integral

$$\iint_R \sqrt{1 - x^2}\, dA$$

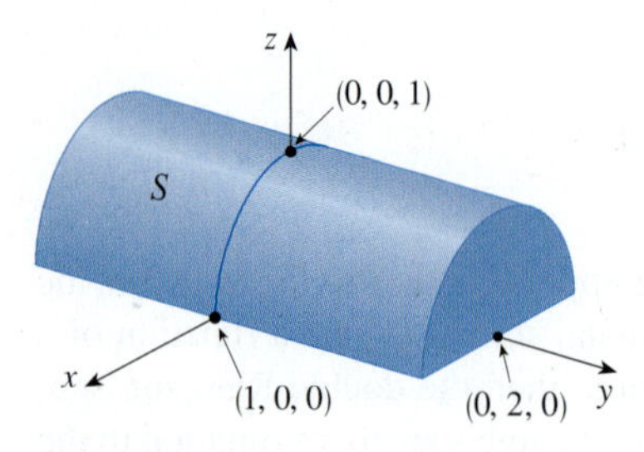

FIGURE 9

SOLUTION It would be very difficult to evaluate this integral directly from Definition 5 but, because $\sqrt{1 - x^2} \geqslant 0$, we can compute the integral by interpreting it as a volume. If $z = \sqrt{1 - x^2}$, then $x^2 + z^2 = 1$ and $z \geqslant 0$, so the given double integral represents the volume of the solid S that lies below the circular cylinder $x^2 + z^2 = 1$ and above the rectangle R. (See Figure 9.) The volume of S is the area of a semicircle with radius 1 times the length of the cylinder. Thus

$$\iint_R \sqrt{1 - x^2}\, dA = \tfrac{1}{2}\pi(1)^2 \times 4 = 2\pi$$ ■

The Midpoint Rule

The methods that we used for approximating single integrals (the Midpoint Rule, the Trapezoidal Rule, Simpson's Rule) all have counterparts for double integrals. Here we consider only the Midpoint Rule for double integrals. This means that we use a double Riemann sum to approximate the double integral, where the sample point (x_{ij}^*, y_{ij}^*) in R_{ij} is chosen to be the center $(\bar{x}_i, \bar{y}_j)$ of R_{ij}. In other words, $\bar{x}_i$ is the midpoint of $[x_{i-1}, x_i]$ and $\bar{y}_j$ is the midpoint of $[y_{j-1}, y_j]$.

Midpoint Rule for Double Integrals

$$\iint_R f(x, y)\, dA \approx \sum_{i=1}^{m} \sum_{j=1}^{n} f(\bar{x}_i, \bar{y}_j)\, \Delta A$$

where $\bar{x}_i$ is the midpoint of $[x_{i-1}, x_i]$ and $\bar{y}_j$ is the midpoint of $[y_{j-1}, y_j]$.

EXAMPLE 3 Use the Midpoint Rule with $m = n = 2$ to estimate the value of the integral $\iint_R (x - 3y^2)\, dA$, where $R = \{(x, y) \mid 0 \leqslant x \leqslant 2,\ 1 \leqslant y \leqslant 2\}$.

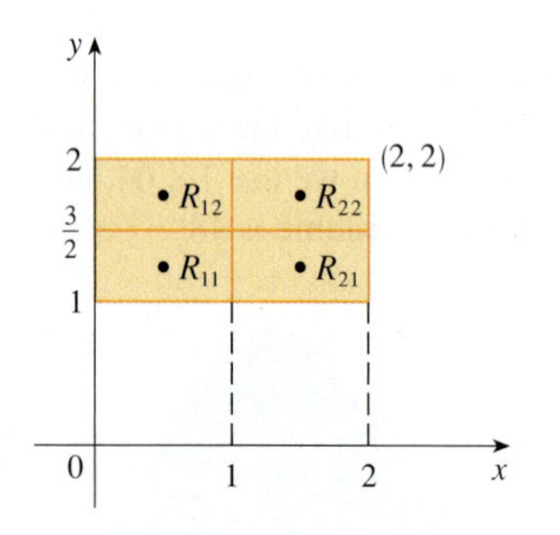

FIGURE 10

SOLUTION In using the Midpoint Rule with $m = n = 2$, we evaluate $f(x, y) = x - 3y^2$ at the centers of the four subrectangles shown in Figure 10. So $\bar{x}_1 = \frac{1}{2}$, $\bar{x}_2 = \frac{3}{2}$, $\bar{y}_1 = \frac{5}{4}$, and $\bar{y}_2 = \frac{7}{4}$. The area of each subrectangle is $\Delta A = \frac{1}{2}$. Thus

$$\begin{aligned}
\iint_R (x - 3y^2)\, dA &\approx \sum_{i=1}^{2} \sum_{j=1}^{2} f(\bar{x}_i, \bar{y}_j)\, \Delta A \\
&= f(\bar{x}_1, \bar{y}_1)\, \Delta A + f(\bar{x}_1, \bar{y}_2)\, \Delta A + f(\bar{x}_2, \bar{y}_1)\, \Delta A + f(\bar{x}_2, \bar{y}_2)\, \Delta A \\
&= f(\tfrac{1}{2}, \tfrac{5}{4})\, \Delta A + f(\tfrac{1}{2}, \tfrac{7}{4})\, \Delta A + f(\tfrac{3}{2}, \tfrac{5}{4})\, \Delta A + f(\tfrac{3}{2}, \tfrac{7}{4})\, \Delta A \\
&= (-\tfrac{67}{16})\tfrac{1}{2} + (-\tfrac{139}{16})\tfrac{1}{2} + (-\tfrac{51}{16})\tfrac{1}{2} + (-\tfrac{123}{16})\tfrac{1}{2} \\
&= -\tfrac{95}{8} = -11.875
\end{aligned}$$

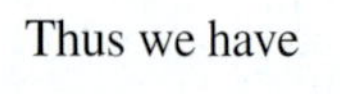

Thus we have
$$\iint_R (x - 3y^2)\, dA \approx -11.875$$ ■

Number of subrectangles	Midpoint Rule approximation
1	−11.5000
4	−11.8750
16	−11.9687
64	−11.9922
256	−11.9980
1024	−11.9995

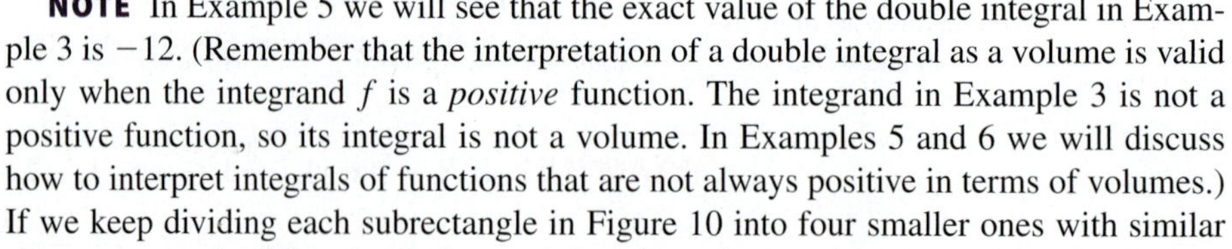

NOTE In Example 5 we will see that the exact value of the double integral in Example 3 is -12. (Remember that the interpretation of a double integral as a volume is valid only when the integrand f is a *positive* function. The integrand in Example 3 is not a positive function, so its integral is not a volume. In Examples 5 and 6 we will discuss how to interpret integrals of functions that are not always positive in terms of volumes.) If we keep dividing each subrectangle in Figure 10 into four smaller ones with similar shape, we get the Midpoint Rule approximations displayed in the chart in the margin. Notice how these approximations approach the exact value of the double integral, -12.

Iterated Integrals

Recall that it is usually difficult to evaluate single integrals directly from the definition of an integral, but the Fundamental Theorem of Calculus provides a much easier method. The evaluation of double integrals from first principles is even more difficult, but here we see how to express a double integral as an iterated integral, which can then be evaluated by calculating two single integrals.

Suppose that f is a function of two variables that is integrable on the rectangle $R = [a, b] \times [c, d]$. We use the notation $\int_c^d f(x, y)\,dy$ to mean that x is held fixed and $f(x, y)$ is integrated with respect to y from $y = c$ to $y = d$. This procedure is called *partial integration with respect to y*. (Notice its similarity to partial differentiation.) Now $\int_c^d f(x, y)\,dy$ is a number that depends on the value of x, so it defines a function of x:

$$A(x) = \int_c^d f(x, y)\,dy$$

If we now integrate the function A with respect to x from $x = a$ to $x = b$, we get

$$\boxed{7} \qquad \int_a^b A(x)\,dx = \int_a^b \left[\int_c^d f(x, y)\,dy \right] dx$$

The integral on the right side of Equation 7 is called an **iterated integral**. Usually the brackets are omitted. Thus

$$\boxed{8} \qquad \int_a^b \int_c^d f(x, y)\,dy\,dx = \int_a^b \left[\int_c^d f(x, y)\,dy \right] dx$$

means that we first integrate with respect to y from c to d and then with respect to x from a to b.

Similarly, the iterated integral

$$\boxed{9} \qquad \int_c^d \int_a^b f(x, y)\,dx\,dy = \int_c^d \left[\int_a^b f(x, y)\,dx \right] dy$$

means that we first integrate with respect to x (holding y fixed) from $x = a$ to $x = b$ and then we integrate the resulting function of y with respect to y from $y = c$ to $y = d$. Notice that in both Equations 8 and 9 we work *from the inside out*.

EXAMPLE 4 Evaluate the iterated integrals.

(a) $\int_0^3 \int_1^2 x^2 y\,dy\,dx$ (b) $\int_1^2 \int_0^3 x^2 y\,dx\,dy$

SOLUTION

(a) Regarding x as a constant, we obtain

$$\int_1^2 x^2 y\,dy = \left[x^2 \frac{y^2}{2} \right]_{y=1}^{y=2} = x^2\left(\frac{2^2}{2}\right) - x^2\left(\frac{1^2}{2}\right) = \tfrac{3}{2}x^2$$

Thus the function A in the preceding discussion is given by $A(x) = \frac{3}{2}x^2$ in this example. We now integrate this function of x from 0 to 3:

$$\int_0^3 \int_1^2 x^2 y\,dy\,dx = \int_0^3 \left[\int_1^2 x^2 y\,dy \right] dx$$

$$= \int_0^3 \tfrac{3}{2}x^2\,dx = \frac{x^3}{2}\bigg]_0^3 = \frac{27}{2}$$

(b) Here we first integrate with respect to x:

$$\int_1^2 \int_0^3 x^2 y \, dx \, dy = \int_1^2 \left[\int_0^3 x^2 y \, dx \right] dy = \int_1^2 \left[\frac{x^3}{3} y \right]_{x=0}^{x=3} dy$$

$$= \int_1^2 9y \, dy = 9 \frac{y^2}{2} \bigg]_1^2 = \frac{27}{2}$$

Notice that in Example 4 we obtained the same answer whether we integrated with respect to y or x first. In general, it turns out (see Theorem 10) that the two iterated integrals in Equations 8 and 9 are always equal; that is, the order of integration does not matter. (This is similar to Clairaut's Theorem on the equality of the mixed partial derivatives.)

The following theorem gives a practical method for evaluating a double integral by expressing it as an iterated integral (in either order).

Theorem 10 is named after the Italian mathematician Guido Fubini (1879–1943), who proved a very general version of this theorem in 1907. But the version for continuous functions was known to the French mathematician Augustin-Louis Cauchy almost a century earlier.

10 Fubini's Theorem If f is continuous on the rectangle $R = \{(x, y) \mid a \leqslant x \leqslant b, c \leqslant y \leqslant d\}$, then

$$\iint_R f(x, y) \, dA = \int_a^b \int_c^d f(x, y) \, dy \, dx = \int_c^d \int_a^b f(x, y) \, dx \, dy$$

More generally, this is true if we assume that f is bounded on R, f is discontinuous only on a finite number of smooth curves, and the iterated integrals exist.

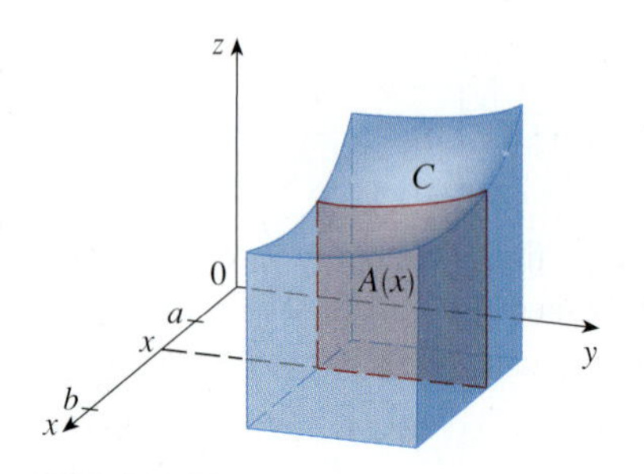

FIGURE 11

The proof of Fubini's Theorem is too difficult to include in this book, but we can at least give an intuitive indication of why it is true for the case where $f(x, y) \geqslant 0$. Recall that if f is positive, then we can interpret the double integral $\iint_R f(x, y) \, dA$ as the volume V of the solid S that lies above R and under the surface $z = f(x, y)$. But we have another formula that we used for volume in Chapter 5, namely,

$$V = \int_a^b A(x) \, dx$$

where $A(x)$ is the area of a cross-section of S in the plane through x perpendicular to the x-axis. From Figure 11 you can see that $A(x)$ is the area under the curve C whose equation is $z = f(x, y)$, where x is held constant and $c \leqslant y \leqslant d$. Therefore

TEC Visual 15.1 illustrates Fubini's Theorem by showing an animation of Figures 11 and 12.

$$A(x) = \int_c^d f(x, y) \, dy$$

and we have

$$\iint_R f(x, y) \, dA = V = \int_a^b A(x) \, dx = \int_a^b \int_c^d f(x, y) \, dy \, dx$$

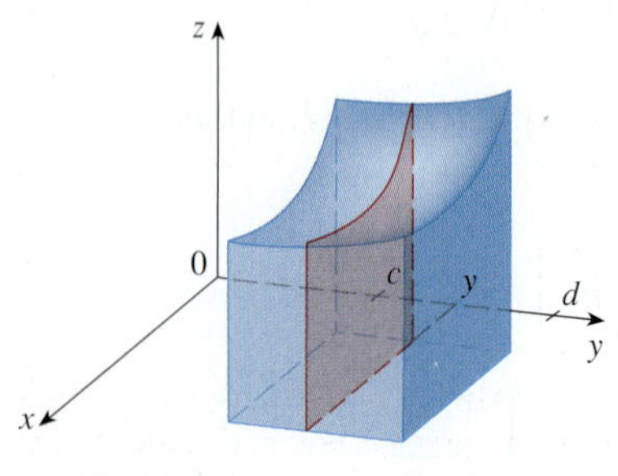

FIGURE 12

A similar argument, using cross-sections perpendicular to the y-axis as in Figure 12, shows that

$$\iint_R f(x, y) \, dA = \int_c^d \int_a^b f(x, y) \, dx \, dy$$

EXAMPLE 5 Evaluate the double integral $\iint_R (x - 3y^2)\,dA$, where $R = \{(x, y) \mid 0 \leqslant x \leqslant 2,\ 1 \leqslant y \leqslant 2\}$. (Compare with Example 3.)

SOLUTION 1 Fubini's Theorem gives

$$\iint_R (x - 3y^2)\,dA = \int_0^2 \int_1^2 (x - 3y^2)\,dy\,dx = \int_0^2 \Big[xy - y^3\Big]_{y=1}^{y=2} dx$$

$$= \int_0^2 (x - 7)\,dx = \frac{x^2}{2} - 7x\Big]_0^2 = -12$$

SOLUTION 2 Again applying Fubini's Theorem, but this time integrating with respect to x first, we have

$$\iint_R (x - 3y^2)\,dA = \int_1^2 \int_0^2 (x - 3y^2)\,dx\,dy$$

$$= \int_1^2 \left[\frac{x^2}{2} - 3xy^2\right]_{x=0}^{x=2} dy$$

$$= \int_1^2 (2 - 6y^2)\,dy = 2y - 2y^3\Big]_1^2 = -12$$ ■

Notice the negative answer in Example 5; nothing is wrong with that. The function f is not a positive function, so its integral doesn't represent a volume. From Figure 13 we see that f is always negative on R, so the value of the integral is the *negative* of the volume that lies *above* the graph of f and *below* R.

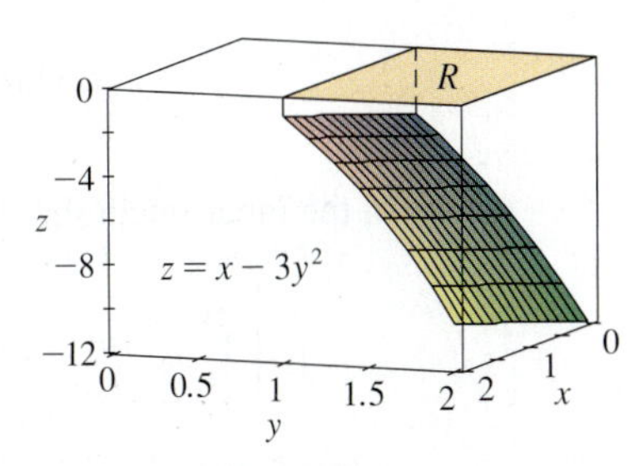

FIGURE 13

EXAMPLE 6 Evaluate $\iint_R y\sin(xy)\,dA$, where $R = [1, 2] \times [0, \pi]$.

SOLUTION If we first integrate with respect to x, we get

$$\iint_R y\sin(xy)\,dA = \int_0^\pi \int_1^2 y\sin(xy)\,dx\,dy$$

$$= \int_0^\pi \Big[-\cos(xy)\Big]_{x=1}^{x=2} dy$$

$$= \int_0^\pi (-\cos 2y + \cos y)\,dy$$

$$= -\tfrac{1}{2}\sin 2y + \sin y\Big]_0^\pi = 0$$ ■

For a function f that takes on both positive and negative values, $\iint_R f(x, y)\,dA$ is a difference of volumes: $V_1 - V_2$, where V_1 is the volume above R and below the graph of f, and V_2 is the volume below R and above the graph. The fact that the integral in Example 6 is 0 means that these two volumes V_1 and V_2 are equal. (See Figure 14.)

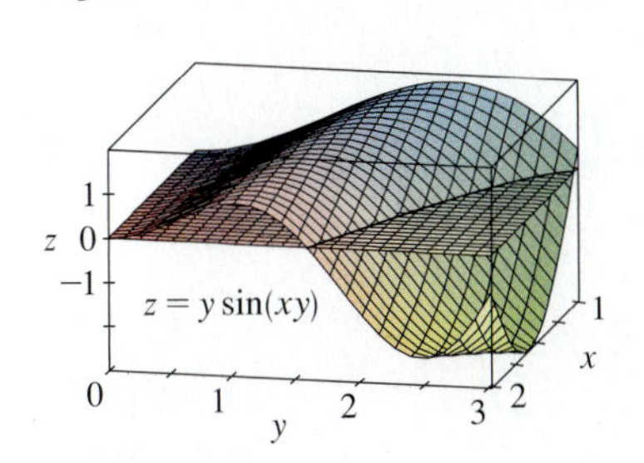

FIGURE 14

NOTE If we reverse the order of integration and first integrate with respect to y in Example 6, we get

$$\iint_R y\sin(xy)\,dA = \int_1^2 \int_0^\pi y\sin(xy)\,dy\,dx$$

but this order of integration is much more difficult than the method given in the example because it involves integration by parts twice. Therefore, when we evaluate double integrals it is wise to choose the order of integration that gives simpler integrals.

EXAMPLE 7 Find the volume of the solid S that is bounded by the elliptic paraboloid $x^2 + 2y^2 + z = 16$, the planes $x = 2$ and $y = 2$, and the three coordinate planes.

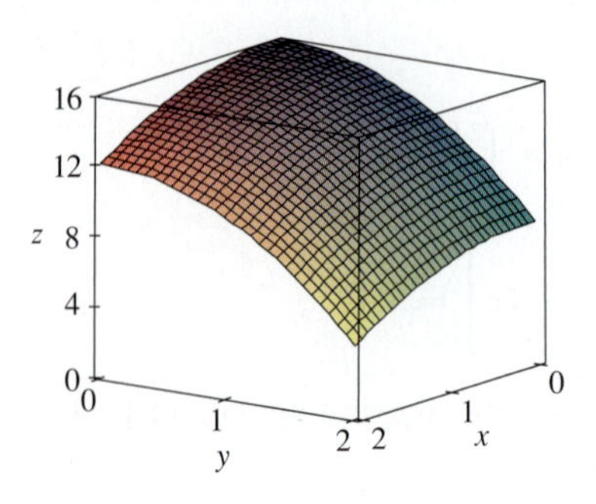

FIGURE 15

SOLUTION We first observe that S is the solid that lies under the surface $z = 16 - x^2 - 2y^2$ and above the square $R = [0, 2] \times [0, 2]$. (See Figure 15.) This solid was considered in Example 1, but we are now in a position to evaluate the double integral using Fubini's Theorem. Therefore

$$V = \iint_R (16 - x^2 - 2y^2)\, dA = \int_0^2 \int_0^2 (16 - x^2 - 2y^2)\, dx\, dy$$

$$= \int_0^2 \Big[16x - \tfrac{1}{3}x^3 - 2y^2 x\Big]_{x=0}^{x=2} dy$$

$$= \int_0^2 \left(\tfrac{88}{3} - 4y^2\right) dy = \Big[\tfrac{88}{3}y - \tfrac{4}{3}y^3\Big]_0^2 = 48$$

In the special case where $f(x, y)$ can be factored as the product of a function of x only and a function of y only, the double integral of f can be written in a particularly simple form. To be specific, suppose that $f(x, y) = g(x)h(y)$ and $R = [a, b] \times [c, d]$. Then Fubini's Theorem gives

$$\iint_R f(x, y)\, dA = \int_c^d \int_a^b g(x)h(y)\, dx\, dy = \int_c^d \left[\int_a^b g(x)h(y)\, dx\right] dy$$

In the inner integral, y is a constant, so $h(y)$ is a constant and we can write

$$\int_c^d \left[\int_a^b g(x)h(y)\, dx\right] dy = \int_c^d \left[h(y)\left(\int_a^b g(x)\, dx\right)\right] dy = \int_a^b g(x)\, dx \int_c^d h(y)\, dy$$

since $\int_a^b g(x)\, dx$ is a constant. Therefore, in this case the double integral of f can be written as the product of two single integrals:

11 $$\iint_R g(x)h(y)\, dA = \int_a^b g(x)\, dx \int_c^d h(y)\, dy \qquad \text{where } R = [a, b] \times [c, d]$$

EXAMPLE 8 If $R = [0, \pi/2] \times [0, \pi/2]$, then, by Equation 11,

$$\iint_R \sin x \cos y\, dA = \int_0^{\pi/2} \sin x\, dx \int_0^{\pi/2} \cos y\, dy$$

$$= \Big[-\cos x\Big]_0^{\pi/2} \Big[\sin y\Big]_0^{\pi/2} = 1 \cdot 1 = 1$$

The function $f(x, y) = \sin x \cos y$ in Example 8 is positive on R, so the integral represents the volume of the solid that lies above R and below the graph of f shown in Figure 16.

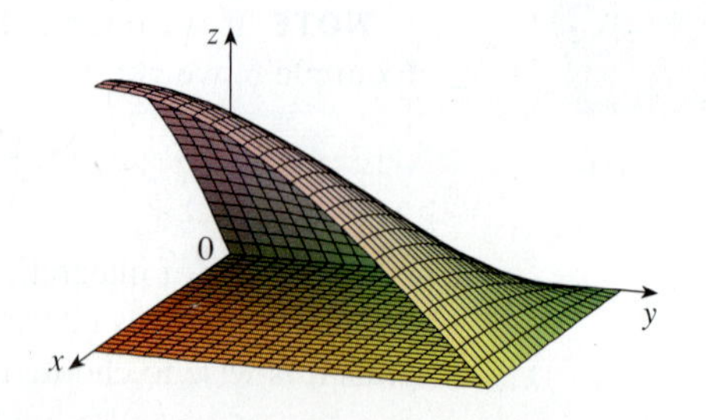

FIGURE 16

Average Value

Recall from Section 5.5 that the average value of a function f of one variable defined on an interval $[a, b]$ is

$$f_{\text{ave}} = \frac{1}{b-a}\int_a^b f(x)\,dx$$

In a similar fashion we define the **average value** of a function f of two variables defined on a rectangle R to be

$$f_{\text{ave}} = \frac{1}{A(R)}\iint_R f(x, y)\,dA$$

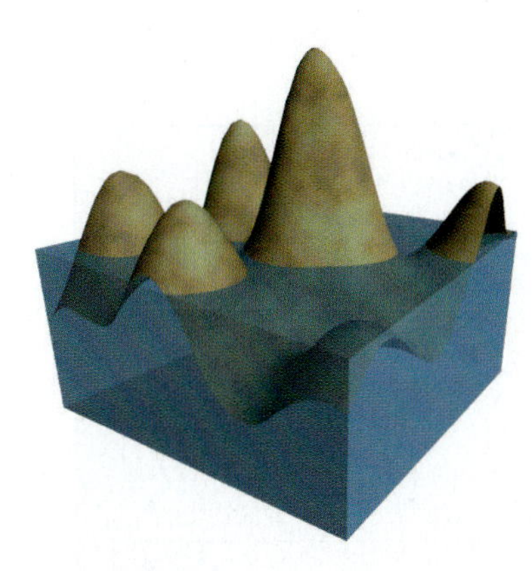

FIGURE 17

where $A(R)$ is the area of R.

If $f(x, y) \geqslant 0$, the equation

$$A(R) \times f_{\text{ave}} = \iint_R f(x, y)\,dA$$

says that the box with base R and height f_{ave} has the same volume as the solid that lies under the graph of f. [If $z = f(x, y)$ describes a mountainous region and you chop off the tops of the mountains at height f_{ave}, then you can use them to fill in the valleys so that the region becomes completely flat. See Figure 17.]

EXAMPLE 9 The contour map in Figure 18 shows the snowfall, in inches, that fell on the state of Colorado on December 20 and 21, 2006. (The state is in the shape of a rectangle that measures 388 mi west to east and 276 mi south to north.) Use the contour map to estimate the average snowfall for the entire state of Colorado on those days.

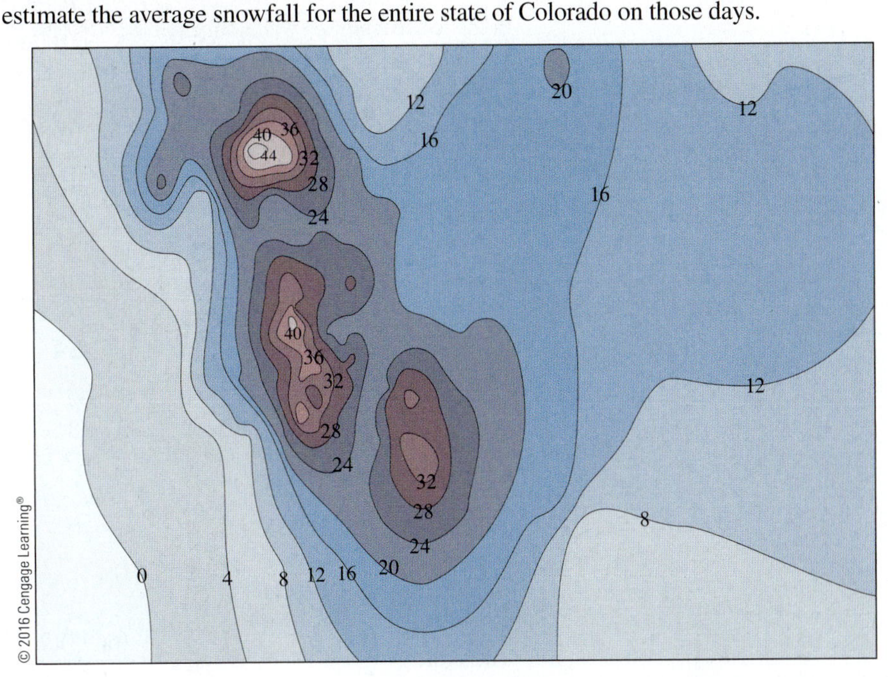

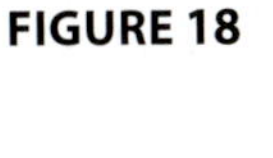

FIGURE 18

SOLUTION Let's place the origin at the southwest corner of the state. Then $0 \leqslant x \leqslant 388$, $0 \leqslant y \leqslant 276$, and $f(x, y)$ is the snowfall, in inches, at a location x miles to the east and y miles to the north of the origin. If R is the rectangle that represents Colorado, then the average snowfall for the state on December 20–21 was

$$f_{\text{ave}} = \frac{1}{A(R)}\iint_R f(x, y)\,dA$$

where $A(R) = 388 \cdot 276$. To estimate the value of this double integral, let's use the Midpoint Rule with $m = n = 4$. In other words, we divide R into 16 subrectangles of equal size, as in Figure 19. The area of each subrectangle is

$$\Delta A = \tfrac{1}{16}(388)(276) = 6693 \text{ mi}^2$$

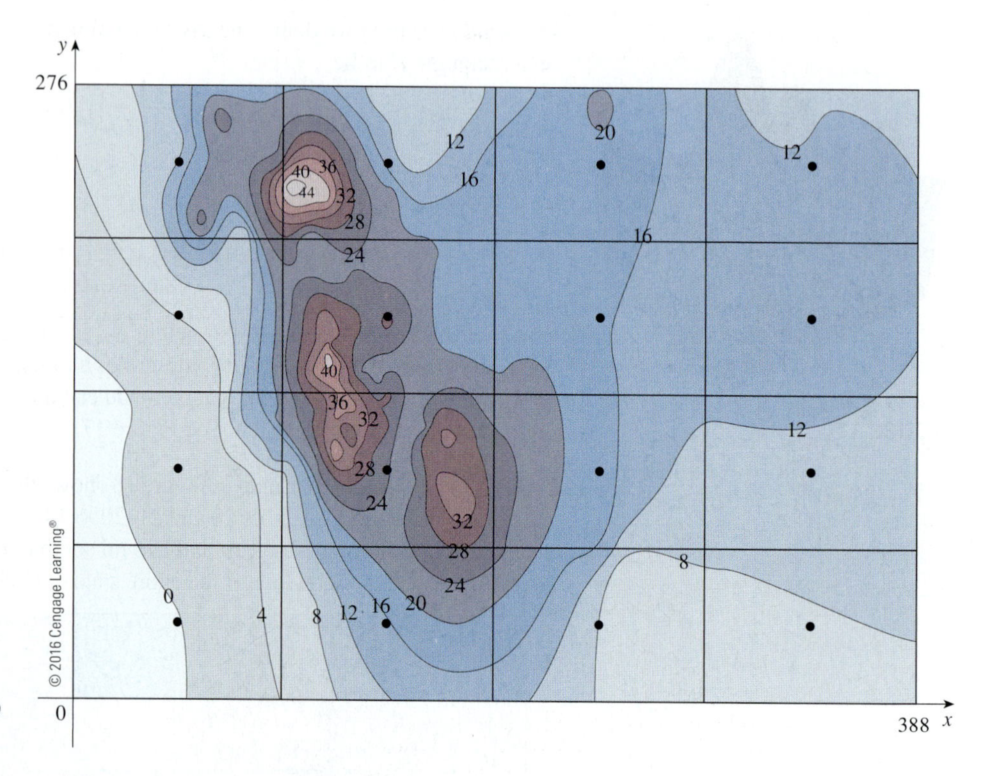

FIGURE 19

Using the contour map to estimate the value of f at the center of each subrectangle, we get

$$\iint_R f(x, y)\, dA \approx \sum_{i=1}^{4} \sum_{j=1}^{4} f(\bar{x}_i, \bar{y}_j)\, \Delta A$$

$$\approx \Delta A[0 + 15 + 8 + 7 + 2 + 25 + 18.5 + 11$$

$$+ 4.5 + 28 + 17 + 13.5 + 12 + 15 + 17.5 + 13]$$

$$= (6693)(207)$$

Therefore

$$f_{\text{ave}} \approx \frac{(6693)(207)}{(388)(276)} \approx 12.9$$

On December 20–21, 2006, Colorado received an average of approximately 13 inches of snow. ■

15.1 EXERCISES

1. (a) Estimate the volume of the solid that lies below the surface $z = xy$ and above the rectangle
$$R = \{(x, y) \mid 0 \leq x \leq 6, 0 \leq y \leq 4\}$$
Use a Riemann sum with $m = 3$, $n = 2$, and take the sample point to be the upper right corner of each square.
(b) Use the Midpoint Rule to estimate the volume of the solid in part (a).

2. If $R = [0, 4] \times [-1, 2]$, use a Riemann sum with $m = 2$, $n = 3$ to estimate the value of $\iint_R (1 - xy^2)\, dA$. Take the sample points to be (a) the lower right corners and (b) the upper left corners of the rectangles.

3. (a) Use a Riemann sum with $m = n = 2$ to estimate the value of $\iint_R xe^{-xy}\, dA$, where $R = [0, 2] \times [0, 1]$. Take the sample points to be upper right corners.
(b) Use the Midpoint Rule to estimate the integral in part (a).

4. (a) Estimate the volume of the solid that lies below the surface $z = 1 + x^2 + 3y$ and above the rectangle $R = [1, 2] \times [0, 3]$. Use a Riemann sum with $m = n = 2$ and choose the sample points to be lower left corners.
(b) Use the Midpoint Rule to estimate the volume in part (a).

5. Let V be the volume of the solid that lies under the graph of $f(x, y) = \sqrt{52 - x^2 - y^2}$ and above the rectangle given by $2 \leq x \leq 4$, $2 \leq y \leq 6$. Use the lines $x = 3$ and $y = 4$ to divide R into subrectangles. Let L and U be the Riemann sums computed using lower left corners and upper right corners, respectively. Without calculating the numbers V, L, and U, arrange them in increasing order and explain your reasoning.

6. A 20-ft-by-30-ft swimming pool is filled with water. The depth is measured at 5-ft intervals, starting at one corner of the pool, and the values are recorded in the table. Estimate the volume of water in the pool.

	0	5	10	15	20	25	30
0	2	3	4	6	7	8	8
5	2	3	4	7	8	10	8
10	2	4	6	8	10	12	10
15	2	3	4	5	6	8	7
20	2	2	2	2	3	4	4

7. A contour map is shown for a function f on the square $R = [0, 4] \times [0, 4]$.
(a) Use the Midpoint Rule with $m = n = 2$ to estimate the value of $\iint_R f(x, y)\, dA$.
(b) Estimate the average value of f.

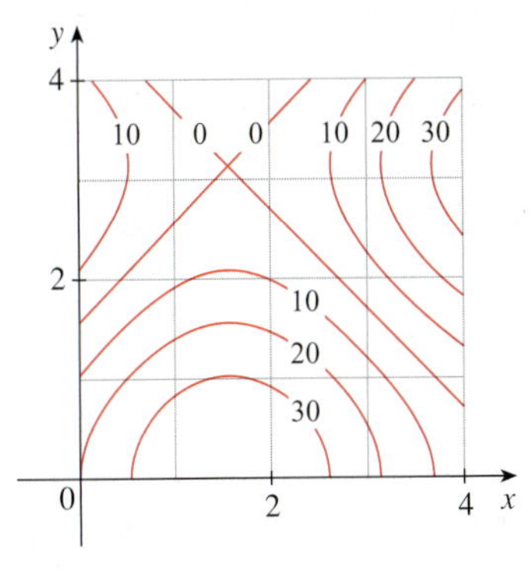

8. The contour map shows the temperature, in degrees Fahrenheit, at 4:00 PM on February 26, 2007, in Colorado. (The state measures 388 mi west to east and 276 mi south to north.) Use the Midpoint Rule with $m = n = 4$ to estimate the average temperature in Colorado at that time.

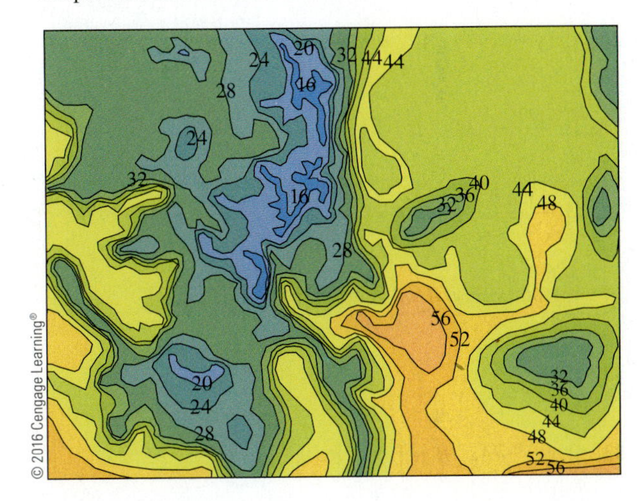

9–11 Evaluate the double integral by first identifying it as the volume of a solid.

9. $\iint_R \sqrt{2}\, dA$, $R = \{(x, y) \mid 2 \leq x \leq 6, -1 \leq y \leq 5\}$

10. $\iint_R (2x + 1)\, dA$, $R = \{(x, y) \mid 0 \leq x \leq 2, 0 \leq y \leq 4\}$

11. $\iint_R (4 - 2y)\, dA$, $R = [0, 1] \times [0, 1]$

12. The integral $\iint_R \sqrt{9 - y^2}\, dA$, where $R = [0, 4] \times [0, 2]$, represents the volume of a solid. Sketch the solid.

13–14 Find $\int_0^2 f(x, y)\, dx$ and $\int_0^3 f(x, y)\, dy$

13. $f(x, y) = x + 3x^2y^2$

14. $f(x, y) = y\sqrt{x + 2}$

15–26 Calculate the iterated integral.

15. $\int_1^4 \int_0^2 (6x^2y - 2x)\, dy\, dx$

16. $\int_0^1 \int_0^1 (x + y)^2\, dx\, dy$

17. $\int_0^1 \int_1^2 (x + e^{-y})\, dx\, dy$

18. $\int_0^{\pi/6} \int_0^{\pi/2} (\sin x + \sin y)\, dy\, dx$

19. $\int_{-3}^3 \int_0^{\pi/2} (y + y^2 \cos x)\, dx\, dy$

20. $\int_1^3 \int_1^5 \frac{\ln y}{xy}\, dy\, dx$

21. $\int_1^4 \int_1^2 \left(\frac{x}{y} + \frac{y}{x}\right) dy\, dx$

22. $\int_0^1 \int_0^2 ye^{x-y}\, dx\, dy$

23. $\int_0^3 \int_0^{\pi/2} t^2 \sin^3\phi\, d\phi\, dt$

24. $\int_0^1 \int_0^1 xy\sqrt{x^2 + y^2}\, dy\, dx$

25. $\int_0^1 \int_0^1 v(u + v^2)^4\, du\, dv$

26. $\int_0^1 \int_0^1 \sqrt{s + t}\, ds\, dt$

27–34 Calculate the double integral.

27. $\iint_R x \sec^2 y\, dA, \quad R = \{(x, y) \mid 0 \leqslant x \leqslant 2, 0 \leqslant y \leqslant \pi/4\}$

28. $\iint_R (y + xy^{-2})\, dA, \quad R = \{(x, y) \mid 0 \leqslant x \leqslant 2, 1 \leqslant y \leqslant 2\}$

29. $\iint_R \frac{xy^2}{x^2 + 1}\, dA, \quad R = \{(x, y) \mid 0 \leqslant x \leqslant 1, -3 \leqslant y \leqslant 3\}$

30. $\iint_R \frac{\tan \theta}{\sqrt{1 - t^2}}\, dA, \quad R = \{(\theta, t) \mid 0 \leqslant \theta \leqslant \pi/3, 0 \leqslant t \leqslant \frac{1}{2}\}$

31. $\iint_R x \sin(x + y)\, dA, \quad R = [0, \pi/6] \times [0, \pi/3]$

32. $\iint_R \frac{x}{1 + xy}\, dA, \quad R = [0, 1] \times [0, 1]$

33. $\iint_R ye^{-xy}\, dA, \quad R = [0, 2] \times [0, 3]$

34. $\iint_R \frac{1}{1 + x + y}\, dA, \quad R = [1, 3] \times [1, 2]$

35–36 Sketch the solid whose volume is given by the iterated integral.

35. $\int_0^1 \int_0^1 (4 - x - 2y)\, dx\, dy$

36. $\int_0^1 \int_0^1 (2 - x^2 - y^2)\, dy\, dx$

37. Find the volume of the solid that lies under the plane $4x + 6y - 2z + 15 = 0$ and above the rectangle $R = \{(x, y) \mid -1 \leqslant x \leqslant 2, -1 \leqslant y \leqslant 1\}$.

38. Find the volume of the solid that lies under the hyperbolic paraboloid $z = 3y^2 - x^2 + 2$ and above the rectangle $R = [-1, 1] \times [1, 2]$.

39. Find the volume of the solid lying under the elliptic paraboloid $x^2/4 + y^2/9 + z = 1$ and above the rectangle $R = [-1, 1] \times [-2, 2]$.

40. Find the volume of the solid enclosed by the surface $z = x^2 + xy^2$ and the planes $z = 0$, $x = 0$, $x = 5$, and $y = \pm 2$.

41. Find the volume of the solid enclosed by the surface $z = 1 + x^2ye^y$ and the planes $z = 0$, $x = \pm 1$, $y = 0$, and $y = 1$.

42. Find the volume of the solid in the first octant bounded by the cylinder $z = 16 - x^2$ and the plane $y = 5$.

43. Find the volume of the solid enclosed by the paraboloid $z = 2 + x^2 + (y - 2)^2$ and the planes $z = 1$, $x = 1$, $x = -1$, $y = 0$, and $y = 4$.

44. Graph the solid that lies between the surface $z = 2xy/(x^2 + 1)$ and the plane $z = x + 2y$ and is bounded by the planes $x = 0$, $x = 2$, $y = 0$, and $y = 4$. Then find its volume.

CAS **45.** Use a computer algebra system to find the exact value of the integral $\iint_R x^5y^3e^{xy}\, dA$, where $R = [0, 1] \times [0, 1]$. Then use the CAS to draw the solid whose volume is given by the integral.

CAS **46.** Graph the solid that lies between the surfaces $z = e^{-x^2} \cos(x^2 + y^2)$ and $z = 2 - x^2 - y^2$ for $|x| \leqslant 1$, $|y| \leqslant 1$. Use a computer algebra system to approximate the volume of this solid correct to four decimal places.

47–48 Find the average value of f over the given rectangle.

47. $f(x, y) = x^2y$, R has vertices $(-1, 0)$, $(-1, 5)$, $(1, 5)$, $(1, 0)$

48. $f(x, y) = e^y\sqrt{x + e^y}, \quad R = [0, 4] \times [0, 1]$

49–50 Use symmetry to evaluate the double integral.

49. $\iint_R \frac{xy}{1 + x^4}\, dA, \quad R = \{(x, y) \mid -1 \leqslant x \leqslant 1, 0 \leqslant y \leqslant 1\}$

50. $\iint_R (1 + x^2 \sin y + y^2 \sin x)\, dA, \quad R = [-\pi, \pi] \times [-\pi, \pi]$

CAS **51.** Use a CAS to compute the iterated integrals

$$\int_0^1 \int_0^1 \frac{x - y}{(x + y)^3}\, dy\, dx \quad \text{and} \quad \int_0^1 \int_0^1 \frac{x - y}{(x + y)^3}\, dx\, dy$$

Do the answers contradict Fubini's Theorem? Explain what is happening.

52. (a) In what way are the theorems of Fubini and Clairaut similar?

(b) If $f(x, y)$ is continuous on $[a, b] \times [c, d]$ and

$$g(x, y) = \int_a^x \int_c^y f(s, t)\, dt\, ds$$

for $a < x < b$, $c < y < d$, show that $g_{xy} = g_{yx} = f(x, y)$.

15.2 Double Integrals over General Regions

For single integrals, the region over which we integrate is always an interval. But for double integrals, we want to be able to integrate a function f not just over rectangles but also over regions D of more general shape, such as the one illustrated in Figure 1. We suppose that D is a bounded region, which means that D can be enclosed in a rectangular region R as in Figure 2. Then we define a new function F with domain R by

1
$$F(x, y) = \begin{cases} f(x, y) & \text{if } (x, y) \text{ is in } D \\ 0 & \text{if } (x, y) \text{ is in } R \text{ but not in } D \end{cases}$$

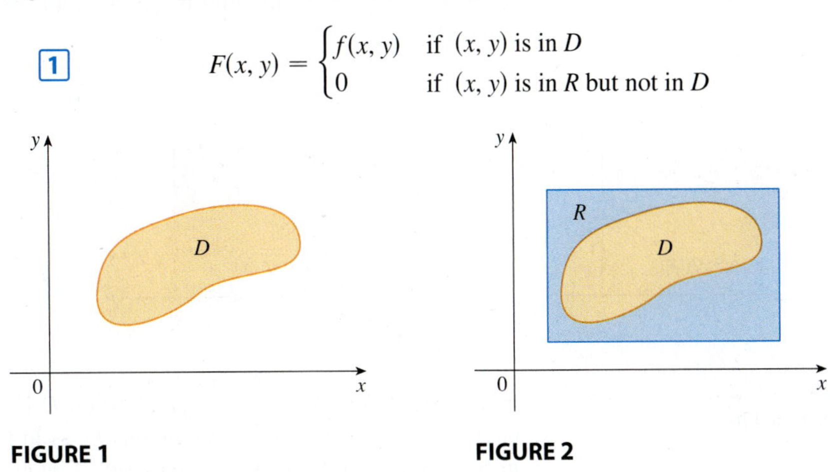

FIGURE 1 **FIGURE 2**

If F is integrable over R, then we define the **double integral of f over D** by

2
$$\iint_D f(x, y)\, dA = \iint_R F(x, y)\, dA \qquad \text{where } F \text{ is given by Equation 1}$$

Definition 2 makes sense because R is a rectangle and so $\iint_R F(x, y)\, dA$ has been previously defined in Section 15.1. The procedure that we have used is reasonable because the values of $F(x, y)$ are 0 when (x, y) lies outside D and so they contribute nothing to the integral. This means that it doesn't matter what rectangle R we use as long as it contains D.

In the case where $f(x, y) \geqslant 0$, we can still interpret $\iint_D f(x, y)\, dA$ as the volume of the solid that lies above D and under the surface $z = f(x, y)$ (the graph of f). You can see that this is reasonable by comparing the graphs of f and F in Figures 3 and 4 and remembering that $\iint_R F(x, y)\, dA$ is the volume under the graph of F.

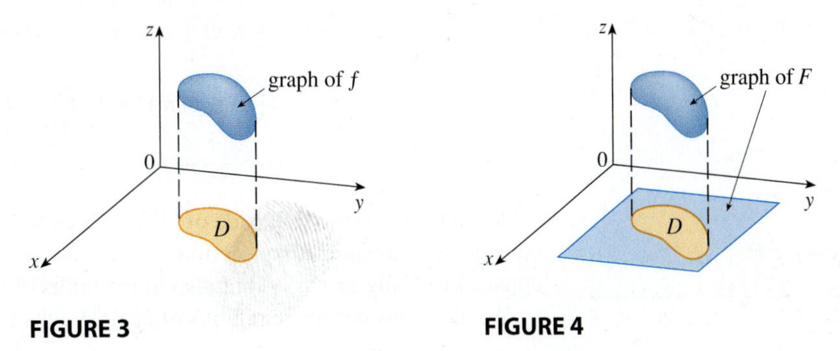

FIGURE 3 **FIGURE 4**

Figure 4 also shows that F is likely to have discontinuities at the boundary points of D. Nonetheless, if f is continuous on D and the boundary curve of D is "well behaved" (in a sense outside the scope of this book), then it can be shown that $\iint_R F(x, y)\, dA$ exists and therefore $\iint_D f(x, y)\, dA$ exists. In particular, this is the case for the following two types of regions.

A plane region D is said to be of **type I** if it lies between the graphs of two continuous functions of x, that is,

$$D = \{(x, y) \mid a \leqslant x \leqslant b,\ g_1(x) \leqslant y \leqslant g_2(x)\}$$

where g_1 and g_2 are continuous on $[a, b]$. Some examples of type I regions are shown in Figure 5.

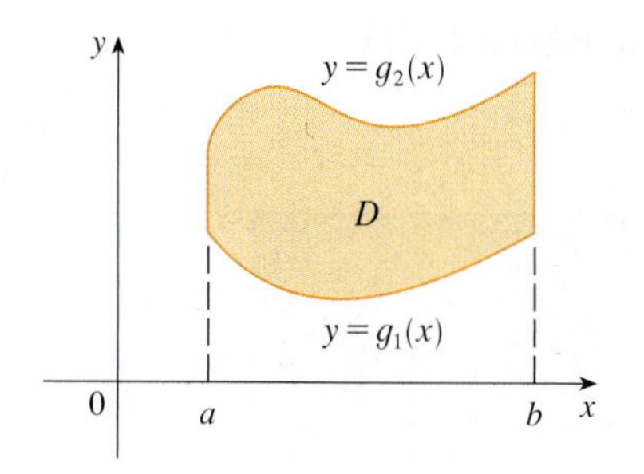

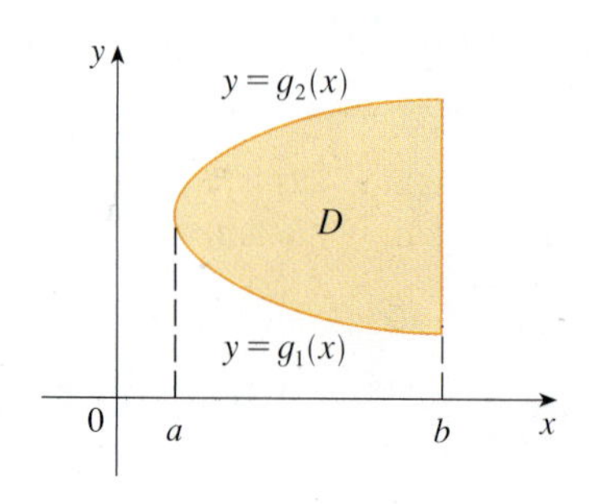

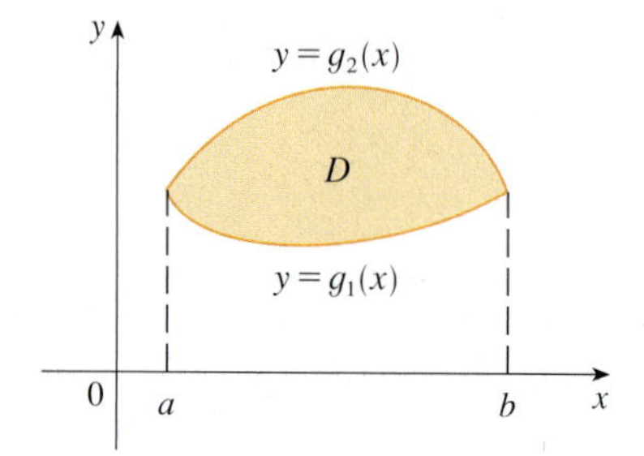

FIGURE 5
Some type I regions

In order to evaluate $\iint_D f(x, y)\, dA$ when D is a region of type I, we choose a rectangle $R = [a, b] \times [c, d]$ that contains D, as in Figure 6, and we let F be the function given by Equation 1; that is, F agrees with f on D and F is 0 outside D. Then, by Fubini's Theorem,

$$\iint\limits_D f(x, y)\, dA = \iint\limits_R F(x, y)\, dA = \int_a^b \int_c^d F(x, y)\, dy\, dx$$

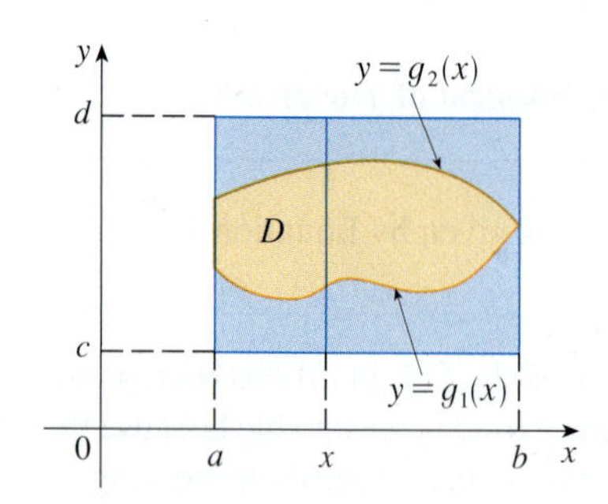

FIGURE 6

Observe that $F(x, y) = 0$ if $y < g_1(x)$ or $y > g_2(x)$ because (x, y) then lies outside D. Therefore

$$\int_c^d F(x, y)\, dy = \int_{g_1(x)}^{g_2(x)} F(x, y)\, dy = \int_{g_1(x)}^{g_2(x)} f(x, y)\, dy$$

because $F(x, y) = f(x, y)$ when $g_1(x) \leqslant y \leqslant g_2(x)$. Thus we have the following formula that enables us to evaluate the double integral as an iterated integral.

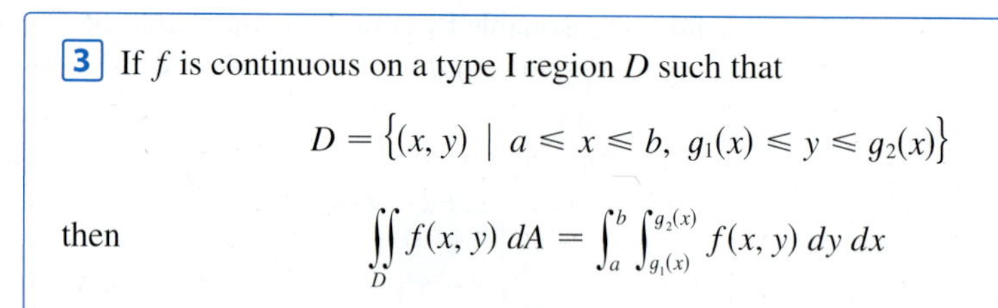

3 If f is continuous on a type I region D such that

$$D = \{(x, y) \mid a \leqslant x \leqslant b,\ g_1(x) \leqslant y \leqslant g_2(x)\}$$

then
$$\iint\limits_D f(x, y)\, dA = \int_a^b \int_{g_1(x)}^{g_2(x)} f(x, y)\, dy\, dx$$

The integral on the right side of (3) is an iterated integral that is similar to the ones we considered in the preceding section, except that in the inner integral we regard x as being constant not only in $f(x, y)$ but also in the limits of integration, $g_1(x)$ and $g_2(x)$.

We also consider plane regions of **type II**, which can be expressed as

4
$$D = \{(x, y) \mid c \leqslant y \leqslant d,\ h_1(y) \leqslant x \leqslant h_2(y)\}$$

where h_1 and h_2 are continuous. Two such regions are illustrated in Figure 7.

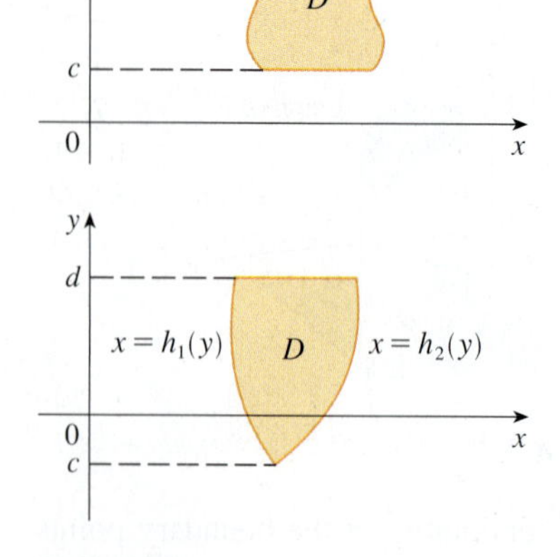

FIGURE 7
Some type II regions

Using the same methods that were used in establishing (3), we can show that

5
$$\iint_D f(x, y)\, dA = \int_c^d \int_{h_1(y)}^{h_2(y)} f(x, y)\, dx\, dy$$

where D is a type II region given by Equation 4.

EXAMPLE 1 Evaluate $\iint_D (x + 2y)\, dA$, where D is the region bounded by the parabolas $y = 2x^2$ and $y = 1 + x^2$.

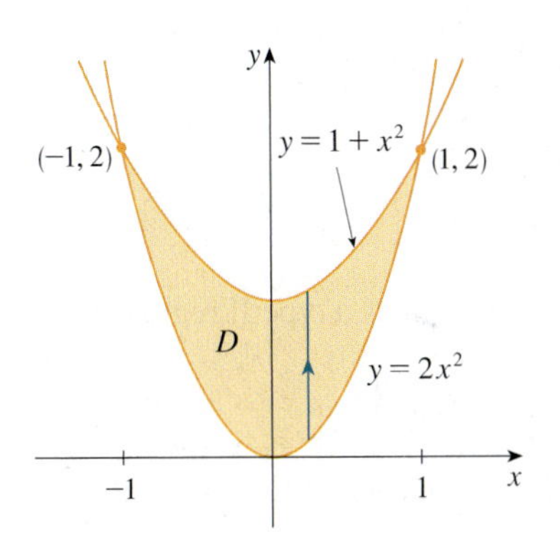

FIGURE 8

SOLUTION The parabolas intersect when $2x^2 = 1 + x^2$, that is, $x^2 = 1$, so $x = \pm 1$. We note that the region D, sketched in Figure 8, is a type I region but not a type II region and we can write

$$D = \{(x, y) \mid -1 \leqslant x \leqslant 1,\ 2x^2 \leqslant y \leqslant 1 + x^2\}$$

Since the lower boundary is $y = 2x^2$ and the upper boundary is $y = 1 + x^2$, Equation 3 gives

$$\iint_D (x + 2y)\, dA = \int_{-1}^{1} \int_{2x^2}^{1+x^2} (x + 2y)\, dy\, dx$$

$$= \int_{-1}^{1} \Big[xy + y^2 \Big]_{y=2x^2}^{y=1+x^2} dx$$

$$= \int_{-1}^{1} [x(1 + x^2) + (1 + x^2)^2 - x(2x^2) - (2x^2)^2]\, dx$$

$$= \int_{-1}^{1} (-3x^4 - x^3 + 2x^2 + x + 1)\, dx$$

$$= -3\frac{x^5}{5} - \frac{x^4}{4} + 2\frac{x^3}{3} + \frac{x^2}{2} + x \Big]_{-1}^{1} = \frac{32}{15}$$ ■

NOTE When we set up a double integral as in Example 1, it is essential to draw a diagram. Often it is helpful to draw a vertical arrow as in Figure 8. Then the limits of integration for the *inner* integral can be read from the diagram as follows: The arrow starts at the lower boundary $y = g_1(x)$, which gives the lower limit in the integral, and the arrow ends at the upper boundary $y = g_2(x)$, which gives the upper limit of integration. For a type II region the arrow is drawn horizontally from the left boundary to the right boundary.

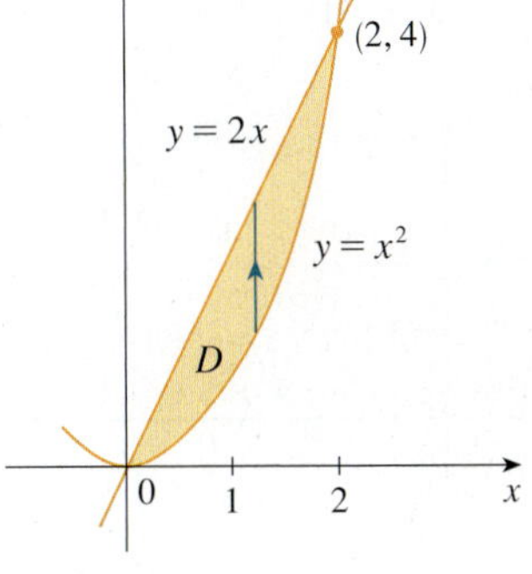

FIGURE 9
D as a type I region

EXAMPLE 2 Find the volume of the solid that lies under the paraboloid $z = x^2 + y^2$ and above the region D in the xy-plane bounded by the line $y = 2x$ and the parabola $y = x^2$.

SOLUTION 1 From Figure 9 we see that D is a type I region and

$$D = \{(x, y) \mid 0 \leqslant x \leqslant 2,\ x^2 \leqslant y \leqslant 2x\}$$

Figure 10 shows the solid whose volume is calculated in Example 2. It lies above the xy-plane, below the paraboloid $z = x^2 + y^2$, and between the plane $y = 2x$ and the parabolic cylinder $y = x^2$.

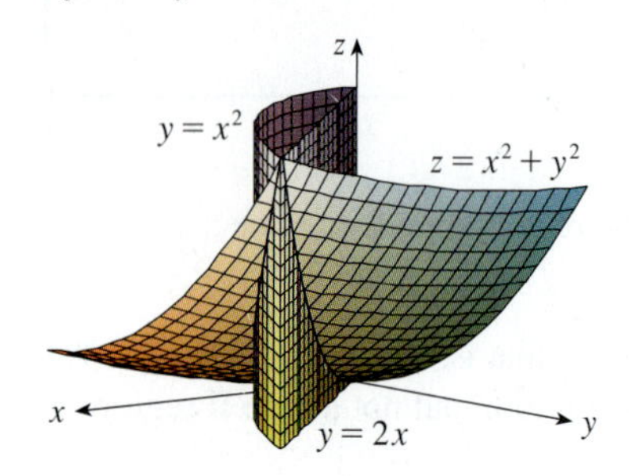

FIGURE 10

Therefore the volume under $z = x^2 + y^2$ and above D is

$$V = \iint_D (x^2 + y^2)\, dA = \int_0^2 \int_{x^2}^{2x} (x^2 + y^2)\, dy\, dx$$

$$= \int_0^2 \left[x^2 y + \frac{y^3}{3} \right]_{y=x^2}^{y=2x} dx$$

$$= \int_0^2 \left[x^2(2x) + \frac{(2x)^3}{3} - x^2 x^2 - \frac{(x^2)^3}{3} \right] dx$$

$$= \int_0^2 \left(-\frac{x^6}{3} - x^4 + \frac{14x^3}{3} \right) dx$$

$$= -\frac{x^7}{21} - \frac{x^5}{5} + \frac{7x^4}{6} \bigg]_0^2 = \frac{216}{35}$$

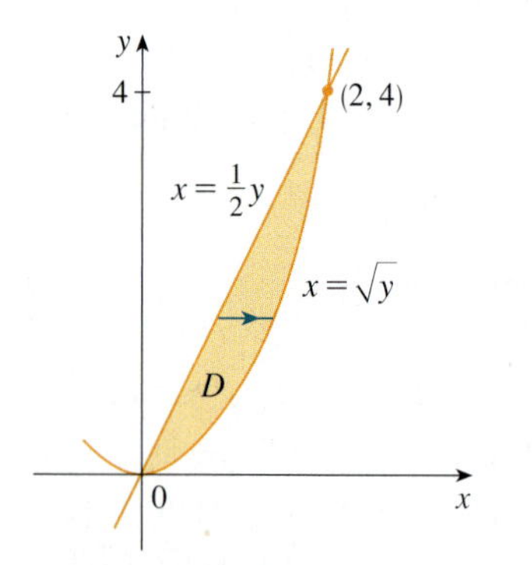

FIGURE 11
D as a type II region

SOLUTION 2 From Figure 11 we see that D can also be written as a type II region:

$$D = \{(x, y) \mid 0 \leqslant y \leqslant 4,\ \tfrac{1}{2}y \leqslant x \leqslant \sqrt{y}\}$$

Therefore another expression for V is

$$V = \iint_D (x^2 + y^2)\, dA = \int_0^4 \int_{\frac{1}{2}y}^{\sqrt{y}} (x^2 + y^2)\, dx\, dy$$

$$= \int_0^4 \left[\frac{x^3}{3} + y^2 x \right]_{x=\frac{1}{2}y}^{x=\sqrt{y}} dy = \int_0^4 \left(\frac{y^{3/2}}{3} + y^{5/2} - \frac{y^3}{24} - \frac{y^3}{2} \right) dy$$

$$= \tfrac{2}{15}y^{5/2} + \tfrac{2}{7}y^{7/2} - \tfrac{13}{96}y^4 \Big]_0^4 = \tfrac{216}{35}$$ ■

EXAMPLE 3 Evaluate $\iint_D xy\, dA$, where D is the region bounded by the line $y = x - 1$ and the parabola $y^2 = 2x + 6$.

SOLUTION The region D is shown in Figure 12. Again D is both type I and type II, but the description of D as a type I region is more complicated because the lower boundary consists of two parts. Therefore we prefer to express D as a type II region:

$$D = \{(x, y) \mid -2 \leqslant y \leqslant 4,\ \tfrac{1}{2}y^2 - 3 \leqslant x \leqslant y + 1\}$$

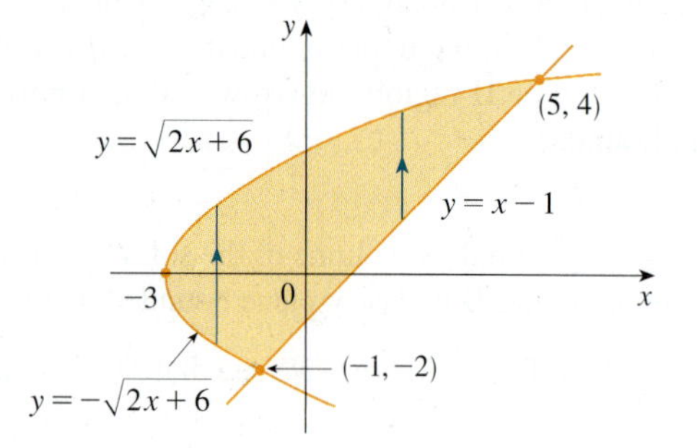

FIGURE 12

(a) D as a type I region

(b) D as a type II region

Then (5) gives

$$\iint_D xy\,dA = \int_{-2}^{4}\int_{\frac{1}{2}y^2-3}^{y+1} xy\,dx\,dy = \int_{-2}^{4}\left[\frac{x^2}{2}y\right]_{x=\frac{1}{2}y^2-3}^{x=y+1} dy$$

$$= \tfrac{1}{2}\int_{-2}^{4} y\left[(y+1)^2 - \left(\tfrac{1}{2}y^2-3\right)^2\right] dy$$

$$= \tfrac{1}{2}\int_{-2}^{4}\left(-\frac{y^5}{4} + 4y^3 + 2y^2 - 8y\right) dy$$

$$= \frac{1}{2}\left[-\frac{y^6}{24} + y^4 + 2\frac{y^3}{3} - 4y^2\right]_{-2}^{4} = 36$$

If we had expressed D as a type I region using Figure 12(a), then we would have obtained

$$\iint_D xy\,dA = \int_{-3}^{-1}\int_{-\sqrt{2x+6}}^{\sqrt{2x+6}} xy\,dy\,dx + \int_{-1}^{5}\int_{x-1}^{\sqrt{2x+6}} xy\,dy\,dx$$

but this would have involved more work than the other method. ■

EXAMPLE 4 Find the volume of the tetrahedron bounded by the planes $x + 2y + z = 2$, $x = 2y$, $x = 0$, and $z = 0$.

SOLUTION In a question such as this, it's wise to draw two diagrams: one of the three-dimensional solid and another of the plane region D over which it lies. Figure 13 shows the tetrahedron T bounded by the coordinate planes $x = 0$, $z = 0$, the vertical plane $x = 2y$, and the plane $x + 2y + z = 2$. Since the plane $x + 2y + z = 2$ intersects the xy-plane (whose equation is $z = 0$) in the line $x + 2y = 2$, we see that T lies above the triangular region D in the xy-plane bounded by the lines $x = 2y$, $x + 2y = 2$, and $x = 0$. (See Figure 14.)

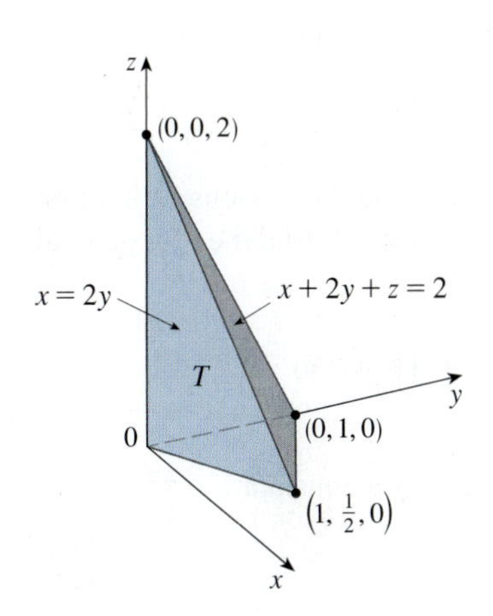

FIGURE 13

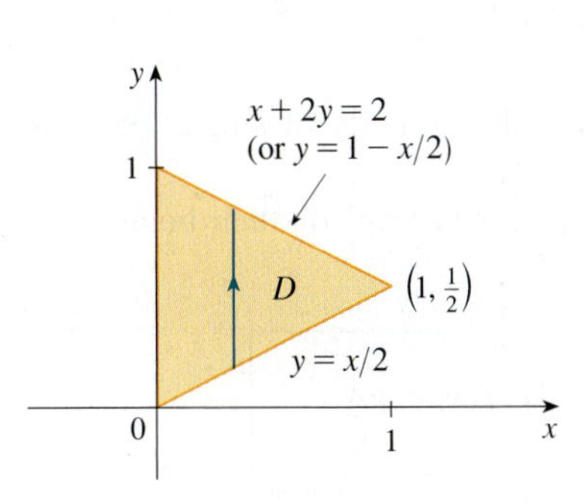

FIGURE 14

The plane $x + 2y + z = 2$ can be written as $z = 2 - x - 2y$, so the required volume lies under the graph of the function $z = 2 - x - 2y$ and above

$$D = \{(x, y) \mid 0 \leqslant x \leqslant 1,\, x/2 \leqslant y \leqslant 1 - x/2\}$$

Therefore

$$V = \iint_D (2 - x - 2y)\,dA$$

$$= \int_0^1\int_{x/2}^{1-x/2} (2 - x - 2y)\,dy\,dx$$

$$= \int_0^1 \left[2y - xy - y^2\right]_{y=x/2}^{y=1-x/2} dx$$

$$= \int_0^1 \left[2 - x - x\left(1 - \frac{x}{2}\right) - \left(1 - \frac{x}{2}\right)^2 - x + \frac{x^2}{2} + \frac{x^2}{4}\right] dx$$

$$= \int_0^1 (x^2 - 2x + 1)\,dx = \frac{x^3}{3} - x^2 + x\bigg]_0^1 = \frac{1}{3}$$ ■

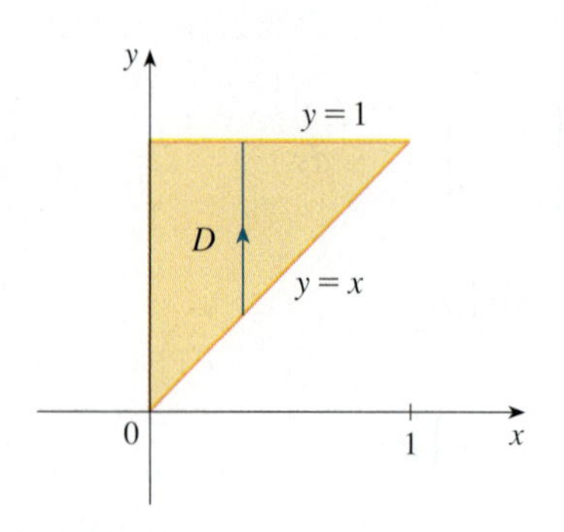

FIGURE 15
D as a type I region

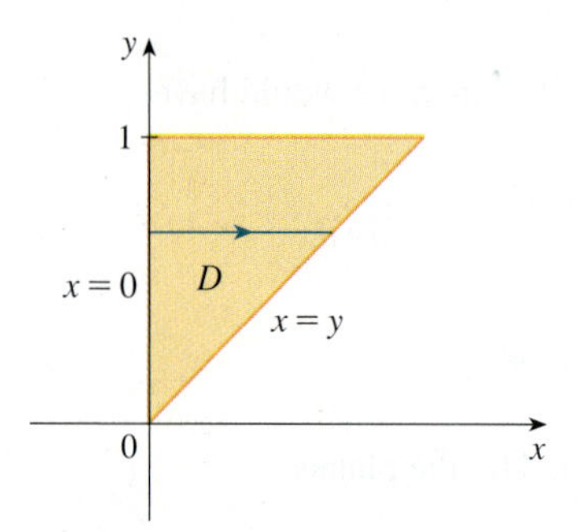

FIGURE 16
D as a type II region

EXAMPLE 5 Evaluate the iterated integral $\int_0^1 \int_x^1 \sin(y^2)\,dy\,dx$.

SOLUTION If we try to evaluate the integral as it stands, we are faced with the task of first evaluating $\int \sin(y^2)\,dy$. But it's impossible to do so in finite terms since $\int \sin(y^2)\,dy$ is not an elementary function. (See the end of Section 7.5.) So we must change the order of integration. This is accomplished by first expressing the given iterated integral as a double integral. Using (3) backward, we have

$$\int_0^1 \int_x^1 \sin(y^2)\,dy\,dx = \iint_D \sin(y^2)\,dA$$

where $$D = \{(x, y) \mid 0 \leqslant x \leqslant 1,\ x \leqslant y \leqslant 1\}$$

We sketch this region D in Figure 15. Then from Figure 16 we see that an alternative description of D is

$$D = \{(x, y) \mid 0 \leqslant y \leqslant 1,\ 0 \leqslant x \leqslant y\}$$

This enables us to use (5) to express the double integral as an iterated integral in the reverse order:

$$\begin{aligned}\int_0^1 \int_x^1 \sin(y^2)\,dy\,dx &= \iint_D \sin(y^2)\,dA \\ &= \int_0^1 \int_0^y \sin(y^2)\,dx\,dy = \int_0^1 \Big[x\sin(y^2)\Big]_{x=0}^{x=y} dy \\ &= \int_0^1 y\sin(y^2)\,dy = -\tfrac{1}{2}\cos(y^2)\Big]_0^1 = \tfrac{1}{2}(1 - \cos 1)\end{aligned}$$

Properties of Double Integrals

We assume that all of the following integrals exist. For rectangular regions D the first three properties can be proved in the same manner as in Section 4.2. And then for general regions the properties follow from Definition 2.

6 $$\iint_D [f(x, y) + g(x, y)]\,dA = \iint_D f(x, y)\,dA + \iint_D g(x, y)\,dA$$

7 $$\iint_D cf(x, y)\,dA = c\iint_D f(x, y)\,dA \qquad \text{where } c \text{ is a constant}$$

If $f(x, y) \geqslant g(x, y)$ for all (x, y) in D, then

8 $$\iint_D f(x, y)\,dA \geqslant \iint_D g(x, y)\,dA$$

The next property of double integrals is similar to the property of single integrals given by the equation $\int_a^b f(x)\,dx = \int_a^c f(x)\,dx + \int_c^b f(x)\,dx$.

If $D = D_1 \cup D_2$, where D_1 and D_2 don't overlap except perhaps on their boundaries (see Figure 17), then

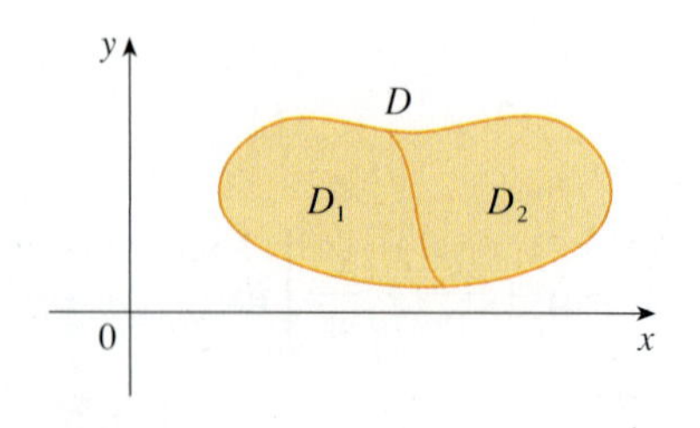

FIGURE 17

9 $$\iint_D f(x, y)\,dA = \iint_{D_1} f(x, y)\,dA + \iint_{D_2} f(x, y)\,dA$$

Property 9 can be used to evaluate double integrals over regions D that are neither type I nor type II but can be expressed as a union of regions of type I or type II. Figure 18 illustrates this procedure. (See Exercises 57 and 58.)

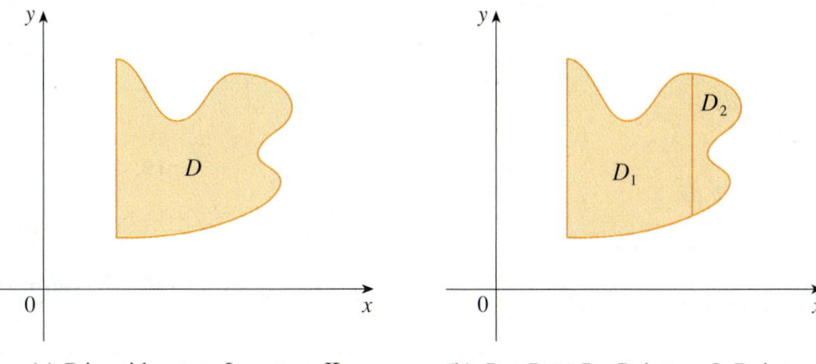

FIGURE 18 (a) D is neither type I nor type II. (b) $D = D_1 \cup D_2$, D_1 is type I, D_2 is type II.

The next property of integrals says that if we integrate the constant function $f(x, y) = 1$ over a region D, we get the area of D:

10
$$\iint_D 1\, dA = A(D)$$

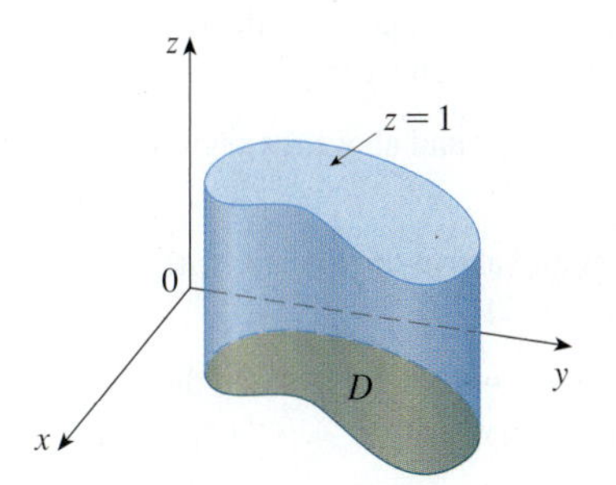

FIGURE 19
Cylinder with base D and height 1

Figure 19 illustrates why Equation 10 is true: A solid cylinder whose base is D and whose height is 1 has volume $A(D) \cdot 1 = A(D)$, but we know that we can also write its volume as $\iint_D 1\, dA$.

Finally, we can combine Properties 7, 8, and 10 to prove the following property. (See Exercise 63.)

11 If $m \leqslant f(x, y) \leqslant M$ for all (x, y) in D, then

$$mA(D) \leqslant \iint_D f(x, y)\, dA \leqslant MA(D)$$

EXAMPLE 6 Use Property 11 to estimate the integral $\iint_D e^{\sin x \cos y}\, dA$, where D is the disk with center the origin and radius 2.

SOLUTION Since $-1 \leqslant \sin x \leqslant 1$ and $-1 \leqslant \cos y \leqslant 1$, we have $-1 \leqslant \sin x \cos y \leqslant 1$ and therefore

$$e^{-1} \leqslant e^{\sin x \cos y} \leqslant e^1 = e$$

Thus, using $m = e^{-1} = 1/e$, $M = e$, and $A(D) = \pi(2)^2$ in Property 11, we obtain

$$\frac{4\pi}{e} \leqslant \iint_D e^{\sin x \cos y}\, dA \leqslant 4\pi e$$

■

15.2 EXERCISES

1–6 Evaluate the iterated integral.

1. $\int_1^5 \int_0^x (8x - 2y)\, dy\, dx$

2. $\int_0^2 \int_0^{y^2} x^2 y\, dx\, dy$

3. $\int_0^1 \int_0^y x e^{y^3}\, dx\, dy$

4. $\int_0^{\pi/2} \int_0^x x \sin y\, dy\, dx$

5. $\int_0^1 \int_0^{s^2} \cos(s^3)\, dt\, ds$

6. $\int_0^1 \int_0^{e^v} \sqrt{1 + e^v}\, dw\, dv$

7–10 Evaluate the double integral.

7. $\iint_D \frac{y}{x^2 + 1}\, dA$, $D = \{(x, y) \mid 0 \leqslant x \leqslant 4, 0 \leqslant y \leqslant \sqrt{x}\}$

8. $\iint_D (2x + y)\, dA$, $D = \{(x, y) \mid 1 \leqslant y \leqslant 2, y - 1 \leqslant x \leqslant 1\}$

9. $\iint_D e^{-y^2}\, dA$, $D = \{(x, y) \mid 0 \leqslant y \leqslant 3, 0 \leqslant x \leqslant y\}$

10. $\iint_D y\sqrt{x^2 - y^2}\, dA$, $D = \{(x, y) \mid 0 \leqslant x \leqslant 2, 0 \leqslant y \leqslant x\}$

11. Draw an example of a region that is
(a) type I but not type II
(b) type II but not type I

12. Draw an example of a region that is
(a) both type I and type II
(b) neither type I nor type II

13–14 Express D as a region of type I and also as a region of type II. Then evaluate the double integral in two ways.

13. $\iint_D x\, dA$, D is enclosed by the lines $y = x$, $y = 0$, $x = 1$

14. $\iint_D xy\, dA$, D is enclosed by the curves $y = x^2$, $y = 3x$

15–16 Set up iterated integrals for both orders of integration. Then evaluate the double integral using the easier order and explain why it's easier.

15. $\iint_D y\, dA$, D is bounded by $y = x - 2$, $x = y^2$

16. $\iint_D y^2 e^{xy}\, dA$, D is bounded by $y = x$, $y = 4$, $x = 0$

17–22 Evaluate the double integral.

17. $\iint_D x \cos y\, dA$, D is bounded by $y = 0$, $y = x^2$, $x = 1$

18. $\iint_D (x^2 + 2y)\, dA$, D is bounded by $y = x$, $y = x^3$, $x \geqslant 0$

19. $\iint_D y^2\, dA$,
D is the triangular region with vertices (0, 1), (1, 2), (4, 1)

20. $\iint_D xy\, dA$, D is enclosed by the quarter-circle $y = \sqrt{1 - x^2}$, $x \geqslant 0$, and the axes

21. $\iint_D (2x - y)\, dA$,
D is bounded by the circle with center the origin and radius 2

22. $\iint_D y\, dA$, D is the triangular region with vertices (0, 0), (1, 1), and (4, 0)

23–32 Find the volume of the given solid.

23. Under the plane $3x + 2y - z = 0$ and above the region enclosed by the parabolas $y = x^2$ and $x = y^2$

24. Under the surface $z = 1 + x^2y^2$ and above the region enclosed by $x = y^2$ and $x = 4$

25. Under the surface $z = xy$ and above the triangle with vertices (1, 1), (4, 1), and (1, 2)

26. Enclosed by the paraboloid $z = x^2 + y^2 + 1$ and the planes $x = 0$, $y = 0$, $z = 0$, and $x + y = 2$

27. The tetrahedron enclosed by the coordinate planes and the plane $2x + y + z = 4$

28. Bounded by the planes $z = x$, $y = x$, $x + y = 2$, and $z = 0$

29. Enclosed by the cylinders $z = x^2$, $y = x^2$ and the planes $z = 0$, $y = 4$

30. Bounded by the cylinder $y^2 + z^2 = 4$ and the planes $x = 2y$, $x = 0$, $z = 0$ in the first octant

31. Bounded by the cylinder $x^2 + y^2 = 1$ and the planes $y = z$, $x = 0$, $z = 0$ in the first octant

32. Bounded by the cylinders $x^2 + y^2 = r^2$ and $y^2 + z^2 = r^2$

33. Use a graphing calculator or computer to estimate the x-coordinates of the points of intersection of the curves $y = x^4$ and $y = 3x - x^2$. If D is the region bounded by these curves, estimate $\iint_D x\, dA$.

34. Find the approximate volume of the solid in the first octant that is bounded by the planes $y = x$, $z = 0$, and $z = x$ and the cylinder $y = \cos x$. (Use a graphing device to estimate the points of intersection.)

35–38 Find the volume of the solid by subtracting two volumes.

35. The solid enclosed by the parabolic cylinders $y = 1 - x^2$, $y = x^2 - 1$ and the planes $x + y + z = 2$, $2x + 2y - z + 10 = 0$

36. The solid enclosed by the parabolic cylinder $y = x^2$ and the planes $z = 3y$, $z = 2 + y$

37. The solid under the plane $z = 3$, above the plane $z = y$, and between the parabolic cylinders $y = x^2$ and $y = 1 - x^2$

38. The solid in the first octant under the plane $z = x + y$, above the surface $z = xy$, and enclosed by the surfaces $x = 0$, $y = 0$, and $x^2 + y^2 = 4$

39–40 Sketch the solid whose volume is given by the iterated integral.

39. $\int_0^1 \int_0^{1-x} (1 - x - y)\, dy\, dx$ **40.** $\int_0^1 \int_0^{1-x^2} (1 - x)\, dy\, dx$

CAS **41–44** Use a computer algebra system to find the exact volume of the solid.

41. Under the surface $z = x^3y^4 + xy^2$ and above the region bounded by the curves $y = x^3 - x$ and $y = x^2 + x$ for $x \geqslant 0$

42. Between the paraboloids $z = 2x^2 + y^2$ and $z = 8 - x^2 - 2y^2$ and inside the cylinder $x^2 + y^2 = 1$

43. Enclosed by $z = 1 - x^2 - y^2$ and $z = 0$

44. Enclosed by $z = x^2 + y^2$ and $z = 2y$

45–50 Sketch the region of integration and change the order of integration.

45. $\int_0^1 \int_0^y f(x, y)\, dx\, dy$ **46.** $\int_0^2 \int_{x^2}^4 f(x, y)\, dy\, dx$

47. $\int_0^{\pi/2} \int_0^{\cos x} f(x, y)\, dy\, dx$ **48.** $\int_{-2}^2 \int_0^{\sqrt{4-y^2}} f(x, y)\, dx\, dy$

49. $\int_1^2 \int_0^{\ln x} f(x, y)\, dy\, dx$ **50.** $\int_0^1 \int_{\arctan x}^{\pi/4} f(x, y)\, dy\, dx$

51–56 Evaluate the integral by reversing the order of integration.

51. $\int_0^1 \int_{3y}^3 e^{x^2}\, dx\, dy$ **52.** $\int_0^1 \int_{x^2}^1 \sqrt{y} \sin y\, dy\, dx$

53. $\int_0^1 \int_{\sqrt{x}}^1 \sqrt{y^3 + 1}\, dy\, dx$

54. $\int_0^2 \int_{y/2}^1 y \cos(x^3 - 1)\, dx\, dy$

55. $\int_0^1 \int_{\arcsin y}^{\pi/2} \cos x \sqrt{1 + \cos^2 x}\, dx\, dy$

56. $\int_0^8 \int_{\sqrt[3]{y}}^2 e^{x^4}\, dx\, dy$

57–58 Express D as a union of regions of type I or type II and evaluate the integral.

57. $\iint_D x^2\, dA$ **58.** $\iint_D y\, dA$

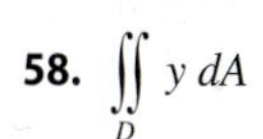

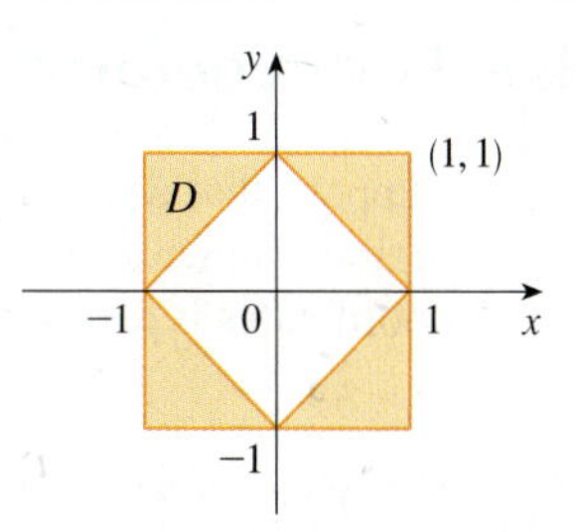

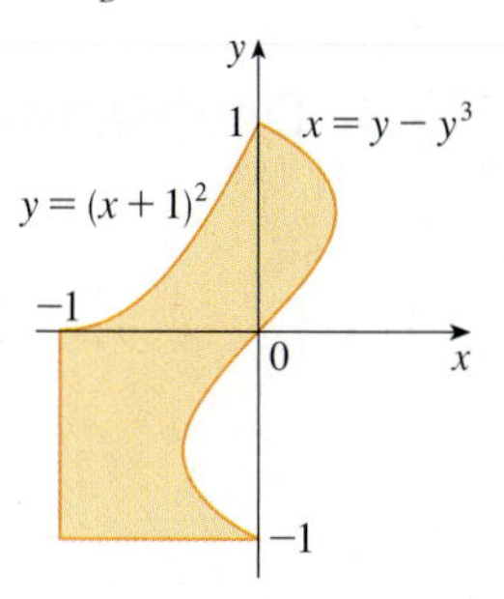

59–60 Use Property 11 to estimate the value of the integral.

59. $\iint_S \sqrt{4 - x^2y^2}\, dA$, $S = \{(x, y) \mid x^2 + y^2 \leqslant 1, x \geqslant 0\}$

60. $\iint_T \sin^4(x + y)\, dA$, T is the triangle enclosed by the lines $y = 0$, $y = 2x$, and $x = 1$

61–62 Find the averge value of f over the region D.

61. $f(x, y) = xy$, D is the triangle with vertices $(0, 0)$, $(1, 0)$, and $(1, 3)$

62. $f(x, y) = x \sin y$, D is enclosed by the curves $y = 0$, $y = x^2$, and $x = 1$

63. Prove Property 11.

64. In evaluating a double integral over a region D, a sum of iterated integrals was obtained as follows:

$$\iint_D f(x, y)\, dA = \int_0^1 \int_0^{2y} f(x, y)\, dx\, dy + \int_1^3 \int_0^{3-y} f(x, y)\, dx\, dy$$

Sketch the region D and express the double integral as an iterated integral with reversed order of integration.

65–69 Use geometry or symmetry, or both, to evaluate the double integral.

65. $\iint_D (x + 2)\, dA$,

$D = \{(x, y) \mid 0 \leqslant y \leqslant \sqrt{9 - x^2}\}$

66. $\iint_D \sqrt{R^2 - x^2 - y^2}\, dA$,

D is the disk with center the origin and radius R

67. $\iint_D (2x + 3y)\, dA$,

D is the rectangle $0 \leqslant x \leqslant a$, $0 \leqslant y \leqslant b$

68. $\iint_D (2 + x^2y^3 - y^2 \sin x)\, dA$,

$D = \{(x, y) \mid |x| + |y| \leqslant 1\}$

69. $\iint_D \left(ax^3 + by^3 + \sqrt{a^2 - x^2}\right) dA$,

$D = [-a, a] \times [-b, b]$

CAS **70.** Graph the solid bounded by the plane $x + y + z = 1$ and the paraboloid $z = 4 - x^2 - y^2$ and find its exact volume. (Use your CAS to do the graphing, to find the equations of the boundary curves of the region of integration, and to evaluate the double integral.)

15.3 Double Integrals in Polar Coordinates

Suppose that we want to evaluate a double integral $\iint_R f(x, y)\, dA$, where R is one of the regions shown in Figure 1. In either case the description of R in terms of rectangular coordinates is rather complicated, but R is easily described using polar coordinates.

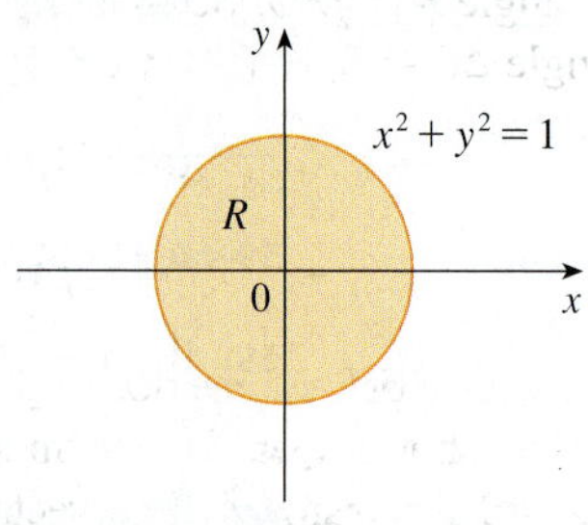

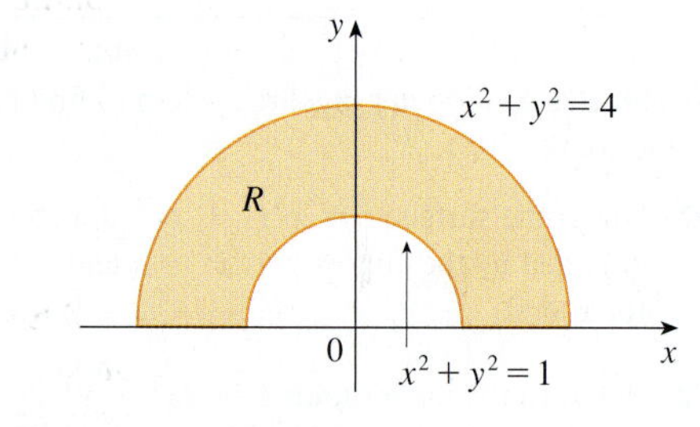

FIGURE 1

(a) $R = \{(r, \theta) \mid 0 \leqslant r \leqslant 1, 0 \leqslant \theta \leqslant 2\pi\}$

(b) $R = \{(r, \theta) \mid 1 \leqslant r \leqslant 2, 0 \leqslant \theta \leqslant \pi\}$

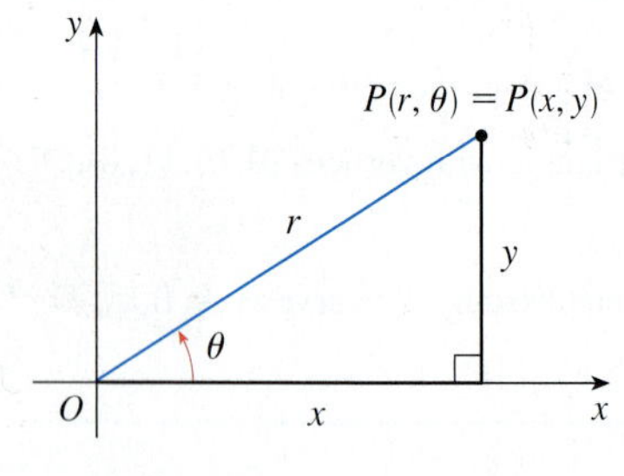

FIGURE 2

Recall from Figure 2 that the polar coordinates (r, θ) of a point are related to the rectangular coordinates (x, y) by the equations

$$r^2 = x^2 + y^2 \qquad x = r\cos\theta \qquad y = r\sin\theta$$

(See Section 10.3.)

The regions in Figure 1 are special cases of a **polar rectangle**

$$R = \{(r, \theta) \mid a \leqslant r \leqslant b, \alpha \leqslant \theta \leqslant \beta\}$$

which is shown in Figure 3. In order to compute the double integral $\iint_R f(x, y)\, dA$, where R is a polar rectangle, we divide the interval $[a, b]$ into m subintervals $[r_{i-1}, r_i]$ of equal width $\Delta r = (b - a)/m$ and we divide the interval $[\alpha, \beta]$ into n subintervals $[\theta_{j-1}, \theta_j]$ of equal width $\Delta\theta = (\beta - \alpha)/n$. Then the circles $r = r_i$ and the rays $\theta = \theta_j$ divide the polar rectangle R into the small polar rectangles R_{ij} shown in Figure 4.

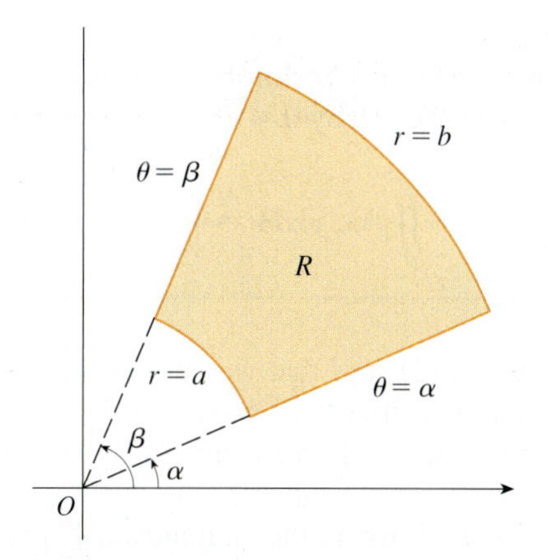

FIGURE 3 Polar rectangle

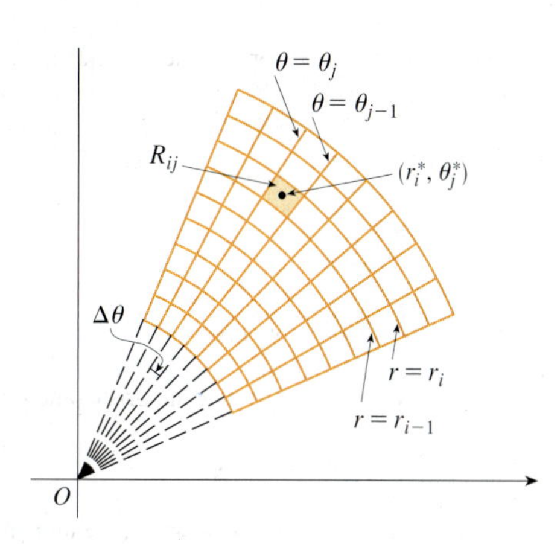

FIGURE 4 Dividing R into polar subrectangles

The "center" of the polar subrectangle

$$R_{ij} = \{(r, \theta) \mid r_{i-1} \leqslant r \leqslant r_i,\ \theta_{j-1} \leqslant \theta \leqslant \theta_j\}$$

has polar coordinates

$$r_i^* = \tfrac{1}{2}(r_{i-1} + r_i) \qquad \theta_j^* = \tfrac{1}{2}(\theta_{j-1} + \theta_j)$$

We compute the area of R_{ij} using the fact that the area of a sector of a circle with radius r and central angle θ is $\frac{1}{2}r^2\theta$. Subtracting the areas of two such sectors, each of which has central angle $\Delta\theta = \theta_j - \theta_{j-1}$, we find that the area of R_{ij} is

$$\Delta A_i = \tfrac{1}{2}r_i^2\,\Delta\theta - \tfrac{1}{2}r_{i-1}^2\,\Delta\theta = \tfrac{1}{2}(r_i^2 - r_{i-1}^2)\,\Delta\theta$$

$$= \tfrac{1}{2}(r_i + r_{i-1})(r_i - r_{i-1})\,\Delta\theta = r_i^*\,\Delta r\,\Delta\theta$$

Although we have defined the double integral $\iint_R f(x, y)\,dA$ in terms of ordinary rectangles, it can be shown that, for continuous functions f, we always obtain the same answer using polar rectangles. The rectangular coordinates of the center of R_{ij} are $(r_i^* \cos\theta_j^*, r_i^* \sin\theta_j^*)$, so a typical Riemann sum is

$$\boxed{1} \quad \sum_{i=1}^{m}\sum_{j=1}^{n} f(r_i^* \cos\theta_j^*, r_i^* \sin\theta_j^*)\,\Delta A_i = \sum_{i=1}^{m}\sum_{j=1}^{n} f(r_i^* \cos\theta_j^*, r_i^* \sin\theta_j^*)\,r_i^*\,\Delta r\,\Delta\theta$$

If we write $g(r, \theta) = rf(r\cos\theta, r\sin\theta)$, then the Riemann sum in Equation 1 can be written as

$$\sum_{i=1}^{m}\sum_{j=1}^{n} g(r_i^*, \theta_j^*)\,\Delta r\,\Delta\theta$$

which is a Riemann sum for the double integral

$$\int_\alpha^\beta \int_a^b g(r, \theta)\,dr\,d\theta$$

Therefore we have

$$\iint_R f(x, y)\,dA = \lim_{m,n\to\infty} \sum_{i=1}^{m}\sum_{j=1}^{n} f(r_i^* \cos\theta_j^*, r_i^* \sin\theta_j^*)\,\Delta A_i$$

$$= \lim_{m,n\to\infty} \sum_{i=1}^{m}\sum_{j=1}^{n} g(r_i^*, \theta_j^*)\,\Delta r\,\Delta\theta = \int_\alpha^\beta \int_a^b g(r, \theta)\,dr\,d\theta$$

$$= \int_\alpha^\beta \int_a^b f(r\cos\theta, r\sin\theta)\,r\,dr\,d\theta$$

2 Change to Polar Coodinates in a Double Integral If f is continuous on a polar rectangle R given by $0 \leqslant a \leqslant r \leqslant b$, $\alpha \leqslant \theta \leqslant \beta$, where $0 \leqslant \beta - \alpha \leqslant 2\pi$, then

$$\iint_R f(x, y)\, dA = \int_\alpha^\beta \int_a^b f(r\cos\theta, r\sin\theta)\, r\, dr\, d\theta$$

The formula in (2) says that we convert from rectangular to polar coordinates in a double integral by writing $x = r\cos\theta$ and $y = r\sin\theta$, using the appropriate limits of integration for r and θ, and replacing dA by $r\,dr\,d\theta$. Be careful not to forget the additional factor r on the right side of Formula 2. A classical method for remembering this is shown in Figure 5, where the "infinitesimal" polar rectangle can be thought of as an ordinary rectangle with dimensions $r\,d\theta$ and dr and therefore has "area" $dA = r\,dr\,d\theta$.

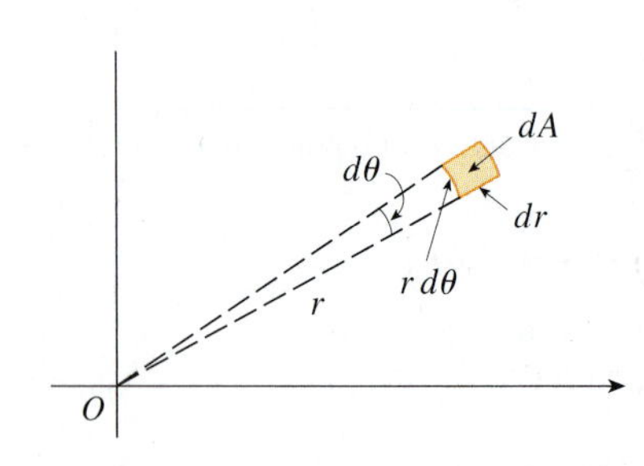

FIGURE 5

EXAMPLE 1 Evaluate $\iint_R (3x + 4y^2)\, dA$, where R is the region in the upper half-plane bounded by the circles $x^2 + y^2 = 1$ and $x^2 + y^2 = 4$.

SOLUTION The region R can be described as

$$R = \{(x, y) \mid y \geqslant 0,\ 1 \leqslant x^2 + y^2 \leqslant 4\}$$

It is the half-ring shown in Figure 1(b), and in polar coordinates it is given by $1 \leqslant r \leqslant 2$, $0 \leqslant \theta \leqslant \pi$. Therefore, by Formula 2,

$$\iint_R (3x + 4y^2)\, dA = \int_0^\pi \int_1^2 (3r\cos\theta + 4r^2\sin^2\theta)\, r\, dr\, d\theta$$

$$= \int_0^\pi \int_1^2 (3r^2\cos\theta + 4r^3\sin^2\theta)\, dr\, d\theta$$

$$= \int_0^\pi \Big[r^3\cos\theta + r^4\sin^2\theta\Big]_{r=1}^{r=2} d\theta = \int_0^\pi (7\cos\theta + 15\sin^2\theta)\, d\theta$$

$$= \int_0^\pi \Big[7\cos\theta + \tfrac{15}{2}(1 - \cos 2\theta)\Big]\, d\theta$$

$$= 7\sin\theta + \frac{15\theta}{2} - \frac{15}{4}\sin 2\theta\Big]_0^\pi = \frac{15\pi}{2}$$

Here we use the trigonometric identity

$$\sin^2\theta = \tfrac{1}{2}(1 - \cos 2\theta)$$

See Section 7.2 for advice on integrating trigonometric functions.

EXAMPLE 2 Find the volume of the solid bounded by the plane $z = 0$ and the paraboloid $z = 1 - x^2 - y^2$.

SOLUTION If we put $z = 0$ in the equation of the paraboloid, we get $x^2 + y^2 = 1$. This means that the plane intersects the paraboloid in the circle $x^2 + y^2 = 1$, so the solid lies under the paraboloid and above the circular disk D given by $x^2 + y^2 \leqslant 1$ [see Figures 6 and 1(a)]. In polar coordinates D is given by $0 \leqslant r \leqslant 1$, $0 \leqslant \theta \leqslant 2\pi$. Since $1 - x^2 - y^2 = 1 - r^2$, the volume is

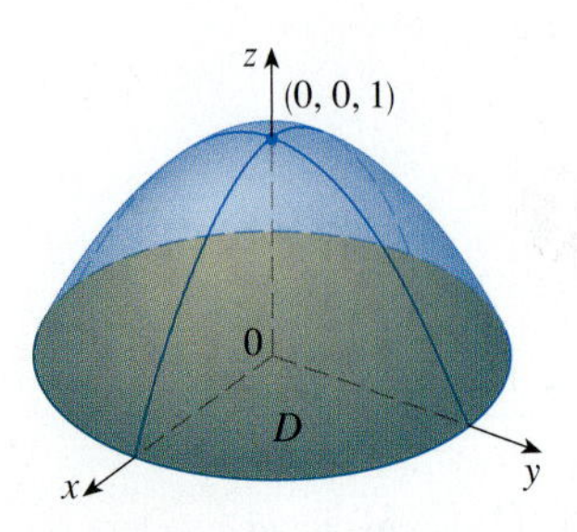

FIGURE 6

$$V = \iint_D (1 - x^2 - y^2)\, dA = \int_0^{2\pi} \int_0^1 (1 - r^2)\, r\, dr\, d\theta$$

$$= \int_0^{2\pi} d\theta \int_0^1 (r - r^3)\, dr = 2\pi\left[\frac{r^2}{2} - \frac{r^4}{4}\right]_0^1 = \frac{\pi}{2}$$

If we had used rectangular coordinates instead of polar coordinates, then we would have obtained

$$V = \iint_D (1 - x^2 - y^2)\, dA = \int_{-1}^{1} \int_{-\sqrt{1-x^2}}^{\sqrt{1-x^2}} (1 - x^2 - y^2)\, dy\, dx$$

which is not easy to evaluate because it involves finding $\int (1 - x^2)^{3/2}\, dx$. ■

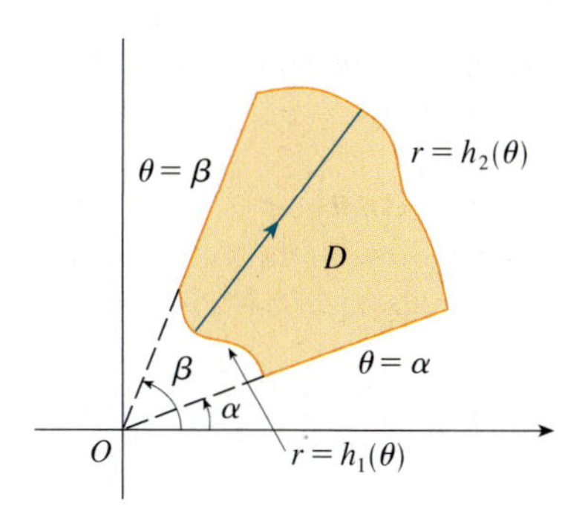

FIGURE 7
$D = \{(r, \theta) \mid \alpha \leqslant \theta \leqslant \beta, h_1(\theta) \leqslant r \leqslant h_2(\theta)\}$

What we have done so far can be extended to the more complicated type of region shown in Figure 7. It's similar to the type II rectangular regions considered in Section 15.2. In fact, by combining Formula 2 in this section with Formula 15.2.5, we obtain the following formula.

3 If f is continuous on a polar region of the form

$$D = \{(r, \theta) \mid \alpha \leqslant \theta \leqslant \beta,\ h_1(\theta) \leqslant r \leqslant h_2(\theta)\}$$

then $$\iint_D f(x, y)\, dA = \int_\alpha^\beta \int_{h_1(\theta)}^{h_2(\theta)} f(r\cos\theta, r\sin\theta)\, r\, dr\, d\theta$$

In particular, taking $f(x, y) = 1$, $h_1(\theta) = 0$, and $h_2(\theta) = h(\theta)$ in this formula, we see that the area of the region D bounded by $\theta = \alpha$, $\theta = \beta$, and $r = h(\theta)$ is

$$A(D) = \iint_D 1\, dA = \int_\alpha^\beta \int_0^{h(\theta)} r\, dr\, d\theta$$

$$= \int_\alpha^\beta \left[\frac{r^2}{2}\right]_0^{h(\theta)} d\theta = \int_\alpha^\beta \tfrac{1}{2}[h(\theta)]^2\, d\theta$$

and this agrees with Formula 10.4.3.

EXAMPLE 3 Use a double integral to find the area enclosed by one loop of the four-leaved rose $r = \cos 2\theta$.

SOLUTION From the sketch of the curve in Figure 8, we see that a loop is given by the region

$$D = \{(r, \theta) \mid -\pi/4 \leqslant \theta \leqslant \pi/4, 0 \leqslant r \leqslant \cos 2\theta\}$$

So the area is

$$A(D) = \iint_D dA = \int_{-\pi/4}^{\pi/4} \int_0^{\cos 2\theta} r\, dr\, d\theta$$

$$= \int_{-\pi/4}^{\pi/4} \left[\tfrac{1}{2} r^2\right]_0^{\cos 2\theta} d\theta = \tfrac{1}{2} \int_{-\pi/4}^{\pi/4} \cos^2 2\theta\, d\theta$$

$$= \tfrac{1}{4} \int_{-\pi/4}^{\pi/4} (1 + \cos 4\theta)\, d\theta = \tfrac{1}{4}\left[\theta + \tfrac{1}{4}\sin 4\theta\right]_{-\pi/4}^{\pi/4} = \frac{\pi}{8}$$ ■

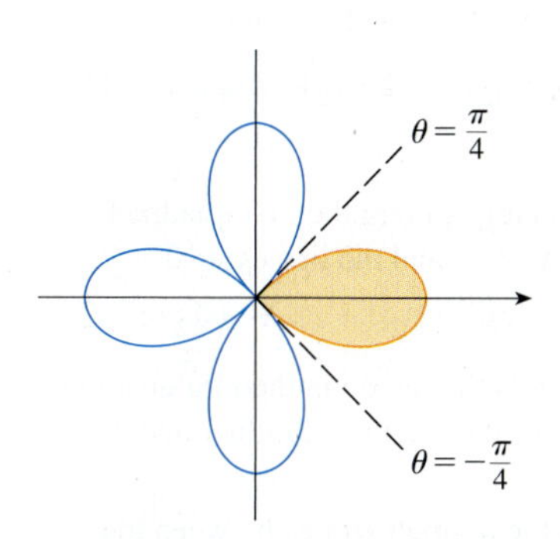

FIGURE 8

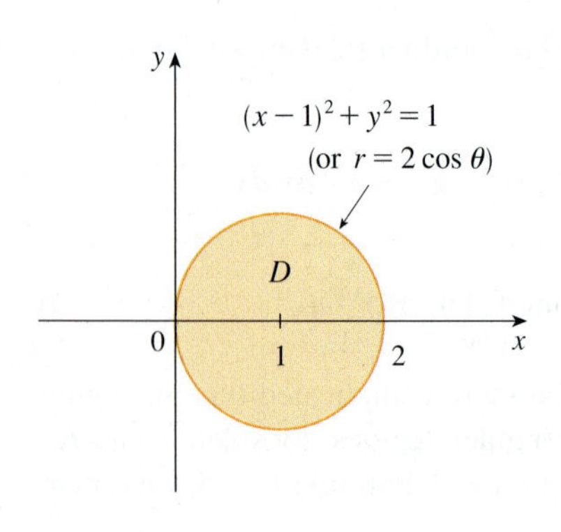

FIGURE 9

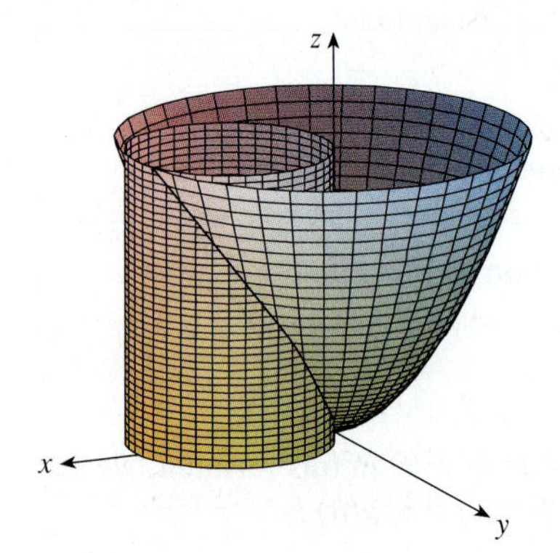

FIGURE 10

EXAMPLE 4 Find the volume of the solid that lies under the paraboloid $z = x^2 + y^2$, above the xy-plane, and inside the cylinder $x^2 + y^2 = 2x$.

SOLUTION The solid lies above the disk D whose boundary circle has equation $x^2 + y^2 = 2x$ or, after completing the square,

$$(x - 1)^2 + y^2 = 1$$

(See Figures 9 and 10.)

In polar coordinates we have $x^2 + y^2 = r^2$ and $x = r\cos\theta$, so the boundary circle becomes $r^2 = 2r\cos\theta$, or $r = 2\cos\theta$. Thus the disk D is given by

$$D = \{(r, \theta) \mid -\pi/2 \leqslant \theta \leqslant \pi/2, 0 \leqslant r \leqslant 2\cos\theta\}$$

and, by Formula 3, we have

$$V = \iint_D (x^2 + y^2)\, dA = \int_{-\pi/2}^{\pi/2} \int_0^{2\cos\theta} r^2\, r\, dr\, d\theta = \int_{-\pi/2}^{\pi/2} \left[\frac{r^4}{4}\right]_0^{2\cos\theta} d\theta$$

$$= 4\int_{-\pi/2}^{\pi/2} \cos^4\theta\, d\theta = 8\int_0^{\pi/2} \cos^4\theta\, d\theta = 8\int_0^{\pi/2} \left(\frac{1 + \cos 2\theta}{2}\right)^2 d\theta$$

$$= 2\int_0^{\pi/2} \left[1 + 2\cos 2\theta + \tfrac{1}{2}(1 + \cos 4\theta)\right] d\theta$$

$$= 2\left[\tfrac{3}{2}\theta + \sin 2\theta + \tfrac{1}{8}\sin 4\theta\right]_0^{\pi/2} = 2\left(\frac{3}{2}\right)\left(\frac{\pi}{2}\right) = \frac{3\pi}{2}$$ ■

15.3 EXERCISES

1–4 A region R is shown. Decide whether to use polar coordinates or rectangular coordinates and write $\iint_R f(x, y)\, dA$ as an iterated integral, where f is an arbitrary continuous function on R.

1.

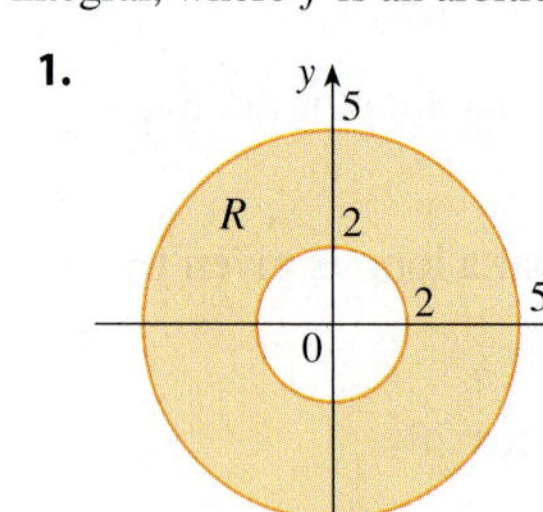

2.

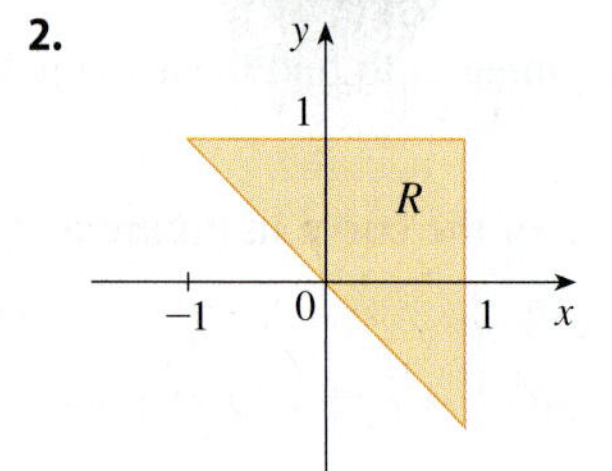

3.

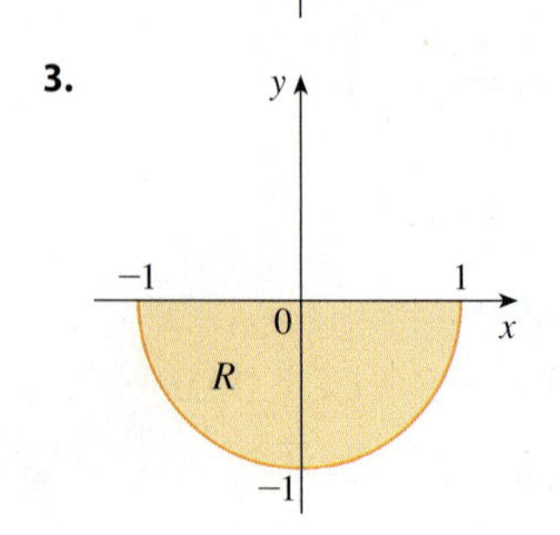

4.

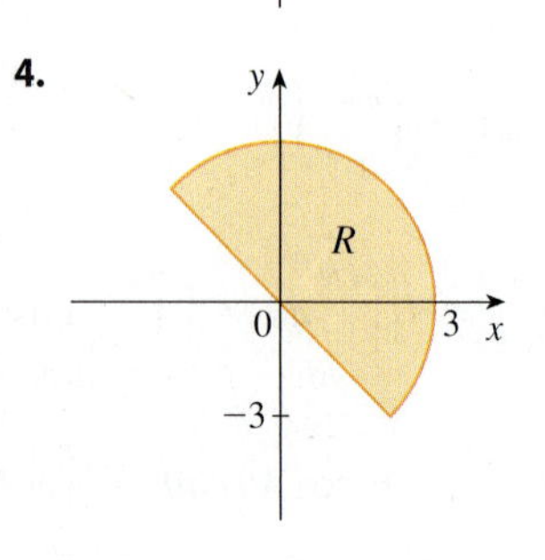

5–6 Sketch the region whose area is given by the integral and evaluate the integral.

5. $\displaystyle\int_{\pi/4}^{3\pi/4} \int_1^2 r\, dr\, d\theta$

6. $\displaystyle\int_{\pi/2}^{\pi} \int_0^{2\sin\theta} r\, dr\, d\theta$

7–14 Evaluate the given integral by changing to polar coordinates.

7. $\iint_D x^2 y\, dA$, where D is the top half of the disk with center the origin and radius 5

8. $\iint_R (2x - y)\, dA$, where R is the region in the first quadrant enclosed by the circle $x^2 + y^2 = 4$ and the lines $x = 0$ and $y = x$

9. $\iint_R \sin(x^2 + y^2)\, dA$, where R is the region in the first quadrant between the circles with center the origin and radii 1 and 3

10. $\displaystyle\iint_R \frac{y^2}{x^2 + y^2}\, dA$, where R is the region that lies between the circles $x^2 + y^2 = a^2$ and $x^2 + y^2 = b^2$ with $0 < a < b$

11. $\iint_D e^{-x^2-y^2}\, dA$, where D is the region bounded by the semicircle $x = \sqrt{4 - y^2}$ and the y-axis

12. $\iint_D \cos\sqrt{x^2 + y^2}\, dA$, where D is the disk with center the origin and radius 2

13. $\iint_R \arctan(y/x)\, dA$, where $R = \{(x, y) \mid 1 \leqslant x^2 + y^2 \leqslant 4,\ 0 \leqslant y \leqslant x\}$

14. $\iint_D x\, dA$, where D is the region in the first quadrant that lies between the circles $x^2 + y^2 = 4$ and $x^2 + y^2 = 2x$

15–18 Use a double integral to find the area of the region.

15. One loop of the rose $r = \cos 3\theta$

16. The region enclosed by both of the cardioids $r = 1 + \cos\theta$ and $r = 1 - \cos\theta$

17. The region inside the circle $(x - 1)^2 + y^2 = 1$ and outside the circle $x^2 + y^2 = 1$

18. The region inside the cardioid $r = 1 + \cos\theta$ and outside the circle $r = 3\cos\theta$

19–27 Use polar coordinates to find the volume of the given solid.

19. Under the paraboloid $z = x^2 + y^2$ and above the disk $x^2 + y^2 \leqslant 25$

20. Below the cone $z = \sqrt{x^2 + y^2}$ and above the ring $1 \leqslant x^2 + y^2 \leqslant 4$

21. Below the plane $2x + y + z = 4$ and above the disk $x^2 + y^2 \leqslant 1$

22. Inside the sphere $x^2 + y^2 + z^2 = 16$ and outside the cylinder $x^2 + y^2 = 4$

23. A sphere of radius a

24. Bounded by the paraboloid $z = 1 + 2x^2 + 2y^2$ and the plane $z = 7$ in the first octant

25. Above the cone $z = \sqrt{x^2 + y^2}$ and below the sphere $x^2 + y^2 + z^2 = 1$

26. Bounded by the paraboloids $z = 6 - x^2 - y^2$ and $z = 2x^2 + 2y^2$

27. Inside both the cylinder $x^2 + y^2 = 4$ and the ellipsoid $4x^2 + 4y^2 + z^2 = 64$

28. (a) A cylindrical drill with radius r_1 is used to bore a hole through the center of a sphere of radius r_2. Find the volume of the ring-shaped solid that remains.

(b) Express the volume in part (a) in terms of the height h of the ring. Notice that the volume depends only on h, not on r_1 or r_2.

29–32 Evaluate the iterated integral by converting to polar coordinates.

29. $\displaystyle\int_0^2 \int_0^{\sqrt{4-x^2}} e^{-x^2-y^2}\, dy\, dx$

30. $\displaystyle\int_0^a \int_{-\sqrt{a^2-y^2}}^{\sqrt{a^2-y^2}} (2x + y)\, dx\, dy$

31. $\displaystyle\int_0^{1/2} \int_{\sqrt{3}\,y}^{\sqrt{1-y^2}} xy^2\, dx\, dy$

32. $\displaystyle\int_0^2 \int_0^{\sqrt{2x-x^2}} \sqrt{x^2 + y^2}\, dy\, dx$

33–34 Express the double integral in terms of a single integral with respect to r. Then use your calculator to evaluate the integral correct to four decimal places.

33. $\iint_D e^{(x^2+y^2)^2}\, dA$, where D is the disk with center the origin and radius 1

34. $\iint_D xy\sqrt{1 + x^2 + y^2}\, dA$, where D is the portion of the disk $x^2 + y^2 \leqslant 1$ that lies in the first quadrant

35. A swimming pool is circular with a 40-ft diameter. The depth is constant along east-west lines and increases linearly from 2 ft at the south end to 7 ft at the north end. Find the volume of water in the pool.

36. An agricultural sprinkler distributes water in a circular pattern of radius 100 ft. It supplies water to a depth of e^{-r} feet per hour at a distance of r feet from the sprinkler.

(a) If $0 < R \leqslant 100$, what is the total amount of water supplied per hour to the region inside the circle of radius R centered at the sprinkler?

(b) Determine an expression for the average amount of water per hour per square foot supplied to the region inside the circle of radius R.

37. Find the average value of the function $f(x, y) = 1/\sqrt{x^2 + y^2}$ on the annular region $a^2 \leqslant x^2 + y^2 \leqslant b^2$, where $0 < a < b$.

38. Let D be the disk with center the origin and radius a. What is the average distance from points in D to the origin?

39. Use polar coordinates to combine the sum

$$\int_{1/\sqrt{2}}^{1} \int_{\sqrt{1-x^2}}^{x} xy\, dy\, dx + \int_1^{\sqrt{2}} \int_0^x xy\, dy\, dx + \int_{\sqrt{2}}^{2} \int_0^{\sqrt{4-x^2}} xy\, dy\, dx$$

into one double integral. Then evaluate the double integral.

40. (a) We define the improper integral (over the entire plane $\mathbb{R}^2$)

$$I = \iint_{\mathbb{R}^2} e^{-(x^2+y^2)}\, dA$$

$$= \int_{-\infty}^{\infty} \int_{-\infty}^{\infty} e^{-(x^2+y^2)}\, dy\, dx$$

$$= \lim_{a\to\infty} \iint_{D_a} e^{-(x^2+y^2)}\, dA$$

where D_a is the disk with radius a and center the origin. Show that

$$\int_{-\infty}^{\infty} \int_{-\infty}^{\infty} e^{-(x^2+y^2)}\, dA = \pi$$

(b) An equivalent definition of the improper integral in part (a) is

$$\iint_{\mathbb{R}^2} e^{-(x^2+y^2)}\,dA = \lim_{a\to\infty} \iint_{S_a} e^{-(x^2+y^2)}\,dA$$

where S_a is the square with vertices $(\pm a, \pm a)$. Use this to show that

$$\int_{-\infty}^{\infty} e^{-x^2}\,dx \int_{-\infty}^{\infty} e^{-y^2}\,dy = \pi$$

(c) Deduce that

$$\int_{-\infty}^{\infty} e^{-x^2}\,dx = \sqrt{\pi}$$

(d) By making the change of variable $t = \sqrt{2}\,x$, show that

$$\int_{-\infty}^{\infty} e^{-x^2/2}\,dx = \sqrt{2\pi}$$

(This is a fundamental result for probability and statistics.)

41. Use the result of Exercise 40 part (c) to evaluate the following integrals.

(a) $\int_0^{\infty} x^2 e^{-x^2}\,dx$ (b) $\int_0^{\infty} \sqrt{x}\, e^{-x}\,dx$

15.4 Applications of Double Integrals

We have already seen one application of double integrals: computing volumes. Another geometric application is finding areas of surfaces and this will be done in the next section. In this section we explore physical applications such as computing mass, electric charge, center of mass, and moment of inertia. We will see that these physical ideas are also important when applied to probability density functions of two random variables.

Density and Mass

In Section 8.3 we were able to use single integrals to compute moments and the center of mass of a thin plate or lamina with constant density. But now, equipped with the double integral, we can consider a lamina with variable density. Suppose the lamina occupies a region D of the xy-plane and its **density** (in units of mass per unit area) at a point (x, y) in D is given by $\rho(x, y)$, where ρ is a continuous function on D. This means that

$$\rho(x, y) = \lim \frac{\Delta m}{\Delta A}$$

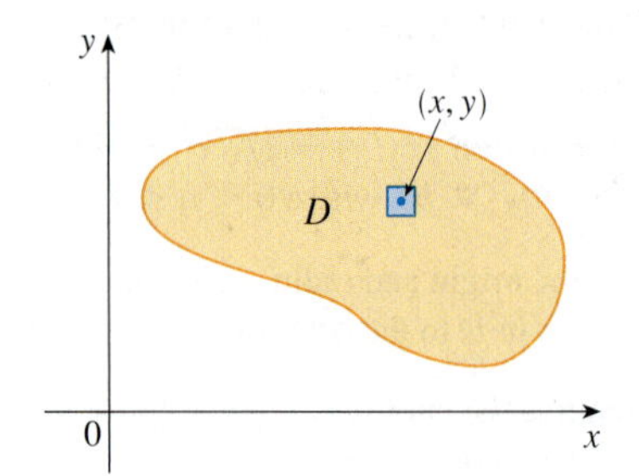

FIGURE 1

where Δm and ΔA are the mass and area of a small rectangle that contains (x, y) and the limit is taken as the dimensions of the rectangle approach 0. (See Figure 1.)

To find the total mass m of the lamina we divide a rectangle R containing D into subrectangles R_{ij} of the same size (as in Figure 2) and consider $\rho(x, y)$ to be 0 outside D. If we choose a point (x_{ij}^*, y_{ij}^*) in R_{ij}, then the mass of the part of the lamina that occupies R_{ij} is approximately $\rho(x_{ij}^*, y_{ij}^*)\,\Delta A$, where ΔA is the area of R_{ij}. If we add all such masses, we get an approximation to the total mass:

$$m \approx \sum_{i=1}^{k} \sum_{j=1}^{l} \rho(x_{ij}^*, y_{ij}^*)\,\Delta A$$

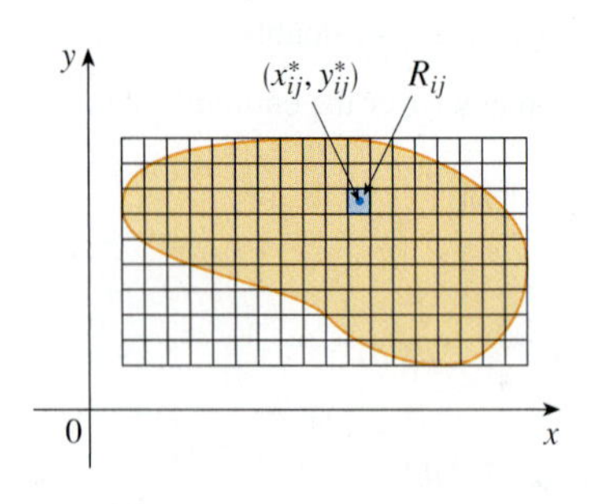

FIGURE 2

If we now increase the number of subrectangles, we obtain the total mass m of the lamina as the limiting value of the approximations:

1
$$m = \lim_{k,\,l\to\infty} \sum_{i=1}^{k} \sum_{j=1}^{l} \rho(x_{ij}^*, y_{ij}^*)\,\Delta A = \iint_D \rho(x, y)\,dA$$

Physicists also consider other types of density that can be treated in the same manner. For example, if an electric charge is distributed over a region D and the charge density (in units of charge per unit area) is given by $\sigma(x, y)$ at a point (x, y) in D, then the total charge Q is given by

2
$$Q = \iint_D \sigma(x, y)\, dA$$

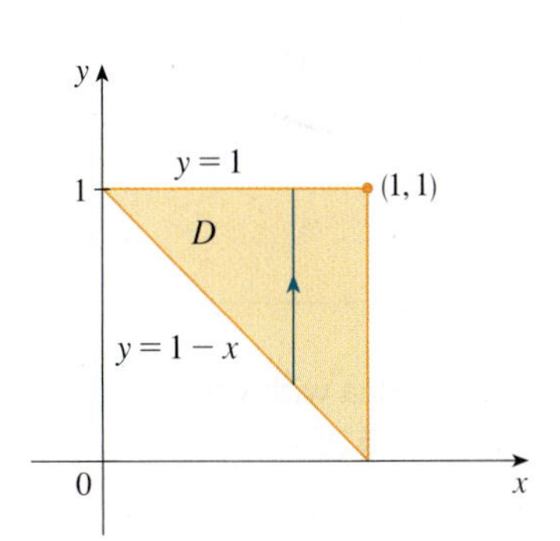

FIGURE 3

EXAMPLE 1 Charge is distributed over the triangular region D in Figure 3 so that the charge density at (x, y) is $\sigma(x, y) = xy$, measured in coulombs per square meter (C/m^2). Find the total charge.

SOLUTION From Equation 2 and Figure 3 we have

$$Q = \iint_D \sigma(x, y)\, dA = \int_0^1 \int_{1-x}^1 xy\, dy\, dx$$

$$= \int_0^1 \left[x \frac{y^2}{2} \right]_{y=1-x}^{y=1} dx = \int_0^1 \frac{x}{2}[1^2 - (1 - x)^2]\, dx$$

$$= \frac{1}{2} \int_0^1 (2x^2 - x^3)\, dx = \frac{1}{2}\left[\frac{2x^3}{3} - \frac{x^4}{4} \right]_0^1 = \frac{5}{24}$$

Thus the total charge is $\frac{5}{24}$ C. ■

Moments and Centers of Mass

In Section 8.3 we found the center of mass of a lamina with constant density; here we consider a lamina with variable density. Suppose the lamina occupies a region D and has density function $\rho(x, y)$. Recall from Chapter 8 that we defined the moment of a particle about an axis as the product of its mass and its directed distance from the axis. We divide D into small rectangles as in Figure 2. Then the mass of R_{ij} is approximately $\rho(x_{ij}^*, y_{ij}^*)\, \Delta A$, so we can approximate the moment of R_{ij} with respect to the x-axis by

$$[\rho(x_{ij}^*, y_{ij}^*)\, \Delta A]\, y_{ij}^*$$

If we now add these quantities and take the limit as the number of subrectangles becomes large, we obtain the **moment** of the entire lamina **about the *x*-axis**:

3
$$M_x = \lim_{m, n \to \infty} \sum_{i=1}^m \sum_{j=1}^n y_{ij}^*\, \rho(x_{ij}^*, y_{ij}^*)\, \Delta A = \iint_D y\rho(x, y)\, dA$$

Similarly, the **moment about the *y*-axis** is

4
$$M_y = \lim_{m, n \to \infty} \sum_{i=1}^m \sum_{j=1}^n x_{ij}^*\, \rho(x_{ij}^*, y_{ij}^*)\, \Delta A = \iint_D x\rho(x, y)\, dA$$

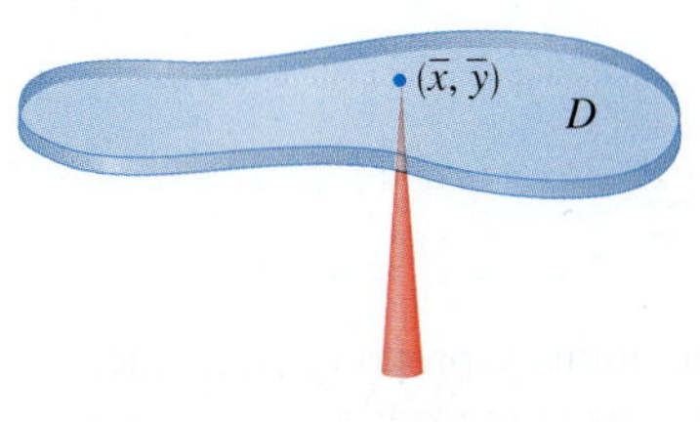

FIGURE 4

As before, we define the center of mass $(\bar{x}, \bar{y})$ so that $m\bar{x} = M_y$ and $m\bar{y} = M_x$. The physical significance is that the lamina behaves as if its entire mass is concentrated at its center of mass. Thus the lamina balances horizontally when supported at its center of mass (see Figure 4).

5 The coordinates $(\bar{x}, \bar{y})$ of the center of mass of a lamina occupying the region D and having density function $\rho(x, y)$ are

$$\bar{x} = \frac{M_y}{m} = \frac{1}{m}\iint_D x\rho(x, y)\,dA \qquad \bar{y} = \frac{M_x}{m} = \frac{1}{m}\iint_D y\rho(x, y)\,dA$$

where the mass m is given by

$$m = \iint_D \rho(x, y)\,dA$$

EXAMPLE 2 Find the mass and center of mass of a triangular lamina with vertices $(0, 0)$, $(1, 0)$, and $(0, 2)$ if the density function is $\rho(x, y) = 1 + 3x + y$.

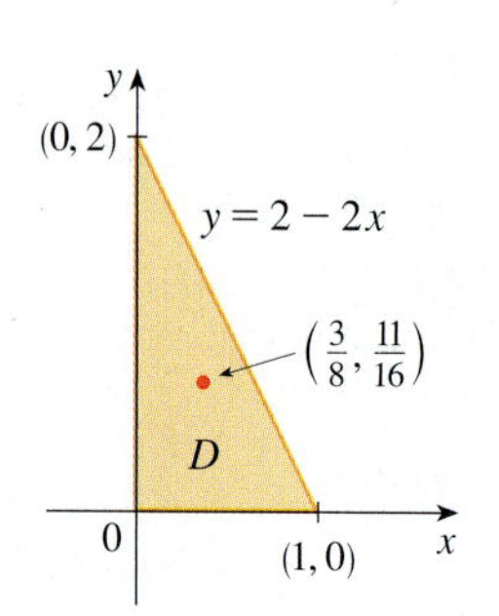

FIGURE 5

SOLUTION The triangle is shown in Figure 5. (Note that the equation of the upper boundary is $y = 2 - 2x$.) The mass of the lamina is

$$m = \iint_D \rho(x, y)\,dA = \int_0^1 \int_0^{2-2x} (1 + 3x + y)\,dy\,dx$$

$$= \int_0^1 \left[y + 3xy + \frac{y^2}{2} \right]_{y=0}^{y=2-2x} dx$$

$$= 4\int_0^1 (1 - x^2)\,dx = 4\left[x - \frac{x^3}{3} \right]_0^1 = \frac{8}{3}$$

Then the formulas in (5) give

$$\bar{x} = \frac{1}{m}\iint_D x\rho(x, y)\,dA = \tfrac{3}{8}\int_0^1 \int_0^{2-2x} (x + 3x^2 + xy)\,dy\,dx$$

$$= \frac{3}{8}\int_0^1 \left[xy + 3x^2y + x\frac{y^2}{2} \right]_{y=0}^{y=2-2x} dx$$

$$= \frac{3}{2}\int_0^1 (x - x^3)\,dx = \frac{3}{2}\left[\frac{x^2}{2} - \frac{x^4}{4} \right]_0^1 = \frac{3}{8}$$

$$\bar{y} = \frac{1}{m}\iint_D y\rho(x, y)\,dA = \tfrac{3}{8}\int_0^1 \int_0^{2-2x} (y + 3xy + y^2)\,dy\,dx$$

$$= \frac{3}{8}\int_0^1 \left[\frac{y^2}{2} + 3x\frac{y^2}{2} + \frac{y^3}{3} \right]_{y=0}^{y=2-2x} dx = \tfrac{1}{4}\int_0^1 (7 - 9x - 3x^2 + 5x^3)\,dx$$

$$= \frac{1}{4}\left[7x - 9\frac{x^2}{2} - x^3 + 5\frac{x^4}{4} \right]_0^1 = \frac{11}{16}$$

The center of mass is at the point $\left(\tfrac{3}{8}, \tfrac{11}{16}\right)$. ■

EXAMPLE 3 The density at any point on a semicircular lamina is proportional to the distance from the center of the circle. Find the center of mass of the lamina.

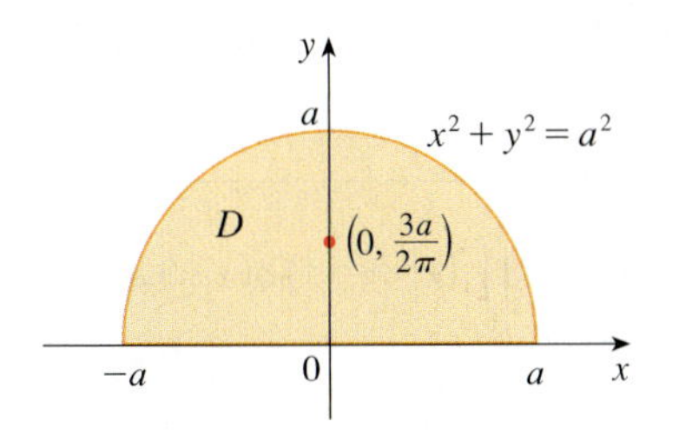

FIGURE 6

SOLUTION Let's place the lamina as the upper half of the circle $x^2 + y^2 = a^2$. (See Figure 6.) Then the distance from a point (x, y) to the center of the circle (the origin) is $\sqrt{x^2 + y^2}$. Therefore the density function is

$$\rho(x, y) = K\sqrt{x^2 + y^2}$$

where K is some constant. Both the density function and the shape of the lamina suggest that we convert to polar coordinates. Then $\sqrt{x^2 + y^2} = r$ and the region D is given by $0 \leqslant r \leqslant a$, $0 \leqslant \theta \leqslant \pi$. Thus the mass of the lamina is

$$m = \iint_D \rho(x, y)\, dA = \iint_D K\sqrt{x^2 + y^2}\, dA$$

$$= \int_0^\pi \int_0^a (Kr)\, r\, dr\, d\theta = K \int_0^\pi d\theta \int_0^a r^2\, dr$$

$$= K\pi \frac{r^3}{3}\bigg]_0^a = \frac{K\pi a^3}{3}$$

Both the lamina and the density function are symmetric with respect to the y-axis, so the center of mass must lie on the y-axis, that is, $\bar{x} = 0$. The y-coordinate is given by

Compare the location of the center of mass in Example 3 with Example 8.3.4, where we found that the center of mass of a lamina with the same shape but uniform density is located at the point $(0, 4a/(3\pi))$.

$$\bar{y} = \frac{1}{m}\iint_D y\rho(x, y)\, dA = \frac{3}{K\pi a^3}\int_0^\pi \int_0^a r\sin\theta\,(Kr)\, r\, dr\, d\theta$$

$$= \frac{3}{\pi a^3}\int_0^\pi \sin\theta\, d\theta \int_0^a r^3\, dr = \frac{3}{\pi a^3}\Big[-\cos\theta\Big]_0^\pi \left[\frac{r^4}{4}\right]_0^a$$

$$= \frac{3}{\pi a^3}\,\frac{2a^4}{4} = \frac{3a}{2\pi}$$

Therefore the center of mass is located at the point $(0, 3a/(2\pi))$. ■

Moment of Inertia

The **moment of inertia** (also called the **second moment**) of a particle of mass m about an axis is defined to be mr^2, where r is the distance from the particle to the axis. We extend this concept to a lamina with density function $\rho(x, y)$ and occupying a region D by proceeding as we did for ordinary moments. We divide D into small rectangles, approximate the moment of inertia of each subrectangle about the x-axis, and take the limit of the sum as the number of subrectangles becomes large. The result is the **moment of inertia** of the lamina **about the *x*-axis**:

6
$$I_x = \lim_{m, n \to \infty} \sum_{i=1}^{m} \sum_{j=1}^{n} (y_{ij}^*)^2 \rho(x_{ij}^*, y_{ij}^*)\, \Delta A = \iint_D y^2 \rho(x, y)\, dA$$

Similarly, the **moment of inertia about the *y*-axis** is

7
$$I_y = \lim_{m, n \to \infty} \sum_{i=1}^{m} \sum_{j=1}^{n} (x_{ij}^*)^2 \rho(x_{ij}^*, y_{ij}^*)\, \Delta A = \iint_D x^2 \rho(x, y)\, dA$$

It is also of interest to consider the **moment of inertia about the origin**, also called the **polar moment of inertia**:

$$\boxed{8} \quad I_0 = \lim_{m,n\to\infty} \sum_{i=1}^{m} \sum_{j=1}^{n} \left[(x_{ij}^*)^2 + (y_{ij}^*)^2\right] \rho(x_{ij}^*, y_{ij}^*)\, \Delta A = \iint_D (x^2 + y^2)\,\rho(x, y)\, dA$$

Note that $I_0 = I_x + I_y$.

EXAMPLE 4 Find the moments of inertia I_x, I_y, and I_0 of a homogeneous disk D with density $\rho(x, y) = \rho$, center the origin, and radius a.

SOLUTION The boundary of D is the circle $x^2 + y^2 = a^2$ and in polar coordinates D is described by $0 \leqslant \theta \leqslant 2\pi$, $0 \leqslant r \leqslant a$. Let's compute I_0 first:

$$I_0 = \iint_D (x^2 + y^2)\rho\, dA = \rho \int_0^{2\pi} \int_0^a r^2\, r\, dr\, d\theta$$

$$= \rho \int_0^{2\pi} d\theta \int_0^a r^3\, dr = 2\pi\rho \left[\frac{r^4}{4}\right]_0^a = \frac{\pi\rho a^4}{2}$$

Instead of computing I_x and I_y directly, we use the facts that $I_x + I_y = I_0$ and $I_x = I_y$ (from the symmetry of the problem). Thus

$$I_x = I_y = \frac{I_0}{2} = \frac{\pi\rho a^4}{4}$$ ■

In Example 4 notice that the mass of the disk is

$$m = \text{density} \times \text{area} = \rho(\pi a^2)$$

so the moment of inertia of the disk about the origin (like a wheel about its axle) can be written as

$$I_0 = \frac{\pi\rho a^4}{2} = \tfrac{1}{2}(\rho\pi a^2)a^2 = \tfrac{1}{2}ma^2$$

Thus if we increase the mass or the radius of the disk, we thereby increase the moment of inertia. In general, the moment of inertia plays much the same role in rotational motion that mass plays in linear motion. The moment of inertia of a wheel is what makes it difficult to start or stop the rotation of the wheel, just as the mass of a car is what makes it difficult to start or stop the motion of the car.

The **radius of gyration of a lamina about an axis** is the number R such that

$$\boxed{9} \qquad mR^2 = I$$

where m is the mass of the lamina and I is the moment of inertia about the given axis. Equation 9 says that if the mass of the lamina were concentrated at a distance R from the axis, then the moment of inertia of this "point mass" would be the same as the moment of inertia of the lamina.

In particular, the radius of gyration $\bar{\bar{y}}$ with respect to the x-axis and the radius of gyration $\bar{\bar{x}}$ with respect to the y-axis are given by the equations

$$\boxed{10} \qquad m\bar{\bar{y}}^2 = I_x \qquad m\bar{\bar{x}}^2 = I_y$$

Thus $(\overline{\overline{x}}, \overline{\overline{y}})$ is the point at which the mass of the lamina can be concentrated without changing the moments of inertia with respect to the coordinate axes. (Note the analogy with the center of mass.)

EXAMPLE 5 Find the radius of gyration about the x-axis of the disk in Example 4.

SOLUTION As noted, the mass of the disk is $m = \rho\pi a^2$, so from Equations 10 we have

$$\overline{\overline{y}}^2 = \frac{I_x}{m} = \frac{\frac{1}{4}\pi\rho a^4}{\rho\pi a^2} = \frac{a^2}{4}$$

Therefore the radius of gyration about the x-axis is $\overline{\overline{y}} = \frac{1}{2}a$, which is half the radius of the disk. ■

Probability

In Section 8.5 we considered the *probability density function* f of a continuous random variable X. This means that $f(x) \geqslant 0$ for all x, $\int_{-\infty}^{\infty} f(x)\,dx = 1$, and the probability that X lies between a and b is found by integrating f from a to b:

$$P(a \leqslant X \leqslant b) = \int_a^b f(x)\,dx$$

Now we consider a pair of continuous random variables X and Y, such as the lifetimes of two components of a machine or the height and weight of an adult female chosen at random. The **joint density function** of X and Y is a function f of two variables such that the probability that (X, Y) lies in a region D is

$$P((X, Y) \in D) = \iint_D f(x, y)\,dA$$

In particular, if the region is a rectangle, the probability that X lies between a and b and Y lies between c and d is

$$P(a \leqslant X \leqslant b,\ c \leqslant Y \leqslant d) = \int_a^b \int_c^d f(x, y)\,dy\,dx$$

(See Figure 7.)

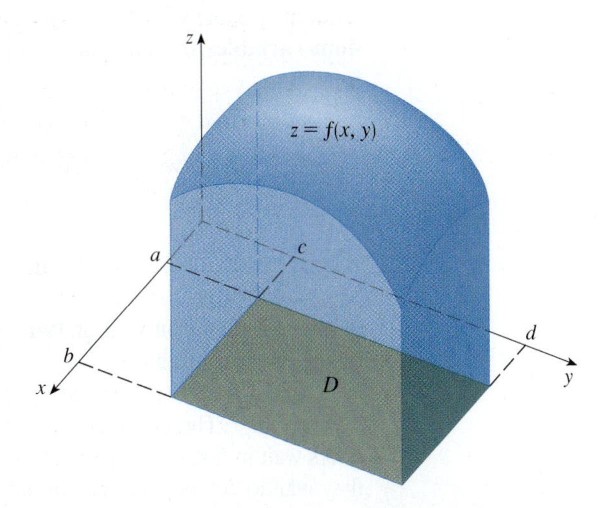

FIGURE 7
The probability that X lies between a and b and Y lies between c and d is the volume that lies above the rectangle $D = [a, b] \times [c, d]$ and below the graph of the joint density function.

Because probabilities aren't negative and are measured on a scale from 0 to 1, the joint density function has the following properties:

$$f(x, y) \geqslant 0 \qquad \iint_{\mathbb{R}^2} f(x, y)\, dA = 1$$

As in Exercise 15.3.40, the double integral over $\mathbb{R}^2$ is an improper integral defined as the limit of double integrals over expanding circles or squares, and we can write

$$\iint_{\mathbb{R}^2} f(x, y)\, dA = \int_{-\infty}^{\infty} \int_{-\infty}^{\infty} f(x, y)\, dx\, dy = 1$$

EXAMPLE 6 If the joint density function for X and Y is given by

$$f(x, y) = \begin{cases} C(x + 2y) & \text{if } 0 \leqslant x \leqslant 10,\ 0 \leqslant y \leqslant 10 \\ 0 & \text{otherwise} \end{cases}$$

find the value of the constant C. Then find $P(X \leqslant 7, Y \geqslant 2)$.

SOLUTION We find the value of C by ensuring that the double integral of f is equal to 1. Because $f(x, y) = 0$ outside the rectangle $[0, 10] \times [0, 10]$, we have

$$\int_{-\infty}^{\infty} \int_{-\infty}^{\infty} f(x, y)\, dy\, dx = \int_0^{10} \int_0^{10} C(x + 2y)\, dy\, dx = C \int_0^{10} \Big[xy + y^2 \Big]_{y=0}^{y=10} dx$$

$$= C \int_0^{10} (10x + 100)\, dx = 1500C$$

Therefore $1500C = 1$ and so $C = \frac{1}{1500}$.

Now we can compute the probability that X is at most 7 and Y is at least 2:

$$P(X \leqslant 7, Y \geqslant 2) = \int_{-\infty}^{7} \int_2^{\infty} f(x, y)\, dy\, dx = \int_0^7 \int_2^{10} \tfrac{1}{1500}(x + 2y)\, dy\, dx$$

$$= \tfrac{1}{1500} \int_0^7 \Big[xy + y^2 \Big]_{y=2}^{y=10} dx = \tfrac{1}{1500} \int_0^7 (8x + 96)\, dx$$

$$= \tfrac{868}{1500} \approx 0.5787$$

Suppose X is a random variable with probability density function $f_1(x)$ and Y is a random variable with density function $f_2(y)$. Then X and Y are called **independent random variables** if their joint density function is the product of their individual density functions:

$$f(x, y) = f_1(x) f_2(y)$$

In Section 8.5 we modeled waiting times by using exponential density functions

$$f(t) = \begin{cases} 0 & \text{if } t < 0 \\ \mu^{-1} e^{-t/\mu} & \text{if } t \geqslant 0 \end{cases}$$

where μ is the mean waiting time. In the next example we consider a situation with two independent waiting times.

EXAMPLE 7 The manager of a movie theater determines that the average time moviegoers wait in line to buy a ticket for this week's film is 10 minutes and the average time they wait to buy popcorn is 5 minutes. Assuming that the waiting times are independent, find the probability that a moviegoer waits a total of less than 20 minutes before taking his or her seat.

SOLUTION Assuming that both the waiting time X for the ticket purchase and the waiting time Y in the refreshment line are modeled by exponential probability density functions, we can write the individual density functions as

$$f_1(x) = \begin{cases} 0 & \text{if } x < 0 \\ \frac{1}{10}e^{-x/10} & \text{if } x \geqslant 0 \end{cases} \qquad f_2(y) = \begin{cases} 0 & \text{if } y < 0 \\ \frac{1}{5}e^{-y/5} & \text{if } y \geqslant 0 \end{cases}$$

Since X and Y are independent, the joint density function is the product:

$$f(x, y) = f_1(x)f_2(y) = \begin{cases} \frac{1}{50}e^{-x/10}e^{-y/5} & \text{if } x \geqslant 0,\ y \geqslant 0 \\ 0 & \text{otherwise} \end{cases}$$

We are asked for the probability that $X + Y < 20$:

$$P(X + Y < 20) = P((X, Y) \in D)$$

where D is the triangular region shown in Figure 8. Thus

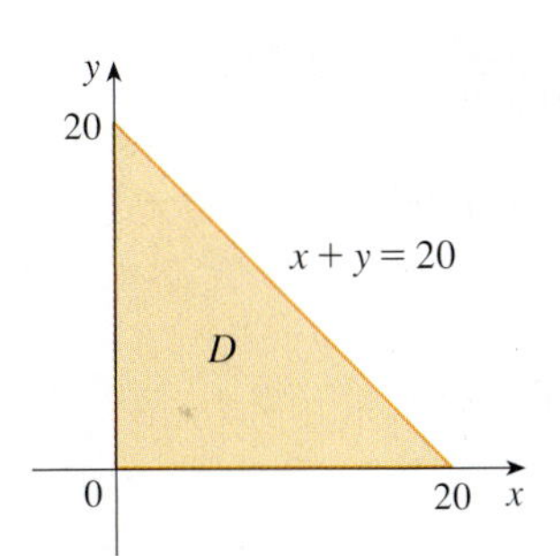

FIGURE 8

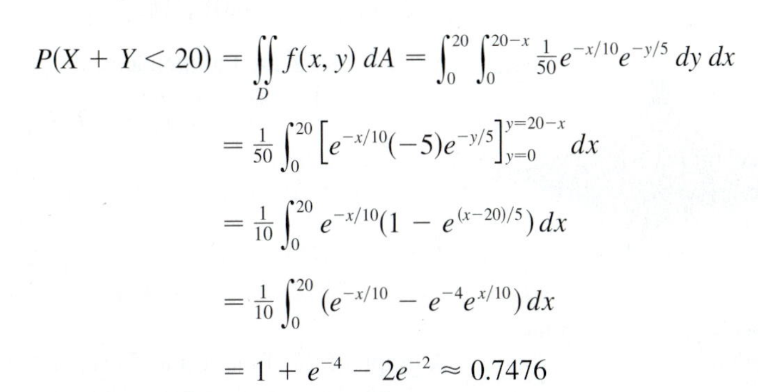

$$\begin{aligned} P(X + Y < 20) &= \iint_D f(x, y)\, dA = \int_0^{20} \int_0^{20-x} \frac{1}{50}e^{-x/10}e^{-y/5}\, dy\, dx \\ &= \frac{1}{50}\int_0^{20} \Big[e^{-x/10}(-5)e^{-y/5}\Big]_{y=0}^{y=20-x} dx \\ &= \frac{1}{10}\int_0^{20} e^{-x/10}(1 - e^{(x-20)/5})\, dx \\ &= \frac{1}{10}\int_0^{20} (e^{-x/10} - e^{-4}e^{x/10})\, dx \\ &= 1 + e^{-4} - 2e^{-2} \approx 0.7476 \end{aligned}$$

This means that about 75% of the moviegoers wait less than 20 minutes before taking their seats. ■

Expected Values

Recall from Section 8.5 that if X is a random variable with probability density function f, then its *mean* is

$$\mu = \int_{-\infty}^{\infty} x f(x)\, dx$$

Now if X and Y are random variables with joint density function f, we define the ***X*-mean** and ***Y*-mean**, also called the **expected values** of X and Y, to be

11 $$\mu_1 = \iint_{\mathbb{R}^2} x f(x, y)\, dA \qquad \mu_2 = \iint_{\mathbb{R}^2} y f(x, y)\, dA$$

Notice how closely the expressions for μ_1 and μ_2 in (11) resemble the moments M_x and M_y of a lamina with density function ρ in Equations 3 and 4. In fact, we can think of probability as being like continuously distributed mass. We calculate probability the way we calculate mass—by integrating a density function. And because the total "probability mass" is 1, the expressions for $\bar{x}$ and $\bar{y}$ in (5) show that we can think of the expected values of X and Y, μ_1 and μ_2, as the coordinates of the "center of mass" of the probability distribution.

In the next example we deal with normal distributions. As in Section 8.5, a single random variable is *normally distributed* if its probability density function is of the form

$$f(x) = \frac{1}{\sigma\sqrt{2\pi}} e^{-(x-\mu)^2/(2\sigma^2)}$$

where μ is the mean and σ is the standard deviation.

EXAMPLE 8 A factory produces (cylindrically shaped) roller bearings that are sold as having diameter 4.0 cm and length 6.0 cm. In fact, the diameters X are normally distributed with mean 4.0 cm and standard deviation 0.01 cm while the lengths Y are normally distributed with mean 6.0 cm and standard deviation 0.01 cm. Assuming that X and Y are independent, write the joint density function and graph it. Find the probability that a bearing randomly chosen from the production line has either length or diameter that differs from the mean by more than 0.02 cm.

SOLUTION We are given that X and Y are normally distributed with $\mu_1 = 4.0$, $\mu_2 = 6.0$, and $\sigma_1 = \sigma_2 = 0.01$. So the individual density functions for X and Y are

$$f_1(x) = \frac{1}{0.01\sqrt{2\pi}} e^{-(x-4)^2/0.0002} \qquad f_2(y) = \frac{1}{0.01\sqrt{2\pi}} e^{-(y-6)^2/0.0002}$$

Since X and Y are independent, the joint density function is the product:

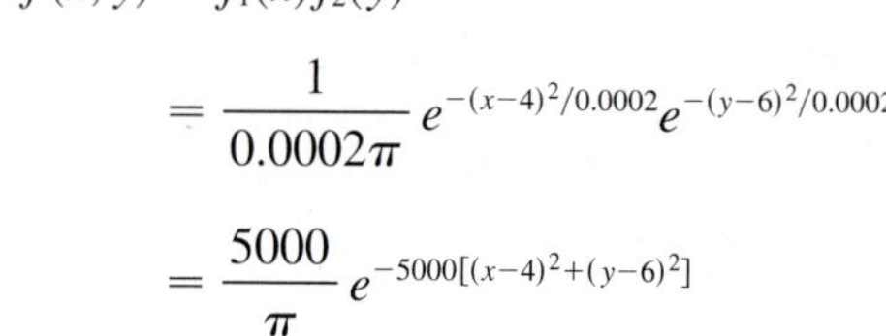

$$f(x, y) = f_1(x)f_2(y)$$

$$= \frac{1}{0.0002\pi} e^{-(x-4)^2/0.0002} e^{-(y-6)^2/0.0002}$$

$$= \frac{5000}{\pi} e^{-5000[(x-4)^2+(y-6)^2]}$$

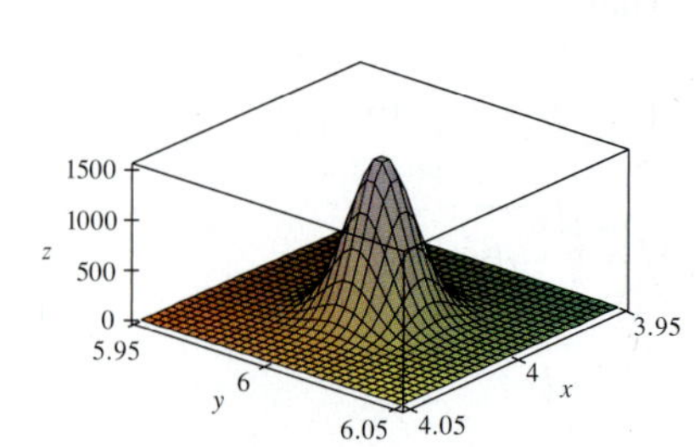

FIGURE 9
Graph of the bivariate normal joint density function in Example 8

A graph of this function is shown in Figure 9.

Let's first calculate the probability that both X and Y differ from their means by less than 0.02 cm. Using a calculator or computer to estimate the integral, we have

$$P(3.98 < X < 4.02,\, 5.98 < Y < 6.02) = \int_{3.98}^{4.02}\int_{5.98}^{6.02} f(x, y)\, dy\, dx$$

$$= \frac{5000}{\pi}\int_{3.98}^{4.02}\int_{5.98}^{6.02} e^{-5000[(x-4)^2+(y-6)^2]}\, dy\, dx$$

$$\approx 0.91$$

Then the probability that either X or Y differs from its mean by more than 0.02 cm is approximately

$$1 - 0.91 = 0.09$$

■

15.4 EXERCISES

1. Electric charge is distributed over the rectangle $0 \leqslant x \leqslant 5$, $2 \leqslant y \leqslant 5$ so that the charge density at (x, y) is $\sigma(x, y) = 2x + 4y$ (measured in coulombs per square meter). Find the total charge on the rectangle.

2. Electric charge is distributed over the disk $x^2 + y^2 \leqslant 1$ so that the charge density at (x, y) is $\sigma(x, y) = \sqrt{x^2 + y^2}$ (measured in coulombs per square meter). Find the total charge on the disk.

3–10 Find the mass and center of mass of the lamina that occupies the region D and has the given density function ρ.

3. $D = \{(x, y) \mid 1 \leqslant x \leqslant 3,\, 1 \leqslant y \leqslant 4\}$; $\rho(x, y) = ky^2$

4. $D = \{(x, y) \mid 0 \leqslant x \leqslant a, 0 \leqslant y \leqslant b\}$;
$\rho(x, y) = 1 + x^2 + y^2$

5. D is the triangular region with vertices $(0, 0)$, $(2, 1)$, $(0, 3)$;
$\rho(x, y) = x + y$

6. D is the triangular region enclosed by the lines $y = 0$, $y = 2x$, and $x + 2y = 1$; $\rho(x, y) = x$

7. D is bounded by $y = 1 - x^2$ and $y = 0$; $\rho(x, y) = ky$

8. D is bounded by $y = x + 2$ and $y = x^2$; $\rho(x, y) = kx^2$

9. D is bounded by the curves $y = e^{-x}$, $y = 0$, $x = 0$, $x = 1$;
$\rho(x, y) = xy$

10. D is enclosed by the curves $y = 0$ and $y = \cos x$,
$-\pi/2 \leqslant x \leqslant \pi/2$; $\rho(x, y) = y$

---

11. A lamina occupies the part of the disk $x^2 + y^2 \leqslant 1$ in the first quadrant. Find its center of mass if the density at any point is proportional to its distance from the x-axis.

12. Find the center of mass of the lamina in Exercise 11 if the density at any point is proportional to the square of its distance from the origin.

13. The boundary of a lamina consists of the semicircles $y = \sqrt{1 - x^2}$ and $y = \sqrt{4 - x^2}$ together with the portions of the x-axis that join them. Find the center of mass of the lamina if the density at any point is proportional to its distance from the origin.

14. Find the center of mass of the lamina in Exercise 13 if the density at any point is inversely proportional to its distance from the origin.

15. Find the center of mass of a lamina in the shape of an isosceles right triangle with equal sides of length a if the density at any point is proportional to the square of the distance from the vertex opposite the hypotenuse.

16. A lamina occupies the region inside the circle $x^2 + y^2 = 2y$ but outside the circle $x^2 + y^2 = 1$. Find the center of mass if the density at any point is inversely proportional to its distance from the origin.

17. Find the moments of inertia I_x, I_y, I_0 for the lamina of Exercise 3.

18. Find the moments of inertia I_x, I_y, I_0 for the lamina of Exercise 6.

19. Find the moments of inertia I_x, I_y, I_0 for the lamina of Exercise 15.

20. Consider a square fan blade with sides of length 2 and the lower left corner placed at the origin. If the density of the blade is $\rho(x, y) = 1 + 0.1x$, is it more difficult to rotate the blade about the x-axis or the y-axis?

21–24 A lamina with constant density $\rho(x, y) = \rho$ occupies the given region. Find the moments of inertia I_x and I_y and the radii of gyration $\bar{\bar{x}}$ and $\bar{\bar{y}}$.

21. The rectangle $0 \leqslant x \leqslant b$, $0 \leqslant y \leqslant h$

22. The triangle with vertices $(0, 0)$, $(b, 0)$, and $(0, h)$

23. The part of the disk $x^2 + y^2 \leqslant a^2$ in the first quadrant

24. The region under the curve $y = \sin x$ from $x = 0$ to $x = \pi$

CAS **25–26** Use a computer algebra system to find the mass, center of mass, and moments of inertia of the lamina that occupies the region D and has the given density function.

25. D is enclosed by the right loop of the four-leaved rose
$r = \cos 2\theta$; $\rho(x, y) = x^2 + y^2$

26. $D = \{(x, y) \mid 0 \leqslant y \leqslant xe^{-x}, 0 \leqslant x \leqslant 2\}$; $\rho(x, y) = x^2y^2$

---

27. The joint density function for a pair of random variables X and Y is

$$f(x, y) = \begin{cases} Cx(1 + y) & \text{if } 0 \leqslant x \leqslant 1,\ 0 \leqslant y \leqslant 2 \\ 0 & \text{otherwise} \end{cases}$$

(a) Find the value of the constant C.
(b) Find $P(X \leqslant 1, Y \leqslant 1)$.
(c) Find $P(X + Y \leqslant 1)$.

28. (a) Verify that

$$f(x, y) = \begin{cases} 4xy & \text{if } 0 \leqslant x \leqslant 1,\ 0 \leqslant y \leqslant 1 \\ 0 & \text{otherwise} \end{cases}$$

is a joint density function.
(b) If X and Y are random variables whose joint density function is the function f in part (a), find
(i) $P\left(X \geqslant \frac{1}{2}\right)$ (ii) $P\left(X \geqslant \frac{1}{2}, Y \leqslant \frac{1}{2}\right)$
(c) Find the expected values of X and Y.

29. Suppose X and Y are random variables with joint density function

$$f(x, y) = \begin{cases} 0.1e^{-(0.5x+0.2y)} & \text{if } x \geqslant 0,\ y \geqslant 0 \\ 0 & \text{otherwise} \end{cases}$$

(a) Verify that f is indeed a joint density function.
(b) Find the following probabilities.
(i) $P(Y \geqslant 1)$ (ii) $P(X \leqslant 2, Y \leqslant 4)$
(c) Find the expected values of X and Y.

30. (a) A lamp has two bulbs, each of a type with average lifetime 1000 hours. Assuming that we can model the probability of failure of a bulb by an exponential density function with mean $\mu = 1000$, find the probability that both of the lamp's bulbs fail within 1000 hours.
(b) Another lamp has just one bulb of the same type as in part (a). If one bulb burns out and is replaced by a bulb of the same type, find the probability that the two bulbs fail within a total of 1000 hours.

CAS **31.** Suppose that X and Y are independent random variables, where X is normally distributed with mean 45 and standard deviation 0.5 and Y is normally distributed with mean 20 and standard deviation 0.1.

(a) Find $P(40 \leqslant X \leqslant 50,\ 20 \leqslant Y \leqslant 25)$.

(b) Find $P(4(X - 45)^2 + 100(Y - 20)^2 \leqslant 2)$.

32. Xavier and Yolanda both have classes that end at noon and they agree to meet every day after class. They arrive at the coffee shop independently. Xavier's arrival time is X and Yolanda's arrival time is Y, where X and Y are measured in minutes after noon. The individual density functions are

$$f_1(x) = \begin{cases} e^{-x} & \text{if } x \geqslant 0 \\ 0 & \text{if } x < 0 \end{cases} \qquad f_2(y) = \begin{cases} \frac{1}{50}y & \text{if } 0 \leqslant y \leqslant 10 \\ 0 & \text{otherwise} \end{cases}$$

(Xavier arrives sometime after noon and is more likely to arrive promptly than late. Yolanda always arrives by 12:10 PM and is more likely to arrive late than promptly.) After Yolanda arrives, she'll wait for up to half an hour for Xavier, but he won't wait for her. Find the probability that they meet.

33. When studying the spread of an epidemic, we assume that the probability that an infected individual will spread the disease to an uninfected individual is a function of the distance between them. Consider a circular city of radius 10 miles in which the population is uniformly distributed. For an uninfected individual at a fixed point $A(x_0, y_0)$, assume that the probability function is given by

$$f(P) = \tfrac{1}{20}[20 - d(P, A)]$$

where $d(P, A)$ denotes the distance between points P and A.

(a) Suppose the exposure of a person to the disease is the sum of the probabilities of catching the disease from all members of the population. Assume that the infected people are uniformly distributed throughout the city, with k infected individuals per square mile. Find a double integral that represents the exposure of a person residing at A.

(b) Evaluate the integral for the case in which A is the center of the city and for the case in which A is located on the edge of the city. Where would you prefer to live?

15.5 Surface Area

In Section 16.6 we will deal with areas of more general surfaces, called parametric surfaces, and so this section need not be covered if that later section will be covered.

In this section we apply double integrals to the problem of computing the area of a surface. In Section 8.2 we found the area of a very special type of surface—a surface of revolution—by the methods of single-variable calculus. Here we compute the area of a surface with equation $z = f(x, y)$, the graph of a function of two variables.

Let S be a surface with equation $z = f(x, y)$, where f has continuous partial derivatives. For simplicity in deriving the surface area formula, we assume that $f(x, y) \geqslant 0$ and the domain D of f is a rectangle. We divide D into small rectangles R_{ij} with area $\Delta A = \Delta x\, \Delta y$. If (x_i, y_j) is the corner of R_{ij} closest to the origin, let $P_{ij}(x_i, y_j, f(x_i, y_j))$ be the point on S directly above it (see Figure 1). The tangent plane to S at P_{ij} is an approximation to S near P_{ij}. So the area ΔT_{ij} of the part of this tangent plane (a parallelogram) that lies directly above R_{ij} is an approximation to the area ΔS_{ij} of the part of S that lies directly above R_{ij}. Thus the sum $\Sigma\Sigma\, \Delta T_{ij}$ is an approximation to the total area of S, and this approximation appears to improve as the number of rectangles increases. Therefore we define the **surface area** of S to be

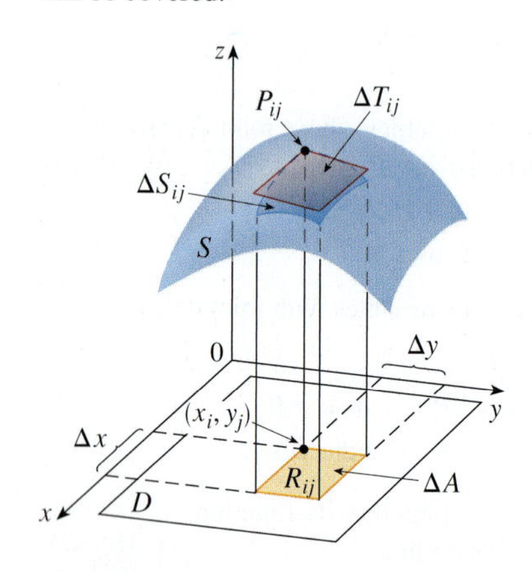

FIGURE 1

1
$$A(S) = \lim_{m,\, n \to \infty} \sum_{i=1}^{m} \sum_{j=1}^{n} \Delta T_{ij}$$

To find a formula that is more convenient than Equation 1 for computational purposes, we let **a** and **b** be the vectors that start at P_{ij} and lie along the sides of the parallelogram with area ΔT_{ij}. (See Figure 2.) Then $\Delta T_{ij} = |\mathbf{a} \times \mathbf{b}|$. Recall from Section 14.3 that $f_x(x_i, y_j)$ and $f_y(x_i, y_j)$ are the slopes of the tangent lines through P_{ij} in the directions of **a** and **b**. Therefore

$$\mathbf{a} = \Delta x\, \mathbf{i} + f_x(x_i, y_j)\, \Delta x\, \mathbf{k}$$

$$\mathbf{b} = \Delta y\, \mathbf{j} + f_y(x_i, y_j)\, \Delta y\, \mathbf{k}$$

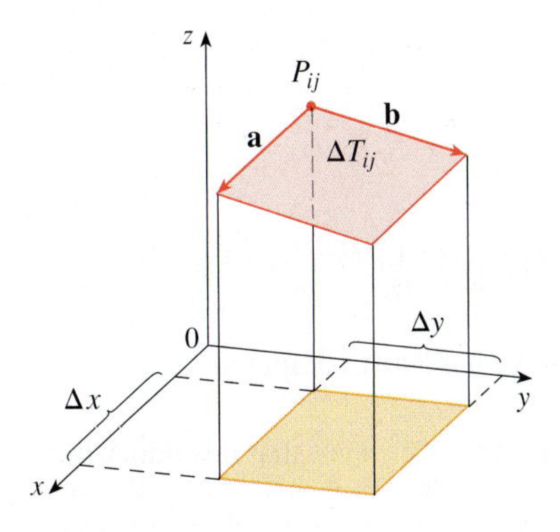

FIGURE 2

and

$$\mathbf{a} \times \mathbf{b} = \begin{vmatrix} \mathbf{i} & \mathbf{j} & \mathbf{k} \\ \Delta x & 0 & f_x(x_i, y_j)\,\Delta x \\ 0 & \Delta y & f_y(x_i, y_j)\,\Delta y \end{vmatrix}$$

$$= -f_x(x_i, y_j)\,\Delta x\,\Delta y\,\mathbf{i} - f_y(x_i, y_j)\,\Delta x\,\Delta y\,\mathbf{j} + \Delta x\,\Delta y\,\mathbf{k}$$

$$= [-f_x(x_i, y_j)\mathbf{i} - f_y(x_i, y_j)\mathbf{j} + \mathbf{k}]\,\Delta A$$

Thus $$\Delta T_{ij} = |\mathbf{a} \times \mathbf{b}| = \sqrt{[f_x(x_i, y_j)]^2 + [f_y(x_i, y_j)]^2 + 1}\,\Delta A$$

From Definition 1 we then have

$$A(S) = \lim_{m,n\to\infty} \sum_{i=1}^{m} \sum_{j=1}^{n} \Delta T_{ij}$$

$$= \lim_{m,n\to\infty} \sum_{i=1}^{m} \sum_{j=1}^{n} \sqrt{[f_x(x_i, y_j)]^2 + [f_y(x_i, y_j)]^2 + 1}\,\Delta A$$

and by the definition of a double integral we get the following formula.

2 The area of the surface with equation $z = f(x, y)$, $(x, y) \in D$, where f_x and f_y are continuous, is

$$A(S) = \iint_D \sqrt{[f_x(x, y)]^2 + [f_y(x, y)]^2 + 1}\; dA$$

We will verify in Section 16.6 that this formula is consistent with our previous formula for the area of a surface of revolution. If we use the alternative notation for partial derivatives, we can rewrite Formula 2 as follows:

3 $$A(S) = \iint_D \sqrt{1 + \left(\frac{\partial z}{\partial x}\right)^2 + \left(\frac{\partial z}{\partial y}\right)^2}\; dA$$

Notice the similarity between the surface area formula in Equation 3 and the arc length formula from Section 8.1:

$$L = \int_a^b \sqrt{1 + \left(\frac{dy}{dx}\right)^2}\; dx$$

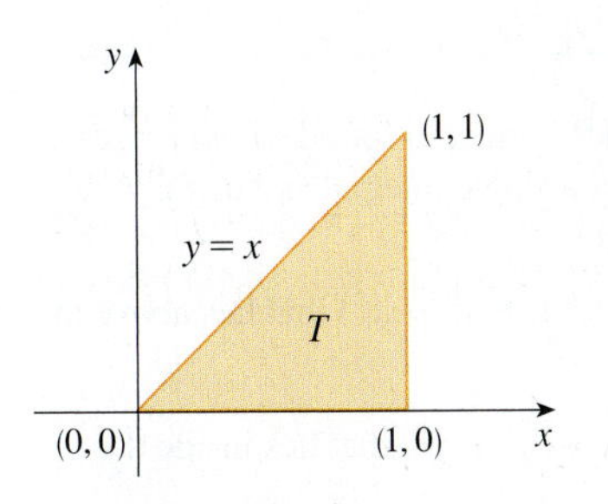

FIGURE 3

EXAMPLE 1 Find the surface area of the part of the surface $z = x^2 + 2y$ that lies above the triangular region T in the xy-plane with vertices (0, 0), (1, 0), and (1, 1).

SOLUTION The region T is shown in Figure 3 and is described by

$$T = \{(x, y) \mid 0 \leqslant x \leqslant 1,\ 0 \leqslant y \leqslant x\}$$

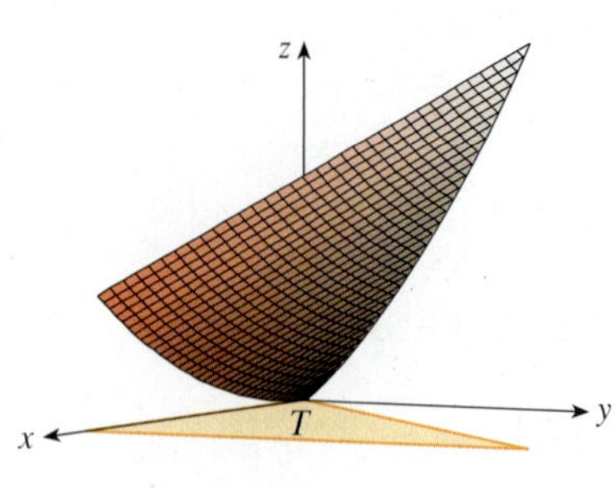

FIGURE 4

Using Formula 2 with $f(x, y) = x^2 + 2y$, we get

$$A = \iint_T \sqrt{(2x)^2 + (2)^2 + 1}\, dA = \int_0^1 \int_0^x \sqrt{4x^2 + 5}\, dy\, dx$$

$$= \int_0^1 x\sqrt{4x^2 + 5}\, dx = \tfrac{1}{8} \cdot \tfrac{2}{3}(4x^2 + 5)^{3/2}\Big]_0^1 = \tfrac{1}{12}\left(27 - 5\sqrt{5}\right)$$

Figure 4 shows the portion of the surface whose area we have just computed. ■

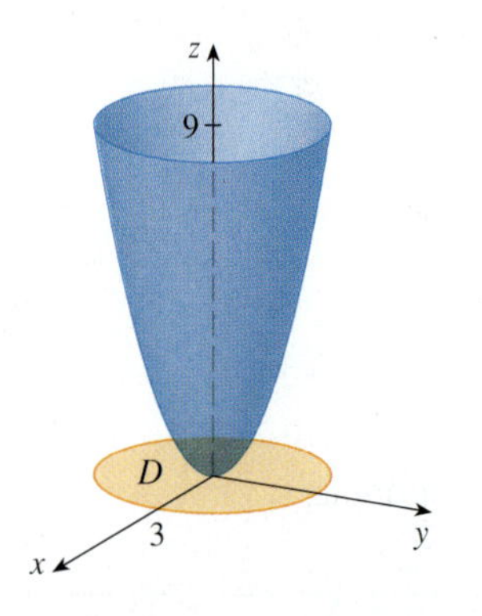

FIGURE 5

EXAMPLE 2 Find the area of the part of the paraboloid $z = x^2 + y^2$ that lies under the plane $z = 9$.

SOLUTION The plane intersects the paraboloid in the circle $x^2 + y^2 = 9$, $z = 9$. Therefore the given surface lies above the disk D with center the origin and radius 3. (See Figure 5.) Using Formula 3, we have

$$A = \iint_D \sqrt{1 + \left(\frac{\partial z}{\partial x}\right)^2 + \left(\frac{\partial z}{\partial y}\right)^2}\, dA = \iint_D \sqrt{1 + (2x)^2 + (2y)^2}\, dA$$

$$= \iint_D \sqrt{1 + 4(x^2 + y^2)}\, dA$$

Converting to polar coordinates, we obtain

$$A = \int_0^{2\pi} \int_0^3 \sqrt{1 + 4r^2}\, r\, dr\, d\theta = \int_0^{2\pi} d\theta \int_0^3 \tfrac{1}{8}\sqrt{1 + 4r^2}\, (8r)\, dr$$

$$= 2\pi\left(\tfrac{1}{8}\right)\tfrac{2}{3}(1 + 4r^2)^{3/2}\Big]_0^3 = \frac{\pi}{6}\left(37\sqrt{37} - 1\right)$$ ■

15.5 EXERCISES

1–12 Find the area of the surface.

1. The part of the plane $5x + 3y - z + 6 = 0$ that lies above the rectangle $[1, 4] \times [2, 6]$

2. The part of the plane $6x + 4y + 2z = 1$ that lies inside the cylinder $x^2 + y^2 = 25$

3. The part of the plane $3x + 2y + z = 6$ that lies in the first octant

4. The part of the surface $2y + 4z - x^2 = 5$ that lies above the triangle with vertices $(0, 0)$, $(2, 0)$, and $(2, 4)$

5. The part of the paraboloid $z = 1 - x^2 - y^2$ that lies above the plane $z = -2$

6. The part of the cylinder $x^2 + z^2 = 4$ that lies above the square with vertices $(0, 0)$, $(1, 0)$, $(0, 1)$, and $(1, 1)$

7. The part of the hyperbolic paraboloid $z = y^2 - x^2$ that lies between the cylinders $x^2 + y^2 = 1$ and $x^2 + y^2 = 4$

8. The surface $z = \frac{2}{3}(x^{3/2} + y^{3/2})$, $0 \leqslant x \leqslant 1$, $0 \leqslant y \leqslant 1$

9. The part of the surface $z = xy$ that lies within the cylinder $x^2 + y^2 = 1$

10. The part of the sphere $x^2 + y^2 + z^2 = 4$ that lies above the plane $z = 1$

11. The part of the sphere $x^2 + y^2 + z^2 = a^2$ that lies within the cylinder $x^2 + y^2 = ax$ and above the xy-plane

12. The part of the sphere $x^2 + y^2 + z^2 = 4z$ that lies inside the paraboloid $z = x^2 + y^2$

13–14 Find the area of the surface correct to four decimal places by expressing the area in terms of a single integral and using your calculator to estimate the integral.

13. The part of the surface $z = 1/(1 + x^2 + y^2)$ that lies above the disk $x^2 + y^2 \leqslant 1$

14. The part of the surface $z = \cos(x^2 + y^2)$ that lies inside the cylinder $x^2 + y^2 = 1$

15. (a) Use the Midpoint Rule for double integrals (see Section 15.1) with four squares to estimate the surface area of the portion of the paraboloid $z = x^2 + y^2$ that lies above the square $[0, 1] \times [0, 1]$.

CAS (b) Use a computer algebra system to approximate the surface area in part (a) to four decimal places. Compare with the answer to part (a).

16. (a) Use the Midpoint Rule for double integrals with $m = n = 2$ to estimate the area of the surface $z = xy + x^2 + y^2$, $0 \leq x \leq 2$, $0 \leq y \leq 2$.

CAS (b) Use a computer algebra system to approximate the surface area in part (a) to four decimal places. Compare with the answer to part (a).

CAS **17.** Find the exact area of the surface $z = 1 + 2x + 3y + 4y^2$, $1 \leq x \leq 4, 0 \leq y \leq 1$.

CAS **18.** Find the exact area of the surface

$$z = 1 + x + y + x^2 \qquad -2 \leq x \leq 1 \quad -1 \leq y \leq 1$$

Illustrate by graphing the surface.

CAS **19.** Find, to four decimal places, the area of the part of the surface $z = 1 + x^2y^2$ that lies above the disk $x^2 + y^2 \leq 1$.

CAS **20.** Find, to four decimal places, the area of the part of the surface $z = (1 + x^2)/(1 + y^2)$ that lies above the square $|x| + |y| \leq 1$. Illustrate by graphing this part of the surface.

21. Show that the area of the part of the plane $z = ax + by + c$ that projects onto a region D in the xy-plane with area $A(D)$ is $\sqrt{a^2 + b^2 + 1}\, A(D)$.

22. If you attempt to use Formula 2 to find the area of the top half of the sphere $x^2 + y^2 + z^2 = a^2$, you have a slight problem because the double integral is improper. In fact, the integrand has an infinite discontinuity at every point of the boundary circle $x^2 + y^2 = a^2$. However, the integral can be computed as the limit of the integral over the disk $x^2 + y^2 \leq t^2$ as $t \to a^-$. Use this method to show that the area of a sphere of radius a is $4\pi a^2$.

23. Find the area of the finite part of the paraboloid $y = x^2 + z^2$ cut off by the plane $y = 25$. [*Hint:* Project the surface onto the xz-plane.]

24. The figure shows the surface created when the cylinder $y^2 + z^2 = 1$ intersects the cylinder $x^2 + z^2 = 1$. Find the area of this surface.

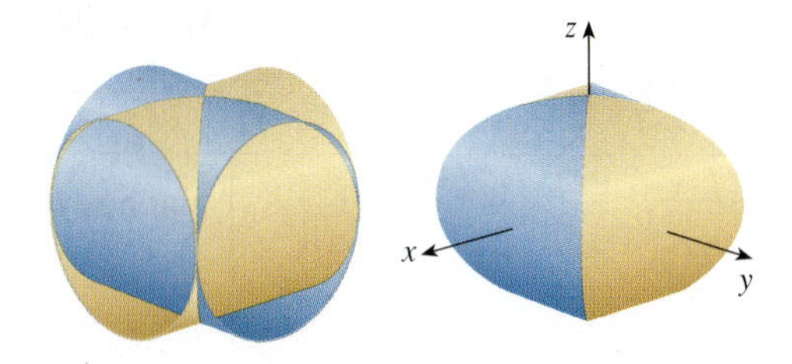

15.6 Triple Integrals

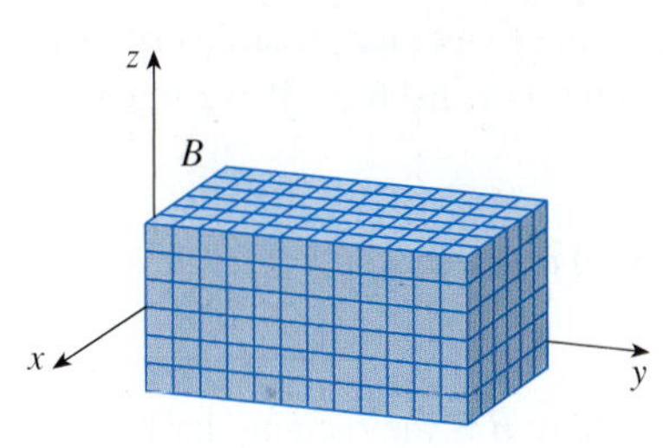

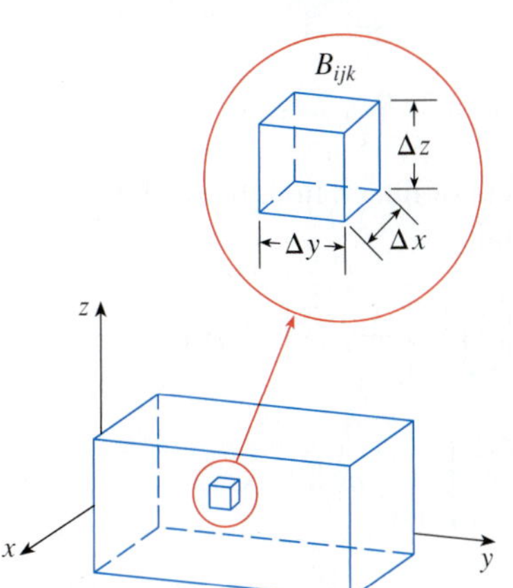

FIGURE 1

Just as we defined single integrals for functions of one variable and double integrals for functions of two variables, so we can define triple integrals for functions of three variables. Let's first deal with the simplest case where f is defined on a rectangular box:

1 $$B = \{(x, y, z) \mid a \leq x \leq b,\ c \leq y \leq d,\ r \leq z \leq s\}$$

The first step is to divide B into sub-boxes. We do this by dividing the interval $[a, b]$ into l subintervals $[x_{i-1}, x_i]$ of equal width Δx, dividing $[c, d]$ into m subintervals of width Δy, and dividing $[r, s]$ into n subintervals of width Δz. The planes through the endpoints of these subintervals parallel to the coordinate planes divide the box B into lmn sub-boxes

$$B_{ijk} = [x_{i-1}, x_i] \times [y_{j-1}, y_j] \times [z_{k-1}, z_k]$$

which are shown in Figure 1. Each sub-box has volume $\Delta V = \Delta x\, \Delta y\, \Delta z$.

Then we form the **triple Riemann sum**

2 $$\sum_{i=1}^{l} \sum_{j=1}^{m} \sum_{k=1}^{n} f(x_{ijk}^*, y_{ijk}^*, z_{ijk}^*)\, \Delta V$$

where the sample point $(x_{ijk}^*, y_{ijk}^*, z_{ijk}^*)$ is in B_{ijk}. By analogy with the definition of a double integral (15.1.5), we define the triple integral as the limit of the triple Riemann sums in (2).

3 Definition The **triple integral** of f over the box B is

$$\iiint_B f(x, y, z)\, dV = \lim_{l, m, n \to \infty} \sum_{i=1}^{l} \sum_{j=1}^{m} \sum_{k=1}^{n} f(x_{ijk}^*, y_{ijk}^*, z_{ijk}^*)\, \Delta V$$

if this limit exists.

Again, the triple integral always exists if f is continuous. We can choose the sample point to be any point in the sub-box, but if we choose it to be the point (x_i, y_j, z_k) we get a simpler-looking expression for the triple integral:

$$\iiint_B f(x, y, z)\, dV = \lim_{l, m, n \to \infty} \sum_{i=1}^{l} \sum_{j=1}^{m} \sum_{k=1}^{n} f(x_i, y_j, z_k)\, \Delta V$$

Just as for double integrals, the practical method for evaluating triple integrals is to express them as iterated integrals as follows.

4 Fubini's Theorem for Triple Integrals If f is continuous on the rectangular box $B = [a, b] \times [c, d] \times [r, s]$, then

$$\iiint_B f(x, y, z)\, dV = \int_r^s \int_c^d \int_a^b f(x, y, z)\, dx\, dy\, dz$$

The iterated integral on the right side of Fubini's Theorem means that we integrate first with respect to x (keeping y and z fixed), then we integrate with respect to y (keeping z fixed), and finally we integrate with respect to z. There are five other possible orders in which we can integrate, all of which give the same value. For instance, if we integrate with respect to y, then z, and then x, we have

$$\iiint_B f(x, y, z)\, dV = \int_a^b \int_r^s \int_c^d f(x, y, z)\, dy\, dz\, dx$$

EXAMPLE 1 Evaluate the triple integral $\iiint_B xyz^2\, dV$, where B is the rectangular box given by

$$B = \{(x, y, z) \mid 0 \leqslant x \leqslant 1,\ -1 \leqslant y \leqslant 2,\ 0 \leqslant z \leqslant 3\}$$

SOLUTION We could use any of the six possible orders of integration. If we choose to integrate with respect to x, then y, and then z, we obtain

$$\iiint_B xyz^2\, dV = \int_0^3 \int_{-1}^2 \int_0^1 xyz^2\, dx\, dy\, dz = \int_0^3 \int_{-1}^2 \left[\frac{x^2yz^2}{2}\right]_{x=0}^{x=1} dy\, dz$$

$$= \int_0^3 \int_{-1}^2 \frac{yz^2}{2}\, dy\, dz = \int_0^3 \left[\frac{y^2z^2}{4}\right]_{y=-1}^{y=2} dz$$

$$= \int_0^3 \frac{3z^2}{4}\, dz = \frac{z^3}{4}\bigg]_0^3 = \frac{27}{4}$$ ■

Now we define the **triple integral over a general bounded region E** in three-dimensional space (a solid) by much the same procedure that we used for double integrals (15.2.2). We enclose E in a box B of the type given by Equation 1. Then we define F so that it agrees with f on E but is 0 for points in B that are outside E. By definition,

$$\iiint_E f(x, y, z)\, dV = \iiint_B F(x, y, z)\, dV$$

This integral exists if f is continuous and the boundary of E is "reasonably smooth." The triple integral has essentially the same properties as the double integral (Properties 6–9 in Section 15.2).

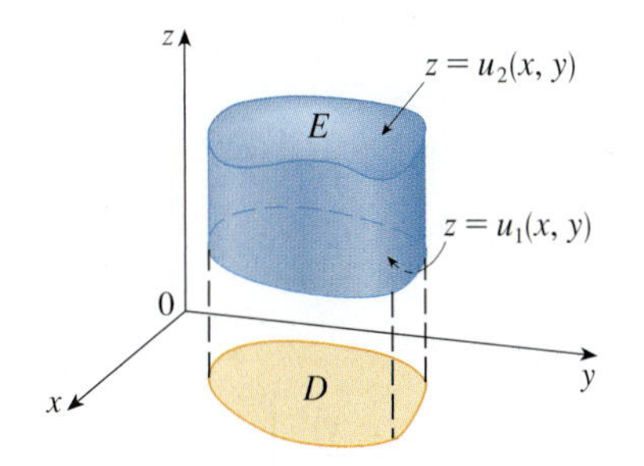

FIGURE 2
A type 1 solid region

We restrict our attention to continuous functions f and to certain simple types of regions. A solid region E is said to be of **type 1** if it lies between the graphs of two continuous functions of x and y, that is,

5 $$E = \{(x, y, z) \mid (x, y) \in D,\ u_1(x, y) \leqslant z \leqslant u_2(x, y)\}$$

where D is the projection of E onto the xy-plane as shown in Figure 2. Notice that the upper boundary of the solid E is the surface with equation $z = u_2(x, y)$, while the lower boundary is the surface $z = u_1(x, y)$.

By the same sort of argument that led to (15.2.3), it can be shown that if E is a type 1 region given by Equation 5, then

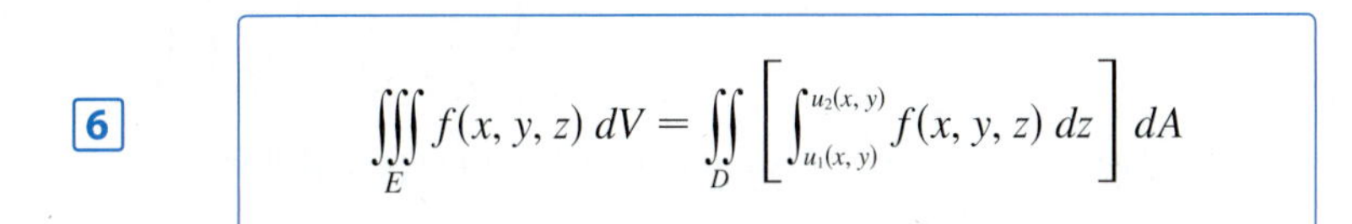

6 $$\iiint_E f(x, y, z)\, dV = \iint_D \left[\int_{u_1(x, y)}^{u_2(x, y)} f(x, y, z)\, dz \right] dA$$

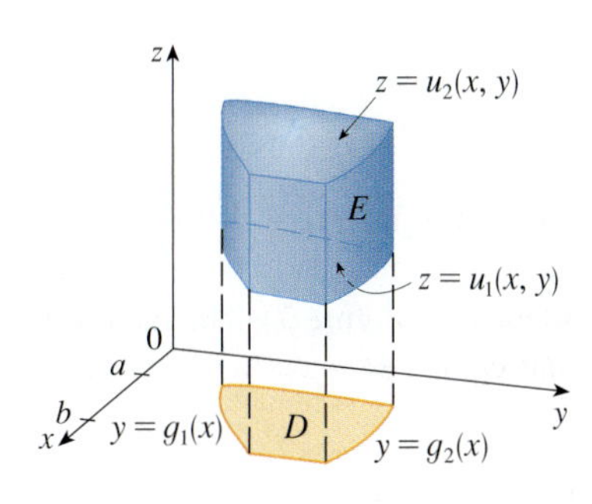

FIGURE 3
A type 1 solid region where the projection D is a type I plane region

The meaning of the inner integral on the right side of Equation 6 is that x and y are held fixed, and therefore $u_1(x, y)$ and $u_2(x, y)$ are regarded as constants, while $f(x, y, z)$ is integrated with respect to z.

In particular, if the projection D of E onto the xy-plane is a type I plane region (as in Figure 3), then

$$E = \{(x, y, z) \mid a \leqslant x \leqslant b,\ g_1(x) \leqslant y \leqslant g_2(x),\ u_1(x, y) \leqslant z \leqslant u_2(x, y)\}$$

and Equation 6 becomes

7 $$\iiint_E f(x, y, z)\, dV = \int_a^b \int_{g_1(x)}^{g_2(x)} \int_{u_1(x, y)}^{u_2(x, y)} f(x, y, z)\, dz\, dy\, dx$$

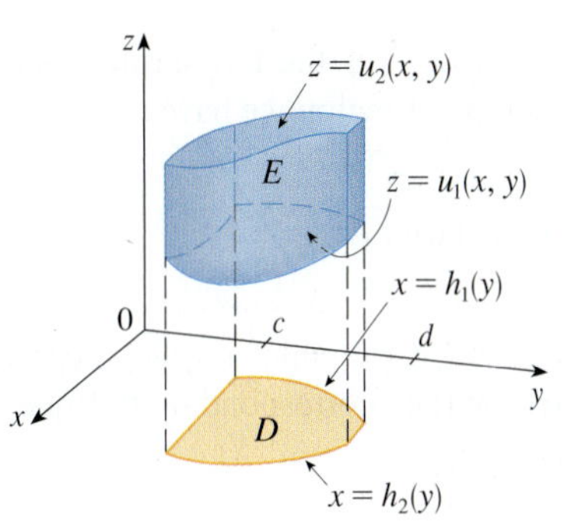

FIGURE 4
A type 1 solid region with a type II projection

If, on the other hand, D is a type II plane region (as in Figure 4), then

$$E = \{(x, y, z) \mid c \leqslant y \leqslant d,\ h_1(y) \leqslant x \leqslant h_2(y),\ u_1(x, y) \leqslant z \leqslant u_2(x, y)\}$$

and Equation 6 becomes

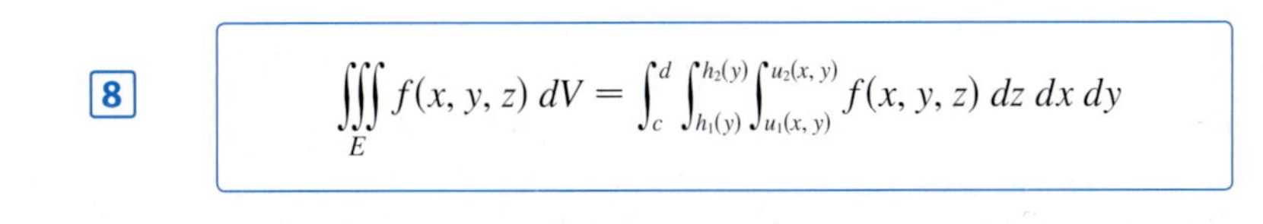

8 $$\iiint_E f(x, y, z)\, dV = \int_c^d \int_{h_1(y)}^{h_2(y)} \int_{u_1(x, y)}^{u_2(x, y)} f(x, y, z)\, dz\, dx\, dy$$

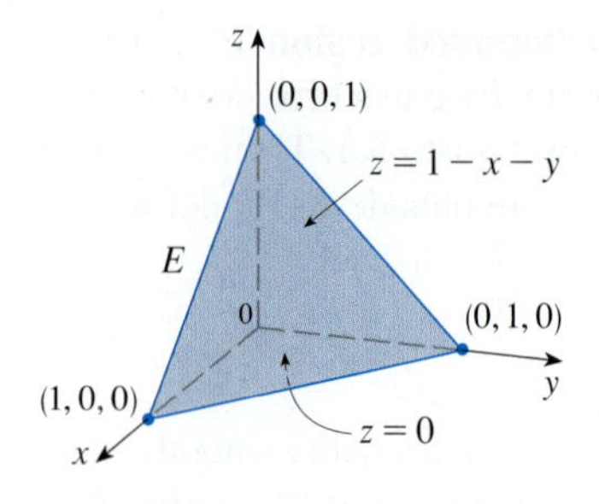

FIGURE 5

EXAMPLE 2 Evaluate $\iiint_E z\, dV$, where E is the solid tetrahedron bounded by the four planes $x = 0$, $y = 0$, $z = 0$, and $x + y + z = 1$.

SOLUTION When we set up a triple integral it's wise to draw *two* diagrams: one of the solid region E (see Figure 5) and one of its projection D onto the xy-plane (see Figure 6). The lower boundary of the tetrahedron is the plane $z = 0$ and the upper boundary is the plane $x + y + z = 1$ (or $z = 1 - x - y$), so we use $u_1(x, y) = 0$ and $u_2(x, y) = 1 - x - y$ in Formula 7. Notice that the planes $x + y + z = 1$ and $z = 0$ intersect in the line $x + y = 1$ (or $y = 1 - x$) in the xy-plane. So the projection of E is the triangular region shown in Figure 6, and we have

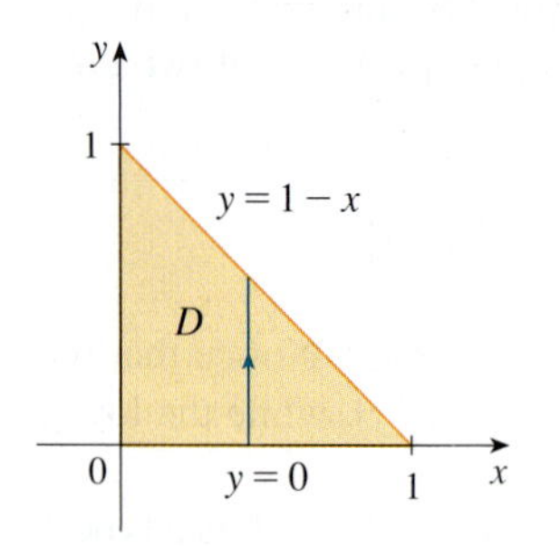

FIGURE 6

$$\boxed{9} \qquad E = \{(x, y, z) \mid 0 \leqslant x \leqslant 1,\ 0 \leqslant y \leqslant 1 - x,\ 0 \leqslant z \leqslant 1 - x - y\}$$

This description of E as a type 1 region enables us to evaluate the integral as follows:

$$\iiint_E z\, dV = \int_0^1 \int_0^{1-x} \int_0^{1-x-y} z\, dz\, dy\, dx = \int_0^1 \int_0^{1-x} \left[\frac{z^2}{2}\right]_{z=0}^{z=1-x-y} dy\, dx$$

$$= \tfrac{1}{2} \int_0^1 \int_0^{1-x} (1 - x - y)^2\, dy\, dx = \tfrac{1}{2} \int_0^1 \left[-\frac{(1 - x - y)^3}{3}\right]_{y=0}^{y=1-x} dx$$

$$= \tfrac{1}{6} \int_0^1 (1 - x)^3\, dx = \frac{1}{6}\left[-\frac{(1 - x)^4}{4}\right]_0^1 = \frac{1}{24}$$ ■

A solid region E is of **type 2** if it is of the form

$$E = \{(x, y, z) \mid (y, z) \in D,\ u_1(y, z) \leqslant x \leqslant u_2(y, z)\}$$

where, this time, D is the projection of E onto the yz-plane (see Figure 7). The back surface is $x = u_1(y, z)$, the front surface is $x = u_2(y, z)$, and we have

$$\boxed{10} \qquad \iiint_E f(x, y, z)\, dV = \iint_D \left[\int_{u_1(y, z)}^{u_2(y, z)} f(x, y, z)\, dx\right] dA$$

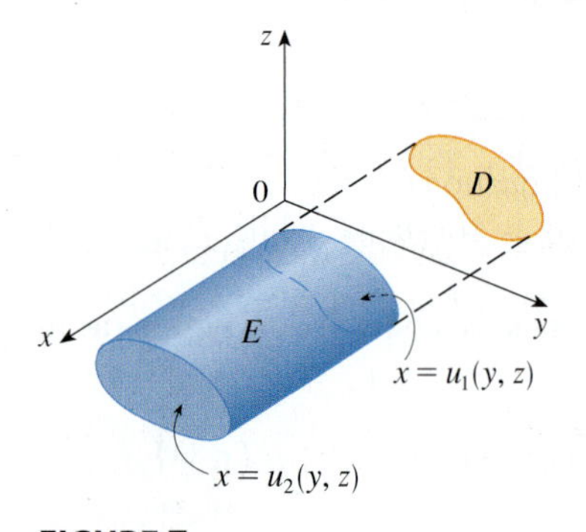

FIGURE 7
A type 2 region

Finally, a **type 3** region is of the form

$$E = \{(x, y, z) \mid (x, z) \in D,\ u_1(x, z) \leqslant y \leqslant u_2(x, z)\}$$

where D is the projection of E onto the xz-plane, $y = u_1(x, z)$ is the left surface, and $y = u_2(x, z)$ is the right surface (see Figure 8). For this type of region we have

$$\boxed{11} \qquad \iiint_E f(x, y, z)\, dV = \iint_D \left[\int_{u_1(x, z)}^{u_2(x, z)} f(x, y, z)\, dy\right] dA$$

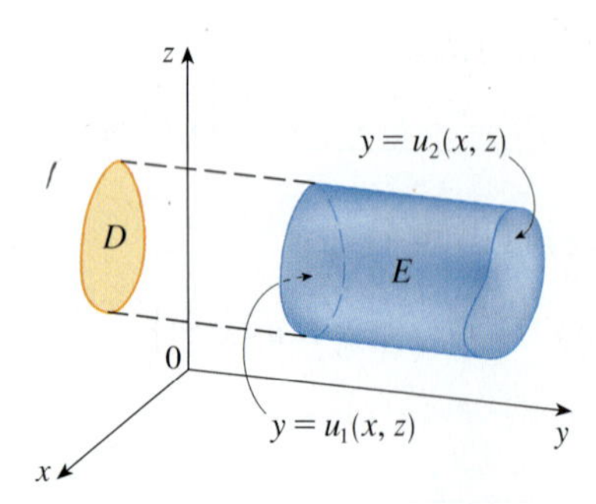

FIGURE 8
A type 3 region

In each of Equations 10 and 11 there may be two possible expressions for the integral depending on whether D is a type I or type II plane region (and corresponding to Equations 7 and 8).

EXAMPLE 3 Evaluate $\iiint_E \sqrt{x^2 + z^2}\, dV$, where E is the region bounded by the paraboloid $y = x^2 + z^2$ and the plane $y = 4$.

SOLUTION The solid E is shown in Figure 9. If we regard it as a type 1 region, then we need to consider its projection D_1 onto the xy-plane, which is the parabolic region in Figure 10. (The trace of $y = x^2 + z^2$ in the plane $z = 0$ is the parabola $y = x^2$.)

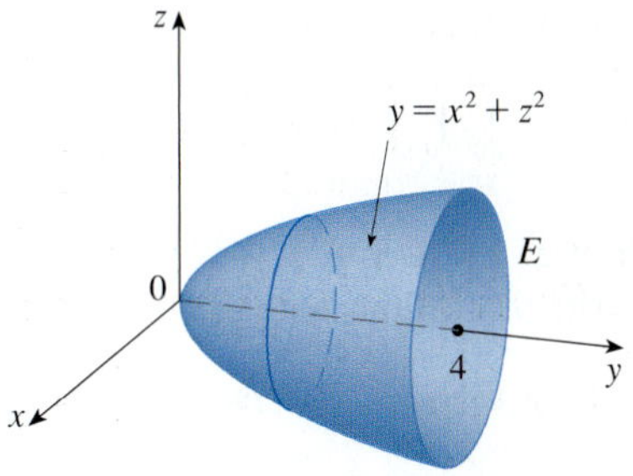

FIGURE 9
Region of integration

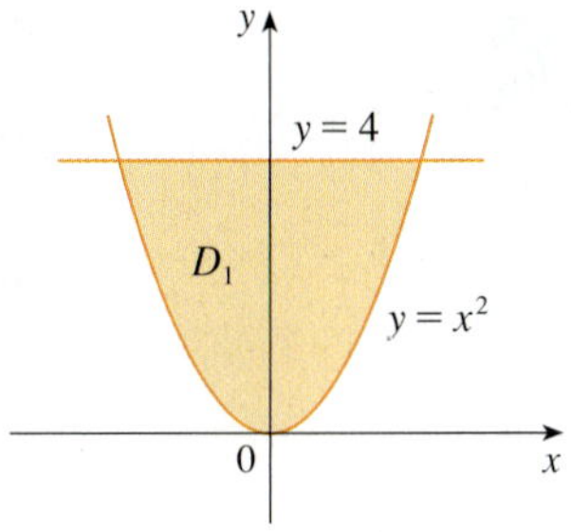

FIGURE 10
Projection onto xy-plane

TEC Visual 15.6 illustrates how solid regions (including the one in Figure 9) project onto coordinate planes.

From $y = x^2 + z^2$ we obtain $z = \pm\sqrt{y - x^2}$, so the lower boundary surface of E is $z = -\sqrt{y - x^2}$ and the upper surface is $z = \sqrt{y - x^2}$. Therefore the description of E as a type 1 region is

$$E = \left\{(x, y, z) \mid -2 \leqslant x \leqslant 2,\ x^2 \leqslant y \leqslant 4,\ -\sqrt{y - x^2} \leqslant z \leqslant \sqrt{y - x^2}\right\}$$

and so we obtain

$$\iiint_E \sqrt{x^2 + z^2}\, dV = \int_{-2}^{2} \int_{x^2}^{4} \int_{-\sqrt{y-x^2}}^{\sqrt{y-x^2}} \sqrt{x^2 + z^2}\, dz\, dy\, dx$$

Although this expression is correct, it is extremely difficult to evaluate. So let's instead consider E as a type 3 region. As such, its projection D_3 onto the xz-plane is the disk $x^2 + z^2 \leqslant 4$ shown in Figure 11.

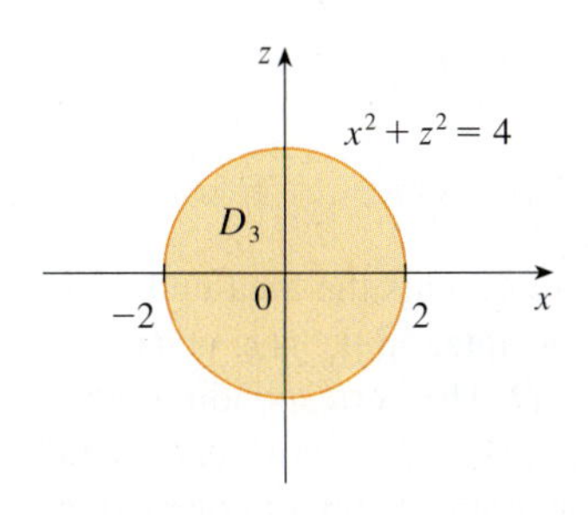

FIGURE 11
Projection onto xz-plane

Then the left boundary of E is the paraboloid $y = x^2 + z^2$ and the right boundary is the plane $y = 4$, so taking $u_1(x, z) = x^2 + z^2$ and $u_2(x, z) = 4$ in Equation 11, we have

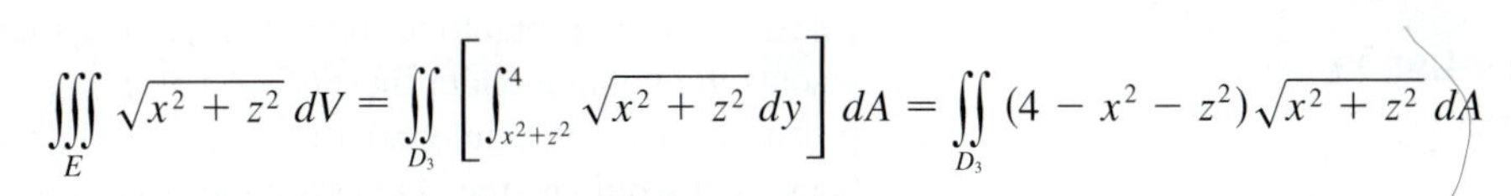

$$\iiint_E \sqrt{x^2 + z^2}\, dV = \iint_{D_3} \left[\int_{x^2+z^2}^{4} \sqrt{x^2 + z^2}\, dy\right] dA = \iint_{D_3} (4 - x^2 - z^2)\sqrt{x^2 + z^2}\, dA$$

Although this integral could be written as

$$\int_{-2}^{2} \int_{-\sqrt{4-x^2}}^{\sqrt{4-x^2}} (4 - x^2 - z^2)\sqrt{x^2 + z^2}\, dz\, dx$$

it's easier to convert to polar coordinates in the xz-plane: $x = r\cos\theta$, $z = r\sin\theta$. This gives

$$\iiint_E \sqrt{x^2 + z^2}\, dV = \iint_{D_3} (4 - x^2 - z^2)\sqrt{x^2 + z^2}\, dA$$

$$= \int_0^{2\pi} \int_0^2 (4 - r^2) r\, r\, dr\, d\theta = \int_0^{2\pi} d\theta \int_0^2 (4r^2 - r^4)\, dr$$

$$= 2\pi\left[\frac{4r^3}{3} - \frac{r^5}{5}\right]_0^2 = \frac{128\pi}{15}$$ ■

⊘ The most difficult step in evaluating a triple integral is setting up an expression for the region of integration (such as Equation 9 in Example 2). Remember that the limits of integration in the inner integral contain at most two variables, the limits of integration in the middle integral contain at most one variable, and the limits of integration in the outer integral must be constants.

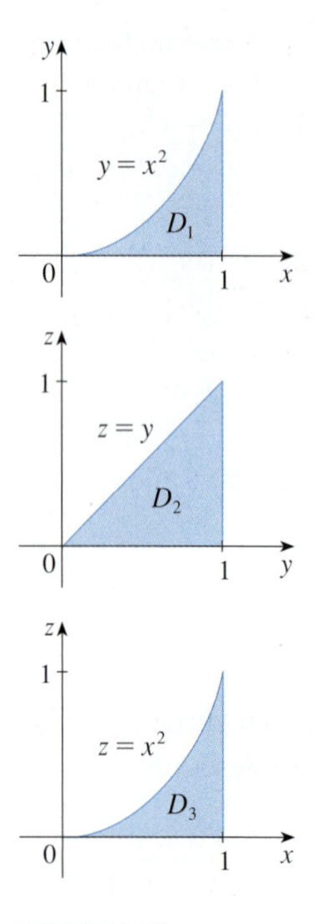

FIGURE 12
Projections of E

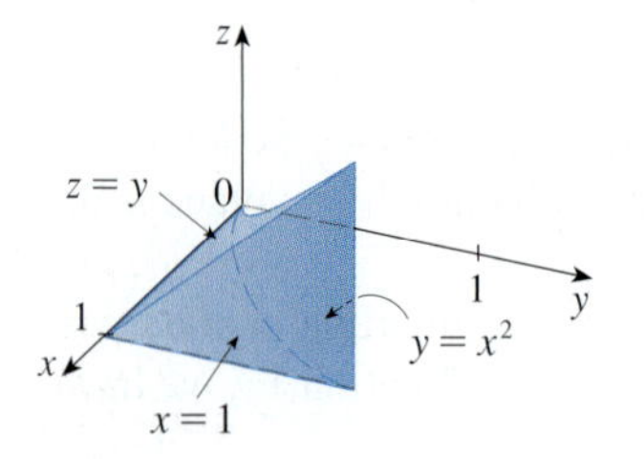

FIGURE 13
The solid E

EXAMPLE 4 Express the iterated integral $\int_0^1 \int_0^{x^2} \int_0^y f(x, y, z)\, dz\, dy\, dx$ as a triple integral and then rewrite it as an iterated integral in a different order, integrating first with respect to x, then z, and then y.

SOLUTION We can write

$$\int_0^1 \int_0^{x^2} \int_0^y f(x, y, z)\, dz\, dy\, dx = \iiint_E f(x, y, z)\, dV$$

where $E = \{(x, y, z) \mid 0 \leqslant x \leqslant 1, 0 \leqslant y \leqslant x^2, 0 \leqslant z \leqslant y\}$. This description of E enables us to write projections onto the three coordinate planes as follows:

$$\text{on the } xy\text{-plane:} \qquad D_1 = \{(x, y) \mid 0 \leqslant x \leqslant 1, 0 \leqslant y \leqslant x^2\}$$

$$= \left\{(x, y) \mid 0 \leqslant y \leqslant 1, \sqrt{y} \leqslant x \leqslant 1\right\}$$

$$\text{on the } yz\text{-plane:} \qquad D_2 = \{(y, z) \mid 0 \leqslant y \leqslant 1, 0 \leqslant z \leqslant y\}$$

$$\text{on the } xz\text{-plane:} \qquad D_3 = \{(x, z) \mid 0 \leqslant x \leqslant 1, 0 \leqslant z \leqslant x^2\}$$

From the resulting sketches of the projections in Figure 12 we sketch the solid E in Figure 13. We see that it is the solid enclosed by the planes $z = 0$, $x = 1$, $y = z$ and the parabolic cylinder $y = x^2$ $\left(\text{or } x = \sqrt{y}\right)$.

If we integrate first with respect to x, then z, and then y, we use an alternate description of E:

$$E = \left\{(x, y, z) \mid 0 \leqslant y \leqslant 1, 0 \leqslant z \leqslant y, \sqrt{y} \leqslant x \leqslant 1\right\}$$

Thus

$$\iiint_E f(x, y, z)\, dV = \int_0^1 \int_0^y \int_{\sqrt{y}}^1 f(x, y, z)\, dx\, dz\, dy$$

Applications of Triple Integrals

Recall that if $f(x) \geqslant 0$, then the single integral $\int_a^b f(x)\, dx$ represents the area under the curve $y = f(x)$ from a to b, and if $f(x, y) \geqslant 0$, then the double integral $\iint_D f(x, y)\, dA$ represents the volume under the surface $z = f(x, y)$ and above D. The corresponding interpretation of a triple integral $\iiint_E f(x, y, z)\, dV$, where $f(x, y, z) \geqslant 0$, is not very useful because it would be the "hypervolume" of a four-dimensional object and, of course, that is very difficult to visualize. (Remember that E is just the *domain* of the function f; the graph of f lies in four-dimensional space.) Nonetheless, the triple integral $\iiint_E f(x, y, z)\, dV$ can be interpreted in different ways in different physical situations, depending on the physical interpretations of x, y, z, and $f(x, y, z)$.

Let's begin with the special case where $f(x, y, z) = 1$ for all points in E. Then the triple integral does represent the volume of E:

$$\boxed{12} \qquad V(E) = \iiint_E dV$$

For example, you can see this in the case of a type 1 region by putting $f(x, y, z) = 1$ in Formula 6:

$$\iiint_E 1\, dV = \iint_D \left[\int_{u_1(x, y)}^{u_2(x, y)} dz\right] dA = \iint_D [u_2(x, y) - u_1(x, y)]\, dA$$

and from Section 15.2 we know this represents the volume that lies between the surfaces $z = u_1(x, y)$ and $z = u_2(x, y)$.

EXAMPLE 5 Use a triple integral to find the volume of the tetrahedron T bounded by the planes $x + 2y + z = 2$, $x = 2y$, $x = 0$, and $z = 0$.

SOLUTION The tetrahedron T and its projection D onto the xy-plane are shown in Figures 14 and 15. The lower boundary of T is the plane $z = 0$ and the upper boundary is the plane $x + 2y + z = 2$, that is, $z = 2 - x - 2y$.

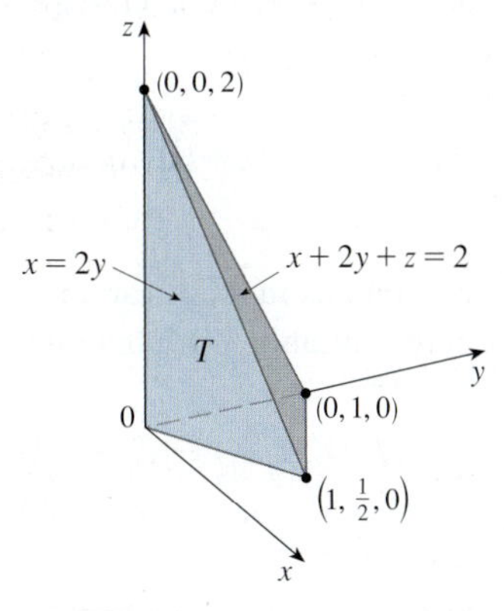

FIGURE 14

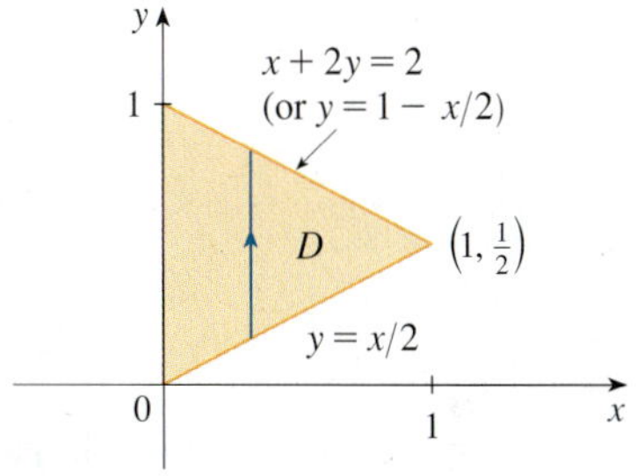

FIGURE 15

Therefore we have

$$V(T) = \iiint_T dV = \int_0^1 \int_{x/2}^{1-x/2} \int_0^{2-x-2y} dz\, dy\, dx$$

$$= \int_0^1 \int_{x/2}^{1-x/2} (2 - x - 2y)\, dy\, dx = \tfrac{1}{3}$$

by the same calculation as in Example 15.2.4.

(Notice that it is not necessary to use triple integrals to compute volumes. They simply give an alternative method for setting up the calculation.) ■

All the applications of double integrals in Section 15.4 can be immediately extended to triple integrals. For example, if the density function of a solid object that occupies the region E is $\rho(x, y, z)$, in units of mass per unit volume, at any given point (x, y, z), then its **mass** is

13
$$m = \iiint_E \rho(x, y, z)\, dV$$

and its **moments** about the three coordinate planes are

14
$$M_{yz} = \iiint_E x\rho(x, y, z)\, dV \qquad M_{xz} = \iiint_E y\rho(x, y, z)\, dV$$

$$M_{xy} = \iiint_E z\rho(x, y, z)\, dV$$

The **center of mass** is located at the point $(\bar{x}, \bar{y}, \bar{z})$, where

15
$$\bar{x} = \frac{M_{yz}}{m} \qquad \bar{y} = \frac{M_{xz}}{m} \qquad \bar{z} = \frac{M_{xy}}{m}$$

If the density is constant, the center of mass of the solid is called the **centroid** of E. The **moments of inertia** about the three coordinate axes are

16
$$I_x = \iiint_E (y^2 + z^2)\rho(x, y, z)\, dV \qquad I_y = \iiint_E (x^2 + z^2)\rho(x, y, z)\, dV$$
$$I_z = \iiint_E (x^2 + y^2)\rho(x, y, z)\, dV$$

As in Section 15.4, the total **electric charge** on a solid object occupying a region E and having charge density $\sigma(x, y, z)$ is

$$Q = \iiint_E \sigma(x, y, z)\, dV$$

If we have three continuous random variables X, Y, and Z, their **joint density function** is a function of three variables such that the probability that (X, Y, Z) lies in E is

$$P((X, Y, Z) \in E) = \iiint_E f(x, y, z)\, dV$$

In particular,

$$P(a \leqslant X \leqslant b,\ c \leqslant Y \leqslant d,\ r \leqslant Z \leqslant s) = \int_a^b \int_c^d \int_r^s f(x, y, z)\, dz\, dy\, dx$$

The joint density function satisfies

$$f(x, y, z) \geqslant 0 \qquad \int_{-\infty}^{\infty} \int_{-\infty}^{\infty} \int_{-\infty}^{\infty} f(x, y, z)\, dz\, dy\, dx = 1$$

EXAMPLE 6 Find the center of mass of a solid of constant density that is bounded by the parabolic cylinder $x = y^2$ and the planes $x = z$, $z = 0$, and $x = 1$.

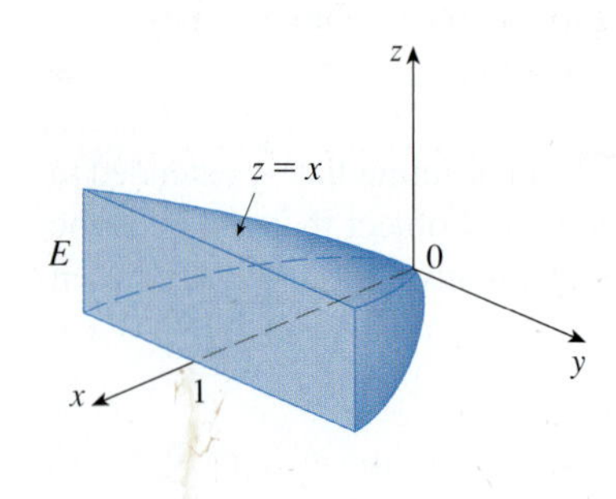

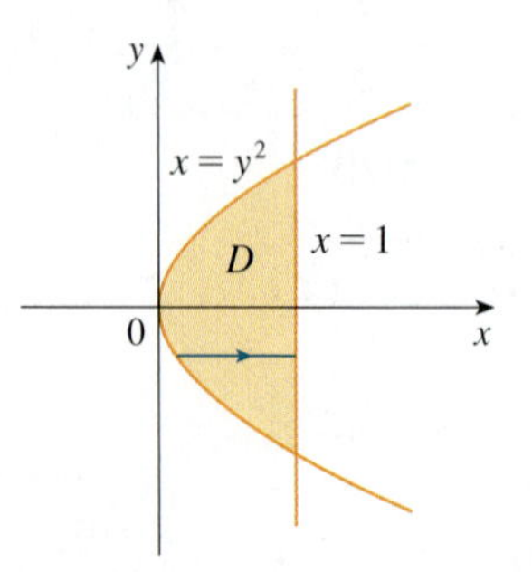

FIGURE 16

SOLUTION The solid E and its projection onto the xy-plane are shown in Figure 16. The lower and upper surfaces of E are the planes $z = 0$ and $z = x$, so we describe E as a type 1 region:

$$E = \{(x, y, z) \mid -1 \leqslant y \leqslant 1,\ y^2 \leqslant x \leqslant 1,\ 0 \leqslant z \leqslant x\}$$

Then, if the density is $\rho(x, y, z) = \rho$, the mass is

$$m = \iiint_E \rho\, dV = \int_{-1}^{1} \int_{y^2}^{1} \int_0^x \rho\, dz\, dx\, dy$$

$$= \rho \int_{-1}^{1} \int_{y^2}^{1} x\, dx\, dy = \rho \int_{-1}^{1} \left[\frac{x^2}{2}\right]_{x=y^2}^{x=1} dy$$

$$= \frac{\rho}{2} \int_{-1}^{1} (1 - y^4)\, dy = \rho \int_0^1 (1 - y^4)\, dy$$

$$= \rho\left[y - \frac{y^5}{5}\right]_0^1 = \frac{4\rho}{5}$$

Because of the symmetry of E and ρ about the xz-plane, we can immediately say that $M_{xz} = 0$ and therefore $\bar{y} = 0$. The other moments are

$$M_{yz} = \iiint_E x\rho \, dV = \int_{-1}^{1} \int_{y^2}^{1} \int_0^x x\rho \, dz \, dx \, dy$$

$$= \rho \int_{-1}^{1} \int_{y^2}^{1} x^2 \, dx \, dy = \rho \int_{-1}^{1} \left[\frac{x^3}{3} \right]_{x=y^2}^{x=1} dy$$

$$= \frac{2\rho}{3} \int_0^1 (1 - y^6) \, dy = \frac{2\rho}{3} \left[y - \frac{y^7}{7} \right]_0^1 = \frac{4\rho}{7}$$

$$M_{xy} = \iiint_E z\rho \, dV = \int_{-1}^{1} \int_{y^2}^{1} \int_0^x z\rho \, dz \, dx \, dy$$

$$= \rho \int_{-1}^{1} \int_{y^2}^{1} \left[\frac{z^2}{2} \right]_{z=0}^{z=x} dx \, dy = \frac{\rho}{2} \int_{-1}^{1} \int_{y^2}^{1} x^2 \, dx \, dy$$

$$= \frac{\rho}{3} \int_0^1 (1 - y^6) \, dy = \frac{2\rho}{7}$$

Therefore the center of mass is

$$(\bar{x}, \bar{y}, \bar{z}) = \left(\frac{M_{yz}}{m}, \frac{M_{xz}}{m}, \frac{M_{xy}}{m} \right) = \left(\tfrac{5}{7}, 0, \tfrac{5}{14} \right)$$ ■

15.6 EXERCISES

1. Evaluate the integral in Example 1, integrating first with respect to y, then z, and then x.

2. Evaluate the integral $\iiint_E (xy + z^2) \, dV$, where

$$E = \{(x, y, z) \mid 0 \leqslant x \leqslant 2, 0 \leqslant y \leqslant 1, 0 \leqslant z \leqslant 3\}$$

using three different orders of integration.

3–8 Evaluate the iterated integral.

3. $\displaystyle\int_0^2 \int_0^{z^2} \int_0^{y-z} (2x - y) \, dx \, dy \, dz$

4. $\displaystyle\int_0^1 \int_y^{2y} \int_0^{x+y} 6xy \, dz \, dx \, dy$

5. $\displaystyle\int_1^2 \int_0^{2z} \int_0^{\ln x} xe^{-y} \, dy \, dx \, dz$

6. $\displaystyle\int_0^1 \int_0^1 \int_0^{\sqrt{1-z^2}} \frac{z}{y+1} \, dx \, dz \, dy$

7. $\displaystyle\int_0^{\pi} \int_0^1 \int_0^{\sqrt{1-z^2}} z \sin x \, dy \, dz \, dx$

8. $\displaystyle\int_0^1 \int_0^1 \int_0^{2-x^2-y^2} xye^z \, dz \, dy \, dx$

9–18 Evaluate the triple integral.

9. $\iiint_E y \, dV$, where
$E = \{(x, y, z) \mid 0 \leqslant x \leqslant 3, 0 \leqslant y \leqslant x, x - y \leqslant z \leqslant x + y\}$

10. $\iiint_E e^{z/y} \, dV$, where
$E = \{(x, y, z) \mid 0 \leqslant y \leqslant 1, y \leqslant x \leqslant 1, 0 \leqslant z \leqslant xy\}$

11. $\displaystyle\iiint_E \frac{z}{x^2 + z^2} \, dV$, where
$E = \{(x, y, z) \mid 1 \leqslant y \leqslant 4, y \leqslant z \leqslant 4, 0 \leqslant x \leqslant z\}$

12. $\iiint_E \sin y \, dV$, where E lies below the plane $z = x$ and above the triangular region with vertices $(0, 0, 0)$, $(\pi, 0, 0)$, and $(0, \pi, 0)$

13. $\iiint_E 6xy \, dV$, where E lies under the plane $z = 1 + x + y$ and above the region in the xy-plane bounded by the curves $y = \sqrt{x}$, $y = 0$, and $x = 1$

14. $\iiint_E (x - y) \, dV$, where E is enclosed by the surfaces $z = x^2 - 1$, $z = 1 - x^2$, $y = 0$, and $y = 2$

15. $\iiint_T y^2 \, dV$, where T is the solid tetrahedron with vertices $(0, 0, 0)$, $(2, 0, 0)$, $(0, 2, 0)$, and $(0, 0, 2)$

16. $\iiint_T xz\, dV$, where T is the solid tetrahedron with vertices $(0, 0, 0)$, $(1, 0, 1)$, $(0, 1, 1)$, and $(0, 0, 1)$

17. $\iiint_E x\, dV$, where E is bounded by the paraboloid $x = 4y^2 + 4z^2$ and the plane $x = 4$

18. $\iiint_E z\, dV$, where E is bounded by the cylinder $y^2 + z^2 = 9$ and the planes $x = 0$, $y = 3x$, and $z = 0$ in the first octant

19–22 Use a triple integral to find the volume of the given solid.

19. The tetrahedron enclosed by the coordinate planes and the plane $2x + y + z = 4$

20. The solid enclosed by the paraboloids $y = x^2 + z^2$ and $y = 8 - x^2 - z^2$

21. The solid enclosed by the cylinder $y = x^2$ and the planes $z = 0$ and $y + z = 1$

22. The solid enclosed by the cylinder $x^2 + z^2 = 4$ and the planes $y = -1$ and $y + z = 4$

23. (a) Express the volume of the wedge in the first octant that is cut from the cylinder $y^2 + z^2 = 1$ by the planes $y = x$ and $x = 1$ as a triple integral.

CAS (b) Use either the Table of Integrals (on Reference Pages 6–10) or a computer algebra system to find the exact value of the triple integral in part (a).

24. (a) In the **Midpoint Rule for triple integrals** we use a triple Riemann sum to approximate a triple integral over a box B, where $f(x, y, z)$ is evaluated at the center $(\bar{x}_i, \bar{y}_j, \bar{z}_k)$ of the box B_{ijk}. Use the Midpoint Rule to estimate $\iiint_B \sqrt{x^2 + y^2 + z^2}\, dV$, where B is the cube defined by $0 \leqslant x \leqslant 4$, $0 \leqslant y \leqslant 4$, $0 \leqslant z \leqslant 4$. Divide B into eight cubes of equal size.

CAS (b) Use a computer algebra system to approximate the integral in part (a) correct to the nearest integer. Compare with the answer to part (a).

25–26 Use the Midpoint Rule for triple integrals (Exercise 24) to estimate the value of the integral. Divide B into eight sub-boxes of equal size.

25. $\iiint_B \cos(xyz)\, dV$, where
$B = \{(x, y, z) \mid 0 \leqslant x \leqslant 1,\ 0 \leqslant y \leqslant 1,\ 0 \leqslant z \leqslant 1\}$

26. $\iiint_B \sqrt{x}\, e^{xyz}\, dV$, where
$B = \{(x, y, z) \mid 0 \leqslant x \leqslant 4,\ 0 \leqslant y \leqslant 1,\ 0 \leqslant z \leqslant 2\}$

27–28 Sketch the solid whose volume is given by the iterated integral.

27. $\int_0^1 \int_0^{1-x} \int_0^{2-2z} dy\, dz\, dx$

28. $\int_0^2 \int_0^{2-y} \int_0^{4-y^2} dx\, dz\, dy$

29–32 Express the integral $\iiint_E f(x, y, z)\, dV$ as an iterated integral in six different ways, where E is the solid bounded by the given surfaces.

29. $y = 4 - x^2 - 4z^2$, $y = 0$

30. $y^2 + z^2 = 9$, $x = -2$, $x = 2$

31. $y = x^2$, $z = 0$, $y + 2z = 4$

32. $x = 2$, $y = 2$, $z = 0$, $x + y - 2z = 2$

33. The figure shows the region of integration for the integral

$$\int_0^1 \int_{\sqrt{x}}^1 \int_0^{1-y} f(x, y, z)\, dz\, dy\, dx$$

Rewrite this integral as an equivalent iterated integral in the five other orders.

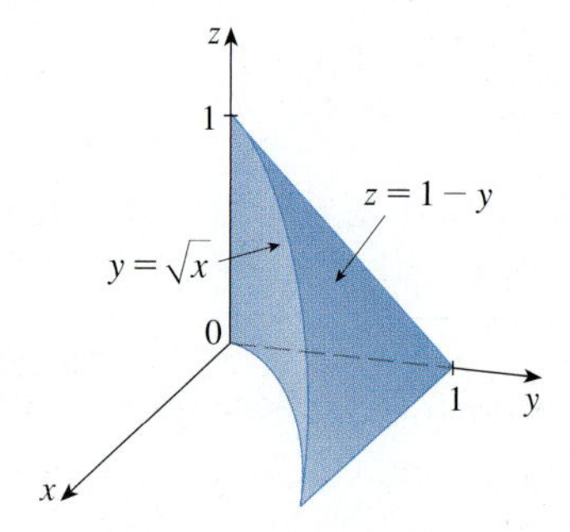

34. The figure shows the region of integration for the integral

$$\int_0^1 \int_0^{1-x^2} \int_0^{1-x} f(x, y, z)\, dy\, dz\, dx$$

Rewrite this integral as an equivalent iterated integral in the five other orders.

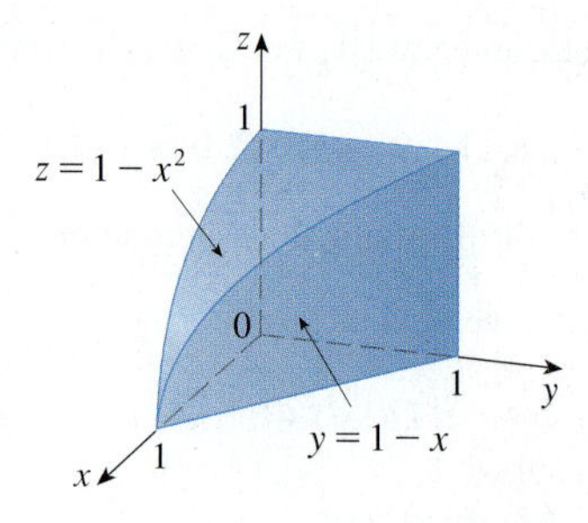

35–36 Write five other iterated integrals that are equal to the given iterated integral.

35. $\int_0^1 \int_y^1 \int_0^y f(x, y, z)\, dz\, dx\, dy$

36. $\int_0^1 \int_y^1 \int_0^z f(x, y, z)\, dx\, dz\, dy$

37–38 Evaluate the triple integral using only geometric interpretation and symmetry.

37. $\iiint_C (4 + 5x^2yz^2)\,dV$, where C is the cylindrical region $x^2 + y^2 \leqslant 4$, $-2 \leqslant z \leqslant 2$

38. $\iiint_B (z^3 + \sin y + 3)\,dV$, where B is the unit ball $x^2 + y^2 + z^2 \leqslant 1$

39–42 Find the mass and center of mass of the solid E with the given density function ρ.

39. E lies above the xy-plane and below the paraboloid $z = 1 - x^2 - y^2$; $\rho(x, y, z) = 3$

40. E is bounded by the parabolic cylinder $z = 1 - y^2$ and the planes $x + z = 1$, $x = 0$, and $z = 0$; $\rho(x, y, z) = 4$

41. E is the cube given by $0 \leqslant x \leqslant a$, $0 \leqslant y \leqslant a$, $0 \leqslant z \leqslant a$; $\rho(x, y, z) = x^2 + y^2 + z^2$

42. E is the tetrahedron bounded by the planes $x = 0$, $y = 0$, $z = 0$, $x + y + z = 1$; $\rho(x, y, z) = y$

43–46 Assume that the solid has constant density k.

43. Find the moments of inertia for a cube with side length L if one vertex is located at the origin and three edges lie along the coordinate axes.

44. Find the moments of inertia for a rectangular brick with dimensions a, b, and c and mass M if the center of the brick is situated at the origin and the edges are parallel to the coordinate axes.

45. Find the moment of inertia about the z-axis of the solid cylinder $x^2 + y^2 \leqslant a^2$, $0 \leqslant z \leqslant h$.

46. Find the moment of inertia about the z-axis of the solid cone $\sqrt{x^2 + y^2} \leqslant z \leqslant h$.

47–48 Set up, but do not evaluate, integral expressions for (a) the mass, (b) the center of mass, and (c) the moment of inertia about the z-axis.

47. The solid of Exercise 21; $\rho(x, y, z) = \sqrt{x^2 + y^2}$

48. The hemisphere $x^2 + y^2 + z^2 \leqslant 1$, $z \geqslant 0$; $\rho(x, y, z) = \sqrt{x^2 + y^2 + z^2}$

CAS **49.** Let E be the solid in the first octant bounded by the cylinder $x^2 + y^2 = 1$ and the planes $y = z$, $x = 0$, and $z = 0$ with the density function $\rho(x, y, z) = 1 + x + y + z$. Use a computer algebra system to find the exact values of the following quantities for E.
(a) The mass
(b) The center of mass
(c) The moment of inertia about the z-axis

CAS **50.** If E is the solid of Exercise 18 with density function $\rho(x, y, z) = x^2 + y^2$, find the following quantities, correct to three decimal places.
(a) The mass
(b) The center of mass
(c) The moment of inertia about the z-axis

51. The joint density function for random variables X, Y, and Z is $f(x, y, z) = Cxyz$ if $0 \leqslant x \leqslant 2$, $0 \leqslant y \leqslant 2$, $0 \leqslant z \leqslant 2$, and $f(x, y, z) = 0$ otherwise.
(a) Find the value of the constant C.
(b) Find $P(X \leqslant 1, Y \leqslant 1, Z \leqslant 1)$.
(c) Find $P(X + Y + Z \leqslant 1)$.

52. Suppose X, Y, and Z are random variables with joint density function $f(x, y, z) = Ce^{-(0.5x+0.2y+0.1z)}$ if $x \geqslant 0$, $y \geqslant 0$, $z \geqslant 0$, and $f(x, y, z) = 0$ otherwise.
(a) Find the value of the constant C.
(b) Find $P(X \leqslant 1, Y \leqslant 1)$.
(c) Find $P(X \leqslant 1, Y \leqslant 1, Z \leqslant 1)$.

53–54 The **average value** of a function $f(x, y, z)$ over a solid region E is defined to be

$$f_{\text{ave}} = \frac{1}{V(E)} \iiint_E f(x, y, z)\,dV$$

where $V(E)$ is the volume of E. For instance, if ρ is a density function, then ρ_{ave} is the average density of E.

53. Find the average value of the function $f(x, y, z) = xyz$ over the cube with side length L that lies in the first octant with one vertex at the origin and edges parallel to the coordinate axes.

54. Find the average height of the points in the solid hemisphere $x^2 + y^2 + z^2 \leqslant 1$, $z \geqslant 0$.

55. (a) Find the region E for which the triple integral

$$\iiint_E (1 - x^2 - 2y^2 - 3z^2)\,dV$$

is a maximum.

CAS (b) Use a computer algebra system to calculate the exact maximum value of the triple integral in part (a).

DISCOVERY PROJECT VOLUMES OF HYPERSPHERES

In this project we find formulas for the volume enclosed by a hypersphere in n-dimensional space.

1. Use a double integral and trigonometric substitution, together with Formula 64 in the Table of Integrals, to find the area of a circle with radius r.

2. Use a triple integral and trigonometric substitution to find the volume of a sphere with radius r.

3. Use a quadruple integral to find the (4-dimensional) volume enclosed by the hypersphere $x^2 + y^2 + z^2 + w^2 = r^2$ in $\mathbb{R}^4$. (Use only trigonometric substitution and the reduction formulas for $\int \sin^n x\, dx$ or $\int \cos^n x\, dx$.)

4. Use an n-tuple integral to find the volume enclosed by a hypersphere of radius r in n-dimensional space $\mathbb{R}^n$. [*Hint:* The formulas are different for n even and n odd.]

15.7 Triple Integrals in Cylindrical Coordinates

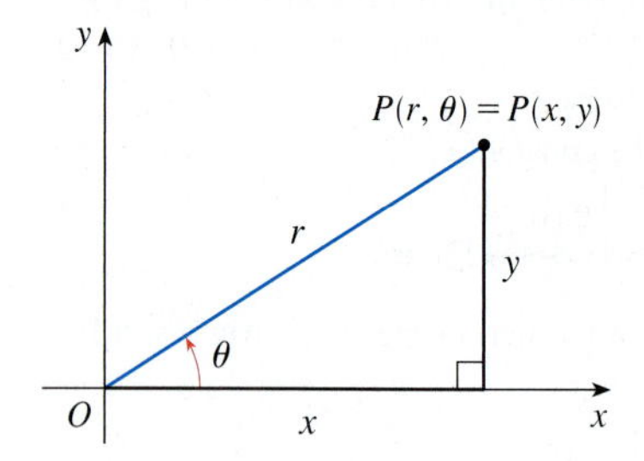

FIGURE 1

In plane geometry the polar coordinate system is used to give a convenient description of certain curves and regions. (See Section 10.3.) Figure 1 enables us to recall the connection between polar and Cartesian coordinates. If the point P has Cartesian coordinates (x, y) and polar coordinates (r, θ), then, from the figure,

$$x = r\cos\theta \qquad y = r\sin\theta$$

$$r^2 = x^2 + y^2 \qquad \tan\theta = \frac{y}{x}$$

In three dimensions there is a coordinate system, called *cylindrical coordinates*, that is similar to polar coordinates and gives convenient descriptions of some commonly occurring surfaces and solids. As we will see, some triple integrals are much easier to evaluate in cylindrical coordinates.

Cylindrical Coordinates

In the **cylindrical coordinate system**, a point P in three-dimensional space is represented by the ordered triple (r, θ, z), where r and θ are polar coordinates of the projection of P onto the xy-plane and z is the directed distance from the xy-plane to P. (See Figure 2.)

To convert from cylindrical to rectangular coordinates, we use the equations

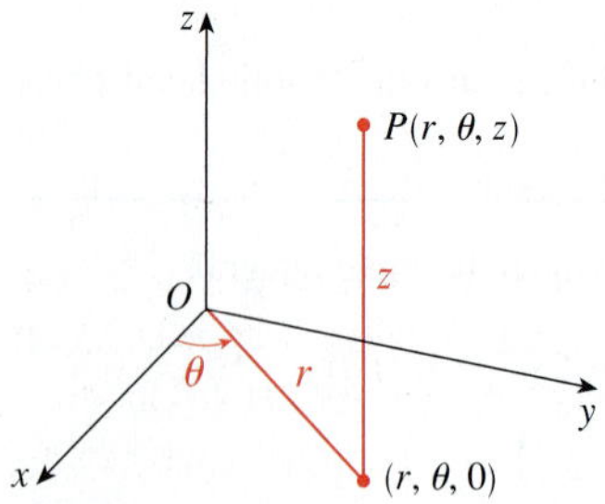

FIGURE 2
The cylindrical coordinates of a point

$$\boxed{1} \qquad x = r\cos\theta \qquad y = r\sin\theta \qquad z = z$$

whereas to convert from rectangular to cylindrical coordinates, we use

$$\boxed{2} \qquad r^2 = x^2 + y^2 \qquad \tan\theta = \frac{y}{x} \qquad z = z$$

EXAMPLE 1

(a) Plot the point with cylindrical coordinates $(2, 2\pi/3, 1)$ and find its rectangular coordinates.

(b) Find cylindrical coordinates of the point with rectangular coordinates $(3, -3, -7)$.

SOLUTION

(a) The point with cylindrical coordinates $(2, 2\pi/3, 1)$ is plotted in Figure 3. From Equations 1, its rectangular coordinates are

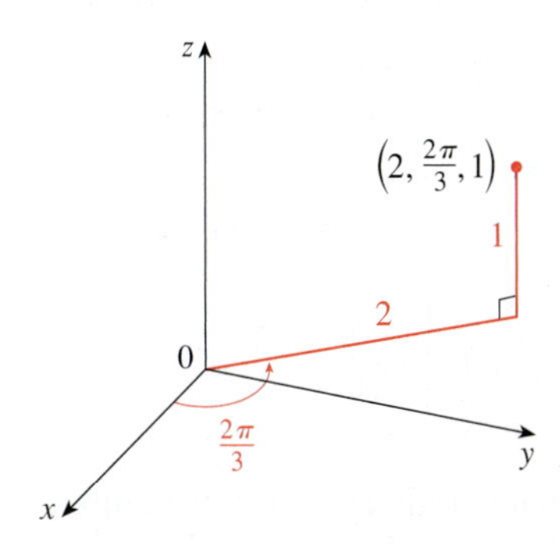

FIGURE 3

$$x = 2\cos\frac{2\pi}{3} = 2\left(-\frac{1}{2}\right) = -1$$

$$y = 2\sin\frac{2\pi}{3} = 2\left(\frac{\sqrt{3}}{2}\right) = \sqrt{3}$$

$$z = 1$$

So the point is $\left(-1, \sqrt{3}, 1\right)$ in rectangular coordinates.

(b) From Equations 2 we have

$$r = \sqrt{3^2 + (-3)^2} = 3\sqrt{2}$$

$$\tan\theta = \frac{-3}{3} = -1 \qquad \text{so} \qquad \theta = \frac{7\pi}{4} + 2n\pi$$

$$z = -7$$

Therefore one set of cylindrical coordinates is $\left(3\sqrt{2}, 7\pi/4, -7\right)$. Another is $\left(3\sqrt{2}, -\pi/4, -7\right)$. As with polar coordinates, there are infinitely many choices. ■

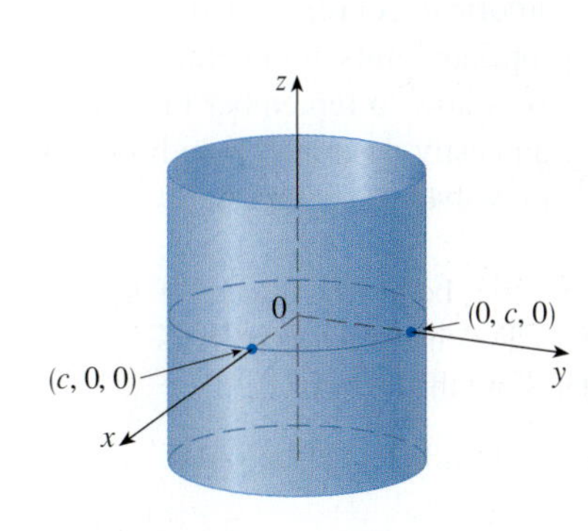

FIGURE 4

$r = c$, a cylinder

Cylindrical coordinates are useful in problems that involve symmetry about an axis, and the z-axis is chosen to coincide with this axis of symmetry. For instance, the axis of the circular cylinder with Cartesian equation $x^2 + y^2 = c^2$ is the z-axis. In cylindrical coordinates this cylinder has the very simple equation $r = c$. (See Figure 4.) This is the reason for the name "cylindrical" coordinates.

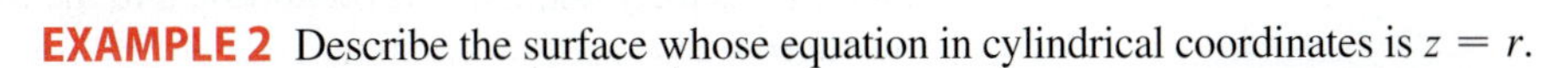

EXAMPLE 2 Describe the surface whose equation in cylindrical coordinates is $z = r$.

SOLUTION The equation says that the z-value, or height, of each point on the surface is the same as r, the distance from the point to the z-axis. Because θ doesn't appear, it can vary. So any horizontal trace in the plane $z = k$ $(k > 0)$ is a circle of radius k. These traces suggest that the surface is a cone. This prediction can be confirmed by converting the equation into rectangular coordinates. From the first equation in (2) we have

$$z^2 = r^2 = x^2 + y^2$$

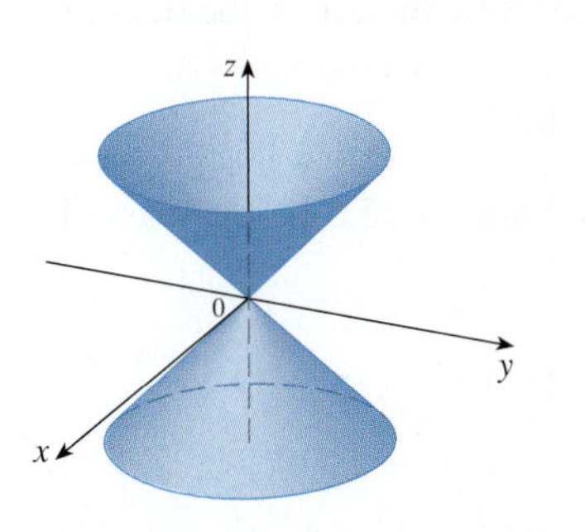

FIGURE 5

$z = r$, a cone

We recognize the equation $z^2 = x^2 + y^2$ (by comparison with Table 1 in Section 12.6) as being a circular cone whose axis is the z-axis (see Figure 5). ■

Evaluating Triple Integrals with Cylindrical Coordinates

Suppose that E is a type 1 region whose projection D onto the xy-plane is conveniently described in polar coordinates (see Figure 6). In particular, suppose that f is continuous and

$$E = \{(x, y, z) \mid (x, y) \in D,\ u_1(x, y) \le z \le u_2(x, y)\}$$

where D is given in polar coordinates by

$$D = \{(r, \theta) \mid \alpha \le \theta \le \beta,\ h_1(\theta) \le r \le h_2(\theta)\}$$

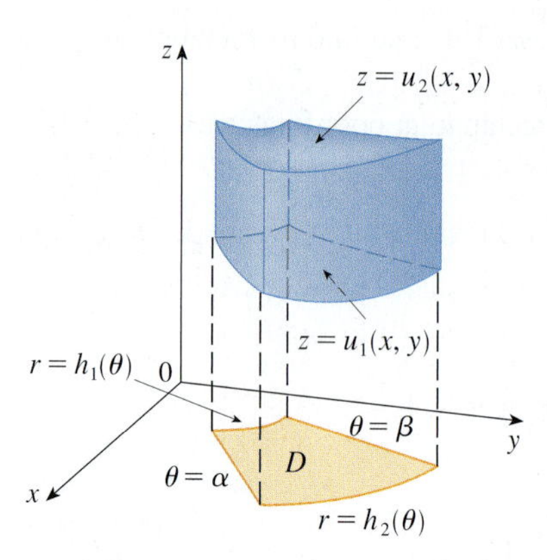

FIGURE 6

We know from Equation 15.6.6 that

3 $$\iiint_E f(x, y, z)\, dV = \iint_D \left[\int_{u_1(x, y)}^{u_2(x, y)} f(x, y, z)\, dz \right] dA$$

But we also know how to evaluate double integrals in polar coordinates. In fact, combining Equation 3 with Equation 15.3.3, we obtain

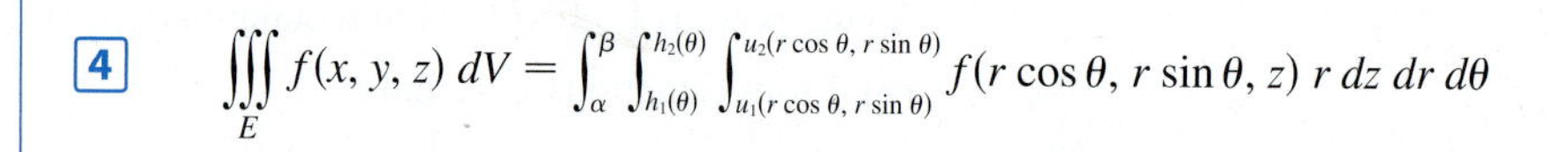

4 $$\iiint_E f(x, y, z)\, dV = \int_\alpha^\beta \int_{h_1(\theta)}^{h_2(\theta)} \int_{u_1(r\cos\theta,\, r\sin\theta)}^{u_2(r\cos\theta,\, r\sin\theta)} f(r\cos\theta, r\sin\theta, z)\, r\, dz\, dr\, d\theta$$

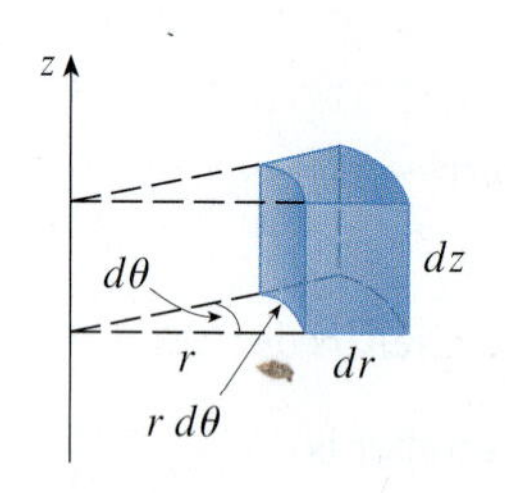

FIGURE 7
Volume element in cylindrical coordinates: $dV = r\, dz\, dr\, d\theta$

Formula 4 is the **formula for triple integration in cylindrical coordinates**. It says that we convert a triple integral from rectangular to cylindrical coordinates by writing $x = r\cos\theta$, $y = r\sin\theta$, leaving z as it is, using the appropriate limits of integration for z, r, and θ, and replacing dV by $r\, dz\, dr\, d\theta$. (Figure 7 shows how to remember this.) It is worthwhile to use this formula when E is a solid region easily described in cylindrical coordinates, and especially when the function $f(x, y, z)$ involves the expression $x^2 + y^2$.

EXAMPLE 3 A solid E lies within the cylinder $x^2 + y^2 = 1$, below the plane $z = 4$, and above the paraboloid $z = 1 - x^2 - y^2$. (See Figure 8.) The density at any point is proportional to its distance from the axis of the cylinder. Find the mass of E.

SOLUTION In cylindrical coordinates the cylinder is $r = 1$ and the paraboloid is $z = 1 - r^2$, so we can write

$$E = \{(r, \theta, z) \mid 0 \le \theta \le 2\pi,\ 0 \le r \le 1,\ 1 - r^2 \le z \le 4\}$$

Since the density at (x, y, z) is proportional to the distance from the z-axis, the density function is

$$f(x, y, z) = K\sqrt{x^2 + y^2} = Kr$$

where K is the proportionality constant. Therefore, from Formula 15.6.13, the mass of E is

$$m = \iiint_E K\sqrt{x^2 + y^2}\, dV = \int_0^{2\pi} \int_0^1 \int_{1-r_2}^4 (Kr)\, r\, dz\, dr\, d\theta$$

$$= \int_0^{2\pi} \int_0^1 Kr^2[4 - (1 - r^2)]\, dr\, d\theta = K \int_0^{2\pi} d\theta \int_0^1 (3r^2 + r^4)\, dr$$

$$= 2\pi K \left[r^3 + \frac{r^5}{5} \right]_0^1 = \frac{12\pi K}{5}$$

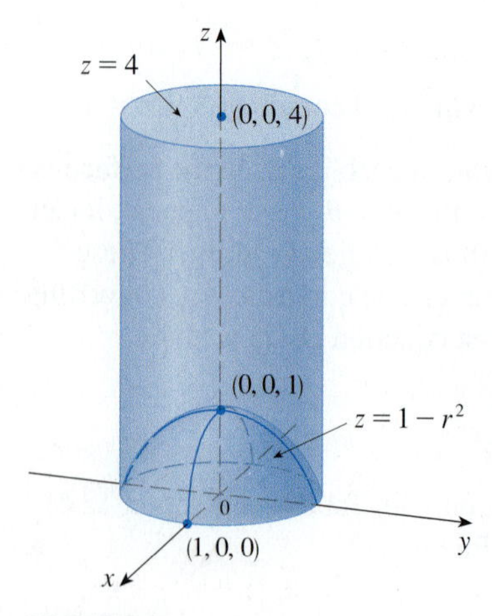

FIGURE 8

EXAMPLE 4 Evaluate $\int_{-2}^{2} \int_{-\sqrt{4-x^2}}^{\sqrt{4-x^2}} \int_{\sqrt{x^2+y^2}}^{2} (x^2 + y^2)\, dz\, dy\, dx$.

SOLUTION This iterated integral is a triple integral over the solid region

$$E = \left\{(x, y, z) \mid -2 \leqslant x \leqslant 2,\ -\sqrt{4 - x^2} \leqslant y \leqslant \sqrt{4 - x^2},\ \sqrt{x^2 + y^2} \leqslant z \leqslant 2\right\}$$

and the projection of E onto the xy-plane is the disk $x^2 + y^2 \leqslant 4$. The lower surface of E is the cone $z = \sqrt{x^2 + y^2}$ and its upper surface is the plane $z = 2$. (See Figure 9.) This region has a much simpler description in cylindrical coordinates:

$$E = \left\{(r, \theta, z) \mid 0 \leqslant \theta \leqslant 2\pi,\ 0 \leqslant r \leqslant 2,\ r \leqslant z \leqslant 2\right\}$$

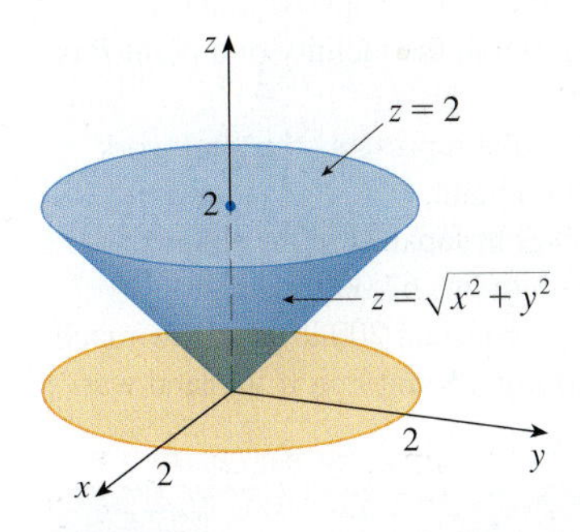

FIGURE 9

Therefore we have

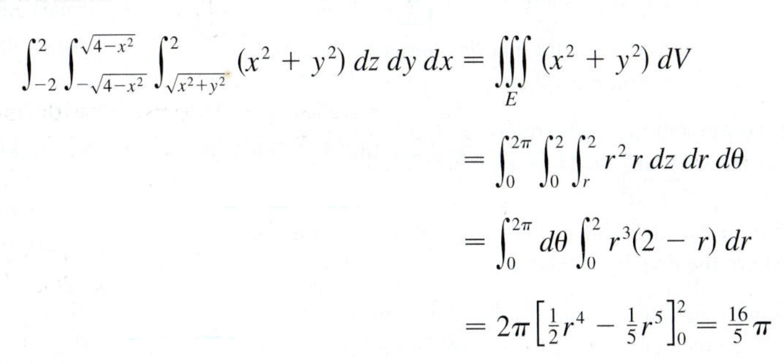

$$\int_{-2}^{2} \int_{-\sqrt{4-x^2}}^{\sqrt{4-x^2}} \int_{\sqrt{x^2+y^2}}^{2} (x^2 + y^2)\, dz\, dy\, dx = \iiint_E (x^2 + y^2)\, dV$$

$$= \int_0^{2\pi} \int_0^2 \int_r^2 r^2\, r\, dz\, dr\, d\theta$$

$$= \int_0^{2\pi} d\theta \int_0^2 r^3(2 - r)\, dr$$

$$= 2\pi\left[\tfrac{1}{2}r^4 - \tfrac{1}{5}r^5\right]_0^2 = \tfrac{16}{5}\pi$$

15.7 EXERCISES

1–2 Plot the point whose cylindrical coordinates are given. Then find the rectangular coordinates of the point.

1. (a) $(4, \pi/3, -2)$ (b) $(2, -\pi/2, 1)$

2. (a) $(\sqrt{2}, 3\pi/4, 2)$ (b) $(1, 1, 1)$

3–4 Change from rectangular to cylindrical coordinates.

3. (a) $(-1, 1, 1)$ (b) $(-2, 2\sqrt{3}, 3)$

4. (a) $(-\sqrt{2}, \sqrt{2}, 1)$ (b) $(2, 2, 2)$

5–6 Describe in words the surface whose equation is given.

5. $r = 2$ **6.** $\theta = \pi/6$

7–8 Identify the surface whose equation is given.

7. $r^2 + z^2 = 4$ **8.** $r = 2\sin\theta$

9–10 Write the equations in cylindrical coordinates.

9. (a) $x^2 - x + y^2 + z^2 = 1$ (b) $z = x^2 - y^2$

10. (a) $2x^2 + 2y^2 - z^2 = 4$ (b) $2x - y + z = 1$

11–12 Sketch the solid described by the given inequalities.

11. $r^2 \leqslant z \leqslant 8 - r^2$

12. $0 \leqslant \theta \leqslant \pi/2,\quad r \leqslant z \leqslant 2$

13. A cylindrical shell is 20 cm long, with inner radius 6 cm and outer radius 7 cm. Write inequalities that describe the shell in an appropriate coordinate system. Explain how you have positioned the coordinate system with respect to the shell.

14. Use a graphing device to draw the solid enclosed by the paraboloids $z = x^2 + y^2$ and $z = 5 - x^2 - y^2$.

15–16 Sketch the solid whose volume is given by the integral and evaluate the integral.

15. $\int_{-\pi/2}^{\pi/2} \int_0^2 \int_0^{r^2} r\, dz\, dr\, d\theta$ **16.** $\int_0^2 \int_0^{2\pi} \int_0^r r\, dz\, d\theta\, dr$

17–28 Use cylindrical coordinates.

17. Evaluate $\iiint_E \sqrt{x^2 + y^2}\, dV$, where E is the region that lies inside the cylinder $x^2 + y^2 = 16$ and between the planes $z = -5$ and $z = 4$.

18. Evaluate $\iiint_E z\, dV$, where E is enclosed by the paraboloid $z = x^2 + y^2$ and the plane $z = 4$.

19. Evaluate $\iiint_E (x + y + z)\,dV$, where E is the solid in the first octant that lies under the paraboloid $z = 4 - x^2 - y^2$.

20. Evaluate $\iiint_E (x - y)\,dV$, where E is the solid that lies between the cylinders $x^2 + y^2 = 1$ and $x^2 + y^2 = 16$, above the xy-plane, and below the plane $z = y + 4$.

21. Evaluate $\iiint_E x^2\,dV$, where E is the solid that lies within the cylinder $x^2 + y^2 = 1$, above the plane $z = 0$, and below the cone $z^2 = 4x^2 + 4y^2$.

22. Find the volume of the solid that lies within both the cylinder $x^2 + y^2 = 1$ and the sphere $x^2 + y^2 + z^2 = 4$.

23. Find the volume of the solid that is enclosed by the cone $z = \sqrt{x^2 + y^2}$ and the sphere $x^2 + y^2 + z^2 = 2$.

24. Find the volume of the solid that lies between the paraboloid $z = x^2 + y^2$ and the sphere $x^2 + y^2 + z^2 = 2$.

25. (a) Find the volume of the region E that lies between the paraboloid $z = 24 - x^2 - y^2$ and the cone $z = 2\sqrt{x^2 + y^2}$.
(b) Find the centroid of E (the center of mass in the case where the density is constant).

26. (a) Find the volume of the solid that the cylinder $r = a\cos\theta$ cuts out of the sphere of radius a centered at the origin.
(b) Illustrate the solid of part (a) by graphing the sphere and the cylinder on the same screen.

27. Find the mass and center of mass of the solid S bounded by the paraboloid $z = 4x^2 + 4y^2$ and the plane $z = a$ $(a > 0)$ if S has constant density K.

28. Find the mass of a ball B given by $x^2 + y^2 + z^2 \leqslant a^2$ if the density at any point is proportional to its distance from the z-axis.

29–30 Evaluate the integral by changing to cylindrical coordinates.

29. $\displaystyle\int_{-2}^{2} \int_{-\sqrt{4-y^2}}^{\sqrt{4-y^2}} \int_{\sqrt{x^2+y^2}}^{2} xz\,dz\,dx\,dy$

30. $\displaystyle\int_{-3}^{3} \int_{0}^{\sqrt{9-x^2}} \int_{0}^{9-x^2-y^2} \sqrt{x^2 + y^2}\,dz\,dy\,dx$

31. When studying the formation of mountain ranges, geologists estimate the amount of work required to lift a mountain from sea level. Consider a mountain that is essentially in the shape of a right circular cone. Suppose that the weight density of the material in the vicinity of a point P is $g(P)$ and the height is $h(P)$.
(a) Find a definite integral that represents the total work done in forming the mountain.
(b) Assume that Mount Fuji in Japan is in the shape of a right circular cone with radius 62,000 ft, height 12,400 ft, and density a constant 200 lb/ft^3. How much work was done in forming Mount Fuji if the land was initially at sea level?

DISCOVERY PROJECT THE INTERSECTION OF THREE CYLINDERS

The figure shows the solid enclosed by three circular cylinders with the same diameter that intersect at right angles. In this project we compute its volume and determine how its shape changes if the cylinders have different diameters.

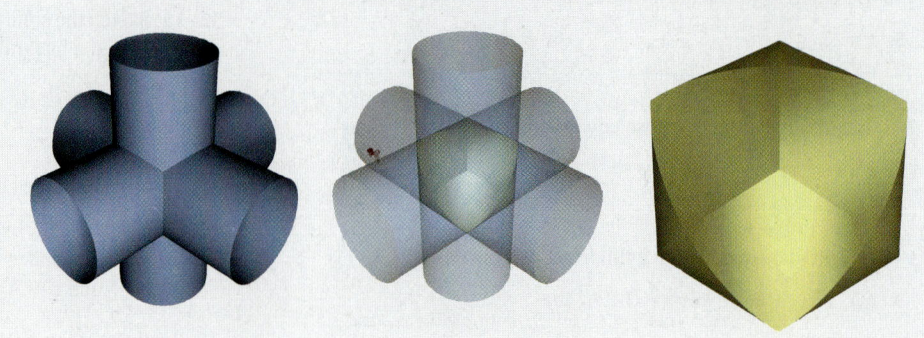

1. Sketch carefully the solid enclosed by the three cylinders $x^2 + y^2 = 1$, $x^2 + z^2 = 1$, and $y^2 + z^2 = 1$. Indicate the positions of the coordinate axes and label the faces with the equations of the corresponding cylinders.

2. Find the volume of the solid in Problem 1.

CAS 3. Use a computer algebra system to draw the edges of the solid.

4. What happens to the solid in Problem 1 if the radius of the first cylinder is different from 1? Illustrate with a hand-drawn sketch or a computer graph.

5. If the first cylinder is $x^2 + y^2 = a^2$, where $a < 1$, set up, but do not evaluate, a double integral for the volume of the solid. What if $a > 1$?

15.8 Triple Integrals in Spherical Coordinates

Another useful coordinate system in three dimensions is the *spherical coordinate system.* It simplifies the evaluation of triple integrals over regions bounded by spheres or cones.

Spherical Coordinates

The **spherical coordinates** (ρ, θ, ϕ) of a point P in space are shown in Figure 1, where $\rho = |OP|$ is the distance from the origin to P, θ is the same angle as in cylindrical coordinates, and ϕ is the angle between the positive z-axis and the line segment OP. Note that

$$\rho \geqslant 0 \qquad 0 \leqslant \phi \leqslant \pi$$

FIGURE 1
The spherical coordinates of a point

The spherical coordinate system is especially useful in problems where there is symmetry about a point, and the origin is placed at this point. For example, the sphere with center the origin and radius c has the simple equation $\rho = c$ (see Figure 2); this is the reason for the name "spherical" coordinates. The graph of the equation $\theta = c$ is a vertical half-plane (see Figure 3), and the equation $\phi = c$ represents a half-cone with the z-axis as its axis (see Figure 4).

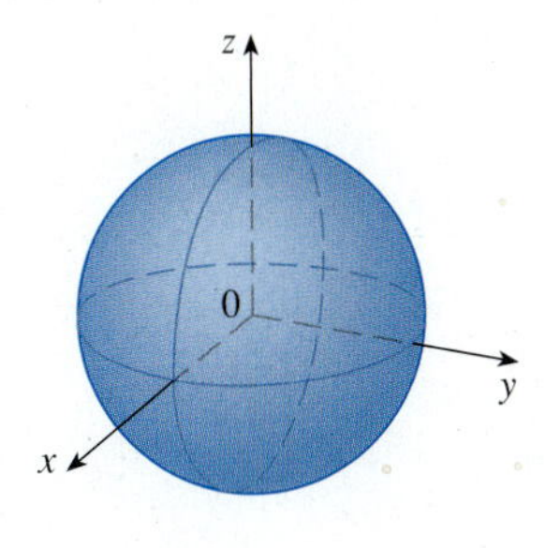

FIGURE 2 $\rho = c$, a sphere

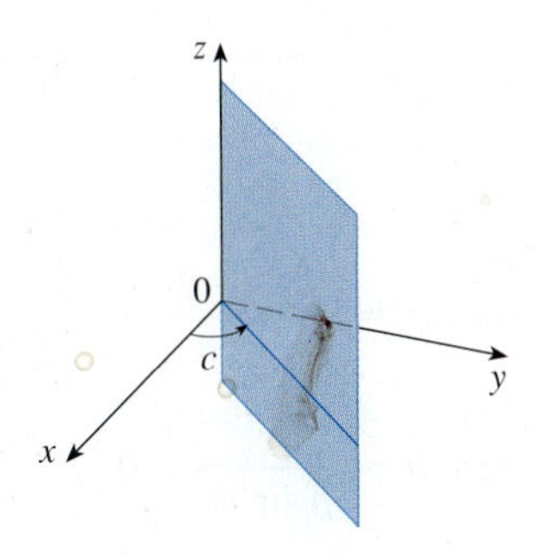

FIGURE 3 $\theta = c$, a half-plane

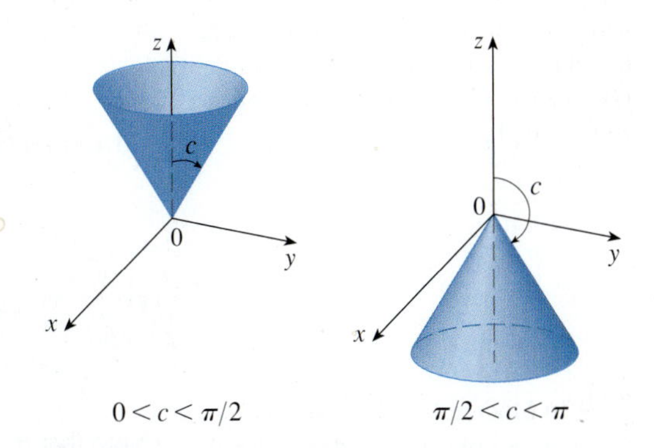

FIGURE 4 $\phi = c$, a half-cone

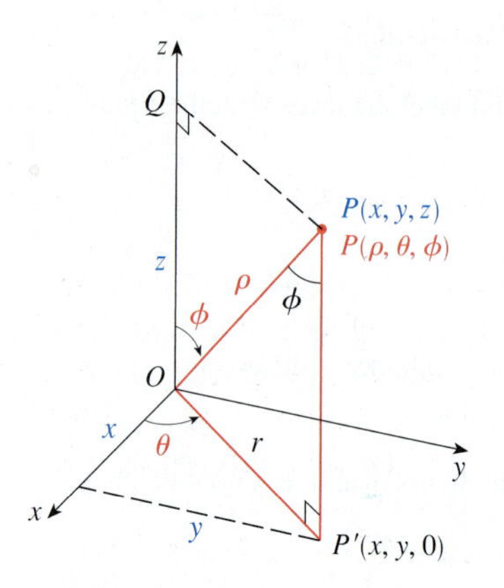

FIGURE 5

The relationship between rectangular and spherical coordinates can be seen from Figure 5. From triangles OPQ and OPP' we have

$$z = \rho \cos \phi \qquad r = \rho \sin \phi$$

But $x = r \cos \theta$ and $y = r \sin \theta$, so to convert from spherical to rectangular coordinates, we use the equations

1
$$x = \rho \sin \phi \cos \theta \qquad y = \rho \sin \phi \sin \theta \qquad z = \rho \cos \phi$$

Also, the distance formula shows that

2
$$\rho^2 = x^2 + y^2 + z^2$$

We use this equation in converting from rectangular to spherical coordinates.

EXAMPLE 1 The point $(2, \pi/4, \pi/3)$ is given in spherical coordinates. Plot the point and find its rectangular coordinates.

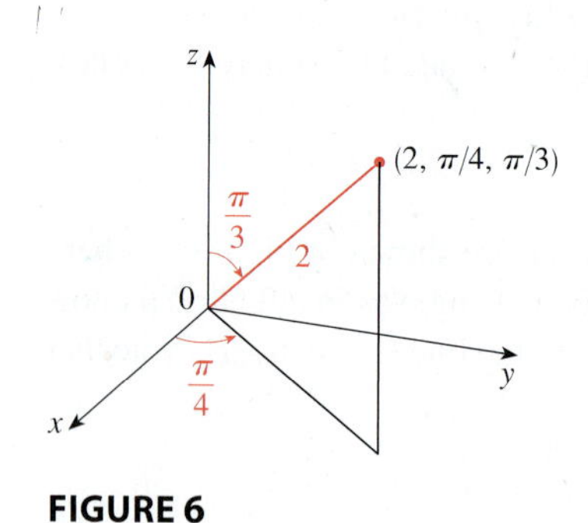

FIGURE 6

SOLUTION We plot the point in Figure 6. From Equations 1 we have

$$x = \rho \sin \phi \cos \theta = 2 \sin \frac{\pi}{3} \cos \frac{\pi}{4} = 2\left(\frac{\sqrt{3}}{2}\right)\left(\frac{1}{\sqrt{2}}\right) = \sqrt{\frac{3}{2}}$$

$$y = \rho \sin \phi \sin \theta = 2 \sin \frac{\pi}{3} \sin \frac{\pi}{4} = 2\left(\frac{\sqrt{3}}{2}\right)\left(\frac{1}{\sqrt{2}}\right) = \sqrt{\frac{3}{2}}$$

$$z = \rho \cos \phi = 2 \cos \frac{\pi}{3} = 2\left(\tfrac{1}{2}\right) = 1$$

Thus the point $(2, \pi/4, \pi/3)$ is $\left(\sqrt{3/2}, \sqrt{3/2}, 1\right)$ in rectangular coordinates. ■

EXAMPLE 2 The point $\left(0, 2\sqrt{3}, -2\right)$ is given in rectangular coordinates. Find spherical coordinates for this point.

WARNING There is not universal agreement on the notation for spherical coordinates. Most books on physics reverse the meanings of θ and ϕ and use r in place of ρ.

SOLUTION From Equation 2 we have

$$\rho = \sqrt{x^2 + y^2 + z^2} = \sqrt{0 + 12 + 4} = 4$$

and so Equations 1 give

$$\cos \phi = \frac{z}{\rho} = \frac{-2}{4} = -\frac{1}{2} \qquad \phi = \frac{2\pi}{3}$$

$$\cos \theta = \frac{x}{\rho \sin \phi} = 0 \qquad \theta = \frac{\pi}{2}$$

TEC In Module 15.8 you can investigate families of surfaces in cylindrical and spherical coordinates.

(Note that $\theta \neq 3\pi/2$ because $y = 2\sqrt{3} > 0$.) Therefore spherical coordinates of the given point are $(4, \pi/2, 2\pi/3)$. ■

Evaluating Triple Integrals with Spherical Coordinates

In the spherical coordinate system the counterpart of a rectangular box is a **spherical wedge**

$$E = \{(\rho, \theta, \phi) \mid a \leqslant \rho \leqslant b,\ \alpha \leqslant \theta \leqslant \beta,\ c \leqslant \phi \leqslant d\}$$

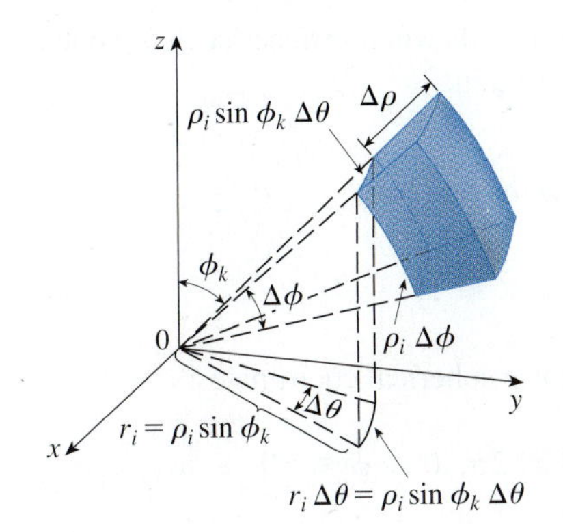

FIGURE 7

where $a \geqslant 0$ and $\beta - \alpha \leqslant 2\pi$, and $d - c \leqslant \pi$. Although we defined triple integrals by dividing solids into small boxes, it can be shown that dividing a solid into small spherical wedges always gives the same result. So we divide E into smaller spherical wedges E_{ijk} by means of equally spaced spheres $\rho = \rho_i$, half-planes $\theta = \theta_j$, and half-cones $\phi = \phi_k$. Figure 7 shows that E_{ijk} is approximately a rectangular box with dimensions $\Delta\rho$, $\rho_i\,\Delta\phi$ (arc of a circle with radius ρ_i, angle $\Delta\phi$), and $\rho_i \sin\phi_k\,\Delta\theta$ (arc of a circle with radius $\rho_i \sin\phi_k$, angle $\Delta\theta$). So an approximation to the volume of E_{ijk} is given by

$$\Delta V_{ijk} \approx (\Delta\rho)(\rho_i\,\Delta\phi)(\rho_i \sin\phi_k\,\Delta\theta) = \rho_i^2 \sin\phi_k\,\Delta\rho\,\Delta\theta\,\Delta\phi$$

In fact, it can be shown, with the aid of the Mean Value Theorem (Exercise 49), that the volume of E_{ijk} is given exactly by

$$\Delta V_{ijk} = \tilde{\rho}_i^2 \sin\tilde{\phi}_k\,\Delta\rho\,\Delta\theta\,\Delta\phi$$

where $(\tilde{\rho}_i, \tilde{\theta}_j, \tilde{\phi}_k)$ is some point in E_{ijk}. Let $(x_{ijk}^*, y_{ijk}^*, z_{ijk}^*)$ be the rectangular coordinates of this point. Then

$$\iiint_E f(x, y, z)\,dV = \lim_{l, m, n \to \infty} \sum_{i=1}^{l} \sum_{j=1}^{m} \sum_{k=1}^{n} f(x_{ijk}^*, y_{ijk}^*, z_{ijk}^*)\,\Delta V_{ijk}$$

$$= \lim_{l, m, n \to \infty} \sum_{i=1}^{l} \sum_{j=1}^{m} \sum_{k=1}^{n} f(\tilde{\rho}_i \sin\tilde{\phi}_k \cos\tilde{\theta}_j, \tilde{\rho}_i \sin\tilde{\phi}_k \sin\tilde{\theta}_j, \tilde{\rho}_i \cos\tilde{\phi}_k)\,\tilde{\rho}_i^2 \sin\tilde{\phi}_k\,\Delta\rho\Delta\theta\Delta\phi$$

But this sum is a Riemann sum for the function

$$F(\rho, \theta, \phi) = f(\rho \sin\phi\ \cos\theta, \rho \sin\phi\ \sin\theta, \rho \cos\phi)\,\rho^2 \sin\phi$$

Consequently, we have arrived at the following **formula for triple integration in spherical coordinates**.

3
$$\iiint_E f(x, y, z)\,dV$$
$$= \int_c^d \int_\alpha^\beta \int_a^b f(\rho \sin\phi\ \cos\theta, \rho \sin\phi\ \sin\theta, \rho \cos\phi)\,\rho^2 \sin\phi\ d\rho\ d\theta\ d\phi$$

where E is a spherical wedge given by

$$E = \{(\rho, \theta, \phi) \mid a \leqslant \rho \leqslant b,\ \alpha \leqslant \theta \leqslant \beta,\ c \leqslant \phi \leqslant d\}$$

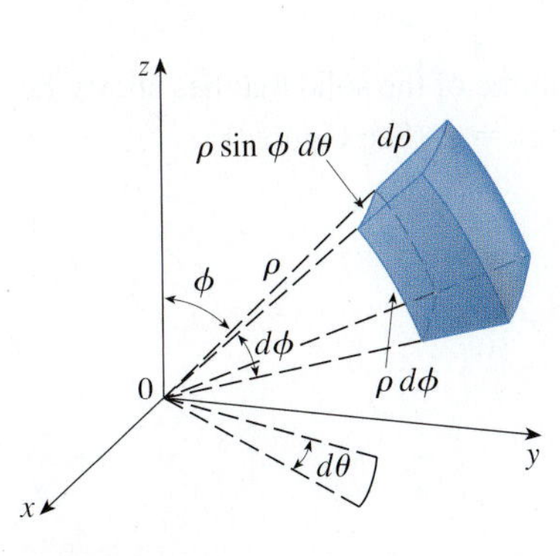

FIGURE 8

Volume element in spherical coordinates: $dV = \rho^2 \sin\phi\ d\rho\,d\theta\,d\phi$

Formula 3 says that we convert a triple integral from rectangular coordinates to spherical coordinates by writing

$$x = \rho \sin\phi\ \cos\theta \qquad y = \rho \sin\phi\ \sin\theta \qquad z = \rho \cos\phi$$

using the appropriate limits of integration, and replacing dV by $\rho^2 \sin\phi\ d\rho\ d\theta\ d\phi$. This is illustrated in Figure 8.

This formula can be extended to include more general spherical regions such as

$$E = \{(\rho, \theta, \phi) \mid \alpha \leqslant \theta \leqslant \beta,\ c \leqslant \phi \leqslant d,\ g_1(\theta, \phi) \leqslant \rho \leqslant g_2(\theta, \phi)\}$$

In this case the formula is the same as in (3) except that the limits of integration for ρ are $g_1(\theta, \phi)$ and $g_2(\theta, \phi)$.

Usually, spherical coordinates are used in triple integrals when surfaces such as cones and spheres form the boundary of the region of integration.

EXAMPLE 3 Evaluate $\iiint_B e^{(x^2+y^2+z^2)^{3/2}}\, dV$, where B is the unit ball:

$$B = \{(x, y, z) \mid x^2 + y^2 + z^2 \leqslant 1\}$$

SOLUTION Since the boundary of B is a sphere, we use spherical coordinates:

$$B = \{(\rho, \theta, \phi) \mid 0 \leqslant \rho \leqslant 1,\ 0 \leqslant \theta \leqslant 2\pi,\ 0 \leqslant \phi \leqslant \pi\}$$

In addition, spherical coordinates are appropriate because

$$x^2 + y^2 + z^2 = \rho^2$$

Thus (3) gives

$$\iiint_B e^{(x^2+y^2+z^2)^{3/2}}\, dV = \int_0^{\pi} \int_0^{2\pi} \int_0^1 e^{(\rho^2)^{3/2}} \rho^2 \sin\phi \, d\rho \, d\theta \, d\phi$$

$$= \int_0^{\pi} \sin\phi \, d\phi \int_0^{2\pi} d\theta \int_0^1 \rho^2 e^{\rho^3} d\rho$$

$$= \Big[-\cos\phi\Big]_0^{\pi} (2\pi) \Big[\tfrac{1}{3} e^{\rho^3}\Big]_0^1 = \tfrac{4}{3}\pi(e - 1)$$

■

NOTE It would have been extremely awkward to evaluate the integral in Example 3 without spherical coordinates. In rectangular coordinates the iterated integral would have been

$$\int_{-1}^{1} \int_{-\sqrt{1-x^2}}^{\sqrt{1-x^2}} \int_{-\sqrt{1-x^2-y^2}}^{\sqrt{1-x^2-y^2}} e^{(x^2+y^2+z^2)^{3/2}}\, dz\, dy\, dx$$

EXAMPLE 4 Use spherical coordinates to find the volume of the solid that lies above the cone $z = \sqrt{x^2 + y^2}$ and below the sphere $x^2 + y^2 + z^2 = z$. (See Figure 9.)

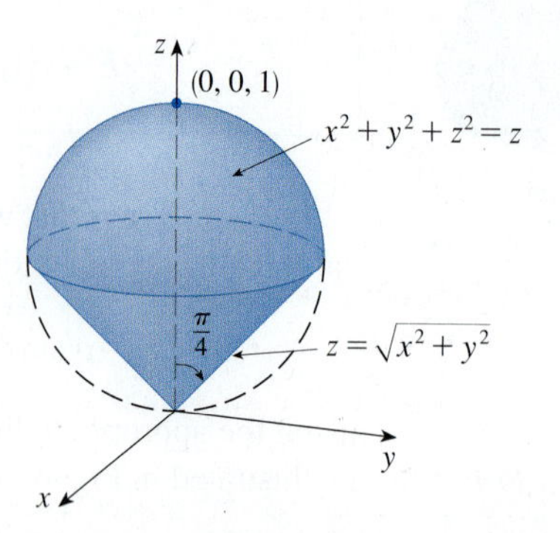

FIGURE 9

Figure 10 gives another look (this time drawn by Maple) at the solid of Example 4.

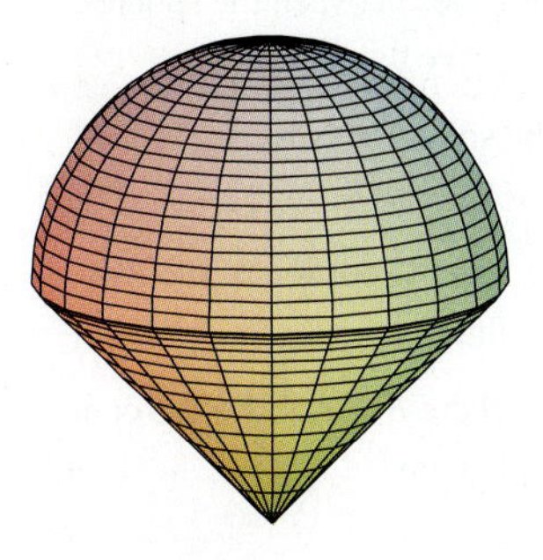

FIGURE 10

SOLUTION Notice that the sphere passes through the origin and has center $\left(0, 0, \frac{1}{2}\right)$. We write the equation of the sphere in spherical coordinates as

$$\rho^2 = \rho \cos \phi \qquad \text{or} \qquad \rho = \cos \phi$$

The equation of the cone can be written as

$$\rho \cos \phi = \sqrt{\rho^2 \sin^2\phi \, \cos^2\theta + \rho^2 \sin^2\phi \, \sin^2\theta} = \rho \sin \phi$$

This gives $\sin \phi = \cos \phi$, or $\phi = \pi/4$. Therefore the description of the solid E in spherical coordinates is

$$E = \left\{(\rho, \theta, \phi) \mid 0 \leqslant \theta \leqslant 2\pi, \ 0 \leqslant \phi \leqslant \pi/4, \ 0 \leqslant \rho \leqslant \cos \phi\right\}$$

Figure 11 shows how E is swept out if we integrate first with respect to ρ, then ϕ, and then θ. The volume of E is

$$V(E) = \iiint_E dV = \int_0^{2\pi} \int_0^{\pi/4} \int_0^{\cos\phi} \rho^2 \sin \phi \, d\rho \, d\phi \, d\theta$$

$$= \int_0^{2\pi} d\theta \int_0^{\pi/4} \sin \phi \left[\frac{\rho^3}{3}\right]_{\rho=0}^{\rho=\cos\phi} d\phi$$

$$= \frac{2\pi}{3} \int_0^{\pi/4} \sin \phi \cos^3\phi \, d\phi = \frac{2\pi}{3}\left[-\frac{\cos^4\phi}{4}\right]_0^{\pi/4} = \frac{\pi}{8}$$

TEC Visual 15.8 shows an animation of Figure 11.

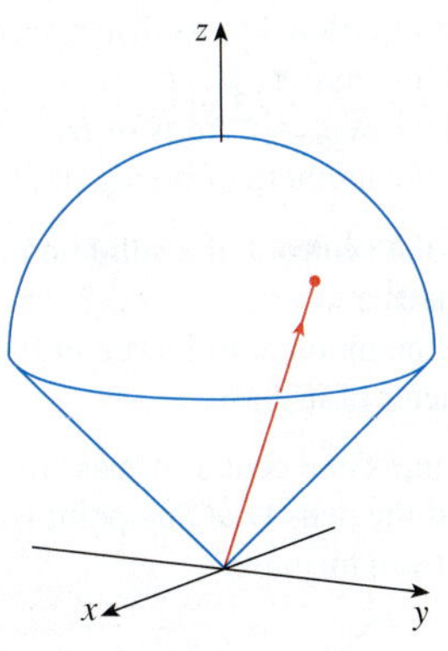

ρ varies from 0 to $\cos\phi$ while ϕ and θ are constant.

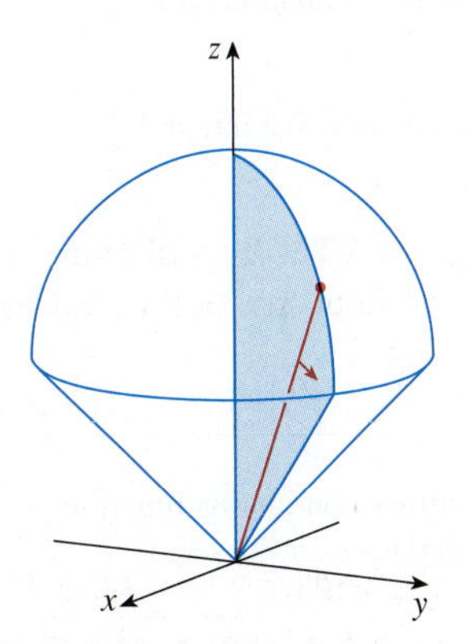

ϕ varies from 0 to $\pi/4$ while θ is constant.

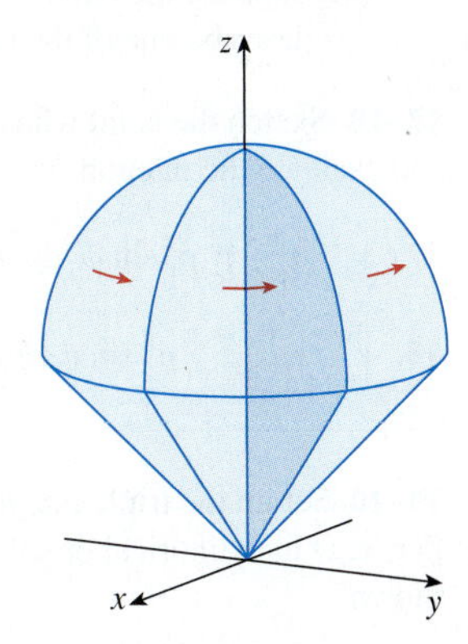

θ varies from 0 to 2π. ■

FIGURE 11

15.8 EXERCISES

1–2 Plot the point whose spherical coordinates are given. Then find the rectangular coordinates of the point.

1. (a) $(6, \pi/3, \pi/6)$ (b) $(3, \pi/2, 3\pi/4)$

2. (a) $(2, \pi/2, \pi/2)$ (b) $(4, -\pi/4, \pi/3)$

3–4 Change from rectangular to spherical coordinates.

3. (a) $(0, -2, 0)$ (b) $\left(-1, 1, -\sqrt{2}\right)$

4. (a) $\left(1, 0, \sqrt{3}\right)$ (b) $\left(\sqrt{3}, -1, 2\sqrt{3}\right)$

5–6 Describe in words the surface whose equation is given.

5. $\phi = \pi/3$ **6.** $\rho^2 - 3\rho + 2 = 0$

7–8 Identify the surface whose equation is given.

7. $\rho \cos \phi = 1$ **8.** $\rho = \cos \phi$

9–10 Write the equation in spherical coordinates.

9. (a) $x^2 + y^2 + z^2 = 9$ (b) $x^2 - y^2 - z^2 = 1$

10. (a) $z = x^2 + y^2$ (b) $z = x^2 - y^2$

11–14 Sketch the solid described by the given inequalities.

11. $\rho \leqslant 1, \quad 0 \leqslant \phi \leqslant \pi/6, \quad 0 \leqslant \theta \leqslant \pi$

12. $1 \leqslant \rho \leqslant 2, \quad \pi/2 \leqslant \phi \leqslant \pi$

13. $2 \leqslant \rho \leqslant 4, \quad 0 \leqslant \phi \leqslant \pi/3, \quad 0 \leqslant \theta \leqslant \pi$

14. $\rho \leqslant 2, \quad \rho \leqslant \csc \phi$

15. A solid lies above the cone $z = \sqrt{x^2 + y^2}$ and below the sphere $x^2 + y^2 + z^2 = z$. Write a description of the solid in terms of inequalities involving spherical coordinates.

16. (a) Find inequalities that describe a hollow ball with diameter 30 cm and thickness 0.5 cm. Explain how you have positioned the coordinate system that you have chosen.
(b) Suppose the ball is cut in half. Write inequalities that describe one of the halves.

17–18 Sketch the solid whose volume is given by the integral and evaluate the integral.

17. $\displaystyle\int_0^{\pi/6} \int_0^{\pi/2} \int_0^3 \rho^2 \sin \phi \, d\rho \, d\theta \, d\phi$

18. $\displaystyle\int_0^{\pi/4} \int_0^{2\pi} \int_0^{\sec \phi} \rho^2 \sin \phi \, d\rho \, d\theta \, d\phi$

19–20 Set up the triple integral of an arbitrary continuous function $f(x, y, z)$ in cylindrical or spherical coordinates over the solid shown.

19.

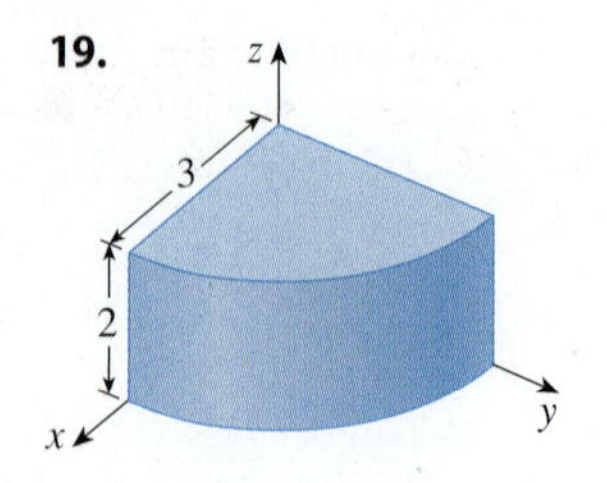

20.

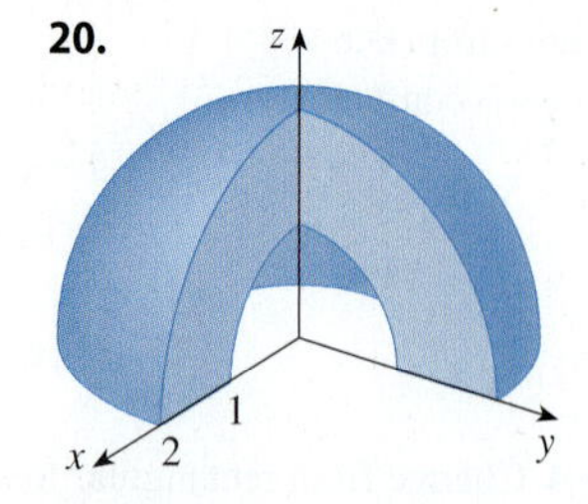

21–34 Use spherical coordinates.

21. Evaluate $\iiint_B (x^2 + y^2 + z^2)^2 \, dV$, where B is the ball with center the origin and radius 5.

22. Evaluate $\iiint_E y^2 z^2 \, dV$, where E lies above the cone $\phi = \pi/3$ and below the sphere $\rho = 1$.

23. Evaluate $\iiint_E (x^2 + y^2) \, dV$, where E lies between the spheres $x^2 + y^2 + z^2 = 4$ and $x^2 + y^2 + z^2 = 9$.

24. Evaluate $\iiint_E y^2 \, dV$, where E is the solid hemisphere $x^2 + y^2 + z^2 \leqslant 9$, $y \geqslant 0$.

25. Evaluate $\iiint_E x e^{x^2+y^2+z^2} \, dV$, where E is the portion of the unit ball $x^2 + y^2 + z^2 \leqslant 1$ that lies in the first octant.

26. Evaluate $\iiint_E \sqrt{x^2 + y^2 + z^2} \, dV$, where E lies above the cone $z = \sqrt{x^2 + y^2}$ and between the spheres $x^2 + y^2 + z^2 = 1$ and $x^2 + y^2 + z^2 = 4$.

27. Find the volume of the part of the ball $\rho \leqslant a$ that lies between the cones $\phi = \pi/6$ and $\phi = \pi/3$.

28. Find the average distance from a point in a ball of radius a to its center.

29. (a) Find the volume of the solid that lies above the cone $\phi = \pi/3$ and below the sphere $\rho = 4 \cos \phi$.
(b) Find the centroid of the solid in part (a).

30. Find the volume of the solid that lies within the sphere $x^2 + y^2 + z^2 = 4$, above the xy-plane, and below the cone $z = \sqrt{x^2 + y^2}$.

31. (a) Find the centroid of the solid in Example 4. (Assume constant density K.)
(b) Find the moment of inertia about the z-axis for this solid.

32. Let H be a solid hemisphere of radius a whose density at any point is proportional to its distance from the center of the base.
(a) Find the mass of H.
(b) Find the center of mass of H.
(c) Find the moment of inertia of H about its axis.

33. (a) Find the centroid of a solid homogeneous hemisphere of radius a.
(b) Find the moment of inertia of the solid in part (a) about a diameter of its base.

34. Find the mass and center of mass of a solid hemisphere of radius a if the density at any point is proportional to its distance from the base.

35–40 Use cylindrical or spherical coordinates, whichever seems more appropriate.

35. Find the volume and centroid of the solid E that lies above the cone $z = \sqrt{x^2 + y^2}$ and below the sphere $x^2 + y^2 + z^2 = 1$.

36. Find the volume of the smaller wedge cut from a sphere of radius a by two planes that intersect along a diameter at an angle of $\pi/6$.

37. A solid cylinder with constant density has base radius a and height h.
(a) Find the moment of inertia of the cylinder about its axis.
(b) Find the moment of inertia of the cylinder about a diameter of its base.

38. A solid right circular cone with constant density has base radius a and height h.
(a) Find the moment of inertia of the cone about its axis.
(b) Find the moment of inertia of the cone about a diameter of its base.

CAS **39.** Evaluate $\iiint_E z\,dV$, where E lies above the paraboloid $z = x^2 + y^2$ and below the plane $z = 2y$. Use either the Table of Integrals (on Reference Pages 6–10) or a computer algebra system to evaluate the integral.

CAS **40.** (a) Find the volume enclosed by the torus $\rho = \sin\phi$.
(b) Use a computer to draw the torus.

41–43 Evaluate the integral by changing to spherical coordinates.

41. $\displaystyle\int_0^1 \int_0^{\sqrt{1-x^2}} \int_{\sqrt{x^2+y^2}}^{\sqrt{2-x^2-y^2}} xy\,dz\,dy\,dx$

42. $\displaystyle\int_{-a}^{a} \int_{-\sqrt{a^2-y^2}}^{\sqrt{a^2-y^2}} \int_{-\sqrt{a^2-x^2-y^2}}^{\sqrt{a^2-x^2-y^2}} (x^2z + y^2z + z^3)\,dz\,dx\,dy$

43. $\displaystyle\int_{-2}^{2} \int_{-\sqrt{4-x^2}}^{\sqrt{4-x^2}} \int_{2-\sqrt{4-x^2-y^2}}^{2+\sqrt{4-x^2-y^2}} (x^2 + y^2 + z^2)^{3/2}\,dz\,dy\,dx$

44. A model for the density δ of the earth's atmosphere near its surface is

$$\delta = 619.09 - 0.000097\rho$$

where ρ (the distance from the center of the earth) is measured in meters and δ is measured in kilograms per cubic meter. If we take the surface of the earth to be a sphere with radius 6370 km, then this model is a reasonable one for $6.370 \times 10^6 \leqslant \rho \leqslant 6.375 \times 10^6$. Use this model to estimate the mass of the atmosphere between the ground and an altitude of 5 km.

45. Use a graphing device to draw a silo consisting of a cylinder with radius 3 and height 10 surmounted by a hemisphere.

46. The latitude and longitude of a point P in the Northern Hemisphere are related to spherical coordinates ρ, θ, ϕ as follows. We take the origin to be the center of the earth and the positive z-axis to pass through the North Pole. The positive x-axis passes through the point where the prime meridian (the meridian through Greenwich, England) intersects the equator. Then the latitude of P is $\alpha = 90° - \phi°$ and the longitude is $\beta = 360° - \theta°$. Find the great-circle distance from Los Angeles (lat. 34.06° N, long. 118.25° W) to Montréal (lat. 45.50° N, long. 73.60° W). Take the radius of the earth to be 3960 mi. (A *great circle* is the circle of intersection of a sphere and a plane through the center of the sphere.)

CAS **47.** The surfaces $\rho = 1 + \frac{1}{5}\sin m\theta \sin n\phi$ have been used as models for tumors. The "bumpy sphere" with $m = 6$ and $n = 5$ is shown. Use a computer algebra system to find the volume it encloses.

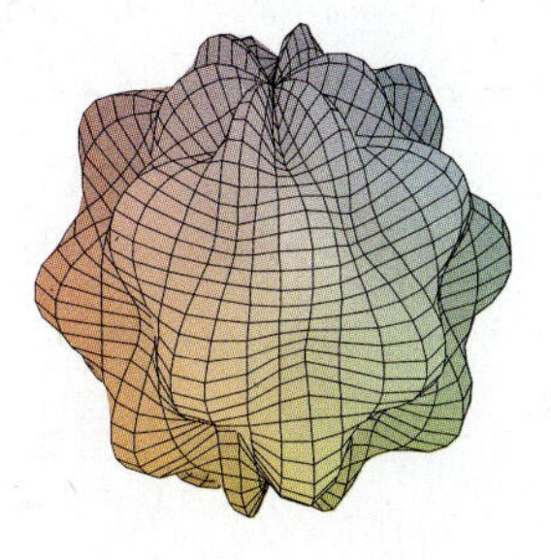

48. Show that

$$\int_{-\infty}^{\infty} \int_{-\infty}^{\infty} \int_{-\infty}^{\infty} \sqrt{x^2 + y^2 + z^2}\, e^{-(x^2+y^2+z^2)}\,dx\,dy\,dz = 2\pi$$

(The improper triple integral is defined as the limit of a triple integral over a solid sphere as the radius of the sphere increases indefinitely.)

49. (a) Use cylindrical coordinates to show that the volume of the solid bounded above by the sphere $r^2 + z^2 = a^2$ and below by the cone $z = r\cot\phi_0$ (or $\phi = \phi_0$), where $0 < \phi_0 < \pi/2$, is

$$V = \frac{2\pi a^3}{3}(1 - \cos\phi_0)$$

(b) Deduce that the volume of the spherical wedge given by $\rho_1 \leqslant \rho \leqslant \rho_2$, $\theta_1 \leqslant \theta \leqslant \theta_2$, $\phi_1 \leqslant \phi \leqslant \phi_2$ is

$$\Delta V = \frac{\rho_2^3 - \rho_1^3}{3}(\cos\phi_1 - \cos\phi_2)(\theta_2 - \theta_1)$$

(c) Use the Mean Value Theorem to show that the volume in part (b) can be written as

$$\Delta V = \tilde{\rho}^2 \sin\tilde{\phi}\,\Delta\rho\,\Delta\theta\,\Delta\phi$$

where $\tilde{\rho}$ lies between ρ_1 and ρ_2, $\tilde{\phi}$ lies between ϕ_1 and ϕ_2, $\Delta\rho = \rho_2 - \rho_1$, $\Delta\theta = \theta_2 - \theta_1$, and $\Delta\phi = \phi_2 - \phi_1$.

APPLIED PROJECT ROLLER DERBY

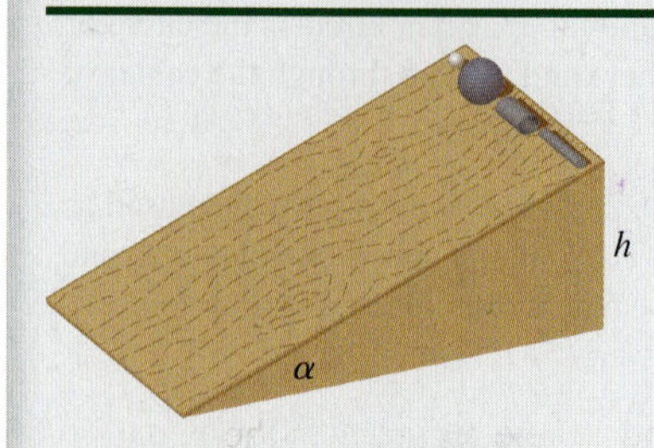

Suppose that a solid ball (a marble), a hollow ball (a squash ball), a solid cylinder (a steel bar), and a hollow cylinder (a lead pipe) roll down a slope. Which of these objects reaches the bottom first? (Make a guess before proceeding.)

To answer this question, we consider a ball or cylinder with mass m, radius r, and moment of inertia I (about the axis of rotation). If the vertical drop is h, then the potential energy at the top is mgh. Suppose the object reaches the bottom with velocity v and angular velocity ω, so $v = \omega r$. The kinetic energy at the bottom consists of two parts: $\frac{1}{2}mv^2$ from translation (moving down the slope) and $\frac{1}{2}I\omega^2$ from rotation. If we assume that energy loss from rolling friction is negligible, then conservation of energy gives

$$mgh = \tfrac{1}{2}mv^2 + \tfrac{1}{2}I\omega^2$$

1. Show that

$$v^2 = \frac{2gh}{1 + I^*} \qquad \text{where } I^* = \frac{I}{mr^2}$$

2. If $y(t)$ is the vertical distance traveled at time t, then the same reasoning as used in Problem 1 shows that $v^2 = 2gy/(1 + I^*)$ at any time t. Use this result to show that y satisfies the differential equation

$$\frac{dy}{dt} = \sqrt{\frac{2g}{1 + I^*}}\,(\sin \alpha)\sqrt{y}$$

where α is the angle of inclination of the plane.

3. By solving the differential equation in Problem 2, show that the total travel time is

$$T = \sqrt{\frac{2h(1 + I^*)}{g \sin^2\alpha}}$$

This shows that the object with the smallest value of I^* wins the race.

4. Show that $I^* = \frac{1}{2}$ for a solid cylinder and $I^* = 1$ for a hollow cylinder.

5. Calculate I^* for a partly hollow ball with inner radius a and outer radius r. Express your answer in terms of $b = a/r$. What happens as $a \to 0$ and as $a \to r$?

6. Show that $I^* = \frac{2}{5}$ for a solid ball and $I^* = \frac{2}{3}$ for a hollow ball. Thus the objects finish in the following order: solid ball, solid cylinder, hollow ball, hollow cylinder.

15.9 Change of Variables in Multiple Integrals

In one-dimensional calculus we often use a change of variable (a substitution) to simplify an integral. By reversing the roles of x and u, we can write the Substitution Rule (4.5.5) as

1
$$\int_a^b f(x)\,dx = \int_c^d f(g(u))\,g'(u)\,du$$

where $x = g(u)$ and $a = g(c)$, $b = g(d)$. Another way of writing Formula 1 is as follows:

2
$$\int_a^b f(x)\,dx = \int_c^d f(x(u))\,\frac{dx}{du}\,du$$

A change of variables can also be useful in double integrals. We have already seen one example of this: conversion to polar coordinates. The new variables r and θ are related to the old variables x and y by the equations

$$x = r\cos\theta \qquad y = r\sin\theta$$

and the change of variables formula (15.3.2) can be written as

$$\iint_R f(x, y)\, dA = \iint_S f(r\cos\theta, r\sin\theta)\, r\, dr\, d\theta$$

where S is the region in the $r\theta$-plane that corresponds to the region R in the xy-plane.

More generally, we consider a change of variables that is given by a **transformation** T from the uv-plane to the xy-plane:

$$T(u, v) = (x, y)$$

where x and y are related to u and v by the equations

3 $$x = g(u, v) \qquad y = h(u, v)$$

or, as we sometimes write,

$$x = x(u, v) \qquad y = y(u, v)$$

We usually assume that T is a C^1 **transformation**, which means that g and h have continuous first-order partial derivatives.

A transformation T is really just a function whose domain and range are both subsets of $\mathbb{R}^2$. If $T(u_1, v_1) = (x_1, y_1)$, then the point (x_1, y_1) is called the **image** of the point (u_1, v_1). If no two points have the same image, T is called **one-to-one**. Figure 1 shows the effect of a transformation T on a region S in the uv-plane. T transforms S into a region R in the xy-plane called the **image of S**, consisting of the images of all points in S.

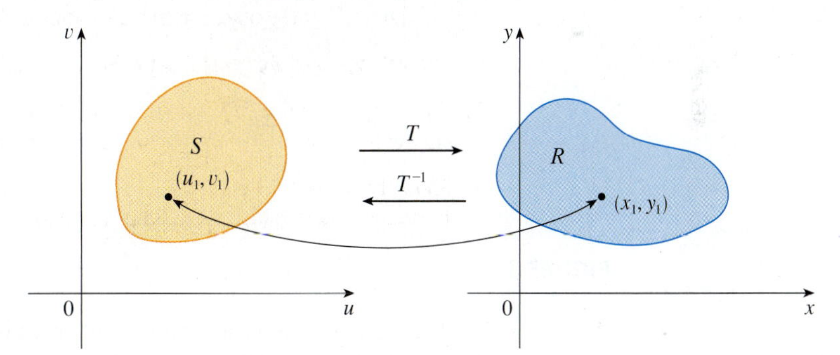

FIGURE 1

If T is a one-to-one transformation, then it has an **inverse transformation** T^{-1} from the xy-plane to the uv-plane and it may be possible to solve Equations 3 for u and v in terms of x and y:

$$u = G(x, y) \qquad v = H(x, y)$$

EXAMPLE 1 A transformation is defined by the equations

$$x = u^2 - v^2 \qquad y = 2uv$$

Find the image of the square $S = \{(u, v) \mid 0 \leqslant u \leqslant 1,\ 0 \leqslant v \leqslant 1\}$.

SOLUTION The transformation maps the boundary of S into the boundary of the image. So we begin by finding the images of the sides of S. The first side, S_1, is given by $v = 0$

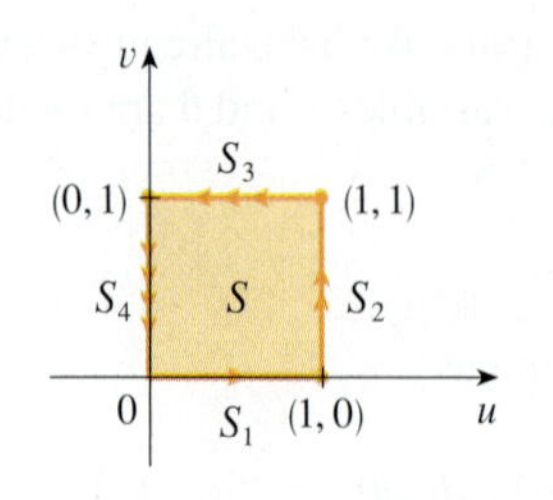

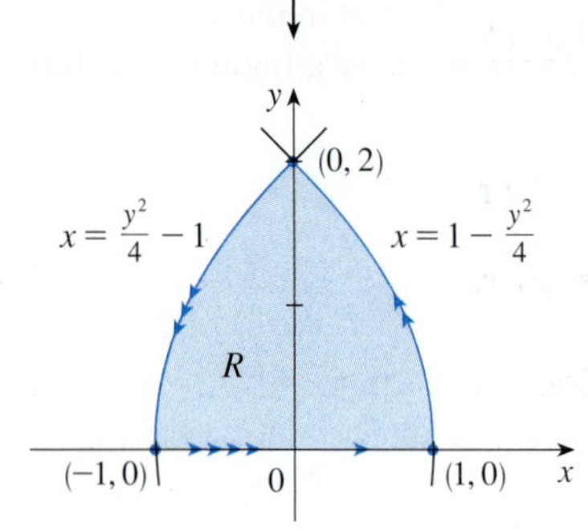

FIGURE 2

($0 \leqslant u \leqslant 1$). (See Figure 2.) From the given equations we have $x = u^2$, $y = 0$, and so $0 \leqslant x \leqslant 1$. Thus S_1 is mapped into the line segment from $(0, 0)$ to $(1, 0)$ in the xy-plane. The second side, S_2, is $u = 1$ ($0 \leqslant v \leqslant 1$) and, putting $u = 1$ in the given equations, we get

$$x = 1 - v^2 \qquad y = 2v$$

Eliminating v, we obtain

$$\boxed{4} \qquad x = 1 - \frac{y^2}{4} \qquad 0 \leqslant x \leqslant 1$$

which is part of a parabola. Similarly, S_3 is given by $v = 1$ ($0 \leqslant u \leqslant 1$), whose image is the parabolic arc

$$\boxed{5} \qquad x = \frac{y^2}{4} - 1 \qquad -1 \leqslant x \leqslant 0$$

Finally, S_4 is given by $u = 0$ ($0 \leqslant v \leqslant 1$) whose image is $x = -v^2$, $y = 0$, that is, $-1 \leqslant x \leqslant 0$. (Notice that as we move around the square in the counterclockwise direction, we also move around the parabolic region in the counterclockwise direction.) The image of S is the region R (shown in Figure 2) bounded by the x-axis and the parabolas given by Equations 4 and 5. ■

Now let's see how a change of variables affects a double integral. We start with a small rectangle S in the uv-plane whose lower left corner is the point (u_0, v_0) and whose dimensions are Δu and Δv. (See Figure 3.)

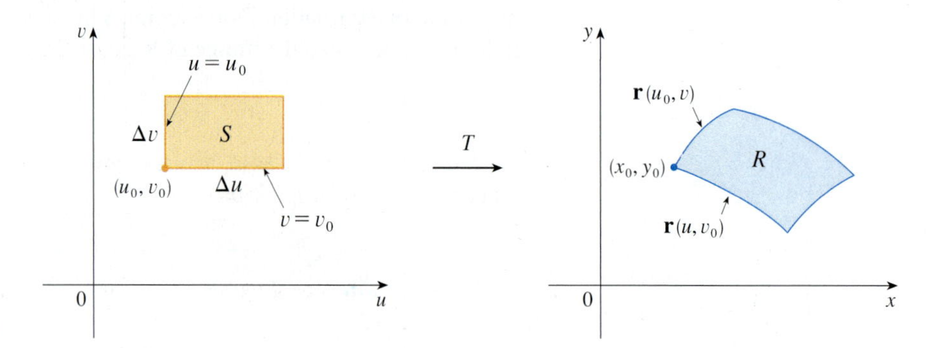

FIGURE 3

The image of S is a region R in the xy-plane, one of whose boundary points is $(x_0, y_0) = T(u_0, v_0)$. The vector

$$\mathbf{r}(u, v) = g(u, v)\,\mathbf{i} + h(u, v)\,\mathbf{j}$$

is the position vector of the image of the point (u, v). The equation of the lower side of S is $v = v_0$, whose image curve is given by the vector function $\mathbf{r}(u, v_0)$. The tangent vector at (x_0, y_0) to this image curve is

$$\mathbf{r}_u = g_u(u_0, v_0)\,\mathbf{i} + h_u(u_0, v_0)\,\mathbf{j} = \frac{\partial x}{\partial u}\,\mathbf{i} + \frac{\partial y}{\partial u}\,\mathbf{j}$$

Similarly, the tangent vector at (x_0, y_0) to the image curve of the left side of S (namely,

$u = u_0$) is

$$\mathbf{r}_v = g_v(u_0, v_0)\,\mathbf{i} + h_v(u_0, v_0)\,\mathbf{j} = \frac{\partial x}{\partial v}\mathbf{i} + \frac{\partial y}{\partial v}\mathbf{j}$$

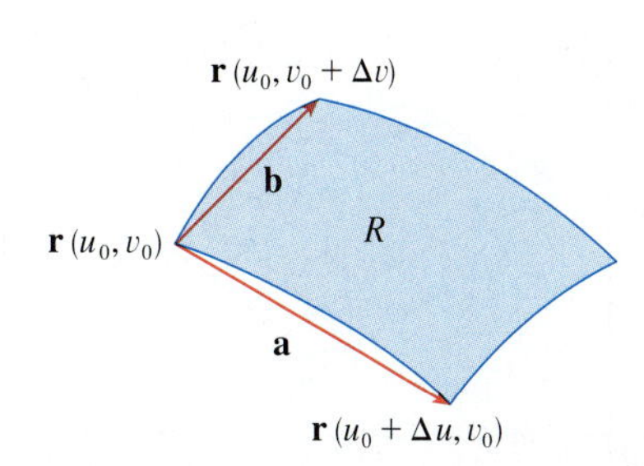

FIGURE 4

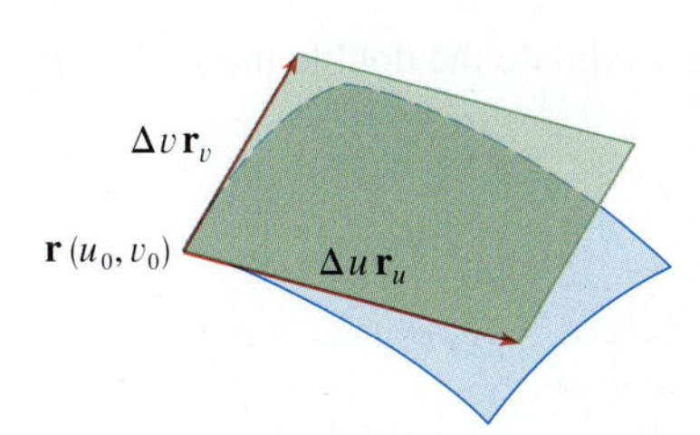

FIGURE 5

We can approximate the image region $R = T(S)$ by a parallelogram determined by the secant vectors

$$\mathbf{a} = \mathbf{r}(u_0 + \Delta u, v_0) - \mathbf{r}(u_0, v_0) \qquad \mathbf{b} = \mathbf{r}(u_0, v_0 + \Delta v) - \mathbf{r}(u_0, v_0)$$

shown in Figure 4. But

$$\mathbf{r}_u = \lim_{\Delta u \to 0} \frac{\mathbf{r}(u_0 + \Delta u, v_0) - \mathbf{r}(u_0, v_0)}{\Delta u}$$

and so $\qquad \mathbf{r}(u_0 + \Delta u, v_0) - \mathbf{r}(u_0, v_0) \approx \Delta u\,\mathbf{r}_u$

Similarly $\qquad \mathbf{r}(u_0, v_0 + \Delta v) - \mathbf{r}(u_0, v_0) \approx \Delta v\,\mathbf{r}_v$

This means that we can approximate R by a parallelogram determined by the vectors $\Delta u\,\mathbf{r}_u$ and $\Delta v\,\mathbf{r}_v$. (See Figure 5.) Therefore we can approximate the area of R by the area of this parallelogram, which, from Section 12.4, is

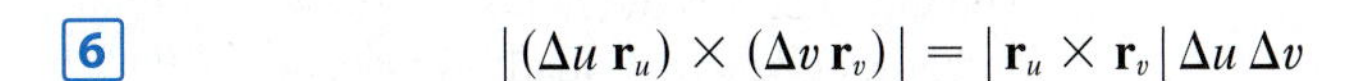

6 $$|(\Delta u\,\mathbf{r}_u) \times (\Delta v\,\mathbf{r}_v)| = |\mathbf{r}_u \times \mathbf{r}_v|\,\Delta u\,\Delta v$$

Computing the cross product, we obtain

$$\mathbf{r}_u \times \mathbf{r}_v = \begin{vmatrix} \mathbf{i} & \mathbf{j} & \mathbf{k} \\ \dfrac{\partial x}{\partial u} & \dfrac{\partial y}{\partial u} & 0 \\ \dfrac{\partial x}{\partial v} & \dfrac{\partial y}{\partial v} & 0 \end{vmatrix} = \begin{vmatrix} \dfrac{\partial x}{\partial u} & \dfrac{\partial y}{\partial u} \\ \dfrac{\partial x}{\partial v} & \dfrac{\partial y}{\partial v} \end{vmatrix} \mathbf{k} = \begin{vmatrix} \dfrac{\partial x}{\partial u} & \dfrac{\partial x}{\partial v} \\ \dfrac{\partial y}{\partial u} & \dfrac{\partial y}{\partial v} \end{vmatrix} \mathbf{k}$$

The determinant that arises in this calculation is called the *Jacobian* of the transformation and is given a special notation.

The Jacobian is named after the German mathematician Carl Gustav Jacob Jacobi (1804–1851). Although the French mathematician Cauchy first used these special determinants involving partial derivatives, Jacobi developed them into a method for evaluating multiple integrals.

7 Definition The **Jacobian** of the transformation T given by $x = g(u, v)$ and $y = h(u, v)$ is

$$\frac{\partial(x, y)}{\partial(u, v)} = \begin{vmatrix} \dfrac{\partial x}{\partial u} & \dfrac{\partial x}{\partial v} \\ \dfrac{\partial y}{\partial u} & \dfrac{\partial y}{\partial v} \end{vmatrix} = \frac{\partial x}{\partial u}\frac{\partial y}{\partial v} - \frac{\partial x}{\partial v}\frac{\partial y}{\partial u}$$

With this notation we can use Equation 6 to give an approximation to the area ΔA of R:

8 $$\Delta A \approx \left| \frac{\partial(x, y)}{\partial(u, v)} \right| \Delta u\,\Delta v$$

where the Jacobian is evaluated at (u_0, v_0).

Next we divide a region S in the uv-plane into rectangles S_{ij} and call their images in the xy-plane R_{ij}. (See Figure 6.)

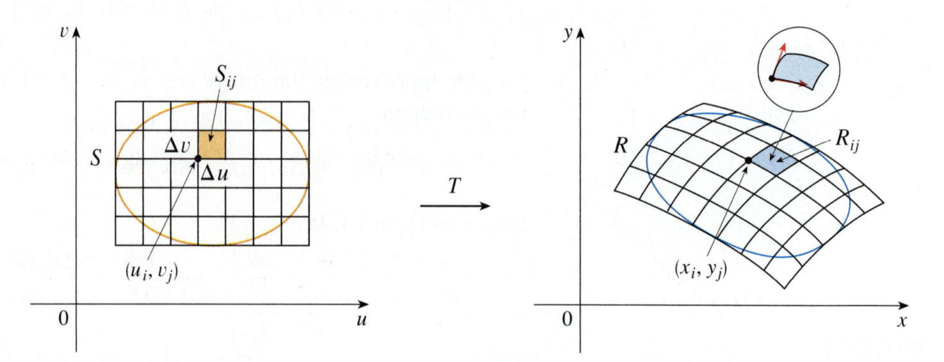

FIGURE 6

Applying the approximation (8) to each R_{ij}, we approximate the double integral of f over R as follows:

$$\iint_R f(x, y)\, dA \approx \sum_{i=1}^{m} \sum_{j=1}^{n} f(x_i, y_j)\, \Delta A$$

$$\approx \sum_{i=1}^{m} \sum_{j=1}^{n} f(g(u_i, v_j), h(u_i, v_j)) \left| \frac{\partial(x, y)}{\partial(u, v)} \right| \Delta u\, \Delta v$$

where the Jacobian is evaluated at (u_i, v_j). Notice that this double sum is a Riemann sum for the integral

$$\iint_S f(g(u, v), h(u, v)) \left| \frac{\partial(x, y)}{\partial(u, v)} \right| du\, dv$$

The foregoing argument suggests that the following theorem is true. (A full proof is given in books on advanced calculus.)

9 Change of Variables in a Double Integral Suppose that T is a C^1 transformation whose Jacobian is nonzero and that T maps a region S in the uv-plane onto a region R in the xy-plane. Suppose that f is continuous on R and that R and S are type I or type II plane regions. Suppose also that T is one-to-one, except perhaps on the boundary of S. Then

$$\iint_R f(x, y)\, dA = \iint_S f(x(u, v), y(u, v)) \left| \frac{\partial(x, y)}{\partial(u, v)} \right| du\, dv$$

Theorem 9 says that we change from an integral in x and y to an integral in u and v by expressing x and y in terms of u and v and writing

$$dA = \left| \frac{\partial(x, y)}{\partial(u, v)} \right| du\, dv$$

Notice the similarity between Theorem 9 and the one-dimensional formula in Equation 2. Instead of the derivative dx/du, we have the absolute value of the Jacobian, that is, $|\partial(x, y)/\partial(u, v)|$.

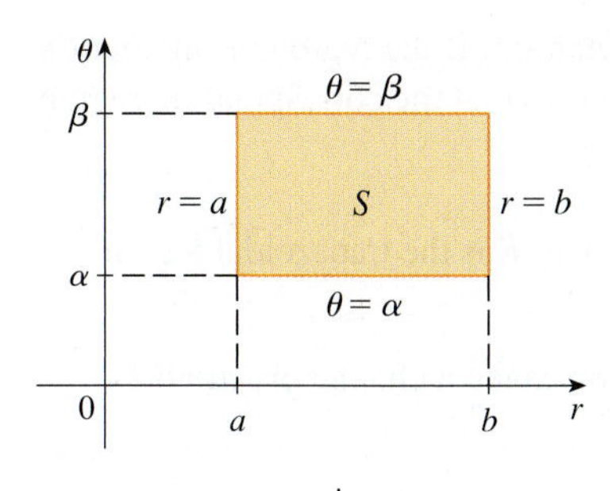

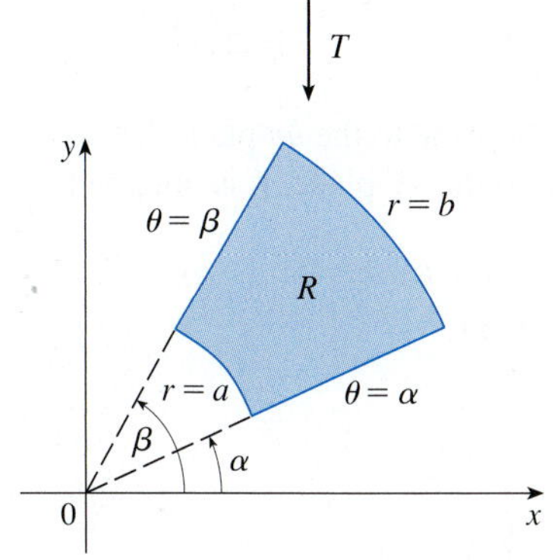

FIGURE 7
The polar coordinate transformation

As a first illustration of Theorem 9, we show that the formula for integration in polar coordinates is just a special case. Here the transformation T from the $r\theta$-plane to the xy-plane is given by

$$x = g(r, \theta) = r\cos\theta \qquad y = h(r, \theta) = r\sin\theta$$

and the geometry of the transformation is shown in Figure 7. T maps an ordinary rectangle in the $r\theta$-plane to a polar rectangle in the xy-plane. The Jacobian of T is

$$\frac{\partial(x, y)}{\partial(r, \theta)} = \begin{vmatrix} \dfrac{\partial x}{\partial r} & \dfrac{\partial x}{\partial \theta} \\ \dfrac{\partial y}{\partial r} & \dfrac{\partial y}{\partial \theta} \end{vmatrix} = \begin{vmatrix} \cos\theta & -r\sin\theta \\ \sin\theta & r\cos\theta \end{vmatrix} = r\cos^2\theta + r\sin^2\theta = r > 0$$

Thus Theorem 9 gives

$$\iint_R f(x, y)\, dx\, dy = \iint_S f(r\cos\theta, r\sin\theta) \left| \frac{\partial(x, y)}{\partial(r, \theta)} \right| dr\, d\theta$$

$$= \int_\alpha^\beta \int_a^b f(r\cos\theta, r\sin\theta)\, r\, dr\, d\theta$$

which is the same as Formula 15.3.2.

EXAMPLE 2 Use the change of variables $x = u^2 - v^2$, $y = 2uv$ to evaluate the integral $\iint_R y\, dA$, where R is the region bounded by the x-axis and the parabolas $y^2 = 4 - 4x$ and $y^2 = 4 + 4x$, $y \geqslant 0$.

SOLUTION The region R is pictured in Figure 2 (on page 1094). In Example 1 we discovered that $T(S) = R$, where S is the square $[0, 1] \times [0, 1]$. Indeed, the reason for making the change of variables to evaluate the integral is that S is a much simpler region than R. First we need to compute the Jacobian:

$$\frac{\partial(x, y)}{\partial(u, v)} = \begin{vmatrix} \dfrac{\partial x}{\partial u} & \dfrac{\partial x}{\partial v} \\ \dfrac{\partial y}{\partial u} & \dfrac{\partial y}{\partial v} \end{vmatrix} = \begin{vmatrix} 2u & -2v \\ 2v & 2u \end{vmatrix} = 4u^2 + 4v^2 > 0$$

Therefore, by Theorem 9,

$$\iint_R y\, dA = \iint_S 2uv \left| \frac{\partial(x, y)}{\partial(u, v)} \right| dA = \int_0^1 \int_0^1 (2uv)4(u^2 + v^2)\, du\, dv$$

$$= 8 \int_0^1 \int_0^1 (u^3 v + uv^3)\, du\, dv = 8 \int_0^1 \left[\tfrac{1}{4}u^4 v + \tfrac{1}{2}u^2 v^3 \right]_{u=0}^{u=1} dv$$

$$= \int_0^1 (2v + 4v^3)\, dv = \left[v^2 + v^4 \right]_0^1 = 2$$

NOTE Example 2 was not a very difficult problem to solve because we were given a suitable change of variables. If we are not supplied with a transformation, then the first step is to think of an appropriate change of variables. If $f(x, y)$ is difficult to integrate, then the form of $f(x, y)$ may suggest a transformation. If the region of integration R is awkward, then the transformation should be chosen so that the corresponding region S in the uv-plane has a convenient description.

EXAMPLE 3 Evaluate the integral $\iint_R e^{(x+y)/(x-y)}\, dA$, where R is the trapezoidal region with vertices $(1, 0)$, $(2, 0)$, $(0, -2)$, and $(0, -1)$.

SOLUTION Since it isn't easy to integrate $e^{(x+y)/(x-y)}$, we make a change of variables suggested by the form of this function:

$$u = x + y \qquad v = x - y \tag{10}$$

These equations define a transformation T^{-1} from the xy-plane to the uv-plane. Theorem 9 talks about a transformation T from the uv-plane to the xy-plane. It is obtained by solving Equations 10 for x and y:

$$x = \tfrac{1}{2}(u + v) \qquad y = \tfrac{1}{2}(u - v) \tag{11}$$

The Jacobian of T is

$$\frac{\partial(x, y)}{\partial(u, v)} = \begin{vmatrix} \dfrac{\partial x}{\partial u} & \dfrac{\partial x}{\partial v} \\ \dfrac{\partial y}{\partial u} & \dfrac{\partial y}{\partial v} \end{vmatrix} = \begin{vmatrix} \frac{1}{2} & \frac{1}{2} \\ \frac{1}{2} & -\frac{1}{2} \end{vmatrix} = -\frac{1}{2}$$

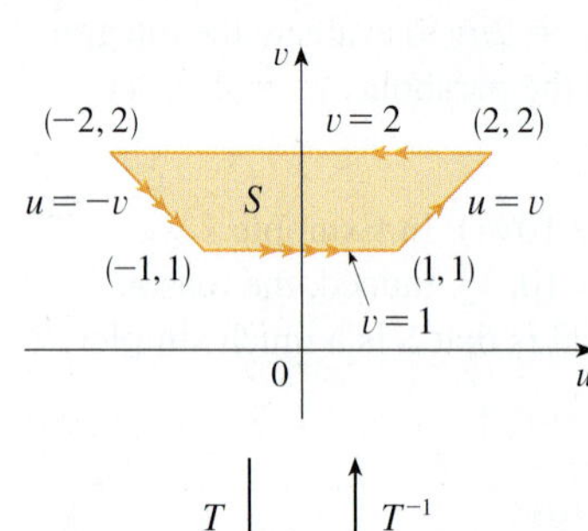

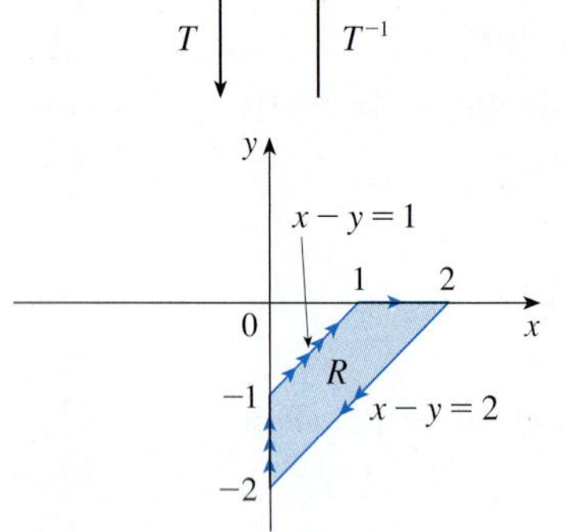

FIGURE 8

To find the region S in the uv-plane corresponding to R, we note that the sides of R lie on the lines

$$y = 0 \qquad x - y = 2 \qquad x = 0 \qquad x - y = 1$$

and, from either Equations 10 or Equations 11, the image lines in the uv-plane are

$$u = v \qquad v = 2 \qquad u = -v \qquad v = 1$$

Thus the region S is the trapezoidal region with vertices $(1, 1)$, $(2, 2)$, $(-2, 2)$, and $(-1, 1)$ shown in Figure 8. Since

$$S = \{(u, v) \mid 1 \leqslant v \leqslant 2,\ -v \leqslant u \leqslant v\}$$

Theorem 9 gives

$$\iint\limits_R e^{(x+y)/(x-y)}\, dA = \iint\limits_S e^{u/v} \left| \frac{\partial(x, y)}{\partial(u, v)} \right| du\, dv$$

$$= \int_1^2 \int_{-v}^{v} e^{u/v}\left(\tfrac{1}{2}\right) du\, dv = \tfrac{1}{2} \int_1^2 \left[ve^{u/v}\right]_{u=-v}^{u=v} dv$$

$$= \tfrac{1}{2} \int_1^2 (e - e^{-1}) v\, dv = \tfrac{3}{4}(e - e^{-1})$$

Triple Integrals

There is a similar change of variables formula for triple integrals. Let T be a transformation that maps a region S in uvw-space onto a region R in xyz-space by means of the equations

$$x = g(u, v, w) \qquad y = h(u, v, w) \qquad z = k(u, v, w)$$

The **Jacobian** of T is the following 3×3 determinant:

12
$$\frac{\partial(x, y, z)}{\partial(u, v, w)} = \begin{vmatrix} \dfrac{\partial x}{\partial u} & \dfrac{\partial x}{\partial v} & \dfrac{\partial x}{\partial w} \\ \dfrac{\partial y}{\partial u} & \dfrac{\partial y}{\partial v} & \dfrac{\partial y}{\partial w} \\ \dfrac{\partial z}{\partial u} & \dfrac{\partial z}{\partial v} & \dfrac{\partial z}{\partial w} \end{vmatrix}$$

Under hypotheses similar to those in Theorem 9, we have the following formula for triple integrals:

13
$$\iiint_R f(x, y, z)\, dV = \iiint_S f(x(u, v, w), y(u, v, w), z(u, v, w)) \left| \frac{\partial(x, y, z)}{\partial(u, v, w)} \right| du\, dv\, dw$$

EXAMPLE 4 Use Formula 13 to derive the formula for triple integration in spherical coordinates.

SOLUTION Here the change of variables is given by

$$x = \rho \sin\phi \cos\theta \qquad y = \rho \sin\phi \sin\theta \qquad z = \rho \cos\phi$$

We compute the Jacobian as follows:

$$\frac{\partial(x, y, z)}{\partial(\rho, \theta, \phi)} = \begin{vmatrix} \sin\phi\cos\theta & -\rho\sin\phi\sin\theta & \rho\cos\phi\cos\theta \\ \sin\phi\sin\theta & \rho\sin\phi\cos\theta & \rho\cos\phi\sin\theta \\ \cos\phi & 0 & -\rho\sin\phi \end{vmatrix}$$

$$= \cos\phi \begin{vmatrix} -\rho\sin\phi\sin\theta & \rho\cos\phi\cos\theta \\ \rho\sin\phi\cos\theta & \rho\cos\phi\sin\theta \end{vmatrix} - \rho\sin\phi \begin{vmatrix} \sin\phi\cos\theta & -\rho\sin\phi\sin\theta \\ \sin\phi\sin\theta & \rho\sin\phi\cos\theta \end{vmatrix}$$

$$= \cos\phi\,(-\rho^2\sin\phi\cos\phi\sin^2\theta - \rho^2\sin\phi\cos\phi\cos^2\theta)$$

$$- \rho\sin\phi\,(\rho\sin^2\phi\cos^2\theta + \rho\sin^2\phi\sin^2\theta)$$

$$= -\rho^2\sin\phi\cos^2\phi - \rho^2\sin\phi\sin^2\phi = -\rho^2\sin\phi$$

Since $0 \leqslant \phi \leqslant \pi$, we have $\sin\phi \geqslant 0$. Therefore

$$\left| \frac{\partial(x, y, z)}{\partial(\rho, \theta, \phi)} \right| = |-\rho^2\sin\phi| = \rho^2\sin\phi$$

and Formula 13 gives

$$\iiint_R f(x, y, z)\, dV = \iiint_S f(\rho\sin\phi\cos\theta, \rho\sin\phi\sin\theta, \rho\cos\phi)\, \rho^2\sin\phi\, d\rho\, d\theta\, d\phi$$

which is equivalent to Formula 15.8.3. ■

15.9 EXERCISES

1–6 Find the Jacobian of the transformation.

1. $x = 2u + v, \quad y = 4u - v$

2. $x = u^2 + uv, \quad y = uv^2$

3. $x = s \cos t, \quad y = s \sin t$

4. $x = pe^q, \quad y = qe^p$

5. $x = uv, \quad y = vw, \quad z = wu$

6. $x = u + vw, \quad y = v + wu, \quad z = w + uv$

7–10 Find the image of the set S under the given transformation.

7. $S = \{(u, v) \mid 0 \leqslant u \leqslant 3,\ 0 \leqslant v \leqslant 2\}$;
$x = 2u + 3v,\ y = u - v$

8. S is the square bounded by the lines $u = 0$, $u = 1$, $v = 0$, $v = 1$; $\quad x = v,\ y = u(1 + v^2)$

9. S is the triangular region with vertices $(0, 0)$, $(1, 1)$, $(0, 1)$; $x = u^2,\ y = v$

10. S is the disk given by $u^2 + v^2 \leqslant 1$; $\quad x = au,\ y = bv$

11–14 A region R in the xy-plane is given. Find equations for a transformation T that maps a rectangular region S in the uv-plane onto R, where the sides of S are parallel to the u- and v-axes.

11. R is bounded by $y = 2x - 1$, $y = 2x + 1$, $y = 1 - x$, $y = 3 - x$

12. R is the parallelogram with vertices $(0, 0)$, $(4, 3)$, $(2, 4)$, $(-2, 1)$

13. R lies between the circles $x^2 + y^2 = 1$ and $x^2 + y^2 = 2$ in the first quadrant

14. R is bounded by the hyperbolas $y = 1/x$, $y = 4/x$ and the lines $y = x$, $y = 4x$ in the first quadrant

---

15–20 Use the given transformation to evaluate the integral.

15. $\iint_R (x - 3y)\, dA$, where R is the triangular region with vertices $(0, 0)$, $(2, 1)$, and $(1, 2)$; $\quad x = 2u + v,\ y = u + 2v$

16. $\iint_R (4x + 8y)\, dA$, where R is the parallelogram with vertices $(-1, 3)$, $(1, -3)$, $(3, -1)$, and $(1, 5)$; $x = \frac{1}{4}(u + v),\ y = \frac{1}{4}(v - 3u)$

17. $\iint_R x^2\, dA$, where R is the region bounded by the ellipse $9x^2 + 4y^2 = 36$; $\quad x = 2u,\ y = 3v$

18. $\iint_R (x^2 - xy + y^2)\, dA$, where R is the region bounded by the ellipse $x^2 - xy + y^2 = 2$;
$x = \sqrt{2}\, u - \sqrt{2/3}\, v,\ y = \sqrt{2}\, u + \sqrt{2/3}\, v$

19. $\iint_R xy\, dA$, where R is the region in the first quadrant bounded by the lines $y = x$ and $y = 3x$ and the hyperbolas $xy = 1$, $xy = 3$; $\quad x = u/v,\ y = v$

20. $\iint_R y^2\, dA$, where R is the region bounded by the curves $xy = 1$, $xy = 2$, $xy^2 = 1$, $xy^2 = 2$; $\quad u = xy,\ v = xy^2$. Illustrate by using a graphing calculator or computer to draw R.

---

21. (a) Evaluate $\iiint_E dV$, where E is the solid enclosed by the ellipsoid $x^2/a^2 + y^2/b^2 + z^2/c^2 = 1$. Use the transformation $x = au,\ y = bv,\ z = cw$.

(b) The earth is not a perfect sphere; rotation has resulted in flattening at the poles. So the shape can be approximated by an ellipsoid with $a = b = 6378$ km and $c = 6356$ km. Use part (a) to estimate the volume of the earth.

(c) If the solid of part (a) has constant density k, find its moment of inertia about the z-axis.

22. An important problem in thermodynamics is to find the work done by an ideal Carnot engine. A cycle consists of alternating expansion and compression of gas in a piston. The work done by the engine is equal to the area of the region R enclosed by two isothermal curves $xy = a$, $xy = b$ and two adiabatic curves $xy^{1.4} = c$, $xy^{1.4} = d$, where $0 < a < b$ and $0 < c < d$. Compute the work done by determining the area of R.

23–27 Evaluate the integral by making an appropriate change of variables.

23. $\displaystyle\iint_R \frac{x - 2y}{3x - y}\, dA$, where R is the parallelogram enclosed by the lines $x - 2y = 0$, $x - 2y = 4$, $3x - y = 1$, and $3x - y = 8$

24. $\iint_R (x + y)e^{x^2 - y^2}\, dA$, where R is the rectangle enclosed by the lines $x - y = 0$, $x - y = 2$, $x + y = 0$, and $x + y = 3$

25. $\displaystyle\iint_R \cos\left(\frac{y - x}{y + x}\right) dA$, where R is the trapezoidal region with vertices $(1, 0)$, $(2, 0)$, $(0, 2)$, and $(0, 1)$

26. $\iint_R \sin(9x^2 + 4y^2)\, dA$, where R is the region in the first quadrant bounded by the ellipse $9x^2 + 4y^2 = 1$

27. $\iint_R e^{x+y}\, dA$, where R is given by the inequality $|x| + |y| \leqslant 1$

---

28. Let f be continuous on $[0, 1]$ and let R be the triangular region with vertices $(0, 0)$, $(1, 0)$, and $(0, 1)$. Show that

$$\iint_R f(x + y)\, dA = \int_0^1 u f(u)\, du$$

15 REVIEW

CONCEPT CHECK

Answers to the Concept Check can be found on the back endpapers.

1. Suppose f is a continuous function defined on a rectangle $R = [a, b] \times [c, d]$.
 (a) Write an expression for a double Riemann sum of f. If $f(x, y) \geqslant 0$, what does the sum represent?
 (b) Write the definition of $\iint_R f(x, y)\, dA$ as a limit.
 (c) What is the geometric interpretation of $\iint_R f(x, y)\, dA$ if $f(x, y) \geqslant 0$? What if f takes on both positive and negative values?
 (d) How do you evaluate $\iint_R f(x, y)\, dA$?
 (e) What does the Midpoint Rule for double integrals say?
 (f) Write an expression for the average value of f.

2. (a) How do you define $\iint_D f(x, y)\, dA$ if D is a bounded region that is not a rectangle?
 (b) What is a type I region? How do you evaluate $\iint_D f(x, y)\, dA$ if D is a type I region?
 (c) What is a type II region? How do you evaluate $\iint_D f(x, y)\, dA$ if D is a type II region?
 (d) What properties do double integrals have?

3. How do you change from rectangular coordinates to polar coordinates in a double integral? Why would you want to make the change?

4. If a lamina occupies a plane region D and has density function $\rho(x, y)$, write expressions for each of the following in terms of double integrals.
 (a) The mass
 (b) The moments about the axes
 (c) The center of mass
 (d) The moments of inertia about the axes and the origin

5. Let f be a joint density function of a pair of continuous random variables X and Y.
 (a) Write a double integral for the probability that X lies between a and b and Y lies between c and d.
 (b) What properties does f possess?
 (c) What are the expected values of X and Y?

6. Write an expression for the area of a surface with equation $z = f(x, y)$, $(x, y) \in D$.

7. (a) Write the definition of the triple integral of f over a rectangular box B.
 (b) How do you evaluate $\iiint_B f(x, y, z)\, dV$?
 (c) How do you define $\iiint_E f(x, y, z)\, dV$ if E is a bounded solid region that is not a box?
 (d) What is a type 1 solid region? How do you evaluate $\iiint_E f(x, y, z)\, dV$ if E is such a region?
 (e) What is a type 2 solid region? How do you evaluate $\iiint_E f(x, y, z)\, dV$ if E is such a region?
 (f) What is a type 3 solid region? How do you evaluate $\iiint_E f(x, y, z)\, dV$ if E is such a region?

8. Suppose a solid object occupies the region E and has density function $\rho(x, y, z)$. Write expressions for each of the following.
 (a) The mass
 (b) The moments about the coordinate planes
 (c) The coordinates of the center of mass
 (d) The moments of inertia about the axes

9. (a) How do you change from rectangular coordinates to cylindrical coordinates in a triple integral?
 (b) How do you change from rectangular coordinates to spherical coordinates in a triple integral?
 (c) In what situations would you change to cylindrical or spherical coordinates?

10. (a) If a transformation T is given by $x = g(u, v)$, $y = h(u, v)$, what is the Jacobian of T?
 (b) How do you change variables in a double integral?
 (c) How do you change variables in a triple integral?

TRUE-FALSE QUIZ

Determine whether the statement is true or false. If it is true, explain why. If it is false, explain why or give an example that disproves the statement.

1. $\displaystyle\int_{-1}^{2}\int_{0}^{6} x^2 \sin(x - y)\, dx\, dy = \int_{0}^{6}\int_{-1}^{2} x^2 \sin(x - y)\, dy\, dx$

2. $\displaystyle\int_{0}^{1}\int_{0}^{x} \sqrt{x + y^2}\, dy\, dx = \int_{0}^{x}\int_{0}^{1} \sqrt{x + y^2}\, dx\, dy$

3. $\displaystyle\int_{1}^{2}\int_{3}^{4} x^2 e^y\, dy\, dx = \int_{1}^{2} x^2\, dx \int_{3}^{4} e^y\, dy$

4. $\displaystyle\int_{-1}^{1}\int_{0}^{1} e^{x^2+y^2} \sin y\, dx\, dy = 0$

5. If f is continuous on $[0, 1]$, then

$$\int_0^1\int_0^1 f(x)f(y)\, dy\, dx = \left[\int_0^1 f(x)\, dx\right]^2$$

6. $\displaystyle\int_{1}^{4}\int_{0}^{1} \left(x^2 + \sqrt{y}\right) \sin(x^2y^2)\, dx\, dy \leqslant 9$

7. If D is the disk given by $x^2 + y^2 \leqslant 4$, then

$$\iint_D \sqrt{4 - x^2 - y^2}\, dA = \tfrac{16}{3}\pi$$

8. The integral $\iiint_E kr^3\, dz\, dr\, d\theta$ represents the moment of inertia about the z-axis of a solid E with constant density k.

9. The integral

$$\int_0^{2\pi}\int_0^2\int_r^2 dz\, dr\, d\theta$$

represents the volume enclosed by the cone $z = \sqrt{x^2 + y^2}$ and the plane $z = 2$.

EXERCISES

1. A contour map is shown for a function f on the square $R = [0, 3] \times [0, 3]$. Use a Riemann sum with nine terms to estimate the value of $\iint_R f(x, y)\, dA$. Take the sample points to be the upper right corners of the squares.

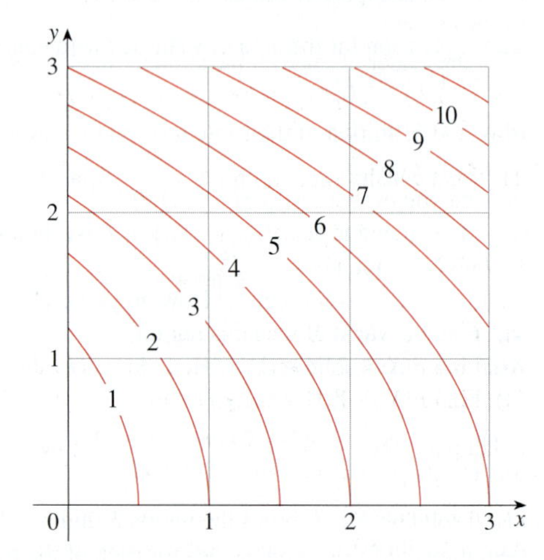

2. Use the Midpoint Rule to estimate the integral in Exercise 1.

3–8 Calculate the iterated integral.

3. $\int_1^2 \int_0^2 (y + 2xe^y)\, dx\, dy$

4. $\int_0^1 \int_0^1 ye^{xy}\, dx\, dy$

5. $\int_0^1 \int_0^x \cos(x^2)\, dy\, dx$

6. $\int_0^1 \int_x^{e^x} 3xy^2\, dy\, dx$

7. $\int_0^{\pi} \int_0^1 \int_0^{\sqrt{1-y^2}} y \sin x\, dz\, dy\, dx$

8. $\int_0^1 \int_0^y \int_x^1 6xyz\, dz\, dx\, dy$

9–10 Write $\iint_R f(x, y)\, dA$ as an iterated integral, where R is the region shown and f is an arbitrary continuous function on R.

9.

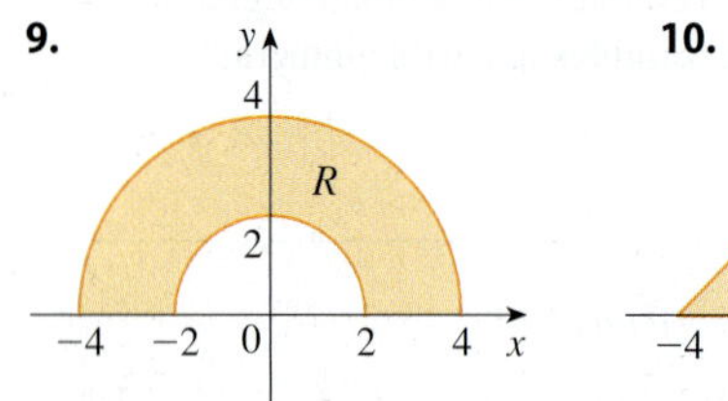

10.

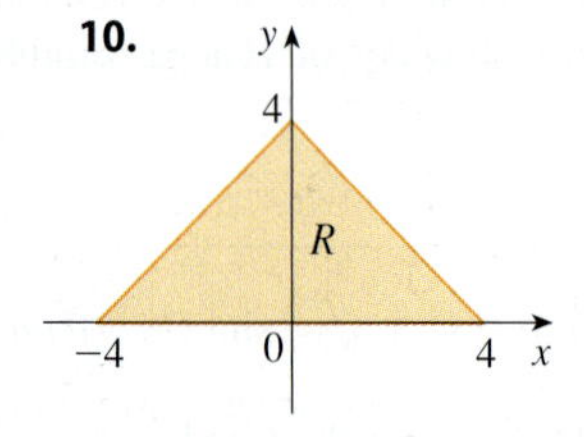

11. The cylindrical coordinates of a point are $(2\sqrt{3}, \pi/3, 2)$. Find the rectangular and spherical coordinates of the point.

12. The rectangular coordinates of a point are $(2, 2, -1)$. Find the cylindrical and spherical coordinates of the point.

13. The spherical coordinates of a point are $(8, \pi/4, \pi/6)$. Find the rectangular and cylindrical coordinates of the point.

14. Identify the surfaces whose equations are given.
(a) $\theta = \pi/4$ (b) $\phi = \pi/4$

15. Write the equation in cylindrical coordinates and in spherical coordinates.
(a) $x^2 + y^2 + z^2 = 4$ (b) $x^2 + y^2 = 4$

16. Sketch the solid consisting of all points with spherical coordinates (ρ, θ, ϕ) such that $0 \leqslant \theta \leqslant \pi/2$, $0 \leqslant \phi \leqslant \pi/6$, and $0 \leqslant \rho \leqslant 2 \cos \phi$.

17. Describe the region whose area is given by the integral

$$\int_0^{\pi/2} \int_0^{\sin 2\theta} r\, dr\, d\theta$$

18. Describe the solid whose volume is given by the integral

$$\int_0^{\pi/2} \int_0^{\pi/2} \int_1^2 \rho^2 \sin \phi\, d\rho\, d\phi\, d\theta$$

and evaluate the integral.

19–20 Calculate the iterated integral by first reversing the order of integration.

19. $\int_0^1 \int_x^1 \cos(y^2)\, dy\, dx$

20. $\int_0^1 \int_{\sqrt{y}}^1 \frac{ye^{x^2}}{x^3}\, dx\, dy$

21–34 Calculate the value of the multiple integral.

21. $\iint_R ye^{xy}\, dA$, where $R = \{(x, y) \mid 0 \leqslant x \leqslant 2,\ 0 \leqslant y \leqslant 3\}$

22. $\iint_D xy\, dA$, where $D = \{(x, y) \mid 0 \leqslant y \leqslant 1,\ y^2 \leqslant x \leqslant y + 2\}$

23. $\displaystyle\iint_D \frac{y}{1 + x^2}\, dA$, where D is bounded by $y = \sqrt{x}$, $y = 0$, $x = 1$

24. $\displaystyle\iint_D \frac{1}{1 + x^2}\, dA$, where D is the triangular region with vertices $(0, 0)$, $(1, 1)$, and $(0, 1)$

25. $\iint_D y\, dA$, where D is the region in the first quadrant bounded by the parabolas $x = y^2$ and $x = 8 - y^2$

26. $\iint_D y\, dA$, where D is the region in the first quadrant that lies above the hyperbola $xy = 1$ and the line $y = x$ and below the line $y = 2$

27. $\iint_D (x^2 + y^2)^{3/2}\, dA$, where D is the region in the first quadrant bounded by the lines $y = 0$ and $y = \sqrt{3}\,x$ and the circle $x^2 + y^2 = 9$

28. $\iint_D x\, dA$, where D is the region in the first quadrant that lies between the circles $x^2 + y^2 = 1$ and $x^2 + y^2 = 2$

29. $\iiint_E xy\, dV$, where
$E = \{(x, y, z) \mid 0 \leqslant x \leqslant 3,\ 0 \leqslant y \leqslant x,\ 0 \leqslant z \leqslant x + y\}$

30. $\iiint_T xy\, dV$, where T is the solid tetrahedron with vertices $(0, 0, 0)$, $(\frac{1}{3}, 0, 0)$, $(0, 1, 0)$, and $(0, 0, 1)$

31. $\iiint_E y^2z^2\, dV$, where E is bounded by the paraboloid $x = 1 - y^2 - z^2$ and the plane $x = 0$

32. $\iiint_E z\, dV$, where E is bounded by the planes $y = 0$, $z = 0$, $x + y = 2$ and the cylinder $y^2 + z^2 = 1$ in the first octant

33. $\iiint_E yz\, dV$, where E lies above the plane $z = 0$, below the plane $z = y$, and inside the cylinder $x^2 + y^2 = 4$

34. $\iiint_H z^3\sqrt{x^2 + y^2 + z^2}\, dV$, where H is the solid hemisphere that lies above the xy-plane and has center the origin and radius 1

35–40 Find the volume of the given solid.

35. Under the paraboloid $z = x^2 + 4y^2$ and above the rectangle $R = [0, 2] \times [1, 4]$

36. Under the surface $z = x^2 y$ and above the triangle in the xy-plane with vertices $(1, 0)$, $(2, 1)$, and $(4, 0)$

37. The solid tetrahedron with vertices $(0, 0, 0)$, $(0, 0, 1)$, $(0, 2, 0)$, and $(2, 2, 0)$

38. Bounded by the cylinder $x^2 + y^2 = 4$ and the planes $z = 0$ and $y + z = 3$

39. One of the wedges cut from the cylinder $x^2 + 9y^2 = a^2$ by the planes $z = 0$ and $z = mx$

40. Above the paraboloid $z = x^2 + y^2$ and below the half-cone $z = \sqrt{x^2 + y^2}$

41. Consider a lamina that occupies the region D bounded by the parabola $x = 1 - y^2$ and the coordinate axes in the first quadrant with density function $\rho(x, y) = y$.
(a) Find the mass of the lamina.
(b) Find the center of mass.
(c) Find the moments of inertia and radii of gyration about the x- and y-axes.

42. A lamina occupies the part of the disk $x^2 + y^2 \leqslant a^2$ that lies in the first quadrant.
(a) Find the centroid of the lamina.
(b) Find the center of mass of the lamina if the density function is $\rho(x, y) = xy^2$.

43. (a) Find the centroid of a solid right circular cone with height h and base radius a. (Place the cone so that its base is in the xy-plane with center the origin and its axis along the positive z-axis.)
(b) If the cone has density function $\rho(x, y, z) = \sqrt{x^2 + y^2}$, find the moment of inertia of the cone about its axis (the z-axis).

44. Find the area of the part of the cone $z^2 = a^2(x^2 + y^2)$ between the planes $z = 1$ and $z = 2$.

45. Find the area of the part of the surface $z = x^2 + y$ that lies above the triangle with vertices $(0, 0)$, $(1, 0)$, and $(0, 2)$.

CAS **46.** Graph the surface $z = x \sin y$, $-3 \leqslant x \leqslant 3$, $-\pi \leqslant y \leqslant \pi$, and find its surface area correct to four decimal places.

47. Use polar coordinates to evaluate

$$\int_0^3 \int_{-\sqrt{9-x^2}}^{\sqrt{9-x^2}} (x^3 + xy^2)\, dy\, dx$$

48. Use spherical coordinates to evaluate

$$\int_{-2}^{2} \int_0^{\sqrt{4-y^2}} \int_{-\sqrt{4-x^2-y^2}}^{\sqrt{4-x^2-y^2}} y^2\sqrt{x^2 + y^2 + z^2}\, dz\, dx\, dy$$

49. If D is the region bounded by the curves $y = 1 - x^2$ and $y = e^x$, find the approximate value of the integral $\iint_D y^2\, dA$. (Use a graphing device to estimate the points of intersection of the curves.)

CAS **50.** Find the center of mass of the solid tetrahedron with vertices $(0, 0, 0)$, $(1, 0, 0)$, $(0, 2, 0)$, $(0, 0, 3)$ and density function $\rho(x, y, z) = x^2 + y^2 + z^2$.

51. The joint density function for random variables X and Y is

$$f(x, y) = \begin{cases} C(x + y) & \text{if } 0 \leqslant x \leqslant 3,\ 0 \leqslant y \leqslant 2 \\ 0 & \text{otherwise} \end{cases}$$

(a) Find the value of the constant C.
(b) Find $P(X \leqslant 2, Y \geqslant 1)$.
(c) Find $P(X + Y \leqslant 1)$.

52. A lamp has three bulbs, each of a type with average lifetime 800 hours. If we model the probability of failure of a bulb by an exponential density function with mean 800, find the probability that all three bulbs fail within a total of 1000 hours.

53. Rewrite the integral

$$\int_{-1}^{1} \int_{x^2}^{1} \int_0^{1-y} f(x, y, z)\, dz\, dy\, dx$$

as an iterated integral in the order $dx\, dy\, dz$.

54. Give five other iterated integrals that are equal to

$$\int_0^2 \int_0^{y^3} \int_0^{y^2} f(x, y, z)\, dz\, dx\, dy$$

55. Use the transformation $u = x - y$, $v = x + y$ to evaluate

$$\iint_R \frac{x - y}{x + y}\, dA$$

where R is the square with vertices $(0, 2)$, $(1, 1)$, $(2, 2)$, and $(1, 3)$.

56. Use the transformation $x = u^2$, $y = v^2$, $z = w^2$ to find the volume of the region bounded by the surface $\sqrt{x} + \sqrt{y} + \sqrt{z} = 1$ and the coordinate planes.

57. Use the change of variables formula and an appropriate transformation to evaluate $\iint_R xy\, dA$, where R is the square with vertices $(0, 0)$, $(1, 1)$, $(2, 0)$, and $(1, -1)$.

58. The **Mean Value Theorem for double integrals** says that if f is a continuous function on a plane region D that is of type I or II, then there exists a point (x_0, y_0) in D such that

$$\iint_D f(x, y)\, dA = f(x_0, y_0)A(D)$$

Use the Extreme Value Theorem (14.7.8) and Property 15.2.11 of integrals to prove this theorem. (Use the proof of the single-variable version in Section 5.5 as a guide.)

59. Suppose that f is continuous on a disk that contains the point (a, b). Let D_r be the closed disk with center (a, b) and radius r. Use the Mean Value Theorem for double integrals (see Exercise 58) to show that

$$\lim_{r \to 0} \frac{1}{\pi r^2} \iint_{D_r} f(x, y)\, dA = f(a, b)$$

60. (a) Evaluate $\displaystyle\iint_D \frac{1}{(x^2 + y^2)^{n/2}}\, dA$, where n is an integer and D is the region bounded by the circles with center the origin and radii r and R, $0 < r < R$.

(b) For what values of n does the integral in part (a) have a limit as $r \to 0^+$?

(c) Find $\displaystyle\iiint_E \frac{1}{(x^2 + y^2 + z^2)^{n/2}}\, dV$, where E is the region bounded by the spheres with center the origin and radii r and R, $0 < r < R$.

(d) For what values of n does the integral in part (c) have a limit as $r \to 0^+$?

Problems Plus

1. If $[\![x]\!]$ denotes the greatest integer in x, evaluate the integral

$$\iint_R [\![x + y]\!] \, dA$$

where $R = \{(x, y) \mid 1 \leqslant x \leqslant 3,\ 2 \leqslant y \leqslant 5\}$.

2. Evaluate the integral

$$\int_0^1 \int_0^1 e^{\max\{x^2, y^2\}} \, dy \, dx$$

where $\max\{x^2, y^2\}$ means the larger of the numbers x^2 and y^2.

3. Find the average value of the function $f(x) = \int_x^1 \cos(t^2) \, dt$ on the interval $[0, 1]$.

4. If $\mathbf{a}$, $\mathbf{b}$, and $\mathbf{c}$ are constant vectors, $\mathbf{r}$ is the position vector $x\,\mathbf{i} + y\,\mathbf{j} + z\,\mathbf{k}$, and E is given by the inequalities $0 \leqslant \mathbf{a} \cdot \mathbf{r} \leqslant \alpha,\ 0 \leqslant \mathbf{b} \cdot \mathbf{r} \leqslant \beta,\ 0 \leqslant \mathbf{c} \cdot \mathbf{r} \leqslant \gamma$, show that

$$\iiint_E (\mathbf{a} \cdot \mathbf{r})(\mathbf{b} \cdot \mathbf{r})(\mathbf{c} \cdot \mathbf{r}) \, dV = \frac{(\alpha\beta\gamma)^2}{8\,|\mathbf{a} \cdot (\mathbf{b} \times \mathbf{c})|}$$

5. The double integral $\int_0^1 \int_0^1 \dfrac{1}{1 - xy} \, dx \, dy$ is an improper integral and could be defined as the limit of double integrals over the rectangle $[0, t] \times [0, t]$ as $t \to 1^-$. But if we expand the integrand as a geometric series, we can express the integral as the sum of an infinite series. Show that

$$\int_0^1 \int_0^1 \frac{1}{1 - xy} \, dx \, dy = \sum_{n=1}^{\infty} \frac{1}{n^2}$$

6. Leonhard Euler was able to find the exact sum of the series in Problem 5. In 1736 he proved that

$$\sum_{n=1}^{\infty} \frac{1}{n^2} = \frac{\pi^2}{6}$$

In this problem we ask you to prove this fact by evaluating the double integral in Problem 5. Start by making the change of variables

$$x = \frac{u - v}{\sqrt{2}} \qquad y = \frac{u + v}{\sqrt{2}}$$

This gives a rotation about the origin through the angle $\pi/4$. You will need to sketch the corresponding region in the uv-plane.

[*Hint:* If, in evaluating the integral, you encounter either of the expressions $(1 - \sin\theta)/\cos\theta$ or $(\cos\theta)/(1 + \sin\theta)$, you might like to use the identity $\cos\theta = \sin((\pi/2) - \theta)$ and the corresponding identity for $\sin\theta$.]

7. (a) Show that

$$\int_0^1 \int_0^1 \int_0^1 \frac{1}{1 - xyz} \, dx \, dy \, dz = \sum_{n=1}^{\infty} \frac{1}{n^3}$$

(Nobody has ever been able to find the exact value of the sum of this series.)

(b) Show that

$$\int_0^1 \int_0^1 \int_0^1 \frac{1}{1 + xyz} \, dx \, dy \, dz = \sum_{n=1}^{\infty} \frac{(-1)^{n-1}}{n^3}$$

Use this equation to evaluate the triple integral correct to two decimal places.

8. Show that

$$\int_0^\infty \frac{\arctan \pi x - \arctan x}{x}\, dx = \frac{\pi}{2} \ln \pi$$

by first expressing the integral as an iterated integral.

9. (a) Show that when Laplace's equation

$$\frac{\partial^2 u}{\partial x^2} + \frac{\partial^2 u}{\partial y^2} + \frac{\partial^2 u}{\partial z^2} = 0$$

is written in cylindrical coordinates, it becomes

$$\frac{\partial^2 u}{\partial r^2} + \frac{1}{r}\frac{\partial u}{\partial r} + \frac{1}{r^2}\frac{\partial^2 u}{\partial \theta^2} + \frac{\partial^2 u}{\partial z^2} = 0$$

(b) Show that when Laplace's equation is written in spherical coordinates, it becomes

$$\frac{\partial^2 u}{\partial \rho^2} + \frac{2}{\rho}\frac{\partial u}{\partial \rho} + \frac{\cot \phi}{\rho^2}\frac{\partial u}{\partial \phi} + \frac{1}{\rho^2}\frac{\partial^2 u}{\partial \phi^2} + \frac{1}{\rho^2 \sin^2\phi}\frac{\partial^2 u}{\partial \theta^2} = 0$$

10. (a) A lamina has constant density ρ and takes the shape of a disk with center the origin and radius R. Use Newton's Law of Gravitation (see Section 13.4) to show that the magnitude of the force of attraction that the lamina exerts on a body with mass m located at the point $(0, 0, d)$ on the positive z-axis is

$$F = 2\pi G m \rho d \left(\frac{1}{d} - \frac{1}{\sqrt{R^2 + d^2}} \right)$$

[*Hint:* Divide the disk as in Figure 15.3.4 and first compute the vertical component of the force exerted by the polar subrectangle R_{ij}.]

(b) Show that the magnitude of the force of attraction of a lamina with density ρ that occupies an entire plane on an object with mass m located at a distance d from the plane is

$$F = 2\pi G m \rho$$

Notice that this expression does not depend on d.

11. If f is continuous, show that

$$\int_0^x \int_0^y \int_0^z f(t)\, dt\, dz\, dy = \tfrac{1}{2} \int_0^x (x - t)^2 f(t)\, dt$$

12. Evaluate $\displaystyle\lim_{n\to\infty} n^{-2} \sum_{i=1}^{n} \sum_{j=1}^{n^2} \frac{1}{\sqrt{n^2 + ni + j}}$.

13. The plane

$$\frac{x}{a} + \frac{y}{b} + \frac{z}{c} = 1 \qquad a > 0, \quad b > 0, \quad c > 0$$

cuts the solid ellipsoid

$$\frac{x^2}{a^2} + \frac{y^2}{b^2} + \frac{z^2}{c^2} \leqslant 1$$

into two pieces. Find the volume of the smaller piece.

16 Vector Calculus

Parametric surfaces, which are studied in Section 16.6, are frequently used by programmers in creating the sophisticated software used in the development of computer animated films like the *Shrek* series. The software employs parametric and other types of surfaces to create 3D models of the characters and objects in a scene. Color, texture, and lighting is then rendered to bring the scene to life.

Everett Collection / Glow Images

IN THIS CHAPTER WE STUDY the calculus of vector fields. (These are functions that assign vectors to points in space.) In particular we define line integrals (which can be used to find the work done by a force field in moving an object along a curve). Then we define surface integrals (which can be used to find the rate of fluid flow across a surface). The connections between these new types of integrals and the single, double, and triple integrals that we have already met are given by the higher-dimensional versions of the Fundamental Theorem of Calculus: Green's Theorem, Stokes' Theorem, and the Divergence Theorem.

16.1 Vector Fields

The vectors in Figure 1 are air velocity vectors that indicate the wind speed and direction at points 10 m above the surface elevation in the San Francisco Bay area. We see at a glance from the largest arrows in part (a) that the greatest wind speeds at that time occurred as the winds entered the bay across the Golden Gate Bridge. Part (b) shows the very different wind pattern 12 hours earlier. Associated with every point in the air we can imagine a wind velocity vector. This is an example of a *velocity vector field.*

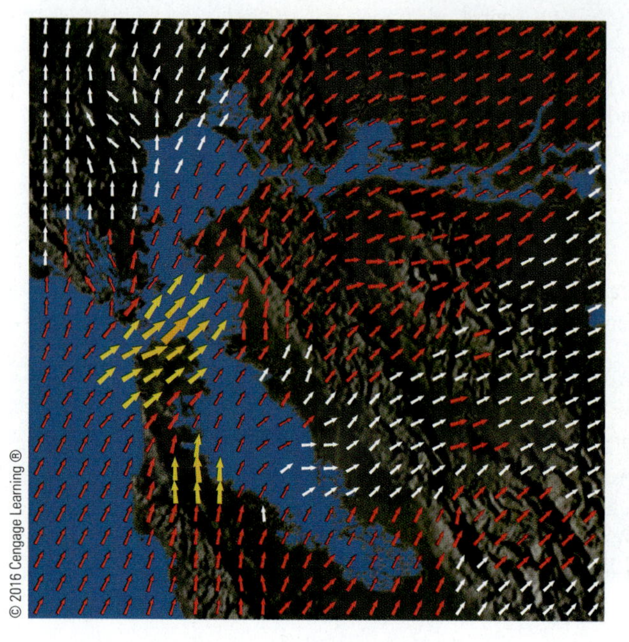

(a) 6:00 PM, March 1, 2010

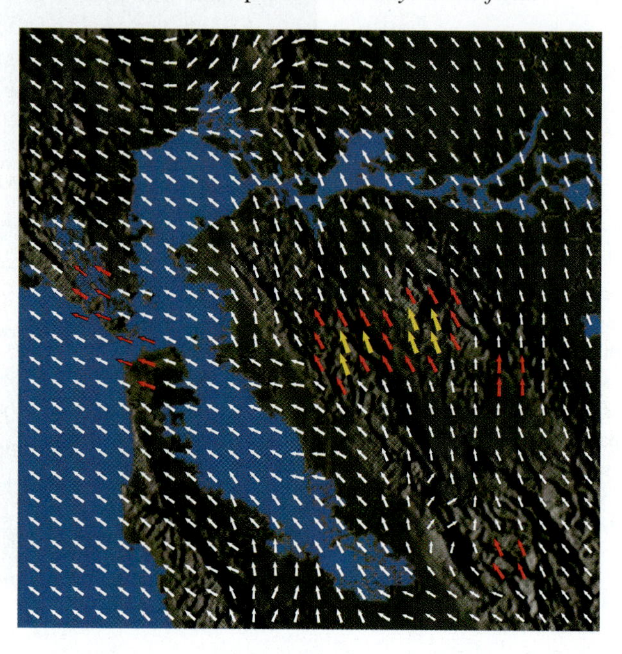

(b) 6:00 AM, March 1, 2010

FIGURE 1 Velocity vector fields showing San Francisco Bay wind patterns

Other examples of velocity vector fields are illustrated in Figure 2: ocean currents and flow past an airfoil.

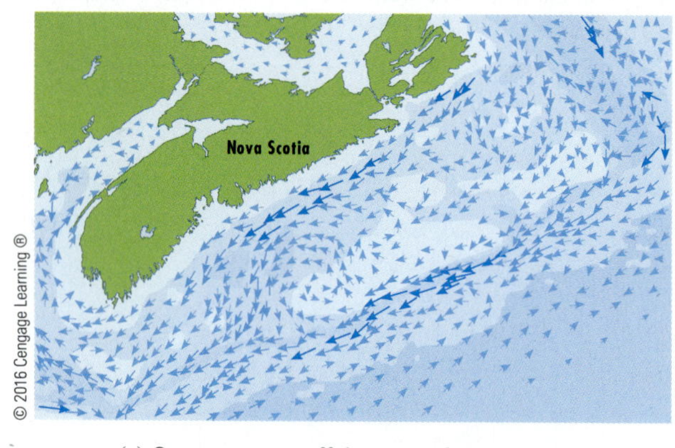

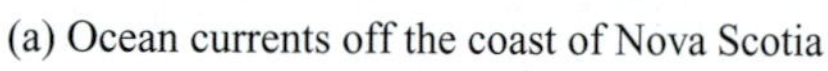

(a) Ocean currents off the coast of Nova Scotia

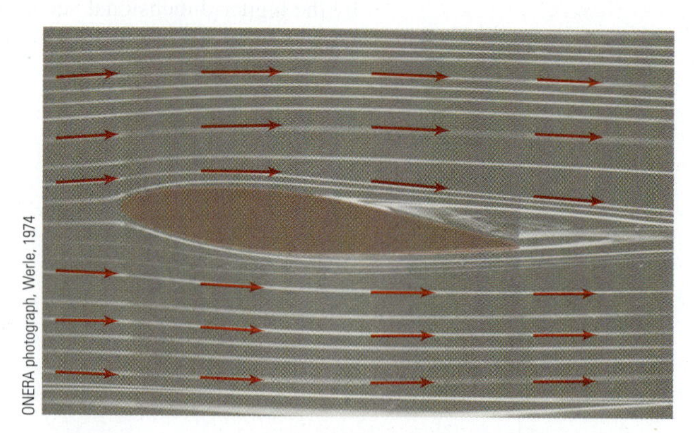

ONERA photograph, Werle, 1974

(b) Airflow past an inclined airfoil

FIGURE 2
Velocity vector fields

Another type of vector field, called a *force field,* associates a force vector with each point in a region. An example is the gravitational force field that we will look at in Example 4.

In general, a vector field is a function whose domain is a set of points in $\mathbb{R}^2$ (or $\mathbb{R}^3$) and whose range is a set of vectors in V_2 (or V_3).

1 Definition Let D be a set in $\mathbb{R}^2$ (a plane region). A **vector field on** $\mathbb{R}^2$ is a function $\mathbf{F}$ that assigns to each point (x, y) in D a two-dimensional vector $\mathbf{F}(x, y)$.

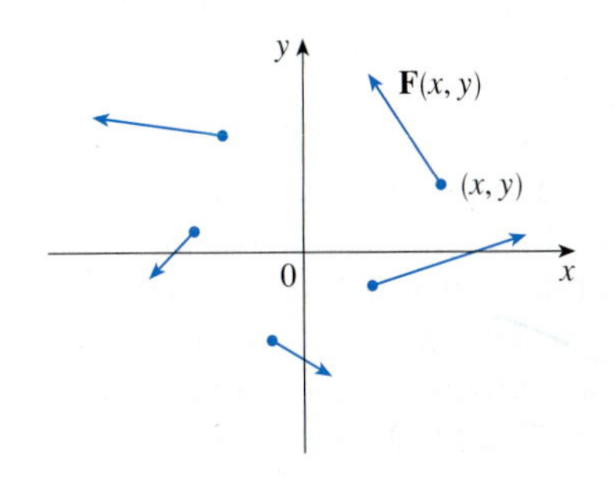

FIGURE 3
Vector field on $\mathbb{R}^2$

The best way to picture a vector field is to draw the arrow representing the vector $\mathbf{F}(x, y)$ starting at the point (x, y). Of course, it's impossible to do this for all points (x, y), but we can gain a reasonable impression of $\mathbf{F}$ by doing it for a few representative points in D as in Figure 3. Since $\mathbf{F}(x, y)$ is a two-dimensional vector, we can write it in terms of its **component functions** P and Q as follows:

$$\mathbf{F}(x, y) = P(x, y)\,\mathbf{i} + Q(x, y)\,\mathbf{j} = \langle P(x, y), Q(x, y)\rangle$$

or, for short,

$$\mathbf{F} = P\,\mathbf{i} + Q\,\mathbf{j}$$

Notice that P and Q are scalar functions of two variables and are sometimes called **scalar fields** to distinguish them from vector fields.

2 Definition Let E be a subset of $\mathbb{R}^3$. A **vector field on** $\mathbb{R}^3$ is a function $\mathbf{F}$ that assigns to each point (x, y, z) in E a three-dimensional vector $\mathbf{F}(x, y, z)$.

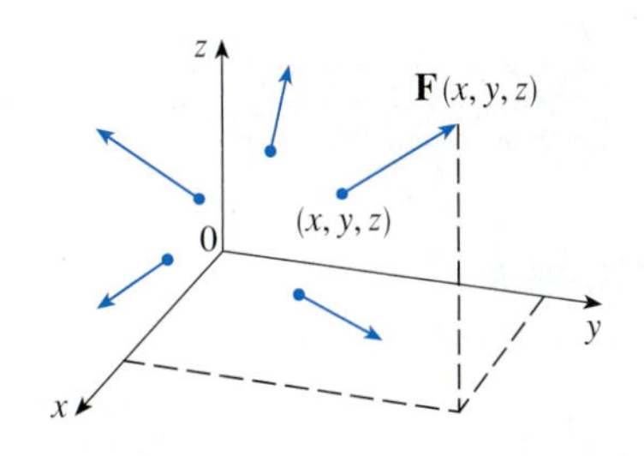

FIGURE 4
Vector field on $\mathbb{R}^3$

A vector field $\mathbf{F}$ on $\mathbb{R}^3$ is pictured in Figure 4. We can express it in terms of its component functions P, Q, and R as

$$\mathbf{F}(x, y, z) = P(x, y, z)\,\mathbf{i} + Q(x, y, z)\,\mathbf{j} + R(x, y, z)\,\mathbf{k}$$

As with the vector functions in Section 13.1, we can define continuity of vector fields and show that $\mathbf{F}$ is continuous if and only if its component functions P, Q, and R are continuous.

We sometimes identify a point (x, y, z) with its position vector $\mathbf{x} = \langle x, y, z\rangle$ and write $\mathbf{F}(\mathbf{x})$ instead of $\mathbf{F}(x, y, z)$. Then $\mathbf{F}$ becomes a function that assigns a vector $\mathbf{F}(\mathbf{x})$ to a vector $\mathbf{x}$.

EXAMPLE 1 A vector field on $\mathbb{R}^2$ is defined by $\mathbf{F}(x, y) = -y\,\mathbf{i} + x\,\mathbf{j}$. Describe $\mathbf{F}$ by sketching some of the vectors $\mathbf{F}(x, y)$ as in Figure 3.

SOLUTION Since $\mathbf{F}(1, 0) = \mathbf{j}$, we draw the vector $\mathbf{j} = \langle 0, 1\rangle$ starting at the point $(1, 0)$ in Figure 5. Since $\mathbf{F}(0, 1) = -\mathbf{i}$, we draw the vector $\langle -1, 0\rangle$ with starting point $(0, 1)$. Continuing in this way, we calculate several other representative values of $\mathbf{F}(x, y)$ in the table and draw the corresponding vectors to represent the vector field in Figure 5.

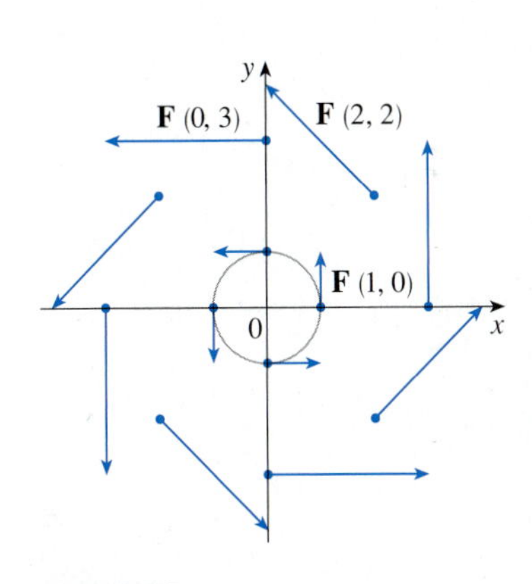

FIGURE 5
$\mathbf{F}(x, y) = -y\,\mathbf{i} + x\,\mathbf{j}$

(x, y)	$\mathbf{F}(x, y)$	(x, y)	$\mathbf{F}(x, y)$
$(1, 0)$	$\langle 0, 1\rangle$	$(-1, 0)$	$\langle 0, -1\rangle$
$(2, 2)$	$\langle -2, 2\rangle$	$(-2, -2)$	$\langle 2, -2\rangle$
$(3, 0)$	$\langle 0, 3\rangle$	$(-3, 0)$	$\langle 0, -3\rangle$
$(0, 1)$	$\langle -1, 0\rangle$	$(0, -1)$	$\langle 1, 0\rangle$
$(-2, 2)$	$\langle -2, -2\rangle$	$(2, -2)$	$\langle 2, 2\rangle$
$(0, 3)$	$\langle -3, 0\rangle$	$(0, -3)$	$\langle 3, 0\rangle$

It appears from Figure 5 that each arrow is tangent to a circle with center the origin. To confirm this, we take the dot product of the position vector $\mathbf{x} = x\,\mathbf{i} + y\,\mathbf{j}$ with the vector $\mathbf{F}(\mathbf{x}) = \mathbf{F}(x, y)$:

$$\mathbf{x} \cdot \mathbf{F}(\mathbf{x}) = (x\,\mathbf{i} + y\,\mathbf{j}) \cdot (-y\,\mathbf{i} + x\,\mathbf{j}) = -xy + yx = 0$$

This shows that $\mathbf{F}(x, y)$ is perpendicular to the position vector $\langle x, y \rangle$ and is therefore tangent to a circle with center the origin and radius $|\mathbf{x}| = \sqrt{x^2 + y^2}$. Notice also that

$$|\mathbf{F}(x, y)| = \sqrt{(-y)^2 + x^2} = \sqrt{x^2 + y^2} = |\mathbf{x}|$$

so the magnitude of the vector $\mathbf{F}(x, y)$ is equal to the radius of the circle. ■

Some computer algebra systems are capable of plotting vector fields in two or three dimensions. They give a better impression of the vector field than is possible by hand because the computer can plot a large number of representative vectors. Figure 6 shows a computer plot of the vector field in Example 1; Figures 7 and 8 show two other vector fields. Notice that the computer scales the lengths of the vectors so they are not too long and yet are proportional to their true lengths.

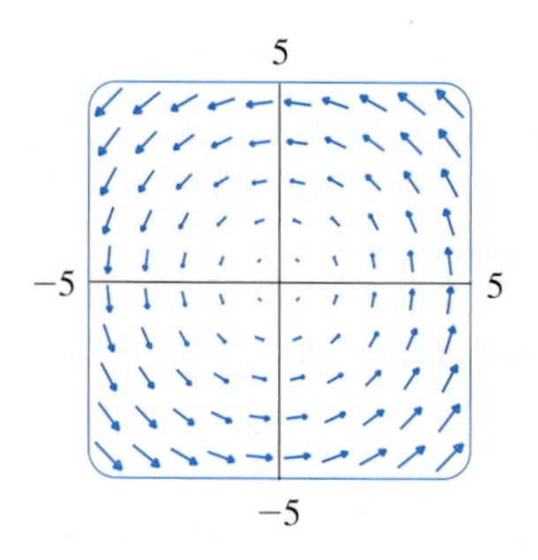

FIGURE 6
$\mathbf{F}(x, y) = \langle -y, x \rangle$

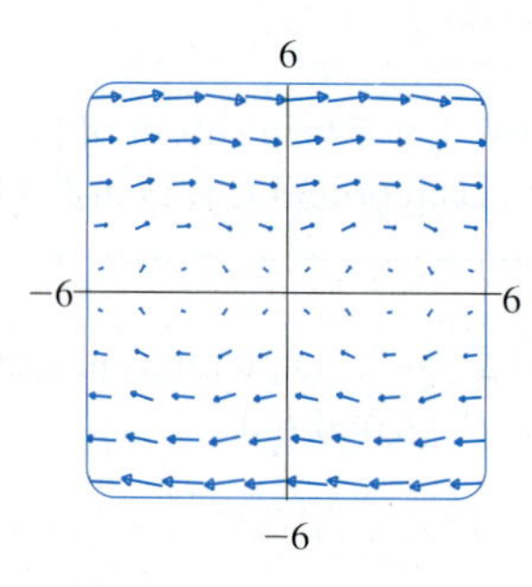

FIGURE 7
$\mathbf{F}(x, y) = \langle y, \sin x \rangle$

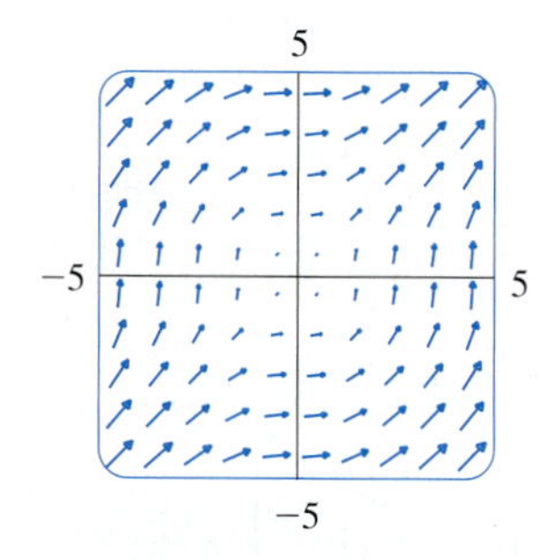

FIGURE 8
$\mathbf{F}(x, y) = \langle \ln(1 + y^2), \ln(1 + x^2) \rangle$

EXAMPLE 2 Sketch the vector field on $\mathbb{R}^3$ given by $\mathbf{F}(x, y, z) = z\,\mathbf{k}$.

SOLUTION The sketch is shown in Figure 9. Notice that all vectors are vertical and point upward above the xy-plane or downward below it. The magnitude increases with the distance from the xy-plane.

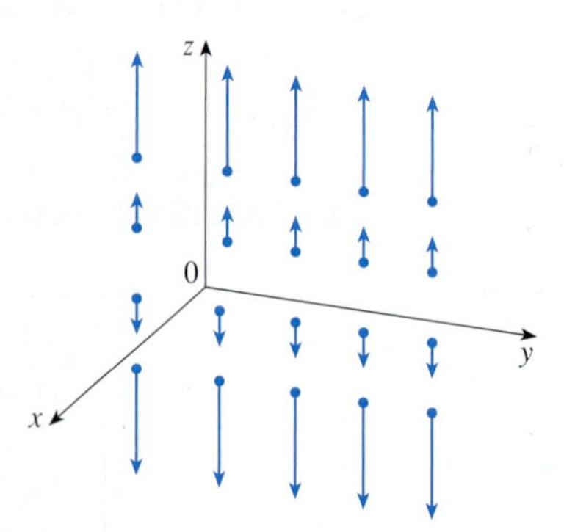

FIGURE 9
$\mathbf{F}(x, y, z) = z\,\mathbf{k}$

■

We were able to draw the vector field in Example 2 by hand because of its particularly simple formula. Most three-dimensional vector fields, however, are virtually impossible to sketch by hand and so we need to resort to computer software. Examples are shown in Figures 10, 11, and 12. Notice that the vector fields in Figures 10 and 11 have similar formulas, but all the vectors in Figure 11 point in the general direction of the negative y-axis because their y-components are all -2. If the vector field in Figure 12 represents a velocity field, then a particle would be swept upward and would spiral around the z-axis in the clockwise direction as viewed from above.

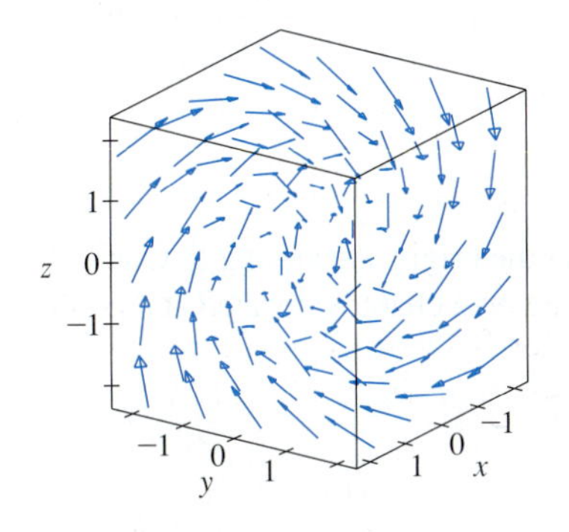

FIGURE 10
$\mathbf{F}(x, y, z) = y\,\mathbf{i} + z\,\mathbf{j} + x\,\mathbf{k}$

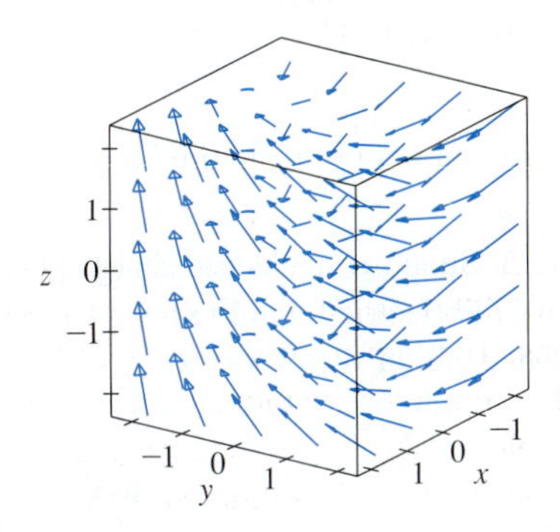

FIGURE 11
$\mathbf{F}(x, y, z) = y\,\mathbf{i} - 2\,\mathbf{j} + x\,\mathbf{k}$

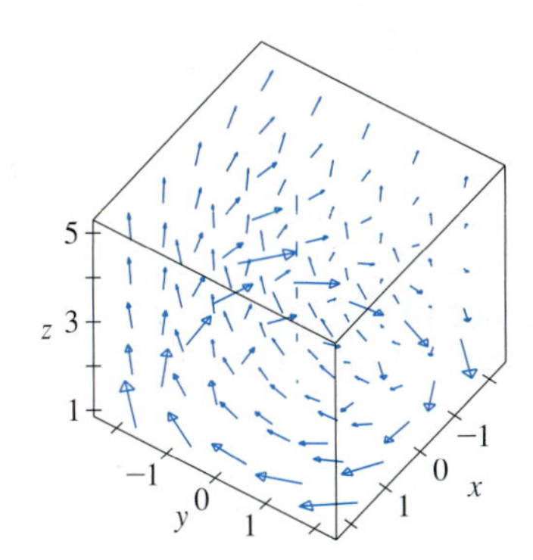

FIGURE 12
$\mathbf{F}(x, y, z) = \dfrac{y}{z}\,\mathbf{i} - \dfrac{x}{z}\,\mathbf{j} + \dfrac{z}{4}\,\mathbf{k}$

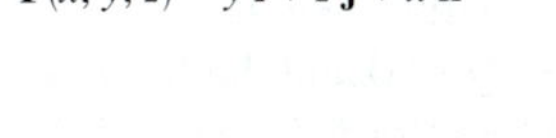
TEC In Visual 16.1 you can rotate the vector fields in Figures 10–12 as well as additional fields.

EXAMPLE 3 Imagine a fluid flowing steadily along a pipe and let $\mathbf{V}(x, y, z)$ be the velocity vector at a point (x, y, z). Then $\mathbf{V}$ assigns a vector to each point (x, y, z) in a certain domain E (the interior of the pipe) and so $\mathbf{V}$ is a vector field on $\mathbb{R}^3$ called a **velocity field**. A possible velocity field is illustrated in Figure 13. The speed at any given point is indicated by the length of the arrow.

Velocity fields also occur in other areas of physics. For instance, the vector field in Example 1 could be used as the velocity field describing the counterclockwise rotation of a wheel. We have seen other examples of velocity fields in Figures 1 and 2. ■

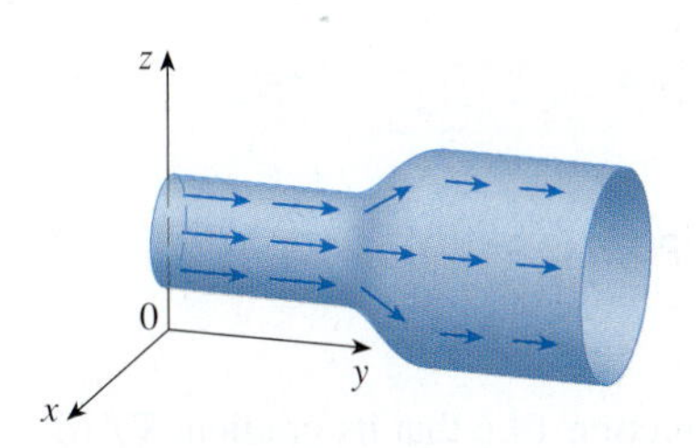

FIGURE 13
Velocity field in fluid flow

EXAMPLE 4 Newton's Law of Gravitation states that the magnitude of the gravitational force between two objects with masses m and M is

$$|\mathbf{F}| = \frac{mMG}{r^2}$$

where r is the distance between the objects and G is the gravitational constant. (This is an example of an inverse square law.) Let's assume that the object with mass M is located at the origin in $\mathbb{R}^3$. (For instance, M could be the mass of the earth and the origin would be at its center.) Let the position vector of the object with mass m be $\mathbf{x} = \langle x, y, z \rangle$. Then $r = |\mathbf{x}|$, so $r^2 = |\mathbf{x}|^2$. The gravitational force exerted on this second object acts toward the origin, and the unit vector in this direction is

$$-\frac{\mathbf{x}}{|\mathbf{x}|}$$

Therefore the gravitational force acting on the object at $\mathbf{x} = \langle x, y, z \rangle$ is

$$\boxed{3} \qquad \mathbf{F}(\mathbf{x}) = -\frac{mMG}{|\mathbf{x}|^3}\,\mathbf{x}$$

[Physicists often use the notation $\mathbf{r}$ instead of $\mathbf{x}$ for the position vector, so you may see Formula 3 written in the form $\mathbf{F} = -(mMG/r^3)\mathbf{r}$.] The function given by Equation 3 is

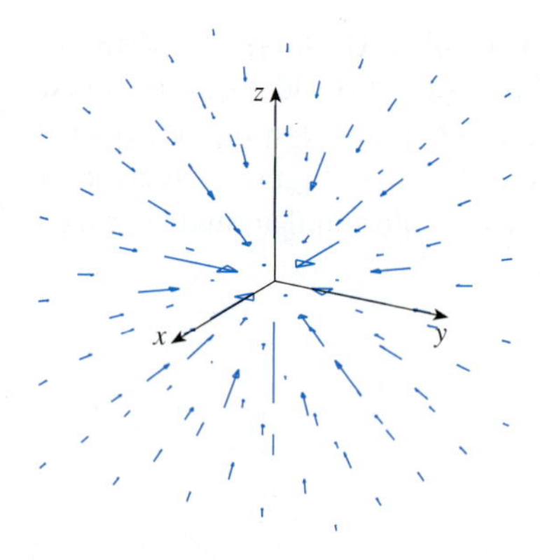

FIGURE 14
Gravitational force field

an example of a vector field, called the **gravitational field**, because it associates a vector [the force $\mathbf{F}(\mathbf{x})$] with every point $\mathbf{x}$ in space.

Formula 3 is a compact way of writing the gravitational field, but we can also write it in terms of its component functions by using the facts that $\mathbf{x} = x\,\mathbf{i} + y\,\mathbf{j} + z\,\mathbf{k}$ and $|\mathbf{x}| = \sqrt{x^2 + y^2 + z^2}$:

$$\mathbf{F}(x, y, z) = \frac{-mMGx}{(x^2 + y^2 + z^2)^{3/2}}\,\mathbf{i} + \frac{-mMGy}{(x^2 + y^2 + z^2)^{3/2}}\,\mathbf{j} + \frac{-mMGz}{(x^2 + y^2 + z^2)^{3/2}}\,\mathbf{k}$$

The gravitational field $\mathbf{F}$ is pictured in Figure 14. ■

EXAMPLE 5 Suppose an electric charge Q is located at the origin. According to Coulomb's Law, the electric force $\mathbf{F}(\mathbf{x})$ exerted by this charge on a charge q located at a point (x, y, z) with position vector $\mathbf{x} = \langle x, y, z \rangle$ is

4
$$\mathbf{F}(\mathbf{x}) = \frac{\varepsilon q Q}{|\mathbf{x}|^3}\,\mathbf{x}$$

where ε is a constant (that depends on the units used). For like charges, we have $qQ > 0$ and the force is repulsive; for unlike charges, we have $qQ < 0$ and the force is attractive. Notice the similarity between Formulas 3 and 4. Both vector fields are examples of **force fields**.

Instead of considering the electric force $\mathbf{F}$, physicists often consider the force per unit charge:

$$\mathbf{E}(\mathbf{x}) = \frac{1}{q}\,\mathbf{F}(\mathbf{x}) = \frac{\varepsilon Q}{|\mathbf{x}|^3}\,\mathbf{x}$$

Then $\mathbf{E}$ is a vector field on $\mathbb{R}^3$ called the **electric field** of Q. ■

Gradient Fields

If f is a scalar function of two variables, recall from Section 14.6 that its gradient ∇f (or grad f) is defined by

$$\nabla f(x, y) = f_x(x, y)\,\mathbf{i} + f_y(x, y)\,\mathbf{j}$$

Therefore ∇f is really a vector field on $\mathbb{R}^2$ and is called a **gradient vector field**. Likewise, if f is a scalar function of three variables, its gradient is a vector field on $\mathbb{R}^3$ given by

$$\nabla f(x, y, z) = f_x(x, y, z)\,\mathbf{i} + f_y(x, y, z)\,\mathbf{j} + f_z(x, y, z)\,\mathbf{k}$$

EXAMPLE 6 Find the gradient vector field of $f(x, y) = x^2y - y^3$. Plot the gradient vector field together with a contour map of f. How are they related?

SOLUTION The gradient vector field is given by

$$\nabla f(x, y) = \frac{\partial f}{\partial x}\,\mathbf{i} + \frac{\partial f}{\partial y}\,\mathbf{j} = 2xy\,\mathbf{i} + (x^2 - 3y^2)\,\mathbf{j}$$

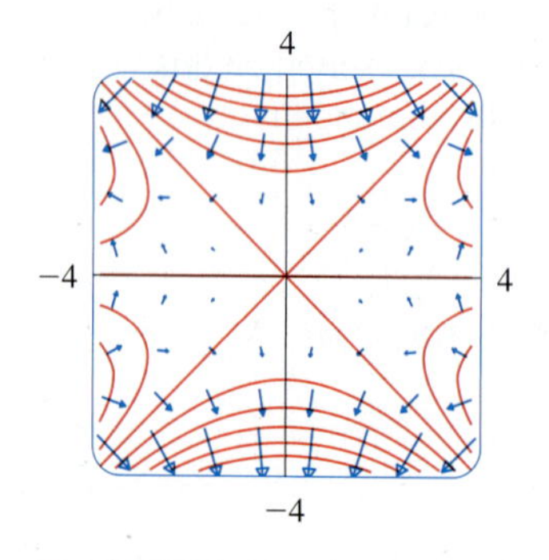

FIGURE 15

Figure 15 shows a contour map of f with the gradient vector field. Notice that the gradient vectors are perpendicular to the level curves, as we would expect from

Section 14.6. Notice also that the gradient vectors are long where the level curves are close to each other and short where the curves are farther apart. That's because the length of the gradient vector is the value of the directional derivative of f and closely spaced level curves indicate a steep graph. ■

A vector field **F** is called a **conservative vector field** if it is the gradient of some scalar function, that is, if there exists a function f such that $\mathbf{F} = \nabla f$. In this situation f is called a **potential function** for **F**.

Not all vector fields are conservative, but such fields do arise frequently in physics. For example, the gravitational field **F** in Example 4 is conservative because if we define

$$f(x, y, z) = \frac{mMG}{\sqrt{x^2 + y^2 + z^2}}$$

then

$$\begin{aligned}\nabla f(x, y, z) &= \frac{\partial f}{\partial x}\mathbf{i} + \frac{\partial f}{\partial y}\mathbf{j} + \frac{\partial f}{\partial z}\mathbf{k} \\ &= \frac{-mMGx}{(x^2 + y^2 + z^2)^{3/2}}\mathbf{i} + \frac{-mMGy}{(x^2 + y^2 + z^2)^{3/2}}\mathbf{j} + \frac{-mMGz}{(x^2 + y^2 + z^2)^{3/2}}\mathbf{k} \\ &= \mathbf{F}(x, y, z)\end{aligned}$$

In Sections 16.3 and 16.5 we will learn how to tell whether or not a given vector field is conservative.

16.1 EXERCISES

1–10 Sketch the vector field **F** by drawing a diagram like Figure 5 or Figure 9.

1. $\mathbf{F}(x, y) = 0.3\,\mathbf{i} - 0.4\,\mathbf{j}$

2. $\mathbf{F}(x, y) = \frac{1}{2}x\,\mathbf{i} + y\,\mathbf{j}$

3. $\mathbf{F}(x, y) = -\frac{1}{2}\,\mathbf{i} + (y - x)\,\mathbf{j}$

4. $\mathbf{F}(x, y) = y\,\mathbf{i} + (x + y)\,\mathbf{j}$

5. $\mathbf{F}(x, y) = \dfrac{y\,\mathbf{i} + x\,\mathbf{j}}{\sqrt{x^2 + y^2}}$

6. $\mathbf{F}(x, y) = \dfrac{y\,\mathbf{i} - x\,\mathbf{j}}{\sqrt{x^2 + y^2}}$

7. $\mathbf{F}(x, y, z) = \mathbf{i}$

8. $\mathbf{F}(x, y, z) = z\,\mathbf{i}$

9. $\mathbf{F}(x, y, z) = -y\,\mathbf{i}$

10. $\mathbf{F}(x, y, z) = \mathbf{i} + \mathbf{k}$

11–14 Match the vector fields **F** with the plots labeled I–IV. Give reasons for your choices.

11. $\mathbf{F}(x, y) = \langle x, -y \rangle$

12. $\mathbf{F}(x, y) = \langle y, x - y \rangle$

13. $\mathbf{F}(x, y) = \langle y, y + 2 \rangle$

14. $\mathbf{F}(x, y) = \langle \cos(x + y), x \rangle$

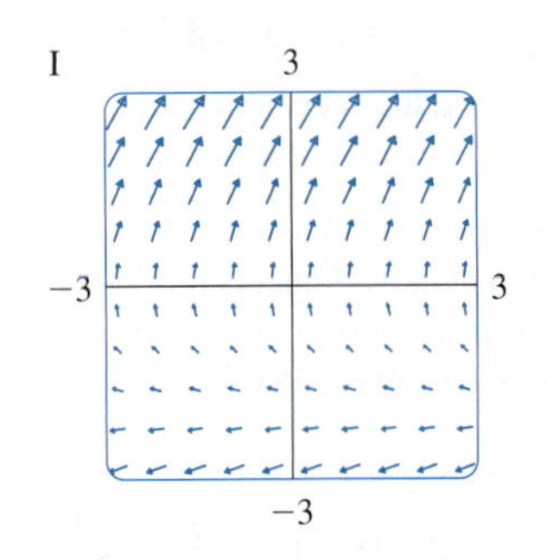

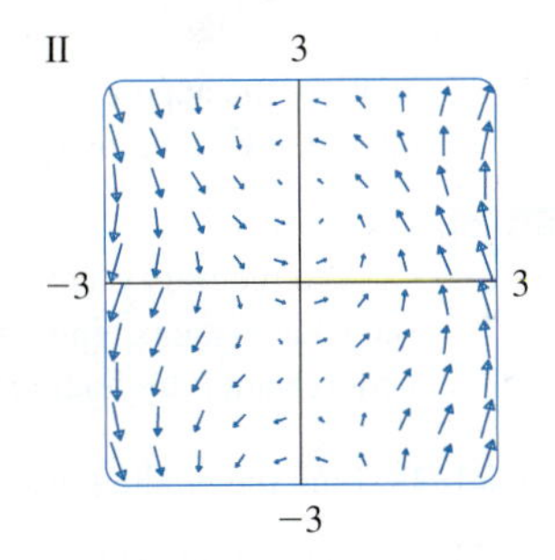

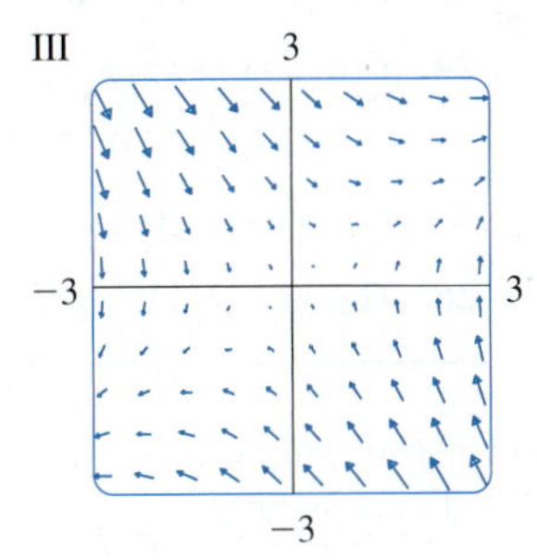

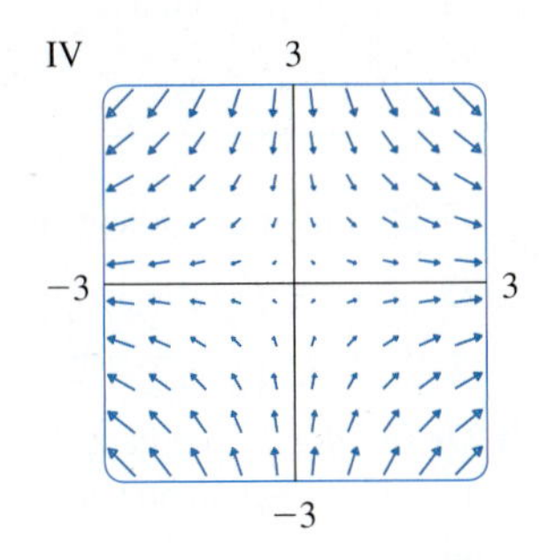

15–18 Match the vector fields $\mathbf{F}$ on $\mathbb{R}^3$ with the plots labeled I–IV. Give reasons for your choices.

15. $\mathbf{F}(x, y, z) = \mathbf{i} + 2\,\mathbf{j} + 3\,\mathbf{k}$

16. $\mathbf{F}(x, y, z) = \mathbf{i} + 2\,\mathbf{j} + z\,\mathbf{k}$

17. $\mathbf{F}(x, y, z) = x\,\mathbf{i} + y\,\mathbf{j} + 3\,\mathbf{k}$

18. $\mathbf{F}(x, y, z) = x\,\mathbf{i} + y\,\mathbf{j} + z\,\mathbf{k}$

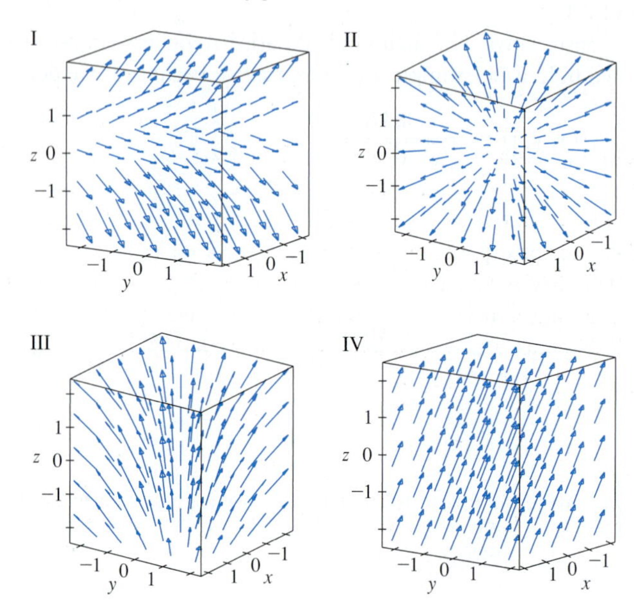

CAS 19. If you have a CAS that plots vector fields (the command is `fieldplot` in Maple and `PlotVectorField` or `VectorPlot` in Mathematica), use it to plot

$$\mathbf{F}(x, y) = (y^2 - 2xy)\,\mathbf{i} + (3xy - 6x^2)\,\mathbf{j}$$

Explain the appearance by finding the set of points (x, y) such that $\mathbf{F}(x, y) = \mathbf{0}$.

CAS 20. Let $\mathbf{F}(\mathbf{x}) = (r^2 - 2r)\mathbf{x}$, where $\mathbf{x} = \langle x, y \rangle$ and $r = |\mathbf{x}|$. Use a CAS to plot this vector field in various domains until you can see what is happening. Describe the appearance of the plot and explain it by finding the points where $\mathbf{F}(\mathbf{x}) = \mathbf{0}$.

21–24 Find the gradient vector field of f.

21. $f(x, y) = y \sin(xy)$

22. $f(s, t) = \sqrt{2s + 3t}$

23. $f(x, y, z) = \sqrt{x^2 + y^2 + z^2}$

24. $f(x, y, z) = x^2 y e^{y/z}$

25–26 Find the gradient vector field ∇f of f and sketch it.

25. $f(x, y) = \frac{1}{2}(x - y)^2$

26. $f(x, y) = \frac{1}{2}(x^2 - y^2)$

CAS 27–28 Plot the gradient vector field of f together with a contour map of f. Explain how they are related to each other.

27. $f(x, y) = \ln(1 + x^2 + 2y^2)$

28. $f(x, y) = \cos x - 2 \sin y$

29–32 Match the functions f with the plots of their gradient vector fields labeled I–IV. Give reasons for your choices.

29. $f(x, y) = x^2 + y^2$

30. $f(x, y) = x(x + y)$

31. $f(x, y) = (x + y)^2$

32. $f(x, y) = \sin\sqrt{x^2 + y^2}$

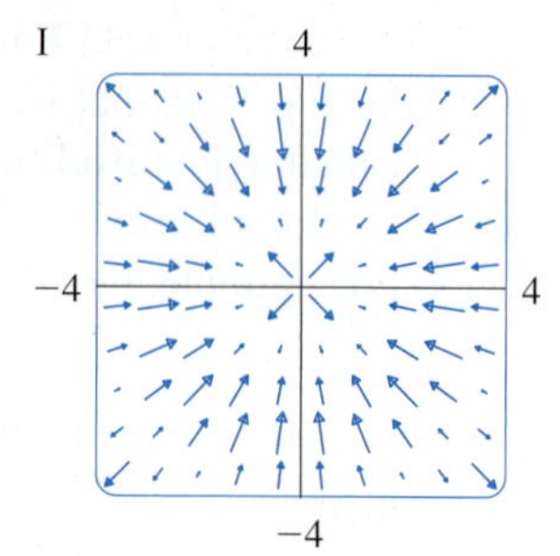

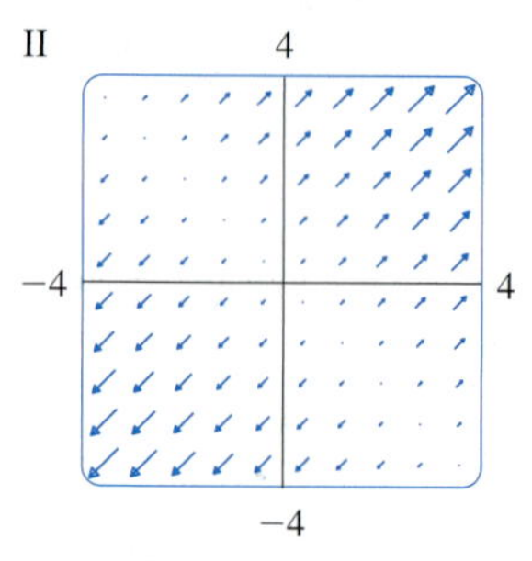

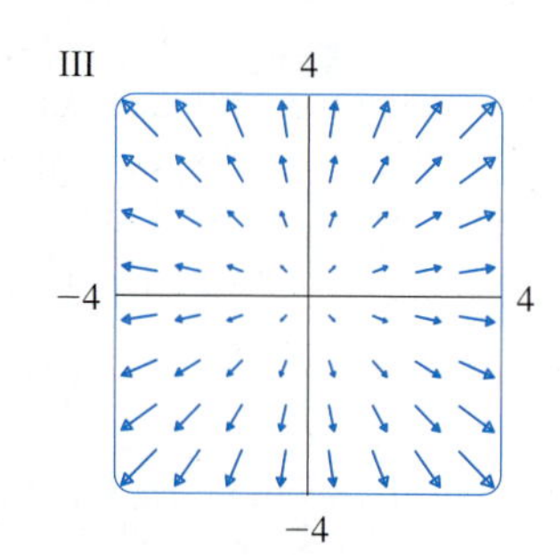

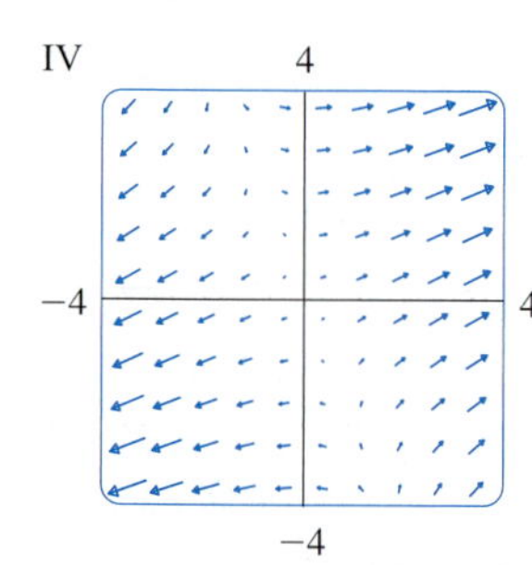

33. A particle moves in a velocity field $\mathbf{V}(x, y) = \langle x^2, x + y^2 \rangle$. If it is at position $(2, 1)$ at time $t = 3$, estimate its location at time $t = 3.01$.

34. At time $t = 1$, a particle is located at position $(1, 3)$. If it moves in a velocity field

$$\mathbf{F}(x, y) = \langle xy - 2, y^2 - 10 \rangle$$

find its approximate location at time $t = 1.05$.

35. The **flow lines** (or **streamlines**) of a vector field are the paths followed by a particle whose velocity field is the given vector field. Thus the vectors in a vector field are tangent to the flow lines.

(a) Use a sketch of the vector field $\mathbf{F}(x, y) = x\,\mathbf{i} - y\,\mathbf{j}$ to draw some flow lines. From your sketches, can you guess the equations of the flow lines?

(b) If parametric equations of a flow line are $x = x(t)$, $y = y(t)$, explain why these functions satisfy the differential equations $dx/dt = x$ and $dy/dt = -y$. Then solve the differential equations to find an equation of the flow line that passes through the point $(1, 1)$.

36. (a) Sketch the vector field $\mathbf{F}(x, y) = \mathbf{i} + x\,\mathbf{j}$ and then sketch some flow lines. What shape do these flow lines appear to have?

(b) If parametric equations of the flow lines are $x = x(t)$, $y = y(t)$, what differential equations do these functions satisfy? Deduce that $dy/dx = x$.

(c) If a particle starts at the origin in the velocity field given by $\mathbf{F}$, find an equation of the path it follows.

16.2 Line Integrals

In this section we define an integral that is similar to a single integral except that instead of integrating over an interval $[a, b]$, we integrate over a curve C. Such integrals are called *line integrals*, although "curve integrals" would be better terminology. They were invented in the early 19th century to solve problems involving fluid flow, forces, electricity, and magnetism.

We start with a plane curve C given by the parametric equations

$$\boxed{1} \qquad x = x(t) \qquad y = y(t) \qquad a \leqslant t \leqslant b$$

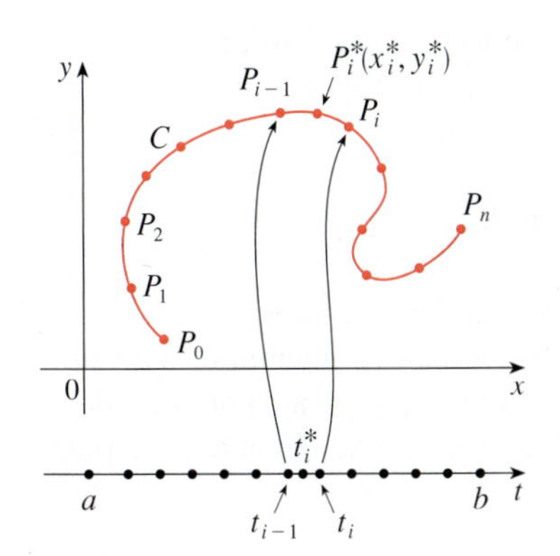

FIGURE 1

or, equivalently, by the vector equation $\mathbf{r}(t) = x(t)\,\mathbf{i} + y(t)\,\mathbf{j}$, and we assume that C is a smooth curve. [This means that $\mathbf{r}'$ is continuous and $\mathbf{r}'(t) \neq \mathbf{0}$. See Section 13.3.] If we divide the parameter interval $[a, b]$ into n subintervals $[t_{i-1}, t_i]$ of equal width and we let $x_i = x(t_i)$ and $y_i = y(t_i)$, then the corresponding points $P_i(x_i, y_i)$ divide C into n subarcs with lengths $\Delta s_1, \Delta s_2, \ldots, \Delta s_n$. (See Figure 1.) We choose any point $P_i^*(x_i^*, y_i^*)$ in the ith subarc. (This corresponds to a point t_i^* in $[t_{i-1}, t_i]$.) Now if f is any function of two variables whose domain includes the curve C, we evaluate f at the point (x_i^*, y_i^*), multiply by the length Δs_i of the subarc, and form the sum

$$\sum_{i=1}^{n} f(x_i^*, y_i^*)\,\Delta s_i$$

which is similar to a Riemann sum. Then we take the limit of these sums and make the following definition by analogy with a single integral.

2 Definition If f is defined on a smooth curve C given by Equations 1, then the **line integral of f along C** is

$$\int_C f(x, y)\,ds = \lim_{n \to \infty} \sum_{i=1}^{n} f(x_i^*, y_i^*)\,\Delta s_i$$

if this limit exists.

In Section 10.2 we found that the length of C is

$$L = \int_a^b \sqrt{\left(\frac{dx}{dt}\right)^2 + \left(\frac{dy}{dt}\right)^2}\,dt$$

A similar type of argument can be used to show that if f is a continuous function, then the limit in Definition 2 always exists and the following formula can be used to evaluate the line integral:

$$\boxed{3} \qquad \int_C f(x, y)\,ds = \int_a^b f(x(t), y(t))\sqrt{\left(\frac{dx}{dt}\right)^2 + \left(\frac{dy}{dt}\right)^2}\,dt$$

The value of the line integral does not depend on the parametrization of the curve, provided that the curve is traversed exactly once as t increases from a to b.

The arc length function s is discussed in Section 13.3.

If $s(t)$ is the length of C between $\mathbf{r}(a)$ and $\mathbf{r}(t)$, then

$$\frac{ds}{dt} = \sqrt{\left(\frac{dx}{dt}\right)^2 + \left(\frac{dy}{dt}\right)^2}$$

So the way to remember Formula 3 is to express everything in terms of the parameter t: Use the parametric equations to express x and y in terms of t and write ds as

$$ds = \sqrt{\left(\frac{dx}{dt}\right)^2 + \left(\frac{dy}{dt}\right)^2}\, dt$$

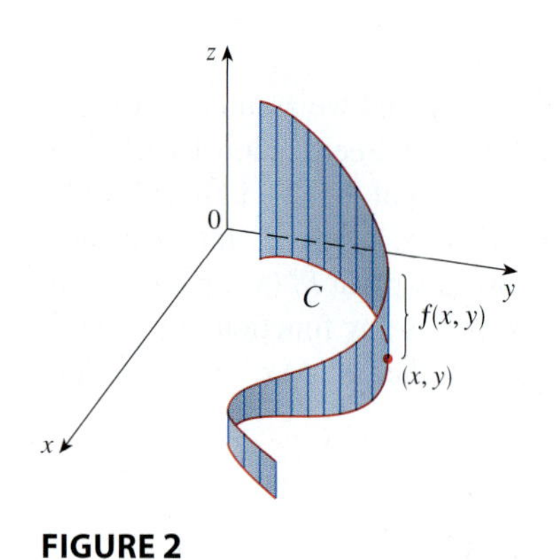

FIGURE 2

In the special case where C is the line segment that joins $(a, 0)$ to $(b, 0)$, using x as the parameter, we can write the parametric equations of C as follows: $x = x$, $y = 0$, $a \leqslant x \leqslant b$. Formula 3 then becomes

$$\int_C f(x, y)\, ds = \int_a^b f(x, 0)\, dx$$

and so the line integral reduces to an ordinary single integral in this case.

Just as for an ordinary single integral, we can interpret the line integral of a *positive* function as an area. In fact, if $f(x, y) \geqslant 0$, $\int_C f(x, y)\, ds$ represents the area of one side of the "fence" or "curtain" in Figure 2, whose base is C and whose height above the point (x, y) is $f(x, y)$.

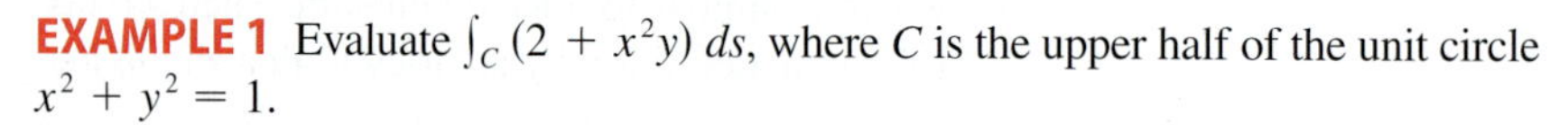

EXAMPLE 1 Evaluate $\int_C (2 + x^2y)\, ds$, where C is the upper half of the unit circle $x^2 + y^2 = 1$.

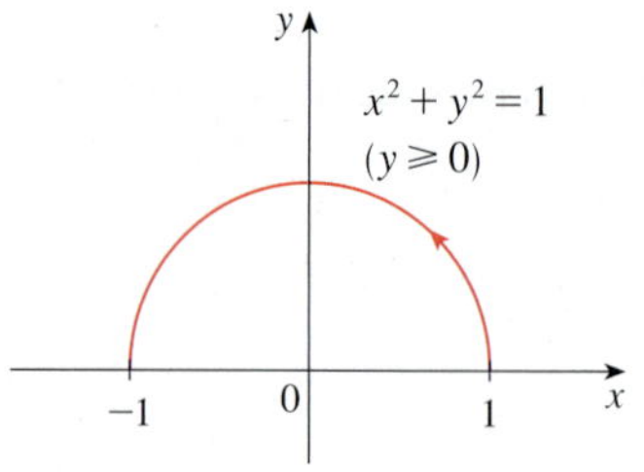

SOLUTION In order to use Formula 3, we first need parametric equations to represent C. Recall that the unit circle can be parametrized by means of the equations

$$x = \cos t \qquad y = \sin t$$

and the upper half of the circle is described by the parameter interval $0 \leqslant t \leqslant \pi$. (See Figure 3.) Therefore Formula 3 gives

FIGURE 3

$$\begin{aligned}\int_C (2 + x^2y)\, ds &= \int_0^\pi (2 + \cos^2 t \sin t)\sqrt{\left(\frac{dx}{dt}\right)^2 + \left(\frac{dy}{dt}\right)^2}\, dt \\ &= \int_0^\pi (2 + \cos^2 t \sin t)\sqrt{\sin^2 t + \cos^2 t}\, dt \\ &= \int_0^\pi (2 + \cos^2 t \sin t)\, dt = \left[2t - \frac{\cos^3 t}{3}\right]_0^\pi \\ &= 2\pi + \tfrac{2}{3}\end{aligned}$$

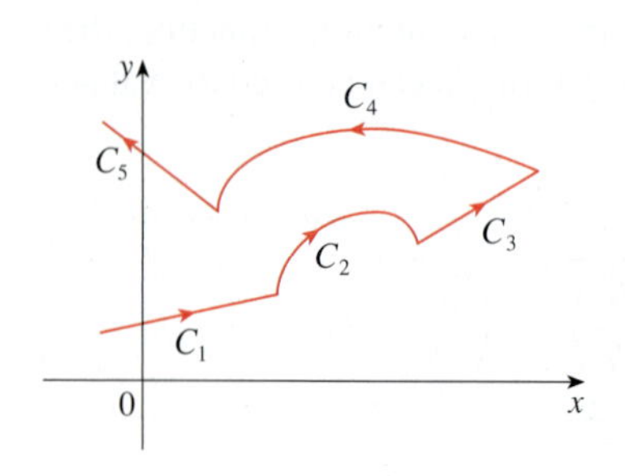

FIGURE 4
A piecewise-smooth curve

Suppose now that C is a **piecewise-smooth curve**; that is, C is a union of a finite number of smooth curves $C_1, C_2, \ldots, C_n$, where, as illustrated in Figure 4, the initial point of C_{i+1} is the terminal point of C_i. Then we define the integral of f along C as the sum of the integrals of f along each of the smooth pieces of C:

$$\int_C f(x, y)\, ds = \int_{C_1} f(x, y)\, ds + \int_{C_2} f(x, y)\, ds + \cdots + \int_{C_n} f(x, y)\, ds$$

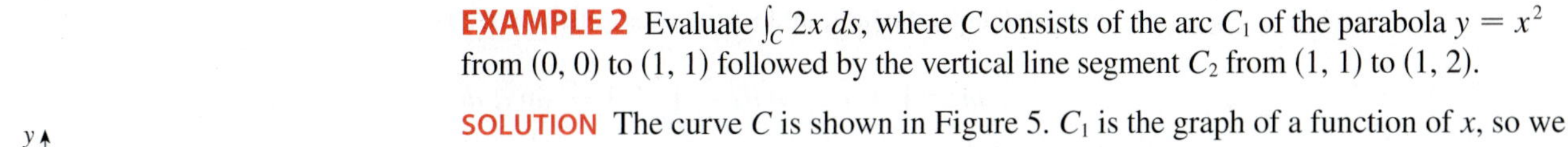

EXAMPLE 2 Evaluate $\int_C 2x\, ds$, where C consists of the arc C_1 of the parabola $y = x^2$ from $(0, 0)$ to $(1, 1)$ followed by the vertical line segment C_2 from $(1, 1)$ to $(1, 2)$.

SOLUTION The curve C is shown in Figure 5. C_1 is the graph of a function of x, so we can choose x as the parameter and the equations for C_1 become

$$x = x \qquad y = x^2 \qquad 0 \leqslant x \leqslant 1$$

Therefore

$$\int_{C_1} 2x\, ds = \int_0^1 2x \sqrt{\left(\frac{dx}{dx}\right)^2 + \left(\frac{dy}{dx}\right)^2}\, dx = \int_0^1 2x\sqrt{1 + 4x^2}\, dx$$

$$= \tfrac{1}{4} \cdot \tfrac{2}{3}(1 + 4x^2)^{3/2}\Big]_0^1 = \frac{5\sqrt{5} - 1}{6}$$

FIGURE 5

$C = C_1 \cup C_2$

On C_2 we choose y as the parameter, so the equations of C_2 are

$$x = 1 \qquad y = y \qquad 1 \leqslant y \leqslant 2$$

and
$$\int_{C_2} 2x\, ds = \int_1^2 2(1)\sqrt{\left(\frac{dx}{dy}\right)^2 + \left(\frac{dy}{dy}\right)^2}\, dy = \int_1^2 2\, dy = 2$$

Thus
$$\int_C 2x\, ds = \int_{C_1} 2x\, ds + \int_{C_2} 2x\, ds = \frac{5\sqrt{5} - 1}{6} + 2$$
■

Any physical interpretation of a line integral $\int_C f(x, y)\, ds$ depends on the physical interpretation of the function f. Suppose that $\rho(x, y)$ represents the linear density at a point (x, y) of a thin wire shaped like a curve C. Then the mass of the part of the wire from P_{i-1} to P_i in Figure 1 is approximately $\rho(x_i^*, y_i^*)\, \Delta s_i$ and so the total mass of the wire is approximately $\Sigma\, \rho(x_i^*, y_i^*)\, \Delta s_i$. By taking more and more points on the curve, we obtain the **mass** m of the wire as the limiting value of these approximations:

$$m = \lim_{n \to \infty} \sum_{i=1}^{n} \rho(x_i^*, y_i^*)\, \Delta s_i = \int_C \rho(x, y)\, ds$$

[For example, if $f(x, y) = 2 + x^2 y$ represents the density of a semicircular wire, then the integral in Example 1 would represent the mass of the wire.] The **center of mass** of the wire with density function ρ is located at the point $(\bar{x}, \bar{y})$, where

$$\boxed{4} \qquad \bar{x} = \frac{1}{m}\int_C x\rho(x, y)\, ds \qquad \bar{y} = \frac{1}{m}\int_C y\rho(x, y)\, ds$$

Other physical interpretations of line integrals will be discussed later in this chapter.

EXAMPLE 3 A wire takes the shape of the semicircle $x^2 + y^2 = 1$, $y \geqslant 0$, and is thicker near its base than near the top. Find the center of mass of the wire if the linear density at any point is proportional to its distance from the line $y = 1$.

SOLUTION As in Example 1 we use the parametrization $x = \cos t$, $y = \sin t$, $0 \leqslant t \leqslant \pi$, and find that $ds = dt$. The linear density is

$$\rho(x, y) = k(1 - y)$$

where k is a constant, and so the mass of the wire is

$$m = \int_C k(1 - y)\, ds = \int_0^{\pi} k(1 - \sin t)\, dt = k\Big[t + \cos t\Big]_0^{\pi} = k(\pi - 2)$$

From Equations 4 we have

$$\bar{y} = \frac{1}{m}\int_C y\rho(x, y)\, ds = \frac{1}{k(\pi - 2)}\int_C y\, k(1 - y)\, ds$$

$$= \frac{1}{\pi - 2}\int_0^{\pi} (\sin t - \sin^2 t)\, dt = \frac{1}{\pi - 2}\Big[-\cos t - \tfrac{1}{2}t + \tfrac{1}{4}\sin 2t\Big]_0^{\pi}$$

$$= \frac{4 - \pi}{2(\pi - 2)}$$

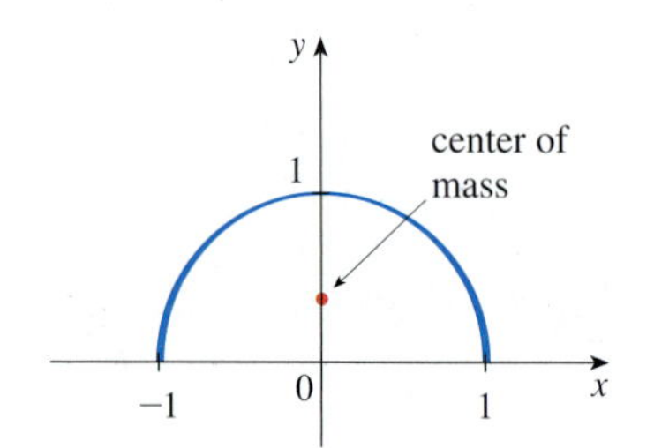

FIGURE 6

By symmetry we see that $\bar{x} = 0$, so the center of mass is

$$\left(0, \frac{4 - \pi}{2(\pi - 2)}\right) \approx (0, 0.38)$$

See Figure 6. ■

Two other line integrals are obtained by replacing Δs_i by either $\Delta x_i = x_i - x_{i-1}$ or $\Delta y_i = y_i - y_{i-1}$ in Definition 2. They are called the **line integrals of f along C with respect to x and y**:

5 $$\int_C f(x, y)\, dx = \lim_{n \to \infty} \sum_{i=1}^{n} f(x_i^*, y_i^*)\, \Delta x_i$$

6 $$\int_C f(x, y)\, dy = \lim_{n \to \infty} \sum_{i=1}^{n} f(x_i^*, y_i^*)\, \Delta y_i$$

When we want to distinguish the original line integral $\int_C f(x, y)\, ds$ from those in Equations 5 and 6, we call it the **line integral with respect to arc length**.

The following formulas say that line integrals with respect to x and y can also be evaluated by expressing everything in terms of t: $x = x(t)$, $y = y(t)$, $dx = x'(t)\, dt$, $dy = y'(t)\, dt$.

7 $$\int_C f(x, y)\, dx = \int_a^b f(x(t), y(t))\, x'(t)\, dt$$

$$\int_C f(x, y)\, dy = \int_a^b f(x(t), y(t))\, y'(t)\, dt$$

It frequently happens that line integrals with respect to x and y occur together. When this happens, it's customary to abbreviate by writing

$$\int_C P(x, y)\, dx + \int_C Q(x, y)\, dy = \int_C P(x, y)\, dx + Q(x, y)\, dy$$

When we are setting up a line integral, sometimes the most difficult thing is to think of a parametric representation for a curve whose geometric description is given. In particular, we often need to parametrize a line segment, so it's useful to remember that a vector representation of the line segment that starts at $\mathbf{r}_0$ and ends at $\mathbf{r}_1$ is given by

$$\boxed{8} \qquad \mathbf{r}(t) = (1 - t)\mathbf{r}_0 + t\,\mathbf{r}_1 \qquad 0 \leqslant t \leqslant 1$$

(See Equation 12.5.4.)

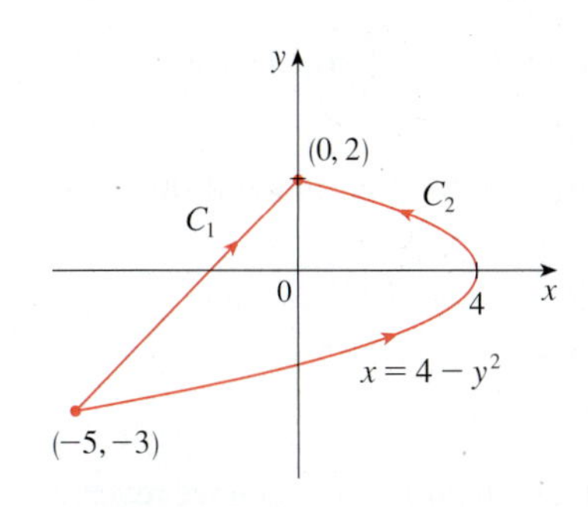

FIGURE 7

EXAMPLE 4 Evaluate $\int_C y^2\,dx + x\,dy$, where (a) $C = C_1$ is the line segment from $(-5, -3)$ to $(0, 2)$ and (b) $C = C_2$ is the arc of the parabola $x = 4 - y^2$ from $(-5, -3)$ to $(0, 2)$. (See Figure 7.)

SOLUTION

(a) A parametric representation for the line segment is

$$x = 5t - 5 \qquad y = 5t - 3 \qquad 0 \leqslant t \leqslant 1$$

(Use Equation 8 with $\mathbf{r}_0 = \langle -5, -3 \rangle$ and $\mathbf{r}_1 = \langle 0, 2 \rangle$.) Then $dx = 5\,dt$, $dy = 5\,dt$, and Formulas 7 give

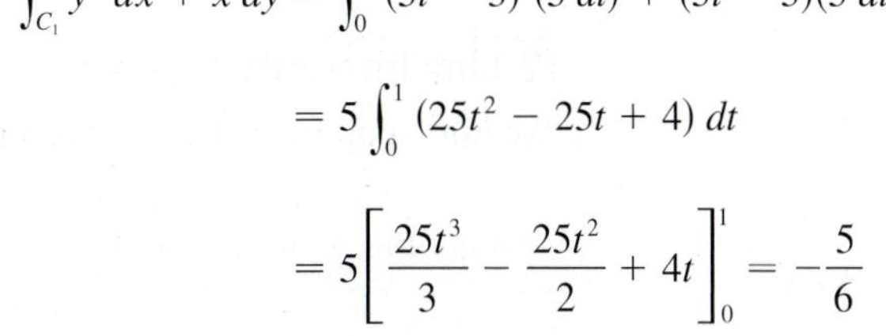

$$\int_{C_1} y^2\,dx + x\,dy = \int_0^1 (5t - 3)^2(5\,dt) + (5t - 5)(5\,dt)$$

$$= 5\int_0^1 (25t^2 - 25t + 4)\,dt$$

$$= 5\left[\frac{25t^3}{3} - \frac{25t^2}{2} + 4t\right]_0^1 = -\frac{5}{6}$$

(b) Since the parabola is given as a function of y, let's take y as the parameter and write C_2 as

$$x = 4 - y^2 \qquad y = y \qquad -3 \leqslant y \leqslant 2$$

Then $dx = -2y\,dy$ and by Formulas 7 we have

$$\int_{C_2} y^2\,dx + x\,dy = \int_{-3}^{2} y^2(-2y)\,dy + (4 - y^2)\,dy$$

$$= \int_{-3}^{2} (-2y^3 - y^2 + 4)\,dy$$

$$= \left[-\frac{y^4}{2} - \frac{y^3}{3} + 4y\right]_{-3}^{2} = 40\tfrac{5}{6} \qquad \blacksquare$$

Notice that we got different answers in parts (a) and (b) of Example 4 even though the two curves had the same endpoints. Thus, in general, the value of a line integral depends not just on the endpoints of the curve but also on the path. (But see Section 16.3 for conditions under which the integral is independent of the path.)

Notice also that the answers in Example 4 depend on the direction, or orientation, of the curve. If $-C_1$ denotes the line segment from $(0, 2)$ to $(-5, -3)$, you can verify, using the parametrization

$$x = -5t \qquad y = 2 - 5t \qquad 0 \leqslant t \leqslant 1$$

that

$$\int_{-C_1} y^2\,dx + x\,dy = \tfrac{5}{6}$$

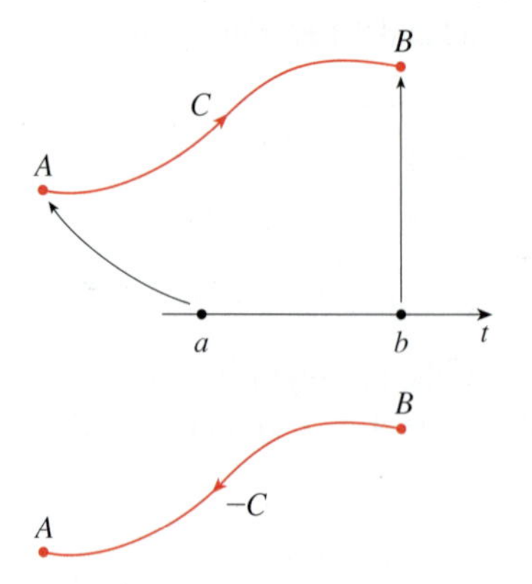

FIGURE 8

In general, a given parametrization $x = x(t)$, $y = y(t)$, $a \leqslant t \leqslant b$, determines an **orientation** of a curve C, with the positive direction corresponding to increasing values of the parameter t. (See Figure 8, where the initial point A corresponds to the parameter value a and the terminal point B corresponds to $t = b$.)

If $-C$ denotes the curve consisting of the same points as C but with the opposite orientation (from initial point B to terminal point A in Figure 8), then we have

$$\int_{-C} f(x, y)\, dx = -\int_C f(x, y)\, dx \qquad \int_{-C} f(x, y)\, dy = -\int_C f(x, y)\, dy$$

But if we integrate with respect to arc length, the value of the line integral does *not* change when we reverse the orientation of the curve:

$$\int_{-C} f(x, y)\, ds = \int_C f(x, y)\, ds$$

This is because Δs_i is always positive, whereas Δx_i and Δy_i change sign when we reverse the orientation of C.

Line Integrals in Space

We now suppose that C is a smooth space curve given by the parametric equations

$$x = x(t) \qquad y = y(t) \qquad z = z(t) \qquad a \leqslant t \leqslant b$$

or by a vector equation $\mathbf{r}(t) = x(t)\,\mathbf{i} + y(t)\,\mathbf{j} + z(t)\,\mathbf{k}$. If f is a function of three variables that is continuous on some region containing C, then we define the **line integral of f along C** (with respect to arc length) in a manner similar to that for plane curves:

$$\int_C f(x, y, z)\, ds = \lim_{n \to \infty} \sum_{i=1}^{n} f(x_i^*, y_i^*, z_i^*)\, \Delta s_i$$

We evaluate it using a formula similar to Formula 3:

$$\boxed{9} \quad \int_C f(x, y, z)\, ds = \int_a^b f(x(t), y(t), z(t)) \sqrt{\left(\frac{dx}{dt}\right)^2 + \left(\frac{dy}{dt}\right)^2 + \left(\frac{dz}{dt}\right)^2}\, dt$$

Observe that the integrals in both Formulas 3 and 9 can be written in the more compact vector notation

$$\int_a^b f(\mathbf{r}(t))\, |\mathbf{r}'(t)|\, dt$$

For the special case $f(x, y, z) = 1$, we get

$$\int_C ds = \int_a^b |\mathbf{r}'(t)|\, dt = L$$

where L is the length of the curve C (see Formula 13.3.3).

Line integrals along C with respect to x, y, and z can also be defined. For example,

$$\int_C f(x, y, z)\, dz = \lim_{n \to \infty} \sum_{i=1}^{n} f(x_i^*, y_i^*, z_i^*)\, \Delta z_i$$

$$= \int_a^b f(x(t), y(t), z(t))\, z'(t)\, dt$$

Therefore, as with line integrals in the plane, we evaluate integrals of the form

10 $$\int_C P(x, y, z)\, dx + Q(x, y, z)\, dy + R(x, y, z)\, dz$$

by expressing everything (x, y, z, dx, dy, dz) in terms of the parameter t.

EXAMPLE 5 Evaluate $\int_C y \sin z\, ds$, where C is the circular helix given by the equations $x = \cos t$, $y = \sin t$, $z = t$, $0 \leqslant t \leqslant 2\pi$. (See Figure 9.)

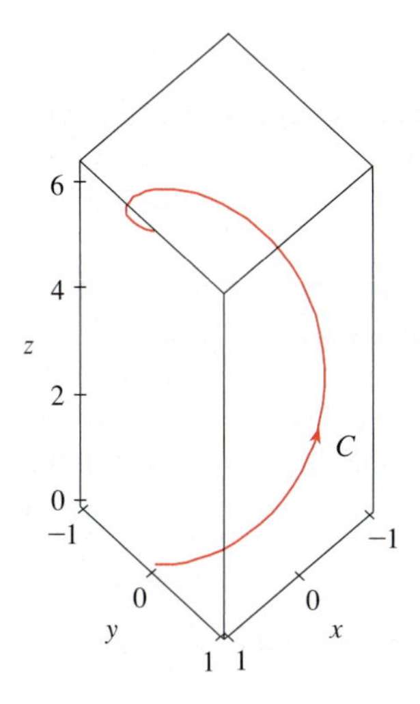

FIGURE 9

SOLUTION Formula 9 gives

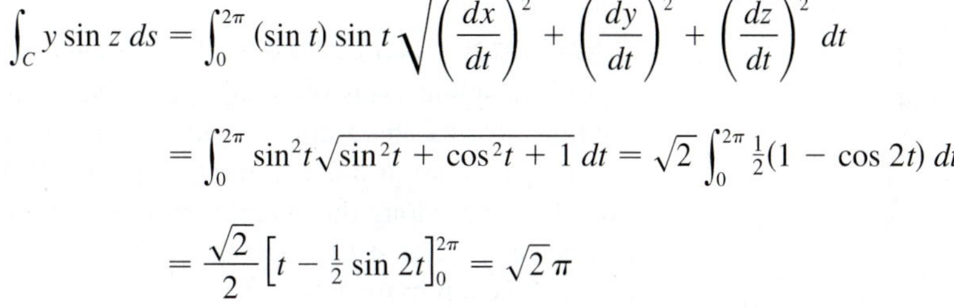

$$\int_C y \sin z\, ds = \int_0^{2\pi} (\sin t) \sin t \sqrt{\left(\frac{dx}{dt}\right)^2 + \left(\frac{dy}{dt}\right)^2 + \left(\frac{dz}{dt}\right)^2}\, dt$$

$$= \int_0^{2\pi} \sin^2 t \sqrt{\sin^2 t + \cos^2 t + 1}\, dt = \sqrt{2} \int_0^{2\pi} \tfrac{1}{2}(1 - \cos 2t)\, dt$$

$$= \frac{\sqrt{2}}{2} \Big[t - \tfrac{1}{2} \sin 2t\Big]_0^{2\pi} = \sqrt{2}\, \pi$$ ■

EXAMPLE 6 Evaluate $\int_C y\, dx + z\, dy + x\, dz$, where C consists of the line segment C_1 from $(2, 0, 0)$ to $(3, 4, 5)$, followed by the vertical line segment C_2 from $(3, 4, 5)$ to $(3, 4, 0)$.

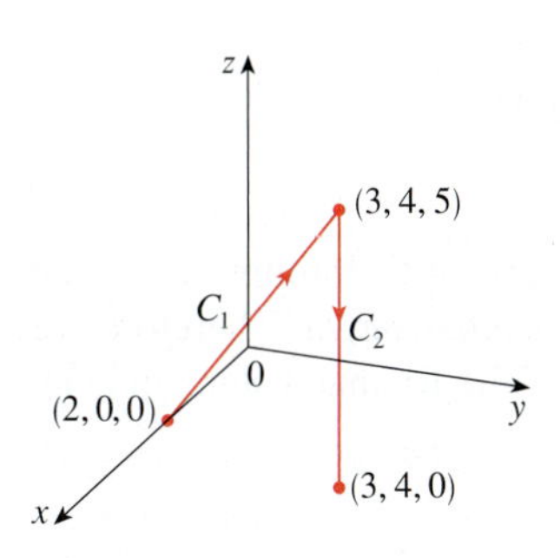

FIGURE 10

SOLUTION The curve C is shown in Figure 10. Using Equation 8, we write C_1 as

$$\mathbf{r}(t) = (1 - t)\langle 2, 0, 0\rangle + t\langle 3, 4, 5\rangle = \langle 2 + t, 4t, 5t\rangle$$

or, in parametric form, as

$$x = 2 + t \qquad y = 4t \qquad z = 5t \qquad 0 \leqslant t \leqslant 1$$

Thus

$$\int_{C_1} y\, dx + z\, dy + x\, dz = \int_0^1 (4t)\, dt + (5t)4\, dt + (2 + t)5\, dt$$

$$= \int_0^1 (10 + 29t)\, dt = 10t + 29\frac{t^2}{2}\Big]_0^1 = 24.5$$

Likewise, C_2 can be written in the form

$$\mathbf{r}(t) = (1 - t)\langle 3, 4, 5\rangle + t\langle 3, 4, 0\rangle = \langle 3, 4, 5 - 5t\rangle$$

or $$x = 3 \qquad y = 4 \qquad z = 5 - 5t \qquad 0 \leqslant t \leqslant 1$$

Then $dx = 0 = dy$, so

$$\int_{C_2} y\,dx + z\,dy + x\,dz = \int_0^1 3(-5)\,dt = -15$$

Adding the values of these integrals, we obtain

$$\int_C y\,dx + z\,dy + x\,dz = 24.5 - 15 = 9.5$$

Line Integrals of Vector Fields

Recall from Section 5.4 that the work done by a variable force $f(x)$ in moving a particle from a to b along the x-axis is $W = \int_a^b f(x)\,dx$. Then in Section 12.3 we found that the work done by a constant force $\mathbf{F}$ in moving an object from a point P to another point Q in space is $W = \mathbf{F} \cdot \mathbf{D}$, where $\mathbf{D} = \overrightarrow{PQ}$ is the displacement vector.

Now suppose that $\mathbf{F} = P\,\mathbf{i} + Q\,\mathbf{j} + R\,\mathbf{k}$ is a continuous force field on $\mathbb{R}^3$, such as the gravitational field of Example 16.1.4 or the electric force field of Example 16.1.5. (A force field on $\mathbb{R}^2$ could be regarded as a special case where $R = 0$ and P and Q depend only on x and y.) We wish to compute the work done by this force in moving a particle along a smooth curve C.

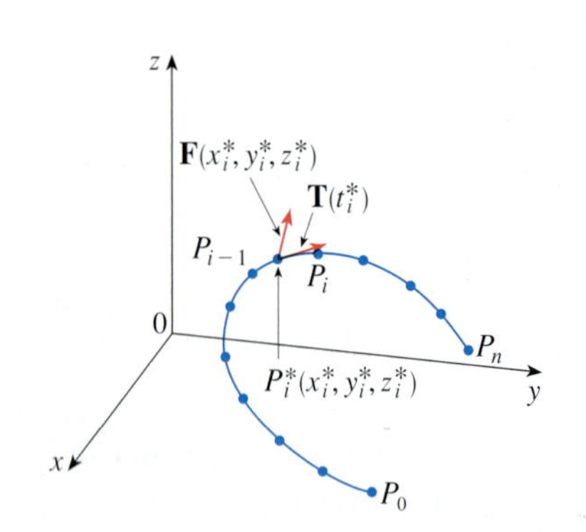

FIGURE 11

We divide C into subarcs $P_{i-1}P_i$ with lengths Δs_i by dividing the parameter interval $[a, b]$ into subintervals of equal width. (See Figure 1 for the two-dimensional case or Figure 11 for the three-dimensional case.) Choose a point $P_i^*(x_i^*, y_i^*, z_i^*)$ on the ith subarc corresponding to the parameter value t_i^*. If Δs_i is small, then as the particle moves from P_{i-1} to P_i along the curve, it proceeds approximately in the direction of $\mathbf{T}(t_i^*)$, the unit tangent vector at P_i^*. Thus the work done by the force $\mathbf{F}$ in moving the particle from P_{i-1} to P_i is approximately

$$\mathbf{F}(x_i^*, y_i^*, z_i^*) \cdot [\Delta s_i\,\mathbf{T}(t_i^*)] = [\mathbf{F}(x_i^*, y_i^*, z_i^*) \cdot \mathbf{T}(t_i^*)]\,\Delta s_i$$

and the total work done in moving the particle along C is approximately

$$\boxed{11} \qquad \sum_{i=1}^{n} [\mathbf{F}(x_i^*, y_i^*, z_i^*) \cdot \mathbf{T}(x_i^*, y_i^*, z_i^*)]\,\Delta s_i$$

where $\mathbf{T}(x, y, z)$ is the unit tangent vector at the point (x, y, z) on C. Intuitively, we see that these approximations ought to become better as n becomes larger. Therefore we define the **work** W done by the force field $\mathbf{F}$ as the limit of the Riemann sums in (11), namely,

$$\boxed{12} \qquad W = \int_C \mathbf{F}(x, y, z) \cdot \mathbf{T}(x, y, z)\,ds = \int_C \mathbf{F} \cdot \mathbf{T}\,ds$$

Equation 12 says that *work is the line integral with respect to arc length of the tangential component of the force.*

If the curve C is given by the vector equation $\mathbf{r}(t) = x(t)\,\mathbf{i} + y(t)\,\mathbf{j} + z(t)\,\mathbf{k}$, then $\mathbf{T}(t) = \mathbf{r}'(t)/|\mathbf{r}'(t)|$, so using Equation 9 we can rewrite Equation 12 in the form

$$W = \int_a^b \left[\mathbf{F}(\mathbf{r}(t)) \cdot \frac{\mathbf{r}'(t)}{|\mathbf{r}'(t)|}\right] |\mathbf{r}'(t)|\,dt = \int_a^b \mathbf{F}(\mathbf{r}(t)) \cdot \mathbf{r}'(t)\,dt$$

This integral is often abbreviated as $\int_C \mathbf{F} \cdot d\mathbf{r}$ and occurs in other areas of physics as well. Therefore we make the following definition for the line integral of *any* continuous vector field.

13 Definition Let $\mathbf{F}$ be a continuous vector field defined on a smooth curve C given by a vector function $\mathbf{r}(t)$, $a \leqslant t \leqslant b$. Then the **line integral of F along C** is

$$\int_C \mathbf{F} \cdot d\mathbf{r} = \int_a^b \mathbf{F}(\mathbf{r}(t)) \cdot \mathbf{r}'(t)\, dt = \int_C \mathbf{F} \cdot \mathbf{T}\, ds$$

When using Definition 13, bear in mind that $\mathbf{F}(\mathbf{r}(t))$ is just an abbreviation for the vector field $\mathbf{F}(x(t), y(t), z(t))$, so we evaluate $\mathbf{F}(\mathbf{r}(t))$ simply by putting $x = x(t)$, $y = y(t)$, and $z = z(t)$ in the expression for $\mathbf{F}(x, y, z)$. Notice also that we can formally write $d\mathbf{r} = \mathbf{r}'(t)\, dt$.

Figure 12 shows the force field and the curve in Example 7. The work done is negative because the field impedes movement along the curve.

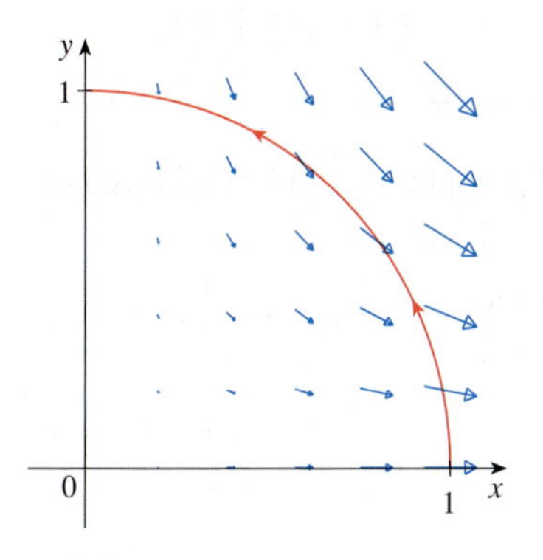

FIGURE 12

EXAMPLE 7 Find the work done by the force field $\mathbf{F}(x, y) = x^2\, \mathbf{i} - xy\, \mathbf{j}$ in moving a particle along the quarter-circle $\mathbf{r}(t) = \cos t\, \mathbf{i} + \sin t\, \mathbf{j}$, $0 \leqslant t \leqslant \pi/2$.

SOLUTION Since $x = \cos t$ and $y = \sin t$, we have

$$\mathbf{F}(\mathbf{r}(t)) = \cos^2 t\, \mathbf{i} - \cos t \sin t\, \mathbf{j}$$

and
$$\mathbf{r}'(t) = -\sin t\, \mathbf{i} + \cos t\, \mathbf{j}$$

Therefore the work done is

$$\int_C \mathbf{F} \cdot d\mathbf{r} = \int_0^{\pi/2} \mathbf{F}(\mathbf{r}(t)) \cdot \mathbf{r}'(t)\, dt = \int_0^{\pi/2} (-2\cos^2 t \sin t)\, dt$$

$$= 2\, \frac{\cos^3 t}{3}\bigg]_0^{\pi/2} = -\frac{2}{3}$$

NOTE Even though $\int_C \mathbf{F} \cdot d\mathbf{r} = \int_C \mathbf{F} \cdot \mathbf{T}\, ds$ and integrals with respect to arc length are unchanged when orientation is reversed, it is still true that

$$\int_{-C} \mathbf{F} \cdot d\mathbf{r} = -\int_C \mathbf{F} \cdot d\mathbf{r}$$

because the unit tangent vector $\mathbf{T}$ is replaced by its negative when C is replaced by $-C$.

Figure 13 shows the twisted cubic C in Example 8 and some typical vectors acting at three points on C.

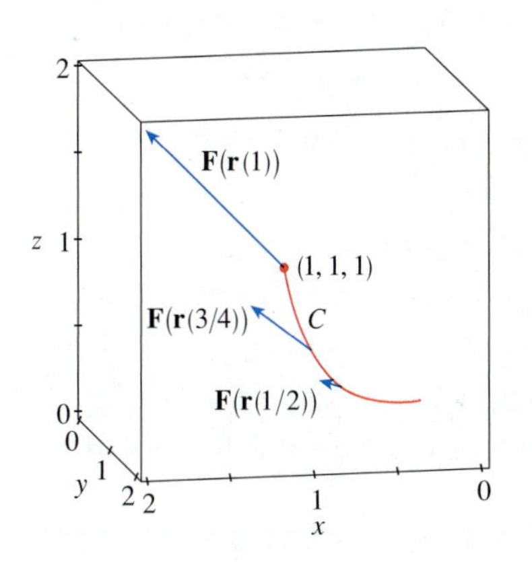

FIGURE 13

EXAMPLE 8 Evaluate $\int_C \mathbf{F} \cdot d\mathbf{r}$, where $\mathbf{F}(x, y, z) = xy\, \mathbf{i} + yz\, \mathbf{j} + zx\, \mathbf{k}$ and C is the twisted cubic given by

$$x = t \qquad y = t^2 \qquad z = t^3 \qquad 0 \leqslant t \leqslant 1$$

SOLUTION We have

$$\mathbf{r}(t) = t\, \mathbf{i} + t^2\, \mathbf{j} + t^3\, \mathbf{k}$$

$$\mathbf{r}'(t) = \mathbf{i} + 2t\, \mathbf{j} + 3t^2\, \mathbf{k}$$

$$\mathbf{F}(\mathbf{r}(t)) = t^3\, \mathbf{i} + t^5\, \mathbf{j} + t^4\, \mathbf{k}$$

Thus
$$\int_C \mathbf{F} \cdot d\mathbf{r} = \int_0^1 \mathbf{F}(\mathbf{r}(t)) \cdot \mathbf{r}'(t)\, dt$$
$$= \int_0^1 (t^3 + 5t^6)\, dt = \frac{t^4}{4} + \frac{5t^7}{7}\Big]_0^1 = \frac{27}{28}$$

Finally, we note the connection between line integrals of vector fields and line integrals of scalar fields. Suppose the vector field $\mathbf{F}$ on $\mathbb{R}^3$ is given in component form by the equation $\mathbf{F} = P\,\mathbf{i} + Q\,\mathbf{j} + R\,\mathbf{k}$. We use Definition 13 to compute its line integral along C:

$$\int_C \mathbf{F} \cdot d\mathbf{r} = \int_a^b \mathbf{F}(\mathbf{r}(t)) \cdot \mathbf{r}'(t)\, dt$$
$$= \int_a^b (P\,\mathbf{i} + Q\,\mathbf{j} + R\,\mathbf{k}) \cdot (x'(t)\,\mathbf{i} + y'(t)\,\mathbf{j} + z'(t)\,\mathbf{k})\, dt$$
$$= \int_a^b \big[P(x(t), y(t), z(t))\, x'(t) + Q(x(t), y(t), z(t))\, y'(t) + R(x(t), y(t), z(t))\, z'(t)\big]\, dt$$

But this last integral is precisely the line integral in (10). Therefore we have

$$\int_C \mathbf{F} \cdot d\mathbf{r} = \int_C P\, dx + Q\, dy + R\, dz \qquad \text{where } \mathbf{F} = P\,\mathbf{i} + Q\,\mathbf{j} + R\,\mathbf{k}$$

For example, the integral $\int_C y\, dx + z\, dy + x\, dz$ in Example 6 could be expressed as $\int_C \mathbf{F} \cdot d\mathbf{r}$ where

$$\mathbf{F}(x, y, z) = y\,\mathbf{i} + z\,\mathbf{j} + x\,\mathbf{k}$$

16.2 EXERCISES

1–16 Evaluate the line integral, where C is the given curve.

1. $\int_C y\, ds$, $C: x = t^2,\ y = 2t,\ 0 \leqslant t \leqslant 3$
2. $\int_C (x/y)\, ds$, $C: x = t^3,\ y = t^4,\ 1 \leqslant t \leqslant 2$
3. $\int_C xy^4\, ds$, C is the right half of the circle $x^2 + y^2 = 16$
4. $\int_C xe^y\, ds$, C is the line segment from $(2, 0)$ to $(5, 4)$
5. $\int_C (x^2y + \sin x)\, dy$, C is the arc of the parabola $y = x^2$ from $(0, 0)$ to (π, π^2)
6. $\int_C e^x\, dx$, C is the arc of the curve $x = y^3$ from $(-1, -1)$ to $(1, 1)$
7. $\int_C (x + 2y)\, dx + x^2\, dy$, C consists of line segments from $(0, 0)$ to $(2, 1)$ and from $(2, 1)$ to $(3, 0)$
8. $\int_C x^2\, dx + y^2\, dy$, C consists of the arc of the circle $x^2 + y^2 = 4$ from $(2, 0)$ to $(0, 2)$ followed by the line segment from $(0, 2)$ to $(4, 3)$
9. $\int_C x^2y\, ds$, $C: x = \cos t,\ y = \sin t,\ z = t,\ 0 \leqslant t \leqslant \pi/2$
10. $\int_C y^2z\, ds$, C is the line segment from $(3, 1, 2)$ to $(1, 2, 5)$
11. $\int_C xe^{yz}\, ds$, C is the line segment from $(0, 0, 0)$ to $(1, 2, 3)$
12. $\int_C (x^2 + y^2 + z^2)\, ds$, $C: x = t,\ y = \cos 2t,\ z = \sin 2t,\ 0 \leqslant t \leqslant 2\pi$
13. $\int_C xye^{yz}\, dy$, $C: x = t,\ y = t^2,\ z = t^3,\ 0 \leqslant t \leqslant 1$
14. $\int_C y\, dx + z\, dy + x\, dz$, $C: x = \sqrt{t},\ y = t,\ z = t^2,\ 1 \leqslant t \leqslant 4$
15. $\int_C z^2\, dx + x^2\, dy + y^2\, dz$, C is the line segment from $(1, 0, 0)$ to $(4, 1, 2)$
16. $\int_C (y + z)\, dx + (x + z)\, dy + (x + y)\, dz$, C consists of line segments from $(0, 0, 0)$ to $(1, 0, 1)$ and from $(1, 0, 1)$ to $(0, 1, 2)$

17. Let $\mathbf{F}$ be the vector field shown in the figure.

(a) If C_1 is the vertical line segment from $(-3, -3)$ to $(-3, 3)$, determine whether $\int_{C_1} \mathbf{F} \cdot d\mathbf{r}$ is positive, negative, or zero.

(b) If C_2 is the counterclockwise-oriented circle with radius 3 and center the origin, determine whether $\int_{C_2} \mathbf{F} \cdot d\mathbf{r}$ is positive, negative, or zero.

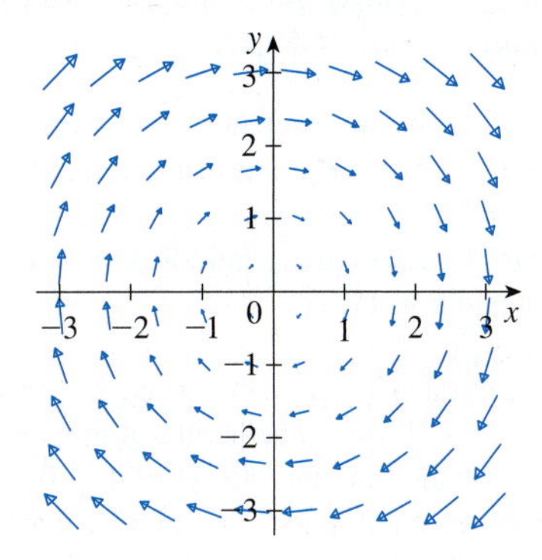

18. The figure shows a vector field $\mathbf{F}$ and two curves C_1 and C_2. Are the line integrals of $\mathbf{F}$ over C_1 and C_2 positive, negative, or zero? Explain.

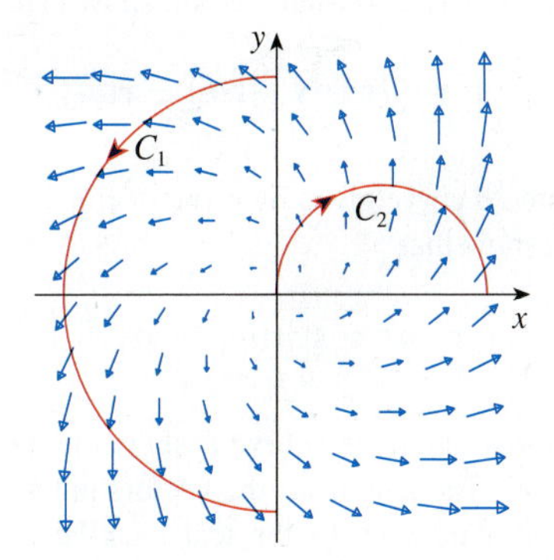

19–22 Evaluate the line integral $\int_C \mathbf{F} \cdot d\mathbf{r}$, where C is given by the vector function $\mathbf{r}(t)$.

19. $\mathbf{F}(x, y) = xy^2\,\mathbf{i} - x^2\,\mathbf{j}$,
$\mathbf{r}(t) = t^3\,\mathbf{i} + t^2\,\mathbf{j}, \quad 0 \leqslant t \leqslant 1$

20. $\mathbf{F}(x, y, z) = (x + y^2)\,\mathbf{i} + xz\,\mathbf{j} + (y + z)\,\mathbf{k}$,
$\mathbf{r}(t) = t^2\,\mathbf{i} + t^3\,\mathbf{j} - 2t\,\mathbf{k}, \quad 0 \leqslant t \leqslant 2$

21. $\mathbf{F}(x, y, z) = \sin x\,\mathbf{i} + \cos y\,\mathbf{j} + xz\,\mathbf{k}$,
$\mathbf{r}(t) = t^3\,\mathbf{i} - t^2\,\mathbf{j} + t\,\mathbf{k}, \quad 0 \leqslant t \leqslant 1$

22. $\mathbf{F}(x, y, z) = x\,\mathbf{i} + y\,\mathbf{j} + xy\,\mathbf{k}$,
$\mathbf{r}(t) = \cos t\,\mathbf{i} + \sin t\,\mathbf{j} + t\,\mathbf{k}, \quad 0 \leqslant t \leqslant \pi$

23–26 Use a calculator to evaluate the line integral correct to four decimal places.

23. $\int_C \mathbf{F} \cdot d\mathbf{r}$, where $\mathbf{F}(x, y) = \sqrt{x + y}\,\mathbf{i} + (y/x)\,\mathbf{j}$ and $\mathbf{r}(t) = \sin^2 t\,\mathbf{i} + \sin t \cos t\,\mathbf{j}, \quad \pi/6 \leqslant t \leqslant \pi/3$

24. $\int_C \mathbf{F} \cdot d\mathbf{r}$, where $\mathbf{F}(x, y, z) = yze^x\,\mathbf{i} + zxe^y\,\mathbf{j} + xye^z\,\mathbf{k}$ and $\mathbf{r}(t) = \sin t\,\mathbf{i} + \cos t\,\mathbf{j} + \tan t\,\mathbf{k}, \quad 0 \leqslant t \leqslant \pi/4$

25. $\int_C xy \arctan z\, ds$, where C has parametric equations $x = t^2, y = t^3, z = \sqrt{t}, \quad 1 \leqslant t \leqslant 2$

26. $\int_C z \ln(x + y)\, ds$, where C has parametric equations $x = 1 + 3t, y = 2 + t^2, z = t^4, \quad -1 \leqslant t \leqslant 1$

CAS **27–28** Use a graph of the vector field $\mathbf{F}$ and the curve C to guess whether the line integral of $\mathbf{F}$ over C is positive, negative, or zero. Then evaluate the line integral.

27. $\mathbf{F}(x, y) = (x - y)\,\mathbf{i} + xy\,\mathbf{j}$,
C is the arc of the circle $x^2 + y^2 = 4$ traversed counterclockwise from $(2, 0)$ to $(0, -2)$

28. $\mathbf{F}(x, y) = \dfrac{x}{\sqrt{x^2 + y^2}}\,\mathbf{i} + \dfrac{y}{\sqrt{x^2 + y^2}}\,\mathbf{j}$,
C is the parabola $y = 1 + x^2$ from $(-1, 2)$ to $(1, 2)$

29. (a) Evaluate the line integral $\int_C \mathbf{F} \cdot d\mathbf{r}$, where $\mathbf{F}(x, y) = e^{x-1}\,\mathbf{i} + xy\,\mathbf{j}$ and C is given by $\mathbf{r}(t) = t^2\,\mathbf{i} + t^3\,\mathbf{j}, \quad 0 \leqslant t \leqslant 1$.

(b) Illustrate part (a) by using a graphing calculator or computer to graph C and the vectors from the vector field corresponding to $t = 0, 1/\sqrt{2}$, and 1 (as in Figure 13).

30. (a) Evaluate the line integral $\int_C \mathbf{F} \cdot d\mathbf{r}$, where $\mathbf{F}(x, y, z) = x\,\mathbf{i} - z\,\mathbf{j} + y\,\mathbf{k}$ and C is given by $\mathbf{r}(t) = 2t\,\mathbf{i} + 3t\,\mathbf{j} - t^2\,\mathbf{k}, \quad -1 \leqslant t \leqslant 1$.

(b) Illustrate part (a) by using a computer to graph C and the vectors from the vector field corresponding to $t = \pm 1$ and $\pm\frac{1}{2}$ (as in Figure 13).

CAS **31.** Find the exact value of $\int_C x^3 y^2 z\, ds$, where C is the curve with parametric equations $x = e^{-t} \cos 4t$, $y = e^{-t} \sin 4t$, $z = e^{-t}$, $0 \leqslant t \leqslant 2\pi$.

32. (a) Find the work done by the force field $\mathbf{F}(x, y) = x^2\,\mathbf{i} + xy\,\mathbf{j}$ on a particle that moves once around the circle $x^2 + y^2 = 4$ oriented in the counterclockwise direction.

CAS (b) Use a computer algebra system to graph the force field and circle on the same screen. Use the graph to explain your answer to part (a).

33. A thin wire is bent into the shape of a semicircle $x^2 + y^2 = 4$, $x \geqslant 0$. If the linear density is a constant k, find the mass and center of mass of the wire.

34. A thin wire has the shape of the first-quadrant part of the circle with center the origin and radius a. If the density function is $\rho(x, y) = kxy$, find the mass and center of mass of the wire.

35. (a) Write the formulas similar to Equations 4 for the center of mass $(\bar{x}, \bar{y}, \bar{z})$ of a thin wire in the shape of a space curve C if the wire has density function $\rho(x, y, z)$.

(b) Find the center of mass of a wire in the shape of the helix $x = 2 \sin t$, $y = 2 \cos t$, $z = 3t$, $0 \leqslant t \leqslant 2\pi$, if the density is a constant k.

36. Find the mass and center of mass of a wire in the shape of the helix $x = t$, $y = \cos t$, $z = \sin t$, $0 \leqslant t \leqslant 2\pi$, if the density at any point is equal to the square of the distance from the origin.

37. If a wire with linear density $\rho(x, y)$ lies along a plane curve C, its **moments of inertia** about the x- and y-axes are defined as

$$I_x = \int_C y^2 \rho(x, y)\, ds \qquad I_y = \int_C x^2 \rho(x, y)\, ds$$

Find the moments of inertia for the wire in Example 3.

38. If a wire with linear density $\rho(x, y, z)$ lies along a space curve C, its **moments of inertia** about the x-, y-, and z-axes are defined as

$$I_x = \int_C (y^2 + z^2)\rho(x, y, z)\, ds$$

$$I_y = \int_C (x^2 + z^2)\rho(x, y, z)\, ds$$

$$I_z = \int_C (x^2 + y^2)\rho(x, y, z)\, ds$$

Find the moments of inertia for the wire in Exercise 35.

39. Find the work done by the force field

$$\mathbf{F}(x, y) = x\,\mathbf{i} + (y + 2)\,\mathbf{j}$$

in moving an object along an arch of the cycloid

$$\mathbf{r}(t) = (t - \sin t)\,\mathbf{i} + (1 - \cos t)\,\mathbf{j} \qquad 0 \leqslant t \leqslant 2\pi$$

40. Find the work done by the force field $\mathbf{F}(x, y) = x^2\,\mathbf{i} + ye^x\,\mathbf{j}$ on a particle that moves along the parabola $x = y^2 + 1$ from $(1, 0)$ to $(2, 1)$.

41. Find the work done by the force field

$$\mathbf{F}(x, y, z) = \langle x - y^2, y - z^2, z - x^2 \rangle$$

on a particle that moves along the line segment from $(0, 0, 1)$ to $(2, 1, 0)$.

42. The force exerted by an electric charge at the origin on a charged particle at a point (x, y, z) with position vector $\mathbf{r} = \langle x, y, z \rangle$ is $\mathbf{F}(\mathbf{r}) = K\mathbf{r}/|\mathbf{r}|^3$ where K is a constant. (See Example 16.1.5.) Find the work done as the particle moves along a straight line from $(2, 0, 0)$ to $(2, 1, 5)$.

43. The position of an object with mass m at time t is $\mathbf{r}(t) = at^2\,\mathbf{i} + bt^3\,\mathbf{j}$, $0 \leqslant t \leqslant 1$.

(a) What is the force acting on the object at time t?

(b) What is the work done by the force during the time interval $0 \leqslant t \leqslant 1$?

44. An object with mass m moves with position function $\mathbf{r}(t) = a \sin t\,\mathbf{i} + b \cos t\,\mathbf{j} + ct\,\mathbf{k}$, $0 \leqslant t \leqslant \pi/2$. Find the work done on the object during this time period.

45. A 160-lb man carries a 25-lb can of paint up a helical staircase that encircles a silo with a radius of 20 ft. If the silo is 90 ft high and the man makes exactly three complete revolutions climbing to the top, how much work is done by the man against gravity?

46. Suppose there is a hole in the can of paint in Exercise 45 and 9 lb of paint leaks steadily out of the can during the man's ascent. How much work is done?

47. (a) Show that a constant force field does zero work on a particle that moves once uniformly around the circle $x^2 + y^2 = 1$.

(b) Is this also true for a force field $\mathbf{F}(\mathbf{x}) = k\mathbf{x}$, where k is a constant and $\mathbf{x} = \langle x, y \rangle$?

48. The base of a circular fence with radius 10 m is given by $x = 10 \cos t$, $y = 10 \sin t$. The height of the fence at position (x, y) is given by the function $h(x, y) = 4 + 0.01(x^2 - y^2)$, so the height varies from 3 m to 5 m. Suppose that 1 L of paint covers 100 m^2. Sketch the fence and determine how much paint you will need if you paint both sides of the fence.

49. If C is a smooth curve given by a vector function $\mathbf{r}(t)$, $a \leqslant t \leqslant b$, and $\mathbf{v}$ is a constant vector, show that

$$\int_C \mathbf{v} \cdot d\mathbf{r} = \mathbf{v} \cdot [\mathbf{r}(b) - \mathbf{r}(a)]$$

50. If C is a smooth curve given by a vector function $\mathbf{r}(t)$, $a \leqslant t \leqslant b$, show that

$$\int_C \mathbf{r} \cdot d\mathbf{r} = \tfrac{1}{2}\left[|\mathbf{r}(b)|^2 - |\mathbf{r}(a)|^2\right]$$

51. An object moves along the curve C shown in the figure from $(1, 2)$ to $(9, 8)$. The lengths of the vectors in the force field $\mathbf{F}$ are measured in newtons by the scales on the axes. Estimate the work done by $\mathbf{F}$ on the object.

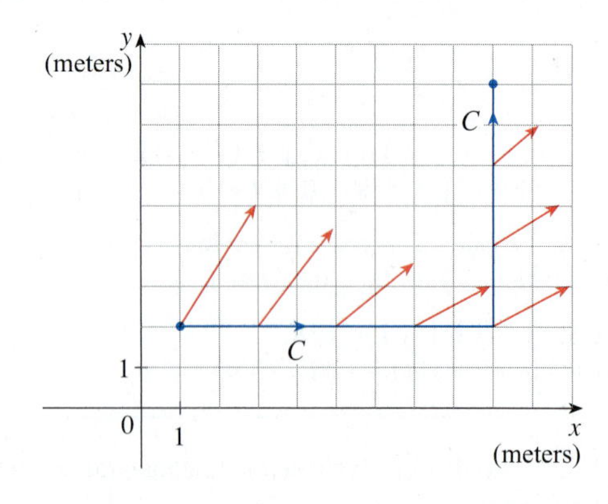

52. Experiments show that a steady current I in a long wire produces a magnetic field $\mathbf{B}$ that is tangent to any circle that lies in the plane perpendicular to the wire and whose center is the axis of the wire (as in the figure). *Ampère's Law* relates the electric current to its magnetic effects and states that

$$\int_C \mathbf{B} \cdot d\mathbf{r} = \mu_0 I$$

where I is the net current that passes through any surface bounded by a closed curve C, and μ_0 is a constant called the permeability of free space. By taking C to be a circle with radius r, show that the magnitude $B = |\mathbf{B}|$ of the magnetic field at a distance r from the center of the wire is

$$B = \frac{\mu_0 I}{2\pi r}$$

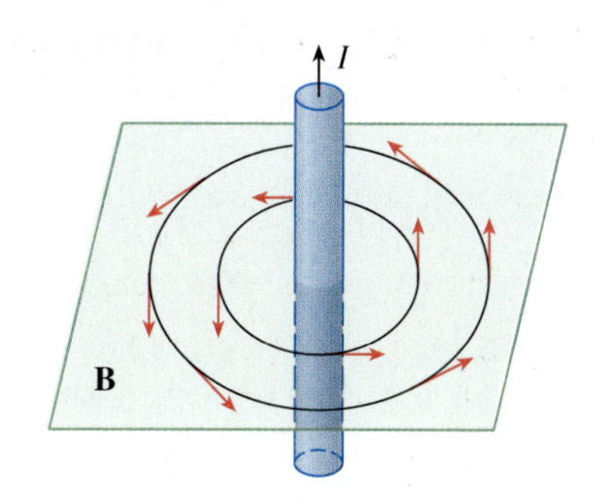

16.3 The Fundamental Theorem for Line Integrals

Recall from Section 4.3 that Part 2 of the Fundamental Theorem of Calculus can be written as

$$\boxed{1} \qquad \int_a^b F'(x)\,dx = F(b) - F(a)$$

where F' is continuous on $[a, b]$. We also called Equation 1 the Net Change Theorem: The integral of a rate of change is the net change.

If we think of the gradient vector ∇f of a function f of two or three variables as a sort of derivative of f, then the following theorem can be regarded as a version of the Fundamental Theorem for line integrals.

2 Theorem Let C be a smooth curve given by the vector function $\mathbf{r}(t)$, $a \leqslant t \leqslant b$. Let f be a differentiable function of two or three variables whose gradient vector ∇f is continuous on C. Then

$$\int_C \nabla f \cdot d\mathbf{r} = f(\mathbf{r}(b)) - f(\mathbf{r}(a))$$

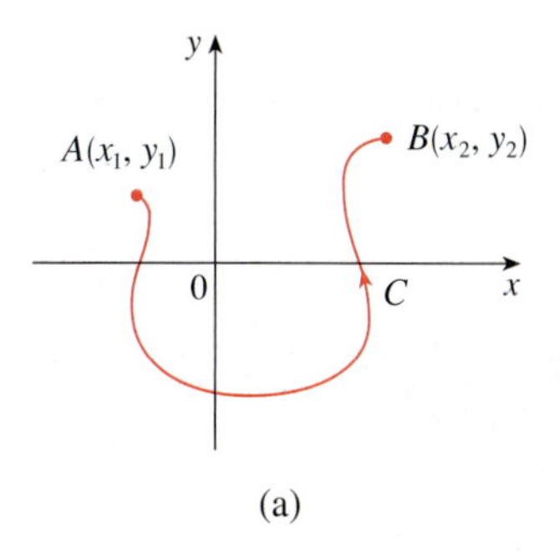

(a)

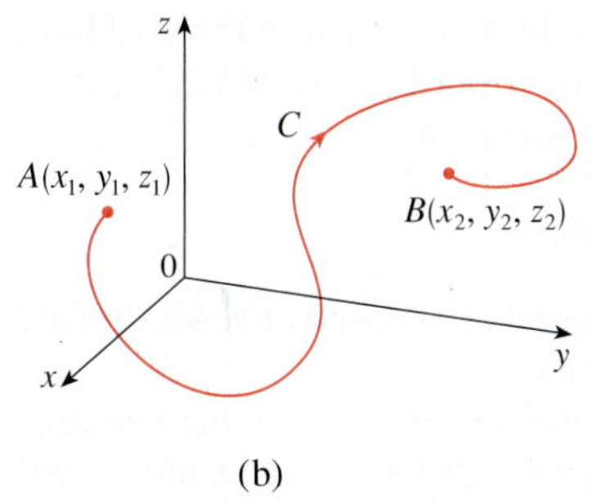

(b)

FIGURE 1

NOTE Theorem 2 says that we can evaluate the line integral of a conservative vector field (the gradient vector field of the potential function f) simply by knowing the value of f at the endpoints of C. In fact, Theorem 2 says that the line integral of ∇f is the net change in f. If f is a function of two variables and C is a plane curve with initial point $A(x_1, y_1)$ and terminal point $B(x_2, y_2)$, as in Figure 1(a), then Theorem 2 becomes

$$\int_C \nabla f \cdot d\mathbf{r} = f(x_2, y_2) - f(x_1, y_1)$$

If f is a function of three variables and C is a space curve joining the point $A(x_1, y_1, z_1)$ to the point $B(x_2, y_2, z_2)$, as in Figure 1(b), then we have

$$\int_C \nabla f \cdot d\mathbf{r} = f(x_2, y_2, z_2) - f(x_1, y_1, z_1)$$

Let's prove Theorem 2 for this case.

PROOF OF THEOREM 2 Using Definition 16.2.13, we have

$$\begin{aligned}\int_C \nabla f \cdot d\mathbf{r} &= \int_a^b \nabla f(\mathbf{r}(t)) \cdot \mathbf{r}'(t)\, dt \\ &= \int_a^b \left(\frac{\partial f}{\partial x}\frac{dx}{dt} + \frac{\partial f}{\partial y}\frac{dy}{dt} + \frac{\partial f}{\partial z}\frac{dz}{dt} \right) dt \\ &= \int_a^b \frac{d}{dt} f(\mathbf{r}(t))\, dt \qquad \text{(by the Chain Rule)} \\ &= f(\mathbf{r}(b)) - f(\mathbf{r}(a))\end{aligned}$$

The last step follows from the Fundamental Theorem of Calculus (Equation 1). ■

Although we have proved Theorem 2 for smooth curves, it is also true for piecewise-smooth curves. This can be seen by subdividing C into a finite number of smooth curves and adding the resulting integrals.

EXAMPLE 1 Find the work done by the gravitational field

$$\mathbf{F}(\mathbf{x}) = -\frac{mMG}{|\mathbf{x}|^3}\mathbf{x}$$

in moving a particle with mass m from the point $(3, 4, 12)$ to the point $(2, 2, 0)$ along a piecewise-smooth curve C. (See Example 16.1.4.)

SOLUTION From Section 16.1 we know that $\mathbf{F}$ is a conservative vector field and, in fact, $\mathbf{F} = \nabla f$, where

$$f(x, y, z) = \frac{mMG}{\sqrt{x^2 + y^2 + z^2}}$$

Therefore, by Theorem 2, the work done is

$$\begin{aligned}W &= \int_C \mathbf{F} \cdot d\mathbf{r} = \int_C \nabla f \cdot d\mathbf{r} \\ &= f(2, 2, 0) - f(3, 4, 12) \\ &= \frac{mMG}{\sqrt{2^2 + 2^2}} - \frac{mMG}{\sqrt{3^2 + 4^2 + 12^2}} = mMG\left(\frac{1}{2\sqrt{2}} - \frac{1}{13}\right)\end{aligned}$$ ■

Independence of Path

Suppose C_1 and C_2 are two piecewise-smooth curves (which are called **paths**) that have the same initial point A and terminal point B. We know from Example 16.2.4 that, in general, $\int_{C_1} \mathbf{F} \cdot d\mathbf{r} \neq \int_{C_2} \mathbf{F} \cdot d\mathbf{r}$. But one implication of Theorem 2 is that

$$\int_{C_1} \nabla f \cdot d\mathbf{r} = \int_{C_2} \nabla f \cdot d\mathbf{r}$$

whenever ∇f is continuous. In other words, the line integral of a *conservative* vector field depends only on the initial point and terminal point of a curve.

In general, if $\mathbf{F}$ is a continuous vector field with domain D, we say that the line integral $\int_C \mathbf{F} \cdot d\mathbf{r}$ is **independent of path** if $\int_{C_1} \mathbf{F} \cdot d\mathbf{r} = \int_{C_2} \mathbf{F} \cdot d\mathbf{r}$ for any two paths C_1 and C_2 in D that have the same initial points and the same terminal points. With this terminology we can say that *line integrals of conservative vector fields are independent of path.*

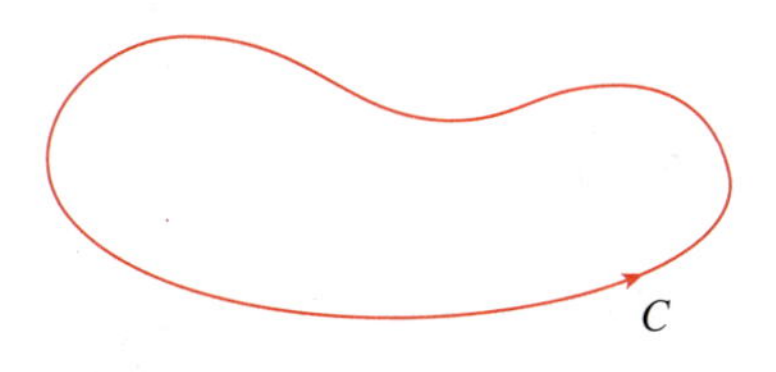

FIGURE 2
A closed curve

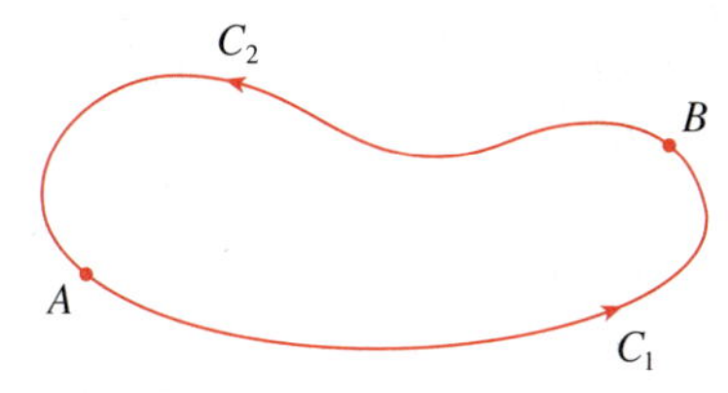

FIGURE 3

A curve is called **closed** if its terminal point coincides with its initial point, that is, $\mathbf{r}(b) = \mathbf{r}(a)$. (See Figure 2.) If $\int_C \mathbf{F} \cdot d\mathbf{r}$ is independent of path in D and C is any closed path in D, we can choose any two points A and B on C and regard C as being composed of the path C_1 from A to B followed by the path C_2 from B to A. (See Figure 3.) Then

$$\int_C \mathbf{F} \cdot d\mathbf{r} = \int_{C_1} \mathbf{F} \cdot d\mathbf{r} + \int_{C_2} \mathbf{F} \cdot d\mathbf{r} = \int_{C_1} \mathbf{F} \cdot d\mathbf{r} - \int_{-C_2} \mathbf{F} \cdot d\mathbf{r} = 0$$

since C_1 and $-C_2$ have the same initial and terminal points.

Conversely, if it is true that $\int_C \mathbf{F} \cdot d\mathbf{r} = 0$ whenever C is a closed path in D, then we demonstrate independence of path as follows. Take any two paths C_1 and C_2 from A to B in D and define C to be the curve consisting of C_1 followed by $-C_2$. Then

$$0 = \int_C \mathbf{F} \cdot d\mathbf{r} = \int_{C_1} \mathbf{F} \cdot d\mathbf{r} + \int_{-C_2} \mathbf{F} \cdot d\mathbf{r} = \int_{C_1} \mathbf{F} \cdot d\mathbf{r} - \int_{C_2} \mathbf{F} \cdot d\mathbf{r}$$

and so $\int_{C_1} \mathbf{F} \cdot d\mathbf{r} = \int_{C_2} \mathbf{F} \cdot d\mathbf{r}$. Thus we have proved the following theorem.

3 Theorem $\int_C \mathbf{F} \cdot d\mathbf{r}$ is independent of path in D if and only if $\int_C \mathbf{F} \cdot d\mathbf{r} = 0$ for every closed path C in D.

Since we know that the line integral of any conservative vector field $\mathbf{F}$ is independent of path, it follows that $\int_C \mathbf{F} \cdot d\mathbf{r} = 0$ for any closed path. The physical interpretation is that the work done by a conservative force field (such as the gravitational or electric field in Section 16.1) as it moves an object around a closed path is 0.

The following theorem says that the *only* vector fields that are independent of path are conservative. It is stated and proved for plane curves, but there is a similar version for space curves. We assume that D is **open**, which means that for every point P in D there is a disk with center P that lies entirely in D. (So D doesn't contain any of its boundary points.) In addition, we assume that D is **connected**: this means that any two points in D can be joined by a path that lies in D.

4 Theorem Suppose $\mathbf{F}$ is a vector field that is continuous on an open connected region D. If $\int_C \mathbf{F} \cdot d\mathbf{r}$ is independent of path in D, then $\mathbf{F}$ is a conservative vector field on D; that is, there exists a function f such that $\nabla f = \mathbf{F}$.

PROOF Let $A(a, b)$ be a fixed point in D. We construct the desired potential function f by defining

$$f(x, y) = \int_{(a, b)}^{(x, y)} \mathbf{F} \cdot d\mathbf{r}$$

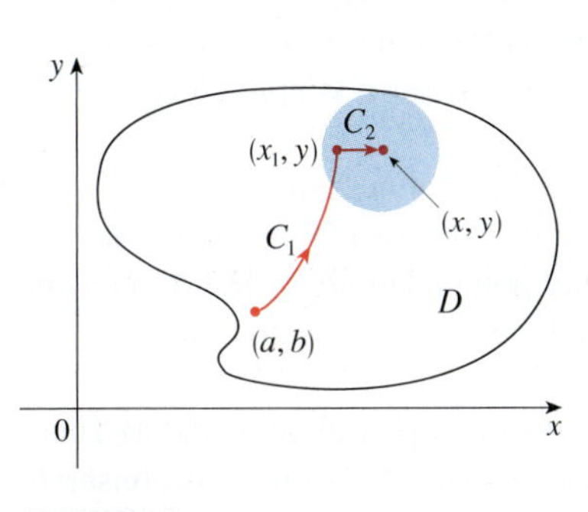

FIGURE 4

for any point (x, y) in D. Since $\int_C \mathbf{F} \cdot d\mathbf{r}$ is independent of path, it does not matter which path C from (a, b) to (x, y) is used to evaluate $f(x, y)$. Since D is open, there exists a disk contained in D with center (x, y). Choose any point (x_1, y) in the disk with $x_1 < x$ and let C consist of any path C_1 from (a, b) to (x_1, y) followed by the horizontal line segment C_2 from (x_1, y) to (x, y). (See Figure 4.) Then

$$f(x, y) = \int_{C_1} \mathbf{F} \cdot d\mathbf{r} + \int_{C_2} \mathbf{F} \cdot d\mathbf{r} = \int_{(a, b)}^{(x_1, y)} \mathbf{F} \cdot d\mathbf{r} + \int_{C_2} \mathbf{F} \cdot d\mathbf{r}$$

Notice that the first of these integrals does not depend on x, so

$$\frac{\partial}{\partial x} f(x, y) = 0 + \frac{\partial}{\partial x} \int_{C_2} \mathbf{F} \cdot d\mathbf{r}$$

If we write $\mathbf{F} = P\,\mathbf{i} + Q\,\mathbf{j}$, then

$$\int_{C_2} \mathbf{F} \cdot d\mathbf{r} = \int_{C_2} P\,dx + Q\,dy$$

On C_2, y is constant, so $dy = 0$. Using t as the parameter, where $x_1 \leqslant t \leqslant x$, we have

$$\frac{\partial}{\partial x} f(x, y) = \frac{\partial}{\partial x} \int_{C_2} P\,dx + Q\,dy = \frac{\partial}{\partial x} \int_{x_1}^{x} P(t, y)\,dt = P(x, y)$$

by Part 1 of the Fundamental Theorem of Calculus (see Section 4.3). A similar argument, using a vertical line segment (see Figure 5), shows that

$$\frac{\partial}{\partial y} f(x, y) = \frac{\partial}{\partial y} \int_{C_2} P\,dx + Q\,dy = \frac{\partial}{\partial y} \int_{y_1}^{y} Q(x, t)\,dt = Q(x, y)$$

FIGURE 5

Thus
$$\mathbf{F} = P\,\mathbf{i} + Q\,\mathbf{j} = \frac{\partial f}{\partial x}\mathbf{i} + \frac{\partial f}{\partial y}\mathbf{j} = \nabla f$$

which says that $\mathbf{F}$ is conservative. ■

The question remains: how is it possible to determine whether or not a vector field $\mathbf{F}$ is conservative? Suppose it is known that $\mathbf{F} = P\,\mathbf{i} + Q\,\mathbf{j}$ is conservative, where P and Q have continuous first-order partial derivatives. Then there is a function f such that $\mathbf{F} = \nabla f$, that is,

$$P = \frac{\partial f}{\partial x} \qquad \text{and} \qquad Q = \frac{\partial f}{\partial y}$$

Therefore, by Clairaut's Theorem,

$$\frac{\partial P}{\partial y} = \frac{\partial^2 f}{\partial y\,\partial x} = \frac{\partial^2 f}{\partial x\,\partial y} = \frac{\partial Q}{\partial x}$$

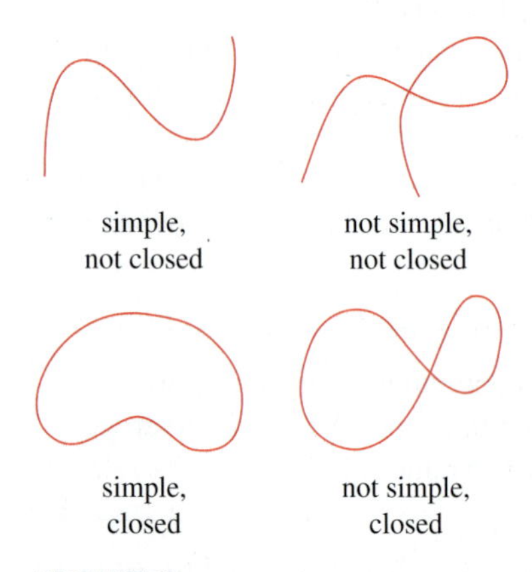

FIGURE 6
Types of curves

> **5 Theorem** If $\mathbf{F}(x, y) = P(x, y)\,\mathbf{i} + Q(x, y)\,\mathbf{j}$ is a conservative vector field, where P and Q have continuous first-order partial derivatives on a domain D, then throughout D we have
>
> $$\frac{\partial P}{\partial y} = \frac{\partial Q}{\partial x}$$

The converse of Theorem 5 is true only for a special type of region. To explain this, we first need the concept of a **simple curve**, which is a curve that doesn't intersect itself anywhere between its endpoints. [See Figure 6; $\mathbf{r}(a) = \mathbf{r}(b)$ for a simple closed curve, but $\mathbf{r}(t_1) \neq \mathbf{r}(t_2)$ when $a < t_1 < t_2 < b$.]

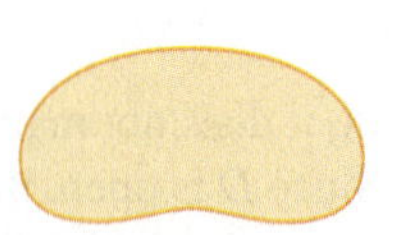

simply-connected region

In Theorem 4 we needed an open connected region. For the next theorem we need a stronger condition. A **simply-connected region** in the plane is a connected region D such that every simple closed curve in D encloses only points that are in D. Notice from Figure 7 that, intuitively speaking, a simply-connected region contains no hole and can't consist of two separate pieces.

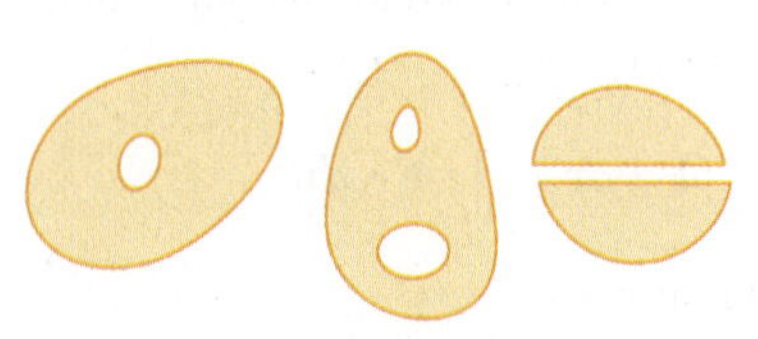

regions that are not simply-connected

FIGURE 7

In terms of simply-connected regions, we can now state a partial converse to Theorem 5 that gives a convenient method for verifying that a vector field on $\mathbb{R}^2$ is conservative. The proof will be sketched in the next section as a consequence of Green's Theorem.

6 Theorem Let $\mathbf{F} = P\,\mathbf{i} + Q\,\mathbf{j}$ be a vector field on an open simply-connected region D. Suppose that P and Q have continuous first-order partial derivatives and

$$\frac{\partial P}{\partial y} = \frac{\partial Q}{\partial x} \qquad \text{throughout } D$$

Then $\mathbf{F}$ is conservative.

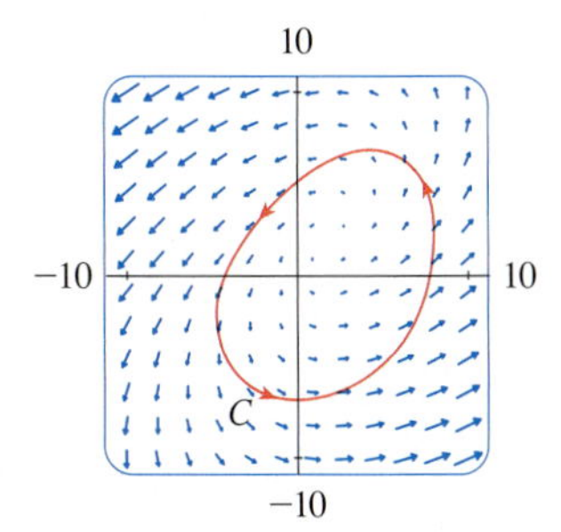

FIGURE 8

Figures 8 and 9 show the vector fields in Examples 2 and 3, respectively. The vectors in Figure 8 that start on the closed curve C all appear to point in roughly the same direction as C. So it looks as if $\int_C \mathbf{F} \cdot d\mathbf{r} > 0$ and therefore $\mathbf{F}$ is not conservative. The calculation in Example 2 confirms this impression. Some of the vectors near the curves C_1 and C_2 in Figure 9 point in approximately the same direction as the curves, whereas others point in the opposite direction. So it appears plausible that line integrals around all closed paths are 0. Example 3 shows that $\mathbf{F}$ is indeed conservative.

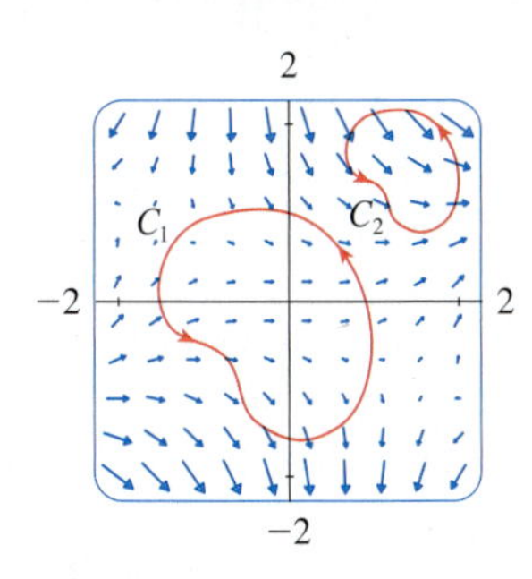

FIGURE 9

EXAMPLE 2 Determine whether or not the vector field

$$\mathbf{F}(x, y) = (x - y)\,\mathbf{i} + (x - 2)\,\mathbf{j}$$

is conservative.

SOLUTION Let $P(x, y) = x - y$ and $Q(x, y) = x - 2$. Then

$$\frac{\partial P}{\partial y} = -1 \qquad \frac{\partial Q}{\partial x} = 1$$

Since $\partial P/\partial y \neq \partial Q/\partial x$, $\mathbf{F}$ is not conservative by Theorem 5. ■

EXAMPLE 3 Determine whether or not the vector field

$$\mathbf{F}(x, y) = (3 + 2xy)\,\mathbf{i} + (x^2 - 3y^2)\,\mathbf{j}$$

is conservative.

SOLUTION Let $P(x, y) = 3 + 2xy$ and $Q(x, y) = x^2 - 3y^2$. Then

$$\frac{\partial P}{\partial y} = 2x = \frac{\partial Q}{\partial x}$$

Also, the domain of $\mathbf{F}$ is the entire plane ($D = \mathbb{R}^2$), which is open and simply-connected. Therefore we can apply Theorem 6 and conclude that $\mathbf{F}$ is conservative. ■

In Example 3, Theorem 6 told us that $\mathbf{F}$ is conservative, but it did not tell us how to find the (potential) function f such that $\mathbf{F} = \nabla f$. The proof of Theorem 4 gives us a clue as to how to find f. We use "partial integration" as in the following example.

EXAMPLE 4

(a) If $\mathbf{F}(x, y) = (3 + 2xy)\,\mathbf{i} + (x^2 - 3y^2)\,\mathbf{j}$, find a function f such that $\mathbf{F} = \nabla f$.

(b) Evaluate the line integral $\int_C \mathbf{F} \cdot d\mathbf{r}$, where C is the curve given by

$$\mathbf{r}(t) = e^t \sin t\,\mathbf{i} + e^t \cos t\,\mathbf{j} \qquad 0 \leqslant t \leqslant \pi$$

SOLUTION

(a) From Example 3 we know that $\mathbf{F}$ is conservative and so there exists a function f with $\nabla f = \mathbf{F}$, that is,

$$\boxed{7} \qquad f_x(x, y) = 3 + 2xy$$

$$\boxed{8} \qquad f_y(x, y) = x^2 - 3y^2$$

Integrating (7) with respect to x, we obtain

$$\boxed{9} \qquad f(x, y) = 3x + x^2y + g(y)$$

Notice that the constant of integration is a constant with respect to x, that is, a function of y, which we have called $g(y)$. Next we differentiate both sides of (9) with respect to y:

$$\boxed{10} \qquad f_y(x, y) = x^2 + g'(y)$$

Comparing (8) and (10), we see that

$$g'(y) = -3y^2$$

Integrating with respect to y, we have

$$g(y) = -y^3 + K$$

where K is a constant. Putting this in (9), we have

$$f(x, y) = 3x + x^2y - y^3 + K$$

as the desired potential function.

(b) To use Theorem 2 all we have to know are the initial and terminal points of C, namely, $\mathbf{r}(0) = (0, 1)$ and $\mathbf{r}(\pi) = (0, -e^{\pi})$. In the expression for $f(x, y)$ in part (a), any value of the constant K will do, so let's choose $K = 0$. Then we have

$$\int_C \mathbf{F} \cdot d\mathbf{r} = \int_C \nabla f \cdot d\mathbf{r} = f(0, -e^{\pi}) - f(0, 1) = e^{3\pi} - (-1) = e^{3\pi} + 1$$

This method is much shorter than the straightforward method for evaluating line integrals that we learned in Section 16.2. ■

A criterion for determining whether or not a vector field $\mathbf{F}$ on $\mathbb{R}^3$ is conservative is given in Section 16.5. Meanwhile, the next example shows that the technique for finding the potential function is much the same as for vector fields on $\mathbb{R}^2$.

EXAMPLE 5 If $\mathbf{F}(x, y, z) = y^2\,\mathbf{i} + (2xy + e^{3z})\,\mathbf{j} + 3ye^{3z}\,\mathbf{k}$, find a function f such that $\nabla f = \mathbf{F}$.

SOLUTION If there is such a function f, then

$$\boxed{11} \qquad f_x(x, y, z) = y^2$$

$$\boxed{12} \qquad f_y(x, y, z) = 2xy + e^{3z}$$

$$\boxed{13} \qquad f_z(x, y, z) = 3ye^{3z}$$

Integrating (11) with respect to x, we get

$$\boxed{14} \qquad f(x, y, z) = xy^2 + g(y, z)$$

where $g(y, z)$ is a constant with respect to x. Then differentiating (14) with respect to y, we have

$$f_y(x, y, z) = 2xy + g_y(y, z)$$

and comparison with (12) gives

$$g_y(y, z) = e^{3z}$$

Thus $g(y, z) = ye^{3z} + h(z)$ and we rewrite (14) as

$$f(x, y, z) = xy^2 + ye^{3z} + h(z)$$

Finally, differentiating with respect to z and comparing with (13), we obtain $h'(z) = 0$ and therefore $h(z) = K$, a constant. The desired function is

$$f(x, y, z) = xy^2 + ye^{3z} + K$$

It is easily verified that $\nabla f = \mathbf{F}$. ■

Conservation of Energy

Let's apply the ideas of this chapter to a continuous force field $\mathbf{F}$ that moves an object along a path C given by $\mathbf{r}(t)$, $a \leqslant t \leqslant b$, where $\mathbf{r}(a) = A$ is the initial point and $\mathbf{r}(b) = B$ is the terminal point of C. According to Newton's Second Law of Motion (see Section 13.4), the force $\mathbf{F}(\mathbf{r}(t))$ at a point on C is related to the acceleration $\mathbf{a}(t) = \mathbf{r}''(t)$ by the equation

$$\mathbf{F}(\mathbf{r}(t)) = m\mathbf{r}''(t)$$

So the work done by the force on the object is

$$\begin{aligned} W &= \int_C \mathbf{F} \cdot d\mathbf{r} = \int_a^b \mathbf{F}(\mathbf{r}(t)) \cdot \mathbf{r}'(t)\, dt = \int_a^b m\mathbf{r}''(t) \cdot \mathbf{r}'(t)\, dt \\ &= \frac{m}{2} \int_a^b \frac{d}{dt} [\mathbf{r}'(t) \cdot \mathbf{r}'(t)]\, dt \qquad \text{(Theorem 13.2.3, Formula 4)} \\ &= \frac{m}{2} \int_a^b \frac{d}{dt} |\mathbf{r}'(t)|^2\, dt = \frac{m}{2} \Big[|\mathbf{r}'(t)|^2 \Big]_a^b \qquad \text{(Fundamental Theorem of Calculus)} \\ &= \frac{m}{2} \left(|\mathbf{r}'(b)|^2 - |\mathbf{r}'(a)|^2 \right) \end{aligned}$$

Therefore

$$\boxed{15} \qquad W = \tfrac{1}{2}m|\mathbf{v}(b)|^2 - \tfrac{1}{2}m|\mathbf{v}(a)|^2$$

where $\mathbf{v} = \mathbf{r}'$ is the velocity.

The quantity $\frac{1}{2}m|\mathbf{v}(t)|^2$, that is, half the mass times the square of the speed, is called the **kinetic energy** of the object. Therefore we can rewrite Equation 15 as

$$\boxed{16} \qquad W = K(B) - K(A)$$

which says that the work done by the force field along C is equal to the change in kinetic energy at the endpoints of C.

Now let's further assume that $\mathbf{F}$ is a conservative force field; that is, we can write $\mathbf{F} = \nabla f$. In physics, the **potential energy** of an object at the point (x, y, z) is defined as $P(x, y, z) = -f(x, y, z)$, so we have $\mathbf{F} = -\nabla P$. Then by Theorem 2 we have

$$W = \int_C \mathbf{F} \cdot d\mathbf{r} = -\int_C \nabla P \cdot d\mathbf{r} = -[P(\mathbf{r}(b)) - P(\mathbf{r}(a))] = P(A) - P(B)$$

Comparing this equation with Equation 16, we see that

$$P(A) + K(A) = P(B) + K(B)$$

which says that if an object moves from one point A to another point B under the influence of a conservative force field, then the sum of its potential energy and its kinetic energy remains constant. This is called the **Law of Conservation of Energy** and it is the reason the vector field is called *conservative*.

16.3 EXERCISES

1. The figure shows a curve C and a contour map of a function f whose gradient is continuous. Find $\int_C \nabla f \cdot d\mathbf{r}$.

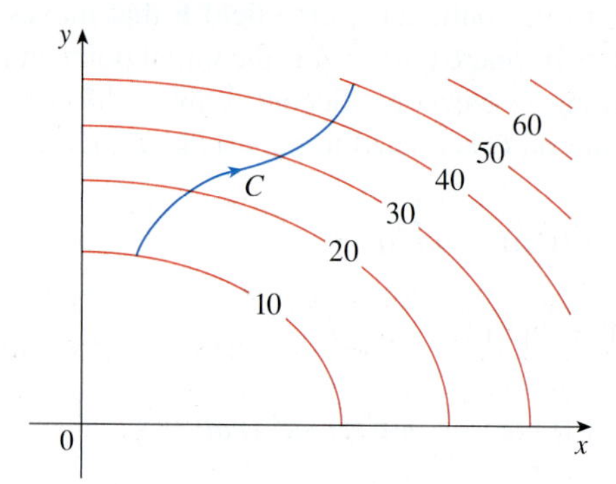

2. A table of values of a function f with continuous gradient is given. Find $\int_C \nabla f \cdot d\mathbf{r}$, where C has parametric equations

$$x = t^2 + 1 \qquad y = t^3 + t \qquad 0 \leqslant t \leqslant 1$$

x \ y	0	1	2
0	1	6	4
1	3	5	7
2	8	2	9

3–10 Determine whether or not $\mathbf{F}$ is a conservative vector field. If it is, find a function f such that $\mathbf{F} = \nabla f$.

3. $\mathbf{F}(x, y) = (xy + y^2)\,\mathbf{i} + (x^2 + 2xy)\,\mathbf{j}$

4. $\mathbf{F}(x, y) = (y^2 - 2x)\,\mathbf{i} + 2xy\,\mathbf{j}$

5. $\mathbf{F}(x, y) = y^2e^{xy}\,\mathbf{i} + (1 + xy)e^{xy}\,\mathbf{j}$

6. $\mathbf{F}(x, y) = ye^x\,\mathbf{i} + (e^x + e^y)\,\mathbf{j}$

7. $\mathbf{F}(x, y) = (ye^x + \sin y)\,\mathbf{i} + (e^x + x\cos y)\,\mathbf{j}$

8. $\mathbf{F}(x, y) = (2xy + y^{-2})\,\mathbf{i} + (x^2 - 2xy^{-3})\,\mathbf{j}, \quad y > 0$

9. $\mathbf{F}(x, y) = (y^2 \cos x + \cos y)\,\mathbf{i} + (2y \sin x - x \sin y)\,\mathbf{j}$

10. $\mathbf{F}(x, y) = (\ln y + y/x)\,\mathbf{i} + (\ln x + x/y)\,\mathbf{j}$

11. The figure shows the vector field $\mathbf{F}(x, y) = \langle 2xy, x^2 \rangle$ and three curves that start at (1, 2) and end at (3, 2).
(a) Explain why $\int_C \mathbf{F} \cdot d\mathbf{r}$ has the same value for all three curves.
(b) What is this common value?

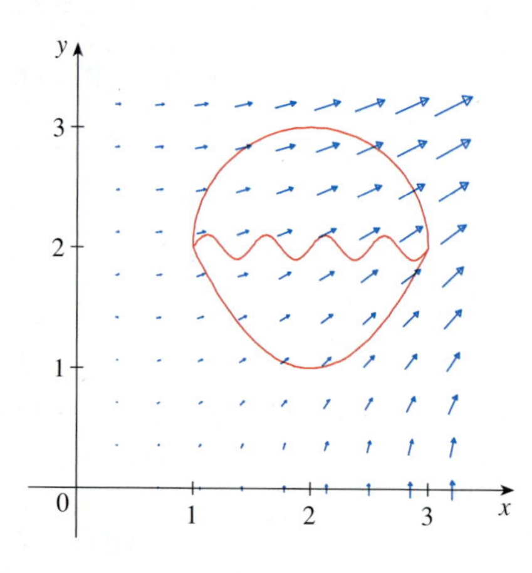

12–18 (a) Find a function f such that $\mathbf{F} = \nabla f$ and (b) use part (a) to evaluate $\int_C \mathbf{F} \cdot d\mathbf{r}$ along the given curve C.

12. $\mathbf{F}(x, y) = (3 + 2xy^2)\,\mathbf{i} + 2x^2y\,\mathbf{j}$,
C is the arc of the hyperbola $y = 1/x$ from (1, 1) to $\left(4, \frac{1}{4}\right)$

13. $\mathbf{F}(x, y) = x^2y^3\,\mathbf{i} + x^3y^2\,\mathbf{j}$,
C: $\mathbf{r}(t) = \langle t^3 - 2t, t^3 + 2t \rangle, \quad 0 \leqslant t \leqslant 1$

14. $\mathbf{F}(x, y) = (1 + xy)e^{xy}\,\mathbf{i} + x^2e^{xy}\,\mathbf{j}$,
C: $\mathbf{r}(t) = \cos t\,\mathbf{i} + 2\sin t\,\mathbf{j}, \quad 0 \leqslant t \leqslant \pi/2$

15. $\mathbf{F}(x, y, z) = yz\,\mathbf{i} + xz\,\mathbf{j} + (xy + 2z)\,\mathbf{k}$,
C is the line segment from $(1, 0, -2)$ to $(4, 6, 3)$

16. $\mathbf{F}(x, y, z) = (y^2z + 2xz^2)\,\mathbf{i} + 2xyz\,\mathbf{j} + (xy^2 + 2x^2z)\,\mathbf{k}$,
C: $x = \sqrt{t}$, $y = t + 1$, $z = t^2$, $0 \leqslant t \leqslant 1$

17. $\mathbf{F}(x, y, z) = yze^{xz}\,\mathbf{i} + e^{xz}\,\mathbf{j} + xye^{xz}\,\mathbf{k}$,
C: $\mathbf{r}(t) = (t^2 + 1)\,\mathbf{i} + (t^2 - 1)\,\mathbf{j} + (t^2 - 2t)\,\mathbf{k}$,
$0 \leqslant t \leqslant 2$

18. $\mathbf{F}(x, y, z) = \sin y\,\mathbf{i} + (x \cos y + \cos z)\,\mathbf{j} - y \sin z\,\mathbf{k}$,
C: $\mathbf{r}(t) = \sin t\,\mathbf{i} + t\,\mathbf{j} + 2t\,\mathbf{k}$, $0 \leqslant t \leqslant \pi/2$

19–20 Show that the line integral is independent of path and evaluate the integral.

19. $\int_C 2xe^{-y}\,dx + (2y - x^2e^{-y})\,dy$,
C is any path from $(1, 0)$ to $(2, 1)$

20. $\int_C \sin y\,dx + (x \cos y - \sin y)\,dy$,
C is any path from $(2, 0)$ to $(1, \pi)$

21. Suppose you're asked to determine the curve that requires the least work for a force field **F** to move a particle from one point to another point. You decide to check first whether **F** is conservative, and indeed it turns out that it is. How would you reply to the request?

22. Suppose an experiment determines that the amount of work required for a force field **F** to move a particle from the point $(1, 2)$ to the point $(5, -3)$ along a curve C_1 is 1.2 J and the work done by **F** in moving the particle along another curve C_2 between the same two points is 1.4 J. What can you say about **F**? Why?

23–24 Find the work done by the force field **F** in moving an object from P to Q.

23. $\mathbf{F}(x, y) = x^3\,\mathbf{i} + y^3\,\mathbf{j}$; $P(1, 0)$, $Q(2, 2)$

24. $\mathbf{F}(x, y) = (2x + y)\,\mathbf{i} + x\,\mathbf{j}$; $P(1, 1)$, $Q(4, 3)$

25–26 Is the vector field shown in the figure conservative? Explain.

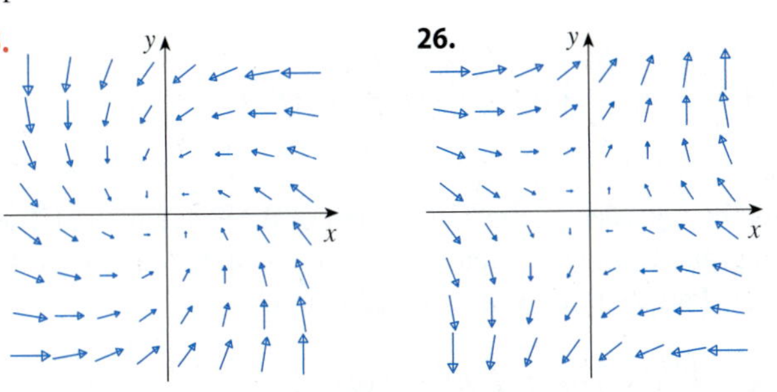

27. (CAS) If $\mathbf{F}(x, y) = \sin y\,\mathbf{i} + (1 + x \cos y)\,\mathbf{j}$, use a plot to guess whether **F** is conservative. Then determine whether your guess is correct.

28. Let $\mathbf{F} = \nabla f$, where $f(x, y) = \sin(x - 2y)$. Find curves C_1 and C_2 that are not closed and satisfy the equation.

(a) $\int_{C_1} \mathbf{F} \cdot d\mathbf{r} = 0$ (b) $\int_{C_2} \mathbf{F} \cdot d\mathbf{r} = 1$

29. Show that if the vector field $\mathbf{F} = P\,\mathbf{i} + Q\,\mathbf{j} + R\,\mathbf{k}$ is conservative and P, Q, R have continuous first-order partial derivatives, then

$$\frac{\partial P}{\partial y} = \frac{\partial Q}{\partial x} \qquad \frac{\partial P}{\partial z} = \frac{\partial R}{\partial x} \qquad \frac{\partial Q}{\partial z} = \frac{\partial R}{\partial y}$$

30. Use Exercise 29 to show that the line integral $\int_C y\,dx + x\,dy + xyz\,dz$ is not independent of path.

31–34 Determine whether or not the given set is (a) open, (b) connected, and (c) simply-connected.

31. $\{(x, y) \mid 0 < y < 3\}$

32. $\{(x, y) \mid 1 < |x| < 2\}$

33. $\{(x, y) \mid 1 \leqslant x^2 + y^2 \leqslant 4,\ y \geqslant 0\}$

34. $\{(x, y) \mid (x, y) \neq (2, 3)\}$

35. Let $\mathbf{F}(x, y) = \dfrac{-y\,\mathbf{i} + x\,\mathbf{j}}{x^2 + y^2}$.

(a) Show that $\partial P/\partial y = \partial Q/\partial x$.

(b) Show that $\int_C \mathbf{F} \cdot d\mathbf{r}$ is not independent of path. [*Hint:* Compute $\int_{C_1} \mathbf{F} \cdot d\mathbf{r}$ and $\int_{C_2} \mathbf{F} \cdot d\mathbf{r}$, where C_1 and C_2 are the upper and lower halves of the circle $x^2 + y^2 = 1$ from $(1, 0)$ to $(-1, 0)$.] Does this contradict Theorem 6?

36. (a) Suppose that **F** is an inverse square force field, that is,

$$\mathbf{F}(\mathbf{r}) = \frac{c\mathbf{r}}{|\mathbf{r}|^3}$$

for some constant c, where $\mathbf{r} = x\,\mathbf{i} + y\,\mathbf{j} + z\,\mathbf{k}$. Find the work done by **F** in moving an object from a point P_1 along a path to a point P_2 in terms of the distances d_1 and d_2 from these points to the origin.

(b) An example of an inverse square field is the gravitational field $\mathbf{F} = -(mMG)\mathbf{r}/|\mathbf{r}|^3$ discussed in Example 16.1.4. Use part (a) to find the work done by the gravitational field when the earth moves from aphelion (at a maximum distance of 1.52×10^8 km from the sun) to perihelion (at a minimum distance of 1.47×10^8 km). (Use the values $m = 5.97 \times 10^{24}$ kg, $M = 1.99 \times 10^{30}$ kg, and $G = 6.67 \times 10^{-11}\ \text{N}\cdot\text{m}^2/\text{kg}^2$.)

(c) Another example of an inverse square field is the electric force field $\mathbf{F} = \varepsilon qQ\mathbf{r}/|\mathbf{r}|^3$ discussed in Example 16.1.5. Suppose that an electron with a charge of -1.6×10^{-19} C is located at the origin. A positive unit charge is positioned a distance 10^{-12} m from the electron and moves to a position half that distance from the electron. Use part (a) to find the work done by the electric force field. (Use the value $\varepsilon = 8.985 \times 10^9$.)

16.4 Green's Theorem

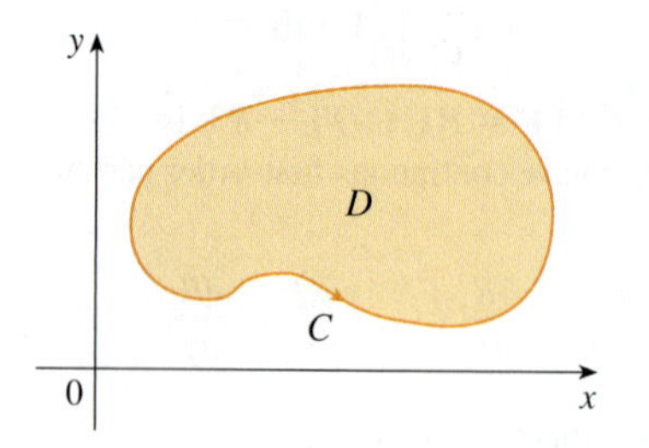

FIGURE 1

Green's Theorem gives the relationship between a line integral around a simple closed curve C and a double integral over the plane region D bounded by C. (See Figure 1. We assume that D consists of all points inside C as well as all points on C.) In stating Green's Theorem we use the convention that the **positive orientation** of a simple closed curve C refers to a single *counterclockwise* traversal of C. Thus if C is given by the vector function $\mathbf{r}(t)$, $a \leqslant t \leqslant b$, then the region D is always on the left as the point $\mathbf{r}(t)$ traverses C. (See Figure 2.)

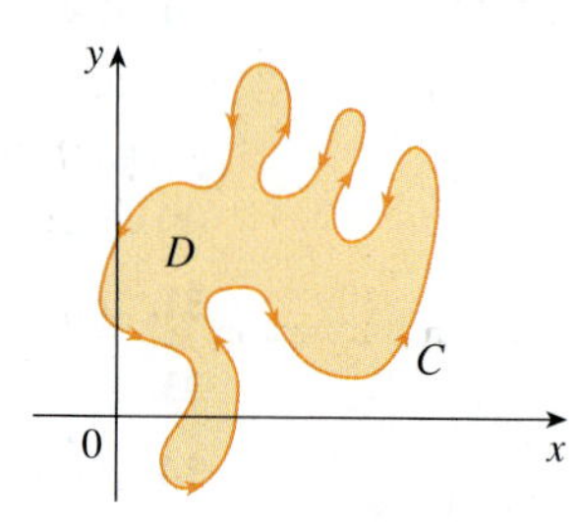

(a) Positive orientation

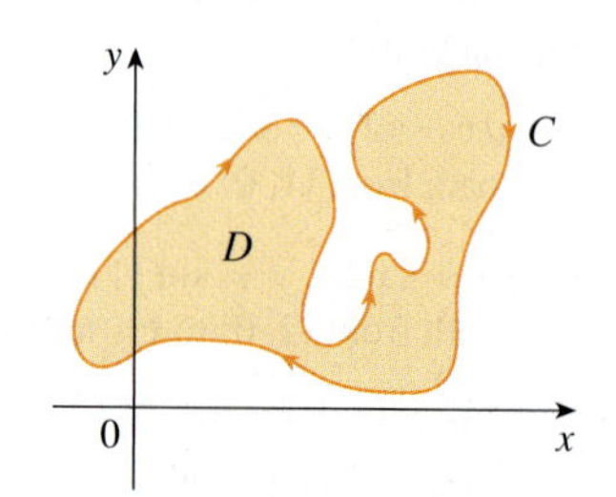

(b) Negative orientation

FIGURE 2

Green's Theorem Let C be a positively oriented, piecewise-smooth, simple closed curve in the plane and let D be the region bounded by C. If P and Q have continuous partial derivatives on an open region that contains D, then

$$\int_C P\,dx + Q\,dy = \iint_D \left(\frac{\partial Q}{\partial x} - \frac{\partial P}{\partial y} \right) dA$$

Recall that the left side of this equation is another way of writing $\int_C \mathbf{F} \cdot d\mathbf{r}$, where $\mathbf{F} = P\,\mathbf{i} + Q\,\mathbf{j}$.

NOTE The notation

$$\oint_C P\,dx + Q\,dy \qquad \text{or} \qquad \circlearrowleft\!\!\!\!\int_C P\,dx + Q\,dy$$

is sometimes used to indicate that the line integral is calculated using the positive orientation of the closed curve C. Another notation for the positively oriented boundary curve of D is ∂D, so the equation in Green's Theorem can be written as

$$\boxed{1} \qquad \iint_D \left(\frac{\partial Q}{\partial x} - \frac{\partial P}{\partial y} \right) dA = \int_{\partial D} P\,dx + Q\,dy$$

Green's Theorem should be regarded as the counterpart of the Fundamental Theorem of Calculus for double integrals. Compare Equation 1 with the statement of the Fundamental Theorem of Calculus, Part 2, in the following equation:

$$\int_a^b F'(x)\,dx = F(b) - F(a)$$

In both cases there is an integral involving derivatives (F', $\partial Q/\partial x$, and $\partial P/\partial y$) on the left side of the equation. And in both cases the right side involves the values of the original functions (F, Q, and P) only on the *boundary* of the domain. (In the one-dimensional case, the domain is an interval $[a, b]$ whose boundary consists of just two points, a and b.)

Green's Theorem is not easy to prove in general, but we can give a proof for the special case where the region is both type I and type II (see Section 15.2). Let's call such regions **simple regions**.

George Green

Green's Theorem is named after the self-taught English scientist George Green (1793–1841). He worked full-time in his father's bakery from the age of nine and taught himself mathematics from library books. In 1828 he published privately *An Essay on the Application of Mathematical Analysis to the Theories of Electricity and Magnetism*, but only 100 copies were printed and most of those went to his friends. This pamphlet contained a theorem that is equivalent to what we know as Green's Theorem, but it didn't become widely known at that time. Finally, at age 40, Green entered Cambridge University as an undergraduate but died four years after graduation. In 1846 William Thomson (Lord Kelvin) located a copy of Green's essay, realized its significance, and had it reprinted. Green was the first person to try to formulate a mathematical theory of electricity and magnetism. His work was the basis for the subsequent electromagnetic theories of Thomson, Stokes, Rayleigh, and Maxwell.

PROOF OF GREEN'S THEOREM FOR THE CASE IN WHICH D IS A SIMPLE REGION Notice that Green's Theorem will be proved if we can show that

$$\boxed{2} \qquad \int_C P\,dx = -\iint_D \frac{\partial P}{\partial y}\,dA$$

and

$$\boxed{3} \qquad \int_C Q\,dy = \iint_D \frac{\partial Q}{\partial x}\,dA$$

We prove Equation 2 by expressing D as a type I region:

$$D = \{(x, y) \mid a \leqslant x \leqslant b,\ g_1(x) \leqslant y \leqslant g_2(x)\}$$

where g_1 and g_2 are continuous functions. This enables us to compute the double integral on the right side of Equation 2 as follows:

$$\boxed{4} \qquad \iint_D \frac{\partial P}{\partial y}\,dA = \int_a^b \int_{g_1(x)}^{g_2(x)} \frac{\partial P}{\partial y}(x, y)\,dy\,dx = \int_a^b [P(x, g_2(x)) - P(x, g_1(x))]\,dx$$

where the last step follows from the Fundamental Theorem of Calculus.

Now we compute the left side of Equation 2 by breaking up C as the union of the four curves C_1, C_2, C_3, and C_4 shown in Figure 3. On C_1 we take x as the parameter and write the parametric equations as $x = x$, $y = g_1(x)$, $a \leqslant x \leqslant b$. Thus

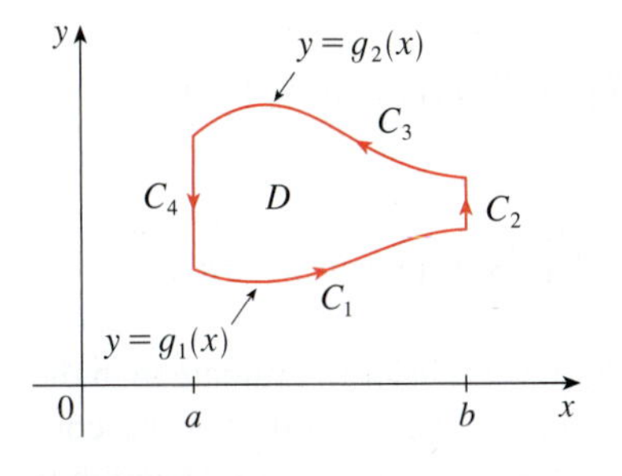

FIGURE 3

$$\int_{C_1} P(x, y)\,dx = \int_a^b P(x, g_1(x))\,dx$$

Observe that C_3 goes from right to left but $-C_3$ goes from left to right, so we can write the parametric equations of $-C_3$ as $x = x$, $y = g_2(x)$, $a \leqslant x \leqslant b$. Therefore

$$\int_{C_3} P(x, y)\,dx = -\int_{-C_3} P(x, y)\,dx = -\int_a^b P(x, g_2(x))\,dx$$

On C_2 or C_4 (either of which might reduce to just a single point), x is constant, so $dx = 0$ and

$$\int_{C_2} P(x, y)\,dx = 0 = \int_{C_4} P(x, y)\,dx$$

Hence

$$\int_C P(x, y)\,dx = \int_{C_1} P(x, y)\,dx + \int_{C_2} P(x, y)\,dx + \int_{C_3} P(x, y)\,dx + \int_{C_4} P(x, y)\,dx$$

$$= \int_a^b P(x, g_1(x))\,dx - \int_a^b P(x, g_2(x))\,dx$$

Comparing this expression with the one in Equation 4, we see that

$$\int_C P(x, y)\, dx = -\iint_D \frac{\partial P}{\partial y}\, dA$$

Equation 3 can be proved in much the same way by expressing D as a type II region (see Exercise 30). Then, by adding Equations 2 and 3, we obtain Green's Theorem. ■

EXAMPLE 1 Evaluate $\int_C x^4\, dx + xy\, dy$, where C is the triangular curve consisting of the line segments from $(0, 0)$ to $(1, 0)$, from $(1, 0)$ to $(0, 1)$, and from $(0, 1)$ to $(0, 0)$.

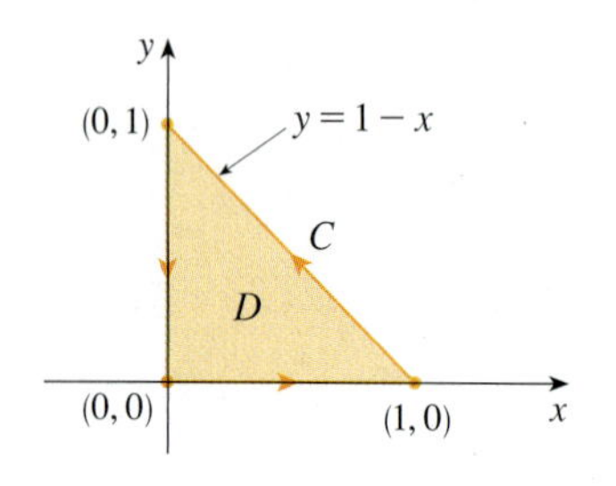

FIGURE 4

SOLUTION Although the given line integral could be evaluated as usual by the methods of Section 16.2, that would involve setting up three separate integrals along the three sides of the triangle, so let's use Green's Theorem instead. Notice that the region D enclosed by C is simple and C has positive orientation (see Figure 4). If we let $P(x, y) = x^4$ and $Q(x, y) = xy$, then we have

$$\int_C x^4\, dx + xy\, dy = \iint_D \left(\frac{\partial Q}{\partial x} - \frac{\partial P}{\partial y} \right) dA = \int_0^1 \int_0^{1-x} (y - 0)\, dy\, dx$$

$$= \int_0^1 \left[\tfrac{1}{2}y^2 \right]_{y=0}^{y=1-x} dx = \tfrac{1}{2} \int_0^1 (1 - x)^2\, dx$$

$$= -\tfrac{1}{6}(1 - x)^3 \Big]_0^1 = \tfrac{1}{6}$$ ■

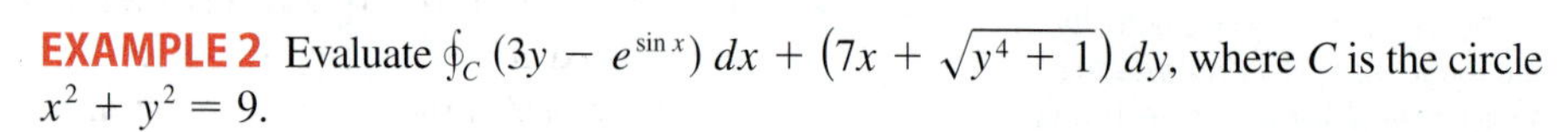

EXAMPLE 2 Evaluate $\oint_C (3y - e^{\sin x})\, dx + \left(7x + \sqrt{y^4 + 1}\right) dy$, where C is the circle $x^2 + y^2 = 9$.

SOLUTION The region D bounded by C is the disk $x^2 + y^2 \leqslant 9$, so let's change to polar coordinates after applying Green's Theorem:

$$\oint_C (3y - e^{\sin x})\, dx + \left(7x + \sqrt{y^4 + 1}\right) dy$$

$$= \iint_D \left[\frac{\partial}{\partial x}\left(7x + \sqrt{y^4 + 1}\right) - \frac{\partial}{\partial y}(3y - e^{\sin x}) \right] dA$$

$$= \int_0^{2\pi} \int_0^3 (7 - 3)\, r\, dr\, d\theta = 4 \int_0^{2\pi} d\theta \int_0^3 r\, dr = 36\pi$$ ■

Instead of using polar coordinates, we could simply use the fact that D is a disk of radius 3 and write

$$\iint_D 4\, dA = 4 \cdot \pi(3)^2 = 36\pi$$

In Examples 1 and 2 we found that the double integral was easier to evaluate than the line integral. (Try setting up the line integral in Example 2 and you'll soon be convinced!) But sometimes it's easier to evaluate the line integral, and Green's Theorem is used in the reverse direction. For instance, if it is known that $P(x, y) = Q(x, y) = 0$ on the curve C, then Green's Theorem gives

$$\iint_D \left(\frac{\partial Q}{\partial x} - \frac{\partial P}{\partial y} \right) dA = \int_C P\, dx + Q\, dy = 0$$

no matter what values P and Q assume in the region D.

Another application of the reverse direction of Green's Theorem is in computing areas. Since the area of D is $\iint_D 1\, dA$, we wish to choose P and Q so that

$$\frac{\partial Q}{\partial x} - \frac{\partial P}{\partial y} = 1$$

There are several possibilities:

$$P(x, y) = 0 \qquad P(x, y) = -y \qquad P(x, y) = -\tfrac{1}{2}y$$

$$Q(x, y) = x \qquad Q(x, y) = 0 \qquad Q(x, y) = \tfrac{1}{2}x$$

Then Green's Theorem gives the following formulas for the area of D:

5
$$A = \oint_C x\,dy = -\oint_C y\,dx = \tfrac{1}{2}\oint_C x\,dy - y\,dx$$

EXAMPLE 3 Find the area enclosed by the ellipse $\dfrac{x^2}{a^2} + \dfrac{y^2}{b^2} = 1$.

SOLUTION The ellipse has parametric equations $x = a\cos t$ and $y = b\sin t$, where $0 \leqslant t \leqslant 2\pi$. Using the third formula in Equation 5, we have

$$\begin{aligned} A &= \tfrac{1}{2}\int_C x\,dy - y\,dx \\ &= \tfrac{1}{2}\int_0^{2\pi} (a\cos t)(b\cos t)\,dt - (b\sin t)(-a\sin t)\,dt \\ &= \frac{ab}{2}\int_0^{2\pi} dt = \pi ab \end{aligned}$$

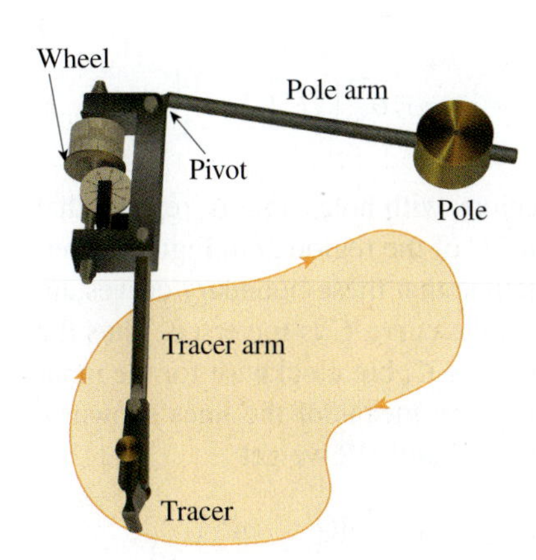

FIGURE 5
A Keuffel and Esser polar planimeter

Formula 5 can be used to explain how planimeters work. A **planimeter** is a mechanical instrument used for measuring the area of a region by tracing its boundary curve. These devices are useful in all the sciences: in biology for measuring the area of leaves or wings, in medicine for measuring the size of cross-sections of organs or tumors, in forestry for estimating the size of forested regions from photographs.

Figure 5 shows the operation of a polar planimeter: the pole is fixed and, as the tracer is moved along the boundary curve of the region, the wheel partly slides and partly rolls perpendicular to the tracer arm. The planimeter measures the distance that the wheel rolls and this is proportional to the area of the enclosed region. The explanation as a consequence of Formula 5 can be found in the following articles:

- R. W. Gatterman, "The planimeter as an example of Green's Theorem" *Amer. Math. Monthly,* Vol. 88 (1981), pp. 701–4.
- Tanya Leise, "As the planimeter wheel turns" *College Math. Journal,* Vol. 38 (2007), pp. 24–31.

Extended Versions of Green's Theorem

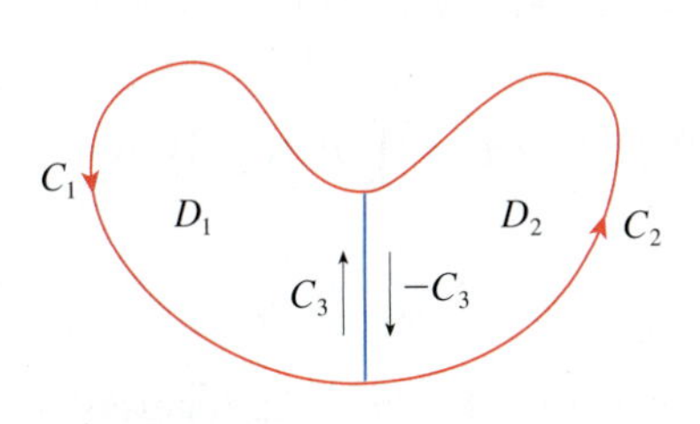

FIGURE 6

Although we have proved Green's Theorem only for the case where D is simple, we can now extend it to the case where D is a finite union of simple regions. For example, if D is the region shown in Figure 6, then we can write $D = D_1 \cup D_2$, where D_1 and D_2 are both simple. The boundary of D_1 is $C_1 \cup C_3$ and the boundary of D_2 is $C_2 \cup (-C_3)$ so, applying Green's Theorem to D_1 and D_2 separately, we get

$$\int_{C_1 \cup C_3} P\,dx + Q\,dy = \iint_{D_1} \left(\frac{\partial Q}{\partial x} - \frac{\partial P}{\partial y}\right) dA$$

$$\int_{C_2 \cup (-C_3)} P\,dx + Q\,dy = \iint_{D_2} \left(\frac{\partial Q}{\partial x} - \frac{\partial P}{\partial y}\right) dA$$

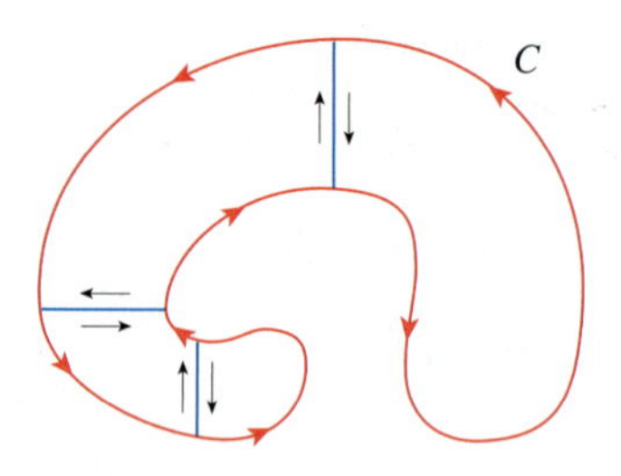

FIGURE 7

If we add these two equations, the line integrals along C_3 and $-C_3$ cancel, so we get

$$\int_{C_1 \cup C_2} P\,dx + Q\,dy = \iint_D \left(\frac{\partial Q}{\partial x} - \frac{\partial P}{\partial y} \right) dA$$

which is Green's Theorem for $D = D_1 \cup D_2$, since its boundary is $C = C_1 \cup C_2$.

The same sort of argument allows us to establish Green's Theorem for any finite union of nonoverlapping simple regions (see Figure 7).

EXAMPLE 4 Evaluate $\oint_C y^2\,dx + 3xy\,dy$, where C is the boundary of the semiannular region D in the upper half-plane between the circles $x^2 + y^2 = 1$ and $x^2 + y^2 = 4$.

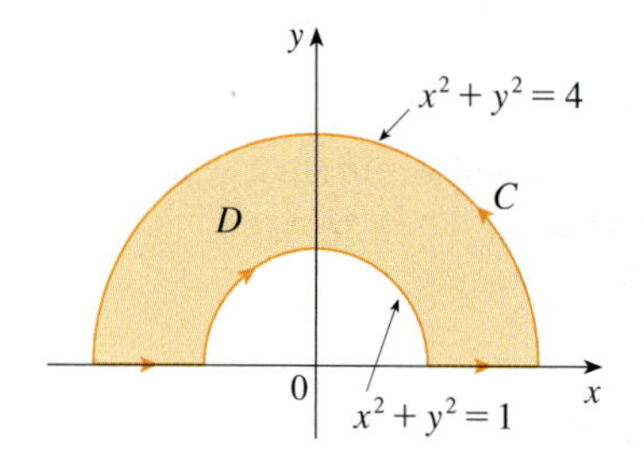

FIGURE 8

SOLUTION Notice that although D is not simple, the y-axis divides it into two simple regions (see Figure 8). In polar coordinates we can write

$$D = \{(r, \theta) \mid 1 \leqslant r \leqslant 2,\ 0 \leqslant \theta \leqslant \pi\}$$

Therefore Green's Theorem gives

$$\oint_C y^2\,dx + 3xy\,dy = \iint_D \left[\frac{\partial}{\partial x}(3xy) - \frac{\partial}{\partial y}(y^2) \right] dA$$

$$= \iint_D y\,dA = \int_0^{\pi} \int_1^2 (r \sin\theta)\, r\,dr\,d\theta$$

$$= \int_0^{\pi} \sin\theta\,d\theta \int_1^2 r^2\,dr = \Big[-\cos\theta\Big]_0^{\pi} \Big[\tfrac{1}{3}r^3\Big]_1^2 = \frac{14}{3}$$

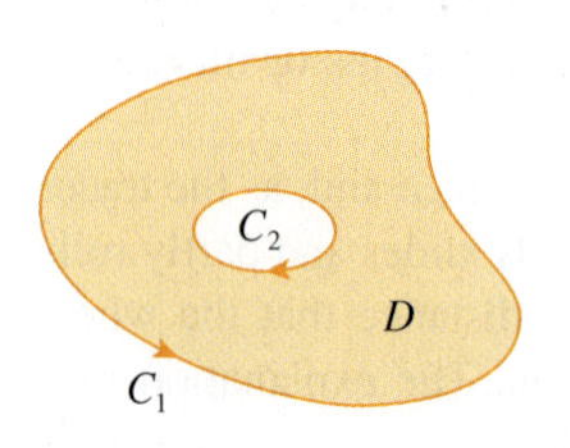

FIGURE 9

Green's Theorem can be extended to apply to regions with holes, that is, regions that are not simply-connected. Observe that the boundary C of the region D in Figure 9 consists of two simple closed curves C_1 and C_2. We assume that these boundary curves are oriented so that the region D is always on the left as the curve C is traversed. Thus the positive direction is counterclockwise for the outer curve C_1 but clockwise for the inner curve C_2. If we divide D into two regions D' and D'' by means of the lines shown in Figure 10 and then apply Green's Theorem to each of D' and D'', we get

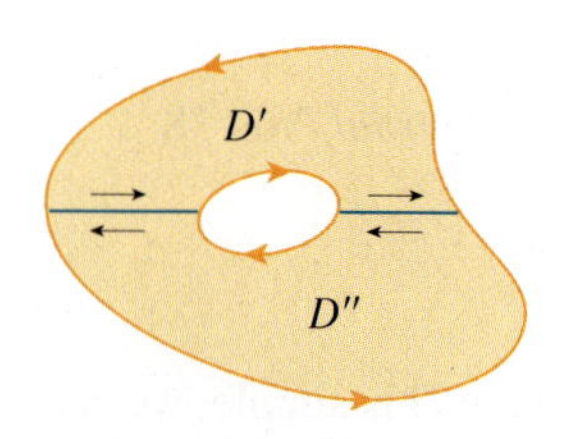

FIGURE 10

$$\iint_D \left(\frac{\partial Q}{\partial x} - \frac{\partial P}{\partial y} \right) dA = \iint_{D'} \left(\frac{\partial Q}{\partial x} - \frac{\partial P}{\partial y} \right) dA + \iint_{D''} \left(\frac{\partial Q}{\partial x} - \frac{\partial P}{\partial y} \right) dA$$

$$= \int_{\partial D'} P\,dx + Q\,dy + \int_{\partial D''} P\,dx + Q\,dy$$

Since the line integrals along the common boundary lines are in opposite directions, they cancel and we get

$$\iint_D \left(\frac{\partial Q}{\partial x} - \frac{\partial P}{\partial y} \right) dA = \int_{C_1} P\,dx + Q\,dy + \int_{C_2} P\,dx + Q\,dy = \int_C P\,dx + Q\,dy$$

which is Green's Theorem for the region D.

EXAMPLE 5 If $\mathbf{F}(x, y) = (-y\,\mathbf{i} + x\,\mathbf{j})/(x^2 + y^2)$, show that $\int_C \mathbf{F} \cdot d\mathbf{r} = 2\pi$ for every positively oriented simple closed path that encloses the origin.

SOLUTION Since C is an *arbitrary* closed path that encloses the origin, it's difficult to compute the given integral directly. So let's consider a counterclockwise-oriented circle C'

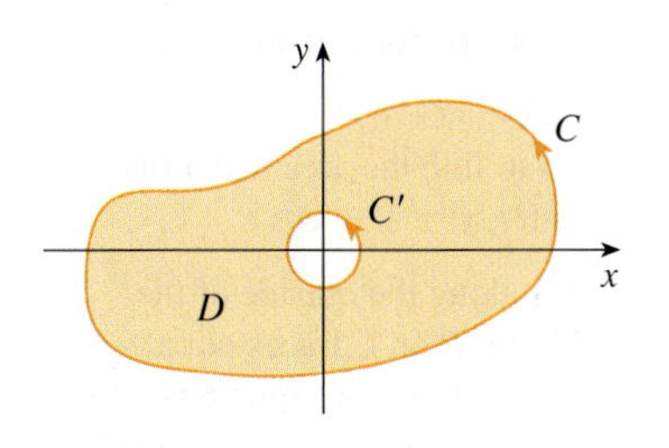

FIGURE 11

with center the origin and radius a, where a is chosen to be small enough that C' lies inside C. (See Figure 11.) Let D be the region bounded by C and C'. Then its positively oriented boundary is $C \cup (-C')$ and so the general version of Green's Theorem gives

$$\int_C P\,dx + Q\,dy + \int_{-C'} P\,dx + Q\,dy = \iint_D \left(\frac{\partial Q}{\partial x} - \frac{\partial P}{\partial y} \right) dA$$

$$= \iint_D \left[\frac{y^2 - x^2}{(x^2 + y^2)^2} - \frac{y^2 - x^2}{(x^2 + y^2)^2} \right] dA = 0$$

Therefore
$$\int_C P\,dx + Q\,dy = \int_{C'} P\,dx + Q\,dy$$

that is,
$$\int_C \mathbf{F} \cdot d\mathbf{r} = \int_{C'} \mathbf{F} \cdot d\mathbf{r}$$

We now easily compute this last integral using the parametrization given by $\mathbf{r}(t) = a\cos t\,\mathbf{i} + a\sin t\,\mathbf{j}$, $0 \leqslant t \leqslant 2\pi$. Thus

$$\int_C \mathbf{F} \cdot d\mathbf{r} = \int_{C'} \mathbf{F} \cdot d\mathbf{r} = \int_0^{2\pi} \mathbf{F}(\mathbf{r}(t)) \cdot \mathbf{r}'(t)\,dt$$

$$= \int_0^{2\pi} \frac{(-a\sin t)(-a\sin t) + (a\cos t)(a\cos t)}{a^2\cos^2 t + a^2\sin^2 t}\,dt = \int_0^{2\pi} dt = 2\pi$$

We end this section by using Green's Theorem to discuss a result that was stated in the preceding section.

SKETCH OF PROOF OF THEOREM 16.3.6 We're assuming that $\mathbf{F} = P\,\mathbf{i} + Q\,\mathbf{j}$ is a vector field on an open simply-connected region D, that P and Q have continuous first-order partial derivatives, and that

$$\frac{\partial P}{\partial y} = \frac{\partial Q}{\partial x} \qquad \text{throughout } D$$

If C is any simple closed path in D and R is the region that C encloses, then Green's Theorem gives

$$\oint_C \mathbf{F} \cdot d\mathbf{r} = \oint_C P\,dx + Q\,dy = \iint_R \left(\frac{\partial Q}{\partial x} - \frac{\partial P}{\partial y} \right) dA = \iint_R 0\,dA = 0$$

A curve that is not simple crosses itself at one or more points and can be broken up into a number of simple curves. We have shown that the line integrals of $\mathbf{F}$ around these simple curves are all 0 and, adding these integrals, we see that $\int_C \mathbf{F} \cdot d\mathbf{r} = 0$ for any closed curve C. Therefore $\int_C \mathbf{F} \cdot d\mathbf{r}$ is independent of path in D by Theorem 16.3.3. It follows that $\mathbf{F}$ is a conservative vector field.

16.4 EXERCISES

1–4 Evaluate the line integral by two methods: (a) directly and (b) using Green's Theorem.

1. $\oint_C y^2\,dx + x^2y\,dy$,
C is the rectangle with vertices (0, 0), (5, 0), (5, 4), and (0, 4)

2. $\oint_C y\,dx - x\,dy$,
C is the circle with center the origin and radius 4

3. $\oint_C xy\,dx + x^2y^3\,dy$,
C is the triangle with vertices (0, 0), (1, 0), and (1, 2)

4. $\oint_C x^2y^2\,dx + xy\,dy$, C consists of the arc of the parabola $y = x^2$ from $(0, 0)$ to $(1, 1)$ and the line segments from $(1, 1)$ to $(0, 1)$ and from $(0, 1)$ to $(0, 0)$

5–10 Use Green's Theorem to evaluate the line integral along the given positively oriented curve.

5. $\int_C ye^x\,dx + 2e^x\,dy$,
C is the rectangle with vertices $(0, 0)$, $(3, 0)$, $(3, 4)$, and $(0, 4)$

6. $\int_C (x^2 + y^2)\,dx + (x^2 - y^2)\,dy$,
C is the triangle with vertices $(0, 0)$, $(2, 1)$, and $(0, 1)$

7. $\int_C \left(y + e^{\sqrt{x}}\right)dx + (2x + \cos y^2)\,dy$,
C is the boundary of the region enclosed by the parabolas $y = x^2$ and $x = y^2$

8. $\int_C y^4\,dx + 2xy^3\,dy$, C is the ellipse $x^2 + 2y^2 = 2$

9. $\int_C y^3\,dx - x^3\,dy$, C is the circle $x^2 + y^2 = 4$

10. $\int_C (1 - y^3)\,dx + (x^3 + e^{y^2})\,dy$, C is the boundary of the region between the circles $x^2 + y^2 = 4$ and $x^2 + y^2 = 9$

11–14 Use Green's Theorem to evaluate $\int_C \mathbf{F} \cdot d\mathbf{r}$. (Check the orientation of the curve before applying the theorem.)

11. $\mathbf{F}(x, y) = \langle y\cos x - xy\sin x,\, xy + x\cos x\rangle$,
C is the triangle from $(0, 0)$ to $(0, 4)$ to $(2, 0)$ to $(0, 0)$

12. $\mathbf{F}(x, y) = \langle e^{-x} + y^2,\, e^{-y} + x^2\rangle$,
C consists of the arc of the curve $y = \cos x$ from $(-\pi/2, 0)$ to $(\pi/2, 0)$ and the line segment from $(\pi/2, 0)$ to $(-\pi/2, 0)$

13. $\mathbf{F}(x, y) = \langle y - \cos y,\, x\sin y\rangle$,
C is the circle $(x - 3)^2 + (y + 4)^2 = 4$ oriented clockwise

14. $\mathbf{F}(x, y) = \left\langle \sqrt{x^2 + 1},\, \tan^{-1}x\right\rangle$, C is the triangle from $(0, 0)$ to $(1, 1)$ to $(0, 1)$ to $(0, 0)$

CAS **15–16** Verify Green's Theorem by using a computer algebra system to evaluate both the line integral and the double integral.

15. $P(x, y) = x^3y^4$, $Q(x, y) = x^5y^4$,
C consists of the line segment from $(-\pi/2, 0)$ to $(\pi/2, 0)$ followed by the arc of the curve $y = \cos x$ from $(\pi/2, 0)$ to $(-\pi/2, 0)$

16. $P(x, y) = 2x - x^3y^5$, $Q(x, y) = x^3y^8$,
C is the ellipse $4x^2 + y^2 = 4$

17. Use Green's Theorem to find the work done by the force $\mathbf{F}(x, y) = x(x + y)\,\mathbf{i} + xy^2\,\mathbf{j}$ in moving a particle from the origin along the x-axis to $(1, 0)$, then along the line segment to $(0, 1)$, and then back to the origin along the y-axis.

18. A particle starts at the origin, moves along the x-axis to $(5, 0)$, then along the quarter-circle $x^2 + y^2 = 25$, $x \geqslant 0$, $y \geqslant 0$ to the point $(0, 5)$, and then down the y-axis back to the origin. Use Green's Theorem to find the work done on this particle by the force field $\mathbf{F}(x, y) = \left\langle \sin x,\, \sin y + xy^2 + \frac{1}{3}x^3\right\rangle$.

19. Use one of the formulas in (5) to find the area under one arch of the cycloid $x = t - \sin t$, $y = 1 - \cos t$.

20. If a circle C with radius 1 rolls along the outside of the circle $x^2 + y^2 = 16$, a fixed point P on C traces out a curve called an *epicycloid*, with parametric equations $x = 5\cos t - \cos 5t$, $y = 5\sin t - \sin 5t$. Graph the epicycloid and use (5) to find the area it encloses.

21. (a) If C is the line segment connecting the point (x_1, y_1) to the point (x_2, y_2), show that

$$\int_C x\,dy - y\,dx = x_1y_2 - x_2y_1$$

(b) If the vertices of a polygon, in counterclockwise order, are $(x_1, y_1), (x_2, y_2), \ldots, (x_n, y_n)$, show that the area of the polygon is

$$A = \tfrac{1}{2}[(x_1y_2 - x_2y_1) + (x_2y_3 - x_3y_2) + \cdots + (x_{n-1}y_n - x_ny_{n-1}) + (x_ny_1 - x_1y_n)]$$

(c) Find the area of the pentagon with vertices $(0, 0)$, $(2, 1)$, $(1, 3)$, $(0, 2)$, and $(-1, 1)$.

22. Let D be a region bounded by a simple closed path C in the xy-plane. Use Green's Theorem to prove that the coordinates of the centroid $(\bar{x}, \bar{y})$ of D are

$$\bar{x} = \frac{1}{2A}\oint_C x^2\,dy \qquad \bar{y} = -\frac{1}{2A}\oint_C y^2\,dx$$

where A is the area of D.

23. Use Exercise 22 to find the centroid of a quarter-circular region of radius a.

24. Use Exercise 22 to find the centroid of the triangle with vertices $(0, 0)$, $(a, 0)$, and (a, b), where $a > 0$ and $b > 0$.

25. A plane lamina with constant density $\rho(x, y) = \rho$ occupies a region in the xy-plane bounded by a simple closed path C. Show that its moments of inertia about the axes are

$$I_x = -\frac{\rho}{3}\oint_C y^3\,dx \qquad I_y = \frac{\rho}{3}\oint_C x^3\,dy$$

26. Use Exercise 25 to find the moment of inertia of a circular disk of radius a with constant density ρ about a diameter. (Compare with Example 15.4.4.)

27. Use the method of Example 5 to calculate $\int_C \mathbf{F} \cdot d\mathbf{r}$, where

$$\mathbf{F}(x, y) = \frac{2xy\,\mathbf{i} + (y^2 - x^2)\,\mathbf{j}}{(x^2 + y^2)^2}$$

and C is any positively oriented simple closed curve that encloses the origin.

28. Calculate $\int_C \mathbf{F} \cdot d\mathbf{r}$, where $\mathbf{F}(x, y) = \langle x^2 + y,\, 3x - y^2\rangle$ and C is the positively oriented boundary curve of a region D that has area 6.

29. If $\mathbf{F}$ is the vector field of Example 5, show that $\int_C \mathbf{F} \cdot d\mathbf{r} = 0$ for every simple closed path that does not pass through or enclose the origin.

30. Complete the proof of the special case of Green's Theorem by proving Equation 3.

31. Use Green's Theorem to prove the change of variables formula for a double integral (Formula 15.9.9) for the case where $f(x, y) = 1$:

$$\iint_R dx\,dy = \iint_S \left| \frac{\partial(x, y)}{\partial(u, v)} \right| du\,dv$$

Here R is the region in the xy-plane that corresponds to the region S in the uv-plane under the transformation given by $x = g(u, v)$, $y = h(u, v)$.

[*Hint:* Note that the left side is $A(R)$ and apply the first part of Equation 5. Convert the line integral over ∂R to a line integral over ∂S and apply Green's Theorem in the uv-plane.]

16.5 Curl and Divergence

In this section we define two operations that can be performed on vector fields and that play a basic role in the applications of vector calculus to fluid flow and electricity and magnetism. Each operation resembles differentiation, but one produces a vector field whereas the other produces a scalar field.

Curl

If $\mathbf{F} = P\,\mathbf{i} + Q\,\mathbf{j} + R\,\mathbf{k}$ is a vector field on $\mathbb{R}^3$ and the partial derivatives of P, Q, and R all exist, then the **curl** of $\mathbf{F}$ is the vector field on $\mathbb{R}^3$ defined by

1
$$\operatorname{curl} \mathbf{F} = \left(\frac{\partial R}{\partial y} - \frac{\partial Q}{\partial z}\right)\mathbf{i} + \left(\frac{\partial P}{\partial z} - \frac{\partial R}{\partial x}\right)\mathbf{j} + \left(\frac{\partial Q}{\partial x} - \frac{\partial P}{\partial y}\right)\mathbf{k}$$

As an aid to our memory, let's rewrite Equation 1 using operator notation. We introduce the vector differential operator ∇ ("del") as

$$\nabla = \mathbf{i}\frac{\partial}{\partial x} + \mathbf{j}\frac{\partial}{\partial y} + \mathbf{k}\frac{\partial}{\partial z}$$

It has meaning when it operates on a scalar function to produce the gradient of f:

$$\nabla f = \mathbf{i}\frac{\partial f}{\partial x} + \mathbf{j}\frac{\partial f}{\partial y} + \mathbf{k}\frac{\partial f}{\partial z} = \frac{\partial f}{\partial x}\mathbf{i} + \frac{\partial f}{\partial y}\mathbf{j} + \frac{\partial f}{\partial z}\mathbf{k}$$

If we think of ∇ as a vector with components $\partial/\partial x$, $\partial/\partial y$, and $\partial/\partial z$, we can also consider the formal cross product of ∇ with the vector field $\mathbf{F}$ as follows:

$$\begin{aligned}
\nabla \times \mathbf{F} &= \begin{vmatrix} \mathbf{i} & \mathbf{j} & \mathbf{k} \\ \dfrac{\partial}{\partial x} & \dfrac{\partial}{\partial y} & \dfrac{\partial}{\partial z} \\ P & Q & R \end{vmatrix} \\
&= \left(\frac{\partial R}{\partial y} - \frac{\partial Q}{\partial z}\right)\mathbf{i} + \left(\frac{\partial P}{\partial z} - \frac{\partial R}{\partial x}\right)\mathbf{j} + \left(\frac{\partial Q}{\partial x} - \frac{\partial P}{\partial y}\right)\mathbf{k} \\
&= \operatorname{curl} \mathbf{F}
\end{aligned}$$

So the easiest way to remember Definition 1 is by means of the symbolic expression

2
$$\operatorname{curl} \mathbf{F} = \nabla \times \mathbf{F}$$

EXAMPLE 1 If $\mathbf{F}(x, y, z) = xz\,\mathbf{i} + xyz\,\mathbf{j} - y^2\,\mathbf{k}$, find curl $\mathbf{F}$.

SOLUTION Using Equation 2, we have

$$\text{curl}\,\mathbf{F} = \nabla \times \mathbf{F} = \begin{vmatrix} \mathbf{i} & \mathbf{j} & \mathbf{k} \\ \dfrac{\partial}{\partial x} & \dfrac{\partial}{\partial y} & \dfrac{\partial}{\partial z} \\ xz & xyz & -y^2 \end{vmatrix}$$

$$= \left[\frac{\partial}{\partial y}(-y^2) - \frac{\partial}{\partial z}(xyz)\right]\mathbf{i} - \left[\frac{\partial}{\partial x}(-y^2) - \frac{\partial}{\partial z}(xz)\right]\mathbf{j}$$

$$+ \left[\frac{\partial}{\partial x}(xyz) - \frac{\partial}{\partial y}(xz)\right]\mathbf{k}$$

$$= (-2y - xy)\,\mathbf{i} - (0 - x)\,\mathbf{j} + (yz - 0)\,\mathbf{k}$$

$$= -y(2 + x)\,\mathbf{i} + x\,\mathbf{j} + yz\,\mathbf{k}$$

CAS Most computer algebra systems have commands that compute the curl and divergence of vector fields. If you have access to a CAS, use these commands to check the answers to the examples and exercises in this section.

Recall that the gradient of a function f of three variables is a vector field on $\mathbb{R}^3$ and so we can compute its curl. The following theorem says that the curl of a gradient vector field is $\mathbf{0}$.

3 Theorem If f is a function of three variables that has continuous second-order partial derivatives, then

$$\text{curl}(\nabla f) = \mathbf{0}$$

PROOF We have

$$\text{curl}(\nabla f) = \nabla \times (\nabla f) = \begin{vmatrix} \mathbf{i} & \mathbf{j} & \mathbf{k} \\ \dfrac{\partial}{\partial x} & \dfrac{\partial}{\partial y} & \dfrac{\partial}{\partial z} \\ \dfrac{\partial f}{\partial x} & \dfrac{\partial f}{\partial y} & \dfrac{\partial f}{\partial z} \end{vmatrix}$$

$$= \left(\frac{\partial^2 f}{\partial y\,\partial z} - \frac{\partial^2 f}{\partial z\,\partial y}\right)\mathbf{i} + \left(\frac{\partial^2 f}{\partial z\,\partial x} - \frac{\partial^2 f}{\partial x\,\partial z}\right)\mathbf{j} + \left(\frac{\partial^2 f}{\partial x\,\partial y} - \frac{\partial^2 f}{\partial y\,\partial x}\right)\mathbf{k}$$

$$= 0\,\mathbf{i} + 0\,\mathbf{j} + 0\,\mathbf{k} = \mathbf{0}$$

by Clairaut's Theorem.

Notice the similarity to what we know from Section 12.4: $\mathbf{a} \times \mathbf{a} = \mathbf{0}$ for every three-dimensional vector $\mathbf{a}$.

Since a conservative vector field is one for which $\mathbf{F} = \nabla f$, Theorem 3 can be rephrased as follows:

If $\mathbf{F}$ is conservative, then curl $\mathbf{F} = \mathbf{0}$.

Compare this with Exercise 16.3.29.

This gives us a way of verifying that a vector field is not conservative.

EXAMPLE 2 Show that the vector field $\mathbf{F}(x, y, z) = xz\,\mathbf{i} + xyz\,\mathbf{j} - y^2\,\mathbf{k}$ is not conservative.

SOLUTION In Example 1 we showed that

$$\text{curl } \mathbf{F} = -y(2 + x)\,\mathbf{i} + x\,\mathbf{j} + yz\,\mathbf{k}$$

This shows that curl $\mathbf{F} \neq \mathbf{0}$ and so, by the remarks preceding this example, $\mathbf{F}$ is not conservative. ■

The converse of Theorem 3 is not true in general, but the following theorem says the converse is true if $\mathbf{F}$ is defined everywhere. (More generally it is true if the domain is simply-connected, that is, "has no hole.") Theorem 4 is the three-dimensional version of Theorem 16.3.6. Its proof requires Stokes' Theorem and is sketched at the end of Section 16.8.

4 Theorem If $\mathbf{F}$ is a vector field defined on all of $\mathbb{R}^3$ whose component functions have continuous partial derivatives and curl $\mathbf{F} = \mathbf{0}$, then $\mathbf{F}$ is a conservative vector field.

EXAMPLE 3
(a) Show that

$$\mathbf{F}(x, y, z) = y^2z^3\,\mathbf{i} + 2xyz^3\,\mathbf{j} + 3xy^2z^2\,\mathbf{k}$$

is a conservative vector field.
(b) Find a function f such that $\mathbf{F} = \nabla f$.

SOLUTION
(a) We compute the curl of $\mathbf{F}$:

$$\text{curl } \mathbf{F} = \nabla \times \mathbf{F} = \begin{vmatrix} \mathbf{i} & \mathbf{j} & \mathbf{k} \\ \dfrac{\partial}{\partial x} & \dfrac{\partial}{\partial y} & \dfrac{\partial}{\partial z} \\ y^2z^3 & 2xyz^3 & 3xy^2z^2 \end{vmatrix}$$

$$= (6xyz^2 - 6xyz^2)\,\mathbf{i} - (3y^2z^2 - 3y^2z^2)\,\mathbf{j} + (2yz^3 - 2yz^3)\,\mathbf{k}$$

$$= \mathbf{0}$$

Since curl $\mathbf{F} = \mathbf{0}$ and the domain of $\mathbf{F}$ is $\mathbb{R}^3$, $\mathbf{F}$ is a conservative vector field by Theorem 4.

(b) The technique for finding f was given in Section 16.3. We have

$$f_x(x, y, z) = y^2z^3 \tag{5}$$

$$f_y(x, y, z) = 2xyz^3 \tag{6}$$

$$f_z(x, y, z) = 3xy^2z^2 \tag{7}$$

Integrating (5) with respect to x, we obtain

$$f(x, y, z) = xy^2z^3 + g(y, z) \tag{8}$$

Differentiating (8) with respect to y, we get $f_y(x, y, z) = 2xyz^3 + g_y(y, z)$, so comparison with (6) gives $g_y(y, z) = 0$. Thus $g(y, z) = h(z)$ and

$$f_z(x, y, z) = 3xy^2z^2 + h'(z)$$

Then (7) gives $h'(z) = 0$. Therefore

$$f(x, y, z) = xy^2z^3 + K$$ ■

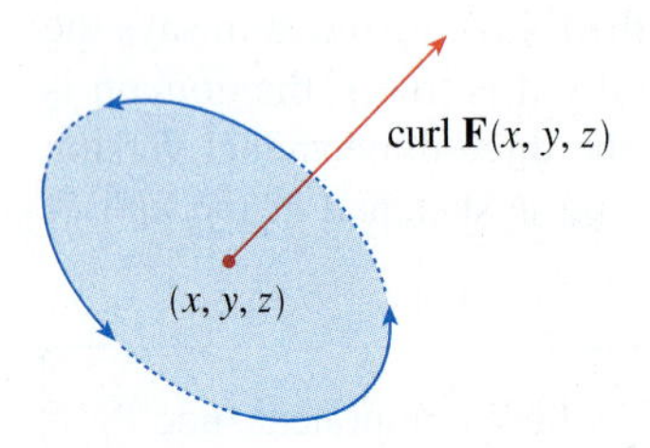

FIGURE 1

The reason for the name *curl* is that the curl vector is associated with rotations. One connection is explained in Exercise 37. Another occurs when **F** represents the velocity field in fluid flow (see Example 16.1.3). Particles near (x, y, z) in the fluid tend to rotate about the axis that points in the direction of curl $\mathbf{F}(x, y, z)$, and the length of this curl vector is a measure of how quickly the particles move around the axis (see Figure 1). If curl $\mathbf{F} = \mathbf{0}$ at a point P, then the fluid is free from rotations at P and **F** is called **irrotational** at P. In other words, there is no whirlpool or eddy at P. If curl $\mathbf{F} = \mathbf{0}$, then a tiny paddle wheel moves with the fluid but doesn't rotate about its axis. If curl $\mathbf{F} \neq \mathbf{0}$, the paddle wheel rotates about its axis. We give a more detailed explanation in Section 16.8 as a consequence of Stokes' Theorem.

Divergence

If $\mathbf{F} = P\,\mathbf{i} + Q\,\mathbf{j} + R\,\mathbf{k}$ is a vector field on $\mathbb{R}^3$ and $\partial P/\partial x$, $\partial Q/\partial y$, and $\partial R/\partial z$ exist, then the **divergence of F** is the function of three variables defined by

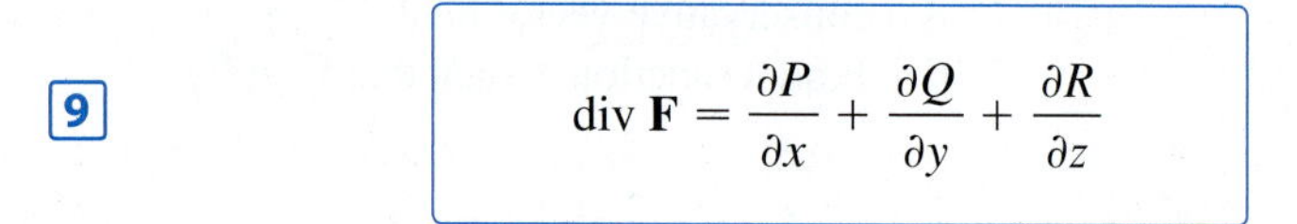

9 $$\text{div}\,\mathbf{F} = \frac{\partial P}{\partial x} + \frac{\partial Q}{\partial y} + \frac{\partial R}{\partial z}$$

Observe that curl **F** is a vector field but div **F** is a scalar field. In terms of the gradient operator $\nabla = (\partial/\partial x)\,\mathbf{i} + (\partial/\partial y)\,\mathbf{j} + (\partial/\partial z)\,\mathbf{k}$, the divergence of **F** can be written symbolically as the dot product of ∇ and **F**:

10 $$\text{div}\,\mathbf{F} = \nabla \cdot \mathbf{F}$$

EXAMPLE 4 If $\mathbf{F}(x, y, z) = xz\,\mathbf{i} + xyz\,\mathbf{j} - y^2\,\mathbf{k}$, find div **F**.

SOLUTION By the definition of divergence (Equation 9 or 10) we have

$$\text{div}\,\mathbf{F} = \nabla \cdot \mathbf{F} = \frac{\partial}{\partial x}(xz) + \frac{\partial}{\partial y}(xyz) + \frac{\partial}{\partial z}(-y^2) = z + xz$$ ■

If **F** is a vector field on $\mathbb{R}^3$, then curl **F** is also a vector field on $\mathbb{R}^3$. As such, we can compute its divergence. The next theorem shows that the result is 0.

11 Theorem If $\mathbf{F} = P\,\mathbf{i} + Q\,\mathbf{j} + R\,\mathbf{k}$ is a vector field on $\mathbb{R}^3$ and P, Q, and R have continuous second-order partial derivatives, then

$$\text{div curl}\,\mathbf{F} = 0$$

PROOF Using the definitions of divergence and curl, we have

Note the analogy with the scalar triple product: $\mathbf{a} \cdot (\mathbf{a} \times \mathbf{b}) = 0$.

$$\begin{aligned}
\text{div curl } \mathbf{F} &= \nabla \cdot (\nabla \times \mathbf{F}) \\
&= \frac{\partial}{\partial x}\left(\frac{\partial R}{\partial y} - \frac{\partial Q}{\partial z}\right) + \frac{\partial}{\partial y}\left(\frac{\partial P}{\partial z} - \frac{\partial R}{\partial x}\right) + \frac{\partial}{\partial z}\left(\frac{\partial Q}{\partial x} - \frac{\partial P}{\partial y}\right) \\
&= \frac{\partial^2 R}{\partial x\,\partial y} - \frac{\partial^2 Q}{\partial x\,\partial z} + \frac{\partial^2 P}{\partial y\,\partial z} - \frac{\partial^2 R}{\partial y\,\partial x} + \frac{\partial^2 Q}{\partial z\,\partial x} - \frac{\partial^2 P}{\partial z\,\partial y} \\
&= 0
\end{aligned}$$

because the terms cancel in pairs by Clairaut's Theorem. ■

EXAMPLE 5 Show that the vector field $\mathbf{F}(x, y, z) = xz\,\mathbf{i} + xyz\,\mathbf{j} - y^2\,\mathbf{k}$ can't be written as the curl of another vector field, that is, $\mathbf{F} \neq \text{curl } \mathbf{G}$.

SOLUTION In Example 4 we showed that

$$\text{div } \mathbf{F} = z + xz$$

and therefore $\text{div } \mathbf{F} \neq 0$. If it were true that $\mathbf{F} = \text{curl } \mathbf{G}$, then Theorem 11 would give

$$\text{div } \mathbf{F} = \text{div curl } \mathbf{G} = 0$$

which contradicts $\text{div } \mathbf{F} \neq 0$. Therefore $\mathbf{F}$ is not the curl of another vector field. ■

The reason for this interpretation of div **F** will be explained at the end of Section 16.9 as a consequence of the Divergence Theorem.

Again, the reason for the name *divergence* can be understood in the context of fluid flow. If $\mathbf{F}(x, y, z)$ is the velocity of a fluid (or gas), then $\text{div } \mathbf{F}(x, y, z)$ represents the net rate of change (with respect to time) of the mass of fluid (or gas) flowing from the point (x, y, z) per unit volume. In other words, $\text{div } \mathbf{F}(x, y, z)$ measures the tendency of the fluid to diverge from the point (x, y, z). If $\text{div } \mathbf{F} = 0$, then $\mathbf{F}$ is said to be **incompressible**.

Another differential operator occurs when we compute the divergence of a gradient vector field ∇f. If f is a function of three variables, we have

$$\text{div}(\nabla f) = \nabla \cdot (\nabla f) = \frac{\partial^2 f}{\partial x^2} + \frac{\partial^2 f}{\partial y^2} + \frac{\partial^2 f}{\partial z^2}$$

and this expression occurs so often that we abbreviate it as $\nabla^2 f$. The operator

$$\nabla^2 = \nabla \cdot \nabla$$

is called the **Laplace operator** because of its relation to **Laplace's equation**

$$\nabla^2 f = \frac{\partial^2 f}{\partial x^2} + \frac{\partial^2 f}{\partial y^2} + \frac{\partial^2 f}{\partial z^2} = 0$$

We can also apply the Laplace operator ∇^2 to a vector field

$$\mathbf{F} = P\,\mathbf{i} + Q\,\mathbf{j} + R\,\mathbf{k}$$

in terms of its components:

$$\nabla^2 \mathbf{F} = \nabla^2 P\,\mathbf{i} + \nabla^2 Q\,\mathbf{j} + \nabla^2 R\,\mathbf{k}$$

Vector Forms of Green's Theorem

The curl and divergence operators allow us to rewrite Green's Theorem in versions that will be useful in our later work. We suppose that the plane region D, its boundary curve C, and the functions P and Q satisfy the hypotheses of Green's Theorem. Then we consider the vector field $\mathbf{F} = P\,\mathbf{i} + Q\,\mathbf{j}$. Its line integral is

$$\oint_C \mathbf{F} \cdot d\mathbf{r} = \oint_C P\,dx + Q\,dy$$

and, regarding $\mathbf{F}$ as a vector field on $\mathbb{R}^3$ with third component 0, we have

$$\text{curl } \mathbf{F} = \begin{vmatrix} \mathbf{i} & \mathbf{j} & \mathbf{k} \\ \dfrac{\partial}{\partial x} & \dfrac{\partial}{\partial y} & \dfrac{\partial}{\partial z} \\ P(x, y) & Q(x, y) & 0 \end{vmatrix} = \left(\frac{\partial Q}{\partial x} - \frac{\partial P}{\partial y} \right) \mathbf{k}$$

Therefore

$$(\text{curl } \mathbf{F}) \cdot \mathbf{k} = \left(\frac{\partial Q}{\partial x} - \frac{\partial P}{\partial y} \right) \mathbf{k} \cdot \mathbf{k} = \frac{\partial Q}{\partial x} - \frac{\partial P}{\partial y}$$

and we can now rewrite the equation in Green's Theorem in the vector form

12
$$\oint_C \mathbf{F} \cdot d\mathbf{r} = \iint_D (\text{curl } \mathbf{F}) \cdot \mathbf{k}\, dA$$

Equation 12 expresses the line integral of the tangential component of $\mathbf{F}$ along C as the double integral of the vertical component of curl $\mathbf{F}$ over the region D enclosed by C. We now derive a similar formula involving the *normal* component of $\mathbf{F}$.

If C is given by the vector equation

$$\mathbf{r}(t) = x(t)\,\mathbf{i} + y(t)\,\mathbf{j} \qquad a \leqslant t \leqslant b$$

then the unit tangent vector (see Section 13.2) is

$$\mathbf{T}(t) = \frac{x'(t)}{|\mathbf{r}'(t)|}\,\mathbf{i} + \frac{y'(t)}{|\mathbf{r}'(t)|}\,\mathbf{j}$$

You can verify that the outward unit normal vector to C is given by

$$\mathbf{n}(t) = \frac{y'(t)}{|\mathbf{r}'(t)|}\,\mathbf{i} - \frac{x'(t)}{|\mathbf{r}'(t)|}\,\mathbf{j}$$

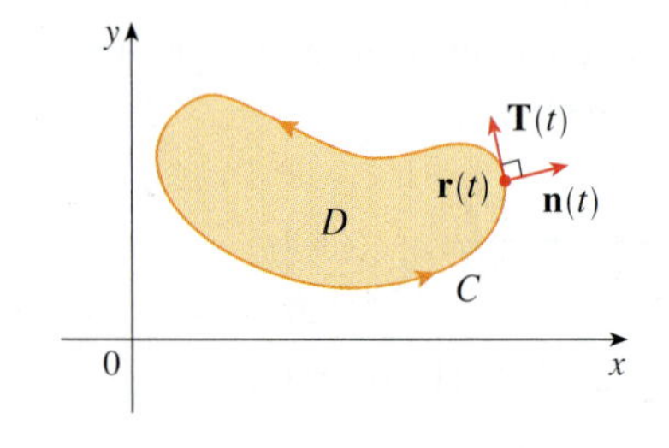

FIGURE 2

(See Figure 2.) Then, from Equation 16.2.3, we have

$$\oint_C \mathbf{F} \cdot \mathbf{n}\, ds = \int_a^b (\mathbf{F} \cdot \mathbf{n})(t)\, |\mathbf{r}'(t)|\, dt$$

$$= \int_a^b \left[\frac{P(x(t), y(t))\, y'(t)}{|\mathbf{r}'(t)|} - \frac{Q(x(t), y(t))\, x'(t)}{|\mathbf{r}'(t)|} \right] |\mathbf{r}'(t)|\, dt$$

$$= \int_a^b P(x(t), y(t))\, y'(t)\, dt - Q(x(t), y(t))\, x'(t)\, dt$$

$$= \int_C P\, dy - Q\, dx = \iint_D \left(\frac{\partial P}{\partial x} + \frac{\partial Q}{\partial y} \right) dA$$

by Green's Theorem. But the integrand in this double integral is just the divergence of **F**. So we have a second vector form of Green's Theorem.

13
$$\oint_C \mathbf{F} \cdot \mathbf{n}\, ds = \iint_D \operatorname{div} \mathbf{F}(x, y)\, dA$$

This version says that the line integral of the normal component of **F** along C is equal to the double integral of the divergence of **F** over the region D enclosed by C.

16.5 EXERCISES

1–8 Find (a) the curl and (b) the divergence of the vector field.

1. $\mathbf{F}(x, y, z) = xy^2z^2\,\mathbf{i} + x^2yz^2\,\mathbf{j} + x^2y^2z\,\mathbf{k}$
2. $\mathbf{F}(x, y, z) = x^3yz^2\,\mathbf{j} + y^4z^3\,\mathbf{k}$
3. $\mathbf{F}(x, y, z) = xye^z\,\mathbf{i} + yze^x\,\mathbf{k}$
4. $\mathbf{F}(x, y, z) = \sin yz\,\mathbf{i} + \sin zx\,\mathbf{j} + \sin xy\,\mathbf{k}$
5. $\mathbf{F}(x, y, z) = \dfrac{\sqrt{x}}{1+z}\,\mathbf{i} + \dfrac{\sqrt{y}}{1+x}\,\mathbf{j} + \dfrac{\sqrt{z}}{1+y}\,\mathbf{k}$
6. $\mathbf{F}(x, y, z) = \ln(2y + 3z)\,\mathbf{i} + \ln(x + 3z)\,\mathbf{j} + \ln(x + 2y)\,\mathbf{k}$
7. $\mathbf{F}(x, y, z) = \langle e^x \sin y,\, e^y \sin z,\, e^z \sin x\rangle$
8. $\mathbf{F}(x, y, z) = \langle \arctan(xy),\, \arctan(yz),\, \arctan(zx)\rangle$

9–11 The vector field **F** is shown in the xy-plane and looks the same in all other horizontal planes. (In other words, **F** is independent of z and its z-component is 0.)
(a) Is div **F** positive, negative, or zero? Explain.
(b) Determine whether curl **F** = **0**. If not, in which direction does curl **F** point?

9.
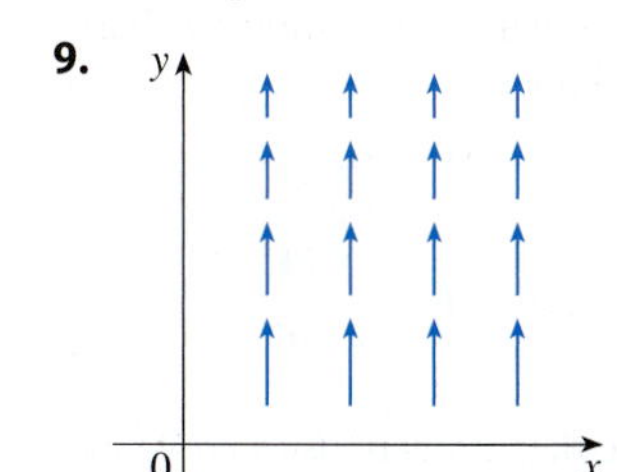

10.
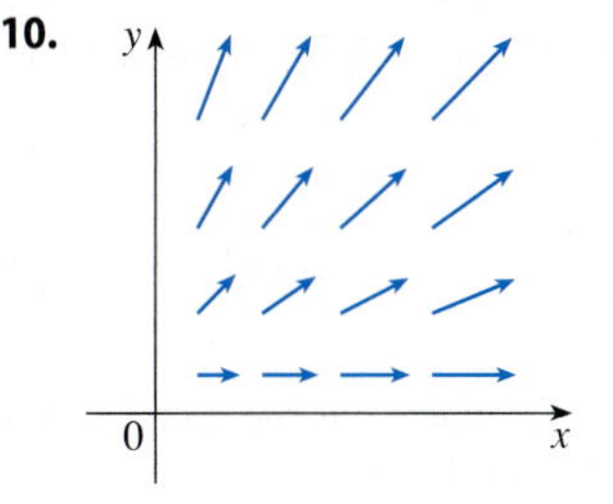

11.
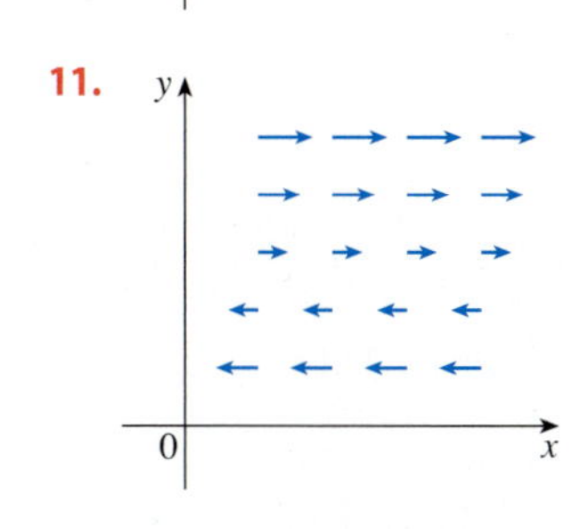

12. Let f be a scalar field and **F** a vector field. State whether each expression is meaningful. If not, explain why. If so, state whether it is a scalar field or a vector field.
(a) curl f (b) grad f
(c) div **F** (d) curl(grad f)
(e) grad **F** (f) grad(div **F**)
(g) div(grad f) (h) grad(div f)
(i) curl(curl **F**) (j) div(div **F**)
(k) (grad f) × (div **F**) (l) div(curl(grad f))

13–18 Determine whether or not the vector field is conservative. If it is conservative, find a function f such that $\mathbf{F} = \nabla f$.

13. $\mathbf{F}(x, y, z) = y^2z^3\,\mathbf{i} + 2xyz^3\,\mathbf{j} + 3xy^2z^2\,\mathbf{k}$
14. $\mathbf{F}(x, y, z) = xyz^4\,\mathbf{i} + x^2z^4\,\mathbf{j} + 4x^2yz^3\,\mathbf{k}$
15. $\mathbf{F}(x, y, z) = z\cos y\,\mathbf{i} + xz\sin y\,\mathbf{j} + x\cos y\,\mathbf{k}$
16. $\mathbf{F}(x, y, z) = \mathbf{i} + \sin z\,\mathbf{j} + y\cos z\,\mathbf{k}$
17. $\mathbf{F}(x, y, z) = e^{yz}\,\mathbf{i} + xze^{yz}\,\mathbf{j} + xye^{yz}\,\mathbf{k}$
18. $\mathbf{F}(x, y, z) = e^x \sin yz\,\mathbf{i} + ze^x\cos yz\,\mathbf{j} + ye^x\cos yz\,\mathbf{k}$

19. Is there a vector field **G** on $\mathbb{R}^3$ such that curl $\mathbf{G} = \langle x\sin y, \cos y, z - xy\rangle$? Explain.

20. Is there a vector field **G** on $\mathbb{R}^3$ such that curl $\mathbf{G} = \langle x, y, z\rangle$? Explain.

21. Show that any vector field of the form
$$\mathbf{F}(x, y, z) = f(x)\,\mathbf{i} + g(y)\,\mathbf{j} + h(z)\,\mathbf{k}$$
where f, g, h are differentiable functions, is irrotational.

22. Show that any vector field of the form
$$\mathbf{F}(x, y, z) = f(y, z)\,\mathbf{i} + g(x, z)\,\mathbf{j} + h(x, y)\,\mathbf{k}$$
is incompressible.

23–29 Prove the identity, assuming that the appropriate partial derivatives exist and are continuous. If f is a scalar field and $\mathbf{F}$, $\mathbf{G}$ are vector fields, then $f\mathbf{F}$, $\mathbf{F}\cdot\mathbf{G}$, and $\mathbf{F}\times\mathbf{G}$ are defined by

$$(f\mathbf{F})(x, y, z) = f(x, y, z)\,\mathbf{F}(x, y, z)$$

$$(\mathbf{F}\cdot\mathbf{G})(x, y, z) = \mathbf{F}(x, y, z)\cdot\mathbf{G}(x, y, z)$$

$$(\mathbf{F}\times\mathbf{G})(x, y, z) = \mathbf{F}(x, y, z)\times\mathbf{G}(x, y, z)$$

23. $\operatorname{div}(\mathbf{F}+\mathbf{G}) = \operatorname{div}\mathbf{F} + \operatorname{div}\mathbf{G}$

24. $\operatorname{curl}(\mathbf{F}+\mathbf{G}) = \operatorname{curl}\mathbf{F} + \operatorname{curl}\mathbf{G}$

25. $\operatorname{div}(f\mathbf{F}) = f\operatorname{div}\mathbf{F} + \mathbf{F}\cdot\nabla f$

26. $\operatorname{curl}(f\mathbf{F}) = f\operatorname{curl}\mathbf{F} + (\nabla f)\times\mathbf{F}$

27. $\operatorname{div}(\mathbf{F}\times\mathbf{G}) = \mathbf{G}\cdot\operatorname{curl}\mathbf{F} - \mathbf{F}\cdot\operatorname{curl}\mathbf{G}$

28. $\operatorname{div}(\nabla f\times\nabla g) = 0$

29. $\operatorname{curl}(\operatorname{curl}\mathbf{F}) = \operatorname{grad}(\operatorname{div}\mathbf{F}) - \nabla^2\mathbf{F}$

30–32 Let $\mathbf{r} = x\,\mathbf{i} + y\,\mathbf{j} + z\,\mathbf{k}$ and $r = |\mathbf{r}|$.

30. Verify each identity.

(a) $\nabla\cdot\mathbf{r} = 3$ (b) $\nabla\cdot(r\mathbf{r}) = 4r$

(c) $\nabla^2 r^3 = 12r$

31. Verify each identity.

(a) $\nabla r = \mathbf{r}/r$ (b) $\nabla\times\mathbf{r} = \mathbf{0}$

(c) $\nabla(1/r) = -\mathbf{r}/r^3$ (d) $\nabla\ln r = \mathbf{r}/r^2$

32. If $\mathbf{F} = \mathbf{r}/r^p$, find $\operatorname{div}\mathbf{F}$. Is there a value of p for which $\operatorname{div}\mathbf{F} = 0$?

33. Use Green's Theorem in the form of Equation 13 to prove **Green's first identity**:

$$\iint_D f\nabla^2 g\,dA = \oint_C f(\nabla g)\cdot\mathbf{n}\,ds - \iint_D \nabla f\cdot\nabla g\,dA$$

where D and C satisfy the hypotheses of Green's Theorem and the appropriate partial derivatives of f and g exist and are continuous. (The quantity $\nabla g\cdot\mathbf{n} = D_{\mathbf{n}}\,g$ occurs in the line integral. This is the directional derivative in the direction of the normal vector $\mathbf{n}$ and is called the **normal derivative** of g.)

34. Use Green's first identity (Exercise 33) to prove **Green's second identity**:

$$\iint_D (f\nabla^2 g - g\nabla^2 f)\,dA = \oint_C (f\nabla g - g\nabla f)\cdot\mathbf{n}\,ds$$

where D and C satisfy the hypotheses of Green's Theorem and the appropriate partial derivatives of f and g exist and are continuous.

35. Recall from Section 14.3 that a function g is called *harmonic* on D if it satisfies Laplace's equation, that is, $\nabla^2 g = 0$ on D. Use Green's first identity (with the same hypotheses as in Exercise 33) to show that if g is harmonic on D, then $\oint_C D_{\mathbf{n}}\,g\,ds = 0$. Here $D_{\mathbf{n}}\,g$ is the normal derivative of g defined in Exercise 33.

36. Use Green's first identity to show that if f is harmonic on D, and if $f(x, y) = 0$ on the boundary curve C, then $\iint_D |\nabla f|^2\,dA = 0$. (Assume the same hypotheses as in Exercise 33.)

37. This exercise demonstrates a connection between the curl vector and rotations. Let B be a rigid body rotating about the z-axis. The rotation can be described by the vector $\mathbf{w} = \omega\,\mathbf{k}$, where ω is the angular speed of B, that is, the tangential speed of any point P in B divided by the distance d from the axis of rotation. Let $\mathbf{r} = \langle x, y, z\rangle$ be the position vector of P.

(a) By considering the angle θ in the figure, show that the velocity field of B is given by $\mathbf{v} = \mathbf{w}\times\mathbf{r}$.

(b) Show that $\mathbf{v} = -\omega y\,\mathbf{i} + \omega x\,\mathbf{j}$.

(c) Show that $\operatorname{curl}\mathbf{v} = 2\mathbf{w}$.

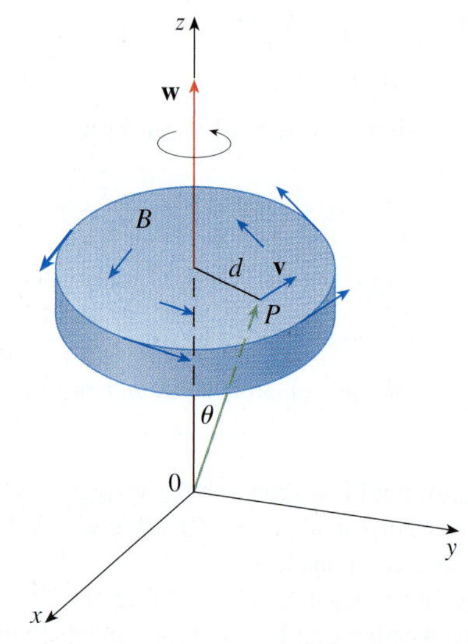

38. Maxwell's equations relating the electric field $\mathbf{E}$ and magnetic field $\mathbf{H}$ as they vary with time in a region containing no charge and no current can be stated as follows:

$$\operatorname{div}\mathbf{E} = 0 \qquad \operatorname{div}\mathbf{H} = 0$$

$$\operatorname{curl}\mathbf{E} = -\frac{1}{c}\frac{\partial\mathbf{H}}{\partial t} \qquad \operatorname{curl}\mathbf{H} = \frac{1}{c}\frac{\partial\mathbf{E}}{\partial t}$$

where c is the speed of light. Use these equations to prove the following:

(a) $\nabla\times(\nabla\times\mathbf{E}) = -\dfrac{1}{c^2}\dfrac{\partial^2\mathbf{E}}{\partial t^2}$

(b) $\nabla\times(\nabla\times\mathbf{H}) = -\dfrac{1}{c^2}\dfrac{\partial^2\mathbf{H}}{\partial t^2}$

(c) $\nabla^2\mathbf{E} = \dfrac{1}{c^2}\dfrac{\partial^2\mathbf{E}}{\partial t^2}$ [*Hint:* Use Exercise 29.]

(d) $\nabla^2\mathbf{H} = \dfrac{1}{c^2}\dfrac{\partial^2\mathbf{H}}{\partial t^2}$

39. We have seen that all vector fields of the form $\mathbf{F} = \nabla g$ satisfy the equation $\text{curl } \mathbf{F} = \mathbf{0}$ and that all vector fields of the form $\mathbf{F} = \text{curl } \mathbf{G}$ satisfy the equation $\text{div } \mathbf{F} = 0$ (assuming continuity of the appropriate partial derivatives). This suggests the question: are there any equations that all functions of the form $f = \text{div } \mathbf{G}$ must satisfy? Show that the answer to this question is "No" by proving that *every* continuous function f on $\mathbb{R}^3$ is the divergence of some vector field.
[*Hint:* Let $\mathbf{G}(x, y, z) = \langle g(x, y, z), 0, 0 \rangle$, where $g(x, y, z) = \int_0^x f(t, y, z)\, dt$.]

16.6 Parametric Surfaces and Their Areas

So far we have considered special types of surfaces: cylinders, quadric surfaces, graphs of functions of two variables, and level surfaces of functions of three variables. Here we use vector functions to describe more general surfaces, called *parametric surfaces*, and compute their areas. Then we take the general surface area formula and see how it applies to special surfaces.

Parametric Surfaces

In much the same way that we describe a space curve by a vector function $\mathbf{r}(t)$ of a single parameter t, we can describe a surface by a vector function $\mathbf{r}(u, v)$ of two parameters u and v. We suppose that

$$\boxed{1} \qquad \mathbf{r}(u, v) = x(u, v)\,\mathbf{i} + y(u, v)\,\mathbf{j} + z(u, v)\,\mathbf{k}$$

is a vector-valued function defined on a region D in the uv-plane. So x, y, and z, the component functions of $\mathbf{r}$, are functions of the two variables u and v with domain D. The set of all points (x, y, z) in $\mathbb{R}^3$ such that

$$\boxed{2} \qquad x = x(u, v) \qquad y = y(u, v) \qquad z = z(u, v)$$

and (u, v) varies throughout D, is called a **parametric surface** S and Equations 2 are called **parametric equations** of S. Each choice of u and v gives a point on S; by making all choices, we get all of S. In other words, the surface S is traced out by the tip of the position vector $\mathbf{r}(u, v)$ as (u, v) moves throughout the region D. (See Figure 1.)

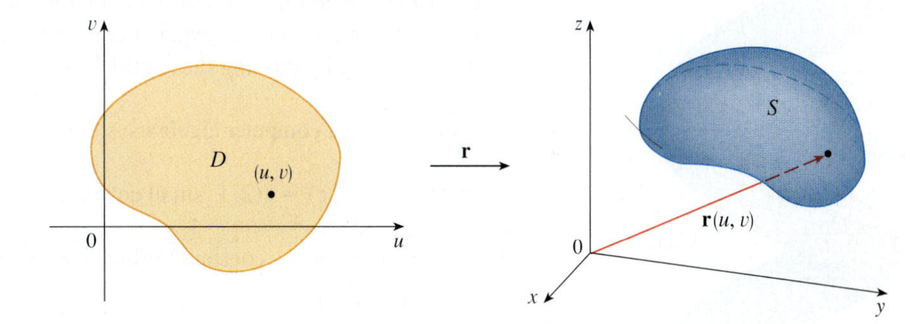

FIGURE 1
A parametric surface

EXAMPLE 1 Identify and sketch the surface with vector equation

$$\mathbf{r}(u, v) = 2 \cos u\,\mathbf{i} + v\,\mathbf{j} + 2 \sin u\,\mathbf{k}$$

SOLUTION The parametric equations for this surface are

$$x = 2 \cos u \qquad y = v \qquad z = 2 \sin u$$

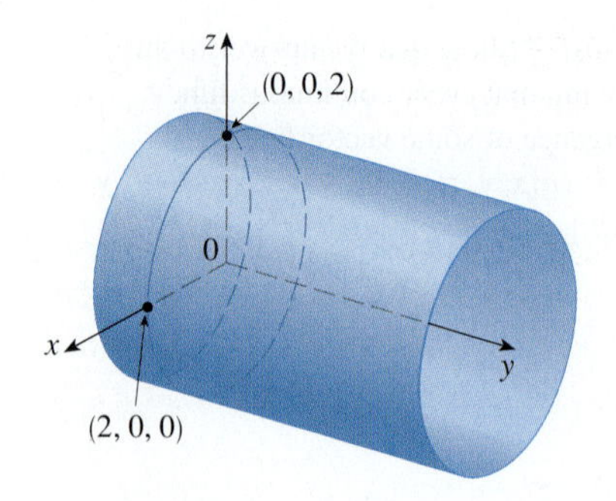

FIGURE 2

So for any point (x, y, z) on the surface, we have

$$x^2 + z^2 = 4\cos^2 u + 4\sin^2 u = 4$$

This means that vertical cross-sections parallel to the xz-plane (that is, with y constant) are all circles with radius 2. Since $y = v$ and no restriction is placed on v, the surface is a circular cylinder with radius 2 whose axis is the y-axis (see Figure 2). ■

In Example 1 we placed no restrictions on the parameters u and v and so we obtained the entire cylinder. If, for instance, we restrict u and v by writing the parameter domain as

$$0 \leqslant u \leqslant \pi/2 \qquad 0 \leqslant v \leqslant 3$$

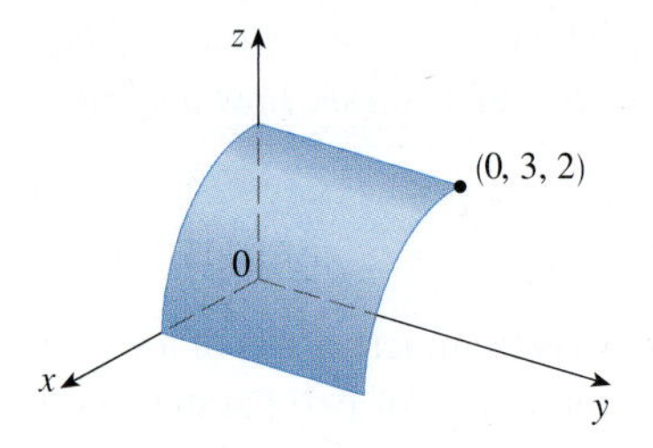

FIGURE 3

then $x \geqslant 0$, $z \geqslant 0$, $0 \leqslant y \leqslant 3$, and we get the quarter-cylinder with length 3 illustrated in Figure 3.

If a parametric surface S is given by a vector function $\mathbf{r}(u, v)$, then there are two useful families of curves that lie on S, one family with u constant and the other with v constant. These families correspond to vertical and horizontal lines in the uv-plane. If we keep u constant by putting $u = u_0$, then $\mathbf{r}(u_0, v)$ becomes a vector function of the single parameter v and defines a curve C_1 lying on S. (See Figure 4.)

TEC Visual 16.6 shows animated versions of Figures 4 and 5, with moving grid curves, for several parametric surfaces.

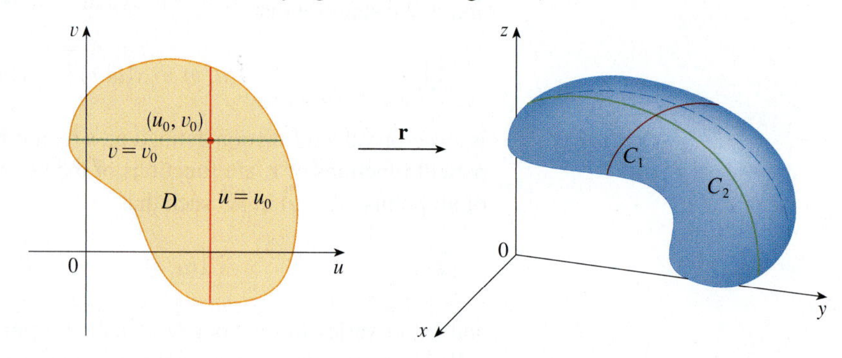

FIGURE 4

Similarly, if we keep v constant by putting $v = v_0$, we get a curve C_2 given by $\mathbf{r}(u, v_0)$ that lies on S. We call these curves **grid curves**. (In Example 1, for instance, the grid curves obtained by letting u be constant are horizontal lines whereas the grid curves with v constant are circles.) In fact, when a computer graphs a parametric surface, it usually depicts the surface by plotting these grid curves, as we see in the following example.

EXAMPLE 2 Use a computer algebra system to graph the surface

$$\mathbf{r}(u, v) = \langle (2 + \sin v)\cos u,\ (2 + \sin v)\sin u,\ u + \cos v \rangle$$

Which grid curves have u constant? Which have v constant?

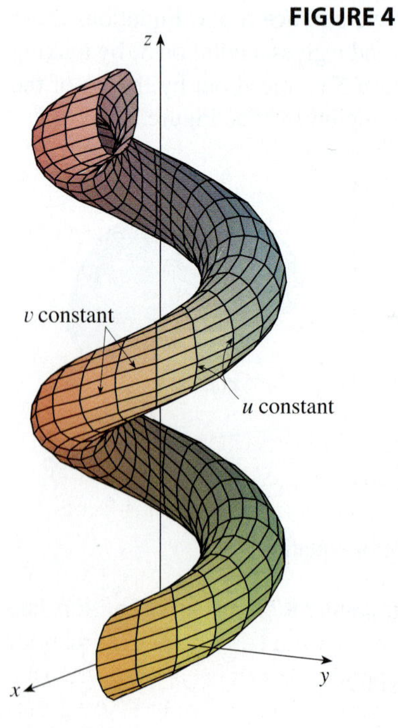

FIGURE 5

SOLUTION We graph the portion of the surface with parameter domain $0 \leqslant u \leqslant 4\pi$, $0 \leqslant v \leqslant 2\pi$ in Figure 5. It has the appearance of a spiral tube. To identify the grid curves, we write the corresponding parametric equations:

$$x = (2 + \sin v)\cos u \qquad y = (2 + \sin v)\sin u \qquad z = u + \cos v$$

If v is constant, then $\sin v$ and $\cos v$ are constant, so the parametric equations resemble those of the helix in Example 13.1.4. Thus the grid curves with v constant are the spiral curves in Figure 5. We deduce that the grid curves with u constant must be the curves that look like circles in the figure. Further evidence for this assertion is that if u is kept constant, $u = u_0$, then the equation $z = u_0 + \cos v$ shows that the z-values vary from $u_0 - 1$ to $u_0 + 1$. ■

In Examples 1 and 2 we were given a vector equation and asked to graph the corresponding parametric surface. In the following examples, however, we are given the more challenging problem of finding a vector function to represent a given surface. In the rest of this chapter we will often need to do exactly that.

EXAMPLE 3 Find a vector function that represents the plane that passes through the point P_0 with position vector $\mathbf{r}_0$ and that contains two nonparallel vectors $\mathbf{a}$ and $\mathbf{b}$.

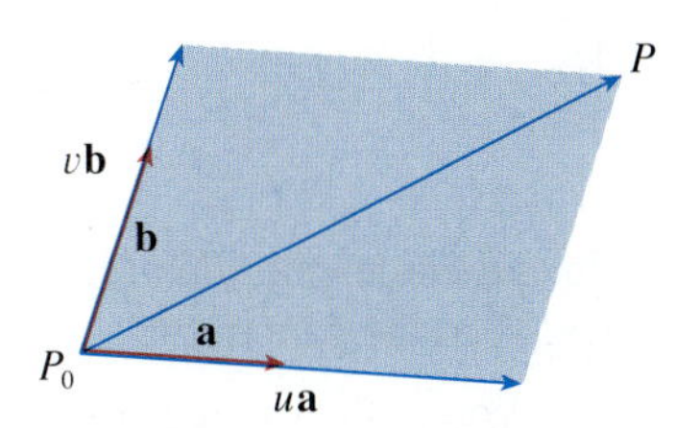

FIGURE 6

SOLUTION If P is any point in the plane, we can get from P_0 to P by moving a certain distance in the direction of $\mathbf{a}$ and another distance in the direction of $\mathbf{b}$. So there are scalars u and v such that $\overrightarrow{P_0P} = u\mathbf{a} + v\mathbf{b}$. (Figure 6 illustrates how this works, by means of the Parallelogram Law, for the case where u and v are positive. See also Exercise 12.2.46.) If $\mathbf{r}$ is the position vector of P, then

$$\mathbf{r} = \overrightarrow{OP_0} + \overrightarrow{P_0P} = \mathbf{r}_0 + u\mathbf{a} + v\mathbf{b}$$

So the vector equation of the plane can be written as

$$\mathbf{r}(u, v) = \mathbf{r}_0 + u\mathbf{a} + v\mathbf{b}$$

where u and v are real numbers.

If we write $\mathbf{r} = \langle x, y, z \rangle$, $\mathbf{r}_0 = \langle x_0, y_0, z_0 \rangle$, $\mathbf{a} = \langle a_1, a_2, a_3 \rangle$, and $\mathbf{b} = \langle b_1, b_2, b_3 \rangle$, then we can write the parametric equations of the plane through the point (x_0, y_0, z_0) as follows:

$$x = x_0 + ua_1 + vb_1 \qquad y = y_0 + ua_2 + vb_2 \qquad z = z_0 + ua_3 + vb_3$$ ■

EXAMPLE 4 Find a parametric representation of the sphere

$$x^2 + y^2 + z^2 = a^2$$

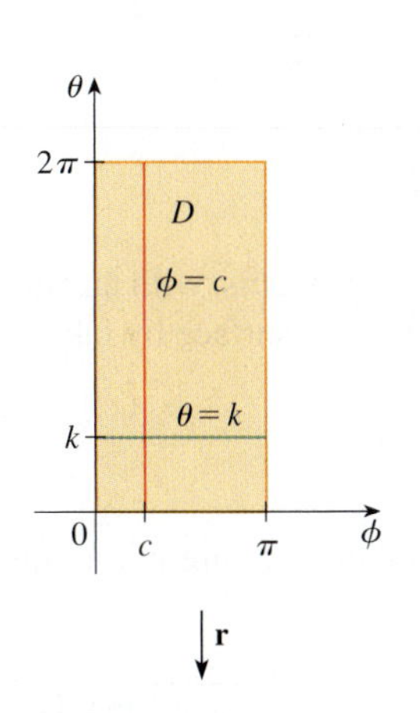

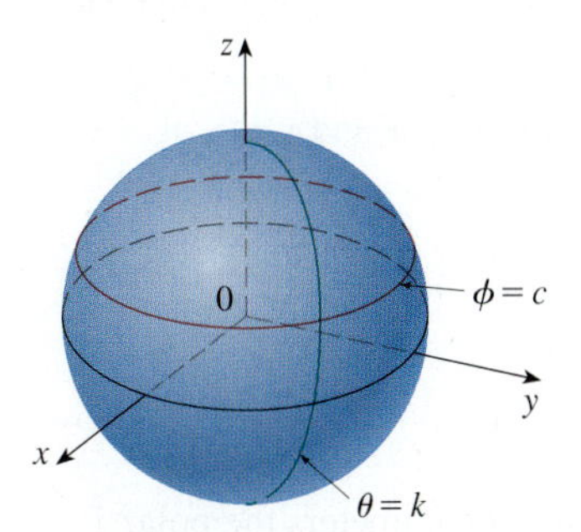

FIGURE 7

SOLUTION The sphere has a simple representation $\rho = a$ in spherical coordinates, so let's choose the angles ϕ and θ in spherical coordinates as the parameters (see Section 15.8). Then, putting $\rho = a$ in the equations for conversion from spherical to rectangular coordinates (Equations 15.8.1), we obtain

$$x = a \sin \phi \cos \theta \qquad y = a \sin \phi \sin \theta \qquad z = a \cos \phi$$

as the parametric equations of the sphere. The corresponding vector equation is

$$\mathbf{r}(\phi, \theta) = a \sin \phi \cos \theta \, \mathbf{i} + a \sin \phi \sin \theta \, \mathbf{j} + a \cos \phi \, \mathbf{k}$$

We have $0 \leqslant \phi \leqslant \pi$ and $0 \leqslant \theta \leqslant 2\pi$, so the parameter domain is the rectangle $D = [0, \pi] \times [0, 2\pi]$. The grid curves with ϕ constant are the circles of constant latitude (including the equator). The grid curves with θ constant are the meridians (semicircles), which connect the north and south poles (see Figure 7). ■

NOTE We saw in Example 4 that the grid curves for a sphere are curves of constant latitude or constant longitude. For a general parametric surface we are really making a map and the grid curves are similar to lines of latitude and longitude. Describing a point on a parametric surface (like the one in Figure 5) by giving specific values of u and v is like giving the latitude and longitude of a point.

One of the uses of parametric surfaces is in computer graphics. Figure 8 shows the result of trying to graph the sphere $x^2 + y^2 + z^2 = 1$ by solving the equation for z and graphing the top and bottom hemispheres separately. Part of the sphere appears to be missing because of the rectangular grid system used by the computer. The much better picture in Figure 9 was produced by a computer using the parametric equations found in Example 4.

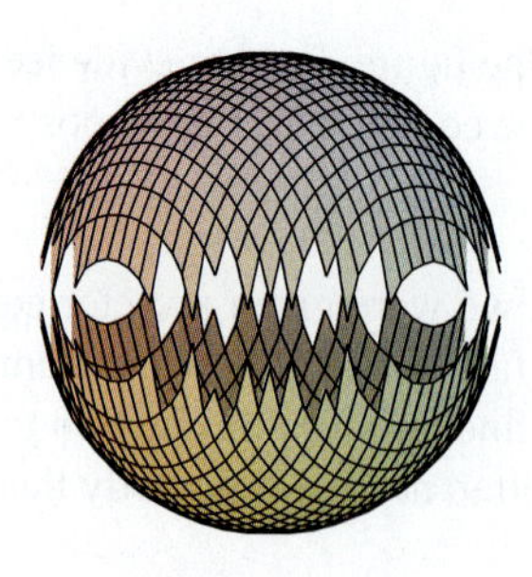

FIGURE 8

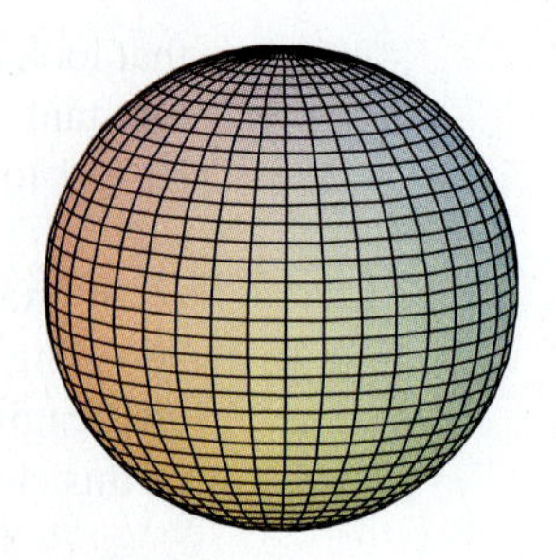

FIGURE 9

EXAMPLE 5 Find a parametric representation for the cylinder

$$x^2 + y^2 = 4 \qquad 0 \leqslant z \leqslant 1$$

SOLUTION The cylinder has a simple representation $r = 2$ in cylindrical coordinates, so we choose as parameters θ and z in cylindrical coordinates. Then the parametric equations of the cylinder are

$$x = 2\cos\theta \qquad y = 2\sin\theta \qquad z = z$$

where $0 \leqslant \theta \leqslant 2\pi$ and $0 \leqslant z \leqslant 1$. ■

EXAMPLE 6 Find a vector function that represents the elliptic paraboloid $z = x^2 + 2y^2$.

SOLUTION If we regard x and y as parameters, then the parametric equations are simply

$$x = x \qquad y = y \qquad z = x^2 + 2y^2$$

and the vector equation is

$$\mathbf{r}(x, y) = x\,\mathbf{i} + y\,\mathbf{j} + (x^2 + 2y^2)\,\mathbf{k}$$ ■

TEC In Module 16.6 you can investigate several families of parametric surfaces.

In general, a surface given as the graph of a function of x and y, that is, with an equation of the form $z = f(x, y)$, can always be regarded as a parametric surface by taking x and y as parameters and writing the parametric equations as

$$x = x \qquad y = y \qquad z = f(x, y)$$

Parametric representations (also called parametrizations) of surfaces are not unique. The next example shows two ways to parametrize a cone.

EXAMPLE 7 Find a parametric representation for the surface $z = 2\sqrt{x^2 + y^2}$, that is, the top half of the cone $z^2 = 4x^2 + 4y^2$.

SOLUTION 1 One possible representation is obtained by choosing x and y as parameters:

$$x = x \qquad y = y \qquad z = 2\sqrt{x^2 + y^2}$$

So the vector equation is

$$\mathbf{r}(x, y) = x\,\mathbf{i} + y\,\mathbf{j} + 2\sqrt{x^2 + y^2}\,\mathbf{k}$$

SOLUTION 2 Another representation results from choosing as parameters the polar coordinates r and θ. A point (x, y, z) on the cone satisfies $x = r\cos\theta$, $y = r\sin\theta$, and

$z = 2\sqrt{x^2 + y^2} = 2r$. So a vector equation for the cone is

$$\mathbf{r}(r, \theta) = r\cos\theta\,\mathbf{i} + r\sin\theta\,\mathbf{j} + 2r\,\mathbf{k}$$

where $r \geqslant 0$ and $0 \leqslant \theta \leqslant 2\pi$. ■

For some purposes the parametric representations in Solutions 1 and 2 are equally good, but Solution 2 might be preferable in certain situations. If we are interested only in the part of the cone that lies below the plane $z = 1$, for instance, all we have to do in Solution 2 is change the parameter domain to

$$0 \leqslant r \leqslant \tfrac{1}{2} \qquad 0 \leqslant \theta \leqslant 2\pi$$

Surfaces of Revolution

Surfaces of revolution can be represented parametrically and thus graphed using a computer. For instance, let's consider the surface S obtained by rotating the curve $y = f(x)$, $a \leqslant x \leqslant b$, about the x-axis, where $f(x) \geqslant 0$. Let θ be the angle of rotation as shown in Figure 10. If (x, y, z) is a point on S, then

3 $$x = x \qquad y = f(x)\cos\theta \qquad z = f(x)\sin\theta$$

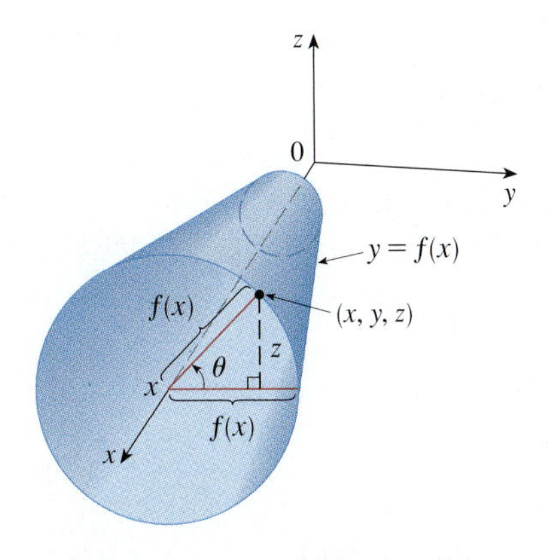

FIGURE 10

Therefore we take x and θ as parameters and regard Equations 3 as parametric equations of S. The parameter domain is given by $a \leqslant x \leqslant b$, $0 \leqslant \theta \leqslant 2\pi$.

EXAMPLE 8 Find parametric equations for the surface generated by rotating the curve $y = \sin x$, $0 \leqslant x \leqslant 2\pi$, about the x-axis. Use these equations to graph the surface of revolution.

SOLUTION From Equations 3, the parametric equations are

$$x = x \qquad y = \sin x\cos\theta \qquad z = \sin x\sin\theta$$

and the parameter domain is $0 \leqslant x \leqslant 2\pi$, $0 \leqslant \theta \leqslant 2\pi$. Using a computer to plot these equations and and then rotating the image, we obtain the graph in Figure 11. ■

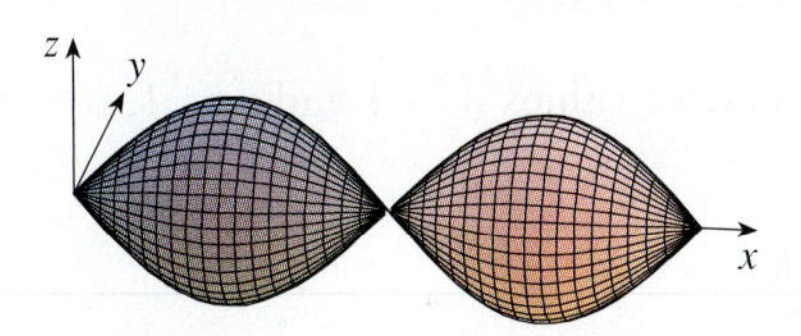

FIGURE 11

We can adapt Equations 3 to represent a surface obtained through revolution about the y- or z-axis (see Exercise 30).

Tangent Planes

We now find the tangent plane to a parametric surface S traced out by a vector function

$$\mathbf{r}(u, v) = x(u, v)\,\mathbf{i} + y(u, v)\,\mathbf{j} + z(u, v)\,\mathbf{k}$$

at a point P_0 with position vector $\mathbf{r}(u_0, v_0)$. If we keep u constant by putting $u = u_0$, then $\mathbf{r}(u_0, v)$ becomes a vector function of the single parameter v and defines a grid curve C_1 lying on S. (See Figure 12.) The tangent vector to C_1 at P_0 is obtained by taking the partial derivative of $\mathbf{r}$ with respect to v:

4 $$\mathbf{r}_v = \frac{\partial x}{\partial v}(u_0, v_0)\,\mathbf{i} + \frac{\partial y}{\partial v}(u_0, v_0)\,\mathbf{j} + \frac{\partial z}{\partial v}(u_0, v_0)\,\mathbf{k}$$

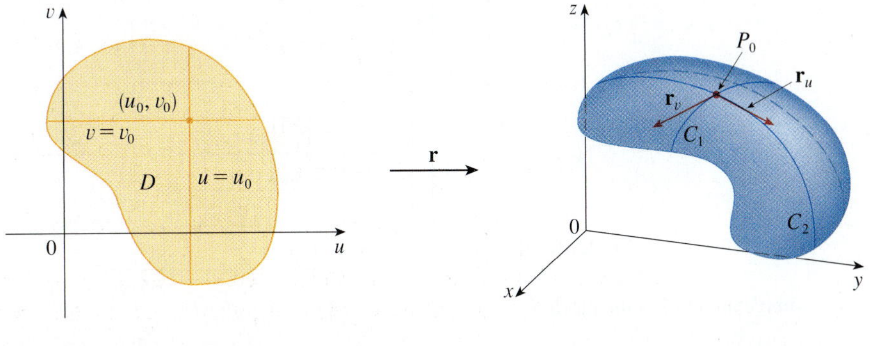

FIGURE 12

Similarly, if we keep v constant by putting $v = v_0$, we get a grid curve C_2 given by $\mathbf{r}(u, v_0)$ that lies on S, and its tangent vector at P_0 is

$$\boxed{5} \qquad \mathbf{r}_u = \frac{\partial x}{\partial u}(u_0, v_0)\,\mathbf{i} + \frac{\partial y}{\partial u}(u_0, v_0)\,\mathbf{j} + \frac{\partial z}{\partial u}(u_0, v_0)\,\mathbf{k}$$

If $\mathbf{r}_u \times \mathbf{r}_v$ is not $\mathbf{0}$, then the surface S is called **smooth** (it has no "corners"). For a smooth surface, the **tangent plane** is the plane that contains the tangent vectors $\mathbf{r}_u$ and $\mathbf{r}_v$, and the vector $\mathbf{r}_u \times \mathbf{r}_v$ is a normal vector to the tangent plane.

Figure 13 shows the self-intersecting surface in Example 9 and its tangent plane at (1, 1, 3).

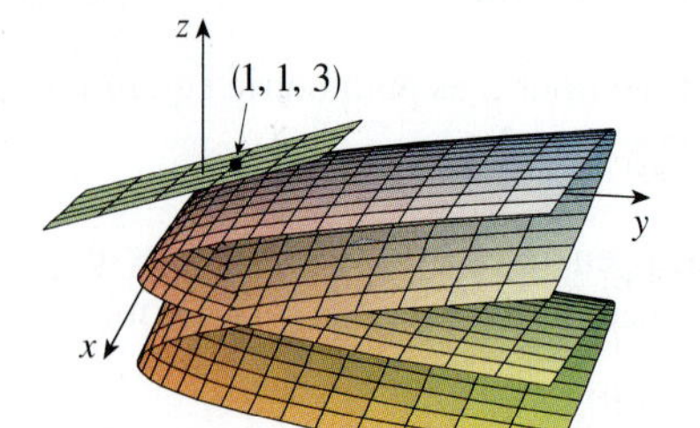

FIGURE 13

EXAMPLE 9 Find the tangent plane to the surface with parametric equations $x = u^2$, $y = v^2$, $z = u + 2v$ at the point (1, 1, 3).

SOLUTION We first compute the tangent vectors:

$$\mathbf{r}_u = \frac{\partial x}{\partial u}\mathbf{i} + \frac{\partial y}{\partial u}\mathbf{j} + \frac{\partial z}{\partial u}\mathbf{k} = 2u\,\mathbf{i} + \mathbf{k}$$

$$\mathbf{r}_v = \frac{\partial x}{\partial v}\mathbf{i} + \frac{\partial y}{\partial v}\mathbf{j} + \frac{\partial z}{\partial v}\mathbf{k} = 2v\,\mathbf{j} + 2\,\mathbf{k}$$

Thus a normal vector to the tangent plane is

$$\mathbf{r}_u \times \mathbf{r}_v = \begin{vmatrix} \mathbf{i} & \mathbf{j} & \mathbf{k} \\ 2u & 0 & 1 \\ 0 & 2v & 2 \end{vmatrix} = -2v\,\mathbf{i} - 4u\,\mathbf{j} + 4uv\,\mathbf{k}$$

Notice that the point (1, 1, 3) corresponds to the parameter values $u = 1$ and $v = 1$, so the normal vector there is

$$-2\,\mathbf{i} - 4\,\mathbf{j} + 4\,\mathbf{k}$$

Therefore an equation of the tangent plane at (1, 1, 3) is

$$-2(x - 1) - 4(y - 1) + 4(z - 3) = 0$$

or

$$x + 2y - 2z + 3 = 0$$

■

Surface Area

Now we define the surface area of a general parametric surface given by Equation 1. For simplicity we start by considering a surface whose parameter domain D is a rectangle, and we divide it into subrectangles R_{ij}. Let's choose (u_i^*, v_j^*) to be the lower left corner of R_{ij}. (See Figure 14.)

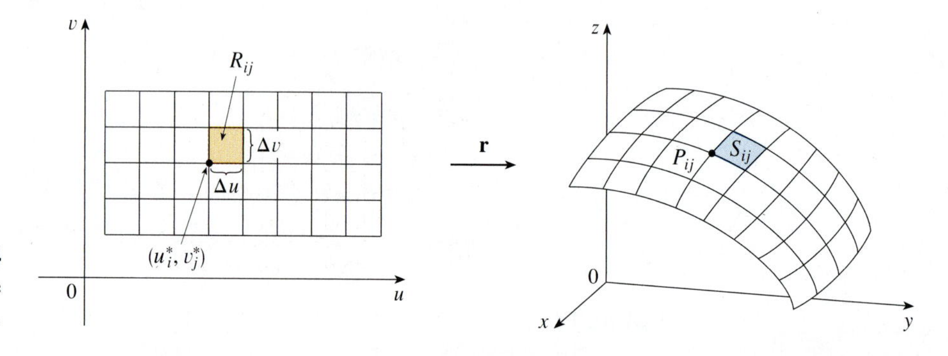

FIGURE 14
The image of the subrectangle R_{ij} is the patch S_{ij}.

The part S_{ij} of the surface S that corresponds to R_{ij} is called a *patch* and has the point P_{ij} with position vector $\mathbf{r}(u_i^*, v_j^*)$ as one of its corners. Let

$$\mathbf{r}_u^* = \mathbf{r}_u(u_i^*, v_j^*) \qquad \text{and} \qquad \mathbf{r}_v^* = \mathbf{r}_v(u_i^*, v_j^*)$$

be the tangent vectors at P_{ij} as given by Equations 5 and 4.

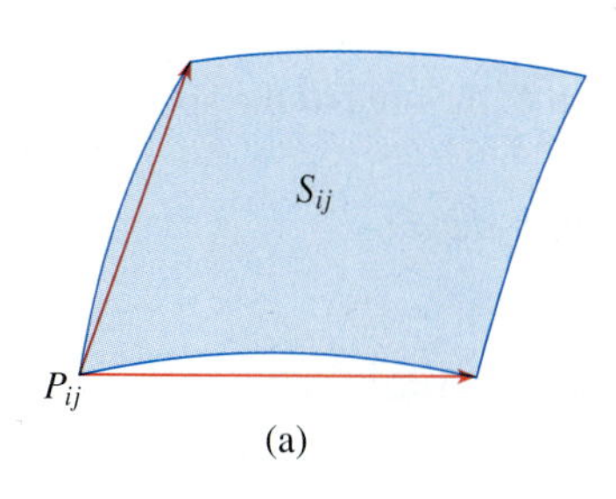

(a)

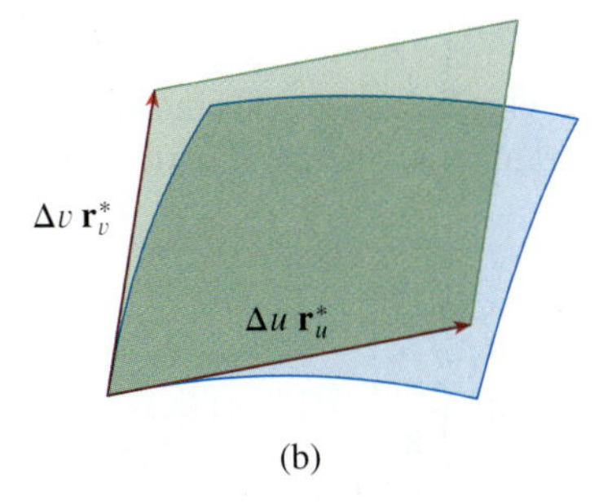

(b)

FIGURE 15
Approximating a patch by a parallelogram

Figure 15(a) shows how the two edges of the patch that meet at P_{ij} can be approximated by vectors. These vectors, in turn, can be approximated by the vectors $\Delta u\, \mathbf{r}_u^*$ and $\Delta v\, \mathbf{r}_v^*$ because partial derivatives can be approximated by difference quotients. So we approximate S_{ij} by the parallelogram determined by the vectors $\Delta u\, \mathbf{r}_u^*$ and $\Delta v\, \mathbf{r}_v^*$. This parallelogram is shown in Figure 15(b) and lies in the tangent plane to S at P_{ij}. The area of this parallelogram is

$$|(\Delta u\, \mathbf{r}_u^*) \times (\Delta v\, \mathbf{r}_v^*)| = |\mathbf{r}_u^* \times \mathbf{r}_v^*|\, \Delta u\, \Delta v$$

and so an approximation to the area of S is

$$\sum_{i=1}^{m} \sum_{j=1}^{n} |\mathbf{r}_u^* \times \mathbf{r}_v^*|\, \Delta u\, \Delta v$$

Our intuition tells us that this approximation gets better as we increase the number of subrectangles, and we recognize the double sum as a Riemann sum for the double integral $\iint_D |\mathbf{r}_u \times \mathbf{r}_v|\, du\, dv$. This motivates the following definition.

6 Definition If a smooth parametric surface S is given by the equation

$$\mathbf{r}(u, v) = x(u, v)\,\mathbf{i} + y(u, v)\,\mathbf{j} + z(u, v)\,\mathbf{k} \qquad (u, v) \in D$$

and S is covered just once as (u, v) ranges throughout the parameter domain D, then the **surface area** of S is

$$A(S) = \iint_D |\mathbf{r}_u \times \mathbf{r}_v|\, dA$$

where $\mathbf{r}_u = \dfrac{\partial x}{\partial u}\mathbf{i} + \dfrac{\partial y}{\partial u}\mathbf{j} + \dfrac{\partial z}{\partial u}\mathbf{k} \qquad \mathbf{r}_v = \dfrac{\partial x}{\partial v}\mathbf{i} + \dfrac{\partial y}{\partial v}\mathbf{j} + \dfrac{\partial z}{\partial v}\mathbf{k}$

EXAMPLE 10 Find the surface area of a sphere of radius a.

SOLUTION In Example 4 we found the parametric representation

$$x = a \sin\phi \cos\theta \qquad y = a \sin\phi \sin\theta \qquad z = a \cos\phi$$

where the parameter domain is

$$D = \{(\phi, \theta) \mid 0 \leqslant \phi \leqslant \pi,\ 0 \leqslant \theta \leqslant 2\pi\}$$

We first compute the cross product of the tangent vectors:

$$\mathbf{r}_\phi \times \mathbf{r}_\theta = \begin{vmatrix} \mathbf{i} & \mathbf{j} & \mathbf{k} \\ \dfrac{\partial x}{\partial \phi} & \dfrac{\partial y}{\partial \phi} & \dfrac{\partial z}{\partial \phi} \\ \dfrac{\partial x}{\partial \theta} & \dfrac{\partial y}{\partial \theta} & \dfrac{\partial z}{\partial \theta} \end{vmatrix} = \begin{vmatrix} \mathbf{i} & \mathbf{j} & \mathbf{k} \\ a\cos\phi\cos\theta & a\cos\phi\sin\theta & -a\sin\phi \\ -a\sin\phi\sin\theta & a\sin\phi\cos\theta & 0 \end{vmatrix}$$

$$= a^2\sin^2\phi\cos\theta\,\mathbf{i} + a^2\sin^2\phi\sin\theta\,\mathbf{j} + a^2\sin\phi\cos\phi\,\mathbf{k}$$

Thus

$$|\mathbf{r}_\phi \times \mathbf{r}_\theta| = \sqrt{a^4 \sin^4\phi \cos^2\theta + a^4 \sin^4\phi \sin^2\theta + a^4 \sin^2\phi \cos^2\phi}$$

$$= \sqrt{a^4 \sin^4\phi + a^4 \sin^2\phi \cos^2\phi} = a^2\sqrt{\sin^2\phi} = a^2 \sin\phi$$

since $\sin\phi \geqslant 0$ for $0 \leqslant \phi \leqslant \pi$. Therefore, by Definition 6, the area of the sphere is

$$A = \iint_D |\mathbf{r}_\phi \times \mathbf{r}_\theta|\, dA = \int_0^{2\pi} \int_0^{\pi} a^2 \sin\phi \, d\phi \, d\theta$$

$$= a^2 \int_0^{2\pi} d\theta \int_0^{\pi} \sin\phi \, d\phi = a^2(2\pi)2 = 4\pi a^2$$

Surface Area of the Graph of a Function

For the special case of a surface S with equation $z = f(x, y)$, where (x, y) lies in D and f has continuous partial derivatives, we take x and y as parameters. The parametric equations are

$$x = x \qquad y = y \qquad z = f(x, y)$$

so

$$\mathbf{r}_x = \mathbf{i} + \left(\frac{\partial f}{\partial x}\right)\mathbf{k} \qquad \mathbf{r}_y = \mathbf{j} + \left(\frac{\partial f}{\partial y}\right)\mathbf{k}$$

and

$$\boxed{7} \qquad \mathbf{r}_x \times \mathbf{r}_y = \begin{vmatrix} \mathbf{i} & \mathbf{j} & \mathbf{k} \\ 1 & 0 & \dfrac{\partial f}{\partial x} \\ 0 & 1 & \dfrac{\partial f}{\partial y} \end{vmatrix} = -\frac{\partial f}{\partial x}\mathbf{i} - \frac{\partial f}{\partial y}\mathbf{j} + \mathbf{k}$$

Thus we have

$$\boxed{8} \qquad |\mathbf{r}_x \times \mathbf{r}_y| = \sqrt{\left(\frac{\partial f}{\partial x}\right)^2 + \left(\frac{\partial f}{\partial y}\right)^2 + 1} = \sqrt{1 + \left(\frac{\partial z}{\partial x}\right)^2 + \left(\frac{\partial z}{\partial y}\right)^2}$$

and the surface area formula in Definition 6 becomes

$$\boxed{9} \qquad A(S) = \iint_D \sqrt{1 + \left(\frac{\partial z}{\partial x}\right)^2 + \left(\frac{\partial z}{\partial y}\right)^2}\, dA$$

Notice the similarity between the surface area formula in Equation 9 and the arc length formula

$$L = \int_a^b \sqrt{1 + \left(\frac{dy}{dx}\right)^2}\, dx$$

from Section 8.1.

EXAMPLE 11 Find the area of the part of the paraboloid $z = x^2 + y^2$ that lies under the plane $z = 9$.

SOLUTION The plane intersects the paraboloid in the circle $x^2 + y^2 = 9$, $z = 9$. Therefore the given surface lies above the disk D with center the origin and radius 3. (See

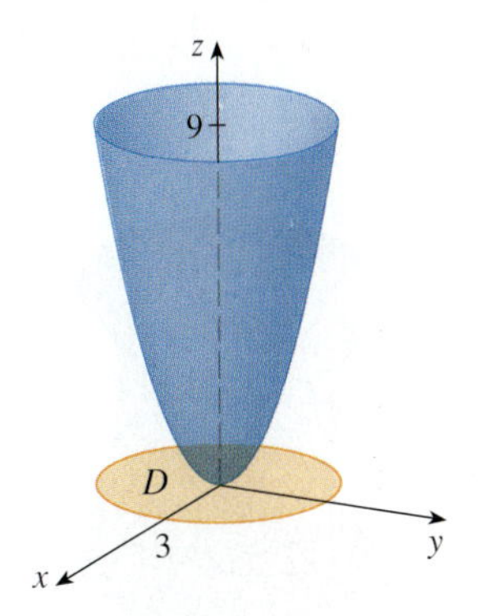

FIGURE 16

Figure 16.) Using Formula 9, we have

$$A = \iint_D \sqrt{1 + \left(\frac{\partial z}{\partial x}\right)^2 + \left(\frac{\partial z}{\partial y}\right)^2}\, dA$$

$$= \iint_D \sqrt{1 + (2x)^2 + (2y)^2}\, dA$$

$$= \iint_D \sqrt{1 + 4(x^2 + y^2)}\, dA$$

Converting to polar coordinates, we obtain

$$A = \int_0^{2\pi} \int_0^3 \sqrt{1 + 4r^2}\, r\, dr\, d\theta = \int_0^{2\pi} d\theta \int_0^3 r\sqrt{1 + 4r^2}\, dr$$

$$= 2\pi\left(\tfrac{1}{8}\right)\tfrac{2}{3}(1 + 4r^2)^{3/2}\Big]_0^3 = \frac{\pi}{6}\left(37\sqrt{37} - 1\right) \qquad \blacksquare$$

The question remains whether our definition of surface area (6) is consistent with the surface area formula from single-variable calculus (8.2.4).

We consider the surface S obtained by rotating the curve $y = f(x)$, $a \leqslant x \leqslant b$, about the x-axis, where $f(x) \geqslant 0$ and f' is continuous. From Equations 3 we know that parametric equations of S are

$$x = x \qquad y = f(x)\cos\theta \qquad z = f(x)\sin\theta \qquad a \leqslant x \leqslant b \qquad 0 \leqslant \theta \leqslant 2\pi$$

To compute the surface area of S we need the tangent vectors

$$\mathbf{r}_x = \mathbf{i} + f'(x)\cos\theta\,\mathbf{j} + f'(x)\sin\theta\,\mathbf{k}$$

$$\mathbf{r}_\theta = -f(x)\sin\theta\,\mathbf{j} + f(x)\cos\theta\,\mathbf{k}$$

Thus

$$\mathbf{r}_x \times \mathbf{r}_\theta = \begin{vmatrix} \mathbf{i} & \mathbf{j} & \mathbf{k} \\ 1 & f'(x)\cos\theta & f'(x)\sin\theta \\ 0 & -f(x)\sin\theta & f(x)\cos\theta \end{vmatrix}$$

$$= f(x)f'(x)\,\mathbf{i} - f(x)\cos\theta\,\mathbf{j} - f(x)\sin\theta\,\mathbf{k}$$

and so

$$|\mathbf{r}_x \times \mathbf{r}_\theta| = \sqrt{[f(x)]^2[f'(x)]^2 + [f(x)]^2\cos^2\theta + [f(x)]^2\sin^2\theta}$$

$$= \sqrt{[f(x)]^2[1 + [f'(x)]^2]} = f(x)\sqrt{1 + [f'(x)]^2}$$

because $f(x) \geqslant 0$. Therefore the area of S is

$$A = \iint_D |\mathbf{r}_x \times \mathbf{r}_\theta|\, dA$$

$$= \int_0^{2\pi} \int_a^b f(x)\sqrt{1 + [f'(x)]^2}\, dx\, d\theta$$

$$= 2\pi \int_a^b f(x)\sqrt{1 + [f'(x)]^2}\, dx$$

This is precisely the formula that was used to define the area of a surface of revolution in single-variable calculus (8.2.4).

16.6 EXERCISES

1–2 Determine whether the points P and Q lie on the given surface.

1. $\mathbf{r}(u, v) = \langle u + v, u - 2v, 3 + u - v \rangle$
$P(4, -5, 1)$, $Q(0, 4, 6)$

2. $\mathbf{r}(u, v) = \langle 1 + u - v, u + v^2, u^2 - v^2 \rangle$
$P(1, 2, 1)$, $Q(2, 3, 3)$

3–6 Identify the surface with the given vector equation.

3. $\mathbf{r}(u, v) = (u + v)\,\mathbf{i} + (3 - v)\,\mathbf{j} + (1 + 4u + 5v)\,\mathbf{k}$

4. $\mathbf{r}(u, v) = u^2\,\mathbf{i} + u \cos v\,\mathbf{j} + u \sin v\,\mathbf{k}$

5. $\mathbf{r}(s, t) = \langle s \cos t, s \sin t, s \rangle$

6. $\mathbf{r}(s, t) = \langle 3 \cos t, s, \sin t \rangle, \quad -1 \leqslant s \leqslant 1$

7–12 Use a computer to graph the parametric surface. Get a printout and indicate on it which grid curves have u constant and which have v constant.

7. $\mathbf{r}(u, v) = \langle u^2, v^2, u + v \rangle$,
$-1 \leqslant u \leqslant 1, \ -1 \leqslant v \leqslant 1$

8. $\mathbf{r}(u, v) = \langle u, v^3, -v \rangle$,
$-2 \leqslant u \leqslant 2, \ -2 \leqslant v \leqslant 2$

9. $\mathbf{r}(u, v) = \langle u^3, u \sin v, u \cos v \rangle$,
$-1 \leqslant u \leqslant 1, \ 0 \leqslant v \leqslant 2\pi$

10. $\mathbf{r}(u, v) = \langle u, \sin(u + v), \sin v \rangle$,
$-\pi \leqslant u \leqslant \pi, \ -\pi \leqslant v \leqslant \pi$

11. $x = \sin v, \quad y = \cos u \sin 4v, \quad z = \sin 2u \sin 4v$,
$0 \leqslant u \leqslant 2\pi, \ -\pi/2 \leqslant v \leqslant \pi/2$

12. $x = \cos u, \quad y = \sin u \sin v, \quad z = \cos v$,
$0 \leqslant u \leqslant 2\pi, \ 0 \leqslant v \leqslant 2\pi$

13–18 Match the equations with the graphs labeled I–VI and give reasons for your answers. Determine which families of grid curves have u constant and which have v constant.

13. $\mathbf{r}(u, v) = u \cos v\,\mathbf{i} + u \sin v\,\mathbf{j} + v\,\mathbf{k}$

14. $\mathbf{r}(u, v) = uv^2\,\mathbf{i} + u^2 v\,\mathbf{j} + (u^2 - v^2)\,\mathbf{k}$

15. $\mathbf{r}(u, v) = (u^3 - u)\,\mathbf{i} + v^2\,\mathbf{j} + u^2\,\mathbf{k}$

16. $x = (1 - u)(3 + \cos v) \cos 4\pi u$,

$y = (1 - u)(3 + \cos v) \sin 4\pi u$,

$z = 3u + (1 - u) \sin v$

17. $x = \cos^3 u \cos^3 v, \quad y = \sin^3 u \cos^3 v, \quad z = \sin^3 v$

18. $x = \sin u, \quad y = \cos u \sin v, \quad z = \sin v$

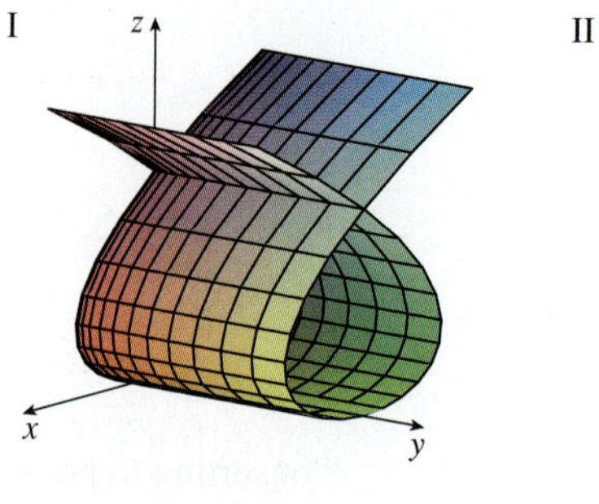

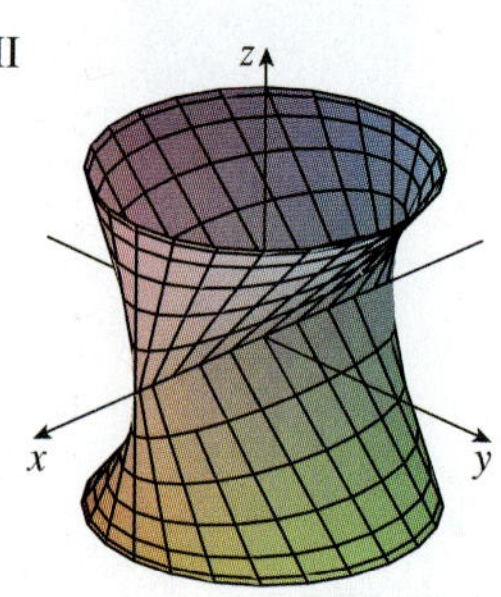

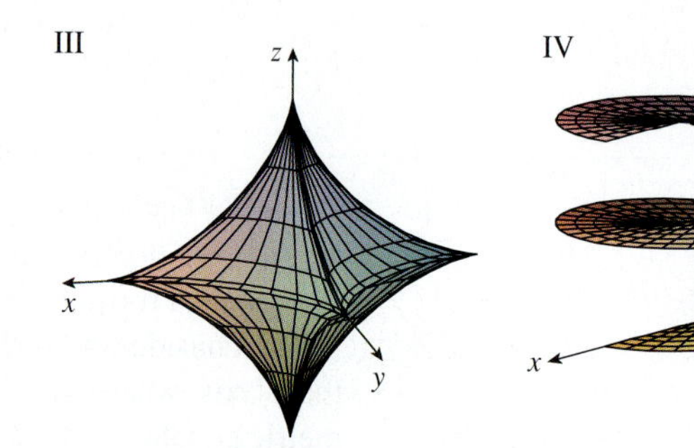

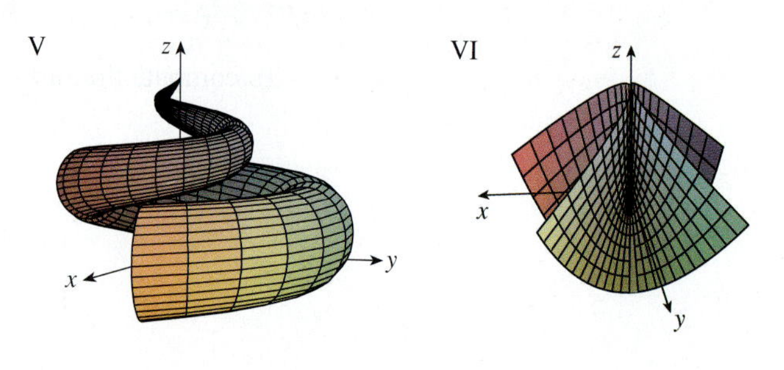

19–26 Find a parametric representation for the surface.

19. The plane through the origin that contains the vectors $\mathbf{i} - \mathbf{j}$ and $\mathbf{j} - \mathbf{k}$

20. The plane that passes through the point $(0, -1, 5)$ and contains the vectors $\langle 2, 1, 4 \rangle$ and $\langle -3, 2, 5 \rangle$

21. The part of the hyperboloid $4x^2 - 4y^2 - z^2 = 4$ that lies in front of the yz-plane

22. The part of the ellipsoid $x^2 + 2y^2 + 3z^2 = 1$ that lies to the left of the xz-plane

23. The part of the sphere $x^2 + y^2 + z^2 = 4$ that lies above the cone $z = \sqrt{x^2 + y^2}$

24. The part of the cylinder $x^2 + z^2 = 9$ that lies above the xy-plane and between the planes $y = -4$ and $y = 4$

25. The part of the sphere $x^2 + y^2 + z^2 = 36$ that lies between the planes $z = 0$ and $z = 3\sqrt{3}$

26. The part of the plane $z = x + 3$ that lies inside the cylinder $x^2 + y^2 = 1$

27–28 Use a graphing device to produce a graph that looks like the given one.

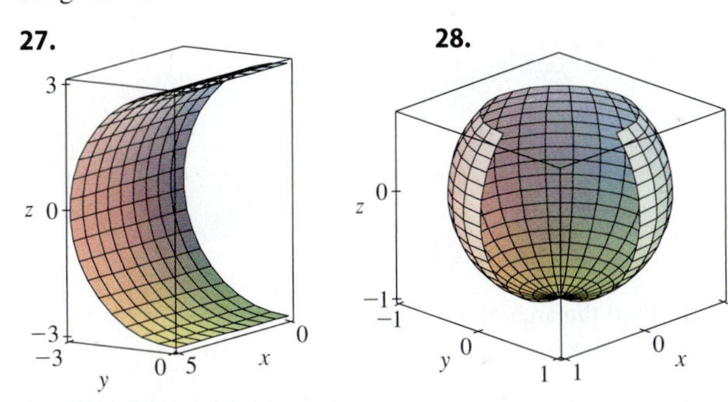

29. Find parametric equations for the surface obtained by rotating the curve $y = 1/(1 + x^2)$, $-2 \leqslant x \leqslant 2$, about the x-axis and use them to graph the surface.

30. Find parametric equations for the surface obtained by rotating the curve $x = 1/y$, $y \geqslant 1$, about the y-axis and use them to graph the surface.

31. (a) What happens to the spiral tube in Example 2 (see Figure 5) if we replace $\cos u$ by $\sin u$ and $\sin u$ by $\cos u$?
(b) What happens if we replace $\cos u$ by $\cos 2u$ and $\sin u$ by $\sin 2u$?

32. The surface with parametric equations

$$x = 2\cos\theta + r\cos(\theta/2)$$
$$y = 2\sin\theta + r\cos(\theta/2)$$
$$z = r\sin(\theta/2)$$

where $-\frac{1}{2} \leqslant r \leqslant \frac{1}{2}$ and $0 \leqslant \theta \leqslant 2\pi$, is called a **Möbius strip**. Graph this surface with several viewpoints. What is unusual about it?

33–36 Find an equation of the tangent plane to the given parametric surface at the specified point.

33. $x = u + v, \quad y = 3u^2, \quad z = u - v; \quad (2, 3, 0)$

34. $x = u^2 + 1, \quad y = v^3 + 1, \quad z = u + v; \quad (5, 2, 3)$

35. $\mathbf{r}(u, v) = u\cos v\,\mathbf{i} + u\sin v\,\mathbf{j} + v\,\mathbf{k}; \quad u = 1, \ v = \pi/3$

36. $\mathbf{r}(u, v) = \sin u\,\mathbf{i} + \cos u\sin v\,\mathbf{j} + \sin v\,\mathbf{k};$
$u = \pi/6, \ v = \pi/6$

CAS **37–38** Find an equation of the tangent plane to the given parametric surface at the specified point. Graph the surface and the tangent plane.

37. $\mathbf{r}(u, v) = u^2\mathbf{i} + 2u\sin v\,\mathbf{j} + u\cos v\,\mathbf{k}; \quad u = 1, \ v = 0$

38. $\mathbf{r}(u, v) = (1 - u^2 - v^2)\,\mathbf{i} - v\,\mathbf{j} - u\,\mathbf{k}; \quad (-1, -1, -1)$

39–50 Find the area of the surface.

39. The part of the plane $3x + 2y + z = 6$ that lies in the first octant

40. The part of the plane with vector equation $\mathbf{r}(u, v) = \langle u + v, 2 - 3u, 1 + u - v\rangle$ that is given by $0 \leqslant u \leqslant 2$, $-1 \leqslant v \leqslant 1$

41. The part of the plane $x + 2y + 3z = 1$ that lies inside the cylinder $x^2 + y^2 = 3$

42. The part of the cone $z = \sqrt{x^2 + y^2}$ that lies between the plane $y = x$ and the cylinder $y = x^2$

43. The surface $z = \frac{2}{3}(x^{3/2} + y^{3/2})$, $0 \leqslant x \leqslant 1$, $0 \leqslant y \leqslant 1$

44. The part of the surface $z = 4 - 2x^2 + y$ that lies above the triangle with vertices $(0, 0)$, $(1, 0)$, and $(1, 1)$

45. The part of the surface $z = xy$ that lies within the cylinder $x^2 + y^2 = 1$

46. The part of the surface $x = z^2 + y$ that lies between the planes $y = 0$, $y = 2$, $z = 0$, and $z = 2$

47. The part of the paraboloid $y = x^2 + z^2$ that lies within the cylinder $x^2 + z^2 = 16$

48. The helicoid (or spiral ramp) with vector equation $\mathbf{r}(u, v) = u\cos v\,\mathbf{i} + u\sin v\,\mathbf{j} + v\,\mathbf{k}$, $0 \leqslant u \leqslant 1$, $0 \leqslant v \leqslant \pi$

49. The surface with parametric equations $x = u^2$, $y = uv$, $z = \frac{1}{2}v^2$, $0 \leqslant u \leqslant 1$, $0 \leqslant v \leqslant 2$

50. The part of the sphere $x^2 + y^2 + z^2 = b^2$ that lies inside the cylinder $x^2 + y^2 = a^2$, where $0 < a < b$

51. If the equation of a surface S is $z = f(x, y)$, where $x^2 + y^2 \leqslant R^2$, and you know that $|f_x| \leqslant 1$ and $|f_y| \leqslant 1$, what can you say about $A(S)$?

52–53 Find the area of the surface correct to four decimal places by expressing the area in terms of a single integral and using your calculator to estimate the integral.

52. The part of the surface $z = \cos(x^2 + y^2)$ that lies inside the cylinder $x^2 + y^2 = 1$

53. The part of the surface $z = \ln(x^2 + y^2 + 2)$ that lies above the disk $x^2 + y^2 \leqslant 1$

CAS **54.** Find, to four decimal places, the area of the part of the surface $z = (1 + x^2)/(1 + y^2)$ that lies above the square $|x| + |y| \leqslant 1$. Illustrate by graphing this part of the surface.

55. (a) Use the Midpoint Rule for double integrals (see Section 15.1) with six squares to estimate the area of the surface $z = 1/(1 + x^2 + y^2)$, $0 \leqslant x \leqslant 6$, $0 \leqslant y \leqslant 4$.
CAS (b) Use a computer algebra system to approximate the surface area in part (a) to four decimal places. Compare with the answer to part (a).

CAS 56. Find the area of the surface with vector equation $\mathbf{r}(u, v) = \langle \cos^3 u \cos^3 v, \sin^3 u \cos^3 v, \sin^3 v \rangle$, $0 \leqslant u \leqslant \pi$, $0 \leqslant v \leqslant 2\pi$. State your answer correct to four decimal places.

CAS 57. Find the exact area of the surface $z = 1 + 2x + 3y + 4y^2$, $1 \leqslant x \leqslant 4$, $0 \leqslant y \leqslant 1$.

58. (a) Set up, but do not evaluate, a double integral for the area of the surface with parametric equations $x = au \cos v$, $y = bu \sin v$, $z = u^2$, $0 \leqslant u \leqslant 2$, $0 \leqslant v \leqslant 2\pi$.

(b) Eliminate the parameters to show that the surface is an elliptic paraboloid and set up another double integral for the surface area.

(c) Use the parametric equations in part (a) with $a = 2$ and $b = 3$ to graph the surface.

CAS (d) For the case $a = 2$, $b = 3$, use a computer algebra system to find the surface area correct to four decimal places.

59. (a) Show that the parametric equations $x = a \sin u \cos v$, $y = b \sin u \sin v$, $z = c \cos u$, $0 \leqslant u \leqslant \pi$, $0 \leqslant v \leqslant 2\pi$, represent an ellipsoid.

(b) Use the parametric equations in part (a) to graph the ellipsoid for the case $a = 1$, $b = 2$, $c = 3$.

(c) Set up, but do not evaluate, a double integral for the surface area of the ellipsoid in part (b).

60. (a) Show that the parametric equations $x = a \cosh u \cos v$, $y = b \cosh u \sin v$, $z = c \sinh u$, represent a hyperboloid of one sheet.

(b) Use the parametric equations in part (a) to graph the hyperboloid for the case $a = 1$, $b = 2$, $c = 3$.

(c) Set up, but do not evaluate, a double integral for the surface area of the part of the hyperboloid in part (b) that lies between the planes $z = -3$ and $z = 3$.

61. Find the area of the part of the sphere $x^2 + y^2 + z^2 = 4z$ that lies inside the paraboloid $z = x^2 + y^2$.

62. The figure shows the surface created when the cylinder $y^2 + z^2 = 1$ intersects the cylinder $x^2 + z^2 = 1$. Find the area of this surface.

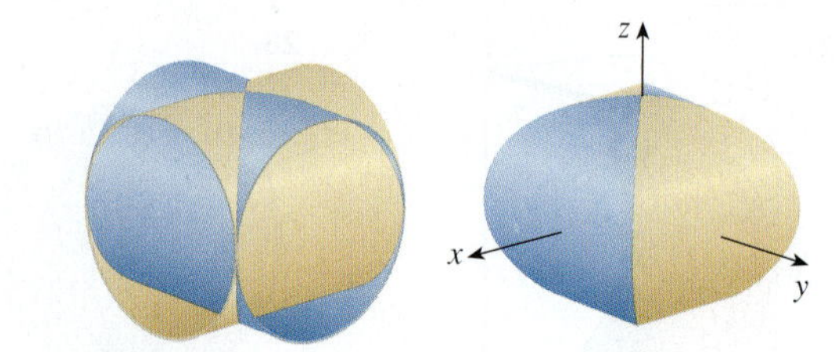

63. Find the area of the part of the sphere $x^2 + y^2 + z^2 = a^2$ that lies inside the cylinder $x^2 + y^2 = ax$.

64. (a) Find a parametric representation for the torus obtained by rotating about the z-axis the circle in the xz-plane with center $(b, 0, 0)$ and radius $a < b$. [*Hint:* Take as parameters the angles θ and α shown in the figure.]

(b) Use the parametric equations found in part (a) to graph the torus for several values of a and b.

(c) Use the parametric representation from part (a) to find the surface area of the torus.

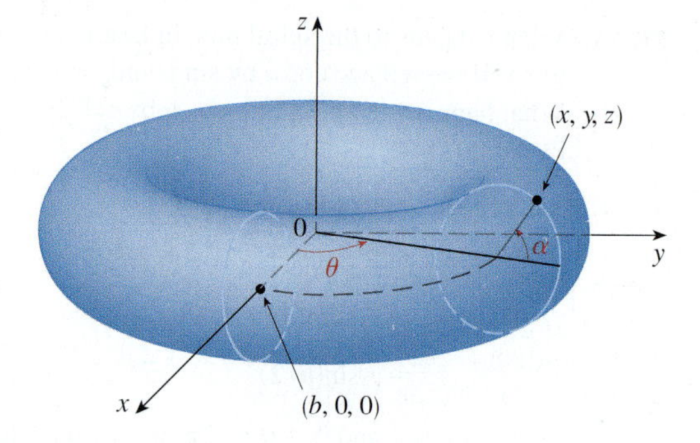

16.7 Surface Integrals

The relationship between surface integrals and surface area is much the same as the relationship between line integrals and arc length. Suppose f is a function of three variables whose domain includes a surface S. We will define the surface integral of f over S in such a way that, in the case where $f(x, y, z) = 1$, the value of the surface integral is equal to the surface area of S. We start with parametric surfaces and then deal with the special case where S is the graph of a function of two variables.

Parametric Surfaces

Suppose that a surface S has a vector equation

$$\mathbf{r}(u, v) = x(u, v)\,\mathbf{i} + y(u, v)\,\mathbf{j} + z(u, v)\,\mathbf{k} \qquad (u, v) \in D$$

We first assume that the parameter domain D is a rectangle and we divide it into subrect-

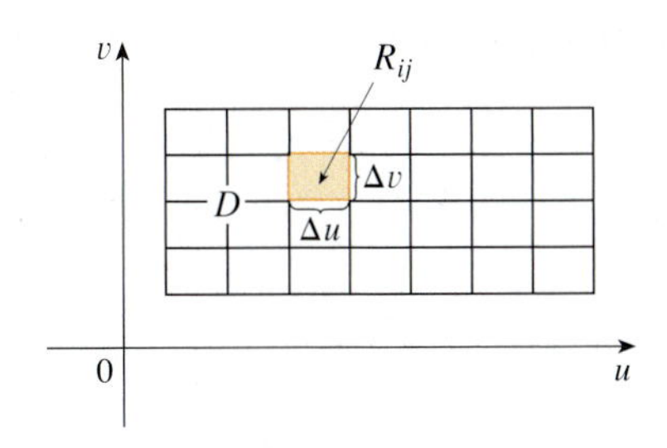

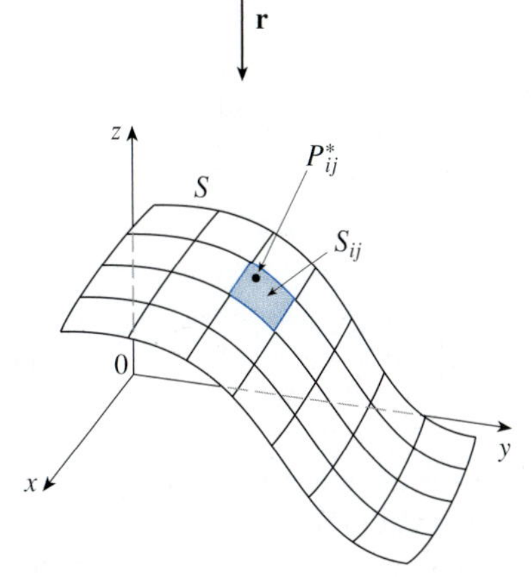

FIGURE 1

angles R_{ij} with dimensions Δu and Δv. Then the surface S is divided into corresponding patches S_{ij} as in Figure 1. We evaluate f at a point P_{ij}^* in each patch, multiply by the area ΔS_{ij} of the patch, and form the Riemann sum

$$\sum_{i=1}^{m} \sum_{j=1}^{n} f(P_{ij}^*)\, \Delta S_{ij}$$

Then we take the limit as the number of patches increases and define the **surface integral of f over the surface S** as

1
$$\iint_S f(x, y, z)\, dS = \lim_{m, n \to \infty} \sum_{i=1}^{m} \sum_{j=1}^{n} f(P_{ij}^*)\, \Delta S_{ij}$$

Notice the analogy with the definition of a line integral (16.2.2) and also the analogy with the definition of a double integral (15.1.5).

To evaluate the surface integral in Equation 1 we approximate the patch area ΔS_{ij} by the area of an approximating parallelogram in the tangent plane. In our discussion of surface area in Section 16.6 we made the approximation

$$\Delta S_{ij} \approx |\mathbf{r}_u \times \mathbf{r}_v|\, \Delta u\, \Delta v$$

where $$\mathbf{r}_u = \frac{\partial x}{\partial u}\mathbf{i} + \frac{\partial y}{\partial u}\mathbf{j} + \frac{\partial z}{\partial u}\mathbf{k} \qquad \mathbf{r}_v = \frac{\partial x}{\partial v}\mathbf{i} + \frac{\partial y}{\partial v}\mathbf{j} + \frac{\partial z}{\partial v}\mathbf{k}$$

are the tangent vectors at a corner of S_{ij}. If the components are continuous and $\mathbf{r}_u$ and $\mathbf{r}_v$ are nonzero and nonparallel in the interior of D, it can be shown from Definition 1, even when D is not a rectangle, that

We assume that the surface is covered only once as (u, v) ranges throughout D. The value of the surface integral does not depend on the parametrization that is used.

2
$$\iint_S f(x, y, z)\, dS = \iint_D f(\mathbf{r}(u, v))\, |\mathbf{r}_u \times \mathbf{r}_v|\, dA$$

This should be compared with the formula for a line integral:

$$\int_C f(x, y, z)\, ds = \int_a^b f(\mathbf{r}(t))\, |\mathbf{r}'(t)|\, dt$$

Observe also that

$$\iint_S 1\, dS = \iint_D |\mathbf{r}_u \times \mathbf{r}_v|\, dA = A(S)$$

Formula 2 allows us to compute a surface integral by converting it into a double integral over the parameter domain D. When using this formula, remember that $f(\mathbf{r}(u, v))$ is evaluated by writing $x = x(u, v)$, $y = y(u, v)$, and $z = z(u, v)$ in the formula for $f(x, y, z)$.

EXAMPLE 1 Compute the surface integral $\iint_S x^2\, dS$, where S is the unit sphere $x^2 + y^2 + z^2 = 1$.

SOLUTION As in Example 16.6.4, we use the parametric representation

$$x = \sin\phi\, \cos\theta \qquad y = \sin\phi\, \sin\theta \qquad z = \cos\phi \qquad 0 \leqslant \phi \leqslant \pi \qquad 0 \leqslant \theta \leqslant 2\pi$$

that is, $$\mathbf{r}(\phi, \theta) = \sin\phi \cos\theta\, \mathbf{i} + \sin\phi \sin\theta\, \mathbf{j} + \cos\phi\, \mathbf{k}$$

As in Example 16.6.10, we can compute that

$$|\mathbf{r}_\phi \times \mathbf{r}_\theta| = \sin\phi$$

Therefore, by Formula 2,

$$\iint_S x^2\, dS = \iint_D (\sin\phi \cos\theta)^2 |\mathbf{r}_\phi \times \mathbf{r}_\theta|\, dA$$

$$= \int_0^{2\pi} \int_0^{\pi} \sin^2\phi \cos^2\theta \sin\phi\, d\phi\, d\theta = \int_0^{2\pi} \cos^2\theta\, d\theta \int_0^{\pi} \sin^3\phi\, d\phi$$

$$= \int_0^{2\pi} \tfrac{1}{2}(1 + \cos 2\theta)\, d\theta \int_0^{\pi} (\sin\phi - \sin\phi \cos^2\phi)\, d\phi$$

$$= \tfrac{1}{2}\left[\theta + \tfrac{1}{2}\sin 2\theta\right]_0^{2\pi} \left[-\cos\phi + \tfrac{1}{3}\cos^3\phi\right]_0^{\pi} = \frac{4\pi}{3}$$ ■

Here we use the identities

$$\cos^2\theta = \tfrac{1}{2}(1 + \cos 2\theta)$$
$$\sin^2\phi = 1 - \cos^2\phi$$

Instead, we could use Formulas 64 and 67 in the Table of Integrals.

Surface integrals have applications similar to those for the integrals we have previously considered. For example, if a thin sheet (say, of aluminum foil) has the shape of a surface S and the density (mass per unit area) at the point (x, y, z) is $\rho(x, y, z)$, then the total **mass** of the sheet is

$$m = \iint_S \rho(x, y, z)\, dS$$

and the **center of mass** is $(\bar{x}, \bar{y}, \bar{z})$, where

$$\bar{x} = \frac{1}{m} \iint_S x\rho(x, y, z)\, dS \qquad \bar{y} = \frac{1}{m} \iint_S y\rho(x, y, z)\, dS \qquad \bar{z} = \frac{1}{m} \iint_S z\rho(x, y, z)\, dS$$

Moments of inertia can also be defined as before (see Exercise 41).

Graphs of Functions

Any surface S with equation $z = g(x, y)$ can be regarded as a parametric surface with parametric equations

$$x = x \qquad y = y \qquad z = g(x, y)$$

and so we have $$\mathbf{r}_x = \mathbf{i} + \left(\frac{\partial g}{\partial x}\right)\mathbf{k} \qquad \mathbf{r}_y = \mathbf{j} + \left(\frac{\partial g}{\partial y}\right)\mathbf{k}$$

Thus

3 $$\mathbf{r}_x \times \mathbf{r}_y = -\frac{\partial g}{\partial x}\mathbf{i} - \frac{\partial g}{\partial y}\mathbf{j} + \mathbf{k}$$

and $$|\mathbf{r}_x \times \mathbf{r}_y| = \sqrt{\left(\frac{\partial z}{\partial x}\right)^2 + \left(\frac{\partial z}{\partial y}\right)^2 + 1}$$

Therefore, in this case, Formula 2 becomes

$$\boxed{4} \quad \iint_S f(x, y, z)\, dS = \iint_D f(x, y, g(x, y)) \sqrt{\left(\frac{\partial z}{\partial x}\right)^2 + \left(\frac{\partial z}{\partial y}\right)^2 + 1}\, dA$$

Similar formulas apply when it is more convenient to project S onto the yz-plane or xz-plane. For instance, if S is a surface with equation $y = h(x, z)$ and D is its projection onto the xz-plane, then

$$\iint_S f(x, y, z)\, dS = \iint_D f(x, h(x, z), z) \sqrt{\left(\frac{\partial y}{\partial x}\right)^2 + \left(\frac{\partial y}{\partial z}\right)^2 + 1}\, dA$$

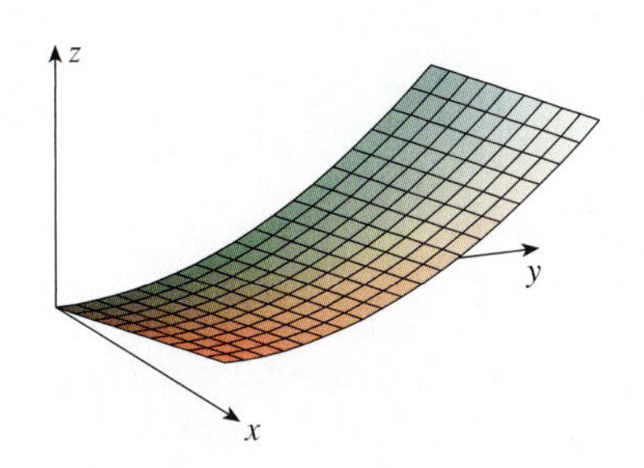

FIGURE 2

EXAMPLE 2 Evaluate $\iint_S y\, dS$, where S is the surface $z = x + y^2$, $0 \leqslant x \leqslant 1$, $0 \leqslant y \leqslant 2$. (See Figure 2.)

SOLUTION Since

$$\frac{\partial z}{\partial x} = 1 \qquad \text{and} \qquad \frac{\partial z}{\partial y} = 2y$$

Formula 4 gives

$$\iint_S y\, dS = \iint_D y \sqrt{1 + \left(\frac{\partial z}{\partial x}\right)^2 + \left(\frac{\partial z}{\partial y}\right)^2}\, dA$$

$$= \int_0^1 \int_0^2 y\sqrt{1 + 1 + 4y^2}\, dy\, dx$$

$$= \int_0^1 dx\, \sqrt{2} \int_0^2 y\sqrt{1 + 2y^2}\, dy$$

$$= \sqrt{2}\left(\tfrac{1}{4}\right)\tfrac{2}{3}(1 + 2y^2)^{3/2}\Big]_0^2 = \frac{13\sqrt{2}}{3}$$

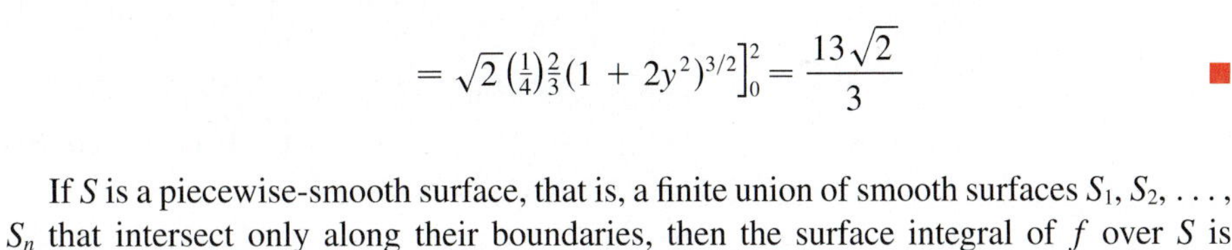

If S is a piecewise-smooth surface, that is, a finite union of smooth surfaces $S_1, S_2, \ldots, S_n$ that intersect only along their boundaries, then the surface integral of f over S is defined by

$$\iint_S f(x, y, z)\, dS = \iint_{S_1} f(x, y, z)\, dS + \cdots + \iint_{S_n} f(x, y, z)\, dS$$

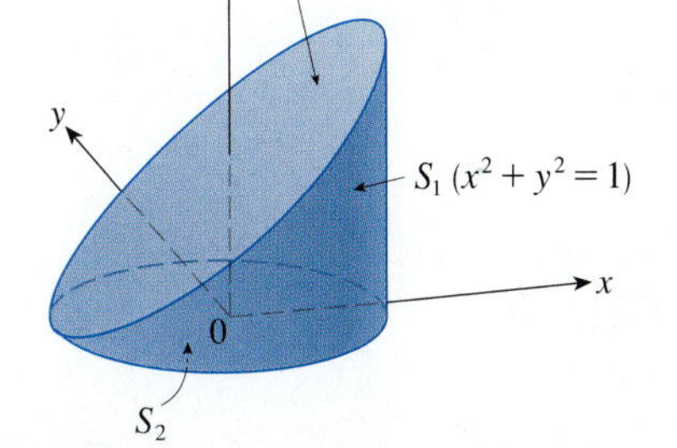

FIGURE 3

EXAMPLE 3 Evaluate $\iint_S z\, dS$, where S is the surface whose sides S_1 are given by the cylinder $x^2 + y^2 = 1$, whose bottom S_2 is the disk $x^2 + y^2 \leqslant 1$ in the plane $z = 0$, and whose top S_3 is the part of the plane $z = 1 + x$ that lies above S_2.

SOLUTION The surface S is shown in Figure 3. (We have changed the usual position of the axes to get a better look at S.) For S_1 we use θ and z as parameters (see Example 16.6.5) and write its parametric equations as

$$x = \cos\theta \qquad y = \sin\theta \qquad z = z$$

where

$$0 \leqslant \theta \leqslant 2\pi \qquad \text{and} \qquad 0 \leqslant z \leqslant 1 + x = 1 + \cos\theta$$

Therefore

$$\mathbf{r}_\theta \times \mathbf{r}_z = \begin{vmatrix} \mathbf{i} & \mathbf{j} & \mathbf{k} \\ -\sin\theta & \cos\theta & 0 \\ 0 & 0 & 1 \end{vmatrix} = \cos\theta\,\mathbf{i} + \sin\theta\,\mathbf{j}$$

and $$|\mathbf{r}_\theta \times \mathbf{r}_z| = \sqrt{\cos^2\theta + \sin^2\theta} = 1$$

Thus the surface integral over S_1 is

$$\iint_{S_1} z\,dS = \iint_D z\,|\mathbf{r}_\theta \times \mathbf{r}_z|\,dA$$

$$= \int_0^{2\pi}\int_0^{1+\cos\theta} z\,dz\,d\theta = \int_0^{2\pi} \tfrac{1}{2}(1+\cos\theta)^2\,d\theta$$

$$= \tfrac{1}{2}\int_0^{2\pi}\left[1 + 2\cos\theta + \tfrac{1}{2}(1+\cos 2\theta)\right]d\theta$$

$$= \tfrac{1}{2}\left[\tfrac{3}{2}\theta + 2\sin\theta + \tfrac{1}{4}\sin 2\theta\right]_0^{2\pi} = \frac{3\pi}{2}$$

Since S_2 lies in the plane $z = 0$, we have

$$\iint_{S_2} z\,dS = \iint_{S_2} 0\,dS = 0$$

The top surface S_3 lies above the unit disk D and is part of the plane $z = 1 + x$. So, taking $g(x, y) = 1 + x$ in Formula 4 and converting to polar coordinates, we have

$$\iint_{S_3} z\,dS = \iint_D (1 + x)\sqrt{1 + \left(\frac{\partial z}{\partial x}\right)^2 + \left(\frac{\partial z}{\partial y}\right)^2}\,dA$$

$$= \int_0^{2\pi}\int_0^1 (1 + r\cos\theta)\sqrt{1 + 1 + 0}\;r\,dr\,d\theta$$

$$= \sqrt{2}\int_0^{2\pi}\int_0^1 (r + r^2\cos\theta)\,dr\,d\theta$$

$$= \sqrt{2}\int_0^{2\pi}\left(\tfrac{1}{2} + \tfrac{1}{3}\cos\theta\right)d\theta$$

$$= \sqrt{2}\left[\frac{\theta}{2} + \frac{\sin\theta}{3}\right]_0^{2\pi} = \sqrt{2}\,\pi$$

Therefore

$$\iint_S z\,dS = \iint_{S_1} z\,dS + \iint_{S_2} z\,dS + \iint_{S_3} z\,dS$$

$$= \frac{3\pi}{2} + 0 + \sqrt{2}\,\pi = \left(\tfrac{3}{2} + \sqrt{2}\right)\pi$$

■

Oriented Surfaces

To define surface integrals of vector fields, we need to rule out nonorientable surfaces such as the Möbius strip shown in Figure 4. [It is named after the German geometer August Möbius (1790–1868).] You can construct one for yourself by taking a long rectangular strip of paper, giving it a half-twist, and taping the short edges together as in Figure 5. If an ant were to crawl along the Möbius strip starting at a point P, it would end up on the "other side" of the strip (that is, with its upper side pointing in the opposite direction). Then, if the ant continued to crawl in the same direction, it would end up back at the same point P without ever having crossed an edge. (If you have constructed a Möbius strip, try drawing a pencil line down the middle.) Therefore a Möbius strip really has only one side. You can graph the Möbius strip using the parametric equations in Exercise 16.6.32.

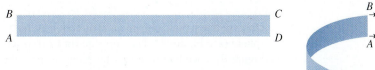

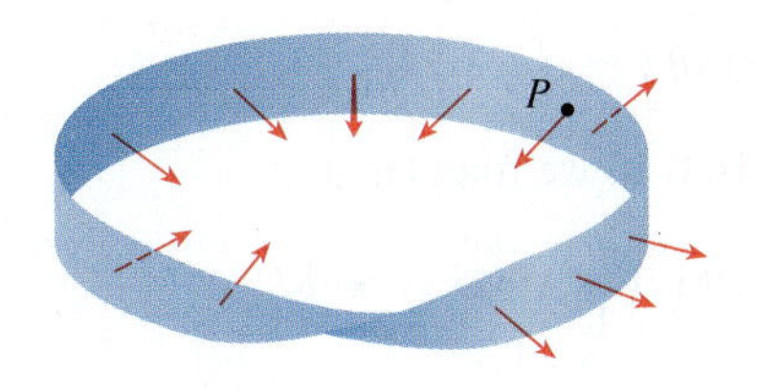

FIGURE 4
A Möbius strip

TEC Visual 16.7 shows a Möbius strip with a normal vector that can be moved along the surface.

FIGURE 5
Constructing a Möbius strip

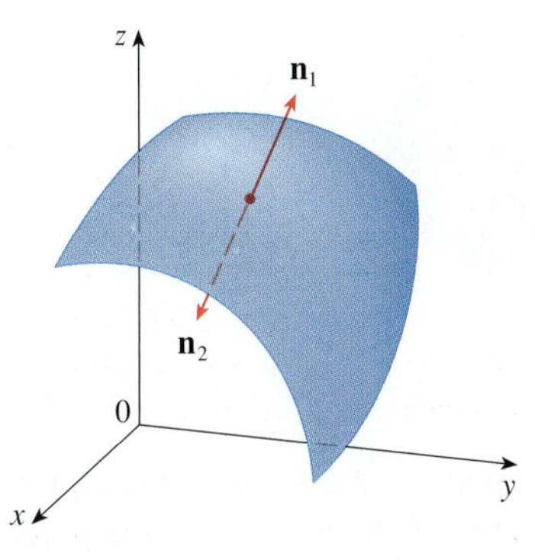

FIGURE 6

From now on we consider only orientable (two-sided) surfaces. We start with a surface S that has a tangent plane at every point (x, y, z) on S (except at any boundary point). There are two unit normal vectors $\mathbf{n}_1$ and $\mathbf{n}_2 = -\mathbf{n}_1$ at (x, y, z). (See Figure 6.)

If it is possible to choose a unit normal vector $\mathbf{n}$ at every such point (x, y, z) so that $\mathbf{n}$ varies continuously over S, then S is called an **oriented surface** and the given choice of $\mathbf{n}$ provides S with an **orientation**. There are two possible orientations for any orientable surface (see Figure 7).

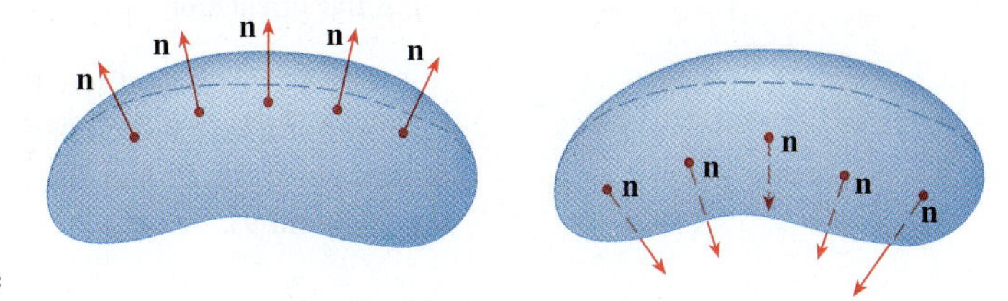

FIGURE 7
The two orientations of an orientable surface

For a surface $z = g(x, y)$ given as the graph of g, we use Equation 3 to associate with the surface a natural orientation given by the unit normal vector

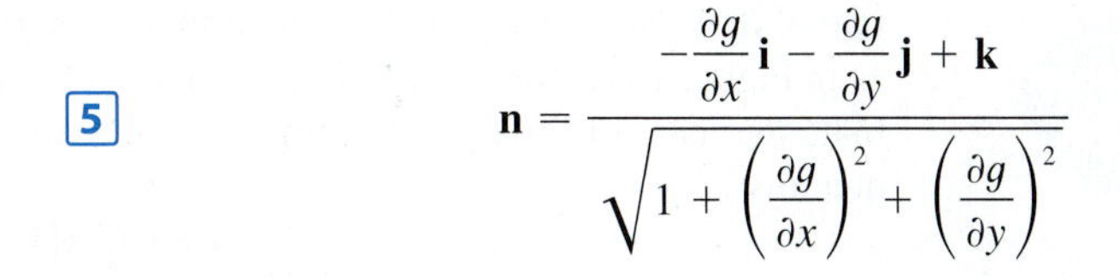

$$\boxed{5} \qquad \mathbf{n} = \frac{-\dfrac{\partial g}{\partial x}\mathbf{i} - \dfrac{\partial g}{\partial y}\mathbf{j} + \mathbf{k}}{\sqrt{1 + \left(\dfrac{\partial g}{\partial x}\right)^2 + \left(\dfrac{\partial g}{\partial y}\right)^2}}$$

Since the $\mathbf{k}$-component is positive, this gives the *upward* orientation of the surface.

If S is a smooth orientable surface given in parametric form by a vector function $\mathbf{r}(u, v)$, then it is automatically supplied with the orientation of the unit normal vector

$$\boxed{6} \qquad \mathbf{n} = \frac{\mathbf{r}_u \times \mathbf{r}_v}{|\mathbf{r}_u \times \mathbf{r}_v|}$$

and the opposite orientation is given by $-\mathbf{n}$. For instance, in Example 16.6.4 we found the parametric representation

$$\mathbf{r}(\phi, \theta) = a \sin\phi \cos\theta\,\mathbf{i} + a \sin\phi \sin\theta\,\mathbf{j} + a \cos\phi\,\mathbf{k}$$

for the sphere $x^2 + y^2 + z^2 = a^2$. Then in Example 16.6.10 we found that

$$\mathbf{r}_\phi \times \mathbf{r}_\theta = a^2 \sin^2\phi \cos\theta\,\mathbf{i} + a^2 \sin^2\phi \sin\theta\,\mathbf{j} + a^2 \sin\phi \cos\phi\,\mathbf{k}$$

and
$$|\mathbf{r}_\phi \times \mathbf{r}_\theta| = a^2 \sin\phi$$

So the orientation induced by $\mathbf{r}(\phi, \theta)$ is defined by the unit normal vector

$$\mathbf{n} = \frac{\mathbf{r}_\phi \times \mathbf{r}_\theta}{|\mathbf{r}_\phi \times \mathbf{r}_\theta|} = \sin\phi \cos\theta\,\mathbf{i} + \sin\phi \sin\theta\,\mathbf{j} + \cos\phi\,\mathbf{k} = \frac{1}{a}\mathbf{r}(\phi, \theta)$$

Observe that $\mathbf{n}$ points in the same direction as the position vector, that is, outward from the sphere (see Figure 8). The opposite (inward) orientation would have been obtained (see Figure 9) if we had reversed the order of the parameters because $\mathbf{r}_\theta \times \mathbf{r}_\phi = -\mathbf{r}_\phi \times \mathbf{r}_\theta$.

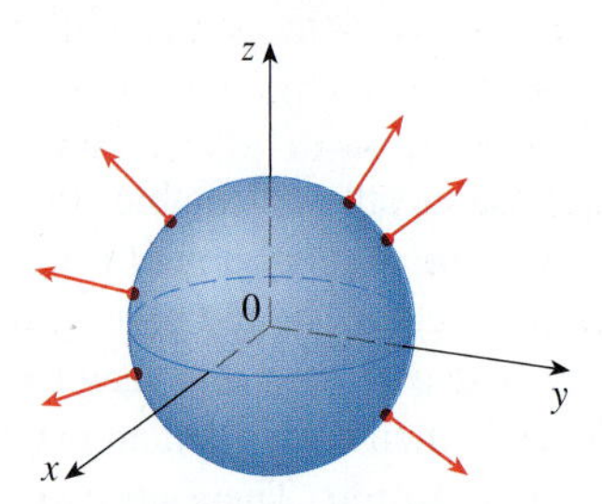

FIGURE 8
Positive orientation

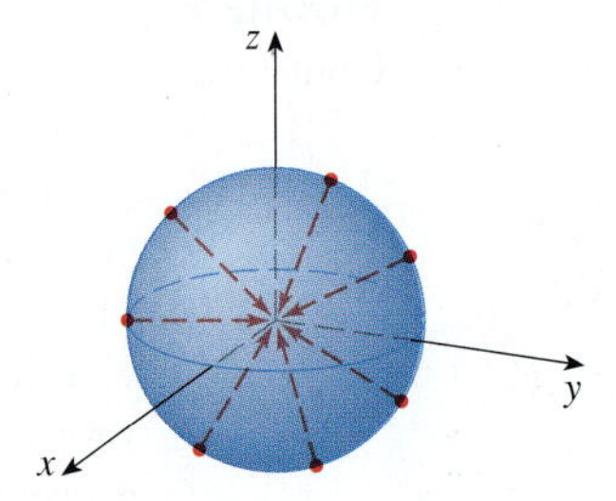

FIGURE 9
Negative orientation

For a **closed surface**, that is, a surface that is the boundary of a solid region E, the convention is that the **positive orientation** is the one for which the normal vectors point *outward* from E, and inward-pointing normals give the negative orientation (see Figures 8 and 9).

Surface Integrals of Vector Fields

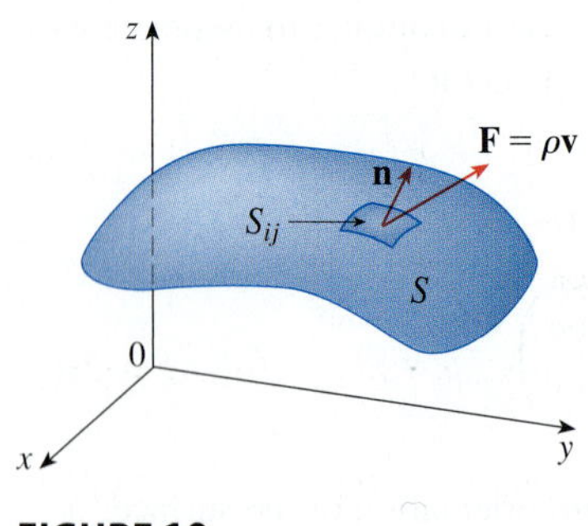

FIGURE 10

Suppose that S is an oriented surface with unit normal vector $\mathbf{n}$, and imagine a fluid with density $\rho(x, y, z)$ and velocity field $\mathbf{v}(x, y, z)$ flowing through S. (Think of S as an imaginary surface that doesn't impede the fluid flow, like a fishing net across a stream.) Then the rate of flow (mass per unit time) per unit area is $\rho\mathbf{v}$. If we divide S into small patches S_{ij}, as in Figure 10 (compare with Figure 1), then S_{ij} is nearly planar and so we can approximate the mass of fluid per unit time crossing S_{ij} in the direction of the normal $\mathbf{n}$ by the quantity

$$(\rho\mathbf{v} \cdot \mathbf{n})A(S_{ij})$$

where ρ, $\mathbf{v}$, and $\mathbf{n}$ are evaluated at some point on S_{ij}. (Recall that the component of the vector $\rho\mathbf{v}$ in the direction of the unit vector $\mathbf{n}$ is $\rho\mathbf{v} \cdot \mathbf{n}$.) By summing these quantities and taking the limit we get, according to Definition 1, the surface integral of the function $\rho\mathbf{v} \cdot \mathbf{n}$ over S:

7
$$\iint_S \rho\mathbf{v} \cdot \mathbf{n}\, dS = \iint_S \rho(x, y, z)\mathbf{v}(x, y, z) \cdot \mathbf{n}(x, y, z)\, dS$$

and this is interpreted physically as the rate of flow through S.

If we write $\mathbf{F} = \rho\mathbf{v}$, then $\mathbf{F}$ is also a vector field on $\mathbb{R}^3$ and the integral in Equation 7 becomes

$$\iint_S \mathbf{F} \cdot \mathbf{n}\, dS$$

A surface integral of this form occurs frequently in physics, even when $\mathbf{F}$ is not $\rho\mathbf{v}$, and is called the *surface integral* (or *flux integral*) of $\mathbf{F}$ over S.

8 Definition If $\mathbf{F}$ is a continuous vector field defined on an oriented surface S with unit normal vector $\mathbf{n}$, then the **surface integral of F over S** is

$$\iint_S \mathbf{F} \cdot d\mathbf{S} = \iint_S \mathbf{F} \cdot \mathbf{n}\, dS$$

This integral is also called the **flux** of $\mathbf{F}$ across S.

In words, Definition 8 says that the surface integral of a vector field over S is equal to the surface integral of its normal component over S (as previously defined).

If S is given by a vector function $\mathbf{r}(u, v)$, then $\mathbf{n}$ is given by Equation 6, and from Definition 8 and Equation 2 we have

$$\iint_S \mathbf{F} \cdot d\mathbf{S} = \iint_S \mathbf{F} \cdot \frac{\mathbf{r}_u \times \mathbf{r}_v}{|\mathbf{r}_u \times \mathbf{r}_v|}\, dS$$

$$= \iint_D \left[\mathbf{F}(\mathbf{r}(u, v)) \cdot \frac{\mathbf{r}_u \times \mathbf{r}_v}{|\mathbf{r}_u \times \mathbf{r}_v|} \right] |\mathbf{r}_u \times \mathbf{r}_v|\, dA$$

where D is the parameter domain. Thus we have

Compare Equation 9 to the similar expression for evaluating line integrals of vector fields in Definition 16.2.13:

$$\int_C \mathbf{F} \cdot d\mathbf{r} = \int_a^b \mathbf{F}(\mathbf{r}(t)) \cdot \mathbf{r}'(t)\, dt$$

9

$$\iint_S \mathbf{F} \cdot d\mathbf{S} = \iint_D \mathbf{F} \cdot (\mathbf{r}_u \times \mathbf{r}_v)\, dA$$

EXAMPLE 4 Find the flux of the vector field $\mathbf{F}(x, y, z) = z\,\mathbf{i} + y\,\mathbf{j} + x\,\mathbf{k}$ across the unit sphere $x^2 + y^2 + z^2 = 1$.

Figure 11 shows the vector field $\mathbf{F}$ in Example 4 at points on the unit sphere.

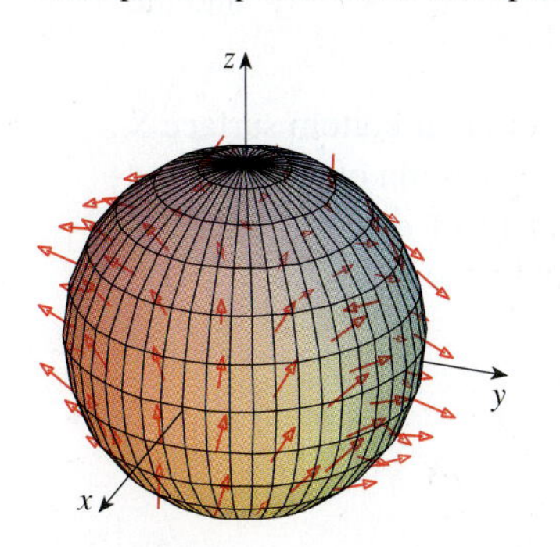

FIGURE 11

SOLUTION As in Example 1, we use the parametric representation

$$\mathbf{r}(\phi, \theta) = \sin\phi\, \cos\theta\, \mathbf{i} + \sin\phi\, \sin\theta\, \mathbf{j} + \cos\phi\, \mathbf{k} \qquad 0 \leqslant \phi \leqslant \pi \qquad 0 \leqslant \theta \leqslant 2\pi$$

Then $\qquad \mathbf{F}(\mathbf{r}(\phi, \theta)) = \cos\phi\, \mathbf{i} + \sin\phi\, \sin\theta\, \mathbf{j} + \sin\phi\, \cos\theta\, \mathbf{k}$

and, from Example 16.6.10,

$$\mathbf{r}_\phi \times \mathbf{r}_\theta = \sin^2\phi\, \cos\theta\, \mathbf{i} + \sin^2\phi\, \sin\theta\, \mathbf{j} + \sin\phi\, \cos\phi\, \mathbf{k}$$

Therefore

$$\mathbf{F}(\mathbf{r}(\phi, \theta)) \cdot (\mathbf{r}_\phi \times \mathbf{r}_\theta) = \cos\phi\, \sin^2\phi\, \cos\theta + \sin^3\phi\, \sin^2\theta + \sin^2\phi\, \cos\phi\, \cos\theta$$

and, by Formula 9, the flux is

$$\iint_S \mathbf{F} \cdot d\mathbf{S} = \iint_D \mathbf{F} \cdot (\mathbf{r}_\phi \times \mathbf{r}_\theta)\, dA$$

$$= \int_0^{2\pi} \int_0^{\pi} (2 \sin^2\phi \cos\phi \cos\theta + \sin^3\phi \sin^2\theta)\, d\phi\, d\theta$$

$$= 2 \int_0^{\pi} \sin^2\phi \cos\phi\, d\phi \int_0^{2\pi} \cos\theta\, d\theta + \int_0^{\pi} \sin^3\phi\, d\phi \int_0^{2\pi} \sin^2\theta\, d\theta$$

$$= 0 + \int_0^{\pi} \sin^3\phi\, d\phi \int_0^{2\pi} \sin^2\theta\, d\theta \qquad \left(\text{since } \int_0^{2\pi} \cos\theta\, d\theta = 0\right)$$

$$= \frac{4\pi}{3}$$

by the same calculation as in Example 1. ■

If, for instance, the vector field in Example 4 is a velocity field describing the flow of a fluid with density 1, then the answer, $4\pi/3$, represents the rate of flow through the unit sphere in units of mass per unit time.

In the case of a surface S given by a graph $z = g(x, y)$, we can think of x and y as parameters and use Equation 3 to write

$$\mathbf{F} \cdot (\mathbf{r}_x \times \mathbf{r}_y) = (P\,\mathbf{i} + Q\,\mathbf{j} + R\,\mathbf{k}) \cdot \left(-\frac{\partial g}{\partial x}\mathbf{i} - \frac{\partial g}{\partial y}\mathbf{j} + \mathbf{k}\right)$$

Thus Formula 9 becomes

10
$$\iint_S \mathbf{F} \cdot d\mathbf{S} = \iint_D \left(-P\frac{\partial g}{\partial x} - Q\frac{\partial g}{\partial y} + R\right) dA$$

This formula assumes the upward orientation of S; for a downward orientation we multiply by -1. Similar formulas can be worked out if S is given by $y = h(x, z)$ or $x = k(y, z)$. (See Exercises 37 and 38.)

EXAMPLE 5 Evaluate $\iint_S \mathbf{F} \cdot d\mathbf{S}$, where $\mathbf{F}(x, y, z) = y\,\mathbf{i} + x\,\mathbf{j} + z\,\mathbf{k}$ and S is the boundary of the solid region E enclosed by the paraboloid $z = 1 - x^2 - y^2$ and the plane $z = 0$.

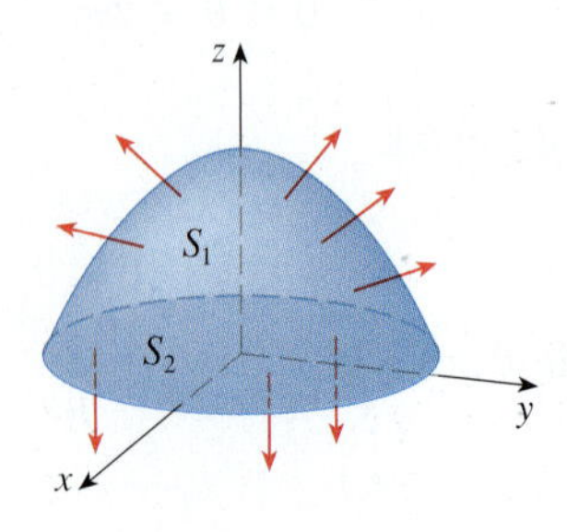

FIGURE 12

SOLUTION S consists of a parabolic top surface S_1 and a circular bottom surface S_2. (See Figure 12.) Since S is a closed surface, we use the convention of positive (outward) orientation. This means that S_1 is oriented upward and we can use Equation 10 with D being the projection of S_1 onto the xy-plane, namely, the disk $x^2 + y^2 \leqslant 1$. Since

$$P(x, y, z) = y \qquad Q(x, y, z) = x \qquad R(x, y, z) = z = 1 - x^2 - y^2$$

on S_1 and
$$\frac{\partial g}{\partial x} = -2x \qquad \frac{\partial g}{\partial y} = -2y$$

we have

$$\iint_{S_1} \mathbf{F} \cdot d\mathbf{S} = \iint_D \left(-P\frac{\partial g}{\partial x} - Q\frac{\partial g}{\partial y} + R \right) dA$$

$$= \iint_D [-y(-2x) - x(-2y) + 1 - x^2 - y^2]\, dA$$

$$= \iint_D (1 + 4xy - x^2 - y^2)\, dA$$

$$= \int_0^{2\pi} \int_0^1 (1 + 4r^2 \cos\theta \sin\theta - r^2)\, r\, dr\, d\theta$$

$$= \int_0^{2\pi} \int_0^1 (r - r^3 + 4r^3 \cos\theta \sin\theta)\, dr\, d\theta$$

$$= \int_0^{2\pi} \left(\tfrac{1}{4} + \cos\theta \sin\theta\right) d\theta = \tfrac{1}{4}(2\pi) + 0 = \frac{\pi}{2}$$

The disk S_2 is oriented downward, so its unit normal vector is $\mathbf{n} = -\mathbf{k}$ and we have

$$\iint_{S_2} \mathbf{F} \cdot d\mathbf{S} = \iint_{S_2} \mathbf{F} \cdot (-\mathbf{k})\, dS = \iint_D (-z)\, dA = \iint_D 0\, dA = 0$$

since $z = 0$ on S_2. Finally, we compute, by definition, $\iint_S \mathbf{F} \cdot d\mathbf{S}$ as the sum of the surface integrals of $\mathbf{F}$ over the pieces S_1 and S_2:

$$\iint_S \mathbf{F} \cdot d\mathbf{S} = \iint_{S_1} \mathbf{F} \cdot d\mathbf{S} + \iint_{S_2} \mathbf{F} \cdot d\mathbf{S} = \frac{\pi}{2} + 0 = \frac{\pi}{2}$$

Although we motivated the surface integral of a vector field using the example of fluid flow, this concept also arises in other physical situations. For instance, if $\mathbf{E}$ is an electric field (see Example 16.1.5), then the surface integral

$$\iint_S \mathbf{E} \cdot d\mathbf{S}$$

is called the **electric flux** of $\mathbf{E}$ through the surface S. One of the important laws of electrostatics is **Gauss's Law**, which says that the net charge enclosed by a closed surface S is

$$Q = \varepsilon_0 \iint_S \mathbf{E} \cdot d\mathbf{S} \tag{11}$$

where ε_0 is a constant (called the permittivity of free space) that depends on the units used. (In the SI system, $\varepsilon_0 \approx 8.8542 \times 10^{-12}\ \mathrm{C^2/N \cdot m^2}$.) Therefore, if the vector field $\mathbf{F}$ in Example 4 represents an electric field, we can conclude that the charge enclosed by S is $Q = \frac{4}{3}\pi\varepsilon_0$.

Another application of surface integrals occurs in the study of heat flow. Suppose the temperature at a point (x, y, z) in a body is $u(x, y, z)$. Then the **heat flow** is defined as the vector field

$$\mathbf{F} = -K\,\nabla u$$

where K is an experimentally determined constant called the **conductivity** of the substance. The rate of heat flow across the surface S in the body is then given by the surface integral

$$\iint_S \mathbf{F} \cdot d\mathbf{S} = -K \iint_S \nabla u \cdot d\mathbf{S}$$

EXAMPLE 6 The temperature u in a metal ball is proportional to the square of the distance from the center of the ball. Find the rate of heat flow across a sphere S of radius a with center at the center of the ball.

SOLUTION Taking the center of the ball to be at the origin, we have

$$u(x, y, z) = C(x^2 + y^2 + z^2)$$

where C is the proportionality constant. Then the heat flow is

$$\mathbf{F}(x, y, z) = -K\,\nabla u = -KC(2x\,\mathbf{i} + 2y\,\mathbf{j} + 2z\,\mathbf{k})$$

where K is the conductivity of the metal. Instead of using the usual parametrization of the sphere as in Example 4, we observe that the outward unit normal to the sphere $x^2 + y^2 + z^2 = a^2$ at the point (x, y, z) is

$$\mathbf{n} = \frac{1}{a}(x\,\mathbf{i} + y\,\mathbf{j} + z\,\mathbf{k})$$

and so

$$\mathbf{F} \cdot \mathbf{n} = -\frac{2KC}{a}(x^2 + y^2 + z^2)$$

But on S we have $x^2 + y^2 + z^2 = a^2$, so $\mathbf{F} \cdot \mathbf{n} = -2aKC$. Therefore the rate of heat flow across S is

$$\iint_S \mathbf{F} \cdot d\mathbf{S} = \iint_S \mathbf{F} \cdot \mathbf{n}\, dS = -2aKC \iint_S dS$$

$$= -2aKCA(S) = -2aKC(4\pi a^2) = -8KC\pi a^3$$

■

16.7 EXERCISES

1. Let S be the surface of the box enclosed by the planes $x = \pm 1$, $y = \pm 1$, $z = \pm 1$. Approximate $\iint_S \cos(x + 2y + 3z)\, dS$ by using a Riemann sum as in Definition 1, taking the patches S_{ij} to be the squares that are the faces of the box S and the points P_{ij}^* to be the centers of the squares.

2. A surface S consists of the cylinder $x^2 + y^2 = 1$, $-1 \leqslant z \leqslant 1$, together with its top and bottom disks. Suppose you know that f is a continuous function with

$$f(\pm 1, 0, 0) = 2 \qquad f(0, \pm 1, 0) = 3 \qquad f(0, 0, \pm 1) = 4$$

Estimate the value of $\iint_S f(x, y, z)\, dS$ by using a Riemann sum, taking the patches S_{ij} to be four quarter-cylinders and the top and bottom disks.

3. Let H be the hemisphere $x^2 + y^2 + z^2 = 50$, $z \geqslant 0$, and suppose f is a continuous function with $f(3, 4, 5) = 7$, $f(3, -4, 5) = 8$, $f(-3, 4, 5) = 9$, and $f(-3, -4, 5) = 12$. By dividing H into four patches, estimate the value of $\iint_H f(x, y, z)\, dS$.

4. Suppose that $f(x, y, z) = g\left(\sqrt{x^2 + y^2 + z^2}\right)$, where g is a function of one variable such that $g(2) = -5$. Evaluate $\iint_S f(x, y, z)\, dS$, where S is the sphere $x^2 + y^2 + z^2 = 4$.

5–20 Evaluate the surface integral.

5. $\iint_S (x + y + z)\, dS$,
S is the parallelogram with parametric equations $x = u + v$, $y = u - v$, $z = 1 + 2u + v$, $0 \leqslant u \leqslant 2$, $0 \leqslant v \leqslant 1$

6. $\iint_S xyz\, dS$,
S is the cone with parametric equations $x = u \cos v$, $y = u \sin v$, $z = u$, $0 \leqslant u \leqslant 1$, $0 \leqslant v \leqslant \pi/2$

7. $\iint_S y\, dS$, S is the helicoid with vector equation $\mathbf{r}(u, v) = \langle u \cos v, u \sin v, v \rangle$, $0 \leqslant u \leqslant 1$, $0 \leqslant v \leqslant \pi$

8. $\iint_S (x^2 + y^2)\, dS$,
S is the surface with vector equation $\mathbf{r}(u, v) = \langle 2uv, u^2 - v^2, u^2 + v^2 \rangle$, $u^2 + v^2 \leqslant 1$

9. $\iint_S x^2yz\, dS$,
S is the part of the plane $z = 1 + 2x + 3y$ that lies above the rectangle $[0, 3] \times [0, 2]$

10. $\iint_S xz\, dS$,
S is the part of the plane $2x + 2y + z = 4$ that lies in the first octant

11. $\iint_S x\, dS$,
S is the triangular region with vertices $(1, 0, 0)$, $(0, -2, 0)$, and $(0, 0, 4)$

12. $\iint_S y\, dS$,
S is the surface $z = \frac{2}{3}(x^{3/2} + y^{3/2})$, $0 \leqslant x \leqslant 1$, $0 \leqslant y \leqslant 1$

13. $\iint_S z^2\, dS$,
S is the part of the paraboloid $x = y^2 + z^2$ given by $0 \leqslant x \leqslant 1$

14. $\iint_S y^2z^2\, dS$,
S is the part of the cone $y = \sqrt{x^2 + z^2}$ given by $0 \leqslant y \leqslant 5$

15. $\iint_S x\, dS$,
S is the surface $y = x^2 + 4z$, $0 \leqslant x \leqslant 1$, $0 \leqslant z \leqslant 1$

16. $\iint_S y^2\, dS$,
S is the part of the sphere $x^2 + y^2 + z^2 = 1$ that lies above the cone $z = \sqrt{x^2 + y^2}$

17. $\iint_S (x^2z + y^2z)\, dS$,
S is the hemisphere $x^2 + y^2 + z^2 = 4$, $z \geqslant 0$

18. $\iint_S (x + y + z)\, dS$,
S is the part of the half-cylinder $x^2 + z^2 = 1$, $z \geqslant 0$, that lies between the planes $y = 0$ and $y = 2$

19. $\iint_S xz\, dS$,
S is the boundary of the region enclosed by the cylinder $y^2 + z^2 = 9$ and the planes $x = 0$ and $x + y = 5$

20. $\iint_S (x^2 + y^2 + z^2)\, dS$,
S is the part of the cylinder $x^2 + y^2 = 9$ between the planes $z = 0$ and $z = 2$, together with its top and bottom disks

---

21–32 Evaluate the surface integral $\iint_S \mathbf{F} \cdot d\mathbf{S}$ for the given vector field $\mathbf{F}$ and the oriented surface S. In other words, find the flux of $\mathbf{F}$ across S. For closed surfaces, use the positive (outward) orientation.

21. $\mathbf{F}(x, y, z) = ze^{xy}\, \mathbf{i} - 3ze^{xy}\, \mathbf{j} + xy\, \mathbf{k}$,
S is the parallelogram of Exercise 5 with upward orientation

22. $\mathbf{F}(x, y, z) = z\, \mathbf{i} + y\, \mathbf{j} + x\, \mathbf{k}$,
S is the helicoid of Exercise 7 with upward orientation

23. $\mathbf{F}(x, y, z) = xy\, \mathbf{i} + yz\, \mathbf{j} + zx\, \mathbf{k}$, S is the part of the paraboloid $z = 4 - x^2 - y^2$ that lies above the square $0 \leqslant x \leqslant 1$, $0 \leqslant y \leqslant 1$, and has upward orientation

24. $\mathbf{F}(x, y, z) = -x\, \mathbf{i} - y\, \mathbf{j} + z^3\, \mathbf{k}$, S is the part of the cone $z = \sqrt{x^2 + y^2}$ between the planes $z = 1$ and $z = 3$ with downward orientation

25. $\mathbf{F}(x, y, z) = x\, \mathbf{i} + y\, \mathbf{j} + z^2\, \mathbf{k}$, S is the sphere with radius 1 and center the origin

26. $\mathbf{F}(x, y, z) = y\, \mathbf{i} - x\, \mathbf{j} + 2z\, \mathbf{k}$, S is the hemisphere $x^2 + y^2 + z^2 = 4$, $z \geqslant 0$, oriented downward

27. $\mathbf{F}(x, y, z) = y\, \mathbf{j} - z\, \mathbf{k}$,
S consists of the paraboloid $y = x^2 + z^2$, $0 \leqslant y \leqslant 1$, and the disk $x^2 + z^2 \leqslant 1$, $y = 1$

28. $\mathbf{F}(x, y, z) = yz\, \mathbf{i} + zx\, \mathbf{j} + xy\, \mathbf{k}$,
S is the surface $z = x \sin y$, $0 \leqslant x \leqslant 2$, $0 \leqslant y \leqslant \pi$, with upward orientation

29. $\mathbf{F}(x, y, z) = x\, \mathbf{i} + 2y\, \mathbf{j} + 3z\, \mathbf{k}$,
S is the cube with vertices $(\pm 1, \pm 1, \pm 1)$

30. $\mathbf{F}(x, y, z) = x\, \mathbf{i} + y\, \mathbf{j} + 5\, \mathbf{k}$, S is the boundary of the region enclosed by the cylinder $x^2 + z^2 = 1$ and the planes $y = 0$ and $x + y = 2$

31. $\mathbf{F}(x, y, z) = x^2\, \mathbf{i} + y^2\, \mathbf{j} + z^2\, \mathbf{k}$, S is the boundary of the solid half-cylinder $0 \leqslant z \leqslant \sqrt{1 - y^2}$, $0 \leqslant x \leqslant 2$

32. $\mathbf{F}(x, y, z) = y\, \mathbf{i} + (z - y)\, \mathbf{j} + x\, \mathbf{k}$,
S is the surface of the tetrahedron with vertices $(0, 0, 0)$, $(1, 0, 0)$, $(0, 1, 0)$, and $(0, 0, 1)$

CAS **33.** Evaluate $\iint_S (x^2 + y^2 + z^2)\, dS$ correct to four decimal places, where S is the surface $z = xe^y$, $0 \leqslant x \leqslant 1$, $0 \leqslant y \leqslant 1$.

CAS **34.** Find the exact value of $\iint_S xyz\, dS$, where S is the surface $z = x^2y^2$, $0 \leqslant x \leqslant 1$, $0 \leqslant y \leqslant 2$.

CAS **35.** Find the value of $\iint_S x^2y^2z^2\, dS$ correct to four decimal places, where S is the part of the paraboloid $z = 3 - 2x^2 - y^2$ that lies above the xy-plane.

CAS **36.** Find the flux of

$$\mathbf{F}(x, y, z) = \sin(xyz)\, \mathbf{i} + x^2y\, \mathbf{j} + z^2e^{x/5}\, \mathbf{k}$$

across the part of the cylinder $4y^2 + z^2 = 4$ that lies above the xy-plane and between the planes $x = -2$ and $x = 2$ with upward orientation. Illustrate by using a computer algebra system to draw the cylinder and the vector field on the same screen.

37. Find a formula for $\iint_S \mathbf{F} \cdot d\mathbf{S}$ similar to Formula 10 for the case where S is given by $y = h(x, z)$ and $\mathbf{n}$ is the unit normal that points toward the left.

38. Find a formula for $\iint_S \mathbf{F} \cdot d\mathbf{S}$ similar to Formula 10 for the case where S is given by $x = k(y, z)$ and $\mathbf{n}$ is the unit normal that points forward (that is, toward the viewer when the axes are drawn in the usual way).

39. Find the center of mass of the hemisphere $x^2 + y^2 + z^2 = a^2$, $z \geqslant 0$, if it has constant density.

40. Find the mass of a thin funnel in the shape of a cone $z = \sqrt{x^2 + y^2}$, $1 \leqslant z \leqslant 4$, if its density function is $\rho(x, y, z) = 10 - z$.

41. (a) Give an integral expression for the moment of inertia I_z about the z-axis of a thin sheet in the shape of a surface S if the density function is ρ.
(b) Find the moment of inertia about the z-axis of the funnel in Exercise 40.

42. Let S be the part of the sphere $x^2 + y^2 + z^2 = 25$ that lies above the plane $z = 4$. If S has constant density k, find (a) the center of mass and (b) the moment of inertia about the z-axis.

43. A fluid has density 870 kg/m^3 and flows with velocity $\mathbf{v} = z\,\mathbf{i} + y^2\,\mathbf{j} + x^2\,\mathbf{k}$, where x, y, and z are measured in meters and the components of $\mathbf{v}$ in meters per second. Find the rate of flow outward through the cylinder $x^2 + y^2 = 4$, $0 \leqslant z \leqslant 1$.

44. Seawater has density 1025 kg/m^3 and flows in a velocity field $\mathbf{v} = y\,\mathbf{i} + x\,\mathbf{j}$, where x, y, and z are measured in meters and the components of $\mathbf{v}$ in meters per second. Find the rate of flow outward through the hemisphere $x^2 + y^2 + z^2 = 9$, $z \geqslant 0$.

45. Use Gauss's Law to find the charge contained in the solid hemisphere $x^2 + y^2 + z^2 \leqslant a^2$, $z \geqslant 0$, if the electric field is

$$\mathbf{E}(x, y, z) = x\,\mathbf{i} + y\,\mathbf{j} + 2z\,\mathbf{k}$$

46. Use Gauss's Law to find the charge enclosed by the cube with vertices $(\pm 1, \pm 1, \pm 1)$ if the electric field is

$$\mathbf{E}(x, y, z) = x\,\mathbf{i} + y\,\mathbf{j} + z\,\mathbf{k}$$

47. The temperature at the point (x, y, z) in a substance with conductivity $K = 6.5$ is $u(x, y, z) = 2y^2 + 2z^2$. Find the rate of heat flow inward across the cylindrical surface $y^2 + z^2 = 6$, $0 \leqslant x \leqslant 4$.

48. The temperature at a point in a ball with conductivity K is inversely proportional to the distance from the center of the ball. Find the rate of heat flow across a sphere S of radius a with center at the center of the ball.

49. Let $\mathbf{F}$ be an inverse square field, that is, $\mathbf{F}(\mathbf{r}) = c\mathbf{r}/|\,\mathbf{r}\,|^3$ for some constant c, where $\mathbf{r} = x\,\mathbf{i} + y\,\mathbf{j} + z\,\mathbf{k}$. Show that the flux of $\mathbf{F}$ across a sphere S with center the origin is independent of the radius of S.

16.8 Stokes' Theorem

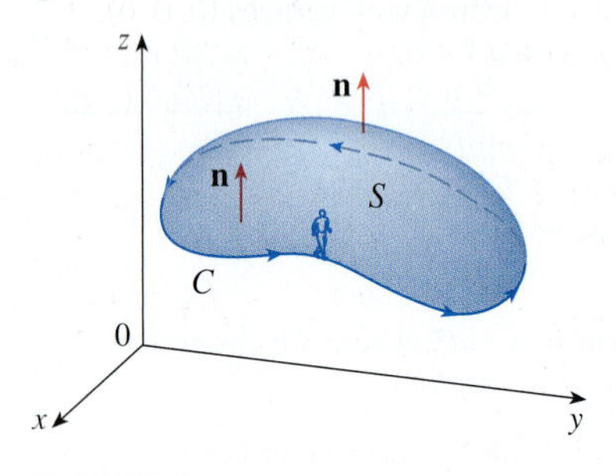

FIGURE 1

Stokes' Theorem can be regarded as a higher-dimensional version of Green's Theorem. Whereas Green's Theorem relates a double integral over a plane region D to a line integral around its plane boundary curve, Stokes' Theorem relates a surface integral over a surface S to a line integral around the boundary curve of S (which is a space curve). Figure 1 shows an oriented surface with unit normal vector $\mathbf{n}$. The orientation of S induces the **positive orientation of the boundary curve *C*** shown in the figure. This means that if you walk in the positive direction around C with your head pointing in the direction of $\mathbf{n}$, then the surface will always be on your left.

Stokes' Theorem Let S be an oriented piecewise-smooth surface that is bounded by a simple, closed, piecewise-smooth boundary curve C with positive orientation. Let $\mathbf{F}$ be a vector field whose components have continuous partial derivatives on an open region in $\mathbb{R}^3$ that contains S. Then

$$\int_C \mathbf{F} \cdot d\mathbf{r} = \iint_S \text{curl}\,\mathbf{F} \cdot d\mathbf{S}$$

Since

$$\int_C \mathbf{F} \cdot d\mathbf{r} = \int_C \mathbf{F} \cdot \mathbf{T}\,ds \qquad \text{and} \qquad \iint_S \text{curl}\,\mathbf{F} \cdot d\mathbf{S} = \iint_S \text{curl}\,\mathbf{F} \cdot \mathbf{n}\,dS$$

Stokes' Theorem says that the line integral around the boundary curve of S of the tangential component of $\mathbf{F}$ is equal to the surface integral over S of the normal component of the curl of $\mathbf{F}$.

The positively oriented boundary curve of the oriented surface S is often written as ∂S, so Stokes' Theorem can be expressed as

$$\boxed{1} \qquad \iint_S \operatorname{curl} \mathbf{F} \cdot d\mathbf{S} = \int_{\partial S} \mathbf{F} \cdot d\mathbf{r}$$

George Stokes

Stokes' Theorem is named after the Irish mathematical physicist Sir George Stokes (1819–1903). Stokes was a professor at Cambridge University (in fact he held the same position as Newton, Lucasian Professor of Mathematics) and was especially noted for his studies of fluid flow and light. What we call Stokes' Theorem was actually discovered by the Scottish physicist Sir William Thomson (1824–1907, known as Lord Kelvin). Stokes learned of this theorem in a letter from Thomson in 1850 and asked students to prove it on an examination at Cambridge University in 1854. We don't know if any of those students was able to do so.

There is an analogy among Stokes' Theorem, Green's Theorem, and the Fundamental Theorem of Calculus. As before, there is an integral involving derivatives on the left side of Equation 1 (recall that curl $\mathbf{F}$ is a sort of derivative of $\mathbf{F}$) and the right side involves the values of $\mathbf{F}$ only on the *boundary* of S.

In fact, in the special case where the surface S is flat and lies in the xy-plane with upward orientation, the unit normal is $\mathbf{k}$, the surface integral becomes a double integral, and Stokes' Theorem becomes

$$\int_C \mathbf{F} \cdot d\mathbf{r} = \iint_S \operatorname{curl} \mathbf{F} \cdot d\mathbf{S} = \iint_S (\operatorname{curl} \mathbf{F}) \cdot \mathbf{k}\, dA$$

This is precisely the vector form of Green's Theorem given in Equation 16.5.12. Thus we see that Green's Theorem is really a special case of Stokes' Theorem.

Although Stokes' Theorem is too difficult for us to prove in its full generality, we can give a proof when S is a graph and $\mathbf{F}$, S, and C are well behaved.

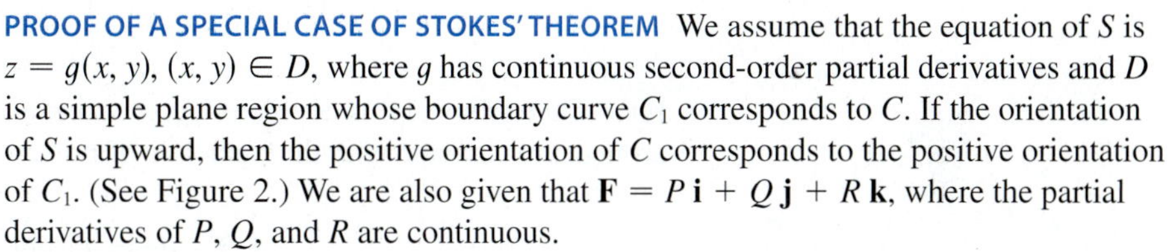

PROOF OF A SPECIAL CASE OF STOKES' THEOREM We assume that the equation of S is $z = g(x, y)$, $(x, y) \in D$, where g has continuous second-order partial derivatives and D is a simple plane region whose boundary curve C_1 corresponds to C. If the orientation of S is upward, then the positive orientation of C corresponds to the positive orientation of C_1. (See Figure 2.) We are also given that $\mathbf{F} = P\,\mathbf{i} + Q\,\mathbf{j} + R\,\mathbf{k}$, where the partial derivatives of P, Q, and R are continuous.

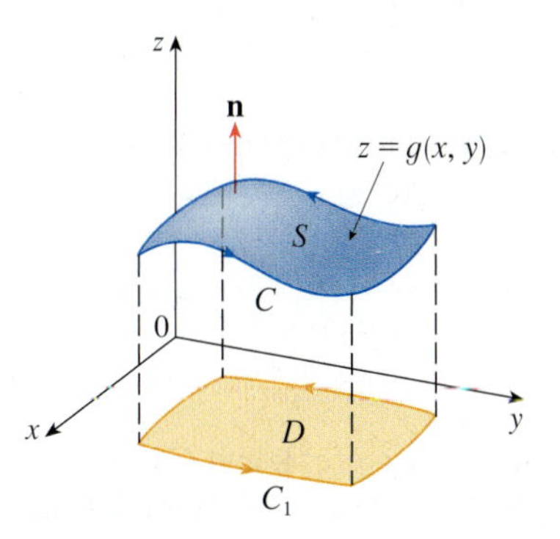

FIGURE 2

Since S is a graph of a function, we can apply Formula 16.7.10 with $\mathbf{F}$ replaced by curl $\mathbf{F}$. The result is

$$\boxed{2} \qquad \iint_S \operatorname{curl} \mathbf{F} \cdot d\mathbf{S}$$

$$= \iint_D \left[-\left(\frac{\partial R}{\partial y} - \frac{\partial Q}{\partial z} \right) \frac{\partial z}{\partial x} - \left(\frac{\partial P}{\partial z} - \frac{\partial R}{\partial x} \right) \frac{\partial z}{\partial y} + \left(\frac{\partial Q}{\partial x} - \frac{\partial P}{\partial y} \right) \right] dA$$

where the partial derivatives of P, Q, and R are evaluated at $(x, y, g(x, y))$. If

$$x = x(t) \qquad y = y(t) \qquad a \leqslant t \leqslant b$$

is a parametric representation of C_1, then a parametric representation of C is

$$x = x(t) \qquad y = y(t) \qquad z = g(x(t), y(t)) \qquad a \leqslant t \leqslant b$$

This allows us, with the aid of the Chain Rule, to evaluate the line integral as follows:

$$\begin{aligned}
\int_C \mathbf{F} \cdot d\mathbf{r} &= \int_a^b \left(P \frac{dx}{dt} + Q \frac{dy}{dt} + R \frac{dz}{dt} \right) dt \\
&= \int_a^b \left[P \frac{dx}{dt} + Q \frac{dy}{dt} + R \left(\frac{\partial z}{\partial x} \frac{dx}{dt} + \frac{\partial z}{\partial y} \frac{dy}{dt} \right) \right] dt \\
&= \int_a^b \left[\left(P + R \frac{\partial z}{\partial x} \right) \frac{dx}{dt} + \left(Q + R \frac{\partial z}{\partial y} \right) \frac{dy}{dt} \right] dt \\
&= \int_{C_1} \left(P + R \frac{\partial z}{\partial x} \right) dx + \left(Q + R \frac{\partial z}{\partial y} \right) dy \\
&= \iint_D \left[\frac{\partial}{\partial x} \left(Q + R \frac{\partial z}{\partial y} \right) - \frac{\partial}{\partial y} \left(P + R \frac{\partial z}{\partial x} \right) \right] dA
\end{aligned}$$

where we have used Green's Theorem in the last step. Then, using the Chain Rule again and remembering that P, Q, and R are functions of x, y, and z and that z is itself a function of x and y, we get

$$\begin{aligned}
\int_C \mathbf{F} \cdot d\mathbf{r} = \iint_D \Bigg[&\left(\frac{\partial Q}{\partial x} + \frac{\partial Q}{\partial z} \frac{\partial z}{\partial x} + \frac{\partial R}{\partial x} \frac{\partial z}{\partial y} + \frac{\partial R}{\partial z} \frac{\partial z}{\partial x} \frac{\partial z}{\partial y} + R \frac{\partial^2 z}{\partial x \, \partial y} \right) \\
&- \left(\frac{\partial P}{\partial y} + \frac{\partial P}{\partial z} \frac{\partial z}{\partial y} + \frac{\partial R}{\partial y} \frac{\partial z}{\partial x} + \frac{\partial R}{\partial z} \frac{\partial z}{\partial y} \frac{\partial z}{\partial x} + R \frac{\partial^2 z}{\partial y \, \partial x} \right) \Bigg] dA
\end{aligned}$$

Four of the terms in this double integral cancel and the remaining six terms can be arranged to coincide with the right side of Equation 2. Therefore

$$\int_C \mathbf{F} \cdot d\mathbf{r} = \iint_S \text{curl } \mathbf{F} \cdot d\mathbf{S}$$

■

EXAMPLE 1 Evaluate $\int_C \mathbf{F} \cdot d\mathbf{r}$, where $\mathbf{F}(x, y, z) = -y^2 \, \mathbf{i} + x \, \mathbf{j} + z^2 \, \mathbf{k}$ and C is the curve of intersection of the plane $y + z = 2$ and the cylinder $x^2 + y^2 = 1$. (Orient C to be counterclockwise when viewed from above.)

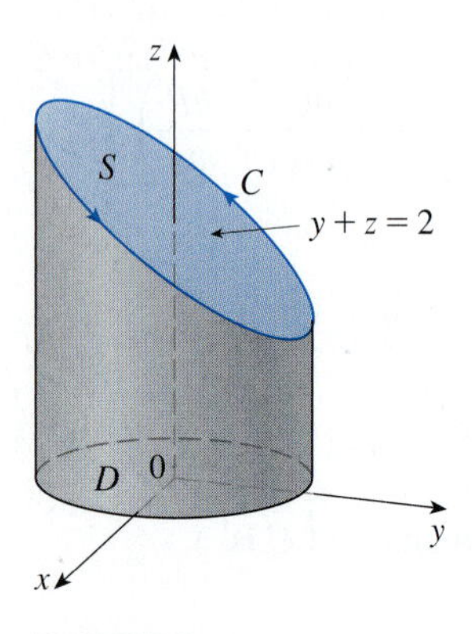

FIGURE 3

SOLUTION The curve C (an ellipse) is shown in Figure 3. Although $\int_C \mathbf{F} \cdot d\mathbf{r}$ could be evaluated directly, it's easier to use Stokes' Theorem. We first compute

$$\text{curl } \mathbf{F} = \begin{vmatrix} \mathbf{i} & \mathbf{j} & \mathbf{k} \\ \dfrac{\partial}{\partial x} & \dfrac{\partial}{\partial y} & \dfrac{\partial}{\partial z} \\ -y^2 & x & z^2 \end{vmatrix} = (1 + 2y) \, \mathbf{k}$$

Although there are many surfaces with boundary C, the most convenient choice is the elliptical region S in the plane $y + z = 2$ that is bounded by C. If we orient S upward, then C has the induced positive orientation. The projection D of S onto the xy-plane is the disk $x^2 + y^2 \leqslant 1$ and so using Equation 16.7.10 with $z = g(x, y) = 2 - y$, we have

$$\int_C \mathbf{F} \cdot d\mathbf{r} = \iint_S \text{curl } \mathbf{F} \cdot d\mathbf{S} = \iint_D (1 + 2y)\, dA$$

$$= \int_0^{2\pi} \int_0^1 (1 + 2r \sin\theta)\, r\, dr\, d\theta$$

$$= \int_0^{2\pi} \left[\frac{r^2}{2} + 2\frac{r^3}{3} \sin\theta \right]_0^1 d\theta = \int_0^{2\pi} \left(\tfrac{1}{2} + \tfrac{2}{3} \sin\theta \right) d\theta$$

$$= \tfrac{1}{2}(2\pi) + 0 = \pi$$

■

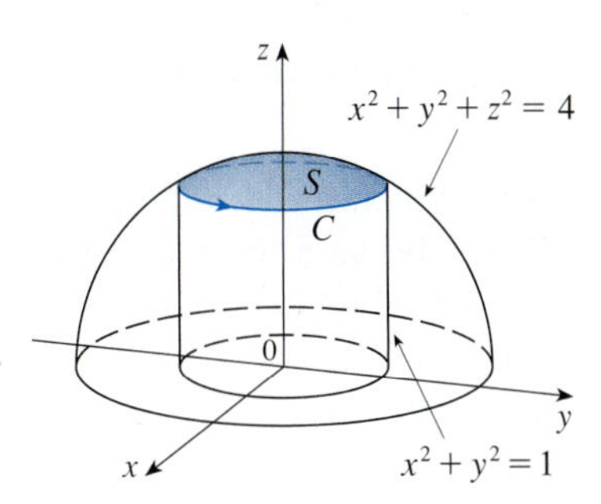

FIGURE 4

EXAMPLE 2 Use Stokes' Theorem to compute the integral $\iint_S \text{curl } \mathbf{F} \cdot d\mathbf{S}$, where $\mathbf{F}(x, y, z) = xz\,\mathbf{i} + yz\,\mathbf{j} + xy\,\mathbf{k}$ and S is the part of the sphere $x^2 + y^2 + z^2 = 4$ that lies inside the cylinder $x^2 + y^2 = 1$ and above the xy-plane. (See Figure 4.)

SOLUTION To find the boundary curve C we solve the equations $x^2 + y^2 + z^2 = 4$ and $x^2 + y^2 = 1$. Subtracting, we get $z^2 = 3$ and so $z = \sqrt{3}$ (since $z > 0$). Thus C is the circle given by the equations $x^2 + y^2 = 1$, $z = \sqrt{3}$. A vector equation of C is

$$\mathbf{r}(t) = \cos t\,\mathbf{i} + \sin t\,\mathbf{j} + \sqrt{3}\,\mathbf{k} \qquad 0 \leqslant t \leqslant 2\pi$$

so

$$\mathbf{r}'(t) = -\sin t\,\mathbf{i} + \cos t\,\mathbf{j}$$

Also, we have

$$\mathbf{F}(\mathbf{r}(t)) = \sqrt{3} \cos t\,\mathbf{i} + \sqrt{3} \sin t\,\mathbf{j} + \cos t \sin t\,\mathbf{k}$$

Therefore, by Stokes' Theorem,

$$\iint_S \text{curl } \mathbf{F} \cdot d\mathbf{S} = \int_C \mathbf{F} \cdot d\mathbf{r} = \int_0^{2\pi} \mathbf{F}(\mathbf{r}(t)) \cdot \mathbf{r}'(t)\, dt$$

$$= \int_0^{2\pi} \left(-\sqrt{3} \cos t \sin t + \sqrt{3} \sin t \cos t \right) dt$$

$$= \sqrt{3} \int_0^{2\pi} 0\, dt = 0$$

■

Note that in Example 2 we computed a surface integral simply by knowing the values of $\mathbf{F}$ on the boundary curve C. This means that if we have another oriented surface with the same boundary curve C, then we get exactly the same value for the surface integral!

In general, if S_1 and S_2 are oriented surfaces with the same oriented boundary curve C and both satisfy the hypotheses of Stokes' Theorem, then

3 $$\iint_{S_1} \text{curl } \mathbf{F} \cdot d\mathbf{S} = \int_C \mathbf{F} \cdot d\mathbf{r} = \iint_{S_2} \text{curl } \mathbf{F} \cdot d\mathbf{S}$$

This fact is useful when it is difficult to integrate over one surface but easy to integrate over the other.

We now use Stokes' Theorem to throw some light on the meaning of the curl vector. Suppose that C is an oriented closed curve and $\mathbf{v}$ represents the velocity field in fluid flow. Consider the line integral

$$\int_C \mathbf{v} \cdot d\mathbf{r} = \int_C \mathbf{v} \cdot \mathbf{T}\, ds$$

and recall that $\mathbf{v} \cdot \mathbf{T}$ is the component of $\mathbf{v}$ in the direction of the unit tangent vector $\mathbf{T}$. This means that the closer the direction of $\mathbf{v}$ is to the direction of $\mathbf{T}$, the larger the value of $\mathbf{v} \cdot \mathbf{T}$. Thus $\int_C \mathbf{v} \cdot d\mathbf{r}$ is a measure of the tendency of the fluid to move around C and is called the **circulation** of $\mathbf{v}$ around C. (See Figure 5.)

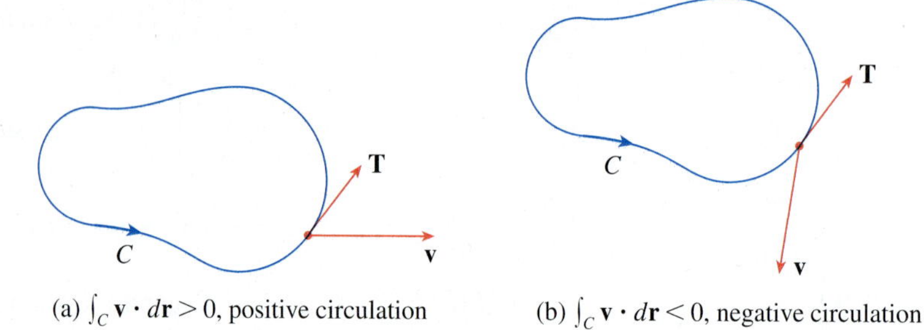

FIGURE 5 (a) $\int_C \mathbf{v} \cdot d\mathbf{r} > 0$, positive circulation (b) $\int_C \mathbf{v} \cdot d\mathbf{r} < 0$, negative circulation

Now let $P_0(x_0, y_0, z_0)$ be a point in the fluid and let S_a be a small disk with radius a and center P_0. Then $(\text{curl } \mathbf{F})(P) \approx (\text{curl } \mathbf{F})(P_0)$ for all points P on S_a because curl $\mathbf{F}$ is continuous. Thus, by Stokes' Theorem, we get the following approximation to the circulation around the boundary circle C_a:

$$\int_{C_a} \mathbf{v} \cdot d\mathbf{r} = \iint_{S_a} \text{curl } \mathbf{v} \cdot d\mathbf{S} = \iint_{S_a} \text{curl } \mathbf{v} \cdot \mathbf{n}\, dS$$

$$\approx \iint_{S_a} \text{curl } \mathbf{v}(P_0) \cdot \mathbf{n}(P_0)\, dS = \text{curl } \mathbf{v}(P_0) \cdot \mathbf{n}(P_0)\pi a^2$$

Imagine a tiny paddle wheel placed in the fluid at a point P, as in Figure 6; the paddle wheel rotates fastest when its axis is parallel to curl $\mathbf{v}$.

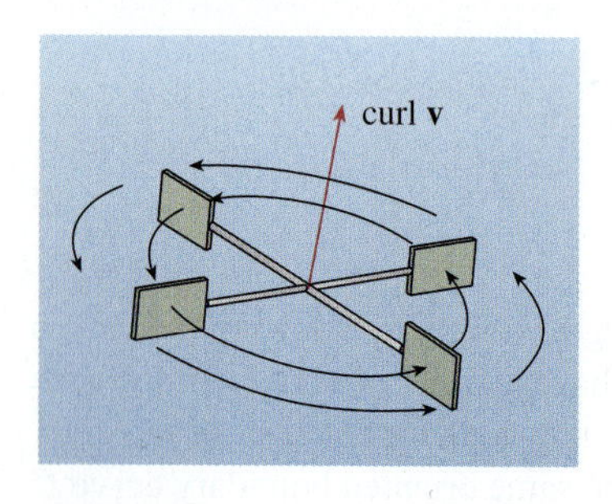

FIGURE 6

This approximation becomes better as $a \to 0$ and we have

4 $$\text{curl } \mathbf{v}(P_0) \cdot \mathbf{n}(P_0) = \lim_{a \to 0} \frac{1}{\pi a^2} \int_{C_a} \mathbf{v} \cdot d\mathbf{r}$$

Equation 4 gives the relationship between the curl and the circulation. It shows that curl $\mathbf{v} \cdot \mathbf{n}$ is a measure of the rotating effect of the fluid about the axis $\mathbf{n}$. The curling effect is greatest about the axis parallel to curl $\mathbf{v}$.

Finally, we mention that Stokes' Theorem can be used to prove Theorem 16.5.4 (which states that if curl $\mathbf{F} = \mathbf{0}$ on all of $\mathbb{R}^3$, then $\mathbf{F}$ is conservative). From our previous work (Theorems 16.3.3 and 16.3.4), we know that $\mathbf{F}$ is conservative if $\int_C \mathbf{F} \cdot d\mathbf{r} = 0$ for every closed path C. Given C, suppose we can find an orientable surface S whose boundary is C. (This can be done, but the proof requires advanced techniques.) Then Stokes' Theorem gives

$$\int_C \mathbf{F} \cdot d\mathbf{r} = \iint_S \text{curl } \mathbf{F} \cdot d\mathbf{S} = \iint_S \mathbf{0} \cdot d\mathbf{S} = 0$$

A curve that is not simple can be broken into a number of simple curves, and the integrals around these simple curves are all 0. Adding these integrals, we obtain $\int_C \mathbf{F} \cdot d\mathbf{r} = 0$ for any closed curve C.

16.8 EXERCISES

1. A hemisphere H and a portion P of a paraboloid are shown. Suppose $\mathbf{F}$ is a vector field on $\mathbb{R}^3$ whose components have continuous partial derivatives. Explain why

$$\iint_H \text{curl } \mathbf{F} \cdot d\mathbf{S} = \iint_P \text{curl } \mathbf{F} \cdot d\mathbf{S}$$

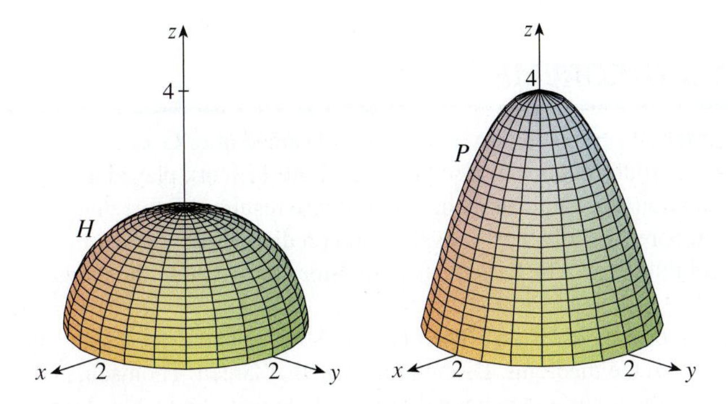

2–6 Use Stokes' Theorem to evaluate $\iint_S \text{curl } \mathbf{F} \cdot d\mathbf{S}$.

2. $\mathbf{F}(x, y, z) = x^2 \sin z \, \mathbf{i} + y^2 \, \mathbf{j} + xy \, \mathbf{k}$,
S is the part of the paraboloid $z = 1 - x^2 - y^2$ that lies above the xy-plane, oriented upward

3. $\mathbf{F}(x, y, z) = ze^y \, \mathbf{i} + x \cos y \, \mathbf{j} + xz \sin y \, \mathbf{k}$,
S is the hemisphere $x^2 + y^2 + z^2 = 16$, $y \geqslant 0$, oriented in the direction of the positive y-axis

4. $\mathbf{F}(x, y, z) = \tan^{-1}(x^2yz^2) \, \mathbf{i} + x^2y \, \mathbf{j} + x^2z^2 \, \mathbf{k}$,
S is the cone $x = \sqrt{y^2 + z^2}$, $0 \leqslant x \leqslant 2$, oriented in the direction of the positive x-axis

5. $\mathbf{F}(x, y, z) = xyz \, \mathbf{i} + xy \, \mathbf{j} + x^2yz \, \mathbf{k}$,
S consists of the top and the four sides (but not the bottom) of the cube with vertices $(\pm 1, \pm 1, \pm 1)$, oriented outward

6. $\mathbf{F}(x, y, z) = e^{xy} \, \mathbf{i} + e^{xz} \, \mathbf{j} + x^2z \, \mathbf{k}$,
S is the half of the ellipsoid $4x^2 + y^2 + 4z^2 = 4$ that lies to the right of the xz-plane, oriented in the direction of the positive y-axis

7–10 Use Stokes' Theorem to evaluate $\int_C \mathbf{F} \cdot d\mathbf{r}$. In each case C is oriented counterclockwise as viewed from above.

7. $\mathbf{F}(x, y, z) = (x + y^2) \, \mathbf{i} + (y + z^2) \, \mathbf{j} + (z + x^2) \, \mathbf{k}$,
C is the triangle with vertices $(1, 0, 0)$, $(0, 1, 0)$, and $(0, 0, 1)$

8. $\mathbf{F}(x, y, z) = \mathbf{i} + (x + yz) \, \mathbf{j} + (xy - \sqrt{z}) \, \mathbf{k}$,
C is the boundary of the part of the plane $3x + 2y + z = 1$ in the first octant

9. $\mathbf{F}(x, y, z) = xy \, \mathbf{i} + yz \, \mathbf{j} + zx \, \mathbf{k}$, C is the boundary of the part of the paraboloid $z = 1 - x^2 - y^2$ in the first octant

10. $\mathbf{F}(x, y, z) = 2y \, \mathbf{i} + xz \, \mathbf{j} + (x + y) \, \mathbf{k}$, C is the curve of intersection of the plane $z = y + 2$ and the cylinder $x^2 + y^2 = 1$

11. (a) Use Stokes' Theorem to evaluate $\int_C \mathbf{F} \cdot d\mathbf{r}$, where

$$\mathbf{F}(x, y, z) = x^2z \, \mathbf{i} + xy^2 \, \mathbf{j} + z^2 \, \mathbf{k}$$

and C is the curve of intersection of the plane $x + y + z = 1$ and the cylinder $x^2 + y^2 = 9$, oriented counterclockwise as viewed from above.

(b) Graph both the plane and the cylinder with domains chosen so that you can see the curve C and the surface that you used in part (a).

(c) Find parametric equations for C and use them to graph C.

12. (a) Use Stokes' Theorem to evaluate $\int_C \mathbf{F} \cdot d\mathbf{r}$, where $\mathbf{F}(x, y, z) = x^2y \, \mathbf{i} + \frac{1}{3}x^3 \, \mathbf{j} + xy \, \mathbf{k}$ and C is the curve of intersection of the hyperbolic paraboloid $z = y^2 - x^2$ and the cylinder $x^2 + y^2 = 1$, oriented counterclockwise as viewed from above.

(b) Graph both the hyperbolic paraboloid and the cylinder with domains chosen so that you can see the curve C and the surface that you used in part (a).

(c) Find parametric equations for C and use them to graph C.

13–15 Verify that Stokes' Theorem is true for the given vector field $\mathbf{F}$ and surface S.

13. $\mathbf{F}(x, y, z) = -y \, \mathbf{i} + x \, \mathbf{j} - 2 \, \mathbf{k}$,
S is the cone $z^2 = x^2 + y^2$, $0 \leqslant z \leqslant 4$, oriented downward

14. $\mathbf{F}(x, y, z) = -2yz \, \mathbf{i} + y \, \mathbf{j} + 3x \, \mathbf{k}$,
S is the part of the paraboloid $z = 5 - x^2 - y^2$ that lies above the plane $z = 1$, oriented upward

15. $\mathbf{F}(x, y, z) = y \, \mathbf{i} + z \, \mathbf{j} + x \, \mathbf{k}$,
S is the hemisphere $x^2 + y^2 + z^2 = 1$, $y \geqslant 0$, oriented in the direction of the positive y-axis

16. Let C be a simple closed smooth curve that lies in the plane $x + y + z = 1$. Show that the line integral

$$\int_C z \, dx - 2x \, dy + 3y \, dz$$

depends only on the area of the region enclosed by C and not on the shape of C or its location in the plane.

17. A particle moves along line segments from the origin to the points $(1, 0, 0)$, $(1, 2, 1)$, $(0, 2, 1)$, and back to the origin under the influence of the force field

$$\mathbf{F}(x, y, z) = z^2 \, \mathbf{i} + 2xy \, \mathbf{j} + 4y^2 \, \mathbf{k}$$

Find the work done.

18. Evaluate

$$\int_C (y + \sin x)\,dx + (z^2 + \cos y)\,dy + x^3\,dz$$

where C is the curve $\mathbf{r}(t) = \langle \sin t, \cos t, \sin 2t \rangle$, $0 \leqslant t \leqslant 2\pi$. [*Hint:* Observe that C lies on the surface $z = 2xy$.]

19. If S is a sphere and $\mathbf{F}$ satisfies the hypotheses of Stokes' Theorem, show that $\iint_S \operatorname{curl} \mathbf{F} \cdot d\mathbf{S} = 0$.

20. Suppose S and C satisfy the hypotheses of Stokes' Theorem and f, g have continuous second-order partial derivatives. Use Exercises 24 and 26 in Section 16.5 to show the following.

(a) $\int_C (f\,\nabla g) \cdot d\mathbf{r} = \iint_S (\nabla f \times \nabla g) \cdot d\mathbf{S}$

(b) $\int_C (f\,\nabla f) \cdot d\mathbf{r} = 0$

(c) $\int_C (f\,\nabla g + g\,\nabla f) \cdot d\mathbf{r} = 0$

WRITING PROJECT — THREE MEN AND TWO THEOREMS

The photograph shows a stained-glass window at Cambridge University in honor of George Green.

Courtesy of the Masters and Fellows of Gonville and Caius College, Cambridge University, England

Although two of the most important theorems in vector calculus are named after George Green and George Stokes, a third man, William Thomson (also known as Lord Kelvin), played a large role in the formulation, dissemination, and application of both of these results. All three men were interested in how the two theorems could help to explain and predict physical phenomena in electricity and magnetism and fluid flow. The basic facts of the story are given in the margin notes on pages 1137 and 1175.

Write a report on the historical origins of Green's Theorem and Stokes' Theorem. Explain the similarities and relationship between the theorems. Discuss the roles that Green, Thomson, and Stokes played in discovering these theorems and making them widely known. Show how both theorems arose from the investigation of electricity and magnetism and were later used to study a variety of physical problems.

The dictionary edited by Gillispie [2] is a good source for both biographical and scientific information. The book by Hutchinson [5] gives an account of Stokes' life and the book by Thompson [8] is a biography of Lord Kelvin. The articles by Grattan-Guinness [3] and Gray [4] and the book by Cannell [1] give background on the extraordinary life and works of Green. Additional historical and mathematical information is found in the books by Katz [6] and Kline [7].

1. D. M. Cannell, *George Green, Mathematician and Physicist 1793–1841: The Background to His Life and Work* (Philadelphia: Society for Industrial and Applied Mathematics, 2001).

2. C. C. Gillispie, ed., *Dictionary of Scientific Biography* (New York: Scribner's, 1974). See the article on Green by P. J. Wallis in Volume XV and the articles on Thomson by Jed Buchwald and on Stokes by E. M. Parkinson in Volume XIII.

3. I. Grattan-Guinness, "Why Did George Green Write his Essay of 1828 on Electricity and Magnetism?" *Amer. Math. Monthly,* Vol. 102 (1995), pp. 387–96.

4. J. Gray, "There Was a Jolly Miller." *The New Scientist,* Vol. 139 (1993), pp. 24–27.

5. G. E. Hutchinson, *The Enchanted Voyage and Other Studies* (Westport, CT: Greenwood Press, 1978).

6. Victor Katz, *A History of Mathematics: An Introduction* (New York: HarperCollins, 1993), pp. 678–80.

7. Morris Kline, *Mathematical Thought from Ancient to Modern Times* (New York: Oxford University Press, 1972), pp. 683–85.

8. Sylvanus P. Thompson, *The Life of Lord Kelvin* (New York: Chelsea, 1976).

16.9 The Divergence Theorem

In Section 16.5 we rewrote Green's Theorem in a vector version as

$$\int_C \mathbf{F} \cdot \mathbf{n}\, ds = \iint_D \operatorname{div} \mathbf{F}(x, y)\, dA$$

where C is the positively oriented boundary curve of the plane region D. If we were seeking to extend this theorem to vector fields on $\mathbb{R}^3$, we might make the guess that

$$\boxed{1} \qquad \iint_S \mathbf{F} \cdot \mathbf{n}\, dS = \iiint_E \operatorname{div} \mathbf{F}(x, y, z)\, dV$$

where S is the boundary surface of the solid region E. It turns out that Equation 1 is true, under appropriate hypotheses, and is called the Divergence Theorem. Notice its similarity to Green's Theorem and Stokes' Theorem in that it relates the integral of a derivative of a function (div $\mathbf{F}$ in this case) over a region to the integral of the original function $\mathbf{F}$ over the boundary of the region.

At this stage you may wish to review the various types of regions over which we were able to evaluate triple integrals in Section 15.6. We state and prove the Divergence Theorem for regions E that are simultaneously of types 1, 2, and 3 and we call such regions **simple solid regions**. (For instance, regions bounded by ellipsoids or rectangular boxes are simple solid regions.) The boundary of E is a closed surface, and we use the convention, introduced in Section 16.7, that the positive orientation is outward; that is, the unit normal vector $\mathbf{n}$ is directed outward from E.

The Divergence Theorem is sometimes called Gauss's Theorem after the great German mathematician Karl Friedrich Gauss (1777–1855), who discovered this theorem during his investigation of electrostatics. In Eastern Europe the Divergence Theorem is known as Ostrogradsky's Theorem after the Russian mathematician Mikhail Ostrogradsky (1801–1862), who published this result in 1826.

The Divergence Theorem Let E be a simple solid region and let S be the boundary surface of E, given with positive (outward) orientation. Let $\mathbf{F}$ be a vector field whose component functions have continuous partial derivatives on an open region that contains E. Then

$$\iint_S \mathbf{F} \cdot d\mathbf{S} = \iiint_E \operatorname{div} \mathbf{F}\, dV$$

Thus the Divergence Theorem states that, under the given conditions, the flux of $\mathbf{F}$ across the boundary surface of E is equal to the triple integral of the divergence of $\mathbf{F}$ over E.

PROOF Let $\mathbf{F} = P\,\mathbf{i} + Q\,\mathbf{j} + R\,\mathbf{k}$. Then

$$\operatorname{div} \mathbf{F} = \frac{\partial P}{\partial x} + \frac{\partial Q}{\partial y} + \frac{\partial R}{\partial z}$$

so

$$\iiint_E \operatorname{div} \mathbf{F}\, dV = \iiint_E \frac{\partial P}{\partial x}\, dV + \iiint_E \frac{\partial Q}{\partial y}\, dV + \iiint_E \frac{\partial R}{\partial z}\, dV$$

If $\mathbf{n}$ is the unit outward normal of S, then the surface integral on the left side of the Divergence Theorem is

$$\iint_S \mathbf{F} \cdot d\mathbf{S} = \iint_S \mathbf{F} \cdot \mathbf{n}\, dS = \iint_S (P\,\mathbf{i} + Q\,\mathbf{j} + R\,\mathbf{k}) \cdot \mathbf{n}\, dS$$

$$= \iint_S P\,\mathbf{i} \cdot \mathbf{n}\, dS + \iint_S Q\,\mathbf{j} \cdot \mathbf{n}\, dS + \iint_S R\,\mathbf{k} \cdot \mathbf{n}\, dS$$

Therefore, to prove the Divergence Theorem, it suffices to prove the following three equations:

2 $$\iint_S P\,\mathbf{i}\cdot\mathbf{n}\,dS = \iiint_E \frac{\partial P}{\partial x}\,dV$$

3 $$\iint_S Q\,\mathbf{j}\cdot\mathbf{n}\,dS = \iiint_E \frac{\partial Q}{\partial y}\,dV$$

4 $$\iint_S R\,\mathbf{k}\cdot\mathbf{n}\,dS = \iiint_E \frac{\partial R}{\partial z}\,dV$$

To prove Equation 4 we use the fact that E is a type 1 region:

$$E = \{(x, y, z) \mid (x, y) \in D,\ u_1(x, y) \leqslant z \leqslant u_2(x, y)\}$$

where D is the projection of E onto the xy-plane. By Equation 15.6.6, we have

$$\iiint_E \frac{\partial R}{\partial z}\,dV = \iint_D \left[\int_{u_1(x,y)}^{u_2(x,y)} \frac{\partial R}{\partial z}(x, y, z)\,dz\right] dA$$

and therefore, by the Fundamental Theorem of Calculus,

5 $$\iiint_E \frac{\partial R}{\partial z}\,dV = \iint_D \big[R(x, y, u_2(x, y)) - R(x, y, u_1(x, y))\big]\,dA$$

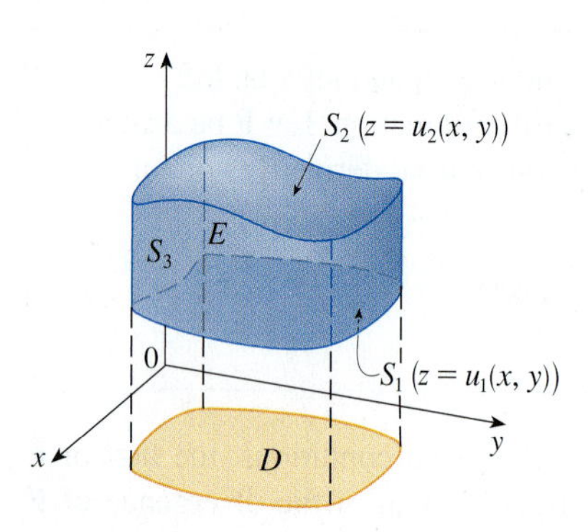

FIGURE 1

The boundary surface S consists of three pieces: the bottom surface S_1, the top surface S_2, and possibly a vertical surface S_3, which lies above the boundary curve of D. (See Figure 1. It might happen that S_3 doesn't appear, as in the case of a sphere.) Notice that on S_3 we have $\mathbf{k}\cdot\mathbf{n} = 0$, because $\mathbf{k}$ is vertical and $\mathbf{n}$ is horizontal, and so

$$\iint_{S_3} R\,\mathbf{k}\cdot\mathbf{n}\,dS = \iint_{S_3} 0\,dS = 0$$

Thus, regardless of whether there is a vertical surface, we can write

6 $$\iint_S R\,\mathbf{k}\cdot\mathbf{n}\,dS = \iint_{S_1} R\,\mathbf{k}\cdot\mathbf{n}\,dS + \iint_{S_2} R\,\mathbf{k}\cdot\mathbf{n}\,dS$$

The equation of S_2 is $z = u_2(x, y)$, $(x, y) \in D$, and the outward normal $\mathbf{n}$ points upward, so from Equation 16.7.10 (with $\mathbf{F}$ replaced by $R\,\mathbf{k}$) we have

$$\iint_{S_2} R\,\mathbf{k}\cdot\mathbf{n}\,dS = \iint_D R(x, y, u_2(x, y))\,dA$$

On S_1 we have $z = u_1(x, y)$, but here the outward normal $\mathbf{n}$ points downward, so we multiply by -1:

$$\iint_{S_1} R\,\mathbf{k}\cdot\mathbf{n}\,dS = -\iint_D R(x, y, u_1(x, y))\,dA$$

Therefore Equation 6 gives

$$\iint_S R\,\mathbf{k}\cdot\mathbf{n}\,dS = \iint_D \big[R(x, y, u_2(x, y)) - R(x, y, u_1(x, y))\big]\,dA$$

Comparison with Equation 5 shows that

$$\iint_S R\,\mathbf{k}\cdot\mathbf{n}\,dS = \iiint_E \frac{\partial R}{\partial z}\,dV$$

Notice that the method of proof of the Divergence Theorem is very similar to that of Green's Theorem.

Equations 2 and 3 are proved in a similar manner using the expressions for E as a type 2 or type 3 region, respectively. ■

EXAMPLE 1 Find the flux of the vector field $\mathbf{F}(x, y, z) = z\,\mathbf{i} + y\,\mathbf{j} + x\,\mathbf{k}$ over the unit sphere $x^2 + y^2 + z^2 = 1$.

SOLUTION First we compute the divergence of $\mathbf{F}$:

$$\text{div}\,\mathbf{F} = \frac{\partial}{\partial x}(z) + \frac{\partial}{\partial y}(y) + \frac{\partial}{\partial z}(x) = 1$$

The unit sphere S is the boundary of the unit ball B given by $x^2 + y^2 + z^2 \leqslant 1$. Thus the Divergence Theorem gives the flux as

The solution in Example 1 should be compared with the solution in Example 16.7.4.

$$\iint_S \mathbf{F}\cdot d\mathbf{S} = \iiint_B \text{div}\,\mathbf{F}\,dV = \iiint_B 1\,dV = V(B) = \tfrac{4}{3}\pi(1)^3 = \frac{4\pi}{3}$$ ■

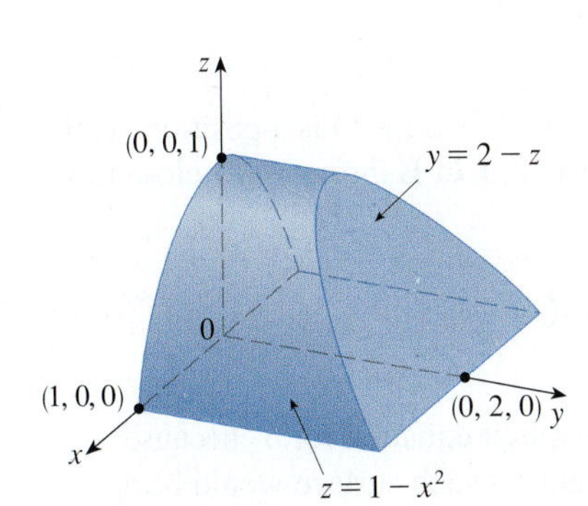

FIGURE 2

EXAMPLE 2 Evaluate $\iint_S \mathbf{F}\cdot d\mathbf{S}$, where

$$\mathbf{F}(x, y, z) = xy\,\mathbf{i} + \left(y^2 + e^{xz^2}\right)\mathbf{j} + \sin(xy)\,\mathbf{k}$$

and S is the surface of the region E bounded by the parabolic cylinder $z = 1 - x^2$ and the planes $z = 0$, $y = 0$, and $y + z = 2$. (See Figure 2.)

SOLUTION It would be extremely difficult to evaluate the given surface integral directly. (We would have to evaluate four surface integrals corresponding to the four pieces of S.) Furthermore, the divergence of $\mathbf{F}$ is much less complicated than $\mathbf{F}$ itself:

$$\text{div}\,\mathbf{F} = \frac{\partial}{\partial x}(xy) + \frac{\partial}{\partial y}\left(y^2 + e^{xz^2}\right) + \frac{\partial}{\partial z}(\sin xy) = y + 2y = 3y$$

Therefore we use the Divergence Theorem to transform the given surface integral into a triple integral. The easiest way to evaluate the triple integral is to express E as a type 3 region:

$$E = \left\{(x, y, z) \mid -1 \leqslant x \leqslant 1,\ 0 \leqslant z \leqslant 1 - x^2,\ 0 \leqslant y \leqslant 2 - z\right\}$$

Then we have

$$\iint_S \mathbf{F}\cdot d\mathbf{S} = \iiint_E \text{div}\,\mathbf{F}\,dV = \iiint_E 3y\,dV$$

$$= 3\int_{-1}^{1}\int_0^{1-x^2}\int_0^{2-z} y\,dy\,dz\,dx = 3\int_{-1}^{1}\int_0^{1-x^2}\frac{(2-z)^2}{2}\,dz\,dx$$

$$= \frac{3}{2}\int_{-1}^{1}\left[-\frac{(2-z)^3}{3}\right]_0^{1-x^2} dx = -\tfrac{1}{2}\int_{-1}^{1}\left[(x^2+1)^3 - 8\right]dx$$

$$= -\int_0^1 (x^6 + 3x^4 + 3x^2 - 7)\,dx = \frac{184}{35}$$ ■

Although we have proved the Divergence Theorem only for simple solid regions, it can be proved for regions that are finite unions of simple solid regions. (The procedure is similar to the one we used in Section 16.4 to extend Green's Theorem.)

For example, let's consider the region E that lies between the closed surfaces S_1 and S_2, where S_1 lies inside S_2. Let $\mathbf{n}_1$ and $\mathbf{n}_2$ be outward normals of S_1 and S_2. Then the boundary surface of E is $S = S_1 \cup S_2$ and its normal $\mathbf{n}$ is given by $\mathbf{n} = -\mathbf{n}_1$ on S_1 and $\mathbf{n} = \mathbf{n}_2$ on S_2. (See Figure 3.) Applying the Divergence Theorem to S, we get

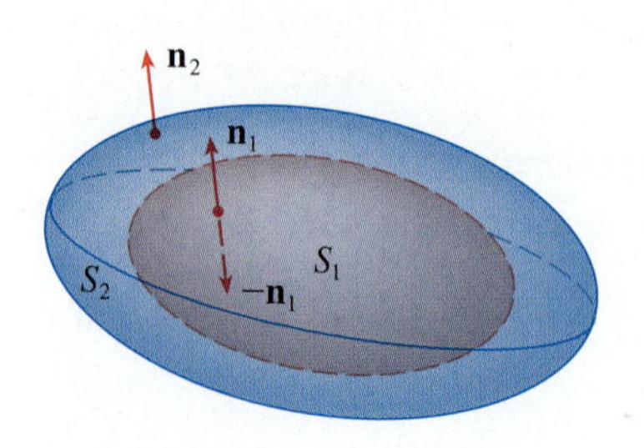

FIGURE 3

7
$$\begin{aligned} \iiint_E \operatorname{div} \mathbf{F}\, dV &= \iint_S \mathbf{F} \cdot d\mathbf{S} = \iint_S \mathbf{F} \cdot \mathbf{n}\, dS \\ &= \iint_{S_1} \mathbf{F} \cdot (-\mathbf{n}_1)\, dS + \iint_{S_2} \mathbf{F} \cdot \mathbf{n}_2\, dS \\ &= -\iint_{S_1} \mathbf{F} \cdot d\mathbf{S} + \iint_{S_2} \mathbf{F} \cdot d\mathbf{S} \end{aligned}$$

EXAMPLE 3 In Example 16.1.5 we considered the electric field

$$\mathbf{E}(\mathbf{x}) = \frac{\varepsilon Q}{|\mathbf{x}|^3}\mathbf{x}$$

where the electric charge Q is located at the origin and $\mathbf{x} = \langle x, y, z \rangle$ is a position vector. Use the Divergence Theorem to show that the electric flux of $\mathbf{E}$ through any closed surface S_2 that encloses the origin is

$$\iint_{S_2} \mathbf{E} \cdot d\mathbf{S} = 4\pi\varepsilon Q$$

SOLUTION The difficulty is that we don't have an explicit equation for S_2 because it is *any* closed surface enclosing the origin. The simplest such surface would be a sphere, so we let S_1 be a small sphere with radius a and center the origin. You can verify that $\operatorname{div} \mathbf{E} = 0$. (See Exercise 23.) Therefore Equation 7 gives

$$\iint_{S_2} \mathbf{E} \cdot d\mathbf{S} = \iint_{S_1} \mathbf{E} \cdot d\mathbf{S} + \iiint_E \operatorname{div} \mathbf{E}\, dV = \iint_{S_1} \mathbf{E} \cdot d\mathbf{S} = \iint_{S_1} \mathbf{E} \cdot \mathbf{n}\, dS$$

The point of this calculation is that we can compute the surface integral over S_1 because S_1 is a sphere. The normal vector at $\mathbf{x}$ is $\mathbf{x}/|\mathbf{x}|$. Therefore

$$\mathbf{E} \cdot \mathbf{n} = \frac{\varepsilon Q}{|\mathbf{x}|^3}\mathbf{x} \cdot \left(\frac{\mathbf{x}}{|\mathbf{x}|}\right) = \frac{\varepsilon Q}{|\mathbf{x}|^4}\mathbf{x} \cdot \mathbf{x} = \frac{\varepsilon Q}{|\mathbf{x}|^2} = \frac{\varepsilon Q}{a^2}$$

since the equation of S_1 is $|\mathbf{x}| = a$. Thus we have

$$\iint_{S_2} \mathbf{E} \cdot d\mathbf{S} = \iint_{S_1} \mathbf{E} \cdot \mathbf{n}\, dS = \frac{\varepsilon Q}{a^2} \iint_{S_1} dS = \frac{\varepsilon Q}{a^2} A(S_1) = \frac{\varepsilon Q}{a^2} 4\pi a^2 = 4\pi\varepsilon Q$$

This shows that the electric flux of $\mathbf{E}$ is $4\pi\varepsilon Q$ through *any* closed surface S_2 that contains the origin. [This is a special case of Gauss's Law (Equation 16.7.11) for a single charge. The relationship between ε and ε_0 is $\varepsilon = 1/(4\pi\varepsilon_0)$.] ■

Another application of the Divergence Theorem occurs in fluid flow. Let $\mathbf{v}(x, y, z)$ be the velocity field of a fluid with constant density ρ. Then $\mathbf{F} = \rho\mathbf{v}$ is the rate of flow per unit area. If $P_0(x_0, y_0, z_0)$ is a point in the fluid and B_a is a ball with center P_0 and very small radius a, then $\text{div}\,\mathbf{F}(P) \approx \text{div}\,\mathbf{F}(P_0)$ for all points P in B_a since $\text{div}\,\mathbf{F}$ is continuous. We approximate the flux over the boundary sphere S_a as follows:

$$\iint_{S_a} \mathbf{F} \cdot d\mathbf{S} = \iiint_{B_a} \text{div}\,\mathbf{F}\,dV \approx \iiint_{B_a} \text{div}\,\mathbf{F}(P_0)\,dV = \text{div}\,\mathbf{F}(P_0)V(B_a)$$

This approximation becomes better as $a \to 0$ and suggests that

8
$$\text{div}\,\mathbf{F}(P_0) = \lim_{a \to 0} \frac{1}{V(B_a)} \iint_{S_a} \mathbf{F} \cdot d\mathbf{S}$$

Equation 8 says that $\text{div}\,\mathbf{F}(P_0)$ is the net rate of outward flux per unit volume at P_0. (This is the reason for the name *divergence*.) If $\text{div}\,\mathbf{F}(P) > 0$, the net flow is outward near P and P is called a **source**. If $\text{div}\,\mathbf{F}(P) < 0$, the net flow is inward near P and P is called a **sink**.

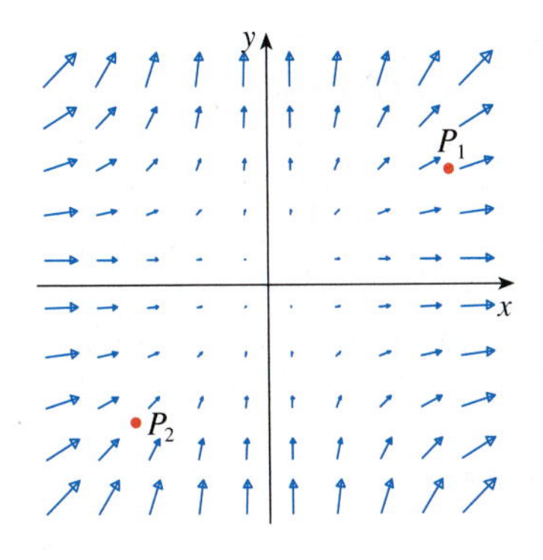

FIGURE 4
The vector field $\mathbf{F} = x^2\,\mathbf{i} + y^2\,\mathbf{j}$

For the vector field in Figure 4, it appears that the vectors that end near P_1 are shorter than the vectors that start near P_1. Thus the net flow is outward near P_1, so $\text{div}\,\mathbf{F}(P_1) > 0$ and P_1 is a source. Near P_2, on the other hand, the incoming arrows are longer than the outgoing arrows. Here the net flow is inward, so $\text{div}\,\mathbf{F}(P_2) < 0$ and P_2 is a sink. We can use the formula for $\mathbf{F}$ to confirm this impression. Since $\mathbf{F} = x^2\,\mathbf{i} + y^2\,\mathbf{j}$, we have $\text{div}\,\mathbf{F} = 2x + 2y$, which is positive when $y > -x$. So the points above the line $y = -x$ are sources and those below are sinks.

16.9 EXERCISES

1–4 Verify that the Divergence Theorem is true for the vector field $\mathbf{F}$ on the region E.

1. $\mathbf{F}(x, y, z) = 3x\,\mathbf{i} + xy\,\mathbf{j} + 2xz\,\mathbf{k}$,
   E is the cube bounded by the planes $x = 0$, $x = 1$, $y = 0$, $y = 1$, $z = 0$, and $z = 1$

2. $\mathbf{F}(x, y, z) = y^2z^3\,\mathbf{i} + 2yz\,\mathbf{j} + 4z^2\,\mathbf{k}$,
   E is the solid enclosed by the paraboloid $z = x^2 + y^2$ and the plane $z = 9$

3. $\mathbf{F}(x, y, z) = \langle z, y, x \rangle$,
   E is the solid ball $x^2 + y^2 + z^2 \leq 16$

4. $\mathbf{F}(x, y, z) = \langle x^2, -y, z \rangle$,
   E is the solid cylinder $y^2 + z^2 \leq 9$, $0 \leq x \leq 2$

5–15 Use the Divergence Theorem to calculate the surface integral $\iint_S \mathbf{F} \cdot d\mathbf{S}$; that is, calculate the flux of $\mathbf{F}$ across S.

5. $\mathbf{F}(x, y, z) = xye^z\,\mathbf{i} + xy^2z^3\,\mathbf{j} - ye^z\,\mathbf{k}$,
   S is the surface of the box bounded by the coordinate planes and the planes $x = 3$, $y = 2$, and $z = 1$

6. $\mathbf{F}(x, y, z) = x^2yz\,\mathbf{i} + xy^2z\,\mathbf{j} + xyz^2\,\mathbf{k}$,
   S is the surface of the box enclosed by the planes $x = 0$, $x = a$, $y = 0$, $y = b$, $z = 0$, and $z = c$, where a, b, and c are positive numbers

7. $\mathbf{F}(x, y, z) = 3xy^2\,\mathbf{i} + xe^z\,\mathbf{j} + z^3\,\mathbf{k}$,
   S is the surface of the solid bounded by the cylinder $y^2 + z^2 = 1$ and the planes $x = -1$ and $x = 2$

8. $\mathbf{F}(x, y, z) = (x^3 + y^3)\,\mathbf{i} + (y^3 + z^3)\,\mathbf{j} + (z^3 + x^3)\,\mathbf{k}$,
   S is the sphere with center the origin and radius 2

9. $\mathbf{F}(x, y, z) = xe^y\,\mathbf{i} + (z - e^y)\,\mathbf{j} - xy\,\mathbf{k}$,
   S is the ellipsoid $x^2 + 2y^2 + 3z^2 = 4$

10. $\mathbf{F}(x, y, z) = z\,\mathbf{i} + y\,\mathbf{j} + zx\,\mathbf{k}$,
    S is the surface of the tetrahedron enclosed by the coordinate planes and the plane
    $$\frac{x}{a} + \frac{y}{b} + \frac{z}{c} = 1$$
    where a, b, and c are positive numbers

11. $\mathbf{F}(x, y, z) = (2x^3 + y^3)\,\mathbf{i} + (y^3 + z^3)\,\mathbf{j} + 3y^2z\,\mathbf{k}$,
    S is the surface of the solid bounded by the paraboloid $z = 1 - x^2 - y^2$ and the xy-plane

12. $\mathbf{F}(x, y, z) = (xy + 2xz)\,\mathbf{i} + (x^2 + y^2)\,\mathbf{j} + (xy - z^2)\,\mathbf{k}$,
    S is the surface of the solid bounded by the cylinder $x^2 + y^2 = 4$ and the planes $z = y - 2$ and $z = 0$

13. $\mathbf{F} = |\mathbf{r}|\,\mathbf{r}$, where $\mathbf{r} = x\,\mathbf{i} + y\,\mathbf{j} + z\,\mathbf{k}$,
    S consists of the hemisphere $z = \sqrt{1 - x^2 - y^2}$ and the disk $x^2 + y^2 \leq 1$ in the xy-plane

14. $\mathbf{F} = |\mathbf{r}|^2\mathbf{r}$, where $\mathbf{r} = x\,\mathbf{i} + y\,\mathbf{j} + z\,\mathbf{k}$,
S is the sphere with radius R and center the origin

CAS **15.** $\mathbf{F}(x, y, z) = e^y \tan z\,\mathbf{i} + y\sqrt{3 - x^2}\,\mathbf{j} + x \sin y\,\mathbf{k}$,
S is the surface of the solid that lies above the xy-plane and below the surface $z = 2 - x^4 - y^4$, $-1 \leqslant x \leqslant 1$, $-1 \leqslant y \leqslant 1$

CAS **16.** Use a computer algebra system to plot the vector field $\mathbf{F}(x, y, z) = \sin x \cos^2 y\,\mathbf{i} + \sin^3 y \cos^4 z\,\mathbf{j} + \sin^5 z \cos^6 x\,\mathbf{k}$ in the cube cut from the first octant by the planes $x = \pi/2$, $y = \pi/2$, and $z = \pi/2$. Then compute the flux across the surface of the cube.

17. Use the Divergence Theorem to evaluate $\iint_S \mathbf{F} \cdot d\mathbf{S}$, where
$\mathbf{F}(x, y, z) = z^2 x\,\mathbf{i} + \left(\frac{1}{3}y^3 + \tan z\right)\mathbf{j} + (x^2 z + y^2)\,\mathbf{k}$
and S is the top half of the sphere $x^2 + y^2 + z^2 = 1$. [*Hint:* Note that S is not a closed surface. First compute integrals over S_1 and S_2, where S_1 is the disk $x^2 + y^2 \leqslant 1$, oriented downward, and $S_2 = S \cup S_1$.]

18. Let $\mathbf{F}(x, y, z) = z \tan^{-1}(y^2)\,\mathbf{i} + z^3 \ln(x^2 + 1)\,\mathbf{j} + z\,\mathbf{k}$. Find the flux of $\mathbf{F}$ across the part of the paraboloid $x^2 + y^2 + z = 2$ that lies above the plane $z = 1$ and is oriented upward.

19. A vector field $\mathbf{F}$ is shown. Use the interpretation of divergence derived in this section to determine whether div $\mathbf{F}$ is positive or negative at P_1 and at P_2.

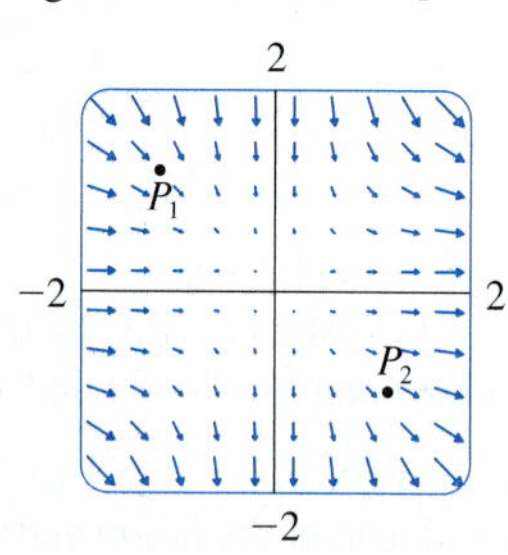

20. (a) Are the points P_1 and P_2 sources or sinks for the vector field $\mathbf{F}$ shown in the figure? Give an explanation based solely on the picture.
(b) Given that $\mathbf{F}(x, y) = \langle x, y^2 \rangle$, use the definition of divergence to verify your answer to part (a).

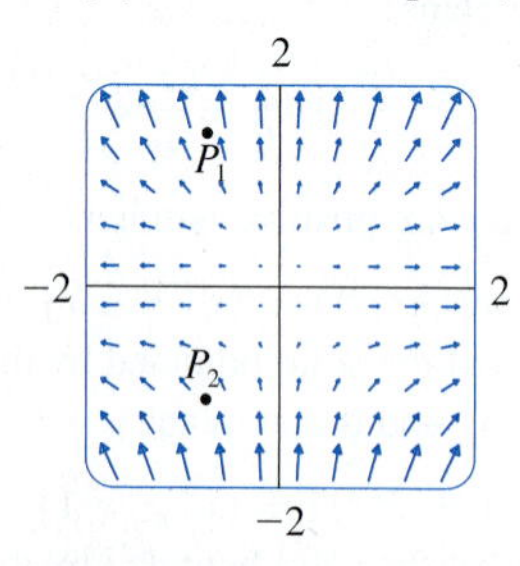

CAS **21–22** Plot the vector field and guess where div $\mathbf{F} > 0$ and where div $\mathbf{F} < 0$. Then calculate div $\mathbf{F}$ to check your guess.

21. $\mathbf{F}(x, y) = \langle xy, x + y^2 \rangle$ **22.** $\mathbf{F}(x, y) = \langle x^2, y^2 \rangle$

---

23. Verify that div $\mathbf{E} = 0$ for the electric field $\mathbf{E}(\mathbf{x}) = \dfrac{\varepsilon Q}{|\mathbf{x}|^3}\mathbf{x}$.

24. Use the Divergence Theorem to evaluate

$$\iint_S (2x + 2y + z^2)\, dS$$

where S is the sphere $x^2 + y^2 + z^2 = 1$.

25–30 Prove each identity, assuming that S and E satisfy the conditions of the Divergence Theorem and the scalar functions and components of the vector fields have continuous second-order partial derivatives.

25. $\displaystyle\iint_S \mathbf{a} \cdot \mathbf{n}\, dS = 0$, where $\mathbf{a}$ is a constant vector

26. $V(E) = \frac{1}{3}\displaystyle\iint_S \mathbf{F} \cdot d\mathbf{S}$, where $\mathbf{F}(x, y, z) = x\,\mathbf{i} + y\,\mathbf{j} + z\,\mathbf{k}$

27. $\displaystyle\iint_S \operatorname{curl} \mathbf{F} \cdot d\mathbf{S} = 0$ **28.** $\displaystyle\iint_S D_{\mathbf{n}} f\, dS = \iiint_E \nabla^2 f\, dV$

29. $\displaystyle\iint_S (f\nabla g) \cdot \mathbf{n}\, dS = \iiint_E (f\nabla^2 g + \nabla f \cdot \nabla g)\, dV$

30. $\displaystyle\iint_S (f\nabla g - g\nabla f) \cdot \mathbf{n}\, dS = \iiint_E (f\nabla^2 g - g\nabla^2 f)\, dV$

---

31. Suppose S and E satisfy the conditions of the Divergence Theorem and f is a scalar function with continuous partial derivatives. Prove that

$$\iint_S f\mathbf{n}\, dS = \iiint_E \nabla f\, dV$$

These surface and triple integrals of vector functions are vectors defined by integrating each component function. [*Hint:* Start by applying the Divergence Theorem to $\mathbf{F} = f\mathbf{c}$, where $\mathbf{c}$ is an arbitrary constant vector.]

32. A solid occupies a region E with surface S and is immersed in a liquid with constant density ρ. We set up a coordinate system so that the xy-plane coincides with the surface of the liquid, and positive values of z are measured downward into the liquid. Then the pressure at depth z is $p = \rho g z$, where g is the acceleration due to gravity (see Section 8.3). The total buoyant force on the solid due to the pressure distribution is given by the surface integral

$$\mathbf{F} = -\iint_S p\mathbf{n}\, dS$$

where $\mathbf{n}$ is the outer unit normal. Use the result of Exercise 31 to show that $\mathbf{F} = -W\mathbf{k}$, where W is the weight of the liquid displaced by the solid. (Note that $\mathbf{F}$ is directed upward because z is directed downward.) The result is *Archimedes' Principle:* The buoyant force on an object equals the weight of the displaced liquid.

16.10 Summary

The main results of this chapter are all higher-dimensional versions of the Fundamental Theorem of Calculus. To help you remember them, we collect them together here (without hypotheses) so that you can see more easily their essential similarity. Notice that in each case we have an integral of a "derivative" over a region on the left side, and the right side involves the values of the original function only on the *boundary* of the region.

Fundamental Theorem of Calculus

$$\int_a^b F'(x)\,dx = F(b) - F(a)$$

Fundamental Theorem for Line Integrals

$$\int_C \nabla f \cdot d\mathbf{r} = f(\mathbf{r}(b)) - f(\mathbf{r}(a))$$

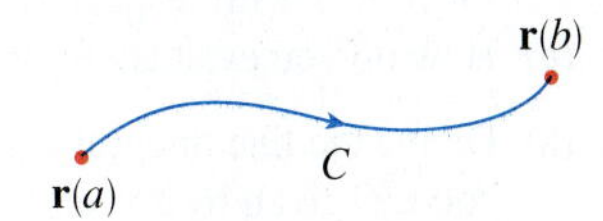

Green's Theorem

$$\iint_D \left(\frac{\partial Q}{\partial x} - \frac{\partial P}{\partial y}\right) dA = \int_C P\,dx + Q\,dy$$

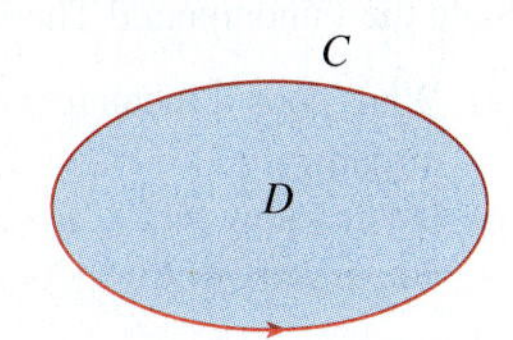

Stokes' Theorem

$$\iint_S \operatorname{curl} \mathbf{F} \cdot d\mathbf{S} = \int_C \mathbf{F} \cdot d\mathbf{r}$$

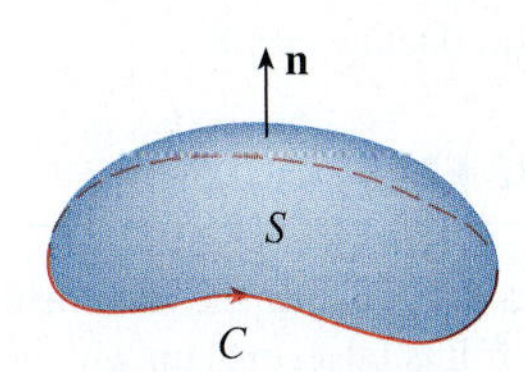

Divergence Theorem

$$\iiint_E \operatorname{div} \mathbf{F}\,dV = \iint_S \mathbf{F} \cdot d\mathbf{S}$$

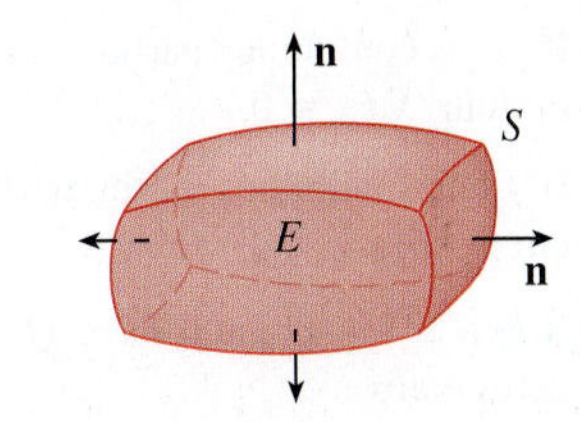

16 REVIEW

CONCEPT CHECK

Answers to the Concept Check can be found on the back endpapers.

1. What is a vector field? Give three examples that have physical meaning.

2. (a) What is a conservative vector field?
(b) What is a potential function?

3. (a) Write the definition of the line integral of a scalar function f along a smooth curve C with respect to arc length.
(b) How do you evaluate such a line integral?
(c) Write expressions for the mass and center of mass of a thin wire shaped like a curve C if the wire has linear density function $\rho(x, y)$.
(d) Write the definitions of the line integrals along C of a scalar function f with respect to x, y, and z.
(e) How do you evaluate these line integrals?

4. (a) Define the line integral of a vector field $\mathbf{F}$ along a smooth curve C given by a vector function $\mathbf{r}(t)$.
(b) If $\mathbf{F}$ is a force field, what does this line integral represent?
(c) If $\mathbf{F} = \langle P, Q, R \rangle$, what is the connection between the line integral of $\mathbf{F}$ and the line integrals of the component functions P, Q, and R?

5. State the Fundamental Theorem for Line Integrals.

6. (a) What does it mean to say that $\int_C \mathbf{F} \cdot d\mathbf{r}$ is independent of path?
(b) If you know that $\int_C \mathbf{F} \cdot d\mathbf{r}$ is independent of path, what can you say about $\mathbf{F}$?

7. State Green's Theorem.

8. Write expressions for the area enclosed by a curve C in terms of line integrals around C.

9. Suppose $\mathbf{F}$ is a vector field on $\mathbb{R}^3$.
(a) Define curl $\mathbf{F}$. (b) Define div $\mathbf{F}$.
(c) If $\mathbf{F}$ is a velocity field in fluid flow, what are the physical interpretations of curl $\mathbf{F}$ and div $\mathbf{F}$?

10. If $\mathbf{F} = P\,\mathbf{i} + Q\,\mathbf{j}$, how do you determine whether $\mathbf{F}$ is conservative? What if $\mathbf{F}$ is a vector field on $\mathbb{R}^3$?

11. (a) What is a parametric surface? What are its grid curves?
(b) Write an expression for the area of a parametric surface.
(c) What is the area of a surface given by an equation $z = g(x, y)$?

12. (a) Write the definition of the surface integral of a scalar function f over a surface S.
(b) How do you evaluate such an integral if S is a parametric surface given by a vector function $\mathbf{r}(u, v)$?
(c) What if S is given by an equation $z = g(x, y)$?
(d) If a thin sheet has the shape of a surface S, and the density at (x, y, z) is $\rho(x, y, z)$, write expressions for the mass and center of mass of the sheet.

13. (a) What is an oriented surface? Give an example of a non-orientable surface.
(b) Define the surface integral (or flux) of a vector field $\mathbf{F}$ over an oriented surface S with unit normal vector $\mathbf{n}$.
(c) How do you evaluate such an integral if S is a parametric surface given by a vector function $\mathbf{r}(u, v)$?
(d) What if S is given by an equation $z = g(x, y)$?

14. State Stokes' Theorem.

15. State the Divergence Theorem.

16. In what ways are the Fundamental Theorem for Line Integrals, Green's Theorem, Stokes' Theorem, and the Divergence Theorem similar?

TRUE-FALSE QUIZ

Determine whether the statement is true or false. If it is true, explain why. If it is false, explain why or give an example that disproves the statement.

1. If $\mathbf{F}$ is a vector field, then div $\mathbf{F}$ is a vector field.

2. If $\mathbf{F}$ is a vector field, then curl $\mathbf{F}$ is a vector field.

3. If f has continuous partial derivatives of all orders on $\mathbb{R}^3$, then $\text{div}(\text{curl}\, \nabla f) = 0$.

4. If f has continuous partial derivatives on $\mathbb{R}^3$ and C is any circle, then $\int_C \nabla f \cdot d\mathbf{r} = 0$.

5. If $\mathbf{F} = P\,\mathbf{i} + Q\,\mathbf{j}$ and $P_y = Q_x$ in an open region D, then $\mathbf{F}$ is conservative.

6. $\int_{-C} f(x, y)\, ds = -\int_C f(x, y)\, ds$

7. If $\mathbf{F}$ and $\mathbf{G}$ are vector fields and div $\mathbf{F}$ = div $\mathbf{G}$, then $\mathbf{F} = \mathbf{G}$.

8. The work done by a conservative force field in moving a particle around a closed path is zero.

9. If $\mathbf{F}$ and $\mathbf{G}$ are vector fields, then

$$\text{curl}(\mathbf{F} + \mathbf{G}) = \text{curl}\, \mathbf{F} + \text{curl}\, \mathbf{G}$$

10. If $\mathbf{F}$ and $\mathbf{G}$ are vector fields, then

$$\text{curl}(\mathbf{F} \cdot \mathbf{G}) = \text{curl}\, \mathbf{F} \cdot \text{curl}\, \mathbf{G}$$

11. If S is a sphere and $\mathbf{F}$ is a constant vector field, then $\iint_S \mathbf{F} \cdot d\mathbf{S} = 0$.

12. There is a vector field $\mathbf{F}$ such that

$$\text{curl}\, \mathbf{F} = x\,\mathbf{i} + y\,\mathbf{j} + z\,\mathbf{k}$$

13. The area of the region bounded by the positively oriented, piecewise smooth, simple closed curve C is $A = \oint_C y\, dx$.

EXERCISES

1. A vector field **F**, a curve C, and a point P are shown.
(a) Is $\int_C \mathbf{F} \cdot d\mathbf{r}$ positive, negative, or zero? Explain.
(b) Is div $\mathbf{F}(P)$ positive, negative, or zero? Explain.

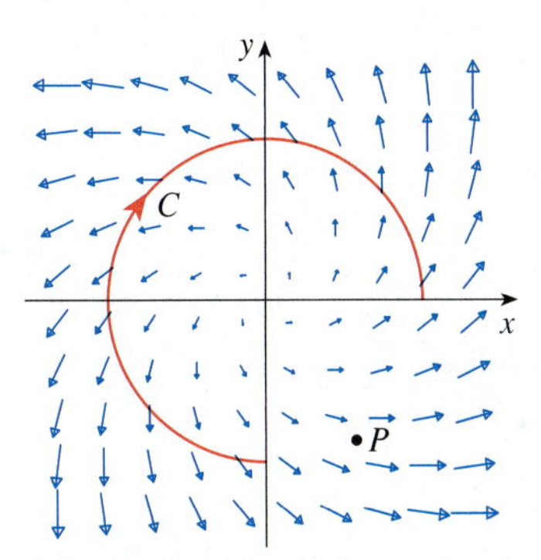

2–9 Evaluate the line integral.

2. $\int_C x\, ds$,
C is the arc of the parabola $y = x^2$ from (0, 0) to (1, 1)

3. $\int_C yz \cos x\, ds$,
C: $x = t,\ y = 3 \cos t,\ z = 3 \sin t,\ 0 \leqslant t \leqslant \pi$

4. $\int_C y\, dx + (x + y^2)\, dy$, C is the ellipse $4x^2 + 9y^2 = 36$ with counterclockwise orientation

5. $\int_C y^3\, dx + x^2\, dy$, C is the arc of the parabola $x = 1 - y^2$ from $(0, -1)$ to $(0, 1)$

6. $\int_C \sqrt{xy}\, dx + e^y\, dy + xz\, dz$,
C is given by $\mathbf{r}(t) = t^4\, \mathbf{i} + t^2\, \mathbf{j} + t^3\, \mathbf{k}, 0 \leqslant t \leqslant 1$

7. $\int_C xy\, dx + y^2\, dy + yz\, dz$,
C is the line segment from $(1, 0, -1)$, to $(3, 4, 2)$

8. $\int_C \mathbf{F} \cdot d\mathbf{r}$, where $\mathbf{F}(x, y) = xy\, \mathbf{i} + x^2\, \mathbf{j}$ and C is given by $\mathbf{r}(t) = \sin t\, \mathbf{i} + (1 + t)\, \mathbf{j}, 0 \leqslant t \leqslant \pi$

9. $\int_C \mathbf{F} \cdot d\mathbf{r}$, where $\mathbf{F}(x, y, z) = e^z\, \mathbf{i} + xz\, \mathbf{j} + (x + y)\, \mathbf{k}$ and C is given by $\mathbf{r}(t) = t^2\, \mathbf{i} + t^3\, \mathbf{j} - t\, \mathbf{k}, 0 \leqslant t \leqslant 1$

10. Find the work done by the force field

$$\mathbf{F}(x, y, z) = z\, \mathbf{i} + x\, \mathbf{j} + y\, \mathbf{k}$$

in moving a particle from the point (3, 0, 0) to the point $(0, \pi/2, 3)$ along
(a) a straight line
(b) the helix $x = 3 \cos t,\ y = t,\ z = 3 \sin t$

11–12 Show that **F** is a conservative vector field. Then find a function f such that $\mathbf{F} = \nabla f$.

11. $\mathbf{F}(x, y) = (1 + xy)e^{xy}\, \mathbf{i} + (e^y + x^2 e^{xy})\, \mathbf{j}$

12. $\mathbf{F}(x, y, z) = \sin y\, \mathbf{i} + x \cos y\, \mathbf{j} - \sin z\, \mathbf{k}$

13–14 Show that **F** is conservative and use this fact to evaluate $\int_C \mathbf{F} \cdot d\mathbf{r}$ along the given curve.

13. $\mathbf{F}(x, y) = (4x^3y^2 - 2xy^3)\, \mathbf{i} + (2x^4y - 3x^2y^2 + 4y^3)\, \mathbf{j}$,
C: $\mathbf{r}(t) = (t + \sin \pi t)\, \mathbf{i} + (2t + \cos \pi t)\, \mathbf{j},\ 0 \leqslant t \leqslant 1$

14. $\mathbf{F}(x, y, z) = e^y\, \mathbf{i} + (xe^y + e^z)\, \mathbf{j} + ye^z\, \mathbf{k}$,
C is the line segment from (0, 2, 0) to (4, 0, 3)

15. Verify that Green's Theorem is true for the line integral $\int_C xy^2\, dx - x^2y\, dy$, where C consists of the parabola $y = x^2$ from $(-1, 1)$ to $(1, 1)$ and the line segment from $(1, 1)$ to $(-1, 1)$.

16. Use Green's Theorem to evaluate

$$\int_C \sqrt{1 + x^3}\, dx + 2xy\, dy$$

where C is the triangle with vertices (0, 0), (1, 0), and (1, 3).

17. Use Green's Theorem to evaluate $\int_C x^2y\, dx - xy^2\, dy$, where C is the circle $x^2 + y^2 = 4$ with counterclockwise orientation.

18. Find curl **F** and div **F** if

$$\mathbf{F}(x, y, z) = e^{-x} \sin y\, \mathbf{i} + e^{-y} \sin z\, \mathbf{j} + e^{-z} \sin x\, \mathbf{k}$$

19. Show that there is no vector field **G** such that

$$\text{curl}\, \mathbf{G} = 2x\, \mathbf{i} + 3yz\, \mathbf{j} - xz^2\, \mathbf{k}$$

20. If **F** and **G** are vector fields whose component functions have continuous first partial derivatives, show that

$$\text{curl}(\mathbf{F} \times \mathbf{G}) = \mathbf{F}\, \text{div}\, \mathbf{G} - \mathbf{G}\, \text{div}\, \mathbf{F} + (\mathbf{G} \cdot \nabla)\mathbf{F} - (\mathbf{F} \cdot \nabla)\mathbf{G}$$

21. If C is any piecewise-smooth simple closed plane curve and f and g are differentiable functions, show that $\int_C f(x)\, dx + g(y)\, dy = 0$.

22. If f and g are twice differentiable functions, show that

$$\nabla^2(fg) = f\nabla^2 g + g\nabla^2 f + 2\nabla f \cdot \nabla g$$

23. If f is a harmonic function, that is, $\nabla^2 f = 0$, show that the line integral $\int f_y\, dx - f_x\, dy$ is independent of path in any simple region D.

24. (a) Sketch the curve C with parametric equations

$$x = \cos t \qquad y = \sin t \qquad z = \sin t \qquad 0 \leqslant t \leqslant 2\pi$$

(b) Find $\int_C 2xe^{2y}\, dx + (2x^2e^{2y} + 2y \cot z)\, dy - y^2 \csc^2 z\, dz$.

25. Find the area of the part of the surface $z = x^2 + 2y$ that lies above the triangle with vertices $(0, 0)$, $(1, 0)$, and $(1, 2)$.

26. (a) Find an equation of the tangent plane at the point $(4, -2, 1)$ to the parametric surface S given by

$$\mathbf{r}(u, v) = v^2\,\mathbf{i} - uv\,\mathbf{j} + u^2\,\mathbf{k} \qquad 0 \leqslant u \leqslant 3,\ -3 \leqslant v \leqslant 3$$

(b) Use a computer to graph the surface S and the tangent plane found in part (a).

(c) Set up, but do not evaluate, an integral for the surface area of S.

CAS (d) If

$$\mathbf{F}(x, y, z) = \frac{z^2}{1 + x^2}\,\mathbf{i} + \frac{x^2}{1 + y^2}\,\mathbf{j} + \frac{y^2}{1 + z^2}\,\mathbf{k}$$

find $\iint_S \mathbf{F} \cdot d\mathbf{S}$ correct to four decimal places.

27–30 Evaluate the surface integral.

27. $\iint_S z\,dS$, where S is the part of the paraboloid $z = x^2 + y^2$ that lies under the plane $z = 4$

28. $\iint_S (x^2 z + y^2 z)\,dS$, where S is the part of the plane $z = 4 + x + y$ that lies inside the cylinder $x^2 + y^2 = 4$

29. $\iint_S \mathbf{F} \cdot d\mathbf{S}$, where $\mathbf{F}(x, y, z) = xz\,\mathbf{i} - 2y\,\mathbf{j} + 3x\,\mathbf{k}$ and S is the sphere $x^2 + y^2 + z^2 = 4$ with outward orientation

30. $\iint_S \mathbf{F} \cdot d\mathbf{S}$, where $\mathbf{F}(x, y, z) = x^2\,\mathbf{i} + xy\,\mathbf{j} + z\,\mathbf{k}$ and S is the part of the paraboloid $z = x^2 + y^2$ below the plane $z = 1$ with upward orientation

---

31. Verify that Stokes' Theorem is true for the vector field $\mathbf{F}(x, y, z) = x^2\,\mathbf{i} + y^2\,\mathbf{j} + z^2\,\mathbf{k}$, where S is the part of the paraboloid $z = 1 - x^2 - y^2$ that lies above the xy-plane and S has upward orientation.

32. Use Stokes' Theorem to evaluate $\iint_S \operatorname{curl} \mathbf{F} \cdot d\mathbf{S}$, where $\mathbf{F}(x, y, z) = x^2yz\,\mathbf{i} + yz^2\,\mathbf{j} + z^3e^{xy}\,\mathbf{k}$, S is the part of the sphere $x^2 + y^2 + z^2 = 5$ that lies above the plane $z = 1$, and S is oriented upward.

33. Use Stokes' Theorem to evaluate $\int_C \mathbf{F} \cdot d\mathbf{r}$, where $\mathbf{F}(x, y, z) = xy\,\mathbf{i} + yz\,\mathbf{j} + zx\,\mathbf{k}$, and C is the triangle with vertices $(1, 0, 0)$, $(0, 1, 0)$, and $(0, 0, 1)$, oriented counterclockwise as viewed from above.

34. Use the Divergence Theorem to calculate the surface integral $\iint_S \mathbf{F} \cdot d\mathbf{S}$, where $\mathbf{F}(x, y, z) = x^3\,\mathbf{i} + y^3\,\mathbf{j} + z^3\,\mathbf{k}$ and S is the surface of the solid bounded by the cylinder $x^2 + y^2 = 1$ and the planes $z = 0$ and $z = 2$.

35. Verify that the Divergence Theorem is true for the vector field $\mathbf{F}(x, y, z) = x\,\mathbf{i} + y\,\mathbf{j} + z\,\mathbf{k}$, where E is the unit ball $x^2 + y^2 + z^2 \leqslant 1$.

36. Compute the outward flux of

$$\mathbf{F}(x, y, z) = \frac{x\,\mathbf{i} + y\,\mathbf{j} + z\,\mathbf{k}}{(x^2 + y^2 + z^2)^{3/2}}$$

through the ellipsoid $4x^2 + 9y^2 + 6z^2 = 36$.

37. Let

$$\mathbf{F}(x, y, z) = (3x^2yz - 3y)\,\mathbf{i} + (x^3z - 3x)\,\mathbf{j} + (x^3y + 2z)\,\mathbf{k}$$

Evaluate $\int_C \mathbf{F} \cdot d\mathbf{r}$, where C is the curve with initial point $(0, 0, 2)$ and terminal point $(0, 3, 0)$ shown in the figure.

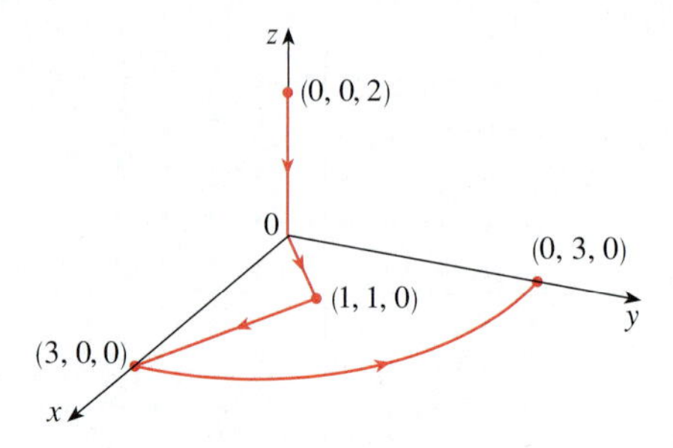

38. Let

$$\mathbf{F}(x, y) = \frac{(2x^3 + 2xy^2 - 2y)\,\mathbf{i} + (2y^3 + 2x^2y + 2x)\,\mathbf{j}}{x^2 + y^2}$$

Evaluate $\oint_C \mathbf{F} \cdot d\mathbf{r}$, where C is shown in the figure.

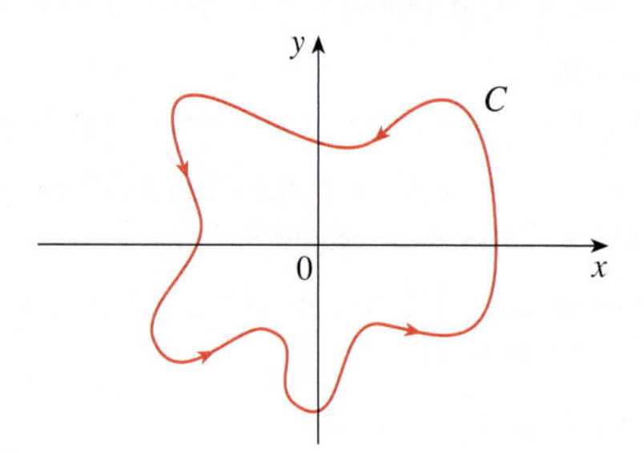

39. Find $\iint_S \mathbf{F} \cdot \mathbf{n}\,dS$, where $\mathbf{F}(x, y, z) = x\,\mathbf{i} + y\,\mathbf{j} + z\,\mathbf{k}$ and S is the outwardly oriented surface shown in the figure (the boundary surface of a cube with a unit corner cube removed).

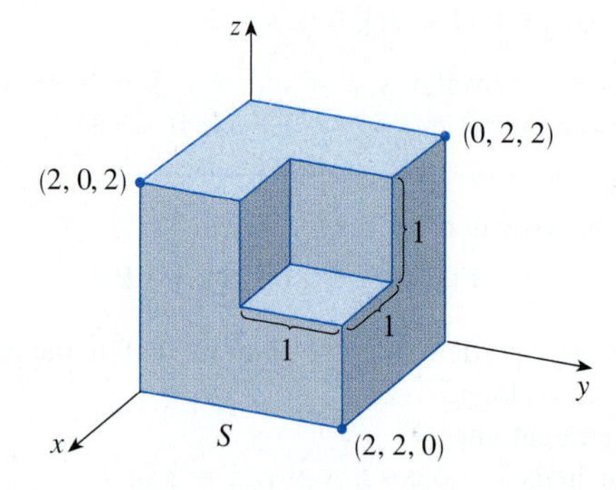

40. If the components of $\mathbf{F}$ have continuous second partial derivatives and S is the boundary surface of a simple solid region, show that $\iint_S \operatorname{curl} \mathbf{F} \cdot d\mathbf{S} = 0$.

41. If $\mathbf{a}$ is a constant vector, $\mathbf{r} = x\,\mathbf{i} + y\,\mathbf{j} + z\,\mathbf{k}$, and S is an oriented, smooth surface with a simple, closed, smooth, positively oriented boundary curve C, show that

$$\iint_S 2\mathbf{a} \cdot d\mathbf{S} = \int_C (\mathbf{a} \times \mathbf{r}) \cdot d\mathbf{r}$$

Problems Plus

1. Let S be a smooth parametric surface and let P be a point such that each line that starts at P intersects S at most once. The **solid angle** $\Omega(S)$ subtended by S at P is the set of lines starting at P and passing through S. Let $S(a)$ be the intersection of $\Omega(S)$ with the surface of the sphere with center P and radius a. Then the measure of the solid angle (in *steradians*) is defined to be

$$|\Omega(S)| = \frac{\text{area of } S(a)}{a^2}$$

Apply the Divergence Theorem to the part of $\Omega(S)$ between $S(a)$ and S to show that

$$|\Omega(S)| = \iint_S \frac{\mathbf{r} \cdot \mathbf{n}}{r^3}\, dS$$

where $\mathbf{r}$ is the radius vector from P to any point on S, $r = |\mathbf{r}|$, and the unit normal vector $\mathbf{n}$ is directed away from P.

This shows that the definition of the measure of a solid angle is independent of the radius a of the sphere. Thus the measure of the solid angle is equal to the area subtended on a *unit* sphere. (Note the analogy with the definition of radian measure.) The total solid angle subtended by a sphere at its center is thus 4π steradians.

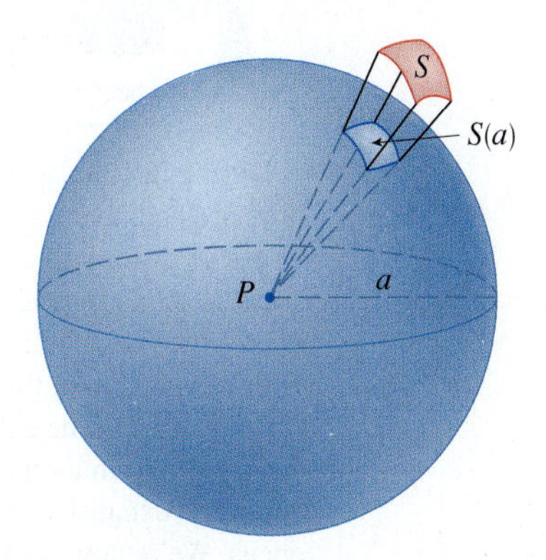

2. Find the positively oriented simple closed curve C for which the value of the line integral

$$\int_C (y^3 - y)\, dx - 2x^3\, dy$$

is a maximum.

3. Let C be a simple closed piecewise-smooth space curve that lies in a plane with unit normal vector $\mathbf{n} = \langle a, b, c \rangle$ and has positive orientation with respect to $\mathbf{n}$. Show that the plane area enclosed by C is

$$\tfrac{1}{2} \int_C (bz - cy)\, dx + (cx - az)\, dy + (ay - bx)\, dz$$

4. Investigate the shape of the surface with parametric equations $x = \sin u$, $y = \sin v$, $z = \sin(u + v)$. Start by graphing the surface from several points of view. Explain the appearance of the graphs by determining the traces in the horizontal planes $z = 0$, $z = \pm 1$, and $z = \pm \frac{1}{2}$.

5. Prove the following identity:

$$\nabla(\mathbf{F} \cdot \mathbf{G}) = (\mathbf{F} \cdot \nabla)\mathbf{G} + (\mathbf{G} \cdot \nabla)\mathbf{F} + \mathbf{F} \times \operatorname{curl} \mathbf{G} + \mathbf{G} \times \operatorname{curl} \mathbf{F}$$

6. The figure depicts the sequence of events in each cylinder of a four-cylinder internal combustion engine. Each piston moves up and down and is connected by a pivoted arm to a rotating crankshaft. Let $P(t)$ and $V(t)$ be the pressure and volume within a cylinder at time t, where $a \leqslant t \leqslant b$ gives the time required for a complete cycle. The graph shows how P and V vary through one cycle of a four-stroke engine.

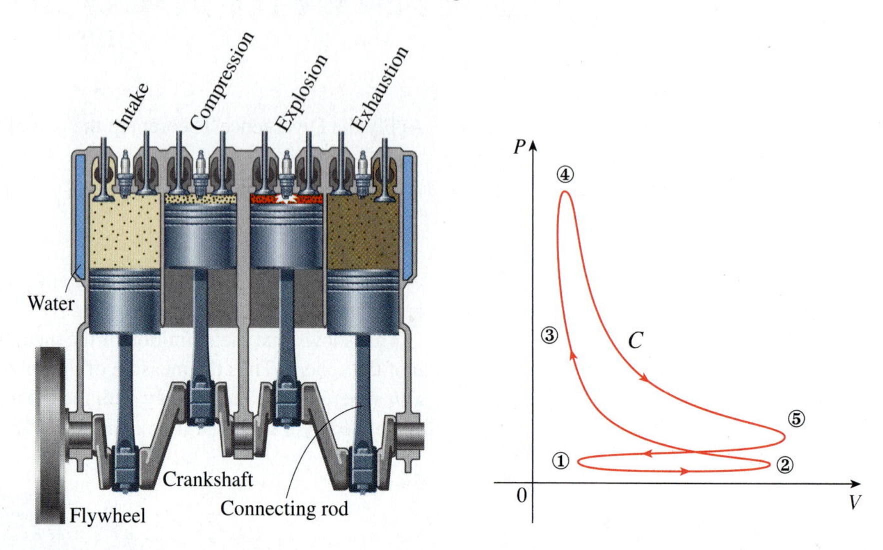

During the intake stroke (from ① to ②) a mixture of air and gasoline at atmospheric pressure is drawn into a cylinder through the intake valve as the piston moves downward. Then the piston rapidly compresses the mix with the valves closed in the compression stroke (from ② to ③) during which the pressure rises and the volume decreases. At ③ the sparkplug ignites the fuel, raising the temperature and pressure at almost constant volume to ④. Then, with valves closed, the rapid expansion forces the piston downward during the power stroke (from ④ to ⑤). The exhaust valve opens, temperature and pressure drop, and mechanical energy stored in a rotating flywheel pushes the piston upward, forcing the waste products out of the exhaust valve in the exhaust stroke. The exhaust valve closes and the intake valve opens. We're now back at ① and the cycle starts again.

(a) Show that the work done on the piston during one cycle of a four-stroke engine is $W = \int_C P\, dV$, where C is the curve in the PV-plane shown in the figure.

[*Hint:* Let $x(t)$ be the distance from the piston to the top of the cylinder and note that the force on the piston is $\mathbf{F} = AP(t)\,\mathbf{i}$, where A is the area of the top of the piston. Then $W = \int_{C_1} \mathbf{F} \cdot d\mathbf{r}$, where C_1 is given by $\mathbf{r}(t) = x(t)\,\mathbf{i}$, $a \leqslant t \leqslant b$. An alternative approach is to work directly with Riemann sums.]

(b) Use Formula 16.4.5 to show that the work is the difference of the areas enclosed by the two loops of C.

17 Second-Order Differential Equations

The motion of a shock absorber in a motorcycle is described by the differential equations that we solve in Section 17.3.

THE BASIC IDEAS OF DIFFERENTIAL equations were explained in Chapter 9; there we concentrated on first-order equations. In this chapter we study second-order linear differential equations and learn how they can be applied to solve problems concerning the vibrations of springs and the analysis of electric circuits. We will also see how infinite series can be used to solve differential equations.

17.1 Second-Order Linear Equations

A **second-order linear differential equation** has the form

$$\boxed{1} \qquad P(x)\frac{d^2y}{dx^2} + Q(x)\frac{dy}{dx} + R(x)y = G(x)$$

where P, Q, R, and G are continuous functions. We saw in Section 9.1 that equations of this type arise in the study of the motion of a spring. In Section 17.3 we will further pursue this application as well as the application to electric circuits.

In this section we study the case where $G(x) = 0$, for all x, in Equation 1. Such equations are called **homogeneous** linear equations. Thus the form of a second-order linear homogeneous differential equation is

$$\boxed{2} \qquad P(x)\frac{d^2y}{dx^2} + Q(x)\frac{dy}{dx} + R(x)y = 0$$

If $G(x) \neq 0$ for some x, Equation 1 is **nonhomogeneous** and is discussed in Section 17.2.

Two basic facts enable us to solve homogeneous linear equations. The first of these says that if we know two solutions y_1 and y_2 of such an equation, then the **linear combination** $y = c_1y_1 + c_2y_2$ is also a solution.

3 Theorem If $y_1(x)$ and $y_2(x)$ are both solutions of the linear homogeneous equation (2) and c_1 and c_2 are any constants, then the function

$$y(x) = c_1y_1(x) + c_2y_2(x)$$

is also a solution of Equation 2.

PROOF Since y_1 and y_2 are solutions of Equation 2, we have

$$P(x)y_1'' + Q(x)y_1' + R(x)y_1 = 0$$

and

$$P(x)y_2'' + Q(x)y_2' + R(x)y_2 = 0$$

Therefore, using the basic rules for differentiation, we have

$$\begin{aligned}
P(x)y'' + Q(x)y' + R(x)y \\
&= P(x)(c_1y_1 + c_2y_2)'' + Q(x)(c_1y_1 + c_2y_2)' + R(x)(c_1y_1 + c_2y_2) \\
&= P(x)(c_1y_1'' + c_2y_2'') + Q(x)(c_1y_1' + c_2y_2') + R(x)(c_1y_1 + c_2y_2) \\
&= c_1[P(x)y_1'' + Q(x)y_1' + R(x)y_1] + c_2[P(x)y_2'' + Q(x)y_2' + R(x)y_2] \\
&= c_1(0) + c_2(0) = 0
\end{aligned}$$

Thus $y = c_1y_1 + c_2y_2$ is a solution of Equation 2. ■

The other fact we need is given by the following theorem, which is proved in more advanced courses. It says that the general solution is a linear combination of two **linearly independent** solutions y_1 and y_2. This means that neither y_1 nor y_2 is a constant multiple of the other. For instance, the functions $f(x) = x^2$ and $g(x) = 5x^2$ are linearly dependent, but $f(x) = e^x$ and $g(x) = xe^x$ are linearly independent.

4 Theorem If y_1 and y_2 are linearly independent solutions of Equation 2 on an interval, and $P(x)$ is never 0, then the general solution is given by

$$y(x) = c_1 y_1(x) + c_2 y_2(x)$$

where c_1 and c_2 are arbitrary constants.

Theorem 4 is very useful because it says that if we know *two* particular linearly independent solutions, then we know *every* solution.

In general, it's not easy to discover particular solutions to a second-order linear equation. But it is always possible to do so if the coefficient functions P, Q, and R are constant functions, that is, if the differential equation has the form

$$ay'' + by' + cy = 0 \tag{5}$$

where a, b, and c are constants and $a \neq 0$.

It's not hard to think of some likely candidates for particular solutions of Equation 5 if we state the equation verbally. We are looking for a function y such that a constant times its second derivative y'' plus another constant times y' plus a third constant times y is equal to 0. We know that the exponential function $y = e^{rx}$ (where r is a constant) has the property that its derivative is a constant multiple of itself: $y' = re^{rx}$. Furthermore, $y'' = r^2e^{rx}$. If we substitute these expressions into Equation 5, we see that $y = e^{rx}$ is a solution if

$$ar^2e^{rx} + bre^{rx} + ce^{rx} = 0$$

or

$$(ar^2 + br + c)e^{rx} = 0$$

But e^{rx} is never 0. Thus $y = e^{rx}$ is a solution of Equation 5 if r is a root of the equation

$$ar^2 + br + c = 0 \tag{6}$$

Equation 6 is called the **auxiliary equation** (or **characteristic equation**) of the differential equation $ay'' + by' + cy = 0$. Notice that it is an algebraic equation that is obtained from the differential equation by replacing y'' by r^2, y' by r, and y by 1.

Sometimes the roots r_1 and r_2 of the auxiliary equation can be found by factoring. In other cases they are found by using the quadratic formula:

$$r_1 = \frac{-b + \sqrt{b^2 - 4ac}}{2a} \qquad r_2 = \frac{-b - \sqrt{b^2 - 4ac}}{2a} \tag{7}$$

We distinguish three cases according to the sign of the discriminant $b^2 - 4ac$.

CASE I $b^2 - 4ac > 0$

In this case the roots r_1 and r_2 of the auxiliary equation are real and distinct, so $y_1 = e^{r_1 x}$ and $y_2 = e^{r_2 x}$ are two linearly independent solutions of Equation 5. (Note that $e^{r_2 x}$ is not a constant multiple of $e^{r_1 x}$.) Therefore, by Theorem 4, we have the following fact.

8 If the roots r_1 and r_2 of the auxiliary equation $ar^2 + br + c = 0$ are real and unequal, then the general solution of $ay'' + by' + cy = 0$ is

$$y = c_1 e^{r_1 x} + c_2 e^{r_2 x}$$

In Figure 1 the graphs of the basic solutions $f(x) = e^{2x}$ and $g(x) = e^{-3x}$ of the differential equation in Example 1 are shown in blue and red, respectively. Some of the other solutions, linear combinations of f and g, are shown in black.

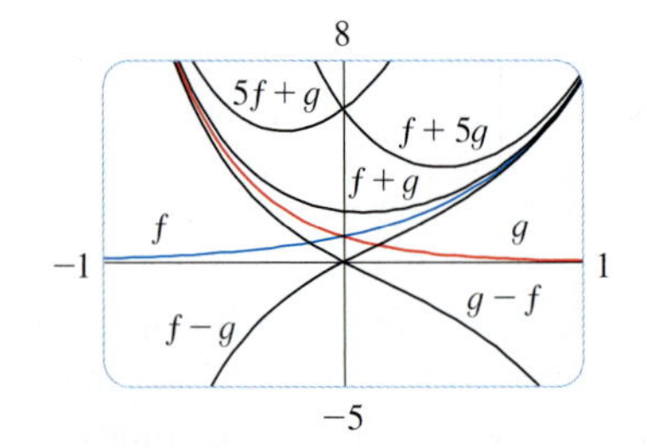

FIGURE 1

EXAMPLE 1 Solve the equation $y'' + y' - 6y = 0$.

SOLUTION The auxiliary equation is

$$r^2 + r - 6 = (r - 2)(r + 3) = 0$$

whose roots are $r = 2, -3$. Therefore, by (8), the general solution of the given differential equation is

$$y = c_1 e^{2x} + c_2 e^{-3x}$$

We could verify that this is indeed a solution by differentiating and substituting into the differential equation. ■

EXAMPLE 2 Solve $3\dfrac{d^2y}{dx^2} + \dfrac{dy}{dx} - y = 0$.

SOLUTION To solve the auxiliary equation $3r^2 + r - 1 = 0$, we use the quadratic formula:

$$r = \frac{-1 \pm \sqrt{13}}{6}$$

Since the roots are real and distinct, the general solution is

$$y = c_1 e^{(-1+\sqrt{13})x/6} + c_2 e^{(-1-\sqrt{13})x/6}$$ ■

CASE II $b^2 - 4ac = 0$

In this case $r_1 = r_2$; that is, the roots of the auxiliary equation are real and equal. Let's denote by r the common value of r_1 and r_2. Then, from Equations 7, we have

9 $$r = -\frac{b}{2a} \quad \text{so} \quad 2ar + b = 0$$

We know that $y_1 = e^{rx}$ is one solution of Equation 5. We now verify that $y_2 = xe^{rx}$ is also a solution:

$$\begin{aligned} ay_2'' + by_2' + cy_2 &= a(2re^{rx} + r^2xe^{rx}) + b(e^{rx} + rxe^{rx}) + cxe^{rx} \\ &= (2ar + b)e^{rx} + (ar^2 + br + c)xe^{rx} \\ &= 0(e^{rx}) + 0(xe^{rx}) = 0 \end{aligned}$$

In the first term, $2ar + b = 0$ by Equations 9; in the second term, $ar^2 + br + c = 0$ because r is a root of the auxiliary equation. Since $y_1 = e^{rx}$ and $y_2 = xe^{rx}$ are linearly independent solutions, Theorem 4 provides us with the general solution.

10 If the auxiliary equation $ar^2 + br + c = 0$ has only one real root r, then the general solution of $ay'' + by' + cy = 0$ is

$$y = c_1 e^{rx} + c_2 x e^{rx}$$

Figure 2 shows the basic solutions $f(x) = e^{-3x/2}$ and $g(x) = xe^{-3x/2}$ in Example 3 and some other members of the family of solutions. Notice that all of them approach 0 as $x \to \infty$.

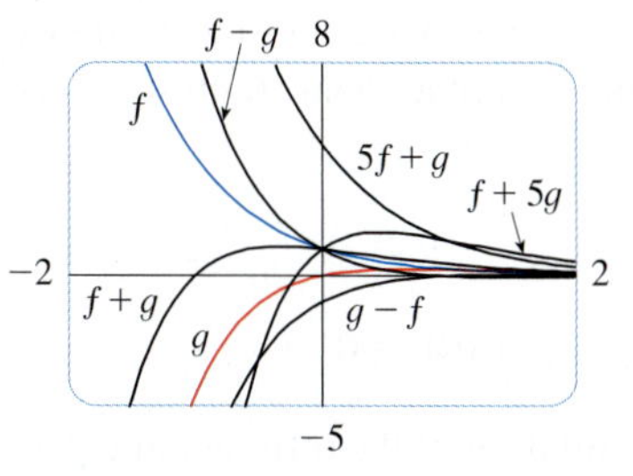

FIGURE 2

EXAMPLE 3 Solve the equation $4y'' + 12y' + 9y = 0$.

SOLUTION The auxiliary equation $4r^2 + 12r + 9 = 0$ can be factored as

$$(2r + 3)^2 = 0$$

so the only root is $r = -\frac{3}{2}$. By (10) the general solution is

$$y = c_1 e^{-3x/2} + c_2 x e^{-3x/2}$$

CASE III $b^2 - 4ac < 0$

In this case the roots r_1 and r_2 of the auxiliary equation are complex numbers. (See Appendix H for information about complex numbers.) We can write

$$r_1 = \alpha + i\beta \qquad r_2 = \alpha - i\beta$$

where α and β are real numbers. [In fact, $\alpha = -b/(2a)$, $\beta = \sqrt{4ac - b^2}/(2a)$.] Then, using Euler's equation

$$e^{i\theta} = \cos\theta + i\sin\theta$$

from Appendix H, we write the solution of the differential equation as

$$\begin{aligned} y &= C_1 e^{r_1 x} + C_2 e^{r_2 x} = C_1 e^{(\alpha + i\beta)x} + C_2 e^{(\alpha - i\beta)x} \\ &= C_1 e^{\alpha x}(\cos\beta x + i\sin\beta x) + C_2 e^{\alpha x}(\cos\beta x - i\sin\beta x) \\ &= e^{\alpha x}[(C_1 + C_2)\cos\beta x + i(C_1 - C_2)\sin\beta x] \\ &= e^{\alpha x}(c_1\cos\beta x + c_2\sin\beta x) \end{aligned}$$

where $c_1 = C_1 + C_2$, $c_2 = i(C_1 - C_2)$. This gives all solutions (real or complex) of the differential equation. The solutions are real when the constants c_1 and c_2 are real. We summarize the discussion as follows.

11 If the roots of the auxiliary equation $ar^2 + br + c = 0$ are the complex numbers $r_1 = \alpha + i\beta$, $r_2 = \alpha - i\beta$, then the general solution of $ay'' + by' + cy = 0$ is

$$y = e^{\alpha x}(c_1\cos\beta x + c_2\sin\beta x)$$

Figure 3 shows the graphs of the solutions in Example 4, $f(x) = e^{3x} \cos 2x$ and $g(x) = e^{3x} \sin 2x$, together with some linear combinations. All solutions approach 0 as $x \to -\infty$.

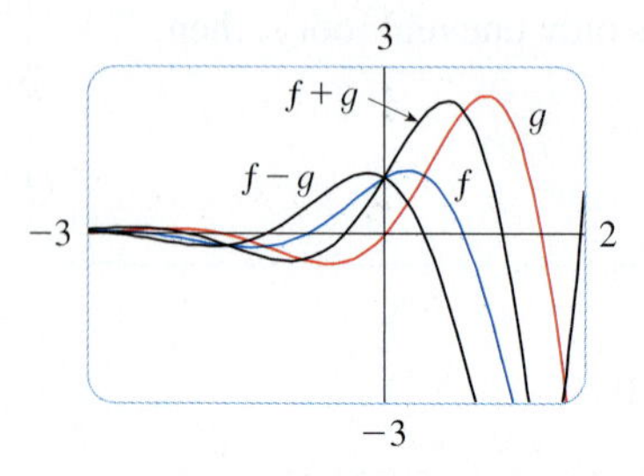

FIGURE 3

EXAMPLE 4 Solve the equation $y'' - 6y' + 13y = 0$.

SOLUTION The auxiliary equation is $r^2 - 6r + 13 = 0$. By the quadratic formula, the roots are

$$r = \frac{6 \pm \sqrt{36 - 52}}{2} = \frac{6 \pm \sqrt{-16}}{2} = 3 \pm 2i$$

By (11), the general solution of the differential equation is

$$y = e^{3x}(c_1 \cos 2x + c_2 \sin 2x)$$

■

Initial-Value and Boundary-Value Problems

An **initial-value problem** for the second-order Equation 1 or 2 consists of finding a solution y of the differential equation that also satisfies initial conditions of the form

$$y(x_0) = y_0 \qquad y'(x_0) = y_1$$

where y_0 and y_1 are given constants. If P, Q, R, and G are continuous on an interval and $P(x) \neq 0$ there, then a theorem found in more advanced books guarantees the existence and uniqueness of a solution to this initial-value problem. Examples 5 and 6 illustrate the technique for solving such a problem.

EXAMPLE 5 Solve the initial-value problem

$$y'' + y' - 6y = 0 \qquad y(0) = 1 \qquad y'(0) = 0$$

SOLUTION From Example 1 we know that the general solution of the differential equation is

$$y(x) = c_1 e^{2x} + c_2 e^{-3x}$$

Figure 4 shows the graph of the solution of the initial-value problem in Example 5. Compare with Figure 1.

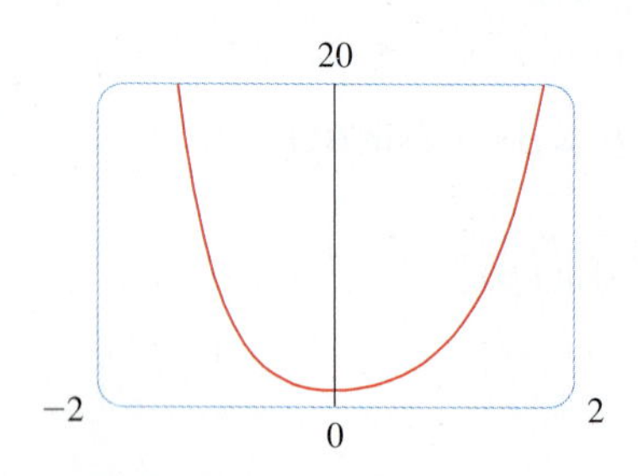

FIGURE 4

Differentiating this solution, we get

$$y'(x) = 2c_1 e^{2x} - 3c_2 e^{-3x}$$

To satisfy the initial conditions we require that

$$\textbf{12} \qquad y(0) = c_1 + c_2 = 1$$

$$\textbf{13} \qquad y'(0) = 2c_1 - 3c_2 = 0$$

From (13), we have $c_2 = \frac{2}{3}c_1$ and so (12) gives

$$c_1 + \tfrac{2}{3}c_1 = 1 \qquad c_1 = \tfrac{3}{5} \qquad c_2 = \tfrac{2}{5}$$

Thus the required solution of the initial-value problem is

$$y = \tfrac{3}{5}e^{2x} + \tfrac{2}{5}e^{-3x}$$

■

EXAMPLE 6 Solve the initial-value problem

$$y'' + y = 0 \qquad y(0) = 2 \qquad y'(0) = 3$$

SOLUTION The auxiliary equation is $r^2 + 1 = 0$, or $r^2 = -1$, whose roots are $\pm i$. Thus $\alpha = 0$, $\beta = 1$, and since $e^{0x} = 1$, the general solution is

$$y(x) = c_1 \cos x + c_2 \sin x$$

Since

$$y'(x) = -c_1 \sin x + c_2 \cos x$$

The solution to Example 6 is graphed in Figure 5. It appears to be a shifted sine curve and, indeed, you can verify that another way of writing the solution is

$y = \sqrt{13}\sin(x + \phi)$ where $\tan\phi = \frac{2}{3}$

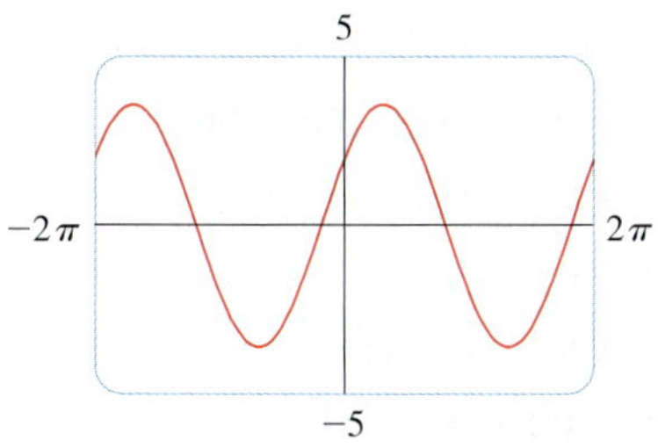
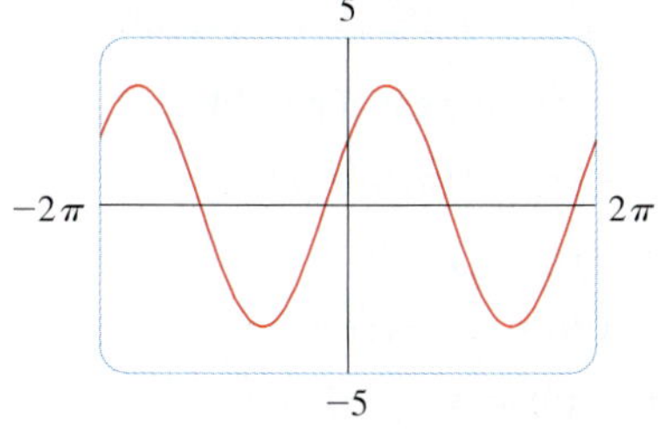

FIGURE 5

the initial conditions become

$$y(0) = c_1 = 2 \qquad y'(0) = c_2 = 3$$

Therefore the solution of the initial-value problem is

$$y(x) = 2\cos x + 3\sin x$$

■

A **boundary-value problem** for Equation 1 or 2 consists of finding a solution y of the differential equation that also satisfies boundary conditions of the form

$$y(x_0) = y_0 \qquad y(x_1) = y_1$$

In contrast with the situation for initial-value problems, a boundary-value problem does not always have a solution. The method is illustrated in Example 7.

EXAMPLE 7 Solve the boundary-value problem

$$y'' + 2y' + y = 0 \qquad y(0) = 1 \qquad y(1) = 3$$

SOLUTION The auxiliary equation is

$$r^2 + 2r + 1 = 0 \qquad \text{or} \qquad (r + 1)^2 = 0$$

whose only root is $r = -1$. Therefore the general solution is

$$y(x) = c_1 e^{-x} + c_2 x e^{-x}$$

Figure 6 shows the graph of the solution of the boundary-value problem in Example 7.

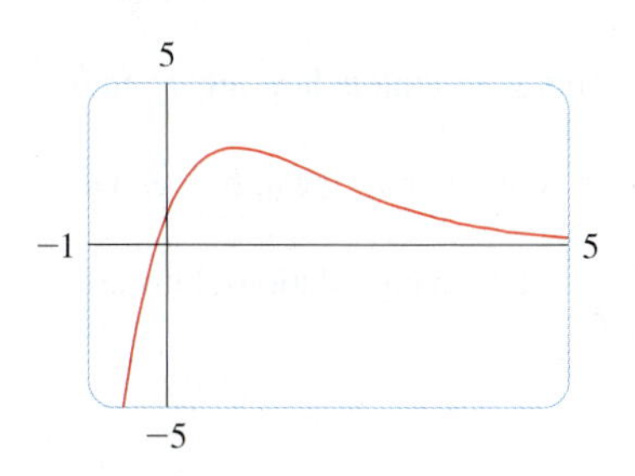

FIGURE 6

The boundary conditions are satisfied if

$$y(0) = c_1 = 1$$

$$y(1) = c_1 e^{-1} + c_2 e^{-1} = 3$$

The first condition gives $c_1 = 1$, so the second condition becomes

$$e^{-1} + c_2 e^{-1} = 3$$

Solving this equation for c_2 by first multiplying through by e, we get

$$1 + c_2 = 3e \qquad \text{so} \qquad c_2 = 3e - 1$$

Thus the solution of the boundary-value problem is

$$y = e^{-x} + (3e - 1)xe^{-x}$$

■

Summary: Solutions of $ay'' + by' + c = 0$

Roots of $ar^2 + br + c = 0$	General solution
r_1, r_2 real and distinct	$y = c_1 e^{r_1 x} + c_2 e^{r_2 x}$
$r_1 = r_2 = r$	$y = c_1 e^{rx} + c_2 x e^{rx}$
r_1, r_2 complex: $\alpha \pm i\beta$	$y = e^{\alpha x}(c_1 \cos\beta x + c_2 \sin\beta x)$

17.1 EXERCISES

1–13 Solve the differential equation.

1. $y'' - y' - 6y = 0$

2. $y'' - 6y' + 9y = 0$

3. $y'' + 2y = 0$

4. $y'' + y' - 12y = 0$

5. $4y'' + 4y' + y = 0$

6. $9y'' + 4y = 0$

7. $3y'' = 4y'$

8. $y = y''$

9. $y'' - 4y' + 13y = 0$

10. $3y'' + 4y' - 3y = 0$

11. $2\dfrac{d^2y}{dt^2} + 2\dfrac{dy}{dt} - y = 0$

12. $\dfrac{d^2R}{dt^2} + 6\dfrac{dR}{dt} + 34R = 0$

13. $3\dfrac{d^2V}{dt^2} + 4\dfrac{dV}{dt} + 3V = 0$

14–16 Graph the two basic solutions along with several other solutions of the differential equation. What features do the solutions have in common?

14. $4\dfrac{d^2y}{dx^2} - 4\dfrac{dy}{dx} + y = 0$

15. $\dfrac{d^2y}{dx^2} + 2\dfrac{dy}{dx} + 2y = 0$

16. $2\dfrac{d^2y}{dx^2} + \dfrac{dy}{dx} - y = 0$

17–24 Solve the initial-value problem.

17. $y'' + 3y = 0, \quad y(0) = 1, \quad y'(0) = 3$

18. $y'' - 2y' - 3y = 0, \quad y(0) = 2, \quad y'(0) = 2$

19. $9y'' + 12y' + 4y = 0, \quad y(0) = 1, \quad y'(0) = 0$

20. $3y'' - 2y' - y = 0, \quad y(0) = 0, \quad y'(0) = -4$

21. $y'' - 6y' + 10y = 0, \quad y(0) = 2, \quad y'(0) = 3$

22. $4y'' - 20y' + 25y = 0, \quad y(0) = 2, \quad y'(0) = -3$

23. $y'' - y' - 12y = 0, \quad y(1) = 0, \quad y'(1) = 1$

24. $4y'' + 4y' + 3y = 0, \quad y(0) = 0, \quad y'(0) = 1$

25–32 Solve the boundary-value problem, if possible.

25. $y'' + 16y = 0, \quad y(0) = -3, \quad y(\pi/8) = 2$

26. $y'' + 6y' = 0, \quad y(0) = 1, \quad y(1) = 0$

27. $y'' + 4y' + 4y = 0, \quad y(0) = 2, \quad y(1) = 0$

28. $y'' - 8y' + 17y = 0, \quad y(0) = 3, \quad y(\pi) = 2$

29. $y'' = y', \quad y(0) = 1, \quad y(1) = 2$

30. $4y'' - 4y' + y = 0, \quad y(0) = 4, \quad y(2) = 0$

31. $y'' + 4y' + 20y = 0, \quad y(0) = 1, \quad y(\pi) = 2$

32. $y'' + 4y' + 20y = 0, \quad y(0) = 1, \quad y(\pi) = e^{-2\pi}$

33. Let L be a nonzero real number.

(a) Show that the boundary-value problem $y'' + \lambda y = 0$, $y(0) = 0$, $y(L) = 0$ has only the trivial solution $y = 0$ for the cases $\lambda = 0$ and $\lambda < 0$.

(b) For the case $\lambda > 0$, find the values of λ for which this problem has a nontrivial solution and give the corresponding solution.

34. If a, b, and c are all positive constants and $y(x)$ is a solution of the differential equation $ay'' + by' + cy = 0$, show that $\lim_{x\to\infty} y(x) = 0$.

35. Consider the boundary-value problem $y'' - 2y' + 2y = 0$, $y(a) = c$, $y(b) = d$.

(a) If this problem has a unique solution, how are a and b related?

(b) If this problem has no solution, how are a, b, c, and d related?

(c) If this problem has infinitely many solutions, how are a, b, c, and d related?

17.2 Nonhomogeneous Linear Equations

In this section we learn how to solve second-order nonhomogeneous linear differential equations with constant coefficients, that is, equations of the form

$$ay'' + by' + cy = G(x) \tag{1}$$

where a, b, and c are constants and G is a continuous function. The related homogeneous equation

$$ay'' + by' + cy = 0 \tag{2}$$

is called the **complementary equation** and plays an important role in the solution of the original nonhomogeneous equation (1).

3 Theorem The general solution of the nonhomogeneous differential equation (1) can be written as

$$y(x) = y_p(x) + y_c(x)$$

where y_p is a particular solution of Equation 1 and y_c is the general solution of the complementary Equation 2.

PROOF We verify that if y is any solution of Equation 1, then $y - y_p$ is a solution of the complementary Equation 2. Indeed

$$\begin{aligned} a(y - y_p)'' + b(y - y_p)' + c(y - y_p) &= ay'' - ay_p'' + by' - by_p' + cy - cy_p \\ &= (ay'' + by' + cy) - (ay_p'' + by_p' + cy_p) \\ &= G(x) - G(x) = 0 \end{aligned}$$

This shows that every solution is of the form $y(x) = y_p(x) + y_c(x)$. It is easy to check that every function of this form is a solution. ■

We know from Section 17.1 how to solve the complementary equation. (Recall that the solution is $y_c = c_1y_1 + c_2y_2$, where y_1 and y_2 are linearly independent solutions of Equation 2.) Therefore Theorem 3 says that we know the general solution of the nonhomogeneous equation as soon as we know a particular solution y_p. There are two methods for finding a particular solution: The method of undetermined coefficients is straightforward but works only for a restricted class of functions G. The method of variation of parameters works for every function G but is usually more difficult to apply in practice.

The Method of Undetermined Coefficients

We first illustrate the method of undetermined coefficients for the equation

$$ay'' + by' + cy = G(x)$$

where $G(x)$ is a polynomial. It is reasonable to guess that there is a particular solution y_p that is a polynomial of the same degree as G because if y is a polynomial, then $ay'' + by' + cy$ is also a polynomial. We therefore substitute $y_p(x) =$ a polynomial (of the same degree as G) into the differential equation and determine the coefficients.

EXAMPLE 1 Solve the equation $y'' + y' - 2y = x^2$.

SOLUTION The auxiliary equation of $y'' + y' - 2y = 0$ is

$$r^2 + r - 2 = (r - 1)(r + 2) = 0$$

with roots $r = 1, -2$. So the solution of the complementary equation is

$$y_c = c_1e^x + c_2e^{-2x}$$

Since $G(x) = x^2$ is a polynomial of degree 2, we seek a particular solution of the form

$$y_p(x) = Ax^2 + Bx + C$$

Then $y_p' = 2Ax + B$ and $y_p'' = 2A$ so, substituting into the given differential equation, we have

$$(2A) + (2Ax + B) - 2(Ax^2 + Bx + C) = x^2$$

or

$$-2Ax^2 + (2A - 2B)x + (2A + B - 2C) = x^2$$

Figure 1 shows four solutions of the differential equation in Example 1 in terms of the particular solution y_p and the functions $f(x) = e^x$ and $g(x) = e^{-2x}$.

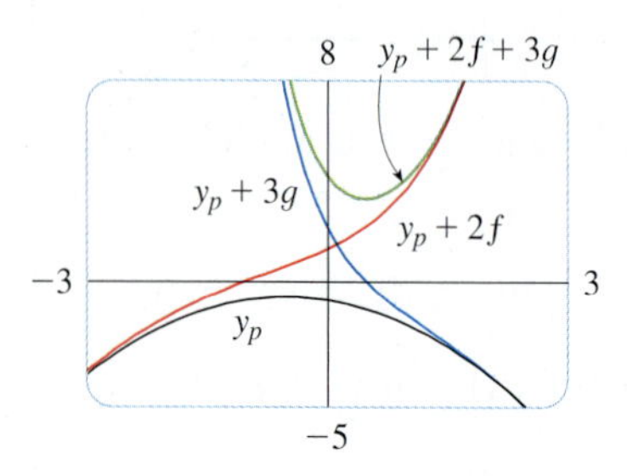

FIGURE 1

Polynomials are equal when their coefficients are equal. Thus

$$-2A = 1 \qquad 2A - 2B = 0 \qquad 2A + B - 2C = 0$$

The solution of this system of equations is

$$A = -\tfrac{1}{2} \qquad B = -\tfrac{1}{2} \qquad C = -\tfrac{3}{4}$$

A particular solution is therefore

$$y_p(x) = -\tfrac{1}{2}x^2 - \tfrac{1}{2}x - \tfrac{3}{4}$$

and, by Theorem 3, the general solution is

$$y = y_c + y_p = c_1 e^x + c_2 e^{-2x} - \tfrac{1}{2}x^2 - \tfrac{1}{2}x - \tfrac{3}{4}$$ ■

If $G(x)$ (the right side of Equation 1) is of the form Ce^{kx}, where C and k are constants, then we take as a trial solution a function of the same form, $y_p(x) = Ae^{kx}$, because the derivatives of e^{kx} are constant multiples of e^{kx}.

EXAMPLE 2 Solve $y'' + 4y = e^{3x}$.

Figure 2 shows solutions of the differential equation in Example 2 in terms of y_p and the functions $f(x) = \cos 2x$ and $g(x) = \sin 2x$. Notice that all solutions approach ∞ as $x \to \infty$ and all solutions (except y_p) resemble sine functions when x is negative.

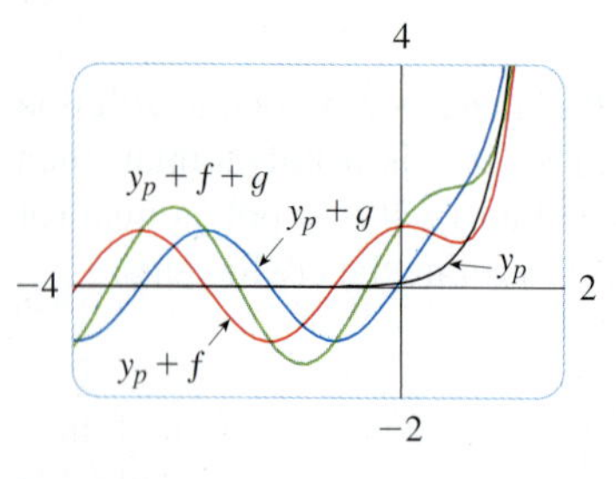

FIGURE 2

SOLUTION The auxiliary equation is $r^2 + 4 = 0$ with roots $\pm 2i$, so the solution of the complementary equation is

$$y_c(x) = c_1 \cos 2x + c_2 \sin 2x$$

For a particular solution we try $y_p(x) = Ae^{3x}$. Then $y_p' = 3Ae^{3x}$ and $y_p'' = 9Ae^{3x}$. Substituting into the differential equation, we have

$$9Ae^{3x} + 4(Ae^{3x}) = e^{3x}$$

so $13Ae^{3x} = e^{3x}$ and $A = \frac{1}{13}$. Thus a particular solution is

$$y_p(x) = \tfrac{1}{13}e^{3x}$$

and the general solution is

$$y(x) = c_1 \cos 2x + c_2 \sin 2x + \tfrac{1}{13}e^{3x}$$ ■

If $G(x)$ is either $C \cos kx$ or $C \sin kx$, then, because of the rules for differentiating the sine and cosine functions, we take as a trial particular solution a function of the form

$$y_p(x) = A \cos kx + B \sin kx$$

EXAMPLE 3 Solve $y'' + y' - 2y = \sin x$.

SOLUTION We try a particular solution

$$y_p(x) = A \cos x + B \sin x$$

Then $\quad y_p' = -A\sin x + B\cos x \qquad y_p'' = -A\cos x - B\sin x$

so substitution in the differential equation gives

$$(-A\cos x - B\sin x) + (-A\sin x + B\cos x) - 2(A\cos x + B\sin x) = \sin x$$

or

$$(-3A + B)\cos x + (-A - 3B)\sin x = \sin x$$

This is true if

$$-3A + B = 0 \qquad \text{and} \qquad -A - 3B = 1$$

The solution of this system is

$$A = -\tfrac{1}{10} \qquad B = -\tfrac{3}{10}$$

so a particular solution is

$$y_p(x) = -\tfrac{1}{10}\cos x - \tfrac{3}{10}\sin x$$

In Example 1 we determined that the solution of the complementary equation is $y_c = c_1e^x + c_2e^{-2x}$. Thus the general solution of the given equation is

$$y(x) = c_1e^x + c_2e^{-2x} - \tfrac{1}{10}(\cos x + 3\sin x)$$ ■

If $G(x)$ is a product of functions of the preceding types, then we take the trial solution to be a product of functions of the same type. For instance, in solving the differential equation

$$y'' + 2y' + 4y = x\cos 3x$$

we would try

$$y_p(x) = (Ax + B)\cos 3x + (Cx + D)\sin 3x$$

If $G(x)$ is a sum of functions of these types, we use the easily verified *principle of superposition*, which says that if y_{p_1} and y_{p_2} are solutions of

$$ay'' + by' + cy = G_1(x) \qquad ay'' + by' + cy = G_2(x)$$

respectively, then $y_{p_1} + y_{p_2}$ is a solution of

$$ay'' + by' + cy = G_1(x) + G_2(x)$$

EXAMPLE 4 Solve $y'' - 4y = xe^x + \cos 2x$.

SOLUTION The auxiliary equation is $r^2 - 4 = 0$ with roots ± 2, so the solution of the complementary equation is $y_c(x) = c_1e^{2x} + c_2e^{-2x}$. For the equation $y'' - 4y = xe^x$ we try

$$y_{p_1}(x) = (Ax + B)e^x$$

Then $y_{p_1}' = (Ax + A + B)e^x$, $y_{p_1}'' = (Ax + 2A + B)e^x$, so substitution in the equation gives

$$(Ax + 2A + B)e^x - 4(Ax + B)e^x = xe^x$$

or

$$(-3Ax + 2A - 3B)e^x = xe^x$$

Thus $-3A = 1$ and $2A - 3B = 0$, so $A = -\frac{1}{3}$, $B = -\frac{2}{9}$, and

$$y_{p_1}(x) = \left(-\tfrac{1}{3}x - \tfrac{2}{9}\right)e^x$$

For the equation $y'' - 4y = \cos 2x$, we try

$$y_{p_2}(x) = C\cos 2x + D\sin 2x$$

In Figure 3 we show the particular solution $y_p = y_{p_1} + y_{p_2}$ of the differential equation in Example 4. The other solutions are given in terms of $f(x) = e^{2x}$ and $g(x) = e^{-2x}$.

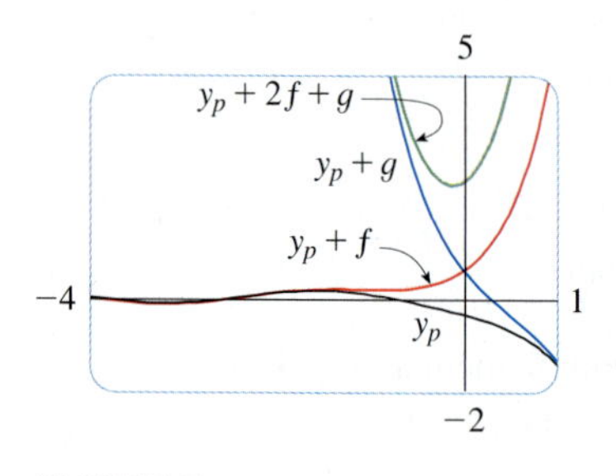

FIGURE 3

Substitution gives

$$-4C\cos 2x - 4D\sin 2x - 4(C\cos 2x + D\sin 2x) = \cos 2x$$

or

$$-8C\cos 2x - 8D\sin 2x = \cos 2x$$

Therefore $-8C = 1$, $-8D = 0$, and

$$y_{p_2}(x) = -\tfrac{1}{8}\cos 2x$$

By the superposition principle, the general solution is

$$y = y_c + y_{p_1} + y_{p_2} = c_1e^{2x} + c_2e^{-2x} - \left(\tfrac{1}{3}x + \tfrac{2}{9}\right)e^x - \tfrac{1}{8}\cos 2x$$ ■

Finally we note that the recommended trial solution y_p sometimes turns out to be a solution of the complementary equation and therefore can't be a solution of the nonhomogeneous equation. In such cases we multiply the recommended trial solution by x (or by x^2 if necessary) so that no term in $y_p(x)$ is a solution of the complementary equation.

EXAMPLE 5 Solve $y'' + y = \sin x$.

SOLUTION The auxiliary equation is $r^2 + 1 = 0$ with roots $\pm i$, so the solution of the complementary equation is

$$y_c(x) = c_1\cos x + c_2\sin x$$

Ordinarily, we would use the trial solution

$$y_p(x) = A\cos x + B\sin x$$

but we observe that it is a solution of the complementary equation, so instead we try

$$y_p(x) = Ax\cos x + Bx\sin x$$

Then

$$y_p'(x) = A\cos x - Ax\sin x + B\sin x + Bx\cos x$$

$$y_p''(x) = -2A\sin x - Ax\cos x + 2B\cos x - Bx\sin x$$

Substitution in the differential equation gives

$$y_p'' + y_p = -2A\sin x + 2B\cos x = \sin x$$

The graphs of four solutions of the differential equation in Example 5 are shown in Figure 4.

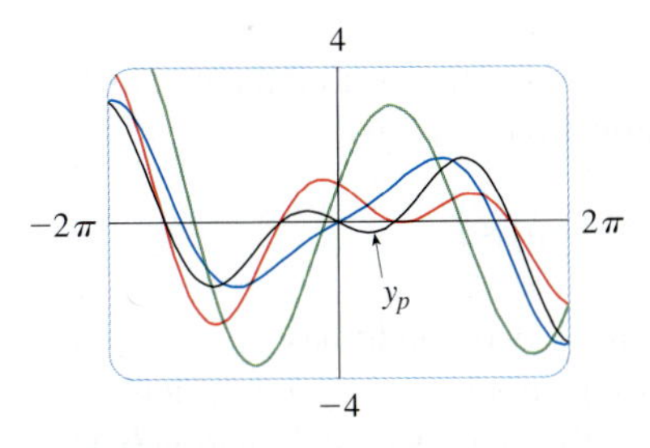

FIGURE 4

so $A = -\frac{1}{2}$, $B = 0$, and

$$y_p(x) = -\tfrac{1}{2}x \cos x$$

The general solution is

$$y(x) = c_1 \cos x + c_2 \sin x - \tfrac{1}{2}x \cos x$$

We summarize the method of undetermined coefficients as follows:

Summary of the Method of Undetermined Coefficients

1. If $G(x) = e^{kx}P(x)$, where P is a polynomial of degree n, then try $y_p(x) = e^{kx}Q(x)$, where $Q(x)$ is an nth-degree polynomial (whose coefficients are determined by substituting in the differential equation).

2. If $G(x) = e^{kx}P(x) \cos mx$ or $G(x) = e^{kx}P(x) \sin mx$, where P is an nth-degree polynomial, then try

$$y_p(x) = e^{kx}Q(x) \cos mx + e^{kx}R(x) \sin mx$$

where Q and R are nth-degree polynomials.

Modification: If any term of y_p is a solution of the complementary equation, multiply y_p by x (or by x^2 if necessary).

EXAMPLE 6 Determine the form of the trial solution for the differential equation $y'' - 4y' + 13y = e^{2x} \cos 3x$.

SOLUTION Here $G(x)$ has the form of part 2 of the summary, where $k = 2$, $m = 3$, and $P(x) = 1$. So, at first glance, the form of the trial solution would be

$$y_p(x) = e^{2x}(A \cos 3x + B \sin 3x)$$

But the auxiliary equation is $r^2 - 4r + 13 = 0$, with roots $r = 2 \pm 3i$, so the solution of the complementary equation is

$$y_c(x) = e^{2x}(c_1 \cos 3x + c_2 \sin 3x)$$

This means that we have to multiply the suggested trial solution by x. So, instead, we use

$$y_p(x) = xe^{2x}(A \cos 3x + B \sin 3x)$$

The Method of Variation of Parameters

Suppose we have already solved the homogeneous equation $ay'' + by' + cy = 0$ and written the solution as

4
$$y(x) = c_1y_1(x) + c_2y_2(x)$$

where y_1 and y_2 are linearly independent solutions. Let's replace the constants (or parameters) c_1 and c_2 in Equation 4 by arbitrary functions $u_1(x)$ and $u_2(x)$. We look for a particular solution of the nonhomogeneous equation $ay'' + by' + cy = G(x)$ of the form

5 $$y_p(x) = u_1(x)\, y_1(x) + u_2(x)\, y_2(x)$$

(This method is called **variation of parameters** because we have varied the parameters c_1 and c_2 to make them functions.) Differentiating Equation 5, we get

6 $$y_p' = (u_1' y_1 + u_2' y_2) + (u_1 y_1' + u_2 y_2')$$

Since u_1 and u_2 are arbitrary functions, we can impose two conditions on them. One condition is that y_p is a solution of the differential equation; we can choose the other condition so as to simplify our calculations. In view of the expression in Equation 6, let's impose the condition that

7 $$u_1' y_1 + u_2' y_2 = 0$$

Then $$y_p'' = u_1' y_1' + u_2' y_2' + u_1 y_1'' + u_2 y_2''$$

Substituting in the differential equation, we get

$$a(u_1' y_1' + u_2' y_2' + u_1 y_1'' + u_2 y_2'') + b(u_1 y_1' + u_2 y_2') + c(u_1 y_1 + u_2 y_2) = G$$

or

8 $$u_1(ay_1'' + by_1' + cy_1) + u_2(ay_2'' + by_2' + cy_2) + a(u_1' y_1' + u_2' y_2') = G$$

But y_1 and y_2 are solutions of the complementary equation, so

$$ay_1'' + by_1' + cy_1 = 0 \qquad \text{and} \qquad ay_2'' + by_2' + cy_2 = 0$$

and Equation 8 simplifies to

9 $$a(u_1' y_1' + u_2' y_2') = G$$

Equations 7 and 9 form a system of two equations in the unknown functions u_1' and u_2'. After solving this system we may be able to integrate to find u_1 and u_2 and then the particular solution is given by Equation 5.

EXAMPLE 7 Solve the equation $y'' + y = \tan x$, $0 < x < \pi/2$.

SOLUTION The auxiliary equation is $r^2 + 1 = 0$ with roots $\pm i$, so the solution of $y'' + y = 0$ is $y(x) = c_1 \sin x + c_2 \cos x$. Using variation of parameters, we seek a solution of the form

$$y_p(x) = u_1(x) \sin x + u_2(x) \cos x$$

Then $$y_p' = (u_1' \sin x + u_2' \cos x) + (u_1 \cos x - u_2 \sin x)$$

Set

10 $$u_1' \sin x + u_2' \cos x = 0$$

Then $$y_p'' = u_1' \cos x - u_2' \sin x - u_1 \sin x - u_2 \cos x$$

For y_p to be a solution we must have

$$y_p'' + y_p = u_1' \cos x - u_2' \sin x = \tan x \tag{11}$$

Solving Equations 10 and 11, we get

$$u_1'(\sin^2 x + \cos^2 x) = \cos x \tan x$$

$$u_1' = \sin x \qquad u_1(x) = -\cos x$$

(We seek a particular solution, so we don't need a constant of integration here.) Then, from Equation 10, we obtain

$$u_2' = -\frac{\sin x}{\cos x} u_1' = -\frac{\sin^2 x}{\cos x} = \frac{\cos^2 x - 1}{\cos x} = \cos x - \sec x$$

So $$u_2(x) = \sin x - \ln(\sec x + \tan x)$$

(Note that $\sec x + \tan x > 0$ for $0 < x < \pi/2$.) Therefore

$$y_p(x) = -\cos x \sin x + [\sin x - \ln(\sec x + \tan x)] \cos x$$
$$= -\cos x \ln(\sec x + \tan x)$$

and the general solution is

$$y(x) = c_1 \sin x + c_2 \cos x - \cos x \ln(\sec x + \tan x)$$

Figure 5 shows four solutions of the differential equation in Example 7.

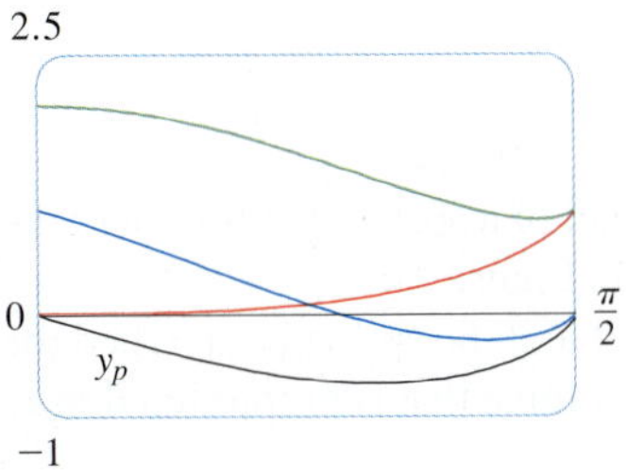

2.5
0
$\frac{\pi}{2}$
y_p
−1

FIGURE 5

17.2 EXERCISES

1–10 Solve the differential equation or initial-value problem using the method of undetermined coefficients.

1. $y'' + 2y' - 8y = 1 - 2x^2$
2. $y'' - 3y' = \sin 2x$
3. $9y'' + y = e^{2x}$
4. $y'' - 2y' + 2y = x + e^x$
5. $y'' - 4y' + 5y = e^{-x}$
6. $y'' - 4y' + 4y = x - \sin x$
7. $y'' - 2y' + 5y = \sin x, \quad y(0) = 1, \quad y'(0) = 1$
8. $y'' - y = xe^{2x}, \quad y(0) = 0, \quad y'(0) = 1$
9. $y'' - y' = xe^x, \quad y(0) = 2, \quad y'(0) = 1$
10. $y'' + y' - 2y = x + \sin 2x, \quad y(0) = 1, \quad y'(0) = 0$

11–12 Graph the particular solution and several other solutions. What characteristics do these solutions have in common?

11. $y'' + 3y' + 2y = \cos x$
12. $y'' + 4y = e^{-x}$

13–18 Write a trial solution for the method of undetermined coefficients. Do not determine the coefficients.

13. $y'' - y' - 2y = xe^x \cos x$
14. $y'' + 4y = \cos 4x + \cos 2x$
15. $y'' - 3y' + 2y = e^x + \sin x$
16. $y'' + 3y' - 4y = (x^3 + x)e^x$
17. $y'' + 2y' + 10y = x^2 e^{-x} \cos 3x$
18. $y'' + 4y = e^{3x} + x \sin 2x$

19–22 Solve the differential equation using (a) undetermined coefficients and (b) variation of parameters.

19. $4y'' + y = \cos x$

20. $y'' - 2y' - 3y = x + 2$

21. $y'' - 2y' + y = e^{2x}$

22. $y'' - y' = e^x$

23–28 Solve the differential equation using the method of variation of parameters.

23. $y'' + y = \sec^2 x, \ 0 < x < \pi/2$

24. $y'' + y = \sec^3 x, \ 0 < x < \pi/2$

25. $y'' - 3y' + 2y = \dfrac{1}{1 + e^{-x}}$

26. $y'' + 3y' + 2y = \sin(e^x)$

27. $y'' - 2y' + y = \dfrac{e^x}{1 + x^2}$

28. $y'' + 4y' + 4y = \dfrac{e^{-2x}}{x^3}$

17.3 Applications of Second-Order Differential Equations

Second-order linear differential equations have a variety of applications in science and engineering. In this section we explore two of them: the vibration of springs and electric circuits.

Vibrating Springs

We consider the motion of an object with mass m at the end of a spring that is either vertical (as in Figure 1) or horizontal on a level surface (as in Figure 2).

In Section 5.4 we discussed Hooke's Law, which says that if the spring is stretched (or compressed) x units from its natural length, then it exerts a force that is proportional to x:

$$\text{restoring force} = -kx$$

where k is a positive constant (called the **spring constant**). If we ignore any external resisting forces (due to air resistance or friction) then, by Newton's Second Law (force equals mass times acceleration), we have

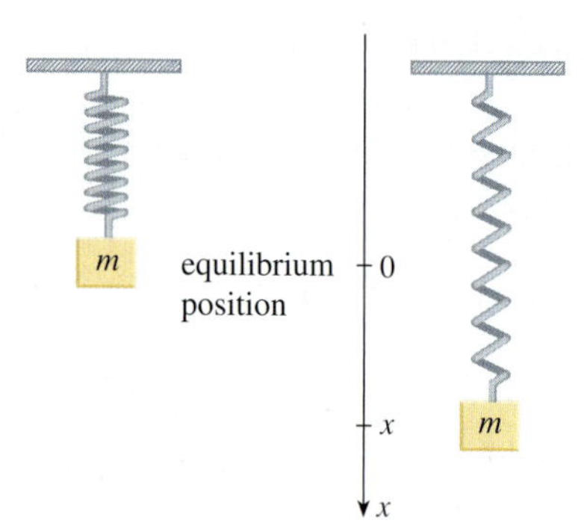

FIGURE 1

1
$$m\frac{d^2x}{dt^2} = -kx \qquad \text{or} \qquad m\frac{d^2x}{dt^2} + kx = 0$$

This is a second-order linear differential equation. Its auxiliary equation is $mr^2 + k = 0$ with roots $r = \pm\omega i$, where $\omega = \sqrt{k/m}$. Thus the general solution is

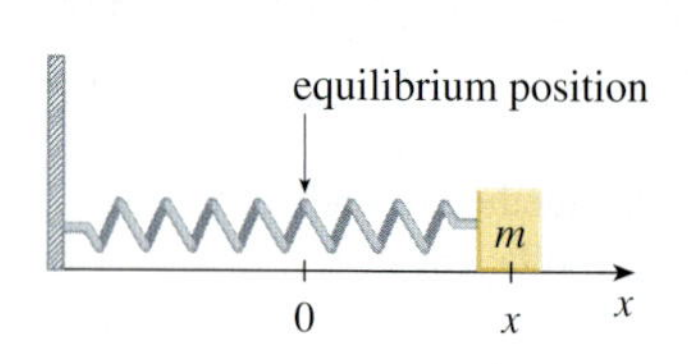

FIGURE 2

$$x(t) = c_1 \cos \omega t + c_2 \sin \omega t$$

which can also be written as

$$x(t) = A \cos(\omega t + \delta)$$

where

$$\omega = \sqrt{k/m} \quad \text{(frequency)}$$

$$A = \sqrt{c_1^2 + c_2^2} \quad \text{(amplitude)}$$

$$\cos \delta = \frac{c_1}{A} \qquad \sin \delta = -\frac{c_2}{A} \quad (\delta \text{ is the phase angle})$$

(See Exercise 17.) This type of motion is called **simple harmonic motion**.

EXAMPLE 1 A spring with a mass of 2 kg has natural length 0.5 m. A force of 25.6 N is required to maintain it stretched to a length of 0.7 m. If the spring is stretched to a length of 0.7 m and then released with initial velocity 0, find the position of the mass at any time t.

SOLUTION From Hooke's Law, the force required to stretch the spring is

$$k(0.2) = 25.6$$

so $k = 25.6/0.2 = 128$. Using this value of the spring constant k, together with $m = 2$ in Equation 1, we have

$$2\frac{d^2x}{dt^2} + 128x = 0$$

As in the earlier general discussion, the solution of this equation is

$$x(t) = c_1 \cos 8t + c_2 \sin 8t \tag{2}$$

We are given the initial condition that $x(0) = 0.2$. But, from Equation 2, $x(0) = c_1$. Therefore $c_1 = 0.2$. Differentiating Equation 2, we get

$$x'(t) = -8c_1 \sin 8t + 8c_2 \cos 8t$$

Since the initial velocity is given as $x'(0) = 0$, we have $c_2 = 0$ and so the solution is

$$x(t) = 0.2 \cos 8t$$

Damped Vibrations

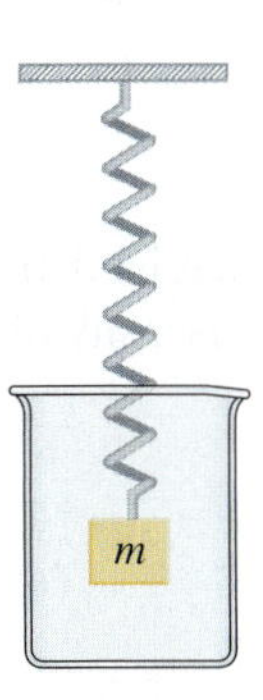

FIGURE 3

We next consider the motion of a spring that is subject to a frictional force (in the case of the horizontal spring of Figure 2) or a damping force (in the case where a vertical spring moves through a fluid as in Figure 3). An example is the damping force supplied by a shock absorber in a car or a bicycle.

We assume that the damping force is proportional to the velocity of the mass and acts in the direction opposite to the motion. (This has been confirmed, at least approximately, by some physical experiments.) Thus

$$\text{damping force} = -c\frac{dx}{dt}$$

where c is a positive constant, called the **damping constant**. Thus, in this case, Newton's Second Law gives

$$m\frac{d^2x}{dt^2} = \text{restoring force} + \text{damping force} = -kx - c\frac{dx}{dt}$$

or

$$m\frac{d^2x}{dt^2} + c\frac{dx}{dt} + kx = 0 \tag{3}$$

Equation 3 is a second-order linear differential equation and its auxiliary equation is $mr^2 + cr + k = 0$. The roots are

$$\boxed{4} \qquad r_1 = \frac{-c + \sqrt{c^2 - 4mk}}{2m} \qquad r_2 = \frac{-c - \sqrt{c^2 - 4mk}}{2m}$$

According to Section 17.1 we need to discuss three cases.

CASE I $c^2 - 4mk > 0$ **(overdamping)**
In this case r_1 and r_2 are distinct real roots and

$$x = c_1 e^{r_1 t} + c_2 e^{r_2 t}$$

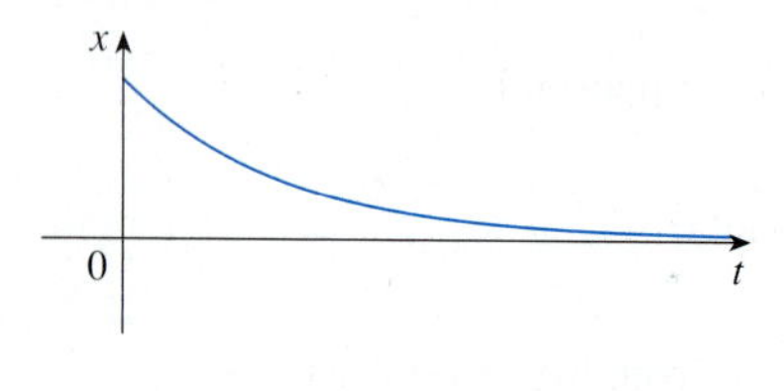

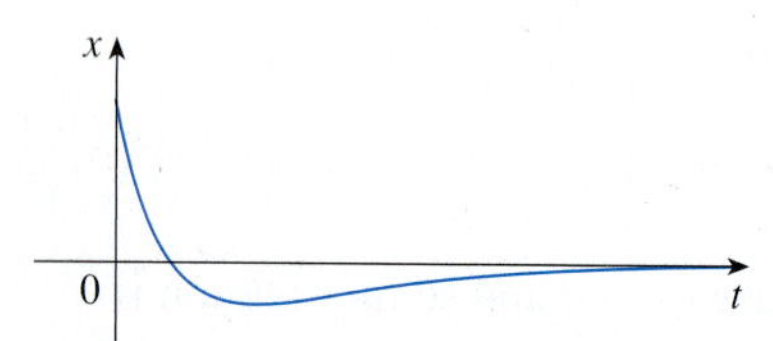

FIGURE 4
Overdamping

Since c, m, and k are all positive, we have $\sqrt{c^2 - 4mk} < c$, so the roots r_1 and r_2 given by Equations 4 must both be negative. This shows that $x \to 0$ as $t \to \infty$. Typical graphs of x as a function of t are shown in Figure 4. Notice that oscillations do not occur. (It's possible for the mass to pass through the equilibrium position once, but only once.) This is because $c^2 > 4mk$ means that there is a strong damping force (high-viscosity oil or grease) compared with a weak spring or small mass.

CASE II $c^2 - 4mk = 0$ **(critical damping)**
This case corresponds to equal roots

$$r_1 = r_2 = -\frac{c}{2m}$$

and the solution is given by

$$x = (c_1 + c_2 t)e^{-(c/2m)t}$$

It is similar to Case I, and typical graphs resemble those in Figure 4 (see Exercise 12), but the damping is just sufficient to suppress vibrations. Any decrease in the viscosity of the fluid leads to the vibrations of the following case.

CASE III $c^2 - 4mk < 0$ **(underdamping)**
Here the roots are complex:

$$\left.\begin{matrix} r_1 \\ r_2 \end{matrix}\right\} = -\frac{c}{2m} \pm \omega i$$

where

$$\omega = \frac{\sqrt{4mk - c^2}}{2m}$$

The solution is given by

$$x = e^{-(c/2m)t}(c_1 \cos \omega t + c_2 \sin \omega t)$$

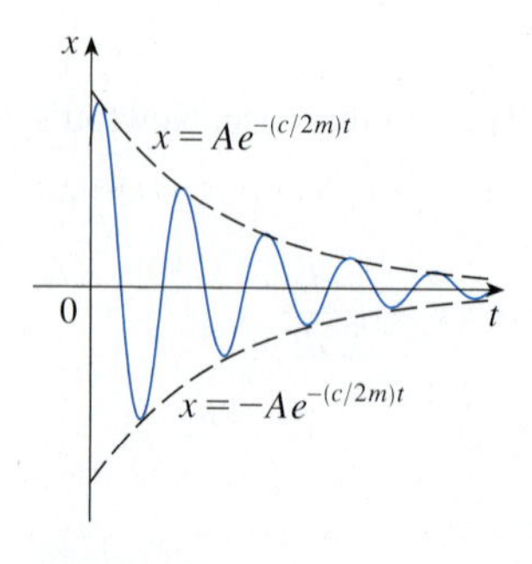

FIGURE 5
Underdamping

We see that there are oscillations that are damped by the factor $e^{-(c/2m)t}$. Since $c > 0$ and $m > 0$, we have $-(c/2m) < 0$ so $e^{-(c/2m)t} \to 0$ as $t \to \infty$. This implies that $x \to 0$ as $t \to \infty$; that is, the motion decays to 0 as time increases. A typical graph is shown in Figure 5.

EXAMPLE 2 Suppose that the spring of Example 1 is immersed in a fluid with damping constant $c = 40$. Find the position of the mass at any time t if it starts from the equilibrium position and is given a push to start it with an initial velocity of 0.6 m/s.

SOLUTION From Example 1, the mass is $m = 2$ and the spring constant is $k = 128$, so the differential equation (3) becomes

$$2\frac{d^2x}{dt^2} + 40\frac{dx}{dt} + 128x = 0$$

or

$$\frac{d^2x}{dt^2} + 20\frac{dx}{dt} + 64x = 0$$

The auxiliary equation is $r^2 + 20r + 64 = (r + 4)(r + 16) = 0$ with roots -4 and -16, so the motion is overdamped and the solution is

Figure 6 shows the graph of the position function for the overdamped motion in Example 2.

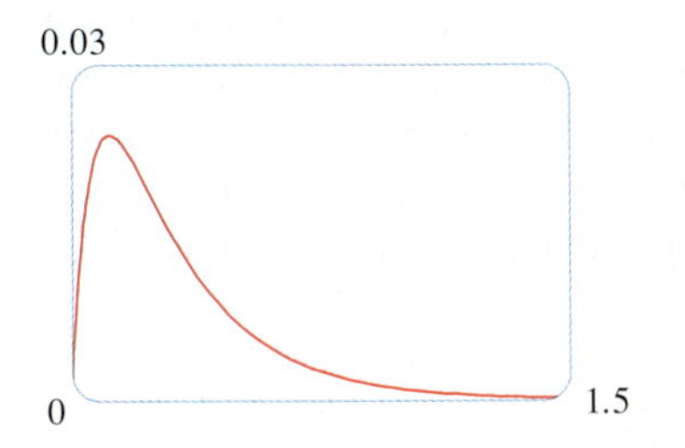

FIGURE 6

$$x(t) = c_1e^{-4t} + c_2e^{-16t}$$

We are given that $x(0) = 0$, so $c_1 + c_2 = 0$. Differentiating, we get

$$x'(t) = -4c_1e^{-4t} - 16c_2e^{-16t}$$

so

$$x'(0) = -4c_1 - 16c_2 = 0.6$$

Since $c_2 = -c_1$, this gives $12c_1 = 0.6$ or $c_1 = 0.05$. Therefore

$$x = 0.05(e^{-4t} - e^{-16t})$$

Forced Vibrations

Suppose that, in addition to the restoring force and the damping force, the motion of the spring is affected by an external force $F(t)$. Then Newton's Second Law gives

$$m\frac{d^2x}{dt^2} = \text{restoring force} + \text{damping force} + \text{external force}$$

$$= -kx - c\frac{dx}{dt} + F(t)$$

Thus, instead of the homogeneous equation (3), the motion of the spring is now governed by the following nonhomogeneous differential equation:

5
$$m\frac{d^2x}{dt^2} + c\frac{dx}{dt} + kx = F(t)$$

The motion of the spring can be determined by the methods of Section 17.2.

A commonly occurring type of external force is a periodic force function

$$F(t) = F_0 \cos \omega_0 t \qquad \text{where} \qquad \omega_0 \neq \omega = \sqrt{k/m}$$

In this case, and in the absence of a damping force ($c = 0$), you are asked in Exercise 9 to use the method of undetermined coefficients to show that

$$\boxed{6} \qquad x(t) = c_1 \cos \omega t + c_2 \sin \omega t + \frac{F_0}{m(\omega^2 - \omega_0^2)} \cos \omega_0 t$$

If $\omega_0 = \omega$, then the applied frequency reinforces the natural frequency and the result is vibrations of large amplitude. This is the phenomenon of **resonance** (see Exercise 10).

Electric Circuits

In Sections 9.3 and 9.5 we were able to use first-order separable and linear equations to analyze electric circuits that contain a resistor and inductor (see Figure 9.3.5 or Figure 9.5.4) or a resistor and capacitor (see Exercise 9.5.29). Now that we know how to solve second-order linear equations, we are in a position to analyze the circuit shown in Figure 7. It contains an electromotive force E (supplied by a battery or generator), a resistor R, an inductor L, and a capacitor C, in series. If the charge on the capacitor at time t is $Q = Q(t)$, then the current is the rate of change of Q with respect to t: $I = dQ/dt$. As in Section 9.5, it is known from physics that the voltage drops across the resistor, inductor, and capacitor are

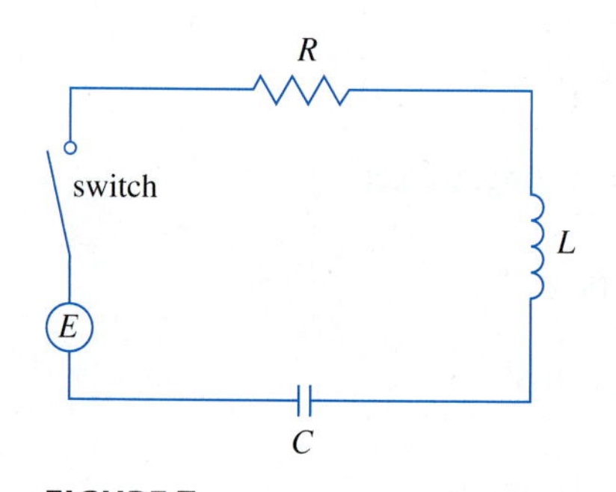

FIGURE 7

$$RI \qquad L\frac{dI}{dt} \qquad \frac{Q}{C}$$

respectively. Kirchhoff's voltage law says that the sum of these voltage drops is equal to the supplied voltage:

$$L\frac{dI}{dt} + RI + \frac{Q}{C} = E(t)$$

Since $I = dQ/dt$, this equation becomes

$$\boxed{7} \qquad \boxed{L\frac{d^2Q}{dt^2} + R\frac{dQ}{dt} + \frac{1}{C}Q = E(t)}$$

which is a second-order linear differential equation with constant coefficients. If the charge Q_0 and the current I_0 are known at time 0, then we have the initial conditions

$$Q(0) = Q_0 \qquad Q'(0) = I(0) = I_0$$

and the initial-value problem can be solved by the methods of Section 17.2.

A differential equation for the current can be obtained by differentiating Equation 7 with respect to t and remembering that $I = dQ/dt$:

$$L\frac{d^2I}{dt^2} + R\frac{dI}{dt} + \frac{1}{C}I = E'(t)$$

EXAMPLE 3 Find the charge and current at time t in the circuit of Figure 7 if $R = 40\ \Omega$, $L = 1$ H, $C = 16 \times 10^{-4}$ F, $E(t) = 100\cos 10t$, and the initial charge and current are both 0.

SOLUTION With the given values of L, R, C, and $E(t)$, Equation 7 becomes

$$\frac{d^2Q}{dt^2} + 40\frac{dQ}{dt} + 625Q = 100\cos 10t \tag{8}$$

The auxiliary equation is $r^2 + 40r + 625 = 0$ with roots

$$r = \frac{-40 \pm \sqrt{-900}}{2} = -20 \pm 15i$$

so the solution of the complementary equation is

$$Q_c(t) = e^{-20t}(c_1\cos 15t + c_2\sin 15t)$$

For the method of undetermined coefficients we try the particular solution

$$Q_p(t) = A\cos 10t + B\sin 10t$$

Then $$Q_p'(t) = -10A\sin 10t + 10B\cos 10t$$

$$Q_p''(t) = -100A\cos 10t - 100B\sin 10t$$

Substituting into Equation 8, we have

$$(-100A\cos 10t - 100B\sin 10t) + 40(-10A\sin 10t + 10B\cos 10t)$$
$$+ 625(A\cos 10t + B\sin 10t) = 100\cos 10t$$

or $$(525A + 400B)\cos 10t + (-400A + 525B)\sin 10t = 100\cos 10t$$

Equating coefficients, we have

$$525A + 400B = 100 \qquad \text{or} \qquad 21A + 16B = 4$$
$$-400A + 525B = 0 \qquad \text{or} \qquad -16A + 21B = 0$$

The solution of this system is $A = \frac{84}{697}$ and $B = \frac{64}{697}$, so a particular solution is

$$Q_p(t) = \tfrac{1}{697}(84\cos 10t + 64\sin 10t)$$

and the general solution is

$$Q(t) = Q_c(t) + Q_p(t)$$

$$= e^{-20t}(c_1 \cos 15t + c_2 \sin 15t) + \tfrac{4}{697}(21 \cos 10t + 16 \sin 10t)$$

Imposing the initial condition $Q(0) = 0$, we get

$$Q(0) = c_1 + \tfrac{84}{697} = 0 \qquad c_1 = -\tfrac{84}{697}$$

To impose the other initial condition, we first differentiate to find the current:

$$I = \frac{dQ}{dt} = e^{-20t}[(-20c_1 + 15c_2)\cos 15t + (-15c_1 - 20c_2)\sin 15t]$$

$$+ \tfrac{40}{697}(-21 \sin 10t + 16 \cos 10t)$$

$$I(0) = -20c_1 + 15c_2 + \tfrac{640}{697} = 0 \qquad c_2 = -\tfrac{464}{2091}$$

Thus the formula for the charge is

$$Q(t) = \frac{4}{697}\left[\frac{e^{-20t}}{3}(-63 \cos 15t - 116 \sin 15t) + (21 \cos 10t + 16 \sin 10t)\right]$$

and the expression for the current is

$$I(t) = \tfrac{1}{2091}[e^{-20t}(-1920 \cos 15t + 13{,}060 \sin 15t) + 120(-21 \sin 10t + 16 \cos 10t)]$$

■

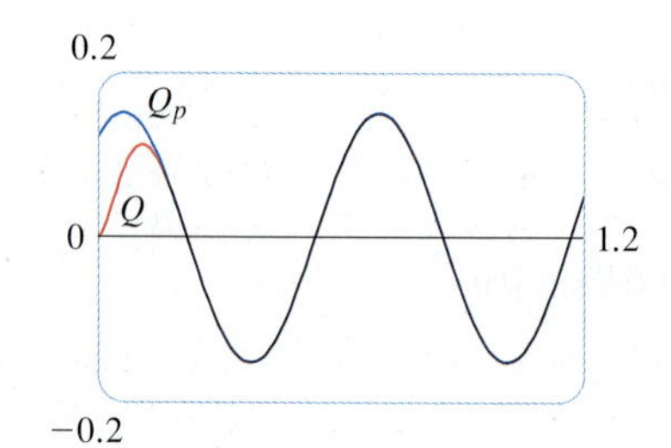

FIGURE 8

NOTE 1 In Example 3 the solution for $Q(t)$ consists of two parts. Since $e^{-20t} \to 0$ as $t \to \infty$ and both $\cos 15t$ and $\sin 15t$ are bounded functions,

$$Q_c(t) = \tfrac{4}{2091}e^{-20t}(-63 \cos 15t - 116 \sin 15t) \to 0 \qquad \text{as } t \to \infty$$

So, for large values of t,

$$Q(t) \approx Q_p(t) = \tfrac{4}{697}(21 \cos 10t + 16 \sin 10t)$$

and, for this reason, $Q_p(t)$ is called the **steady state solution**. Figure 8 shows how the graph of the steady state solution compares with the graph of Q in this case.

$$\boxed{5} \quad m\frac{d^2x}{dt^2} + c\frac{dx}{dt} + kx = F(t)$$

$$\boxed{7} \quad L\frac{d^2Q}{dt^2} + R\frac{dQ}{dt} + \frac{1}{C}Q = E(t)$$

NOTE 2 Comparing Equations 5 and 7, we see that mathematically they are identical. This suggests the analogies given in the following chart between physical situations that, at first glance, are very different.

Spring system		Electric circuit	
x	displacement	Q	charge
dx/dt	velocity	$I = dQ/dt$	current
m	mass	L	inductance
c	damping constant	R	resistance
k	spring constant	$1/C$	elastance
$F(t)$	external force	$E(t)$	electromotive force

We can also transfer other ideas from one situation to the other. For instance, the steady state solution discussed in Note 1 makes sense in the spring system. And the phenomenon of resonance in the spring system can be usefully carried over to electric circuits as electrical resonance.

17.3 EXERCISES

1. A spring has natural length 0.75 m and a 5-kg mass. A force of 25 N is needed to keep the spring stretched to a length of 1 m. If the spring is stretched to a length of 1.1 m and then released with velocity 0, find the position of the mass after t seconds.

2. A spring with an 8-kg mass is kept stretched 0.4 m beyond its natural length by a force of 32 N. The spring starts at its equilibrium position and is given an initial velocity of 1 m/s. Find the position of the mass at any time t.

3. A spring with a mass of 2 kg has damping constant 14, and a force of 6 N is required to keep the spring stretched 0.5 m beyond its natural length. The spring is stretched 1 m beyond its natural length and then released with zero velocity. Find the position of the mass at any time t.

4. A force of 13 N is needed to keep a spring with a 2-kg mass stretched 0.25 m beyond its natural length. The damping constant of the spring is $c = 8$.

(a) If the mass starts at the equilibrium position with a velocity of 0.5 m/s, find its position at time t.

(b) Graph the position function of the mass.

5. For the spring in Exercise 3, find the mass that would produce critical damping.

6. For the spring in Exercise 4, find the damping constant that would produce critical damping.

7. A spring has a mass of 1 kg and its spring constant is $k = 100$. The spring is released at a point 0.1 m above its equilibrium position. Graph the position function for the following values of the damping constant c: 10, 15, 20, 25, 30. What type of damping occurs in each case?

8. A spring has a mass of 1 kg and its damping constant is $c = 10$. The spring starts from its equilibrium position with a velocity of 1 m/s. Graph the position function for the following values of the spring constant k: 10, 20, 25, 30, 40. What type of damping occurs in each case?

9. Suppose a spring has mass m and spring constant k and let $\omega = \sqrt{k/m}$. Suppose that the damping constant is so small that the damping force is negligible. If an external force $F(t) = F_0 \cos \omega_0 t$ is applied, where $\omega_0 \neq \omega$, use the method of undetermined coefficients to show that the motion of the mass is described by Equation 6.

10. As in Exercise 9, consider a spring with mass m, spring constant k, and damping constant $c = 0$, and let $\omega = \sqrt{k/m}$. If an external force $F(t) = F_0 \cos \omega t$ is applied (the applied frequency equals the natural frequency), use the method of undetermined coefficients to show that the motion of the mass is given by

$$x(t) = c_1 \cos \omega t + c_2 \sin \omega t + \frac{F_0}{2m\omega} t \sin \omega t$$

11. Show that if $\omega_0 \neq \omega$, but ω/ω_0 is a rational number, then the motion described by Equation 6 is periodic.

12. Consider a spring subject to a frictional or damping force.

(a) In the critically damped case, the motion is given by $x = c_1 e^{rt} + c_2 t e^{rt}$. Show that the graph of x crosses the t-axis whenever c_1 and c_2 have opposite signs.

(b) In the overdamped case, the motion is given by $x = c_1 e^{r_1 t} + c_2 e^{r_2 t}$, where $r_1 > r_2$. Determine a condition on the relative magnitudes of c_1 and c_2 under which the graph of x crosses the t-axis at a positive value of t.

13. A series circuit consists of a resistor with $R = 20\ \Omega$, an inductor with $L = 1$ H, a capacitor with $C = 0.002$ F, and a 12-V battery. If the initial charge and current are both 0, find the charge and current at time t.

14. A series circuit contains a resistor with $R = 24\ \Omega$, an inductor with $L = 2$ H, a capacitor with $C = 0.005$ F, and a 12-V battery. The initial charge is $Q = 0.001$ C and the initial current is 0.

(a) Find the charge and current at time t.

(b) Graph the charge and current functions.

15. The battery in Exercise 13 is replaced by a generator producing a voltage of $E(t) = 12 \sin 10t$. Find the charge at time t.

16. The battery in Exercise 14 is replaced by a generator producing a voltage of $E(t) = 12 \sin 10t$.

(a) Find the charge at time t.

(b) Graph the charge function.

17. Verify that the solution to Equation 1 can be written in the form $x(t) = A \cos(\omega t + \delta)$.

18. The figure shows a pendulum with length L and the angle θ from the vertical to the pendulum. It can be shown that θ, as a function of time, satisfies the nonlinear differential equation

$$\frac{d^2\theta}{dt^2} + \frac{g}{L}\sin\theta = 0$$

where g is the acceleration due to gravity. For small values of θ we can use the linear approximation $\sin\theta \approx \theta$ and then the differential equation becomes linear.

(a) Find the equation of motion of a pendulum with length 1 m if θ is initially 0.2 rad and the initial angular velocity is $d\theta/dt = 1$ rad/s.

(b) What is the maximum angle from the vertical?

(c) What is the period of the pendulum (that is, the time to complete one back-and-forth swing)?

(d) When will the pendulum first be vertical?

(e) What is the angular velocity when the pendulum is vertical?

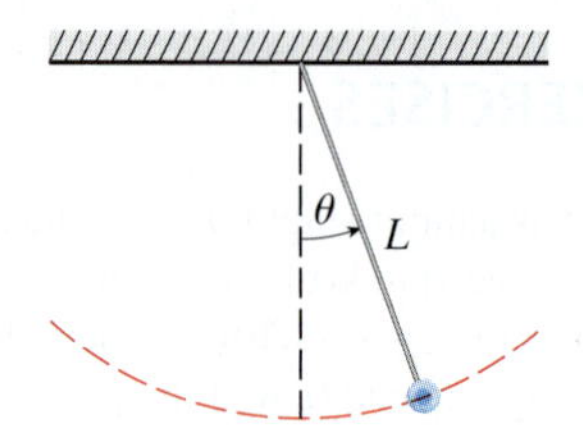

17.4 Series Solutions

Many differential equations can't be solved explicitly in terms of finite combinations of simple familiar functions. This is true even for a simple-looking equation like

$$\boxed{1} \qquad y'' - 2xy' + y = 0$$

But it is important to be able to solve equations such as Equation 1 because they arise from physical problems and, in particular, in connection with the Schrödinger equation in quantum mechanics. In such a case we use the method of power series; that is, we look for a solution of the form

$$y = f(x) = \sum_{n=0}^{\infty} c_n x^n = c_0 + c_1 x + c_2 x^2 + c_3 x^3 + \cdots$$

The method is to substitute this expression into the differential equation and determine the values of the coefficients $c_0, c_1, c_2, \ldots$. This technique resembles the method of undetermined coefficients discussed in Section 17.2.

Before using power series to solve Equation 1, we illustrate the method on the simpler equation $y'' + y = 0$ in Example 1. It's true that we already know how to solve this equation by the techniques of Section 17.1, but it's easier to understand the power series method when it is applied to this simpler equation.

EXAMPLE 1 Use power series to solve the equation $y'' + y = 0$.

SOLUTION We assume there is a solution of the form

$$\boxed{2} \qquad y = c_0 + c_1 x + c_2 x^2 + c_3 x^3 + \cdots = \sum_{n=0}^{\infty} c_n x^n$$

We can differentiate power series term by term, so

$$y' = c_1 + 2c_2 x + 3c_3 x^2 + \cdots = \sum_{n=1}^{\infty} n c_n x^{n-1}$$

$$\boxed{3} \qquad y'' = 2c_2 + 2 \cdot 3c_3 x + \cdots = \sum_{n=2}^{\infty} n(n-1) c_n x^{n-2}$$

In order to compare the expressions for y and y'' more easily, we rewrite y'' as follows:

By writing out the first few terms of (4), you can see that it is the same as (3). To obtain (4), we replaced n by $n + 2$ and began the summation at 0 instead of 2.

$$y'' = \sum_{n=0}^{\infty} (n + 2)(n + 1)c_{n+2}x^n \tag{4}$$

Substituting the expressions in Equations 2 and 4 into the differential equation, we obtain

$$\sum_{n=0}^{\infty} (n + 2)(n + 1)c_{n+2}x^n + \sum_{n=0}^{\infty} c_n x^n = 0$$

or

$$\sum_{n=0}^{\infty} [(n + 2)(n + 1)c_{n+2} + c_n]x^n = 0 \tag{5}$$

If two power series are equal, then the corresponding coefficients must be equal. Therefore the coefficients of x^n in Equation 5 must be 0:

$$(n + 2)(n + 1)c_{n+2} + c_n = 0$$

$$c_{n+2} = -\frac{c_n}{(n + 1)(n + 2)} \qquad n = 0, 1, 2, 3, \ldots \tag{6}$$

Equation 6 is called a *recursion relation.* If c_0 and c_1 are known, this equation allows us to determine the remaining coefficients recursively by putting $n = 0, 1, 2, 3, \ldots$ in succession.

Put $n = 0$: $\quad c_2 = -\dfrac{c_0}{1 \cdot 2}$

Put $n = 1$: $\quad c_3 = -\dfrac{c_1}{2 \cdot 3}$

Put $n = 2$: $\quad c_4 = -\dfrac{c_2}{3 \cdot 4} = \dfrac{c_0}{1 \cdot 2 \cdot 3 \cdot 4} = \dfrac{c_0}{4!}$

Put $n = 3$: $\quad c_5 = -\dfrac{c_3}{4 \cdot 5} = \dfrac{c_1}{2 \cdot 3 \cdot 4 \cdot 5} = \dfrac{c_1}{5!}$

Put $n = 4$: $\quad c_6 = -\dfrac{c_4}{5 \cdot 6} = -\dfrac{c_0}{4!\,5 \cdot 6} = -\dfrac{c_0}{6!}$

Put $n = 5$: $\quad c_7 = -\dfrac{c_5}{6 \cdot 7} = -\dfrac{c_1}{5!\,6 \cdot 7} = -\dfrac{c_1}{7!}$

By now we see the pattern:

For the even coefficients, $c_{2n} = (-1)^n \dfrac{c_0}{(2n)!}$

For the odd coefficients, $c_{2n+1} = (-1)^n \dfrac{c_1}{(2n + 1)!}$

Putting these values back into Equation 2, we write the solution as

$$y = c_0 + c_1 x + c_2 x^2 + c_3 x^3 + c_4 x^4 + c_5 x^5 + \cdots$$

$$= c_0\left(1 - \frac{x^2}{2!} + \frac{x^4}{4!} - \frac{x^6}{6!} + \cdots + (-1)^n \frac{x^{2n}}{(2n)!} + \cdots\right)$$

$$+ c_1\left(x - \frac{x^3}{3!} + \frac{x^5}{5!} - \frac{x^7}{7!} + \cdots + (-1)^n \frac{x^{2n+1}}{(2n+1)!} + \cdots\right)$$

$$= c_0 \sum_{n=0}^{\infty} (-1)^n \frac{x^{2n}}{(2n)!} + c_1 \sum_{n=0}^{\infty} (-1)^n \frac{x^{2n+1}}{(2n+1)!}$$

Notice that there are two arbitrary constants, c_0 and c_1. ■

NOTE 1 We recognize the series obtained in Example 1 as being the Maclaurin series for $\cos x$ and $\sin x$. (See Equations 11.10.16 and 11.10.15.) Therefore we could write the solution as

$$y(x) = c_0 \cos x + c_1 \sin x$$

But we are not usually able to express power series solutions of differential equations in terms of known functions.

EXAMPLE 2 Solve $y'' - 2xy' + y = 0$.

SOLUTION We assume there is a solution of the form

$$y = \sum_{n=0}^{\infty} c_n x^n$$

Then $$y' = \sum_{n=1}^{\infty} n c_n x^{n-1}$$

and $$y'' = \sum_{n=2}^{\infty} n(n-1) c_n x^{n-2} = \sum_{n=0}^{\infty} (n+2)(n+1) c_{n+2} x^n$$

as in Example 1. Substituting in the differential equation, we get

$$\sum_{n=0}^{\infty} (n+2)(n+1) c_{n+2} x^n - 2x \sum_{n=1}^{\infty} n c_n x^{n-1} + \sum_{n=0}^{\infty} c_n x^n = 0$$

$$\sum_{n=0}^{\infty} (n+2)(n+1) c_{n+2} x^n - \sum_{n=1}^{\infty} 2n c_n x^n + \sum_{n=0}^{\infty} c_n x^n = 0$$

$$\sum_{n=1}^{\infty} 2n c_n x^n = \sum_{n=0}^{\infty} 2n c_n x^n$$

$$\sum_{n=0}^{\infty} [(n+2)(n+1) c_{n+2} - (2n-1) c_n] x^n = 0$$

This equation is true if the coefficients of x^n are 0:

$$(n+2)(n+1) c_{n+2} - (2n-1) c_n = 0$$

7 $$c_{n+2} = \frac{2n-1}{(n+1)(n+2)} c_n \qquad n = 0, 1, 2, 3, \ldots$$

We solve this recursion relation by putting $n = 0, 1, 2, 3, \ldots$ successively in Equation 7:

$$\text{Put } n = 0: \qquad c_2 = \frac{-1}{1 \cdot 2} c_0$$

$$\text{Put } n = 1: \qquad c_3 = \frac{1}{2 \cdot 3} c_1$$

$$\text{Put } n = 2: \qquad c_4 = \frac{3}{3 \cdot 4} c_2 = -\frac{3}{1 \cdot 2 \cdot 3 \cdot 4} c_0 = -\frac{3}{4!} c_0$$

$$\text{Put } n = 3: \qquad c_5 = \frac{5}{4 \cdot 5} c_3 = \frac{1 \cdot 5}{2 \cdot 3 \cdot 4 \cdot 5} c_1 = \frac{1 \cdot 5}{5!} c_1$$

$$\text{Put } n = 4: \qquad c_6 = \frac{7}{5 \cdot 6} c_4 = -\frac{3 \cdot 7}{4!\, 5 \cdot 6} c_0 = -\frac{3 \cdot 7}{6!} c_0$$

$$\text{Put } n = 5: \qquad c_7 = \frac{9}{6 \cdot 7} c_5 = \frac{1 \cdot 5 \cdot 9}{5!\, 6 \cdot 7} c_1 = \frac{1 \cdot 5 \cdot 9}{7!} c_1$$

$$\text{Put } n = 6: \qquad c_8 = \frac{11}{7 \cdot 8} c_6 = -\frac{3 \cdot 7 \cdot 11}{8!} c_0$$

$$\text{Put } n = 7: \qquad c_9 = \frac{13}{8 \cdot 9} c_7 = \frac{1 \cdot 5 \cdot 9 \cdot 13}{9!} c_1$$

In general, the even coefficients are given by

$$c_{2n} = -\frac{3 \cdot 7 \cdot 11 \cdot \cdots \cdot (4n - 5)}{(2n)!} c_0$$

and the odd coefficients are given by

$$c_{2n+1} = \frac{1 \cdot 5 \cdot 9 \cdot \cdots \cdot (4n - 3)}{(2n + 1)!} c_1$$

The solution is

$$\begin{aligned} y &= c_0 + c_1 x + c_2 x^2 + c_3 x^3 + c_4 x^4 + \cdots \\ &= c_0\left(1 - \frac{1}{2!}x^2 - \frac{3}{4!}x^4 - \frac{3 \cdot 7}{6!}x^6 - \frac{3 \cdot 7 \cdot 11}{8!}x^8 - \cdots\right) \\ &\quad + c_1\left(x + \frac{1}{3!}x^3 + \frac{1 \cdot 5}{5!}x^5 + \frac{1 \cdot 5 \cdot 9}{7!}x^7 + \frac{1 \cdot 5 \cdot 9 \cdot 13}{9!}x^9 + \cdots\right) \end{aligned}$$

or

$$\boxed{8} \qquad \begin{aligned} y &= c_0\left(1 - \frac{1}{2!}x^2 - \sum_{n=2}^{\infty} \frac{3 \cdot 7 \cdot \cdots \cdot (4n - 5)}{(2n)!} x^{2n}\right) \\ &\quad + c_1\left(x + \sum_{n=1}^{\infty} \frac{1 \cdot 5 \cdot 9 \cdot \cdots \cdot (4n - 3)}{(2n + 1)!} x^{2n+1}\right) \end{aligned}$$

■

NOTE 2 In Example 2 we had to *assume* that the differential equation had a series solution. But now we could verify directly that the function given by Equation 8 is indeed a solution.

NOTE 3 Unlike the situation of Example 1, the power series that arise in the solution of Example 2 do not define elementary functions. The functions

$$y_1(x) = 1 - \frac{1}{2!}x^2 - \sum_{n=2}^{\infty} \frac{3 \cdot 7 \cdot \cdots \cdot (4n - 5)}{(2n)!} x^{2n}$$

and

$$y_2(x) = x + \sum_{n=1}^{\infty} \frac{1 \cdot 5 \cdot 9 \cdot \cdots \cdot (4n - 3)}{(2n + 1)!} x^{2n+1}$$

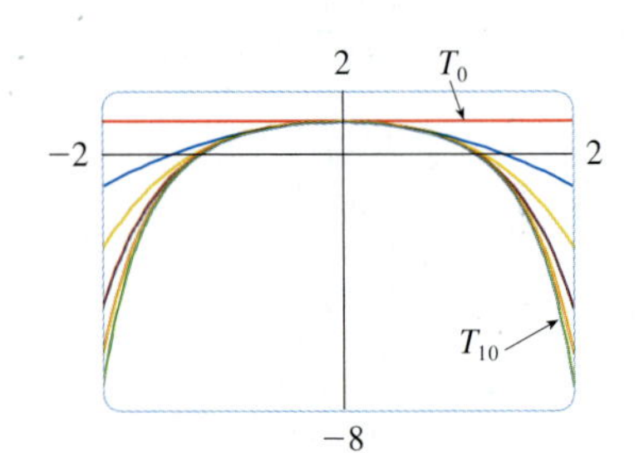

FIGURE 1

are perfectly good functions but they can't be expressed in terms of familiar functions. We can use these power series expressions for y_1 and y_2 to compute approximate values of the functions and even to graph them. Figure 1 shows the first few partial sums $T_0, T_2, T_4, \ldots$ (Taylor polynomials) for $y_1(x)$, and we see how they converge to y_1. In this way we can graph both y_1 and y_2 as in Figure 2.

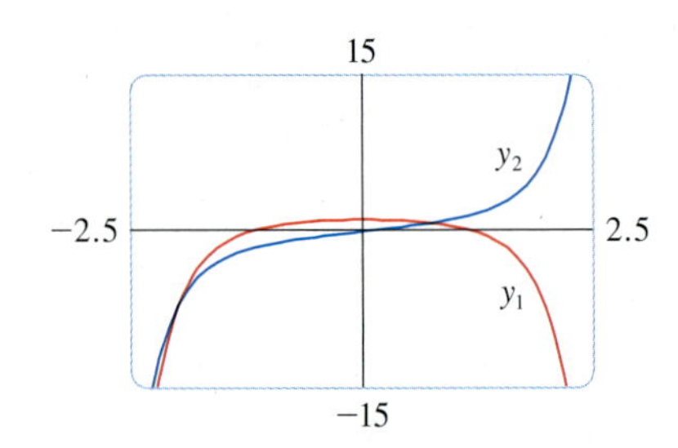

FIGURE 2

NOTE 4 If we were asked to solve the initial-value problem

$$y'' - 2xy' + y = 0 \qquad y(0) = 0 \qquad y'(0) = 1$$

we would observe from Theorem 11.10.5 that

$$c_0 = y(0) = 0 \qquad c_1 = y'(0) = 1$$

This would simplify the calculations in Example 2, since all of the even coefficients would be 0. The solution to the initial-value problem is

$$y(x) = x + \sum_{n=1}^{\infty} \frac{1 \cdot 5 \cdot 9 \cdot \cdots \cdot (4n - 3)}{(2n + 1)!} x^{2n+1}$$

17.4 EXERCISES

1–11 Use power series to solve the differential equation.

1. $y' - y = 0$

2. $y' = xy$

3. $y' = x^2 y$

4. $(x - 3)y' + 2y = 0$

5. $y'' + xy' + y = 0$

6. $y'' = y$

7. $(x - 1)y'' + y' = 0$

8. $y'' = xy$

9. $y'' - xy' - y = 0, \quad y(0) = 1, \quad y'(0) = 0$

10. $y'' + x^2 y = 0, \quad y(0) = 1, \quad y'(0) = 0$

11. $y'' + x^2 y' + xy = 0, \quad y(0) = 0, \quad y'(0) = 1$

12. The solution of the initial-value problem

$$x^2 y'' + xy' + x^2 y = 0 \qquad y(0) = 1 \qquad y'(0) = 0$$

is called a Bessel function of order 0.

(a) Solve the initial-value problem to find a power series expansion for the Bessel function.

(b) Graph several Taylor polynomials until you reach one that looks like a good approximation to the Bessel function on the interval $[-5, 5]$.

17 REVIEW

CONCEPT CHECK

Answers to the Concept Check can be found on the back endpapers.

1. (a) Write the general form of a second-order homogeneous linear differential equation with constant coefficients.
(b) Write the auxiliary equation.
(c) How do you use the roots of the auxiliary equation to solve the differential equation? Write the form of the solution for each of the three cases that can occur.

2. (a) What is an initial-value problem for a second-order differential equation?
(b) What is a boundary-value problem for such an equation?

3. (a) Write the general form of a second-order nonhomogeneous linear differential equation with constant coefficients.
(b) What is the complementary equation? How does it help solve the original differential equation?
(c) Explain how the method of undetermined coefficients works.
(d) Explain how the method of variation of parameters works.

4. Discuss two applications of second-order linear differential equations.

5. How do you use power series to solve a differential equation?

TRUE-FALSE QUIZ

Determine whether the statement is true or false. If it is true, explain why. If it is false, explain why or give an example that disproves the statement.

1. If y_1 and y_2 are solutions of $y'' + y = 0$, then $y_1 + y_2$ is also a solution of the equation.

2. If y_1 and y_2 are solutions of $y'' + 6y' + 5y = x$, then $c_1 y_1 + c_2 y_2$ is also a solution of the equation.

3. The general solution of $y'' - y = 0$ can be written as

$$y = c_1 \cosh x + c_2 \sinh x$$

4. The equation $y'' - y = e^x$ has a particular solution of the form

$$y_p = Ae^x$$

EXERCISES

1–10 Solve the differential equation.

1. $4y'' - y = 0$

2. $y'' - 2y' + 10y = 0$

3. $y'' + 3y = 0$

4. $y'' + 8y' + 16y = 0$

5. $\dfrac{d^2y}{dx^2} - 4\dfrac{dy}{dx} + 5y = e^{2x}$

6. $\dfrac{d^2y}{dx^2} + \dfrac{dy}{dx} - 2y = x^2$

7. $\dfrac{d^2y}{dx^2} - 2\dfrac{dy}{dx} + y = x \cos x$

8. $\dfrac{d^2y}{dx^2} + 4y = \sin 2x$

9. $\dfrac{d^2y}{dx^2} - \dfrac{dy}{dx} - 6y = 1 + e^{-2x}$

10. $\dfrac{d^2y}{dx^2} + y = \csc x, \quad 0 < x < \pi/2$

11–14 Solve the initial-value problem.

11. $y'' + 6y' = 0, \quad y(1) = 3, \quad y'(1) = 12$

12. $y'' - 6y' + 25y = 0, \quad y(0) = 2, \quad y'(0) = 1$

13. $y'' - 5y' + 4y = 0, \quad y(0) = 0, \quad y'(0) = 1$

14. $9y'' + y = 3x + e^{-x}, \quad y(0) = 1, \quad y'(0) = 2$

15–16 Solve the boundary-value problem, if possible.

15. $y'' + 4y' + 29y = 0, \quad y(0) = 1, \quad y(\pi) = -1$

16. $y'' + 4y' + 29y = 0, \quad y(0) = 1, \quad y(\pi) = -e^{-2\pi}$

17. Use power series to solve the initial-value problem

$$y'' + xy' + y = 0 \qquad y(0) = 0 \qquad y'(0) = 1$$

18. Use power series to solve the differential equation

$$y'' - xy' - 2y = 0$$

19. A series circuit contains a resistor with $R = 40\ \Omega$, an inductor with $L = 2$ H, a capacitor with $C = 0.0025$ F, and a 12-V battery. The initial charge is $Q = 0.01$ C and the initial current is 0. Find the charge at time t.

20. A spring with a mass of 2 kg has damping constant 16, and a force of 12.8 N keeps the spring stretched 0.2 m beyond its natural length. Find the position of the mass at time t if it starts at the equilibrium position with a velocity of 2.4 m/s.

21. Assume that the earth is a solid sphere of uniform density with mass M and radius $R = 3960$ mi. For a particle of mass m within the earth at a distance r from the earth's center, the gravitational force attracting the particle to the center is

$$F_r = \frac{-GM_r m}{r^2}$$

where G is the gravitational constant and M_r is the mass of the earth within the sphere of radius r.

(a) Show that $F_r = \dfrac{-GMm}{R^3} r$.

(b) Suppose a hole is drilled through the earth along a diameter. Show that if a particle of mass m is dropped from rest at the surface, into the hole, then the distance $y = y(t)$ of the particle from the center of the earth at time t is given by

$$y''(t) = -k^2 y(t)$$

where $k^2 = GM/R^3 = g/R$.

(c) Conclude from part (b) that the particle undergoes simple harmonic motion. Find the period T.

(d) With what speed does the particle pass through the center of the earth?

Appendixes

F Proofs of Theorems

In this appendix we present proofs of several theorems that are stated in the main body of the text. The sections in which they occur are indicated in the margin.

Section 11.8

In order to prove Theorem 11.8.4, we first need the following results.

Theorem

1. If a power series $\Sigma c_n x^n$ converges when $x = b$ (where $b \neq 0$), then it converges whenever $|x| < |b|$.
2. If a power series $\Sigma c_n x^n$ diverges when $x = d$ (where $d \neq 0$), then it diverges whenever $|x| > |d|$.

PROOF OF 1 Suppose that $\Sigma c_n b^n$ converges. Then, by Theorem 11.2.6, we have $\lim_{n \to \infty} c_n b^n = 0$. According to Definition 11.1.2 with $\varepsilon = 1$, there is a positive integer N such that $|c_n b^n| < 1$ whenever $n \geqslant N$. Thus, for $n \geqslant N$, we have

$$|c_n x^n| = \left| \frac{c_n b^n x^n}{b^n} \right| = |c_n b^n| \left| \frac{x}{b} \right|^n < \left| \frac{x}{b} \right|^n$$

If $|x| < |b|$, then $|x/b| < 1$, so $\Sigma |x/b|^n$ is a convergent geometric series. Therefore, by the Comparison Test, the series $\Sigma_{n=N}^{\infty} |c_n x^n|$ is convergent. Thus the series $\Sigma c_n x^n$ is absolutely convergent and therefore convergent. ■

PROOF OF 2 Suppose that $\Sigma c_n d^n$ diverges. If x is any number such that $|x| > |d|$, then $\Sigma c_n x^n$ cannot converge because, by part 1, the convergence of $\Sigma c_n x^n$ would imply the convergence of $\Sigma c_n d^n$. Therefore $\Sigma c_n x^n$ diverges whenever $|x| > |d|$. ■

Theorem For a power series $\Sigma c_n x^n$ there are only three possibilities:

1. The series converges only when $x = 0$.
2. The series converges for all x.
3. There is a positive number R such that the series converges if $|x| < R$ and diverges if $|x| > R$.

PROOF Suppose that neither case 1 nor case 2 is true. Then there are nonzero numbers b and d such that $\Sigma c_n x^n$ converges for $x = b$ and diverges for $x = d$. Therefore the set $S = \{x \mid \Sigma c_n x^n \text{ converges}\}$ is not empty. By the preceding theorem, the series diverges if $|x| > |d|$, so $|x| \leqslant |d|$ for all $x \in S$. This says that $|d|$ is an upper bound for the set S. Thus, by the Completeness Axiom (see Section 11.1), S has a least upper bound R. If $|x| > R$, then $x \notin S$, so $\Sigma c_n x^n$ diverges. If $|x| < R$, then $|x|$ is not an upper bound for S and so there exists $b \in S$ such that $b > |x|$. Since $b \in S$, $\Sigma c_n x^n$ converges, so by the preceding theorem $\Sigma c_n x^n$ converges. ■

4 Theorem For a power series $\Sigma\, c_n(x - a)^n$ there are only three possibilities:

1. The series converges only when $x = a$.
2. The series converges for all x.
3. There is a positive number R such that the series converges if $|x - a| < R$ and diverges if $|x - a| > R$.

PROOF If we make the change of variable $u = x - a$, then the power series becomes $\Sigma\, c_n u^n$ and we can apply the preceding theorem to this series. In case 3 we have convergence for $|u| < R$ and divergence for $|u| > R$. Thus we have convergence for $|x - a| < R$ and divergence for $|x - a| > R$. ■

Section 14.3

Clairaut's Theorem Suppose f is defined on a disk D that contains the point (a, b). If the functions f_{xy} and f_{yx} are both continuous on D, then $f_{xy}(a, b) = f_{yx}(a, b)$.

PROOF For small values of h, $h \neq 0$, consider the difference

$$\Delta(h) = [f(a + h, b + h) - f(a + h, b)] - [f(a, b + h) - f(a, b)]$$

Notice that if we let $g(x) = f(x, b + h) - f(x, b)$, then

$$\Delta(h) = g(a + h) - g(a)$$

By the Mean Value Theorem, there is a number c between a and $a + h$ such that

$$g(a + h) - g(a) = g'(c)h = h[f_x(c, b + h) - f_x(c, b)]$$

Applying the Mean Value Theorem again, this time to f_x, we get a number d between b and $b + h$ such that

$$f_x(c, b + h) - f_x(c, b) = f_{xy}(c, d)h$$

Combining these equations, we obtain

$$\Delta(h) = h^2 f_{xy}(c, d)$$

If $h \to 0$, then $(c, d) \to (a, b)$, so the continuity of f_{xy} at (a, b) gives

$$\lim_{h \to 0} \frac{\Delta(h)}{h^2} = \lim_{(c, d) \to (a, b)} f_{xy}(c, d) = f_{xy}(a, b)$$

Similarly, by writing

$$\Delta(h) = [f(a + h, b + h) - f(a, b + h)] - [f(a + h, b) - f(a, b)]$$

and using the Mean Value Theorem twice and the continuity of f_{yx} at (a, b), we obtain

$$\lim_{h \to 0} \frac{\Delta(h)}{h^2} = f_{yx}(a, b)$$

It follows that $f_{xy}(a, b) = f_{yx}(a, b)$. ■

Section 14.4

8 Theorem If the partial derivatives f_x and f_y exist near (a, b) and are continuous at (a, b), then f is differentiable at (a, b).

PROOF Let

$$\Delta z = f(a + \Delta x, b + \Delta y) - f(a, b)$$

According to (14.4.7), to prove that f is differentiable at (a, b) we have to show that we can write Δz in the form

$$\Delta z = f_x(a, b)\,\Delta x + f_y(a, b)\,\Delta y + \varepsilon_1\,\Delta x + \varepsilon_2\,\Delta y$$

where ε_1 and $\varepsilon_2 \to 0$ as $(\Delta x, \Delta y) \to (0, 0)$.

Referring to Figure 4, we write

$$\boxed{1} \quad \Delta z = [f(a + \Delta x, b + \Delta y) - f(a, b + \Delta y)] + [f(a, b + \Delta y) - f(a, b)]$$

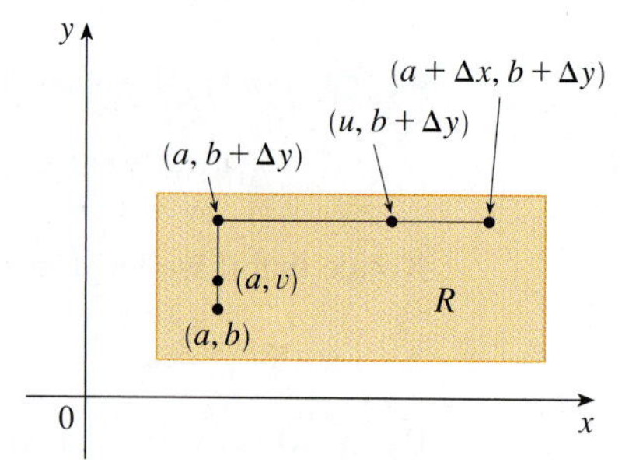

FIGURE 4

Observe that the function of a single variable

$$g(x) = f(x, b + \Delta y)$$

is defined on the interval $[a, a + \Delta x]$ and $g'(x) = f_x(x, b + \Delta y)$. If we apply the Mean Value Theorem to g, we get

$$g(a + \Delta x) - g(a) = g'(u)\,\Delta x$$

where u is some number between a and $a + \Delta x$. In terms of f, this equation becomes

$$f(a + \Delta x, b + \Delta y) - f(a, b + \Delta y) = f_x(u, b + \Delta y)\,\Delta x$$

This gives us an expression for the first part of the right side of Equation 1. For the second part we let $h(y) = f(a, y)$. Then h is a function of a single variable defined on the interval $[b, b + \Delta y]$ and $h'(y) = f_y(a, y)$. A second application of the Mean Value Theorem then gives

$$h(b + \Delta y) - h(b) = h'(v)\,\Delta y$$

where v is some number between b and $b + \Delta y$. In terms of f, this becomes

$$f(a, b + \Delta y) - f(a, b) = f_y(a, v)\,\Delta y$$

We now substitute these expressions into Equation 1 and obtain

$$\begin{aligned}\Delta z &= f_x(u, b + \Delta y)\,\Delta x + f_y(a, v)\,\Delta y \\ &= f_x(a, b)\,\Delta x + [f_x(u, b + \Delta y) - f_x(a, b)]\,\Delta x + f_y(a, b)\,\Delta y \\ &\qquad + [f_y(a, v) - f_y(a, b)]\,\Delta y \\ &= f_x(a, b)\,\Delta x + f_y(a, b)\,\Delta y + \varepsilon_1\,\Delta x + \varepsilon_2\,\Delta y\end{aligned}$$

where
$$\varepsilon_1 = f_x(u, b + \Delta y) - f_x(a, b)$$
$$\varepsilon_2 = f_y(a, v) - f_y(a, b)$$

Since $(u, b + \Delta y) \to (a, b)$ and $(a, v) \to (a, b)$ as $(\Delta x, \Delta y) \to (0, 0)$ and since f_x and f_y are continuous at (a, b), we see that $\varepsilon_1 \to 0$ and $\varepsilon_2 \to 0$ as $(\Delta x, \Delta y) \to (0, 0)$.

Therefore f is differentiable at (a, b). ■

G Complex Numbers

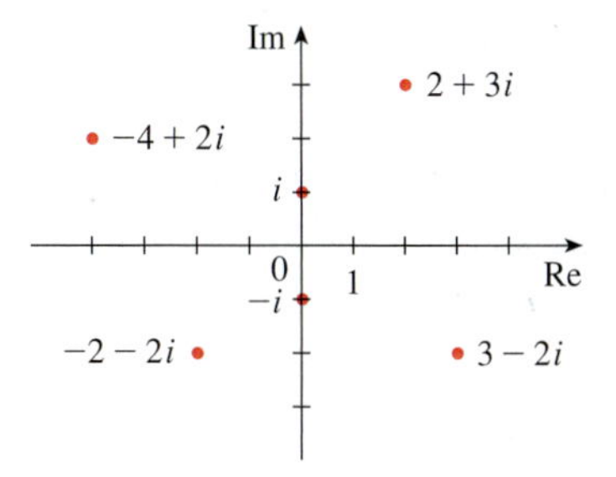

FIGURE 1
Complex numbers as points in the Argand plane

A **complex number** can be represented by an expression of the form $a + bi$, where a and b are real numbers and i is a symbol with the property that $i^2 = -1$. The complex number $a + bi$ can also be represented by the ordered pair (a, b) and plotted as a point in a plane (called the Argand plane) as in Figure 1. Thus the complex number $i = 0 + 1 \cdot i$ is identified with the point $(0, 1)$.

The **real part** of the complex number $a + bi$ is the real number a and the **imaginary part** is the real number b. Thus the real part of $4 - 3i$ is 4 and the imaginary part is -3. Two complex numbers $a + bi$ and $c + di$ are **equal** if $a = c$ and $b = d$, that is, their real parts are equal and their imaginary parts are equal. In the Argand plane the horizontal axis is called the real axis and the vertical axis is called the imaginary axis.

The sum and difference of two complex numbers are defined by adding or subtracting their real parts and their imaginary parts:

$$(a + bi) + (c + di) = (a + c) + (b + d)i$$
$$(a + bi) - (c + di) = (a - c) + (b - d)i$$

For instance,
$$(1 - i) + (4 + 7i) = (1 + 4) + (-1 + 7)i = 5 + 6i$$

The product of complex numbers is defined so that the usual commutative and distributive laws hold:

$$\begin{aligned}(a + bi)(c + di) &= a(c + di) + (bi)(c + di) \\ &= ac + adi + bci + bdi^2\end{aligned}$$

Since $i^2 = -1$, this becomes

$$(a + bi)(c + di) = (ac - bd) + (ad + bc)i$$

EXAMPLE 1

$$\begin{aligned}(-1 + 3i)(2 - 5i) &= (-1)(2 - 5i) + 3i(2 - 5i) \\ &= -2 + 5i + 6i - 15(-1) = 13 + 11i\end{aligned}$$ ■

Division of complex numbers is much like rationalizing the denominator of a rational expression. For the complex number $z = a + bi$, we define its **complex conjugate** to be $\overline{z} = a - bi$. To find the quotient of two complex numbers we multiply numerator and denominator by the complex conjugate of the denominator.

EXAMPLE 2 Express the number $\dfrac{-1 + 3i}{2 + 5i}$ in the form $a + bi$.

SOLUTION We multiply numerator and denominator by the complex conjugate of $2 + 5i$, namely, $2 - 5i$, and we take advantage of the result of Example 1:

$$\frac{-1 + 3i}{2 + 5i} = \frac{-1 + 3i}{2 + 5i} \cdot \frac{2 - 5i}{2 - 5i} = \frac{13 + 11i}{2^2 + 5^2} = \frac{13}{29} + \frac{11}{29}i$$

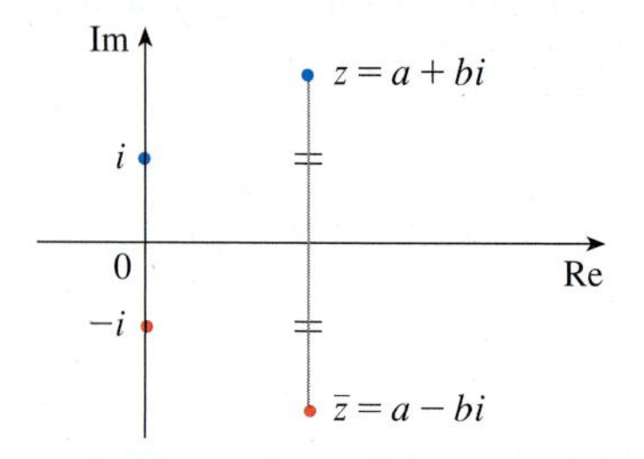

FIGURE 2

The geometric interpretation of the complex conjugate is shown in Figure 2: $\overline{z}$ is the reflection of z in the real axis. We list some of the properties of the complex conjugate in the following box. The proofs follow from the definition and are requested in Exercise 18.

Properties of Conjugates

$$\overline{z + w} = \overline{z} + \overline{w} \qquad \overline{zw} = \overline{z}\,\overline{w} \qquad \overline{z^n} = \overline{z}^n$$

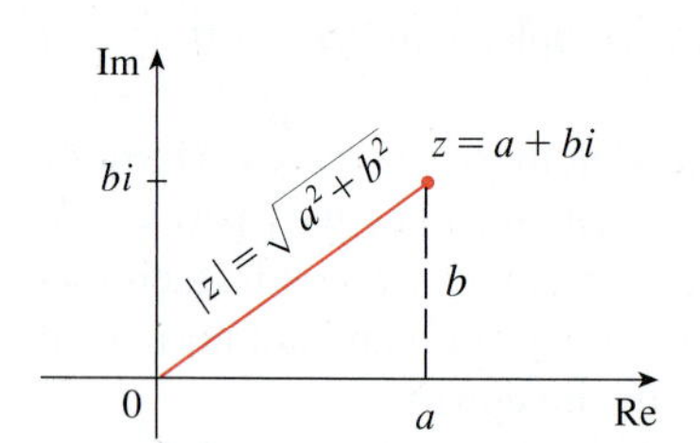

FIGURE 3

The **modulus**, or **absolute value**, $|z|$ of a complex number $z = a + bi$ is its distance from the origin. From Figure 3 we see that if $z = a + bi$, then

$$|z| = \sqrt{a^2 + b^2}$$

Notice that

$$z\overline{z} = (a + bi)(a - bi) = a^2 + abi - abi - b^2i^2 = a^2 + b^2$$

and so

$$z\overline{z} = |z|^2$$

This explains why the division procedure in Example 2 works in general:

$$\frac{z}{w} = \frac{z\overline{w}}{w\overline{w}} = \frac{z\overline{w}}{|w|^2}$$

Since $i^2 = -1$, we can think of i as a square root of -1. But notice that we also have $(-i)^2 = i^2 = -1$ and so $-i$ is also a square root of -1. We say that i is the **principal square root** of -1 and write $\sqrt{-1} = i$. In general, if c is any positive number, we write

$$\sqrt{-c} = \sqrt{c}\, i$$

With this convention, the usual derivation and formula for the roots of the quadratic equation $ax^2 + bx + c = 0$ are valid even when $b^2 - 4ac < 0$:

$$x = \frac{-b \pm \sqrt{b^2 - 4ac}}{2a}$$

EXAMPLE 3 Find the roots of the equation $x^2 + x + 1 = 0$.

SOLUTION Using the quadratic formula, we have

$$x = \frac{-1 \pm \sqrt{1^2 - 4 \cdot 1}}{2} = \frac{-1 \pm \sqrt{-3}}{2} = \frac{-1 \pm \sqrt{3}\, i}{2}$$

We observe that the solutions of the equation in Example 3 are complex conjugates of each other. In general, the solutions of any quadratic equation $ax^2 + bx + c = 0$ with real coefficients a, b, and c are always complex conjugates. (If z is real, $\bar{z} = z$, so z is its own conjugate.)

We have seen that if we allow complex numbers as solutions, then every quadratic equation has a solution. More generally, it is true that every polynomial equation

$$a_n x^n + a_{n-1} x^{n-1} + \cdots + a_1 x + a_0 = 0$$

of degree at least one has a solution among the complex numbers. This fact is known as the Fundamental Theorem of Algebra and was proved by Gauss.

Polar Form

We know that any complex number $z = a + bi$ can be considered as a point (a, b) and that any such point can be represented by polar coordinates (r, θ) with $r \geqslant 0$. In fact,

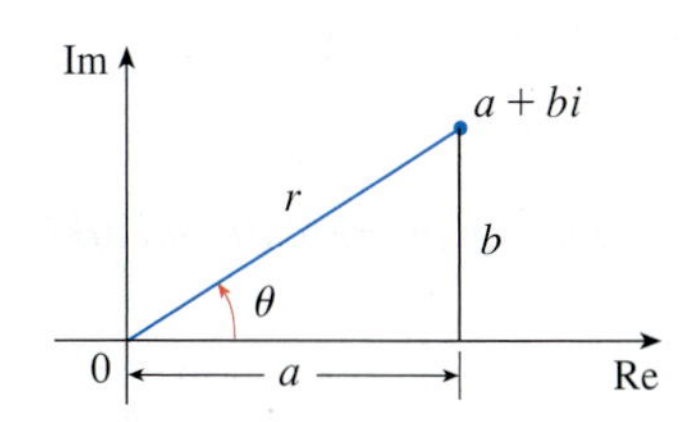

FIGURE 4

$$a = r\cos\theta \qquad b = r\sin\theta$$

as in Figure 4. Therefore we have

$$z = a + bi = (r\cos\theta) + (r\sin\theta)i$$

Thus we can write any complex number z in the form

$$z = r(\cos\theta + i\sin\theta)$$

where $$r = |z| = \sqrt{a^2 + b^2} \qquad \text{and} \qquad \tan\theta = \frac{b}{a}$$

The angle θ is called the **argument** of z and we write $\theta = \arg(z)$. Note that $\arg(z)$ is not unique; any two arguments of z differ by an integer multiple of 2π.

EXAMPLE 4 Write the following numbers in polar form.

(a) $z = 1 + i$ (b) $w = \sqrt{3} - i$

SOLUTION

(a) We have $r = |z| = \sqrt{1^2 + 1^2} = \sqrt{2}$ and $\tan\theta = 1$, so we can take $\theta = \pi/4$. Therefore the polar form is

$$z = \sqrt{2}\left(\cos\frac{\pi}{4} + i\sin\frac{\pi}{4}\right)$$

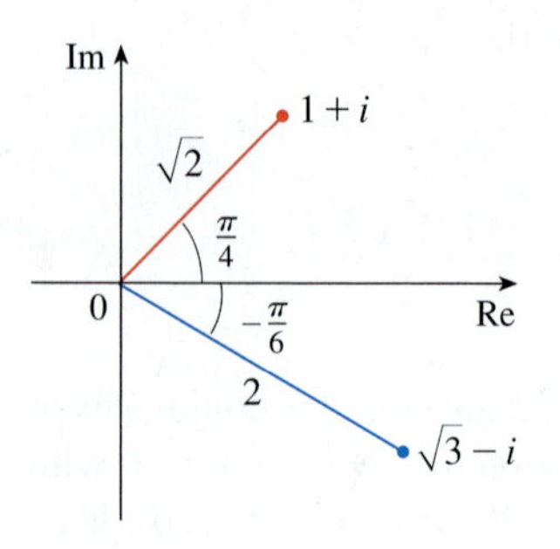

FIGURE 5

(b) Here we have $r = |w| = \sqrt{3 + 1} = 2$ and $\tan \theta = -1/\sqrt{3}$. Since w lies in the fourth quadrant, we take $\theta = -\pi/6$ and

$$w = 2\left[\cos\left(-\frac{\pi}{6}\right) + i \sin\left(-\frac{\pi}{6}\right)\right]$$

The numbers z and w are shown in Figure 5. ■

The polar form of complex numbers gives insight into multiplication and division. Let

$$z_1 = r_1(\cos \theta_1 + i \sin \theta_1) \qquad z_2 = r_2(\cos \theta_2 + i \sin \theta_2)$$

be two complex numbers written in polar form. Then

$$\begin{aligned} z_1 z_2 &= r_1 r_2(\cos \theta_1 + i \sin \theta_1)(\cos \theta_2 + i \sin \theta_2) \\ &= r_1 r_2[(\cos \theta_1 \cos \theta_2 - \sin \theta_1 \sin \theta_2) + i(\sin \theta_1 \cos \theta_2 + \cos \theta_1 \sin \theta_2)] \end{aligned}$$

Therefore, using the addition formulas for cosine and sine, we have

$$\boxed{1} \qquad z_1 z_2 = r_1 r_2[\cos(\theta_1 + \theta_2) + i \sin(\theta_1 + \theta_2)]$$

This formula says that *to multiply two complex numbers we multiply the moduli and add the arguments*. (See Figure 6.)

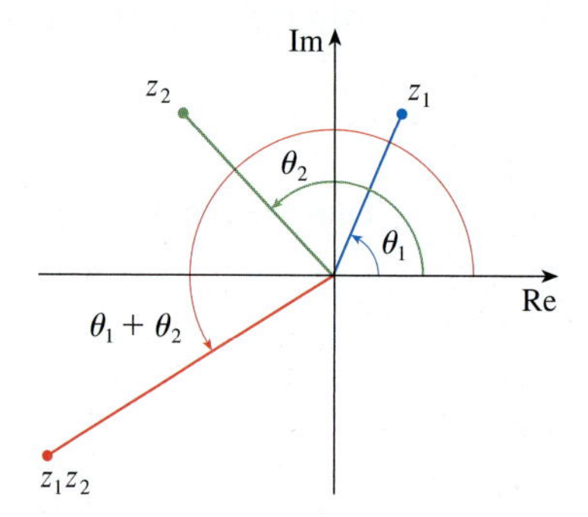

FIGURE 6

A similar argument using the subtraction formulas for sine and cosine shows that *to divide two complex numbers we divide the moduli and subtract the arguments.*

$$\frac{z_1}{z_2} = \frac{r_1}{r_2}[\cos(\theta_1 - \theta_2) + i \sin(\theta_1 - \theta_2)] \qquad z_2 \neq 0$$

In particular, taking $z_1 = 1$ and $z_2 = z$ (and therefore $\theta_1 = 0$ and $\theta_2 = \theta$), we have the following, which is illustrated in Figure 7.

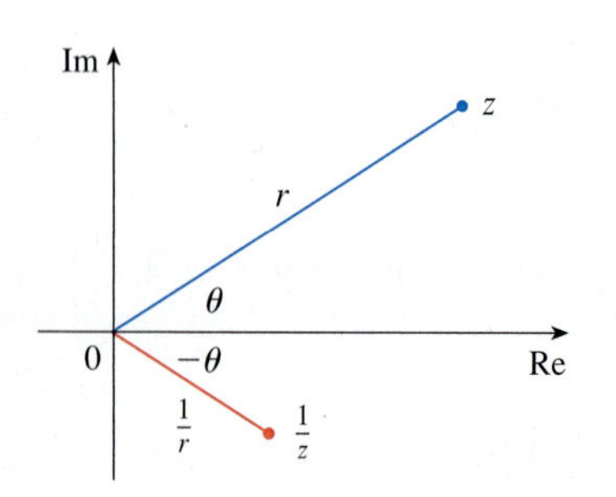

FIGURE 7

$$\text{If} \quad z = r(\cos \theta + i \sin \theta), \quad \text{then} \quad \frac{1}{z} = \frac{1}{r}(\cos \theta - i \sin \theta).$$

EXAMPLE 5 Find the product of the complex numbers $1 + i$ and $\sqrt{3} - i$ in polar form.

SOLUTION From Example 4 we have

$$1 + i = \sqrt{2}\left(\cos\frac{\pi}{4} + i\sin\frac{\pi}{4}\right)$$

and
$$\sqrt{3} - i = 2\left[\cos\left(-\frac{\pi}{6}\right) + i\sin\left(-\frac{\pi}{6}\right)\right]$$

So, by Equation 1,

$$\begin{aligned}(1 + i)(\sqrt{3} - i) &= 2\sqrt{2}\left[\cos\left(\frac{\pi}{4} - \frac{\pi}{6}\right) + i\sin\left(\frac{\pi}{4} - \frac{\pi}{6}\right)\right] \\ &= 2\sqrt{2}\left(\cos\frac{\pi}{12} + i\sin\frac{\pi}{12}\right)\end{aligned}$$

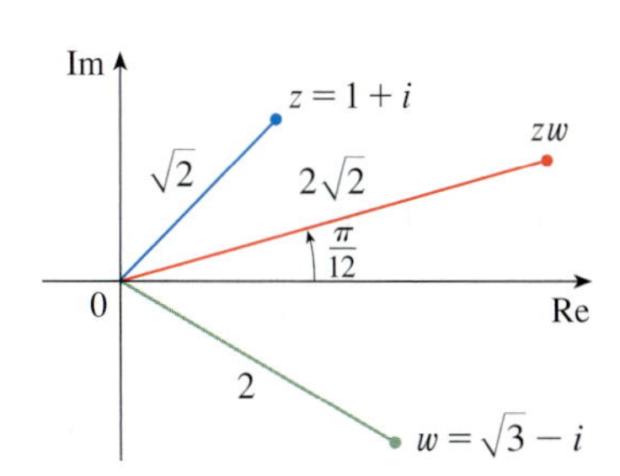

FIGURE 8

This is illustrated in Figure 8. ■

Repeated use of Formula 1 shows how to compute powers of a complex number. If

$$z = r(\cos\theta + i\sin\theta)$$

then
$$z^2 = r^2(\cos 2\theta + i\sin 2\theta)$$

and
$$z^3 = zz^2 = r^3(\cos 3\theta + i\sin 3\theta)$$

In general, we obtain the following result, which is named after the French mathematician Abraham De Moivre (1667–1754).

> **2 De Moivre's Theorem** If $z = r(\cos\theta + i\sin\theta)$ and n is a positive integer, then
>
> $$z^n = [r(\cos\theta + i\sin\theta)]^n = r^n(\cos n\theta + i\sin n\theta)$$

This says that *to take the nth power of a complex number we take the nth power of the modulus and multiply the argument by n.*

EXAMPLE 6 Find $\left(\frac{1}{2} + \frac{1}{2}i\right)^{10}$.

SOLUTION Since $\frac{1}{2} + \frac{1}{2}i = \frac{1}{2}(1 + i)$, it follows from Example 4(a) that $\frac{1}{2} + \frac{1}{2}i$ has the polar form

$$\frac{1}{2} + \frac{1}{2}i = \frac{\sqrt{2}}{2}\left(\cos\frac{\pi}{4} + i\sin\frac{\pi}{4}\right)$$

So by De Moivre's Theorem,

$$\left(\frac{1}{2}+\frac{1}{2}i\right)^{10} = \left(\frac{\sqrt{2}}{2}\right)^{10}\left(\cos\frac{10\pi}{4} + i\sin\frac{10\pi}{4}\right)$$

$$= \frac{2^5}{2^{10}}\left(\cos\frac{5\pi}{2} + i\sin\frac{5\pi}{2}\right) = \frac{1}{32}i$$

De Moivre's Theorem can also be used to find the nth roots of complex numbers. An nth root of the complex number z is a complex number w such that

$$w^n = z$$

Writing these two numbers in trigonometric form as

$$w = s(\cos\phi + i\sin\phi) \qquad \text{and} \qquad z = r(\cos\theta + i\sin\theta)$$

and using De Moivre's Theorem, we get

$$s^n(\cos n\phi + i\sin n\phi) = r(\cos\theta + i\sin\theta)$$

The equality of these two complex numbers shows that

$$s^n = r \qquad \text{or} \qquad s = r^{1/n}$$

and

$$\cos n\phi = \cos\theta \qquad \text{and} \qquad \sin n\phi = \sin\theta$$

From the fact that sine and cosine have period 2π, it follows that

$$n\phi = \theta + 2k\pi \qquad \text{or} \qquad \phi = \frac{\theta + 2k\pi}{n}$$

Thus

$$w = r^{1/n}\left[\cos\left(\frac{\theta + 2k\pi}{n}\right) + i\sin\left(\frac{\theta + 2k\pi}{n}\right)\right]$$

Since this expression gives a different value of w for $k = 0, 1, 2, \ldots, n - 1$, we have the following.

3 Roots of a Complex Number Let $z = r(\cos\theta + i\sin\theta)$ and let n be a positive integer. Then z has the n distinct nth roots

$$w_k = r^{1/n}\left[\cos\left(\frac{\theta + 2k\pi}{n}\right) + i\sin\left(\frac{\theta + 2k\pi}{n}\right)\right]$$

where $k = 0, 1, 2, \ldots, n - 1$.

Notice that each of the nth roots of z has modulus $|w_k| = r^{1/n}$. Thus all the nth roots of z lie on the circle of radius $r^{1/n}$ in the complex plane. Also, since the argument of each successive nth root exceeds the argument of the previous root by $2\pi/n$, we see that the nth roots of z are equally spaced on this circle.

EXAMPLE 7 Find the six sixth roots of $z = -8$ and graph these roots in the complex plane.

SOLUTION In trigonometric form, $z = 8(\cos \pi + i \sin \pi)$. Applying Equation 3 with $n = 6$, we get

$$w_k = 8^{1/6}\left(\cos \frac{\pi + 2k\pi}{6} + i \sin \frac{\pi + 2k\pi}{6}\right)$$

We get the six sixth roots of -8 by taking $k = 0, 1, 2, 3, 4, 5$ in this formula:

$$w_0 = 8^{1/6}\left(\cos \frac{\pi}{6} + i \sin \frac{\pi}{6}\right) = \sqrt{2}\left(\frac{\sqrt{3}}{2} + \frac{1}{2}i\right)$$

$$w_1 = 8^{1/6}\left(\cos \frac{\pi}{2} + i \sin \frac{\pi}{2}\right) = \sqrt{2}\, i$$

$$w_2 = 8^{1/6}\left(\cos \frac{5\pi}{6} + i \sin \frac{5\pi}{6}\right) = \sqrt{2}\left(-\frac{\sqrt{3}}{2} + \frac{1}{2}i\right)$$

$$w_3 = 8^{1/6}\left(\cos \frac{7\pi}{6} + i \sin \frac{7\pi}{6}\right) = \sqrt{2}\left(-\frac{\sqrt{3}}{2} - \frac{1}{2}i\right)$$

$$w_4 = 8^{1/6}\left(\cos \frac{3\pi}{2} + i \sin \frac{3\pi}{2}\right) = -\sqrt{2}\, i$$

$$w_5 = 8^{1/6}\left(\cos \frac{11\pi}{6} + i \sin \frac{11\pi}{6}\right) = \sqrt{2}\left(\frac{\sqrt{3}}{2} - \frac{1}{2}i\right)$$

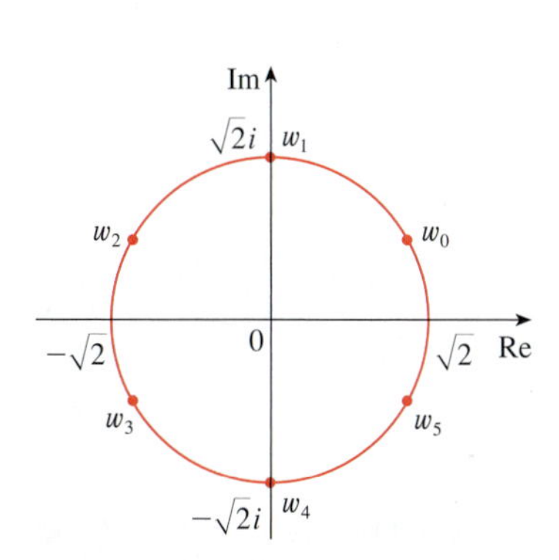

FIGURE 9
The six sixth roots of $z = -8$

All these points lie on the circle of radius $\sqrt{2}$ as shown in Figure 9. ■

Complex Exponentials

We also need to give a meaning to the expression e^z when $z = x + iy$ is a complex number. The theory of infinite series as developed in Chapter 11 can be extended to the case where the terms are complex numbers. Using the Taylor series for e^x (11.10.11) as our guide, we define

4 $$e^z = \sum_{n=0}^{\infty} \frac{z^n}{n!} = 1 + z + \frac{z^2}{2!} + \frac{z^3}{3!} + \cdots$$

and it turns out that this complex exponential function has the same properties as the real exponential function. In particular, it is true that

5 $$e^{z_1 + z_2} = e^{z_1} e^{z_2}$$

If we put $z = iy$, where y is a real number, in Equation 4, and use the facts that

$$i^2 = -1, \quad i^3 = i^2 i = -i, \quad i^4 = 1, \quad i^5 = i, \quad \ldots$$

we get $$e^{iy} = 1 + iy + \frac{(iy)^2}{2!} + \frac{(iy)^3}{3!} + \frac{(iy)^4}{4!} + \frac{(iy)^5}{5!} + \cdots$$
$$= 1 + iy - \frac{y^2}{2!} - i\frac{y^3}{3!} + \frac{y^4}{4!} + i\frac{y^5}{5!} + \cdots$$
$$= \left(1 - \frac{y^2}{2!} + \frac{y^4}{4!} - \frac{y^6}{6!} + \cdots\right) + i\left(y - \frac{y^3}{3!} + \frac{y^5}{5!} - \cdots\right)$$
$$= \cos y + i \sin y$$

Here we have used the Taylor series for $\cos y$ and $\sin y$ (Equations 11.10.16 and 11.10.15). The result is a famous formula called **Euler's formula**:

6 $$e^{iy} = \cos y + i \sin y$$

Combining Euler's formula with Equation 5, we get

7 $$e^{x+iy} = e^x e^{iy} = e^x(\cos y + i \sin y)$$

EXAMPLE 8 Evaluate: (a) $e^{i\pi}$ (b) $e^{-1+i\pi/2}$

SOLUTION

(a) From Euler's equation (6) we have

$$e^{i\pi} = \cos \pi + i \sin \pi = -1 + i(0) = -1$$

We could write the result of Example 8(a) as

$$e^{i\pi} + 1 = 0$$

This equation relates the five most famous numbers in all of mathematics: 0, 1, e, i, and π.

(b) Using Equation 7 we get

$$e^{-1+i\pi/2} = e^{-1}\left(\cos \frac{\pi}{2} + i \sin \frac{\pi}{2}\right) = \frac{1}{e}[0 + i(1)] = \frac{i}{e}$$ ■

Finally, we note that Euler's equation provides us with an easier method of proving De Moivre's Theorem:

$$[r(\cos \theta + i \sin \theta)]^n = (re^{i\theta})^n = r^n e^{in\theta} = r^n(\cos n\theta + i \sin n\theta)$$

G EXERCISES

1–14 Evaluate the expression and write your answer in the form $a + bi$.

1. $(5 - 6i) + (3 + 2i)$

2. $\left(4 - \frac{1}{2}i\right) - \left(9 + \frac{5}{2}i\right)$

3. $(2 + 5i)(4 - i)$

4. $(1 - 2i)(8 - 3i)$

5. $\overline{12 + 7i}$

6. $\overline{2i\left(\frac{1}{2} - i\right)}$

7. $\dfrac{1 + 4i}{3 + 2i}$

8. $\dfrac{3 + 2i}{1 - 4i}$

9. $\dfrac{1}{1 + i}$

10. $\dfrac{3}{4 - 3i}$

11. i^3

12. i^{100}

13. $\sqrt{-25}$

14. $\sqrt{-3}\sqrt{-12}$

15–17 Find the complex conjugate and the modulus of the number.

15. $12 - 5i$

16. $-1 + 2\sqrt{2}\,i$

17. $-4i$

18. Prove the following properties of complex numbers.
(a) $\overline{z + w} = \bar{z} + \bar{w}$ (b) $\overline{zw} = \bar{z}\,\bar{w}$
(c) $\overline{z^n} = \bar{z}^n$, where n is a positive integer
[*Hint:* Write $z = a + bi$, $w = c + di$.]

19–24 Find all solutions of the equation.

19. $4x^2 + 9 = 0$

20. $x^4 = 1$

21. $x^2 + 2x + 5 = 0$

22. $2x^2 - 2x + 1 = 0$

23. $z^2 + z + 2 = 0$

24. $z^2 + \frac{1}{2}z + \frac{1}{4} = 0$

25–28 Write the number in polar form with argument between 0 and 2π.

25. $-3 + 3i$

26. $1 - \sqrt{3}\,i$

27. $3 + 4i$

28. $8i$

29–32 Find polar forms for zw, z/w, and $1/z$ by first putting z and w into polar form.

29. $z = \sqrt{3} + i, \quad w = 1 + \sqrt{3}\,i$

30. $z = 4\sqrt{3} - 4i, \quad w = 8i$

31. $z = 2\sqrt{3} - 2i, \quad w = -1 + i$

32. $z = 4(\sqrt{3} + i), \quad w = -3 - 3i$

33–36 Find the indicated power using De Moivre's Theorem.

33. $(1 + i)^{20}$

34. $(1 - \sqrt{3}\,i)^5$

35. $(2\sqrt{3} + 2i)^5$

36. $(1 - i)^8$

37–40 Find the indicated roots. Sketch the roots in the complex plane.

37. The eighth roots of 1

38. The fifth roots of 32

39. The cube roots of i

40. The cube roots of $1 + i$

41–46 Write the number in the form $a + bi$.

41. $e^{i\pi/2}$

42. $e^{2\pi i}$

43. $e^{i\pi/3}$

44. $e^{-i\pi}$

45. $e^{2+i\pi}$

46. $e^{\pi+i}$

47. Use De Moivre's Theorem with $n = 3$ to express $\cos 3\theta$ and $\sin 3\theta$ in terms of $\cos\theta$ and $\sin\theta$.

48. Use Euler's formula to prove the following formulas for $\cos x$ and $\sin x$:

$$\cos x = \frac{e^{ix} + e^{-ix}}{2} \qquad \sin x = \frac{e^{ix} - e^{-ix}}{2i}$$

49. If $u(x) = f(x) + ig(x)$ is a complex-valued function of a real variable x and the real and imaginary parts $f(x)$ and $g(x)$ are differentiable functions of x, then the derivative of u is defined to be $u'(x) = f'(x) + ig'(x)$. Use this together with Equation 7 to prove that if $F(x) = e^{rx}$, then $F'(x) = re^{rx}$ when $r = a + bi$ is a complex number.

50. (a) If u is a complex-valued function of a real variable, its indefinite integral $\int u(x)\,dx$ is an antiderivative of u. Evaluate

$$\int e^{(1+i)x}\,dx$$

(b) By considering the real and imaginary parts of the integral in part (a), evaluate the real integrals

$$\int e^x \cos x\,dx \qquad \text{and} \qquad \int e^x \sin x\,dx$$

(c) Compare with the method used in Example 7.1.4.

H Answers to Odd-Numbered Exercises

CHAPTER 10

EXERCISES 10.1 ■ PAGE 685

1.

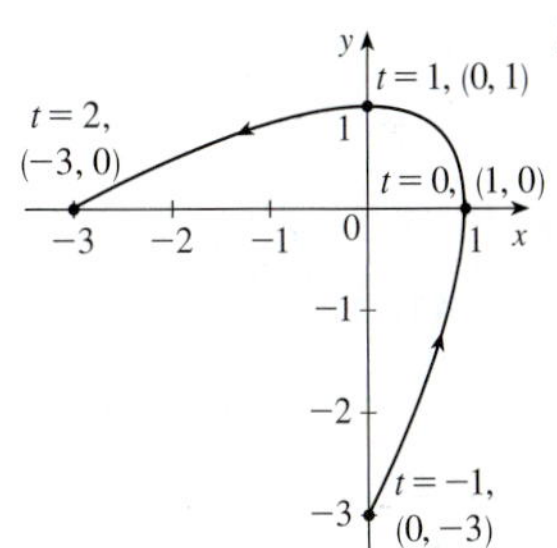

3.

5. (a)

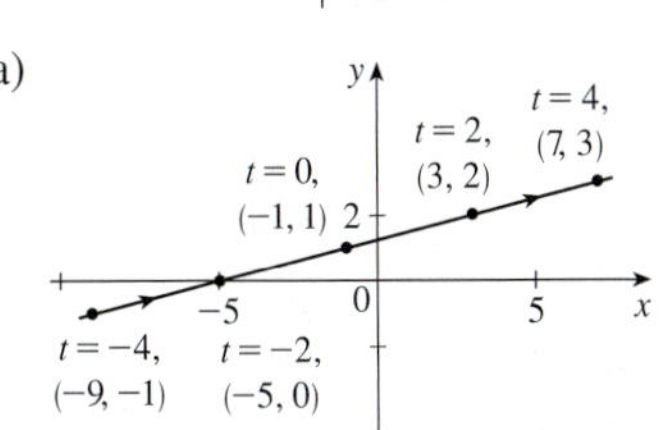

(b) $y = \frac{1}{4}x + \frac{5}{4}$

7. (a)

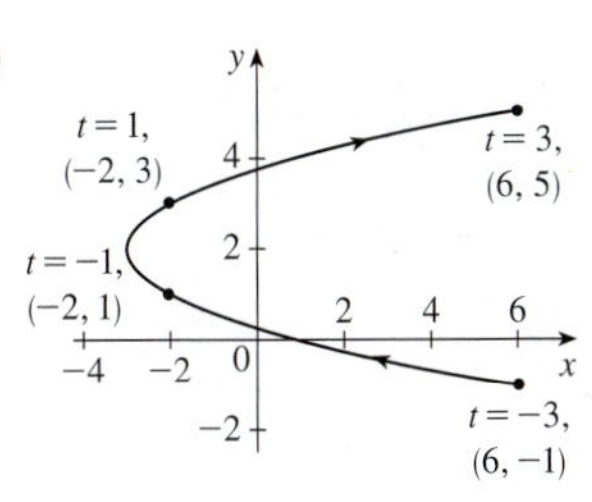

(b) $x = y^2 - 4y + 1$, $-1 \leqslant y \leqslant 5$

9. (a)

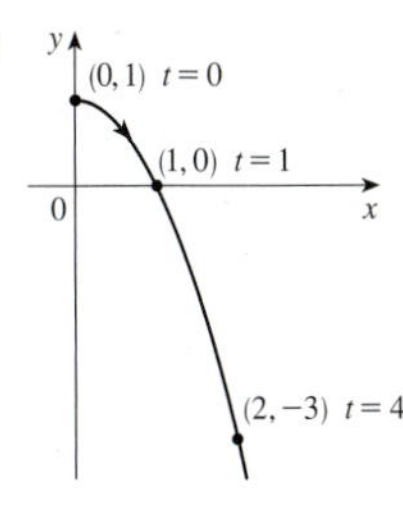

(b) $y = 1 - x^2, x \geqslant 0$

11. (a) $x^2 + y^2 = 1, y \geqslant 0$ (b)

13. (a) $y = 1/x, y > 1$ (b)

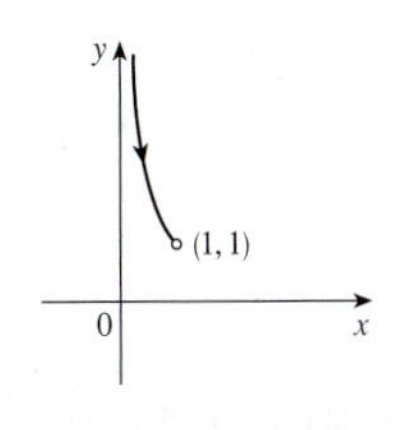

15. (a) $x = e^{2y}$ (b)

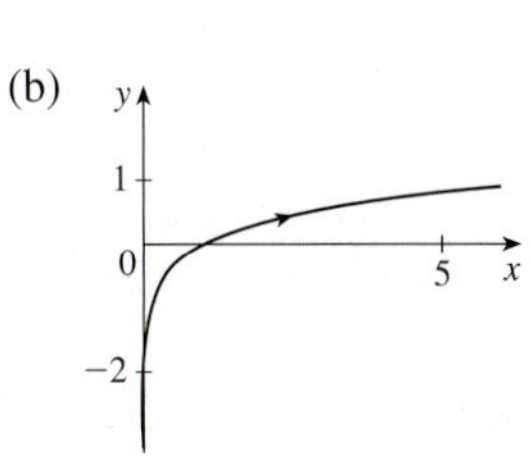

17. (a) $y^2 - x^2 = 1, y \geqslant 1$ (b)

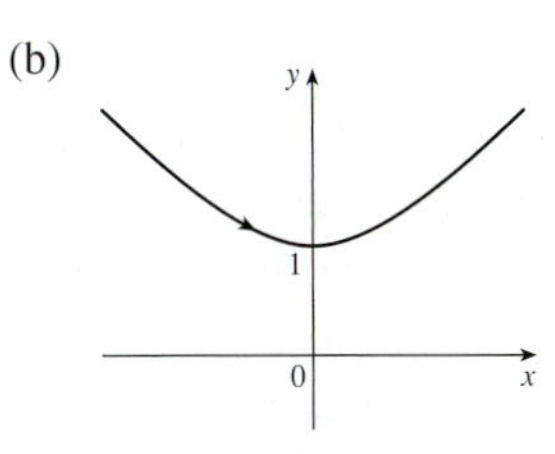

19. Moves counterclockwise along the circle
$$\left(\frac{x-5}{2}\right)^2 + \left(\frac{y-3}{2}\right)^2 = 1 \text{ from } (3, 3) \text{ to } (7, 3)$$

21. Moves 3 times clockwise around the ellipse $(x^2/25) + (y^2/4) = 1$, starting and ending at $(0, -2)$

23. It is contained in the rectangle described by $1 \leqslant x \leqslant 4$ and $2 \leqslant y \leqslant 3$.

25.

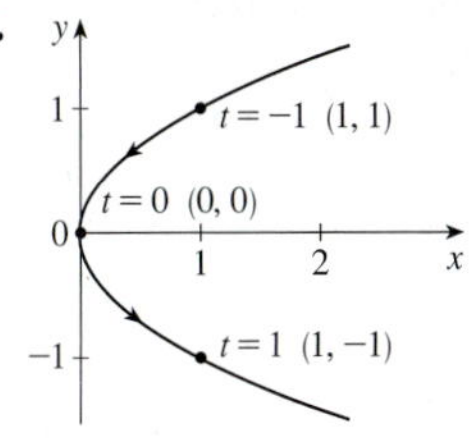

27.

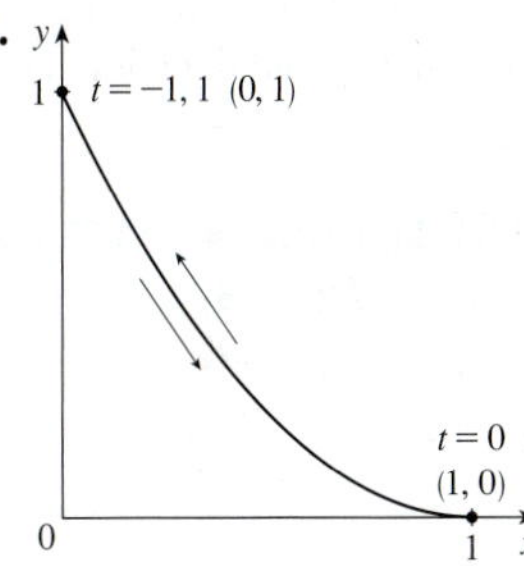

29.

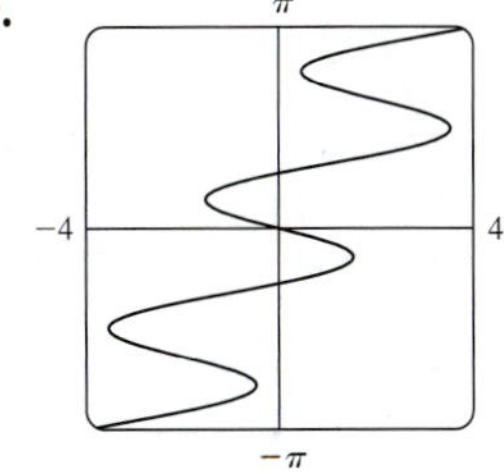

31. (b) $x = -2 + 5t, y = 7 - 8t, 0 \leqslant t \leqslant 1$

33. (a) $x = 2 \cos t, y = 1 - 2 \sin t, 0 \leqslant t \leqslant 2\pi$
(b) $x = 2 \cos t, y = 1 + 2 \sin t, 0 \leqslant t \leqslant 6\pi$
(c) $x = 2 \cos t, y = 1 + 2 \sin t, \pi/2 \leqslant t \leqslant 3\pi/2$

37. The curve $y = x^{2/3}$ is generated in (a). In (b), only the portion with $x \geqslant 0$ is generated, and in (c) we get only the portion with $x > 0$.

41. $x = a \cos \theta, y = b \sin \theta$; $(x^2/a^2) + (y^2/b^2) = 1$, ellipse

43.

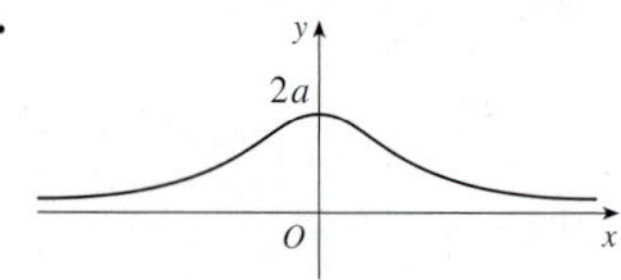

45. (a) Two points of intersection

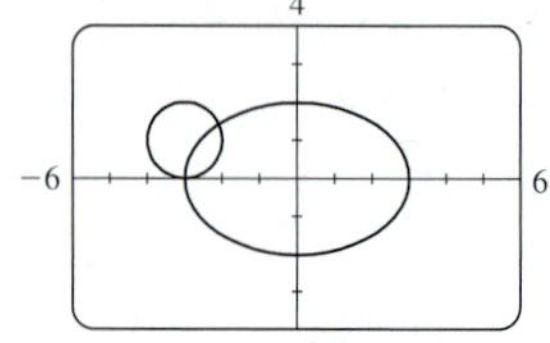

(b) One collision point at $(-3, 0)$ when $t = 3\pi/2$
(c) There are still two intersection points, but no collision point.

47. For $c = 0$, there is a cusp; for $c > 0$, there is a loop whose size increases as c increases.

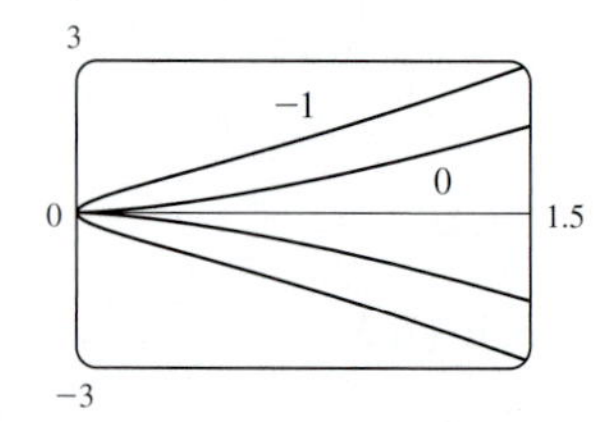

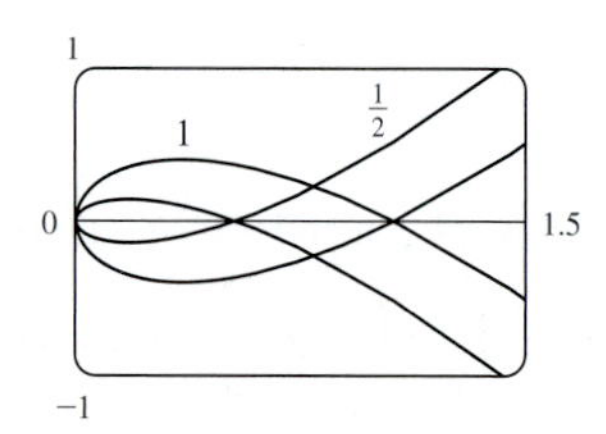

49. The curves roughly follow the line $y = x$, and they start having loops when a is between 1.4 and 1.6 (more precisely, when $a > \sqrt{2}$). The loops increase in size as a increases.
51. As n increases, the number of oscillations increases; a and b determine the width and height.

EXERCISES 10.2 ■ PAGE 695

1. $\frac{1}{2}(1 + t)^{3/2}$ **3.** $y = -x$ **5.** $y = \pi x + \pi^2$
7. $y = 2x + 1$
9. $y = 3x + 3$

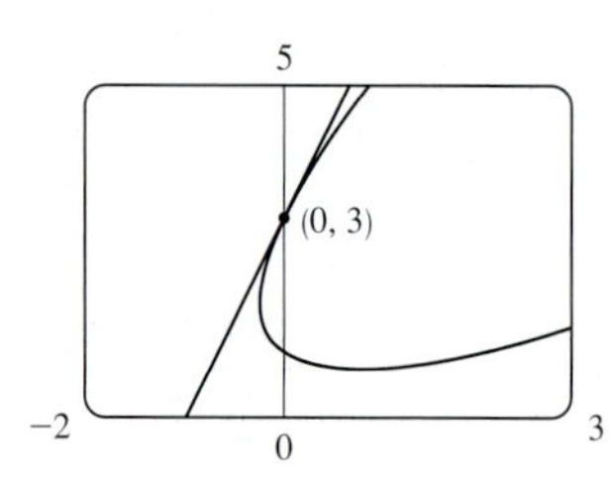

11. $\dfrac{2t + 1}{2t}, -\dfrac{1}{4t^3}, t < 0$ **13.** $e^{-2t}(1 - t), e^{-3t}(2t - 3), t > \frac{3}{2}$
15. $\dfrac{t + 1}{t - 1}, \dfrac{-2t}{(t - 1)^3}, 0 < t < 1$
17. Horizontal at $(0, -3)$, vertical at $(\pm 2, -2)$
19. Horizontal at $\left(\frac{1}{2}, -1\right)$ and $\left(-\frac{1}{2}, 1\right)$, no vertical
21. $(0.6, 2)$; $\left(5 \cdot 6^{-6/5}, e^{6^{-1/5}}\right)$
23.

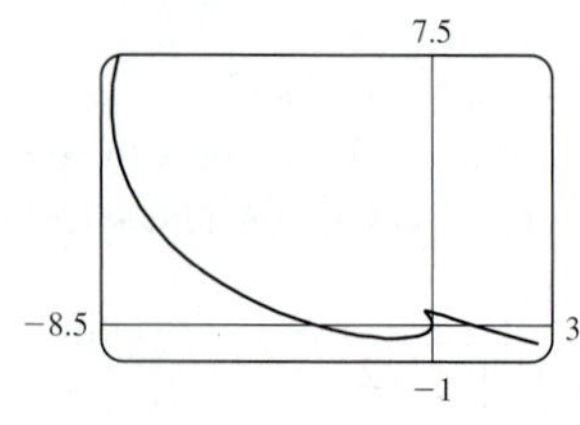

25. $y = x, y = -x$

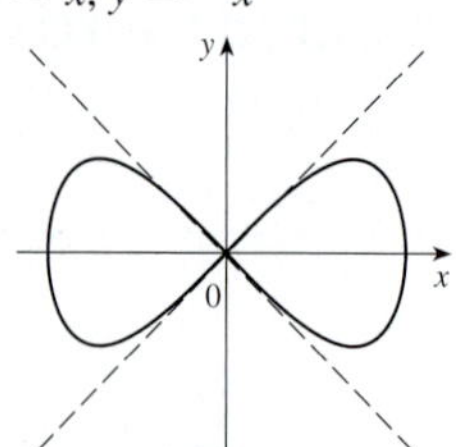

27. (a) $d \sin \theta/(r - d \cos \theta)$ **29.** $(4, 0)$ **31.** πab
33. $\frac{24}{5}$ **35.** $2\pi r^2 + \pi d^2$ **37.** $\int_0^2 \sqrt{2 + 2e^{-2t}}\, dt \approx 3.1416$
39. $\int_0^{4\pi} \sqrt{5 - 4 \cos t}\, dt \approx 26.7298$ **41.** $4\sqrt{2} - 2$
43. $\frac{1}{2}\sqrt{2} + \frac{1}{2}\ln\left(1 + \sqrt{2}\right)$
45. $\sqrt{2}\,(e^\pi - 1)$

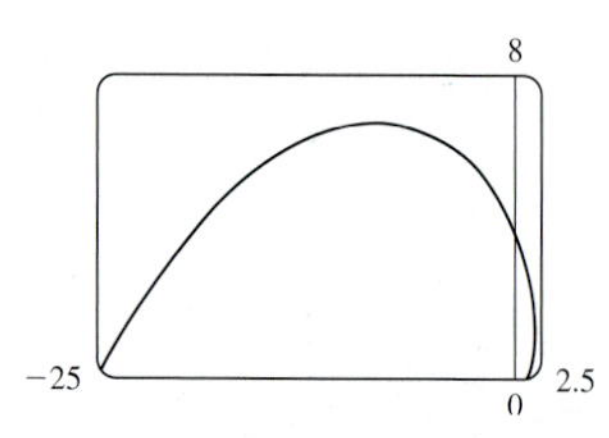

47. 16.7102

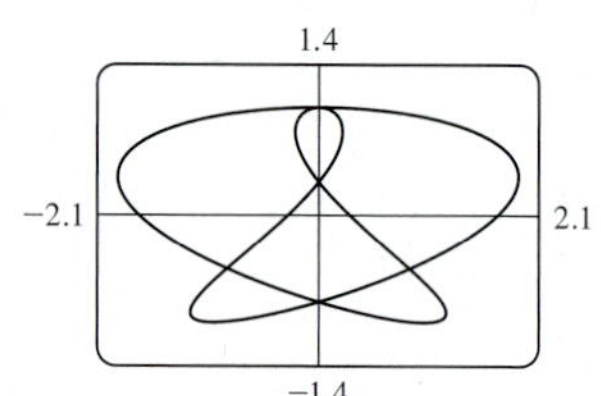

49. 612.3053 **51.** $6\sqrt{2}, \sqrt{2}$
55. (a)

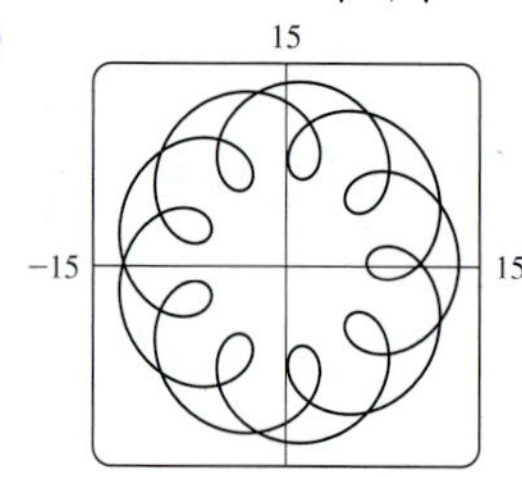

$t \in [0, 4\pi]$

(b) 294
57. $\int_0^{\pi/2} 2\pi t \cos t \sqrt{t^2 + 1}\, dt \approx 4.7394$
59. $\int_0^1 2\pi e^{-t}\sqrt{1 + 2e^t + e^{2t} + e^{-2t}}\, dt \approx 10.6705$
61. $\frac{2}{1215}\pi\left(247\sqrt{3} + 64\right)$ **63.** $\frac{6}{5}\pi a^2$
65. $\frac{24}{5}\pi\left(949\sqrt{26} + 1\right)$ **71.** $\frac{1}{4}$

EXERCISES 10.3 ■ PAGE 706

1. (a)

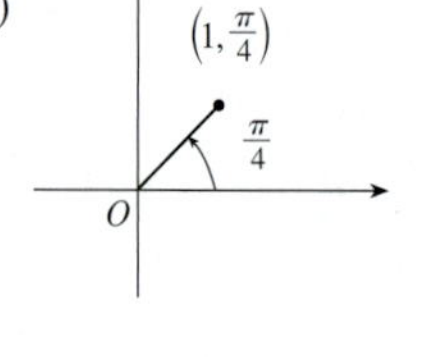

$(1, 9\pi/4), (-1, 5\pi/4)$

(b)

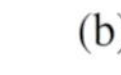

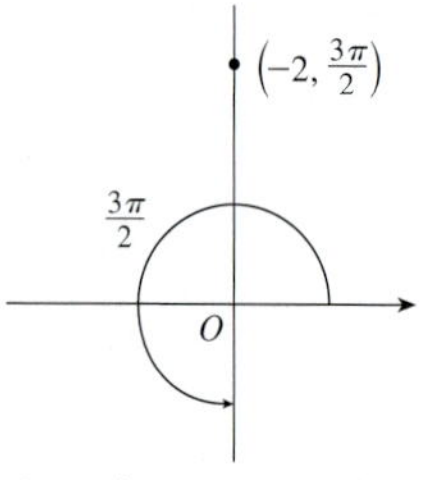

$(2, \pi/2), (-2, 7\pi/2)$

(c)

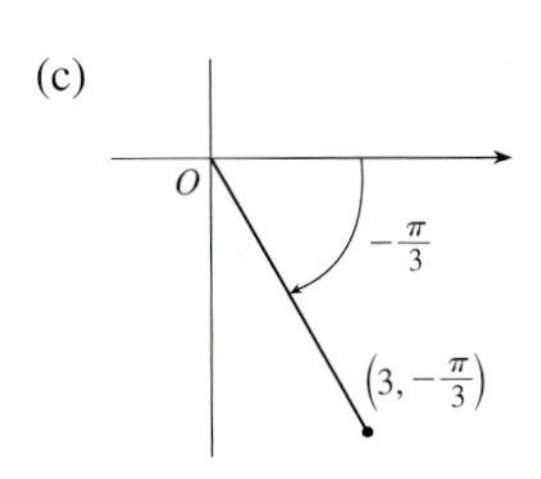

$(3, 5\pi/3), (-3, 2\pi/3)$

3. (a)

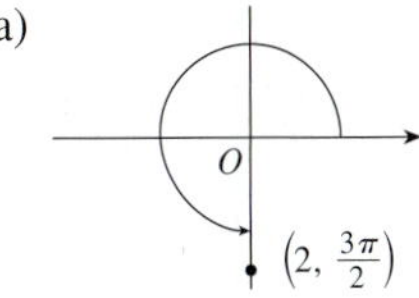

$(0, -2)$

(b) $(1, 1)$

(c)

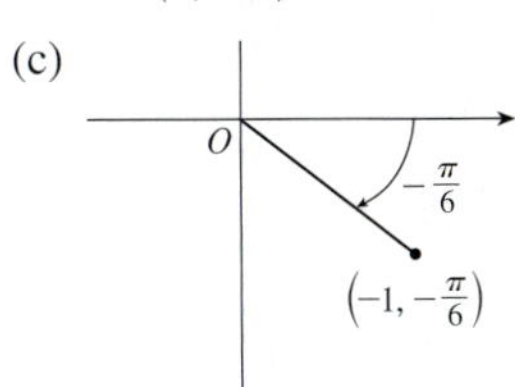

$\left(-\sqrt{3}/2, 1/2\right)$

5. (a) (i) $\left(4\sqrt{2}, 3\pi/4\right)$ (ii) $\left(-4\sqrt{2}, 7\pi/4\right)$
(b) (i) $(6, \pi/3)$ (ii) $(-6, 4\pi/3)$

7.

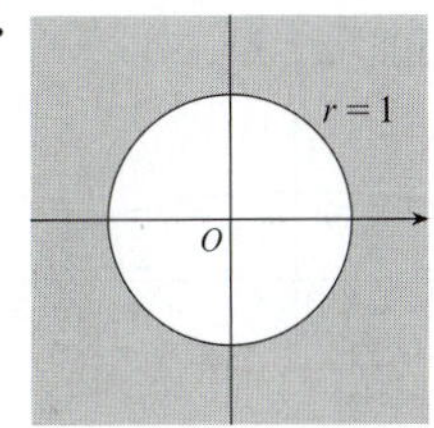

9.

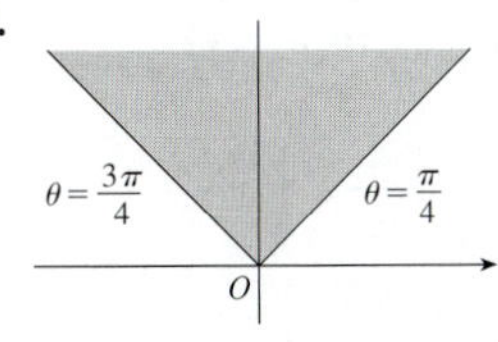

11.

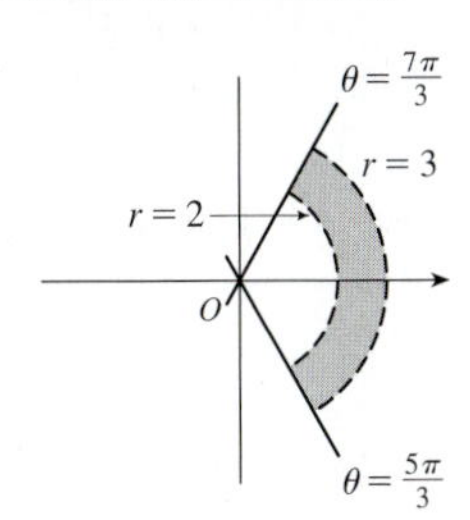

13. $2\sqrt{7}$ **15.** Circle, center O, radius $\sqrt{5}$
17. Circle, center $(5/2, 0)$, radius $5/2$
19. Hyperbola, center O, foci on x-axis
21. $r = 2 \csc\theta$ **23.** $r = 1/(\sin\theta - 3\cos\theta)$
25. $r = 2c\cos\theta$ **27.** (a) $\theta = \pi/6$ (b) $x = 3$

29.

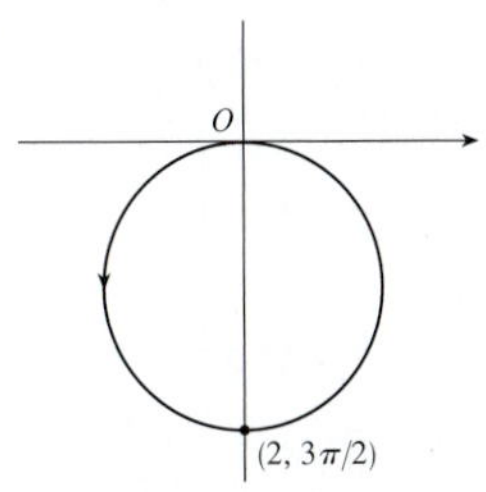

31.

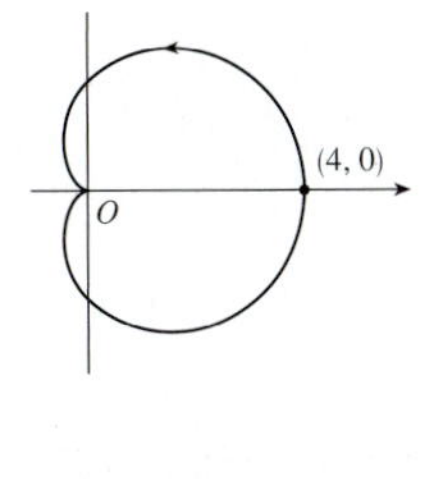

33.

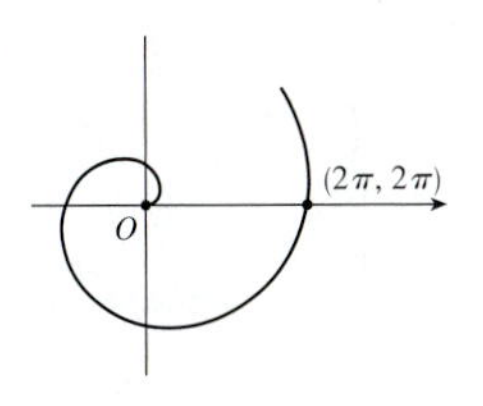

35.

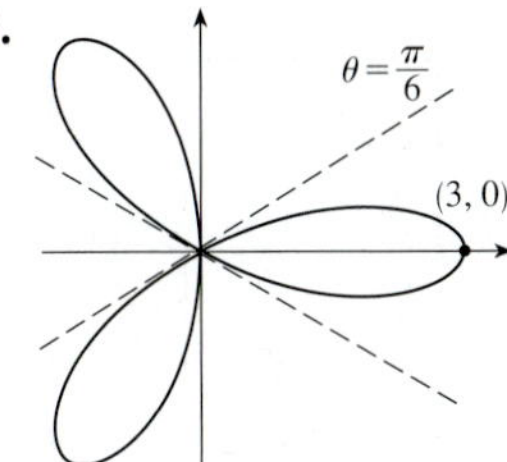

37.

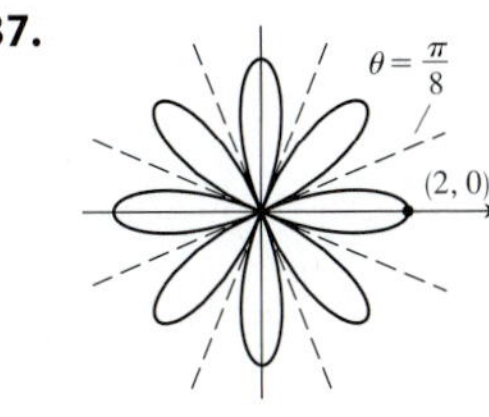

39.

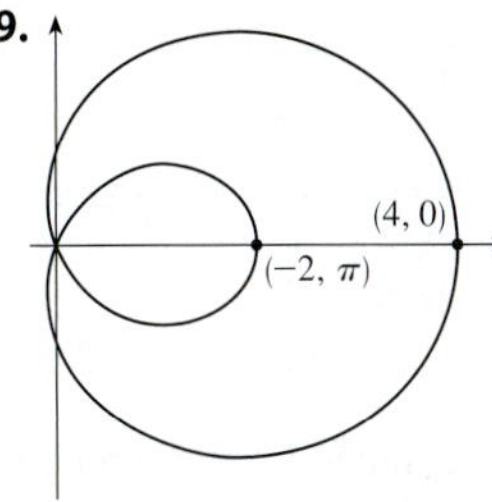

41.

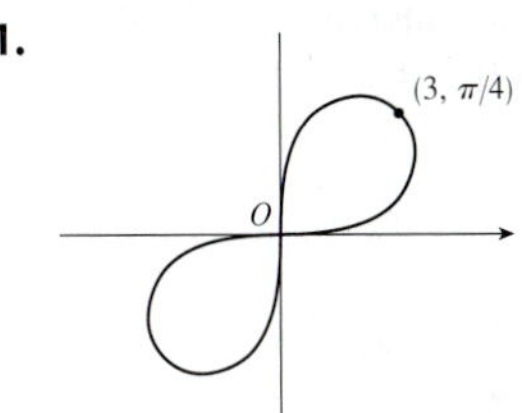

43.

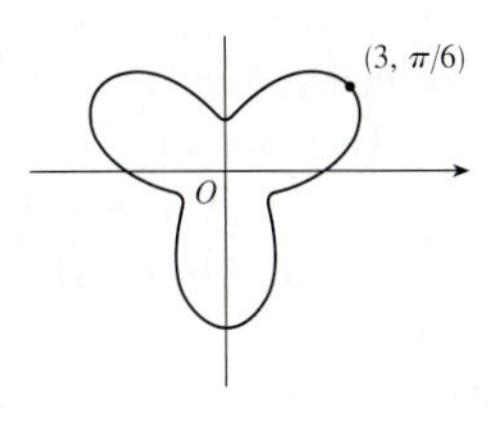

45.

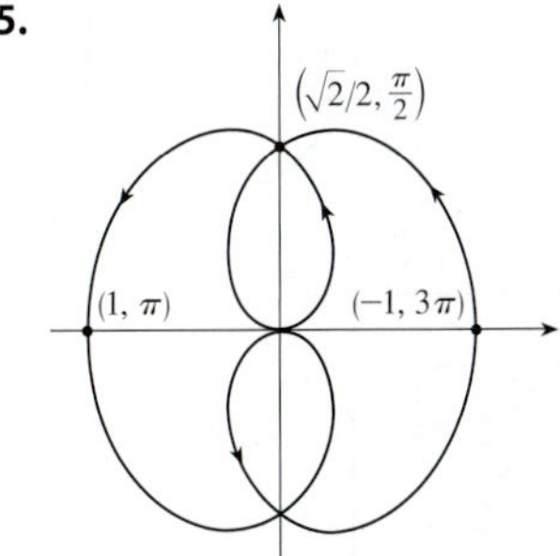

47.

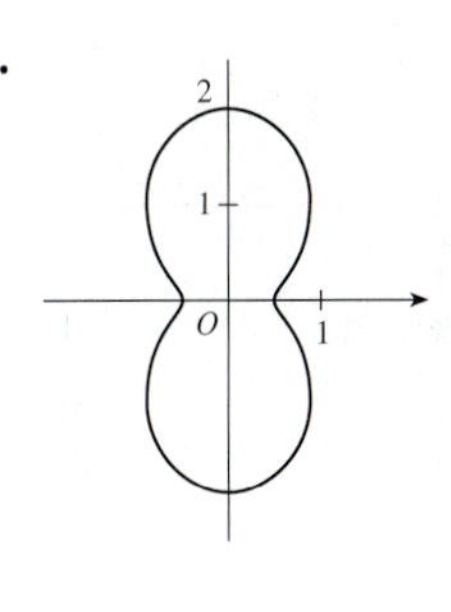

49.

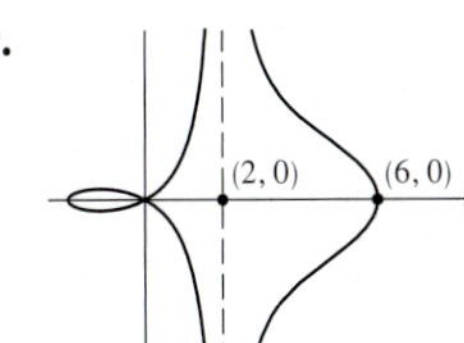

51.

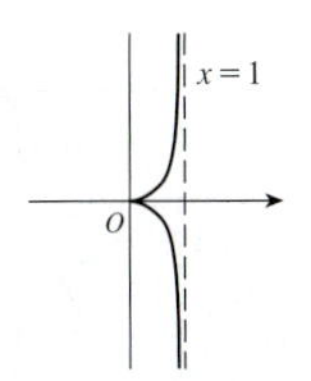

53. (a) For $c < -1$, the inner loop begins at $\theta = \sin^{-1}(-1/c)$ and ends at $\theta = \pi - \sin^{-1}(-1/c)$; for $c > 1$, it begins at $\theta = \pi + \sin^{-1}(1/c)$ and ends at $\theta = 2\pi - \sin^{-1}(1/c)$.
55. $1/\sqrt{3}$ **57.** $-\pi$ **59.** 1
61. Horizontal at $\left(3/\sqrt{2}, \pi/4\right), \left(-3/\sqrt{2}, 3\pi/4\right)$; vertical at $(3, 0), (0, \pi/2)$
63. Horizontal at $\left(\frac{3}{2}, \pi/3\right)$, $(0, \pi)$ [the pole], and $\left(\frac{3}{2}, 5\pi/3\right)$; vertical at $(2, 0)$, $\left(\frac{1}{2}, 2\pi/3\right)$, $\left(\frac{1}{2}, 4\pi/3\right)$
65. Center $(b/2, a/2)$, radius $\sqrt{a^2 + b^2}/2$

67.

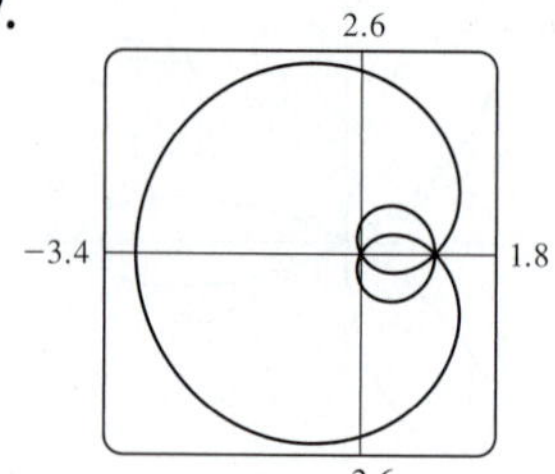

69.

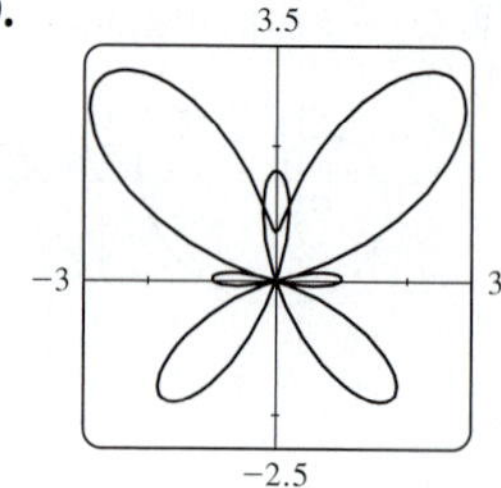

71.

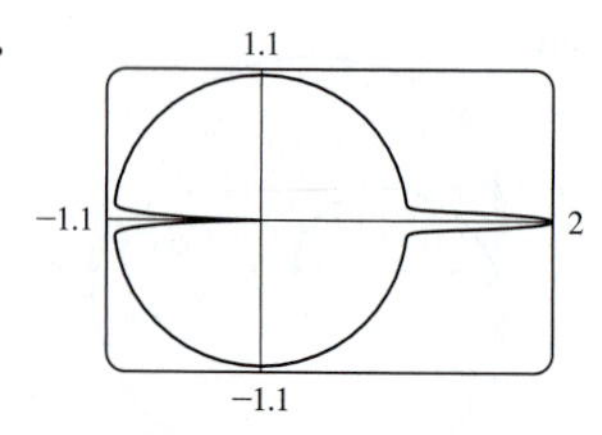

73. By counterclockwise rotation through angle $\pi/6$, $\pi/3$, or α about the origin

75. For $c = 0$, the curve is a circle. As c increases, the left side gets flatter, then has a dimple for $0.5 < c < 1$, a cusp for $c = 1$, and a loop for $c > 1$.

EXERCISES 10.4 ■ PAGE 712

1. $e^{-\pi/4} - e^{-\pi/2}$ **3.** $\pi/2$ **5.** $\frac{1}{2}$ **7.** $\frac{41}{4}\pi$

9. π

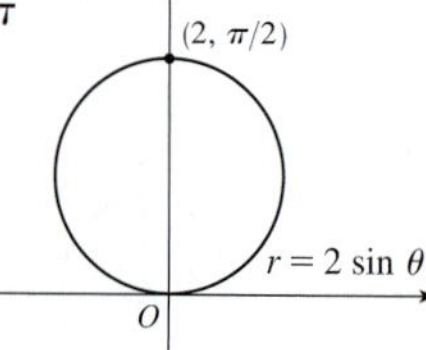

11. 11π

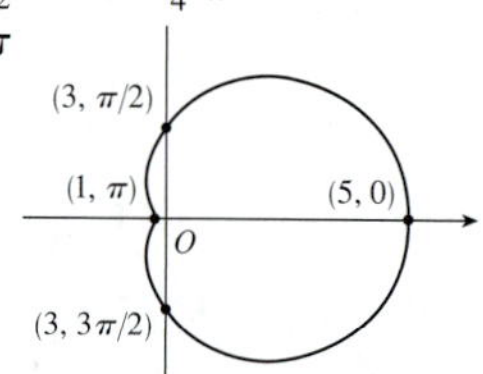

13. $\frac{9}{2}\pi$

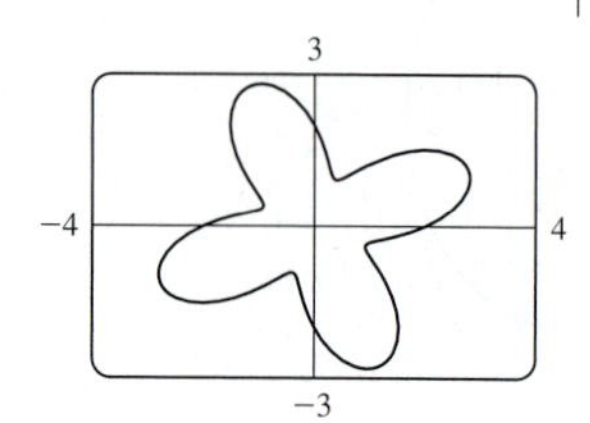

15. $\frac{3}{2}\pi$

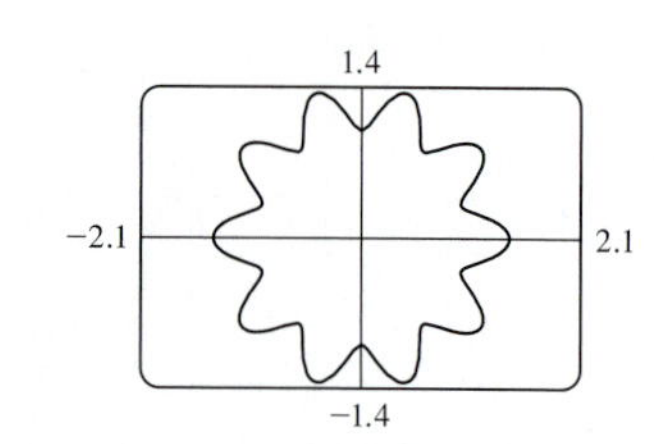

17. $\frac{4}{3}\pi$ **19.** $\frac{1}{16}\pi$ **21.** $\pi - \frac{3}{2}\sqrt{3}$ **23.** $\frac{4}{3}\pi + 2\sqrt{3}$

25. $4\sqrt{3} - \frac{4}{3}\pi$ **27.** π **29.** $\frac{9}{8}\pi - \frac{9}{4}$ **31.** $\frac{1}{2}\pi - 1$

33. $-\sqrt{3} + 2 + \frac{1}{3}\pi$ **35.** $\frac{1}{4}(\pi + 3\sqrt{3})$

37. $(\frac{1}{2}, \pi/6)$, $(\frac{1}{2}, 5\pi/6)$, and the pole

39. $(1, \theta)$ where $\theta = \pi/12, 5\pi/12, 13\pi/12, 17\pi/12$ and $(-1, \theta)$ where $\theta = 7\pi/12, 11\pi/12, 19\pi/12, 23\pi/12$

41. $(\frac{1}{2}\sqrt{3}, \pi/3)$, $(\frac{1}{2}\sqrt{3}, 2\pi/3)$, and the pole

43. Intersection at $\theta \approx 0.89, 2.25$; area ≈ 3.46

45. 2π **47.** $\frac{8}{3}[(\pi^2 + 1)^{3/2} - 1]$ **49.** $\frac{16}{3}$

51. 2.4221 **53.** 8.0091

55. (b) $2\pi(2 - \sqrt{2})$

EXERCISES 10.5 ■ PAGE 720

1. $(0, 0)$, $(0, \frac{3}{2})$, $y = -\frac{3}{2}$

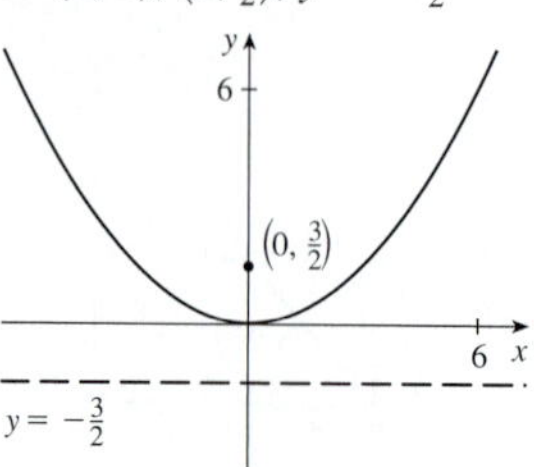

3. $(0, 0)$, $(-\frac{1}{2}, 0)$, $x = \frac{1}{2}$

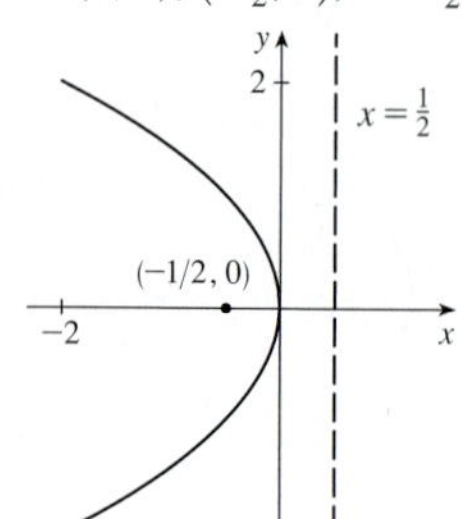

5. $(-2, 3)$ $(-2, 5)$, $y = 1$

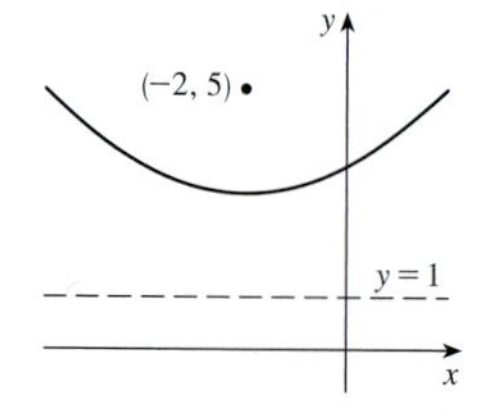

7. $(4, -3)$, $(\frac{7}{2}, -3)$, $x = \frac{9}{2}$

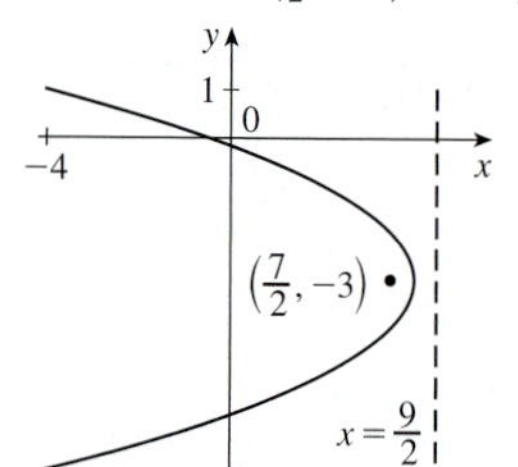

9. $x = -y^2$, focus $(\frac{1}{4}, 0)$, directrix $x = \frac{1}{4}$

11. $(0, \pm 2)$, $(0, \pm\sqrt{2})$

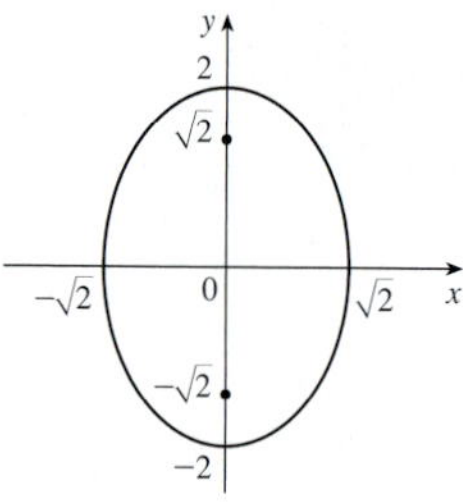

13. $(\pm 3, 0)$, $(\pm 2\sqrt{2}, 0)$

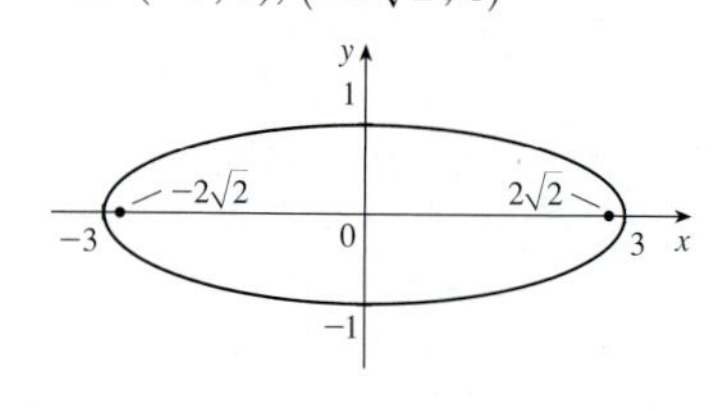

15. $(1, \pm 3)$, $(1, \pm\sqrt{5})$

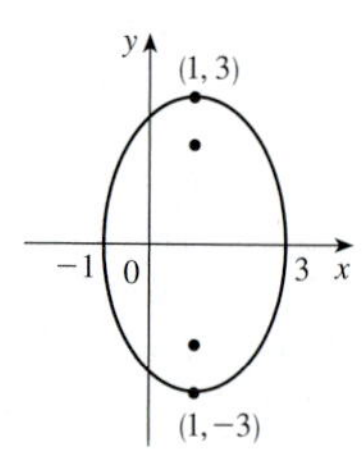

17. $\dfrac{x^2}{4} + \dfrac{y^2}{9} = 1$, foci $(0, \pm\sqrt{5})$

19. $(0, \pm 5)$; $(0, \pm\sqrt{34})$; $y = \pm\frac{5}{3}x$

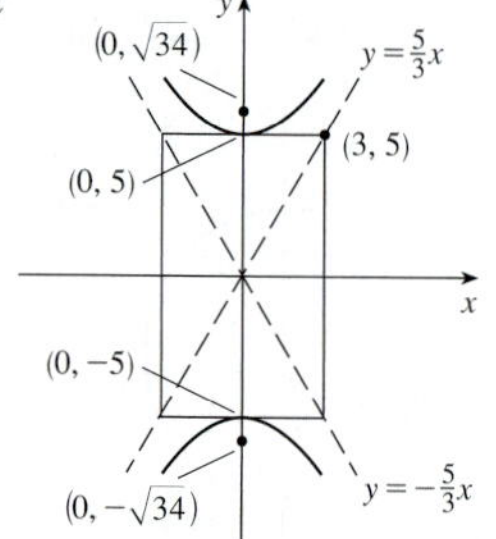

21. $(\pm 10, 0)$, $(\pm 10\sqrt{2}, 0)$, $y = \pm x$

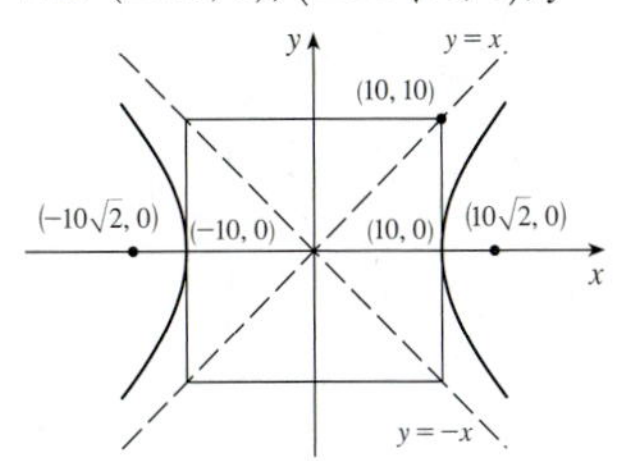

23. $(\pm 1, 1)$, $(\pm\sqrt{2}, 1)$, $y - 1 = \pm x$

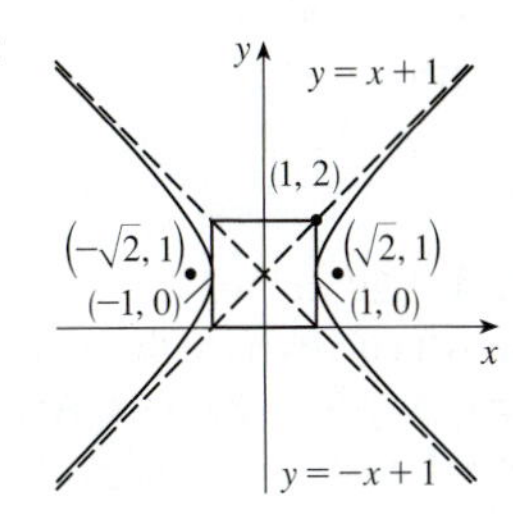

25. Hyperbola, $(\pm 1, 0)$, $(\pm\sqrt{5}, 0)$

27. Ellipse, $(\pm\sqrt{2}, 1)$, $(\pm 1, 1)$

29. Parabola, $(1, -2)$, $\left(1, -\frac{11}{6}\right)$

31. $y^2 = 4x$ **33.** $y^2 = -12(x + 1)$

35. $(y + 1)^2 = -\frac{1}{2}(x - 3)$

37. $\dfrac{x^2}{25} + \dfrac{y^2}{21} = 1$ **39.** $\dfrac{x^2}{12} + \dfrac{(y - 4)^2}{16} = 1$

41. $\dfrac{(x + 1)^2}{12} + \dfrac{(y - 4)^2}{16} = 1$ **43.** $\dfrac{x^2}{9} - \dfrac{y^2}{16} = 1$

45. $\dfrac{(y - 1)^2}{25} - \dfrac{(x + 3)^2}{39} = 1$ **47.** $\dfrac{x^2}{9} - \dfrac{y^2}{36} = 1$

49. $\dfrac{x^2}{3{,}763{,}600} + \dfrac{y^2}{3{,}753{,}196} = 1$

51. (a) $\dfrac{121x^2}{1{,}500{,}625} - \dfrac{121y^2}{3{,}339{,}375} = 1$ (b) ≈ 248 mi

55. (a) Ellipse (b) Hyperbola (c) No curve

59. 15.9

61. $\dfrac{b^2c}{a} + ab\ln\left(\dfrac{a}{b + c}\right)$ where $c^2 = a^2 + b^2$

63. $(0, 4/\pi)$

EXERCISES 10.6 ■ PAGE 728

1. $r = \dfrac{4}{2 + \cos\theta}$ **3.** $r = \dfrac{6}{2 + 3\sin\theta}$

5. $r = \dfrac{10}{3 - 2\cos\theta}$ **7.** $r = \dfrac{6}{1 + \sin\theta}$

9. (a) $\frac{4}{5}$ (b) Ellipse (c) $y = -1$

(d)

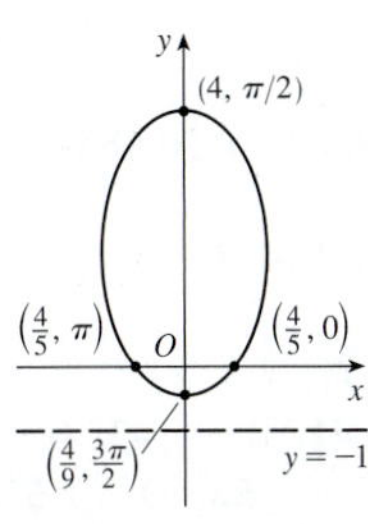

11. (a) 1 (b) Parabola (c) $y = \frac{2}{3}$

(d)

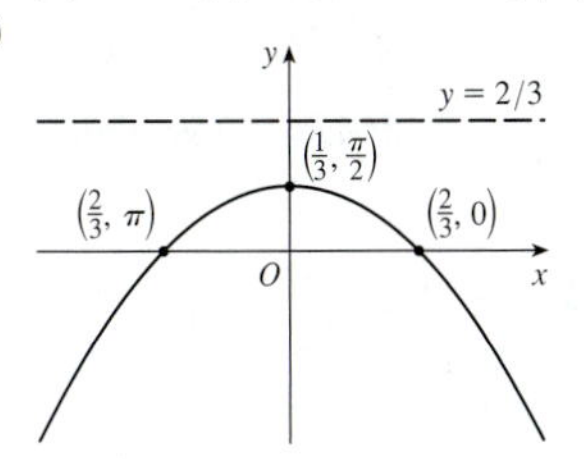

13. (a) $\frac{1}{3}$ (b) Ellipse (c) $x = \frac{9}{2}$

(d)

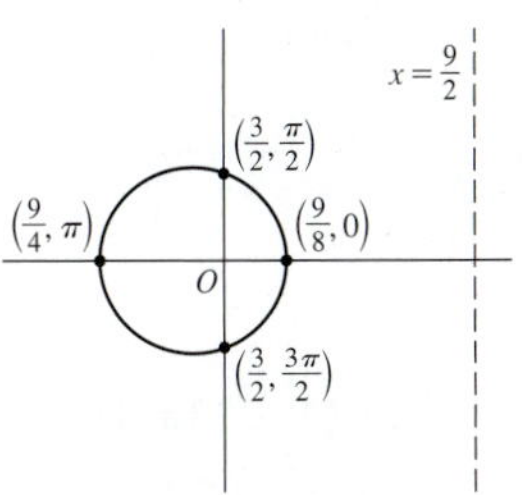

15. (a) 2 (b) Hyperbola (c) $x = -\frac{3}{8}$

(d)

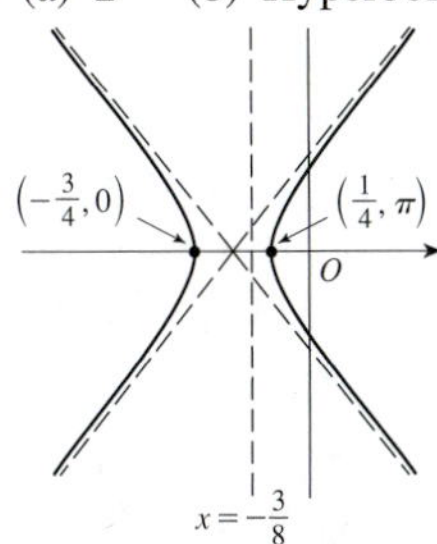

17. (a) $2, y = -\frac{1}{2}$

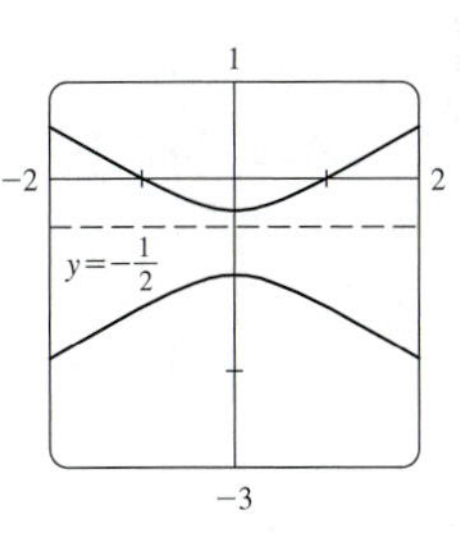

(b) $r = \dfrac{1}{1 - 2\sin(\theta - 3\pi/4)}$

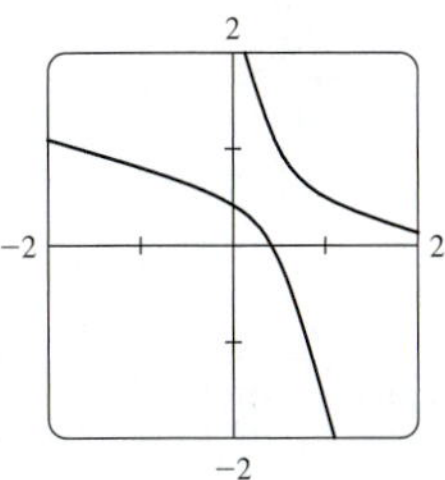

19. The ellipse is nearly circular when e is close to 0 and becomes more elongated as $e \to 1^-$. At $e = 1$, the curve becomes a parabola.

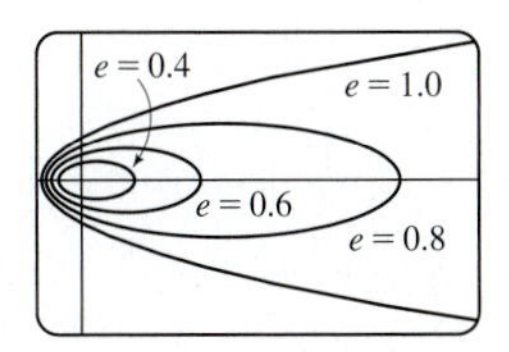

25. $r = \dfrac{2.26 \times 10^8}{1 + 0.093\cos\theta}$ **27.** $r = \dfrac{1.07}{1 + 0.97\cos\theta}$; 35.64 AU

29. 7.0×10^7 km **31.** 3.6×10^8 km

CHAPTER 10 REVIEW ■ PAGE 729

True-False Quiz

1. False **3.** False **5.** True **7.** False **9.** True

Exercises

1. $x = y^2 - 8y + 12$

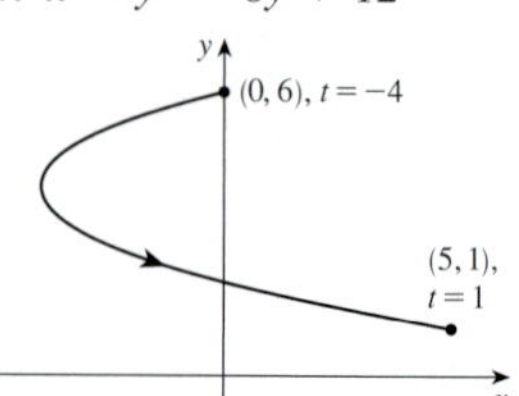

3. $y = 1/x$

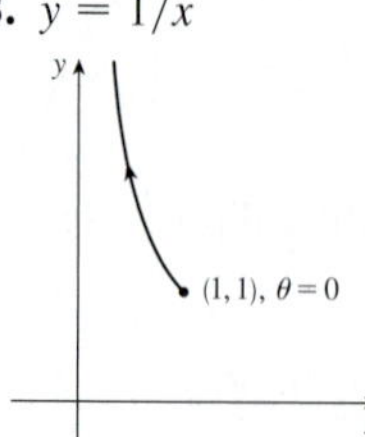

5. $x = t,\ y = \sqrt{t};\ x = t^4,\ y = t^2;$
$x = \tan^2 t,\ y = \tan t,\ 0 \leqslant t < \pi/2$

7. (a)

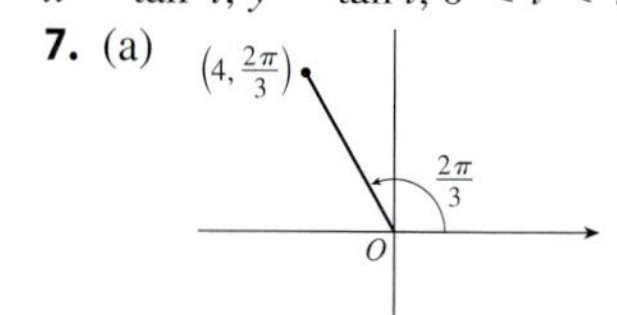

$(-2, 2\sqrt{3})$

(b) $(3\sqrt{2}, 3\pi/4)$, $(-3\sqrt{2}, 7\pi/4)$

9.

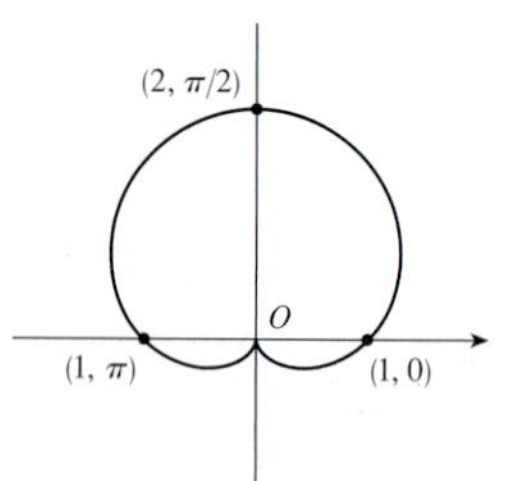

11.

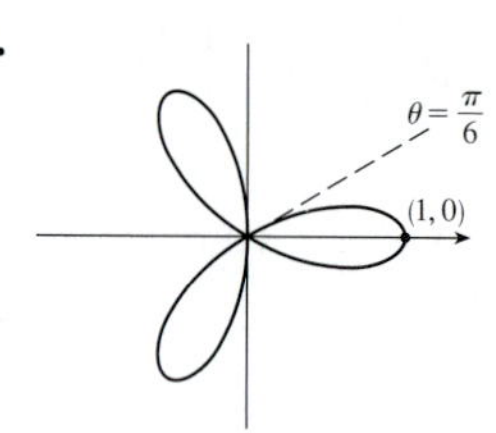

13.

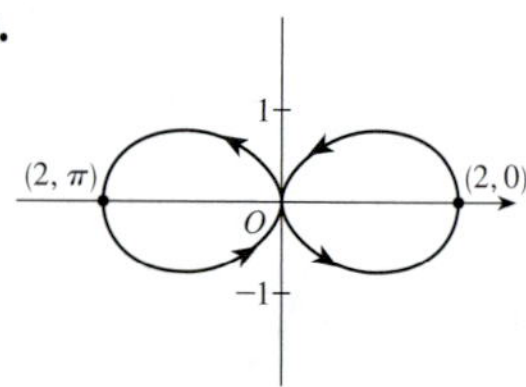

15.

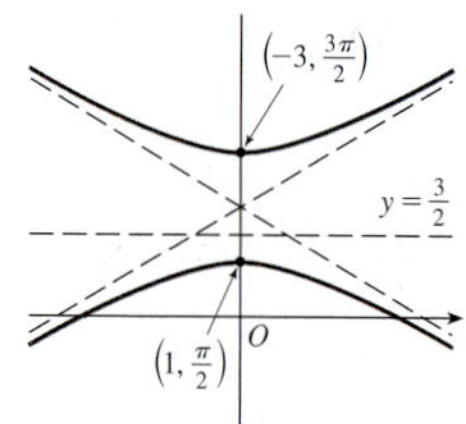

17. $r = \dfrac{2}{\cos\theta + \sin\theta}$

19.

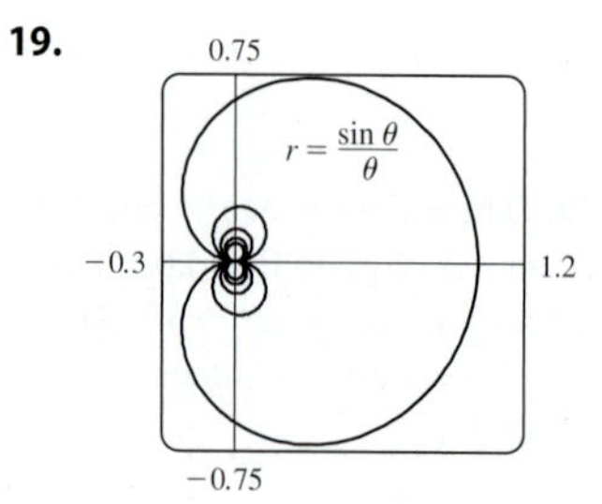

21. 2 **23.** −1

25. $\dfrac{1 + \sin t}{1 + \cos t}, \dfrac{1 + \cos t + \sin t}{(1 + \cos t)^3}$ **27.** $\left(\frac{11}{8}, \frac{3}{4}\right)$

29. Vertical tangent at $\left(\frac{3}{2}a, \pm\frac{1}{2}\sqrt{3}\,a\right), (-3a, 0)$; horizontal tangent at $(a, 0), \left(-\frac{1}{2}a, \pm\frac{3}{2}\sqrt{3}\,a\right)$

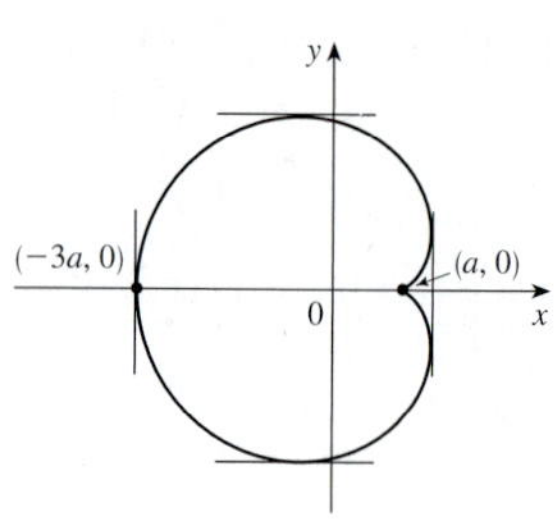

31. 18 **33.** $(2, \pm\pi/3)$ **35.** $\frac{1}{2}(\pi - 1)$

37. $2(5\sqrt{5} - 1)$

39. $\dfrac{2\sqrt{\pi^2 + 1} - \sqrt{4\pi^2 + 1}}{2\pi} + \ln\left(\dfrac{2\pi + \sqrt{4\pi^2 + 1}}{\pi + \sqrt{\pi^2 + 1}}\right)$

41. $471{,}295\pi/1024$

43. All curves have the vertical asymptote $x = 1$. For $c < -1$, the curve bulges to the right. At $c = -1$, the curve is the line $x = 1$. For $-1 < c < 0$, it bulges to the left. At $c = 0$ there is a cusp at $(0, 0)$. For $c > 0$, there is a loop.

45. $(\pm 1, 0), (\pm 3, 0)$

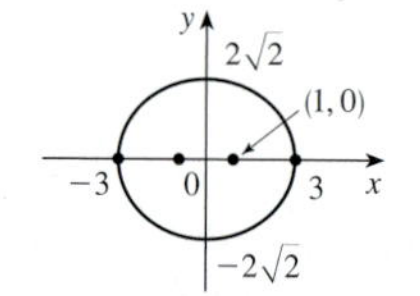

47. $\left(-\frac{25}{24}, 3\right), (-1, 3)$

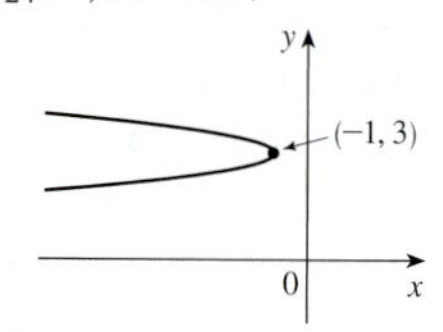

49. $\dfrac{x^2}{25} + \dfrac{y^2}{9} = 1$ **51.** $\dfrac{y^2}{72/5} - \dfrac{x^2}{8/5} = 1$

53. $\dfrac{x^2}{25} + \dfrac{(8y - 399)^2}{160{,}801} = 1$ **55.** $r = \dfrac{4}{3 + \cos\theta}$

57. $x = a(\cot\theta + \sin\theta\cos\theta),\ y = a(1 + \sin^2\theta)$

PROBLEMS PLUS ■ PAGE 732

1. $\ln(\pi/2)$ **3.** $\left[-\frac{3}{4}\sqrt{3}, \frac{3}{4}\sqrt{3}\right] \times [-1, 2]$

CHAPTER 11

EXERCISES 11.1 ■ PAGE 744

Abbreviations: C, convergent; D, divergent

1. (a) A sequence is an ordered list of numbers. It can also be defined as a function whose domain is the set of positive integers.
(b) The terms a_n approach 8 as n becomes large.
(c) The terms a_n become large as n becomes large.

3. $\frac{2}{3}, \frac{4}{5}, \frac{8}{7}, \frac{16}{9}, \frac{32}{11}$ **5.** $\frac{1}{5}, -\frac{1}{25}, \frac{1}{125}, -\frac{1}{625}, \frac{1}{3125}$ **7.** $\frac{1}{2}, \frac{1}{6}, \frac{1}{24}, \frac{1}{120}, \frac{1}{720}$

9. 1, 2, 7, 32, 157 **11.** $2, \frac{2}{3}, \frac{2}{5}, \frac{2}{7}, \frac{2}{9}$ **13.** $a_n = 1/(2n)$

15. $a_n = -3\left(-\frac{2}{3}\right)^{n-1}$ **17.** $a_n = (-1)^{n+1}\dfrac{n^2}{n + 1}$

19. 0.4286, 0.4615, 0.4737, 0.4800, 0.4839, 0.4865, 0.4884, 0.4898, 0.4909, 0.4918; yes; $\frac{1}{2}$

21. 0.5000, 1.2500, 0.8750, 1.0625, 0.9688, 1.0156, 0.9922, 1.0039, 0.9980, 1.0010; yes; 1

23. 5 **25.** D **27.** 0 **29.** 1 **31.** 2

33. D **35.** 0 **37.** 0 **39.** D **41.** 0 **43.** 0

45. 1 **47.** e^2 **49.** ln 2 **51.** $\pi/2$ **53.** D **55.** D

57. D **59.** $\pi/4$ **61.** D **63.** 0
65. (a) 1060, 1123.60, 1191.02, 1262.48, 1338.23 (b) D
67. (b) 5734 **69.** $-1 < r < 1$
71. Convergent by the Monotonic Sequence Theorem; $5 \leqslant L < 8$
73. Decreasing; yes **75.** Not monotonic; no
77. Increasing; yes
79. 2 **81.** $\frac{1}{2}(3 + \sqrt{5})$ **83.** (b) $\frac{1}{2}(1 + \sqrt{5})$
85. (a) 0 (b) 9, 11

EXERCISES 11.2 ■ PAGE 755

1. (a) A sequence is an ordered list of numbers whereas a series is the *sum* of a list of numbers.
(b) A series is convergent if the sequence of partial sums is a convergent sequence. A series is divergent if it is not convergent.
3. 2
5. 0.5, 0.55, 0.5611, 0.5648, 0.5663, 0.5671, 0.5675, 0.5677; C
7. 1, 1.7937, 2.4871, 3.1170, 3.7018, 4.2521, 4.7749, 5.2749; D
9. −2.40000, −1.92000, −2.01600, −1.99680, −2.00064, −1.99987, −2.00003, −1.99999, −2.00000, −2.00000; convergent, sum = −2

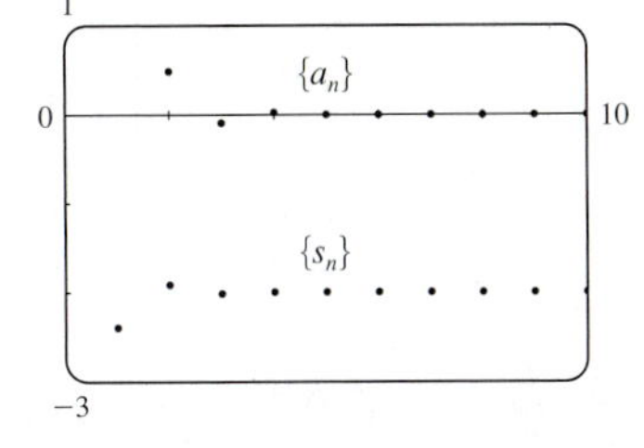

11. 0.44721, 1.15432, 1.98637, 2.88080, 3.80927, 4.75796, 5.71948, 6.68962, 7.66581, 8.64639; divergent

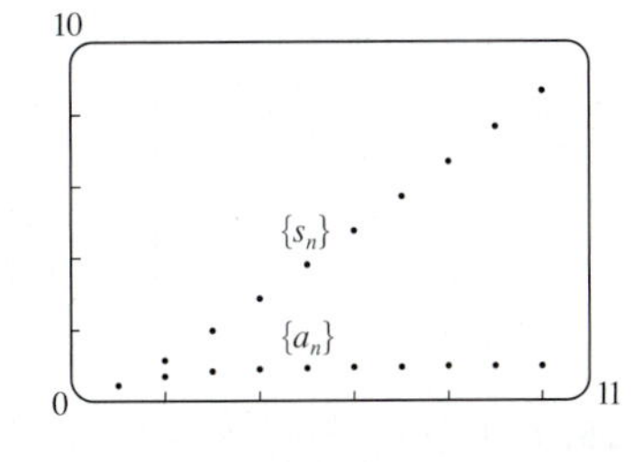

13. 1.00000, 1.33333, 1.50000, 1.60000, 1.66667, 1.71429, 1.75000, 1.77778, 1.80000, 1.81818; convergent, sum = 2

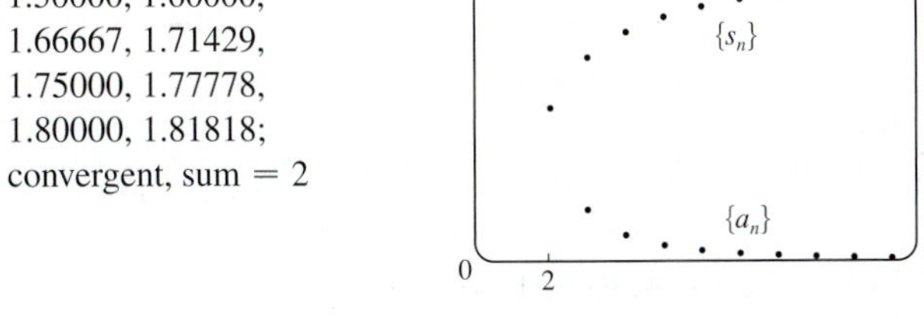

15. (a) Yes (b) No **17.** D **19.** $\frac{25}{3}$ **21.** $\frac{400}{9}$
23. $\frac{1}{7}$ **25.** D **27.** D **29.** D **31.** 9 **33.** D
35. $\dfrac{\sin 100}{1 - \sin 100}$
37. D **39.** D **41.** $e/(e-1)$ **43.** $\frac{3}{2}$ **45.** $\frac{11}{6}$
47. $e - 1$
49. (b) 1 (c) 2 (d) All rational numbers with a terminating decimal representation, except 0
51. $\frac{8}{9}$ **53.** $\frac{838}{333}$ **55.** 45,679/37,000
57. $-\frac{1}{5} < x < \frac{1}{5}$; $\dfrac{-5x}{1 + 5x}$
59. $-1 < x < 5$; $\dfrac{3}{5 - x}$
61. $x > 2$ or $x < -2$; $\dfrac{x}{x - 2}$ **63.** $x < 0$; $\dfrac{1}{1 - e^x}$
65. 1 **67.** $a_1 = 0$, $a_n = \dfrac{2}{n(n+1)}$ for $n > 1$, sum = 1
69. (a) 120 mg; 124 mg
(b) $Q_{n+1} = 100 + 0.20Q_n$ (c) 125 mg
71. (a) 157.875 mg; $\frac{3000}{19}(1 - 0.05^n)$ (b) 157.895 mg
73. (a) $S_n = \dfrac{D(1 - c^n)}{1 - c}$ (b) 5 **75.** $\frac{1}{2}(\sqrt{3} - 1)$
79. $\dfrac{1}{n(n+1)}$ **81.** The series is divergent.
87. $\{s_n\}$ is bounded and increasing.
89. (a) $0, \frac{1}{9}, \frac{2}{9}, \frac{1}{3}, \frac{2}{3}, \frac{7}{9}, \frac{8}{9}, 1$
91. (a) $\frac{1}{2}, \frac{5}{6}, \frac{23}{24}, \frac{119}{120}$; $\dfrac{(n+1)! - 1}{(n+1)!}$ (c) 1

EXERCISES 11.3 ■ PAGE 765

1. C

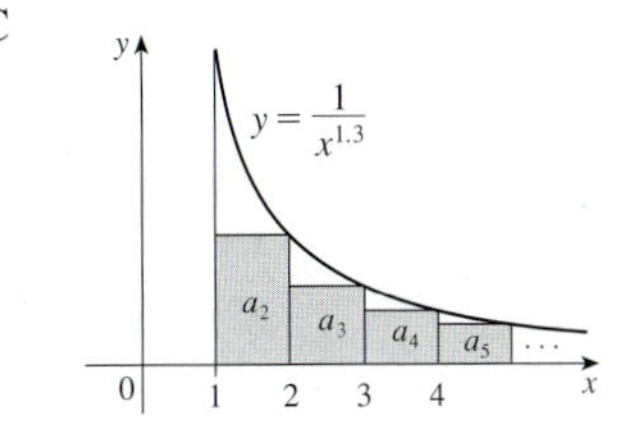

3. C **5.** D **7.** D **9.** C **11.** C **13.** D
15. C **17.** C **19.** D **21.** D **23.** C **25.** C
27. f is neither positive nor decreasing.
29. $p > 1$ **31.** $p < -1$ **33.** $(1, \infty)$
35. (a) $\frac{9}{10}\pi^4$ (b) $\frac{1}{90}\pi^4 - \frac{17}{16}$
37. (a) 1.54977, error $\leqslant$ 0.1 (b) 1.64522, error $\leqslant$ 0.005
(c) 1.64522 compared to 1.64493 (d) $n > 1000$
39. 0.00145 **45.** $b < 1/e$

EXERCISES 11.4 ■ PAGE 771

1. (a) Nothing (b) C **3.** C **5.** D **7.** C **9.** D
11. C **13.** C **15.** D **17.** D **19.** C **21.** D
23. C **25.** D **27.** C **29.** C **31.** D
33. 0.1993, error $< 2.5 \times 10^{-5}$
35. 0.0739, error $< 6.4 \times 10^{-8}$
45. Yes

EXERCISES 11.5 ■ PAGE 776

1. (a) A series whose terms are alternately positive and negative (b) $0 < b_{n+1} \leqslant b_n$ and $\lim_{n\to\infty} b_n = 0$, where $b_n = |a_n|$ (c) $|R_n| \leqslant b_{n+1}$
3. D **5.** C **7.** D **9.** C **11.** C **13.** D
15. C **17.** C **19.** D **21.** −0.5507 **23.** 5
25. 5 **27.** −0.4597 **29.** −0.1050
31. An underestimate
33. p is not a negative integer. **35.** $\{b_n\}$ is not decreasing.

EXERCISES 11.6 ■ PAGE 782

Abbreviations: AC, absolutely convergent; CC, conditionally convergent

1. (a) D (b) C (c) May converge or diverge
3. CC **5.** AC **7.** AC **9.** D **11.** AC
13. AC **15.** D **17.** AC **19.** AC **21.** AC
23. D **25.** AC **27.** AC **29.** D **31.** CC
33. AC **35.** D **37.** AC **39.** D **41.** AC
43. (a) and (d)
47. (a) $\frac{661}{960} \approx 0.68854$, error < 0.00521
(b) $n \geqslant 11$, 0.693109
53. (b) $\sum_{n=2}^{\infty} \frac{(-1)^n}{n \ln n}$; $\sum_{n=1}^{\infty} \frac{(-1)^{n-1}}{n}$

EXERCISES 11.7 ■ PAGE 786

1. D **3.** CC **5.** D **7.** D **9.** C **11.** C
13. C **15.** C **17.** C **19.** C **21.** D **23.** D
25. C **27.** C **29.** C **31.** D
33. C **35.** D **37.** C

EXERCISES 11.8 ■ PAGE 791

1. A series of the form $\sum_{n=0}^{\infty} c_n(x-a)^n$, where x is a variable and a and the c_n's are constants
3. 1, $(-1, 1)$ **5.** 1, $[-1, 1)$
7. ∞, $(-\infty, \infty)$ **9.** 4, $[-4, 4]$
11. $\frac{1}{4}$, $\left(-\frac{1}{4}, \frac{1}{4}\right]$ **13.** 2, $[-2, 2)$
15. 1, $[1, 3]$ **17.** 2, $[-4, 0)$
19. ∞, $(-\infty, \infty)$ **21.** b, $(a-b, a+b)$ **23.** 0, $\left\{\frac{1}{2}\right\}$
25. $\frac{1}{5}$, $\left[\frac{3}{5}, 1\right]$ **27.** ∞, $(-\infty, \infty)$
29. (a) Yes (b) No
31. k^k **33.** No
35. (a) $(-\infty, \infty)$
(b), (c)

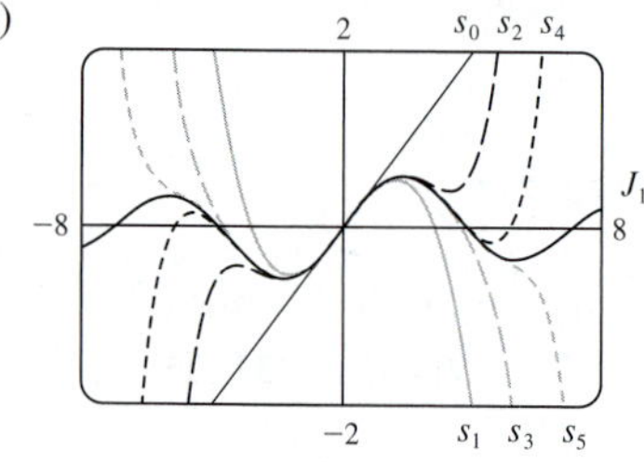

37. $(-1, 1)$, $f(x) = (1 + 2x)/(1 - x^2)$ **41.** 2

EXERCISES 11.9 ■ PAGE 797

1. 10 **3.** $\sum_{n=0}^{\infty} (-1)^n x^n$, $(-1, 1)$ **5.** $2\sum_{n=0}^{\infty} \frac{1}{3^{n+1}} x^n$, $(-3, 3)$

7. $\sum_{n=0}^{\infty} \frac{(-1)^n x^{4n+2}}{2^{4n+4}}$, $(-2, 2)$ **9.** $-\frac{1}{2} - \sum_{n=1}^{\infty} \frac{(-1)^n 3x^n}{2^{n+1}}$, $(-2, 2)$

11. $\sum_{n=0}^{\infty} \left(-1 - \frac{1}{3^{n+1}}\right) x^n$, $(-1, 1)$

13. (a) $\sum_{n=0}^{\infty} (-1)^n (n+1) x^n$, $R = 1$

(b) $\frac{1}{2} \sum_{n=0}^{\infty} (-1)^n (n+2)(n+1) x^n$, $R = 1$

(c) $\frac{1}{2} \sum_{n=2}^{\infty} (-1)^n n(n-1) x^n$, $R = 1$

15. $\ln 5 - \sum_{n=1}^{\infty} \frac{x^n}{n5^n}$, $R = 5$

17. $\sum_{n=0}^{\infty} (-1)^n 4^n (n+1) x^{n+1}$, $R = \frac{1}{4}$

19. $\sum_{n=0}^{\infty} (2n+1) x^n$, $R = 1$

21. $\sum_{n=0}^{\infty} (-1)^n x^{2n+2}$, $R = 1$

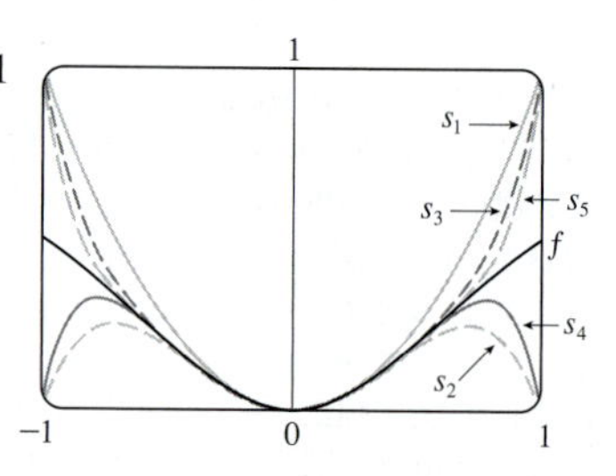

23. $\sum_{n=0}^{\infty} \frac{2x^{2n+1}}{2n+1}$, $R = 1$

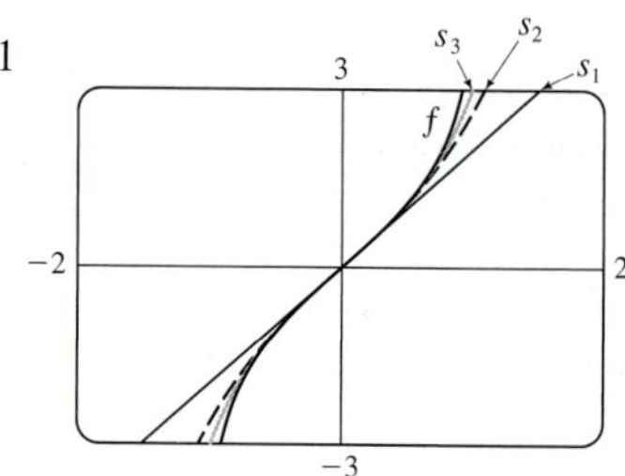

25. $C + \sum_{n=0}^{\infty} \frac{t^{8n+2}}{8n+2}$, $R = 1$

27. $C + \sum_{n=1}^{\infty} (-1)^n \frac{x^{n+3}}{n(n+3)}$, $R = 1$

29. 0.044522 **31.** 0.000395
33. 0.19740
35. (b) 0.920 **39.** $[-1, 1]$, $[-1, 1)$, $(-1, 1)$

EXERCISES 11.10 ■ PAGE 811

1. $b_8 = f^{(8)}(5)/8!$ **3.** $\sum_{n=0}^{\infty} (n+1) x^n$, $R = 1$

5. $x + x^2 + \frac{1}{2}x^3 + \frac{1}{6}x^4$

7. $2 + \frac{1}{12}(x-8) - \frac{1}{288}(x-8)^2 + \frac{5}{20{,}736}(x-8)^3$

9. $\frac{1}{2} + \frac{\sqrt{3}}{2}\left(x - \frac{\pi}{6}\right) - \frac{1}{4}\left(x - \frac{\pi}{6}\right)^2 - \frac{\sqrt{3}}{12}\left(x - \frac{\pi}{6}\right)^3$

11. $\sum_{n=0}^{\infty} (n+1) x^n$, $R = 1$ **13.** $\sum_{n=0}^{\infty} (-1)^n \frac{x^{2n}}{(2n)!}$, $R = \infty$

15. $\sum_{n=0}^{\infty} \frac{(\ln 2)^n}{n!} x^n$, $R = \infty$ **17.** $\sum_{n=0}^{\infty} \frac{x^{2n+1}}{(2n+1)!}$, $R = \infty$

19. $50 + 105(x-2) + 92(x-2)^2 + 42(x-2)^3 + 10(x-2)^4 + (x-2)^5$, $R = \infty$

21. $\ln 2 + \sum_{n=1}^{\infty} (-1)^{n+1} \frac{1}{n 2^n} (x-2)^n$, $R = 2$

23. $\sum_{n=0}^{\infty} \frac{2^n e^6}{n!} (x-3)^n$, $R = \infty$

25. $\sum_{n=0}^{\infty} \frac{(-1)^{n+1}}{(2n+1)!} (x-\pi)^{2n+1}$, $R = \infty$

31. $1 - \frac{1}{4}x - \sum_{n=2}^{\infty} \frac{3 \cdot 7 \cdot \cdots \cdot (4n-5)}{4^n \cdot n!} x^n$, $R = 1$

33. $\sum_{n=0}^{\infty} (-1)^n \frac{(n+1)(n+2)}{2^{n+4}} x^n$, $R = 2$

35. $\sum_{n=0}^{\infty} (-1)^n \frac{1}{2n+1} x^{4n+2}$, $R = 1$

37. $\sum_{n=0}^{\infty} (-1)^n \frac{2^{2n}}{(2n)!} x^{2n+1}$, $R = \infty$

39. $\sum_{n=0}^{\infty} (-1)^n \frac{1}{2^{2n}(2n)!} x^{4n+1}$, $R = \infty$

41. $\frac{1}{2}x + \sum_{n=1}^{\infty} (-1)^n \frac{1 \cdot 3 \cdot 5 \cdot \cdots \cdot (2n-1)}{n!\, 2^{3n+1}} x^{2n+1}$, $R = 2$

43. $\sum_{n=1}^{\infty} (-1)^{n+1} \frac{2^{2n-1}}{(2n)!} x^{2n}$, $R = \infty$

45. $\sum_{n=0}^{\infty} (-1)^n \frac{1}{(2n)!} x^{4n}$, $R = \infty$

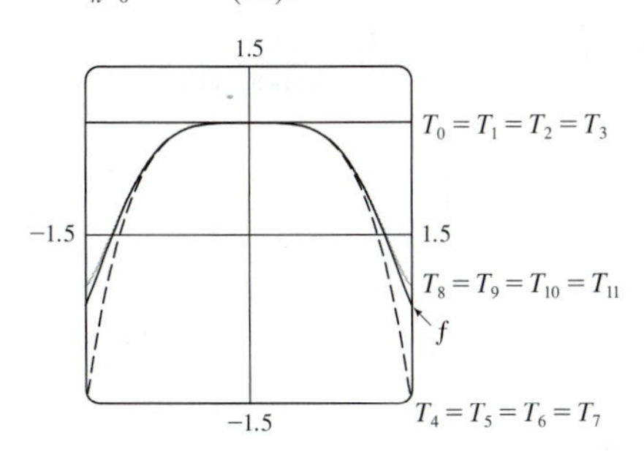

47. $\sum_{n=1}^{\infty} \frac{(-1)^{n-1}}{(n-1)!} x^n$, $R = \infty$

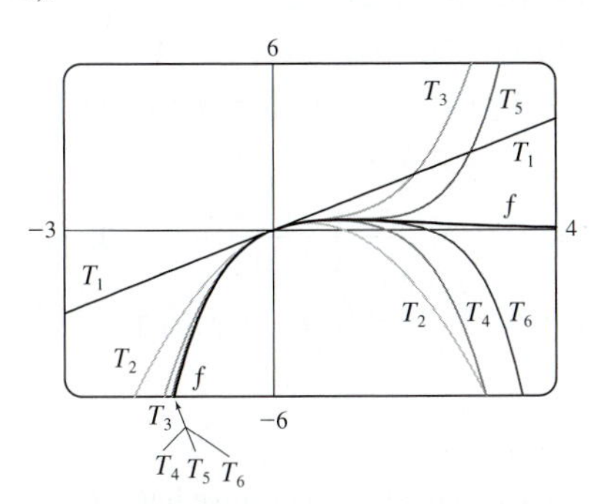

49. 0.99619

51. (a) $1 + \sum_{n=1}^{\infty} \frac{1 \cdot 3 \cdot 5 \cdot \cdots \cdot (2n-1)}{2^n n!} x^{2n}$

(b) $x + \sum_{n=1}^{\infty} \frac{1 \cdot 3 \cdot 5 \cdot \cdots \cdot (2n-1)}{(2n+1)2^n n!} x^{2n+1}$

53. $C + \sum_{n=0}^{\infty} \binom{\frac{1}{2}}{n} \frac{x^{3n+1}}{3n+1}$, $R = 1$

55. $C + \sum_{n=1}^{\infty} (-1)^n \frac{1}{2n(2n)!} x^{2n}$, $R = \infty$

57. 0.0059 **59.** 0.40102 **61.** $\frac{1}{2}$ **63.** $\frac{1}{120}$ **65.** $\frac{3}{5}$

67. $1 - \frac{3}{2}x^2 + \frac{25}{24}x^4$ **69.** $1 + \frac{1}{6}x^2 + \frac{7}{360}x^4$

71. $x - \frac{2}{3}x^4 + \frac{23}{45}x^6$

73. e^{-x^4} **75.** $\ln \frac{8}{5}$

77. $1/\sqrt{2}$ **79.** $e^3 - 1$

EXERCISES 11.11 ■ PAGE 820

1. (a) $T_0(x) = 0$, $T_1(x) = T_2(x) = x$, $T_3(x) = T_4(x) = x - \frac{1}{6}x^3$, $T_5(x) = x - \frac{1}{6}x^3 + \frac{1}{120}x^5$

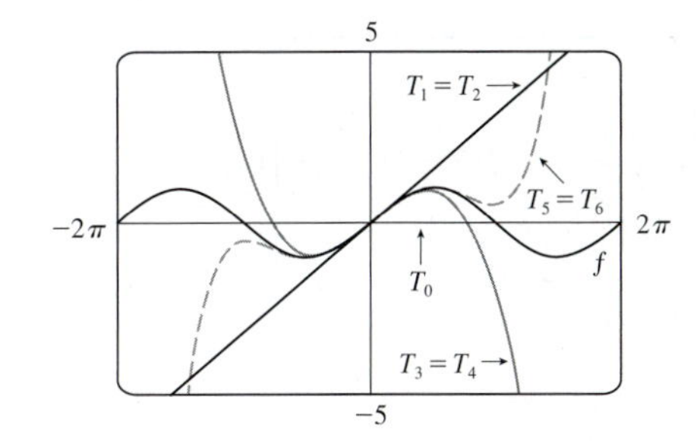

(b)

x	f	T_0	$T_1 = T_2$	$T_3 = T_4$	T_5
$\pi/4$	0.7071	0	0.7854	0.7047	0.7071
$\pi/2$	1	0	1.5708	0.9248	1.0045
π	0	0	3.1416	−2.0261	0.5240

(c) As n increases, $T_n(x)$ is a good approximation to $f(x)$ on a larger and larger interval.

3. $e + e(x-1) + \frac{1}{2}e(x-1)^2 + \frac{1}{6}e(x-1)^3$

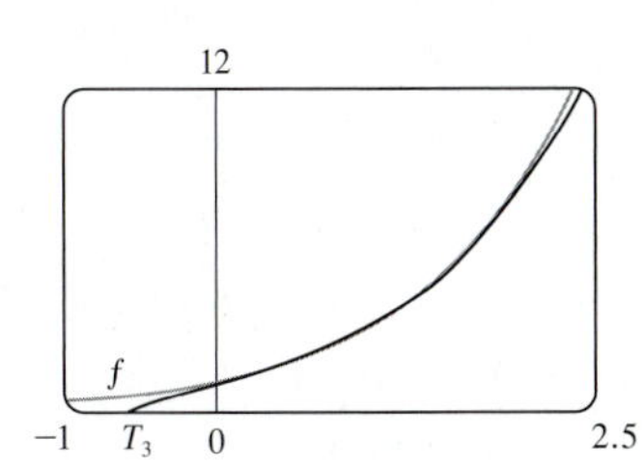

5. $-\left(x-\frac{\pi}{2}\right)+\frac{1}{6}\left(x-\frac{\pi}{2}\right)^3$

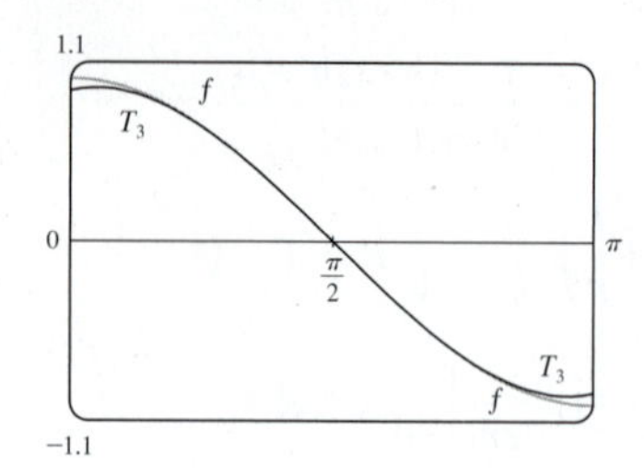

7. $(x-1)-\frac{1}{2}(x-1)^2+\frac{1}{3}(x-1)^3$

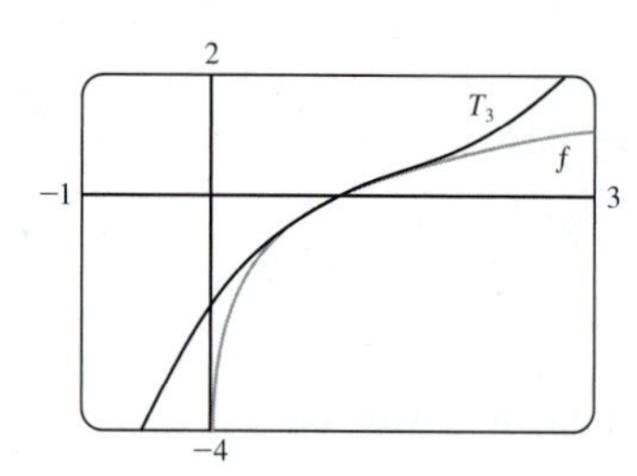

9. $x-2x^2+2x^3$

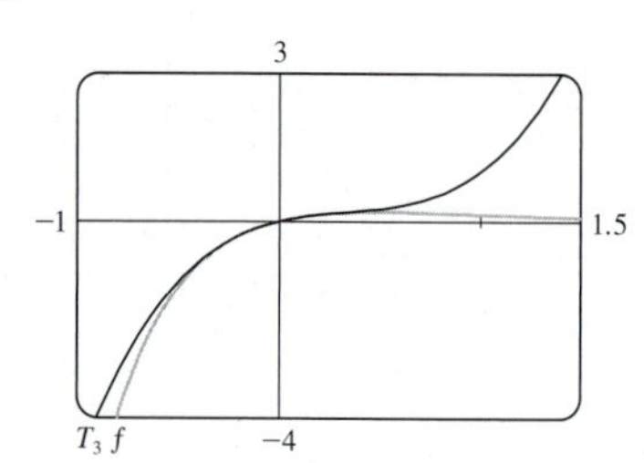

11. $T_5(x)=1-2\left(x-\frac{\pi}{4}\right)+2\left(x-\frac{\pi}{4}\right)^2-\frac{8}{3}\left(x-\frac{\pi}{4}\right)^3$
$\quad+\frac{10}{3}\left(x-\frac{\pi}{4}\right)^4-\frac{64}{15}\left(x-\frac{\pi}{4}\right)^5$

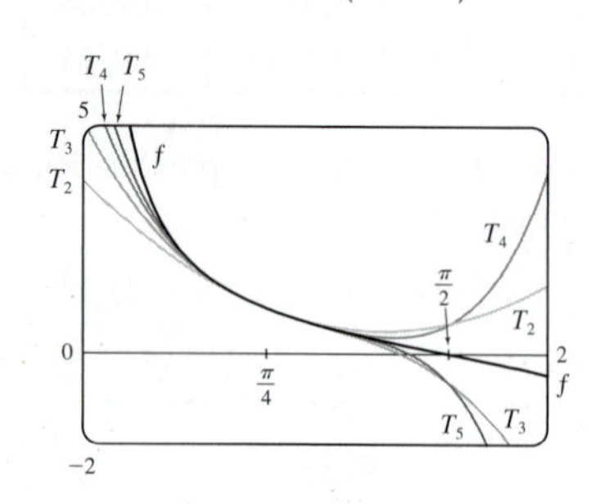

13. (a) $1-(x-1)+(x-1)^2$ (b) 0.006 482 7
15. (a) $1+\frac{2}{3}(x-1)-\frac{1}{9}(x-1)^2+\frac{4}{81}(x-1)^3$ (b) 0.000 097
17. (a) $1+\frac{1}{2}x^2$ (b) 0.0015
19. (a) $1+x^2$ (b) 0.000 06 **21.** (a) $x^2-\frac{1}{6}x^4$ (b) 0.042
23. 0.17365 **25.** Four **27.** $-1.037<x<1.037$
29. $-0.86<x<0.86$ **31.** 21 m, no
37. (c) They differ by about 8×10^{-9} km.

CHAPTER 11 REVIEW ■ PAGE 824

True-False Quiz

1. False **3.** True **5.** False **7.** False **9.** False
11. True **13.** True **15.** False **17.** True
19. True **21.** True

Exercises

1. $\frac{1}{2}$ **3.** D **5.** 0 **7.** e^{12} **9.** 2 **11.** C
13. C **15.** D **17.** C **19.** C **21.** C **23.** CC
25. AC **27.** $\frac{1}{11}$ **29.** $\pi/4$ **31.** e^{-e} **35.** 0.9721
37. 0.189 762 24, error $<6.4\times10^{-7}$
41. 4, $[-6, 2)$ **43.** 0.5, $[2.5, 3.5)$

45. $\frac{1}{2}\sum_{n=0}^{\infty}(-1)^n\left[\frac{1}{(2n)!}\left(x-\frac{\pi}{6}\right)^{2n}+\frac{\sqrt{3}}{(2n+1)!}\left(x-\frac{\pi}{6}\right)^{2n+1}\right]$

47. $\sum_{n=0}^{\infty}(-1)^n x^{n+2}$, $R=1$ **49.** $\ln 4-\sum_{n=1}^{\infty}\frac{x^n}{n4^n}$, $R=4$

51. $\sum_{n=0}^{\infty}(-1)^n\frac{x^{8n+4}}{(2n+1)!}$, $R=\infty$

53. $\frac{1}{2}+\sum_{n=1}^{\infty}\frac{1\cdot5\cdot9\cdot\cdots\cdot(4n-3)}{n!\,2^{6n+1}}x^n$, $R=16$

55. $C+\ln|x|+\sum_{n=1}^{\infty}\frac{x^n}{n\cdot n!}$

57. (a) $1+\frac{1}{2}(x-1)-\frac{1}{8}(x-1)^2+\frac{1}{16}(x-1)^3$
(b) (c) 0.000 006

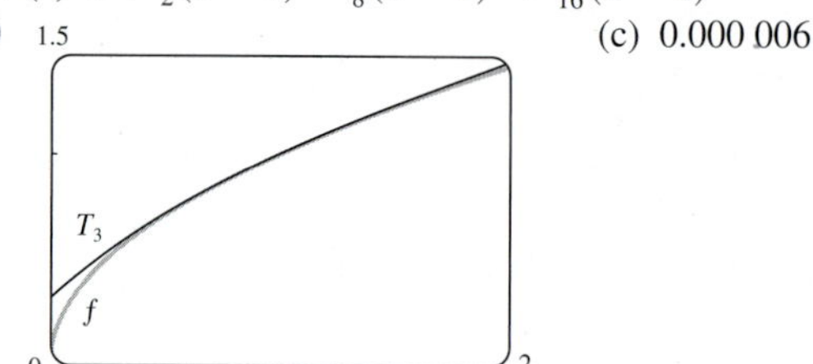

59. $-\frac{1}{6}$

PROBLEMS PLUS ■ PAGE 827

1. $15!/5! = 10{,}897{,}286{,}400$
3. (b) 0 if $x=0$, $(1/x)-\cot x$ if $x\neq k\pi$, k an integer
5. (a) $s_n=3\cdot4^n$, $l_n=1/3^n$, $p_n=4^n/3^{n-1}$ (c) $\frac{2}{5}\sqrt{3}$
9. $\frac{3\pi}{4}$ **11.** $(-1, 1)$, $\frac{x^3+4x^2+x}{(1-x)^4}$ **13.** $\ln\frac{1}{2}$
17. (a) $\frac{250}{101}\pi\left(e^{-(n-1)\pi/5}-e^{-n\pi/5}\right)$ (b) $\frac{250}{101}\pi$
19. $\frac{\pi}{2\sqrt{3}}-1$
21. $-\left(\frac{\pi}{2}-\pi k\right)^2$, where k is a positive integer

CHAPTER 12

EXERCISES 12.1 ■ PAGE 836

1. $(4, 0, -3)$ **3.** C; A

5. A line parallel to the y-axis and 4 units to the right of it; a vertical plane parallel to the yz-plane and 4 units in front of it.

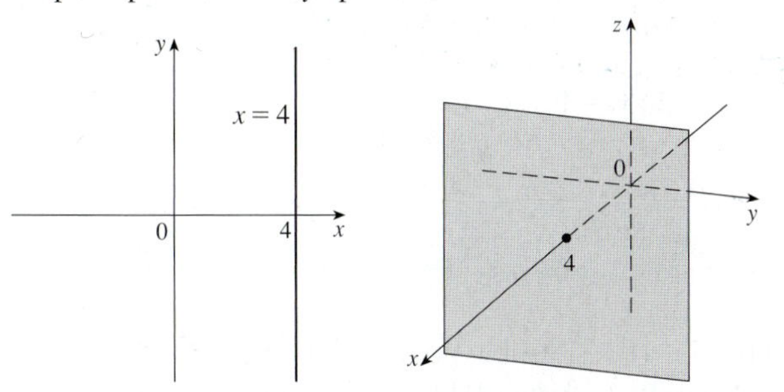

7. A vertical plane that intersects the xy-plane in the line $y = 2 - x$, $z = 0$

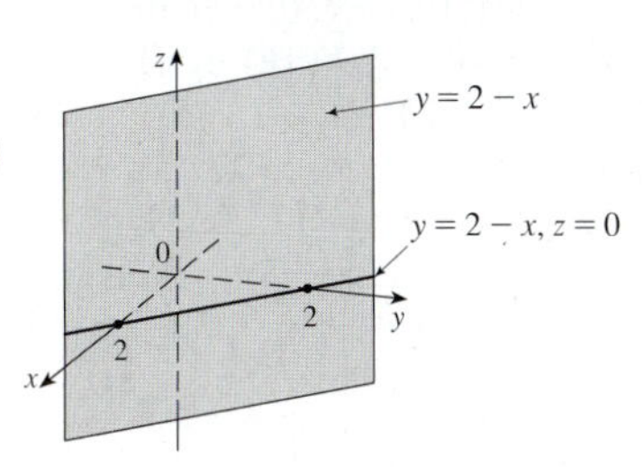

9. (a) $|PQ| = 6$, $|QR| = 2\sqrt{10}$, $|RP| = 6$; isosceles triangle

11. (a) No (b) Yes

13. $(x + 3)^2 + (y - 2)^2 + (z - 5)^2 = 16$; $(y - 2)^2 + (z - 5)^2 = 7$, $x = 0$ (a circle)

15. $(x - 3)^2 + (y - 8)^2 + (z - 1)^2 = 30$

17. $(1, 2, -4)$, 6 **19.** $(2, 0, -6)$, $9/\sqrt{2}$

21. (b) $\frac{5}{2}, \frac{1}{2}\sqrt{94}, \frac{1}{2}\sqrt{85}$

23. (a) $(x - 2)^2 + (y + 3)^2 + (z - 6)^2 = 36$
(b) $(x - 2)^2 + (y + 3)^2 + (z - 6)^2 = 4$
(c) $(x - 2)^2 + (y + 3)^2 + (z - 6)^2 = 9$

25. A plane parallel to the yz-plane and 5 units in front of it

27. A half-space consisting of all points to the left of the plane $y = 8$

29. All points on or between the horizontal planes $z = 0$ and $z = 6$

31. All points on a circle with radius 2 with center on the z-axis that is contained in the plane $z = -1$

33. All point on a sphere with radius 2 and center $(0, 0, 0)$

35. All points on or between spheres with radii 1 and $\sqrt{5}$ and centers $(0, 0, 0)$

37. All points on or inside a circular cylinder of radius 3 with axis the y-axis

39. $0 < x < 5$ **41.** $r^2 < x^2 + y^2 + z^2 < R^2$

43. (a) $(2, 1, 4)$ (b)

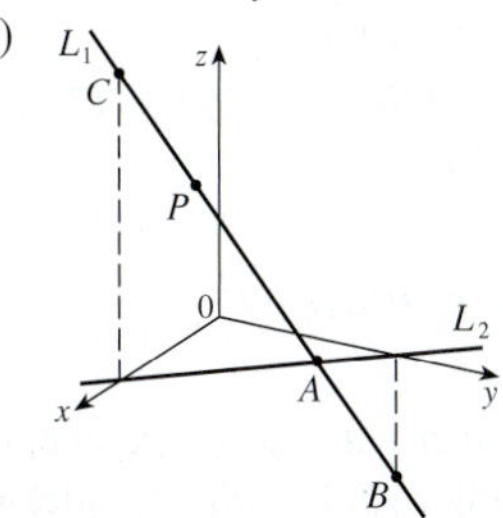

45. $14x - 6y - 10z = 9$, a plane perpendicular to AB

47. $2\sqrt{3} - 3$

EXERCISES 12.2 ■ PAGE 845

1. (a) Scalar (b) Vector (c) Vector (d) Scalar

3. $\overrightarrow{AB} = \overrightarrow{DC}$, $\overrightarrow{DA} = \overrightarrow{CB}$, $\overrightarrow{DE} = \overrightarrow{EB}$, $\overrightarrow{EA} = \overrightarrow{CE}$

5. (a) u + v, v, u (b)

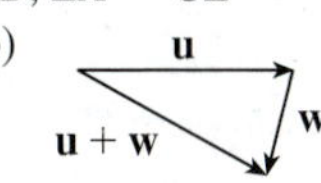

(c)

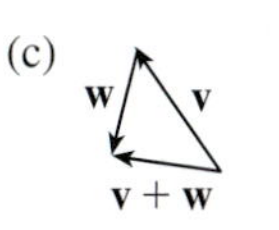

(d)

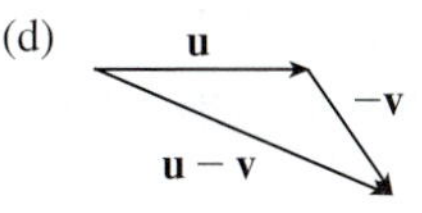

(e)

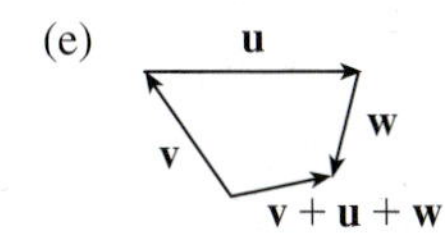

(f)

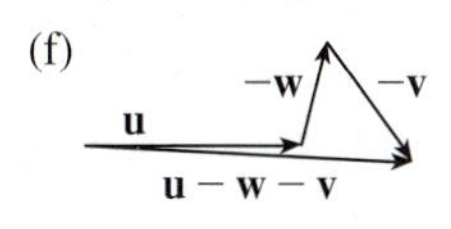

7. $\mathbf{c} = \frac{1}{2}\mathbf{a} + \frac{1}{2}\mathbf{b}$, $\mathbf{d} = \frac{1}{2}\mathbf{b} - \frac{1}{2}\mathbf{a}$

9. $\mathbf{a} = \langle 3, 1 \rangle$

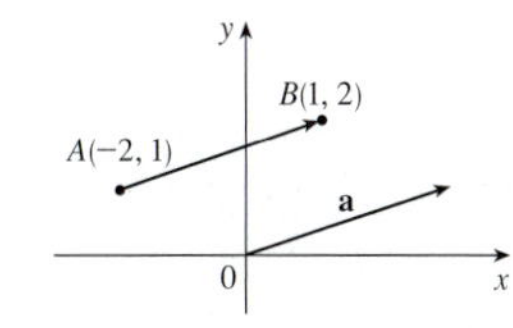

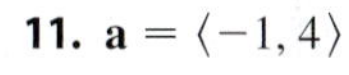

11. $\mathbf{a} = \langle -1, 4 \rangle$

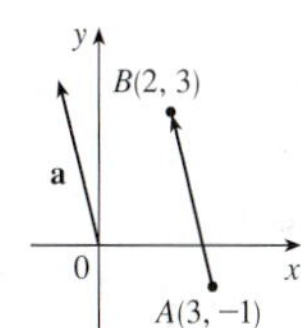

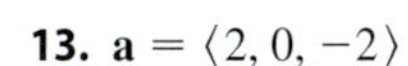

13. $\mathbf{a} = \langle 2, 0, -2 \rangle$

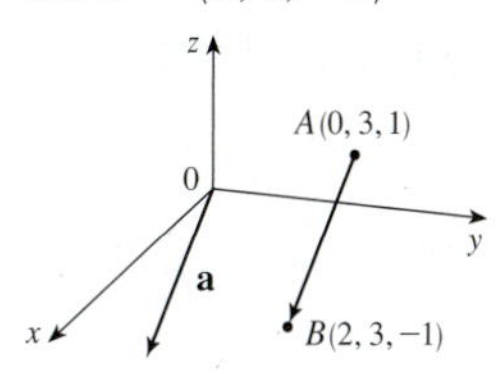

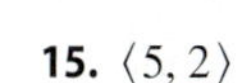

15. $\langle 5, 2 \rangle$

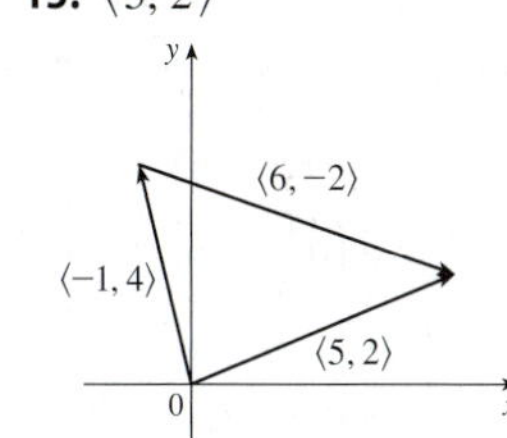

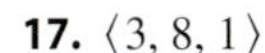

17. $\langle 3, 8, 1 \rangle$

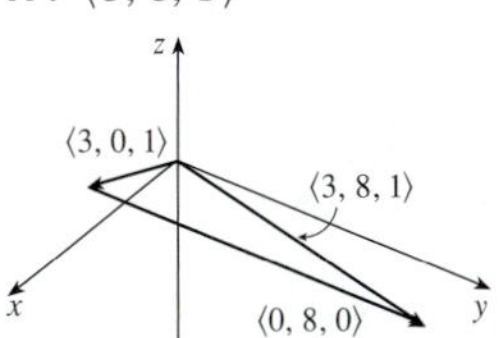

19. $\langle 6, 3 \rangle$, $\langle 6, 14 \rangle$, 5, 13

21. $6\mathbf{i} - 3\mathbf{j} - 2\mathbf{k}$, $20\mathbf{i} - 12\mathbf{j}$, $\sqrt{29}$, 7

23. $\left\langle \dfrac{3}{\sqrt{10}}, -\dfrac{1}{\sqrt{10}} \right\rangle$ **25.** $\frac{8}{9}\mathbf{i} - \frac{1}{9}\mathbf{j} + \frac{4}{9}\mathbf{k}$ **27.** $60°$

29. $\langle 2, 2\sqrt{3} \rangle$ **31.** ≈ 45.96 ft/s, ≈ 38.57 ft/s

33. $100\sqrt{7} \approx 264.6$ N, $\approx 139.1°$

35. $\sqrt{493} \approx 22.2$ mi/h, N8°W

37. $\approx -177.39\,\mathbf{i} + 211.41\,\mathbf{j}$, $\approx 177.39\,\mathbf{i} + 138.59\,\mathbf{j}$; ≈ 275.97 N, ≈ 225.11 N

39. (a) At an angle of 43.4° from the bank, toward upstream
(b) 20.2 min

41. $\pm(\mathbf{i} + 4\,\mathbf{j})/\sqrt{17}$ **43.** $\mathbf{0}$

45. (a), (b)

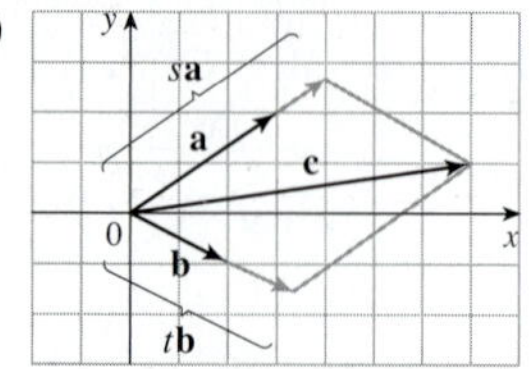

(d) $s = \frac{9}{7}, t = \frac{11}{7}$

47. A sphere with radius 1, centered at (x_0, y_0, z_0)

EXERCISES 12.3 ▪ PAGE 852

1. (b), (c), (d) are meaningful **3.** -3.6 **5.** 19 **7.** 1

9. $14\sqrt{3}$ **11.** $\mathbf{u} \cdot \mathbf{v} = \frac{1}{2}$, $\mathbf{u} \cdot \mathbf{w} = -\frac{1}{2}$

15. $\cos^{-1}\left(\dfrac{1}{\sqrt{5}}\right) \approx 63°$ **17.** $\cos^{-1}\left(-\frac{5}{6}\right) \approx 146°$

19. $\cos^{-1}\left(\dfrac{7}{\sqrt{130}}\right) \approx 52°$ **21.** 48°, 75°, 57°

23. (a) Orthogonal (b) Neither
(c) Parallel (d) Orthogonal

25. Yes **27.** $(\mathbf{i} - \mathbf{j} - \mathbf{k})/\sqrt{3}$ $\left[\text{or } (-\mathbf{i} + \mathbf{j} + \mathbf{k})/\sqrt{3}\right]$

29. 45° **31.** 0° at (0, 0), $\approx 8.1°$ at (1, 1)

33. $\frac{2}{3}, \frac{1}{3}, \frac{2}{3}$; 48°, 71°, 48°

35. $1/\sqrt{14}, -2/\sqrt{14}, -3/\sqrt{14}$; 74°, 122°, 143°

37. $1/\sqrt{3}, 1/\sqrt{3}, 1/\sqrt{3}$; 55°, 55°, 55° **39.** 4, $\left\langle -\frac{20}{13}, \frac{48}{13} \right\rangle$

41. $\frac{1}{9}$, $\left\langle \frac{4}{81}, \frac{7}{81}, -\frac{4}{81} \right\rangle$ **43.** $-7/\sqrt{19}$, $-\frac{21}{19}\mathbf{i} + \frac{21}{19}\mathbf{j} - \frac{7}{19}\mathbf{k}$

47. $\langle 0, 0, -2\sqrt{10} \rangle$ or any vector of the form $\langle s, t, 3s - 2\sqrt{10} \rangle$, $s, t \in \mathbb{R}$

49. 144 J **51.** $2400\cos(40°) \approx 1839$ ft-lb

53. $\frac{13}{5}$ **55.** $\cos^{-1}(1/\sqrt{3}) \approx 55°$

EXERCISES 12.4 ▪ PAGE 861

1. $15\,\mathbf{i} - 10\,\mathbf{j} - 3\,\mathbf{k}$ **3.** $14\,\mathbf{i} + 4\,\mathbf{j} + 2\,\mathbf{k}$

5. $-\frac{3}{2}\mathbf{i} + \frac{7}{4}\mathbf{j} + \frac{2}{3}\mathbf{k}$ **7.** $(1 - t)\,\mathbf{i} + (t^3 - t^2)\,\mathbf{k}$

9. $\mathbf{0}$ **11.** $\mathbf{i} + \mathbf{j} + \mathbf{k}$

13. (a) Scalar (b) Meaningless (c) Vector
(d) Meaningless (e) Meaningless (f) Scalar

15. $96\sqrt{3}$; into the page **17.** $\langle -7, 10, 8 \rangle$, $\langle 7, -10, -8 \rangle$

19. $\left\langle -\dfrac{1}{3\sqrt{3}}, -\dfrac{1}{3\sqrt{3}}, \dfrac{5}{3\sqrt{3}} \right\rangle, \left\langle \dfrac{1}{3\sqrt{3}}, \dfrac{1}{3\sqrt{3}}, -\dfrac{5}{3\sqrt{3}} \right\rangle$

27. 20 **29.** (a) $\langle 0, 18, -9 \rangle$ (b) $\frac{9}{2}\sqrt{5}$

31. (a) $\langle 13, -14, 5 \rangle$ (b) $\frac{1}{2}\sqrt{390}$

33. 9 **35.** 16 **39.** $10.8 \sin 80° \approx 10.6$ N·m

41. ≈ 417 N **43.** 60°

45. (b) $\sqrt{97/3}$ **53.** (a) No (b) No (c) Yes

EXERCISES 12.5 ▪ PAGE 871

1. (a) True (b) False (c) True (d) False
(e) False (f) True (g) False (h) True (i) True
(j) False (k) True

3. $\mathbf{r} = (2\,\mathbf{i} + 2.4\,\mathbf{j} + 3.5\,\mathbf{k}) + t(3\,\mathbf{i} + 2\,\mathbf{j} - \mathbf{k})$;
$x = 2 + 3t, y = 2.4 + 2t, z = 3.5 - t$

5. $\mathbf{r} = (\mathbf{i} + 6\mathbf{k}) + t(\mathbf{i} + 3\,\mathbf{j} + \mathbf{k})$;
$x = 1 + t, y = 3t, z = 6 + t$

7. $x = 2 + 2t, y = 1 + \frac{1}{2}t, z = -3 - 4t$;
$(x - 2)/2 = 2y - 2 = (z + 3)/(-4)$

9. $x = -8 + 11t, y = 1 - 3t, z = 4$; $\dfrac{x + 8}{11} = \dfrac{y - 1}{-3}, z = 4$

11. $x = -6 + 2t, y = 2 + 3t, z = 3 + t$;
$(x + 6)/2 = (y - 2)/3 = z - 3$

13. Yes

15. (a) $(x - 1)/(-1) = (y + 5)/2 = (z - 6)/(-3)$
(b) $(-1, -1, 0)$, $\left(-\frac{3}{2}, 0, -\frac{3}{2}\right)$, $(0, -3, 3)$

17. $\mathbf{r}(t) = (6\mathbf{i} - \mathbf{j} + 9\mathbf{k}) + t(\mathbf{i} + 7\mathbf{j} - 9\mathbf{k}), 0 \leqslant t \leqslant 1$

19. Skew **21.** $(4, -1, -5)$ **23.** $x - 2y + 5z = 0$

25. $x + 4y + z = 4$ **27.** $5x - y - z = 7$

29. $6x + 6y + 6z = 11$ **31.** $x + y + z = 2$

33. $5x - 3y - 8z = -9$ **35.** $8x + y - 2z = 31$

37. $x - 2y - z = -3$ **39.** $3x - 8y - z = -38$

41.

43.

45. $(-2, 6, 3)$ **47.** $\left(\frac{2}{5}, 4, 0\right)$ **49.** $1, 0, -1$

51. Perpendicular **53.** Neither, $\cos^{-1}\left(-\dfrac{1}{\sqrt{6}}\right) \approx 114.1°$

55. Parallel

57. (a) $x = 1, y = -t, z = t$ (b) $\cos^{-1}\left(\dfrac{5}{3\sqrt{3}}\right) \approx 15.8°$

59. $x = 1, y - 2 = -z$

61. $x + 2y + z = 5$

63. $(x/a) + (y/b) + (z/c) = 1$

65. $x = 3t, y = 1 - t, z = 2 - 2t$

67. P_2 and P_3 are parallel, P_1 and P_4 are identical

69. $\sqrt{61/14}$ **71.** $\frac{18}{7}$ **73.** $5/(2\sqrt{14})$

77. $1/\sqrt{6}$ **79.** $13/\sqrt{69}$

81. (a) $x = 325 + 440t, y = 810 - 135t, z = 561 + 38t$, $0 \leqslant t \leqslant 1$ (b) No

EXERCISES 12.6 ▪ PAGE 879

1. (a) Parabola
(b) Parabolic cylinder with rulings parallel to the z-axis
(c) Parabolic cylinder with rulings parallel to the x-axis

3. Circular cylinder

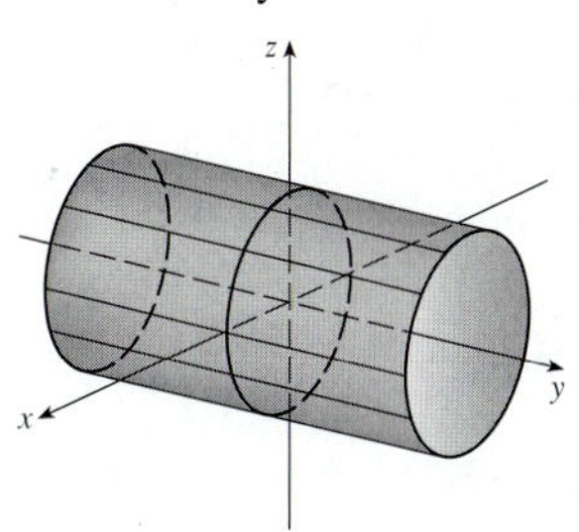

5. Parabolic cylinder

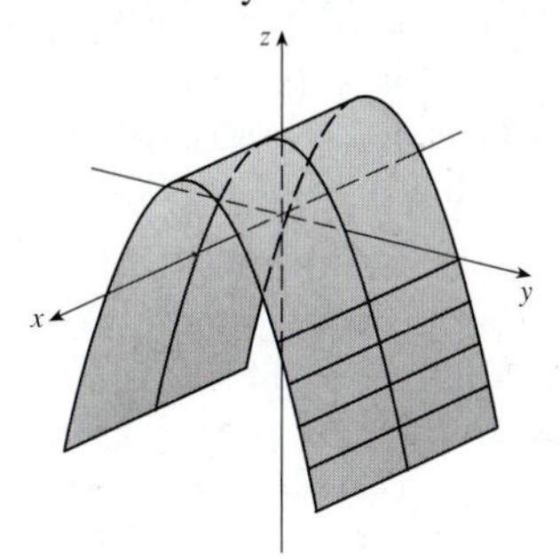

7. Hyperbolic cylinder

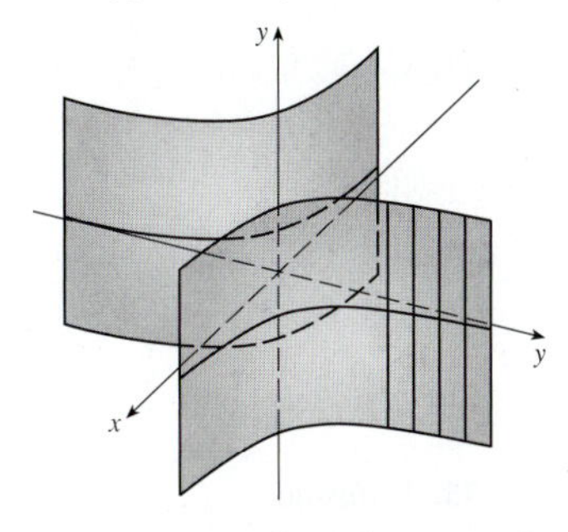

9. (a) $x = k$, $y^2 - z^2 = 1 - k^2$, hyperbola $(k \neq \pm 1)$;
$y = k$, $x^2 - z^2 = 1 - k^2$, hyperbola $(k \neq \pm 1)$;
$z = k$, $x^2 + y^2 = 1 + k^2$, circle
(b) The hyperboloid is rotated so that it has axis the y-axis
(c) The hyperboloid is shifted one unit in the negative y-direction

11. Elliptic paraboloid with axis the x-axis

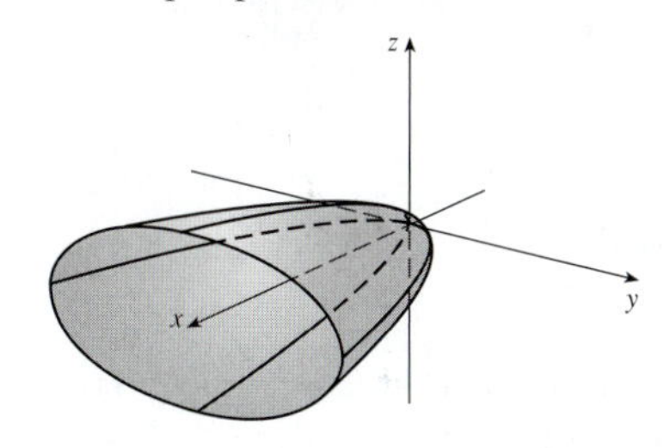

13. Elliptic cone with axis the x-axis

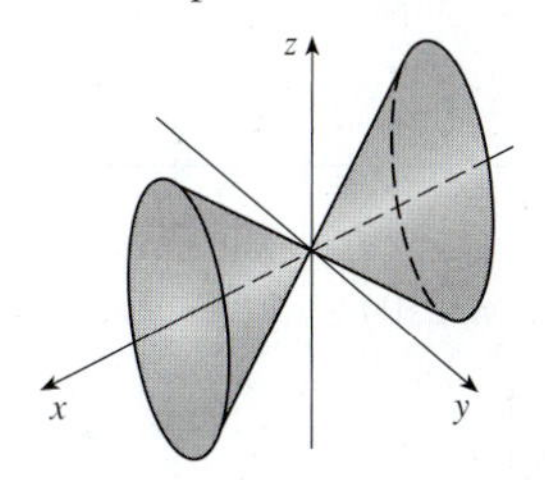

15. Hyperboloid of one sheet with axis the x-axis

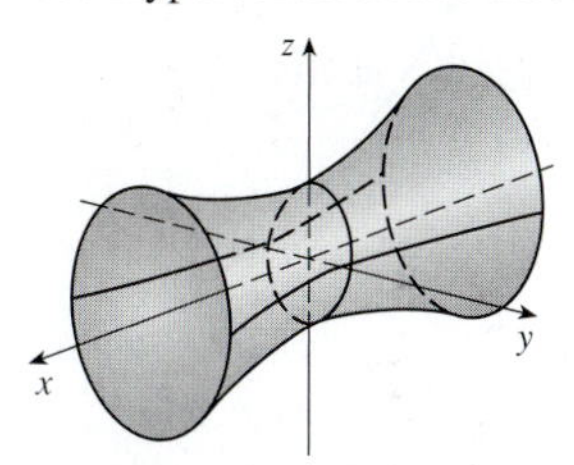

17. Ellipsoid

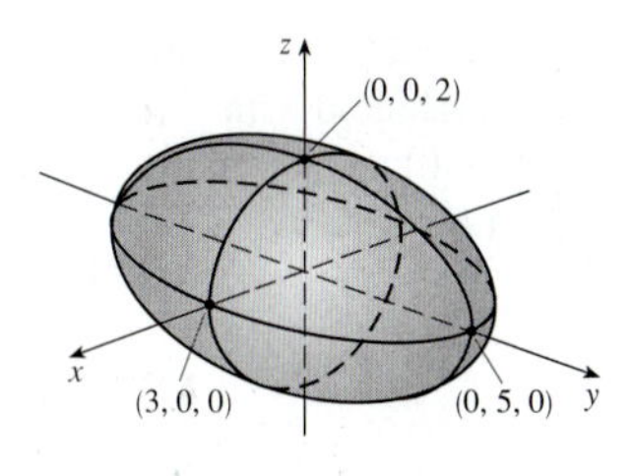

19. Hyperbolic paraboloid

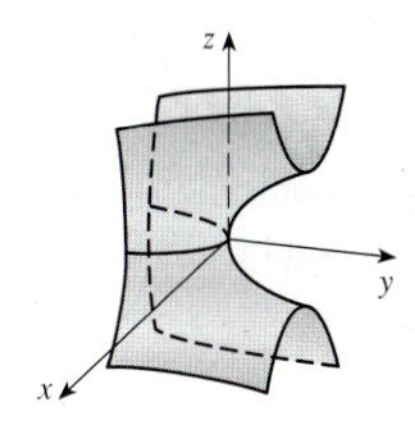

21. VII **23.** II **25.** VI **27.** VIII

29. Circular paraboloid

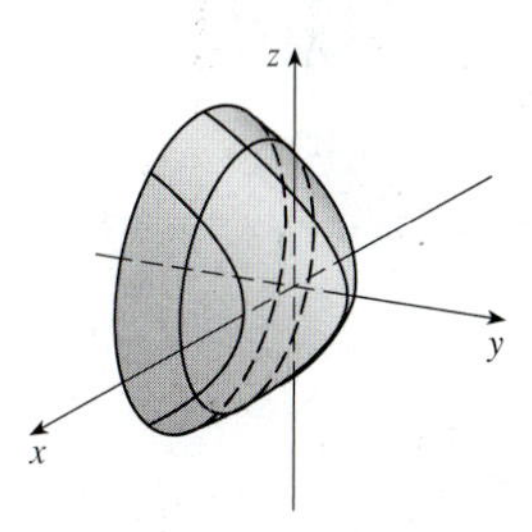

31. $y^2 = x^2 + \dfrac{z^2}{9}$

Elliptic cone with axis the y-axis

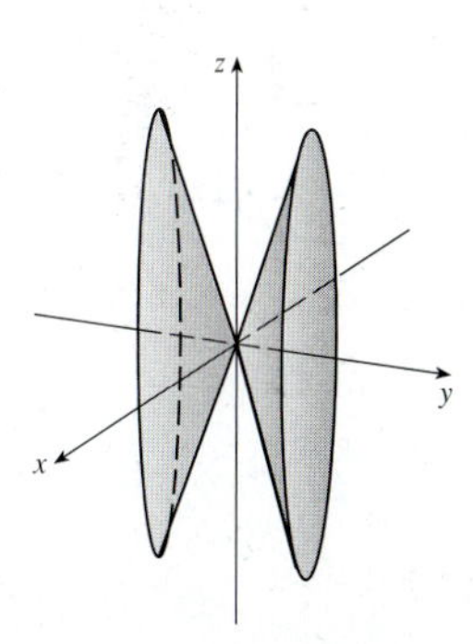

33. $y = z^2 - \dfrac{x^2}{2}$

Hyperbolic paraboloid

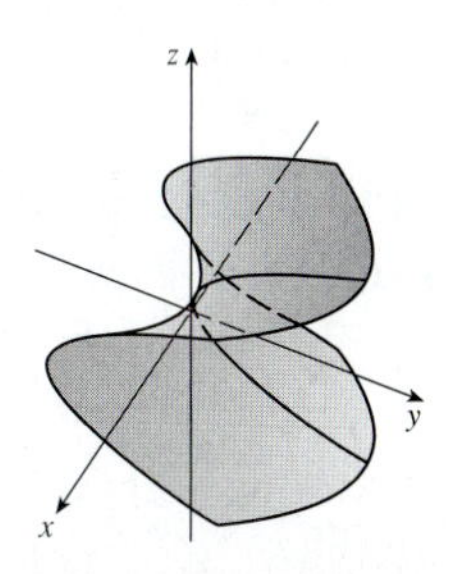

35. $z = (x-1)^2 + (y-3)^2$
Circular paraboloid with vertex $(1, 3, 0)$ and axis the vertical line $x = 1, y = 3$

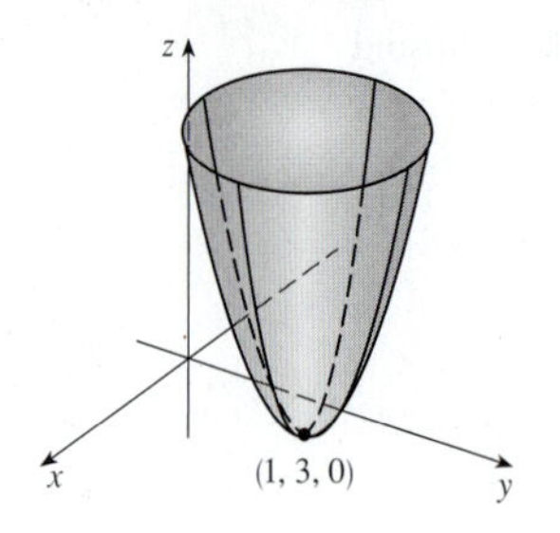

37. $\dfrac{(x-2)^2}{5} - \dfrac{y^2}{5} + \dfrac{(z-1)^2}{5} = 1$
Hyperboloid of one sheet with center $(2, 0, 1)$ and axis the horizontal line $x = 2, z = 1$

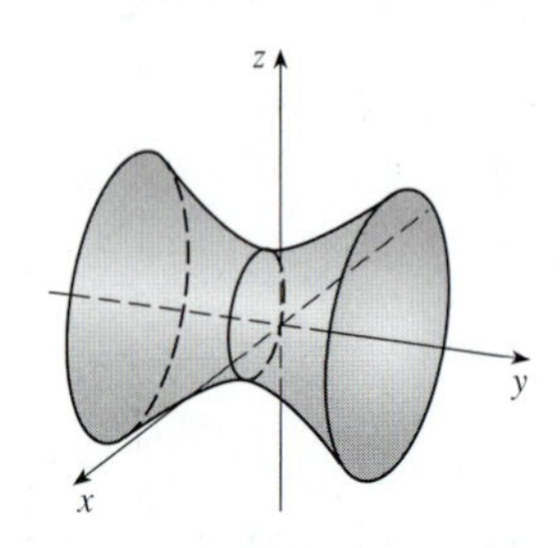

39.

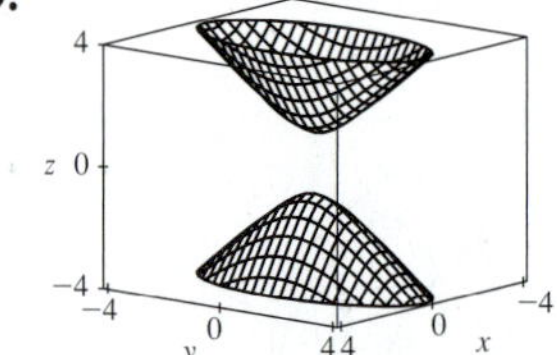

41.

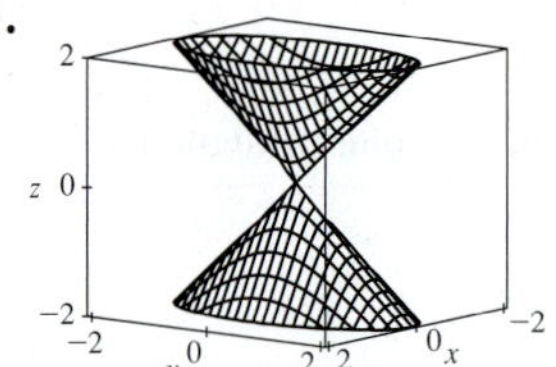

43.

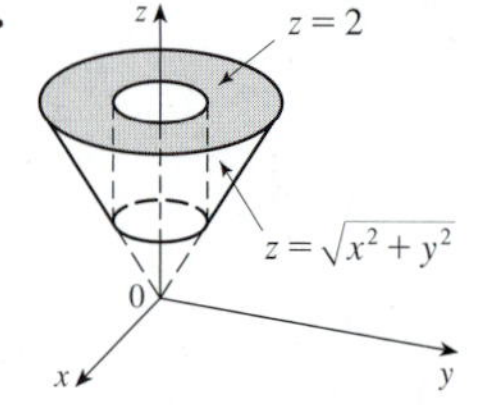

45. $x = y^2 + z^2$ **47.** $-4x = y^2 + z^2$, paraboloid

49. (a) $\dfrac{x^2}{(6378.137)^2} + \dfrac{y^2}{(6378.137)^2} + \dfrac{z^2}{(6356.523)^2} = 1$
(b) Circle (c) Ellipse

53.

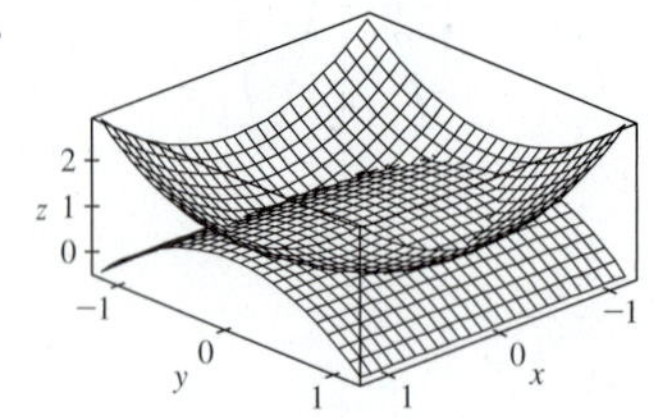

CHAPTER 12 REVIEW ■ PAGE 882

True-False Quiz

1. False **3.** False **5.** True **7.** True **9.** True
11. True **13.** True **15.** False **17.** False
19. False **21.** True

Exercises

1. (a) $(x+1)^2 + (y-2)^2 + (z-1)^2 = 69$
(b) $(y-2)^2 + (z-1)^2 = 68, x = 0$
(c) Center $(4, -1, -3)$, radius 5

3. $\mathbf{u} \cdot \mathbf{v} = 3\sqrt{2}$; $|\mathbf{u} \times \mathbf{v}| = 3\sqrt{2}$; out of the page

5. $-2, -4$ **7.** (a) 2 (b) -2 (c) -2 (d) 0

9. $\cos^{-1}(\frac{1}{3}) \approx 71°$ **11.** (a) $\langle 4, -3, 4 \rangle$ (b) $\sqrt{41}/2$

13. ≈ 166 N, ≈ 114 N

15. $x = 4 - 3t, y = -1 + 2t, z = 2 + 3t$

17. $x = -2 + 2t, y = 2 - t, z = 4 + 5t$

19. $-4x + 3y + z = -14$ **21.** $(1, 4, 4)$ **23.** Skew

25. $x + y + z = 4$ **27.** $22/\sqrt{26}$

29. Plane **31.** Cone

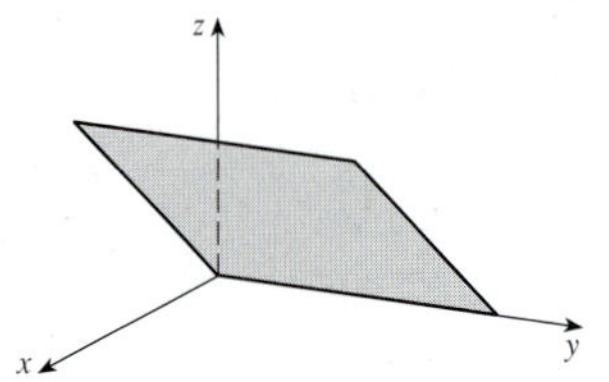

33. Hyperboloid of two sheets **35.** Ellipsoid

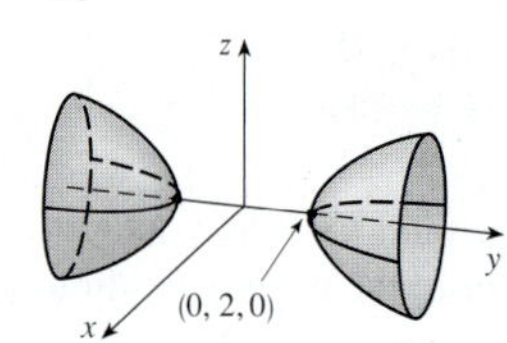

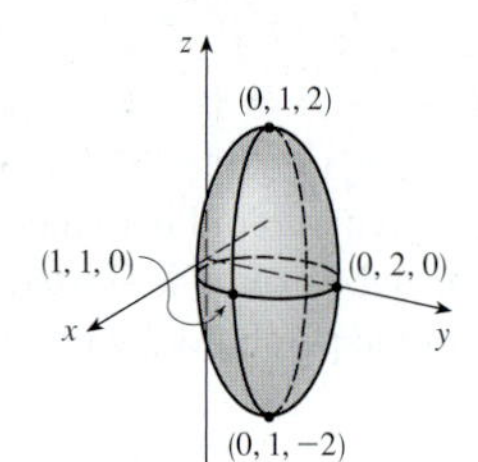

37. $4x^2 + y^2 + z^2 = 16$

PROBLEMS PLUS ■ PAGE 884

1. $\left(\sqrt{3} - \frac{3}{2}\right)$ m

3. (a) $(x+1)/(-2c) = (y-c)/(c^2-1) = (z-c)/(c^2+1)$
(b) $x^2 + y^2 = t^2 + 1, z = t$ (c) $4\pi/3$

5. 20

CHAPTER 13

EXERCISES 13.1 ■ PAGE 893

1. $(-1, 3)$ **3.** $\mathbf{i} + \mathbf{j} + \mathbf{k}$ **5.** $\langle -1, \pi/2, 0 \rangle$

7.

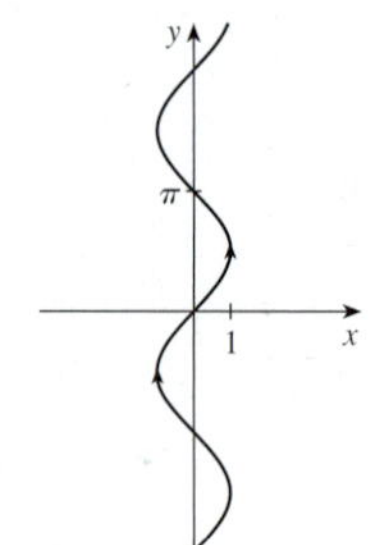

9.

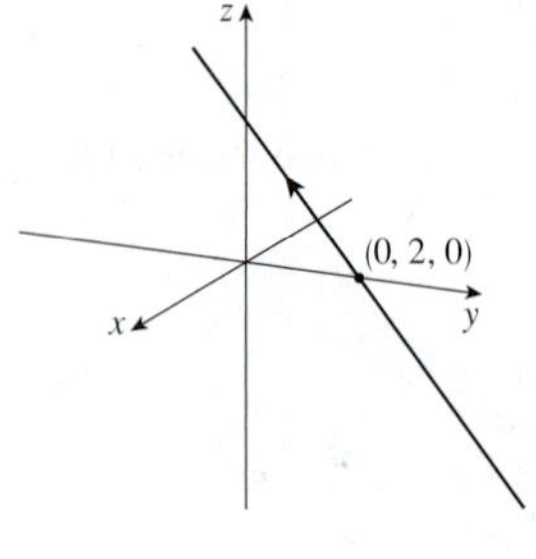

11.

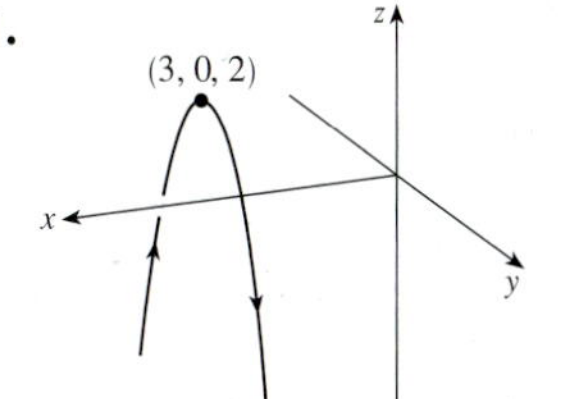

13.

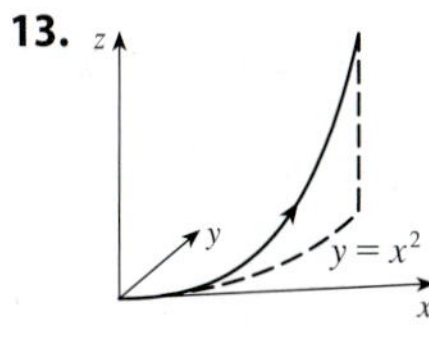

15.

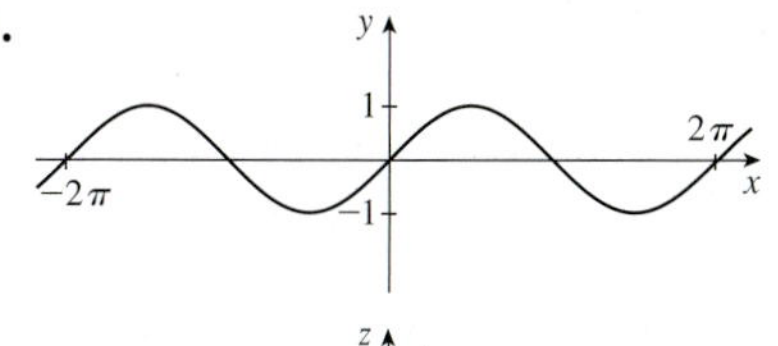

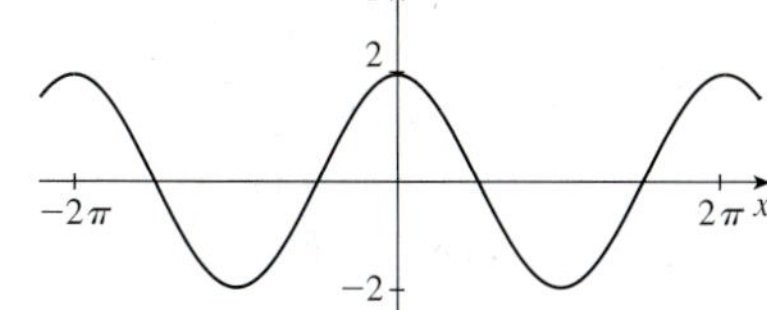

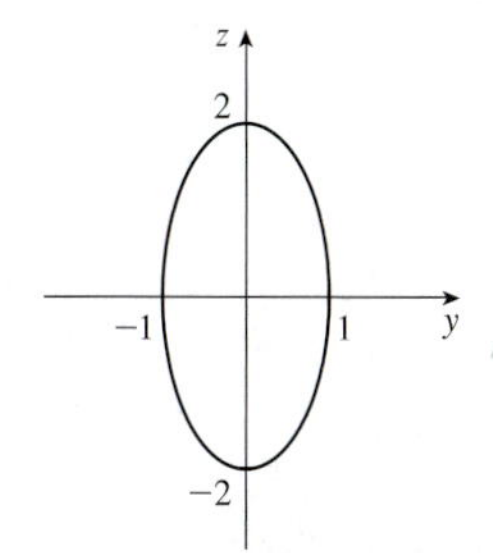

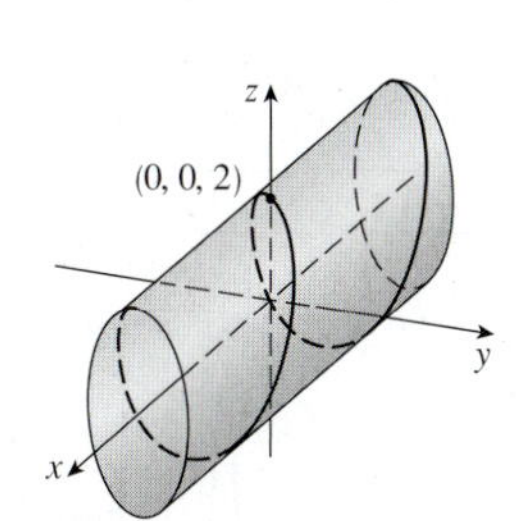

17. $\mathbf{r}(t) = \langle 2 + 4t, 2t, -2t \rangle$, $0 \leqslant t \leqslant 1$;
$x = 2 + 4t$, $y = 2t$, $z = -2t$, $0 \leqslant t \leqslant 1$

19. $\mathbf{r}(t) = \left\langle \frac{1}{2}t, -1 + \frac{4}{3}t, 1 - \frac{3}{4}t \right\rangle$, $0 \leqslant t \leqslant 1$;
$x = \frac{1}{2}t$, $y = -1 + \frac{4}{3}t$, $z = 1 - \frac{3}{4}t$, $0 \leqslant t \leqslant 1$

21. II **23.** V **25.** IV

27.

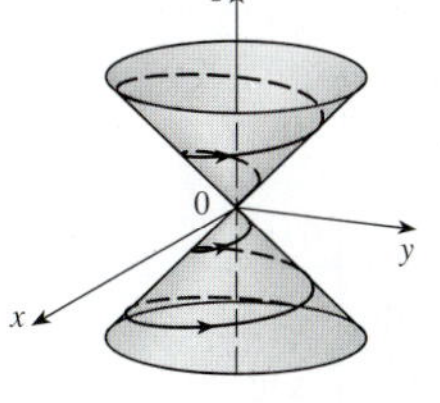

29. $y = e^{x/2}$, $z = e^x$, $z = y^2$

31. (0, 0, 0), (1, 0, 1)

33.

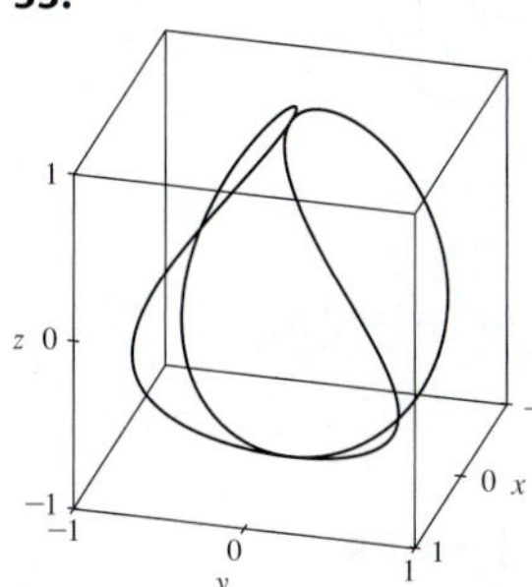

35.

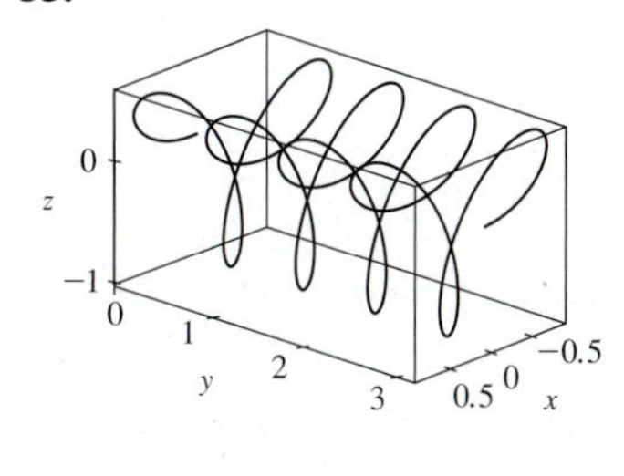

37.

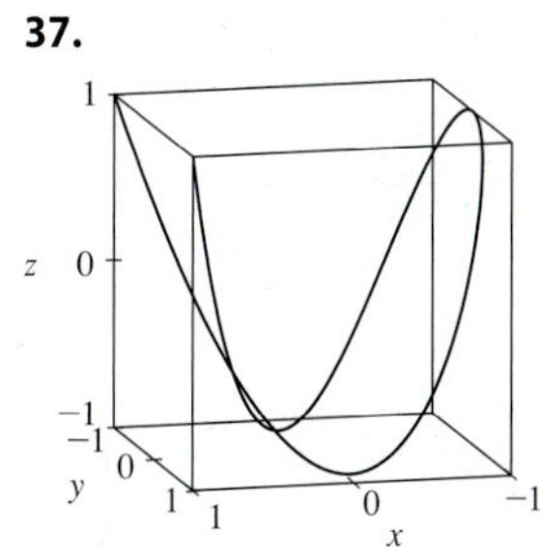

39.

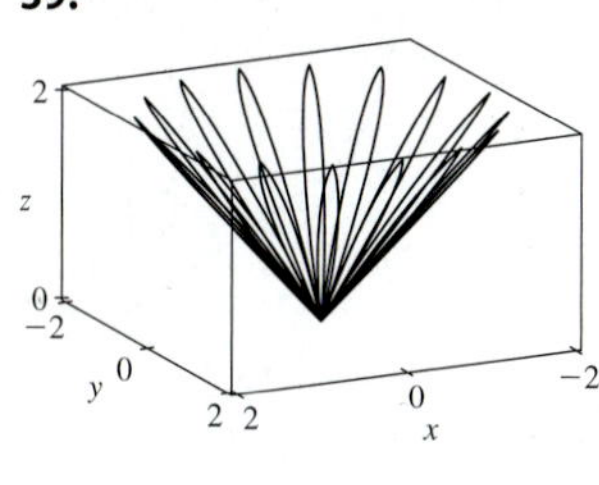

43. $\mathbf{r}(t) = t\,\mathbf{i} + \frac{1}{2}(t^2 - 1)\,\mathbf{j} + \frac{1}{2}(t^2 + 1)\,\mathbf{k}$

45. $\mathbf{r}(t) = \cos t\,\mathbf{i} + \sin t\,\mathbf{j} + \cos 2t\,\mathbf{k}$, $0 \leqslant t \leqslant 2\pi$

47. $x = 2\cos t$, $y = 2\sin t$, $z = 4\cos^2 t$, $0 \leqslant t \leqslant 2\pi$ **49.** Yes

51. (a)

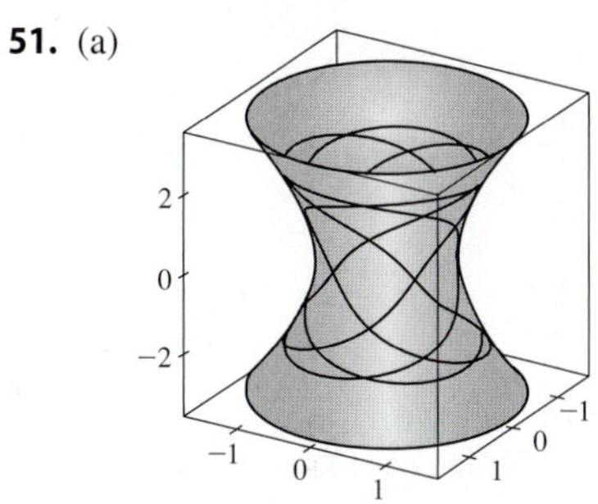

EXERCISES 13.2 ■ PAGE 900

1. (a)

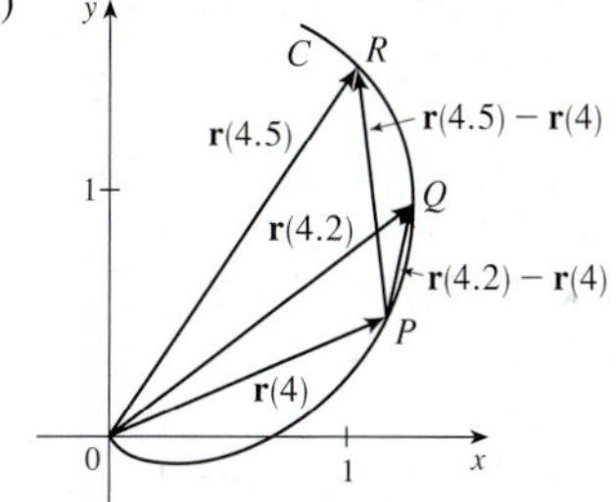

(b), (d)

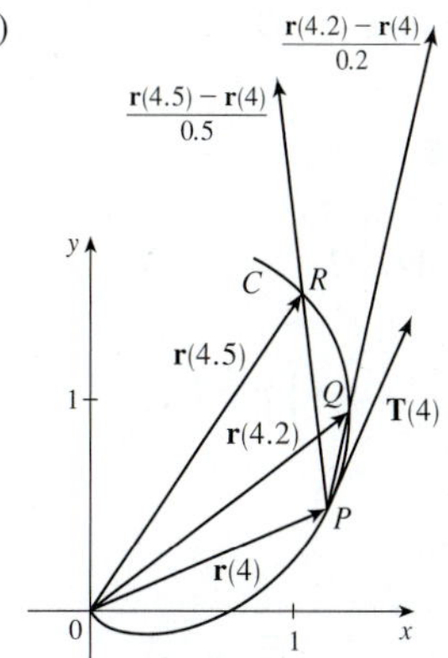

(c) $\mathbf{r}'(4) = \lim\limits_{h \to 0} \dfrac{\mathbf{r}(4 + h) - \mathbf{r}(4)}{h}$; $\mathbf{T}(4) = \dfrac{\mathbf{r}'(4)}{|\mathbf{r}'(4)|}$

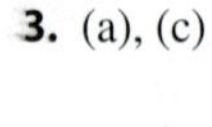

3. (a), (c)

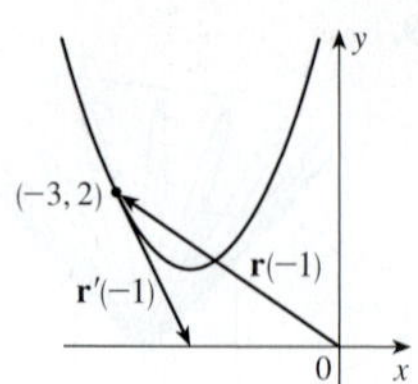

(b) $\mathbf{r}'(t) = \langle 1, 2t \rangle$

5. (a), (c)

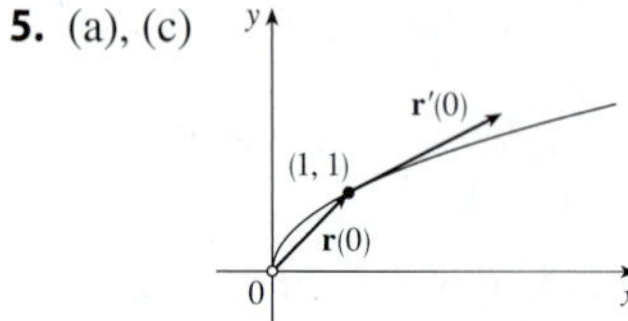

(b) $\mathbf{r}'(t) = 2e^{2t}\,\mathbf{i} + e^t\,\mathbf{j}$

7. (a), (c)

(b) $\mathbf{r}'(t) = 4\cos t\,\mathbf{i} + 2\sin t\,\mathbf{j}$

9. $\mathbf{r}'(t) = \left\langle \dfrac{1}{2\sqrt{t-2}}, 0, -\dfrac{2}{t^3} \right\rangle$

11. $\mathbf{r}'(t) = 2t\,\mathbf{i} - 2t\sin(t^2)\,\mathbf{j} + 2\sin t\cos t\,\mathbf{k}$

13. $\mathbf{r}'(t) = (t\cos t + \sin t)\,\mathbf{i} + e^t(\cos t - \sin t)\,\mathbf{j} + (\cos^2 t - \sin^2 t)\,\mathbf{k}$

15. $\mathbf{r}'(t) = \mathbf{b} + 2t\mathbf{c}$ **17.** $\langle \frac{2}{7}, \frac{3}{7}, \frac{6}{7} \rangle$ **19.** $\frac{3}{5}\mathbf{j} + \frac{4}{5}\mathbf{k}$

21. $\langle 1, 2t, 3t^2 \rangle, \langle 1/\sqrt{14}, 2/\sqrt{14}, 3/\sqrt{14} \rangle, \langle 0, 2, 6t \rangle, \langle 6t^2, -6t, 2 \rangle$

23. $x = 2 + 2t, y = 4 + 2t, z = 1 + t$

25. $x = 1 - t, y = t, z = 1 - t$

27. $\mathbf{r}(t) = (3 - 4t)\,\mathbf{i} + (4 + 3t)\,\mathbf{j} + (2 - 6t)\,\mathbf{k}$

29. $x = t, y = 1 - t, z = 2t$

31. $x = -\pi - t, y = \pi + t, z = -\pi t$

33. 66° **35.** $2\,\mathbf{i} - 4\,\mathbf{j} + 32\,\mathbf{k}$

37. $(\ln 2)\,\mathbf{i} + (\pi/4)\,\mathbf{j} + \frac{1}{2}\ln 2\,\mathbf{k}$

39. $\tan t\,\mathbf{i} + \frac{1}{8}(t^2 + 1)^4\mathbf{j} + \left(\frac{1}{3}t^3\ln t - \frac{1}{9}t^3\right)\mathbf{k} + \mathbf{C}$

41. $t^2\mathbf{i} + t^3\mathbf{j} + \left(\frac{2}{3}t^{3/2} - \frac{2}{3}\right)\mathbf{k}$

47. $2t\cos t + 2\sin t - 2\cos t\sin t$ **49.** 35

EXERCISES 13.3 ■ PAGE 908

1. $10\sqrt{10}$ **3.** $e - e^{-1}$ **5.** $\frac{1}{27}(13^{3/2} - 8)$

7. 18.6833 **9.** 10.3311 **11.** 42

13. (a) $s(t) = \sqrt{26}\,(t - 1)$;

$$\mathbf{r}(t(s)) = \left(4 - \frac{s}{\sqrt{26}}\right)\mathbf{i} + \left(\frac{4s}{\sqrt{26}} + 1\right)\mathbf{j} + \left(\frac{3s}{\sqrt{26}} + 3\right)\mathbf{k}$$

(b) $\left(4 - \dfrac{4}{\sqrt{26}}, \dfrac{16}{\sqrt{26}} + 1, \dfrac{12}{\sqrt{26}} + 3\right)$

15. $(3\sin 1, 4, 3\cos 1)$

17. (a) $\langle 1/\sqrt{10}, (-3/\sqrt{10})\sin t, (3/\sqrt{10})\cos t \rangle$, $\langle 0, -\cos t, -\sin t \rangle$ (b) $\frac{3}{10}$

19. (a) $\dfrac{1}{e^{2t} + 1}\langle \sqrt{2}e^t, e^{2t}, -1 \rangle, \dfrac{1}{e^{2t} + 1}\langle 1 - e^{2t}, \sqrt{2}e^t, \sqrt{2}e^t \rangle$

(b) $\sqrt{2}e^{2t}/(e^{2t} + 1)^2$

21. $6t^2/(9t^4 + 4t^2)^{3/2}$ **23.** $\dfrac{\sqrt{6}}{2(3t^2 + 1)^2}$

25. $\frac{1}{7}\sqrt{\frac{19}{14}}$ **27.** $12x^2/(1 + 16x^6)^{3/2}$

29. $e^x|x + 2|/[1 + (xe^x + e^x)^2]^{3/2}$

31. $\left(-\frac{1}{2}\ln 2, 1/\sqrt{2}\right)$; approaches 0 **33.** (a) P (b) 1.3, 0.7

35.

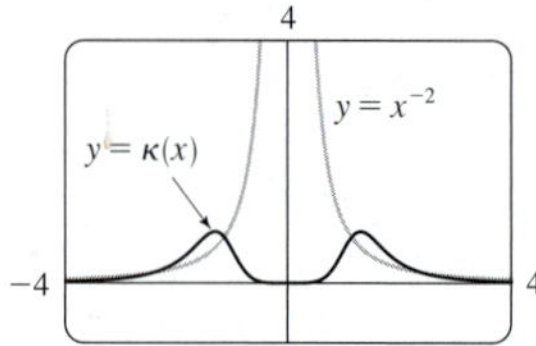

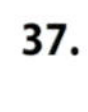

37.

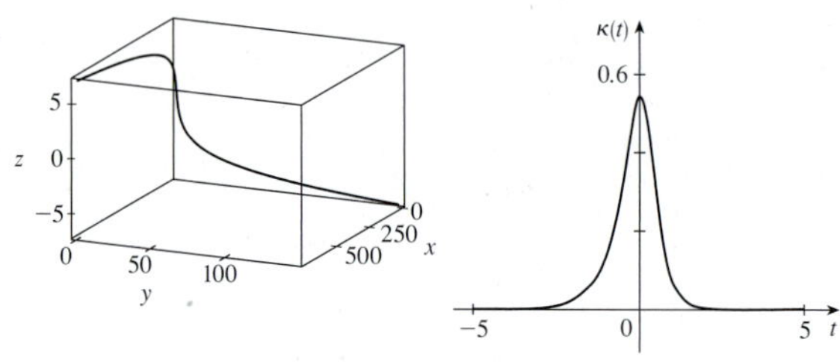

39. a is $y = f(x)$, b is $y = \kappa(x)$

41. $\kappa(t) = \dfrac{6\sqrt{4\cos^2 t - 12\cos t + 13}}{(17 - 12\cos t)^{3/2}}$

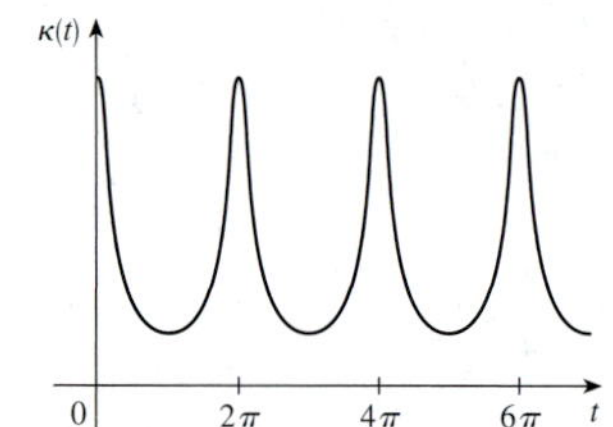

integer multiples of 2π

43. $6t^2/(4t^2 + 9t^4)^{3/2}$

45. $1/(\sqrt{2}e^t)$ **47.** $\langle \frac{2}{3}, \frac{2}{3}, \frac{1}{3} \rangle, \langle -\frac{1}{3}, \frac{2}{3}, -\frac{2}{3} \rangle, \langle -\frac{2}{3}, \frac{1}{3}, \frac{2}{3} \rangle$

49. $x - 2z = -4\pi, 2x + z = 2\pi$

51. $\left(x + \frac{5}{2}\right)^2 + y^2 = \frac{81}{4}, x^2 + \left(y - \frac{5}{3}\right)^2 = \frac{16}{9}$

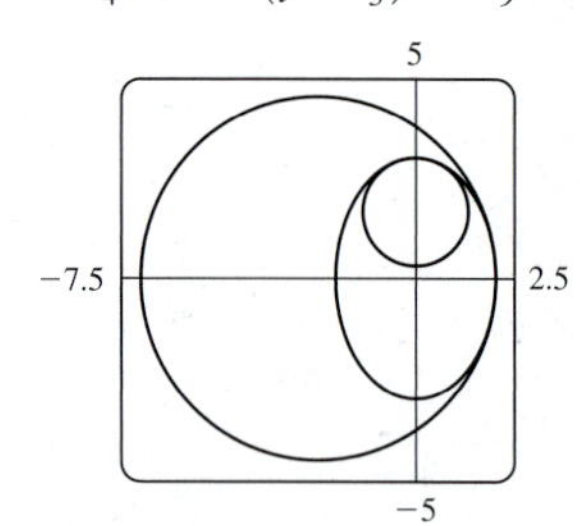

53. $(-1, -3, 1)$

55. $2x + y + 4z = 7, 6x - 8y - z = -3$

65. $2/(t^4 + 4t^2 + 1)$ **67.** 2.07×10^{10} Å ≈ 2 m

EXERCISES 13.4 ■ PAGE 918

1. (a) $1.8\mathbf{i} - 3.8\mathbf{j} - 0.7\mathbf{k}, 2.0\mathbf{i} - 2.4\mathbf{j} - 0.6\mathbf{k}, 2.8\mathbf{i} + 1.8\mathbf{j} - 0.3\mathbf{k}, 2.8\mathbf{i} + 0.8\mathbf{j} - 0.4\mathbf{k}$

(b) $2.4\mathbf{i} - 0.8\mathbf{j} - 0.5\mathbf{k}$, 2.58

3. $\mathbf{v}(t) = \langle -t, 1 \rangle$
$\mathbf{a}(t) = \langle -1, 0 \rangle$
$|\mathbf{v}(t)| = \sqrt{t^2 + 1}$

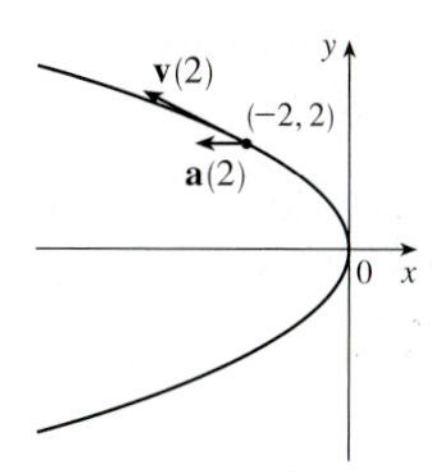

5. $\mathbf{v}(t) = -3 \sin t\, \mathbf{i} + 2 \cos t\, \mathbf{j}$
$\mathbf{a}(t) = -3 \cos t\, \mathbf{i} - 2 \sin t\, \mathbf{j}$
$|\mathbf{v}(t)| = \sqrt{5 \sin^2 t + 4}$

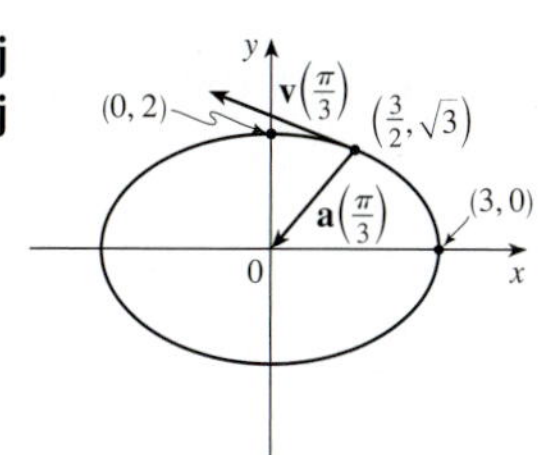

7. $\mathbf{v}(t) = \mathbf{i} + 2t\, \mathbf{j}$
$\mathbf{a}(t) = 2\, \mathbf{j}$
$|\mathbf{v}(t)| = \sqrt{1 + 4t^2}$

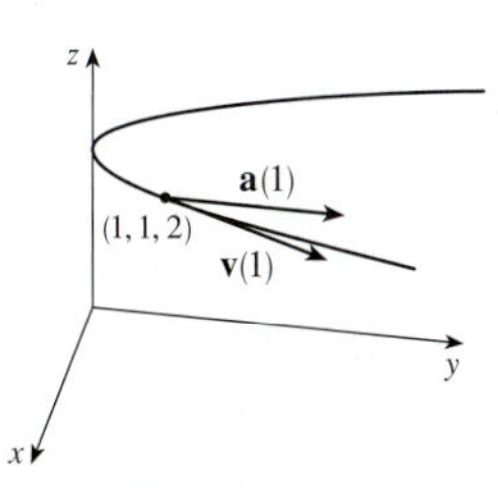

9. $\langle 2t + 1, 2t - 1, 3t^2 \rangle$, $\langle 2, 2, 6t \rangle$, $\sqrt{9t^4 + 8t^2 + 2}$

11. $\sqrt{2}\, \mathbf{i} + e^t \mathbf{j} - e^{-t} \mathbf{k}$, $e^t \mathbf{j} + e^{-t} \mathbf{k}$, $e^t + e^{-t}$

13. $e^t[(\cos t - \sin t)\mathbf{i} + (\sin t + \cos t)\mathbf{j} + (t + 1)\mathbf{k}]$,
$e^t[-2 \sin t\, \mathbf{i} + 2 \cos t\, \mathbf{j} + (t + 2)\mathbf{k}]$, $e^t\sqrt{t^2 + 2t + 3}$

15. $\mathbf{v}(t) = (2t + 3)\, \mathbf{i} - \mathbf{j} + t^2\, \mathbf{k}$,
$\mathbf{r}(t) = (t^2 + 3t)\, \mathbf{i} + (1 - t)\, \mathbf{j} + \left(\frac{1}{3}t^3 + 1\right) \mathbf{k}$

17. (a) $\mathbf{r}(t) = \left(\frac{1}{3}t^3 + t\right) \mathbf{i} + (t - \sin t + 1)\, \mathbf{j} + \left(\frac{1}{4} - \frac{1}{4}\cos 2t\right) \mathbf{k}$
(b)

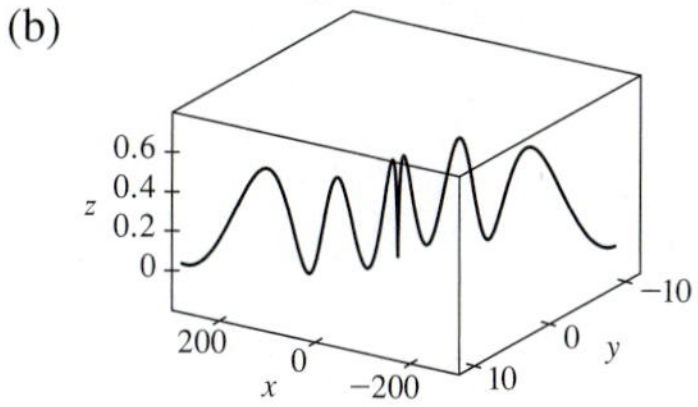

19. $t = 4$

21. $\mathbf{r}(t) = t\, \mathbf{i} - t\, \mathbf{j} + \frac{5}{2}t^2\, \mathbf{k}$, $|\mathbf{v}(t)| = \sqrt{25t^2 + 2}$

23. (a) $\approx$3535 m (b) $\approx$1531 m (c) 200 m/s

25. $\approx$30 m/s **27.** $\approx$ 544 ft/s

29. $13.0° < \theta < 36.0°$, $55.4° < \theta < 85.5°$

31. $(250, -50, 0)$; $10\sqrt{93} \approx 96.4$ ft/s

33. (a) 16 m (b) $\approx$23.6° upstream

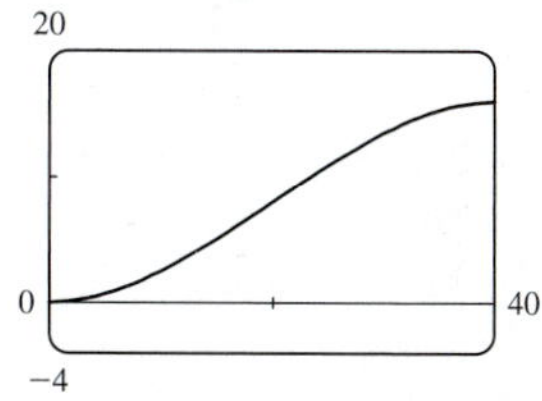

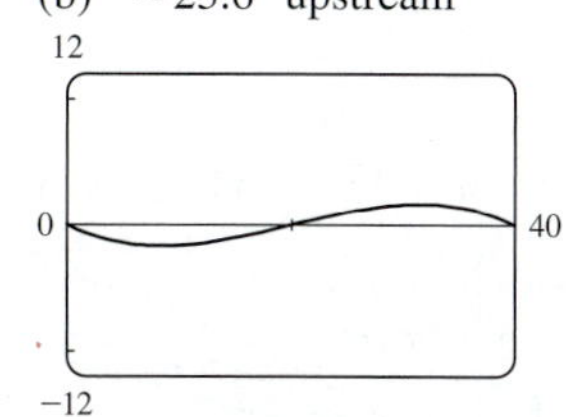

35. The path is contained in a circle that lies in a plane perpendicular to **c** with center on a line through the origin in the direction of **c**.

37. $\dfrac{4 + 18t^2}{\sqrt{4 + 9t^2}}$, $\dfrac{6t}{\sqrt{4 + 9t^2}}$ **39.** 0, 1 **41.** $\dfrac{7}{\sqrt{30}}$, $\sqrt{\dfrac{131}{30}}$

43. 4.5 cm/s², 9.0 cm/s² **45.** $t = 1$

CHAPTER 13 REVIEW ■ PAGE 921

True-False Quiz

1. True **3.** False **5.** False **7.** False
9. True **11.** False **13.** True

Exercises

1. (a)

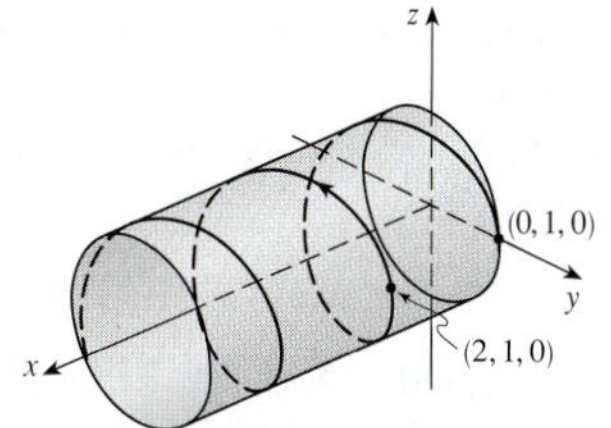

(b) $\mathbf{r}'(t) = \mathbf{i} - \pi \sin \pi t\, \mathbf{j} + \pi \cos \pi t\, \mathbf{k}$,
$\mathbf{r}''(t) = -\pi^2 \cos \pi t\, \mathbf{j} - \pi^2 \sin \pi t\, \mathbf{k}$

3. $\mathbf{r}(t) = 4 \cos t\, \mathbf{i} + 4 \sin t\, \mathbf{j} + (5 - 4 \cos t)\mathbf{k}$, $0 \leqslant t \leqslant 2\pi$

5. $\frac{1}{3}\mathbf{i} - (2/\pi^2)\mathbf{j} + (2/\pi)\mathbf{k}$ **7.** 86.631 **9.** 90°

11. (a) $\dfrac{1}{\sqrt{13}}\langle 3 \sin t, -3 \cos t, 2 \rangle$ (b) $\langle \cos t, \sin t, 0 \rangle$
(c) $\dfrac{1}{\sqrt{13}}\langle -2 \sin t, 2 \cos t, 3 \rangle$
(d) $\dfrac{3}{13 \sin t \cos t}$ or $\dfrac{3}{13} \sec t \csc t$

13. $12/17^{3/2}$ **15.** $x - 2y + 2\pi = 0$

17. $\mathbf{v}(t) = (1 + \ln t)\, \mathbf{i} + \mathbf{j} - e^{-t}\, \mathbf{k}$,
$|\mathbf{v}(t)| = \sqrt{2 + 2 \ln t + (\ln t)^2 + e^{-2t}}$, $\mathbf{a}(t) = (1/t)\mathbf{i} + e^{-t}\, \mathbf{k}$

19. $\mathbf{r}(t) = (t^3 + t)\, \mathbf{i} + (t^4 - t)\, \mathbf{j} + (3t - t^3)\, \mathbf{k}$

21. $\approx$37.3°, $\approx$157.4 m

23. (c) $-2e^{-t}\, \mathbf{v}_d + e^{-t}\, \mathbf{R}$

PROBLEMS PLUS ■ PAGE 924

1. (a) $\mathbf{v} = \omega R(-\sin \omega t\, \mathbf{i} + \cos \omega t\, \mathbf{j})$ (c) $\mathbf{a} = -\omega^2 \mathbf{r}$

3. (a) 90°, $v_0^2/(2g)$

5. (a) $\approx$0.94 ft to the right of the table's edge, $\approx$15 ft/s
(b) $\approx$7.6° (c) $\approx$2.13 ft to the right of the table's edge

7. 56°

9. $(a_2b_3 - a_3b_2)(x - c_1) + (a_3b_1 - a_1b_3)(y - c_2) + (a_1b_2 - a_2b_1)(z - c_3) = 0$

CHAPTER 14

EXERCISES 14.1 ■ PAGE 939

1. (a) −27; a temperature of −15°C with wind blowing at 40 km/h feels equivalent to about −27°C without wind.
(b) When the temperature is −20°C, what wind speed gives a wind chill of −30°C? 20 km/h
(c) With a wind speed of 20 km/h, what temperature gives a wind chill of −49°C? −35°C

(d) A function of wind speed that gives wind-chill values when the temperature is $-5°C$
(e) A function of temperature that gives wind-chill values when the wind speed is 50 km/h
3. ≈ 94.2; the manufacturer's yearly production is valued at \$94.2 million when 120,000 labor hours are spent and \$20 million in capital is invested.
5. (a) ≈ 20.5; the surface area of a person 70 inches tall who weighs 160 pounds is approximately 20.5 square feet.
7. (a) 25; a 40-knot wind blowing in the open sea for 15 h will create waves about 25 ft high.
(b) $f(30, t)$ is a function of t giving the wave heights produced by 30-knot winds blowing for t hours.
(c) $f(v, 30)$ is a function of v giving the wave heights produced by winds of speed v blowing for 30 hours.
9. (a) 1 (b) $\mathbb{R}^2$ (c) $[-1, 1]$
11. (a) 3
(b) $\{(x, y, z) \mid x^2 + y^2 + z^2 < 4, x \geq 0, y \geq 0, z \geq 0\}$, interior of a sphere of radius 2, center the origin, in the first octant
13. $\{(x, y) \mid x \geq 2, y \geq 1\}$

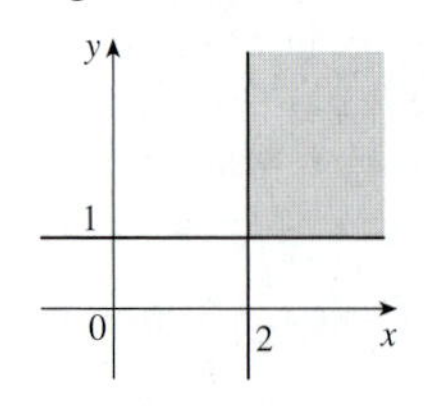

15. $\left\{(x, y) \mid \frac{1}{9}x^2 + y^2 < 1\right\}$

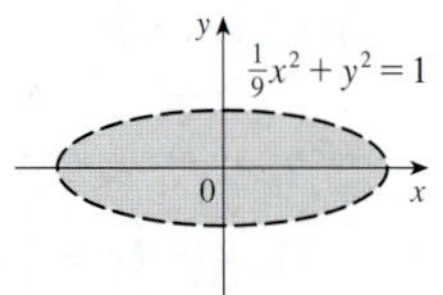

17. $\{(x, y) \mid y \neq -x\}$

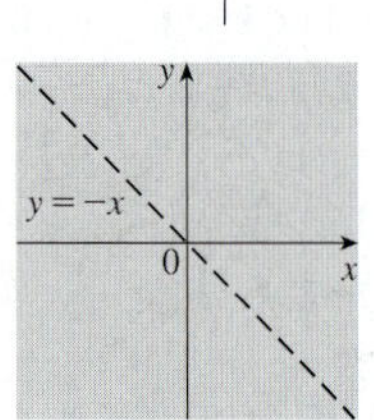

19. $\{(x, y) \mid y \geq x^2, x \neq \pm 1\}$

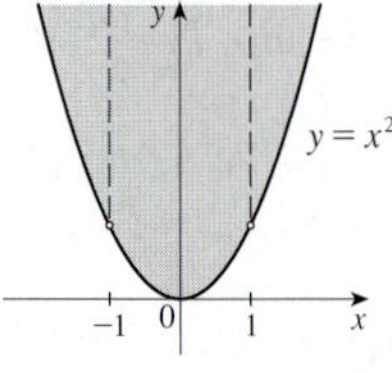

21. $\{(x, y, z) \mid -2 \leq x \leq 2, -3 \leq y \leq 3, -1 \leq z \leq 1\}$

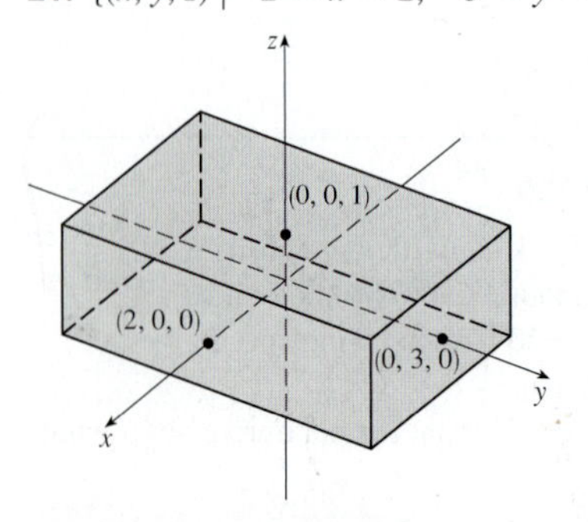

23. $z = y$, plane through the x-axis

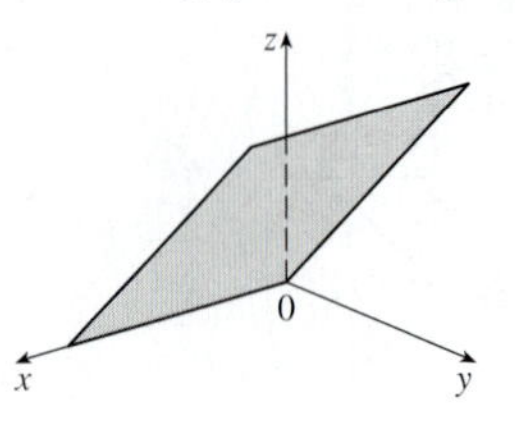

25. $4x + 5y + z = 10$, plane

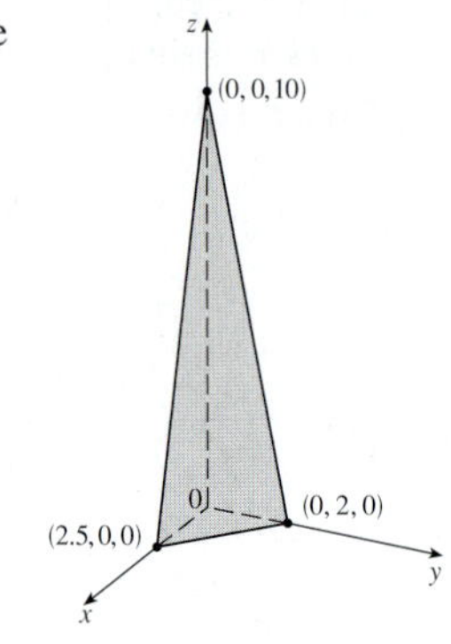

27. $z = \sin x$, cylinder

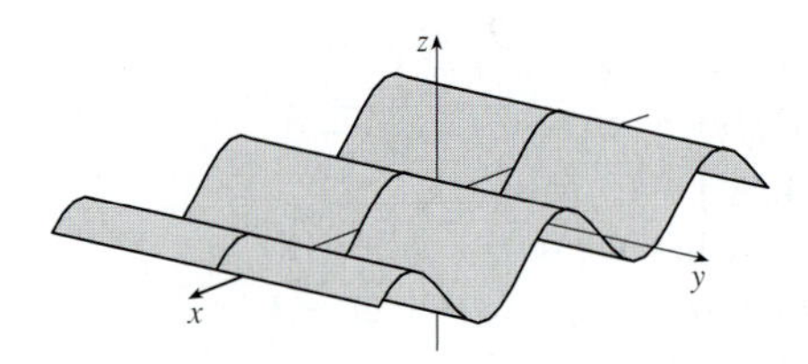

29. $z = x^2 + 4y^2 + 1$, elliptic paraboloid

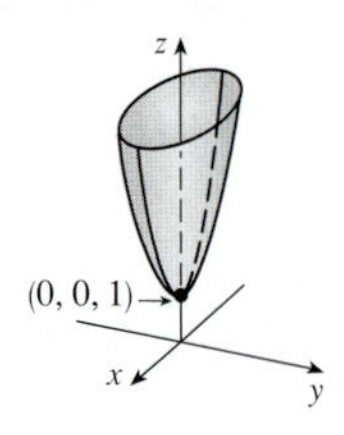

31. $z = \sqrt{4 - 4x^2 - y^2}$, top half of ellipsoid

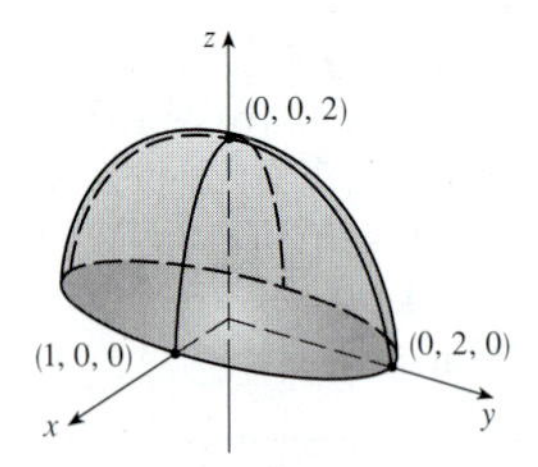

33. ≈ 56, ≈ 35 **35.** 11°C, 19.5°C **37.** Steep; nearly flat
39. h, 2.5, 2, 1.5, 1, 0.5, 0, 20 40 60 80 100 120 140 m; $B(m, h) = k$; $k = 40$, $k = 30$, $k = 25$, $k = 18.5$
No

41.

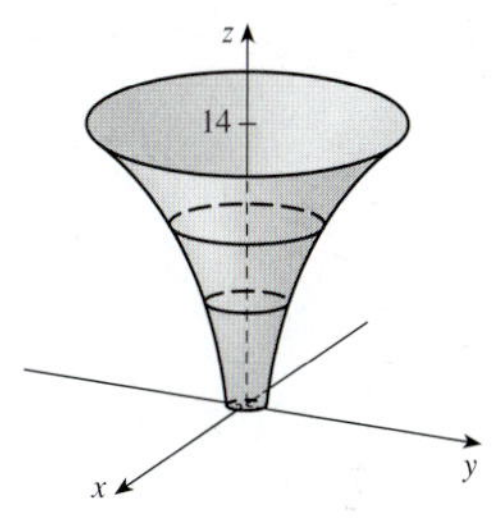

43.

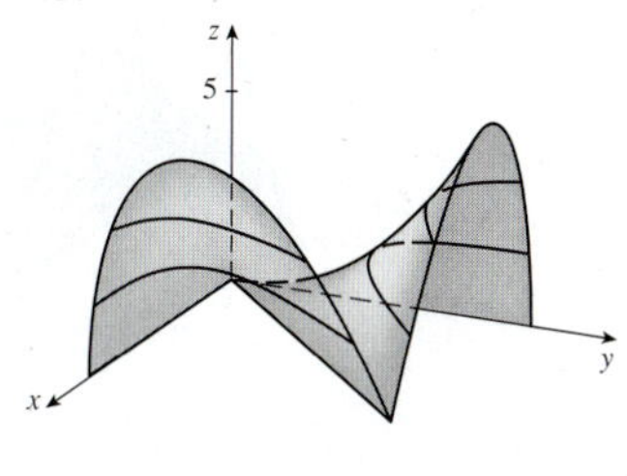

45. $x^2 - y^2 = k$

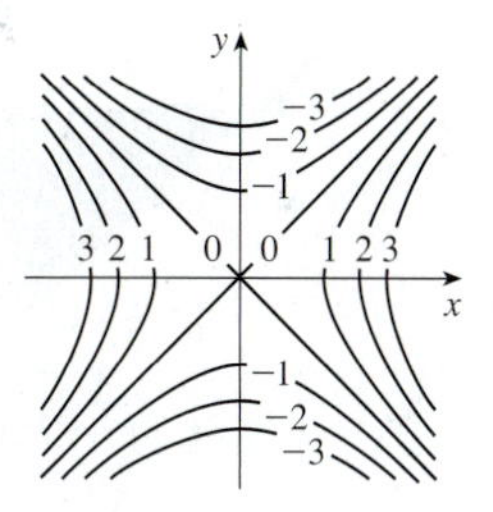

47. $y = -\sqrt{x} + k$

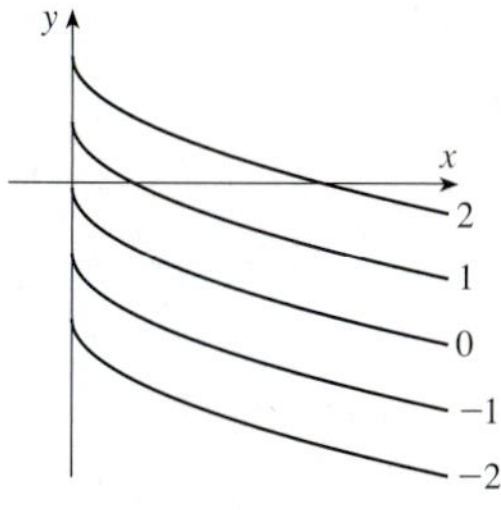

49. $y = ke^{-x}$

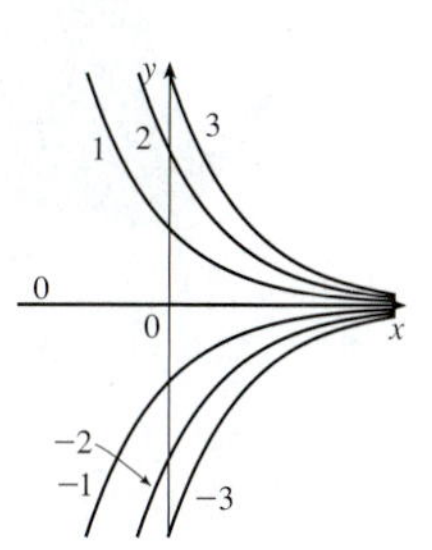

51. $x^2 + y^2 = k^3 \ (k \geqslant 0)$

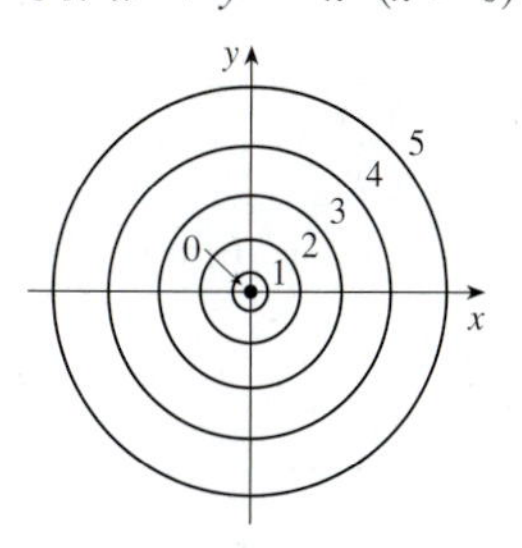

53. $x^2 + 9y^2 = k$

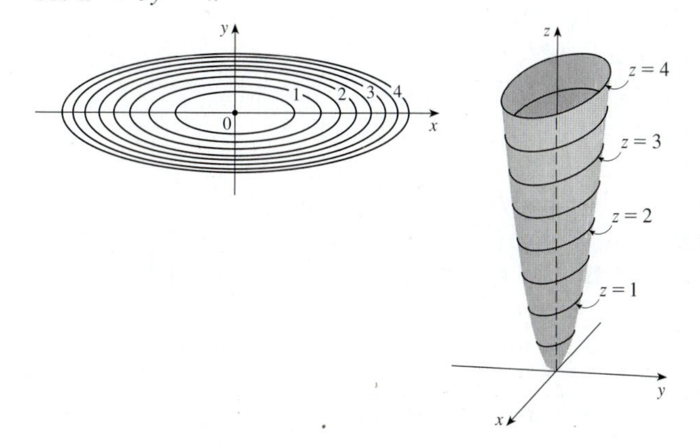

55.

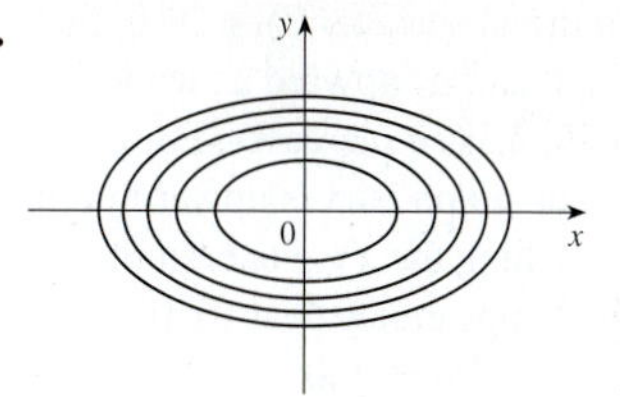

57.

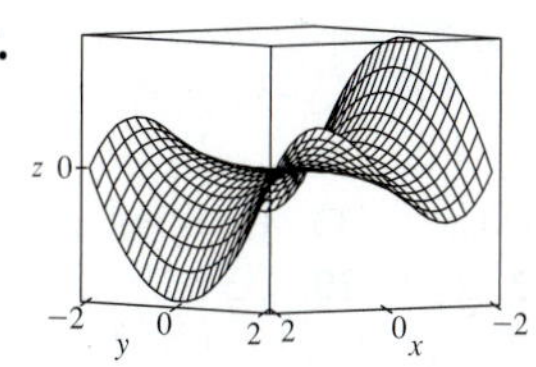

59.

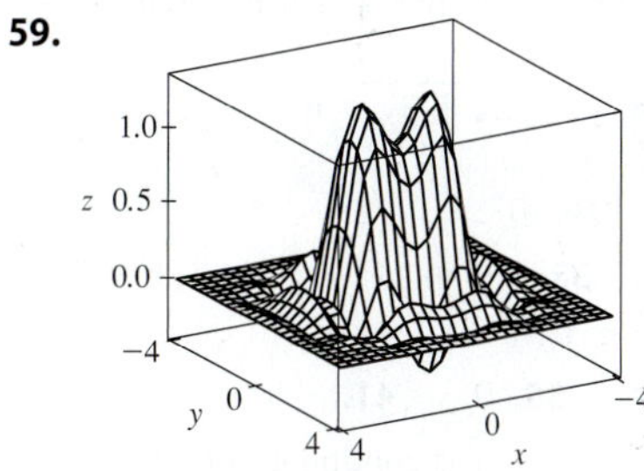

61. (a) C (b) II **63.** (a) F (b) I
65. (a) B (b) VI **67.** Family of parallel planes
69. Family of circular cylinders with axis the x-axis ($k > 0$)
71. (a) Shift the graph of f upward 2 units
(b) Stretch the graph of f vertically by a factor of 2
(c) Reflect the graph of f about the xy-plane
(d) Reflect the graph of f about the xy-plane and then shift it upward 2 units

73.

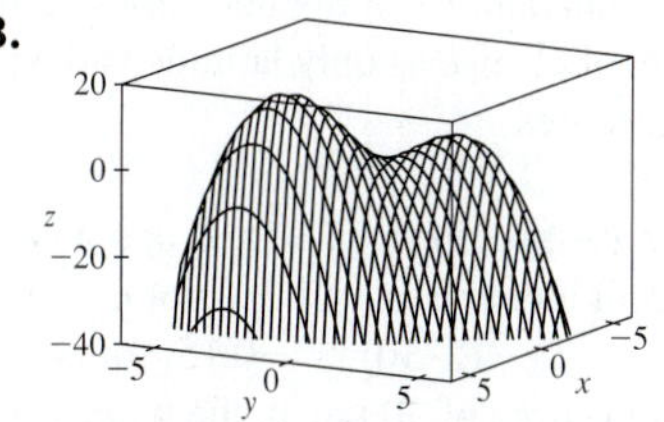

f appears to have a maximum value of about 15. There are two local maximum points but no local minimum point.

75.

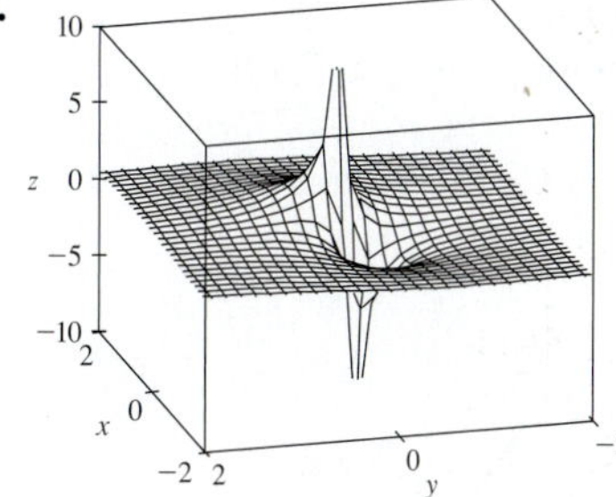

The function values approach 0 as x, y become large; as (x, y) approaches the origin, f approaches $\pm\infty$ or 0, depending on the direction of approach.

77. If $c = 0$, the graph is a cylindrical surface. For $c > 0$, the level curves are ellipses. The graph curves upward as we leave the origin, and the steepness increases as c increases. For $c < 0$, the level curves are hyperbolas. The graph curves upward in the y-direction and downward, approaching the xy-plane, in the x-direction giving a saddle-shaped appearance near $(0, 0, 1)$.
79. $c = -2, 0, 2$ **81.** (b) $y = 0.75x + 0.01$

EXERCISES 14.2 ■ PAGE 950

1. Nothing; if f is continuous, $f(3, 1) = 6$ **3.** $-\frac{5}{2}$
5. 56 **7.** $\pi/2$ **9.** Does not exist **11.** Does not exist
13. 0 **15.** Does not exist **17.** 2 **19.** $\sqrt{3}$
21. Does not exist
23. The graph shows that the function approaches different numbers along different lines.
25. $h(x, y) = (2x + 3y - 6)^2 + \sqrt{2x + 3y - 6}$; $\{(x, y) \mid 2x + 3y \geqslant 6\}$
27. Along the line $y = x$ **29.** $\mathbb{R}^2$
31. $\{(x, y) \mid x^2 + y^2 \neq 1\}$ **33.** $\{(x, y) \mid x^2 + y^2 \leqslant 1, x \geqslant 0\}$
35. $\{(x, y, z) \mid x^2 + y^2 + z^2 \leqslant 1\}$
37. $\{(x, y) \mid (x, y) \neq (0, 0)\}$ **39.** 0 **41.** -1
43.

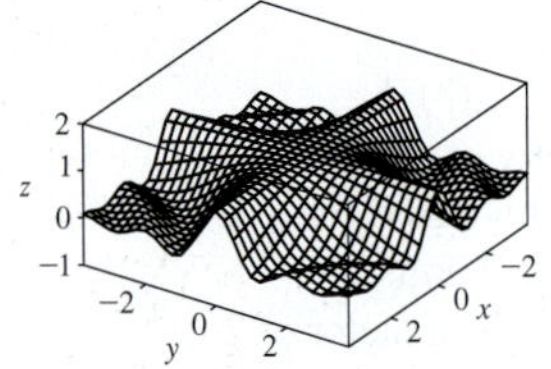

f is continuous on $\mathbb{R}^2$

EXERCISES 14.3 ■ PAGE 963

1. (a) The rate of change of temperature as longitude varies, with latitude and time fixed; the rate of change as only latitude varies; the rate of change as only time varies
(b) Positive, negative, positive
3. (a) $f_T(-15, 30) \approx 1.3$; for a temperature of $-15°C$ and wind speed of 30 km/h, the wind-chill index rises by 1.3°C for each degree the temperature increases. $f_v(-15, 30) \approx -0.15$; for a temperature of $-15°C$ and wind speed of 30 km/h, the wind-chill index decreases by 0.15°C for each km/h the wind speed increases.
(b) Positive, negative (c) 0
5. (a) Positive (b) Negative
7. (a) Positive (b) Negative
9. $c = f, b = f_x, a = f_y$
11. $f_x(1, 2) = -8 =$ slope of C_1, $f_y(1, 2) = -4 =$ slope of C_2

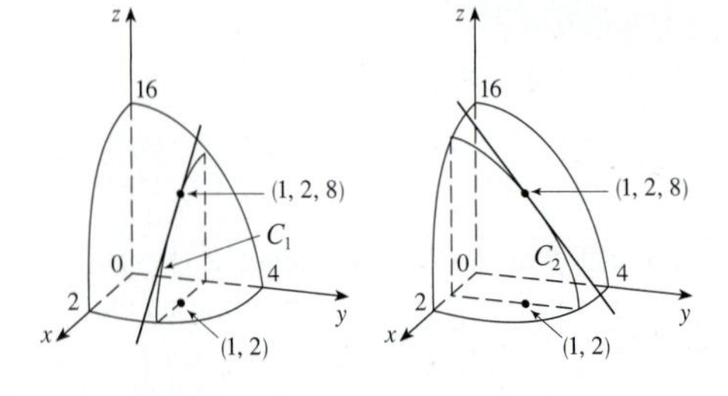

13.

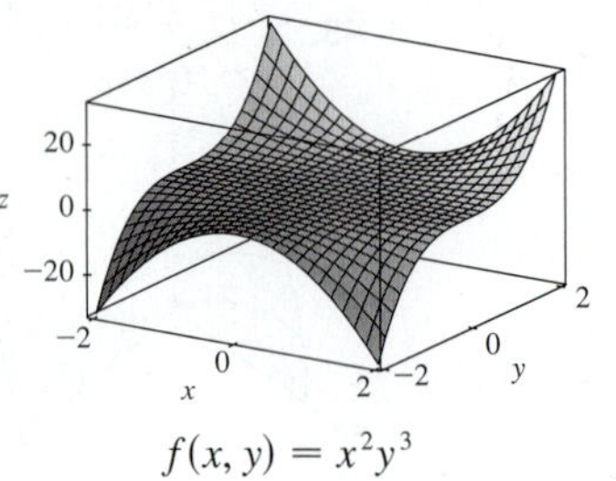

$f(x, y) = x^2y^3$

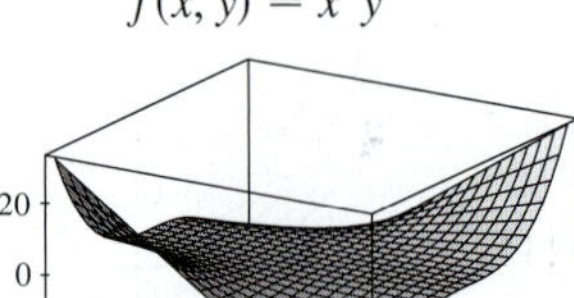

$f_x(x, y) = 2xy^3$

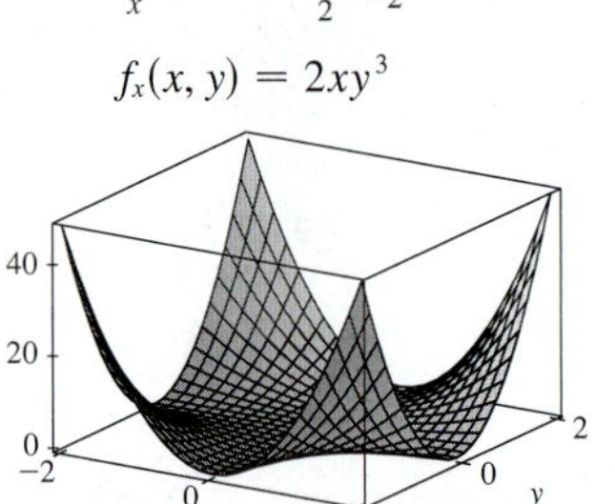

$f_y(x, y) = 3x^2y^2$

15. $f_x(x, y) = 4x^3 + 5y^3$, $f_y(x, y) = 15xy^2$
17. $f_x(x, t) = -t^2e^{-x}$, $f_t(x, t) = 2te^{-x}$
19. $\dfrac{\partial z}{\partial x} = \dfrac{1}{x + t^2}, \dfrac{\partial z}{\partial t} = \dfrac{2t}{x + t^2}$
21. $f_x(x, y) = 1/y$, $f_y(x, y) = -x/y^2$
23. $f_x(x, y) = \dfrac{(ad - bc)y}{(cx + dy)^2}, f_y(x, y) = \dfrac{(bc - ad)x}{(cx + dy)^2}$
25. $g_u(u, v) = 10uv(u^2v - v^3)^4$, $g_v(u, v) = 5(u^2 - 3v^2)(u^2v - v^3)^4$
27. $R_p(p, q) = \dfrac{q^2}{1 + p^2q^4}, R_q(p, q) = \dfrac{2pq}{1 + p^2q^4}$
29. $F_x(x, y) = \cos(e^x)$, $F_y(x, y) = -\cos(e^y)$
31. $f_x = 3x^2yz^2$, $f_y = x^3z^2 + 2z$, $f_z = 2x^3yz + 2y$
33. $\partial w/\partial x = 1/(x + 2y + 3z)$, $\partial w/\partial y = 2/(x + 2y + 3z)$, $\partial w/\partial z = 3/(x + 2y + 3z)$
35. $\partial p/\partial t = 2t^3/\sqrt{t^4 + u^2 \cos v}$, $\partial p/\partial u = u \cos v/\sqrt{t^4 + u^2 \cos v}$, $\partial p/\partial v = -u^2 \sin v/\left(2\sqrt{t^4 + u^2 \cos v}\right)$
37. $h_x = 2xy \cos(z/t)$, $h_y = x^2 \cos(z/t)$, $h_z = (-x^2y/t) \sin(z/t)$, $h_t = (x^2yz/t^2) \sin(z/t)$
39. $\partial u/\partial x_i = x_i/\sqrt{x_1^2 + x_2^2 + \cdots + x_n^2}$
41. 1 **43.** $\frac{1}{6}$ **45.** $f_x(x, y) = y^2 - 3x^2y$, $f_y(x, y) = 2xy - x^3$
47. $\dfrac{\partial z}{\partial x} = -\dfrac{x}{3z}, \dfrac{\partial z}{\partial y} = -\dfrac{2y}{3z}$
49. $\dfrac{\partial z}{\partial x} = \dfrac{yz}{e^z - xy}, \dfrac{\partial z}{\partial y} = \dfrac{xz}{e^z - xy}$

51. (a) $f'(x), g'(y)$ (b) $f'(x+y), f'(x+y)$

53. $f_{xx} = 12x^2y - 12xy^2, f_{xy} = 4x^3 - 12x^2y = f_{yx}, f_{yy} = -4x^3$

55. $z_{xx} = \dfrac{8y}{(2x+3y)^3}, z_{xy} = \dfrac{6y-4x}{(2x+3y)^3} = z_{yx},$

$z_{yy} = -\dfrac{12x}{(2x+3y)^3}$

57. $v_{ss} = 2\cos(s^2-t^2) - 4s^2\sin(s^2-t^2),$
$v_{st} = 4st\sin(s^2-t^2) = v_{ts},$
$v_{tt} = -2\cos(s^2-t^2) - 4t^2\sin(s^2-t^2)$

63. $24xy^2 - 6y, 24x^2y - 6x$ **65.** $(2x^2y^2z^5 + 6xyz^3 + 2z)e^{xyz^2}$

67. $\frac{3}{4}v(u+v^2)^{-5/2}$ **69.** $4/(y+2z)^3, 0$ **71.** $6yz^2$

73. $\approx 12.2, \approx 16.8, \approx 23.25$ **83.** R^2/R_1^2

87. $\dfrac{\partial T}{\partial P} = \dfrac{V-nb}{nR}, \dfrac{\partial P}{\partial V} = \dfrac{2n^2a}{V^3} - \dfrac{nRT}{(V-nb)^2}$

91. (a) ≈ 0.0545; for a person 70 inches tall who weighs 160 pounds, an increase in weight causes the surface area to increase at a rate of about 0.0545 square feet per pound.
(b) ≈ 0.213; for a person 70 inches tall who weighs 160 pounds, an increase in height (with no change in weight) causes the surface area to increase at a rate of about 0.213 square feet per inch of height.

93. $\partial P/\partial v = 3Av^2 - \dfrac{B(mg/x)^2}{v^2}$ is the rate of change of the power needed during flapping mode with respect to the bird's velocity when the mass and fraction of flapping time remain constant;
$\partial P/\partial x = -\dfrac{2Bm^2g^2}{x^3v}$ is the rate at which the power changes when only the fraction of time spent in flapping mode varies;
$\partial P/\partial m = \dfrac{2Bmg^2}{x^2v}$ is the rate of change of the power when only the mass varies.

97. No **99.** $x = 1+t, y = 2, z = 2-2t$ **103.** -2

105. (a)

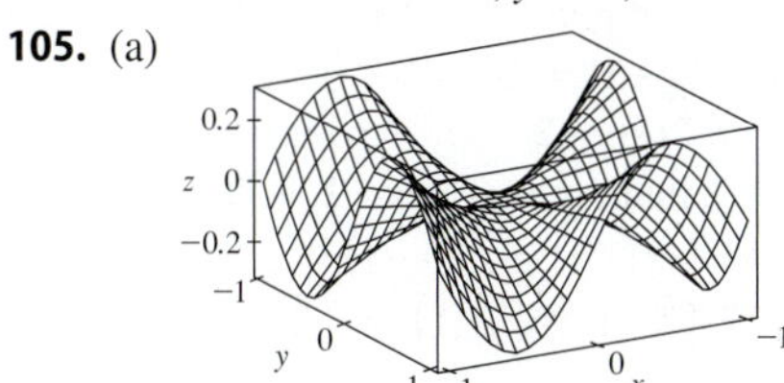

(b) $f_x(x,y) = \dfrac{x^4y + 4x^2y^3 - y^5}{(x^2+y^2)^2}, f_y(x,y) = \dfrac{x^5 - 4x^3y^2 - xy^4}{(x^2+y^2)^2}$

(c) 0, 0 (e) No, since f_{xy} and f_{yx} are not continuous.

EXERCISES 14.4 ■ PAGE 974

1. $z = 4x - y - 6$ **3.** $z = x - y + 1$ **5.** $x + y + z = 0$

7.

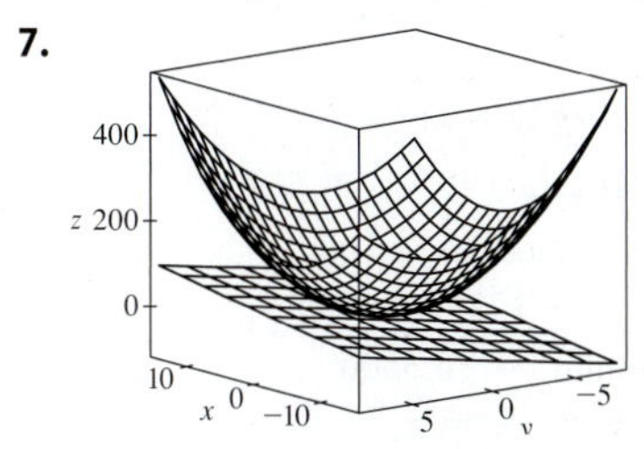

9.

11. $6x + 4y - 23$ **13.** $2x + y - 1$

15. $2x + 2y + \pi - 4$ **19.** 6.3

21. $\frac{3}{7}x + \frac{2}{7}y + \frac{6}{7}z$; 6.9914 **23.** $4T + H - 329$; 129°F

25. $dz = -2e^{-2x}\cos 2\pi t\,dx - 2\pi e^{-2x}\sin 2\pi t\,dt$

27. $dm = 5p^4q^3\,dp + 3p^5q^2\,dq$

29. $dR = \beta^2\cos\gamma\,d\alpha + 2\alpha\beta\cos\gamma\,d\beta - \alpha\beta^2\sin\gamma\,d\gamma$

31. $\Delta z = 0.9225, dz = 0.9$ **33.** 5.4 cm² **35.** 16 cm³

37. $\approx -0.0165mg$; decrease **39.** $\frac{1}{17} \approx 0.059\ \Omega$

41. (a) $0.8264m - 34.56h + 38.02$ (b) 18.801

43. $\varepsilon_1 = \Delta x, \varepsilon_2 = \Delta y$

EXERCISES 14.5 ■ PAGE 983

1. $2t(y^3 - 2xy + 3xy^2 - x^2)$

3. $\dfrac{1}{2\sqrt{t}}\cos x\cos y + \dfrac{1}{t^2}\sin x\sin y$

5. $e^{y/z}[2t - (x/z) - (2xy/z^2)]$

7. $\partial z/\partial s = 5(x-y)^4(2st - t^2), \partial z/\partial t = 5(x-y)^4(s^2 - 2st)$

9. $\dfrac{\partial z}{\partial s} = \dfrac{3\sin t - 2t\sin s}{3x+2y}, \dfrac{\partial z}{\partial t} = \dfrac{3s\cos t + 2\cos s}{3x+2y}$

11. $\dfrac{\partial z}{\partial s} = e^r\left(t\cos\theta - \dfrac{s}{\sqrt{s^2+t^2}}\sin\theta\right),$

$\dfrac{\partial z}{\partial t} = e^r\left(s\cos\theta - \dfrac{t}{\sqrt{s^2+t^2}}\sin\theta\right)$

13. 42 **15.** 7, 2

17. $\dfrac{\partial u}{\partial r} = \dfrac{\partial u}{\partial x}\dfrac{\partial x}{\partial r} + \dfrac{\partial u}{\partial y}\dfrac{\partial y}{\partial r}, \dfrac{\partial u}{\partial s} = \dfrac{\partial u}{\partial x}\dfrac{\partial x}{\partial s} + \dfrac{\partial u}{\partial y}\dfrac{\partial y}{\partial s},$

$\dfrac{\partial u}{\partial t} = \dfrac{\partial u}{\partial x}\dfrac{\partial x}{\partial t} + \dfrac{\partial u}{\partial y}\dfrac{\partial y}{\partial t}$

19. $\dfrac{\partial T}{\partial x} = \dfrac{\partial T}{\partial p}\dfrac{\partial p}{\partial x} + \dfrac{\partial T}{\partial q}\dfrac{\partial q}{\partial x} + \dfrac{\partial T}{\partial r}\dfrac{\partial r}{\partial x},$

$\dfrac{\partial T}{\partial y} = \dfrac{\partial T}{\partial p}\dfrac{\partial p}{\partial y} + \dfrac{\partial T}{\partial q}\dfrac{\partial q}{\partial y} + \dfrac{\partial T}{\partial r}\dfrac{\partial r}{\partial y},$

$\dfrac{\partial T}{\partial z} = \dfrac{\partial T}{\partial p}\dfrac{\partial p}{\partial z} + \dfrac{\partial T}{\partial q}\dfrac{\partial q}{\partial z} + \dfrac{\partial T}{\partial r}\dfrac{\partial r}{\partial z}$

21. 1582, 3164, −700 **23.** $2\pi, -2\pi$

25. $\frac{5}{144}, -\frac{5}{96}, \frac{5}{144}$ **27.** $\dfrac{2x + y\sin x}{\cos x - 2y}$

29. $\dfrac{1 + x^4y^2 + y^2 + x^4y^4 - 2xy}{x^2 - 2xy - 2x^5y^3}$

31. $-\dfrac{x}{3z}, -\dfrac{2y}{3z}$ **33.** $\dfrac{yz}{e^z - xy}, \dfrac{xz}{e^z - xy}$

35. 2°C/s **37.** ≈ -0.33 m/s per minute

39. (a) 6 m³/s (b) 10 m²/s (c) 0 m/s

41. ≈ -0.27 L/s **43.** $-1/(12\sqrt{3})$ rad/s

45. (a) $\partial z/\partial r = (\partial z/\partial x)\cos\theta + (\partial z/\partial y)\sin\theta$, $\partial z/\partial\theta = -(\partial z/\partial x)\, r\sin\theta + (\partial z/\partial y)\, r\cos\theta$
51. $4rs\,\partial^2 z/\partial x^2 + (4r^2 + 4s^2)\partial^2 z/\partial x\,\partial y + 4rs\,\partial^2 z/\partial y^2 + 2\,\partial z/\partial y$

EXERCISES 14.6 ■ PAGE 996

1. ≈ -0.08 mb/km **3.** ≈ 0.778 **5.** $\sqrt{2}/2$
7. (a) $\nabla f(x, y) = (1/y)\mathbf{i} - (x/y^2)\mathbf{j}$ (b) $\mathbf{i} - 2\mathbf{j}$ (c) -1
9. (a) $\langle 2xyz - yz^3, x^2z - xz^3, x^2y - 3xyz^2\rangle$
(b) $\langle -3, 2, 2\rangle$ (c) $\frac{2}{5}$
11. $\dfrac{4 - 3\sqrt{3}}{10}$ **13.** $7/(2\sqrt{5})$ **15.** 1 **17.** $\frac{23}{42}$
19. $\frac{2}{5}$ **21.** $\sqrt{65}, \langle 1, 8\rangle$ **23.** $1, \langle 0, 1\rangle$
25. $\frac{3}{4}, \langle 1, -2, -2\rangle$ **27.** (b) $\langle -12, 92\rangle$
29. All points on the line $y = x + 1$ **31.** (a) $-40/(3\sqrt{3})$
33. (a) $32/\sqrt{3}$ (b) $\langle 38, 6, 12\rangle$ (c) $2\sqrt{406}$
35. $\frac{327}{13}$ **39.** $\frac{774}{25}$
41. (a) $x + y + z = 11$ (b) $x - 3 = y - 3 = z - 5$
43. (a) $x + 2y + 6z = 12$ (b) $x - 2 = \dfrac{y - 2}{2} = \dfrac{z - 1}{6}$
45. (a) $x + y + z = 1$ (b) $x = y = z - 1$
47.

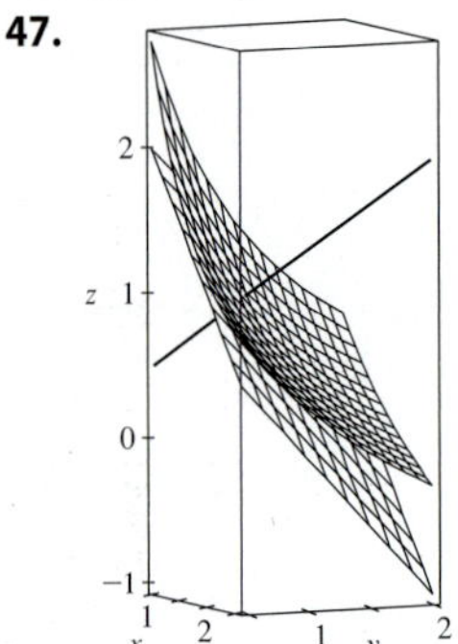

49. $\langle 2, 3\rangle$, $2x + 3y = 12$

55. No **59.** $\left(-\frac{5}{4}, -\frac{5}{4}, \frac{25}{8}\right)$
63. $x = -1 - 10t, y = 1 - 16t, z = 2 - 12t$
65. $(-1, 0, 1)$; $\approx 7.8°$
69. If $\mathbf{u} = \langle a, b\rangle$ and $\mathbf{v} = \langle c, d\rangle$, then $af_x + bf_y$ and $cf_x + df_y$ are known, so we solve linear equations for f_x and f_y.

EXERCISES 14.7 ■ PAGE 1007

1. (a) f has a local minimum at (1, 1).
(b) f has a saddle point at (1, 1).
3. Local minimum at (1, 1), saddle point at (0, 0)
5. Minimum $f\left(\frac{1}{3}, -\frac{2}{3}\right) = -\frac{1}{3}$
7. Saddle points at $(1, 1), (-1, -1)$
9. Minima $f\left(\dfrac{1}{\sqrt{2}}, -\dfrac{1}{\sqrt{2}}\right) = f\left(-\dfrac{1}{\sqrt{2}}, \dfrac{1}{\sqrt{2}}\right) = -\dfrac{1}{4}$, saddle point at (0, 0)
11. Maximum $f(-1, 0) = 2$, minimum $f(1, 0) = -2$, saddle points at $(0, \pm 1)$
13. Maximum $f(0, -1) = 2$, minima $f(\pm 1, 1) = -3$, saddle points at $(0, 1), (\pm 1, -1)$
15. None
17. Minima $f(x, y) = 1$ at all points (x, y) on x- and y-axes
19. Minima $f(0, 1) = f(\pi, -1) = f(2\pi, 1) = -1$, saddle points at $(\pi/2, 0), (3\pi/2, 0)$
23. Minima $f(1, \pm 1) = f(-1, \pm 1) = 3$
25. Maximum $f(\pi/3, \pi/3) = 3\sqrt{3}/2$, minimum $f(5\pi/3, 5\pi/3) = -3\sqrt{3}/2$, saddle point at (π, π)
27. Minima $f(0, -0.794) \approx -1.191$, $f(\pm 1.592, 1.267) \approx -1.310$, saddle points $(\pm 0.720, 0.259)$, lowest points $(\pm 1.592, 1.267, -1.310)$
29. Maximum $f(0.170, -1.215) \approx 3.197$, minima $f(-1.301, 0.549) \approx -3.145$, $f(1.131, 0.549) \approx -0.701$, saddle points $(-1.301, -1.215), (0.170, 0.549), (1.131, -1.215)$, no highest or lowest point
31. Maximum $f(0, \pm 2) = 4$, minimum $f(1, 0) = -1$
33. Maximum $f(\pm 1, 1) = 7$, minimum $f(0, 0) = 4$
35. Maximum $f(0, 3) = f(2, 3) = 7$, minimum $f(1, 1) = -2$
37. Maximum $f(1, 0) = 2$, minimum $f(-1, 0) = -2$
39.

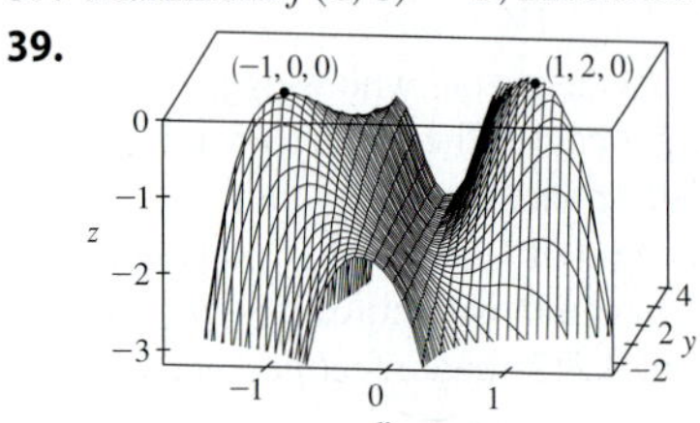

41. $2/\sqrt{3}$ **43.** $(2, 1, \sqrt{5}), (2, 1, -\sqrt{5})$ **45.** $\frac{100}{3}, \frac{100}{3}, \frac{100}{3}$
47. $8r^3/(3\sqrt{3})$ **49.** $\frac{4}{3}$ **51.** Cube, edge length $c/12$
53. Square base of side 40 cm, height 20 cm **55.** $L^3/(3\sqrt{3})$
57. (a) $H = -p_1 \ln p_1 - p_2 \ln p_2 - (1 - p_1 - p_2)\ln(1 - p_1 - p_2)$
(b) $\{(p_1, p_2) \mid 0 < p_1 < 1, p_2 < 1 - p_1\}$
(c) $\ln 3$; $p_1 = p_2 = p_3 = \frac{1}{3}$

EXERCISES 14.8 ■ PAGE 1017

1. $\approx 59, 30$
3. Maximum $f(\pm 1, 0) = 1$, minimum $f(0, \pm 1) = -1$
5. Maximum $f(1, 2) = f(-1, -2) = 2$, minimum $f(1, -2) = f(-1, 2) = -2$
7. Maximum $f(2, 2, 1) = 9$, minimum $f(-2, -2, -1) = -9$
9. Maximum $f(1, \pm\sqrt{2}, 1) = f(-1, \pm\sqrt{2}, -1) = 2$, minimum $f(1, \pm\sqrt{2}, -1) = f(-1, \pm\sqrt{2}, 1) = -2$
11. Maximum $\sqrt{3}$, minimum 1
13. Maximum $f\left(\frac{1}{2}, \frac{1}{2}, \frac{1}{2}, \frac{1}{2}\right) = 2$, minimum $f\left(-\frac{1}{2}, -\frac{1}{2}, -\frac{1}{2}, -\frac{1}{2}\right) = -2$
15. Minimum $f(1, 1) = f(-1, -1) = 2$
17. Maximum $f(0, 1, \sqrt{2}) = 1 + \sqrt{2}$, minimum $f(0, 1, -\sqrt{2}) = 1 - \sqrt{2}$
19. Maximum $\frac{3}{2}$, minimum $\frac{1}{2}$
21. Maximum $f(3/\sqrt{2}, -3/\sqrt{2}) = 9 + 12\sqrt{2}$, minimum $f(-2, 2) = -8$
23. Maximum $f(\pm 1/\sqrt{2}, \mp 1/(2\sqrt{2})) = e^{1/4}$, minimum $f(\pm 1/\sqrt{2}, \pm 1/(2\sqrt{2})) = e^{-1/4}$
31–43. See Exercises 41–55 in Section 14.7.
45. Nearest $\left(\frac{1}{2}, \frac{1}{2}, \frac{1}{2}\right)$, farthest $(-1, -1, 2)$
47. Maximum ≈ 9.7938, minimum ≈ -5.3506
49. (a) c/n (b) When $x_1 = x_2 = \cdots = x_n$

CHAPTER 14 REVIEW ■ PAGE 1022

True-False Quiz

1. True **3.** False **5.** False **7.** True **9.** False
11. True

Exercises

1. $\{(x, y) \mid y > -x - 1\}$

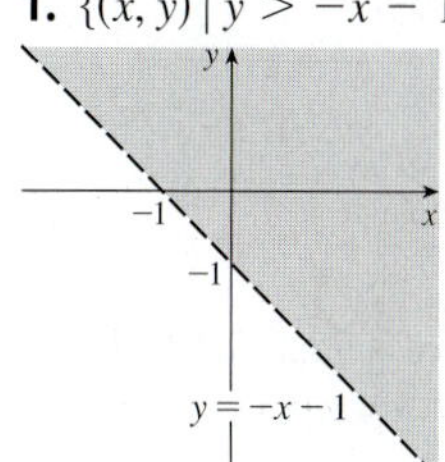

3.

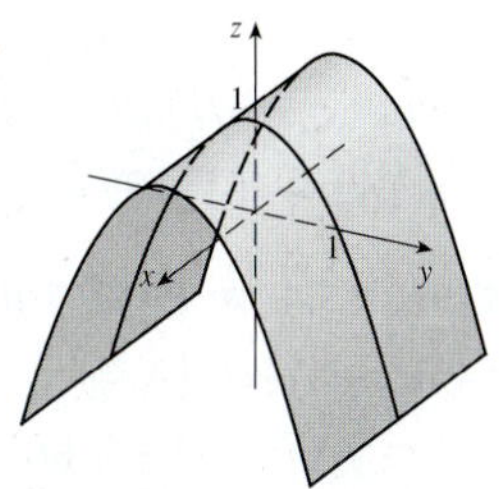

5.

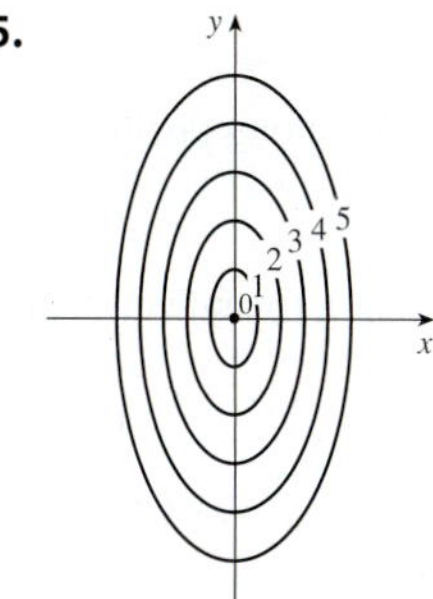

7.

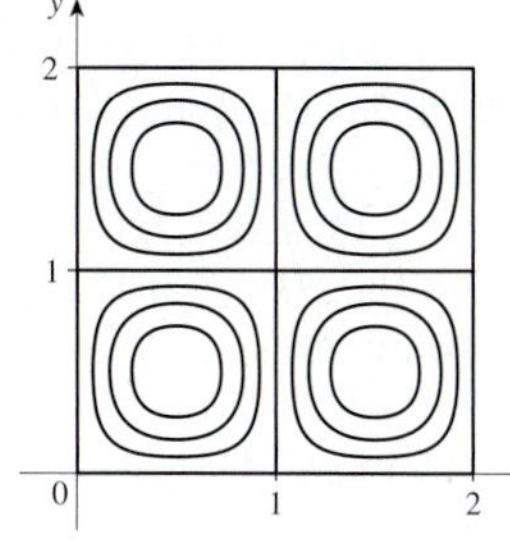

9. $\frac{2}{3}$

11. (a) $\approx 3.5°\text{C/m}, -3.0°\text{C/m}$
(b) $\approx 0.35°\text{C/m}$ by Equation 14.6.9 (Definition 14.6.2 gives $\approx 1.1°\text{C/m}$.)
(c) -0.25

13. $f_x = 32xy(5y^3 + 2x^2y)^7,\ f_y = (16x^2 + 120y^2)(5y^3 + 2x^2y)^7$

15. $F_\alpha = \dfrac{2\alpha^3}{\alpha^2 + \beta^2} + 2\alpha \ln(\alpha^2 + \beta^2),\ F_\beta = \dfrac{2\alpha^2\beta}{\alpha^2 + \beta^2}$

17. $S_u = \arctan(v\sqrt{w}),\ S_v = \dfrac{u\sqrt{w}}{1 + v^2w},\ S_w = \dfrac{uv}{2\sqrt{w}(1 + v^2w)}$

19. $f_{xx} = 24x,\ f_{xy} = -2y = f_{yx},\ f_{yy} = -2x$

21. $f_{xx} = k(k-1)x^{k-2}y^lz^m,\ f_{xy} = klx^{k-1}y^{l-1}z^m = f_{yx}$,
$f_{xz} = kmx^{k-1}y^lz^{m-1} = f_{zx},\ f_{yy} = l(l-1)x^ky^{l-2}z^m$,
$f_{yz} = lmx^ky^{l-1}z^{m-1} = f_{zy},\ f_{zz} = m(m-1)x^ky^lz^{m-2}$

25. (a) $z = 8x + 4y + 1$ (b) $\dfrac{x-1}{8} = \dfrac{y+2}{4} = \dfrac{z-1}{-1}$

27. (a) $2x - 2y - 3z = 3$ (b) $\dfrac{x-2}{4} = \dfrac{y+1}{-4} = \dfrac{z-1}{-6}$

29. (a) $x + 2y + 5z = 0$
(b) $x = 2 + t,\ y = -1 + 2t,\ z = 5t$

31. $\left(2, \frac{1}{2}, -1\right), \left(-2, -\frac{1}{2}, 1\right)$

33. $60x + \frac{24}{5}y + \frac{32}{5}z - 120$; 38.656

35. $2xy^3(1 + 6p) + 3x^2y^2(pe^p + e^p) + 4z^3(p\cos p + \sin p)$

37. $-47, 108$

43. $\langle 2xe^{yz^2}, x^2z^2e^{yz^2}, 2x^2yze^{yz^2}\rangle$ **45.** $-\frac{4}{5}$

47. $\sqrt{145}/2, \left\langle 4, \frac{9}{2}\right\rangle$ **49.** $\approx \frac{5}{8}$ knots/mi

51. Minimum $f(-4, 1) = -11$

53. Maximum $f(1, 1) = 1$; saddle points $(0, 0), (0, 3), (3, 0)$

55. Maximum $f(1, 2) = 4$, minimum $f(2, 4) = -64$

57. Maximum $f(-1, 0) = 2$, minima $f(1, \pm 1) = -3$, saddle points $(-1, \pm 1), (1, 0)$

59. Maximum $f\left(\pm\sqrt{2/3}, 1/\sqrt{3}\right) = 2/(3\sqrt{3})$, minimum $f\left(\pm\sqrt{2/3}, -1/\sqrt{3}\right) = -2/(3\sqrt{3})$

61. Maximum 1, minimum -1

63. $\left(\pm 3^{-1/4}, 3^{-1/4}\sqrt{2}, \pm 3^{1/4}\right), \left(\pm 3^{-1/4}, -3^{-1/4}\sqrt{2}, \pm 3^{1/4}\right)$

65. $P(2 - \sqrt{3}), P(3 - \sqrt{3})/6, P(2\sqrt{3} - 3)/3$

PROBLEMS PLUS ■ PAGE 1025

1. $L^2W^2, \frac{1}{4}L^2W^2$ **3.** (a) $x = w/3$, base $= w/3$ (b) Yes

7. $\sqrt{3/2}, 3/\sqrt{2}$

CHAPTER 15

EXERCISES 15.1 ■ PAGE 1039

1. (a) 288 (b) 144 **3.** (a) 0.990 (b) 1.151

5. $U < V < L$ **7.** (a) ≈ 248 (b) ≈ 15.5

9. $24\sqrt{2}$ **11.** 3 **13.** $2 + 8y^2, 3x + 27x^2$

15. 222 **17.** $\frac{5}{2} - e^{-1}$ **19.** 18

21. $\frac{15}{2}\ln 2 + \frac{3}{2}\ln 4$ or $\frac{21}{2}\ln 2$ **23.** 6

25. $\frac{31}{30}$ **27.** 2 **29.** $9\ln 2$

31. $\frac{1}{2}(\sqrt{3} - 1) - \frac{1}{12}\pi$ **33.** $\frac{1}{2}e^{-6} + \frac{5}{2}$

35.

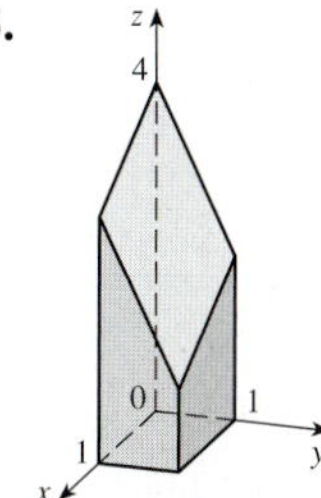

37. 51 **39.** $\frac{166}{27}$ **41.** $\frac{8}{3}$ **43.** $\frac{64}{3}$

45. $21e - 57$

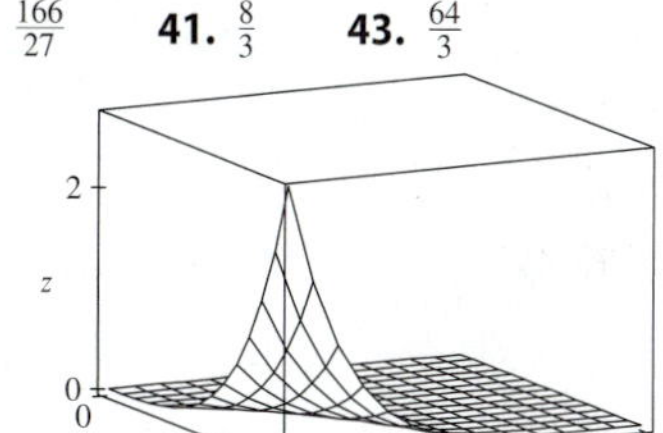

47. $\frac{5}{6}$ **49.** 0

51. Fubini's Theorem does not apply. The integrand has an infinite discontinuity at the origin.

EXERCISES 15.2 ■ PAGE 1048

1. $\frac{868}{3}$ **3.** $\frac{1}{6}(e - 1)$ **5.** $\frac{1}{3}\sin 1$

7. $\frac{1}{4}\ln 17$ **9.** $\frac{1}{2}(1 - e^{-9})$

11. (a) (b)

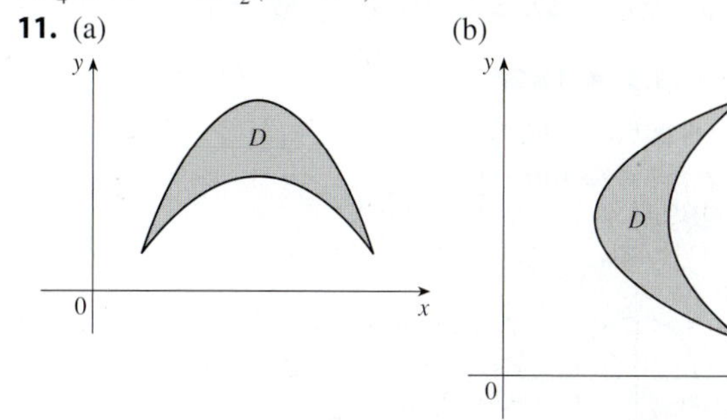

13. Type I: $D = \{(x, y) \mid 0 \leqslant x \leqslant 1, 0 \leqslant y \leqslant x\}$,
type II: $D = \{(x, y) \mid 0 \leqslant y \leqslant 1, y \leqslant x \leqslant 1\}$; $\frac{1}{3}$

15. $\int_0^1 \int_{-\sqrt{x}}^{\sqrt{x}} y\,dy\,dx + \int_1^4 \int_{x-2}^{\sqrt{x}} y\,dy\,dx = \int_{-1}^2 \int_{y^2}^{y+2} y\,dx\,dy = \frac{9}{4}$

17. $\frac{1}{2}(1 - \cos 1)$ **19.** $\frac{11}{3}$ **21.** 0 **23.** $\frac{3}{4}$

25. $\frac{31}{8}$ **27.** $\frac{16}{3}$ **29.** $\frac{128}{15}$ **31.** $\frac{1}{3}$

33. 0, 1.213; 0.713 **35.** $\frac{64}{3}$

37. $\dfrac{10}{3\sqrt{2}}$ or $\dfrac{5\sqrt{2}}{3}$

39.

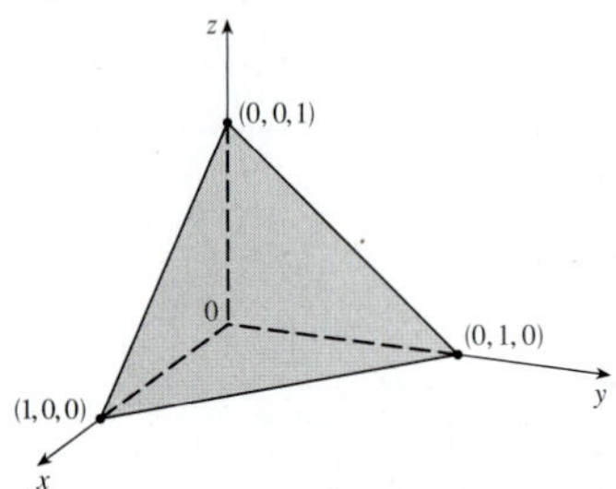

41. 13,984,735,616/14,549,535 **43.** $\pi/2$

45. $\int_0^1 \int_x^1 f(x, y)\,dy\,dx$

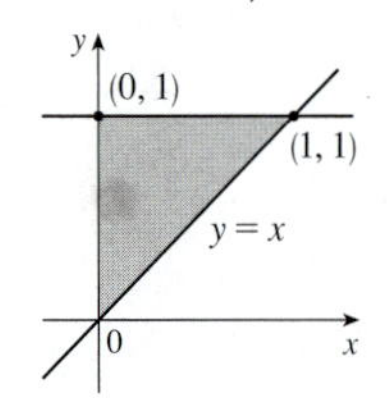

47. $\int_0^1 \int_0^{\cos^{-1} y} f(x, y)\,dx\,dy$

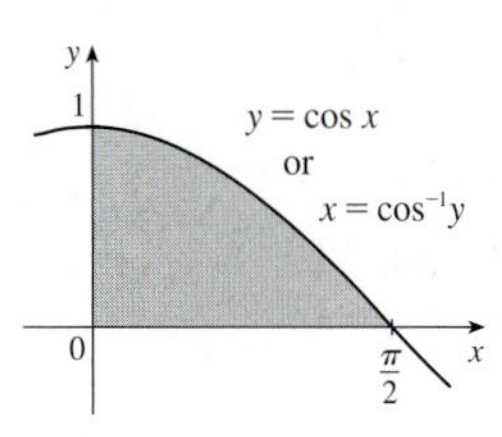

49. $\int_0^{\ln 2} \int_{e^y}^2 f(x, y)\,dx\,dy$

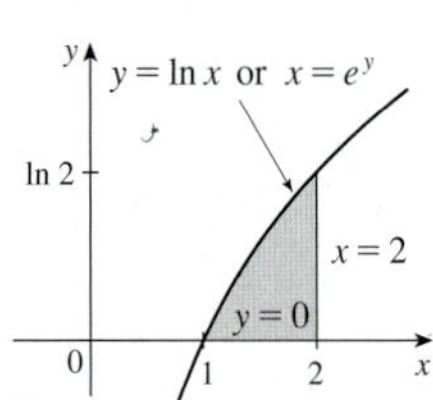

51. $\frac{1}{6}(e^9 - 1)$ **53.** $\frac{2}{9}(2\sqrt{2} - 1)$

55. $\frac{1}{3}(2\sqrt{2} - 1)$ **57.** 1

59. $\dfrac{\sqrt{3}}{2}\pi \leqslant \iint_S \sqrt{4 - x^2y^2}\,dA \leqslant \pi$

61. $\frac{3}{4}$ **65.** 9π **67.** $a^2b + \frac{3}{2}ab^2$ **69.** $\pi a^2 b$

EXERCISES 15.3 ■ PAGE 1054

1. $\int_0^{2\pi} \int_2^5 f(r\cos\theta, r\sin\theta)\,r\,dr\,d\theta$

3. $\int_{\pi}^{2\pi} \int_0^1 f(r\cos\theta, r\sin\theta)\,r\,dr\,d\theta$

5. $\theta = \frac{3\pi}{4}$ $\theta = \frac{\pi}{4}$ $3\pi/4$

y, R, −2, −1, 0, 1, 2, x

7. $\frac{1250}{3}$ **9.** $(\pi/4)(\cos 1 - \cos 9)$

11. $(\pi/2)(1 - e^{-4})$ **13.** $\frac{3}{64}\pi^2$ **15.** $\pi/12$

17. $\dfrac{\pi}{3} + \dfrac{\sqrt{3}}{2}$ **19.** $\frac{625}{2}\pi$ **21.** 4π **23.** $\frac{4}{3}\pi a^3$

25. $(\pi/3)(2 - \sqrt{2})$ **27.** $(8\pi/3)(64 - 24\sqrt{3})$

29. $(\pi/4)(1 - e^{-4})$ **31.** $\frac{1}{120}$ **33.** 4.5951

35. 1800π ft^3 **37.** $2/(a + b)$ **39.** $\frac{15}{16}$

41. (a) $\sqrt{\pi}/4$ (b) $\sqrt{\pi}/2$

EXERCISES 15.4 ■ PAGE 1064

1. 285 C **3.** $42k$, $(2, \frac{85}{28})$ **5.** 6, $(\frac{3}{4}, \frac{3}{2})$ **7.** $\frac{8}{15}k$, $(0, \frac{4}{7})$

9. $\frac{1}{8}(1 - 3e^{-2})$, $\left(\dfrac{e^2 - 5}{e^2 - 3}, \dfrac{8(e^3 - 4)}{27(e^3 - 3e)}\right)$

11. $(\frac{3}{8}, 3\pi/16)$ **13.** $(0, 45/(14\pi))$

15. $(2a/5, 2a/5)$ if vertex is (0, 0) and sides are along positive axes

17. $409.2k$, $182k$, $591.2k$

19. $7ka^6/180$, $7ka^6/180$, $7ka^6/90$ if vertex is (0, 0) and sides are along positive axes

21. $\rho bh^3/3$, $\rho b^3h/3$; $b/\sqrt{3}$, $h/\sqrt{3}$

23. $\rho a^4\pi/16$, $\rho a^4\pi/16$; $a/2$, $a/2$

25. $m = 3\pi/64$, $(\bar{x}, \bar{y}) = \left(\dfrac{16384\sqrt{2}}{10395\pi}, 0\right)$,

$I_x = \dfrac{5\pi}{384} - \dfrac{4}{105}$, $I_y = \dfrac{5\pi}{384} + \dfrac{4}{105}$, $I_0 = \dfrac{5\pi}{192}$

27. (a) $\frac{1}{2}$ (b) 0.375 (c) $\frac{5}{48} \approx 0.1042$

29. (b) (i) $e^{-0.2} \approx 0.8187$
(ii) $1 + e^{-1.8} - e^{-0.8} - e^{-1} \approx 0.3481$ (c) 2, 5

31. (a) ≈ 0.500 (b) ≈ 0.632

33. (a) $\iint_D k\left[1 - \frac{1}{20}\sqrt{(x - x_0)^2 + (y - y_0)^2}\right] dA$, where D is the disk with radius 10 mi centered at the center of the city
(b) $200\pi k/3 \approx 209k$, $200(\pi/2 - \frac{8}{9})k \approx 136k$, on the edge

EXERCISES 15.5 ■ PAGE 1068

1. $12\sqrt{35}$ **3.** $3\sqrt{14}$ **5.** $(\pi/6)(13\sqrt{13} - 1)$

7. $(\pi/6)(17\sqrt{17} - 5\sqrt{5})$ **9.** $(2\pi/3)(2\sqrt{2} - 1)$

11. $a^2(\pi - 2)$ **13.** 3.6258

15. (a) ≈ 1.83 (b) ≈ 1.8616

17. $\frac{45}{8}\sqrt{14} + \frac{15}{16}\ln\left[(11\sqrt{5} + 3\sqrt{70})/(3\sqrt{5} + \sqrt{70})\right]$

19. 3.3213 **23.** $(\pi/6)(101\sqrt{101} - 1)$

EXERCISES 15.6 ■ PAGE 1077

1. $\frac{27}{4}$ **3.** $\frac{16}{15}$ **5.** $\frac{5}{3}$ **7.** $\frac{2}{3}$ **9.** $\frac{27}{2}$ **11.** $9\pi/8$

13. $\frac{65}{28}$ **15.** $\frac{8}{15}$ **17.** $16\pi/3$ **19.** $\frac{16}{3}$ **21.** $\frac{8}{15}$

23. (a) $\int_0^1 \int_0^x \int_0^{\sqrt{1-y^2}} dz\,dy\,dx$ (b) $\frac{1}{4}\pi - \frac{1}{3}$

25. ≈ 0.985 **27.**

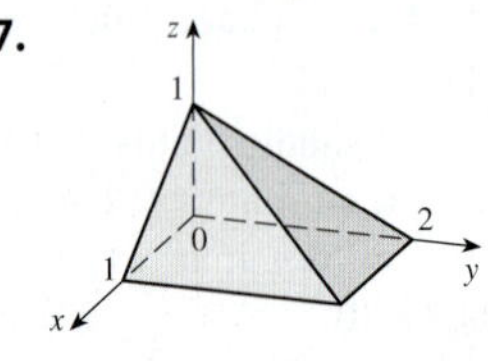

29. $\int_{-2}^{2}\int_{0}^{4-x^2}\int_{-\sqrt{4-x^2-y}/2}^{\sqrt{4-x^2-y}/2} f(x,y,z)\,dz\,dy\,dx$
$= \int_{0}^{4}\int_{-\sqrt{4-y}}^{\sqrt{4-y}}\int_{-\sqrt{4-x^2-y}/2}^{\sqrt{4-x^2-y}/2} f(x,y,z)\,dz\,dx\,dy$
$= \int_{-1}^{1}\int_{0}^{4-4z^2}\int_{-\sqrt{4-y-4z^2}}^{\sqrt{4-y-4z^2}} f(x,y,z)\,dx\,dy\,dz$
$= \int_{0}^{4}\int_{-\sqrt{4-y}/2}^{\sqrt{4-y}/2}\int_{-\sqrt{4-y-4z^2}}^{\sqrt{4-y-4z^2}} f(x,y,z)\,dx\,dz\,dy$
$= \int_{-2}^{2}\int_{-\sqrt{4-x^2}/2}^{\sqrt{4-x^2}/2}\int_{0}^{4-x^2-4z^2} f(x,y,z)\,dy\,dz\,dx$
$= \int_{-1}^{1}\int_{-\sqrt{4-4z^2}}^{\sqrt{4-4z^2}}\int_{0}^{4-x^2-4z^2} f(x,y,z)\,dy\,dx\,dz$

31. $\int_{-2}^{2}\int_{x^2}^{4}\int_{0}^{2-y/2} f(x,y,z)\,dz\,dy\,dx$
$= \int_{0}^{4}\int_{-\sqrt{y}}^{\sqrt{y}}\int_{0}^{2-y/2} f(x,y,z)\,dz\,dx\,dy$
$= \int_{0}^{2}\int_{0}^{4-2z}\int_{-\sqrt{y}}^{\sqrt{y}} f(x,y,z)\,dx\,dy\,dz$
$= \int_{0}^{4}\int_{0}^{2-y/2}\int_{-\sqrt{y}}^{\sqrt{y}} f(x,y,z)\,dx\,dz\,dy$
$= \int_{-2}^{2}\int_{0}^{2-x^2/2}\int_{x^2}^{4-2z} f(x,y,z)\,dy\,dz\,dx$
$= \int_{0}^{2}\int_{-\sqrt{4-2z}}^{\sqrt{4-2z}}\int_{x^2}^{4-2z} f(x,y,z)\,dy\,dx\,dz$

33. $\int_{0}^{1}\int_{\sqrt{x}}^{1}\int_{0}^{1-y} f(x,y,z)\,dz\,dy\,dx = \int_{0}^{1}\int_{0}^{y^2}\int_{0}^{1-y} f(x,y,z)\,dz\,dx\,dy$
$= \int_{0}^{1}\int_{0}^{1-z}\int_{0}^{y^2} f(x,y,z)\,dx\,dy\,dz = \int_{0}^{1}\int_{0}^{1-y}\int_{0}^{y^2} f(x,y,z)\,dx\,dz\,dy$
$= \int_{0}^{1}\int_{0}^{1-\sqrt{x}}\int_{\sqrt{x}}^{1-z} f(x,y,z)\,dy\,dz\,dx = \int_{0}^{1}\int_{0}^{(1-z)^2}\int_{\sqrt{x}}^{1-z} f(x,y,z)\,dy\,dx\,dz$

35. $\int_{0}^{1}\int_{y}^{1}\int_{0}^{y} f(x,y,z)\,dz\,dx\,dy = \int_{0}^{1}\int_{0}^{x}\int_{0}^{y} f(x,y,z)\,dz\,dy\,dx$
$= \int_{0}^{1}\int_{z}^{1}\int_{y}^{1} f(x,y,z)\,dx\,dy\,dz = \int_{0}^{1}\int_{0}^{y}\int_{y}^{1} f(x,y,z)\,dx\,dz\,dy$
$= \int_{0}^{1}\int_{0}^{x}\int_{z}^{x} f(x,y,z)\,dy\,dz\,dx = \int_{0}^{1}\int_{z}^{1}\int_{z}^{x} f(x,y,z)\,dy\,dx\,dz$

37. 64π **39.** $\frac{3}{2}\pi, \left(0, 0, \frac{1}{3}\right)$

41. $a^5, (7a/12, 7a/12, 7a/12)$

43. $I_x = I_y = I_z = \frac{2}{3}kL^5$ **45.** $\frac{1}{2}\pi kha^4$

47. (a) $m = \int_{-1}^{1}\int_{x^2}^{1}\int_{0}^{1-y}\sqrt{x^2+y^2}\,dz\,dy\,dx$
(b) $(\bar{x}, \bar{y}, \bar{z})$, where
$\bar{x} = (1/m)\int_{-1}^{1}\int_{x^2}^{1}\int_{0}^{1-y} x\sqrt{x^2+y^2}\,dz\,dy\,dx$,
$\bar{y} = (1/m)\int_{-1}^{1}\int_{x^2}^{1}\int_{0}^{1-y} y\sqrt{x^2+y^2}\,dz\,dy\,dx$,
and $\bar{z} = (1/m)\int_{-1}^{1}\int_{x^2}^{1}\int_{0}^{1-y} z\sqrt{x^2+y^2}\,dz\,dy\,dx$
(c) $\int_{-1}^{1}\int_{x^2}^{1}\int_{0}^{1-y}(x^2+y^2)^{3/2}\,dz\,dy\,dx$

49. (a) $\frac{3}{32}\pi + \frac{11}{24}$
(b) $\left(\dfrac{28}{9\pi+44}, \dfrac{30\pi+128}{45\pi+220}, \dfrac{45\pi+208}{135\pi+660}\right)$
(c) $\frac{1}{240}(68 + 15\pi)$

51. (a) $\frac{1}{8}$ (b) $\frac{1}{64}$ (c) $\frac{1}{5760}$ **53.** $L^3/8$

55. (a) The region bounded by the ellipsoid $x^2 + 2y^2 + 3z^2 = 1$
(b) $4\sqrt{6}\pi/45$

EXERCISES 15.7 ■ PAGE 1083

1. (a)

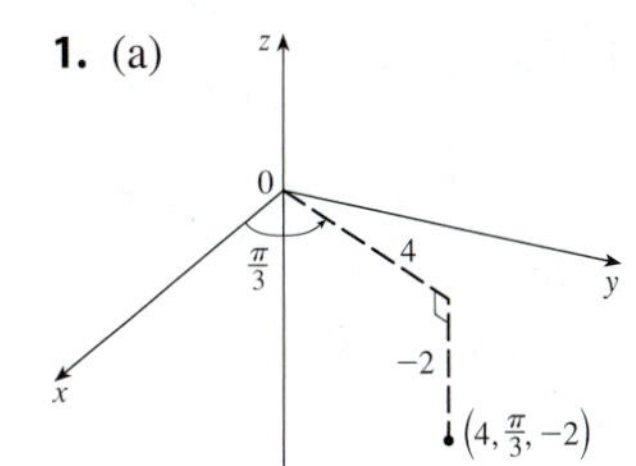

$(2, 2\sqrt{3}, -2)$

(b)

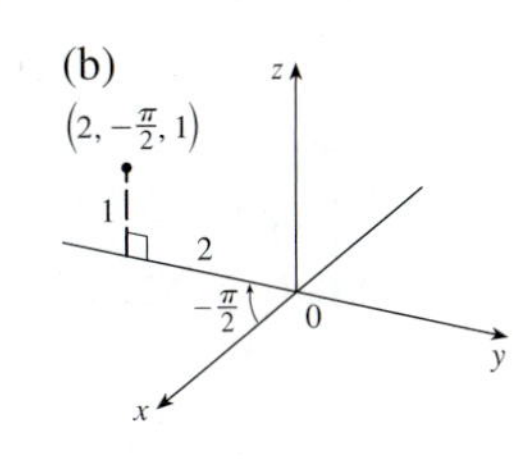

$(0, -2, 1)$

3. (a) $(\sqrt{2}, 3\pi/4, 1)$ (b) $(4, 2\pi/3, 3)$

5. Circular cylinder with radius 2 and axis the z-axis

7. Sphere, radius 2, centered at the origin

9. (a) $z^2 = 1 + r\cos\theta - r^2$ (b) $z = r^2\cos 2\theta$

11.

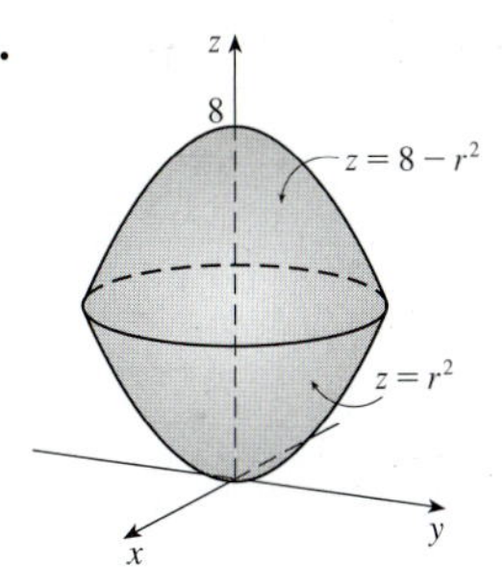

13. Cylindrical coordinates: $6 \leqslant r \leqslant 7, 0 \leqslant \theta \leqslant 2\pi, 0 \leqslant z \leqslant 20$

15. 4π

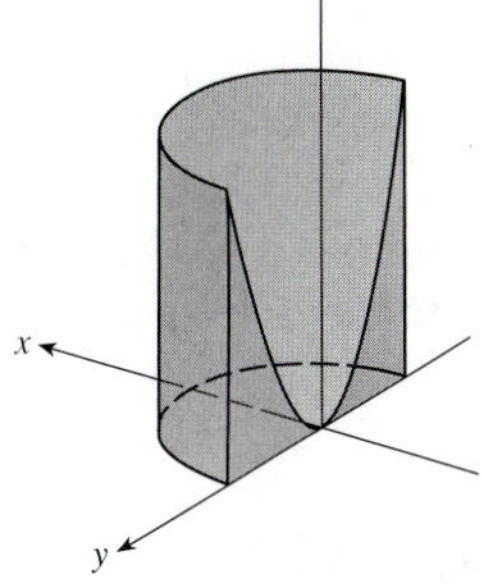

17. 384π **19.** $\frac{8}{3}\pi + \frac{128}{15}$ **21.** $2\pi/5$ **23.** $\frac{4}{3}\pi\left(\sqrt{2} - 1\right)$

25. (a) $\frac{512}{3}\pi$ (b) $\left(0, 0, \frac{23}{2}\right)$

27. $\pi Ka^2/8, (0, 0, 2a/3)$ **29.** 0

31. (a) $\iiint_C h(P)g(P)\,dV$, where C is the cone
(b) $\approx 3.1 \times 10^{19}$ ft-lb

EXERCISES 15.8 ■ PAGE 1089

1. (a)

$\left(\frac{3}{2}, \frac{3\sqrt{3}}{2}, 3\sqrt{3}\right)$

(b)

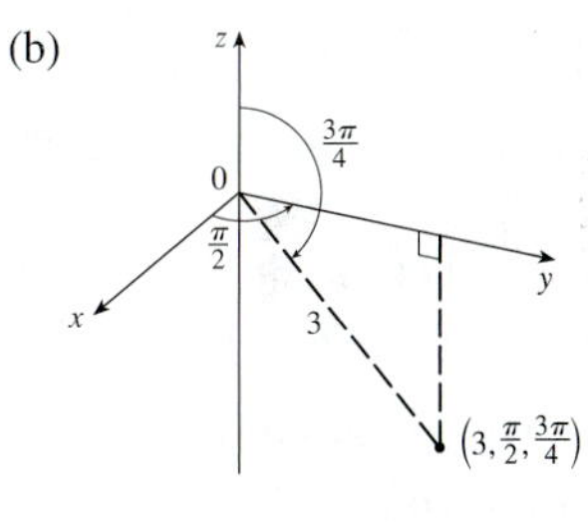

$\left(0, \frac{3\sqrt{2}}{2}, -\frac{3\sqrt{2}}{2}\right)$

3. (a) $(2, 3\pi/2, \pi/2)$ (b) $(2, 3\pi/4, 3\pi/4)$

5. Half-cone **7.** Horizontal plane

9. (a) $\rho = 3$ (b) $\rho^2(\sin^2\phi\cos 2\theta - \cos^2\phi) = 1$

11.

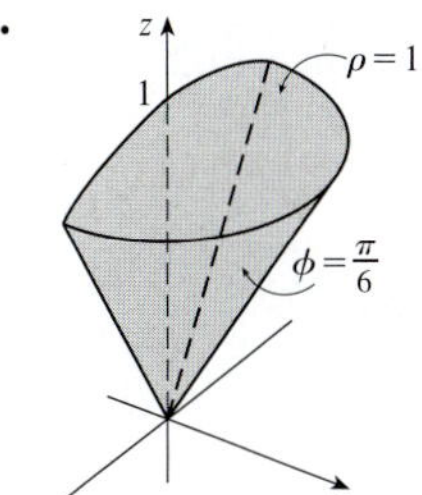

13.

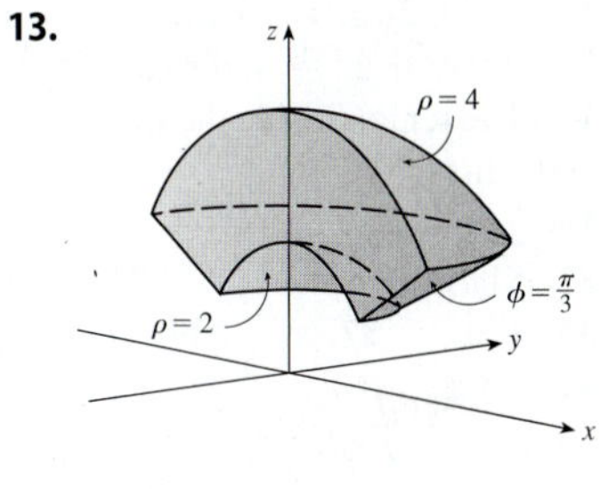

15. $0 \leqslant \phi \leqslant \pi/4, 0 \leqslant \rho \leqslant \cos\phi$

17. 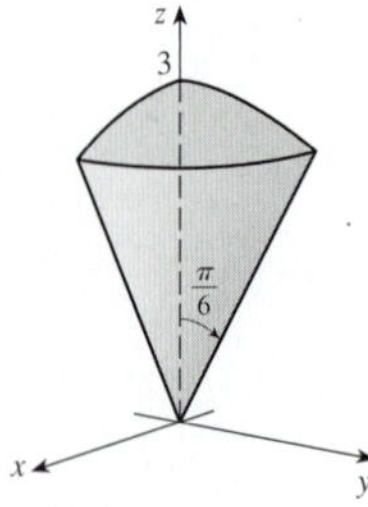 $(9\pi/4)(2 - \sqrt{3})$

19. $\int_0^{\pi/2} \int_0^3 \int_0^2 f(r\cos\theta, r\sin\theta, z)\, r\, dz\, dr\, d\theta$
21. $312{,}500\pi/7$ **23.** $1688\pi/15$ **25.** $\pi/8$
27. $(\sqrt{3} - 1)\pi a^3/3$ **29.** (a) 10π (b) $(0, 0, 2.1)$
31. (a) $(0, 0, \frac{7}{12})$ (b) $11K\pi/960$
33. (a) $(0, 0, \frac{3}{8}a)$ (b) $4K\pi a^5/15$ (K is the density)
35. $\frac{1}{3}\pi(2 - \sqrt{2}), (0, 0, 3/[8(2 - \sqrt{2})])$
37. (a) $\pi K a^4 h/2$ (K is the density) (b) $\pi K a^2 h(3a^2 + 4h^2)/12$
39. $5\pi/6$ **41.** $(4\sqrt{2} - 5)/15$ **43.** $4096\pi/21$
45. **47.** $136\pi/99$

EXERCISES 15.9 ■ PAGE 1100

1. -6 **3.** s **5.** $2uvw$
7. The parallelogram with vertices $(0, 0)$, $(6, 3)$, $(12, 1)$, $(6, -2)$
9. The region bounded by the line $y = 1$, the y-axis, and $y = \sqrt{x}$
11. $x = \frac{1}{3}(v - u)$, $y = \frac{1}{3}(u + 2v)$ is one possible transformation, where $S = \{(u, v) \mid -1 \leqslant u \leqslant 1, 1 \leqslant v \leqslant 3\}$
13. $x = u\cos v$, $y = u\sin v$ is one possible transformation, where $S = \{(u, v) \mid 1 \leqslant u \leqslant \sqrt{2}, 0 \leqslant v \leqslant \pi/2\}$
15. -3 **17.** 6π **19.** $2\ln 3$
21. (a) $\frac{4}{3}\pi abc$ (b) 1.083×10^{12} km^3 (c) $\frac{4}{15}\pi(a^2 + b^2)abck$
23. $\frac{8}{5}\ln 8$ **25.** $\frac{3}{2}\sin 1$ **27.** $e - e^{-1}$

CHAPTER 15 REVIEW ■ PAGE 1101

True-False Quiz

1. True **3.** True **5.** True **7.** True **9.** False

Exercises

1. ≈ 64.0 **3.** $4e^2 - 4e + 3$ **5.** $\frac{1}{2}\sin 1$ **7.** $\frac{2}{3}$
9. $\int_0^{\pi} \int_2^4 f(r\cos\theta, r\sin\theta)\, r\, dr\, d\theta$
11. $(\sqrt{3}, 3, 2)$, $(4, \pi/3, \pi/3)$
13. $(2\sqrt{2}, 2\sqrt{2}, 4\sqrt{3})$, $(4, \pi/4, 4\sqrt{3})$
15. (a) $r^2 + z^2 = 4$, $\rho = 2$ (b) $r = 2$, $\rho\sin\phi = 2$
17. The region inside the loop of the four-leaved rose $r = \sin 2\theta$ in the first quadrant
19. $\frac{1}{2}\sin 1$ **21.** $\frac{1}{2}e^6 - \frac{7}{2}$ **23.** $\frac{1}{4}\ln 2$ **25.** 8
27. $81\pi/5$ **29.** $\frac{81}{2}$ **31.** $\pi/96$ **33.** $\frac{64}{15}$
35. 176 **37.** $\frac{2}{3}$ **39.** $2ma^3/9$
41. (a) $\frac{1}{4}$ (b) $(\frac{1}{3}, \frac{8}{15})$
(c) $I_x = \frac{1}{12}$, $I_y = \frac{1}{24}$; $\bar{\bar{y}} = 1/\sqrt{3}$, $\bar{\bar{x}} = 1/\sqrt{6}$
43. (a) $(0, 0, h/4)$ (b) $\pi a^5 h/15$
45. $\ln(\sqrt{2} + \sqrt{3}) + \sqrt{2}/3$ **47.** $\frac{486}{5}$ **49.** 0.0512

51. (a) $\frac{1}{15}$ (b) $\frac{1}{3}$ (c) $\frac{1}{45}$
53. $\int_0^1 \int_0^{1-z} \int_{-\sqrt{y}}^{\sqrt{y}} f(x, y, z)\, dx\, dy\, dz$ **55.** $-\ln 2$ **57.** 0

PROBLEMS PLUS ■ PAGE 1105

1. 30 **3.** $\frac{1}{2}\sin 1$ **7.** (b) 0.90
13. $abc\pi\left(\frac{2}{3} - \frac{8}{9\sqrt{3}}\right)$

CHAPTER 16

EXERCISES 16.1 ■ PAGE 1113

1.
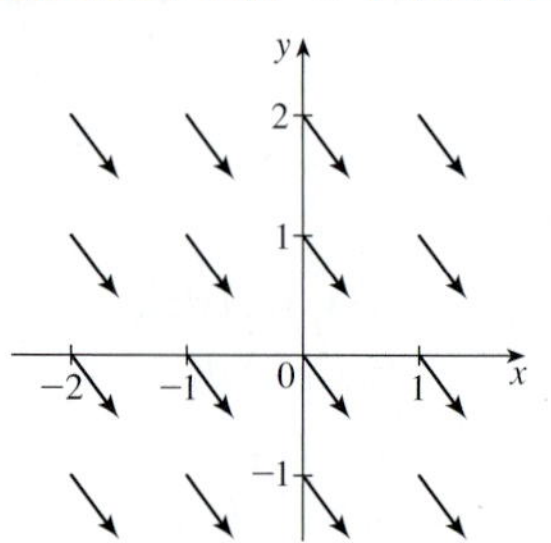

3.
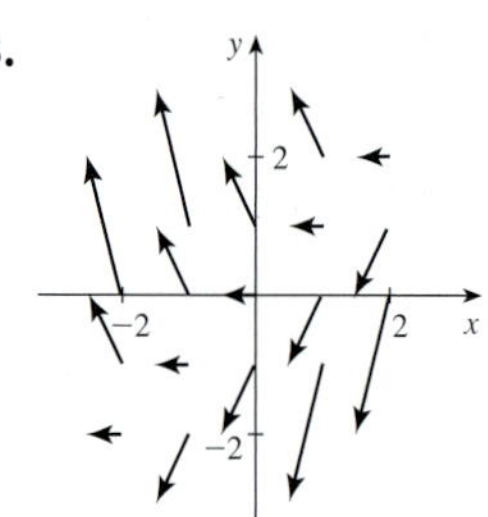

5.
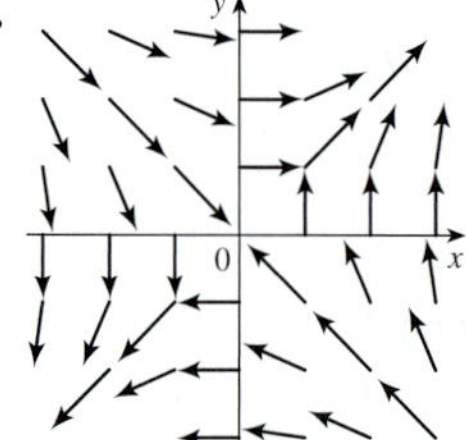

7.
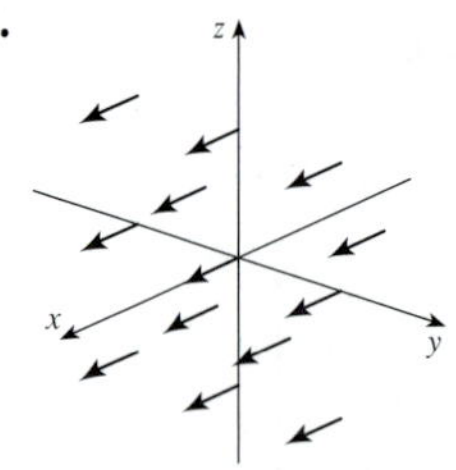

9.
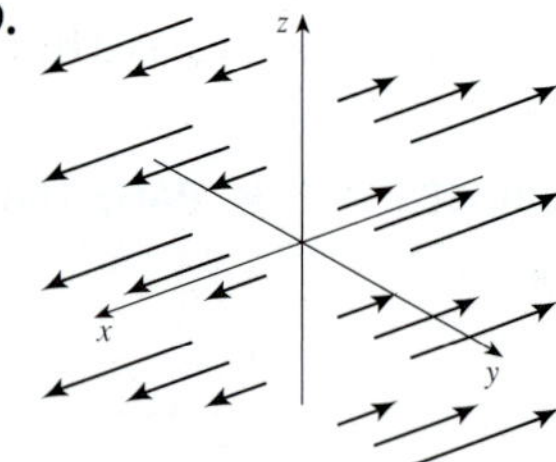

11. IV **13.** I **15.** IV **17.** III
19.
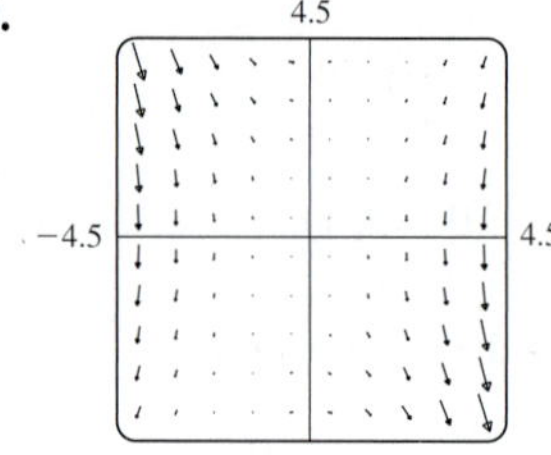
The line $y = 2x$

21. $\nabla f(x, y) = y^2 \cos(xy)\,\mathbf{i} + [xy\cos(xy) + \sin(xy)]\,\mathbf{j}$

23. $\nabla f(x, y, z) = \dfrac{x}{\sqrt{x^2 + y^2 + z^2}}\,\mathbf{i} + \dfrac{y}{\sqrt{x^2 + y^2 + z^2}}\,\mathbf{j} + \dfrac{z}{\sqrt{x^2 + y^2 + z^2}}\,\mathbf{k}$

25. $\nabla f(x, y) = (x - y)\,\mathbf{i} + (y - x)\,\mathbf{j}$

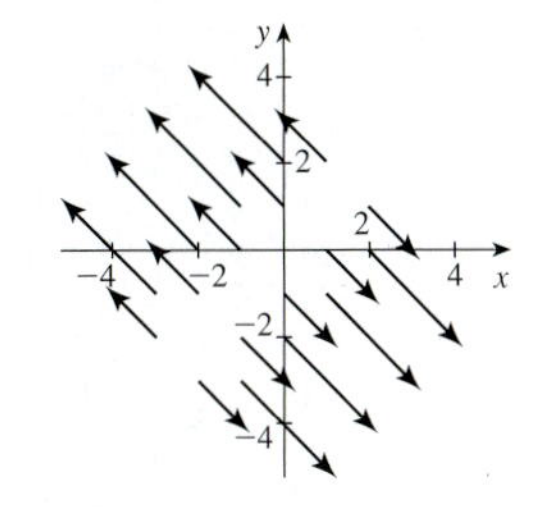

27.

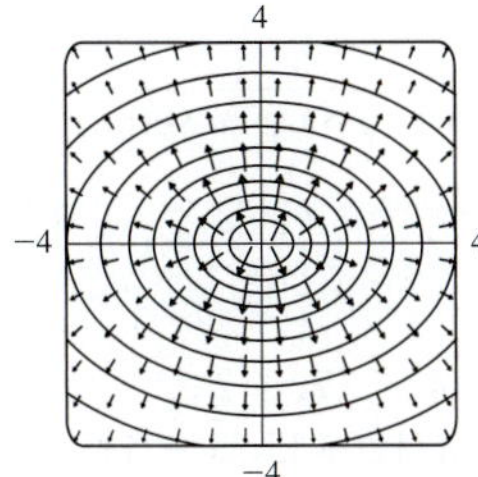

29. III **31.** II **33.** (2.04, 1.03)

35. (a)

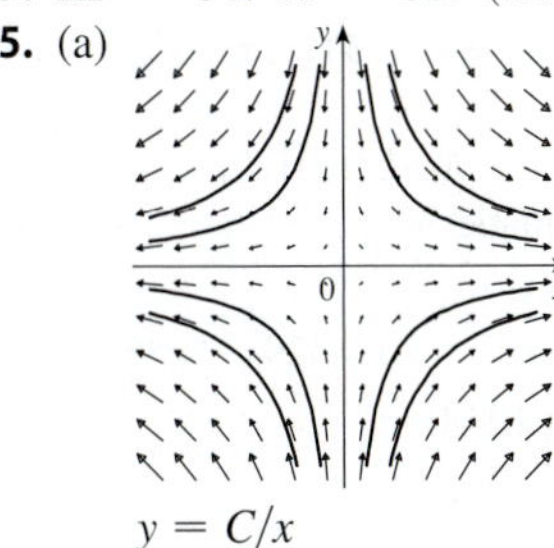

$y = C/x$

(b) $y = 1/x,\ x > 0$

EXERCISES 16.2 ■ PAGE 1124

1. $\frac{4}{3}(10^{3/2} - 1)$ **3.** 1638.4 **5.** $\frac{1}{3}\pi^6 + 2\pi$ **7.** $\frac{5}{2}$

9. $\sqrt{2}/3$ **11.** $\frac{1}{12}\sqrt{14}(e^6 - 1)$ **13.** $\frac{2}{5}(e - 1)$ **15.** $\frac{35}{3}$

17. (a) Positive (b) Negative **19.** $\frac{1}{20}$

21. $\frac{6}{5} - \cos 1 - \sin 1$ **23.** 0.5424 **25.** 94.8231

27. $3\pi + \frac{2}{3}$

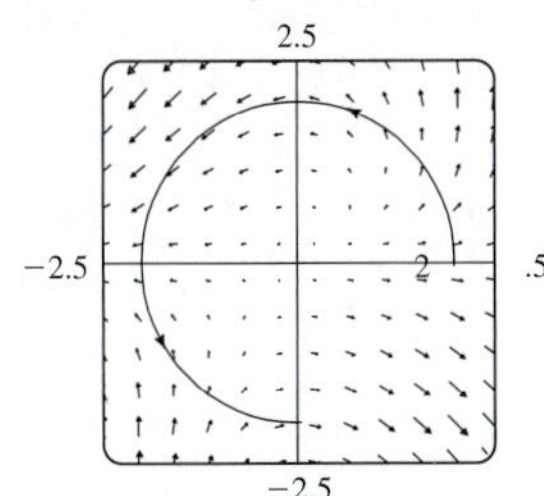

29. (a) $\frac{11}{8} - 1/e$ (b)

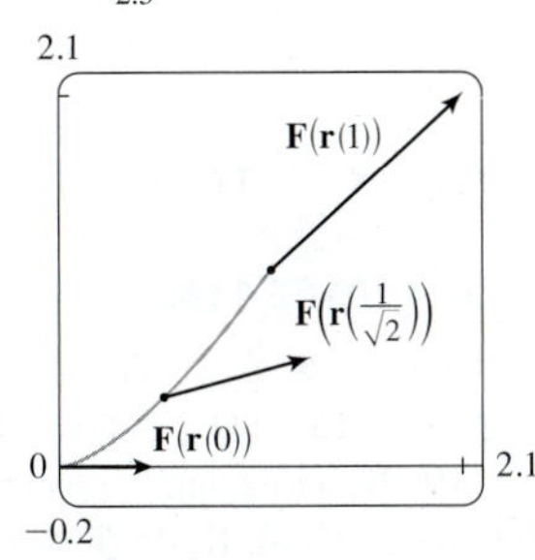

31. $\frac{172{,}704}{5{,}632{,}705}\sqrt{2}(1 - e^{-14\pi})$ **33.** $2\pi k,\ (4/\pi, 0)$

35. (a) $\bar{x} = (1/m)\int_C x\rho(x, y, z)\,ds$,
$\bar{y} = (1/m)\int_C y\rho(x, y, z)\,ds$,
$\bar{z} = (1/m)\int_C z\rho(x, y, z)\,ds$, where $m = \int_C \rho(x, y, z)\,ds$
(b) $(0, 0, 3\pi)$

37. $I_x = k\left(\frac{1}{2}\pi - \frac{4}{3}\right),\ I_y = k\left(\frac{1}{2}\pi - \frac{2}{3}\right)$ **39.** $2\pi^2$ **41.** $\frac{7}{3}$

43. (a) $2ma\,\mathbf{i} + 6mbt\,\mathbf{j},\ 0 \leqslant t \leqslant 1$ (b) $2ma^2 + \frac{9}{2}mb^2$

45. $\approx 1.67 \times 10^4$ ft-lb **47.** (b) Yes **51.** ≈ 22 J

EXERCISES 16.3 ■ PAGE 1134

1. 40 **3.** Not conservative

5. $f(x, y) = ye^{xy} + K$ **7.** $f(x, y) = ye^x + x\sin y + K$

9. $f(x, y) = y^2 \sin x + x\cos y + K$

11. (b) 16 **13.** (a) $f(x, y) = \frac{1}{3}x^3y^3$ (b) -9

15. (a) $f(x, y, z) = xyz + z^2$ (b) 77

17. (a) $f(x, y, z) = ye^{xz}$ (b) 4 **19.** $4/e$

21. It doesn't matter which curve is chosen.

23. $\frac{31}{4}$ **25.** No **27.** Conservative

31. (a) Yes (b) Yes (c) Yes

33. (a) No (b) Yes (c) Yes

EXERCISES 16.4 ■ PAGE 1141

1. 120 **3.** $\frac{2}{3}$ **5.** $4(e^3 - 1)$ **7.** $\frac{1}{3}$

9. -24π **11.** $-\frac{16}{3}$ **13.** 4π

15. $\frac{1}{15}\pi^4 - \frac{4144}{1125}\pi^2 + \frac{7{,}578{,}368}{253{,}125} \approx 0.0779$

17. $-\frac{1}{12}$ **19.** 3π **21.** (c) $\frac{9}{2}$

23. $(4a/3\pi, 4a/3\pi)$ if the region is the portion of the disk $x^2 + y^2 = a^2$ in the first quadrant

27. 0

EXERCISES 16.5 ■ PAGE 1149

1. (a) $\mathbf{0}$ (b) $y^2z^2 + x^2z^2 + x^2y^2$

3. (a) $ze^x\,\mathbf{i} + (xye^z - yze^x)\,\mathbf{j} - xe^z\,\mathbf{k}$ (b) $y(e^z + e^x)$

5. (a) $-\dfrac{\sqrt{z}}{(1 + y)^2}\,\mathbf{i} - \dfrac{\sqrt{x}}{(1 + z)^2}\,\mathbf{j} - \dfrac{\sqrt{y}}{(1 + x)^2}\,\mathbf{k}$

(b) $\dfrac{1}{2\sqrt{x}(1 + z)} + \dfrac{1}{2\sqrt{y}(1 + x)} + \dfrac{1}{2\sqrt{z}(1 + y)}$

7. (a) $\langle -e^y \cos z, -e^z \cos x, -e^x \cos y \rangle$
(b) $e^x \sin y + e^y \sin z + e^z \sin x$

9. (a) Negative (b) curl $\mathbf{F} = \mathbf{0}$

11. (a) Zero (b) curl $\mathbf{F}$ points in the negative z-direction.

13. $f(x, y, z) = xy^2z^3 + K$ **15.** Not conservative

17. $f(x, y, z) = xe^{yz} + K$ **19.** No

EXERCISES 16.6 ■ PAGE 1160

1. P: yes; Q: no

3. Plane through (0, 3, 1) containing vectors $\langle 1, 0, 4 \rangle$, $\langle 1, -1, 5 \rangle$

5. Circular cone with axis the z-axis

7.

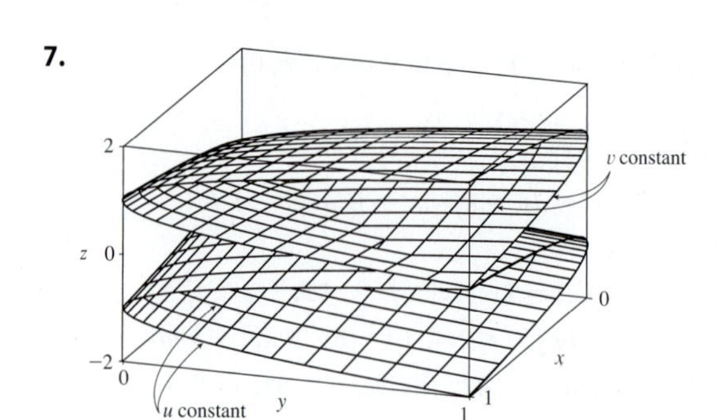

9.

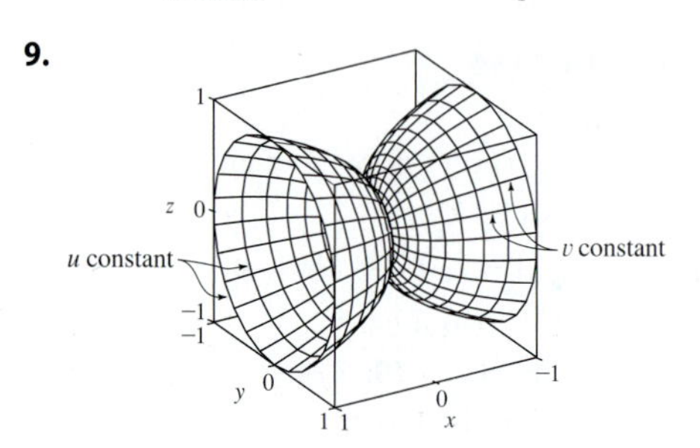

11.

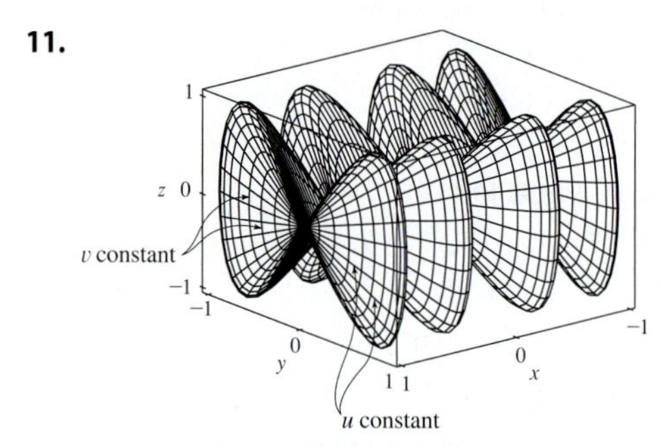

13. IV **15.** I **17.** III

19. $x = u,\ y = v - u,\ z = -v$

21. $y = y,\ z = z,\ x = \sqrt{1 + y^2 + \frac{1}{4}z^2}$

23. $x = 2 \sin\phi \cos\theta,\ y = 2 \sin\phi \sin\theta,$
$z = 2\cos\phi,\ 0 \leqslant \phi \leqslant \pi/4,\ 0 \leqslant \theta \leqslant 2\pi$
$\left[\text{or } x = x,\ y = y,\ z = \sqrt{4 - x^2 - y^2},\ x^2 + y^2 \leqslant 2\right]$

25. $x = 6 \sin\phi \cos\theta,\ y = 6 \sin\phi \sin\theta,\ z = 6\cos\phi,$
$\pi/6 \leqslant \phi \leqslant \pi/2,\ 0 \leqslant \theta \leqslant 2\pi$

29. $x = x,\ y = \dfrac{1}{1 + x^2}\cos\theta,\ y = \dfrac{1}{1 + x^2}\sin\theta,$
$-2 \leqslant x \leqslant 2,\ 0 \leqslant \theta \leqslant 2\pi$

31. (a) Direction reverses (b) Number of coils doubles

33. $3x - y + 3z = 3$ **35.** $\dfrac{\sqrt{3}}{2}x - \dfrac{1}{2}y + z = \dfrac{\pi}{3}$

37. $-x + 2z = 1$ **39.** $3\sqrt{14}$ **41.** $\sqrt{14}\,\pi$

43. $\frac{4}{15}(3^{5/2} - 2^{7/2} + 1)$ **45.** $(2\pi/3)(2\sqrt{2} - 1)$

47. $(\pi/6)(65^{3/2} - 1)$ **49.** 4 **51.** $\pi R^2 \leqslant A(S) \leqslant \sqrt{3}\,\pi R^2$

53. 3.5618 **55.** (a) ≈ 24.2055 (b) 24.2476

57. $\frac{45}{8}\sqrt{14} + \frac{15}{16}\ln\left[(11\sqrt{5} + 3\sqrt{70})/(3\sqrt{5} + \sqrt{70})\right]$

59. (b)

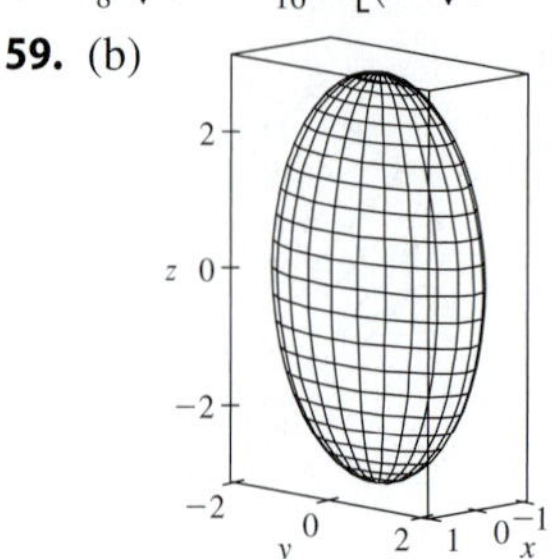

(c) $\int_0^{2\pi}\int_0^{\pi} \sqrt{36 \sin^4 u \cos^2 v + 9 \sin^4 u \sin^2 v + 4 \cos^2 u \sin^2 u}\ du\,dv$

61. 4π **63.** $2a^2(\pi - 2)$

EXERCISES 16.7 ■ PAGE 1172

1. ≈ -6.93 **3.** 900π **5.** $11\sqrt{14}$ **7.** $\frac{2}{3}(2\sqrt{2} - 1)$

9. $171\sqrt{14}$ **11.** $\sqrt{21}/3$ **13.** $(\pi/120)(25\sqrt{5} + 1)$

15. $\frac{7}{4}\sqrt{21} - \frac{17}{12}\sqrt{17}$ **17.** 16π **19.** 0 **21.** 4

23. $\frac{713}{180}$ **25.** $\frac{8}{3}\pi$ **27.** 0 **29.** 48 **31.** $2\pi + \frac{8}{3}$

33. 4.5822 **35.** 3.4895

37. $\iint_S \mathbf{F} \cdot d\mathbf{S} = \iint_D [P(\partial h/\partial x) - Q + R(\partial h/\partial z)]\, dA,$
where D = projection of S onto xz-plane

39. $(0, 0, a/2)$

41. (a) $I_z = \iint_S (x^2 + y^2)\rho(x, y, z)\, dS$ (b) $4329\sqrt{2}\,\pi/5$

43. 0 kg/s **45.** $\frac{8}{3}\pi a^3 \varepsilon_0$ **47.** 1248π

EXERCISES 16.8 ■ PAGE 1179

3. 16π **5.** 0 **7.** -1 **9.** $-\frac{17}{20}$

11. (a) $81\pi/2$ (b)

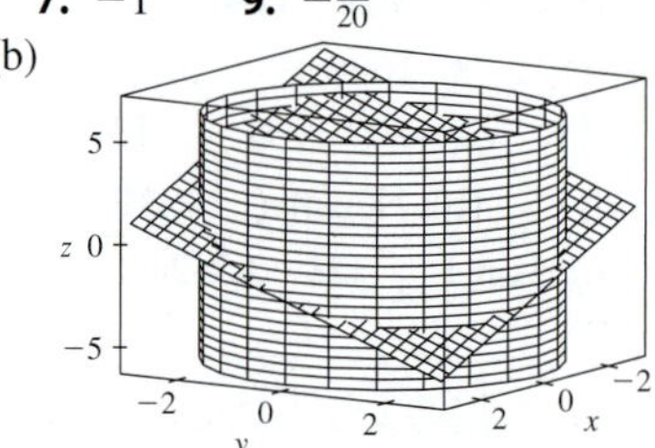

(c) $x = 3\cos t,\ y = 3\sin t,$
$z = 1 - 3(\cos t + \sin t),$
$0 \leqslant t \leqslant 2\pi$

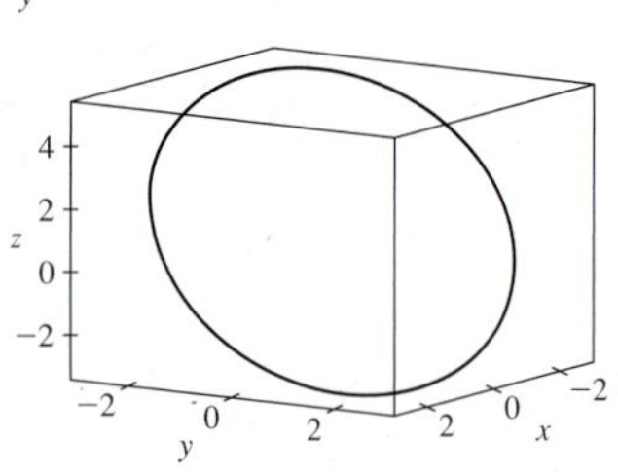

13. -32π **15.** $-\pi$ **17.** 3

EXERCISES 16.9 ■ PAGE 1185

1. $\frac{9}{2}$ **3.** $256\pi/3$ **5.** $\frac{9}{2}$ **7.** $9\pi/2$ **9.** 0

11. π **13.** 2π **15.** $341\sqrt{2}/60 + \frac{81}{20}\arcsin(\sqrt{3}/3)$

17. $13\pi/20$ **19.** Negative at P_1, positive at P_2

21. div $\mathbf{F} > 0$ in quadrants I, II; div $\mathbf{F} < 0$ in quadrants III, IV

CHAPTER 16 REVIEW ■ PAGE 1188

True-False Quiz

1. False **3.** True **5.** False **7.** False
9. True **11.** True **13.** False

Exercises

1. (a) Negative (b) Positive **3.** $6\sqrt{10}$ **5.** $\frac{4}{15}$ **7.** $\frac{110}{3}$
9. $\frac{11}{12} - 4/e$ **11.** $f(x, y) = e^y + xe^{xy} + K$ **13.** 0
15. 0 **17.** -8π **25.** $\frac{1}{6}(27 - 5\sqrt{5})$
27. $(\pi/60)(391\sqrt{17} + 1)$ **29.** $-64\pi/3$ **31.** 0
33. $-\frac{1}{2}$ **35.** 4π **37.** -4 **39.** 21

CHAPTER 17

EXERCISES 17.1 ■ PAGE 1200

1. $y = c_1e^{3x} + c_2e^{-2x}$ **3.** $y = c_1\cos(\sqrt{2}x) + c_2\sin(\sqrt{2}x)$
5. $y = c_1e^{-x/2} + c_2xe^{-x/2}$ **7.** $y = c_1 + c_2e^{4x/3}$
9. $y = e^{2x}(c_1\cos 3x + c_2\sin 3x)$
11. $y = c_1e^{(\sqrt{3}-1)t/2} + c_2e^{-(\sqrt{3}+1)t/2}$
13. $V = e^{-2t/3}\left[c_1\cos\left(\frac{\sqrt{5}}{3}t\right) + c_2\sin\left(\frac{\sqrt{5}}{3}t\right)\right]$
15. $f(x) = e^{-x}\cos x$, $g(x) = e^{-x}\sin x$. All solution curves approach 0 as $x \to \infty$ and oscillate with amplitudes that become arbitrarily large as $x \to -\infty$.

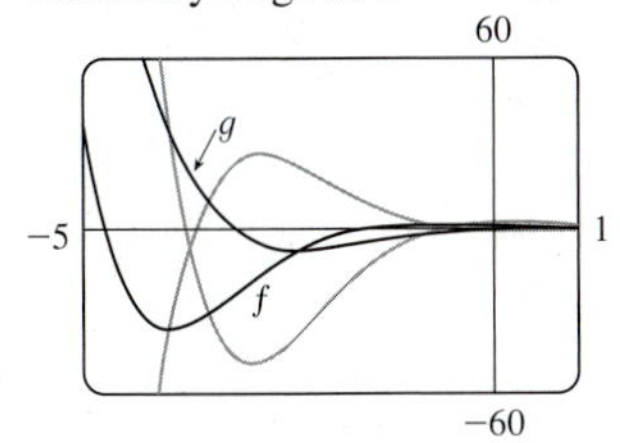

17. $y = \cos(\sqrt{3}x) + \sqrt{3}\sin(\sqrt{3}x)$ **19.** $y = e^{-2x/3} + \frac{2}{3}xe^{-2x/3}$
21. $y = e^{3x}(2\cos x - 3\sin x)$
23. $y = \frac{1}{7}e^{4x-4} - \frac{1}{7}e^{3-3x}$ **25.** $y = -3\cos 4x + 2\sin 4x$
27. $y = 2e^{-2x} - 2xe^{-2x}$ **29.** $y = \dfrac{e-2}{e-1} + \dfrac{e^x}{e-1}$
31. No solution
33. (b) $\lambda = n^2\pi^2/L^2$, n a positive integer; $y = C\sin(n\pi x/L)$
35. (a) $b - a \neq n\pi$, n any integer
(b) $b - a = n\pi$ and $\dfrac{c}{d} \neq e^{a-b}\dfrac{\cos a}{\cos b}$ unless $\cos b = 0$, then $\dfrac{c}{d} \neq e^{a-b}\dfrac{\sin a}{\sin b}$
(c) $b - a = n\pi$ and $\dfrac{c}{d} = e^{a-b}\dfrac{\cos a}{\cos b}$ unless $\cos b = 0$, then $\dfrac{c}{d} = e^{a-b}\dfrac{\sin a}{\sin b}$

EXERCISES 17.2 ■ PAGE 1207

1. $y = c_1e^{2x} + c_2e^{-4x} + \frac{1}{4}x^2 + \frac{1}{8}x - \frac{1}{32}$
3. $y = c_1\cos(\frac{1}{3}x) + c_2\sin(\frac{1}{3}x) + \frac{1}{37}e^{2x}$
5. $y = e^{2x}(c_1\cos x + c_2\sin x) + \frac{1}{10}e^{-x}$
7. $y = e^x(\frac{9}{10}\cos 2x - \frac{1}{20}\sin 2x) + \frac{1}{10}\cos x + \frac{1}{5}\sin x$
9. $y = e^x(\frac{1}{2}x^2 - x + 2)$
11.

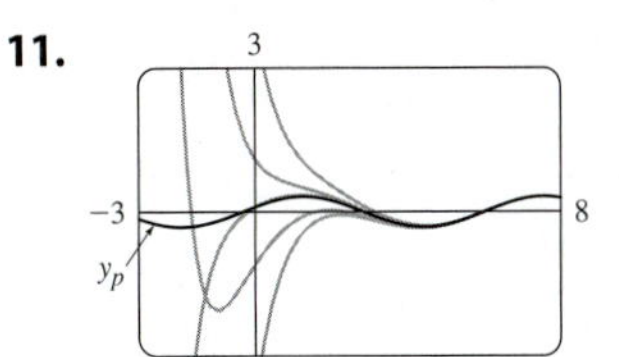

The solutions are all asymptotic to $y_p = \frac{1}{10}\cos x + \frac{3}{10}\sin x$ as $x \to \infty$. Except for y_p, all solutions approach either ∞ or $-\infty$ as $x \to -\infty$.

13. $y_p = (Ax + B)e^x\cos x + (Cx + D)e^x\sin x$
15. $y_p = Axe^x + B\cos x + C\sin x$
17. $y_p = xe^{-x}[(Ax^2 + Bx + C)\cos 3x + (Dx^2 + Ex + F)\sin 3x]$
19. $y = c_1\cos(\frac{1}{2}x) + c_2\sin(\frac{1}{2}x) - \frac{1}{3}\cos x$
21. $y = c_1e^x + c_2xe^x + e^{2x}$
23. $y = c_1\sin x + c_2\cos x + \sin x\ln(\sec x + \tan x) - 1$
25. $y = [c_1 + \ln(1 + e^{-x})]e^x + [c_2 - e^{-x} + \ln(1 + e^{-x})]e^{2x}$
27. $y = e^x\left[c_1 + c_2x - \frac{1}{2}\ln(1 + x^2) + x\tan^{-1}x\right]$

EXERCISES 17.3 ■ PAGE 1215

1. $x = 0.35\cos(2\sqrt{5}\,t)$ **3.** $x = -\frac{1}{5}e^{-6t} + \frac{6}{5}e^{-t}$ **5.** $\frac{49}{12}$ kg
7.

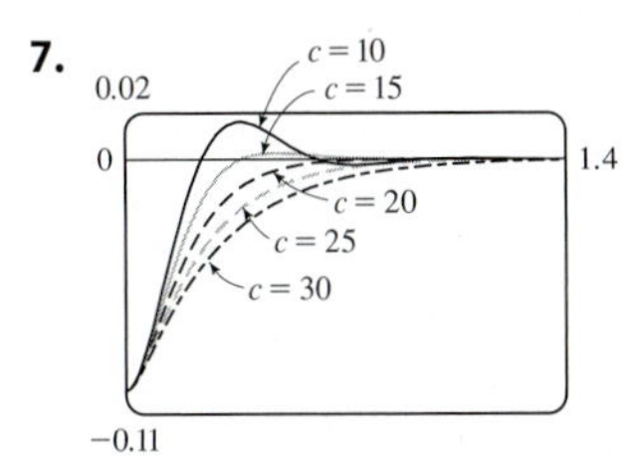

13. $Q(t) = (-e^{-10t}/250)(6\cos 20t + 3\sin 20t) + \frac{3}{125}$,
$I(t) = \frac{3}{5}e^{-10t}\sin 20t$
15. $Q(t) = e^{-10t}\left[\frac{3}{250}\cos 20t - \frac{3}{500}\sin 20t\right] - \frac{3}{250}\cos 10t + \frac{3}{125}\sin 10t$

EXERCISES 17.4 ■ PAGE 1220

1. $c_0\sum_{n=0}^{\infty}\frac{x^n}{n!} = c_0e^x$ **3.** $c_0\sum_{n=0}^{\infty}\frac{x^{3n}}{3^n n!} = c_0e^{x^3/3}$
5. $c_0\sum_{n=0}^{\infty}\frac{(-1)^n}{2^n n!}x^{2n} + c_1\sum_{n=0}^{\infty}\frac{(-2)^n n!}{(2n+1)!}x^{2n+1}$
7. $c_0 + c_1\sum_{n=1}^{\infty}\frac{x^n}{n} = c_0 - c_1\ln(1 - x)$ for $|x| < 1$
9. $\sum_{n=0}^{\infty}\frac{x^{2n}}{2^n n!} = e^{x^2/2}$
11. $x + \sum_{n=1}^{\infty}\frac{(-1)^n 2^2 5^2 \cdot \cdots \cdot (3n-1)^2}{(3n+1)!}x^{3n+1}$

CHAPTER 17 REVIEW ■ PAGE 1221

True-False Quiz

1. True **3.** True

Exercises

1. $y = c_1e^{x/2} + c_2e^{-x/2}$
3. $y = c_1\cos(\sqrt{3}x) + c_2\sin(\sqrt{3}x)$
5. $y = e^{2x}(c_1\cos x + c_2\sin x + 1)$
7. $y = c_1e^x + c_2xe^x - \frac{1}{2}\cos x - \frac{1}{2}(x + 1)\sin x$

9. $y = c_1e^{3x} + c_2e^{-2x} - \frac{1}{6} - \frac{1}{5}xe^{-2x}$
11. $y = 5 - 2e^{-6(x-1)}$ **13.** $y = (e^{4x} - e^x)/3$
15. No solution **17.** $\sum_{n=0}^{\infty} \frac{(-2)^n n!}{(2n+1)!} x^{2n+1}$
19. $Q(t) = -0.02e^{-10t}(\cos 10t + \sin 10t) + 0.03$
21. (c) $2\pi/k \approx 85$ min (d) $\approx$17,600 mi/h

APPENDIXES

EXERCISES G ■ PAGE A12

1. $8 - 4i$ **3.** $13 + 18i$ **5.** $12 - 7i$ **7.** $\frac{11}{13} + \frac{10}{13}i$
9. $\frac{1}{2} - \frac{1}{2}i$ **11.** $-i$ **13.** $5i$ **15.** $12 + 5i$, 13
17. $4i$, 4 **19.** $\pm\frac{3}{2}i$ **21.** $-1 \pm 2i$
23. $-\frac{1}{2} \pm (\sqrt{7}/2)i$ **25.** $3\sqrt{2}\,[\cos(3\pi/4) + i\sin(3\pi/4)]$
27. $5\{\cos[\tan^{-1}(\frac{4}{3})] + i\sin[\tan^{-1}(\frac{4}{3})]\}$
29. $4[\cos(\pi/2) + i\sin(\pi/2)]$, $\cos(-\pi/6) + i\sin(-\pi/6)$, $\frac{1}{2}[\cos(-\pi/6) + i\sin(-\pi/6)]$
31. $4\sqrt{2}\,[\cos(7\pi/12) + i\sin(7\pi/12)]$, $(2\sqrt{2})[\cos(13\pi/12) + i\sin(13\pi/12)]$, $\frac{1}{4}[\cos(\pi/6) + i\sin(\pi/6)]$
33. -1024 **35.** $-512\sqrt{3} + 512i$
37. $\pm1, \pm i, (1/\sqrt{2})(\pm1 \pm i)$ **39.** $\pm(\sqrt{3}/2) + \frac{1}{2}i, -i$

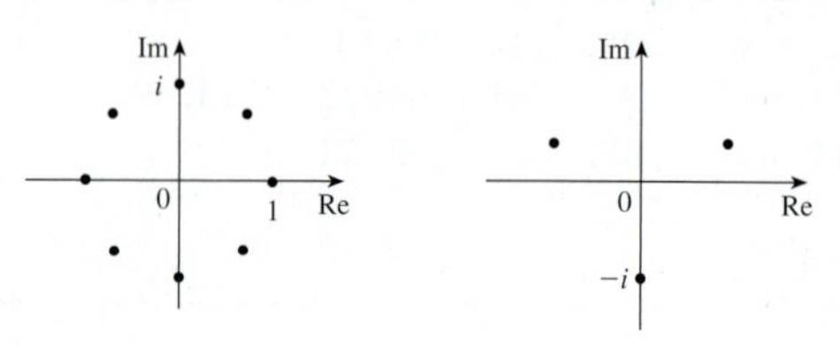

41. i **43.** $\frac{1}{2} + (\sqrt{3}/2)i$ **45.** $-e^2$
47. $\cos 3\theta = \cos^3\theta - 3\cos\theta\sin^2\theta$, $\sin 3\theta = 3\cos^2\theta\sin\theta - \sin^3\theta$

Index

RP denotes Reference Page numbers.

Cut here and keep for reference

DIFFERENTIATION RULES

General Formulas

1. $\dfrac{d}{dx}(c) = 0$

2. $\dfrac{d}{dx}[cf(x)] = cf'(x)$

3. $\dfrac{d}{dx}[f(x) + g(x)] = f'(x) + g'(x)$

4. $\dfrac{d}{dx}[f(x) - g(x)] = f'(x) - g'(x)$

5. $\dfrac{d}{dx}[f(x)g(x)] = f(x)g'(x) + g(x)f'(x)$ (Product Rule)

6. $\dfrac{d}{dx}\left[\dfrac{f(x)}{g(x)}\right] = \dfrac{g(x)f'(x) - f(x)g'(x)}{[g(x)]^2}$ (Quotient Rule)

7. $\dfrac{d}{dx}f(g(x)) = f'(g(x))g'(x)$ (Chain Rule)

8. $\dfrac{d}{dx}(x^n) = nx^{n-1}$ (Power Rule)

Exponential and Logarithmic Functions

9. $\dfrac{d}{dx}(e^x) = e^x$

10. $\dfrac{d}{dx}(b^x) = b^x \ln b$

11. $\dfrac{d}{dx}\ln|x| = \dfrac{1}{x}$

12. $\dfrac{d}{dx}(\log_b x) = \dfrac{1}{x \ln b}$

Trigonometric Functions

13. $\dfrac{d}{dx}(\sin x) = \cos x$

14. $\dfrac{d}{dx}(\cos x) = -\sin x$

15. $\dfrac{d}{dx}(\tan x) = \sec^2 x$

16. $\dfrac{d}{dx}(\csc x) = -\csc x \cot x$

17. $\dfrac{d}{dx}(\sec x) = \sec x \tan x$

18. $\dfrac{d}{dx}(\cot x) = -\csc^2 x$

Inverse Trigonometric Functions

19. $\dfrac{d}{dx}(\sin^{-1}x) = \dfrac{1}{\sqrt{1 - x^2}}$

20. $\dfrac{d}{dx}(\cos^{-1}x) = -\dfrac{1}{\sqrt{1 - x^2}}$

21. $\dfrac{d}{dx}(\tan^{-1}x) = \dfrac{1}{1 + x^2}$

22. $\dfrac{d}{dx}(\csc^{-1}x) = -\dfrac{1}{x\sqrt{x^2 - 1}}$

23. $\dfrac{d}{dx}(\sec^{-1}x) = \dfrac{1}{x\sqrt{x^2 - 1}}$

24. $\dfrac{d}{dx}(\cot^{-1}x) = -\dfrac{1}{1 + x^2}$

Hyperbolic Functions

25. $\dfrac{d}{dx}(\sinh x) = \cosh x$

26. $\dfrac{d}{dx}(\cosh x) = \sinh x$

27. $\dfrac{d}{dx}(\tanh x) = \operatorname{sech}^2 x$

28. $\dfrac{d}{dx}(\operatorname{csch} x) = -\operatorname{csch} x \coth x$

29. $\dfrac{d}{dx}(\operatorname{sech} x) = -\operatorname{sech} x \tanh x$

30. $\dfrac{d}{dx}(\coth x) = -\operatorname{csch}^2 x$

Inverse Hyperbolic Functions

31. $\dfrac{d}{dx}(\sinh^{-1}x) = \dfrac{1}{\sqrt{1 + x^2}}$

32. $\dfrac{d}{dx}(\cosh^{-1}x) = \dfrac{1}{\sqrt{x^2 - 1}}$

33. $\dfrac{d}{dx}(\tanh^{-1}x) = \dfrac{1}{1 - x^2}$

34. $\dfrac{d}{dx}(\operatorname{csch}^{-1}x) = -\dfrac{1}{|x|\sqrt{x^2 + 1}}$

35. $\dfrac{d}{dx}(\operatorname{sech}^{-1}x) = -\dfrac{1}{x\sqrt{1 - x^2}}$

36. $\dfrac{d}{dx}(\coth^{-1}x) = \dfrac{1}{1 - x^2}$

TABLE OF INTEGRALS

Basic Forms

1. $\int u\,dv = uv - \int v\,du$

2. $\int u^n\,du = \frac{u^{n+1}}{n+1} + C, \quad n \neq -1$

3. $\int \frac{du}{u} = \ln|u| + C$

4. $\int e^u\,du = e^u + C$

5. $\int b^u\,du = \frac{b^u}{\ln b} + C$

6. $\int \sin u\,du = -\cos u + C$

7. $\int \cos u\,du = \sin u + C$

8. $\int \sec^2 u\,du = \tan u + C$

9. $\int \csc^2 u\,du = -\cot u + C$

10. $\int \sec u \tan u\,du = \sec u + C$

11. $\int \csc u \cot u\,du = -\csc u + C$

12. $\int \tan u\,du = \ln|\sec u| + C$

13. $\int \cot u\,du = \ln|\sin u| + C$

14. $\int \sec u\,du = \ln|\sec u + \tan u| + C$

15. $\int \csc u\,du = \ln|\csc u - \cot u| + C$

16. $\int \frac{du}{\sqrt{a^2 - u^2}} = \sin^{-1}\frac{u}{a} + C, \quad a > 0$

17. $\int \frac{du}{a^2 + u^2} = \frac{1}{a}\tan^{-1}\frac{u}{a} + C$

18. $\int \frac{du}{u\sqrt{u^2 - a^2}} = \frac{1}{a}\sec^{-1}\frac{u}{a} + C$

19. $\int \frac{du}{a^2 - u^2} = \frac{1}{2a}\ln\left|\frac{u+a}{u-a}\right| + C$

20. $\int \frac{du}{u^2 - a^2} = \frac{1}{2a}\ln\left|\frac{u-a}{u+a}\right| + C$

Forms Involving $\sqrt{a^2 + u^2}$, $a > 0$

21. $\int \sqrt{a^2 + u^2}\,du = \frac{u}{2}\sqrt{a^2 + u^2} + \frac{a^2}{2}\ln\left(u + \sqrt{a^2 + u^2}\right) + C$

22. $\int u^2\sqrt{a^2 + u^2}\,du = \frac{u}{8}(a^2 + 2u^2)\sqrt{a^2 + u^2} - \frac{a^4}{8}\ln\left(u + \sqrt{a^2 + u^2}\right) + C$

23. $\int \frac{\sqrt{a^2 + u^2}}{u}\,du = \sqrt{a^2 + u^2} - a\ln\left|\frac{a + \sqrt{a^2 + u^2}}{u}\right| + C$

24. $\int \frac{\sqrt{a^2 + u^2}}{u^2}\,du = -\frac{\sqrt{a^2 + u^2}}{u} + \ln\left(u + \sqrt{a^2 + u^2}\right) + C$

25. $\int \frac{du}{\sqrt{a^2 + u^2}} = \ln\left(u + \sqrt{a^2 + u^2}\right) + C$

26. $\int \frac{u^2\,du}{\sqrt{a^2 + u^2}} = \frac{u}{2}\sqrt{a^2 + u^2} - \frac{a^2}{2}\ln\left(u + \sqrt{a^2 + u^2}\right) + C$

27. $\int \frac{du}{u\sqrt{a^2 + u^2}} = -\frac{1}{a}\ln\left|\frac{\sqrt{a^2 + u^2} + a}{u}\right| + C$

28. $\int \frac{du}{u^2\sqrt{a^2 + u^2}} = -\frac{\sqrt{a^2 + u^2}}{a^2 u} + C$

29. $\int \frac{du}{(a^2 + u^2)^{3/2}} = \frac{u}{a^2\sqrt{a^2 + u^2}} + C$

Cut here and keep for reference

TABLE OF INTEGRALS

Forms Involving $\sqrt{a^2 - u^2}$, $a > 0$

30. $\displaystyle\int \sqrt{a^2 - u^2}\, du = \frac{u}{2}\sqrt{a^2 - u^2} + \frac{a^2}{2}\sin^{-1}\frac{u}{a} + C$

31. $\displaystyle\int u^2\sqrt{a^2 - u^2}\, du = \frac{u}{8}(2u^2 - a^2)\sqrt{a^2 - u^2} + \frac{a^4}{8}\sin^{-1}\frac{u}{a} + C$

32. $\displaystyle\int \frac{\sqrt{a^2 - u^2}}{u}\, du = \sqrt{a^2 - u^2} - a\ln\left|\frac{a + \sqrt{a^2 - u^2}}{u}\right| + C$

33. $\displaystyle\int \frac{\sqrt{a^2 - u^2}}{u^2}\, du = -\frac{1}{u}\sqrt{a^2 - u^2} - \sin^{-1}\frac{u}{a} + C$

34. $\displaystyle\int \frac{u^2\, du}{\sqrt{a^2 - u^2}} = -\frac{u}{2}\sqrt{a^2 - u^2} + \frac{a^2}{2}\sin^{-1}\frac{u}{a} + C$

35. $\displaystyle\int \frac{du}{u\sqrt{a^2 - u^2}} = -\frac{1}{a}\ln\left|\frac{a + \sqrt{a^2 - u^2}}{u}\right| + C$

36. $\displaystyle\int \frac{du}{u^2\sqrt{a^2 - u^2}} = -\frac{1}{a^2u}\sqrt{a^2 - u^2} + C$

37. $\displaystyle\int (a^2 - u^2)^{3/2}\, du = -\frac{u}{8}(2u^2 - 5a^2)\sqrt{a^2 - u^2} + \frac{3a^4}{8}\sin^{-1}\frac{u}{a} + C$

38. $\displaystyle\int \frac{du}{(a^2 - u^2)^{3/2}} = \frac{u}{a^2\sqrt{a^2 - u^2}} + C$

Forms Involving $\sqrt{u^2 - a^2}$, $a > 0$

39. $\displaystyle\int \sqrt{u^2 - a^2}\, du = \frac{u}{2}\sqrt{u^2 - a^2} - \frac{a^2}{2}\ln\left|u + \sqrt{u^2 - a^2}\right| + C$

40. $\displaystyle\int u^2\sqrt{u^2 - a^2}\, du = \frac{u}{8}(2u^2 - a^2)\sqrt{u^2 - a^2} - \frac{a^4}{8}\ln\left|u + \sqrt{u^2 - a^2}\right| + C$

41. $\displaystyle\int \frac{\sqrt{u^2 - a^2}}{u}\, du = \sqrt{u^2 - a^2} - a\cos^{-1}\frac{a}{|u|} + C$

42. $\displaystyle\int \frac{\sqrt{u^2 - a^2}}{u^2}\, du = -\frac{\sqrt{u^2 - a^2}}{u} + \ln\left|u + \sqrt{u^2 - a^2}\right| + C$

43. $\displaystyle\int \frac{du}{\sqrt{u^2 - a^2}} = \ln\left|u + \sqrt{u^2 - a^2}\right| + C$

44. $\displaystyle\int \frac{u^2\, du}{\sqrt{u^2 - a^2}} = \frac{u}{2}\sqrt{u^2 - a^2} + \frac{a^2}{2}\ln\left|u + \sqrt{u^2 - a^2}\right| + C$

45. $\displaystyle\int \frac{du}{u^2\sqrt{u^2 - a^2}} = \frac{\sqrt{u^2 - a^2}}{a^2u} + C$

46. $\displaystyle\int \frac{du}{(u^2 - a^2)^{3/2}} = -\frac{u}{a^2\sqrt{u^2 - a^2}} + C$

(continued)

TABLE OF INTEGRALS

Forms Involving $a + bu$

47. $$\int \frac{u\,du}{a+bu} = \frac{1}{b^2}\left(a + bu - a\ln|a+bu|\right) + C$$

48. $$\int \frac{u^2\,du}{a+bu} = \frac{1}{2b^3}\left[(a+bu)^2 - 4a(a+bu) + 2a^2\ln|a+bu|\right] + C$$

49. $$\int \frac{du}{u(a+bu)} = \frac{1}{a}\ln\left|\frac{u}{a+bu}\right| + C$$

50. $$\int \frac{du}{u^2(a+bu)} = -\frac{1}{au} + \frac{b}{a^2}\ln\left|\frac{a+bu}{u}\right| + C$$

51. $$\int \frac{u\,du}{(a+bu)^2} = \frac{a}{b^2(a+bu)} + \frac{1}{b^2}\ln|a+bu| + C$$

52. $$\int \frac{du}{u(a+bu)^2} = \frac{1}{a(a+bu)} - \frac{1}{a^2}\ln\left|\frac{a+bu}{u}\right| + C$$

53. $$\int \frac{u^2\,du}{(a+bu)^2} = \frac{1}{b^3}\left(a + bu - \frac{a^2}{a+bu} - 2a\ln|a+bu|\right) + C$$

54. $$\int u\sqrt{a+bu}\,du = \frac{2}{15b^2}(3bu-2a)(a+bu)^{3/2} + C$$

55. $$\int \frac{u\,du}{\sqrt{a+bu}} = \frac{2}{3b^2}(bu-2a)\sqrt{a+bu} + C$$

56. $$\int \frac{u^2\,du}{\sqrt{a+bu}} = \frac{2}{15b^3}(8a^2 + 3b^2u^2 - 4abu)\sqrt{a+bu} + C$$

57. $$\int \frac{du}{u\sqrt{a+bu}} = \frac{1}{\sqrt{a}}\ln\left|\frac{\sqrt{a+bu}-\sqrt{a}}{\sqrt{a+bu}+\sqrt{a}}\right| + C, \quad \text{if } a > 0$$
$$= \frac{2}{\sqrt{-a}}\tan^{-1}\sqrt{\frac{a+bu}{-a}} + C, \qquad \text{if } a < 0$$

58. $$\int \frac{\sqrt{a+bu}}{u}\,du = 2\sqrt{a+bu} + a\int\frac{du}{u\sqrt{a+bu}}$$

59. $$\int \frac{\sqrt{a+bu}}{u^2}\,du = -\frac{\sqrt{a+bu}}{u} + \frac{b}{2}\int\frac{du}{u\sqrt{a+bu}}$$

60. $$\int u^n\sqrt{a+bu}\,du = \frac{2}{b(2n+3)}\left[u^n(a+bu)^{3/2} - na\int u^{n-1}\sqrt{a+bu}\,du\right]$$

61. $$\int \frac{u^n\,du}{\sqrt{a+bu}} = \frac{2u^n\sqrt{a+bu}}{b(2n+1)} - \frac{2na}{b(2n+1)}\int\frac{u^{n-1}\,du}{\sqrt{a+bu}}$$

62. $$\int \frac{du}{u^n\sqrt{a+bu}} = -\frac{\sqrt{a+bu}}{a(n-1)u^{n-1}} - \frac{b(2n-3)}{2a(n-1)}\int\frac{du}{u^{n-1}\sqrt{a+bu}}$$

Cut here and keep for reference

TABLE OF INTEGRALS

Trigonometric Forms

63. $\int \sin^2 u\, du = \frac{1}{2}u - \frac{1}{4}\sin 2u + C$

64. $\int \cos^2 u\, du = \frac{1}{2}u + \frac{1}{4}\sin 2u + C$

65. $\int \tan^2 u\, du = \tan u - u + C$

66. $\int \cot^2 u\, du = -\cot u - u + C$

67. $\int \sin^3 u\, du = -\frac{1}{3}(2 + \sin^2 u)\cos u + C$

68. $\int \cos^3 u\, du = \frac{1}{3}(2 + \cos^2 u)\sin u + C$

69. $\int \tan^3 u\, du = \frac{1}{2}\tan^2 u + \ln|\cos u| + C$

70. $\int \cot^3 u\, du = -\frac{1}{2}\cot^2 u - \ln|\sin u| + C$

71. $\int \sec^3 u\, du = \frac{1}{2}\sec u \tan u + \frac{1}{2}\ln|\sec u + \tan u| + C$

72. $\int \csc^3 u\, du = -\frac{1}{2}\csc u \cot u + \frac{1}{2}\ln|\csc u - \cot u| + C$

73. $\int \sin^n u\, du = -\frac{1}{n}\sin^{n-1}u \cos u + \frac{n-1}{n}\int \sin^{n-2}u\, du$

74. $\int \cos^n u\, du = \frac{1}{n}\cos^{n-1}u \sin u + \frac{n-1}{n}\int \cos^{n-2}u\, du$

75. $\int \tan^n u\, du = \frac{1}{n-1}\tan^{n-1}u - \int \tan^{n-2}u\, du$

76. $\int \cot^n u\, du = \frac{-1}{n-1}\cot^{n-1}u - \int \cot^{n-2}u\, du$

77. $\int \sec^n u\, du = \frac{1}{n-1}\tan u \sec^{n-2}u + \frac{n-2}{n-1}\int \sec^{n-2}u\, du$

78. $\int \csc^n u\, du = \frac{-1}{n-1}\cot u \csc^{n-2}u + \frac{n-2}{n-1}\int \csc^{n-2}u\, du$

79. $\int \sin au \sin bu\, du = \frac{\sin(a-b)u}{2(a-b)} - \frac{\sin(a+b)u}{2(a+b)} + C$

80. $\int \cos au \cos bu\, du = \frac{\sin(a-b)u}{2(a-b)} + \frac{\sin(a+b)u}{2(a+b)} + C$

81. $\int \sin au \cos bu\, du = -\frac{\cos(a-b)u}{2(a-b)} - \frac{\cos(a+b)u}{2(a+b)} + C$

82. $\int u \sin u\, du = \sin u - u\cos u + C$

83. $\int u \cos u\, du = \cos u + u \sin u + C$

84. $\int u^n \sin u\, du = -u^n \cos u + n\int u^{n-1}\cos u\, du$

85. $\int u^n \cos u\, du = u^n \sin u - n\int u^{n-1}\sin u\, du$

86. $\int \sin^n u \cos^m u\, du = -\frac{\sin^{n-1}u \cos^{m+1}u}{n+m} + \frac{n-1}{n+m}\int \sin^{n-2}u \cos^m u\, du$

$= \frac{\sin^{n+1}u \cos^{m-1}u}{n+m} + \frac{m-1}{n+m}\int \sin^n u \cos^{m-2}u\, du$

Inverse Trigonometric Forms

87. $\int \sin^{-1}u\, du = u\sin^{-1}u + \sqrt{1-u^2} + C$

88. $\int \cos^{-1}u\, du = u\cos^{-1}u - \sqrt{1-u^2} + C$

89. $\int \tan^{-1}u\, du = u\tan^{-1}u - \frac{1}{2}\ln(1+u^2) + C$

90. $\int u\sin^{-1}u\, du = \frac{2u^2-1}{4}\sin^{-1}u + \frac{u\sqrt{1-u^2}}{4} + C$

91. $\int u\cos^{-1}u\, du = \frac{2u^2-1}{4}\cos^{-1}u - \frac{u\sqrt{1-u^2}}{4} + C$

92. $\int u\tan^{-1}u\, du = \frac{u^2+1}{2}\tan^{-1}u - \frac{u}{2} + C$

93. $\int u^n \sin^{-1}u\, du = \frac{1}{n+1}\left[u^{n+1}\sin^{-1}u - \int \frac{u^{n+1}\, du}{\sqrt{1-u^2}}\right], \quad n \neq -1$

94. $\int u^n \cos^{-1}u\, du = \frac{1}{n+1}\left[u^{n+1}\cos^{-1}u + \int \frac{u^{n+1}\, du}{\sqrt{1-u^2}}\right], \quad n \neq -1$

95. $\int u^n \tan^{-1}u\, du = \frac{1}{n+1}\left[u^{n+1}\tan^{-1}u - \int \frac{u^{n+1}\, du}{1+u^2}\right], \quad n \neq -1$

(continued)

TABLE OF INTEGRALS

Exponential and Logarithmic Forms

96. $\displaystyle\int ue^{au}\,du = \frac{1}{a^2}(au - 1)e^{au} + C$

97. $\displaystyle\int u^n e^{au}\,du = \frac{1}{a}u^n e^{au} - \frac{n}{a}\int u^{n-1}e^{au}\,du$

98. $\displaystyle\int e^{au}\sin bu\,du = \frac{e^{au}}{a^2 + b^2}(a\sin bu - b\cos bu) + C$

99. $\displaystyle\int e^{au}\cos bu\,du = \frac{e^{au}}{a^2 + b^2}(a\cos bu + b\sin bu) + C$

100. $\displaystyle\int \ln u\,du = u\ln u - u + C$

101. $\displaystyle\int u^n\ln u\,du = \frac{u^{n+1}}{(n+1)^2}[(n+1)\ln u - 1] + C$

102. $\displaystyle\int \frac{1}{u\ln u}\,du = \ln|\ln u| + C$

Hyperbolic Forms

103. $\displaystyle\int \sinh u\,du = \cosh u + C$

104. $\displaystyle\int \cosh u\,du = \sinh u + C$

105. $\displaystyle\int \tanh u\,du = \ln\cosh u + C$

106. $\displaystyle\int \coth u\,du = \ln|\sinh u| + C$

107. $\displaystyle\int \operatorname{sech} u\,du = \tan^{-1}|\sinh u| + C$

108. $\displaystyle\int \operatorname{csch} u\,du = \ln\left|\tanh\tfrac{1}{2}u\right| + C$

109. $\displaystyle\int \operatorname{sech}^2 u\,du = \tanh u + C$

110. $\displaystyle\int \operatorname{csch}^2 u\,du = -\coth u + C$

111. $\displaystyle\int \operatorname{sech} u\tanh u\,du = -\operatorname{sech} u + C$

112. $\displaystyle\int \operatorname{csch} u\coth u\,du = -\operatorname{csch} u + C$

Forms Involving $\sqrt{2au - u^2}$, $a > 0$

113. $\displaystyle\int \sqrt{2au - u^2}\,du = \frac{u-a}{2}\sqrt{2au - u^2} + \frac{a^2}{2}\cos^{-1}\left(\frac{a-u}{a}\right) + C$

114. $\displaystyle\int u\sqrt{2au - u^2}\,du = \frac{2u^2 - au - 3a^2}{6}\sqrt{2au - u^2} + \frac{a^3}{2}\cos^{-1}\left(\frac{a-u}{a}\right) + C$

115. $\displaystyle\int \frac{\sqrt{2au - u^2}}{u}\,du = \sqrt{2au - u^2} + a\cos^{-1}\left(\frac{a-u}{a}\right) + C$

116. $\displaystyle\int \frac{\sqrt{2au - u^2}}{u^2}\,du = -\frac{2\sqrt{2au - u^2}}{u} - \cos^{-1}\left(\frac{a-u}{a}\right) + C$

117. $\displaystyle\int \frac{du}{\sqrt{2au - u^2}} = \cos^{-1}\left(\frac{a-u}{a}\right) + C$

118. $\displaystyle\int \frac{u\,du}{\sqrt{2au - u^2}} = -\sqrt{2au - u^2} + a\cos^{-1}\left(\frac{a-u}{a}\right) + C$

119. $\displaystyle\int \frac{u^2\,du}{\sqrt{2au - u^2}} = -\frac{(u + 3a)}{2}\sqrt{2au - u^2} + \frac{3a^2}{2}\cos^{-1}\left(\frac{a-u}{a}\right) + C$

120. $\displaystyle\int \frac{du}{u\sqrt{2au - u^2}} = -\frac{\sqrt{2au - u^2}}{au} + C$

Cut here and keep for reference

CHAPTER 10 CONCEPT CHECK ANSWERS

1. (a) What is a parametric curve?

A parametric curve is a set of points of the form $(x, y) = (f(t), g(t))$, where f and g are functions of a variable t, the parameter.

(b) How do you sketch a parametric curve?

Sketching a parametric curve, like sketching the graph of a function, is difficult to do in general. We can plot points on the curve by finding $f(t)$ and $g(t)$ for various values of t, either by hand or with a calculator or computer. Sometimes, when f and g are given by formulas, we can eliminate t from the equations $x = f(t)$ and $y = g(t)$ to get a Cartesian equation relating x and y. It may be easier to graph that equation than to work with the original formulas for x and y in terms of t.

2. (a) How do you find the slope of a tangent to a parametric curve?

You can find dy/dx as a function of t by calculating

$$\frac{dy}{dx} = \frac{dy/dt}{dx/dt} \qquad \text{if } dx/dt \neq 0$$

(b) How do you find the area under a parametric curve?

If the curve is traced out once by the parametric equations $x = f(t)$, $y = g(t)$, $\alpha \leqslant t \leqslant \beta$, then the area is

$$A = \int_a^b y\,dx = \int_\alpha^\beta g(t)f'(t)\,dt$$

[or $\int_\beta^\alpha g(t)f'(t)\,dt$ if the leftmost point is $(f(\beta), g(\beta))$ rather than $(f(\alpha), g(\alpha))$].

3. Write an expression for each of the following:

(a) The length of a parametric curve

If the curve is traced out once by the parametric equations $x = f(t)$, $y = g(t)$, $\alpha \leqslant t \leqslant \beta$, then the length is

$$L = \int_\alpha^\beta \sqrt{(dx/dt)^2 + (dy/dt)^2}\,dt$$

$$= \int_\alpha^\beta \sqrt{[f'(t)]^2 + [g'(t)]^2}\,dt$$

(b) The area of the surface obtained by rotating a parametric curve about the x-axis

$$S = \int_\alpha^\beta 2\pi y\sqrt{(dx/dt)^2 + (dy/dt)^2}\,dt$$

$$= \int_\alpha^\beta 2\pi g(t)\sqrt{[f'(t)]^2 + [g'(t)]^2}\,dt$$

4. (a) Use a diagram to explain the meaning of the polar coordinates (r, θ) of a point.

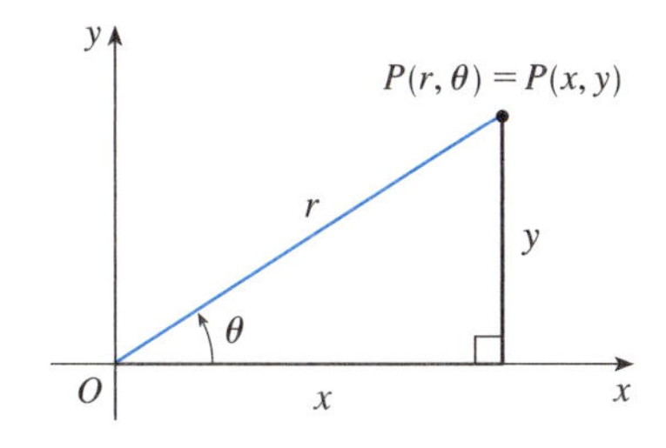

(b) Write equations that express the Cartesian coordinates (x, y) of a point in terms of the polar coordinates.

$$x = r\cos\theta \qquad y = r\sin\theta$$

(c) What equations would you use to find the polar coordinates of a point if you knew the Cartesian coordinates?

To find a polar representation (r, θ) with $r \geqslant 0$ and $0 \leqslant \theta < 2\pi$, first calculate $r = \sqrt{x^2 + y^2}$. Then θ is specified by $\tan\theta = y/x$. Be sure to choose θ so that (r, θ) lies in the correct quadrant.

5. (a) How do you find the slope of a tangent line to a polar curve?

$$\frac{dy}{dx} = \frac{\dfrac{dy}{d\theta}}{\dfrac{dx}{d\theta}} = \frac{\dfrac{d}{d\theta}(y)}{\dfrac{d}{d\theta}(x)} = \frac{\dfrac{d}{d\theta}(r\sin\theta)}{\dfrac{d}{d\theta}(r\cos\theta)}$$

$$= \frac{\left(\dfrac{dr}{d\theta}\right)\sin\theta + r\cos\theta}{\left(\dfrac{dr}{d\theta}\right)\cos\theta - r\sin\theta} \qquad \text{where } r = f(\theta)$$

(b) How do you find the area of a region bounded by a polar curve?

$$A = \int_a^b \tfrac{1}{2}r^2\,d\theta = \int_a^b \tfrac{1}{2}[f(\theta)]^2\,d\theta$$

(c) How do you find the length of a polar curve?

$$L = \int_a^b \sqrt{(dx/d\theta)^2 + (dy/d\theta)^2}\,d\theta$$

$$= \int_a^b \sqrt{r^2 + (dr/d\theta)^2}\,d\theta$$

$$= \int_a^b \sqrt{[f(\theta)]^2 + [f'(\theta)]^2}\,d\theta$$

6. (a) Give a geometric definition of a parabola.

A parabola is a set of points in a plane whose distances from a fixed point F (the focus) and a fixed line l (the directrix) are equal.

(b) Write an equation of a parabola with focus $(0, p)$ and directrix $y = -p$. What if the focus is $(p, 0)$ and the directrix is $x = -p$?

In the first case an equation is $x^2 = 4py$ and in the second case, $y^2 = 4px$.

7. (a) Give a definition of an ellipse in terms of foci.

An ellipse is a set of points in a plane the sum of whose distances from two fixed points (the foci) is a constant.

(b) Write an equation for the ellipse with foci $(\pm c, 0)$ and vertices $(\pm a, 0)$.

$$\frac{x^2}{a^2} + \frac{y^2}{b^2} = 1$$

where $a \geqslant b > 0$ and $c^2 = a^2 - b^2$.

(continued)

8. (a) Give a definition of a hyperbola in terms of foci.

A hyperbola is a set of points in a plane the difference of whose distances from two fixed points (the foci) is a constant. This difference should be interpreted as the larger distance minus the smaller distance.

(b) Write an equation for the hyperbola with foci $(\pm c, 0)$ and vertices $(\pm a, 0)$.

$$\frac{x^2}{a^2} - \frac{y^2}{b^2} = 1$$

where $c^2 = a^2 + b^2$.

(c) Write equations for the asymptotes of the hyperbola in part (b).

$$y = \pm \frac{b}{a} x$$

9. (a) What is the eccentricity of a conic section?

If a conic section has focus F and corresponding directrix l, then the eccentricity e is the fixed ratio $|PF|/|Pl|$ for points P of the conic section.

(b) What can you say about the eccentricity if the conic section is an ellipse? A hyperbola? A parabola?

$e < 1$ for an ellipse; $e > 1$ for a hyperbola; $e = 1$ for a parabola

(c) Write a polar equation for a conic section with eccentricity e and directrix $x = d$. What if the directrix is $x = -d$? $y = d$? $y = -d$?

$$\text{directrix } x = d\text{:} \quad r = \frac{ed}{1 + e\cos\theta}$$

$$x = -d\text{:} \quad r = \frac{ed}{1 - e\cos\theta}$$

$$y = d\text{:} \quad r = \frac{ed}{1 + e\sin\theta}$$

$$y = -d\text{:} \quad r = \frac{ed}{1 - e\sin\theta}$$

CHAPTER 11 CONCEPT CHECK ANSWERS

1. (a) What is a convergent sequence?

A convergent sequence $\{a_n\}$ is an ordered list of numbers where $\lim_{n\to\infty} a_n$ exists.

(b) What is a convergent series?

A series $\Sigma\, a_n$ is the *sum* of a sequence of numbers. It is convergent if the partial sums $s_n = \sum_{i=1}^{n} a_n$ approach a finite value, that is, $\lim_{n\to\infty} s_n$ exists as a real number.

(c) What does $\lim_{n\to\infty} a_n = 3$ mean?

The terms of the sequence $\{a_n\}$ approach 3 as n becomes large.

(d) What does $\Sigma_{n=1}^{\infty} a_n = 3$ mean?

By adding sufficiently many terms of the series, we can make the partial sums as close to 3 as we like.

2. (a) What is a bounded sequence?

A sequence $\{a_n\}$ is bounded if there are numbers m and M such that $m \leqslant a_n \leqslant M$ for all $n \geqslant 1$.

(b) What is a monotonic sequence?

A sequence is monotonic if it is either increasing or decreasing for all $n \geqslant 1$.

(c) What can you say about a bounded monotonic sequence?

Every bounded, monotonic sequence is convergent.

3. (a) What is a geometric series? Under what circumstances is it convergent? What is its sum?

A geometric series is of the form

$$\sum_{n=1}^{\infty} ar^{n-1} = a + ar + ar^2 + \cdots$$

It is convergent if $|r| < 1$ and its sum is $\dfrac{a}{1-r}$.

(b) What is a p-series? Under what circumstances is it convergent?

A p-series is of the form $\displaystyle\sum_{n=1}^{\infty} \frac{1}{n^p}$. It is convergent if $p > 1$.

4. Suppose $\Sigma\, a_n = 3$ and s_n is the nth partial sum of the series. What is $\lim_{n\to\infty} a_n$? What is $\lim_{n\to\infty} s_n$?

If $\Sigma\, a_n = 3$, then $\lim_{n\to\infty} a_n = 0$ and $\lim_{n\to\infty} s_n = 3$.

5. State the following.

(a) The Test for Divergence

If $\lim_{n\to\infty} a_n$ does not exist or if $\lim_{n\to\infty} a_n \neq 0$, then the series $\displaystyle\sum_{n=1}^{\infty} a_n$ is divergent.

(b) The Integral Test

Suppose f is a continuous, positive, decreasing function on $[1, \infty)$ and let $a_n = f(n)$.

- If $\int_1^{\infty} f(x)\,dx$ is convergent, then $\displaystyle\sum_{n=1}^{\infty} a_n$ is convergent.
- If $\int_1^{\infty} f(x)\,dx$ is divergent, then $\displaystyle\sum_{n=1}^{\infty} a_n$ is divergent.

(c) The Comparison Test

Suppose that $\Sigma\, a_n$ and $\Sigma\, b_n$ are series with positive terms.

- If $\Sigma\, b_n$ is convergent and $a_n \leqslant b_n$ for all n, then $\Sigma\, a_n$ is also convergent.
- If $\Sigma\, b_n$ is divergent and $a_n \geqslant b_n$ for all n, then $\Sigma\, a_n$ is also divergent.

(d) The Limit Comparison Test

Suppose that $\Sigma\, a_n$ and $\Sigma\, b_n$ are series with positive terms. If $\lim_{n\to\infty} a_n/b_n = c$, where c is a finite number and $c > 0$, then either both series converge or both diverge.

(e) The Alternating Series Test

If the alternating series

$$\sum_{n=1}^{\infty} (-1)^{n-1} b_n = b_1 - b_2 + b_3 - b_4 + b_5 - b_6 + \cdots$$

where $b_n > 0$ satisfies (i) $b_{n+1} \leqslant b_n$ for all n and (ii) $\lim_{n\to\infty} b_n = 0$, then the series is convergent.

(f) The Ratio Test

- If $\displaystyle\lim_{n\to\infty} \left|\frac{a_{n+1}}{a_n}\right| = L < 1$, then the series $\displaystyle\sum_{n=1}^{\infty} a_n$ is absolutely convergent (and therefore convergent).
- If $\displaystyle\lim_{n\to\infty} \left|\frac{a_{n+1}}{a_n}\right| = L > 1$ or $\displaystyle\lim_{n\to\infty} \left|\frac{a_{n+1}}{a_n}\right| = \infty$, then the series $\displaystyle\sum_{n=1}^{\infty} a_n$ is divergent.
- If $\displaystyle\lim_{n\to\infty} \left|\frac{a_{n+1}}{a_n}\right| = 1$, the Ratio Test is inconclusive.

(g) The Root Test

- If $\lim_{n\to\infty} \sqrt[n]{|a_n|} = L < 1$, then the series $\displaystyle\sum_{n=1}^{\infty} a_n$ is absolutely convergent (and therefore convergent).
- If $\lim_{n\to\infty} \sqrt[n]{|a_n|} = L > 1$ or $\lim_{n\to\infty} \sqrt[n]{|a_n|} = \infty$, then the series $\displaystyle\sum_{n=1}^{\infty} a_n$ is divergent.
- If $\lim_{n\to\infty} \sqrt[n]{|a_n|} = 1$, the Root Test is inconclusive.

6. (a) What is an absolutely convergent series?

A series $\Sigma\, a_n$ is called absolutely convergent if the series of absolute values $\Sigma\, |a_n|$ is convergent.

(b) What can you say about such a series?

If a series $\Sigma\, a_n$ is absolutely convergent, then it is convergent.

(c) What is a conditionally convergent series?

A series $\Sigma\, a_n$ is called conditionally convergent if it is convergent but not absolutely convergent.

(continued)

7. (a) If a series is convergent by the Integral Test, how do you estimate its sum?

The sum s can be estimated by the inequality

$$s_n + \int_{n+1}^{\infty} f(x)\,dx \leqslant s \leqslant s_n + \int_{n}^{\infty} f(x)\,dx$$

where s_n is the nth partial sum.

(b) If a series is convergent by the Comparison Test, how do you estimate its sum?

We first estimate the remainder for the comparison series. This gives an upper bound for the remainder of the original series (as in Example 11.4.5).

(c) If a series is convergent by the Alternating Series Test, how do you estimate its sum?

We can use a partial sum s_n of an alternating series as an approximation to the total sum. The size of the error is guaranteed to be no more than $|a_{n+1}|$, the absolute value of the first neglected term.

8. (a) Write the general form of a power series.

A power series centered at a is

$$\sum_{n=0}^{\infty} c_n(x-a)^n$$

(b) What is the radius of convergence of a power series?

Given the power series $\sum_{n=0}^{\infty} c_n(x-a)^n$, the radius of convergence is:

(i) 0 if the series converges only when $x = a$,

(ii) ∞ if the series converges for all x, or

(iii) a positive number R such that the series converges if $|x-a| < R$ and diverges if $|x-a| > R$.

(c) What is the interval of convergence of a power series?

The interval of convergence of a power series is the interval that consists of all values of x for which the series converges. Corresponding to the cases in part (b), the interval of convergence is (i) the single point $\{a\}$, (ii) $(-\infty, \infty)$, or (iii) an interval with endpoints $a - R$ and $a + R$ that can contain neither, either, or both of the endpoints.

9. Suppose $f(x)$ is the sum of a power series with radius of convergence R.

(a) How do you differentiate f? What is the radius of convergence of the series for f'?

If $f(x) = \sum_{n=0}^{\infty} c_n(x-a)^n$, then $f'(x) = \sum_{n=1}^{\infty} nc_n(x-a)^{n-1}$ with radius of convergence R.

(b) How do you integrate f? What is the radius of convergence of the series for $\int f(x)\,dx$?

$\int f(x)\,dx = C + \sum_{n=0}^{\infty} c_n \dfrac{(x-a)^{n+1}}{n+1}$ with radius of convergence R.

10. (a) Write an expression for the nth-degree Taylor polynomial of f centered at a.

$$T_n(x) = \sum_{i=0}^{n} \frac{f^{(i)}(a)}{i!}(x-a)^i$$

(b) Write an expression for the Taylor series of f centered at a.

$$\sum_{n=0}^{\infty} \frac{f^{(n)}(a)}{n!}(x-a)^n$$

(c) Write an expression for the Maclaurin series of f.

$$\sum_{n=0}^{\infty} \frac{f^{(n)}(0)}{n!}x^n \qquad [a = 0 \text{ in part (b)}]$$

(d) How do you show that $f(x)$ is equal to the sum of its Taylor series?

If $f(x) = T_n(x) + R_n(x)$, where $T_n(x)$ is the nth-degree Taylor polynomial of f and $R_n(x)$ is the remainder of the Taylor series, then we must show that

$$\lim_{n \to \infty} R_n(x) = 0$$

(e) State Taylor's Inequality.

If $|f^{(n+1)}(x)| \leqslant M$ for $|x-a| \leqslant d$, then the remainder $R_n(x)$ of the Taylor series satisfies the inequality

$$|R_n(x)| \leqslant \frac{M}{(n+1)!}|x-a|^{n+1} \qquad \text{for } |x-a| \leqslant d$$

11. Write the Maclaurin series and the interval of convergence for each of the following functions.

(a) $\dfrac{1}{1-x} = \sum_{n=0}^{\infty} x^n, \quad R = 1$

(b) $e^x = \sum_{n=0}^{\infty} \dfrac{x^n}{n!}, \quad R = \infty$

(c) $\sin x = \sum_{n=0}^{\infty} (-1)^n \dfrac{x^{2n+1}}{(2n+1)!}, \quad R = \infty$

(d) $\cos x = \sum_{n=0}^{\infty} (-1)^n \dfrac{x^{2n}}{(2n)!}, \quad R = \infty$

(e) $\tan^{-1} x = \sum_{n=0}^{\infty} (-1)^n \dfrac{x^{2n+1}}{2n+1}, \quad R = 1$

(f) $\ln(1+x) = \sum_{n=1}^{\infty} (-1)^{n-1} \dfrac{x^n}{n}, \quad R = 1$

12. Write the binomial series expansion of $(1+x)^k$. What is the radius of convergence of this series?

If k is any real number and $|x| < 1$, then

$$(1+x)^k = \sum_{n=0}^{\infty} \binom{k}{n} x^n$$

$$= 1 + kx + \frac{k(k-1)}{2!}x^2 + \frac{k(k-1)(k-2)}{3!}x^3 + \cdots$$

The radius of convergence for the binomial series is 1.

CHAPTER 12 CONCEPT CHECK ANSWERS

1. What is the difference between a vector and a scalar?

A scalar is a real number, whereas a vector is a quantity that has both a real-valued magnitude and a direction.

2. How do you add two vectors geometrically? How do you add them algebraically?

To add two vectors geometrically, we can use either the Triangle Law or the Parallelogram Law:

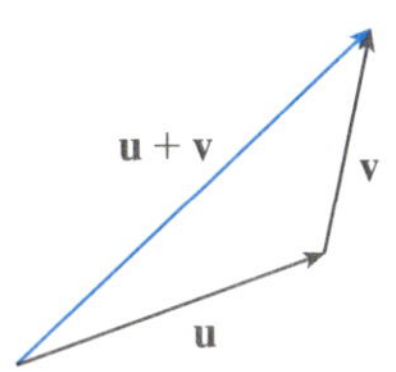

Triangle Law

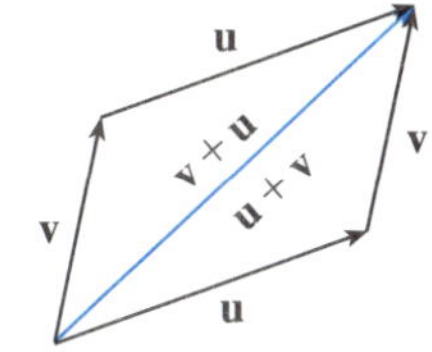

Parallelogram Law

Algebraically, we add the corresponding components of the vectors.

3. If **a** is a vector and c is a scalar, how is $c\mathbf{a}$ related to **a** geometrically? How do you find $c\mathbf{a}$ algebraically?

For $c > 0$, $c\mathbf{a}$ is a vector with the same direction as **a** and length c times the length of **a**. If $c < 0$, $c\mathbf{a}$ points in the direction opposite to **a** and has length $|c|$ times the length of **a**. Algebraically, to find $c\mathbf{a}$ we multiply each component of **a** by c.

4. How do you find the vector from one point to another?

The vector from point $A(x_1, y_1, z_1)$ to point $B(x_2, y_2, z_2)$ is given by

$$\langle x_2 - x_1, y_2 - y_1, z_2 - z_1 \rangle$$

5. How do you find the dot product $\mathbf{a} \cdot \mathbf{b}$ of two vectors if you know their lengths and the angle between them? What if you know their components?

If θ is the angle between the vectors **a** and **b**, then

$$\mathbf{a} \cdot \mathbf{b} = |\mathbf{a}||\mathbf{b}| \cos \theta$$

If $\mathbf{a} = \langle a_1, a_2, a_3 \rangle$ and $\mathbf{b} = \langle b_1, b_2, b_3 \rangle$, then

$$\mathbf{a} \cdot \mathbf{b} = a_1 b_1 + a_2 b_2 + a_3 b_3$$

6. How are dot products useful?

The dot product can be used to find the angle between two vectors. In particular, it can be used to determine whether two vectors are orthogonal. We can also use the dot product to find the scalar projection of one vector onto another. Additionally, if a constant force moves an object, the work done is the dot product of the force and displacement vectors.

7. Write expressions for the scalar and vector projections of **b** onto **a**. Illustrate with diagrams.

Scalar projection of **b** onto **a**: $\quad \text{comp}_{\mathbf{a}}\, \mathbf{b} = \dfrac{\mathbf{a} \cdot \mathbf{b}}{|\mathbf{a}|}$

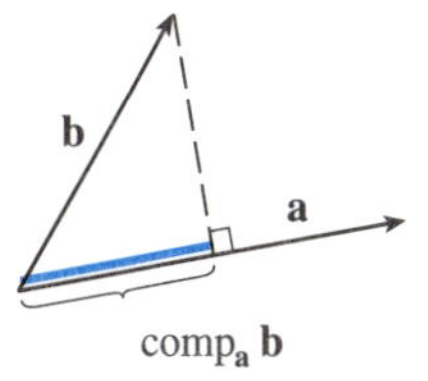

Vector projection of **b** onto **a**:

$$\text{proj}_{\mathbf{a}}\, \mathbf{b} = \left(\frac{\mathbf{a} \cdot \mathbf{b}}{|\mathbf{a}|}\right) \frac{\mathbf{a}}{|\mathbf{a}|} = \frac{\mathbf{a} \cdot \mathbf{b}}{|\mathbf{a}|^2}\, \mathbf{a}$$

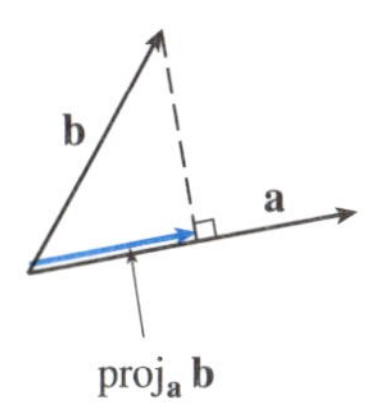

8. How do you find the cross product $\mathbf{a} \times \mathbf{b}$ of two vectors if you know their lengths and the angle between them? What if you know their components?

If θ is the angle between **a** and **b** $(0 \leqslant \theta \leqslant \pi)$, then $\mathbf{a} \times \mathbf{b}$ is the vector with length $|\mathbf{a} \times \mathbf{b}| = |\mathbf{a}||\mathbf{b}| \sin \theta$ and direction orthogonal to both **a** and **b**, as given by the right-hand rule. If

$$\mathbf{a} = \langle a_1, a_2, a_3 \rangle \qquad \text{and} \qquad \mathbf{b} = \langle b_1, b_2, b_3 \rangle$$

then

$$\mathbf{a} \times \mathbf{b} = \begin{vmatrix} \mathbf{i} & \mathbf{j} & \mathbf{k} \\ a_1 & a_2 & a_3 \\ b_1 & b_2 & b_3 \end{vmatrix}$$

$$= \langle a_2 b_3 - a_3 b_2, a_3 b_1 - a_1 b_3, a_1 b_2 - a_2 b_1 \rangle$$

9. How are cross products useful?

The cross product can be used to create a vector orthogonal to two given vectors and it can be used to compute the area of a parallelogram determined by two vectors. Two nonzero vectors are parallel if and only if their cross product is **0**. In addition, if a force acts on a rigid body, then the torque vector is the cross product of the position and force vectors.

(continued)

10. (a) How do you find the area of the parallelogram determined by $\mathbf{a}$ and $\mathbf{b}$?

The area of the parallelogram determined by $\mathbf{a}$ and $\mathbf{b}$ is the length of the cross product: $|\mathbf{a} \times \mathbf{b}|$.

(b) How do you find the volume of the parallelepiped determined by $\mathbf{a}$, $\mathbf{b}$, and $\mathbf{c}$?

The volume of the parallelepiped determined by $\mathbf{a}$, $\mathbf{b}$, and $\mathbf{c}$ is the magnitude of their scalar triple product: $|\mathbf{a} \cdot (\mathbf{b} \times \mathbf{c})|$.

11. How do you find a vector perpendicular to a plane?

If an equation of the plane is known, it can be written in the form $ax + by + cz + d = 0$. A normal vector, which is perpendicular to the plane, is $\langle a, b, c\rangle$ (or any nonzero scalar multiple of $\langle a, b, c\rangle$). If an equation is not known, we can use points on the plane to find two nonparallel vectors that lie in the plane. The cross product of these vectors is a vector perpendicular to the plane.

12. How do you find the angle between two intersecting planes?

The angle between two intersecting planes is defined as the acute angle θ between their normal vectors. If $\mathbf{n}_1$ and $\mathbf{n}_2$ are the normal vectors, then

$$\cos\theta = \frac{\mathbf{n}_1 \cdot \mathbf{n}_2}{|\mathbf{n}_1||\mathbf{n}_2|}$$

13. Write a vector equation, parametric equations, and symmetric equations for a line.

A vector equation for a line that is parallel to a vector $\mathbf{v}$ and that passes through a point with position vector $\mathbf{r}_0$ is $\mathbf{r} = \mathbf{r}_0 + t\mathbf{v}$. Parametric equations for a line through the point (x_0, y_0, z_0) and parallel to the vector $\langle a, b, c\rangle$ are

$$x = x_0 + at \qquad y = y_0 + bt \qquad z = z_0 + ct$$

while symmetric equations are

$$\frac{x - x_0}{a} = \frac{y - y_0}{b} = \frac{z - z_0}{c}$$

14. Write a vector equation and a scalar equation for a plane.

A vector equation of a plane that passes through a point with position vector $\mathbf{r}_0$ and that has normal vector $\mathbf{n}$ (meaning $\mathbf{n}$ is orthogonal to the plane) is $\mathbf{n} \cdot (\mathbf{r} - \mathbf{r}_0) = 0$ or, equivalently, $\mathbf{n} \cdot \mathbf{r} = \mathbf{n} \cdot \mathbf{r}_0$.

A scalar equation of a plane through a point (x_0, y_0, z_0) with normal vector $\mathbf{n} = \langle a, b, c\rangle$ is

$$a(x - x_0) + b(y - y_0) + c(z - z_0) = 0$$

15. (a) How do you tell if two vectors are parallel?

Two (nonzero) vectors are parallel if and only if one is a scalar multiple of the other. In addition, two nonzero vectors are parallel if and only if their cross product is $\mathbf{0}$.

(b) How do you tell if two vectors are perpendicular?

Two vectors are perpendicular if and only if their dot product is 0.

(c) How do you tell if two planes are parallel?

Two planes are parallel if and only if their normal vectors are parallel.

16. (a) Describe a method for determining whether three points P, Q, and R lie on the same line.

Determine the vectors $\overrightarrow{PQ} = \mathbf{a}$ and $\overrightarrow{PR} = \mathbf{b}$. If there is a scalar t such that $\mathbf{a} = t\mathbf{b}$, then the vectors are parallel and the points must all lie on the same line.

Alternatively, if $\overrightarrow{PQ} \times \overrightarrow{PR} = \mathbf{0}$, then $\overrightarrow{PQ}$ and $\overrightarrow{PR}$ are parallel, so P, Q, and R are collinear.

An algebraic method is to determine an equation of the line joining two of the points, and then check whether or not the third point satisfies this equation.

(b) Describe a method for determining whether four points P, Q, R, and S lie in the same plane.

Find the vectors $\overrightarrow{PQ} = \mathbf{a}$, $\overrightarrow{PR} = \mathbf{b}$, $\overrightarrow{PS} = \mathbf{c}$. Then $\mathbf{a} \times \mathbf{b}$ is normal to the plane formed by P, Q, and R, and so S lies on this plane if $\mathbf{a} \times \mathbf{b}$ and $\mathbf{c}$ are orthogonal, that is, if $(\mathbf{a} \times \mathbf{b}) \cdot \mathbf{c} = 0$.

Alternatively, we can check if the volume of the parallelepiped determined by $\mathbf{a}$, $\mathbf{b}$, and $\mathbf{c}$ is 0 (see Example 12.4.5).

An algebraic method is to find an equation of the plane determined by three of the points, and then check whether or not the fourth point satisfies this equation.

17. (a) How do you find the distance from a point to a line?

Let P be a point not on the line L that passes through the points Q and R and let $\mathbf{a} = \overrightarrow{QR}$, $\mathbf{b} = \overrightarrow{QP}$. The distance from the point P to the line L is

$$d = \frac{|\mathbf{a} \times \mathbf{b}|}{|\mathbf{a}|}$$

(b) How do you find the distance from a point to a plane?

Let $P_0(x_0, y_0, z_0)$ be any point in the plane $ax + by + cz + d = 0$ and let $P_1(x_1, y_1, z_1)$ be a point not in the plane. If $\mathbf{b} = \overrightarrow{P_0P_1} = \langle x_1 - x_0, y_1 - y_0, z_1 - z_0\rangle$, then the distance D from P_1 to the plane is equal to the absolute value of the scalar projection of $\mathbf{b}$ onto the plane's normal vector $\mathbf{n} = \langle a, b, c\rangle$:

$$D = |\text{comp}_{\mathbf{n}}\,\mathbf{b}| = \frac{|\mathbf{n} \cdot \mathbf{b}|}{|\mathbf{n}|} = \frac{|ax_1 + by_1 + cz_1 + d|}{\sqrt{a^2 + b^2 + c^2}}$$

(c) How do you find the distance between two lines?

Two skew lines L_1 and L_2 can be viewed as lying on two parallel planes, each with normal vector $\mathbf{n} = \mathbf{v}_1 \times \mathbf{v}_2$, where $\mathbf{v}_1$ and $\mathbf{v}_2$ are the direction vectors of L_1 and L_2. After choosing one point on L_1 and determining the equation of the plane containing L_2, we can proceed as in part (b). (See Example 12.5.10.)

(continued)

CHAPTER 12 CONCEPT CHECK ANSWERS *(continued)*

18. What are the traces of a surface? How do you find them?

The traces of a surface are the curves of intersection of the surface with planes parallel to the coordinate planes. We can find the trace in the plane $x = k$ (parallel to the yz-plane) by setting $x = k$ and determining the curve represented by the resulting equation. Traces in the planes $y = k$ (parallel to the xz-plane) and $z = k$ (parallel to the xy-plane) are found similarly.

19. Write equations in standard form of the six types of quadric surfaces.

Equations for the quadric surfaces symmetric with respect to the z-axis are as follows.

Ellipsoid: $$\frac{x^2}{a^2} + \frac{y^2}{b^2} + \frac{z^2}{c^2} = 1$$

Cone: $$\frac{z^2}{c^2} = \frac{x^2}{a^2} + \frac{y^2}{b^2}$$

Elliptic paraboloid: $$\frac{z}{c} = \frac{x^2}{a^2} + \frac{y^2}{b^2}$$

Hyperboloid of one sheet: $$\frac{x^2}{a^2} + \frac{y^2}{b^2} - \frac{z^2}{c^2} = 1$$

Hyperboloid of two sheets: $$-\frac{x^2}{a^2} - \frac{y^2}{b^2} + \frac{z^2}{c^2} = 1$$

Hyperbolic paraboloid: $$\frac{z}{c} = \frac{x^2}{a^2} - \frac{y^2}{b^2}$$

Cut here and keep for reference

CHAPTER 13 CONCEPT CHECK ANSWERS

1. What is a vector function? How do you find its derivative and its integral?

A vector function is a function whose domain is a set of real numbers and whose range is a set of vectors. To find the derivative or integral, we can differentiate or integrate each component function of the vector function.

2. What is the connection between vector functions and space curves?

A continuous vector function $\mathbf{r}$ defines a space curve that is traced out by the tip of the moving position vector $\mathbf{r}(t)$.

3. How do you find the tangent vector to a smooth curve at a point? How do you find the tangent line? The unit tangent vector?

The tangent vector to a smooth curve at a point P with position vector $\mathbf{r}(t)$ is the vector $\mathbf{r}'(t)$. The tangent line at P is the line through P parallel to the tangent vector $\mathbf{r}'(t)$. The unit tangent vector is $\mathbf{T}(t) = \dfrac{\mathbf{r}'(t)}{|\mathbf{r}'(t)|}$.

4. If $\mathbf{u}$ and $\mathbf{v}$ are differentiable vector functions, c is a scalar, and f is a real-valued function, write the rules for differentiating the following vector functions.

(a) $\mathbf{u}(t) + \mathbf{v}(t)$

$$\frac{d}{dt}[\mathbf{u}(t) + \mathbf{v}(t)] = \mathbf{u}'(t) + \mathbf{v}'(t)$$

(b) $c\mathbf{u}(t)$

$$\frac{d}{dt}[c\mathbf{u}(t)] = c\mathbf{u}'(t)$$

(c) $f(t)\,\mathbf{u}(t)$

$$\frac{d}{dt}[f(t)\,\mathbf{u}(t)] = f'(t)\,\mathbf{u}(t) + f(t)\,\mathbf{u}'(t)$$

(d) $\mathbf{u}(t) \cdot \mathbf{v}(t)$

$$\frac{d}{dt}[\mathbf{u}(t) \cdot \mathbf{v}(t)] = \mathbf{u}'(t) \cdot \mathbf{v}(t) + \mathbf{u}(t) \cdot \mathbf{v}'(t)$$

(e) $\mathbf{u}(t) \times \mathbf{v}(t)$

$$\frac{d}{dt}[\mathbf{u}(t) \times \mathbf{v}(t)] = \mathbf{u}'(t) \times \mathbf{v}(t) + \mathbf{u}(t) \times \mathbf{v}'(t)$$

(f) $\mathbf{u}(f(t))$

$$\frac{d}{dt}[\mathbf{u}(f(t))] = f'(t)\mathbf{u}'(f(t))$$

5. How do you find the length of a space curve given by a vector function $\mathbf{r}(t)$?

If $\mathbf{r}(t) = \langle f(t), g(t), h(t)\rangle$, $a \leqslant t \leqslant b$, and the curve is traversed exactly once as t increases from a to b, then the length is

$$L = \int_a^b |\mathbf{r}'(t)|\,dt = \int_a^b \sqrt{[f'(t)]^2 + [g'(t)]^2 + [h'(t)]^2}\,dt$$

6. (a) What is the definition of curvature?

The curvature of a curve is $\kappa = \left|\dfrac{d\mathbf{T}}{ds}\right|$ where $\mathbf{T}$ is the unit tangent vector.

(b) Write a formula for curvature in terms of $\mathbf{r}'(t)$ and $\mathbf{T}'(t)$.

$$\kappa(t) = \frac{|\mathbf{T}'(t)|}{|\mathbf{r}'(t)|}$$

(c) Write a formula for curvature in terms of $\mathbf{r}'(t)$ and $\mathbf{r}''(t)$.

$$\kappa(t) = \frac{|\mathbf{r}'(t) \times \mathbf{r}''(t)|}{|\mathbf{r}'(t)|^3}$$

(d) Write a formula for the curvature of a plane curve with equation $y = f(x)$.

$$\kappa(x) = \frac{|f''(x)|}{[1 + (f'(x))^2]^{3/2}}$$

7. (a) Write formulas for the unit normal and binormal vectors of a smooth space curve $\mathbf{r}(t)$.

Unit normal vector: $\mathbf{N}(t) = \dfrac{\mathbf{T}'(t)}{|\mathbf{T}'(t)|}$

Binormal vector: $\mathbf{B}(t) = \mathbf{T}(t) \times \mathbf{N}(t)$

(b) What is the normal plane of a curve at a point? What is the osculating plane? What is the osculating circle?

The normal plane of a curve at a point P is the plane determined by the normal and binormal vectors $\mathbf{N}$ and $\mathbf{B}$ at P. The tangent vector $\mathbf{T}$ is orthogonal to the normal plane.

The osculating plane at P is the plane determined by the vectors $\mathbf{T}$ and $\mathbf{N}$. It is the plane that comes closest to containing the part of the curve near P.

The osculating circle at P is the circle that lies in the osculating plane of C at P, has the same tangent as C at P, lies on the concave side of C (toward which $\mathbf{N}$ points), and has radius $\rho = 1/\kappa$ (the reciprocal of the curvature). It is the circle that best describes how C behaves near P; it shares the same tangent, normal, and curvature at P.

(continued)

8. (a) How do you find the velocity, speed, and acceleration of a particle that moves along a space curve?

If $\mathbf{r}(t)$ is the position vector of the particle on the space curve, the velocity vector is $\mathbf{v}(t) = \mathbf{r}'(t)$, the speed is given by $|\mathbf{v}(t)|$, and the acceleration is $\mathbf{a}(t) = \mathbf{v}'(t) = \mathbf{r}''(t)$.

(b) Write the acceleration in terms of its tangential and normal components.

$\mathbf{a} = a_T\mathbf{T} + a_N\mathbf{N}$, where $a_T = v'$ and $a_N = \kappa v^2$ ($v = |\mathbf{v}|$ is speed and κ is the curvature).

9. State Kepler's Laws.

- A planet revolves around the sun in an elliptical orbit with the sun at one focus.
- The line joining the sun to a planet sweeps out equal areas in equal times.
- The square of the period of revolution of a planet is proportional to the cube of the length of the major axis of its orbit.

CHAPTER 14 CONCEPT CHECK ANSWERS

1. (a) What is a function of two variables?

A function f of two variables is a rule that assigns to each ordered pair (x, y) of real numbers in its domain a unique real number denoted by $f(x, y)$.

(b) Describe three methods for visualizing a function of two variables.

One way to visualize a function of two variables is by graphing it, resulting in the surface $z = f(x, y)$. Another method is a contour map, consisting of level curves $f(x, y) = k$ (k a constant), which are horizontal traces of the graph of the function projected onto the xy-plane. Also, we can use an arrow diagram such as the one below.

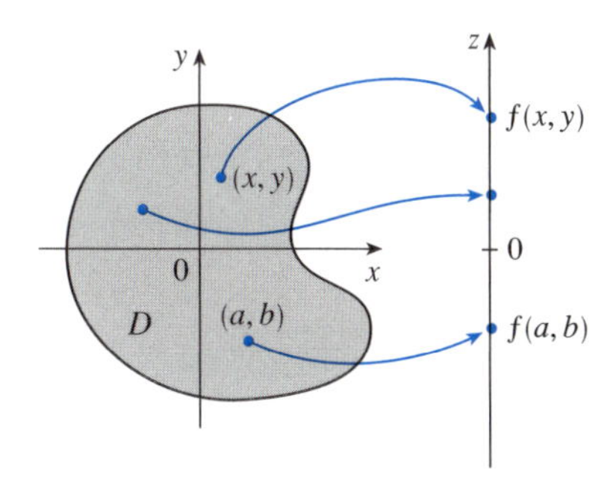

2. What is a function of three variables? How can you visualize such a function?

A function f of three variables is a rule that assigns to each ordered triple (x, y, z) in its domain a unique real number $f(x, y, z)$. We can visualize a function of three variables by examining its level surfaces $f(x, y, z) = k$, where k is a constant.

3. What does

$$\lim_{(x, y)\to(a, b)} f(x, y) = L$$

mean? How can you show that such a limit does not exist?

$\lim_{(x, y)\to(a, b)} f(x, y) = L$ means that the values of $f(x, y)$ approach the number L as the point (x, y) approaches the point (a, b) along any path that is within the domain of f. We can show that a limit at a point does not exist by finding two different paths approaching the point along which $f(x, y)$ has different limits.

4. (a) What does it mean to say that f is continuous at (a, b)?

A function f of two variables is continuous at (a, b) if

$$\lim_{(x, y)\to(a, b)} f(x, y) = f(a, b)$$

(b) If f is continuous on $\mathbb{R}^2$, what can you say about its graph?

If f is continuous on $\mathbb{R}^2$, its graph will appear as a surface without holes or breaks.

5. (a) Write expressions for the partial derivatives $f_x(a, b)$ and $f_y(a, b)$ as limits.

$$f_x(a, b) = \lim_{h\to 0} \frac{f(a + h, b) - f(a, b)}{h}$$

$$f_y(a, b) = \lim_{h\to 0} \frac{f(a, b + h) - f(a, b)}{h}$$

(b) How do you interpret $f_x(a, b)$ and $f_y(a, b)$ geometrically? How do you interpret them as rates of change?

If $f(a, b) = c$, then the point $P(a, b, c)$ lies on the surface S given by $z = f(x, y)$. We can interpret $f_x(a, b)$ as the slope of the tangent line at P to the curve of intersection of the vertical plane $y = b$ and S. In other words, if we restrict ourselves to the path along S through P that is parallel to the xz-plane, then $f_x(a, b)$ is the slope at P looking in the positive x-direction. Similarly, $f_y(a, b)$ is the slope of the tangent line at P to the curve of intersection of the vertical plane $x = a$ and S.

If $z = f(x, y)$, then $f_x(x, y)$ can be interpreted as the rate of change of z with respect to x when y is fixed. Thus $f_x(a, b)$ is the rate of change of z (with respect to x) when y is fixed at b and x is allowed to vary from a. Similarly, $f_y(a, b)$ is the rate of change of z (with respect to y) when x is fixed at a and y is allowed to vary from b.

(c) If $f(x, y)$ is given by a formula, how do you calculate f_x and f_y?

To find f_x, regard y as a constant and differentiate $f(x, y)$ with respect to x. To find f_y, regard x as a constant and differentiate $f(x, y)$ with respect to y.

6. What does Clairaut's Theorem say?

If f is a function of two variables that is defined on a disk D containing the point (a, b) and the functions f_{xy} and f_{yx} are both continuous on D, then Clairaut's Theorem states that $f_{xy}(a, b) = f_{yx}(a, b)$.

7. How do you find a tangent plane to each of the following types of surfaces?

(a) A graph of a function of two variables, $z = f(x, y)$

If f has continuous partial derivatives, an equation of the tangent plane to the surface $z = f(x, y)$ at the point (x_0, y_0, z_0) is

$$z - z_0 = f_x(x_0, y_0)(x - x_0) + f_y(x_0, y_0)(y - y_0)$$

(b) A level surface of a function of three variables, $F(x, y, z) = k$

The tangent plane to the level surface $F(x, y, z) = k$ at $P(x_0, y_0, z_0)$ is the plane that passes through P and has normal vector $\nabla F(x_0, y_0, z_0)$:

$$F_x(x_0, y_0, z_0)(x - x_0) + F_y(x_0, y_0, z_0)(y - y_0) + F_z(x_0, y_0, z_0)(z - z_0) = 0$$

(continued)

8. Define the linearization of f at (a, b). What is the corresponding linear approximation? What is the geometric interpretation of the linear approximation?

The linearization of f at (a, b) is the linear function whose graph is the tangent plane to the surface $z = f(x, y)$ at the point $(a, b, f(a, b))$:

$$L(x, y) = f(a, b) + f_x(a, b)(x - a) + f_y(a, b)(y - b)$$

The linear approximation of f at (a, b) is

$$f(x, y) \approx f(a, b) + f_x(a, b)(x - a) + f_y(a, b)(y - b)$$

Geometrically, the linear approximation says that function values $f(x, y)$ can be approximated by values $L(x, y)$ from the tangent plane to the graph of f at $(a, b, f(a, b))$ when (x, y) is near (a, b).

9. (a) What does it mean to say that f is differentiable at (a, b)?

If $z = f(x, y)$, then f is differentiable at (a, b) if Δz can be expressed in the form

$$\Delta z = f_x(a, b)\,\Delta x + f_y(a, b)\,\Delta y + \varepsilon_1\,\Delta x + \varepsilon_2\,\Delta y$$

where ε_1 and $\varepsilon_2 \to 0$ as $(\Delta x, \Delta y) \to (0, 0)$. In other words, a differentiable function is one for which the linear approximation as stated above is a good approximation when (x, y) is near (a, b).

(b) How do you usually verify that f is differentiable?

If the partial derivatives f_x and f_y exist near (a, b) and are continuous at (a, b), then f is differentiable at (a, b).

10. If $z = f(x, y)$, what are the differentials dx, dy, and dz?

The differentials dx and dy are independent variables that can be given any values. If f is differentiable, the differential dz is then defined by

$$dz = f_x(x, y)\,dx + f_y(x, y)\,dy$$

11. State the Chain Rule for the case where $z = f(x, y)$ and x and y are functions of one variable. What if x and y are functions of two variables?

Suppose that $z = f(x, y)$ is a differentiable function of x and y, where $x = g(t)$ and $y = h(t)$ are both differentiable functions of t. Then z is a differentiable function of t and

$$\frac{dz}{dt} = \frac{\partial f}{\partial x}\frac{dx}{dt} + \frac{\partial f}{\partial y}\frac{dy}{dt}$$

If $z = f(x, y)$ is a differentiable function of x and y, where $x = g(s, t)$ and $y = h(s, t)$ are differentiable functions of s and t, then

$$\frac{\partial z}{\partial s} = \frac{\partial z}{\partial x}\frac{\partial x}{\partial s} + \frac{\partial z}{\partial y}\frac{\partial y}{\partial s} \qquad \frac{\partial z}{\partial t} = \frac{\partial z}{\partial x}\frac{\partial x}{\partial t} + \frac{\partial z}{\partial y}\frac{\partial y}{\partial t}$$

12. If z is defined implicitly as a function of x and y by an equation of the form $F(x, y, z) = 0$, how do you find $\partial z/\partial x$ and $\partial z/\partial y$?

If F is differentiable and $\partial F/\partial z \neq 0$, then

$$\frac{\partial z}{\partial x} = -\frac{\dfrac{\partial F}{\partial x}}{\dfrac{\partial F}{\partial z}} \qquad \frac{\partial z}{\partial y} = -\frac{\dfrac{\partial F}{\partial y}}{\dfrac{\partial F}{\partial z}}$$

13. (a) Write an expression as a limit for the directional derivative of f at (x_0, y_0) in the direction of a unit vector $\mathbf{u} = \langle a, b \rangle$. How do you interpret it as a rate? How do you interpret it geometrically?

The directional derivative of f at (x_0, y_0) in the direction of $\mathbf{u}$ is

$$D_{\mathbf{u}} f(x_0, y_0) = \lim_{h \to 0} \frac{f(x_0 + ha, y_0 + hb) - f(x_0, y_0)}{h}$$

if this limit exists.

We can interpret it as the rate of change of f (with respect to distance) at (x_0, y_0) in the direction of $\mathbf{u}$.

Geometrically, if P is the point $(x_0, y_0, f(x_0, y_0))$ on the graph of f and C is the curve of intersection of the graph of f with the vertical plane that passes through P in the direction of $\mathbf{u}$, then $D_{\mathbf{u}} f(x_0, y_0)$ is the slope of the tangent line to C at P.

(b) If f is differentiable, write an expression for $D_{\mathbf{u}} f(x_0, y_0)$ in terms of f_x and f_y.

$$D_{\mathbf{u}} f(x_0, y_0) = f_x(x_0, y_0)\,a + f_y(x_0, y_0)\,b$$

14. (a) Define the gradient vector ∇f for a function f of two or three variables.

If f is a function of two variables, then

$$\nabla f(x, y) = \langle f_x(x, y), f_y(x, y) \rangle = \frac{\partial f}{\partial x}\mathbf{i} + \frac{\partial f}{\partial y}\mathbf{j}$$

For a function f of three variables,

$$\nabla f(x, y, z) = \langle f_x(x, y, z), f_y(x, y, z), f_z(x, y, z) \rangle$$
$$= \frac{\partial f}{\partial x}\mathbf{i} + \frac{\partial f}{\partial y}\mathbf{j} + \frac{\partial f}{\partial z}\mathbf{k}$$

(b) Express $D_{\mathbf{u}} f$ in terms of ∇f.

$$D_{\mathbf{u}} f(x, y) = \nabla f(x, y) \cdot \mathbf{u}$$

or $$D_{\mathbf{u}} f(x, y, z) = \nabla f(x, y, z) \cdot \mathbf{u}$$

(continued)

Cut here and keep for reference

(c) Explain the geometric significance of the gradient.

The gradient vector of f gives the direction of maximum rate of increase of f. On the graph of $z = f(x, y)$, ∇f points in the direction of steepest ascent. Also, the gradient vector is perpendicular to the level curves or level surfaces of a function.

15. What do the following statements mean?

(a) f has a local maximum at (a, b).

f has a local maximum at (a, b) if $f(x, y) \leqslant f(a, b)$ when (x, y) is near (a, b).

(b) f has an absolute maximum at (a, b).

f has an absolute maximum at (a, b) if $f(x, y) \leqslant f(a, b)$ for all points (x, y) in the domain of f.

(c) f has a local minimum at (a, b).

f has a local minimum at (a, b) if $f(x, y) \geqslant f(a, b)$ when (x, y) is near (a, b).

(d) f has an absolute minimum at (a, b).

f has an absolute minimum at (a, b) if $f(x, y) \geqslant f(a, b)$ for all points (x, y) in the domain of f.

(e) f has a saddle point at (a, b).

f has a saddle point at (a, b) if $f(a, b)$ is a local maximum in one direction but a local minimum in another.

16. (a) If f has a local maximum at (a, b), what can you say about its partial derivatives at (a, b)?

If f has a local maximum at (a, b) and the first-order partial derivatives of f exist there, then $f_x(a, b) = 0$ and $f_y(a, b) = 0$.

(b) What is a critical point of f?

A critical point of f is a point (a, b) such that $f_x(a, b) = 0$ and $f_y(a, b) = 0$ or one of these partial derivatives does not exist.

17. State the Second Derivatives Test.

Suppose the second partial derivatives of f are continuous on a disk with center (a, b), and suppose that $f_x(a, b) = 0$ and $f_y(a, b) = 0$ [that is, (a, b) is a critical point of f]. Let

$$D = D(a, b) = f_{xx}(a, b)\, f_{yy}(a, b) - [f_{xy}(a, b)]^2$$

- If $D > 0$ and $f_{xx}(a, b) > 0$, then $f(a, b)$ is a local minimum.
- If $D > 0$ and $f_{xx}(a, b) < 0$, then $f(a, b)$ is a local maximum.
- If $D < 0$, then $f(a, b)$ is not a local maximum or minimum. The point (a, b) is a saddle point of f.

18. (a) What is a closed set in $\mathbb{R}^2$? What is a bounded set?

A closed set in $\mathbb{R}^2$ is one that contains all its boundary points. If one or more points on the boundary curve are omitted, the set is not closed.

A bounded set is one that is contained within some disk. In other words, it is finite in extent.

(b) State the Extreme Value Theorem for functions of two variables.

If f is continuous on a closed, bounded set D in $\mathbb{R}^2$, then f attains an absolute maximum value $f(x_1, y_1)$ and an absolute minimum value $f(x_2, y_2)$ at some points (x_1, y_1) and (x_2, y_2) in D.

(c) How do you find the values that the Extreme Value Theorem guarantees?

- Find the values of f at the critical points of f in D.
- Find the extreme values of f on the boundary of D.
- The largest of the values from the above steps is the absolute maximum value; the smallest of these values is the absolute minimum value.

19. Explain how the method of Lagrange multipliers works in finding the extreme values of $f(x, y, z)$ subject to the constraint $g(x, y, z) = k$. What if there is a second constraint $h(x, y, z) = c$?

To find the maximum and minimum values of $f(x, y, z)$ subject to the constraint $g(x, y, z) = k$ [assuming that these extreme values exist and $\nabla g \neq \mathbf{0}$ on the surface $g(x, y, z) = k$], we first find all values of x, y, z, and λ where $\nabla f(x, y, z) = \lambda\, \nabla g(x, y, z)$ and $g(x, y, z) = k$. (Thus we are finding the points from the constraint where the gradient vectors ∇f and ∇g are parallel.) Evaluate f at all the resulting points (x, y, z); the largest of these values is the maximum value of f, and the smallest is the minimum value of f.

If there is a second constraint $h(x, y, z) = c$, then we find all values of x, y, z, λ, and μ such that

$$\nabla f(x, y, z) = \lambda\, \nabla g(x, y, z) + \mu\, \nabla h(x, y, z)$$

Again we find the extreme values of f by evaluating f at the resulting points (x, y, z).

Cut here and keep for reference

CHAPTER 15 CONCEPT CHECK ANSWERS

1. Suppose f is a continuous function defined on a rectangle $R = [a, b] \times [c, d]$.

(a) Write an expression for a double Riemann sum of f. If $f(x, y) \geq 0$, what does the sum represent?

A double Riemann sum of f is

$$\sum_{i=1}^{m} \sum_{j=1}^{n} f(x_{ij}^*, y_{ij}^*)\, \Delta A$$

where ΔA is the area of each subrectangle and (x_{ij}^*, y_{ij}^*) is a sample point in each subrectangle. If $f(x, y) \geq 0$, this sum represents an approximation to the volume of the solid that lies above the rectangle R and below the graph of f.

(b) Write the definition of $\iint_R f(x, y)\, dA$ as a limit.

$$\iint_R f(x, y)\, dA = \lim_{m,\, n \to \infty} \sum_{i=1}^{m} \sum_{j=1}^{n} f(x_{ij}^*, y_{ij}^*)\, \Delta A$$

(c) What is the geometric interpretation of $\iint_R f(x, y)\, dA$ if $f(x, y) \geq 0$? What if f takes on both positive and negative values?

If $f(x, y) \geq 0$, $\iint_R f(x, y)\, dA$ represents the volume of the solid that lies above the rectangle R and below the surface $z = f(x, y)$. If f takes on both positive and negative values, then $\iint_R f(x, y)\, dA$ is $V_1 - V_2$, where V_1 is the volume above R and below the surface $z = f(x, y)$, and V_2 is the volume below R and above the surface.

(d) How do you evaluate $\iint_R f(x, y)\, dA$?

We usually evaluate $\iint_R f(x, y)\, dA$ as an iterated integral according to Fubini's Theorem:

$$\iint_R f(x, y)\, dA = \int_a^b \int_c^d f(x, y)\, dy\, dx = \int_c^d \int_a^b f(x, y)\, dx\, dy$$

(e) What does the Midpoint Rule for double integrals say?

The Midpoint Rule for double integrals says that we approximate the double integral $\iint_R f(x, y)\, dA$ by the double Riemann sum $\sum_{i=1}^{m} \sum_{j=1}^{n} f(\bar{x}_i, \bar{y}_j)\, \Delta A$, where the sample points $(\bar{x}_i, \bar{y}_j)$ are the centers of the subrectangles.

(f) Write an expression for the average value of f.

$$f_{\text{ave}} = \frac{1}{A(R)} \iint_R f(x, y)\, dA$$

where $A(R)$ is the area of R.

2. (a) How do you define $\iint_D f(x, y)\, dA$ if D is a bounded region that is not a rectangle?

Since D is bounded, it can be enclosed in a rectangular region R. We define a new function F with domain R by

$$F(x, y) = \begin{cases} f(x, y) & \text{if } (x, y) \text{ is in } D \\ 0 & \text{if } (x, y) \text{ is in } R \text{ but not in } D \end{cases}$$

Then we define

$$\iint_D f(x, y)\, dA = \iint_R F(x, y)\, dA$$

(b) What is a type I region? How do you evaluate $\iint_D f(x, y)\, dA$ if D is a type I region?

A region D is of type I if it lies between the graphs of two continuous functions of x, that is,

$$D = \{(x, y) \mid a \leq x \leq b,\ g_1(x) \leq y \leq g_2(x)\}$$

where g_1 and g_2 are continuous on $[a, b]$. Then

$$\iint_D f(x, y)\, dA = \int_a^b \int_{g_1(x)}^{g_2(x)} f(x, y)\, dy\, dx$$

(c) What is a type II region? How do you evaluate $\iint_D f(x, y)\, dA$ if D is a type II region?

A region D is of type II if it lies between the graphs of two continuous functions of y, that is,

$$D = \{(x, y) \mid c \leq y \leq d,\ h_1(y) \leq x \leq h_2(y)\}$$

where h_1 and h_2 are continuous on $[c, d]$. Then

$$\iint_D f(x, y)\, dA = \int_c^d \int_{h_1(y)}^{h_2(y)} f(x, y)\, dx\, dy$$

(d) What properties do double integrals have?

- $$\iint_D [f(x, y) + g(x, y)]\, dA = \iint_D f(x, y)\, dA + \iint_D g(x, y)\, dA$$
- $$\iint_D cf(x, y)\, dA = c \iint_D f(x, y)\, dA$$
  where c is a constant
- If $f(x, y) \geq g(x, y)$ for all (x, y) in D, then
  $$\iint_D f(x, y)\, dA \geq \iint_D g(x, y)\, dA$$
- If $D = D_1 \cup D_2$, where D_1 and D_2 don't overlap except perhaps on their boundaries, then
  $$\iint_D f(x, y)\, dA = \iint_{D_1} f(x, y)\, dA + \iint_{D_2} f(x, y)\, dA$$
- $\iint_D 1\, dA = A(D)$, the area of D.
- If $m \leq f(x, y) \leq M$ for all (x, y) in D, then
  $$mA(D) \leq \iint_D f(x, y)\, dA \leq MA(D)$$

(continued)

3. How do you change from rectangular coordinates to polar coordinates in a double integral? Why would you want to make the change?

We may want to change from rectangular to polar coordinates in a double integral if the region D of integration is more easily described in polar coordinates:

$$D = \{(r, \theta) \mid \alpha \leqslant \theta \leqslant \beta,\ h_1(\theta) \leqslant r \leqslant h_2(\theta)\}$$

To evaluate $\iint_R f(x, y)\, dA$, we replace x by $r \cos \theta$, y by $r \sin \theta$, and dA by $r\, dr\, d\theta$ (and use appropriate limits of integration):

$$\iint_D f(x, y)\, dA = \int_\alpha^\beta \int_{h_1(\theta)}^{h_2(\theta)} f(r \cos \theta, r \sin \theta)\, r\, dr\, d\theta$$

4. If a lamina occupies a plane region D and has density function $\rho(x, y)$, write expressions for each of the following in terms of double integrals.

(a) The mass: $m = \iint_D \rho(x, y)\, dA$

(b) The moments about the axes:

$$M_x = \iint_D y\rho(x, y)\, dA \qquad M_y = \iint_D x\rho(x, y)\, dA$$

(c) The center of mass:

$$(\bar{x}, \bar{y}), \qquad \text{where } \bar{x} = \frac{M_y}{m} \quad \text{and} \quad \bar{y} = \frac{M_x}{m}$$

(d) The moments of inertia about the axes and the origin:

$$I_x = \iint_D y^2 \rho(x, y)\, dA$$

$$I_y = \iint_D x^2 \rho(x, y)\, dA$$

$$I_0 = \iint_D (x^2 + y^2)\rho(x, y)\, dA$$

5. Let f be a joint density function of a pair of continuous random variables X and Y.

(a) Write a double integral for the probability that X lies between a and b and Y lies between c and d.

$$P(a \leqslant X \leqslant b,\ c \leqslant Y \leqslant d) = \int_a^b \int_c^d f(x, y)\, dy\, dx$$

(b) What properties does f possess?

$$f(x, y) \geqslant 0 \qquad \iint_{\mathbb{R}^2} f(x, y)\, dA = 1$$

(c) What are the expected values of X and Y?

The expected value of X is $\mu_1 = \iint_{\mathbb{R}^2} x f(x, y)\, dA$

The expected value of Y is $\mu_2 = \iint_{\mathbb{R}^2} y f(x, y)\, dA$

6. Write an expression for the area of a surface with equation $z = f(x, y)$, $(x, y) \in D$.

$$A(S) = \iint_D \sqrt{[f_x(x, y)]^2 + [f_y(x, y)]^2 + 1}\, dA$$

(assuming that f_x and f_y are continuous).

7. (a) Write the definition of the triple integral of f over a rectangular box B.

$$\iiint_B f(x, y, z)\, dV = \lim_{l, m, n \to \infty} \sum_{i=1}^{l} \sum_{j=1}^{m} \sum_{k=1}^{n} f(x_{ijk}^*, y_{ijk}^*, z_{ijk}^*)\, \Delta V$$

where ΔV is the volume of each sub-box and $(x_{ijk}^*, y_{ijk}^*, z_{ijk}^*)$ is a sample point in each sub-box.

(b) How do you evaluate $\iiint_B f(x, y, z)\, dV$?

We usually evaluate $\iiint_B f(x, y, z)\, dV$ as an iterated integral according to Fubini's Theorem for Triple Integrals: If f is continuous on $B = [a, b] \times [c, d] \times [r, s]$, then

$$\iiint_B f(x, y, z)\, dV = \int_r^s \int_c^d \int_a^b f(x, y, z)\, dx\, dy\, dz$$

Note that there are five other orders of integration that we can use.

(c) How do you define $\iiint_E f(x, y, z)\, dV$ if E is a bounded solid region that is not a box?

Since E is bounded, it can be enclosed in a box B as described in part (b). We define a new function F with domain B by

$$F(x, y, z) = \begin{cases} f(x, y, z) & \text{if } (x, y, z) \text{ is in } E \\ 0 & \text{if } (x, y, z) \text{ is in } B \text{ but not in } E \end{cases}$$

Then we define

$$\iiint_E f(x, y, z)\, dV = \iiint_B F(x, y, z)\, dV$$

(continued)

Cut here and keep for reference

(d) What is a type 1 solid region? How do you evaluate $\iiint_E f(x, y, z)\, dV$ if E is such a region?

A region E is of type 1 if it lies between the graphs of two continuous functions of x and y, that is,

$$E = \{(x, y, z) \mid (x, y) \in D,\ u_1(x, y) \leqslant z \leqslant u_2(x, y)\}$$

where D is the projection of E onto the xy-plane. Then

$$\iiint_E f(x, y, z)\, dV = \iint_D \left[\int_{u_1(x, y)}^{u_2(x, y)} f(x, y, z)\, dz\right] dA$$

(e) What is a type 2 solid region? How do you evaluate $\iiint_E f(x, y, z)\, dV$ if E is such a region?

A type 2 region is of the form

$$E = \{(x, y, z) \mid (y, z) \in D,\ u_1(y, z) \leqslant x \leqslant u_2(y, z)\}$$

where D is the projection of E onto the yz-plane. Then

$$\iiint_E f(x, y, z)\, dV = \iint_D \left[\int_{u_1(y, z)}^{u_2(y, z)} f(x, y, z)\, dx\right] dA$$

(f) What is a type 3 solid region? How do you evaluate $\iiint_E f(x, y, z)\, dV$ if E is such a region?

A type 3 region is of the form

$$E = \{(x, y, z) \mid (x, z) \in D,\ u_1(x, z) \leqslant y \leqslant u_2(x, z)\}$$

where D is the projection of E onto the xz-plane. Then

$$\iiint_E f(x, y, z)\, dV = \iint_D \left[\int_{u_1(x, z)}^{u_2(x, z)} f(x, y, z)\, dy\right] dA$$

8. Suppose a solid object occupies the region E and has density function $\rho(x, y, z)$. Write expressions for each of the following.

(a) The mass:

$$m = \iiint_E \rho(x, y, z)\, dV$$

(b) The moments about the coordinate planes:

$$M_{yz} = \iiint_E x\rho(x, y, z)\, dV$$

$$M_{xz} = \iiint_E y\rho(x, y, z)\, dV$$

$$M_{xy} = \iiint_E z\rho(x, y, z)\, dV$$

(c) The coordinates of the center of mass:

$$(\bar{x}, \bar{y}, \bar{z}), \text{ where } \bar{x} = \frac{M_{yz}}{m},\ \bar{y} = \frac{M_{xz}}{m},\ \bar{z} = \frac{M_{xy}}{m}$$

(d) The moments of inertia about the axes:

$$I_x = \iiint_E (y^2 + z^2)\rho(x, y, z)\, dV$$

$$I_y = \iiint_E (x^2 + z^2)\rho(x, y, z)\, dV$$

$$I_z = \iiint_E (x^2 + y^2)\rho(x, y, z)\, dV$$

(continued)

9. (a) How do you change from rectangular coordinates to cylindrical coordinates in a triple integral?

$$\iiint_E f(x, y, z)\, dV = \int_\alpha^\beta \int_{h_1(\theta)}^{h_2(\theta)} \int_{u_1(r\cos\theta,\, r\sin\theta)}^{u_2(r\cos\theta,\, r\sin\theta)} f(r\cos\theta, r\sin\theta, z)\, r\, dz\, dr\, d\theta$$

where

$$E = \{(r, \theta, z) \mid \alpha \leqslant \theta \leqslant \beta,\, h_1(\theta) \leqslant r \leqslant h_2(\theta),\, u_1(r\cos\theta, r\sin\theta) \leqslant z \leqslant u_2(r\cos\theta, r\sin\theta)\}$$

Thus we replace x by $r\cos\theta$, y by $r\sin\theta$, dV by $r\, dz\, dr\, d\theta$, and use appropriate limits of integration.

(b) How do you change from rectangular coordinates to spherical coordinates in a triple integral?

$$\iiint_E f(x, y, z)\, dV = \int_c^d \int_\alpha^\beta \int_{g_1(\theta,\phi)}^{g_2(\theta,\phi)} f(\rho\sin\phi\cos\theta, \rho\sin\phi\sin\theta, \rho\cos\phi)\, \rho^2 \sin\phi\, d\rho\, d\theta\, d\phi$$

where

$$E = \{(\rho, \theta, \phi) \mid \alpha \leqslant \theta \leqslant \beta,\ c \leqslant \phi \leqslant d,\ g_1(\theta, \phi) \leqslant \rho \leqslant g_2(\theta, \phi)\}$$

Thus we replace x by $\rho\sin\phi\cos\theta$, y by $\rho\sin\phi\sin\theta$, z by $\rho\cos\phi$, dV by $\rho^2\sin\phi\, d\rho\, d\theta\, d\phi$, and use appropriate limits of integration.

(c) In what situations would you change to cylindrical or spherical coordinates?

We may want to change from rectangular to cylindrical or spherical coordinates in a triple integral if the region E of integration is more easily described in cylindrical or spherical coordinates. Regions that involve symmetry about the z-axis are often simpler to describe using cylindrical coordinates, and regions that are symmetrical about the origin are often simpler in spherical coordinates. Also, sometimes the integrand is easier to integrate using cylindrical or spherical coordinates.

10. (a) If a transformation T is given by $x = g(u, v)$, $y = h(u, v)$, what is the Jacobian of T?

$$\frac{\partial(x, y)}{\partial(u, v)} = \begin{vmatrix} \dfrac{\partial x}{\partial u} & \dfrac{\partial x}{\partial v} \\ \dfrac{\partial y}{\partial u} & \dfrac{\partial y}{\partial v} \end{vmatrix} = \frac{\partial x}{\partial u}\frac{\partial y}{\partial v} - \frac{\partial x}{\partial v}\frac{\partial y}{\partial u}$$

(b) How do you change variables in a double integral?

We change from an integral in x and y to an integral in u and v by expressing x and y in terms of u and v and writing

$$dA = \left|\frac{\partial(x, y)}{\partial(u, v)}\right| du\, dv$$

Thus, under the appropriate conditions,

$$\iint_R f(x, y)\, dA = \iint_S f(x(u, v), y(u, v)) \left|\frac{\partial(x, y)}{\partial(u, v)}\right| du\, dv$$

where R is the image of S under the transformation.

(c) How do you change variables in a triple integral?

Similarly to the case of two variables in part (b),

$$\iiint_R f(x, y, z)\, dV = \iiint_S f(x(u, v, w), y(u, v, w), z(u, v, w)) \left|\frac{\partial(x, y, z)}{\partial(u, v, w)}\right| du\, dv\, dw$$

where

$$\frac{\partial(x, y, z)}{\partial(u, v, w)} = \begin{vmatrix} \dfrac{\partial x}{\partial u} & \dfrac{\partial x}{\partial v} & \dfrac{\partial x}{\partial w} \\ \dfrac{\partial y}{\partial u} & \dfrac{\partial y}{\partial v} & \dfrac{\partial y}{\partial w} \\ \dfrac{\partial z}{\partial u} & \dfrac{\partial z}{\partial v} & \dfrac{\partial z}{\partial w} \end{vmatrix}$$

is the Jacobian.

Cut here and keep for reference

CHAPTER 16 CONCEPT CHECK ANSWERS

1. What is a vector field? Give three examples that have physical meaning.

A vector field is a function that assigns a vector to each point in its domain.

A vector field can represent, for example, the wind velocity at any location in space, the speed and direction of the ocean current at any location, or the force vector of the earth's gravitational field at a location in space.

2. (a) What is a conservative vector field?

A conservative vector field $\mathbf{F}$ is a vector field that is the gradient of some scalar function f, that is, $\mathbf{F} = \nabla f$.

(b) What is a potential function?

The function f in part (a) is called a potential function for $\mathbf{F}$.

3. (a) Write the definition of the line integral of a scalar function f along a smooth curve C with respect to arc length.

If C is given by the parametric equations $x = x(t)$, $y = y(t)$, $a \leqslant t \leqslant b$, we divide the parameter interval $[a, b]$ into n subintervals $[t_{i-1}, t_i]$ of equal width. The ith subinterval corresponds to a subarc of C with length Δs_i. Then

$$\int_C f(x, y)\, ds = \lim_{n\to\infty} \sum_{i=1}^{n} f(x_i^*, y_i^*)\, \Delta s_i$$

where (x_i^*, y_i^*) is any sample point in the ith subarc.

(b) How do you evaluate such a line integral?

$$\int_C f(x, y)\, ds = \int_a^b f(x(t), y(t)) \sqrt{\left(\frac{dx}{dt}\right)^2 + \left(\frac{dy}{dt}\right)^2}\, dt$$

Similarly, if C is a smooth space curve, then

$$\int_C f(x, y, z)\, ds$$

$$= \int_a^b f(x(t), y(t), z(t)) \sqrt{\left(\frac{dx}{dt}\right)^2 + \left(\frac{dy}{dt}\right)^2 + \left(\frac{dz}{dt}\right)^2}\, dt$$

(c) Write expressions for the mass and center of mass of a thin wire shaped like a curve C if the wire has linear density function $\rho(x, y)$.

The mass is $m = \int_C \rho(x, y)\, ds$.

The center of mass is $(\bar{x}, \bar{y})$, where

$$\bar{x} = \frac{1}{m} \int_C x\rho(x, y)\, ds$$

$$\bar{y} = \frac{1}{m} \int_C y\rho(x, y)\, ds$$

(d) Write the definitions of the line integrals along C of a scalar function f with respect to x, y, and z.

$$\int_C f(x, y, z)\, dx = \lim_{n\to\infty} \sum_{i=1}^{n} f(x_i^*, y_i^*, z_i^*)\, \Delta x_i$$

$$\int_C f(x, y, z)\, dy = \lim_{n\to\infty} \sum_{i=1}^{n} f(x_i^*, y_i^*, z_i^*)\, \Delta y_i$$

$$\int_C f(x, y, z)\, dz = \lim_{n\to\infty} \sum_{i=1}^{n} f(x_i^*, y_i^*, z_i^*)\, \Delta z_i$$

(We have similar results when f is a function of two variables.)

(e) How do you evaluate these line integrals?

$$\int_C f(x, y, z)\, dx = \int_a^b f(x(t), y(t), z(t))\, x'(t)\, dt$$

$$\int_C f(x, y, z)\, dy = \int_a^b f(x(t), y(t), z(t))\, y'(t)\, dt$$

$$\int_C f(x, y, z)\, dz = \int_a^b f(x(t), y(t), z(t))\, z'(t)\, dt$$

4. (a) Define the line integral of a vector field $\mathbf{F}$ along a smooth curve C given by a vector function $\mathbf{r}(t)$.

If $\mathbf{F}$ is a continuous vector field and C is given by a vector function $\mathbf{r}(t)$, $a \leqslant t \leqslant b$, then

$$\int_C \mathbf{F} \cdot d\mathbf{r} = \int_a^b \mathbf{F}(\mathbf{r}(t)) \cdot \mathbf{r}'(t)\, dt = \int_C \mathbf{F} \cdot \mathbf{T}\, ds$$

(b) If $\mathbf{F}$ is a force field, what does this line integral represent?

It represents the work done by $\mathbf{F}$ in moving a particle along the curve C.

(c) If $\mathbf{F} = \langle P, Q, R\rangle$, what is the connection between the line integral of $\mathbf{F}$ and the line integrals of the component functions P, Q, and R?

$$\int_C \mathbf{F} \cdot d\mathbf{r} = \int_C P\, dx + Q\, dy + R\, dz$$

5. State the Fundamental Theorem for Line Integrals.

If C is a smooth curve given by $\mathbf{r}(t)$, $a \leqslant t \leqslant b$, and f is a differentiable function whose gradient vector ∇f is continuous on C, then

$$\int_C \nabla f \cdot d\mathbf{r} = f(\mathbf{r}(b)) - f(\mathbf{r}(a))$$

6. (a) What does it mean to say that $\int_C \mathbf{F} \cdot d\mathbf{r}$ is independent of path?

$\int_C \mathbf{F} \cdot d\mathbf{r}$ is independent of path if the line integral has the same value for any two curves that have the same initial points and the same terminal points.

(b) If you know that $\int_C \mathbf{F} \cdot d\mathbf{r}$ is independent of path, what can you say about $\mathbf{F}$?

We know that $\mathbf{F}$ is a conservative vector field, that is, there exists a function f such that $\nabla f = \mathbf{F}$.

(continued)

7. State Green's Theorem.

Let C be a positively oriented, piecewise-smooth, simple closed curve in the plane and let D be the region bounded by C. If P and Q have continuous partial derivatives on an open region that contains D, then

$$\int_C P\,dx + Q\,dy = \iint_D \left(\frac{\partial Q}{\partial x} - \frac{\partial P}{\partial y}\right) dA$$

8. Write expressions for the area enclosed by a curve C in terms of line integrals around C.

$$A = \oint_C x\,dy = -\oint_C y\,dx = \tfrac{1}{2}\oint_C x\,dy - y\,dx$$

9. Suppose $\mathbf{F}$ is a vector field on $\mathbb{R}^3$.

(a) Define curl $\mathbf{F}$.

$$\text{curl } \mathbf{F} = \left(\frac{\partial R}{\partial y} - \frac{\partial Q}{\partial z}\right)\mathbf{i} + \left(\frac{\partial P}{\partial z} - \frac{\partial R}{\partial x}\right)\mathbf{j} + \left(\frac{\partial Q}{\partial x} - \frac{\partial P}{\partial y}\right)\mathbf{k}$$
$$= \nabla \times \mathbf{F}$$

(b) Define div $\mathbf{F}$.

$$\text{div } \mathbf{F} = \frac{\partial P}{\partial x} + \frac{\partial Q}{\partial y} + \frac{\partial R}{\partial z} = \nabla \cdot \mathbf{F}$$

(c) If $\mathbf{F}$ is a velocity field in fluid flow, what are the physical interpretations of curl $\mathbf{F}$ and div $\mathbf{F}$?

At a point in the fluid, the vector curl $\mathbf{F}$ aligns with the axis about which the fluid tends to rotate, and its length measures the speed of rotation; div $\mathbf{F}$ at a point measures the tendency of the fluid to flow away (diverge) from that point.

10. If $\mathbf{F} = P\,\mathbf{i} + Q\,\mathbf{j}$, how do you determine whether $\mathbf{F}$ is conservative? What if $\mathbf{F}$ is a vector field on $\mathbb{R}^3$?

If P and Q have continuous first-order derivatives and $\dfrac{\partial P}{\partial y} = \dfrac{\partial Q}{\partial x}$, then $\mathbf{F}$ is conservative.

If $\mathbf{F}$ is a vector field on $\mathbb{R}^3$ whose component functions have continuous partial derivatives and curl $\mathbf{F} = \mathbf{0}$, then $\mathbf{F}$ is conservative.

11. (a) What is a parametric surface? What are its grid curves?

A parametric surface S is a surface in $\mathbb{R}^3$ described by a vector function

$$\mathbf{r}(u, v) = x(u, v)\,\mathbf{i} + y(u, v)\,\mathbf{j} + z(u, v)\,\mathbf{k}$$

of two parameters u and v. Equivalent parametric equations are

$$x = x(u, v) \qquad y = y(u, v) \qquad z = z(u, v)$$

The grid curves of S are the curves that correspond to holding either u or v constant.

(b) Write an expression for the area of a parametric surface.

If S is a smooth parametric surface given by

$$\mathbf{r}(u, v) = x(u, v)\,\mathbf{i} + y(u, v)\,\mathbf{j} + z(u, v)\,\mathbf{k}$$

where $(u, v) \in D$ and S is covered just once as (u, v) ranges throughout D, then the surface area of S is

$$A(S) = \iint_D |\mathbf{r}_u \times \mathbf{r}_v|\,dA$$

(c) What is the area of a surface given by an equation $z = g(x, y)$?

$$A(S) = \iint_D \sqrt{1 + \left(\frac{\partial z}{\partial x}\right)^2 + \left(\frac{\partial z}{\partial y}\right)^2}\,dA$$

12. (a) Write the definition of the surface integral of a scalar function f over a surface S.

We divide S into "patches" S_{ij}. Then

$$\iint_S f(x, y, z)\,dS = \lim_{m, n \to \infty} \sum_{i=1}^{m} \sum_{j=1}^{n} f(P_{ij}^*)\,\Delta S_{ij}$$

where ΔS_{ij} is the area of the patch S_{ij} and P_{ij}^* is a sample point from the patch. (S is divided into patches in such a way that ensures that $\Delta S_{ij} \to 0$ as $m, n \to \infty$.)

(b) How do you evaluate such an integral if S is a parametric surface given by a vector function $\mathbf{r}(u, v)$?

$$\iint_S f(x, y, z)\,dS = \iint_D f(\mathbf{r}(u, v))\,|\mathbf{r}_u \times \mathbf{r}_v|\,dA$$

where D is the parameter domain of S.

(c) What if S is given by an equation $z = g(x, y)$?

$$\iint_S f(x, y, z)\,dS$$
$$= \iint_D f(x, y, g(x, y))\sqrt{\left(\frac{\partial z}{\partial x}\right)^2 + \left(\frac{\partial z}{\partial y}\right)^2 + 1}\,dA$$

(d) If a thin sheet has the shape of a surface S, and the density at (x, y, z) is $\rho(x, y, z)$, write expressions for the mass and center of mass of the sheet.

The mass is

$$m = \iint_S \rho(x, y, z)\,dS$$

The center of mass is $(\bar{x}, \bar{y}, \bar{z})$, where

$$\bar{x} = \frac{1}{m}\iint_S x\,\rho(x, y, z)\,dS$$

$$\bar{y} = \frac{1}{m}\iint_S y\,\rho(x, y, z)\,dS$$

$$\bar{z} = \frac{1}{m}\iint_S z\,\rho(x, y, z)\,dS$$

(continued)

Cut here and keep for reference

13. (a) What is an oriented surface? Give an example of a nonorientable surface.

An oriented surface S is one for which we can choose a unit normal vector $\mathbf{n}$ at every point so that $\mathbf{n}$ varies continuously over S. The choice of $\mathbf{n}$ provides S with an orientation.

A Möbius strip is a nonorientable surface. (It has only one side.)

(b) Define the surface integral (or flux) of a vector field $\mathbf{F}$ over an oriented surface S with unit normal vector $\mathbf{n}$.

$$\iint_S \mathbf{F} \cdot d\mathbf{S} = \iint_S \mathbf{F} \cdot \mathbf{n}\, dS$$

(c) How do you evaluate such an integral if S is a parametric surface given by a vector function $\mathbf{r}(u, v)$?

$$\iint_S \mathbf{F} \cdot d\mathbf{S} = \iint_D \mathbf{F} \cdot (\mathbf{r}_u \times \mathbf{r}_v)\, dA$$

We multiply by -1 if the opposite orientation of S is desired.

(d) What if S is given by an equation $z = g(x, y)$?

If $\mathbf{F} = \langle P, Q, R \rangle$,

$$\iint_S \mathbf{F} \cdot d\mathbf{S} = \iint_D \left(-P \frac{\partial g}{\partial x} - Q \frac{\partial g}{\partial y} + R \right) dA$$

for the upward orientation of S; we multiply by -1 for the downward orientation.

14. State Stokes' Theorem.

Let S be an oriented piecewise-smooth surface that is bounded by a simple, closed, piecewise-smooth boundary curve C with positive orientation. Let $\mathbf{F}$ be a vector field whose components have continuous partial derivatives on an open region in $\mathbb{R}^3$ that contains S. Then

$$\int_C \mathbf{F} \cdot d\mathbf{r} = \iint_S \operatorname{curl} \mathbf{F} \cdot d\mathbf{S}$$

15. State the Divergence Theorem.

Let E be a simple solid region and let S be the boundary surface of E, given with positive (outward) orientation. Let $\mathbf{F}$ be a vector field whose component functions have continuous partial derivatives on an open region that contains E. Then

$$\iint_S \mathbf{F} \cdot d\mathbf{S} = \iiint_E \operatorname{div} \mathbf{F}\, dV$$

16. In what ways are the Fundamental Theorem for Line Integrals, Green's Theorem, Stokes' Theorem, and the Divergence Theorem similar?

In each theorem, we integrate a "derivative" over a region, and this integral is equal to an expression involving the values of the original function only on the *boundary* of the region.

Cut here and keep for reference

CHAPTER 17 CONCEPT CHECK ANSWERS

1. (a) Write the general form of a second-order homogeneous linear differential equation with constant coefficients.

$$ay'' + by' + cy = 0$$

where a, b, and c are constants and $a \neq 0$.

(b) Write the auxiliary equation.

$$ar^2 + br + c = 0$$

(c) How do you use the roots of the auxiliary equation to solve the differential equation? Write the form of the solution for each of the three cases that can occur.

If the auxiliary equation has two distinct real roots r_1 and r_2, the general solution of the differential equation is

$$y = c_1 e^{r_1 x} + c_2 e^{r_2 x}$$

If the roots are real and equal, the solution is

$$y = c_1 e^{rx} + c_2 x e^{rx}$$

where r is the common root.

If the roots are complex, we can write $r_1 = \alpha + i\beta$ and $r_2 = \alpha - i\beta$, and the solution is

$$y = e^{\alpha x}(c_1 \cos \beta x + c_2 \sin \beta x)$$

2. (a) What is an initial-value problem for a second-order differential equation?

An initial-value problem consists of finding a solution y of the differential equation that also satisfies given conditions $y(x_0) = y_0$ and $y'(x_0) = y_1$, where y_0 and y_1 are constants.

(b) What is a boundary-value problem for such an equation?

A boundary-value problem consists of finding a solution y of the differential equation that also satisfies given boundary conditions $y(x_0) = y_0$ and $y(x_1) = y_1$.

3. (a) Write the general form of a second-order nonhomogeneous linear differential equation with constant coefficients.

$ay'' + by' + cy = G(x)$, where a, b, and c are constants and G is a continuous function.

(b) What is the complementary equation? How does it help solve the original differential equation?

The complementary equation is the related homogeneous equation $ay'' + by' + cy = 0$. If we find the general solution y_c of the complementary equation and y_p is any particular solution of the nonhomogeneous differential equation, then the general solution of the original differential equation is $y(x) = y_p(x) + y_c(x)$.

(c) Explain how the method of undetermined coefficients works.

To determine a particular solution y_p of $ay'' + by' + cy = G(x)$, we make an initial guess that y_p is a general function of the same type as G. If $G(x)$ is a polynomial, choose y_p to be a general polynomial of the same degree. If $G(x)$ is of the form Ce^{kx}, choose $y_p(x) = Ae^{kx}$. If $G(x)$ is $C \cos kx$ or $C \sin kx$, choose $y_p(x) = A \cos kx + B \sin kx$. If $G(x)$ is a product of functions, choose y_p to be a product of functions of the same type. Some examples are:

$G(x)$	$y_p(x)$
x^2	$Ax^2 + Bx + C$
e^{2x}	Ae^{2x}
$\sin 2x$	$A \cos 2x + B \sin 2x$
xe^{-x}	$(Ax + B)e^{-x}$

We then substitute y_p, y_p', and y_p'' into the differential equation and determine the coefficients.

If y_p happens to be a solution of the complementary equation, then multiply the initial trial solution by x (or x^2 if necessary).

If $G(x)$ is a sum of functions, we find a particular solution for each function and then y_p is the sum of these.

The general solution of the differential equation is

$$y(x) = y_p(x) + y_c(x)$$

(d) Explain how the method of variation of parameters works.

We write the solution of the complementary equation $ay'' + by' + cy = 0$ as $y_c(x) = c_1 y_1(x) + c_2 y_2(x)$, where y_1 and y_2 are linearly independent solutions. We then take $y_p(x) = u_1(x)\,y_1(x) + u_2(x)\,y_2(x)$ as a particular solution, where $u_1(x)$ and $u_2(x)$ are arbitrary functions. After computing y_p', we impose the condition that

$$u_1' y_1 + u_2' y_2 = 0 \qquad \textbf{(1)}$$

and then compute y_p''. Substituting y_p, y_p', and y_p'' into the original differential equation gives

$$a(u_1' y_1' + u_2' y_2') = G \qquad \textbf{(2)}$$

We then solve equations **(1)** and **(2)** for the unknown functions u_1' and u_2'. If we are able to integrate these functions, then a particular solution is $y_p(x) = u_1(x)\,y_1(x) + u_2(x)\,y_2(x)$ and the general solution is $y(x) = y_p(x) + y_c(x)$.

4. Discuss two applications of second-order linear differential equations.

The motion of an object with mass m at the end of a spring is an example of simple harmonic motion and is described by the second-order linear differential equation

$$m\frac{d^2x}{dt^2} + kx = 0$$

(continued)

where k is the spring constant and x is the distance the spring is stretched (or compressed) from its natural length. If there are external forces acting on the spring, then the differential equation is modified.

Second-order linear differential equations are also used to analyze electrical circuits involving an electromotive force, a resistor, an inductor, and a capacitor in series.

See the discussion in Section 17.3 for additional details.

5. How do you use power series to solve a differential equation?

We first assume that the differential equation has a power series solution of the form

$$y = \sum_{n=0}^{\infty} c_n x^n = c_0 + c_1 x + c_2 x^2 + c_3 x^3 + \cdots$$

Differentiating gives

$$y' = \sum_{n=1}^{\infty} n c_n x^{n-1} = \sum_{n=0}^{\infty} (n+1) c_{n+1} x^n$$

and

$$y'' = \sum_{n=2}^{\infty} n(n-1) c_n x^{n-2} = \sum_{n=0}^{\infty} (n+2)(n+1) c_{n+2} x^n$$

We substitute these expressions into the differential equation and equate the coefficients of x^n to find a recursion relation involving the constants c_n. Solving the recursion relation gives a formula for c_n and then

$$y = \sum_{n=0}^{\infty} c_n x^n$$

is the solution of the differential equation.